Clinical Applications of Magnetic Nanoparticles

Clinical Applications of Magnetic Nanoparticles

Design to Diagnosis
Manufacturing to Medicine

Edited by
Nguyễn T. K. Thanh

CRC Press is an imprint of the
Taylor & Francis Group, an **informa** business

CRC Press
Taylor & Francis Group
6000 Broken Sound Parkway NW, Suite 300
Boca Raton, FL 33487-2742

Printed on acid-free paper

International Standard Book Number-13: 978-1-138-05155-3 (Hardback)

Library of Congress Cataloging-in-Publication Data

Names: Thanh, Nguyen T. K. (Thi Kim), editor.
Title: Clinical applications of magnetic nanoparticles / [edited by] Nguyen T.K. Thanh.
Description: Boca Raton : Taylor & Francis, 2018. | Includes bibliographical references and index.
Identifiers: LCCN 2017045414 | ISBN 9781138051553 (hardback : alk. paper)
Subjects: | MESH: Magnetite Nanoparticles--therapeutic use | Drug Carriers--therapeutic use | Diagnostic Imaging
Classification: LCC R857.N34 | NLM QT 36.5 | DDC 610.28--dc23
LC record available at https://lccn.loc.gov/2017045414

Visit the Taylor & Francis Web site at
http://www.taylorandfrancis.com

and the CRC Press Web site at
http://www.crcpress.com

This book is dedicated to my dear parents, who encouraged me at a very young age to be independent and allowed me to fly as far as I wish.

Contents

SECTION III In-Vitro Application of MNPs

SECTION IV In-Vivo Application of MNPs

Foreword

This book builds on the demand created by Professor Thanh's earlier volume *Magnetic Nanoparticles: From Fabrication to Clinical Applications*. The editor is a well-known and highly respected research scientist, well versed in many aspects of nanoscience and especially in this particular topic.

In the earlier volume, the foundations for making magnetic nanoparticles were clearly set out and it became apparent that another book would be needed to give scientists and clinicians a guide to what is happening in this field as the subject translates from the laboratory towards clinical practice. Inevitably, there is a very slight overlap with the earlier volume, and the first four chapters of this new book cover some new synthesis routes not covered in the earlier volume, which include methods of varying the particle size and shape, of the creation of chains of nanoparticles and the coating of metal particles by carbon and using bio-inspired methods for synthesis. These chapters, like many others in the book, deal with the potential applications and give very fair appraisals of the strengths and weaknesses of their approaches. Professor Thanh is to be congratulated in urging the many authors to be as objective as possible!

Surface functionalization is the key to most applications of nanoparticles of all types, and this is especially important here. There are several chapters devoted to the topic, and it recurs in many of the specific application chapters. It is very helpful to have available the many approaches that are in use in one book. Magnetic particles have the potential to enhance contrast in MRI images, and this topic receives detailed attention from several authors.

Magnetic particles hold the promise for separation of biomolecules and enhancing biosensing. There are chapters covering all of the current aspects of this very important application, both *in vitro* and *in vivo*. This enables the manipulation and sorting of cells and internal components of cells, such as mitochondria, and it is an area in which we will see increasing utility.

Health issues are always paramount in any use of nanotechnology, and several chapters deal with this with specific regard to the magnetic nanoparticles that are being developed for medical applications. The subject of the fate and possible toxicity of all types of nanoparticles is still debated and we are now getting a much more clear and objective view of these issues.

Magnetic hyperthermia is receiving a lot of attention because it has the possibility of offering a purely physical, nonchemical method of destroying cancerous cells. Added to this is the magnetically guided therapy afforded by directing a drug-loaded particle precisely to a site in the body, and this raises the question about the transport of nanoparticles through porous body tissue, which is dealt with in a complete chapter.

Imaging is important in medicine, and additional modalities can be added to particles designed for magnetic resonance imaging (MRI), and this aspect is described in several chapters and for various future clinical applications.

Finally, the considerations of Good Manufacturing Practice are discussed with respect to magnetic nanoparticles, and this is a useful chapter for all of the contemporary clinical applications of nanoparticles.

Having been asked to write this Foreword, I read the individual chapters and I found that it is very difficult to stop! Each set of specialist authors has produced a fascinating and informative review of their subject matter. This will render the book as a useful tool for learning about almost every aspect of magnetic nanoparticles for clinical application. It is therefore recommended for a wide range of readers, from students to research professors and medical practitioners, and it forms a good companion to Professor Thanh's earlier book.

Peter J. Dobson, OBE
The Queen's College, Oxford

Preface

More than six years ago, when I wrote the preface for the book *Magnetic Nanoparticles: From Fabrication to Clinical Applications* (http://www.crcpress.com/product/isbn/9781439869321), it did not occur to me that I would write another one so soon. However, at every conference I attended, I saw the burgeoning research of magnetic nanoparticles (NPs). I could not help wanting to try to capture the most cutting-edge discovery and ever-expanding research in this field. With the success of the first book, it was a tall order to get this book right as well (e.g. being comprehensive, with 26 chapters, serving a wide audience from early-year research students to professors and being useful not only to practitioners but also to researchers who would like to join the field).

The current book is not overlapping with the first one, but complementary, and covers areas the first book did not, such as the extensive background and development of magnetic NPs (MNPs) as negative and positive contrast agents for magnetic resonance imaging. The 'Fabrication' part of the book covers different synthetic methods for iron oxide MNPs, the mechanism of NP formation, how to control the dimensions and morphology of NPs, which is essential for the optimization of their properties. New magnetic nanostructures, such as nanochains and carbon-coated magnetic metal NPs, as well as bio-inspired synthesis of MNPs are covered.

Very detailed strategies for *biofunctionalization* of MNPs, and their interaction with the biological environment, are beautifully covered. A new framework to experimentally take NP syntheses in the laboratory towards scalable manufacturing, including not only the synthesis but also surface modification, is covered to address the outstanding challenge of creating robust and reproducible syntheses of functionalized NPs. For the biofunctionalization of NPs, polymersomes and multidentate polymers are of particular interest for the stability they provide to the MNPs. Magnetic core–mesoporous silica shell composites with improved drug payloads and the ability to tune the drug release are presented.

In vivo applications, including high- and low-gradient magnetic separations with distinctive separation mechanisms, are discussed for biomedical diagnostics. Magnetic separation in integrated micro-analytical systems and magnetic separation of cellular organelles such as endosomes, exosomes and mitochondria are introduced. MNP-based biosensing with giant magnetoresistance biosensors and Hall sensors are reviewed.

Preceding *in vivo applications*, immunotoxicity and safety considerations for iron oxide NPs and the impact of MNPs on human health should be investigated. It is fascinating to see the idea of using nanorobots to navigate in multiscale complex vascular networks to deliver cancer therapy in addition to the extensive research on magnetic hyperthermia. MNPs used for drug delivery in Alzheimer's disease, as well as for on-demand drug delivery devices based on a magnetic sponge, which are transported in complex media, are presented. When magnetic cores are functionalized with molecules such as nerve growth factors or neuroprotective molecules, multifunctional devices can be developed for neurological diseases, specifically those based on the use of engineered MNPs applied to neuroprotection and neuroregeneration.

For *in vivo applications*, red blood cells were used as carriers for NP-based MRI and magnetic particle imaging contrast agents to prolong their circulation in the bloodstream. The advantages of nuclear imaging of radiolabeled MNPs for biomedical applications and roadmap for developing and imaging radiolabeled NPs are covered. Stimuli-regulated cancer theranostics based on MNPs such as internal stimuli-responsive NPs, including pH, reduction-sensitive NPs and external stimuli-inductive NPs, such as magnetic field- and light-controlled NPs, is also presented.

Finally, establishing large-scale good manufacturing practice (GMP) compliant NPs is the prerequisite to successfully translate the laboratory-scale synthesis to commercial products. The importance of continuous manufacturing methods enables the control of critical quality attributes with adjustment of production parameters, which are also closely monitored with in-process controls, are highlighted.

Similarly with the first book, the chapters were written by world-leading experts in the broad range of disciplines (e.g. physics, chemistry, biochemistry, biology, medicine, engineering and entrepreneurship). They not only present the most cutting-edge research for active scientists in the field but also provide the fundamental knowledge to enable students and other incoming researchers to take steps to translate their technologies to clinics.

Nguyễn T. K. Thanh, FRSC
Biophysics Group
Department of Physics and Astronomy
University College London
London, United Kingdom
Email: ntk.thanh@ucl.ac.uk

Acknowledgements

First and foremost, I would like to thank my husband and wonderful children for allowing me to devote my time to editing this book.

I am privileged to have trust from colleagues in agreeing to spend their precious time in preparing very comprehensive chapters and then to have done their revision after reviewing to produce the highest quality ones.

I am indebted to many volunteer reviewers who provided their insightful comments and made the book valuable not only to people already working on magnetic nanoparticles for clinical application but also to wider communities.

My thanks to Dr. Alejandro Baeza García, Mr. Andreas Sergides, Prof. Andreas Tschope, Dr. Ángel del Pozo, Dr. Ángel Millán Escolano, Prof. Antoine Perreira, Dr. Aristides Bakandritsos, Dr. Beata Kalska-Szostko, Dr. Bernd Baumstümmler, Dr. Boris Polyak, Prof. Carlos Frederico de Gusmão Campos Geraldes, Prof. Christine Ménager, Dr. Christopher Adams, Dr. Claire Wilhelm, Dr. Cordula Grüttner, Prof. Daishun Ling, Dr. Damien Faivre, Prof. Daniel Horak, Dr. Dimitri Stanicki, Dr. Eleni K. Efthimiadou, Prof. Enrico Bergamaschi, Dr. Enza Torino, Prof. Erwann Guenin, Prof. Franca Bigi, Dr. Franck Couillard, Mr. Georgios Kasparis, Dr. Gurvinder Singh, Prof. Ian Prior, Prof. Igor Chourpa, Dr. Inge Katrin Herrmann, Prof. Ivo Safarik, Dr. Jean-Olivier Durand, Prof. Jeff W.M. Bulte, Prof. Jeffrey Chalmers, Dr. Jennifer Hall Grossman, Dr. Jordi Faraudo, Dr. Joseph Bear, Dr. Kazunori Shimizu, Prof. Ladislau Vakas, Dr. Li Zhang, Dr. Liliana Maria Pires Ferreira, Miss Lilin Wang, Dr. Maciej Zborowski, Prof. Manuel Ricardo Ibarra García, Dr. Maria Francesca Casula, Dr. Marijana Mionic, Dr. Marin Tadić, Dr. Nora M. Dempsey, Dr. Oliver Weber, Dr. Olivier Sandre, Dr. Oliviero Gobbo, Prof. Oula Peñate Medina, Miss Panagiota Chondrou, Dr. Penelope Bouziotis, Prof. Peter Dobson, Dr. Quoc Lam Vuong, Dr. Rafael Torres Martin De Rosales, Dr. Ralf Mundkowski, Prof. Randall Erb, Miss Raquel Rodrigues, Dr. Ruxandra Gref, Prof. Samuel D. Bader, Dr. Sofia Lima, Dr. Spriridon V. Spirou, Dr. Stefanos Mourdikoudis, Prof. Sylvio Dutz, Dr. Tanya Prozrorov, Dr. Touraj Ehtezazi, Dr. Vladan Kusigerski, Prof. Vladimir L. Kolesnichenko, Prof. Wilfried Weber and Dr. Yaowu Hao for their useful discussion on the book.

My special thanks to Prof. Jon Preece for helping me in deciding the title of this book to reflect its nature as a complementary volume to the first one. Also thanks to EU COST Action TD1402 Radiomag for many networking opportunities to discuss the book.

I would like to thank Barbara Glunn, Danielle Zarfati, Jonathan Achorn and their colleagues at CRC Press/Taylor & Francis who worked closely with me to publish this timely book.

I am ever thankful for the support from many wonderful scientists, engineers, clinicians and entrepreneurs around the world in this project. Your effort will surely push the research closer to meaningful applications that would one day save many more lives.

Nguyễn T. K. Thanh, FRSC
Biophysics Group
Department of Physics and Astronomy
University College London
London, United Kingdom
Email: ntk.thanh@ucl.ac.uk

Editor

Nguyễn T. K. Thanh, FRSC, MInstP (http://www.ntk-thanh.co.uk) held a prestigious Royal Society University Research Fellowship (2005–2014). She was appointed a full professor in nanomaterials in 2013 at Biophysics Group, Department of Physics and Astronomy, University College London, UK. She leads a very dynamic group conducting cutting-edge *interdisciplinary and innovative research* on the design and synthesis of magnetic and plasmonic nanomaterials for biomedical applications. A very strong feature of her research program is developing new chemical methods and, in collaboration with chemical engineers, producing the next generation of nanoparticles with very high magnetic moment and novel hybrid and multifunctional nanostructures. Detailed mechanistic studies of their formation by sophisticated and advanced analysis of the nanostructure allow tuning of the physical properties at the nanoscale; these can subsequently be exploited for diagnosis and treatment of various diseases. These studies are conducted to provide insight for future material design approaches. It will also help to identify the critical process parameters that can be manipulated in order to obtain the suitable physical properties for the intended applications.

She was the sole editor of the book *Magnetic Nanoparticles: From Fabrication to Clinical Applications* published by CRC Press/Taylor & Francis: http://www.crcpress.com/product/isbn/9781439869321. In 2016, she was a guest editor of the Royal Society Interface Focus on "Multifunctional nanostructures for diagnosis and therapy of diseases."

She has published over 80 peer-reviewed journal articles and book chapters with over 4500 citations so far. She has been a visiting professor at various universities in France, Japan, China and Singapore. She has been an invited speaker at over 200 institutes and scientific meetings. She has a leadership role in professional communities by chairing and organising 30 high-profile international conferences such as the American Chemistry Society symposia in 2018, 2012 and 2010; Royal Society of Chemistry UK Colloids Conferences in 2017, 2014 and 2011; European Material Research Society Symposia in 2016 and 2013; ICMAT Singapore in 2015 and 2013; Faraday Discussions in 2014, and being a member of advisory boards in Europe, the United States and Japan. She served in the Joint Committee of the Royal Society of Chemistry Colloid & Interface Science Group and the Society of Chemical Industry Colloid & Surface Chemistry Group (2008–2017). She is an elected member of The Royal Society of Chemistry Faraday Division Council and is currently serving on the Awards Committee and was a representative member of Joint Colloids Groups (2013–2016). She is a workgroup leader of EU COST Action TD1402 on Multifunctional Nanoparticles for Magnetic Hyperthermia and Indirect Radiation Therapy (RADIOMAG). She is a cochair of the 13th International Conference on the Scientific and Clinical Applications of Magnetic Carriers in June 2020, London, UK.

Foreword Author

Professor Peter J. Dobson, OBE, BSc, MA (Oxon), PhD, C Phys, F Inst P, Member of the ACS, FRCS (The Queen's College, Oxford)

Peter has had a broad career covering a wide range of disciplines, from physics and chemistry to materials science and engineering. He has also worked in industry (Philips) as well as academia (Imperial College and Oxford) and was responsible for creating and building the Begbroke Science Park for Oxford University. He has published over 190 papers and 32 patents. He has founded three companies and advised on the formation of eight more. He was the strategic advisor on nanotechnology to the Research Councils in the UK (2009–2013) and has been on several Engineering and Physical Sciences Research Council (EPSRC) panels and committees. Peter currently sits on the EPSRC Strategic Advisory Board on Quantum Technology. He was awarded the Officer of the Order of the British Empire in 2013 in recognition of his contributions to science and engineering. He is currently a Principal Fellow at Warwick Manufacturing Group and a visiting professor at King's College London and University College London and chairs the Industrial Advisory Board of the Physics Department at Bristol University. He also chairs the Natural Environment Research Council Facility for Environmental Nanoscience Analysis and Characterization at Birmingham University. Peter delivers courses at graduate level in the areas of biosensors, nanotechnology, innovation, entrepreneurship and related topics and advises on innovation.

Contributors

Laurent Adumeau
Centre for BioNano Interactions
School of Chemistry and Chemical Biology
University College Dublin
Dublin, Ireland

Faiz Ul Amin
Department of Biology and Applied Life Science
Gyeongsang National University
Jinju, Republic of Korea

Antonella Antonelli
Department of Biomolecular Sciences
University of Urbino Carlo Bo
Italy

Stephen J. Archibald
Department of Chemistry
and
Positron Emission Tomography Research Centre
University of Hull
Hull, United Kingdom

Laura Asín
Aragon Institute of Materials Science & CIBER-BBN
University of Zaragoza/CSIC
Zaragoza, Spain

Sylvie Bégin-Colin
Institut de Physique et Chimie des Matériaux de Strasbourg
Université de Strasbourg
Strasbourg Cedex, France

Maximilian O. Besenhard
Department of Chemical Engineering
University College London
London, United Kingdom

Maria Eugênia Fortes Brollo
Department of Energy, Environment & Health
Instituto de Ciencia de Materiales de Madrid (ICMM)
Consejo Superior de Investigaciones Científicas, CSIC
Madrid, Spain

Benjamin P. Burke
Department of Chemistry
and
Positron Emission Tomography Research Centre
University of Hull
Hull, United Kingdom

Changqian Cao
France-China Bio-Mineralization and Nano-Structures Laboratory
and
Palaeomagnetism and Geochronology Laboratory
CAS Key Laboratory of Earth and Planetary Physics
Institute of Geology and Geophysics
Chinese Academy of Sciences
Beijing, China

Christopher Cawthorne
School of Life Sciences
and
Positron Emission Tomography Research Centre
University of Hull
Hull, United Kingdom

Mu Chiao
Department of Mechanical Engineering
University of British Columbia
Vancouver, Canada

Sagar Chowdhury
Weinberg Medical Physics, Inc.
North Bethesda, Maryland

Jesús Martínez de la Fuente
Aragon Institute of Materials Science & CIBER-BBN
University of Zaragoza/CSIC
Zaragoza, Spain

María del Puerto Morales
Department of Energy, Environment and Health
Instituto de Ciencia de Materiales de Madrid (ICMM)
Consejo Superior de Investigaciones Científicas, CSIC
Madrid, Spain

Marie-Hélène Delville
Institute for Condensed Matter Chemistry of Bordeaux (ICMCB)
National Center for Scientific Research, CNRS
University of Bordeaux
Pessac, France

Yinglong Feng
The Center for Micromagnetics & Information Technologies (MINT)
Department of Electrical & Computer Engineering
University of Minnesota
Minneapolis, Minnesota

Marc-André Fortin
Laboratory for Biomaterials in Imaging
Department of Mining, Metallurgical and Materials Engineering
and
Centre de Recherche du Centre Hospitalier Universitaire de Québec (CR-CHUQ)
Université Laval
Québec, Canada

Raluca Maria Fratila
Aragon Institute of Materials Science & CIBER-BBN
University of Zaragoza/CSIC
Zaragoza, Spain

Kévin Gagné
Polytechnique Montréal
Department of Computer and Software Engineering
Institute of Biomedical Engineering
Nanorobotics Laboratory
Montréal, Canada

Helena Gavilán
Department of Energy, Environment and Health
Instituto de Ciencia de Materiales de Madrid (ICMM)
Consejo Superior de Investigaciones Científicas, CSIC
Madrid, Spain

Gerardo F. Goya
Institute of Nanoscience of Aragón
Department of Condensed Matter Physics
University of Zaragoza
Zaragoza, Spain

Robert N. Grass
Functional Materials Laboratory
Institute for Chemical and Bioengineering
Department of Chemistry and Applied Biosciences
ETH Zurich
Zurich, Switzerland

Lucía Gutiérrez
Instituto de Nanociencia de Aragón, INA
Universidad de Zaragoza
Zaragoza, Spain

Roxanne Hachani
Department of Physics and Astronomy
University College London
and
UCL Healthcare Biomagnetic and Nanomaterials Laboratories
London, United Kingdom

Adeline Hannecart
Department of General, Organic and Biomedical Chemistry
Laboratory of NMR and Molecular Imaging
University of Mons
Mons, Belgium

Gary Hannon
Laboratory for Biological Characterization of Advanced Materials (LBCAM)
Department of Clinical Medicine,
Trinity Translational Medicine Institute (TTMI)
Trinity College Dublin
Dublin, Ireland

Ryan Hilaman
Weinberg Medical Physics, Inc.
North Bethesda, Maryland

Ingrid Hilger
Department of Experimental Radiology
Institute of Diagnostic and Interventional Radiology
University Hospital Jena
Jena, Germany

Uwe Himmelreich
Biomedical MRI Unit
Department of Imaging and Pathology
Faculty of Biomedical Sciences
Katholieke Universiteit Leuven
Leuven, Belgium

Ali Kafash Hoshiar
Mechanical and Aerospace Engineering and ReCAPT
Gyeongsang National University
Jinju, Gyeongnam, Republic of Korea

and

Industrial and Mechanical Engineering
Islamic Azad University, Qazvin Branch
Qazvin, Iran

Kazunori Hoshino
Department of Biomedical Engineering
University of Connecticut
Storrs, Connecticut

Yanglong Hou
College of Engineering
Peking University
Beijing, China

Sahar Jafari
Weinberg Medical Physics, Inc.
North Bethesda, Maryland

Yanmin Ju
College of Engineering & College of Life Science
Peking University
Beijing, China

Myeong Ok Kim
Department of Biology and Applied Life Science
Gyeongsang National University
Jinju, Republic of Korea

Alec P. LaGrow
Department of Physics and Astronomy
University College London
and
UCL Healthcare Biomagnetic and Nanomaterials Laboratories
London, United Kingdom

Maxime Latulippe
Polytechnique Montréal
Department of Computer and Software Engineering
Institute of Biomedical Engineering
Nanorobotics Laboratory
Montréal, Canada

Sophie Laurent
Department of General, Organic and Biomedical Chemistry
Laboratory of NMR and Molecular Imaging
University of Mons
Mons, Belgium

Tuan Anh Lea
Mechanical and Aerospace Engineering and ReCAPT
Gyeongsang National University
Jinju, Gyeongnam, Republic of Korea

Sim Siong Leong
School of Chemical Engineering
Universiti Sains Malaysia
Nibong Tebal, Penang, Malaysia

Ning Li
Polytechnique Montréal
Department of Computer and Software Engineering
Institute of Biomedical Engineering
Nanorobotics Laboratory
Montréal, Canada

JitKang Lim
School of Chemical Engineering
Universiti Sains Malaysia
Nibong Tebal, Penang, Malaysia

Siew Chun Low
School of Chemical Engineering
Universiti Sains Malaysia
Nibong Tebal, Penang, Malaysia

Shinya Maenosono
School of Materials Science
Japan Advanced Institute of Science and Technology
Nomi, Ishikawa, Japan

Mauro Magnani
Department of Biomolecular Sciences
University of Urbino Carlo Bo
Urbino (PU), Italy

Lamar O. Mair
Weinberg Medical Physics, Inc.
North Bethesda, Maryland

Bella B. Manshian
Biomedical MRI Unit
Department of Imaging and Pathology
Faculty of Biomedical Sciences
Katholieke Universiteit Leuven
Leuven, Belgium

Irena Markovic-Milosevic
Powder Technology Laboratory
Institute of Materials
École Polytechnique Fédérale de Lausanne (EPFL)
Lausanne, Switzerland

Sylvain Martel
Polytechnique Montréal
Department of Computer and Software Engineering
Institute of Biomedical Engineering
Nanorobotics Laboratory
Montréal, Canada

Damien Mertz
Institut de Physique et Chimie des Matériaux de Strasbourg
Centre National de la Recherche Scientique (CNRS)
Université de Strasbourg
Strasbourg Cedex, France

Mahmood Mohammadi
Polytechnique Montréal
Department of Computer and Software Engineering
Institute of Biomedical Engineering
Nanorobotics Laboratory
Montréal, Canada

Rohimah Mohamud
School of Medical Science
Universiti Sains Malaysia
Kubang Kerian, Khota Bharu, Kelantan, Malaysia

Stéphane Mornet
Institute for Condensed Matter Chemistry of Bordeaux (ICMCB)
National Center for Scientific Research, CNRS
University of Bordeaux
Pessac, France

María Moros
Istituto di Scienze Applicate e Sistemi Intelligenti 'Eduardo Caianiello'
Consiglio Nazionale delle Ricerche
Pozzuoli, Italy

Laurence Motte
Laboratory for Vascular Translational Science
Université Paris 13
Sorbonne Paris Cité
Bobigny, France

Robert N. Muller
Department of General, Organic and Biomedical Chemistry
Laboratory of NMR and Molecular Imaging
University of Mons
Mons, Belgium

Aleksandar N. Nacev
Weinberg Medical Physics, Inc.
North Bethesda, Maryland

Jung Kwon (John) Oh
Department of Chemistry and Biochemistry
Concordia University
Montreal, QC, Canada

Yongxin Pan
France-China Bio-Mineralization and Nano-Structures Laboratory
Palaeomagnetism and Geochronology Laboratory
and
CAS Key Laboratory of Earth and Planetary Physics
Institute of Geology and Geophysics
Chinese Academy of Sciences
Beijing, China

Adriele Prina-Mello
Laboratory for Biological Characterization of Advanced Materials (LBCAM)
Centre for Research on Adaptive Nanostructures and Nanodevices (CRANN)
Department of Clinical Medicine
Trinity Translational Medicine Institute (TTMI)
Trinity College Dublin
Dublin, Ireland

Vittoria Raffa
Università di Pisa
Department of Biology
Pisa, Italy

Vincent Russier
Institut de Chimie et des Materiaux Paris-Est-UMR 7182
CNRS and UPEC
Thiais, France

Marie-Louise Saboungi
Institut de Minéralogie, de Physique des Matériaux et de Cosmochimie Sorbonne Universités
UPMC Univ Paris 06
UMR Museum National d'Histoire Naturelle
IRD UMR 206
Paris, France

Ali Shademani
Department of Biomedical Engineering
University of British Columbia
Vancouver, Canada

Benjamin Shapiro
Fischell Department of Bioengineering
Institute for Systems Research
University of Maryland, College Park
College Park, Maryland

Stefaan J. Soenen
Department of Imaging and Pathology
Faculty of Biomedical Sciences
Katholieke Universiteit Leuven
Leuven, Belgium

Dimitri Stanicki
Department of General, Organic and Biomedical Chemistry
Laboratory of NMR and Molecular Imaging
University of Mons
Mons, Belgium

Pavel Stepanov
Weinberg Medical Physics, Inc.
North Bethesda, Maryland

Grazyna Stepien
Institute of Nanoscience of Aragón
University of Zaragoza
Zaragoza, Spain

Diqing Su
The Center for Micromagnetics and Information Technologies (MINT)
Department of Chemical Engineering and Material Science
University of Minnesota
Minneapolis, Minnesota

Maryam S. Tabatabaei
Polytechnique Montréal
Department of Computer and Software Engineering
Institute of Biomedical Engineering
Nanorobotics Laboratory
Montréal, Canada

Mari Takahashi
School of Materials Science
Japan Advanced Institute of Science and Technology
Nomi, Ishikawa, Japan

Nguyễn T. K. Thanh
Department of Physics and Astronomy
University College London
and
UCL Healthcare Biomagnetic and Nanomaterials Laboratories
London, United Kingdom

Shiyan Tong
College of Engineering & College of Life Science
Peking University
Beijing, China

Charles Tremblay
Polytechnique Montréal
Department of Computer and Software Engineering
Institute of Biomedical Engineering
Nanorobotics Laboratory
Montréal, Canada

Akif Emre Türeli
Research and Development Department
MJR PharmJet GmbH
Überherrn, Germany

Nazende Günday Türeli
Research and Development Department
MJR PharmJet GmbH
Überherrn, Germany

Melissa Anne Tutty
Laboratory for Biological Characterization of Advanced Materials (LBCAM)
Department of Clinical Medicine
Trinity Translational Medicine Institute (TTMI)
Trinity College Dublin
Dublin, Ireland

Luce Vander Elst
Department of General, Organic and Biomedical Chemistry
Laboratory of NMR and Molecular Imaging
University of Mons
Mons, Belgium

Sabino Veintemillas-Verdaguer
Department of Energy, Environment and Health
Instituto de Ciencia de Materiales de Madrid (ICMM)
Consejo Superior de Investigaciones Científicas, CSIC
Madrid, Spain

Jian-Ping Wang
The Center for Micromagnetics and Information Technologies (MINT)
Department of Electrical and Computer Engineering
University of Minnesota
Minneapolis, Minnesota

Irving N. Weinberg
Weinberg Medical Physics, Inc.
North Bethesda, Maryland

Kai Wu
The Center for Micromagnetics and Information Technologies (MINT)
Department of Electrical and Computer Engineering
University of Minnesota
Minneapolis, Minnesota

Swee Pin Yeap
Faculty of Engineering, Technology and Built Environment
UCSI University
Kuala Lumpur, Malaysia

Jungwon Yoon
School of Integrated Technology
Gwangju Institute of Science and Technology
Gwangju, Republic of Korea

Martin Zeltner
Functional Materials Laboratory
Institute for Chemical and Bioengineering
Department of Chemistry and Applied Biosciences
ETH Zurich
Zurich, Switzerland

Hongbin Zhang
Department of Mechanical Engineering
University of British Columbia
Vancouver, Canada

Xingming Zhang
School of Naval Architecture and Ocean Engineering
Harbin Institute of Technology at Weihai
Weihai, Shandong, China

Section I

Fabrication, Characterisation of MNPs

1 Controlling the Size and Shape of Uniform Magnetic Iron Oxide Nanoparticles for Biomedical Applications

*Helena Gavilán, Maria Eugênia Fortes Brollo, Lucía Gutiérrez, Sabino Veintemillas-Verdaguer and María del Puerto Morales**

CONTENTS

1.1 STATE OF THE ART: SIZE, SHAPE CONTROL AND SELF-ASSEMBLY PROCESSES

In the last decade, there have been huge advances in the synthesis of magnetic nanoparticles (MNPs) for biomedical applications encompassing requirements in terms of size, surface and colloidal stability under physiological conditions intrinsic to each particular application.[1–4] In some cases MNPs have reached the clinical practice, as it is the case of magnetic resonance imaging (MRI) contrast agents. Nevertheless, for other applications such as hyperthermia treatments, tissue regeneration, magnetically driven transfection of stem cells or delivery of genetic materials, nanoparticles' (NPs') use in an efficient and biocompatible manner remains to be assessed.[5] The scientific and industrial challenges involved in developing MNPs for clinical applications have been recently highlighted.[6] The limitation in many cases comes from the wide size distribution of the NPs, the lack of aggregation control or the poor/weak functionality of the surface.

Typically, MNPs for biomedical applications are composed of a magnetic core, usually a ferrite and most commonly an iron oxide, modified with a biocompatible material resulting in a core–shell structure.[3] The shell acts not only as a hydrophilic layer to render colloidal stability and avoid aggregation but also as a platform for functionalization for specific applications within the biomedical area. To improve their magnetic properties, other metal ions have been chosen to dope into the ferrite spinel structure, such as manganese,[7] zinc and cobalt.[8–10] With the same purpose, metallic NPs[11,12] with higher magnetic moments have also been synthesized and stabilized by different coatings.[13–16] However, magnetite and maghemite are still the most popular materials for biomedical applications because of their biocompatibility and good magnetic properties when properly synthesized with high crystallinity and free of impurities.

This chapter aims to provide an update of recent advances in the synthesis of iron oxide MNPs, consisting of uniform magnetic cores stabilized in water forming biocompatible aqueous colloids (Figure 1.1).

* Corresponding author

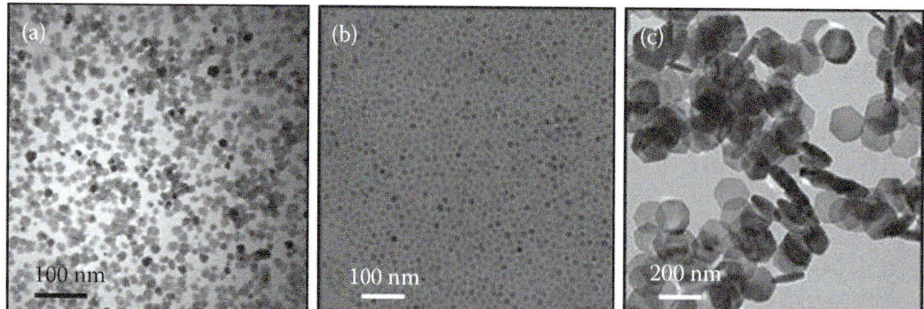

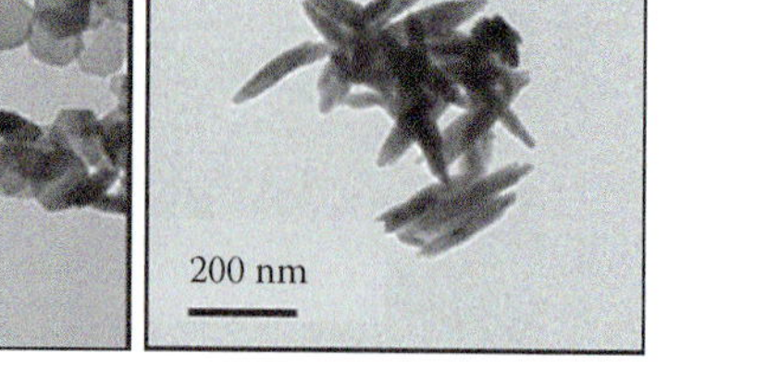

FIGURE 1.1 Transmission electron microscopy (TEM) images of uniform MNPs with well-defined size and shape prepared by different routes: From left to right: (a) polyol process, (b) decomposition in organic media, and (c) and (d) reduction of antiferromagnetic precursors synthesized in aqueous media (hematite and goethite).

Some of the methods described in this chapter were already mentioned in Chapter 2, 'Synthesis and Characterisation of Iron Oxide Ferrite NPs and Ferrite-Based Aqueous Fluids' by Etienne Duguet et al. in the book *Magnetic Nanoparticles: From Fabrication to Clinical Applications* (ISBN 9781439869321),[17] but they have been modified to better control the size, shape or distribution of the NPs. In the case of aqueous synthetic routes, the last advances in NP shape control by the synthesis of antiferromagnetic precursors[18] and the biomineralization of highly ordered magnetic nanostructures[19] are presented in Section 1.2.1. In Sections 1.2.2–1.2.6, alternative synthetic routes to the conventional ones, allowing NP size and aggregation control, and surface functionality tuning are presented. In these sections, we include synthetic routes in organic media (Section 1.2.2) that have recently received a significant amount of effort, the polyol process (1.2.3), microwave-assisted methods (1.2.4) and others such as electrochemistry or plasma techniques (Sections 1.2.5 and 1.2.6).

We will distinguish whether a synthetic route leads to single or multicore particles, taking into account that the term 'core' describes an individual NP, and 'multicore' describes a collection of cores held by a matrix forming a fixed structure.[20] Single-core and multicore particles could further agglomerate as a consequence of weak physical interactions in a reversible process. To differentiate this reversible agglomeration process from stronger irreversible processes, we use the term 'aggregate' to refer to the stronger assemblage that occurs in multicore NPs to generate the discrete entity. These systems may present strong magnetic interactions between the cores, changing the properties of the material if compared to a noninteracting system.[21,22] The need for understanding the different magnetic properties of single-core and multicore particles underlies the importance of reliable synthetic methods to reproduce NP size, shape and structural homogeneity.[23]

Special attention has also been paid to the mechanism of NP formation and self-assembling processes that could be a powerful tool to control the dimensions and morphology of NPs, essential for the optimization of their properties. The typical crystal growth pathway by monomer addition is characterized by a free energy, as a function of the crystal size that exhibits a maximum representing the free energy barrier of nucleation, followed by a progressive decrease as the crystal size increases. Alternatively, less common mechanisms have been discovered in the recent years that may lead to crystals, whose morphologies or internal structures can be very different from the thermodynamically stable final phase, *i.e.* the bulk octahedral crystals of magnetite.[24] Some of these alternative growth pathways, in particular those leading to uniform NPs, involve the oriented aggregation of small subunits in an assembly process driven by the presence of specific molecules or under the action of an external field. This leads to the generation of nanostructures with properties different from those of the discrete NPs from which they derive.[25–28]

In general, the self-assembly processes on the base of the oriented aggregation mechanism – also indicated as oriented attachment, as they are able to generate perfect monocrystals – are difficult to differentiate from the classical growth by molecular incorporation, and there are very few systems in which this mechanism has been unequivocally demonstrated.[24,29,30] An oriented attachment process has been observed for iron oxide MNPs prepared in water or in polyol and it has been related to the biomineralization process.[31–33] In aqueous media, the pH in which magnetite particle growth takes place determines whether or not aggregation of the growing particles occurs. If the pH of the system is close to the isoelectric point of magnetite (pH = 6–7), aggregation of particles will happen readily, whereas at pH values away from that pH, growing particles have a surface charge and electrostatic forces that will prevent aggregation. The effect of the ensemble, *i.e.* the size of the subunits and the degree of orientation, on the magnetic behaviour of magnetite NPs has also been studied.[34–36]

1.2 PROGRESS ON SYNTHESIS ROUTES

1.2.1 Aqueous Synthesis

Synthetic routes to obtain magnetite NPs in aqueous media are currently one of the most commonly studied processes due to the high-availability lab setup and low cost of its reagents, the overall ease of scaling up the process and the low toxicity associated to this route. Conventionally, there have been processes based on the aqueous precipitation of a mixed Fe(II)/Fe(III) solution, the so-called co-precipitation route, and secondly, the partial reduction or oxidation of a Fe(III) or Fe(II) salt, respectively, always in the presence of a base. However, these routes lead to NP populations with broad core size distribution (generally the size range is below 20 nm) and not well-defined morphology, in comparison with other synthetic routes, *i.e.* high-temperature decomposition of organic precursors or polyol-based processes.[3,37,38] Nevertheless, these conventional routes have recently allowed a thorough control over the iron concentration, aging time, counter-ions present in the reaction and the presence of extra organic additives, improving the above-mentioned limitations of aqueous syntheses.

Alternatively, new strategies have emerged as promising aqueous routes to obtain crystalline, monodisperse and well-defined magnetite crystals. These strategies are based on the use of iron hydroxide (ferrihydrite or white/green rust) and iron oxide (hematite or goethite) NPs as starting precursors to obtain magnetite via biomineralization routes.

1.2.1.1 Co-Precipitation of Iron (II) and (III) Salts

The simplest and most straightforward method to obtain magnetite synthetically is the coprecipitation of Fe(III) and Fe(II) in aqueous alkaline conditions, which can be carried out at room temperature under inert atmosphere.[39] The introduction of the acidic Fe(III)/Fe(II) mixture into a highly alkaline solution leads to instant magnetite precipitation according to Equation 1.1. This typically results in small NPs with diameters <20 nm that, due to the limited size of the magnetic domain, have superparamagnetic properties.

$$2Fe^{3+} + Fe^{2+} + 8OH^- \rightarrow Fe_3O_4 + 4H_2O \quad (1.1)$$

Unfortunately, the synthetic procedure provides little means of control over the size distribution and morphology. Indeed, it was shown that the size and shape of the NP can be affected by the type of base and pH value,[40] ionic strength and temperature,[41] iron concentration and aging time[42] and nature of the counteranion.[43] An optimized chemical protocol consisting of an acid treatment postsynthesis reduces the size distribution width (standard deviation <0.25) (Figure 1.2)

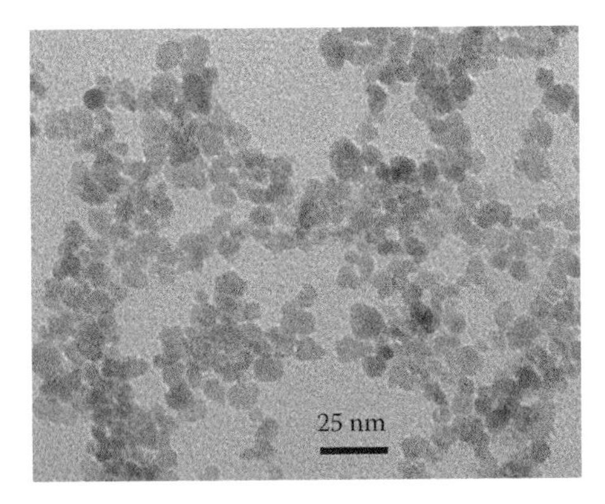

FIGURE 1.2 TEM images of magnetite NPs prepared by the coprecipitation method and subjected to an acid treatment postsynthesis to reduce the size dispersion, as described in Ref. 44.

and improves the magnetic properties and colloidal stability of iron oxide MNPs through the reduction of particle surface disorder mostly in the range of sizes below 15 nm.[44] Recent advances in the coprecipitation method led to the scale-up of the process with continuous monitoring of the magnetization.[45]

Iron oxide NPs synthesised using the coprecipitation method in the presence of oleic acid have been used to develop a drug delivery system to the gastrointestinal (GI) tract with potential for magnetic targeting and tracking, which can release its payload on demand using localised magnetic hyperthermia to trigger release.[46] Gelatin capsules were dip coated into a dispersion of oleic-acid-capped iron oxide NPs in molten eicosane. This renders the capsule impervious to water/acid/base ingress providing a coating that is resistant to the harsh conditions experienced in the GI tract.

An interesting alternative to conventional routes to prepare in one-step hydrophilic magnetite NPs in aqueous media is the hydrothermal route. Recently, a rapid and easily scalable hydrothermal synthesis of single-core magnetite NPs[47] and multicore NPs assembled in a unique flower-shaped structure was reported.[48] Furthermore, surface-modified magnetite NPs were obtained using a one-step continuous hydrothermal process in a countercurrent flow reactor.[49] It remains challenging to obtain Fe_3O_4 (magnetite) NP exclusively starting from a mixture of Fe(III)/Fe(II) under high pressure, temperature and flow of suspension with controlled shape and a narrow size distribution, avoiding the presence of large particles and therefore ensuring the superparamagnetic nature of the NPs.

Another potential solution to narrow the size distribution was found using bioinspired routes. The basic coprecipitation in a hydrogel network offers diffusion control of nucleation and crystal growth by suppressing convection.[50] Living organisms such as chitons and magnetotactic bacteria are also able to form magnetite crystals with well-controlled sizes and shapes using macromolecular templates. Following the same approach, the use of water-soluble (bio)macromolecular control agents, such as magnetosome proteins, has become of great interest.[51] The cationic polypeptide poly(-L-arginine) was used in an ultraslow titration coprecipitation reaction that led to the formation of monodisperse stable single-domain magnetite NPs of 35 ± 5 nm.[51] Interestingly, other magnetosome proteins such as MamJ and MtxAD1–24 have significantly different effects on magnetite coprecipitation, since they strongly inhibit magnetite nucleation. Biomacromolecules such as Mms6 have proven to promote well-defined cubic-octahedral magnetite crystals when used as extra additive in the coprecipitation process. This is due to the amphiphilic nature of the magnetosome protein Mms6 that allows the formation of micelle-like aggregates in aqueous solution.[51]

Based on these results, the authors concluded that 'in the case of magnetite formation proteins, larger complexes, or membrane components promoting the nucleation *in vivo* are likely to expose positively charged residues to a negatively charged crystal surface'.[52] Moreover, the limitations on the production of core sizes above 20 nm (up to 60 nm) and different shapes were overcome by conducting slow coprecipitation of magnetite through a ferrihydrite/Fe (II) precursor in mildly alkaline aqueous medium using NH_3 by the use of different copolypeptides poly(L-aspartic acid) and poly(L-lysine) with varying copolypeptides/Fe ratios as additives.

1.2.1.2 Partial Reduction of Iron (III) Salts

Hydrophilic magnetite NPs have also been obtained in aqueous media in one-step syntheses by the hydrothermal-reduction of Fe(III) salts in an autoclave. Hydrazine citrate, sodium borohydride, carbon monoxide and dimethyl-formamide have been used as reductants, although they are highly reactive and pose potential environmental risks.[53–55] Other mild and nontoxic reducing agents such as ascorbic acid, tartaric acid, aspartic acid and α-D glucose have also been used for the synthesis of iron oxide NP by hydrothermal-reduction method.[56] Tunable size and narrow size distribution can be achieved by choosing an appropriate mixture of solvents and varying parameters such as temperature, pressure and reaction time. The advantages of using a hydrothermal/solvothermal approach for synthesizing MNP include a high degree of product purity, easy control of the size, high crystallinity and uniform morphology of the NP, relatively lower temperatures (in general <200°C) and use of nonspecialized equipment and simple overall process.

1.2.1.3 Partial Oxidation of Iron (II) Salts

Partial oxidation of Fe(II) salt solution in alkaline media is summarized in Equation 1.2.[57,58] An intermediate phase (the green rust) is formed owing to the presence of the base and undergoes dehydroxylation steps leading to the formation of magnetite. Sugimoto and Matijević developed several procedures for the preparation of magnetite particles between 100 and up to 1000 nm using this route.[59]

$$\begin{aligned} &3\,FeCl_2 + 6\,NaOH \rightarrow 3\,Fe_3(OH)_2 + 6\,NaCl \\ &2\,Fe(OH)_2 + 0.5\,O_2 \rightarrow 2\,FeOOH + H_2O \\ &2\,FeOOH + Fe(OH)_2 \rightarrow Fe_3O_4 + 2\,H_2O \end{aligned} \tag{1.2}$$

In the latter work, it was pointed out that a dramatic change in mean particle diameter was observed over a very small concentration range of Fe(II) salt due to a sharp pH change.[60] In this case, the Fe(II) salt was kept oxygen-free by bubbling N_2 in a water bath at a fixed temperature. Both the base and the oxidant (KNO_3 instead of air) were added dropwise. The obtained magnetite particles had cubic shape and were monodisperse, with a size range of 50–200 nm. In past decades, partial oxidation has been less studied than other synthetic routes. Interestingly, this strategy led to magnetite NPs with a size range of 20–100 nm, well-defined size and shape and with few defects on the surface.

The control of the main parameters of the process (the oxidant, iron concentration, pH and temperature), following the guidelines stated by Sugimoto and Matijević,[59] enables the reproducible large-scale synthesis of magnetite nanocrystals.[61] Unfortunately, two iron oxide phases are often formed simultaneously if the conditions are not properly chosen. The origin of such difficulty lies in the complexity of the process that takes place through the gelatinous intermediate (the green rust) that evolves to magnetite by reaction with Fe(II) in solution or to an oxyhydroxide such as lepidocrocite or goethite, depending mostly on the rate of oxidation. The authors emphasized that magnetite formation requires the complete dehydroxylation of the precursor (green rust) prior to their oxidation.[62] If oxidation is too fast and precedes dehydroxylation, lepidocrocite is formed in preference to magnetite. MNPs prepared in this way are characterized by very high saturation magnetization values and have been proven to be useful for biomedical applications such as magnetothermia and magnetically guided drug delivery.[61,63,64]

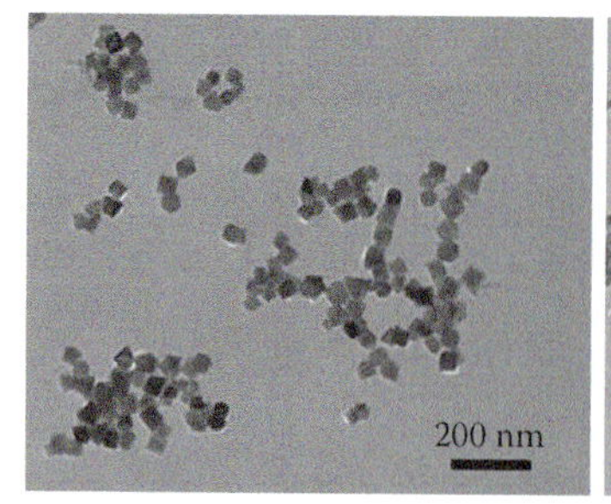

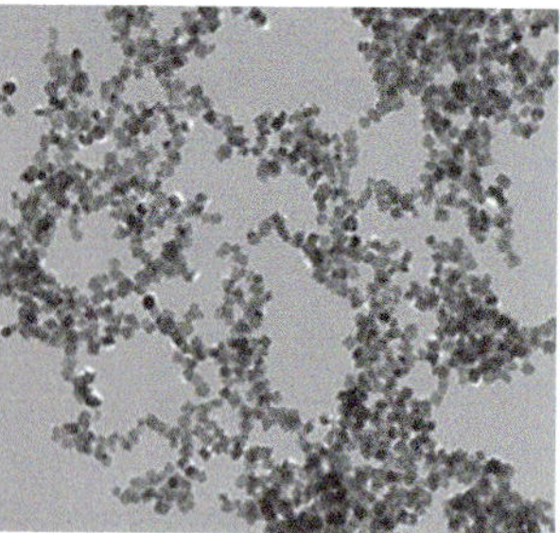

FIGURE 1.3 TEM micrographs at the same magnification of the sodium bearing magnetite nanocrystals obtained by partial oxidation of Fe(II) salts in water (on the left) and ethanol 25% (on the right).

Among the factors that influence the oxidation rate of green rust, counterions are presumably relevant because they affect green rust's structure that consists of positively charged octahedral brucite-type Fe(II)-Fe(III) hydroxyl layers linked by anions.[62] Recently, a detailed study on the effect of different counterions, along with the use of mixed solvents, was carried out.[65] The characteristic stabilities of green rusts as well as the dehydration and oxidation processes of Fe(II) that follow (both facilitated by cosmotropic environments) are responsible for the differences in magnetite particle sizes obtained (Figure 1.3). This is especially relevant because the average particle size of the obtained particles is often close to the superparamagnetic-ferrimagnetic limit of magnetite (close to 20 nm). The range of nanocrystal sizes obtained (20–60 nm) has a profound influence on the magnetic moment per particle.

1.2.1.4 Reduction of Antiferromagnetic Precursor

One of the current challenges in MNP research is the production of large magnetite cores with well-controlled size and shape, large magnetic moment and long-term colloidal stability. An interesting aqueous-based approach to produce single-core magnetite NPs with different morphologies and core sizes above 25 nm using antiferromagnetic precursors has been developed recently. It consists of a three-step process (Figure 1.4) and enables one to obtain uniform magnetite rhombohedra, discs and elongated MNPs.[66] A similar procedure was used to obtain magnetite and iron metal nanorods.[18,67]

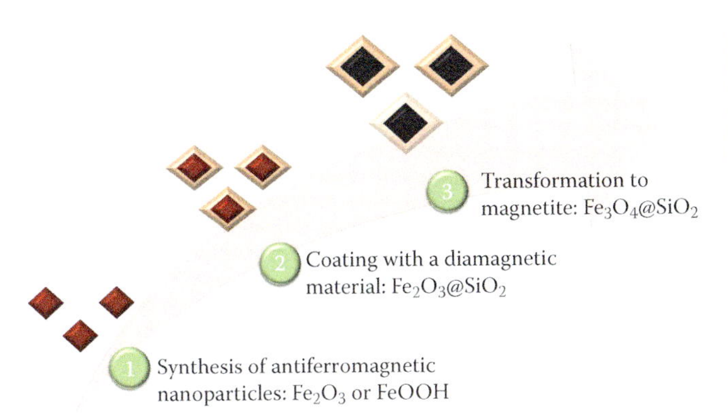

FIGURE 1.4 General scheme for the synthesis of single-core NPs with sizes above 25 nm and different morphologies through the reduction of antiferromagnetic NPs.

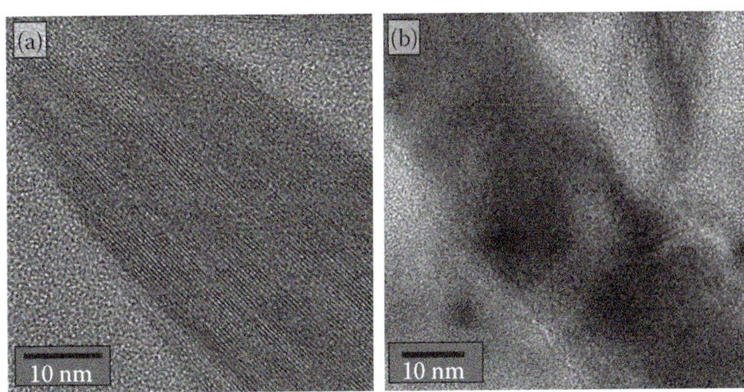

FIGURE 1.6 High-resolution TEM images of the goethite precursor (a) and the resulting needle magnetite NPs (b) after reduction, showing pore formation.

First, an aqueous synthesis route is followed to obtain uniform antiferromagnetic precursors such as hematite or goethite, whose size and shape can be tuned by changing the synthetic conditions, including temperature, pressure and nature and concentration of the salts used (Figure 1.5).[62] Furthermore, this route could be scalable for large production. Then, the antiferromagnetic precursor particles are coated by a silica or alumina layer that prevents their aggregation.[18,68] Finally, the silica-coated antiferromagnetic particles are reduced to magnetite. This can be performed on particles either in powder form, by exposing them to a hydrogen atmosphere at a certain partial pressure (dry reduction),[69] or in liquid form, where oleic acid and an organic solvent are used (wet reduction).[70] In both processes, hydrogen and oleic acid act as the reducing agents.

As a result, exotic morphologies of magnetite such as rhombohedra (70 nm), discs (140 × 20 nm) and needles (180 × 30 nm) were obtained, and both reduction methods transform completely the starting precursor to a single iron oxide phase of magnetite without morphology changes[66] (Equations 1.3 and 1.4 and Figure 1.5). Powder reduction shows fewer defects in the crystal structure, slightly higher values of saturation magnetization and a more homogeneous remanent magnetization state. Pores within the core structure were observed, being more pronounced in the case of using goethite precursor as a consequence of the phase transformation (goethite suffers dehydration to hematite prior transformation to magnetite, Equation 1.3). Moreover, when lower reduction temperatures and shorter times are used, nanorods consisting of clusters of maghemite embedded in an antiferromagnetic hematite matrix are obtained, with very interesting relaxometric properties and potential use as MRI contrast agents.[71]

$$2\,FeOOH \rightarrow Fe_2O_3 + H_2O \tag{1.3}$$

$$3\,Fe_2O_3 + H_2 \rightarrow 2\,Fe_3O_4 + H_2O \tag{1.4}$$

The possibility of generating a discontinuous structure within a particle by forcing the pore formation as those showed in the magnetite needles after reduction (Figure 1.6)[71] is a strategy to develop novel materials for biomedical applications. Moreover, this method has the advantage of covering different particle size ranges and morphologies, opening up new possibilities and the potential interesting magnetic properties.

1.2.1.5 Biomineralization

Biomineralization in a broad sense is all processes that biological systems employ to build the organic–inorganic hybrid

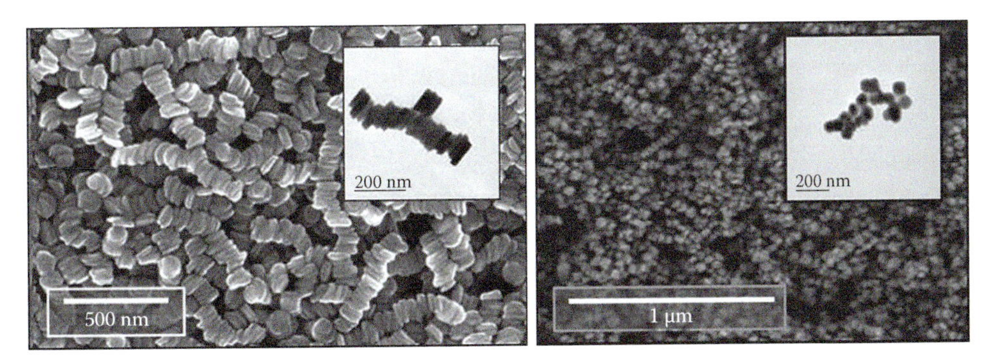

FIGURE 1.5 Scanning electron microscopy (SEM) images of the antiferromagnetic hematite precursors of magnetite NPs with different shape: discs on the left and rhombohedra on the right.

materials present in all living systems, with functions ranging from navigation, mechanical support, photonics, to the protection of the soft parts of the body.[72,73] Often, these biominerals have complex shapes and textures, exceptional structural hierarchy and, in general, are characterized by the highest observed level of control over composition, structure, size and morphology of the constituent mineral components. Examples of iron-based biominerals are the radula teeth of chitons mollusks that contain crystalline iron oxides, such as magnetite (Fe_3O_4) and lepidocrocite (γ-FeO(OH)), and the intracellular chains of magnetite NPs synthesized by magnetotactic bacteria.[74]

Researchers have taken inspiration from nature, aiming to apply the key aspects of biomineralization to more sustainable synthetic methods. Indeed, in particular, mimicking the pathways used by magnetotactic bacteria would open the way to aqueous room temperature synthetic methods that still allow control over the dimension, structure and, as a consequence, magnetic properties of the magnetite synthesized. Recently, the synthesis of magnetite at ambient conditions was followed by using hexagonal ferrihydrite as a precursor, starting from Fe(III) salt, obtaining a gel-like precursor material identified as 6-line ferrihydrite.[75] The latter transformation of ferrihydrite to magnetite was carried out by the addition of Fe(II) salt under an N_2 atmosphere and the subsequent increase in the solution pH by NH_3 diffusion. It was observed that the assembly of 1.5–2.0 nm primary particles into aggregates after a reaction time of ~1.5 hours led to 10–20 nm uniform magnetite NPs after >12 hours. This route was conducted in the presence of random copolymer of glutamic acid, lysine and alanine, producing magnetite with a less polydispersed size distribution.[75]

1.2.2 Organic Synthesis by Thermal Decomposition of an Organic Precursor

Thermal decomposition of organometallic compounds is able to produce MNPs with good crystallinity and high monodispersity.[76] This approach offers two routes to control nucleation and growth processes that occur during the particle synthesis. One procedure is the injection of organometallic compounds into a hot surfactant solution, which results in the formation of nuclei almost instantaneously. The other option is the controlled heating of organometallic compounds in a surfactant solution to generate the nuclei. Once the nucleation has occurred, particles grow at high temperature. Finally, through a rapid decrease in the reaction temperature, the growth of the NPs can be stopped.[77] Control of the NP size, shape and aggregation depends on the number of nuclei initially generated and the presence of surfactant or other molecules hampering the NP growth process.[78] Direct syntheses of magnetite NPs with different sizes and morphologies have been reported, including nanospheres,[78] nanocubes,[79–81] nanowires,[82] nanorods,[83] nanooctahedra,[84] nanoplates[85] and nanoprisms.[86] Examples of particles obtained by thermal decomposition of organometallic compounds showing very narrow core size distributions can be observed in Figure 1.7. Other materials such as cobalt

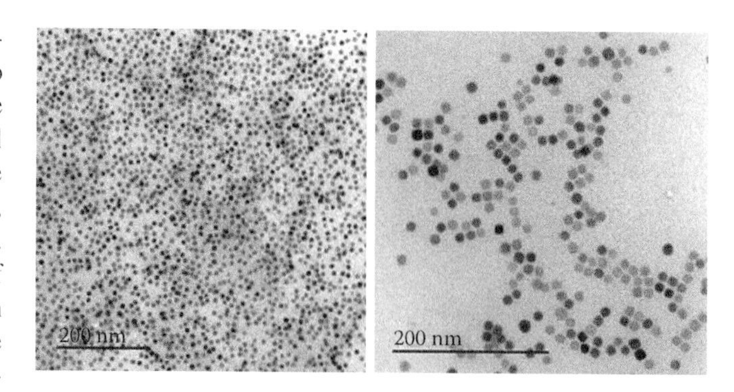

FIGURE 1.7 TEM images of magnetite NPs prepared by thermal decomposition of Fe(III)-oleate complex in 1-octadecene.

substituted ferrite NPs[87] and Ag@Fe_3O_4 core–shell NPs[88] can also be prepared by thermal decomposition of organometallic precursors. In the last case, a temperature pause introduced in a simple single-step thermal decomposition of iron precursors, with the presence of silver seeds formed in the same reaction mixture, gives rise to this novel compact heterostructure.

The control of particle size could be achieved by the use of an extension of the La-Mer mechanism in combination with a careful addition of reactants[89] or by a seeded-growth technique.[90,91] Most commonly, particle sizes can be controlled by varying the precursor concentration, the Fe(III)/oleic acid ratio[92] or the solvents used.[93,94] Single-core iron oxide nanocubes with sizes in the range between 20 nm and 160 nm have been synthetized with Fe(III) acetylacetonate in oleic acid and benzyl ether.[79] Particle shape may be modified from spheres to cubes by adding sodium oleate to the synthesis.[95] Alternative iron precursors include iron pentacarbonyl, leading to smallest NPs,[96] and iron oleate, which allows obtaining very uniform and larger particles.[78]

Nanocrystal shape control has been generally achieved by selective adhesion of surfactant to a particular crystal face and its slow growth along this direction. Without surfactants, NPs can suffer an oriented aggregation into dense or hollow micrometer spheres.[97,98] Uniform anisotropic one-dimensional (1-D) magnetite nanorods with length of from 63 to 140 nm and diameter of ca. 6.5 nm were prepared by solvothermal reaction with iron pentacarbonyl, oleic acid and hexadecylamine as raw materials and 1-octanol as solvent. Oleic acid displaces the carbonyl group to form iron oleate, which could be a precursor of iron oxide. Simultaneously, the easier thermal decomposition of the residual iron pentacarbonyl supplied the growth sites for the incorporation of the magnetite nucleus generated from the iron oleate by hydrolysis driven by the water released by the condensation between oleic acid and hexadecylamine present in the reaction media. The controlled releasing water promotes the formation of 1-D magnetite nanorods. The length of the nanorods could be tuned by changing reaction time and the amount of hexadecylamine.[83] Calorimetric and alternating current (AC) magnetometry experiments performed for the first time on highly crystalline Fe_3O_4 nanorods consistently show large specific absorption rate (SAR) values (862 W/g for an AC field of 800 Oe and

310 kHz), which are superior to spherical and cubic NPs of similar volume (~140 and ~314 W/g, respectively). Increasing the aspect ratio of the nanorods from 6 to 11 improves the SAR by 1.5 times.[99]

In the synthesis of iron oxide NPs by thermal decomposition in organic media, the growth occurs mainly by monomer addition, and the cluster dimension is controlled by the presence of surfactants, usually oleic acid and oleylamine.[100] The importance of the ligands that form the protecting monolayer in controlling the oriented aggregation growth was evidenced by Xue et al.[101] In this frame, the presence of a multidentate ligand on the NP surface would roughen the growing surface, making it prone to aggregation. Good candidates could be calix[n]arenes, synthetic aromatic macrocycles, extensively employed in the host–guest chemistry as platforms for the construction of very efficient and selective hosts.[102] It is also known that they can chelate Fe(III) ions and are able to act as capping agent for iron oxide NPs.[103,104] The *p*-tert-butyl calix[8]arene induces oriented aggregation of iron oxide NPs to obtain homogeneous and monocrystalline magnetite 45-nm nano-octahedra. This process allows a high control degree of product morphology. In particular, stopping the reaction before the aggregation process reaches its completion, it is possible to obtain crystals in the multicore state, with interesting magnetic and hyperthermia properties (Figure 1.8).[105] Application of calixarene derivatives in biotechnology and biomedical researches has been already reviewed.[106]

Although thermal decomposition in organic media has been proven to be an effective approach for preparing uniform, single-crystalline, well-defined and phase pure MNPs, some disadvantages remain, such as the required rigorous and more control of the synthetic conditions, including oxygen-free atmosphere, long reaction time and high reaction temperatures. In addition, many of the reactants and by-products are considered to be toxic. More important, the surfactants that bind to the NP surface render them hydrophobic, and therefore, further steps are needed after the synthesis to stabilize the particles in an aqueous medium for its use in biomedical applications. This transfer to water can lead to irreversible aggregation of the particles.

There are many methods employed to avoid this outcome and that can successfully transfer particles from nonpolar solvents to aqueous media in the form of individual particles

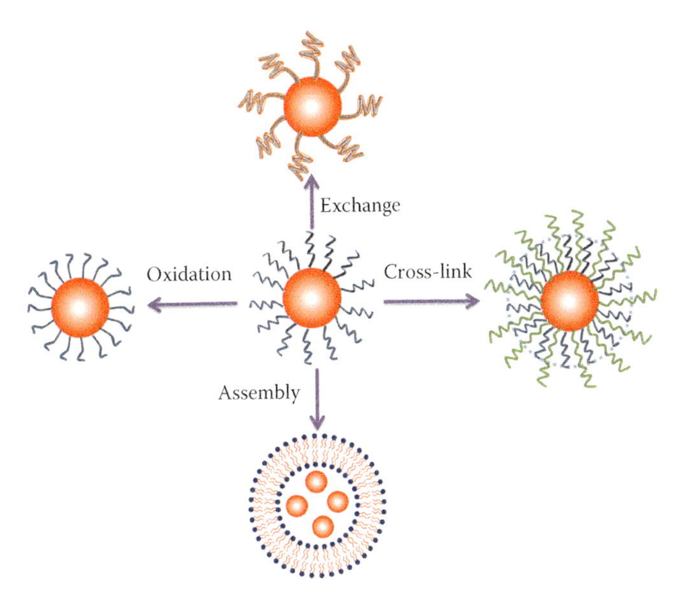

FIGURE 1.9 Different strategies to transfer particles from nonpolar solvents to aqueous media.

using either inorganic or organic coatings (Figure 1.9).[107–110] In this sense, high binding affinity of polyethylene glycol-gallol (PEG-gallol) has been shown and allows freeze drying and redispersion of 9-nm iron oxide cores individually stabilized with approximately 9-nm-thick stealth coatings, yielding particle stability for at least 20 months.[111] PEG-gallol-coated iron oxide nanocubes can be remotely activated with an alternating magnetic field and a near-infrared laser, achieving a very efficient heat conversion at clinical doses.[112]

Also, the transfer process to water has benefits that further functionalities to the NPs can be added. For example, an amphiphilic polymer, poly(maleic anhydride-alt-1-octadecene), was modified with tetramethylrhodamine 5(6)-carboxamide cadaverine and used to obtain a fluorescent magnetic nanosystem that, after further steps of biofunctionalization, offers a great opportunity for the development of active targeting strategies for the early detection and treatment of cancer.[113,114] Using the same approach, different biocompatible tumour-cell-targeting ligands have been investigated to date, such as saccharides, that represent promising molecules for the delivery of such nanoprobes.[115–118] By careful choice of a

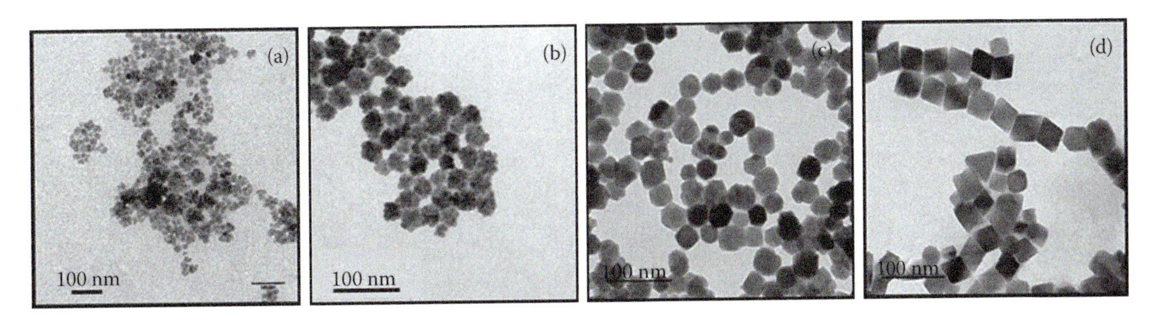

FIGURE 1.8 TEM images showing the evolution of particle aggregation from (a) to (d) with heating for a mixture of iron(III) acetylacetonate and oleic acid in benzyl ether in the presence of calixarenes. (Reproduced with permission of Dr Vita.)

PEG-ylated amphiphilic polymer, it was possible to stabilize hydrophobic magnetite NPs of 27 nm in water with hydrodynamic diameters as low as 54 nm. This colloid was successfully used as a tracer for magnetic particle imaging with a superior performance to Resovist.[119]

Recently, colloidal inorganic NPs have been coated with a thin, cross-linked and functionalized shell containing organic and inorganic layers.[120] The synthesis of the hybrid surface layer takes advantage of the adsorption of amphiphilic polymers to the hydrophobic stabilizing ligands on the colloidal NP surfaces. Commercial poly(styrene-co-maleic anhydride) was adsorbed. Then, the silica precursor, 3-aminopropyltriethoxysilane, was reacted, resulting in a silane being tethered to the polymer. The temperature of the reaction mixture was lowered to further aid this process. The polymerization of the tethered silica precursors occurs resulting in particles that are soluble in dimethyl sulfoxide, dimethylformamide and 0.05 M NaOH.[120]

1.2.3 Polyol Synthesis

The synthesis routes to obtain magnetite NPs mediated by polyols (polyhydric alcohols or etherglycols) are, at present, one of most versatile processes due to the possibility of obtaining particles with very different structures and morphologies in the nanometer size range (Figure 1.10).[121,122] Generally, the polyol is able to dissolve the iron precursor (nitrate, chloride and acetate) due to its high dielectric constant and because of its ability to establish hydrogen bonds. The solution is heated to a given temperature, which can reach the boiling point of the polyol, for a certain period after which the NPs are formed. Among other advantages, the monodispersity of the obtained particles as well as the functionality that the polyol provides (this allows them to remain stable in aqueous media and other polar solvents) are highlighted. Furthermore, great crystallinity is achieved owing to the typical high temperatures used during the synthesis.

The polyol-mediated synthesis was described at first as a novel route for preparing metal powders by reducing the metal ions present in the precursor through the polyols.[122] These can be oxidized to their corresponding aldehydes (Equation 1.5)

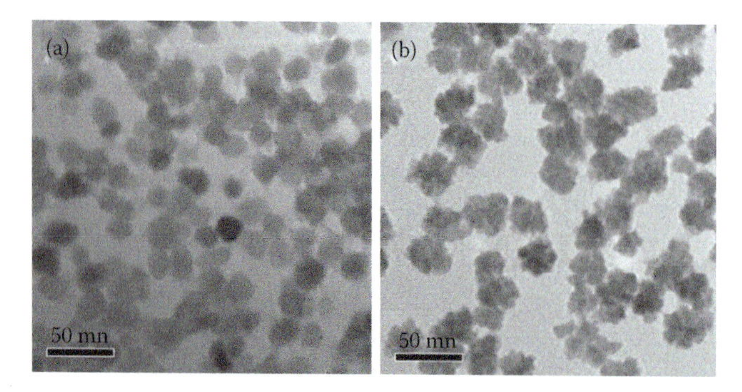

FIGURE 1.10 Representative TEM images of magnetite NPs prepared in a polyol-mediated synthesis by mixing the initial Fe(II) and (III) salts in a DEG/NMDEA (1:1, w/w). (a) single-core particles. (b) Multicore nanoflowers.

and act as mild reducing agents, being capable of reducing some metals easily, such as copper and noble metals.

$$HOCH_2\text{-}CH_2OH + 2/n\, M^{n+} \rightarrow HOCH_2\text{-}CHO + M^{(0)} \quad (1.5)$$

However, in the case of iron and other less reducible metals, the reaction temperature should reach the boiling point of the polyol, resulting in oxidation to the diacetyl, leading to the reduction of the metal. In order to obtain the corresponding metal oxide instead of the metal with zero oxidation state, the presence of oxygen or a certain amount of water is required. Moreover, polyols such as α-diols have chelating properties and act as coordinating solvents, which can form complexes with many metal cations.

Therefore, they can form reactive intermediate species on one hand and, on the other, adsorb onto the surface of the growing particles, preventing aggregation. Iron oxide MNPs (such as magnetite Fe_3O_4 or maghemite γ-Fe_2O_3 phase) have been synthesized by the polyol mediated process.[35] This is due to the capability of producing hydrophilic iron oxide NPs with high magnetization values and zero remanence at room temperature within one step. Some of the above-mentioned morphologies that the crystals can adopt by optimizing the reaction conditions are spherical single-core NPs,[121,123] nanocrystal clusters,[124] compact microspheres,[125] hollow nanospheres[126] and flower-like NPs.[34] In Figure 1.10a, magnetite single-core particles are obtained by mixing the Fe(II) and Fe(III) salts in a 1:1 w/w solution of diethylene glycol (DEG) and *N*-methyl diethanolamine (NMDEA), the addition of a sodium hydroxide solution and a subsequent thermal treatment. By optimizing the thermal treatment that the initial reactants undergo, multicore nanoflowers (Figure 1.10b) can be produced.

These Fe_3O_4 multicore NPs assembled in flower-shape structures have shown enhanced longitudinal and transverse relaxivities for MRI contrast generation and enhanced SAR values for magnetic hyperthermia.[35] Another common polyol-mediated synthesis route starts from an Fe(III) salt and the reaction is performed in an autoclave. In such a way, the reaction solution is held at high temperatures and pressures. This simple route allows obtaining relatively large batches of particles. Water-dispersible iron oxide NPs (6–14 nm) were obtained in such way starting from iron acetylacetonate ($Fe(acac)_3$) and triethylene glycol (TREG).[127] The particles were subsequently coated with carboxylic acid ligands, achieving long-term stability and proving to be effective as MRI contrast agents for applications such as stem cell tracking or cancer cell targeted imaging.

Polyols have diverse physicochemical properties (see Table 1.1) that allow magnetite NPs with different properties to be obtained.

Ethylene glycol (EG) and propylene glycol (PG) were compared as solvents in the reaction. It was found that the morphology evolutions of iron oxide NPs are significantly different in these two polyol processes. Because of their different reductive ability, the formation and growth rate of NPs are different, achieving more aggregated NPs in the case of PG

TABLE 1.1
Principal Physicochemical Properties of Different Polyols and Their Comparison with Water and Ethanol

Solvent	Water	EG	PG	DEG	TREG	Ethanol
ε_r	78.50	38.99	32.0	30.03	19.35	24.30
μ (D)	1.85	2.28	–	2.31	–	1.66
T_{eb} (°C)	100	198	189	245	325	36

Note: ε_r, dielectric constant; μ, electric dipole moment in Debye units; T_{eb}, boiling point.

and resulting in porous single crystals after Ostwald ripening process.[127] Nonetheless, for the cluster formed in EG with a relatively lower formation rate of NPs, most of NPs have enough time to self-assemble along the same orientation in the cluster with retention of its secondary structure (flower-like cluster), owing to the lower surface energy.

Figure 1.11(a)–(f) shows TEM images of flower-shaped magnetite NPs formed by aggregation of primary particles. The TEM images in Figure 1.11 show a reproduced synthesis of magnetite dense and hollow spheres. The hollow spheres (Figure 1.11c and d) are formed by dissolving ferric chloride, polyvinylpyrrolidone (PVP) and an optimized amount of sodium acetate (NaAc) in EG.[126] The mixture is heated in an autoclave and maintained at 200°C for 16 hours. Nanosized hollow magnetite has attracted great interest because of its properties such as low density, selective permeability and large specific area.[126] Dense spheres (Figure 1.11a and b) of 100 nm were obtained using sodium citrate and not PVP as stabilizer.[124] The rest of the experimental conditions and reactants (sodium acetate, EG, and solvothermal crystallization in an autoclave at 200°C for 16 hours) are very similar to those used in the synthesis of hollow spheres. Finally, Figure 1.11e and f shows TEM images of a reproduced synthesis of core/shell structure of magnetite/carbon colloidal NPs with average size about 190 nm prepared via a one-step solvothermal process using ferrocene as a single reactant.[37,128]

Improving the water stability of magnetic NPs produced by the polyol process have been achieved by coating with different hydrophilic polymers, such as poly(hydroxyethyl methacrylate) and poly(methacrylic acid), using atomic transfer radical polymerization.[129] Similar 150-nm superparamagnetic nanocrystal clusters were synthesized by a modified polyol process in the presence of glucose and poly(vinyl-pyrrolidone). At high concentrations due to the thick and uniform coating and size, these systems in presence of magnetic fields form photonic crystals with reflection colours depending on the strength of the magnetic field applied.[130]

On the other hand, manganese-doped iron oxide NP clusters have also been synthesized by a similar approach and showed distinct performance as imaging contrast agents and excellent characteristics as heating mediators in magnetic fluid hyperthermia arising from Mn doping.[131]

The superparamagnetic behaviour, high magnetization and high water dispersibility make nanocrystal clusters or

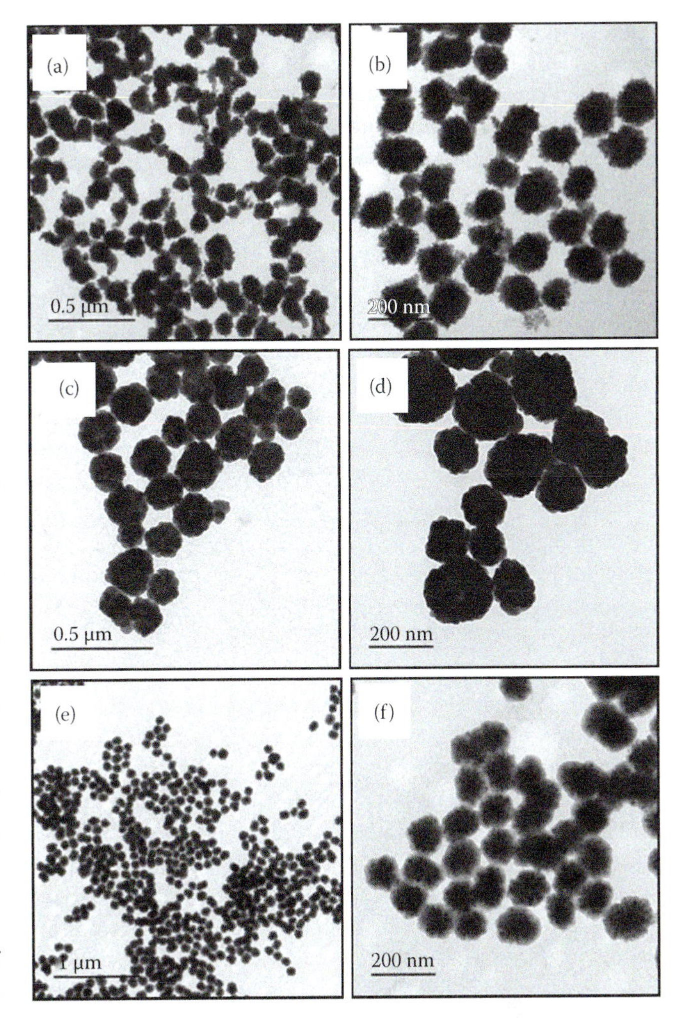

FIGURE 1.11 TEM images of magnetite flower-shaped NPs prepared in a polyol-mediated synthesis from iron (III): (a–b) 100-nm solid spheres. (c–d) 200-nm hollow spheres. (e–f) 100-nm carbon encapsulated flower-like NPs obtained by hydrothermal decomposition of ferrocene in acetone in an autoclave.

flower-shaped NPs below 150 nm ideal candidates for diverse biomedical applications, since they are composed of crystals below 10 nm, and they do not have strong magnetic interactions in dispersion.[132–134] In addition, some of them exhibit low cytotoxicity, good biocompatibility and high capacity for efficient and convenient enrichment of trace peptides.[134] These make them promising candidates for bioapplications in various related fields, such as cell imaging and cell sorting, and for sample preenrichment to analyze trace peptides or proteins in proteomics and in particular those related with diseases and to find biomarkers.[124]

The polyol synthesis of MNPs has two main limitations, the restricted reducing power of polyols and the resulting polyol-functionalized particle surface of the as-prepared particles.

1.2.4 Microwave-Assisted Synthesis

In the past two decades, microwave dielectric heating has gained a lot of attention as a new NP synthesis method.

This is due to its versatile nature in different disciplines, such as polymer chemistry, biomedicine, materials science and nanotechnology. This nonclassical heating method has shown an impressive reduction in synthesis time (from hours to minutes), an increase in product yield and superior material properties as well as reproducibility when compared to the conventional heating (by heat transfer).[135] The first reports on microwave heating in chemistry were published in 1986 by the groups of Gedye and Giguere,[136,137] since the first commercial microwave oven for home use was launched in 1954.

Microwaves are electromagnetic energy with low frequency in the range of 300 to 300,000 MHz. When a sample is irradiated with microwave frequencies, the dipoles tend to align in the direction of the applied electric field. As the field oscillates, the dipoles try to realign along the alternating field streamlines, in such a way energy is lost in the form of heat, through dielectric loss and molecular friction.[138] If the dipole does not have enough time to realign with the applied field, no heating will occur; the same if it reorients too quickly. The frequency of 2450 MHz, corresponding to a wavelength of 12.24 cm, chosen by all commercial systems, is between these two extremes and does not interfere with phone frequencies and telecommunication.

Either the substrate or reagents are likely to be polar, presenting a dielectric property for the reaction, allowing sufficient heating by microwaves. When considering solvents for the microwave reaction, boiling points become less important than in conventional heating under reflux since the pressurized vessels provide reasonable use of solvents with lower boiling points, but the efficiency of the mixture to couple with an applied microwave field becomes an important factor. It can be found in the literature tables of solvents, classifying them by dielectric constant, tan (δ) and dielectric loss,[138] that gives an idea of which solvent is more appropriate for the required NP synthesis. Table 1.2 shows some common organic solvents and their respective values for the dissipative factor $\tan(\delta) = \varepsilon''/\varepsilon'$ (ε'' is the dielectric loss factor related with the energy transformed to heat and ε' is the dielectric constant related with the capacity of the material to store energy in form of electrical potential). If $\tan(\delta) > 0.5$, the solvent is classified as high microwave absorbing; if $0.1 < \tan(\delta) < 0.5$, as medium; and if $\tan(\delta) < 0.1$, as low microwave absorbing.

TABLE 1.2
Dissipative Factors of Some Common Organic Solvents

Solvent	tan(δ)
EG	1.350
Ethanol	0.941
2-Propanol	0.799
Water	0.123
Acetone	0.054
Dichloromethane	0.042
Toluene	0.040
Hexane	0.020

Note: Dissipative factor $\tan(\delta) = \varepsilon''/\varepsilon'$ (ε'' is the dielectric loss factor related with the energy transformed to heat and ε' is the dielectric constant related with the capacity of the material to store energy in form of electrical potential.

Preparation routes for transition-metal oxide nanocrystals and in particular iron oxide NPs rely on nonhydrolytic pathways, in nonaqueous solvents.[139] These routes allow good control over the structure, size and shape of the nanocrystals. Alcohols are classified as high tan(δ) solvents, being convenient as reaction media for this nonaqueous microwave-assisted synthesis, although other solvents like dibenzyl ether or EG have been used.

In organic media, starting from Fe(III)acetylacetonate in benzyl alcohol, Fe_3O_4 nanocrystals of 10 nm were successfully produced by ultrafast reaction, *i.e.* in just a few minutes at 200°C under microwave heating[140] or at 180°C for 10 minutes.[141] Ultrasmall iron oxide NPs of around 3.7 nm with excellent T_1 MRI contrast properties were synthesized using also Fe(III)acetylacetonate in benzyl alcohol, but adding oleylamine and 1,2-dodecanediol.[142] The presence of a polymer like PVP also leads to small iron oxide NPs of around 5 nm whose size can be increased up to 7 nm by changing the heating ramp up to 210°C for 10 hours.[143]

On the contrary, uniform flower-like Fe_3O_4 clusters of a few μm were fabricated in EG with $FeCl_3$, sodium acetate and a surfactant, under microwave irradiation at 160°C for 15 to 60 minutes (Figure 1.12).[144] It was speculated that microwave irradiation set the conditions for creating nanocrystal seeds and accelerated its clustering under the assistance of stabilizers. Using a pressurized microwave reactor, particles in the range of 20 to 130 nm were obtained, dissolving $FeCl_3.6H_2O$ in EG followed by the addition of ammonium acetate, for 30 minutes at 220°C.[145]

Alternatively, room temperature ionic liquids (such as 1-butyl-3-methylimidazolium tetrafluoroborate ([BMIM] $[BF_4]$) are ideal candidates for making a nonpolar solvent

FIGURE 1.12 SEM images of uniform flower-like Fe_3O_4 clusters of a few μm, fabricated in EG with $FeCl_3$ by microwave-assisted synthesis.[144] (Reprinted with permission from Ai, Z.; Deng, K.; Wan, Q.; Zhang, L.; Lee, S., Facile microwave-assisted synthesis and magnetic and gas sensing properties of Fe_3O_4 nanoroses. *The Journal of Physical Chemistry C* 2010, *114* (14), 6237–6242. Copyright 2010 American Chemical Society.)

suitable for microwave heating. Ionic liquids have received a great deal of attention in recent years as novel solvent systems for a range of organic reactions due to their polar nature and attractive properties such as incombustibility, nonvolatility, unique phase behaviour and good solubilizing capacity.[146–148] The ionic character of ionic liquids provides excellent coupling capability with microwave irradiation. Although numerous nanomaterials including iron oxide nanocrystals have been prepared by using ionic liquids as solvents or cosolvents,[140,149,150] only a few reports deal with the synthesis of Fe_3O_4 NPs.[151] To combine the high-temperature solution-phase reaction in an organic solvent like benzyl ether (boiling point = 297°C) and microwave heating, an ionic liquid $[BMIM][BF_4]$ was used in a small proportion (ionic liquid: dibenzyl ether = 1:20 v/v) and magnetite NPs of 6 nm up to 10 nm were obtained in 10 minutes. Benzyl ether is not preferred for use in microwave-assisted high-temperature synthesis due to its substantially lower dielectric constant (e = 3.86). The ionic liquid can be recovered and reused in successive reactions for many times.[151]

In aqueous media, NPs with an average core size from 13 to 17 nm were successfully synthesized by irradiating a mixture of $FeCl_2$ and $FeCl_3$ with sodium carbonate solution, for 10 to 60 minutes at 60°C followed by the addition of citric acid solution. The solution was maintained at 60°C for further 10 to 60 minutes.[152] The particles have a multicore structure with a loose random packing of the constituent core particles. Varying the microwave power from 50 to 300 W or the reaction time from 10 to 60 minutes does not have a major effect on core size. However, the postprecipitation addition of a citric acid solution and microwave treatment lead to a decrease in the particle size from around 150 nm to 50 nm (Figure 1.12). Under similar conditions but starting from Fe(II) salt ($FeSO_4.7H_2O$), spherical NPs with a core size of around 80 nm were synthesized in an alkaline medium at pH 11 in a microwave oven at high power for 1 minute.[153] When starting from $FeCl_3$ and in the presence of citric acid trisodium salt, hydrazine monohydrate is needed. This reaction is allowed to continue in the microwave and stirred at different temperatures from 60°C up to 140°C at 240 W during 10 minutes. It was shown that, contrary to other approaches, citric acid can be incorporated from the beginning to get very small citric-acid-coated NPs that are well dispersed and possess good crystallinity. These properties are mandatory requirements for a good T_1-weighted contrast in MRI.[154]

In general, a higher quality of the microwave-derived NPs was observed in comparison with their convection heated equivalents, regarding increased phase purity, narrower size distribution and lower surface defects. This led to speculation on the so-called 'specific microwave effects.' Using conventional heating, nanocrystals tend to nucleate on the vessel walls first, given its inhomogeneous heating profile. In contrast, microwave produces efficient internal heating, creating numerous 'hot spots,' which could trigger multiple nucleation of seeds throughout the solution, leading to a faster NP development and increasing the product yield. This different nucleation system could explain why NPs synthesized by microwave are smaller than the ones synthesized by conventional heating, as shown in Figure 1.13. Microwave effects on the synthesis are still in debate and are a subject of controversy. It is uncertain whether the unique outcome of NPs irradiated by microwave derives from genuine effects of the dielectric heating mechanisms or by misinterpretation of experimental evidence.[156]

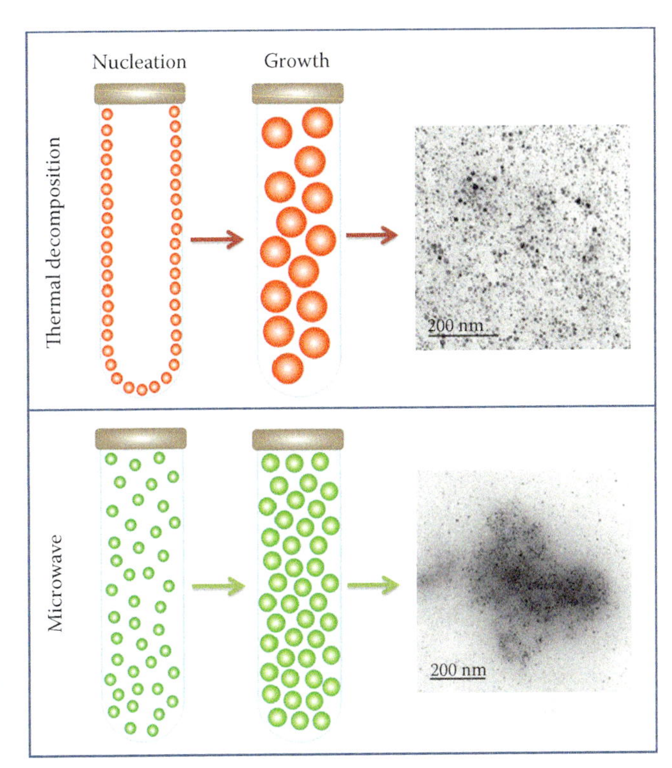

FIGURE 1.13 Different nucleation systems for thermal decomposition and microwave, when the nanocrystals nucleate on the vessel wall first or throughout the entire solution, respectively. The images of TEM show the two syntheses on the same conditions with different heating mechanisms, with mean core sizes of 6 nm and 5 nm, respectively. Temperature gradients in oil-bath heating (up) versus microwave (below) after 1 min of irradiation. The wall temperature in the oil-bath is much higher than the one from microwave, where the whole volume is heated simultaneously.[155]

1.2.5 Electrochemical Synthesis

Anodic oxidation of metal salts in solution or sacrificial metal electrodes allows the preparation of nanometric magnetic materials, in particular iron oxides.[157] Synthetic conditions, such as the nature of the electrolyte, its concentration, pH, current density or potential, play an important role on the powder nature, purity and particle size and distribution. In addition, the synthesis can be carried out in the presence of a surfactant that may provide colloidal stability and active sites at the NP surface for further functionalization.

The advantages of electrochemical synthesis over other methods are the control of the particle size by adjusting experimental conditions such as the imposed electrooxidation current density or potential and the electrolytes present in the reaction media. The generated particles are hydrophilic, and therefore, their surface could be easily modified with biomolecules of interest. The electrochemical synthesis method is an environmentally friendly approach since it is based on an aqueous medium. The yield of this technique is low at approximately 15% because the applied current is also used in other reactions, such as the anodic oxidation of water.[157] Recent works have reported the electrochemical synthesis of Fe_3O_4, $ZnFe_2O_4$, $MnFe_2O_4$ and $CoFe_2O_4$ NPs of different sizes.[158–160] Moderate size and anisotropy of some of these ferrites ensure higher performances for hyperthermia applications. This is the case of cobalt ferrite[161] and manganese ferrite.[159]

Polydopamine-coated magnetite NPs were synthesized in one step by electrooxidation of iron in an aqueous medium in the presence of dopamine.[162] The oxidative conditions and alkaline pH involved in the synthesis favour the self-polymerization of dopamine that adheres at the surface of the MNPs in a simultaneous process. It is shown that the size of the magnetite NPs as well as the polydopamine coating can be controlled by varying the synthetic approach, that is, adding dopamine at the beginning of the electrosynthesis, in the middle or at the end of the process. The particle size of the cores varies between a few nanometres and 25 nm, whilst the shell can reach thicknesses of up to 5 nm.[162]

1.2.6 Other Synthetic Routes

New techniques apart from the chemical routes have been explored for the synthesis of MNPs for biomedicine. In particular, gas aggregation sources offer the possibility to fabricate NPs with controlled size, composition and different structures.[163–165] These nanometric particles are generated in the gas phase and deposited on substrates in vacuum or ultrahigh vacuum conditions. A modified gas aggregation approach allows the one-step generation of well-controlled complex NPs. Thus, it is demonstrated that the atoms of the core and shell of the NPs can be easily inverted, avoiding intrinsic constraints of chemical methods. A fabrication route in ultrahigh vacuum that is compatible with the subsequent dispersion and functionalization of NPs with PEG in aqueous media in one single step was also proposed.[164] The result is the formation of NPs with a structure mainly composed by a metallic iron core and an iron oxide shell, surrounded by a second PEG shell dispersed in aqueous solution. Relaxivity measurements of these PEG functionalized NPs assessed their effectiveness as contrast agents for MRI. Therefore, this new fabrication route is a reliable alternative for the synthesis of NPs for biomedicine, whose toxicity *in vitro* and *in vivo* needs further study.

Multifunctional core–shell NPs consisting of a single-domain metallic Fe core covered with a biocompatible MgO shell can attain a significant increase in the efficiency of magnetic thermal induction compared to conventional superparamagnetic oxides due to interparticle dipolar interactions' substantial influence. Those NPs can be synthesized directly from the gas phase by using a physical vapour deposition technique under inert argon atmosphere. This process has been developed at PROMES facilities in Odeillo-Font Romeu (France) using reactors operating with concentrated sunlight in a solar furnace apparatus.[166] The solar furnace is constituted of a mobile plane mirror that tracks the sun and reflects the radiation on a 2-kW parabolic concentrator (Ø 2 m). The target-material to be melted is placed onto a water-cooled holder in the centre of a glass vacuum chamber. The chamber's pressure is adjusted in the high vacuum region by introducing argon and is maintained by a rotary pump. The target is transferred to the focus of the concentrator and evaporation starts to take place. Shields facilitate the growth rate control by regulating the solar beam. Particles are collected in a cold finger (nanoporous ceramic filter). By this setup, the NP production rate is about 1 g/hour when the beam flux is in the order of 1 kW/m^2. The control of the Fe-to-MgO evaporation ratio and the final particle size was achieved by the variation of the iron-to-magnesium ratio.[166]

MNPs produced via nanoimprint lithography can change the current paradigm of fabrication processes from chemical 'bottom–up' synthesis to 'top–down' fabrication. The combination of controlled nondirectional magnetron sputtering, ethylene tetrafluoroethylene mould, bilayer lift-off and dry etching release can control the shape, size and structure of the fabricated NPs. The resulting MNPs have a novel 'sombrero' shape with complex and unique physical/magnetic properties. Particles around 200–300 nm long by 20–50 nm thick are obtained by this technique.[167,168]

Pulsed laser ablation in liquids has been employed to obtain polymetallic MNPs using a pulsed Nd-Yag laser and a variety of alloys as targets. Due to the presence of liquid, the system affords synthesis and coating in a single step. By changing the target and the solvent, a variety of systems could be obtained.[169] The biomedical applications of the ablation materials are still rare due to the complexity of the equipment required and the polydispersity of the samples, but terbium doped gadolinium oxide obtained using this technique was employed as a dual-contrast agent (fluorescent and T_1 MRI).[170] For the optimized 1% Tb-ion-doped samples, Gd_2O_3:Tb ion is capable of optical labeling, efficient for MRI, and does not cause significant cytotoxic effects.

1.3 PARTICLES' COATING AND POLYMER ENCAPSULATION

In common cases, the synthetic methods described above result in naked MNPs. If biomedical applications are pursued, NPs must be coated with biocompatible molecules or polymers, such as PEG,[171] dextran,[172] chitosan,[173] polyethylenimine[174] and phospholipids.[175] Such coatings stabilize the NPs under physiological conditions. PEG, dextran and chitosan are particularly interesting because they are nontoxic and biocompatible and can prolong NPs lifetime in blood.[176]

Dextran has been widely used to coat superparamagnetic iron oxide NPs because of its polar interactions based on multiple hydrogen bonding that provide high affinity for iron oxide NP surfaces.[177] Many of the commercial ferrofluids are dextran-coated NPs. In the literature, different methods to attach dextran to the NP surface are described; however, covalent bonds instead of hydrogen bonding are preferred for their enhanced stability in physiological conditions.[61,172]

PEG is a biocompatible linear synthetic polyether that can be prepared with a wide range of sizes and terminal functional groups. The uncharged, extremely hydrophilic nature of PEG, combined with its low toxicity and low immunogenicity, renders these PEG-coated NPs 'invisible' to the immune system, making them attractive for biomedical applications.[171] As a result of that, PEG is commonly used in many drug and gene delivery applications.

Chitosan is a unique cationic, hydrophilic polymer that has beneficial properties such as low immunogenicity, excellent biodegradability as well as a high positive charge that easily forms polyelectrolyte complexes with negatively charged entities.[178] Magnetite–chitosan NPs have been obtained by cross-linking chitosan amino groups using glutaraldehyde.[176,179] The disadvantage of this method is the toxicity of this cross-linker. In contrast, ionic gelation (polyionic coacervation) is an interesting technique that uses nontoxic polyanions, such as sodium tripolyphosphate as ionic cross-linker. This procedure is simple and reproducible and NPs are encapsulated in a chitosan shell by ionic interactions.

Various temperature-responsive *N*-isopropylacrylamide-based (NIPAM) functional copolymers have been used for the stabilization of iron oxide NPs.[180] Although this polymer (pNIPAM) is one of the most popular stimulus-responsive polymers for research, it has been demonstrated that the NIPAM monomer is toxic and needs to be completely removed to ensure biocompatibility. Long-term experiments have proved that pNIPAM-coated surfaces were not cytotoxic under some conditions.[181]

When multicore NPs are desired, a common approach is the use of a polymeric matrix to entrap several magnetic cores. Synthetic methods for polymer/NPs hybrid multicore particles, where the polymer is not only intended as a stabilizer for NPs, have been reported in literature and can be divided in to three major approaches: (a) the incorporation of NPs into a forming polymer phase, e.g. polymerization in the presence of the NPs; (b) NP formation from iron salts in an existing polymer particle; and (c) the trapping within a precipitating polymer in the so-called emulsion–solvent–evaporation process (ESE). All of these processes have their advantages and disadvantages.[182]

The ESE process for magnetic particles was first reported to produce hybrid beads in the size range of 125–250 μm. Hamoudeh et al.[183] reported a modified process, yielding magnetite/poly(lactic acid) (PLA) hybrids in the size range between 320 nm and 1.5 μm, based on earlier poly(caprolactone) (PCL) hybrids between 3 and 23 μm.[184] Smaller 90–180 nm hybrid particles can be achieved based on a solvent diffusion process rather than a solvent evaporation process. The major advantage of the ESE process is the wide choice of polymers. Presynthesized polymers can be used, even such that cannot be synthesized in an aqueous environment such as PLA and PCL. Multicore NPs can be synthesised by controlled precipitation within a well-defined oil-in-water emulsion to trap the NPs in a range of polymer matrices of choice, such as poly(styrene), PLA, poly(methyl methacrylate) and PCL.[185] Multicore particles were obtained within the size range of 130 to 340 nm (hydrodynamic diameter determined by dynamic light scattering). With the aim to combine the fast room temperature magnetic relaxation of small individual cores with high magnetization of the ensemble, small-core (<10 nm) NPs were used. The performed synthesis is highly flexible with respect to the choice of polymer and magnetic nanoparticles loading and gives rise to multicore particles with interesting magnetic properties and MRI contrast efficacy.[185]

1.4 FINAL REMARKS

In general, cost-effective, environmentally friendly and large-scale synthesis methods have been pursued keeping good control of size, shape and composition of MNPs, which is a difficult task considering that the difference of only few nanometres in particle size means huge differences in volume, resulting in a functional or failed product.[63] The reproducibility of current synthetic methods, which are able to manufacture high-quality MNPs in large scale, is still a major challenge. On one hand, there is a fundamental and pressing need to develop more sustainable protocols and less toxic nanomaterials in a more efficacious manner.[186] On the other hand, uniformity is a critical point in order to establish the relationships between the physicochemical properties of the nanostructures and their behaviour *in vitro* and *in vivo*, which is currently poorly understood. This is partly due to the complexity of the phenomena of aggregation or transformation that may take place when particles are injected into the body, interact with blood components and furthermore accumulate in different organs.[187–190] Long-term consequences of NPs on human health need further studies in the coming decades.[191] The other problem is the lack of standardization protocols for the characterization of MNP aqueous suspensions for biomedical applications that need to be developed and implemented.

For comparison of synthetic approaches to obtain magnetite NPs, in order to have comparable crystallinity and monodispersity of magnetite, elevated temperature and pressure must be applied, *i.e.* hydrothermal syntheses, using a coprecipitation reaction, a reductive process or an oxidative process. Regarding the thermal decomposition method, it has superior structural properties in terms of controlling size, size distribution and crystallinity. However, organic iron precursors require high temperature in an organic medium containing surfactant stabilizers. This method yields hydrophobic particles stabilized by the surfactants that need further treatments to make them hydrophilic. The polyol method likewise utilizes high-boiling compounds such as EG, DEG or TREG and obtains water dispersible, monodisperse magnetite NPs. Therefore, all these methods require extreme conditions in comparison with biomineralization processes at ambient conditions and, in several cases, the use of harmful organic additives or solvents. Finally, dispersion and stabilization of MNPs in water using nontoxic coatings are important issues and have been the subject of numerous publications[192] and will be discussed more in detail in Chapters 5 and 6 in this book.

ACKNOWLEDGEMENTS

This work was partially supported by the Spanish Ministry of Economy and Competitiveness (Mago project, no. MAT2014-52069-R) and by the European Commission Framework Program 7 (NanoMag project, no. 604448).

REFERENCES

1. Xu, C.; Sun, S., New forms of superparamagnetic nanoparticles for biomedical applications. *Advanced Drug Delivery Reviews* **2013**, *65* (5), 732–743.
2. Bohara, R. A.; Thorat, N. D.; Pawar, S. H., Role of functionalization: Strategies to explore potential nano-bio applications of magnetic nanoparticles. *RSC Advances* **2016**, *6* (50), 43989–44012.
3. Ling, D.; Lee, N.; Hyeon, T., Chemical synthesis and assembly of uniformly sized iron oxide nanoparticles for medical applications. *Accounts of Chemical Research* **2015**, *48* (5), 1276–1285.
4. Blanco-Andujar, C.; Walter, A.; Cotin, G.; Bordeianu, C.; Mertz, D.; Felder-Flesch, D.; Begin-Colin, S., Design of iron oxide-based nanoparticles for MRI and magnetic hyperthermia. *Nanomedicine* **2016**, *11* (14), 1889–1910.
5. Shah, B.; Yin, P. T.; Ghoshal, S.; Lee, K. B., Multimodal magnetic core–shell nanoparticles for effective stem-cell differentiation and imaging. *Angewandte Chemie – International Edition* **2013**, *52* (24), 6190–6195.
6. Wang, Y. X. J.; Idee, J.-M.; Corot, C., Scientific and industrial challenges of developing nanoparticle-based theranostics and multiple-modality contrast agents for clinical application. *Nanoscale* **2015**, *7* (39), 16146–16150.
7. Makridis, A.; Topouridou, K.; Tziomaki, M.; Sakellari, D.; Simeonidis, K.; Angelakeris, M.; Yavropoulou, M. P.; Yovos, J. G.; Kalogirou, O., *In vitro* application of Mn-ferrite nanoparticles as novel magnetic hyperthermia agents. *Journal of Materials Chemistry B* **2014**, *2* (47), 8390–8398.
8. Lee, J. H.; Huh, Y. M.; Jun, Y. W.; Seo, J. W.; Jang, J. T.; Song, H. T.; Kim, S.; Cho, E. J.; Yoon, H. G.; Suh, J. S.; Cheon, J., Artificially engineered magnetic nanoparticles for ultra-sensitive molecular imaging. *Nature Medicine* **2007**, *13* (1), 95–99.
9. Hao, R.; Xing, R.; Xu, Z.; Hou, Y.; Gao, S.; Sun, S., Synthesis, functionalization, and biomedical applications of multifunctional magnetic nanoparticles. *Advanced Materials* **2010**, *22* (25), 2729–2742.
10. Sanpo, N.; Berndt, C. C.; Wen, C.; Wang, J., Transition metal-substituted cobalt ferrite nanoparticles for biomedical applications. *Acta Biomaterialia* **2013**, *9* (3), 5830–5837.
11. Kostevšek, N.; Šturm, S.; Serša, I.; Sepe, A.; Bloemen, M.; Verbiest, T.; Kobe, S.; Rožman, K. Ž., "Single-" and "multi-core" FePt nanoparticles: From controlled synthesis via zwitterionic and silica bio-functionalization to MRI applications. *Journal of Nanoparticle Research* **2015**, *17* (12), 1–15.
12. Lukanov, P.; Anuganti, V. K.; Krupskaya, Y.; Galibert, A. M.; Soula, B.; Tilmaciu, C.; Velders, A. H.; Klingeler, R.; Büchner, B.; Flahaut, E., CCVD synthesis of carbon-encapsulated cobalt nanoparticles for biomedical applications. *Advanced Functional Materials* **2011**, *21* (18), 3583–3588.
13. Branca, M.; Marciello, M.; Ciuculescu-Pradines, D.; Respaud, M.; Morales, M. P.; Serra, R.; Casanove, M. J.; Amiens, C., Towards MRI T2 contrast agents of increased efficiency. *Journal of Magnetism and Magnetic Materials* **2015**, *377*, 348–353.
14. Bruce, I. J.; Sen, T., Surface modification of magnetic nanoparticles with alkoxysilanes and their application in magnetic bioseparations. *Langmuir* **2005**, *21* (15), 7029–7035.
15. Gruar, R. I.; Tighe, C. J.; Southern, P.; Pankhurst, Q. A.; Darr, J. A., A direct and continuous supercritical water process for the synthesis of surface-functionalized nanoparticles. *Industrial & Engineering Chemistry Research* **2015**, *54* (30), 7436–7451.
16. Wei, W.; Zhaohui, W.; Taekyung, Y.; Changzhong, J.; Woo-Sik, K., Recent progress on magnetic iron oxide nanoparticles: Synthesis, surface functional strategies and biomedical applications. *Science and Technology of Advanced Materials* **2015**, *16* (2), 023501.
17. Thanh, N. T. K., *Magnetic nanoparticles: From fabrication to clinical applications: Theory to therapy, chemistry to clinic, bench to bedside*. CRC: Boca Raton, FL, London, 2012.
18. Rebolledo, A. F.; Bomatí-Miguel, O.; Marco, J. F.; Tartaj, P., A facile synthetic route for the preparation of superparamagnetic iron oxide nanorods and nanorices with tunable surface functionality. *Advanced Materials* **2008**, *20* (9), 1760–1765.
19. Kolinko, I.; Lohße, A.; Borg, S.; Raschdorf, O.; Jogler, C.; Tu, Q.; Pósfai, M.; Tompa, É.; Plitzko, J. M.; Brachmann, A.; Wanner, G.; Müller, R.; Zhang, Y.; Schüler, D., Biosynthesis of magnetic nanostructures in a foreign organism by transfer of bacterial magnetosome gene clusters. *Nature Nanotechnology* **2014**, *9* (3), 193–197.
20. Gutiérrez, L.; Costo, R.; Grüttner, C.; Westphal, F.; Gehrke, N.; Heinke, D.; Fornara, A.; Pankhurst, Q. A.; Johansson, C.; Veintemillas-Verdaguer, S.; Morales, M. P., Synthesis methods to prepare single- and multi-core iron oxide nanoparticles for biomedical applications. *Dalton Transactions* **2015**, *44* (7), 2943–2952.
21. Branquinho, L. C.; Carrião, M. S.; Costa, A. S.; Zufelato, N.; Sousa, M. H.; Miotto, R.; Ivkov, R.; Bakuzis, A. F., Effect of magnetic dipolar interactions on nanoparticle heating efficiency: Implications for cancer hyperthermia. *Scientific Reports* **2013**, *3*, 2887.
22. Orozco-Henao, J. M.; Coral, D. F.; Muraca, D.; Moscoso-Londoño, O.; Mendoza Zélis, P.; Fernandez van Raap, M. B.; Sharma, S. K.; Pirota, K. R.; Knobel, M., Effects of nanostructure and dipolar interactions on magnetohyperthermia in iron oxide nanoparticles. *The Journal of Physical Chemistry C* **2016**, *120* (23), 12796–12809.
23. Ludwig, F.; Kazakova, O.; Barquín, L. F.; Fornara, A.; Trahms, L.; Steinhoff, U.; Svedlindh, P.; Wetterskog, E.; Pankhurst, Q. A.; Southern, P.; Morales, M. P.; Hansen, M. F.; Frandsen, C.; Olsson, E.; Gustafsson, S.; Gehrke, N.; Lüdtke-Buzug, K.; Grüttner, C.; Jonasson, C.; Johansson, C., Magnetic, structural, and particle size analysis of single- and multi-core magnetic nanoparticles. *IEEE Transactions on Magnetics* **2014**, *50* (11).
24. De Yoreo, J. J.; Gilbert, P. U. P. A.; Sommerdijk, N. A. J. M.; Penn, R. L.; Whitelam, S.; Joester, D.; Zhang, H.; Rimer, J. D.; Navrotsky, A.; Banfield, J. F.; Wallace, A. F.; Michel, F. M.; Meldrum, F. C.; Cölfen, H.; Dove, P. M., Crystallization by particle attachment in synthetic, biogenic, and geologic environments. *Science* **2015**, *349* (6247).
25. Glotzer, S. C.; Solomon, M. J., Anisotropy of building blocks and their assembly into complex structures. *Nature Materials* **2007**, *6* (8), 557–562.
26. Nie, Z.; Petukhova, A.; Kumacheva, E., Properties and emerging applications of self-assembled structures made from inorganic nanoparticles. *Nature Nanotechnology* **2010**, *5* (1), 15–25.

27. Cheng, K.; Chen, Q.; Wu, Z.; Wang, M.; Wang, H., Colloids of superparamagnetic shell: Synthesis and self-assembly into 3D colloidal crystals with anomalous optical properties. *CrystEngComm* **2011**, *13* (17), 5394–5400.
28. Finney, E. E.; Finke, R. G., Nanocluster nucleation and growth kinetic and mechanistic studies: A review emphasizing transition-metal nanoclusters. *Journal of Colloid and Interface Science* **2008**, *317* (2), 351–374.
29. Frandsen, C.; Legg, B. A.; Comolli, L. R.; Zhang, H.; Gilbert, B.; Johnson, E.; Banfield, J. F., Aggregation-induced growth and transformation of β-FeOOH nanorods to micron-sized α-Fe_2O_3 spindles. *CrystEngComm* **2014**, *16* (8), 1451–1458.
30. Yang, J.; Choi, M. K.; Kim, D. H.; Hyeon, T., Designed assembly and integration of colloidal nanocrystals for device applications. *Advanced Materials* **2016**, *28* (6), 1176–1207.
31. Baumgartner, J.; Dey, A.; Bomans, P. H. H.; Le Coadou, C.; Fratzl, P.; Sommerdijk, N. A. J. M.; Faivre, D., Nucleation and growth of magnetite from solution. *Nature Materials* **2013**, *12* (4), 310–314.
32. Revealed, I.; Yuwono, V. M.; Burrows, N. D.; Soltis, J. A.; Lee Penn, R., Oriented aggregation: Formation and transformation of mesocrystal. *Journal of the American Chemical Society* **2010**, *132* (7), 2163–2165.
33. Vereda, F.; Morales, M. P.; Rodríguez-González, B.; Vicente, J. D.; Hidalgo-Alvarez, R., Control of surface morphology and internal structure in magnetite microparticles: From smooth single crystals to rough polycrystals. *CrystEngComm* **2013**, *15* (26), 5236–5244.
34. Lartigue, L.; Hugounenq, P.; Alloyeau, D.; Clarke, S. P.; Lévy, M.; Bacri, J. C.; Bazzi, R.; Brougham, D. F.; Wilhelm, C.; Gazeau, F., Cooperative organization in iron oxide multicore nanoparticles potentiates their efficiency as heating mediators and MRI contrast agents. *ACS Nano* **2012**, *6* (12), 10935–10949.
35. Hachani, R.; Lowdell, M.; Birchall, M.; Hervault, A.; Mertz, D.; Begin-Colin, S.; Thanh, N. T. B. D. K., Polyol synthesis, functionalisation, and biocompatibility studies of superparamagnetic iron oxide nanoparticles as potential MRI contrast agents. *Nanoscale* **2016**, *8* (6), 3278–3287.
36. Materia, M. E.; Guardia, P.; Sathya, A.; Pernia Leal, M.; Marotta, R.; Di Corato, R.; Pellegrino, T., Mesoscale assemblies of iron oxide nanocubes as heat mediators and image contrast agents. *Langmuir* **2015**, *31* (2), 808–816.
37. Herman, D. A. J.; Cheong-Tilley, S.; McGrath, A. J.; McVey, B. F. P.; Lein, M.; Tilley, R. D., How to choose a precursor for decomposition solution-phase synthesis: The case of iron nanoparticles. *Nanoscale* **2015**, *7* (14), 5951–5954.
38. Ravikumar, C.; Bandyopadhyaya, R., Mechanistic study on magnetite nanoparticle formation by thermal decomposition and coprecipitation routes. *The Journal of Physical Chemistry C* **2011**, *115* (5), 1380–1387.
39. Massart, R., Preparation of aqueous magnetic liquids in alkaline and acidic media. *IEEE Transactions on Magnetics* **1981**, *17* (2), 1247–1248.
40. Gribanov, N. M.; Bibik, E. E.; Buzunov, O. V.; Naumov, V. N., Physico-chemical regularities of obtaining highly dispersed magnetite by the method of chemical condensation. *Journal of Magnetism and Magnetic Materials* **1990**, *85* (1–3), 7–10.
41. Qiu, X. P., Synthesis and characterization of magnetic nanoparticles. *Chinese Journal of Chemistry* **2000**, *18* (6), 834–837.
42. Martínez-Mera, I.; Espinosa-Pesqueira, M. E.; Pérez-Hernández, R.; Arenas-Alatorre, J., Synthesis of magnetite (Fe_3O_4) nanoparticles without surfactants at room temperature. *Materials Letters* **2007**, *61* (23–24), 4447–4451.
43. Iwasaki, T.; Mizutani, N.; Watano, S.; Yanagida, T.; Kawai, T., Size control of magnetite nanoparticles by organic solvent-free chemical coprecipitation at room temperature. *Journal of Experimental Nanoscience* **2010**, *5* (3), 251–262.
44. Costo, R.; Bello, V.; Robic, C.; Port, M.; Marco, J. F.; Morales, M. P.; Veintemillas-Verdaguer, S., Ultrasmall iron oxide nanoparticles for biomedical applications: Improving the colloidal and magnetic properties. *Langmuir* **2012**, *28* (1), 178–185.
45. Milosevic, I.; Warmont, F.; Lalatonne, Y.; Motte, L., Magnetic metrology for iron oxide nanoparticle scaled-up synthesis. *RSC Advances* **2014**, *4* (90), 49086–49089.
46. Che Rose, L.; Bear, J. C.; McNaughter, P. D.; Southern, P.; Piggott, R. B.; Parkin, I. P.; Qi, S.; Mayes, A. G., A SPION-eicosane protective coating for water soluble capsules: Evidence for on-demand drug release triggered by magnetic hyperthermia. *Scientific Reports* **2016**, *6*, 20271.
47. Maurizi, L.; Bouyer, F.; Paris, J.; Demoisson, F.; Saviot, L.; Millot, N., One step continuous hydrothermal synthesis of very fine stabilized superparamagnetic nanoparticles of magnetite. *Chemical Communications* **2011**, *47* (42), 11706–11708.
48. Thomas, G.; Demoisson, F.; Chassagnon, R.; Popova, E.; Millot, N., One-step continuous synthesis of functionalized magnetite nanoflowers. *Nanotechnology* **2016**, *27* (13), 135604.
49. Thomas, G.; Demoisson, F.; Boudon, J.; Millot, N., Efficient functionalization of magnetite nanoparticles with phosphonate using a one-step continuous hydrothermal process. *Dalton Transactions* **2016**, *45* (26), 10821–10829.
50. Heinke, D., Diffusion-controlled synthesis of magnetic nanoparticles. *International Journal on Magnetic Particle Imaging* **2016**.
51. Baumgartner, J.; Antonietta Carillo, M.; Eckes, K. M.; Werner, P.; Faivre, D., Biomimetic magnetite formation: From biocombinatorial approaches to mineralization effects. *Langmuir* **2014**, *30* (8), 2129–2136.
52. Lenders, J. J. M.; Zope, H. R.; Yamagishi, A.; Bomans, P. H. H.; Arakaki, A.; Kros, A.; De With, G.; Sommerdijk, N. A. J. M., Bioinspired magnetite crystallization directed by random copolypeptides. *Advanced Functional Materials* **2015**, *25* (5), 711–719.
53. Park, S., Preparation of iron oxides using ammonium iron citrate precursor: Thin films and nanoparticles. *Journal of Solid State Chemistry* **2009**, *182* (9), 2456–2460.
54. Zheng, Y.-h.; Cheng, Y.; Bao, F.; Wang, Y.-s., Synthesis and magnetic properties of Fe_3O_4 nanoparticles. *Materials Research Bulletin* **2006**, *41* (3), 525–529.
55. Behdadfar, B.; Kermanpur, A.; Sadeghi-Aliabadi, H.; Morales, M. P.; Mozaffari, M., Synthesis of high intrinsic loss power aqueous ferrofluids of iron oxide nanoparticles by citric acid-assisted hydrothermal-reduction route. *Journal of Solid State Chemistry* **2012**, *187*, 20–26.
56. Qu, X.-F.; Zhou, G.-T.; Yao, Q.-Z.; Fu, S.-Q., Aspartic-acid-assisted hydrothermal growth and properties of magnetite octahedrons. *The Journal of Physical Chemistry C* **2010**, *114* (1), 284–289.
57. David, I.; Welch, A. J. E., The oxidation of magnetite and related spinels: Constitution of gamma ferric oxide. *Transactions of the Faraday Society* **1956**, *52*, 1642–1650.
58. Sidhu, P. S.; Gilkes, R. J.; Posner, A. M., Mechanism of the low temperature oxidation of synthetic magnetites. *Journal of Inorganic and Nuclear Chemistry* **1977**, *39* (11), 1953–1958.
59. Sugimoto, T.; Matijević, E., Formation of uniform spherical magnetite particles by crystallization from ferrous hydroxide gels. *Journal of Colloid and Interface Science* **1980**, *74* (1), 227–243.

60. Vergés, M. A.; Costo, R.; Roca, A. G.; Marco, J. F.; Goya, G. F.; Serna, C. J.; Morales, M. P., Uniform and water stable magnetite nanoparticles with diameters around the monodomain-multidomain limit. *Journal of Physics D: Applied Physics* **2008**, *41* (13).
61. Marciello, M.; Connord, V.; Veintemillas-Verdaguer, S.; Vergés, M. A.; Carrey, J.; Respaud, M.; Serna, C. J.; Morales, M. P., Large scale production of biocompatible magnetite nanocrystals with high saturation magnetization values through green aqueous synthesis. *Journal of Materials Chemistry B* **2013**, *1* (43), 5995–6004.
62. Cornell, R. M.; Schwertmann, U., *The iron oxides: Structure, properties, reactions, occurrences and uses.* Weinheim, Germany, Wiley-VCH: 2003.
63. Salas, G.; Veintemillas-Verdaguer, S.; Morales, M. P., Relationship between physico-chemical properties of magnetic fluids and their heating capacity. *International Journal of Hyperthermia* **2013**, *29* (8), 768–776.
64. González-Fernández, M. A.; Torres, T. E.; Andrés-Vergés, M.; Costo, R.; de la Presa, P.; Serna, C. J.; Morales, M. P.; Marquina, C.; Ibarra, M. R.; Goya, G. F., Magnetic nanoparticles for power absorption: Optimizing size, shape and magnetic properties. *Journal of Solid State Chemistry* **2009**, *182* (10), 2779–2784.
65. Luengo, Y.; Morales, M. P.; Gutiérrez, L.; Veintemillas-Verdaguer, S., Counterion and solvent effects on the size of magnetite nanocrystals obtained by oxidative precipitation. *Journal of Materials Chemistry C* **2016**, *4* (40), 9482–9488.
66. Gavilán, H.; Posth, O.; Bogart, L. K.; Steinhoff, U.; Gutiérrez, L.; Morales, M. P., How shape and internal structure affect the magnetic properties of anisometric magnetite nanoparticles. *Acta Materialia* **2017**, *125*, 416–424.
67. Pozas, R.; Ocãa, M.; Morales, M. P.; Serna, C. J., The influence of protective coatings on the magnetic properties of acicular iron nanoparticles. *Nanotechnology* **2006**, *17* (5), 1421–1427.
68. Graf, C.; Vossen, D. L. J.; Imhof, A.; Van Blaaderen, A., A general method to coat colloidal particles with silica. *Langmuir* **2003**, *19* (17), 6693–6700.
69. Morales, M. P.; Pecharroman, C.; Carreñ, T. G.; Serna, C. J., Structural characteristics of uniform γ-Fe2O3 particles with different axial (length/width) ratios. *Journal of Solid State Chemistry* **1994**, *108* (1), 158–163.
70. Yang, Y.; Liu, X.; Ding, J., Synthesis of α-Fe_2O_3 templates via hydrothermal route and Fe_3O_4 particles through subsequent chemical reduction. *Science of Advanced Materials* **2013**, *5* (9), 1199–1207.
71. Rebolledo, A. F.; Laurent, S.; Calero, M.; Villanueva, A.; Knobel, M.; Marco, J. F.; Tartaj, P., Iron oxide nanosized clusters embedded in porous nanorods: A new colloidal design to enhance capabilities of MRI contrast agents. *ACS Nano* **2010**, *4* (4), 2095–2103.
72. Mirabello, G.; Lenders, J. J. M.; Sommerdijk, N. A. J. M., Bioinspired synthesis of magnetite nanoparticles. *Chemical Society Reviews* **2016**, *45* (18), 5085–5106.
73. Lenders, J. J. M.; Mirabello, G.; Sommerdijk, N. A. J. M., Bioinspired magnetite synthesis via solid precursor phases. *Chemical Science* **2016**, *7* (9), 5624–5634.
74. Mann, S., *Biomineralization: Principles and concepts in bioinorganic materials chemistry.* New York, Oxford University Press: 2001; Vol. 5.
75. Dey, A.; Lenders, J. J. M.; Sommerdijk, N. A. J. M., Bioinspired magnetite formation from a disordered ferrihydrite-derived precursor. *Faraday Discussions* **2015**, *179*, 215–225.
76. Hyeon, T.; Su Seong, L.; Park, J.; Chung, Y.; Hyon Bin, N., Synthesis of highly crystalline and monodisperse maghemite nanocrystallites without a size-selection process. *Journal of the American Chemical Society* **2001**, *123* (51), 12798–12801.
77. Kwon, S. G.; Hyeon, T., Formation mechanisms of uniform nanocrystals via hot-injection and heat-up methods. *Small* **2011**, *7* (19), 2685–2702.
78. Salas, G.; Casado, C.; Teran, F. J.; Miranda, R.; Serna, C. J.; Morales, M. P., Controlled synthesis of uniform magnetite nanocrystals with high-quality properties for biomedical applications. *Journal of Materials Chemistry* **2012**, *22* (39), 21065–21075.
79. Kim, D.; Lee, N.; Park, M.; Kim, B. H.; An, K.; Hyeon, T., Synthesis of uniform ferrimagnetic magnetite nanocubes. *Journal of the American Chemical Society* **2009**, *131* (2), 454–455.
80. Guardia, P.; Di Corato, R.; Lartigue, L.; Wilhelm, C.; Espinosa, A.; Garcia-Hernandez, M.; Gazeau, F.; Manna, L.; Pellegrino, T., Water-soluble iron oxide nanocubes with high values of specific absorption rate for cancer cell hyperthermia treatment. *ACS Nano* **2012**, *6* (4), 3080–3091.
81. Kovalenko, M. V.; Bodnarchuk, M. I.; Lechner, R. T.; Hesser, G.; Schäffler, F.; Heiss, W., Fatty acid salts as stabilizers in size- and shape-controlled nanocrystal synthesis: The case of inverse spinel iron oxide. *Journal of the American Chemical Society* **2007**, *129* (20), 6352–6353.
82. Palchoudhury, S.; An, W.; Xu, Y.; Qin, Y.; Zhang, Z.; Chopra, N.; Holler, R. A.; Turner, C. H.; Bao, Y., Synthesis and growth mechanism of iron oxide nanowhiskers. *Nano Letters* **2011**, *11* (3), 1141–1146.
83. Sun, H.; Chen, B.; Jiao, X.; Jiang, Z.; Qin, Z.; Chen, D., Solvothermal synthesis of tunable electroactive magnetite nanorods by controlling the side reaction. *Journal of Physical Chemistry C* **2012**, *116* (9), 5476–5481.
84. Zhang, L.; Wu, J.; Liao, H.; Hou, Y.; Gao, S., Octahedral Fe_3O_4 nanoparticles and their assembled structures. *Chemical Communications* **2009**, *0* (29), 4378–4380.
85. Jian, L.; Xiuling, J.; Dairong, C.; Wei, L., Solvothermal synthesis and characterization of Fe_3O_4 and γ-Fe_2O_3 nanoplates. *Journal of Physical Chemistry C* **2009**, *113* (10), 4012–4017.
86. Zeng, Y.; Hao, R.; Xing, B.; Hou, Y.; Xu, Z., One-pot synthesis of Fe_3O_4 nanoprisms with controlled electrochemical properties. *Chemical Communications* **2010**, *46* (22), 3920–3922.
87. Fantechi, E.; Campo, G.; Carta, D.; Corrias, A.; de Julián Fernández, C.; Gatteschi, D.; Innocenti, C.; Pineider, F.; Rugi, F.; Sangregorio, C., Exploring the effect of Co doping in fine maghemite nanoparticles. *The Journal of Physical Chemistry C* **2012**, *116* (14), 8261–8270.
88. Brollo, M. E. F.; López-Ruiz, R.; Muraca, D.; Figueroa, S. J. A.; Pirota, K. R.; Knobel, M., Compact Ag@Fe_3O_4 core–shell nanoparticles by means of single-step thermal decomposition reaction. *Scientific Reports* **2014**, *4*, 6839.
89. Vreeland, E. C.; Watt, J.; Schober, G. B.; Hance, B. G.; Austin, M. J.; Price, A. D.; Fellows, B. D.; Monson, T. C.; Hudak, N. S.; Maldonado-Camargo, L.; Bohorquez, A. C.; Rinaldi, C.; Huber, D. L., Enhanced nanoparticle size control by extending LaMer's mechanism. *Chemistry of Materials* **2015**, *27* (17), 6059–6066.
90. Castellanos-Rubio, I.; Insausti, M.; Garaio, E.; Gil de Muro, I.; Plazaola, F.; Rojo, T.; Lezama, L., Fe_3O_4 nanoparticles prepared by the seeded-growth route for hyperthermia: Electron magnetic resonance as a key tool to evaluate size distribution in magnetic nanoparticles. *Nanoscale* **2014**, *6* (13), 7542–7552.

91. Salado, J.; Insausti, M.; Lezama, L.; Gil de Muro, I.; Goikolea, E.; Rojo, T., Preparation and Characterization of monodisperse Fe_3O_4 nanoparticles: An electron magnetic resonance study. *Chemistry of Materials* **2011**, *23* (11), 2879–2885.
92. Roca, A. G.; Marco, J. F.; Morales, M. P.; Serna, C. J., Effect of nature and particle size on properties of uniform magnetite and maghemite nanoparticles. *Journal of Physical Chemistry C* **2007**, *111* (50), 18577–18584.
93. Park, J.; Lee, E.; Hwang, N. M.; Kang, M.; Sung, C. K.; Hwang, Y.; Park, J. G.; Noh, H. J.; Kim, J. Y.; Park, J. H.; Hyeon, T., One-nanometer-scale size-controlled synthesis of monodisperse magnetic iron oxide nanoparticles. *Angewandte Chemie – International Edition* **2005**, *44* (19), 2872–2877.
94. Baaziz, W.; Pichon, B. P.; Fleutot, S.; Liu, Y.; Lefevre, C.; Greneche, J. M.; Toumi, M.; Mhiri, T.; Begin-Colin, S., Magnetic iron oxide nanoparticles: Reproducible tuning of the size and nanosized-dependent composition, defects, and spin canting. *Journal of Physical Chemistry C* **2014**, *118* (7), 3795–3810.
95. Wetterskog, E.; Agthe, M.; Mayence, A.; Grins, J.; Wang, D.; Rana, S.; Ahniyaz, A.; Salazar-Alvarez, G.; Bergström, L., Precise control over shape and size of iron oxide nanocrystals suitable for assembly into ordered particle arrays. *Science and Technology of Advanced Materials* **2014**, *15* (5), 055010.
96. Kim, B. H.; Hackett, M. J.; Park, J.; Hyeon, T., Synthesis, characterization, and application of ultrasmall nanoparticles. *Chemistry of Materials* **2014**, *26* (1), 59–71.
97. Guan, N.; Wang, Y.; Sun, D.; Xu, J., A simple one-pot synthesis of single-crystalline magnetite hollow spheres from a single iron precursor. *Nanotechnology* **2009**, *20* (10), 105603.
98. Jia, B.; Gao, L., Morphological transformation of Fe3O4 spherical aggregates from solid to hollow and their self-assembly under an external magnetic field. *Journal of Physical Chemistry C* **2008**, *112* (3), 666–671.
99. Das, R.; Alonso, J.; Nemati Porshokouh, Z.; Kalappattil, V.; Torres, D.; Phan, M. H.; Garaio, E.; García, J. A.; Sanchez Llamazares, J. L.; Srikanth, H., Tunable high aspect ratio iron oxide nanorods for enhanced hyperthermia. *Journal of Physical Chemistry C* **2016**, *120* (18), 10086–10093.
100. Sun, S.; Zeng, H.; Robinson, D. B.; Raoux, S.; Rice, P. M.; Wang, S. X.; Li, G., Monodisperse MFe_2O_4 (M = Fe, Co, Mn) Nanoparticles. *Journal of the American Chemical Society* **2004**, *126* (1), 273–279.
101. Xue, X.; Penn, R. L.; Leite, E. R.; Huang, F.; Lin, Z., Crystal growth by oriented attachment: Kinetic models and control factors. *CrystEngComm* **2014**, *16* (8), 1419–1429.
102. Arbaoui, A.; Redshaw, C.; Elsegood, M. R. J.; Wright, V. E.; Yoshizawa, A.; Yamato, T., Iron(III) and Zinc(II) calixarene complexes: Synthesis, structural studies, and use as procatalysts for ε-caprolactone polymerization. *Chemistry – An Asian Journal* **2010**, *5* (3), 621–633.
103. Ben-Ishay, M. L.; Gedanken, A., Difference in the bonding scheme of calix(6)arene and p-sulfonic calix(6)arene to nanoparticles of Fe_2O_3 and Fe_3O_4. *Langmuir* **2007**, *23* (10), 5238–5242.
104. Khan, L. U.; Brito, H. F.; Hölsä, J.; Pirota, K. R.; Muraca, D.; Felinto, M. C. F. C.; Teotonio, E. E. S.; Malta, O. L., Red-green emitting and superparamagnetic nanomarkers containing Fe_3O_4 functionalized with calixarene and rare earth complexes. *Inorganic Chemistry* **2014**, *53* (24), 12902–12910.
105. Vita, F. V.; Gavilán, H.; Rossi, F.; De Julian Fernandez, C.; Secchi, A.; Arduini, A.; Albertini, F.; Morales, M. P., Tuning morphology and magnetism of magnetite nanoparticles by calixarene-induced oriented aggregation. *CrysEngComm* 2016, 18, 8591–8598.
106. Roman, V. R.; Vyacheslav, I. B.; Vitaly, I. K., Calixarenes in bio-medical researches. *Current Medicinal Chemistry* **2009**, *16* (13), 1630–1655.
107. Palma, S. I. C. J.; Marciello, M.; Carvalho, A.; Veintemillas-Verdaguer, S.; Morales, M. P.; Roque, A. C. A., Effects of phase transfer ligands on monodisperse iron oxide magnetic nanoparticles. *Journal of Colloid and Interface Science* **2015**, *437*, 147–155.
108. Roca, A. G.; Carmona, D.; Miguel-Sancho, N.; Bomatí-Miguel, O.; Balas, F.; Piquer, C.; Santamaría, J., Surface functionalization for tailoring the aggregation and magnetic behaviour of silica-coated iron oxide nanostructures. *Nanotechnology* **2012**, *23* (15), 155603.
109. Fratila, R. M.; Moros, M.; de la Fuente, J. M., Recent advances in biosensing using magnetic glyconanoparticles. *Analytical and Bioanalytical Chemistry* **2016**, *408* (7), 1783–1803.
110. Branca, M.; Ibrahim, M.; Ciuculescu, D.; Philippot, K.; Amiens, C., Water transfer of hydrophobic nanoparticles: Principles and methods. In *Handbook of nanoparticles*, Aliofkhazraei, M., Ed. Cham, Switzerland, Springer International Publishing: 2015; pp 1–26.
111. Amstad, E.; Zurcher, S.; Mashaghi, A.; Wong, J. Y.; Textor, M.; Reimhult, E., Surface functionalization of single superparamagnetic iron oxide nanoparticles for targeted magnetic resonance imaging. *Small* **2009**, *5* (11), 1334–1342.
112. Espinosa, A.; Di Corato, R.; Kolosnjaj-Tabi, J.; Flaud, P.; Pellegrino, T.; Wilhelm, C., Duality of iron oxide nanoparticles in cancer therapy: Amplification of heating efficiency by magnetic hyperthermia and photothermal bimodal treatment. *ACS Nano* **2016**, *10* (2), 2436–2446.
113. Dias, A. M. G. C.; Hussain, A.; Marcos, A. S.; Roque, A. C. A., A biotechnological perspective on the application of iron oxide magnetic colloids modified with polysaccharides. *Biotechnology Advances* **2011**, *29* (1), 142–155.
114. Moros, M.; Delhaes, F.; Puertas, S.; Saez, B.; de la Fuente, J. M.; Grazú, V.; Feracci, H., Surface engineered magnetic nanoparticles for specific immunotargeting of cadherin expressing cells. *Journal of Physics D: Applied Physics* **2015**, *49* (5), 054003.
115. El-Boubbou, K.; Zhu, D. C.; Vasileiou, C.; Borhan, B.; Prosperi, D.; Li, W.; Huang, X., Magnetic glyco-nanoparticles: A tool to detect, differentiate, and unlock the glyco-codes of cancer via magnetic resonance imaging. *Journal of the American Chemical Society* **2010**, *132* (12), 4490–4499.
116. Lartigue, L.; Oumzil, K.; Guari, Y.; Larionova, J.; Guérin, C.; Montero, J.-L.; Barragan-Montero, V.; Sangregorio, C.; Caneschi, A.; Innocenti, C.; Kalaivani, T.; Arosio, P.; Lascialfari, A., Water-soluble rhamnose-coated Fe_3O_4 nanoparticles. *Organic Letters* **2009**, *11* (14), 2992–2995.
117. Lartigue, L.; Innocenti, C.; Kalaivani, T.; Awwad, A.; Sanchez Duque, M. d. M.; Guari, Y.; Larionova, J.; Guérin, C.; Montero, J.-L. G.; Barragan-Montero, V.; Arosio, P.; Lascialfari, A.; Gatteschi, D.; Sangregorio, C., Water-dispersible sugar-coated iron oxide nanoparticles: An evaluation of their relaxometric and magnetic hyperthermia properties. *Journal of the American Chemical Society* **2011**, *133* (27), 10459–10472.
118. Palma, S. I. C. J.; Rodrigues, C. A. V.; Carvalho, A.; Morales, M. P.; Freitas, F.; Fernandes, A. R.; Cabral, J. M. S.; Roque, A. C. A., A value-added exopolysaccharide as a coating agent for MRI nanoprobes. *Nanoscale* **2015**, *7* (34), 14272–14283.
119. Ferguson, R. M.; Khandhar, A. P.; Kemp, S. J.; Arami, H.; Saritas, E. U.; Croft, L. R.; Konkle, J.; Goodwill, P. W.; Halkola, A.; Rahmer, J.; Borgert, J.; Conolly, S. M.;

Krishnan, K. M., Magnetic particle imaging with tailored iron oxide nanoparticle tracers. *IEEE Transactions on Medical Imaging* **2015**, *34* (5), 1077–1084.

120. McNaughter, P. D.; Bear, J. C.; Steytler, D. C.; Mayes, A. G.; Nann, T., A thin silica–polymer shell for functionalizing colloidal inorganic nanoparticles. *Angewandte Chemie – International Edition* **2011**, *50* (44), 10384–10387.
121. Cai, W.; Wan, J., Facile synthesis of superparamagnetic magnetite nanoparticles in liquid polyols. *Journal of Colloid and Interface Science* **2007**, *305* (2), 366–370.
122. Fievet, F.; Lagier, J. P.; Blin, B.; Beaudoin, B.; Figlarz, M., Homogeneous and heterogeneous nucleations in the polyol process for the preparation of micron and submicron size metal particles. *Solid State Ionics* **1989**, *32*, 198–205.
123. Daniela, C.; Gabriel, C.; Charles, J. O. C., Magnetic properties of variable-sized Fe_3O_4 nanoparticles synthesized from non-aqueous homogeneous solutions of polyols. *Journal of Physics D: Applied Physics* **2007**, *40* (19), 5801.
124. Liu, J.; Sun, Z.; Deng, Y.; Zou, Y.; Li, C.; Guo, X.; Xiong, L.; Gao, Y.; Li, F.; Zhao, D., Highly water-dispersible biocompatible magnetite particles with low cytotoxicity stabilized by citrate groups. *Angewandte Chemie – International Edition* **2009**, *48* (32), 5875–5879.
125. Liang, J.; Ma, H.; Luo, W.; Wang, S., Synthesis of magnetite submicrospheres with tunable size and superparamagnetism by a facile polyol process. *Materials Chemistry and Physics* **2013**, *139* (2–3), 383–388.
126. Sun, Q.; Ren, Z.; Wang, R.; Chen, W.; Chen, C., Magnetite hollow spheres: Solution synthesis, phase formation and magnetic property. *Journal of Nanoparticle Research* **2011**, *13* (1), 213–220.
127. Cheng, C.; Xu, F.; Gu, H., Facile synthesis and morphology evolution of magnetic iron oxide nanoparticles in different polyol processes. *New Journal of Chemistry* **2011**, *35* (5), 1072–1079.
128. Wang, H.; Sun, Y.-B.; Chen, Q.-W.; Yu, Y.-F.; Cheng, K., Synthesis of carbon-encapsulated superparamagnetic colloidal nanoparticles with magnetic-responsive photonic crystal property. *Dalton Transactions* **2010**, *39* (40), 9565–9569.
129. Basti, H.; Tahar, L. B.; Smiri, L. S.; Herbst, F.; Nowak, S.; Mangeney, C.; Ammar, S., Surface modification of γ-Fe_2O_3 nanoparticles by grafting from poly-(hydroxyethylmethacrylate) and poly-(methacrylic acid): Qualitative and quantitative analysis of the polymeric coating. *Colloids and Surfaces A: Physicochemical and Engineering Aspects* **2016**, *490*, 222–231.
130. Luo, W.; Ma, H.; Mou, F.; Zhu, M.; Yan, J.; Guan, J., Steric-repulsion-based magnetically responsive photonic crystals. *Advanced Materials* **2014**, *26* (7), 1058–1064.
131. Casula, M. F.; Conca, E.; Bakaimi, I.; Sathya, A.; Materia, M. E.; Casu, A.; Falqui, A.; Sogne, E.; Pellegrino, T.; Kanaras, A. G., Manganese doped-iron oxide nanoparticle clusters and their potential as agents for magnetic resonance imaging and hyperthermia. *Physical Chemistry Chemical Physics* **2016**, *18* (25), 16848–16855.
132. Park, J.; Joo, J.; Soon, G. K.; Jang, Y.; Hyeon, T., Synthesis of monodisperse spherical nanocrystals. *Angewandte Chemie – International Edition* **2007**, *46* (25), 4630–4660.
133. Gu, H.; Xu, K.; Xu, C.; Xu, B., Biofunctional magnetic nanoparticles for protein separation and pathogen detection. *Chemical Communications* **2006**, *0* (9), 941–949.
134. Jun, Y.-w.; Huh, Y.-M.; Choi, J.-s.; Lee, J.-H.; Song, H.-T.; KimKim; Yoon, S.; Kim, K.-S.; Shin, J.-S.; Suh, J.-S.; Cheon, J., Nanoscale size effect of magnetic nanocrystals and their utilization for cancer diagnosis via magnetic resonance imaging. *Journal of the American Chemical Society* **2005**, *127* (16), 5732–5733.
135. Baghbanzadeh, M.; Carbone, L.; Cozzoli, P. D.; Kappe, C. O., Microwave-assisted synthesis of colloidal inorganic nanocrystals. *Angewandte Chemie – International Edition* **2011**, *50* (48), 11312–11359.
136. Gedye, R.; Smith, F.; Westaway, K.; Ali, H.; Baldisera, L.; Laberge, L.; Rousell, J., The use of microwave ovens for rapid organic synthesis. *Tetrahedron Letters* **1986**, *27* (3), 279–282.
137. Giguere, R. J.; Bray, T. L.; Duncan, S. M.; Majetich, G., Application of commercial microwave ovens to organic synthesis. *Tetrahedron Letters* **1986**, *27* (41), 4945–4948.
138. Rana, K. K.; Rana, S., Microwave reactors: A brief review on its fundamental aspects and applications. *Open Access Library Journal* **2014**, *1* (06), 1.
139. Niederberger, M.; Pinna, N., *Metal oxide nanoparticles in organic solvents: Synthesis, formation, assembly and application.* Heidelberg, Germany, Springer Science & Business Media: 2009.
140. Bilecka, I.; Djerdj, I.; Niederberger, M., One-minute synthesis of crystalline binary and ternary metal oxide nanoparticles. *Chemical Communications* **2008**, *0* (7), 886–888.
141. Carenza, E.; Barceló, V.; Morancho, A.; Montaner, J.; Rosell, A.; Roig, A., Rapid synthesis of water-dispersible superparamagnetic iron oxide nanoparticles by a microwave-assisted route for safe labeling of endothelial progenitor cells. *Acta Biomaterialia* **2014**, *10* (8), 3775–3785.
142. Pellico, J.; Lechuga-Vieco, A. V.; Benito, M.; García-Segura, J. M.; Fuster, V.; Ruiz-Cabello, J.; Herranz, F., Microwave-driven synthesis of bisphosphonate nanoparticles allows *in vivo* visualisation of atherosclerotic plaque. *RSC Advances* **2015**, *5* (3), 1661–1665.
143. Yu, S.; Hachtel, J. A.; Chisholm, M. F.; Pantelides, S. T.; Laromaine, A.; Roig, A., Magnetic gold nanotriangles by microwave-assisted polyol synthesis. *Nanoscale* **2015**, *7* (33), 14039–14046.
144. Ai, Z.; Deng, K.; Wan, Q.; Zhang, L.; Lee, S., Facile microwave-assisted synthesis and magnetic and gas sensing properties of Fe_3O_4 nanoroses. *The Journal of Physical Chemistry C* **2010**, *114* (14), 6237–6242.
145. Kozakova, Z.; Kuritka, I.; Kazantseva, N. E.; Babayan, V.; Pastorek, M.; Machovsky, M.; Bazant, P.; Saha, P., The formation mechanism of iron oxide nanoparticles within the microwave-assisted solvothermal synthesis and its correlation with the structural and magnetic properties. *Dalton Transactions* **2015**, *44* (48), 21099–21108.
146. Wilkes, J. S., A short history of ionic liquids-from molten salts to neoteric solvents. *Green Chemistry* **2002**, *4* (2), 73–80.
147. Rogers, R. D.; Seddon, K. R., Ionic liquids – solvents of the future? *Science* **2003**, *302* (5646), 792.
148. Chiappe, C.; Pieraccini, D., Ionic liquids: Solvent properties and organic reactivity. *Journal of Physical Organic Chemistry* **2005**, *18* (4), 275–297.
149. Wasserscheid, P.; Keim, W., Ionic Liquids – New “solutions” for transition metal catalysis. *Angewandte Chemie – International Edition* **2000**, *39* (21), 3772–3789.
150. Jacob, D. S.; Bitton, L.; Grinblat, J.; Felner, I.; Koltypin, Y.; Gedanken, A., Are ionic liquids really a boon for the synthesis of inorganic materials? A general method for the fabrication of nanosized metal fluorides. *Chemistry of Materials* **2006**, *18* (13), 3162–3168.
151. Hu, H.; Yang, H.; Huang, P.; Cui, D.; Peng, Y.; Zhang, J.; Lu, F.; Lian, J.; Shi, D., Unique role of ionic liquid in microwave-assisted synthesis of monodisperse magnetite nanoparticles. *Chemical Communications* **2010**, *46* (22), 3866–3868.

152. Blanco-Andujar, C.; Ortega, D.; Southern, P.; Pankhurst, Q. A.; Thanh, N. T. K., High performance multi-core iron oxide nanoparticles for magnetic hyperthermia: Microwave synthesis, and the role of core-to-core interactions. *Nanoscale* **2015**, *7* (5), 1768–1775.
153. Zheng, B.; Zhang, M.; Xiao, D.; Jin, Y.; Choi, M. M. F., Fast microwave synthesis of Fe_3O_4 and Fe_3O_4/Ag magnetic nanoparticles using Fe^{2+} as precursor. *Inorganic Materials* **2010**, *46* (10), 1106–1111.
154. Pellico, J.; Ruiz-Cabello, J.; Saiz-Alía, M.; del Rosario, G.; Caja, S.; Montoya, M.; Fernández de Manuel, L.; Morales, M. P.; Gutiérrez, L.; Galiana, B.; Enríquez, J. A.; Herranz, F., Fast synthesis and bioconjugation of 68Ga core-doped extremely small iron oxide nanoparticles for PET/MR imaging. *Contrast Media & Molecular Imaging* **2016**, *11* (3), 203–210.
155. Schanche, J.-S., Microwave synthesis solutions from personal chemistry. *Molecular Diversity* **2003**, *7* (2), 291–298.
156. Obermayer, D.; Kappe, C. O., On the importance of simultaneous infrared/fiber-optic temperature monitoring in the microwave-assisted synthesis of ionic liquids. *Organic & Biomolecular Chemistry* **2010**, *8* (1), 114–121.
157. Cabrera, L.; Gutierrez, S.; Menendez, N.; Morales, M. P.; Herrasti, P., Magnetite nanoparticles: Electrochemical synthesis and characterization. *Electrochimica Acta* **2008**, *53* (8), 3436–3441.
158. Mazarío, E.; Herrasti, P.; Morales, M. P.; Menéndez, N., Synthesis and characterization of $CoFe_2O_4$ ferrite nanoparticles obtained by an electrochemical method. *Nanotechnology* **2012**, *23* (35), 355708.
159. Mazarío, E.; Sánchez-Marcos, J.; Menéndez, N.; Cañete, M.; Mayoral, A.; Rivera-Fernández, S.; de la Fuente, J. M.; Herrasti, P., High specific absorption rate and transverse relaxivity effects in manganese ferrite nanoparticles obtained by an electrochemical route. *The Journal of Physical Chemistry C* **2015**, *119* (12), 6828–6834.
160. Rivero, M.; del Campo, A.; Mayoral, A.; Mazario, E.; Sanchez-Marcos, J.; Munoz-Bonilla, A., Synthesis and structural characterization of $ZnxFe_3$-xO_4 ferrite nanoparticles obtained by an electrochemical method. *RSC Advances* **2016**, *6* (46), 40067–40076.
161. Mazario, E.; Menéndez, N.; Herrasti, P.; Cañete, M.; Connord, V.; Carrey, J., Magnetic hyperthermia properties of electrosynthesized cobalt ferrite nanoparticles. *The Journal of Physical Chemistry C* **2013**, *117* (21), 11405–11411.
162. Mazario, E.; Sanchez-Marcos, J.; Menendez, N.; Herrasti, P.; Garcia-Hernandez, M.; Munoz-Bonilla, A., One-pot electrochemical synthesis of polydopamine coated magnetite nanoparticles. *RSC Advances* **2014**, *4* (89), 48353–48361.
163. Llamosa, D.; Ruano, M.; Martinez, L.; Mayoral, A.; Roman, E.; Garcia-Hernandez, M.; Huttel, Y., The ultimate step towards a tailored engineering of core@shell and core@shell@shell nanoparticles. *Nanoscale* **2014**, *6* (22), 13483–13486.
164. Oprea, B.; Martínez, L.; Román, E.; Vanea, E.; Simon, S.; Huttel, Y., Dispersion and functionalization of nanoparticles synthesized by gas aggregation source: Opening new routes toward the fabrication of nanoparticles for biomedicine. *Langmuir* **2015**, *31* (51), 13813–13820.
165. Llamosa, D.; Ruano, M.; Martinez, L.; Mayoral, A.; Roman, E.; Garcia-Hernandez, M.; Huttel, Y., The ultimate step towards a tailored engineering of core@shell and core@shell@shell nanoparticles. *Nanoscale* **2014**, *6* (22), 13483–13486.
166. Martinez-Boubeta, C.; Simeonidis, K.; Serantes, D.; Conde-Leborán, I.; Kazakis, I.; Stefanou, G.; Peña, L.; Galceran, R.; Balcells, L.; Monty, C.; Baldomir, D.; Mitrakas, M.; Angelakeris, M., adjustable hyperthermia response of self-assembled ferromagnetic Fe-MgO core–shell nanoparticles by tuning dipole–dipole interactions. *Advanced Functional Materials* **2012**, *22* (17), 3737–3744.
167. Kwon, B. S.; Zhang, W.; Li, Z.; Krishnan, K. M., Direct release of sombrero-shaped magnetite nanoparticles via nanoimprint lithography. *Advanced Materials Interfaces* **2015**, *2* (3), 1400511-n/a.
168. Hu, W.; Wilson, R. J.; Koh, A.; Fu, A.; Faranesh, A. Z.; Earhart, C. M.; Osterfeld, S. J.; Han, S.-J.; Xu, L.; Guccione, S.; Sinclair, R.; Wang, S. X., High-moment antiferromagnetic nanoparticles with tunable magnetic properties. *Advanced Materials* **2008**, *20* (8), 1479–1483.
169. Wagener, P.; Jakobi, J.; Rehbock, C.; Chakravadhanula, V. S. K.; Thede, C.; Wiedwald, U.; Bartsch, M.; Kienle, L.; Barcikowski, S., Solvent-surface interactions control the phase structure in laser-generated iron-gold core–shell nanoparticles. *Scientific Reports* **2016**, *6*, 23352.
170. Chen, F.; Chen, M.; Yang, C.; Liu, J.; Luo, N.; Yang, G.; Chen, D.; Li, L., Terbium-doped gadolinium oxide nanoparticles prepared by laser ablation in liquid for use as a fluorescence and magnetic resonance imaging dual-modal contrast agent. *Physical Chemistry Chemical Physics* **2015**, *17* (2), 1189–1196.
171. Ruiz, A.; Salas, G.; Calero, M.; Hernández, Y.; Villanueva, A.; Herranz, F.; Veintemillas-Verdaguer, S.; Martínez, E.; Barber, D. F.; Morales, M. P., Short-chain PEG molecules strongly bound to magnetic nanoparticle for MRI long circulating agents. *Acta Biomaterialia* **2013**, *9* (5), 6421–6430.
172. Creixell, M.; Herrera, A. P.; Latorre-Esteves, M.; Ayala, V.; Torres-Lugo, M.; Rinaldi, C., The effect of grafting method on the colloidal stability and *in vitro* cytotoxicity of carboxymethyl dextran coated magnetic nanoparticles. *Journal of Materials Chemistry* **2010**, *20* (39), 8539–8547.
173. Qu, J.; Liu, G.; Wang, Y.; Hong, R., Preparation of Fe_3O_4–chitosan nanoparticles used for hyperthermia. *Advanced Powder Technology* **2010**, *21* (4), 461–467.
174. Mulens-Arias, V.; Rojas, J. M.; Pérez-Yagüe, S.; Morales, M. P.; Barber, D. F., Polyethylenimine-coated SPION exhibits potential intrinsic anti-metastatic properties inhibiting migration and invasion of pancreatic tumor cells. *Journal of Controlled Release* **2015**, *216*, 78–92.
175. Cintra, E. R.; Ferreira, F. S.; Junior, J. L. S.; Campello, J. C.; Socolovsky, L. M.; Lima, E. M.; Bakuzis, A. F., Nanoparticle agglomerates in magnetoliposomes. *Nanotechnology* **2009**, *20* (4), 045103.
176. Zahraei, M.; Marciello, M.; Lazaro-Carrillo, A.; Villanueva, A.; Herranz, F.; Talelli, M.; Costo, R.; Monshi, A.; Shahbazi-Gahrouei, D.; Amirnasr, M.; Behdadfar, B.; Morales, M. P., Versatile theranostics agents designed by coating ferrite nanoparticles with biocompatible polymers. *Nanotechnology* **2016**, *27* (25).
177. Veiseh, O.; Gunn, J. W.; Zhang, M., Design and fabrication of magnetic nanoparticles for targeted drug delivery and imaging. *Advanced Drug Delivery Reviews* **2010**, *62* (3), 284–304.
178. Laurent, S.; Forge, D.; Port, M.; Roch, A.; Robic, C.; Vander Elst, L.; Muller, R. N., Magnetic iron oxide nanoparticles: Synthesis, stabilization, vectorization, physicochemical characterizations, and biological applications. *Chemical Reviews* **2008**, *108* (6), 2064–2110.

179. Patil, R. M.; Shete, P. B.; Thorat, N. D.; Otari, S. V.; Barick, K. C.; Prasad, A.; Ningthoujam, R. S.; Tiwale, B. M.; Pawar, S. H., Superparamagnetic iron oxide/chitosan core/shells for hyperthermia application: Improved colloidal stability and biocompatibility. *Journal of Magnetism and Magnetic Materials* **2014**, *355*, 22–30.
180. Aqil, A.; Vasseur, S.; Duguet, E.; Passirani, C.; Benoit, J. P.; Jerome, R.; Jerome, C., Magnetic nanoparticles coated by temperature responsive copolymers for hyperthermia. *Journal of Materials Chemistry* **2008**, *18* (28), 3352–3360.
181. Cooperstein, M. A.; Canavan, H. E., Assessment of cytotoxicity of (*N*-isopropyl acrylamide) and poly(*N*-isopropyl acrylamide)-coated surfaces. *Biointerphases* **2013**, *8* (1), 19.
182. Rahman, M. M.; Elaissari, A., Organic–inorganic hybrid magnetic latex. In *Hybrid latex particles: Preparation with (mini)emulsion polymerization*, van Herk, A. M.; Landfester, K., Eds. Springer Berlin Heidelberg: Berlin, Heidelberg, 2010; pp. 237–281.
183. Hamoudeh, M.; Faraj, A. A.; Canet-Soulas, E.; Bessueille, F.; Léonard, D.; Fessi, H., Elaboration of PLLA-based superparamagnetic nanoparticles: Characterization, magnetic behaviour study and *in vitro* relaxivity evaluation. *International Journal of Pharmaceutics* **2007**, *338* (1–2), 248–257.
184. Hamoudeh, M.; Fessi, H., Preparation, characterization and surface study of poly-epsilon caprolactone magnetic microparticles. *Journal of Colloid and Interface Science* **2006**, *300* (2), 584–590.
185. Sommertune, J.; Sugunan, A.; Ahniyaz, A.; Bejhed, S. R.; Sarwe, A.; Johansson, C.; Balceris, C.; Ludwig, F.; Posth, O.; Fornara, A., Polymer/iron oxide nanoparticle composites – A straight forward and scalable synthesis approach. *International Journal of Molecular Sciences* **2015**, *16* (8).
186. Hyeon, T.; Manna, L.; Wong, S. S., Sustainable nanotechnology. *Chemical Society Reviews* **2015**, *44* (16), 5755–5757.
187. Mejías, R.; Gutiérrez, L.; Salas, G.; Pérez-Yagüe, S.; Zotes, T. M.; Lázaro, F. J.; Morales, M. P.; Barber, D. F., Long term biotransformation and toxicity of dimercaptosuccinic acid-coated magnetic nanoparticles support their use in biomedical applications. *Journal of Controlled Release* **2013**, *171* (2), 225–233.
188. Mazuel, F.; Espinosa, A.; Luciani, N.; Reffay, M.; Le Borgne, R.; Motte, L.; Desboeufs, K.; Michel, A.; Pellegrino, T.; Lalatonne, Y.; Wilhelm, C., Massive intracellular biodegradation of iron oxide nanoparticles evidenced magnetically at single-endosome and tissue levels. *ACS Nano* **2016**, *10* (8), 7627–7638.
189. Kolosnjaj-Tabi, J.; Di Corato, R.; Lartigue, L.; Marangon, I.; Guardia, P.; Silva, A. K. A.; Luciani, N.; Clément, O.; Flaud, P.; Singh, J. V.; Decuzzi, P.; Pellegrino, T.; Wilhelm, C.; Gazeau, F., Heat-generating iron oxide nanocubes: Subtle "destructurators" of the tumoral microenvironment. *ACS Nano* **2014**, *8* (5), 4268–4283.
190. Ruiz-Molina, D.; Novio, F.; Poscini, C., *Bio- and Bioinspired Nanomaterials*, Weinheim, Germany, Wiley-VCH Verlag GmbH & CO. KGaA: 2014; pp. 139–172.
191. Kendall, M.; Lynch, I., Long-term monitoring for nanomedicine implants and drugs. *Nat Nano* **2016**, *11* (3), 206–210.
192. Kharisov, B. I.; Dias, H. V. R.; Kharissova, O. V.; Vazquez, A.; Pena, Y.; Gomez, I., Solubilization, dispersion and stabilization of magnetic nanoparticles in water and non-aqueous solvents: Recent trends. *RSC Advances* **2014**, *4* (85), 45354–45381.

María del Puerto Morales (E-mail: puerto@icmm.csic.es) has been a senior scientist at the Institute of Material Science in Madrid, Spain (CSIC), since 2008. She earned her degree in chemistry at the University of Salamanca in 1989 and her PhD in material science at the Madrid Autonomous University in 1993. From 1994 to 1996, she worked as a postdoctoral fellow at the School of Electronic Engineering and Computer Systems of the University of Wales (UK). Her research activities are focused on the area of nanotechnology, in particular in the synthesis and characterization of magnetic nanoparticles for biomedicine, including the mechanism of particle formation, dispersion, coating and doping; study of structural, colloidal and magnetic properties; and its performance in biomedical applications such as biomolecule separation, contrast agents for nuclear magnetic resonance imaging, computed tomography, drug delivery and hyperthermia. Major achievements include the description of new mechanisms of nanoparticle formation by aggregation and the effect on the magnetic properties, development of drug magnetic carriers for cancer immunotherapy and a new methodology for the detection, identification and quantification of magnetic nanoparticles in different biosystems that allows nanoparticle biodistribution and long-term degradation studies after intravenous injection.

Helena Gavilán has completed her PhD in Advanced Chemistry in 2017 in the Madrid Complutense University. She has been doing research at the Institute of Material Science in Madrid, Spain (ICMM-CSIC), since 2014. She earned her degree in chemistry in 2013 and her MSc in chemical science and technology at Complutense University of Madrid. Her research activities are concentrated on nanotechnology, in particular on the synthesis and characterization of metal and metal oxide nanoparticles. Her thesis was on the synthesis strategies of single-core and multicore magnetite nanoparticles for biomedical purposes, focusing on the influence of the synthesis conditions on the size, shape and aggregation state of the magnetic particles. She explores dispersion and coating processes to achieve functional materials and studies the structural, colloidal and magnetic properties displayed by the nanoparticles, as well as their performance in biomedical applications such as hyperthermia.

Maria Eugênia Fortes Brollo has been a physics PhD student at the Institute of Material Science in Madrid, Spain (CSIC), since 2015. She earned her degree in physics at the University of Santa Catarina, Brazil, in 2012, and her MSc degree at the University of Campinas, Brazil. Her interest is focused on nanotechnology, magnetic materials and condensed matter physics, in particular on the synthesis and characterization of metal oxide nanoparticles. The objective of her thesis is to achieve functional materials and studies the structural, colloidal and magnetic properties displayed by nanoparticles, as well as their performance in biomedical applications employing liposomes as carriers.

Lucía Gutiérrez is a 'Ramón y Cajal' Research Fellow at the Instituto de Nanociencia de Aragón, from Universidad de Zaragoza, Spain. Including the completion of her PhD in 2008, she has been dedicated to full-time research since 2004. Her work has taken place in four different institutions: Universidad de Zaragoza in Spain, Queen Mary University of London in the United Kingdom, the Materials Science Institute of Madrid (ICMM) in Spain and the University of Western Australia in Australia.

Her work has been focused on the study of iron-containing nanomaterials in diseases, diagnosis and treatments. Included under this topic are iron deposits in the liver or brain as a consequence of different pathologies and the analysis of magnetic nanoparticles for biomedical applications such as contrast agents for magnetic resonance imaging or drug delivery systems.

Her current main research line focuses on the evaluation of magnetic nanoparticle toxicity. Her research aims at developing methods for the *in situ* identification and quantification of magnetic nanoparticles in tissues based on magnetic characterization techniques. Results from her work are fundamental to the evaluation of particle performance and side effects.

Sabino Veintemillas-Verdaguer graduated with a degree in chemistry in 1980 from the University of Barcelona and earned a PhD degree at the Complutense University of Madrid in 1986 with research work on crystal growth from boiling solutions. He joined the Spanish Council of Scientific Research (CSIC) in 1987. His scientific objectives are focused in the development of new methods for the preparation of materials and the study of the physicochemical processes involved in synthesis. Until 1995, he worked at the Materials Research Institute of Barcelona CSIC, in the crystal growth of water-soluble ferroelectric materials such as potassium dihydrogenphosphate (KDP) and L-Arginine Phosphate Monohydrate (LAP) for nonlinear optics. He then moved to the Materials Research Institute of Madrid CSIC to develop the synthesis of magnetic nanoparticles from laser pyrolysis and actually worked in the field of preparation and biomedical uses of magnetic nanoparticles. Dr Veintemillas has published more than 100 papers concerning crystal growth and preparation of materials.

2 Magnetic Nanochains
Properties, Syntheses and Prospects

*Irena Markovic-Milosevic, Vincent Russier, Marie-Louise Saboungi and Laurence Motte**

CONTENTS

2.1 INTRODUCTION

Hierarchical assemblies of magnetic materials arouse considerable interest because of their singular structures, unusual physical properties and potential technological applications.[1,2] In particular, one-dimensional (1-D) magnetic nanoparticle (NP) assemblies present a rich research field from both experimental and theoretical viewpoints. In contrast to individual NPs, *i.e.* 0-D systems, 1-D nanochains (NCs) present enhanced properties and provide surface functionalities because of the alignment possibilities and thus may be suitable for technological applications especially in the medical and environmental fields.

The purpose of the present work is to review the state of the art in 1-D assemblies of magnetic NPs. We first briefly recall the properties of magnetic NPs with an emphasis on modelling systems of particles with dipole–dipole interactions (DDIs). These are responsible for the collective behaviour of magnetic NP assemblies because of their long range, making it possible to form chains. A good understanding of dipolar effects can help to predict this behaviour.

The second part is devoted to a presentation of different experimental strategies developed recently to achieve such organization: dipolar-driven self-organization, external-magnetic-field-induced assembly, template-assisted synthesis, chemical assembly of multifunctionalized particles and physical methods such as electrospinning and microfluidics. Some of these methods can be combined, as is the case for magnetic field application, which can be used to orient the chain formation in all the previously cited methods.

The last part presents the various applications of these NCs, in particular in the fields of life and environmental sciences.

* Corresponding author.

2.2 PROPERTIES AND INTERACTIONS OF MAGNETIC NPs

2.2.1 Magnetic NPs: Magnetic Properties at the Nanometric Scale

The basic feature of nanoscale or nanostructured magnetic materials is to have a length scale smaller than, or on the order of, one of the characteristic magnetic length scales. The latter are determined by competition between two of the mesoscopic magnetic energy components:

(i) The magnetostatic energy due to polarization of the medium;
(ii) The magnetocrystalline anisotropy energy due to the presence of one or three preferential magnetization axes for uniaxial or cubic symmetry, respectively, stemming from interactions at the atomic scale and thus depending on the crystal lattice symmetry; and
(iii) The exchange energy term coupling the moments in a parallel (ferromagnetic [FM] materials) or antiparallel (antiferromagnetic [AFM] materials) way.[3–5]

These energy components are characterized by the saturation magnetization, M_s (or saturation polarization $J_s = \mu_0 M_s$ with μ_0 vacuum permeability), the anisotropy constant, K_1, and the exchange constant A. Dimensional analysis yields, on the one hand, the exchange length

$$l_{ex} = \sqrt{2A/\mu_0 M_s^2} \quad (2.1)$$

corresponding to the coherence length defined as the length over which the moments stay collinear and, on the other hand, the Bloch domain wall thickness

$$\delta_{dw} = \pi\sqrt{(A/K_1)}, \quad (2.2)$$

which represents the minimum distance for local moment reversal between two homogeneous domains. To the domain wall corresponds a surface energy $\gamma = \sqrt{(A.\ K_1)}$.

The relevant intrinsic length scale is either l_{ex} or δ_{dw}, depending on whether the material is magnetically soft or hard (*i.e.* $\mu_0 K_1 < J_s^2$ or $> J_s^2$). We then obtain the critical radius R_c (or critical diameter d_c), which is the value of the NP radius under which a spherical NP is single-domain and homogeneously magnetized. According to the previous discussion, the latter is either the coherence radius directly related to the coherence length defined above, $R_{coh} \approx 5 l_{ex}$, or the single-domain radius, $R_{sd} = 36(A|K_1|)^{1/2}/(\mu_0 M_s^2)$, determined by the competition between the surface energy brought by the introduction of a domain wall and the reduction of the magnetostatic energy. Typical values of R_c are 15 nm for Fe, 35 nm for Co and 35 nm for γ-Fe_2O_3.[5]

2.2.2 Isolated Magnetic NPs

Following from the definitions given earlier, the finite-size effect in magnetic NPs is represented by the critical size under which an NP can be considered as a homogeneous single-domain object.[4–6] For sizes greater than R_c, a hard material presents a multidomain structure,[6–8] and in a soft material, the NP can accommodate a vortex regime[9,10] in which the spin moments form closed loops in order to diminish the magnetostatic energy. In any case, for $R > R_c$, the NP's magnetization and coercivity decrease strongly and even vanish.[7,8]

For sizes lower than R_c, the NPs have a uniform magnetization with a coherent moment reversal under an external field. From a modelling point of view, this provides the basis of effective-one-spin (EOS) models in which each NP carries a certain moment and the local structure of the magnetization inside the NP is ignored. This important simplification makes it possible to model the magnetic properties of NP assemblies with conventional statistical physics methods for systems with only pairwise interactions.[11]

Nevertheless, even in this monodomain regime, the spins are not perfectly collinear, especially close to the NP surface. This spin misorientation at the NP surface defines the concept of spin canting.[12] This spin canting is due either to broken symmetry, crystalline surface defects or chemical bonds with the organic coating of the NP. As a result, the total NP magnetization is smaller than the bulk saturation magnetization, increasingly so as the NP becomes smaller, and the magnetization curve may differ from the usual Langevin curve in the approach to saturation. The simplest version of EOS model, the dipolar-hard-sphere model (DHS), treats spherical particles of volume v bearing at their centre a point dipole of moment $\vec{m} = \hat{m} v M_s$, where $\hat{m}$ is the unit vector in the direction of $\vec{m}$, and exposed to the magnetocrystalline energy, E_K. In the case of uniaxial symmetry with easy axis $\hat{n}$, the latter is a double-well potential:

$$E_K = -K_1 v(\hat{n}.\hat{m})^2. \quad (2.3)$$

Another consequence of the finite size concerns the anisotropy energy, which does not reduce to the magnetocrystalline one of the bulk material but also includes a shape component for particles that are not strictly spherical[6] and a surface component.[13] The shape component has a magnetostatic origin and comes from a demagnetizing effect at the NP scale; the surface component can originate from surface crystalline defects or chemical bonds of the chemisorbed species. Assuming on the one hand that the spin canting effect can be represented by an effective value for the saturation magnetization and on the other hand that the resulting anisotropy is of uniaxial symmetry, the energy of an isolated NP in an external field includes E_K and the Zeeman energy E_Z and is given by

$$E = E_K - \mu_0 (M_s v)\hat{m}.\vec{H} = E_K + E_Z. \quad (2.4)$$

It is thus completely characterized by M_s and the anisotropy constant K_1, which leads to the introduction of the anisotropy field $H_K = 2K_1/J_s$.

In colloidal suspension at room temperature or above the solvent boiling temperature, moment reorientation can result from Brownian rotation of the NP instead of coherent reorientation within the NP through the Néel process. In the Brownian process, the dynamics are determined by both particle size and solvent viscosity η. The two processes are characterized by their specific relaxation times,

$$\tau_N = \tau_0 \exp(K_1 v/k_B T); \tau_B = 3\eta v/(k_B T), \tag{2.5}$$

where τ_0 is the microscopic attempt time, of the order of the ns. These relaxation times characterize the dynamics of the corresponding processes and must be compared with the measurement times, going typically from 10^{-8} to *ca.* 10^2 s for Mössbauer spectroscopy and magnetization measurements, respectively. The important feature of the Néel relaxation is the blocking temperature T_b, widely used to characterize magnetic NPs,[9] which is the temperature beyond which the thermal activation leads to crossing the anisotropy barrier within the time of the measurement.

2.2.3 Interactions between Particles

We recalled earlier the main features of isolated magnetic NPs, in the absence of mutual interactions. When the NP concentration increases or for colloidal suspensions of NPs with a large magnetic moment (*i.e.* materials with large saturation magnetization or large particles), it becomes necessary to take into account the mutual interactions. In most cases, the NPs are coated by an organic layer, which makes them uncoupled by exchange interactions and the pair interactions are of dipolar origin (DDI):

$$E_{dd} = \frac{\mu_0}{4\pi}\frac{(M_s v)^2}{4\pi r_{ij}^3}\left[\widehat{m_i}.\widehat{m_j} - 3\left(\widehat{m_i}\widehat{r_{ij}}\right)\left(\widehat{m_j}\widehat{r_{ij}}\right)\right], \tag{2.6}$$

where r_{ij} and $\widehat{r_{ij}}$ are the distance between particles *i*, *j* and the associated unit vector, respectively. One can thus model the system through the DHS model for which both numerical simulations and theoretical results are available and which has proved useful to predict, at least qualitatively, the DDI-induced behaviour of interacting NP assemblies. In the following, we present some of the major results along these lines.

First, the relevant dipolar coupling parameter is defined as

$$\lambda = \frac{\mu_0 (M_s v)^2}{k_B T d^3}, \tag{2.7}$$

where *d* is the NP diameter, equal to the ratio of the dipolar interaction at contact to the thermal energy. From the available theoretical results, one defines a low- or high-coupling regime according to whether λ is smaller or greater than a threshold value λ = *ca.* 3. The role of the DDI, especially regarding the emergence of collective behaviour, depends strongly on the NP volume fraction $\Phi = Nv/V$, where *N* and *V* are the number of particles and total volume, respectively. The two significant features of the DDI are its long range ($1/r^3$), responsible for collective and demagnetizing effects, and its anisotropy, *i.e.* its dependence on the NP moment and orientation $\widehat{r_{ij}}$ and not only on r_{ij}. This second feature makes possible the spontaneous formation of NP chains, where the moments of neighbouring NPs are oriented parallel along the local chain direction and also leads to the occurrence of ordered phases (either FM or AFM), where the interaction anisotropy couples to the underlying structure.

2.2.4 Dipolar Interactions: Consequences for Orientational Order and Spontaneous Chain Formation

We first recall the features of the DDI via results obtained with the DHS model. In 3-D, in the condensed phase, the moments form an ordered phase at low temperature (*i.e.* in the strong-coupling regime) if the particles are located on a perfect lattice and for a zero or small anisotropy energy. This ordered phase is either FM or AFM according to the lattice symmetry: FM in case of a face-centred cubic (fcc) or body-centred tetragonallattice[14] and AFM for a simple cubic (sc) lattice. These phases correspond to the ground state,[15] *i.e.* at zero temperature, and have been confirmed by Monte Carlo (MC) simulations at finite temperature. The transition threshold (corresponding to the inverse transition temperature) is at $\Phi\lambda = 1.22$ for the fcc and 1.72 for the sc lattice. For an off-lattice DHS in the condensed phase, a similar result is obtained: the FM transition is observed at $\Phi\lambda = 1.75$ ($\Phi = 0.45$) and 1.98 ($\Phi = 0.42$).[16] In order to characterize the ordered phase in 3-D, the nematic order parameter is taken as the largest eigenvalue of the nematic tensor *Q*, the associated eigenvector being the polarization direction

$$\bar{Q} = \frac{1}{N}\sum\left(3\widehat{m_i}\widehat{m_j} - \bar{I}\right) \tag{2.8}$$

For disordered structures with an anisotropy energy and a random distribution of easy axes, one gets in the strong-coupling regime, beyond a threshold value related to the inverse transition temperature, a spin-glass-like collective state (super-spin-glass if one refers to the NP moments as *super spins*).[6,17] This state corresponds to a frozen moment configuration and is characterized by a slowing-down dynamics. Such behaviour has been observed experimentally in multilayers of metallic FeCo NPs embedded in an alumina matrix, iron oxide and Co NP assemblies.[17]

In the case of small volume fractions, $\Phi < 0.10$, MC simulations predict spontaneous chain formation beyond a coupling value $\lambda \approx 3.0$.[18,19] In the absence of an external field, these NP chains show various morphologies: linear chains (not straight), rings and branched chains. In the literature, focus has been put on the study of the link between these morphologies and the relevant physical parameters, namely volume fraction, magnetic moment and temperature.[19–22] The structure of the system is then characterized by the mean value of the chain length and the persistence length replacing the nematic order parameter used at high volume fraction.

In 2-D, *i.e.* for NP monolayers deposited on a surface, the first effect induced by the DDI is to orient the dipoles parallel to the surface.[23] For this reason, some simulations of 2-D systems with strong dipolar coupling are performed by limiting the moments to the surface plane.[24] In the strong-coupling regime, one finds qualitatively the same situation as in 3-D: NP monolayers organized in a lattice with an ordered phase of either FM or AFM character depending on whether the underlying lattice is triangular or square,[25] which can be compared with the *fcc* and *sc* lattices in 3-D. This confirms that for dipolar particle assemblies with a well-organized location, the symmetry of the lattice couples to the anisotropy of the dipolar interaction to form the ordered phase. It is worth mentioning that on perfect lattices, the dipolar system is unfrustrated. For off-lattice systems and at small surface density, the MC simulations show spontaneous chain formation beyond a threshold value for the coupling constant λ (or equivalently, below a critical temperature for a given dipolar interaction). As is the case in 3-D, these chains present various morphologies: linear chains (not straight), rings and branched chains. At small enough surface fraction, the threshold coupling value for the chain formation can be easily deduced from the ratio $-E_d/(\lambda N k_B T)$, where E_d is the dipolar energy and N is the number of NPs, which provides an estimate of the mean number of bonds per particle since this is equal to the number of particles in contact with each particle with an interaction energy $E_{dd} = \lambda k_B T$. For instance, spontaneous chain formation is obtained at $\lambda = 4.75$ in the case of a monodiperse DHS assembly at $\Phi = 0.08$. More precisely, characterization of the structure in terms of chain number and length is based upon a classification of the NP according to either the pair energy compared to some conveniently chosen threshold value or to the number of first, second and third neighbours determined from a distance-of-approach criterion.[21] It should be mentioned that the addition of an isotropic short-range attractive interaction to the dipolar one may have a significant effect on the chain morphology and lead to their aggregation in the form of bands of variable thickness and connected in isotropic patterns in the monolayer.

Finally, in 2-D systems, the effect of an external field is to make the chains straight and oriented in the field direction and moreover to form bands of chains in contact.[22] Interestingly, the chains organize in such a way that the NP configuration corresponds locally to a 2-D hexagonal lattice, the favourable configuration for the FM state in 2-D, since the dipoles are parallel to each other and in the chain direction.

Hence, as a result of DDI, modelled in the framework of the DHS model or more generally of EOS models including or not the anisotropy energy, one expects a singular behaviour, such as an ordered or frozen state, for magnetic NP assemblies when the coupling parameter defined in Equation 2.7 takes a sufficiently high value, of the order of 3, as discussed previously.

2.2.5 Experimental Evidence of Dipolar Behavior

Among the collective states induced by DDI, in the following, we focus on the chain type of organization. Spontaneous chain formation has been obtained experimentally in very diverse assemblies of NPs with dipolar interactions.[23,24] It is then important to determine whether these self-organizations stem from the DDI in the framework of EOS models. For this, we have to estimate both the value of the dipolar coupling parameter and the nature of interactions other than dipolar in experimental situations where such chain organizations are observed.

The origin of the chain-like self-organizations of NPs carrying a permanent dipole, in agreement with what is predicted from EOS models, is illustrated by the general character of such an organization. One of the first experimental evidence of them was the work of Tang et al.[26] dealing with semiconductor NPs carrying an electrostatic permanent dipole, while the general rule is to get dipole-induced chain formation with magnetic NPs. The chain morphologies obtained with depositions on a surface at low surface fractions are qualitatively similar for semiconductors,[27] Fe_3O_4,[27,28] coated Co[29–31] and platinum (Pt)[32] NPs. In the last case, the permanent moment arises from the electronic structure change induced by the finite size. Moreover, these morphologies in the absence of an external field are qualitatively similar to the linear chains and rings deduced from the MC simulations based on EOS models. The value of the dipolar coupling parameter in Tang et al.[26] is of the order of $\lambda = 3.7$–4.0. Wang et al.[27] have found ring formation of Fe_3O_4 and γ-Fe_2O_3 NPs of 40 nm in diameter where, given the value of the NP saturation magnetization ($M_s = 70$ emu/g) and the diameter, the dipolar coupling parameter is estimated to be λ $65/(1 + \Delta/d)^3$, where Δ represents the edge-to-edge distance between NPs, giving $\lambda = 45$ for $\Delta/d = 0.1$. In the work of Jia et al.,[33] assuming that the $FeNi_3$ nanospheres are characterized by an M_s on the order of 55 emu/g,[34] the dipolar coupling parameter takes a value of $\lambda = 12$ for particles 150 nm in diameter, suggesting that the anisotropic clustering leading to chain formation results from the DDI; however, the authors estimated the M_s value to be very much lower,[33] leading to a negligible value for λ. Under this latter hypothesis, the chain formation must result from the synthetic method and/or the growth process.

Work of the group of Pyun[29–31] has shown that polystyrene-coated Co NPs may spontaneously form chains after their synthesis; these are characterized by a coupling parameter, which can be estimated as $\lambda = 2$ ($M_s = 38$ emu/g and $d = 15$ nm)[29]. In subsequent work, λ takes values from 8 to 30 depending on the NP diameter (20 to 50 nm). The direct influence of the NP size and, thus, of λ on the structure in 2-D depositions has been clearly shown by Klokkenburg et al.[28] for magnetite NPs, where chain self-organization was obtained when $d = 20$ nm ($\lambda = 7$), while for smaller sizes, $d = 16$ nm ($\lambda = 2$), the particles aggregate in only isotropic islands. Tripp et al.[35] have also shown the formation of chains with Co NPs with diameter 25 nm, where one can estimate $\lambda = 30$ given the nonmagnetic layer at the NP surface and the value of M_s. Moreover, they have shown that ring formation does not result from any process related to the solvent evaporation after deposition.

Thus, in the examples given previously, chain-like NP self-organizations are obtained in systems with a dipolar coupling value in agreement with the threshold found necessary to get such organizations from simulations with EOS models.

Gao et al.[36] have developed an alternative method for forming NCs made of magnetite NP clusters. The synthesis of the chains is a three-step process: NP synthesis followed by cluster formation in the form of 'flower' structures, which then form chains of NP clusters.

It is difficult in this case to deduce the value of λ since it involves the effective moment of the clusters, which cannot be precisely determined since it depends on the cluster size and structure.

In the earlier discussion, we have mentioned only NC self-organization obtained in the absence of an external field, which necessitates a large value for the coupling parameter λ. In weaker coupling situations, a clustering process takes place and 2-D islands form[36–38]; then NC formation requires an external field during the NP deposition by solvent evaporation from colloidal suspensions. Lalatonne et al.[37,38] have shown experimentally and confirmed from theoretical simulations based upon Brownian dynamics such a behaviour for maghemite NPs. Because of the coating layer of either octanoic acic (C8 chain) or dodecanoic acid (C12 chain) at the NP surface, the coupling parameter takes too small a value ($\lambda \approx 0.7$) to get self-organization during the solvent evaporation. Conversely, by applying an external field during the solvent evaporation, NP bands oriented in the direction of the field can be obtained. The influence of λ is nevertheless important since the formation of NCs under the external field is possible only beyond a minimum NP size value[37,38] or below a maximum thickness of the coating layer,[37] *i.e.* beyond a critical value of λ.

Ge et al.,[39] Hu et al.,[40] Gao et al.,[36] Kralj and Makovec[41] as well as Zhou et al.[42] start from colloidal nanoclusters (CNC) with a size controlled by the synthesis conditions. Then the application of the external field induces magnetization inside the CNCs, which then become the basic magnetized objects, leading to the organization of the latter into NCs through dipolar interactions. These chains can be stabilized by embedding them in a silica layer[40,41] or by formation of an external layer of polydopamine.[42]

2.3 SYNTHETIC STRATEGIES

As described previously, the organization of NPs into chains can be spontaneous due to the strong dipolar interactions between them. Other strategies, involving the application of an external magnetic field, electrospinning and microfluidic methods, will be discussed in the following.

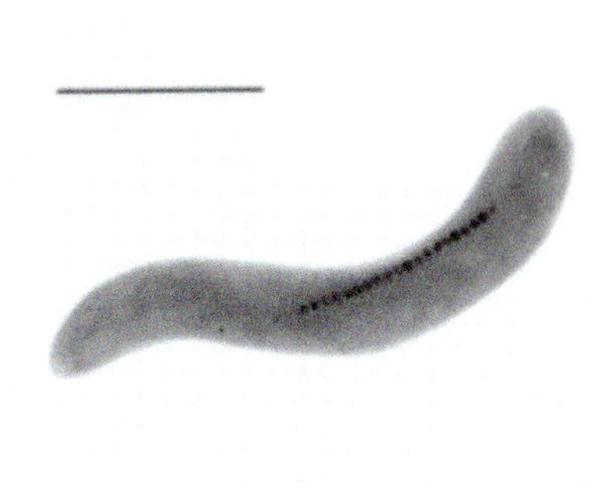

FIGURE 2.1 Magnetosome alignment from *Magnetospirillum magnetotacticum* (scale bar = 1 μm). http://www.calpoly.edu/~rfrankel/mtbphoto.html.

2.3.1 MAGNETOTACTIC BACTERIA

Magnetotactic bacteria, discovered by Blackmore in 1975, are microorganisms that use the Earth's magnetic field to optimize their movements in their natural environment.[43,44] This remarkable ability is based on the alignment of the cytoplasm of specific magnetic organelles called magnetosomes.

Magnetosomes are magnetite (Fe_3O_4) or greigite (Fe_3S_4) particles with a protein–lipid membrane (Figure 2.1), with an average size between 25 and 100 nm and a single magnetic domain characterized by FM behaviour at room temperature. Each particle has a maximum magnetic dipole moment and the alignment of these moments maximizes the total magnetic moment of the cell, leading to alignment with the Earth's magnetic field, and travel toward environments more suitable for their proliferation, usually oxygen-deficient regions. Although widespread in nature, magnetotactic bacteria are difficult to cultivate in the laboratory due to their fastidious lifestyle.[45]

2.3.2 SELF-ASSEMBLY

The transition from a random Brownian fluid to NPs organized in chains in 1-D occurs when the interparticle dipole–dipole potential dominates the thermal fluctuations, corresponding to a dipolar coupling constant λ (defined in Equation 2.7) greater than 3. In this case, the particles spontaneously align in the direction of their magnetic moment. This self-assembly phenomenon was well illustrated by Klokkenburg et al. in 2004.[28] They used different size magnetite NPs and compared the different organizations observed after evaporation of the ferrofluid onto a transmission electron microscopy (TEM) grid (Figure 2.2). NPs with an average size of 16 nm, for which isotropic interactions are dominant ($\lambda \approx 2$), formed a droplet-shaped assembly, while NPs with a diameter of 21 nm self-assembled into a linear organization due to the predominant magnetic interactions ($\lambda \approx 7$).

Grzelczak et al.[46] studied the spontaneous formation of linear organizations of FM nickel NPs with an average size of

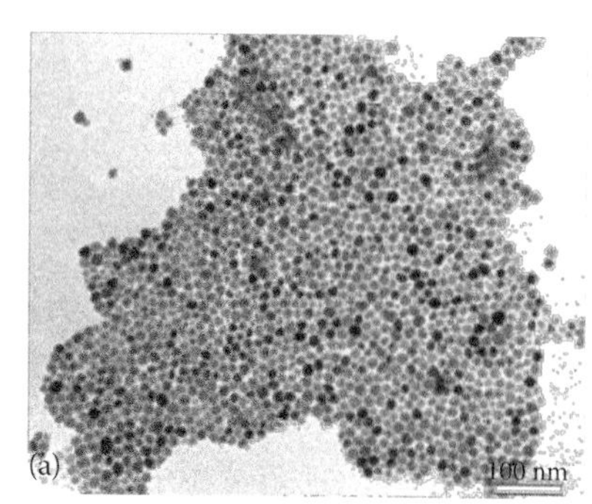

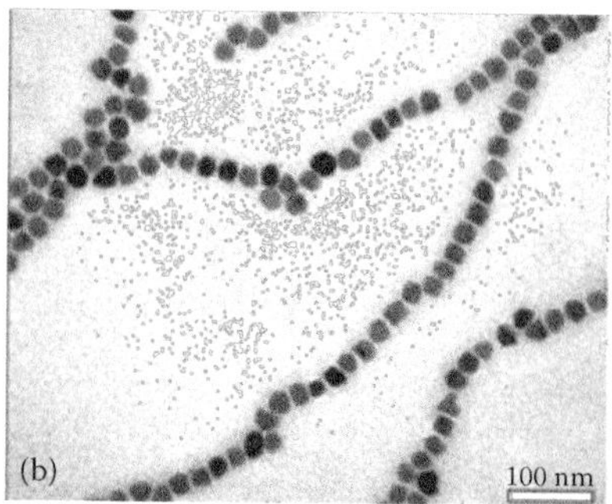

FIGURE 2.2 TEM images showing two types of magnetite NP organization according to their size: (a) 16 nm and (b) 21 nm.[28] (Reprinted with permission from Klokkenburg, M., Vonk, C., Claesson, E.M., Meeldijk, J.D., Erné, B.H., Philipse A.P. 2004. Direct imaging of zero-field dipolar structures in colloidal dispersions of synthetic magnetite. *JACS* 126:16706–7. Copyright 2004 American Chemical Society.)

33 nm after evaporation of the colloidal solution onto a TEM grid. Even in the absence of a magnetic field, linear structures were formed. This was also observed by the *in situ* coating with silica, which results in fossilized NP chains.

Spontaneous linear organization has also been reported for Co particles with sizes between 20 and 30 nm surface functionalized with silica[47] and for 15-nm NPs coated with polystyrene.[29]

In contrast to the 16-nm magnetite NPs,[28] the particles aggregate in isotropic islands.[28] Varon et al.[47] showed that ε-Co NPs with similar size, 15 nm in diameter, self-assemble linearly. For this size and phase, the magnetic dipole moment is strong enough to maintain the magnetic order at room temperature.

Finally, Gao et al.[32] observed NC assembly with 7.5 nm (in diameter) Pt NPs having surface-functionalized with polyvinylpyrrolidone (PVP). As is well known, Pt is non-FM in the bulk state and so the self-organization was attributed to the appearance of ferromagnetism in nanoscale Pt due to electron transfer from the adsorbed PVP molecules.

2.3.3 Self-Assembly Induced by External Forces or Constraints

Self-assembly of magnetic NPs is induced by DDIs guiding their association. This method is valid only for particles with a sufficiently high magnetic moment to be insensitive to temperature fluctuations. This is the case for FM NPs. For superparamagnetic (SPM) NPs, the magnetic moments are randomly oriented due to thermal fluctuations, and so other strategies have to be applied to form linear chains, such as the application of an external magnetic field or the use of polymers or templates to maintain the structures.

2.3.3.1 Application of External Magnetic Field

As described earlier, in the absence of a magnetic field and in the case of weak dipolar coupling, NC organization is not observed. In this case, two configurations predominate, depending on the NP interactions: a random NP dispersion or formation of spherical aggregates (droplet shape). In the latter case, it is possible to form chains by applying an external magnetic field. This has been demonstrated from both theoretical and experimental points of view.[37,38,48,49]

To better understand this phenomenon, we have to consider that two attractive terms contribute to the interactions between particles: magnetic dipolar interactions and van der Waals forces, as well as a repulsive term related to electrostatic interactions (charges on the NP surface) in the case of a hydrophilic ferrofluid or/and that of steric interactions (due to the presence of molecules on the NP surface). When the attractive interactions predominate, formation of spherical aggregates is generally observed. The application of a magnetic field induces a macroscopic moment in each aggregate, sufficiently high that the dipolar coupling between aggregates (λ) induces their organization into chains in the direction of the field. Thus, by changing various parameters such as ionic strength, chain length of the NP coating or the NP volume fraction, it is possible to induce 2-D (chains) and 3-D (cylinder) organizations of SPM NPs. After evaporation of the carrier liquid or by polymer addition, these structures are maintained on a substrate and could be observed with TEM or scanning electron microscopy (SEM).

Lalatonne et al.[37,38,49] studied the influence of a surface passivation agent, notably that of chain length, on the organization of maghemite NPs. Due to their size, these NPs have a λ parameter less than 0.92, and therefore, chain organization is not expected. Figures 2.3 and 2.4 show TEM and SEM images obtained after evaporation of ferrofluids in diluted and concentrated regimes, respectively, with and without an external magnetic field. The formation of chains or cylinders was observed only with NPs forming droplet-shape aggregates in the absence of a magnetic field.

These authors also showed the influence of magnetic field on the 3-D structures (Figure 2.5). Linear organizations in the direction of the applied magnetic field were obtained in all cases, and an increase in cylinder diameter, from 0.8 μm to 3 μm, was observed when the magnetic field intensity increased.

Sheparovych et al.[50] have developed NCs with SPM magnetite NPs, 10–12 nm in diameter, surface-functionalized with citrate ions (negatively charged NPs), by applying a 1-T

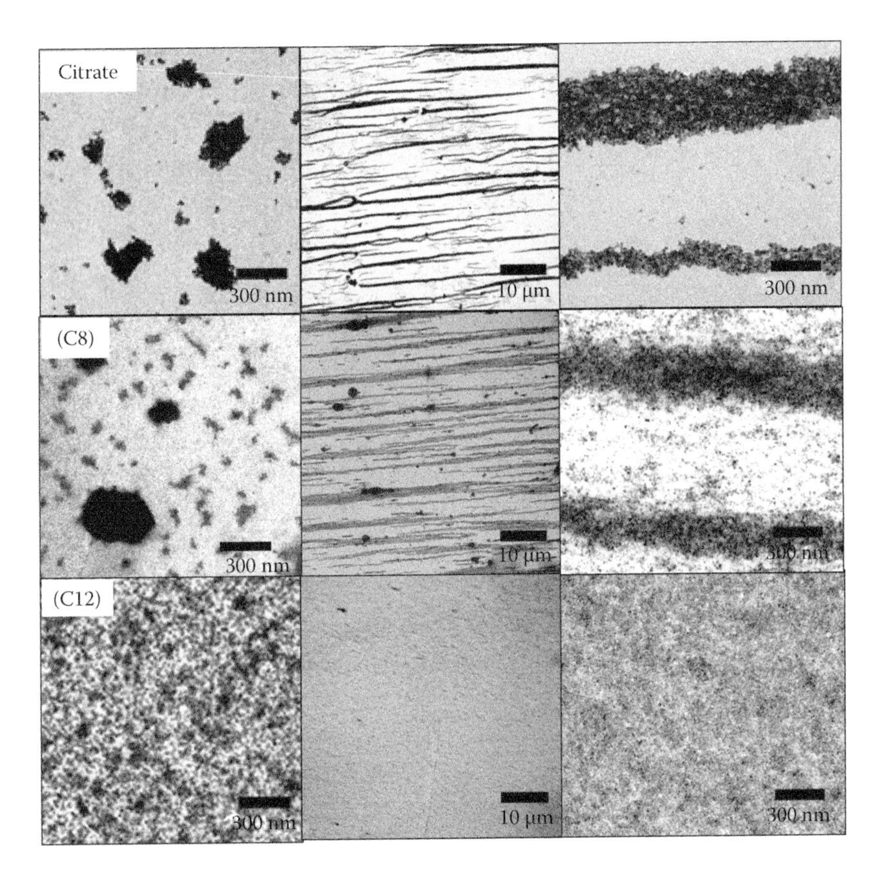

FIGURE 2.3 TEM images obtained at various magnifications for maghemite nanocrystals (diluted solutions) evaporated without (left column) and with (middle and right columns) a magnetic field (0.59 T) applied parallel to the substrate plane and with different coatings: NPs coated with citrate ions and dispersed in water (top), NPs coated with octanoic acid (middle) and dodecanoic acid (bottom) and dispersed in hexane.[37] (Reprinted with permission from Lalatonne, Y., Motte, L., Richardi, J., Pileni, M.P. *Phys. Rev. E.*, 71, 011404, 2005. Copyright 2005 by the American Physical Society.)

magnetic field. Structures were fixed in the carrier fluid in spite of the thermal agitation by using a positively charged polyelectrolyte, poly (2-vinyl *N*-methylpyridinium iodide) (P2VPq) as binding agent. To obtain the structures presented in Figure 2.6B, magnetite NPs dispersed in water were introduced into the bottom of a special setup (Figure 2.6A) under a 1-T magnet and the P2VPq solution was introduced under conditions of slow diffusion.

In 2009, Yan et al.[51] synthesized nanorods made of maghemite NPs (6.7 nm and 8.3 nm in diameter) surface-functionalized with polyacrylic acid (PAA) polymer (negatively charged particles), by applying a magnetic field of 0.3 T and stabilized in solution by the addition of diblock cationic-neutral copolymers: poly (trimethylammonium ethylacrylate) and β-poly (acrylamide), respectively. The process was based on the mixing of NPs and copolymers and slow reduction of the ionic strength of the mixture by dialysis in the presence of a magnetic field (Figure 2.7a). Nanorods of various diameters and lengths were observed (Figure 2.7b). It was possible to generate a morphology diagram of the NP aggregates as a function of ionic strength, dialysis time and initial NP concentration. The authors showed that to form nanorods, it is necessary to work at high NP concentration ($C > 10^{-2}$ wt.%) and low ionic strength ($Is < 0.3$ M). Moreover, they showed that the NC formation mechanism was related to two processes occurring simultaneously: the formation of spherical clusters and the alignment of these clusters induced by the magnetic field, in accordance with the studies of Lalatonne et al.[37,38]

Recently, Singh et al.[52] have shown that, under well-defined experimental conditions, cube-shaped magnetic nanocrystals can self-assemble into helical superstructures (Figure 2.8). The nanocubes (13.4 nm edge, Figure 2.8b), surface-functionalized with oleic acid (OA), are dispersed in hexane under OA excess. A drop of this solution was deposited into a diethylene glycol solution/air interface in the presence of a magnetic field with intensity varying from 0 to 700 G. After hexane evaporation, the structures were transferred to a substrate (Figure 2.8a). The authors showed that the nature of the superstructures was highly dependent on the NP concentration. When the amount of NP was increased, a transition from a 'belt' configuration, with lengths up to 100 μm and a width corresponding to alignment of two or three particles (Figure 2.8b) to single-stranded, double-stranded (Figure 2.8c) and even triple-stranded helices, was observed. The influence of magnetic field intensity was reported only for 'belt' structures: increasing the magnetic field intensity reduced the width of

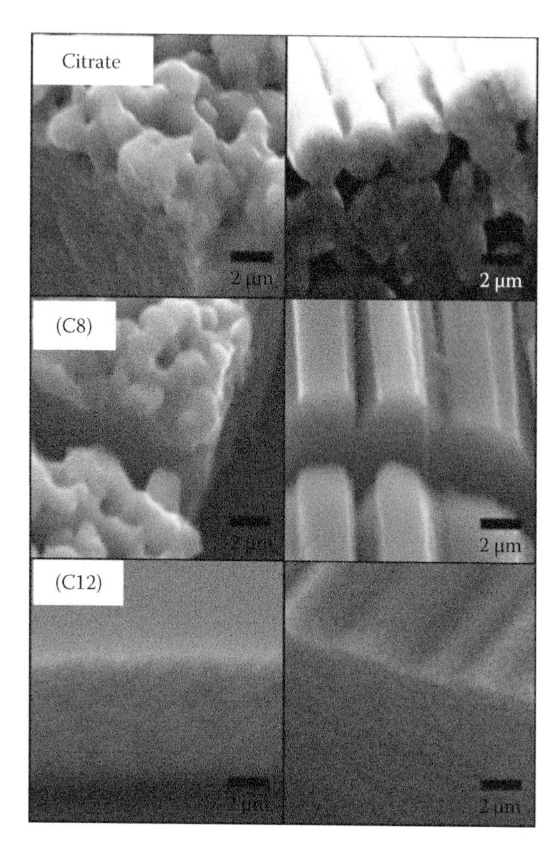

FIGURE 2.4 SEM images obtained at various magnifications for maghemite nanocrystals (concentrated solutions) evaporated without (left column) and with (right column) a magnetic field (0.59 T) applied parallel to the substrate plane and with different coatings: NPs coated with citrate ions and dispersed in water (top), NPs coated with octanoic acid (middle) and dodecanoic acid (bottom) and dispersed in hexane.[38] (Reprinted with permission from Lalatonne, Y., Motte, L., Russier, V., Ngo, A.T., Bonville, P., Pileni, M.P. 2004. Mesoscopic structures of nanocrystals: collective magnetic properties due to the alignment of nanocrystals. *J. Phys. Chem. B.* 108:1848–54. Copyright 2004 American Chemical Society.)

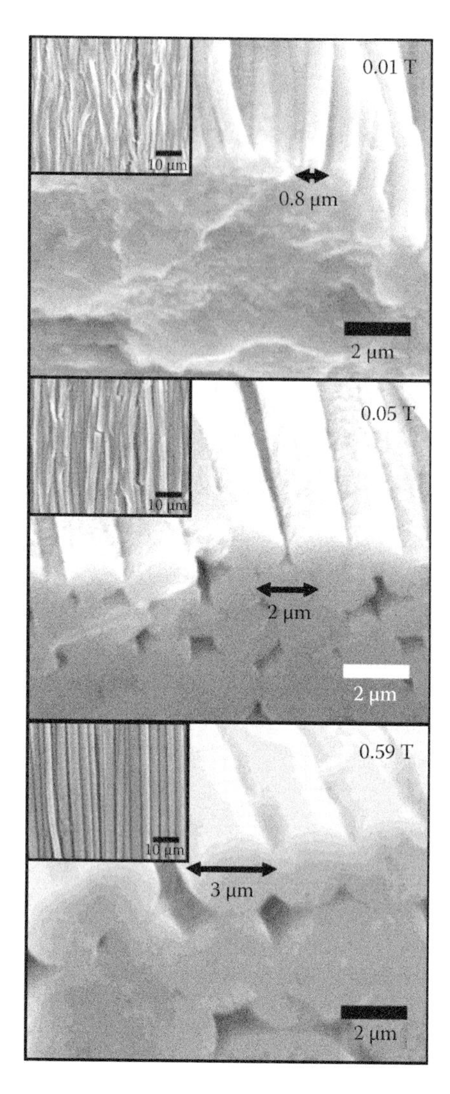

FIGURE 2.5 SEM images obtained at various magnifications for maghemite nanocrystals coated with citrate ions (concentrated solutions) evaporated with a magnetic field: 0.01 T (top), 0.05 T (middle) and 0.59 T (bottom).[38] (Reprinted with permission from Lalatonne, Y., Motte, L., Russier, V., Ngo, A.T., Bonville, P., Pileni, M.P. 2004. Mesoscopic structures of nanocrystals: collective magnetic properties due to the alignment of nanocrystals. *J. Phys. Chem. B.* 108:1848–54. Copyright 2004 American Chemical Society.)

the structures, due to the dipole repulsive interactions between the cubes of two neighbouring chains.

Sone and Stupp[53] showed that it is possible to 'decorate' the peptide-amphiphile (PA) nanofibers with magnetite NPs. The PA molecules self-assemble into cylindrical fibres, 6 to 8 nm in diameter, and act as templates for the nucleation and growth of magnetite. Indeed, the PA (Figure 2.9a) contains three histidine amino acids at its C-terminus that are chelating agents of iron ions through the lone electron pair on the δ-nitrogen. The three histidines are separated from the alkyl tail of the molecule by alanine and glycine residues. An additional aspartic acid residue was added and PA was also terminated by a carboxylic acid at the C-terminus. These two carboxylic acid groups are expected to act also as coordination sites for Fe^{3+}.

In the configuration of the nanofibers, histidine parts are expected to be exposed at the fibre/water interface and the two carboxylic functions act also as iron ion coordination sites. Thus, a supersaturation of iron ions is created around the fibres and the production of iron oxide NPs is induced by an increase in pH by exposing the suspension to ammonia vapour. The nucleation and growth of magnetite NPs are guided by the surface of the nanofibers. The TEM pictures (Figure 2.9b and c) show the final nanofibers that are decorated with magnetite NPs with a diameter of 10–20 nm. The electron diffraction in the inset of Figure 2.9b shows a good match with the pattern expected for magnetite.[53]

Researchers from Dublin[54] used denatured (substantially single-stranded) herring sperm DNA (hs-DNA) both as a template and a surfactant for the preparation and the organization of magnetite NPs. NPs were directly coprecipitated in a DNA solution. The obtained NPs were 9 ± 2 nm in size and assemble linearly yielding materials with a remarkably high relaxivity at low field.

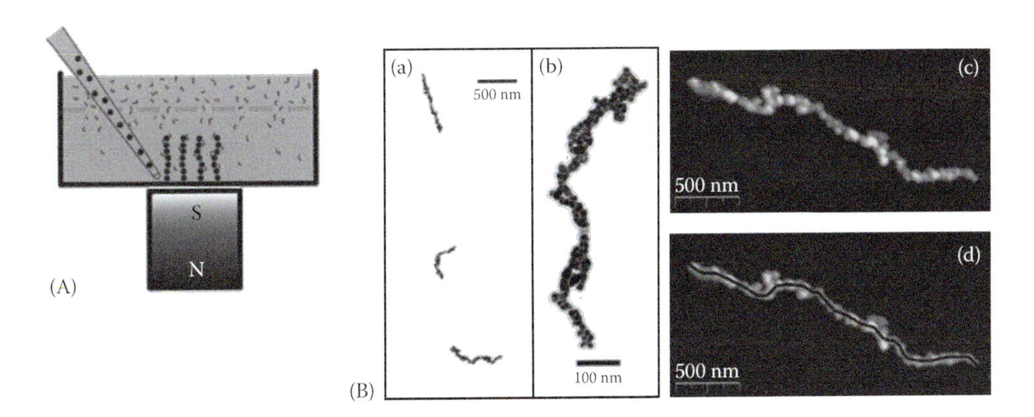

FIGURE 2.6 (A) Schematic representation of the NC synthesis process (A); (B) TEM images of magnetic NCs at different scales (a) and (b) AFM images of the magnetic wire-like structures stabilized by P2VPq, original figure (c), and with a line showing the locus of the cross-section (d).[50] (Reprinted with permission from Sheparovych, R., Sahoo, Y., Motornov, M., Wang, S., Luo, H., Prasad, P.N., Sokolov, I., Minko, S. 2006. Polyelectrolyte stabilized NWs from Fe_3O_4 nanoparticles via magnetic field induced self-assembly. *Chem. Mater.* 18:591–3. Copyright 2006 American Chemical Society.)

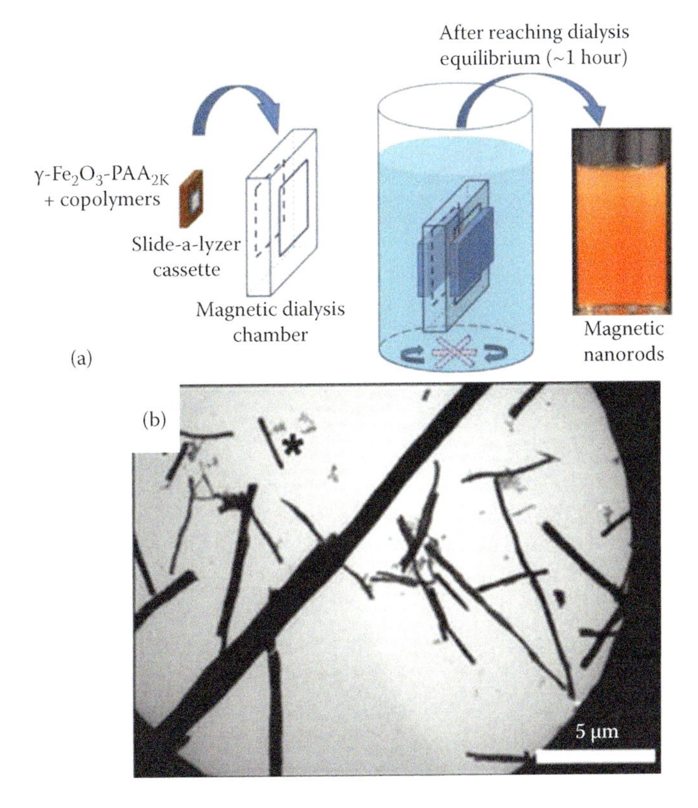

FIGURE 2.7 Schematic representation of the protocol used for the assembly of linear organizations into superparamagnetic nanostructured rods in the presence of an external magnetic field using a slide-a-lyser dialysis cassette (a); TEM images (b).[51] (Adapted from Yan, M., Fresnais, J., Berret, J.F., *Soft Matter*, 6, 1997–2005, 2010.)

In a second paper, these researchers prepared stable suspension of chain-like assemblies in water using polysodium-4-styrene sulfonate (PSSS) electrolyte.[54] The polyelectrolyte solution was added to a predetermined volume of iron solution under argon. The negatively charged polyelectrolyte acts both as a stabilizer, where the positively charged iron ions can accumulate before particle precipitation, and as template for nanowire (NW) formation once the particles are formed. The Fe-PSSS ratio could be tuned in order to obtain different properties (R = Fe/monomer = 1:2 [PSSS-Mag1], 3:1 [PSSS-Mag2] and 6:1 [PSSS-Mag3]). The resulting particle size depended on R: 7 nm (PSSS-Mag1), 10 nm (PSSS-Mag2) and 11 for the largest value of R (PSSS-Mag3). After water evaporation from the suspension and without magnetic field, aggregated structures were observed in the form of necklace or chain (Figure 2.10a). In the presence of a magnetic field, NWs were

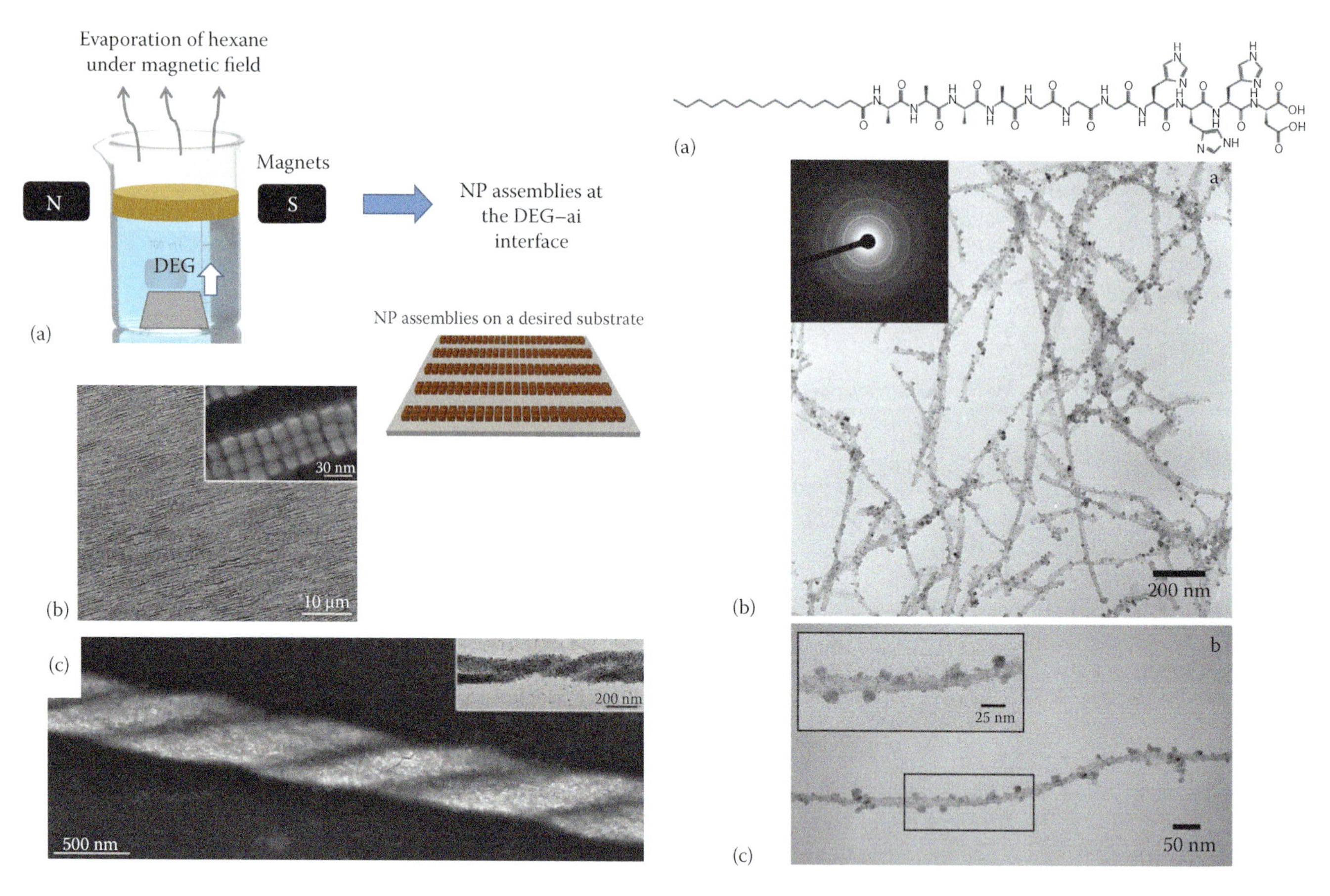

FIGURE 2.8 Schematic representation of the protocol used for the assembly of cubic NPs (a); TEM snapshots of the 'belt' nanocube structures (b); double-helix structure (c).[52] (Adapted from Singh, G., Chan, H., Baskin, B., Gelman, E., Repnin, N., Král, P., Klajn, R., *Science*, 345, 1149–1153, 2014.)

FIGURE 2.9 PA molecule that can form nanofibers (a), TEM micrograph of magnetite nanocrystals on fibers formed from PA molecule A: Network of magnetite-decorated fibers with inset showing electron diffraction pattern (b) and an isolated nanofiber (c).[53] (Reprinted with permission from Sone, E.D., Stupp, S.I. 2011. Bioinspired magnetite mineralization of peptide–amphiphile nanofibers. *Chem. Mater.* 23:2005–7. Copyright 2011 American Chemical Society.)

observed (Figure 2.10b–d), attributed to the interaction of NPs with polyelectrolytes and induced by the interpenetration of polymer chains. The width of the NWs was also dependent on the ratio R: the widths increased from 670 nm for PSSS-MAG1 to 1.5 µm for PSSS-Mag2.

Finally, Lu et al.[55] showed that cetyltrimethylammonium bromide (CTAB) can be used in a simple mediated-assembly method to promote 1-D NCs consisting of crystalline Fe-Ni NPs 90 to 110 nm in size. According to them, the CTAB serves as a cross-linker and also plays an important role in the formation of these NCs. The process corresponds to the adsorption of cationic charged CTAB molecules on the surface of particles, reducing the initial negative surface charge and leading to an anisotropic distribution of the residual surface charges.[55] Thus, an electric dipole moment is created around each particle and is responsible for the linear organization of the particles into chains. Shape-anisotropy induced by the large aspect ratio of NCs is believed to be the main reason for the enhancement of coercivity of this material. As explained by the authors, the strength of shape anisotropy characterized by shape-anisotropy constant Ks [$Ks = 1/2(N_a - N_c)M^2$, with M as the level of magnetization and N_a and N_c the demagnetizing coefficients along the a and c axes] depends on the axial ratio of the specimen. As the demagnetizing field along the long axis of the chains is much smaller than that along the short axis, this results in the pinning of the magnetization in the long axis direction and leads to a harder reverse of the magnetization so that larger values of H_c and M_r/M_s are obtained.

2.3.3.2 Chemical Synthesis

It has been reported that elongated structures called nanoworms were synthesized by coprecipitation of iron salts in the presence of dextran and these iron oxide structures can be obtained simply by increasing the concentrations in metal salts (Fe II and Fe III) and the molecular weight of dextran.[56–60]

Highly crystalline iron oxide nanoworms can also be obtained via a modified 'heat-up' method using iron oleate complex as the precursor in the presence of a mixture of trioctylphosphine oxide (TOPO)/OA.[61] The reaction was carried out at 320°C for 2.5 h. Instead of the usual homogeneous

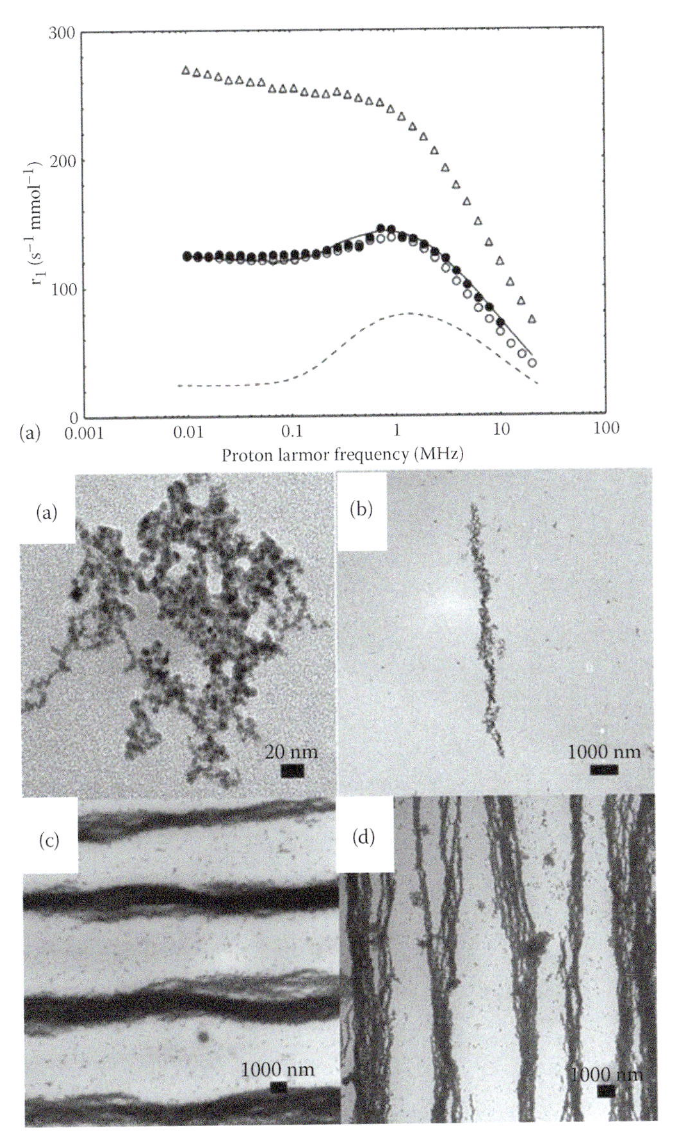

FIGURE 2.10 TEM images of sample PSSS-Mag1 dried in the absence of magnetic field (a), PSSS-Mag1 (b), PSSS-Mag2 (c) and PSSS-Mag3 (d) dried under a 0.5-T magnetic field.[54] (Reprinted with permission from Corr, S.A., Byrne, S.J., Tekoriute, R., Meledandri, C.J., Brougham, D.F., Lynch, M., Kerskens, C., O'Dwyer, L., Gun'ko, Y.K. 2008. Linear assemblies of magnetic nanoparticles as MRI contrast agents. *JACS*. 130:4214–15. Copyright 2008 American Chemical Society.)

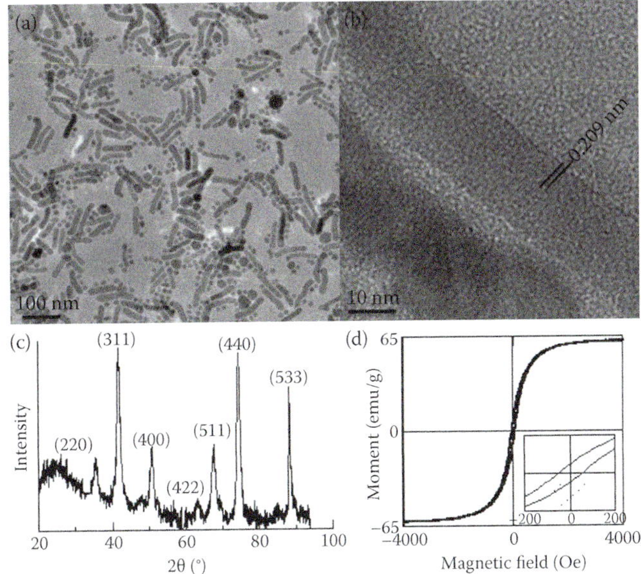

FIGURE 2.11 (a) TEM image of synthesized nanoworms, (b) HRTEM, (c) XRD scan and (d) M-H curve.[61] Reprinted with the permission from Palchoudhury, S., Xu, Y., Goodwin, J., Bao, Y. 2011. Synthesis of iron oxide nanoworms. *J. Appl. Phys.* 109:07E314. Copyright 2011, AIP Publishing.)

spherical particles, the high concentration of TOPO ([TOPO]/[OA] = 6.3/1) induced the aggregation and the coalescence of spherical iron oxide NPs into nanoworms (Figure 2.11). A time study was also conducted to understand the growth mechanism. After 1 h of reaction, the NPs were spherical; at 2.5 h, nanoworms were observed in coexistence with NPs. Spherical particles were no longer observed after 5 h and nanoworms could grow beyond 200 nm in length, while their diameter corresponded to that of the spherical particles (12 nm). Their magnetization curve (Figure 2.11d) showed no longer an SPM behaviour (as for the spherical NPs) but an FM behaviour supported by the appearance of a hysteresis loop with a coercive field (H_c = 50 Oe) and with a saturation magnetization of 60 emu/g. Both the spheres and nanoworms displayed the maghemite crystal structure with an interfringe distance of 0.209 nm very close to the d-spacing of the (400) crystal plane of the maghemite crystal structure. The X-ray diffraction (XRD) pattern also matches with the maghemite structure (Figure 2.11c).

On the other hand, SPM 13 nm iron oxide NPs coated with a mixture of nonanoic acid and 4-phenylbutyric acid had 'rippled' ligand shells with reactive polar point defects that could be selectively modified to form chains of such particles. The chains were formed when 1-(10-carboxydecyldisulfanyl) undecanoic acid was used as a cross-linking agent.[62] Another approach employed by Peiris et al.[63,64] consists of asymmetric surface functionalization of SPM iron oxide NPs. Solid-phase synthesis was used to produce iron oxide nanospheres with asymmetric surface chemistry consisting of two areas with distinct functional group distributions. The as-functionalized NPs carry amine groups on one side of the NPs and thiol groups on the other side, Figure 2.12. Initially, the NPs functionalized with amine groups were linked to a solid support via an amine coupling agent containing a disulfide bond, 3,3'-dithiobis (sulfosuccinimidylpropionate). By cleaving the disulfide bond with a reducing agent, tris[2-carboxyethyl] phosphine asymmetric features of the NPs were obtained in suspension. The last step was to bind the thiol groups to the amine functions. Thus, NCs three to four particles in length were synthesized as shown in Figure 2.12.

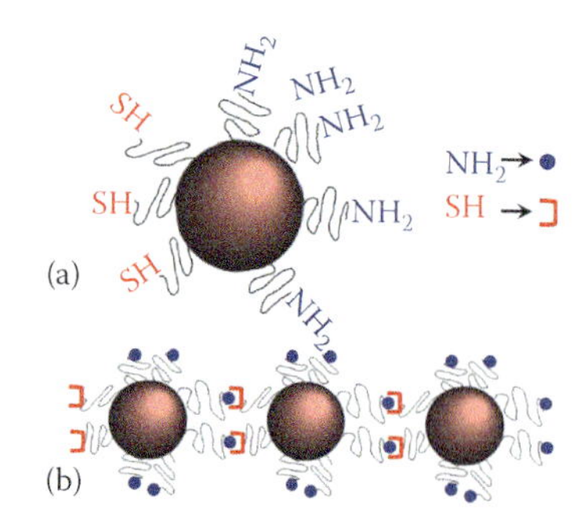

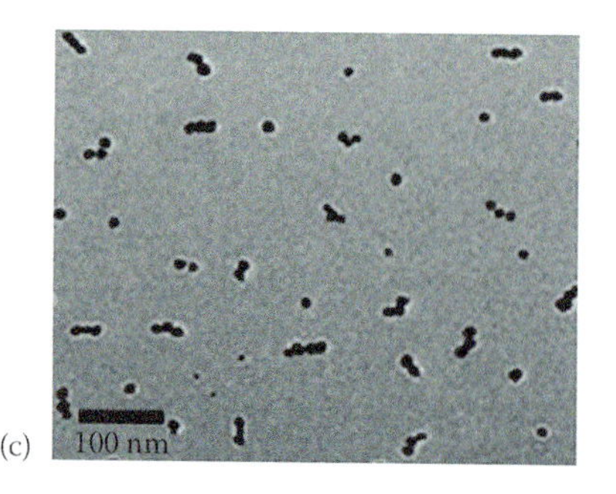

FIGURE 2.12 Illustration of nanospheres with two distinctive faces in terms of chemical functionality and their use as components of nanochains. (a) Nanospheres with asymmetric surface chemistry (ASC) and (b) linear nanochains assembled from spheres with ASC. (c) TEM image of the NPs alignments.[63,64] (Reproduced with permission from Peiris, P.M., Schmidt, E., Calabrese, M., Karathanasis, E., 2011. Assembly of linear nano-chains from iron oxide nanospheres with asymmetric surface chemistry. *PLoS ONE*. 6:e15927. Copyright 2011 American Chemical Society; Peiris, P.M., Bauer, L., Toy, R., Tran, E., Pansky, J., Doolitle, E., Schmidt, E., Hayden, E., Mayer, A., Keri, R.A., Griswold, M.A., Karathanasis, E. 2012. Enhanced delivery of chemotherapy to tumors using a multicomponent nanochain with radio-frequency-tunable drug release. *ACS Nano*. 6:4157–68. Copyright 2012 American Chemical Society.)

An original method is to use niobate nanosheets to encase aligned cobalt NPs: the structures are called nanopeapod.[65] Moreover, cobalt NPs were formed by *in situ* thermal decomposition of cobalt carbonyl in the presence of niobate sheets. When the reaction medium was heated to 100°C–130°C, a very wide distribution of NPs from 2 up to 45 nm was observed, whereas when the temperature is 140°C–150°C, a narrow distribution of 10- to 20-nm-sized NPs was obtained (Figure 2.13). Due to strong dipolar interactions, the NPs aligned in chains on the nanosheet (Figure 2.13a), then the nanosheet wrapped around these alignments to encase the chains (Figure 2.13b) and finally the peapod was detached from the surface (Figure 2.13c).

NCs of $FeNi_3$ were obtained by hydrothermal synthesis under microwave radiations.[33] NCs were synthesized in one step by reducing iron(III)acetylacetonate and nickel(II)acetylacetonate with hydrazine in ethylene glycol solution without any template under a rapid and economical microwave irradiation. Chains with a diameter of 150 to 550 nm and lengths up to several tens of μm consisted of an alignment of nanospheres. These nanospheres resulted from an assembly of NPs. The size of the NC was controlled by the amount of precursors.

Finally, very recently, Situ et al.[65] have synthesized NCs constituted of Fe_3O_4 NPs covered by a carbon layer. The approach involves the carbonization of glucose in the presence of wüstite (FeO) NPs, leading to the fabrication of rapidly acting and potent antibacterial agents based on iron oxide@carbon NCs. By controlling the heating temperature, they control the conversion rate and chain length. The optimal chains in terms of colloidal stability and magnetic properties were obtained after heating for 4 h. These chains had an average length of 0.15 μm and a saturation magnetization of 95.4 emu/g.

2.3.3.3 Magnetic Electrospinning

Alignment of SPM iron oxide NPs can also be obtained by electrospinning of microfibers or magnetic electrospinning.

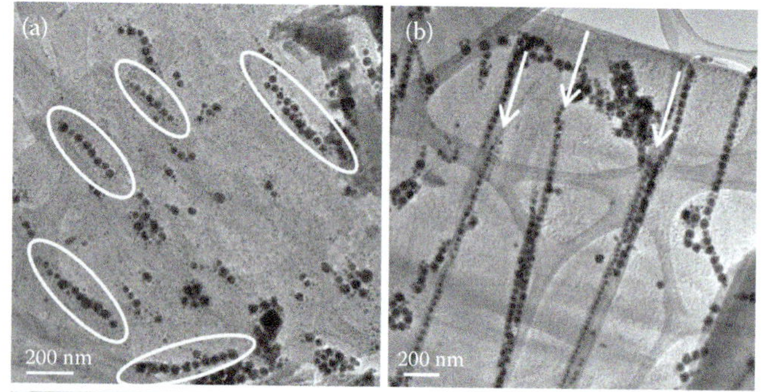

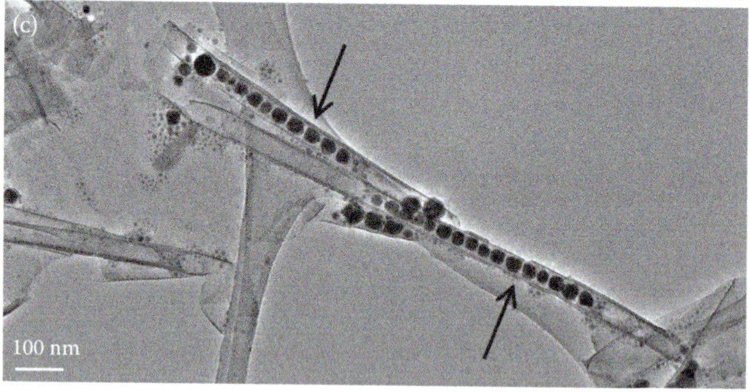

FIGURE 2.13 Free Co NC on a nanosheet (circled in white) (a), aligned peapods on a nanosheet (indicated by white arrows) (b) and free peapod structure in which the nanosheet appears to be curling from two sides to encase two adjacent NC (c).[65] (Reproduced with permission from Yao, Y., Chaubey, G.S., Wiley, J.B. 2012. Fabrication of nanopeapods: scrolling of niobate nanosheets for magnetic nanoparticle chain encapsulation. *JACS* 134:2450–2. Copyright 2012 American Chemical Society.)

Some examples exist in the literature.[66–70] Only two of these items are relevant in our case.

Roskov et al.[68] have used ε-polycaprolactone (PCL) with a molecular weight of 80 kDa. PCL was mixed with known amounts of 18 nm OA-coated iron oxide NPs (0.5, 1 and 2.5 vol%). The experiment was performed in the absence and in the presence of a magnetic field. In the absence of a magnetic field, the NPs were randomly distributed along the PCL fibre, while in the presence of a magnetic field, alignment of the NPs along the fibre was observed.

The method developed by Wang et al.[67] uses 7.5-nm magnetite NPs functionalized with PAA/Jeffamine copolymer and dispersed in a solution of polyethylene oxide (PEO) or poly (vinyl alcohol) (PVA). By electrospinning, magnetic fibres were formed: PEO fibres had a larger diameter (400 nm) compared to PVA/magnetite fibres, which had a diameter of 320 nm. Furthermore, the diameter of the PVA/magnetite fibres could be reduced to 140 nm by adding 1% sodium dodecyl sulfate to the solution. Magnetic measurements showed that the composite fibres have an SPM behaviour at 300 K. TEM analysis showed that in both types of fibres, NPs were aligned in columns parallel to the to the fibre axis direction within the fibre.

2.3.3.4 Microfluidics

The use of microfluidic devices for the synthesis of magnetic nanofibers presents many advantages: rapid prototyping, ease of manufacture and the ability to produce several fibres in parallel.

To our knowledge, only one "microfluidic" approach has been considered for the production of magnetic NP alignment and is the subject of a patent.[71] The main purpose of the invention is the assembly of magnetic particles in the form of permanent colloidal chains for use in the detection and/or analysis of specific species in a fluid. Magnetic microbeads are introduced into a channel and maintained in packing order by applying a magnetic field. The liquid containing the species to be analyzed flows through this stack. Hybridization is then detected by fluorescence. However, the concentrations of analytes used to demonstrate the principle of the method are significantly higher than those actually used in the chips, which suggests that the sensitivity is low.

2.4 APPLICATIONS

2.4.1 Individual Magnetic NPs

The first industrial applications of magnetic NPs date from the 1970s. Ferrofluids were first used in the fabrication of loudspeakers and turbo-molecular pump seals; later on, many other applications were developed, especially in electronics for data storage, the extraction of pollutants from water etc. Currently, thanks to their features, magnetic NPs are particularly exploited for their potential applications in the medical field,[72,73] commonly known as nanomedicine:

- To separate molecules of biological interest from their medium using NPs containing on their surfaces specific ligands by magnetic separation;
- As biosensors for *in vitro* diagnosis;
- As contrast agents for magnetic resonance imaging (MRI) to identify pathology as early as possible;
- As therapeutic agents via targeted drug delivery;
- For the destruction of tumour cells by hyperthermia; and
- In regenerative medicine to regenerate damaged tissues or organs.

Generally, one seeks to synthesize tailored NPs having the highest possible magnetic moment to increase their effectiveness in different applications. This is the reason why magnetic NCs are particularly interesting since their properties are significantly enhanced compared with isolated NPs. Control of the synthesis of these chains is fairly new: it is only in recent years that their potentiality has been explored, particularly in the field of life sciences.

2.4.2 1-D Assemblies – Applications in Life Sciences

2.4.2.1 Biomarkers and MRI Contrast Agents

Byrne et al.[54] studied NW suspensions, formed by assembling iron oxide NPs stabilized by PSSS (see Section 2.3.3.2), by MRI and relaxometry. At 37°C, a significant reduction in the relaxation times was observed for all tested magnetic field strengths and the r_1-r_2 ratios decreased, indicating an improvement in contrast generated by the NCs.[54] Thus, NCs have the potential to become agents that generate stronger contrast than isolated NPs, implying both a reduction of the injected dose and an improvement of the MRI signal for the detection of pathology.

Park et al. showed that when one changes the shape of the NPs from spherical to NWs, one can increase the number of targeting agents that can be functionalized to the surface of the NPs, resulting in a strong growth of cooperative effects with the receptors overexpressed on the surface of the cells (biomarker aspect) (see Section 2.3.3.2).[57] It also increases the value of the transverse relaxivity, leading to a stronger MRI contrast.[58,59,74] Similar effects were reported by Peiris et al.[64] with NCs consisting of an assembly of approximately four iron oxide NPs of 20 nm diameter, with surface functionalized by the cyclic RGD (Arg-Gly-Asp) peptide (cRGD), recognized as a targeting agent of integrin proteins overexpressed on tumour cells. They showed that the transverse relaxivity of the NCs was twice that of nanospheres alone and that the high surface-volume ratio of the NCs, *i.e.* the large number of cRGD coupled to the surface of the NCs, promotes the multivalency effect with respect to the integrins and, thus, early detection of metastatic lesions. Similar effects have been observed by Bolley et al.[75]

2.4.2.2 Therapy: Delivery of Medicines and Hyperthermia

Agrawal et al.[56] developed NWs consisting of five to seven iron oxide NPs, decorated on the surface by fluorophores and derivatives of the dendrimer polyamidoamine (PAMAM, 50–60 PAMAM per NW). These NWs were associated with small RNA (siRNA) by interactions with the PAMAM present on the surface. Thus, the final assembly comprised siRNA (drug aspect) and fluorophores, enabling both double MRI (related to the magnetic cores) and fluorescence in the infrared (thus avoiding autofluorescence of the tissues). The authors showed by *in vitro* experiments that with their NWs, the sustained release of the siRNA was more efficient than with the initial components. The experiments also showed that these NWs were well tolerated and blocked the EGFRvIII transcription factor involved in the proliferation of tumour cells in a model of glioblastoma.

Lu et al.[55] showed that $FeNi_3$ NCs (Section 2.3.3.1) present an effective response to microwave absorption over a wide frequency band and that these materials represent promising tools for generating effective temperature gradients for cancer therapy with hyperthermia.

Peiris et al. (Section 2.3.3.1) used an external stimulus to deliver a drug encapsulated in a liposome connected with arrays of iron oxide NPs 20 nm in diameter. More precisely, the whole structure is composed of three magnetic nanospheres (iron oxide) and one doxorubicin-loaded liposome assembled in a 100-nm-long chain (cf. Figure 2.12 for the assembly of the magnetic nanospheres). They showed that their scaffolding leads to preferential accumulation in the tumour because of the increased permeability of the blood vessels and that the external stimulus leads to effective delivery of medicinal products inside the tumour.[64]

2.4.2.3 Antibacterial Properties

Early studies indicated that fullerenes, single-walled carbon nanotubes and graphene oxide NPs showed potent antibacterial properties.[76] The proposed mechanism is based on a direct interaction of nanotubes with bacterial cells, causing major damage to the membrane and chemical stress-induced *in situ* production of reactive oxygen species. Situ et al. (Section 2.3.3.2) evaluated the antibacterial properties of Fe_3O_4 NCs coated with carbon on pathogenic *Escherichia coli* bacteria.[67] The studies showed that the bacterial effect depends on chain length and colloidal stability. Thus, 100% cell death was observed with shorter chains that also combined the best colloidal stability since the specific surface area offered by well-dispersed NCs enabled better interaction with the cells. The main interest of these magnetic NCs is related to the fact that the application of an external magnetic field makes it possible to extract and recycle them. The authors reported that the NCs could be reused up to five times without any loss of their antibacterial activity, but with the need to carry out a washing step between each cycle to remove cell debris.

2.4.2.4 Regenerative Medicine

Regenerative medicine aims to replace or repair tissues or damaged organs (blood vessels, articular cartilage and muscle).[77,78] Tissue engineering is designed to grow damaged tissues by developing tissue substitutes, making it possible to control the organization of the cells by applying a magnetic field. Depending on the source and the geometry of the magnetic field used, cells can be manipulated to form dense aggregates with a desired shape or to arrange in a monolayer covering a 3-D structure. To our knowledge, only one publication has reported the use of magnetic nanorods for this application. Gil et al.[79] reported that a better internalisation of the spherical NPs in cells leads to an improved response to an external magnetic field for the construction of cellular tissue. However, the authors did not report the magnetic properties of the two types of sample and it is therefore difficult to determine the impact of the shape compared with that of the magnetic properties for this application.

2.5 CONCLUSIONS

At the present time, a large majority of products in our daily life contain nanomaterials. Among the wide spectrum of materials involved in nanoscience and technology, the magnetic ones, either metallic or oxides, play a central role due to their specific properties on the one hand and the possibilities of self- or field-induced assembly provided by the magnetic interactions on the other hand. In parallel, new experimental strategies are being developed to organize these individual NPs into 1-, 2- and 3-D assemblies.

The principal motivation, besides the scientific challenges, is the emergence of new properties due to collective effects and surface-specific shape effects of magnetic NCs. The exploration of potential applications of these systems is only just beginning. Nevertheless, several studies have already indicated substantial promise for medical diagnosis and therapy. First, NCs have demonstrated potential to become agents that generate stronger contrast than isolated NPs, implying both a reduction in the injected dose and an improvement in the MRI signal for the detection of pathology. Second, studies of delivery of chemotherapy to tumours with multicomponent NCs have shown that their scaffolding leads to preferential accumulation in the tumour because of the increased permeability of the blood vessels and that an external stimulus leads to effective delivery of medicinal products inside the tumour. Third, NCs exhibit an effective response to microwave absorption over a wide frequency band and are thus promising tools for generating effective temperature gradients for cancer therapy with hyperthermia. Fourth, Fe_3O_4 NCs coated with carbon have shown highly efficient antibacterial action on pathogenic *E. coli* bacteria: 100% cell death was observed with shorter NCs that also exhibited good colloidal stability. Finally, rod-shaped magnetic NCs have shown an improved response to an external magnetic field for the construction of cellular tissue: here, there is additional interest in magnetic NCs due to the fact that the application of an external magnetic field makes it possible to extract and recycle them.

2.6 FUTURE DIRECTIONS

The importance of linear NCs of magnetic NPs is evident: improved magnetic properties and emergence of new capabilities that open the way to innovative applications. Many achievements have been made in the synthesis of NCs, demonstrating the potential of this kind of structure. The mastery of the synthesis of chains at the nanometric scale still remains, however, a considerable challenge: so far, it has not proved possible to control the number of assembled NPs, the chain length and the collective properties all at the same time. This research topic is extremely interesting in both fundamental and applied aspects. Future work should focus on mastering the assembly of individual SPM NPs and deeper exploration of potential applications of these linear nanosystems in the biomedical, environmental and energy fields.

REFERENCES

1. Le Sage, D., Arai, D., Glenn, D.R., DeVience, S.J., Pham, L.M., Rahn-Lee, L., Lukin, M.D., Yacoby, A., Komeili, A., Walsworth, R.L. 2013. Optical magnetic imaging of living cells. *Nature*. 496:486–9.
2. Faivre, D., Bennet, D. 2016. Magnetic nanoparticles line up. *Nature*. 535:235–6; Jiang, X., Feng, J., Huang, L., Wu, Y., Su, B., Yang, B., Mai, L., Jiang, L. 2016. Bioinspired 1D superparamagnetic magnetite arrays with magnetic field perception. *Adv Mater*. 28:6952–8.
3. de Trémolet de Lacheisserie, E. 1999. *Magnétisme*. EDP Science, Les Ulis, France.
4. Coey, J.M.D. 2010. *Magnetism and Magnetic Materials*. Cambridge University Press, Cambridge, UK.
5. Skomsky, R. 2003. Nanomagnetism. *J. Phys. Condensed Matter*. 15:R841–96.
6. Bedanta, S., Kleemann, W. 2009. Superparamagnetism. *J. Phys. D*. 42:013001.
7. Dormann, J.L., Fiorani, D., Tronc, E. 1997. Magnetic relaxation in fine-particle systems. *Advances in Chemical Physics*, I. Prigogine and S.A. Rice, Ed. Wiley, Hoboken, NJ. 98:283.
8. Hadjipanayis, G.C. 1999. Nanophase hard magnets. *J. Magn. Magn. Mater*. 200:373–91.
9. Hytch, M.J., Dunin-Borkowski, R.E., Scheinfein, M.R., Moulin, J., Duhamel, C., Mazaleyrat, F., Champion, Y. 2003. Vortex flux channeling in magnetic nanoparticle chains. *Phys. Rev. Lett*. 91:257207.
10. Russier, V. 2009. Spherical magnetic nanoparticles: Magnetic structure and interparticle interaction. *J. Appl. Phys*. 105: 073915.
11. Kechrakos, K., Trohidou, K.N. 1998. Magnetic properties of dipolar interacting single-domain particles. *Phys. Rev. B*. 58:12169; Jonsson, P.E., Garcia-Palacios, J.L. 2001. Thermodynamic perturbation theory for dipolar superparamagnets. *Phys. Rev. B*. 64:174416; Holm, C., Weis, J.-J. 2005. The structure of ferrofluids: A status report. *Curr. Opin. Colloid Interface Sci*. 10:133–40; Azeggagh, M., Kachkachi, H. 2007. Effects of dipolar interactions on the zero-field-cooled magnetization of a nanoparticle assembly. *Phys. Rev. B*. 75:174410; Russier, V., de Montferrand, C., Lalatonne, Y., Motte, L. 2013. Magnetization of densely packed interacting magnetic nanoparticles with cubic and uniaxial anisotropies: A Monte Carlo study. *J. Appl. Phys*. 114:143904.
12. Chen, J.P., Sorensen, C.M., Klabunde, K.J., Hadjipanayis, G.C. 1995. Enhanced magnetization of nanoscale colloidal cobalt particles. *Phys. Rev. B*. 51:11527–32; Tronc, E., Fiorani, D., Noguès, M., Testa, A.M., Lucari, F., D'Orazio, F., Greneche, J.M., Wernsdorfer, W., Galvez, N., Chanéac, C., Mailly, D., Jolivet, J.P. 2003. Surface effects in noninteracting and interacting γ-Fe_2O_3 nanoparticles. *J Magn. Magn. Mater*. 262:6–14.
13. Morales, M.P., Serna, C.J., Bodker, F., Morup, S. 1997. Spin canting due to structural disorder in maghemite. *J. Phys. Condens. Matter*. 9:5461–67; Morales, M.P., Veintemillas, S., Montero, M.I., Serna, C.J., Roig, A., Casas, L., Martinez, B., Sandiumenge, F. 1999. Surface and internal spin canting in γ-Fe_2O_3 nanoparticles. *Chem. Mater*. 11:3058–64; Daou, T.J., Pourroy, G., Bégin-Colin, S., Grenèche, J.M., Ulhaq-Bouillet, C., Legaré, P., Bernhardt, P., Leuvrey, C., Rogez, G. 2006. Hydrothermal synthesis of monodisperse magnetite nanoparticles. *Chem. Mater*. 18:4399–404.
14. Levesque, D., Weis, J.-J. 2011. Stability of solid phases in the dipolar hard sphere system. *Mol. Phys*. 109:2747–56.
15. Luttinger, J.M., Tisza, L. 1946. Theory of dipole interaction in crystals. *Phys. Rev*. 70:954–64.
16. Weis, J.-J. 2005. The ferroelectric transition of dipolar hard spheres *J. Chem. Phys*. 123:044503.
17. Hiroi, K., Kura, H., Ogawa, T., Takahashi, M., Sato, T. 2011. Spin-glasslike behavior of magnetic ordered state originating from strong interparticle magnetostatic interaction in α-Fe nanoparticle agglomerate. *Appl. Phys. Lett*. 98:252505; Nakamae, S. 2014. Out-of-equilibrium dynamics in superspin glass state of strongly interacting magnetic nanoparticle assemblies. *J. Magn. Magn. Mater*. 355:225–29.
18. Levesque, D., Weis, J.-J. 1994. Orientational and structural order in strongly interacting dipolar hard spheres. *Phys. Rev. E* 49:5131–40.
19. Weis, J.-J., Levesque, D. 2005. Simple dipolar fluids as generic models for soft matter. *Adv. Polym. Sci*. 165:164.
20. Tavares, J.M., Weis, J.-J., Telo de Gama, M.M. 1999. Strongly dipolar fluids at low densities compared to living polymers. *Phys. Rev. E*. 59:4388–95.
21. Tavares, J.M., Weis, J.-J., Telo de Gama, M.M. 2002. Quasi-two-dimensional dipolar fluid at low densities: Monte Carlo simulations and theory. *Phys. Rev. E*. 65:061201.
22. Weis, J.-J. 2003. Simulation of quasi-two-dimensional dipolar systems. *J. Phys. Condens. Matter*. 15:S1471–95.
23. Lomba, E., Lado, F., Weis, J.-J. 2000. Structure and thermodynamics of a ferrofluid monolayer. *Phys. Rev. E*. 61:3838–49.
24. Weis, J.-J., Tavares, J.M., Telo da Gama, M.M. 2002. Structural and conformational properties of a quasi-two-dimensional dipolar fluid. *J. Phys. Condens. Matter*. 14:9171–86.
25. De'Bell, K., MacIsaac, A.B., Booth, I.N., Whitehead, J.P. 1997. Dipolar-induced planar anisotropy in ultrathin magnetic films. *Phys. Rev. B*. 55:15108–118; Russier, V. 2001. Calculated magnetic properties of two-dimensional arrays of nanoparticles at vanishing temperature. *J. Appl. Phys*. 89:1287–94.
26. Tang, Z., Kotov, N., Giersig, M. 2002. Spontaneous organization of single CdTe nanoparticles into luminescent nanowires. *Science*. 297:237–40.
27. Wang, H., Chen, Q.W., Sun, Y.B., Wang, M.S., Sun, L.X., Yan, W.S. 2010. Synthesis of necklace-like magnetic nanorings. *Langmuir*. 26:5957–62.
28. Klokkenburg, M., Vonk, C., Claesson, E.M., Meeldijk, J.D., Erné, B.H., Philipse A.P. 2004. Direct imaging of zero-field dipolar structures in colloidal dispersions of synthetic magnetite. *JACS*. 126:16706–7.
29. Korth, B.D., Keng, P., Shim, I., Bowles, S.E., Tang, C., Kowalewski, T., Nebesny, K.W., Pyun, J. 2006. Polymer-coated ferromagnetic colloids from well-defined macromolecular surfactants and assembly into nanoparticle chains. *JACS*. 128:6562–3.
30. Hill, L.J., Pyun, J., 2014. Colloidal polymers via dipolar assembly of magnetic nanoparticle monomers. *ACS Appl. Mater. Interfaces*. 6: 6022–32.
31. Hill, L.J., Richey, N., Sung, Y., Dirlan, P., Griebel, J., Lavoie-Higgins, E., Shim, I., Pinna, N., Willinger, M.G., Vogel, W., Benkoski, J.J., Char, K., Pyun, J. 2014. Colloidal polymers from dipolar assembly of cobalt-tipped CdSe@CdS nanorods. *ACS Nano*. 8:3272–84.

32. Gao, M.R., Zhang, S.R., Xu, Y.F., Zheng, Y.-R., Jiang, J., Yu, S.H. 2014. Self-assembled platinum nanochain networks driven by induced magnetic dipoles. *Adv. Funct. Mater.* 24:916–24.
33. Jia, J., Yu, J.C., Wang, Y.X.J., Chan, K.M. 2010. Magnetic nanochains of $FeNi_3$ prepared by a template-free microwave-hydrothermal method. ACS *Appl. Mater. Int.* 2:2579–84.
34. Wang, H., Li, J., Kou, X., Zhang, L. 2008. Synthesis and characterizations of size-controlled $FeNi_3$ nanoplatelets. *J. Cryst. Growth.* 310:3072–6.
35. Tripp, S.L., Pusztay, S.V., Ribbe, A.E., Wei, A. 2002. Self-assembly of cobalt nanoparticle rings. *JACS.* 124:7914–5.
36. Gao, M.R. Zhang, S.R., Jiang, J., Zheng, Y.R., Tao, D.Q., Yu, S.H. 2011. One-pot synthesis of hierarchical magnetite nanochain assemblies with complex building units and their application for water treatment. *J. Mater. Chem.* 21, 16888–92.
37. Lalatonne, Y., Richardi, J., Pileni, M.P. 2004. Van der Waals versus dipolar forces controlling mesoscopic organizations of magnetic nanocrystals. *Nature Mater.* 3:121–5; Lalatonne, Y., Motte, L., Richardi, J., Pileni, M.P. 2005. Influence of short-range interactions on the mesoscopic organization of magnetic nanocrystals. *Phys. Rev. E.* 71:011404.
38. Lalatonne, Y., Motte, L., Russier, V., Ngo, A.T., Bonville, P., Pileni, M.P. 2004. Mesoscopic structures of nanocrystals: Collective magnetic properties due to the alignment of nanocrystals. *J. Phys. Chem. B.* 108:1848–54.
39. Ge, J., Hu, J., Biasini, M., Beyermann, W.P., Yin, Y. 2007. Superparamagnetic magnetite colloidal nanocrystal clusters. *Angew Chem. Int. Ed.* 46: 4342–5.
40. Hu, Y., He, L., Yin, Y. 2011. Magnetically responsive photonic nanochains. *Angew Chem. Int. Ed.* 50:3747–50.
41. Kralj, S., Makovec, D. 2015. Magnetic assembly of superparamagnetic iron oxide nanoparticle clusters into nanochains and nanobundles. *ACS Nano.* 9(10):9700–7.
42. Zhou, J., Wang, P., Messersmith Ph.B., Duan, H., 2015. Multifunctional magnetic nanochains: Exploiting self-polymerization and versatile reactivity of mussel-inspired polydopamine. *Chem. Mater.* 27:3071.
43. Blakemore, R. 1975. Magnetotactic bacteria. *Science.* 190:377–9.
44. Bazylinski, D.A., Frankel, R.B. 2004. Magnetosome formation in prokaryotes. *Nature Rev. Microbiol.* 2:217–30.
45. Postec, A., Tapia, N., Bernadac, A., Joseph, M., Davidson, S., Wu, L.F., Ollivier, B., Pradel, N. 2012. Magnetotactic bacteria in microcosms originating from the French Mediterranean Coast subjected to oil industry activities. *Microb. Ecol.* 63:1–11.
46. Grzelczak, M., Pérez-Juste, J., Rodríguez-González, B., Spasova, M., Barsukov, I., Farle, M., Liz-Marzàn, L.M. 2008. Pt-catalyzed growth of Ni nanoparticles in aqueous CTAB solution. *Chem. Mater.* 20:5399–405.
47. Varon, M., Beleggia, M., Kasama, T., Harrison, R.J., Dunin-Borkowski, R.E., Puntes, V.F., Frandsen, C. 2013. Dipolar magnetism in ordered and disordered low-dimensional nanoparticle assemblies. *Sci. Rep.* 3:1234.
48. Bertoni, G., Torre, B., Falqui, A., Fragouli, D., Athanassiou, A., Cingolani, R. 2011. Nanochains formation of superparamagnetic nanoparticles. *J. Phys. Chem. C.* 115:7249–54.
49. Richardi, J., Motte, L., Pileni, M.P. 2004. Mesoscopic organizations of magnetic nanocrystal: The influence of short-range interactions. *Curr. Opin. Colloid Interface Sci.* 9:185–91.
50. Sheparovych, R., Sahoo, Y., Motornov, M., Wang, S., Luo, H., Prasad, P.N., Sokolov, I., Minko, S. 2006. Polyelectrolyte stabilized nanowires from Fe_3O_4 nanoparticles via magnetic field induced self-assembly. *Chem. Mater.* 18:591–3.
51. Yan, M., Fresnais, J., Berret, J.F. 2010. Growth mechanism of nanostructured superparamagnetic rods obtained by electrostatic co-assembly. *Soft Matter.* 6:1997–2005.
52. Singh, G., Chan, H., Baskin, B., Gelman, E., Repnin, N., Král, P., Klajn, R., 2014. Self-assembly of magnetite nanocubes into helical superstructures. *Science.* 345:1149–53.
53. Sone, E.D., Stupp, S.I. 2011. Bioinspired magnetite mineralization of peptide–amphiphile nanofibers. *Chem. Mater.* 23:2005–7.
54. Byrne, S.J., Corr, S.A., Gun'ko, Y.K., Kelly, J.M., Brougham, D.F., Ghosh, S. 2004. Magnetic nanoparticle assemblies on denatured DNA show unusual magnetic relaxivity and potential applications for MRI. *Chem. Comm.* 22:2560–1; Corr, S.A., Byrne, S.J., Tekoriute, R., Meledandri, C.J., Brougham, D.F., Lynch, M., Kerskens, C., O'Dwyer, L., Gun'ko, Y.K. 2008. Linear assemblies of magnetic nanoparticles as MRI contrast agents. *JACS.* 130:4214–15.
55. Lu, X., Liu, Q., Huo, G., Liang, G., Sun, Q., Song, X. 2012. CTAB-mediated synthesis of iron–nickel alloy nanochains and their magnetic properties. *Coll. Surf. A Physicochem. Eng. Aspects.* 407:23–8.
56. Agrawal, A., Min, D.H., Singh, N., Zhu, H., Birjiniuk, A., von Maltzahn, G., Harris, T.J., Xing, D., Woolfenden, S.D., Sharp, P.A., Charest, A., Bhatia, S. 2009. Functional delivery of siRNA in mice using dendriworms. *ACS Nano.* 3:2495–504.
57. Park, J-H., von Maltzahn, G., Zhang, L., Schwartz, M.P., Ruoslahti, E., Bhatia, S.N., Sailor, M.J. 2008. Magnetic iron oxide nanoworms for tumor targeting and imaging. *Adv. Mater.* 20:1630–5; Park, J.H., von Maltzahn, G., Zhang, L., Derfus, A.M., Simberg, D., Harris, T.J., Ruoslahti, E., Bhatia, S.N., Sailor, M.J. 2009. Systematic surface engineering of magnetic nanoworms for *in vivo* tumor targeting. *Small.* 5:694–700.
58. Lin, K.Y., Kwong, G.A., Warren, A.D., Wood, D.K., Bhatia, S.N. 2013. Nanoparticles that sense thrombin activity as synthetic urinary biomarkers of thrombosis. *ACS Nano.* 7:9001–9.
59. Park, J.H., von Maltzahn, G., Xu, M.J., Fogal, V., Kotamraju, V.R., Ruoslahti, E., Bhatia, S.N., Sailor, M.J. 2010. Cooperative nanomaterial system to sensitize, target, and treat tumors. *Proc. Natl. Acad. Sci.* 107:981–6.
60. Lo, J.H., von Maltzahn, G., Douglass, J., Park, J.H., Sailor, M.J., Ruoslahti, E., Bhatia, S.N. 2013. Nanoparticle amplification via photothermal unveiling of cryptic collagen binding sites. *J. Mater. Chem. B Mater. Biol. Med.* 1:5235–40.
61. Palchoudhury, S., Xu, Y., Goodwin, J., Bao, Y. 2011. Synthesis of iron oxide nanoworms. *J. Appl. Phys.* 109:07E314.
62. Nakata, K., Hu, Y., Uzun, O., Bakr, O., Stellacci, F. 2008. Chains of superparamagnetic nanoparticles. *Adv. Mater.* 20:4294–9.
63. Peiris, P.M., Schmidt, E., Calabrese, M., Karathanasis, E., 2011. Assembly of linear nano-chains from iron oxide nanospheres with asymmetric surface chemistry. *PLoS ONE.* 6:e15927.
64. Peiris, P.M., Bauer, L., Toy, R., Tran, E., Pansky, J., Doolitle, E., Schmidt, E., Hayden, E., Mayer, A., Keri, R.A., Griswold, M.A., Karathanasis, E. 2012. Enhanced delivery of chemotherapy to tumors using a multicomponent nanochain with radio-frequency-tunable drug release. *ACS Nano.* 6:4157–68.
65. Yao, Y., Chaubey, G.S., Wiley, J.B. 2012. Fabrication of nanopeapods: Scrolling of niobate nanosheets for magnetic nanoparticle chain encapsulation. *JACS* 134:2450–2.
66. Situ, S.F., Samia, A.C.S. 2014. Highly efficient antibacterial iron oxide@carbon nanochains from Wüstite precursor nanoparticles. *ACS Appl. Mater. Interfaces.* 6:20154–63.

67. Wang, M., Singh, H., Hatton, T.A., Rutledge, G.C. 2004. Field-responsive superparamagnetic composite nanofibers by electrospinning. *Polymer.* 45:5505–14.
68. Roskov, K.E., Atkinson, J.E., Bronstein, L.M., Spontak, R.J. 2012. Magnetic field-induced alignment of nanoparticles in electrospun microfibers. *RSC Adv.* 2:4603–7.
69. Li, D., Herricks, T., Xia, Y. 2003. Magnetic nanofibers of nickel ferrite prepared by electrospinning. *Appl. Phys. Lett.* 83:4586–8.
70. Yang, D., Lu, B., Zhao, Y., Jiang, X. 2007. Fabrication of aligned fibrous arrays by magnetic electrospinning. *Adv. Mater.* 19:3702–6.
71. Irreversible colloidal chains with recognition sites. Patent WO 2003071276 A1.
72. Thanh, N.T.K. (Ed.) (2012). *Magnetic Nanoparticles: From Fabrication to Clinical Applications.* Boca Raton, London, New York: CRC Press, Taylor & Francis.
73. Pankhurst, Q.A., Thanh, N.T.K., Jones, S.K., Dobson, J. 2009. Progress in applications of magnetic nanoparticles in biomedicine. *J. Phys. D Appl. Phys.* 42:224001.
74. Gossuin, Y., Disch, S., Vuong, Q.L., Gillis, P., Hermann, R.P., Park, J.H., Sailor, M.J. 2010. NMR relaxation and magnetic properties of superparamagnetic nanoworms. *Contrast Media Mol. Imaging.* 5:318–22.
75. Bolley, J., Lalatonne, Y., Haddad, O., Letourneur, D., Soussan, M., Pérard-Viret, J., Motte, L. 2013. Optimized multimodal nanoplatforms for targeting α(v)β3 integrins. *Nanoscale.* 5:11478–89; Bolley, J., Guenin, E., Lievre, N., Lecouvey, M., Soussan, M., Lalatonne, Y., Motte, L. 2013. Carbodiimide versus click chemistry for nanoparticle surface functionalization: A comparative study for the elaboration of multimodal superparamagnetic nanoparticles targeting αvβ3 integrins. *Langmuir.* 29:14639–47.
76. Dizaj, S.M., Mennati, A., Jafari, S., Khezri, K., Adibkia, K. 2015. Antimicrobial activity of carbon-based nanoparticles. *Adv. Pharm. Bull.* 5:19–23.
77. Ito, A., Kamihira, M. 2011. Tissue engineering using magnetite nanoparticles. *Prog. Mol. Biol. Transl. Sci.* 104:355–395.
78. Yamamoto, Y., Ito, A., Fujita, H., Nagamori, E., Kawabe, Y., Kamihira, M. 2011. Functional evaluation of artificial skeletal muscle tissue constructs fabricated by a magnetic force-based tissue engineering technique. *Tissue Eng. Part A.* 17:107–114.
79. Gil, S., Correia, C.R., Mano, J.F. 2015. Magnetically labeled cells with surface-modified Fe_3O_4 spherical and rod-shaped magnetic nanoparticles for tissue engineering applications. *Adv. Healthcare Mater.* 4:883–91.

Laurence Motte (E-mail: laurence.motte@univ-paris13.fr) is a professor at the University of Paris 13 in the Laboratory for Vascular Translational Science. She earned her PhD at the LM2N laboratory (Université Paris 6) on the synthesis and photophysical properties of quantum dots. She was appointed as an assistant professor in the same laboratory and studied the various parameters involved in the self-organization of nanoparticles. She obtained a professor position at the Université Paris 13 in 2005 and, since 2006, has been managing the Nanomaterials group first in CSPBAT laboratory and since 2015 in the LVTS laboratory. The two main expertise areas of the team are the synthesis and characterization of inorganic nanoparticles of different sizes, shapes and compositions and the surface functionalization of these hybrid inorganic nanomaterials for biological and environmental applications and in translational research with industrial (Magnisense, Neelogy) or clinical (CHU Avicenne, Neurospin, CEA) applications.

Irena Markovic-Milosevic conducted her studies in Paris, France, where she earned her BSc in physical chemistry and master's degree (with honours) in materials science at the University of Paris 7, France, in 2006. From 2006 to 2009, she completed her doctoral studies at the University of Orléans, France, with Prof. M.-L. Saboungi. During her thesis, she worked on a thermomagnetic delivery system composed of magnetic nanoparticles and structured emulsions embedded in a hydrogel. In 2010, she joined the group of Prof. L. Motte as a postdoctoral fellow at the University of Sorbonne-Paris Cité, France. She conducted her research on the development (synthesis and surface biofunctionalization) and the evaluation of magnetic nanomaterials for point-of-care diagnostics. She was then hired as a research engineer for 3 years (2011–2014) in a start-up company named Neelogy working on scale-up synthesis of nanoparticles and magnetic nanocomposites for "harsh environment" current sensor. In 2014, after a 5-month stay at Valsem industry (Lachelle, France) as manager in advanced technology, she became a research associate in the University of Paris 13 and lecturer. In 2016, she moved to Switzerland at the Institute of Materials (LTP) at EPFL as scientific collaborator with Prof. H. Hofmann. Her research area includes the synthesis, processing, and characterization of nanostructured materials based on various kinds of nanoparticles, the modification of their surfaces using colloidal surface chemistry and their assembly.

Vincent Russier works in the Institute of Chemistry and Materials at Paris-East, mainly devoted to material sciences. He is a theoretician in condensed matter physics and statistical physics focusing mainly on numerical simulations of magnetic materials properties from Monte Carlo simulations or micromagnetism. Most of his present work concerns the condensed phases of magnetic nanoparticles.

Marie-Louise Saboungi conducts her research at the Pierre et Marie Curie University, Paris, and Institute of Functional Nano & Soft Materials (FUNSOM) in Suzhou, China. She is a fellow of American Physical Society, American Association for the Advancement of Sciences and Alexander von Humboldt Foundation. She served as director of a research institute in Orleans (Centre de la Recherche sur la Matière Divisée) and as director of the Materials, Energy and Geosciences Thrust Area at the University of Orleans. Prior to that, she was a senior scientist at Argonne National Laboratory with close collaborations with the University of Chicago's James Franck Institute and with Cornell University. In the United States, she has served on national committees for the Department of Energy and the National Science Foundation; in France for the Centre National de la Recherche Scientifique and the Agence d'Evaluation de la Recherche et de l'Enseignement Supérieur; in Germany for the Helmholtz Zentrum Berlin; and in the European Community on the Future and Emerging Technologies (FET) Advisory Board for Horizon 2020. She has organized and chaired over 50 international conferences and workshops including two Gordon Research Conferences. She actively participated in the Association of Women in Science and received an Award for Leadership from the Young Women's Christian Association (YWCA) of Metropolitan Chicago. At present, she is involved in investigating complex soft materials with a special interest in the thermal, magnetic and biomedical properties of functionalized nanomaterials with a view to applications in energy and biotechnology.

3 Carbon-Coated Magnetic Metal Nanoparticles for Clinical Applications

*Martin Zeltner and Robert N. Grass**

CONTENTS

3.1 SYNTHESIS OF CARBON-COATED NANOMAGNETS

In contrast to the wide range of liquid phase synthesis procedures for the formation of iron oxide nanoparticles as described in Chapter 1 of this book, the formation of carbon-coated metallic nanoparticles has been so far restricted to the gas phase. Carbon-coated nanoparticles were first observed in carbon arc processes,[1,2] very similar to the processes leading to the original discovery of carbon nanotubes (CNTs).[3] Whereas carbon arc high-temperature processes are still a route to carbon-coated nanomaterials, flame synthesis has evolved as a useful route to manufacture these materials at high qualities and useful quantities.[4,5]

Flame synthesis is a commercial large-scale process, which has its origins in the formation of carbon black (tire rubber additive), silica (e.g. aerosil) and titania (white pigment).[6] Over the last decade, this process has also been adapted for the formation of a wide range of oxide nanomaterials and also metallic nanoparticles.[7,8] For the formation of nonnoble metallic nanomaterials, the flame process has to be performed at well-controlled fuel to oxygen combustion conditions (reducing flame). Additionally, an oxygen-free environment (e.g. in glovebox) is required to prevent the reoxidation of the metal in the off-gas of the flame.

The magnetic metals of nickel, cobalt and iron are accessible by flame synthesis, as well as their alloys.[9] However, metallic nanoparticles of these nonnoble elements are not air stable; if such materials are removed from the protective atmosphere after synthesis, they may ignite (be oxidized) spontaneously and are therefore not useful as magnetic carriers. Only by adapting flame synthesis via injecting acetylene (a carbon forming 'monomer') into the flame atmosphere is it possible to form a thin carbon layer around the individual nanoparticles in a self-limiting process (i.e. as soon as a few layers are present, carbon deposition ceases).[10] This carbon layer consists of two to five layers of 'graphene-like' sp^2 hybridized carbon.[8] Carbon deposition has only been reported on Cu, Co, Fe and Ni nanoparticles, metals that are known from CNTs for their interaction with carbon (either by carbon dissolution or the formation of carbides).[11] Alternative gas-phase methods for the preparation of carbon-coated nanomagnets are high-pressure chemical vapour deposition[12] or hydrogen reduction of cobalt chloride in the presence of ethene.[13]

3.2 PHYSICAL PROPERTIES: HOW METAL NANOMAGNETS ARE DIFFERENT FROM IRON OXIDES

Due to the strong difference in chemical composition, carbon-coated metallic (Co, Fe, Ni) nanoparticles differ considerably from iron oxides in their physical properties. Firstly, the achievable saturation magnetizations are higher due to the higher concentration of magnetic element within the material (no oxygen).[14] Reported mass saturation magnetizations of 158 Am^2/kg for cobalt (30 nm)[10] and 140 Am^2/kg for iron carbide nanoparticles (30 nm)[15] are close to the reported bulk values of these materials (166 Am^2/kg and 140 Am^2/kg, respectively)[16] and are nearly twice as high as the most magnetic iron oxide nanoparticles. As a consequence, metallic nanoparticles can be displaced and collected in field gradients at significantly larger speeds than iron oxide nanoparticles of equal size.[17] This effect is additionally amplified due to field-induced nanoparticle aggregation (individual nanoparticles of higher magnetization have a stronger attraction to each other).[18] This effect allows the rapid separation of magnetic nanoparticles in relatively small field gradients. The magnetic force generated on individual nanoparticles in such fields is usually too small to induce a significant displacement of the particles towards the surface of the magnet. However, experimentally, it can be evidenced that

* Corresponding author.

the particles travel towards the magnet within seconds.[19] This effect can only be explained by cooperative phenomenon of magnetically induced particle aggregation and magnetic attraction of the aggregates.[20] In terms of potential clinical applications utilizing magnetic separation (see Section 3.6), the higher magnetization of the particles in relevant magnetic field gradients has a strong influence on the particle travel velocity. It is expected not only that this property decreases required contact/treatment times but also that a smaller number of particles (if any) can escape separation under a given treatment setup, lowering the possibility of nanoparticle-induced adverse effects if they are to be used in *in vivo* settings.

Also, metallic nanoparticles at obtainable particle sizes (20–40 nm) are ferromagnetic (as opposed to superparamagnetic), displaying a hysteresis with a remanence of ca. 150–300 Oe.[12,15] As a result, dispersions of carbon-coated nanoparticles have a tendency to aggregate due to the enhanced dipole interactions between adjacent particles in the solution. However, by suitable surface modification (see Section 3.4), this attractive interaction can be overcome by steric and electrostatic repulsion, and stable dispersion of metallic nanoparticles can be formed in water and many relevant biological media.[21,22]

A second difference in properties lies within the chemical stability of the nanoparticles. Although metallic nanoparticles as such are highly unstable, the carbon coating gives the underlying metal protection from the environment. This is most directly exemplified by the thermal stability of the material: without carbon coating, it is unstable in air and ignites spontaneously, whereas with carbon coating, it is stable in air up to ~180°C (see Figure 3.1c). At this temperature, the carbon protection breaks down, and the metallic core oxidizes rapidly, resulting in a corresponding weight increase (oxidation of the metal to the metal oxide). Not only does the carbon coating ensure stability in air, but also, carbon-coated metal nanoparticles have extraordinary chemical stabilities: they are near to indestructible in water at a wide range of pH values (Figure 3.1b) and are tolerant to all standard organic solvents.[23] Only very prolonged treatments under oxidizing conditions and elevated temperatures are sufficient to attack the carbon surface and dissolve the metal core.[24] Although it is not obvious, this property enables a greatly increased shelf life of nanoparticle loaded buffer solutions, and a high chemical inertia of these nanoparticles further allows the application of harsh treatment/wash cycles during *in vitro* diagnostic procedures.

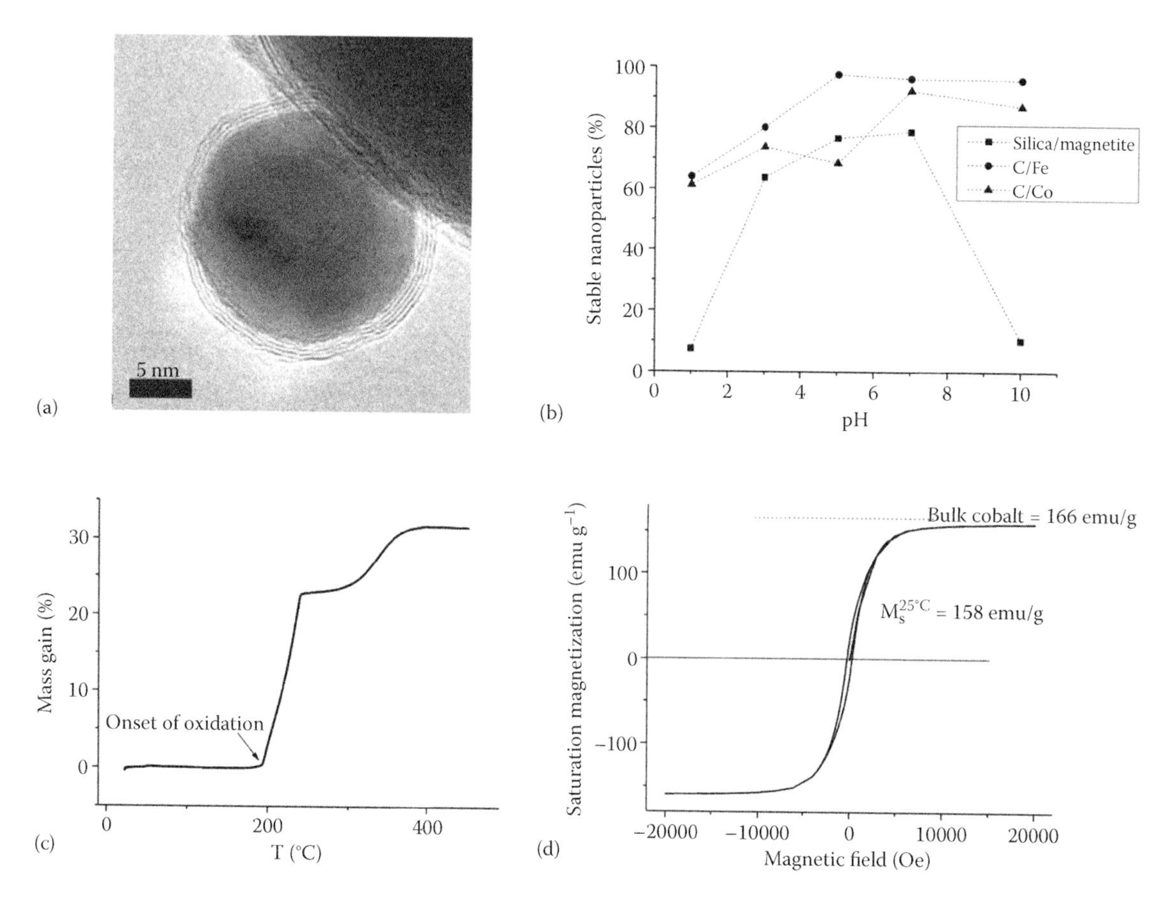

FIGURE 3.1 Carbon-coated cobalt nanoparticle (a). (Reproduced with permission from Grass, R.N., Athanassiou, E.K., Stark, W.J., *Angew Chem. Int. Ed.*, 46, 4909–4912, 2007. Copyright 2007 John Wiley & Sons.) Carbon-coated metallic particles have a high chemical stability under extreme pH conditions not suitable for iron oxide nanoparticles (b), only by heating the particles in air to over 180°C the carbon coating decomposes and the metal ignites resulting in a measurable weight gain due to the formation of metal oxides (c). Magnetic hysteresis loop of carbon-coated cobalt nanoparticles (d) with a saturation magnetization close to the saturation of bulk cobalt.

An application of carbon-coated nanomagnets, resulting directly from the increased magnetization and chemical resistance to a wide range of conditions, is the concentration of gold and palladium from precious metal leaching solutions.[25] In combination with the opportunities of chemical derivatization, it might be possible that these NPs can also be useful for clinical applications (see following chapters in this book).

3.3 INITIAL CHEMICAL DERIVATIZATION OF THE CARBON SURFACE

Resulting from the chemical inertia of carbon surfaces, direct modification of these surfaces by straightforward chemical means has been of limited success. However, it is known from the printing-ink industry (black printing ink = surface modified carbon black) that carbon surfaces are susceptible to radical chemistry, especially to reaction with aryl radicals.[26] A convenient way to surface modify carbon-coated nanoparticles therefore involves the usage of anilines, which can be converted to aryl radicals via diazonium intermediates (Figure 3.2). This chemistry is hassle free as it can be performed in water at room temperature within 1 hour. In contrast to the better known azo-coupling reaction, the reaction of the diazo compound with the particles under acidic conditions involves the formation of an aryl radical, which reacts with the particle surface to form C–C bonds.[27] Confirmation of this reaction route comes from the visible formation of nitrogen gas (bubbling) and the absence of nitrogen in the final material.[10] Standard chemical analytics for these materials involves Fourier-transform infrared spectroscopy for qualitative inspection and elemental microanalysis (C, H, N, S, but also Cl, I and Br) for functional group quantification. This chemistry allows the introduction of a select range of chemical functionalities at a high surface coverage (ca. 0.01 mmol/m^2 = 6 molecules/nm^2), which is close to the theoretical limit (space filling, see Figure 3.2). From here, the next step of conjugation for biomedical application is similar to that of iron oxide nanoparticles, which has been covered by Thanh et al.[28] and in Chapter 5 of this book.

It may be noted that such chemically functionalized particles serve several applications outside of the biomedical space, involving not only water detoxification[29] but also the formation of magnetically actuating polymers,[30] magnetic printing inks,[31] novel analytical tools[32] and magnetic catalysts.[33,34]

3.4 APPLICATION OF SPECIFIC SURFACE FUNCTIONALIZATIONS

Although the introduction of first functionalities allows the attachment of small molecules[35] and proteins[36] to the particle surface, the particles remain hydrophobic and retain their tendency to aggregate in aqueous solutions. As a result, their versatility in biomedical applications remains limited. However,

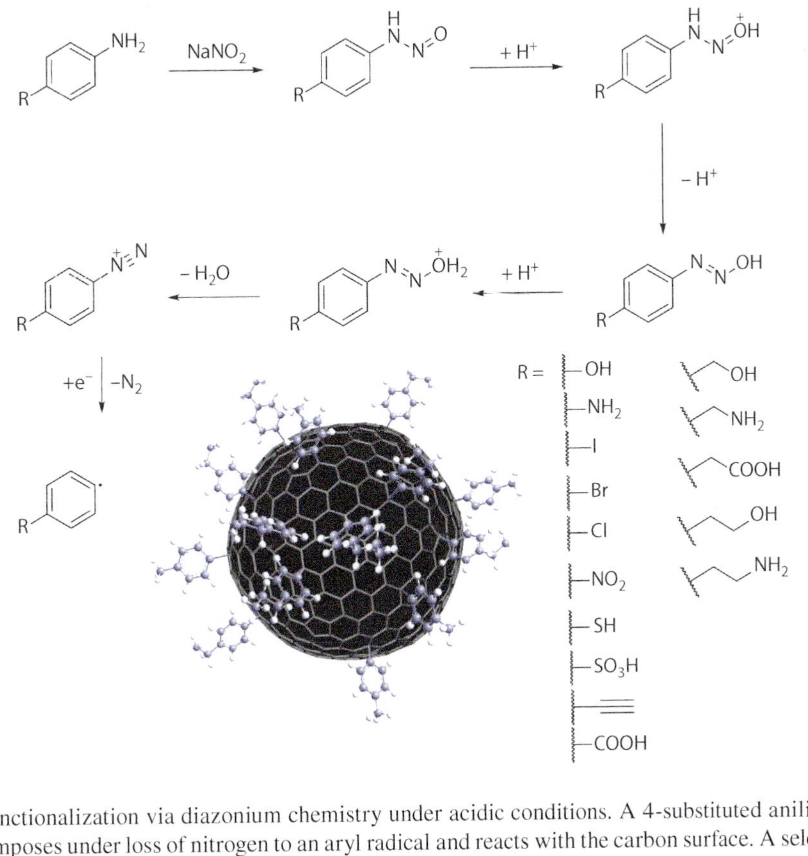

FIGURE 3.2 Particle functionalization via diazonium chemistry under acidic conditions. A 4-substituted aniline is converted to a diazo intermediate, which decomposes under loss of nitrogen to an aryl radical and reacts with the carbon surface. A select range of simple surface functionalities is thereby accessible. As shown in the scheme, the obtainable surface density is relatively high (ca. every 12th surface carbon atom functionalized).

the initial surface functionalities can be utilized as a handle to introduce more complex and oligomeric coatings, by which the surface nature of the particles can be completely altered. Whereas it has been shown that a wide range of chemistry is possible on the surface of the particles (Figure 3.3), the most versatile approaches are the introduction of short polymers (oligos) by surface initiated atom transfer radical polymerization (SI-ATRP)[37] and the straightforward attachment of ligands by click chemistry.[38] The advantages of SI-ATRP come from the wide range of available monomers, which allow the introduction of various functionalities, including cationic and anionic charges,[22] temperature switches[39] and reactive sites.[21] By changing the length of the oligomers, the hydrophobicity (protein fouling) and aggregation stability can be tuned for a wide range of biochemically relevant fluids (see Figure 3.4 and Hofer et al.[21]). In more detail, metallic nanoparticles coated with oligomers of less than six 3-sulfopropyl methacrylate monomer units (i.e. carrying less than ~ 60,000 individual charges per particle) had a tendency to aggregate in water and could be rapidly removed from solution using a standard solid-state magnet taking advantage of the previously described field-induced aggregation. If the particles were coated with oligomers of six or more monomeric building blocks, the particles remained colloidally stable in aqueous solution for at least 2 days and also remained stable in biologically more relevant buffer systems, such as protein loaded phosphate buffered saline (PBS) and Dulbecco's modified Eagle medium (DMEM). As a result, the polymer length can be utilized as a variable to tune the stability of the particles in various aqueous media and to design particles, which are better suited for either magnetic separation applications or magnetic detection applications. By additionally carefully adapting the surface chemistry by choice of functional side group, end group and length of the oligomers, particles can be formed to serve in a specific biomedical application. An example is the formation of a magnetic click-and-release system with which a protein can be specifically bound to the particle surface by copper-free click chemistry.[40] After magnetic collection of the particles, the protein can be decoupled from the particle surface unharmed by bio-orthogonal chemistry.

Alternative approaches for the formation of oligomeric structures on the particle surface to enable an increase in the functional group density are the formation of dendrimers.[41]

3.5 CARBON-COATED METAL NANOMAGNETS FOR DIAGNOSTICS

The usage of magnetic particles in diagnostics can be divided into two parts, one where the particles are utilized for the collection and concentrating of an analyte of interest,[42,43] the other where the magnetic nature of the particles is part of the analytical routine.[44] In terms of concentrating of analytes (cell sorting, immunoprecipitation, nucleic acid purification etc.), metallic particles can be utilized instead of iron oxide particles in nearly any application. The advantages of this transition include a faster nanoparticle recovery (higher

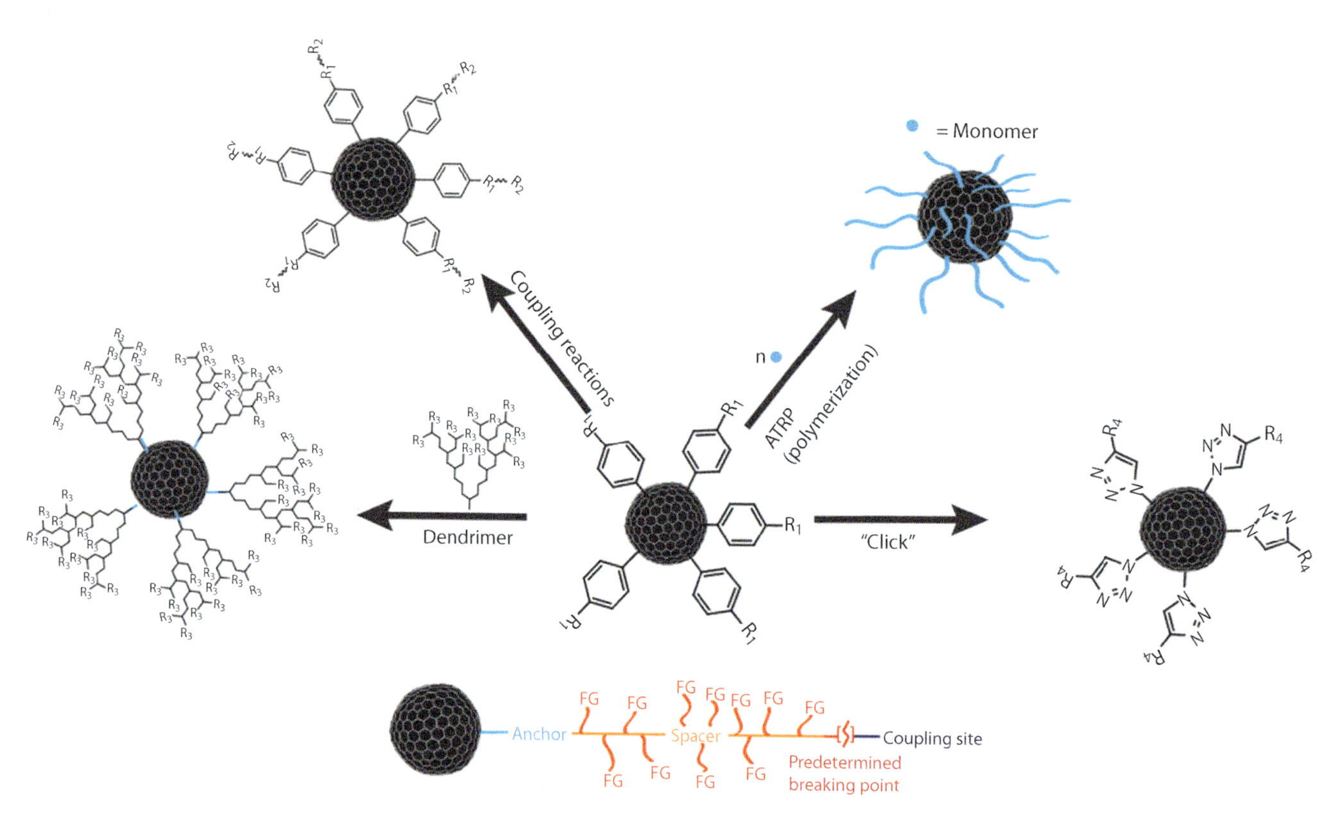

FIGURE 3.3 Initial functionality introduced via diazonium chemistry is used as an anchor to introduce more complex functionalities, using standard organic chemistry and polymer chemistry routines. In a schematic representation (below), this enables the introduction of a functional group loaded spacer and a reactive end group. Also additional functionalities, such as predefined cleavage sites can be introduced.

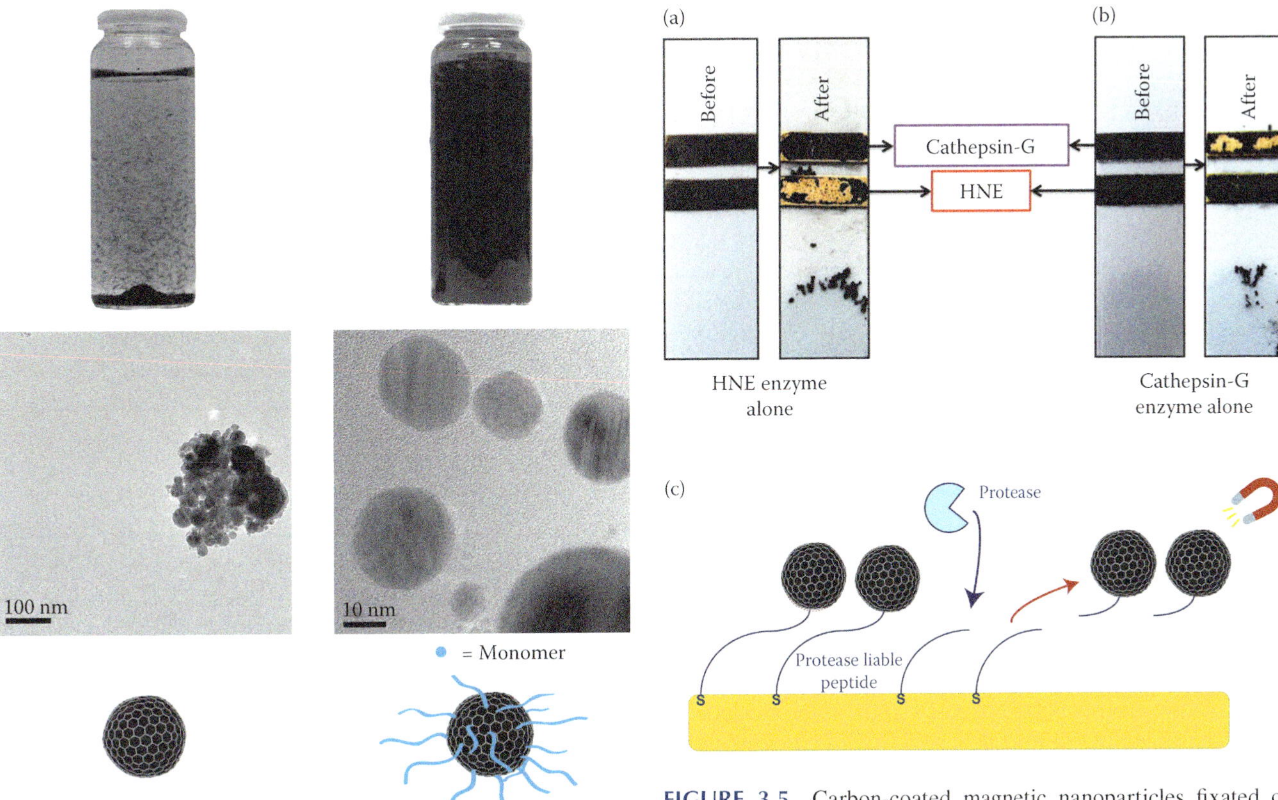

FIGURE 3.4 Unfunctionalized particles (left) sediment rapidly in water due to the hydrophobicity of the carbon surface and the magnetic remanence of the ferromagnetic nanoparticles. By introduction of charged and steric surface oligomers (of at least six monomer units), these effects can be reversed and stable dispersion of ferromagnets can be formed in water and in buffer systems (DMEM, PBS etc.).

FIGURE 3.5 Carbon-coated magnetic nanoparticles fixated on a gold support via a protease sensitive linker are used to detect (a) human neutrophil elastase (HNE) and (b) cathepsin G in an optically readable point of care device (c).[48] (Reprinted with permission from Wignarajah, S., G. A. R. Y. Suaifan, S. Bizzarro, F. J. Bikker, W. E. Kaman, and M. Zourob. 2015. "Colorimetric assay for the detection of typical biomarkers for periodontitis using a magnetic nanoparticle biosensor." *Anal Chem* 87 (24):12161–12168. Copyright 2015 American Chemical Society.)

saturation magnetization), smaller amounts of nanoparticles required (higher surface areas) and a wider range of usable nanoparticle storage buffers (e.g. storage in high-salt buffers instead of isopropanol). It must be added that in every case, this requires the adaption of the nanoparticle surface chemistry to the desired application, which may involve the surface chemistries depicted earlier or nanoparticle surface coating with amorphous silica.[45]

For applications in which the magnetic particles are required for signal generation, metallic nanoparticles have been utilized alongside iron oxide nanoparticles in order to achieve multiplexing by having two distinct magnetic signatures.[46] An innovative analytical method for the detection of proteases was developed by Zourob et al., taking advantage of the relative ease of chemical functionalization, the high magnetic moments of the particles and the high visible opacity of magnetic nanoparticles.[47] As shown in Figure 3.5, magnetic particles are fixed to a gold surface by a protease-specific peptide linker. Upon exposure to the corresponding protease, the particles are cleaved from the surface and automatically removed from the surface by an adjacently placed magnet. Due to the high opacity of the particles and good contrast to the gold surface, even very small changes can be detected visually after just a few minutes. With this method and knowledge of microbe-specific proteases, this method is most suitable as a point-of-care device for the detection of pathogens (e.g. Periodontitis, Listeria, Salmonella and *Staphylococcus aureus*).[48–50] Other important clinical diagnostic methods involving carbon-coated metallic nanoparticles include magnetic resonance imaging (MRI), where the increased saturation magnetization of the particles results in greater hypointensities on T_2-weighted MRI signals.[51,52] As an application example thereof, Balla et al.[53] have detected single native pancreatic islets in mice.

3.6 THE FUTURE OF CARBON-COATED METAL NANOMAGNETS IN BLOOD PURIFICATION

For several medical conditions, the removal of a specific biochemical entity (small molecular compound, protein or microorganism) would bring rapid relief to the patients. Some examples thereof are lead poisoning, drug overdose (e.g. digoxin), inflammation (e.g. interleukin [IL]-6 and IL-1) and bacterial sepsis (endotoxin and lipopolysaccharide). These compounds cannot be removed from blood via hemodialysis, as dialysis is not target specific enough and is only suitable for the removal of small chemical solutes (ions, urea etc.). However, as previously indicated in this chapter, surface functionalized carbon-coated metallic nanoparticles are suitable to remove specific analytes from solution. Herrmann et al. have shown that carbon-coated iron carbide nanoparticles are well suited for this task in whole blood. Specificity is created by chemically attached targets that are specific binding agents to the surface of the nanoparticles.[54] These can be complexating agents (e.g. for lead), antibodies and antigen-binding (Fab) fragments (for a wide range of antigens) as well as antibiotics (for specific bacteria). Covalent linkage of these entities via C–C bonds to the carbon layer of the particles guarantees that the functionality is not lost, even after prolonged usage and storage. The high saturation magnetization of the metallic cores enables the particles to be effectively recovered from a flowing blood stream at flow rates >30 mm s^{-1}, conditions at which comparable iron-oxide-based nanoparticles could no longer be quantitatively collected.[17] Application of an extracorporeal magnetic blood purification device using carbon-coated iron carbide nanoparticles (see Figure 3.6) has been shown for the removal of various toxins from whole blood in laboratory setups and for the rapid removal of lead and digoxin from poisoned rats.[55] Although magnetic blood purification does not require that particles are injected into the body (in contrast to magnetic drug-targeting), detailed studies on the blood compatibility and short- and long-term *in vivo* toxicology of the particles have been preformed. Thereby, it has been shown that the particles do not affect blood coagulation to an extent larger than the intersubject variance,[56] and a high *in vivo* (mouse) compatibility of the carbon-coated particles for a timeframe of 1 year post exposure has been shown,[57] even at relatively high exposure concentrations (up to 60 mg kg^{-1}, equivalent to a ~4 g nanoparticle dose for a 70 kg human).

From the previously described medical conditions in which a toxin should be rapidly removed from a blood stream, the treatment of bacterial sepsis seems to be the most valuable target. Sepsis has been reported to be the most expensive condition treated in the United States in 2008, with nearly 1.1 million hospitalizations and a mortality rate of >30%.[58] More current estimations report 30 million cases a year globally.[59] Although sepsis is a systemic inflammation, the condition is often caused as the result of a bacteremia and the inflammatory system responding to the bacteria themselves as well as to endotoxins (e.g. lipopolysaccharides [LPS] and bacterial decay products). In order to halt and reverse the inflammatory process, bacteria and endotoxins have to be removed from the blood stream as quickly as possible. Magnetic particles carrying a polymyxin-B surface functionality (an antibiotic binding to LPS sites in gram-negative bacteria) on the other hand could be utilized in an extracorporal blood purification device to rapidly remove *Escherichia coli*-derived LPS from whole

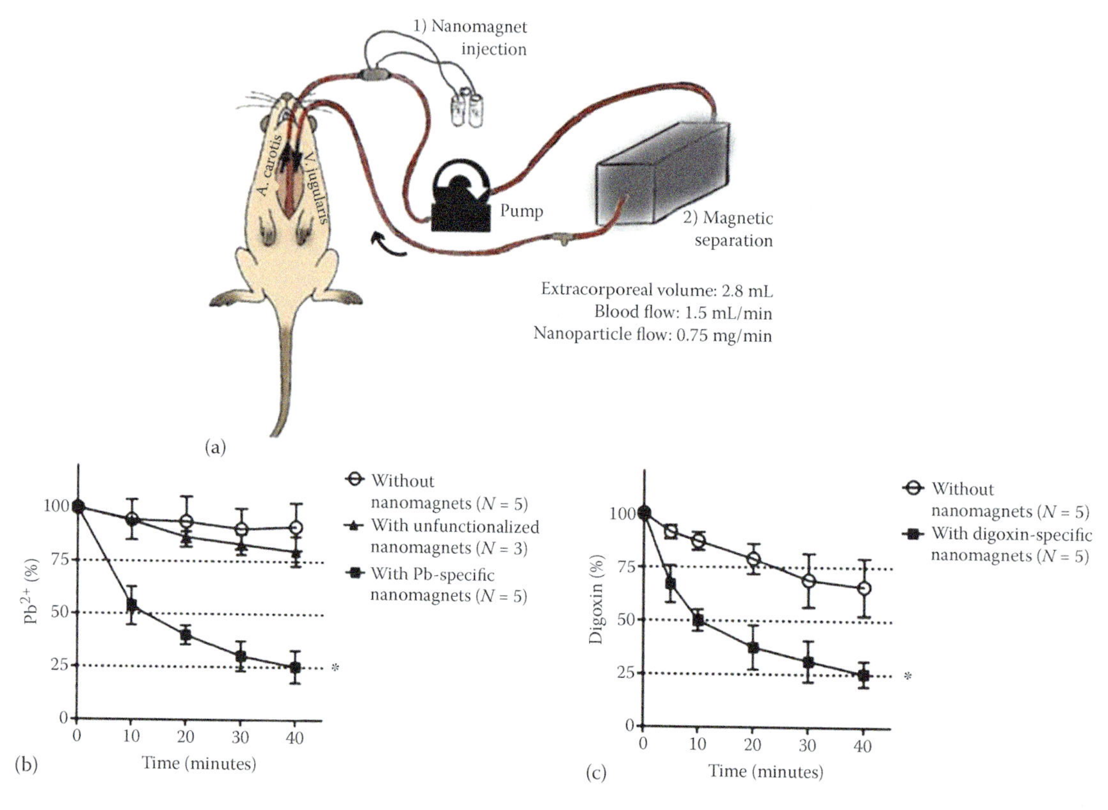

FIGURE 3.6 Extracorporeal magnetic blood purification device (a) enabling the removal of lead (b) and digoxin (c) from poisoned rats.[55] (Reprinted with permission from Herrmann, I. K., A. Schlegel, R. Graf, C. M. Schumacher, N. Senn, M. Hasler, S. Gschwind, A. M. Hirt, D. Gunther, P. A. Clavien, W. J. Stark, and B. Beck-Schimmer. 2013. "Nanomagnet-based removal of lead and digoxin from living rats." *Nanoscale* 5 (18):8718–8723. Copyright 2013 Royal Chemical Society.)

blood.[60] The reversal of the inflammatory cascade was shown by decreased IL-6 and chemokine ligand 1 levels. As there is evidence of a dose–response correlation between LPS levels and the poor prognosis of sepsis patients,[61,62] the possibility of rapidly removing endotoxins from blood utilizing such magnetic nanoparticles may be able to halt/reverse/cure sepsis caused by gram-negative bacteria. Spurred by these results, the ETH start-up Hemotune[63] is aiming to make this technology available for clinical application. While this requires a more controlled safety evaluation of the magnetic particles in order to achieve regulatory approval and first human trials, competing column-based technologies are currently giving additional evidence that endotoxin removal efficiency has a direct effect on sepsis induced mortality.[64]

3.7 CONCLUSION AND FUTURE OUTLOOK

From the data shown in this chapter, it is evident that carbon-coated magnetic nanometals differ from iron oxide nanoparticles in terms of not only magnetic properties but also in other physicochemical aspects. So for every potential application of magnetic nanoparticles in a clinical setting, it might be advisable to list the properties required from the nanoparticles and choose the corresponding material base (e.g. iron oxide, ferrites, silica, gold, polymer or carbon coated metals) and surface functionality to fit the need. This noncomprehensive list of properties may include the following:

- Saturation magnetization, magnetic hardness (i.e. remanence)
- Particle size (available surface area)
- Colloidal stability in media, chemical inertness, surface charge, available surface functionality, antifouling (i.e. repelling biofilm growth, protein adhesion)
- Storage stability (shelf-life)
- Separability, magnetic hyperthermia, MRI and magnetic particle imaging (MPI) efficiency

Whereas the physical properties in terms of material basis cannot be adjusted, surface chemistry modification can be a very powerful tool to achieve/adapt nonmatching requirements. As a result, we see strong efforts in improving nanoparticle surface chemistries and adapting them to novel clinical applications.

REFERENCES

1. Ruoff, R. S., D. C. Lorents, B. Chan, R. Malhotra, and S. Subramoney. 1993. "Single-crystal metals encapsulated in carbon nanoparticles." *Science* 259 (5093):346–348.
2. Majetich, S. A., J. O. Artman, M. E. Mchenry, N. T. Nuhfer, and S. W. Staley. 1993. "Preparation and properties of carbon-coated magnetic nanocrystallites." *Phys Rev B* 48 (22):16845–16848.
3. Iijima, S. 1991. "Helical microtubules of graphitic carbon." *Nature* 354 (6348):56–58.
4. Kammler, H. K., L. Madler, and S. E. Pratsinis. 2001. "Flame synthesis of nanoparticles." *Chem Eng Technol* 24 (6):583–596.
5. Swihart, M. T. 2003. "Vapor-phase synthesis of nanoparticles." *Curr Opin Colloid Interface Sci* 8 (1):127–133.
6. Pratsinis, S. E. 1998. "Flame aerosol synthesis of ceramic powders." *Prog Energ Combust* 24 (3):197–219.
7. Grass, R. N., and W. J. Stark. 2006. "Flame spray synthesis under a non-oxidizing atmosphere: Preparation of metallic bismuth nanoparticles and nanocrystalline bulk bismuth metal." *J Nanopart Res* 8 (5):729–736.
8. Athanassiou, E. K., R. N. Grass, and W. J. Stark. 2006. "Large-scale production of carbon-coated copper nanoparticles for sensor applications." *Nanotechnology* 17 (6):1668–1673.
9. Athanassiou, E. K., R. N. Grass, and W. J. Stark. 2010. "Chemical aerosol engineering as a novel tool for material science: From oxides to salt and metal nanoparticles." *Aerosol Sci Tech* 44 (2):161–172.
10. Grass, R. N., E. K. Athanassiou, and W. J. Stark. 2007. "Covalently functionalized cobalt nanoparticles as a platform for magnetic separations in organic synthesis." *Angew Chem Int Ed* 46 (26):4909–4912.
11. Hoyos-Palacio, L. M., A. G. Garcia, J. F. Perez-Robles, J. Gonzalez, and H. V. Martinez-Tejada. 2014. "Catalytic effect of Fe, Ni, Co and Mo on the CNTs production." *Iop Conf Ser-Mat Sci* 59 (012005):1–8.
12. El-Gendy, A. A., E. M. M. Ibrahim, V. O. Khavrus, Y. Krupskaya, S. Hampel, A. Leonhardt, B. Buchner, and R. Klingeler. 2009. "The synthesis of carbon coated Fe, Co and Ni nanoparticles and an examination of their magnetic properties." *Carbon* 47 (12):2821–2828.
13. Mattila, P., H. Heinonen, K. Loimula, J. Forsman, L. S. Johansson, U. Tapper, R. Mahlberg, H. P. Hentze, A. Auvinen, J. Jokiniemi, and R. Milani. 2014. "Scalable synthesis and functionalization of cobalt nanoparticles for versatile magnetic separation and metal adsorption." *J Nanopart Res* 16 (9) (2606):1–11.
14. Billas, I. M. L., A. Chatelain, and W. A. Deheer. 1994. "Magnetism from the atom to the bulk in iron, cobalt, and nickel clusters." *Science* 265 (5179):1682–1684.
15. Herrmann, I. K., R. N. Grass, D. Mazunin, and W. J. Stark. 2009. "Synthesis and covalent surface functionalization of nonoxidic iron core–shell nanomagnets." *Chem Mater* 21 (14):3275–3281.
16. Hofer, L. J. E., and E. M. Cohn. 1959. "Saturation magnetizations of iron carbides." *J Am Chem Soc* 81 (7):1576–1582.
17. Schumacher, C. M., I. K. Herrmann, S. B. Bubenhofer, S. Gschwind, A. M. Hirt, B. Beck-Schimmer, D. Gunther, and W. J. Stark. 2013. "Quantitative recovery of magnetic nanoparticles from flowing blood: Trace analysis and the role of magnetization." *Adv Funct Mater* 23 (39):4888–4896.
18. Sun, J. F., Y. Zhang, Z. P. Chen, H. Zhou, and N. Gu. 2007. "Fibrous aggregation of magnetite nanoparticles induced by a time varied magnetic field." *Angew Chem Int Ed* 46 (25):4767–4770.
19. Yavuz, C. T., J. T. Mayo, W. W. Yu, A. Prakash, J. C. Falkner, S. Yean, L. L. Cong, H. J. Shipley, A. Kan, M. Tomson, D. Natelson, and V. L. Colvin. 2006. "Low-field magnetic separation of monodisperse Fe_3O_4 nanocrystals." *Science* 314 (5801):964–967.
20. De Las Cuevas, G., J. Faraudo, and J. Camacho. 2008. "Low-gradient magnetophoresis through field-induced reversible aggregation." *J Phys Chem C* 112 (4):945–950.
21. Hofer, C. J., V. Zlateski, P. R. Stoessel, D. Paunescu, E. M. Schneider, R. N. Grass, M. Zeltner, and W. J. Stark. 2015. "Stable dispersions of azide functionalized ferromagnetic metal nanoparticles." *Chem Commun* 51 (10):1826–1829.

22. Zeltner, M., R. N. Grass, A. Schaetz, S. B. Bubenhofer, N. A. Luechinger, and W. J. Stark. 2012. "Stable dispersions of ferromagnetic carbon-coated metal nanoparticles: Preparation via surface initiated atom transfer radical polymerization." *J Mater Chem* 22 (24):12064–12071.
23. Schaetz, A., W. J. Stark, and R. N. Grass. 2012. "Magnetic cobalt[0]-graphene nanospheres." *e-EROS Encyclopedia of Reagents for Organic Synthesis.*
24. Hofer, C. J., R. N. Grass, M. Zeltner, C. A. Mora, F. Krumeich, and W. J. Stark. 2016. "Hollow carbon nanobubbles: Synthesis, chemical functionalization, and container-type behavior in water." *Angew Chem Int Ed* 55 (30):8761–8765.
25. Rossier, M., F. M. Koehler, E. K. Athanassiou, R. N. Grass, B. Aeschlimann, D. Gunther, and W. J. Stark. 2009. "Gold adsorption on the carbon surface of C/Co nanoparticles allows magnetic extraction from extremely diluted aqueous solutions." *J Mater Chem* 19 (43):8239–8243.
26. Belmont, J. A., R. M. Amici, and C. P. Galloway. 1994. Reaction of carbon black with diazonium salts, resultant carbon black products and their uses. US Patent 6042643.
27. Dyke, C. A., M. P. Stewart, F. Maya, and J. M. Tour. 2004. "Diazonium-based functionalization of carbon nanotubes: XPS and GC-MS analysis and mechanistic implications." *Synlett* (1):155–160.
28. Thanh, N. T. K., and L. A. W. Green. 2010. "Functionalisation of nanoparticles for biomedical applications." *Nano Today* 5 (3):213–230.
29. Koehler, F. M., M. Rossier, M. Waelle, E. K. Athanassiou, L. K. Limbach, R. N. Grass, D. Gunther, and W. J. Stark. 2009. "Magnetic EDTA: Coupling heavy metal chelators to metal nanomagnets for rapid removal of cadmium, lead and copper from contaminated water." *Chem Commun* (32):4862–4864.
30. Fuhrer, R., E. K. Athanassiou, N. A. Luechinger, and W. J. Stark. 2009. "Crosslinking metal nanoparticles into the polymer backbone of hydrogels enables preparation of soft, magnetic field-driven actuators with muscle-like flexibility." *Small* 5 (3):383–388.
31. Zeltner, M., L. M. Toedtli, N. Hild, R. Fuhrer, M. Rossier, L. C. Gerber, R. A. Raso, R. N. Grass, and W. J. Stark. 2013. "Ferromagnetic inks facilitate large scale paper recycling and reduce bleach chemical consumption." *Langmuir* 29 (16):5093–5098.
32. Kawasaki, H., K. Nakai, R. Arakawa, E. K. Athanassiou, R. N. Grass, and W. J. Stark. 2012. "Functionalized graphene-coated cobalt nanoparticles for highly efficient surface-assisted laser desorption/ionization mass spectrometry analysis." *Anal Chem* 84 (21):9268–9275.
33. Schatz, A., O. Reiser, and W. J. Stark. 2010. "Nanoparticles as semi-heterogeneous catalyst supports." *Chem-Eur J* 16 (30):8950–8967.
34. Rossi, L. M., N. J. S. Costa, F. P. Silva, and R. Wojcieszak. 2014. "Magnetic nanomaterials in catalysis: Advanced catalysts for magnetic separation and beyond." *Green Chem* 16 (6):2906–2933.
35. Tan, C. G., and R. N. Grass. 2008. "Suzuki cross-coupling reactions on the surface of carbon-coated cobalt: Expanding the applicability of core-shell nano-magnets." *Chem Commun* (36):4297–4299.
36. Zlateski, V., R. Fuhrer, F. M. Koehler, S. Wharry, M. Zeltner, W. J. Stark, T. S. Moody, and R. N. Grass. 2014. "Efficient magnetic recycling of covalently attached enzymes on carbon-coated metallic nanomagnets." *Bioconjugate Chem* 25 (4):677–684.
37. Pyun, J., K. Matyjaszewski, T. Kowalewski, D. Savin, G. Patterson, G. Kickelbick, and N. Huesing. 2001. "Synthesis of well-defined block copolymers tethered to polysilsesquioxane nanoparticles and their nanoscale morphology on surfaces." *J Am Chem Soc* 123 (38):9445–9446.
38. Evans, R. A. 2007. "The rise of azide-alkyne 1,3-dipolar 'click' cycloaddition and its application to polymer science and surface modification." *Aust J Chem* 60 (6):384–395.
39. Zeltner, M., A. Schatz, M. L. Hefti, and W. J. Stark. 2011. "Magnetothermally responsive C/Co@PNIPAM-nanoparticles enable preparation of self-separating phase-switching palladium catalysts." *J Mater Chem* 21 (9):2991–2996.
40. Schneider, E. M., M. Zeltner, V. Zlateski, R. N. Grass, and W. J. Stark. 2016. "Click and release: Fluoride cleavable linker for mild bioorthogonal separation." *Chem Commun* 52 (5):938–941.
41. Kainz, Q. M., and O. Reiser. 2014. "Polymer- and dendrimer-coated magnetic nanoparticles as versatile supports for catalysts, scavengers, and reagents." *Accounts Chem Res* 47 (2):667–677.
42. Borlido, L., A. M. Azevedo, A. C. A. Roque, and M. R. Aires-Barros. 2013. "Magnetic separations in biotechnology." *Biotechnol Adv* 31 (8):1374–1385.
43. Huy, T. Q., P. V. Chung, N. T. Thuy, C. Blanco-Andujar, and N. T. K. Thanh. 2014. "Protein A-conjugated iron oxide nanoparticles for separation of *Vibrio cholerae* from water samples." *Faraday Discuss* 175:73–82.
44. Holzinger, M., A. Le Goff, and S. Cosnier. 2014. "Nanomaterials for biosensing applications: A review." *Front Chem* 2 (63):1–10.
45. Grass, R. N. 2010. Process for manufacturing chemically stable magnetic carriers EP Patent 2244268.
46. Oscarsson, S., K. Eriksson, and P. Svendindh. 2016. Diagnostic assay using particles with magnetic properties. US Patent Application 20160187327.
47. Suaifan, G. A. R. Y., C. Esseghaier, A. Ng, and M. Zourob. 2013. "Ultra-rapid colorimetric assay for protease detection using magnetic nanoparticle-based biosensors." *Analyst* 138 (13):3735–3739.
48. Wignarajah, S., G. A. R. Y. Suaifan, S. Bizzarro, F. J. Bikker, W. E. Kaman, and M. Zourob. 2015. "Colorimetric assay for the detection of typical biomarkers for periodontitis using a magnetic nanoparticle biosensor." *Anal Chem* 87 (24):12161–12168.
49. Alhogail, S., G. A. R. Y. Suaifan, and M. Zourob. 2016. "Rapid colorimetric sensing platform for the detection of *Listeria monocytogenes* foodborne pathogen." *Biosens Bioelectron* 86:1061–1066.
50. Suaifan, G. A., S. Alhogail, and M. Zourob. 2017. "Rapid and low-cost biosensor for the detection of *Staphylococcus aureus*." *Biosens. Bioelectron.* 90:230–237.
51. Yu, J., F. Chen, W. Gao, Y. Ju, X. Chu, S. Che, F. Sheng, and Y. Hou. 2017. "Iron carbide nanoparticles: An innovative nanoplatform for biomedical applications." *Nanoscale Horizons.* Available online.
52. Seo, W. S., J. H. Lee, X. M. Sun, Y. Suzuki, D. Mann, Z. Liu, M. Terashima, P. C. Yang, M. V. McConnell, D. G. Nishimura, and H. J. Dai. 2006. "FeCo/graphitic-shell nanocrystals as advanced magnetic-resonance-imaging and near-infrared agents." *Nat Mater* 5 (12):971–976.
53. Balla, D. Z., S. Gottschalk, G. Shajan, S. Ueberberg, S. Schneider, M. Hardtke-Wolenski, E. Jaeckel, V. Hoerr, C. Faber, K. Scheffler, R. Pohmann, and J. Engelmann. 2013. "*In vivo* visualization of single native pancreatic islets in the mouse." *Contrast Media Mol I* 8 (6):495–504.

54. Herrmann, I. K., M. Urner, F. M. Koehler, M. Hasler, B. Roth-Z'Graggen, R. N. Grass, U. Ziegler, B. Beck-Schimmer, and W. J. Stark. 2010. "Blood purification using functionalized core/shell nanomagnets." *Small* 6 (13):1388–1392.
55. Herrmann, I. K., A. Schlegel, R. Graf, C. M. Schumacher, N. Senn, M. Hasler, S. Gschwind, A. M. Hirt, D. Gunther, P. A. Clavien, W. J. Stark, and B. Beck-Schimmer. 2013. "Nanomagnet-based removal of lead and digoxin from living rats." *Nanoscale* 5 (18):8718–8723.
56. Bircher, L., O. M. Theusinger, S. Locher, P. Eugster, B. Roth-Z'graggen, C. M. Schumacher, J. D. Studt, W. J. Stark, B. Beck-Schimmer, and I. K. Herrmann. 2014. "Characterization of carbon-coated magnetic nanoparticles using clinical blood coagulation assays: Effect of PEG-functionalization and comparison to silica nanoparticles." *J Mater Chem B* 2 (24):3753–3758.
57. Herrmann, I. K., B. Beck-Schimmer, C. M. Schumacher, S. Gschwind, A. Kaech, U. Ziegler, P. A. Clavien, D. Gunther, W. J. Stark, R. Graf, and A. A. Schlegel. 2016. "*In vivo* risk evaluation of carbon-coated iron carbide nanoparticles based on short- and long-term exposure scenarios." *Nanomedicine-Uk* 11 (7):783–796.
58. Daniels, R. 2011. "Surviving the first hours in sepsis: Getting the basics right (an intensivist's perspective)." *J Antimicrob Chemoth* 66:Ii11–Ii23.
59. Dugani, S., J. Veillard, and N. Kissoon. 2017. "Reducing the global burden of sepsis." *Can Med Assoc J* 189 (1):E2–E3.
60. Herrmann, I. K., M. Urner, S. Graf, C. M. Schumacher, B. Roth-Z'graggen, M. Hasler, W. J. Stark, and B. Beck-Schimmer. 2013. "Endotoxin removal by magnetic separation-based blood purification." *Adv Healthc Mater* 2 (6):829–835.
61. Marshall, J. C., D. Foster, J. L. Vincent, D. J. Cook, J. Cohen, R. P. Dellinger, S. Opal, E. Abraham, S. J. Brett, T. Smith, S. Mehta, A. Derzko, and A. Romaschin. 2004. "Diagnostic and prognostic implications of endotoxemia in critical illness: Results of the MEDIC study." *J Infect Dis* 190 (3):527–534.
62. Klein, D. J., D. Foster, C. A. Schorr, K. Kazempour, P. M. Walker, and R. P. Dellinger. 2014. "The EUPHRATES trial (Evaluating the Use of Polymyxin B Hemoperfusion in a Randomized controlled trial of Adults Treated for Endotoxemia and Septic shock): Study protocol for a randomized controlled trial." *Trials* 15 (218):1–15.
63. Tomczak, A. 2016. Magnetic forces purify the blood. https://www.ethz.ch/en/news-and-events/eth-news/news/2016/12/magnetic-forces-purify-the-blood.html (accessed 31.01.2017).
64. Terayama, T., K. Yamakawa, Y. Umemura, M. Aihara, and S. Fujimi. 2017. "Polymyxin B hemoperfusion for sepsis and septic shock: A systematic review and meta-analysis." *Surg Infect (Larchmt)* 18:225–233.

Dr Robert N. Grass (E-mail: robert.grass@chem.ethz.ch) is a senior scientist and Titular Professor in the Functional Materials Laboratory at ETH Zurich. He studied chemical engineering at ETH with a stay at CASE Western Reserve University Cleveland in 2003, after which he pursued a PhD at ETH Zurich.

He has written over 120 research papers, eight patents and three book chapters. Based on the results of his research, he founded the company TurboBeads GmbH in 2007, which makes chemically functionalized magnetic materials commercially available (e.g. via Sigma Aldrich). Aside from his work on magnetic nanomaterials, Robert Grass also has a strong interest in utilizing nucleic acids as information carrying molecules in non-medical applications.

Dr Martin Zeltner has been a senior scientist at the Institute of Chemical and Bioengineering at ETH Zurich since 2015. He earned his degree in chemistry at the Swiss Federal Institute of Technology in 2009 with specialization in polymer chemistry. At the same institute, he also completed his PhD in 2013. His work has strongly focused on the chemistry of carbon surfaces, where he introduced novel functionalities enabling catalysis, phase-transfer and most importantly control of colloidal stability in aqueous solutions. Within his thesis, he worked on the application of polymer-coated magnetic nanoparticles in chemistry. The investigated modification of carbon surfaces essentially uses controlled radical polymerization techniques such as atom transfer radical polymerization, reversible addition-fragmentation chain transfer or nitroxide-mediated polymerization.

4 Bioinspired Magnetic Nanoparticles for Biomedical Applications

Changqian Cao and Yongxin Pan*

CONTENTS

In order to advance cancer treatment using nanotechnology, innovations in synthesis and functionalization of magnetic nanoparticles (MNPs) are required. In the past decades, most research groups have synthesized biocompatible superparamagnetic iron oxide (SPIO) nanoparticles and tuned their size and shape using inorganic techniques, such as coprecipitation,[1,2] thermal decomposition,[3–6] micelle synthesis[7,8] and hydrothermal synthesis.[9–12] These synthesized SPIO nanoparticles need surface coating or grafting with molecules to preserve their stability and function for desired application.

Bioinspired synthesis provides an alternative way to achieve control over the kinetics of nanoparticle crystallization under ambient conditions and in aqueous media. Biogenesis of magnetite (Fe_3O_4) is found in various organisms. For example, magnetotactic bacteria, a group of bacteria that orient and swim along the magnetic field lines of the Earth's magnetic field, mineralize tens to hundreds of membrane encapsulated magnetic crystals in the cell, which are strictly controlled by a large number of biomineralization proteins. The magnetosome membrane consists of lipids and proteins, which allows couplings of bioactive substances to its surface, a characteristic important for many applications. Currently, some of these biomineralization proteins have been successfully employed for bioinspired synthesis of MNPs with defined shape and size distribution.[13–15]

Ferritin is an iron storage protein existing in nearly all organisms. Through biomimetic mineralization, a magnetite core can be mineralized into human H-chain ferritin (HFn) cavity to form magnetoferritin (M-HFn) nanoparticles. Recent studies have indicated that M-HFn nanoparticles have potential in early diagnosis of microscopic (<1–2 mm) tumours at the preangiogenic stage due to their intrinsic tumour targeting ability.[16] Tumour-targeting peptides can also be fused to HFn-based nanoparticles through genetic engineering, which avoids nonspecific linking by current chemical surface modification.

In this chapter, Sections 4.1 and 4.2 are dedicated to biomimetic synthesis of MNPs inspired by biomineralization of ferritin and magnetotactic bacteria. Section 4.3 covers application of bioinspired MNPs for cancer diagnosis and therapy, including *in vitro* staining of tumour tissues, *in vivo* imaging of microscopic tumours and magnetic hyperthermia. Section 4.4 indicates some future directions for bioinspired MNPs.

4.1 MNPs IN LIVING ORGANISMS

The first biogenic magnetite (Fe_3O_4) crystals were found on the radulas of chitons and reported in 1962.[17] Over the past 50 years, biomineralization of ferrimagnetic nanoparticles was found in a wide range of various organisms, such as bacteria, algae, molluscs, insects and vertebrates. Although the biological function of these MNPs was not fully known, biomineralization illustrated exceptional control over the composition, crystallography, morphology and material properties achieved under mild conditions (physiological temperature, pressure and pH).

* Corresponding author.

4.1.1 MNPs in Magnetotactic Bacteria

Magnetotactic bacteria are prokaryotic microorganisms that ubiquitously existed in freshwater and marine habitats. They synthesize chains of nano-sized MNPs that function as a compass needle and allow the microbes to navigate using the Earth's geomagnetic field. These MNPs are synthesized by a specific set of proteins contained within membrane-bound organelles called magnetosomes. The mineral structure of the magnetosome is either magnetite (Fe_3O_4) or greigite (Fe_3S_4); has high purity, narrow distribution size (35–120 nm) and a species-specific crystal morphology; and arranges as chains within the cell.[18] The magnetosomes in uncultured and cultured magnetotactic bacteria are dominated by single-domain magnetite with Verwey transition temperature of about 100 K, distinctly lower than that of stoichiometric magnetite (120–125 K).[19–21] These features indicate that the formation of magnetosomes is precisely controlled by a biomineralization process.

Magnetosome biomineralization was mostly investigated based on two laboratory cultivated magnetotactic bacteria, *Magnetosirillum magneticum* AMB-1 and *Magnetospirillum gryphiswaldense* MSR-1. Figure 4.1a–f illustrates that the mineralization process of magnetosomes can be divided into four steps. First, magnetosome vesicles are formed throughout the cell by invagination of the inner cytoplasmic membrane, and then magnetosome proteins are sorted to the membrane (Figure 4.1a and b). Second, iron is transported into magnetosome vesicles and mineralized as magnetic nanocrystals (Figure 4.1c). Fe^{2+} or Fe^{3+} may enter into the magnetosome membrane using magnetosome-specific transporters and subsequently form magnetite within a narrow redox range (Fe^{3+} and Fe^{2+} are present at a ratio of 2/1). Nucleation of magnetite crystals may occur through either direct coprecipitation of soluble Fe^{2+} and Fe^{3+} under basic condition (>pH 7) or phase transformations of precursor mineral phases such as ferrihydrite and hematite. Third, the magnetosome membrane regulates maturation of the magnetic nanocrystals of defined size and shape. The mature magnetosomes are aligned into chains through the interaction of MamJ (yellow star) with the actin-like MamK filament (green line) (Figure 4.1d). Fourth, cell division is initiated through asymmetric constriction of the FtsZ ring (light gray circle, Figure 4.1e) and the magnetosome chains bend to reduce magnetostatic forces to promote an even segregation of the magnetosome to the daughter cells (Figure 4.1e and f).[22,23]

Each step is strictly controlled by complex machinery that is composed of a large number of genetic determinants. These

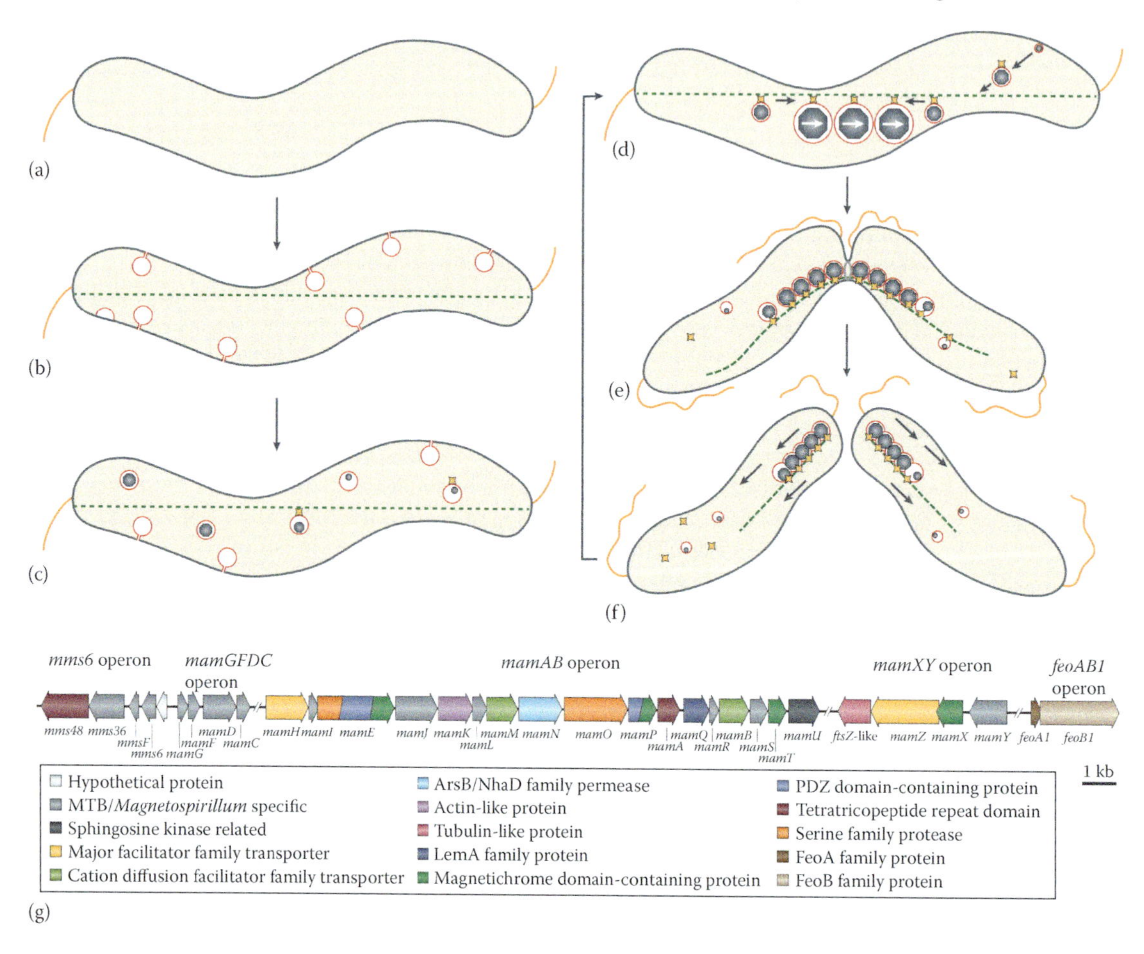

FIGURE 4.1 Mechanisms and genetics of magnetosome biogenesis in *M. gryphiswaldense* MSR-1.[22] (Reprinted by permission from Macmillan Publishers Ltd. *Nature Reviews. Microbiology.* Uebe, R.; Schüler, D., Magnetosome biogenesis in magnetotactic bacteria. *Nat. Rev. Microbiol.* 2016, *14*(10), 621–637. Copyright 2016.)

responsible genes are located mostly within a gnenomic magnetosome island (MAI) that is composed of *mamAB* operon, *felAB1*, *mms6* operon, *mamGFDC* operon and *mamXY* operon (Figure 4.1g).[24–26] These operons are highly conserved in closely related *Magnetospirillum magneticum*. An operon is made up of several structural genes arranged under a common promoter and regulated by a common operator. Usually, these genes encode proteins working together in a metabolic pathway. The large (16–17 kb) *mamAB* operon contains several genes that play key roles in magnetosome biogenesis.[27,28] Comprehensive functional analysis of MAI deletion mutants in *M. magneticum* AMB-1 shows that *mamAB* operons are encoded for factors important for magnetosome membrane biogenesis, for targeting of proteins in this compartment and for several steps of magnetite mineralization. For example, four conserved genes of *mamI*, *mamL*, *mamQ* and *mamB* in *mamAB* gene clusters are significantly important for inner membrane invagination and magnetosome vesicle formation. *mamE* is essential for localization of a subset of proteins in the magnetosome membrane. However, another four small operons encode nonessential proteins, including the putative iron transporters FeoAB1 and MamZ; the tubulin-like protein FtsZm; and the magnetotactic bacteria-specific proteins MamC (also known as Mms13), MamD (also known as Mms7), MamF, MamG, MamX, MamY, Mms6, MmsF, Mms36 and Mms48. These proteins have supplementary roles in regulating the size and shape of biomineralized magnetic nanocrystal.[29–31]

The biomineralization mechanism of many uncultured magnetotactic bacteria remained poorly understood. '*Candidatus* Magnetobacterium bavaricum' was one of the most interesting, uncultivated magnetotactic bacteria first discovered in sediment samples taken from Lake Chiemsee and Lake Ammersee in southern Germany.[32,33] It formed 600 to 1000 bullet-shaped magnetite crystals that arranged into several parallel chains, which were about 20–100 times more than magnetosome crystals found in most other magnetotactic bacteria. This kind of bacterium was affiliated with the independent deep-branching *Nitrospira* phylum and was very large (8–10 μm length, 1.5–2 μm diameter). Fossilized magnetosome crystals of '*Candidatus* Magnetobacterium bavaricum' accounted for a large proportion of the total magnetization in certain aquatic sediments.[20] Recently, a closely related magnetotactic bacterium named '*Candidatus* Magnetobacterium casensis' was identified from Lake Miyun near Beijing, China, and called MYR-1.[34–36] This bacterium was morphologically similar to '*Candidatus* Magnetobacterium bavaricum' and had hundreds of bullet-shaped magnetite magnetosomes arranged into three to five braid-like bundles of chains.[35] The genome sequence suggested a special autotrophic lifestyle using the Wood-Ljungdahl pathway for a CO_2 fixation that was not found in any previously known magnetotactic bacteria.[37] The '*Candidatus* Magnetobacterium casensis' needed multistep crystal growth to form magnetite crystals, which was very different from that of the cubo-octahedral and prismatic magnetosomes in other bacteria. The multiple steps involved initial isotropic growth forming cubo-octahedral particles (less than ~40 nm), followed by anisotropic crystal growth and a systematic elongation along the [001] crystallographic direction.[35,38]

4.1.2 MNPs in Animal

The topic of magnetite nanoparticles is an exciting field of research in animals. Behavioural studies have demonstrated that various birds, fish, reptiles, amphibians and insects use geomagnetic fields for navigation. However, the mechanisms of magnetoreception are still mostly unknown. Currently, two hypotheses have been developed for explanation of magnetoreception in animals.[39–41] The first hypothesis is 'radical-pair hypothesis,' which invokes magnetically sensitive biochemical reactions involving spin-correlated radical pairs that are produced by a light-driven electron transfer reaction of photoreceptor molecules (cryptochrome) in the retina. The magnetic field interacting with the radical pair controls the reaction yield and is transduced into a neuronal stimulus. The second hypothesis is 'magnetite hypothesis,' which assumes that an external magnetic field interacts with magnetite inclusions in tissue that convert the received magnetic energy into a mechanic stimulus to be detected by adjacent mechanoreceptors that eventually generate a nervous signal. A sockeye salmon is well known to respond to magnetic field direction in two different stages, fry and smolts.[42,43] Single-domain magnetite particles that closely resemble those from magnetotactic bacteria have been extracted from the ethmoid tissue of the sockeye salmon with grain sizes between 25 and 60 nm. These MNPs show cubo-octahedral crystal morphology and are assembled into a chain-like structure.[44] For rainbow trout, iron-rich particles in the olfactory epithelium are imaged using confocal laser-scanning microscopy and magnetic force microscopy, and they are identified as chains of single-domain magnetite.[45] Recently, Eder et al.[46] have successfully isolated candidate magnetoreceptor cells that quickly respond to external magnetic fields from trout olfactory epithelium, as shown in Figure 4.2. It is found that μm-sized intracellular structures of magnetic inclusions are firmly coupled with the cell membrane in candidate magnetoreceptor cells.[46] Homing pigeons were the most thoroughly studied subject for magnetoreception. It was first proposed that single-domain magnetite particles existed in the pigeon brain.[40] Later, clusters of superparamagnetic magnetite nanoparticles (1–5 nm) associated with nervous tissue were found on the upper-beak skin.[47,48] However, the 'magnetite hypothesis' was challenged by recent studies, which found that the clusters of iron-rich cells on the upper-beak skin were macrophages, not magnetosensitive neurons.[49,50] The bat was famously known for determining its orientation at night by using echolocation. However, echolocation could work for only a short range, and some bats could use the Earth's magnetic field to migrate hundreds of kilometres between their breeding and wintering roosts each year.[51,52] A recent study from Tian et al. showed that one kind of bat, named the Chinese Noctule, *Nyctalus plancyi*, could sense 1/5 of a present geomagnetic field strength. Such field strengths occurred during geomagnetic field excursions or polarity reversals.[53] Although

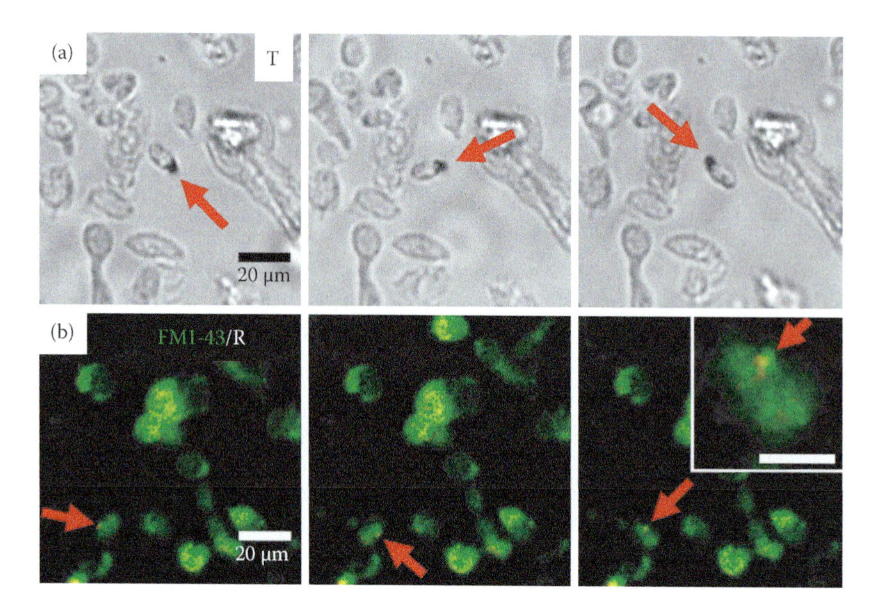

FIGURE 4.2 Candidate magnetoreceptor cells from dissociated trout olfactory epithelium. Transmitted light images (a) and simultaneously recorded dark-field reflection (R) and fluorescence (FM1-43, lipophilic dye) (b), showing an individual cell containing an opaque inclusion (red arrow) that rotates with magnetic field. The rotating cell contains a strong reflective inclusion (red arrow), displayed in the close-up (upper right corner of (b), scale bar 10 μm).[46] (From Eder, S. H.; Cadiou, H.; Muhamad, A.; McNaughton, P. A.; Kirschvink, J. L.; Winklhofer, M., Magnetic characterization of isolated candidate vertebrate magnetoreceptor cells. *Proc. Natl. Acad. Sci. U S A* 2012, *109*(30), 12022–12027. Copyright 2012 National Academy of Sciences U.S.A.)

the mechanism for magnetoreception was unknown, magnetic measurements indicated that the bat's head contained soft magnetic minerals (magnetite/maghemite).[54]

Three models are proposed to explain the magnetite-based magnetoreception. The first model is based on a chain of single-domain magnetite particles anchored on the cell membrane. External magnetic field interacts with the chain to open ion channels, which cause positively charged ions (e.g. Na^{2+} and Ca^{2+}) to cross the membrane and change the receptor potential of the cell.[55] A second possible model postulates that magnetic field induces interacting clusters of superparamagnetic magnetite nanoparticles attractive or repulsive on the membrane, resulting stress transferred to nerve system.[47] The third is a mechanotransduction model of external magnetite based on a mechanism of auditory hair cells. Hair cells connecting with mechanically gated ion channels are specialized neuronal cells that exist in the lateral line of fish and in the ear of vertebrates. In this model, the magnetite is located in a nerve terminal and linked to ion channels through filaments. The molecular motor on actin filaments repositions the ion channel if the channel opened by movement of the magnetite nanoparticles.[56]

4.1.3 Ferritins

Ferritin is an iron storage protein that exists in nearly all organisms. It plays a significant role in detoxification of Fe(II) and O_2 by mineralizing a ferrihydrite core into a protein shell.[57] Figure 4.3 illustrates the secondary structure of

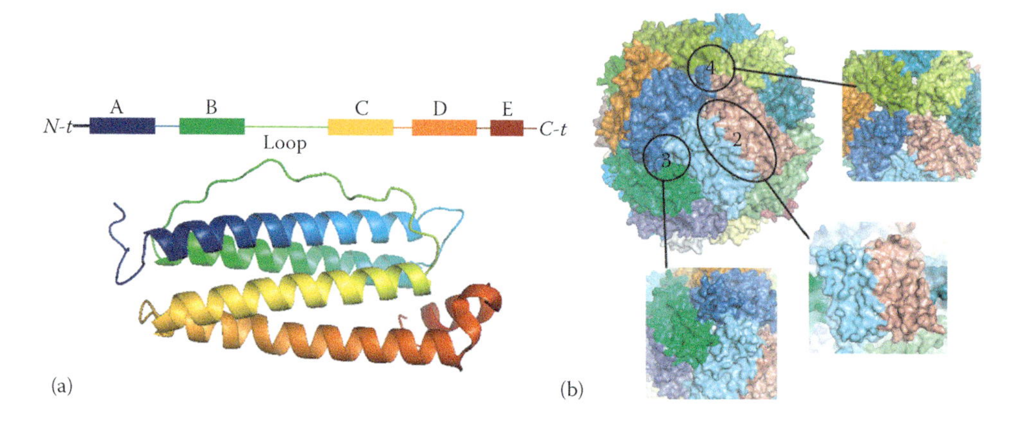

FIGURE 4.3 Secondary structure (a) and three-dimensional structure (b) of ferritin.[58] (Reprinted with permission from Honarmand Ebrahimi, K.; Hagedoorn, P. L.; Hagen, W. R., Unity in the biochemistry of the iron-storage proteins ferritin and bacterioferritin. *Chem. Rev.* 2015, *115*(1), 295–326. Copyright 2015 American Chemical Society.)

the ferritin subunit is composed of four α helices (ABCD) that form a bundle and a short C-terminal α helix (E), and the three-dimensional structure of the ferritin shell is highly conserved with self-assembly of 24 subunits into 4-3-2 symmetry. The outer and inner diameter of ferritin cavity is about 12 and 8 nm, separately, which can accommodate up to 4500 iron atoms. Mammalian ferritin usually has two kinds of subunits, a heavy (H) chain and light (L) chain. For example, the human ferritin H-chain possesses a ferroxidase center to quickly oxidize Fe(II). The ferroxidase center is composed of two iron sites, site A (Glu27, His65, Glu62 and Gln141) and site B (Glu62, Glu61 and Glu107), with Glu62 bridging the two sites. While the L-chain has no ferroxidase center, it possesses more acidic amino acids than the H-chain does in the nucleation site (Glu107, Glu57, Glu60, Glu61, Glu64 and Glu67), which allows a more efficiently formed iron core.[59–61] The H-chain and L-chain can be assembled at any ratio having a cooperative function for the iron uptake mechanism of ferritin. The self-assembled ferritin shell also has eight hydrophilic threefold channels and six hydrophobic fourfold channels. The threefold channel is lined at its narrowest part by Asp131 and Glu134, residues that are conserved in all mammalian ferritins. Mutation of these two residues causes a decreased amount of iron loading and decreased ferroxidase activity and indicates that the threefold channels are responsible for iron entry into the ferritin cavity.[62]

The mineral structure of the ferritin core was characterized as ferrihydrite ($5Fe_2O_3{\cdot}9H_2O$).[63] High-resolution transmission electron microscopy (HRTEM) observations showed that amorphous and crystalline nanoparticles were copresent and the degree of crystallinity improved with increasing particle size.[64] The amorphous structure in the ferrihydrite core probably accounts for its low stability of rapid mobilization of iron from ferritin.[65] However, recent studies showed a multiple mineral phase in the ferritin core. Electron nanodiffraction and HRTEM results demonstrated that physiological ferritin cores were mainly composed of single nanocrystals containing hexagonal ferrihydrite, hematite and little cubic magnetite/maghemite phase, while pathological ferritin cores extracted from patients with neurodegenerative diseases were composed of wüstite and a magnetite-like structure. It was proposed that a magnetite structure in pathological ferritin cores might cause an increase in the concentration of brain ferrous toxic iron.[66–68] When iron was gradually removed from physiological ferritin cores, the proportion of the magnetite phase increased, while the ferrihydrite phase decreased.[69] Magnetic characterization of horse spleen ferritin showed typical antiferromagnetic properties. When below the blocking temperature of $T_b \approx 12$ K, it had a net magnetic moment that came from the uncompensated spins of the nanoparticle surface or spin-canting.[70]

The ferritin cavity strictly controlled the crystal growth of the core. The three-dimensional structure reconstructed from high-angle annular dark field (HAADF) and scanning transmission electron microscopy images is shown in Figure 4.4. It has a hepatic ferritin core with cubic morphology

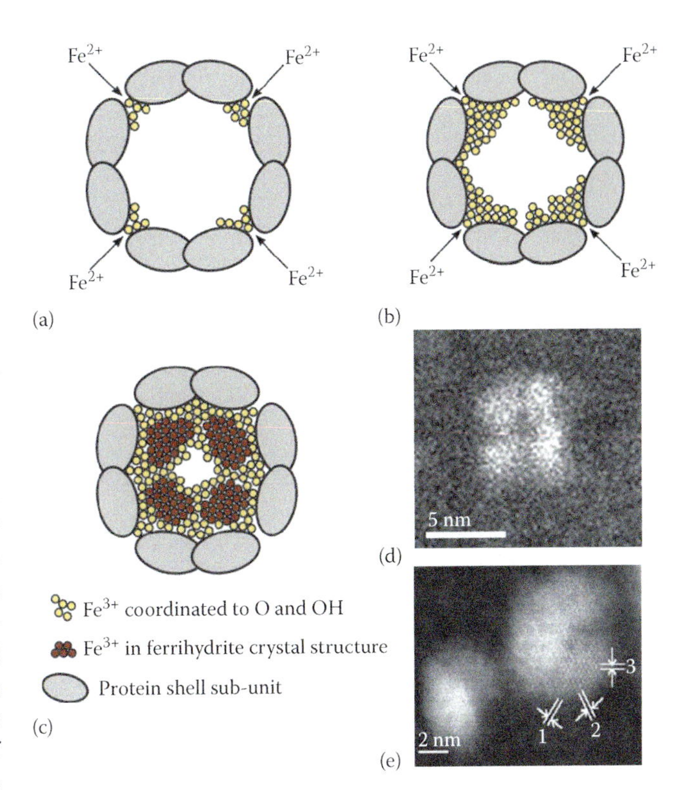

FIGURE 4.4 Mechanism for biomineralization of a ferritin core. (a) Fe^{2+} ions enter the ferritin cavity through the threefold channel and are oxidized at the ferroxidase center. The yellow circle represents oxidized iron (Fe^{3+}). (b) As more iron enters the cavity, Fe^{2+} ions are directly oxidized and deposit on any Fe^{3+} existing near the entry channels and form core subunits. (c) A ferrihydrite core is formed in the cavity with higher iron loading. (d) An HAADF image of a single ferritin core exhibits a cubic-core structure with eight subunits. (e) A high-resolution HAADF image of ferritin core shows that both the crystalline structure and amorphous structure exist in the core.[71] (From Pan, Y. H.; Sader, K.; Powell, J. J.; Bleloch, A.; Gass, M.; Trinick, J.; Warley, A.; Li, A.; Brydson, A.; Brown, A., *J. Struct. Biol.*, 166, 22–31, 2009. Reproduced under a Creative Commons Licence.)

composed of up to eight polycrystalline subunits, corresponding to the eight threefold channels in the protein shell that delivered iron to the central cavity.[71] There was also some phosphate absorbed on the surface of the ferritin core. About 60% of the phosphate present in horse spleen ferritin was released in the initial stage of iron release.[72] *In vitro* reconstitution experiments indicated that phosphate accelerated the rate of iron loading into ferritin.[73] The molecular mechanism underlying this stimulatory effect was that phosphate accelerated displacement of Fe(III) by Fe(II) in the ferroxidase center.[74]

Numerous *in vitro* experiments indicate that iron mineralization in ferritin can be achieved by at least three pathways, corresponding to ferroxidase, the mineral surface and Fe(II) + H_2O_2 detoxification reactions.[75,76] The ferroxidase reaction occurs at low iron loading of ferritin (≤50 Fe(II)/protein) and under aerobic conditions. First, two Fe^{2+} transport through the threefold channels and arrive at the ferroxidase center in the

H subunit of ferritin. Reaction with O_2 results in the reduction of O_2 into H_2O_2 and is seen Equation 4.1.

$$[Fe(II)_{2(FC)}\text{-}P]^{z+4} + O_2 + H_2O \rightarrow [Fe(III)_2O_{(FC)}\text{-}P]^{z+4} + H_2O_2 \quad (4.1)$$

Here, P represents ferritin-bound iron, FC signifies the ferroxidase center and z is the net charge on the protein. This reaction produces an intermediate μ-1,2-peroxodiferric species ($[Fe(II)_2O_{(FC)}\text{-}P]^{z+4}$) with absorption at 650 nm.[77] Secondly, the μ-1,2-peroxodiferric species are not stable at the ferroxidase center and undergo a hydrolysis reaction to form a [2FeOOH] core. The reaction is seen in Equation 4.2.

$$[Fe(II)_{2(FC)}\text{-}P]^{z+4} + O_2 + 4H_2O \rightarrow 2[Fe(III)OOH_{(core)}\text{-}P]^{z} + H_2O_2 + 4H^+ \quad (4.2)$$

However, the mineral surface reaction takes a major role in the formation of the iron core when a large flux of Fe (200 Fe(II)/protein) enters the protein. This mineral surface reaction is similar to Fe(II) autoxidation and hydrolysis and is shown in Equation 4.3.

$$4[Fe(II)_{cavity}\text{-}P]^{z+4} + O_2 + 6H_2O \rightarrow 4[Fe(III)OOH_{(core)}\text{-}P]^{z} + 8H^+ \quad (4.3)$$

In addition to the ferroxidase and mineral surface reaction, the previous observations of ultraviolet (UV) spectroscopy, stopped-flow kinetics and oximetry found that $Fe(II)/O_2$ stoichiometry increases with increasing iron added to the ferritin, especially during intermediate loading of ferritin at 100–500 Fe/protein per addition.[77] This indicates that a third iron oxidation reaction involving H_2O_2 participates in core mineralization (Equation 4.4) because H_2O_2 can be produced through Equations 4.1 and 4.2 on the ferroxidase center of the HFn.

$$2[Fe(II)_{cavity}\text{-}P]^{z+4} + H_2O_2 + 2H_2O \rightarrow 2[Fe(III)OOH_{(core)}\text{-}P]^{z} + 4H^+ \quad (4.4)$$

4.2 BIOINSPIRED SYNTHESIS OF MNPs

4.2.1 Synthesis of MNPs Inspired by Biomineralization of Ferritin

MNPs have great potential in science and technology. Therefore, diverse synthetic approaches have been developed to obtain monodispersed particles and to tune their size distribution, shape and crystallinity. However, conventional approaches using a coprecipitation reaction, a reductive process, an oxidative process, thermal decomposition and surfactant routes may require the use of organic additives or solvents. Rapid chemical synthesis leads to a decreased degree of crystallinity and significant spin misalignment, thereby reducing the total magnetic moment.[13] In contrast, biomimetics allow the use of different functionalities of chemical agents and biomacromolecules (peptides, proteins, viruses, DNA molecules and polymers) to synthesize MNPs under mild conditions. These macromolecules act as matrices, scaffolds or templating agents to control MNP formation and growth at the molecular level.[78]

Demineralized ferritin formed a hollow cavity called an apoferritin that homogeneously dispersed in aqueous media. The protein cage was very stable up to 85°C and tolerated reasonably high levels of urea, guanidinium chloride and many other denaturants at a neutral pH.[79,80] Therefore, it was an ideal nanoplatform for synthesizing MNPs. M-HFn, composed of a magnetite/maghemite core, was first synthesized in 1992.[81] The conventional synthesis of M-HFn needed two steps. First, the native horse spleen ferritin was devoid of its inorganic core by dialysis through thioglycolic acid at a pH of 4.5. Then a magnetite core was reconstituted into the apoferritin by stepwise addition of Fe(II) and air into a solution of apoferritin anaerobically under Ar or N_2 at an elevated pH (8.5) and temperature (60°C–65°C). During the reaction, Fe(II) was added slowly at 20-min intervals to allow partial oxidation and prevent precipitation of iron oxide nanoparticles in the solution. Compared with the blood red colour of native horse spleen ferritin, the produced M-HFn was black with no precipitation observed. Transmission electron microscopy analysis revealed that the majority of the particles synthesized in the apoferritin cavity were discrete, spherical nanometer-sized crystals with the mineral structure of magnetite. The Mössbauer spectra obtained at low temperatures and in large applied magnetic fields clearly showed that the M-HFn core was very similar to mineral maghemite (γ-Fe_2O_3) rather than magnetite (Fe_3O_4), which was quite different to that of native ferritin.[82–84] The M-HFn nanoparticles were superparamagnetic with 13,200 Bohr magnetons per molecule (6 emu/g), and the magnetic moment was sufficient for immunomagnetic isolation of lymphocytes from mononuclear cell preparation.[85,86] The longitudinal (r_1) and transversal (r_2) relaxivity was 8 and 175 $L{\cdot}mM^{-1}\,s^{-1}$, respectively. The high r_2/r_1 ratio of 22 indicated M-HFn was a potential magnetic resonance imaging (MRI) contrast agent to obtain a wide margin of negative signal enhancement.[87] However, *in vivo* studies of M-HFn demonstrated that these nanoparticles were rapidly cleared by the liver and spleen without binding to ferritin receptors, which suggested a limited application for M-HFn in molecular imaging.[87] Furthermore, the conventionally synthesized M-HFn nanoparticles were easy to aggregate and have magnetostatic interactions.[84] This might be caused by protein damage during a long duration (over 200 min) in elevated pH and temperature.

The synthetic procedure was later improved by adding stoichiometric amounts of the oxidant trimethylamine-*N*-oxide (Me_3NO). By limiting the number of stepwise cycles of the $Fe(II)/Me_3NO$ additions, it produced M-HFn with different iron loadings and core sizes. The use of Me_3NO as an oxidant under controlled chemical conditions permitted improved fine-tuning and reproducibility in the synthesis of magnetite/

maghemite cores. The superparamagnetic blocking temperature increased linearly with iron loading.[88] By probing the structure of the M-HFn by small-angle synchrotron X-ray and neutron scattering, it was found that the apoferritin shell underwent structural changes with iron loading above ~150 atoms. This change might be attributed to the effect of iron oxide binding and ordering inside the protein cavity.[89] A detailed experimental study of M-HFn synthesized with Me_3NO as the oxidant revealed a low degree of crystalline in the structure of the core and highly reduced magnetic moments.[90] These properties indicated that the two-step synthesis procedure was unfavourable for synthesis of high-magnetic-moment M-HFn nanoparticles. Recently, Clavijo Jordan et al.[91] developed a simple way to create monodispersed M-HFn using commercial apoferritin that could be used for molecular MRI. This was achieved via stepwise adding of $FeCl_2$ to the protein solution under low O_2 conditions. Subsequent filtration steps allowed for separation of the completely filled M-HFn from the partially filled ferritin. These particles exhibited a high transversal relaxivity r_2 of 130 $mM^{-1} s^{-1}$.

As the human HFn could be genetically engineered and easily overexpressed in *Escherichia coli*, Uchida et al.[92] created a novel way to synthesize M-HFn using genetically engineered HFn and its variants as a biotemplate. The genetically engineered HFn after purification had intact protein cages and nearly no iron oxide core in the ferritin cavity. Therefore, compared with the above-mentioned two-step synthetic procedure, M-HFn synthesis using the human HFn needed only one step to mineralizing an iron oxide core into the ferritin cavity, as shown in Figure 4.5a. Since the synthesis procedure used H_2O_2 as the oxidant and an autotitrator to strictly control pH, the Fe(II) and oxidant could be added to the protein solution continuously and the whole reaction finished within 1 h. Recently, this method was further improved through strictly controlled anaerobic conditions, temperature and pH, in an anaerobic box.[93] The newly synthesized M-HFn nanoparticles were highly monodispersed with each iron oxide core encapsulated by an intact protein shell (Figure 4.5b). Electron energy loss spectroscopy determined whether the structure of the core was stoichiometric magnetite rather than maghemite.[94] The magnetite cores could be tuned with different size distributions by adding different iron atoms, which have a narrow size distribution, a uniformly spherical shape and high crystallinity (Figure 4.5c–f). Even when the core was grown as small as 2 nm, no obvious lattice defects (such as dislocation and stacking fault) existed in the magnetite cores.[95] When the diameter of the M-HFn core was about 4 nm, there was no magnetostatic interaction between the particles. All single-domain particles were unblocked at 15 K and the median unblocking temperature was 8.2 K. The value of saturation remanence to saturation magnetization (M_{rs}/M_s) was nearly 0.5 and remanent coercivity (zero remanence at a field in a direct current demagnetization curve) to coercivity (B_{cr}/B_c) was 1.12, indicating that M-HFn was dominated by uniaxial anisotropy. Therefore, these structural and magnetic properties suggested that M-HFn was an ideal noninteracting model for studying superparamagnetic particles. By measuring the alternating current (AC) susceptibility of M-HFn nanoparticles, the value of preexponential factor f_0 in the Néel-Arrhenius equation was obtained as $(9.2 \pm 7.9) \times 10^{10}$ Hz.[93,94]

The ferritin cage was able to disassemble its subunits at an acidic pH (~2) and reassemble at a basic pH (7–8.5). This assembly and disassembly property was used for loading drugs, imaging agents and inorganic nanoparticles.[96–98] Recently, M-HFn was synthesized in this same way. When chemically synthesized magnetite nanoparticles (4–6 nm) were incubated with disassembled apoferritin at a pH of 2, then increased to a pH at 7, each magnetite nanoparticle could be encapsulated into the ferritin cavity. The ferritin encapsulated magnetite nanoparticle was water-soluble and size-controlled with a structure very similar to M-HFn. Its protein shell could be functionalized with two types of monosaccharides: *N*-acetyl-D-glucosamine and D-mannose vinylsulfone derivatives.[99] However, the approach for magnetite nanoparticles synthesized by disassembly and assembly of ferritin was very different from the biomineralization of ferritin because the structure of ferritin left two hole defects after assembling.[100] The biomineralization and biomimetic synthesis should rely on an intact ferritin cage containing channels, a ferroxidase center and a nucleation site.

4.2.2 Synthesis of MNPs Inspired by Magnetotactic Bacteria

Magnetotactic bacteria produced chain(s) of membrane-enveloped, nanometer-sized structurally perfect magnetite and/or greigite nanocrystals. As mentioned previously, the biogenesis of magnetosomes needed multiple steps and each step was strictly controlled by genes of proteins in magnetotactic bacteria. Therefore, it was of great potential to use macromolecules of this bacterium for biomimetic synthesis of high-quality MNPs. In the context of magnetite biomineralization by magnetotactic bacteria, it was possible to influence the chemical purity and magnetic properties of magnetosomes by controlling the composition of the growth medium. Staniland et al.[101] first achieved *in vivo* cobalt doping in three cultured strains of magnetotactic bacteria: *M. gryphiswaldense* (MSR-1), *Magnetospirillum magnetotacticum* (MS-1) and *M. magneticum* (AMB-1). They found that cobalt doping (0.2%–1.4%) was achieved by simply replacing various quantities of iron quinate with cobalt quinate. Further exploration of *in vivo* doping allowed greater cobalt doping, permitting a larger increase in magnetocrystalline anisotropy and magnetic coercivity.[102] Recently, Li et al. used combined methods of transmission electron microscopy (TEM), energy-dispersive X-ray spectroscopy, bulk magnetic measurements and element- and site-specific X-ray magnetic circular dichroism analysis to study the site occupancy, valence and distribution of cobalt within the magnetosomes of *M. magneticum* AMB-1. It was found that Co^{2+} could be incorporated into octahedral sites through the replacement of Fe^{2+} ions, as shown in Figure 4.6. The Co^{2+} doping in spinel structures of magnetosomes varied in the amount of different particles and was enriched at the

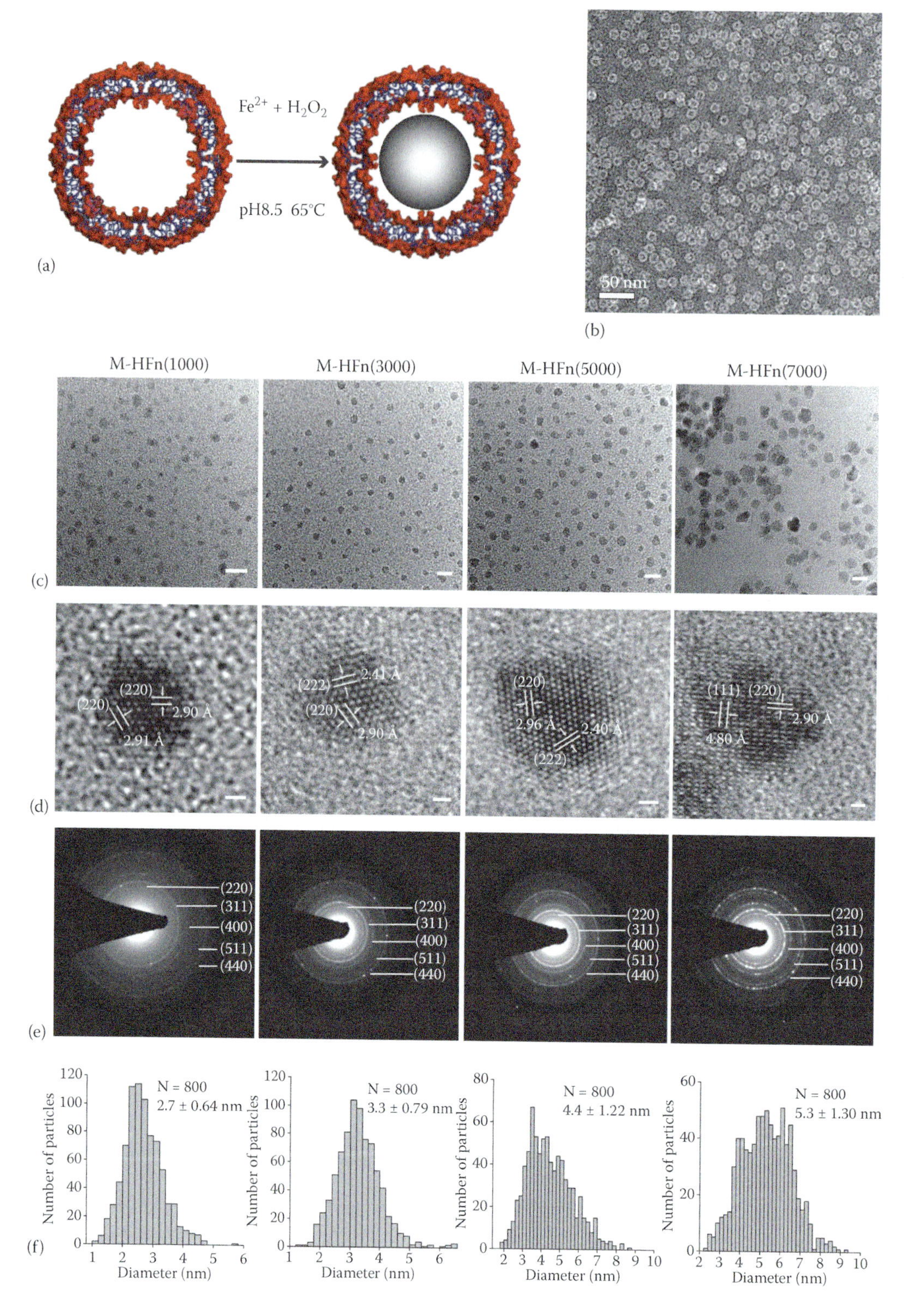

FIGURE 4.5 Biomimetic synthesis of M-HFn nanoparticles using human HFn as a biotemplate: (a) a model depicts a one-step synthesis of M-HFn nanoparticles; (b) a negative staining transmission electron microscopy (TEM) image of M-HFn clearly shows each single core encapsulated by a protein shell; (c) high-resolution TEM images; (d) selected area electron diffraction patterns; (e) and size distribution diagrams; (f) demonstrate single crystalline structure of M-HFn nanoparticles with different iron atoms loading.[95] (Reproduced with permission from Cai, Y.; Cao, C.; He, X.; Yang, C.; Tian, L.; Zhu, R.; Pan, Y., *Int. J. Nanomedicine*, 10, 2619, 2015.)

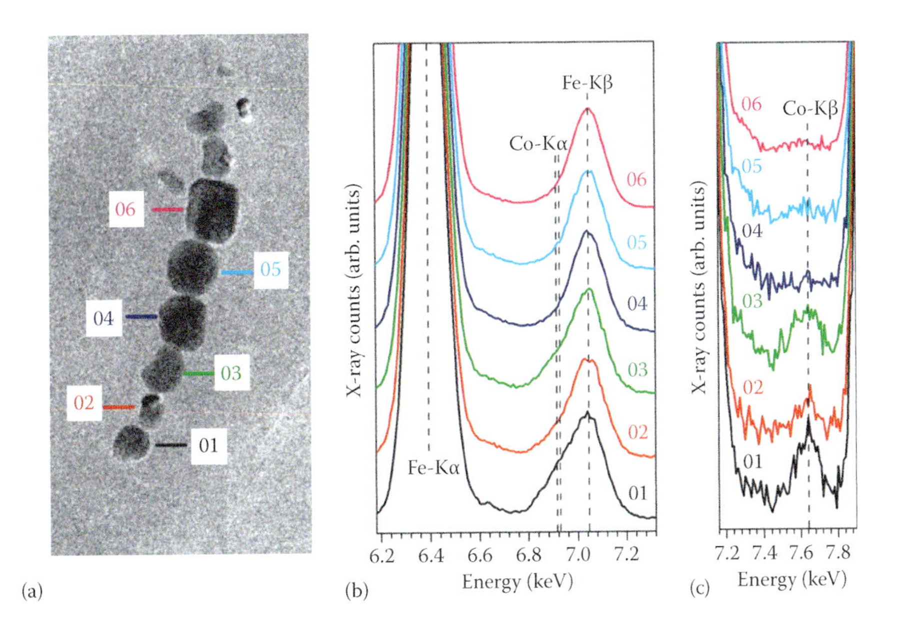

FIGURE 4.6 Six individual magnetic crystals in a cobalt-doped magnetosome chain of *M. magneticum* AMB-1 are analyzed by X-ray absorption spectroscopy (XAS), X-ray magnetic circular dichroism (XMCD) and magnetic characterization, indicating the cobalt ions are heterogeneously doped into O_h sites of magnetite by replacement of Fe^{2+}. (a) TEM image of magnetosome chain; (b, c) EDX spectra of six individual magnetosome particles as indicated by the numbers in (a). (Reproduced with permission from Professor Jinhua Li.)

surface of individual ones.[103] Magnetite crystals produced by magnetotactic bacteria was regarded as chemically pure, but it was reported that uncultured magnetotactic bacteria could incorporate manganese up to 2.8% of the total metal content when manganese chloride was added to microcosms.[104] Manganese doping was also evidenced by a recent study on the cultured magnetotactic bacterium stain MS-1. When manganese, ruthenium, zinc and vanadium were simultaneously added into the growth medium, only manganese incorporated within the magnetosome magnetite crystals.[105]

Although the size, shape and chemical purity of the magnetic crystals can be tuned through genetic engineering or chemical doping, it is difficult to produce large quantities of material, because most of the magnetotactic bacteria are difficult to culture and grow. *In vitro* biomimetic synthesis has become possible now due to the availability of many proteins that regulate the formation of magnetosome. Mms6 represents a class of protein first discovered to tightly bind to magnetic crystals in the *M. magneticum* strain AMB-1. Figure 4.7a illustrates the Mms6 amino acid sequence is

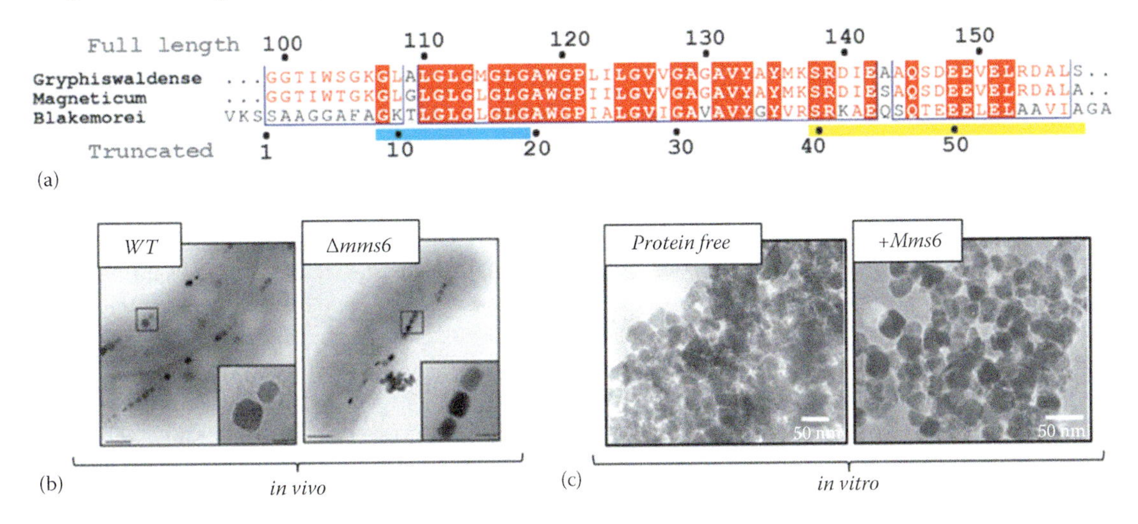

FIGURE 4.7 The sequence of Mms6 and its function demonstrated *in vivo* and *in vitro*. (a) Sequence alignment of the truncated Mms6 from different magnetotactic bacteria species indicates a highly conserved truncated protein. Conserved residues are highlighted in red boxes and similar residues are in red type. The blue bar highlights the glycine-leucine repeating sequence and the yellow bar highlights the hydrophilic, acidic rich, C-terminal amino acid region. (b) *In vivo* activity of Mms6 is demonstrated by knockout *mms6* in AMB-1. The mms6 mutant strain forms smaller magnetosomes with nonregular shape. (c) Mms6 mediates synthesis of highly uniform magnetite nanoparticles *in vitro*.[106] (Reproduced with permission from Staniland, S. S.; Rawlings, A. E., *Biochem. Soc. Trans.*, 44, 883–890, 2016.)

amphiphilic, which consists of an N-terminal hydrophobic region and a C-terminal hydrophilic region containing multiple acidic amino acids. The gene sequence of Mms6 is conserved in several strains of magnetotactic bacteria, particularly in the C-terminal region. In contrast with wild-type *M. magneticum* AMB-1, *in vivo* depletion of the *mms6* gene results in smaller magnetic crystals (27.4 ± 8.9 nm) that are formed with uncommon crystal faces of (210), (211) and (311) (Figure 4.7b).[107] To investigate the key amino acid residues for Mms6-mediated biomineralization, Yamagishi et al. have established and analyzed a series of gene deletion mutants of *M. magneticum* strain AMB-1. Deficiencies in the C-terminal region and repetitive GL region cause elongated morphology of magnetic crystals, indicating that both the N-terminal and C-terminal are essential for protein function, structure formation or localization. Asp123, Glu124 and Glu125 are key amino acid residues that directly control the morphology of magnetic crystal *in vivo*.[108] Although Mms6 plays a key role in controlling the crystal morphology, a recent study of gene deletion mutants indicates that other Mms proteins (Mms5, Mms7 and Mms13) have a coordinated function to regulate the cubo-octahedral magnetite crystals in magnetotactic bacteria.[109]

Numerous *in vitro* experiments have indicated that the Mms6 protein could control both the size and shape of magnetic crystals. It was suggested that coprecipitation of ferrous and ferric ions in the presence of Mms6 produced uniform magnetite nanoparticles with sizes ranging from 20 to 30 nm (Figure 4.7c). These MNPs showed cubo-octahedral morphology similar to that of magnetosomes in the *M. magneticum* strain AMB-1. Without this protein, the magnetic particles formed irregular shapes and sizes (Figure 4.7c).[110] Compared with other iron binding proteins (ferritin, lipocalin and bovine serum albumin), only Mms6 could mediate the formation of uniform *ca.* 30 nm magnetite nanoparticles in the presence of viscous aqueous Pluronic F127 gels. The Pluronic F127 gels created a viscous media to better mimic the synthetic environment, but the entire reaction needed 5 days.[111] Amemiya et al.[112] developed a quick way to partially oxidize ferrous hydroxide to synthesize uniform magnetite nanoparticles in the presence of Mms6 under temperatures as high as 90°C. Further exploration of this protein in biomimetic synthesis allowed room temperature synthesis of 50–80 nm uniform cobalt ferrite ($CoFe_2O4$) nanoparticles through covalently conjugating the full-length Mms6 protein and a synthetic C-terminal of Mms6 protein to the self-assembled Pluronic F127 gels. The $CoFe_2O_4$ nanoparticles in this size range were very difficult to produce using conventional ways.[113] Currently, many different biomimetic methods using Mms6 have been developed to synthesize MNPs with tuning sizes, shape and function. For example, the Mms6 protein could attach to a silicon substrate through hydrophobic interaction with a monolayer of octadecyltrimethoxysilane, which could mediate site-specific formation of magnetite crystals.[114] The Mms6 protein controlled the sizes and shape of not only pure magnetite but also Co-doped magnetite nanoparticles synthesized by the coprecipitation method.[115] Using ferrihydrite as a precursor and optimizing the reaction condition, magnetite with a particle size up to 60 nm could be synthesized in the active sequence DIESAQSDEEVE in the Mms6 protein (M6A).[116] Specifically, the magnetite binding peptide M6A could be fused to the C-terminal of the murine HFn, which enabled *in vivo* magnetite biomineralization in the ferritin cavity. This provided a novel reporter gene for MRI.[117]

To understand the mineralization mechanisms of Mms6, competitive iron binding analysis with other inorganic cations suggested that the C-terminal region was an iron binding site. Prozorov et al.[113] first demonstrated that a synthetic C-terminal domain of Mms6 protein containing 25 amino acids (C25-Mms6) functioned the same as full-length Mms6 to control the size and shape of $CoFe_2O4$ nanoparticles in the presence of Pluronic gel. Later, Arakaki et al. designed short synthetic peptides that mimicked the functions of Mms6 to determine its functional amino acid region for the magnetite synthesis. The study demonstrated that the putative iron binding site in the C-terminal acidic region of Mms6 comprising 12 amino acids (M6A) played a key role in the formation of cubo-octahedral particles. Fluorescence labeling experiments indicated that the M6A peptides localized on the surface of magnetite crystal.[114] The Mms6 self-assembled as a micelle consisting of 20–40 monomers with the hydrophilic C-terminals on the surface and hydrophobic N-terminals buried inside. The C-terminal domain played a key role in forming and maintaining the micellar structure between 200–400 kD.[118] The Mms6 protein also could readily form a stable monomolecular layer at the liquid/vapour interface depending on the pressure and ions in the solution,[119] Self-assembly of Mms6 involved an interlaced structure of intramolecular and intermolecular interactions that resulted in a coordinated structural change with iron binding.[120] Kashyap et al.[121] directly visualized the iron binding on the C-terminal region and nucleation of iron oxide mediated by Mms6 with *in situ* liquid cell TEM. The TEM observations indicated that an amorphous precursor phase formed first on the surface of the protein micelles, then nucleation of iron oxide on the micelles followed.

Since the biogenesis of magnetosomes in magnetotactic bacteria were regulated by several key biomineralization proteins, many other proteins in spite of Mms6 also had potential in biomimetic synthesis of MNPs. The magnetosome associated MamP, a new class of c-type cytochrome, was exclusively found in mangetotactic bacteria. The structure of MamP was composed of a self-plugged PDZ domain fused to two magnetochrome domains, which served as an iron oxidase catalyzing Fe(II) to form a ferrihydrite precursor for growth of magnetite crystal *in vivo*.[122] *In vitro* biochemical experiments demonstrated that purified MamP oxidized Fe(II) to Fe(III) and finally produced mixed valent iron oxides related to magnetite under several different pH.[123] MmsF was a recently characterized magnetosome membrane protein that played a major role in regulating crystal size and morphology. Compared with wild-type magnetic crystals, deletion of the *mmsF* gene from the magnetosome gene island of the *M. magneticum* strain AMB-1 produced overall smaller magnetic crystals, where 77.3% of the crystals were below 35 nm. Addition of *mmsF* to 18 genes of the *mam AB* gene cluster caused a significant enhancement of magnetite biomineralization *in vivo* and cellular magnetic response, which suggested that MmsF was a key biomineralization protein required for maturation of the magnetic crystals.[124] Rawlings et al.[125] overexpressed MmsF in *E. coli* and found that the purified protein could self-assemble into water-soluble proteinosomes with an average size of 100 ± 25 nm determined by dynamic light scattering, but only 36 nm determined by TEM. *In vitro* biomimetic synthesis by coprecipitation of ferrous and ferric ions in the presence of MmsF produced uniform magnetite nanoparticles with a mean length of 56 nm. The saturation magnetization of MmsF-templated magnetite nanoparticles was 129 emu/g, much larger than that of the controlled magnetite synthesis without MmsF.[125]

4.3 BIOINSPIRED MNPs FOR CANCER DIAGNOSIS AND THERAPY

4.3.1 Peroxidase Activity of M-HFN for *In Vitro* Staining of Tumour Cells

Horseradish peroxidase (HRP, EC 1.11.1.7), the haem-containing enzyme, which is by far the most researched peroxidase, is widely used as a detection tool because of its intrinsic catalytic activity that catalyzes oxidation of various substrates in the presence of H_2O_2. For instance, it is used as a reporter enzyme in immunological diagnosis and histochemistry. However, the practical application of HRP is often hampered by the intrinsic drawbacks of natural enzymes such as environmental dependencies of catalytic activity, inherent instability in extreme conditions (high temperature and strong pH) and their expensive preparation and purification. Fe_3O_4 MNPs have been reported to possess intrinsic peroxidase-like activity, which is similar to naturally occurring HRP.[126] In comparison with HRP, the Fe_3O_4 nanoparticles are more stable against denaturation and resistant to high concentration of the substrate. Furthermore, their preparation and storage are relatively simple. However, Fe_3O_4 often needs surface modification to prevent aggregation and linking with targeting ligands (such as antibodies, peptides or small molecules) special for tumour diagnosis. Nevertheless, these modifications are expensive and may severely reduce peroxidase-like activity in practical application.

The peroxidase activity of M-HFn was first found in M-HFn nanoparticles. When the peroxidase substrates 3,3,5,5-tetramethylbenzidine (TMB) and di-azo-aminobenzene (DAB) were added into the M-HFn solution with H_2O_2, the M-HFn nanoparticle could catalyze TMB and DAB to produce a blue colour (Figure 4.8a) and brown colour (Figure 4.8b), respectively.[127] The peroxidase activity of M-HFn nanoparticles was size dependent. With the increase in iron atoms loading into the ferritin cavity, the size of the mineral core and peroxidase activity increased (Figure 4.8a, b).[89,95] Electron spin resonance measurements showed that •OH was produced during the peroxidase-like reaction in the presence of both M-HFn nanoparticles and H_2O_2. Thus, the possible catalyzing mechanism was that H_2O_2 diffused into the ferritin cavity through its hydrophilic channels and interacted with the iron oxide core of M-HFn to generate •OH on the surface of the iron core. The generated •OH catalyzed the peroxidase substrate to give a colour reaction.

Despite the iron oxide core possessing peroxidase activity, the HFn shell of M-HFn nanoparticles also had an intrinsic tumour targeting ability. The HFn shell could specifically and directly bind to the transferrin receptor 1 (TfR1) overexpressed on tumour cells. The TfR1 (also named as CD71) was a type II transmembrane glycoprotein and was originally identified as the receptor for transferrin (Tf).[127,128] Because proliferating cancer cells required more iron uptake for metabolism, it was well known that TfR1 was highly expressed in many kinds of malignant carcinoma cells and was widely used in tumour imaging and therapy.[129,130] Recently, we screened 24 cancer cell lines and found only one type of cancer cell named MX-1 with no TfR1 overexpression. The HFn shell could be specifically endocytosed into tumour cells mediated by TfR1.[16,127]

Based on the finding that M-HFn nanoparticles had intrinsic dual functionality of peroxidase activity and tumour targeting ability, we developed a new technique for tumour detection recently, as shown in Figure 4.8c. This technique simply used M-HFn nanoparticles to visualize various tumour tissues, including ovarian, liver, prostate, lung, breast, pancreas, cervical, thymus, colorectal and oesophageal cancers. The protein shell specifically targeted cancer cells and the iron oxide core catalyzing DAB formed brown precipitate after adding a DAB substrate and H_2O_2. This technique was verified through the screening of 474 clinical specimens, including 247 clinical tumour tissue samples as well as 227 nontumour control samples. The M-HFn nanoparticles could efficiently differentiate tumour tissue and nontumour tissue with a sensitivity of 98% and specificity of 95%. Compared with traditional antibody-based tumour detection, the M-HFn staining technique was easy to use because it needed only one reagent and one step, but traditional antibody-based immunohistochemical staining always used primary antibody, secondary antibody or enzyme-labeled third antibody with multiple steps between each incubation.[127]

Co-doped M-HFn nanoparticles (hereafter termed as M-HFn-$Co_xFe_{3-x}O_4$) could be synthesized by simultaneously adding Fe^{2+} and Co^{2+} with H_2O_2 as the oxidant.[131–133] Our recent study showed that when the starting molar percentage of cobalt (Co/(Co + Fe) × 100) was 0%, 20%, 40% and 60%, kinetic analysis of M-HFn-$Co_xFe_{3-x}O_4$ nanoparticles showed that the K_m values of the M-HFn-$Co_xFe_{3-x}O_4$ nanoparticles were nearly constant by statistical analysis.[133] K_m is the Michaelis constant, which is an indicator of enzyme affinity for its substrate, which means that the doped and undoped M-HFn nanoparticles have similar affinities for H_2O_2 as well as TMB. However, the V_{max} (maximal reaction velocity) of Co_{60} was nearly 2.7 times higher than the undoped M-HFn nanoparticles. It was evident that Co ions had excellent catalytic activity in the H_2O_2 decomposition reaction in place of Fe compounds. The optimal pH and temperature for M-HFn-$Co_xFe_{3-x}O_4$ nanoparticles were 4.5 and 60°C,

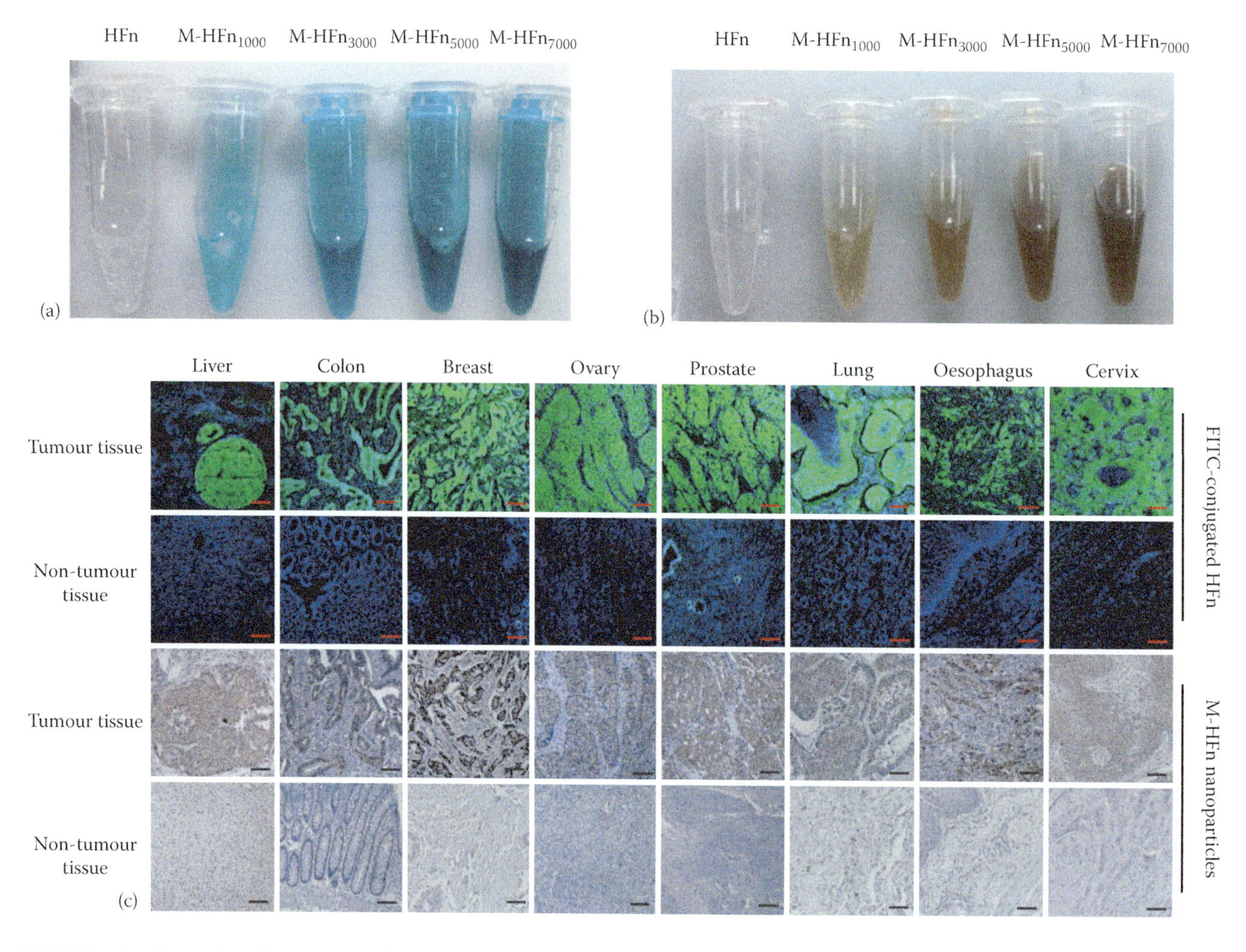

FIGURE 4.8 Peroxidase-like activity of M-HFn nanoparticles for developing a new technique for visualizing tumour tissues. M-HFn nanoparticles catalyze the peroxidase substrates TMB (a) and DAB (b) with H_2O_2 to produce blue and brown colours, respectively. (c) Paraffin-embedded clinical tumour tissues and corresponding normal tissues are stained with FITC-labeled HFn protein shells and M-HFn nanoparticles. The tumour tissues can be specifically stained with either FITC-labeled HFn protein shell (green colour) or M-HFn nanoparticles (brown colour), whereas the normal tissues cannot be stained with HFn shell and M-HFn nanoparticles.[95,127] (Reproduced with permission from Cai, Y., Cao, C., He, X., Yang, C., Tian, L., Zhu, R., Pan, Y., *Int. J. Nanomedicine*, 10, 2619, 2015; Reprinted by permission from Macmillan Publishers Ltd. *Nature Nanotechnology*. Fan, K.; Cao, C.; Pan, Y.; Lu, D.; Yang, D.; Feng, J.; Song, L.; Liang, M.; Yan, X., Magnetoferritin nanoparticles for targeting and visualizing tumour tissues. *Nat. Nanotechnol.* 2012, *7*(7), 459–464. Copyright 2012.)

respectively. Co_{60}-doped M-HFn nanoparticles for staining frozen breast tumour tissue (MDA-MB-231), pancreas tumour tissue (CFPAC-1), colorectal tumour tissue (HCT-116) and stomach tumour tissue (MGC-803) efficiently differentiated tumour tissue from nontumour tissue. The Co_{60} significantly enhanced the staining of tumour tissues compared with tissue treated with undoped M-HFn nanoparticles.

4.3.2 *In Vivo* Targeting and Imaging of Microscopic Tumours

The detection of cancer at its earliest stage is vital for physicians to treat and cure cancer. SPIO is a preferred material to use in MRI of tumours because it has higher relaxivity and lower toxicity than gadolinium-based contrast agents. It can significantly change the spin–spin relaxation times (T_2 and T_2^*), which results in significant intensity reduction on T_2- and T_2^*-weighted MRIs and enables single-cell detection.[134,135] However, commercially available SPIO nanoparticles mainly accumulate in the liver, spleen and bone marrow, all of which are dependent on the reticuloendothelial system. They are nonspecific to cancer cells and therefore are limited to diagnostic applications of the liver and lymph nodes with low sensitivity.[136,137] More specifically, when detecting primary cancer lesions using MRI, SPIO nanoparticles always need complex coatings and conjugation with targeting ligands, such as antibodies, peptides and small molecules.[138] But a major problem with SPIO nanoparticle conjugation with targeting ligands is the targeting ability *in vivo* is unsatisfactory, which leads to insufficient signal amplification and contrast enhancement for the cancer lesion.[139,140] The targeting efficiency of SPIO nanoparticles largely depends on the surface coating and targeting ligand conjugation method.[141–143]

Compared with MNPs synthesized by other methods, the functionalisation and targeting of biomimetic MNPs are covered in Chapters 5 and 6 in this book; biomimetic MNPs can target cancer cells either through indirect targeting with tumour targeted ligands or direct targeting without any modification. For instance, the M-HFn nanoparticles can be genetically engineered with RGD-4C that specifically targets cancer cells via binding to highly expressed integrin molecules ($\alpha_v\beta_3$).[92] Another targeting ligand, epidermal growth factor, can fuse to the N-terminus of the human HFn to form chimeric protein nanoparticles. The chimeric protein can specifically bind to and be taken up by breast cancer MCF-7 cells and MDA-MB-231 cells, but not normal breast epithelial cells.[144] Using the disassembly and reassembly properties of ferritin under different pH conditions, the ferritin heavy chain coupled with Cy5.5, ferritin conjugated with RGD4C and $^{64}CuCl_2$, can mix to form multimodal probes with both positron emission tomography and near-infrared fluorescence (NIRF) functionalities for tumour imaging, as shown in Figure 4.9.[98] Fibroblast activation protein-α (FAP-α) is a membrane-bound serine protease highly expressed by cancer-associated fibroblasts and pericytes in epithelial tumours. Ji et al.[145] have constructed hybrid ferritin probes for optical imaging of tumour microenvironment by coupling a fluorescence tagged peptide that is specifically cleaved by FAP-α. If the HFn is engineered with multiple tumour receptor-binding peptides, it can enhance *in vivo* imaging of tumours by increasing the accumulation of nanoparticles at the tumour site.[146] Recently, two sites of the ferritin monomer (the N-terminus and the loop between the fourth and fifth helices) have been engineered with 24 RGD peptides and 24 AP1 peptides to independently target tumour cells expressing the corresponding receptors, which allows 'super affinity' and bispecificity.[147]

Actually, our recent finding demonstrated that HFn nanoparticles could be directly used for specifically targeting tumours *in vivo*, because TfR1 was highly expressed on various tumour cells.[16] After injecting Cy5.5-labeled HFn nanoparticles, microscopic (<1–2 mm) tumours were sensitively detected by NIRF imaging (Figure 4.10a and d). In particular, the M-HFn nanoparticles exhibited high transversal relaxivity of 224 mM^{-1} s^{-1}. After intravenous injection, the M-HFn nanoparticles were capable of transporting across endothelial, epithelial and blood–brain barriers. It targeted the microscopic breast tumour and was endocytosed by the cancer cells. Although the mechanism of crossing endothelial barriers was unknown, TEM results showed that M-HFn nanoparticles localized in clathrin-coated pits of the endothelial surface; in cytoplasmic sorting endosomes, coated caveolae, multivesicular bodies (200–500 nm), transport vesicles, intercellular tight junctions, subendothelial interstitial space, basement membrane, perivascular space and epithelial caveolae; and exiting the epithelium from the caveolae into the subepithelial space, as shown in Figure 4.10b. These observations indicated that there were multiple transport mechanisms exploited by HFn-based nanoparticles across the endothelium, which included passage through the intercellular tight junctions and transcytosis through clathrin-coated pits and caveolae. Figure 4.10c shows that M-HFn nanoparticles

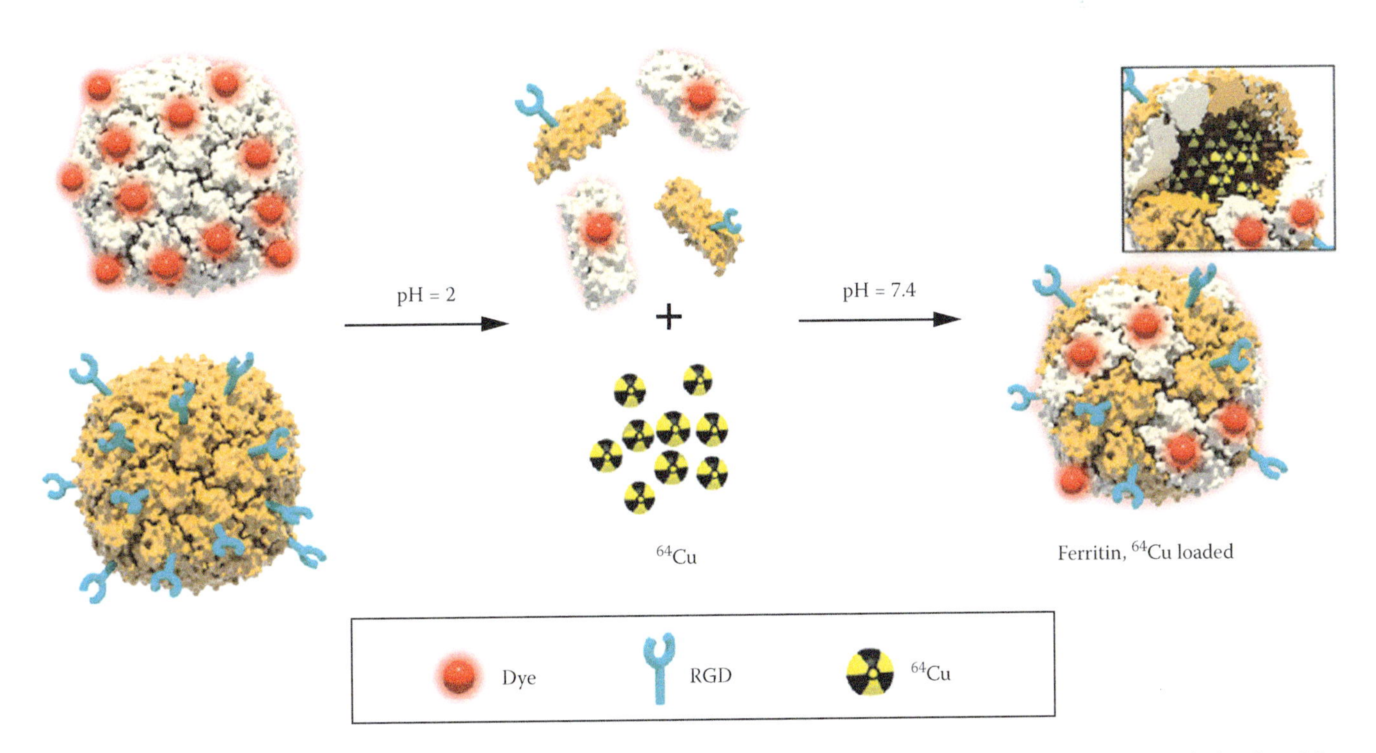

FIGURE 4.9 Construction of hybrid ferritin nanoparticles for multimodal imaging of tumours.[98] (Reproduced with permission from Lin, X.; Xie, J.; Niu, G.; Zhang, F.; Gao, H.; Yang, M.; Quan, Q.; Aronova, M. A.; Zhang, G.; Lee, S.; Leapman, R.; Chen, X., Chimeric ferritin nanocages for multiple function loading and multimodal imaging. *Nano Letters* 2011, *11*(2), 814–819. Copyright 2011 American Chemical Society.)

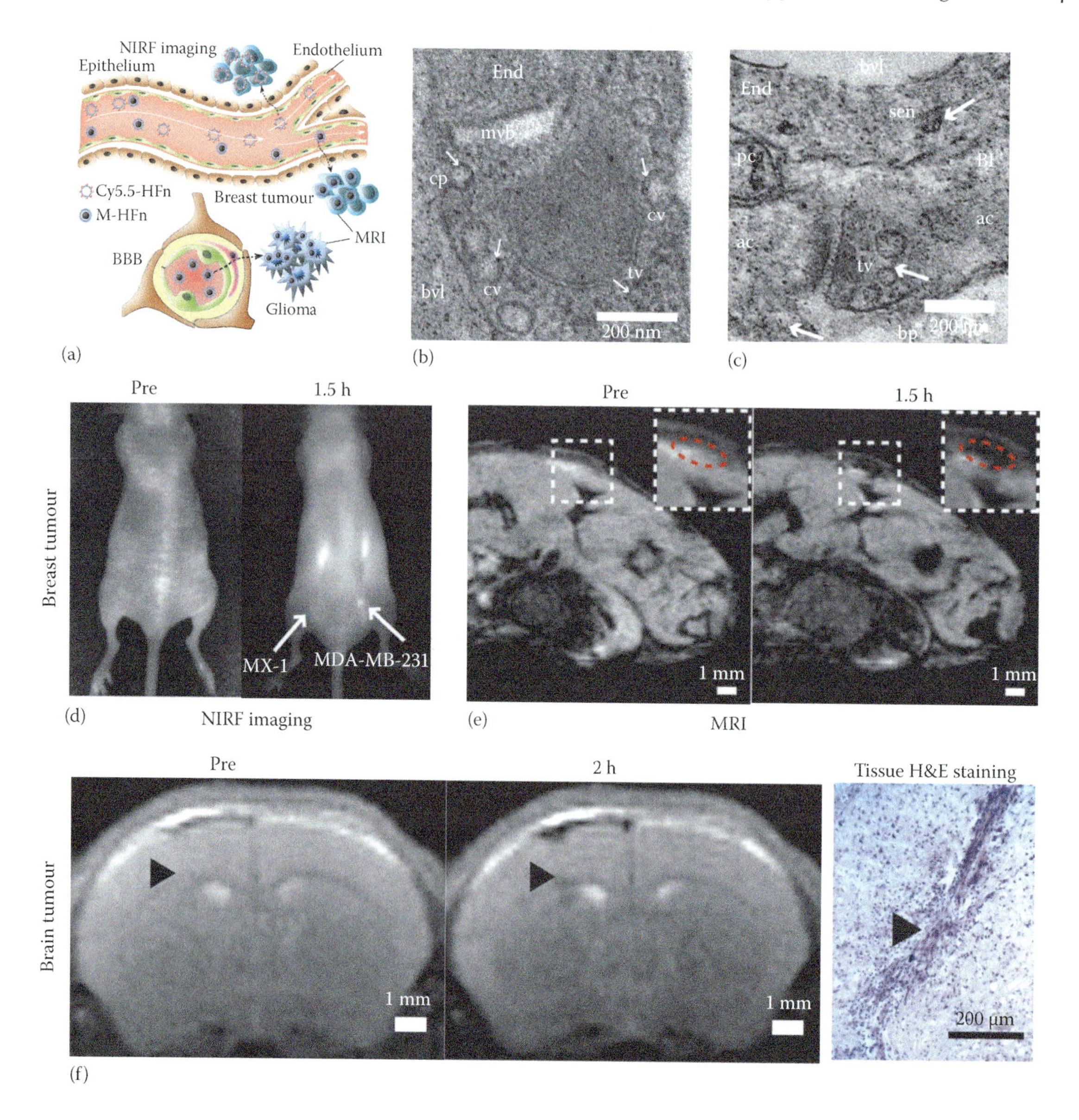

FIGURE 4.10 Imaging of microscopic breast and brain tumours by HFn-based nanoparticles across biological barriers: (a) the upper image depicts that Cy5.5-labeled HFn shell and M-HFn nanoparticles transport across endothelial and epithelial barriers for NIRF imaging and MRI of a microscopic breast tumour. The lower image shows that M-HFn nanoparticles cross the blood–brain barrier for enhanced MRI of a microscopic glioma; (b, c) M-HFn nanoparticles cross endothelial and blood–brain barriers, as shown in TEM images of the ultrastructure of subcutaneous normal blood microvessels (b) and brain microvessels (c) 2 min post injection of M-HFn nanoparticles; (d, e) Detecting microscopic breast tumours using a Cy5.5-labeled HFn shell as an NIRF imaging probe and M-HFn nanoparticle as a MRI probe; (f) Microscopic brain tumour (arrow head) can be sensitively detected by M-HFn nanoparticle enhanced MRI.[16] (Reproduced with permission from Cao, C.; Wang, X.; Cai, Y.; Sun, L.; Tian, L.; Wu, H.; Lei, H.; Liu, W.; Chen, G.; Zhu, R.; Pan, Y., Targeted *in vivo* imaging of microscopic tumors with ferritin-based nanoprobes across biological barriers. *Adv. Mater.*, 26, 2566–2571, 2014. Copyright 2014 John Wiley and Sons.)

transporting across the blood–brain barriers appeared to be similar to transcytosis of Tf mediated by a TfR. Most of the nanoparticles were found to be bound to the clathrin-coated pits and localized to the morphologically tubular sorting endosomes, multivesicular bodies and transport vesicles within the endothelium after 2 min of injection.

Due to a large accumulation of M-HFn nanoparticles in the tumour tissue, a xenografted breast tumour as small as 0.6 mm with no angiogenesis could be ultrasensitively detected by MRI (Figure 4.10e). Also, the M-HFn nanoparticles could be used as a contrast agent for enhanced MRI of microscopic brain tumours as small as 1 mm (Figure 4.10f).[16] Recently, M-HFn nanoparticles were labeled with ^{125}I radionuclide and served as a combination probe for both single-photon emission tomography and MRI. A single-dose injection of ^{125}I-M-HFn nanoparticles provided sufficient spatial and temporal resolutions as well as high sensitivity for imaging of tumours.[148] The intrinsic tumour targeting ability of HFn-based nanoparticles provided a great potential for applications in clinical practice because this kind of nanoparticle bypassed three limitations of current nanoparticle-based targeting and imaging strategies. First, the HFn protein cage provided a biotemplate for one-step nanoparticle synthesis and tumour targeting, thereby eliminating the need of conventional surface coating and targeted ligand modification. Second, the HFn-based nanoparticles could cross biological barriers, such as the endothelium, epithelium and blood–brain barriers, which often severely inhibit most types of nanoparticles from reaching their targets and rely on additional functional groups designed to facilitate barrier permeation. Third, M-HFn-based nanoparticles could be used to sensitively detect a microscopic tumour with no angiogenesis at the earliest stage of tumour growth.

4.3.3 Hyperthermia

Magnetic hyperthermia uses AC field stimulation of MNPs to generate heat for cancer cell destruction. In synergy with chemotherapy and radiotherapy, the heating effect of MNPs can be used for specific targeting and destruction of the tumour, because cancerous cells are selectively damaged when the temperature increases above 42°C. Theoretically, M-HFn is not a good agent to use in hyperthermia because the ferritin cavity limits the magnetite core to less than 8 nm while the best size for magnetic hyperthermia is 15–18 nm. However, a recent study by Fantechi et al.[132] solves this problem with a certain quantity (5%) of cobalt doping of the M-HFn cores. The formed cobalt ferrite nanoparticles significantly increase the temperature up to 5°C, with a specific absorption rate (SAR) value of 2.81 $\pm$ 0.02 W/g compared with undoped M-HFn nanoparticles.[132]

SAR was defined as the overall heating produced by MNPs under an AC field. A large SAR was a key parameter to realize the therapeutic potential of this technique. Interestingly, magnetosomes that were generated by magnetotactic bacteria had the largest reported SAR value. The intact bacterial cell contained chains of magnetosomes that yielded a SAR of 115 $\pm$ 12 W/g_{Fe} at 23 mT and 864 $\pm$ 9 W/g_{Fe} at 88 mT. The individual magnetosomes with membrane removed exhibited a SAR value of 529 $\pm$ 14 W/g_{Fe} at 23 mT and 950 $\pm$ 18 W/g_{Fe} at 88 mT.[149] The hyperthermia effect of magnetosomes also varied with growth conditions. When adding a high amount of Wofe's vitamin solution or ferric quinate into the growth medium of AMB-1 magnetotactic bacteria, the median size of magnetosomes increased from 47 nm (normal condition) up to 52 and 58 nm and caused an increase in coercivity and SAR.[150] To increase magnetic anisotropy, 20 μM cobalt quinate solution was added to the growth condition of AMB-1 magnetotactic bacteria, resulting in an increase in the magnetocrystalline anisotropy constant from 12 kJ/m^3 up to 104 kJ/m^3. This increase in magnetosome magnetic anisotropy led to an increase in the SAR.[102]

However, the antitumour activity resulting from magnetic hyperthermia effect of magnetosomes is still poorly understood. Liu et al.[151] found that magnetosomes isolated from *M. gryphiswaldense* MSR-1 exhibited high heating speed and temperature under an alternative magnetic field (AMF) of 300 kHz. When MCF-7 tumour cells were incubated with magnetosomes and heated under the application of an alternating magnetic field, the temperature could be increased up to 47°C and 80% of the cell proliferation was inhibited.[151] Alphandéry et al.[102] investigated the *in vivo* antitumour effect of magnetosome and found that the extracted chains of magnetosomes resulted in an increase in tumour temperature up to ~43°C, which enabled efficient destruction of cancer cells *in vivo*.[102]

4.4 FUTURE DIRECTIONS

4.4.1 Exploring Biomineralization for Synthesis of High-Quality MNPs

The human HFn contains a di-iron ferroxidase center of A and B binding sites composed of coordinating residues His65, Glu27, Glu61, Glu62 and Glu107 with Glu62 bridging the two sites. According to the protein catalysis model originally proposed by Roman (1978),[152] the ferroxidase center was involved in oxidation of Fe(II) at all stages of core formation. HFn also had a cluster of putative core nucleation site consisting of residues of Glu61, Glu64 and Glu67. Glu61 was a shared ligand between the nucleation site and ferroxidase center. In the past, UV spectroscopy, stopped-flow kinetics and electrode oximetry were usually used to investigate the biomineralization mechanisms of ferritin. However, it was very difficult to use these techniques to study the mineralization and crystal growth of M-HFn because the synthesis of M-HFn was under strictly controlled anaerobic conditions. Little was known about whether the ferroxidase center and nucleation site were involved in the mineralization of the magnetite core. The precursor and mineralization process of M-HFn was also unknown. Moreover, the assembly process of ferritin was not easily monitored. Therefore, the diameter of the cavity limited the size of the core. If the mineralization mechanism and assembly process of M-HFn could be revealed in the future, the size, crystallinity and magnetic properties of MNPs would be further improved.

The scale-up production of M-HFn nanoparticles is possible because the HFn protein cage can be expressed in *E. coli* or yeast at high yield and easily purified by using their heat-resistant property. In addition, the recent synthetic procedures developed for M-HFn offer several important advantageous features over the conventional methods for the synthesis of MNPs. First, this process allows monodisperse magnetite nanoparticles with tumour targeting ability to be obtained directly in the HFn cavity and without a further surface modification. Second, the synthetic process is environmentally friendly and economical, because it is under relatively mild conditions and uses nontoxic and inexpensive reagents such as ferrous ammonium sulfate and H_2O_2.

Magnetotactic bacteria provide a good way to synthesize high-quality MNPs under mild conditions because the shape, size, composition and crystallinity of magnetosomes were strictly controlled by different proteins. However, current biomimetic synthesis of MNPs used only limited magnetosome-related proteins (e.g. Mms6 and MmsF) for synthesizing magnetite and cobalt ferrite nanoparticles. New methods need to be developed to explore more magnetosome-related proteins used for biomimetic synthesis. In addition, more protein structures should be resolved to understand the mechanism of biomineralization and engineered for synthesis of high-quality MNPs.

4.4.2 Genetic Engineering for Functionalization of Bioinspired MNP for Targeted Diagnosis and Therapy

Synthesis of multifunctional MNPs is one of the hottest research areas in advanced materials. Multifunctional surfaces allow rational conjugations of imaging probes, biological and drug molecules, making it possible to achieve a target-specific diagnosis and treatment.[143,153] There are various cross-linkers that can be chosen for modification the surface of MNPs. 1-Ethyl-3-(3-dimethylaminopropyl) carbodiimide (EDC)/*N*-hydroxysuccinimide (NHS) chemistry is most commonly used to conjugate amine and carboxyl functionalities that are present on the surface of MNPs and targeting molecules. This method first needs the carboxyl terminal of the targeting molecules (e.g. monoclonal antibody, peptide, small molecules) to be activated by EDC and form an unstable intermediate stage. Then, the EDC-activated targeting molecules are stabilized by forming NHS esters at the carboxyl terminal. Finally, the NHS-targeting molecule covalently links with the amine functionality of MNP.[154]

Compared with chemically synthesized MNP, the biomimetic nanoparticle possessed protein or peptide molecules with amine and carboyl functionalities on the surface. The protein and peptide located on the surface could not only be modified with various targeting molecules through chemical linking but also could be genetically engineered with targeting molecules. For example, ferritin had three distinctive interfaces that could be utilized for ferritin modification, the inner and outer surfaces as well as the interface between the subunits.[155,156] Chemically linking the outer surface with polyethylene glycol enhanced biomimetic mineralization.[157] The surface interface was genetically engineered with RGD peptides for targeted imaging and therapy.[98,158] It could also be engineered with a PAS polypeptide sequence rich in proline (P), alanine (A) and serine (S), residues that enhance the encapsulation of doxorubicin and prolong the blood half-life for applications in cancer therapy.[159] Recently, a small number of 'silent' amino acid residues were deleted to change the subunit interface and produced a nonnative 48-mer nanocage ferritin.[160] In the future, many other biomineralization proteins should be genetically engineered with various targeting molecules and drugs for targeted imaging and therapy. Last but not the least, several breakthroughs have been recently made on magnetoferritins in tumour staining and MRI, using the unique intrinsic targeting and peroxidase-like property of M-HFns.[16,127,133] Great effort is needed to make the diagnosis and therapy of cancer.

ACKNOWLEDGEMENTS

This work was supported by grants from National Natural Science Foundation of China (41574062, 41330104 and 41621004) and CAS key research program of frontier sciences.

REFERENCES

1. Roth, H. C.; Schwaminger, S. P.; Schindler, M.; Wagner, F. E.; Berensmeier, S., Influencing factors in the CO-precipitation process of superparamagnetic iron oxide nano particles: A model based study. *J. Magn. Magn. Mater.* **2015**, *377*, 81–89.
2. Pušnik, K.; Goršak, T.; Drofenik, M.; Makovec, D., Synthesis of aqueous suspensions of magnetic nanoparticles with the co-precipitation of iron ions in the presence of aspartic acid. *J. Magn. Magn. Mater.* **2016**, *413*: 65–75.
3. Hufschmid, R.; Arami, H.; Ferguson, R. M.; Gonzales, M.; Teeman, E.; Brush, L. N.; Browning, N. D.; Krishnan, K. M., Synthesis of phase-pure and monodisperse iron oxide nanoparticles by thermal decomposition. *Nanoscale* **2015**, *7*(25), 11142–11154.
4. Park, J.; Lee, E.; Hwang, N. M.; Kang, M.; Kim, S. C.; Hwang, Y.; Park, J. G.; Hyeon, T., One-nanometer-scale size-controlled synthesis of monodisperse magnetic Iron oxide nanoparticles. *Angew. Chem.* **2005**, *117*(19), 2932–2937.
5. Park, J.; An, K.; Hwang, Y.; Park, J. G.; Noh, H. J.; Kim, J. Y.; Park J. H.; Hwang N. M.; Hyeon, T., Ultra-large-scale syntheses of monodisperse nanocrystals. *Nat. Mater.* **2004**, *3*(12), 891–895.
6. Unni, M.; Uhl, A. M.; Savliwala, S.; Savitzky, B. H.; Dhavalikar, R.; Garraud, N.; Arnold, D. P.; Kourkoutis, L. F.; Aandrew, J. S,; Rinaldi, C., Thermal decomposition synthesis of iron oxide nanoparticles with diminished magnetic dead layer by controlled addition of oxygen. *ACS Nano*, **2017**, *11*(2), 2284–2303.
7. Unni, M.; Uhl, A. M.; Savliwala, S.; Savitzky, B. H.; Dhavalikar, R.; Garraud, N.; Arnold, D. P.; Kourkoutis, L. F.; Andrew J. S.; Rinaldi, C., Synthesis and characterization of silica-coated iron oxide nanoparticles in microemulsion: The effect of nonionic surfactants. *Langmuir* **2001**, *17*(10), 2900–2906.
8. Vargo, K. B.; Zaki, A. A.; Warden-Rothman, R.; Tsourkas, A.; Hammer, D. A., Superparamagnetic iron oxide nanoparticle micelles stabilized by recombinant oleosin for targeted magnetic resonance imaging. *Small* **2015**, *11*(12), 1409–1413.
9. Wang, X.; Zhuang, J.; Peng, Q.; Li, Y., A general strategy for nanocrystal synthesis. *Nature* **2005**, *437*(7055), 121–124.
10. Gyergyek, S.; Makovec, D.; Jagodič, M.; Drofenik, M.; Schenk, K.; Jordan, O.; Kovač, J.; Dra žič, G.; Hofmann, H., Hydrothermal growth of iron oxide NPs with a uniform size distribution for magnetically induced hyperthermia: Structural, colloidal and magnetic properties. *J. Alloys Compounds* **2017**, *694*, 261–271.
11. Bhavani, P.; Rajababu, C.; Arif, M.; Reddy, I. V. S.; Reddy, N. R., Synthesis and characterization of iron oxide nanoparticles prepared hydrothermally at different reaction temperatures and pH. *Int. J. Mater. Res.* **2016**, *107*(10), 942–947.
12. Cai, H.; An, X.; Cui, J.; Li, J.; Wen, S.; Li, K.; Shen, M.; Zheng, L.; Zhang, G.; Shi, X., Facile hydrothermal synthesis and surface functionalization of polyethyleneimine-coated iron oxide nanoparticles for biomedical applications. *ACS Appl. Mater. Interfaces* **2013**, *5*(5), 1722–1731.
13. Mirabello, G.; Lenders, J. J.; Sommerdijk, N. A., Bioinspired synthesis of magnetite nanoparticles. *Chem. Soc. Rev.* **2016**, *45*(18), 5085–5106.
14. Prozorov, T.; Bazylinski, D. A.; Mallapragada, S. K.; Prozorov, R., Novel magnetic nanomaterials inspired by magnetotactic bacteria: Topical review. *Mater. Sci. Eng. R Rep.* **2013**, *74*(5), 133–172.

15. Peigneux, A.; Valverde-Tercedor, C.; Lopez-Moreno, R.; Pérez-González, T.; Fernandez-Vivas, M. A.; Jiménez-López, C., Learning from magnetotactic bacteria: A review on the synthesis of biomimetic nanoparticles mediated by magnetosome-associated proteins. *J. Struct. Biol.* **2016**, *196*(2), 75–84.
16. Cao, C.; Wang, X.; Cai, Y.; Sun, L.; Tian, L.; Wu, H.; Lei, H.; Liu, W.; Chen, G.; Zhu, R.; Pan, Y., Targeted *in vivo* imaging of microscopic tumors with ferritin-based nanoprobes across biological barriers. *Adv. Mater.* **2014**, *26*(16), 2566–2571.
17. Lowenstam, H. A., Magnetite in denticle capping in recent chitons (Polyplacophora). *Geol. Soc. Am. Bull.* **1962**, *73*(4), 435–438.
18. Bazylinski, D. A.; Frankel, R. B., Magnetosome formation in prokaryotes. *Nat. Rev. Microbiol.* **2004**, *2*(3), 217–230.
19. Pan, Y.; Petersen, N.; Winklhofer, M.; Davila, A. F.; Liu, Q.; Frederichs, T.; Hanzlik, M.; Zhu, R., Rock magnetic properties of uncultured magnetotactic bacteria. *Earth Planet Sci. Lett.* **2005**, *237*(3), 311–325.
20. Pan, Y.; Petersen, N.; Davila, A. F.; Zhang, L.; Winklhofer, M.; Liu, Q.; Hanzlik, M.; Zhu, R. The detection of bacterial magnetite in recent sediments of Lake Chiemsee (southern Germany). *Earth Planet Sci. Lett.* **2005**, *232*(1), 109–123.
21. Li, J.; Pan, Y.; Chen, G.; Liu, Q.; Tian, L.; Lin, W., Magnetite magnetosome and fragmental chain formation of *Magnetospirillum magneticum* AMB-1: Transmission electron microscopy and magnetic observations. *Geophys. J. Int.* **2009**, *177*(1), 33–42.
22. Uebe, R.; Schüler, D., Magnetosome biogenesis in magnetotactic bacteria. *Nat. Rev. Microbiol.* **2016**, *14*(10), 621–637.
23. Lohße, A.; Borg, S.; Raschdorf, O.; Kolinko, I.; Tompa, É.; Pósfai, M.; Faivre, D.; Baumgartner, J.; Schüler, D., Genetic dissection of the mamAB and mms6 operons reveals a gene set essential for magnetosome biogenesis in *Magnetospirillum gryphiswaldense. J. Bacteriol.* **2014**, *196*(14), 2658–2669.
24. Schübbe, S.; Kube, M.; Scheffel, A.; Wawer, C.; Heyen, U.; Meyerdierks, A.; Madkour, M. H.; Mayer, F.; Reinhardt, R.; Schüler, D., Characterization of a spontaneous nonmagnetic mutant of *Magnetospirillum gryphiswaldense* reveals a large deletion comprising a putative magnetosome island. *J. Bacteriol.* **2003**, *185*(19), 5779–5790.
25. Ullrich, S.; Kube, M.; Schübbe, S.; Reinhardt, R.; Schüler, D., A hypervariable 130-kilobase genomic region of *Magnetospirillum gryphiswaldense* comprises a magnetosome island which undergoes frequent rearrangements during stationary growth. *J. Bacteriol.* **2005**, *187*(21), 7176–7184.
26. Grünberg, K.; Wawer, C.; Tebo, B. M.; Schüler, D., A large gene cluster encoding several magnetosome proteins is conserved in different species of magnetotactic bacteria. *Appl. Environ. Microbiol.* **2001**, *67*(10), 4573–4582.
27. Murat, D.; Quinlan, A.; Vali, H.; Komeili, A., Comprehensive genetic dissection of the magnetosome gene island reveals the step-wise assembly of a prokaryotic organelle. *Proc. Natl. Acad. Sci. U S A* **2010**, *107*(12), 5593–5598.
28. Lohße, A.; Ullrich, S.; Katzmann, E.; Borg, S.; Wanner, G.; Richter, M.; Voigt, B.; Schweder, T., Schüler, D., Functional analysis of the magnetosome island in *Magnetospirillum gryphiswaldense*: The mamAB operon is sufficient for magnetite biomineralization. *PLoS One* **2011**, *6*(10), e25561.
29. Scheffel, A.; Gärdes, A.; Grünberg, K.; Wanner, G.; Schüler, D., The major magnetosome proteins MamGFDC are not essential for magnetite biomineralization in *Magnetospirillum gryphiswaldense* but regulate the size of magnetosome crystals. *J. Bacteriol.* **2008**, *190*(1), 377–386.
30. Rong, C.; Zhang, C.; Zhang, Y.; Qi, L.; Yang, J.; Guan, G.; Li, L.; Li, J., FeoB2 functions in magnetosome formation and oxidative stress protection in *Magnetospirillum gryphiswaldense* strain MSR-1. *J. Bacteriol.* **2012**, *194*(15), 3972–3976.
31. Raschdorf, O.; Müller, F. D.; Pósfai, M.; Plitzko, J. M.; Schüler, D., The magnetosome proteins MamX, MamZ and MamH are involved in redox control of magnetite biomineralization in *Magnetospirillum gryphiswaldense. Mol. Microbiol.* **2013**, *89*(5), 872–886.
32. Vali, H.; Förster, O.; Amarantidis, G.; Petersen, N., Magnetotactic bacteria and their magnetofossils in sediments. *Earth Planet Sci. Lett.* **1987**, *86*(2–4), 389–400.
33. Petersen, N.; Weiss, D. G.; Vali, H., Magnetic bacteria in lake sediments. In *Geomagnetism and Palaeomagnetism*. Springer, the Netherlands. Editors: Lowes, F. J.; Collinson, D. W.; Parry, J. H.; Runoorm, S. K.; Tozer, D. C.; Soward, A., **1989**, 231–241.
34. Lin, W.; Li, J.; Schüler, D.; Jogler, C.; Pan, Y., Diversity analysis of magnetotactic bacteria in Lake Miyun, northern China, by restriction fragment length polymorphism. *Syst. Appl. Microbiol.* **2009**, *32*(5), 342–350.
35. Li, J.; Pan, Y.; Liu, Q.; Yu-Zhang, K.; Menguy, N.; Che, R.; Qin, H.; Lin, W.; Wu, W.; Petersen N.; Yang, X. A., Biomineralization, crystallography and magnetic properties of bullet-shaped magnetite magnetosomes in giant rod magnetotactic bacteria. *Earth Planet Sci. Lett.* **2010**, *293*(3), 368–376.
36. Lin, W.; Jogler, C.; Schüler, D.; Pan, Y., Metagenomic analysis reveals unexpected subgenomic diversity of magnetotactic bacteria within the phylum *Nitrospirae. Appl. Environ. Microbiol.* **2011**, *77*(1), 323–326.
37. Lin, W.; Deng, A.; Wang, Z.; Li, Y.; Wen, T.; Wu, L. F.; Wu, M.; Pan, Y., Genomic insights into the uncultured genus '*Candidatus magnetobacterium*' in the phylum *Nitrospirae. ISME J.* **2014**, *8*(12), 2463–2477.
38. Li, J.; Menguy, N.; Gatel, C.; Boureau, V.; Snoeck, E.; Patriarche, G.; Leroy, E.; Pan, Y., Crystal growth of bullet-shaped magnetite in magnetotactic bacteria of the *Nitrospirae* phylum. *J. R. Soc. Interface* **2015**, *12*(103), 20141288.
39. Schulten, K.; Swenberg, C. E.; Weller, A., A biomagnetic sensory mechanism based on magnetic field modulated coherent electron spin motion. *Zeitschrift für Physikalische Chemie.* **1978**, *111*(1), 1–5.
40. Walcott, C.; Gould, J. L.; Kirschvink, J. L., Pigeons have magnets. *Science* **1979**, *205*(4410), 1027–1029.
41. Winklhofer, M., An avian magnetometer. *Science* **2012**, *336*(6084), 991–992.
42. Quinn, T. P., Evidence for celestial and magnetic compass orientation in lake migrating sockeye salmon fry. *J. Comp. Physiol.* **1980**, *137*(3), 243–248.
43. Quinn, T. P.; Brannon, E. L., The use of celestial and magnetic cues by orienting sockeye salmon smolts. *J. Comp. Physiol. A Neuroethol. Sens. Neural Behav. Physiol.* **1982**, *147*(4), 547–552.
44. Mann, S.; Sparks, N. H.; Walker, M. M.; Kirschvink, J. L., Ultrastructure, morphology and organization of biogenic magnetite from sockeye salmon, *Oncorhynchus nerka*: Implications for magnetoreception. *J. Exp. Biol.* **1988**, *140*(1), 35–49.
45. Diebel, C. E.; Proksch, R.; Green, C. R.; Neilson, P.; Walker, M. M., Magnetite defines a vertebrate magnetoreceptor. *Nature* **2000**, *406*(6793), 299–302.
46. Eder, S. H.; Cadiou, H.; Muhamad, A.; McNaughton, P. A.; Kirschvink, J. L.; Winklhofer, M., Magnetic characterization of isolated candidate vertebrate magnetoreceptor cells. *Proc. Natl. Acad. Sci. U S A* **2012**, *109*(30), 12022–12027.

47. Davila, A. F.; Fleissner, G.; Winklhofer, M.; Petersen, N., A new model for a magnetoreceptor in homing pigeons based on interacting clusters of superparamagnetic magnetite. *Phys. Chem. Earth, Parts A/B/C* **2003**, *28*(16), 647–652.
48. Tian, L.; Xiao, B.; Lin, W.; Zhang, S.; Zhu, R.; Pan, Y., Testing for the presence of magnetite in the upper-beak skin of homing pigeons. *BioMetals* **2007**, *20*(2), 197–203.
49. Treiber, C. D.; Salzer, M. C.; Riegler, J.; Edelman, N.; Sugar, C.; Breuss, M.; Pichler, P.; Cadiou H.; Saunders, M.; Lythgoe, M.; Shaw, J.; Keays, D. A., Clusters of iron-rich cells in the upper beak of pigeons are macrophages not magnetosensitive neurons. *Nature* **2012**, *484*(7394), 367–370.
50. Treiber, C. D.; Salzer, M.; Breuss, M.; Ushakova, L.; Lauwers, M.; Edelman, N.; Keays, D. A., High resolution anatomical mapping confirms the absence of a magnetic sense system in the rostral upper beak of pigeons. *Commun. Integr. Biol.* **2013**, *6*(4), e24859.
51. Holland, R. A.; Thorup, K.; Vonhof, M. J.; Cochran, W. W.; Wikelski, M., Navigation: Bat orientation using Earth's magnetic field. *Nature* **2006**, *444*(7120), 702–702.
52. Wang, Y.; Pan, Y.; Parsons, S.; Walker, M.; Zhang, S., Bats respond to polarity of a magnetic field. *Proc. R. Soc. Lond. B Biol. Sci.* **2007**, *274*(1627), 2901–2905.
53. Tian, L.; Pan, Y.; Metzner, W.; Zhang, J.; Zhang, B., Bats respond to very weak magnetic fields. *PloS one* **2015**, *10*(4), e0123205.
54. Tian, L.; Lin, W.; Zhang, S.; Pan, Y., Bat head contains soft magnetic particles: Evidence from magnetism. *Bioelectromagnetics* **2010**, *31*(7), 499–503.
55. Walker, M. M.; Dennis, T. E.; Kirschvink, J. L., The magnetic sense and its use in long-distance navigation by animals. *Curr. Opin. Neurobiol.* **2002**, *12*(6), 735–744.
56. Cadiou, H.; McNaughton, P. A., Avian magnetite-based magnetoreception: A physiologist's perspective. *J. R. Soc. Interface* **2010**, *7*(2), S193–S205.
57. Arosio, P.; Ingrassia, R.; Cavadini, P., Ferritins: A family of molecules for iron storage, antioxidation and more. *Biochim. Biophys. Acta (BBA)—General Subjects.* **2009**, *1790*(7), 589–599.
58. Honarmand Ebrahimi, K.; Hagedoorn, P. L.; Hagen, W. R., Unity in the biochemistry of the iron-storage proteins ferritin and bacterioferritin. *Chem. Rev.* **2015**, *115*(1), 295–326.
59. Levi, S.; Yewdall, S. J.; Harrison, P. M.; Santambrogio, P.; Cozzi, A.; Rovida, E.; Albertini, A.; Arosio, P., Evidence of H-and L-chains have co-operative roles in the iron-uptake mechanism of human ferritin. *Biochem. J.* **1992**, *288*(2), 591–596.
60. Hempstead, P. D.; Yewdall, S. J.; Fernie, A. R.; Lawson, D. M.; Artymiuk, P. J.; Rice, D. W.; Ford, G. C.; Harrison, P. M., Comparison of the three-dimensional structures of recombinant human H and horse L ferritins at high resolution. *J. Mol. Biol.* **1997**, *268*(2), 424–448.
61. Bou-Abdallah, F.; Biasiotto, G.; Arosio, P.; Chasteen, N. D., The putative "nucleation site" in human H-chain ferritin is not required for mineralization of the iron core. *Biochemistry* **2004**, *43*(14), 4332–4337.
62. Sonia, L.; Santambrogio, P.; Corsi, B.; Cozzi, A.; Arosio, P., Evidence that residues exposed on the three-fold channels have active roles in the mechanism of ferritin iron incorporation. *Biochem. J.* **1996**, *317*(2), 467–473.
63. Harrison, P. M.; Fischbach, F. A.; Hoy, T. G., Ferric oxyhydroxide core of ferritin. *Nature* **1967**, *216*:1188–1190.
64. Liu, G.; Debnath, S.; Paul, K. W.; Han, W.; Hausner, D. B.; Hosein, H. A.; Michel, F. M.; Parise, J. B.; Sparks, D. L.; Strongin, D. R., Characterization and surface reactivity of ferrihydrite nanoparticles assembled in ferritin. *Langmuir* **2006**, *22*(22), 9313–9321.
65. Chasteen, N. D.; Harrison, P. M., Mineralization in ferritin: An efficient means of iron storage. *J. Struct. Biol.* **1999**, *126*(3), 182–194.
66. Quintana, C.; Lancin, M.; Marhic, C.; Pérez, M.; Martin-Benito, J.; Avila, J.; Carrascosa, J. L., Initial studies with high resolution TEM and electron energy loss spectroscopy studies of ferritin cores extracted from brains of patients with progressive supranuclear palsy and Alzheimer disease. *Cell Mol. Biol. (Noisy-le-grand)* **2000**, *46*(4), 807–820.
67. Dobson, J., Nanoscale biogenic iron oxides and neurodegenerative disease. *FEBS Lett.* **2001**, *496*(1), 1–5.
68. Quintana, C.; Cowley, J. M.; Marhic, C., Electron nanodiffraction and high-resolution electron microscopy studies of the structure and composition of physiological and pathological ferritin. *J. Struct. Biol.* **2004**, *147*(2), 166–178.
69. Gálvez, N.; Fernández, B.; Sánchez, P.; Cuesta, R.; Ceolín, M.; Clemente-León, M.; Trasobares S.; López-Haro, M.; Calvino, J. J.; Stéphan, O.; Domínguez-Vera, J. M., Comparative structural and chemical studies of ferritin cores with gradual removal of their iron contents. *J. Am. Chem. Soc.* **2008**, *130*(25), 8062–8068.
70. Tian, L.; Cao, C.; Liu, Q.; Pan, Y., Low-temperature magnetic properties of horse spleen ferritin. *Chin. Sci. Bull.* **2010**, *55*(27), 3174–3180.
71. Pan, Y. H.; Sader, K.; Powell, J. J.; Bleloch, A.; Gass, M.; Trinick, J.; Warley, A.; Li, A.; Brydson, A.; Brown, A., 3D morphology of the human hepatic ferritin mineral core: New evidence for a subunit structure revealed by single particle analysis of HAADF-STEM images. *J. Struct. Biol.* **2009**, *166*(1), 22–31.
72. Trefry, A.; Harrison, P. M., Incorporation and release of inorganic phosphate in horse spleen ferritin. *Biochem. J.* **1978**, *171*(2), 313–320.
73. Polanams, J.; Ray, A. D.; Watt, R. K., Nanophase iron phosphate, iron arsenate, iron vanadate, and iron molybdate minerals synthesized within the protein cage of ferritin. *Inorg. Chem.* **2005**, *44*(9), 3203–3209.
74. Honarmand Ebrahimi, K.; Hagedoorn, P. L.; Hagen, W. R., Phosphate accelerates displacement of Fe (III) by Fe (II) in the ferroxidase center of *Pyrococcus furiosus* ferritin. *FEBS Lett.* **2013**, *587*(2), 220–225.
75. Zhao, G.; Bou-Abdallah, F.; Arosio, P.; Levi, S.; Janus-Chandler, C.; Chasteen, N. D., Multiple pathways for mineral core formation in mammalian apoferritin. The role of hydrogen peroxide. *Biochemistry* **2003**, *42*(10), 3142–3150.
76. Bradley, J. M.; Moore, G. R.; Le Brun N. E., Mechanisms of iron mineralization in ferritins: One size does not fit all. *J. Biol. Inorg. Chem.* **2014**, *19*(6), 775–785.
77. Zhao, G.; Su, M.; Chasteen, N. D., μ-1, 2-Peroxo diferric complex formation in horse spleen ferritin. A mixed H/L-subunit heteropolymer. *J. Mol. Biol.* **2005**, *352*(2), 467–477.
78. Klem, M. Willits, T., D.; Solis, D. J., Bio-inspired synthesis of protein-encapsulated CoPt nanoparticles. *Adv. Funct. Mater.* **2005**, *15*(9), 1489–1494.
79. Linder, M. C.; Kakavandi, H. R.; Miller, P.; Wirth, P. L.; Nagel, G. M., Dissociation of ferritins. *Arch. Biochem. Biophys.* **1989**, *269*(2), 485–496.
80. Liu, X.; Jin, W.; Theil, E. C., Opening protein pores with chaotropes enhances Fe reduction and chelation of Fe from the ferritin biomineral. *Proc. Natl. Acad. Sci. U S A* **2003**, *100*(7), 3653–3658.

81. Meldrum, F. C.; Heywood, B. R.; Mann, S., Magnetoferritin: *In vitro* synthesis of a novel magnetic protein. *Science* **1992**, *257*(5069), 522–524.
82. Pankhurst, Q. A.; Betteridge, S.; Dickson, D. P. E.; Douglas, T.; Mann, S.; Frankel, R. B., Mössbauer spectroscopic and magnetic studies of magnetoferritin. *Hyperfine Interact* **1994**, *91*(1), 847–851.
83. Dickson, D. P. E.; Walton, S. A.; Mann, S.; Wong, K., Properties of magnetoferritin: A novel biomagnetic nanoparticle. *NanoStructured Mater.* **1997**, *9*(1), 595–598.
84. Moskowitz, B. M.; Frankel, R. B.; Walton, S. A.; Dickson, D. P.; Wong, K. K. W.; Douglas, T.; Mann, S., Determination of the preexponential frequency factor for superparamagnetic maghemite particles in magnetoferritin. *J. Geophys. Res. Solid Earth* **1997**, *102*(B10), 22671–22680.
85. Bulte, J. W.; Douglas, T.; Mann, S.; Frankel, R. B.; Moskowitz, B. M.; Brooks, R. A.; Baumgarner, B. S.; Vymazal, J.; Strub, M. P.; Frank, J. A., Magnetoferritin: Characterization of a novel superparamagnetic MR contrast agent. *J. Magn. Reson. Imaging* **1994**, *4*(3), 497–505.
86. Zborowski, M.; Fuh, C. B.; Green, R.; Baldwin, N. J.; Reddy, S.; Douglas, T.; Mann, S.; Chalmers, J. J., Immunomagnetic isolation of magnetoferritin-labeled cells in a modified ferrograph. *Cytometry* **1996**, *24*(3), 251–259.
87. Bulte, J. W.; Douglas, T.; Mann, S.; Vymazal, J.; Laughlin, P. G.; Frank, J. A., Initial assessment of magnetoferritin biokinetics and proton relaxation enhancement in rats. *Acad. Radiol.* **1995**, *2*(10), 871–878.
88. Wong, K. K.; Douglas, T.; Gider, S.; Awschalom, D. D.; Mann, S., Biomimetic synthesis and characterization of magnetic proteins (magnetoferritin). *Chem. Mater.* **1998**, *10*(1), 279–285.
89. Melnikova, L.; Pospiskova, K.; Mitroova, Z.; Kopcansky, P.; Safarik, I., Peroxidase-like activity of magnetoferritin. *Microchimica Acta* **2014**, *181*(3–4), 295–301.
90. Martinez-Perez, M. J.; de Miguel, R.; Carbonera, C.; Martinez-Julvez, M.; Lostao, A.; Piquer, C.; Gómez-Moreno, C.; Bartolomé, J.; Luis, F., Size-dependent properties of magnetoferritin. *Nanotechnology* **2010**, *21*(46), 465707.
91. Clavijo Jordan, V., Caplan, M. R., Bennett, K. M., Simplified synthesis and relaxometry of magnetoferritin for magnetic resonance imaging. *Magn. Reson. Med.* **2010**, *64*(5), 1260–1266.
92. Uchida, M.; Flenniken, M. L.; Allen, M.; Willits, D. A.; Crowley, B. E.; Brumfield, S.; Willis, A. F.; Jackiw, L.; Jutila, M.; Young, M. J.; Douglas, T., Targeting of cancer cells with ferrimagnetic ferritin cage nanoparticles. *J. Am. Chem. Soc.* **2006**, *128*(51), 16626–16633.
93. Cao, C.; Tian, L.; Liu, Q.; Liu, W.; Chen, G.; Pan, Y., Magnetic characterization of noninteracting, randomly oriented, nanometer-scale ferrimagnetic particles. *J. Geophys. Res. Solid Earth* **2010**, *115*(B7), B07403.
94. Walls, M. G.; Cao, C.; Yu-Zhang, K.; Li, J.; Che, R.; Pan, Y., Identification of ferrous-ferric Fe_3O_4 nanoparticles in recombinant human ferritin cages. *Microsc. Microanal.* **2013**, *19*(04), 835–841.
95. Cai, Y.; Cao, C.; He, X.; Yang, C.; Tian, L.; Zhu, R.; Pan, Y., Enhanced magnetic resonance imaging and staining of cancer cells using ferrimagnetic H-ferritin nanoparticles with increasing core size. *Int. J. Nanomedicine* **2015**, *10*, 2619.
96. Dalmau, M.; Lim, S.; Wang, S. W., pH-triggered disassembly in a caged protein complex. *Biomacromolecules* **2009**, *10*(12), 3199–3206.
97. Ma-Ham, A.; Wu, H.; Wang, J.; Kang, X.; Zhang, Y.; Lin, Y., Apoferritin-based nanomedicine platform for drug delivery: Equilibrium binding study of daunomycin with DNA. *J. Mater. Chem.* **2011**, *21*(24), 8700–8708.
98. Lin, X.; Xie, J.; Niu, G.; Zhang, F.; Gao, H.; Yang, M.; Quan, Q.; Aronova, M. A.; Zhang, G.; Lee, S.; Leapman, R.; Chen, X., Chimeric ferritin nanocages for multiple function loading and multimodal imaging. *Nano Letters* **2011**, *11*(2), 814–819.
99. Valero, E.; Tambalo, S.; Marzola, P.; Ortega-Munoz, M.; López-Jaramillo, F. J.; Santoyo-González, F.; de Dios López, J.; Delgado, J. J.; Calvino, J. J.; Cuesta, R.; Domínguez-Vera, J. M.; Gálvez, N., Magnetic nanoparticles-templated assembly of protein subunits: A new platform for carbohydrate-based MRI nanoprobes. *J. Am. Chem. Soc.* **2011**, *133*(13), 4889–4895.
100. Kim, M.; Rho, Y.; Jin, K. S.; Ahn, B.; Jung, S.; Kim, H.; Ree, M., pH-dependent structures of ferritin and apoferritin in solution: Disassembly and reassembly. *Biomacromolecules* **2011**, *12*(5), 1629–1640.
101. Staniland, S.; Williams, W. Y. N.; Telling, N.; Van Der Laan, G.; Harrison, A.; Ward, B., Controlled cobalt doping of magnetosomes *in vivo*. *Nat. Nanotechnol.* **2008**, *3*(3), 158–162.
102. Alphandéry, E.; Faure, S.; Seksek, O.; Guyot, F.; Chebbi, I., Chains of magnetosomes extracted from AMB-1 magnetotactic bacteria for application in alternative magnetic field cancer therapy. *ACS Nano* **2011**, *5*(8), 6279–6296.
103. Li, J.; Menguy, N.; Arrio, M. A.; Sainctavit, P.; Juhin, A.; Wang, Y.; Chen, H.; Bunau, O.; Otero, E.; Ohresser, P.; Pan, Y., Controlled cobalt doping in the spinel structure of magnetosome magnetite: New evidences from element-and site-specific X-ray magnetic circular dichroism analyses. *J. R. Soc. Interface* **2016**, *13*(121), 20160355.
104. Keim, C. N.; Lins, U.; Farina, M., Manganese in biogenic magnetite crystals from magnetotactic bacteria. *FEMS Microbiol. Lett.* **2009**, *292*(2), 250–253.
105. Prozorov, T.; Perez-Gonzalez, T.; Valverde-Tercedor, C.; Jimenez-Lopez, C.; Yebra-Rodriguez, A.; Körnig, A.; Faivre, D.; Mallapragada S. K.; Howse, A.; Bazylinski, D. A.; Prozorov, R., Manganese incorporation into the magnetosome magnetite: Magnetic signature of doping. *European Journal of Mineralogy* **2014**, *26*(4), 457–471.
106. Staniland, S. S.; Rawlings, A. E., Crystallizing the function of the magnetosome membrane mineralization protein Mms6. *Biochem. Soc. Trans.* **2016**, *44*(3), 883–890.
107. Tanaka, M.; Mazuyama, E.; Arakaki, A.; Matsunaga, T., Mms6 protein regulates crystal morphology during nano-sized magnetite biomineralization *in vivo*. *J. Biol. Chem.* **2011**, *286*(8), 6386–6392.
108. Yamagishi, A.; Narumiya, K.; Tanaka, M.; Matsunaga, T.; Arakaki, A., Core amino acid residues in the morphology-regulating protein, Mms6, for intracellular magnetite biomineralization. *Scientific Reports* **2016**, *6*.
109. Arakaki, A.; Yamagishi, A.; Fukuyo, A.; Tanaka, M.; Matsunaga, T., Co-ordinated functions of Mms proteins define the surface structure of cubo-octahedral magnetite crystals in magnetotactic bacteria. *Mol. Microbial.* **2014**, *93*(3), 554–567.
110. Arakaki, A.; Webb, J.; Matsunaga, T., A novel protein tightly bound to bacterial magnetic particles in Magnetospirillum magneticum strain AMB-1. *J. Biol. Chem.* **2003**, *278*(10), 8745–8750.
111. Prozorov, T.; Mallapragada, S. K.; Narasimhan, B.; Wang, L.; Palo, P.; Nilsen-Hamilton, M.; Williams, T. J.; Bazylinski, D. A.; Prozorov, P.; Canfield, P. C., Protein-mediated synthesis of uniform superparamagnetic magnetite nanocrystals. *Adv. Funct. Mater.* **2007**, *17*(6), 951–957.
112. Amemiya, Y.; Arakaki, A.; Staniland, S. S.; Tanaka, T.; Matsunaga, T., Controlled formation of magnetite crystal by partial oxidation of ferrous hydroxide in the presence of recombinant magnetotactic bacterial protein Mms6. *Biomaterials* **2007**, *28*(35), 5381–5389.

113. Prozorov, T.; Palo, P.; Wang, L.; Nilsen-Hamilton, M.; Jones, D.; Orr, D.; Mallapragada, S. K.; Narasimhan, B.; Canfield, P. C.; Prozorov, R., Cobalt ferrite nanocrystals: Out-performing magnetotactic bacteria. *ACS Nano* **2007**, *1*(3), 228–233.
114. Arakaki, A., Masuda, F.; Matsunaga, T., Iron oxide crystal formation on a substrate modified with the Mms6 protein from magnetotactic bacteria. *MRS Online Proceedings Library Archive*, **2009**, 1187, 1187-KK03-08.
115. Galloway, J. M.; Arakaki, A.; Masuda, F.; Tanaka, T.; Matsunaga, T.; Staniland, S. S., Magnetic bacterial protein Mms6 controls morphology, crystallinity and magnetism of cobalt-doped magnetite nanoparticles *in vitro*. *J. Mater. Chem.* **2011**, *21*(39), 15244–15254.
116. Lenders, J. J.; Altan, C. L.; Bomans, P. H.; Arakaki, A.; Bucak, S.; de With, G.; Sommerdijk, N. A., A bioinspired coprecipitation method for the controlled synthesis of magnetite nanoparticles. *Cryst. Growth Des.* **2014**, *14*(11), 5561–5568.
117. Radoul, M.; Lewin, L.; Cohen, B.; Oren, R.; Popov, S.; Davidov, G.; Vandsburger, M. H.; Harmelin, A.; Bitton, R.; Greneche, J. M.; Neeman, M.; Zarivach, R., Genetic manipulation of iron biomineralization enhances MR relaxivity in a ferritin-M6A chimeric complex. *Sci. Rep.* **2016**, *6*, 26550.
118. Wang, L.; Prozorov, T.; Palo, P. E.; Liu, X.; Vaknin, D.; Prozorov, R.; Mallapragada, S.; Nilsen-Hamilton, M., Self-assembly and biphasic iron-binding characteristics of Mms6, a bacterial protein that promotes the formation of superparamagnetic magnetite nanoparticles of uniform size and shape. *Biomacromolecules* **2011**, *13*(1), 98–105.
119. Wang, W.; Bu, W.; Wang, L.; Palo, P. E.; Mallapragada, S.; Nilsen-Hamilton, M.; Vaknin, D., Interfacial properties and iron binding to bacterial proteins that promote the growth of magnetite nanocrystals: X-ray reflectivity and surface spectroscopy studies. *Langmuir* **2012**, *28*(9), 4274–4282.
120. Feng, S.; Wang, L.; Palo, P.; Liu, X.; Mallapragada, S. K.; Nilsen-Hamilton, M., Integrated self-assembly of the mms6 magnetosome protein to form an iron-responsive structure. *Int. J. Mol. Sci.* **2013**, *14*(7), 14594–14606.
121. Kashyap, S.; Woehl, T. J.; Liu, X.; Mallapragada, S. K.; Prozorov, T., Nucleation of iron oxide nanoparticles mediated by Mms6 protein *in situ*. *ACS Nano* **2014**, *8*(9), 9097–9106.
122. Siponen, M. I.; Legrand, P.; Widdrat, M.; Jones, S. R.; Zhang, W. J.; Chang, M. C.; Faivre, D.; Arnoux, P.; Pignol, D., Structural insight into magnetochrome-mediated magnetite biomineralization. *Nature* **2013**, *502*(7473),681–684.
123. Jones, S. R.; Wilson, T. D.; Brown, M. E.; Rahn-Lee, L.; Yu, Y.; Fredriksen, L. L.; Ozyamak, E.; Komeili, A.; Chang, M. C., Genetic and biochemical investigations of the role of MamP in redox control of iron biomineralization in *Magnetospirillum magneticum*. *Proc. Natl. Acad. Sci. U S A* **2015**, *112*(13), 3904–3909.
124. Murat, D.; Falahati, V.; Bertinetti, L.; Csencsits, R.; Körnig, A.; Downing, K.; Faivre, D.; Komeili, A., The magnetosome membrane protein, MmsF, is a major regulator of magnetite biomineralization in *Magnetospirillum magneticum* AMB-1. *Mol. Microbial.* **2012**, *85*(4), 684–699.
125. Rawlings, A. E.; Bramble, J. P.; Walker, R.; Bain, J.; Galloway, J. M.; Staniland, S. S., Self-assembled MmsF proteinosomes control magnetite nanoparticle formation *in vitro*. *Proc. Natl. Acad. Sci. U S A* **2014**, *111*(45), 16094–16099.
126. Gao, L.; Zhuang, J.; Nie, L.; Zhang, J.; Zhang, Y.; Gu, N.; Wang, T.; Feng, J.; Yang, D.; Perrett, S.; Yan, X., Intrinsic peroxidase-like activity of ferromagnetic nanoparticles. *Nat. Nanotechnol.* **2007**, *2*(9), 577–583.
127. Fan, K.; Cao, C.; Pan, Y.; Lu, D.; Yang, D.; Feng, J.; Song, L.; Liang, M.; Yan, X., Magnetoferritin nanoparticles for targeting and visualizing tumour tissues. *Nat. Nanotechnol.* **2012**, *7*(7), 459–464.
128. Li, L.; Fang, C. J.; Ryan, J. C.; Niemi, E. C.; Lebrón, J. A.; Björkman, P. J.; Arase, H.; Torti, F. M.; Nakamura, M. C.; Seaman, W. E., Binding and uptake of H-ferritin are mediated by human transferrin receptor-1. *Proc. Natl. Acad. Sci. U S A* **2010**, *107*(8), 3505–3510.
129. Högemann-Savellano, D.; Bos, E.; Blondet, C.; Sato, F.; Abe, T.; Josephson, L.; Weissleder, R.; Gaudet, J.; Sgroi, D.; Peters, P. J.; Basilion, J. P., The transferrin receptor: A potential molecular imaging marker for human cancer. *Neoplasia* **2003**, *5*(6), 495–506.
130. Daniels, T. R.; Delgado, T.; Rodriguez, J. A.; Helguera, G.; Penichet, M. L., The transferrin receptor part I: Biology and targeting with cytotoxic antibodies for the treatment of cancer. *Clin. Immunol.* **2006**, *121*(2), 144–158.
131. Klem, M. T.; Resnick, D. A.; Gilmore, K.; Young, M.; Idzerda, Y. U.; Douglas, T., Synthetic control over magnetic moment and exchange bias in all-oxide materials encapsulated within a spherical protein cage. *J. Am. Chem. Soc.* **2007**, *129*(1), 197–201.
132. Fantechi, E.; Innocenti, C.; Zanardelli, M.; Fittipaldi, M.; Falvo, E.; Carbo, M.; Shullani, V.; Mannelli, L. D. C.; Ghelardini, C.; Ferretti, M.; Ponti, A. A.; Sangregorio, C.; Ceci, P., A smart platform for hyperthermia application in cancer treatment: Cobalt-doped ferrite nanoparticles mineralized in human ferritin cages. *ACS Nano* **2014**, *8*(5), 4705–4719.
133. Zhang, T.; Cao, C.; Tang, X.; Cai, Y.; Yang, C.; Pan, Y., Enhanced peroxidase activity and tumour tissue visualization by cobalt-doped magnetoferritin nanoparticles. *Nanotechnology* **2016**, *28*(4), 045704.
134. Bulte, J. W. M.; Kraitchman, D. L., Iron oxide MR contrast agents for molecular and cellular imaging. *NMR Biomed.* **2004**, *17*(7), 484–499.
135. Shapiro, E. M.; Skrtic, S.; Sharer, K.; Hill, J. M.; Dunbar, C. E.; Koretsky, A. P., MRI detection of single particles for cellular imaging. *Proc. Natl. Acad. Sci. U S A* **2004**, *101*(30), 10901–10906.
136. Mintorovitch, J.; Shamsi, K., Eovist Injection and Resovist Injection: Two new liver-specific contrast agents for MRI. *Oncology (Williston Park, NY)* **2000**, *14* (6 Suppl 3), 37–40.
137. Harisinghani, M. G.; Barentsz, J.; Hahn, P. F.; Deserno, W. M.; Tabatabaei, S.; van de Kaa, C. H.; de la Rosette, J.; Weissleder, R., Noninvasive detection of clinically occult lymph-node metastases in prostate cancer. *N. Engl. J. Med.* **2003**, *348*, 2491–2499.
138. McCarthy, J. R.; Kelly, K. A.; Sun, E. Y.; Weissleder, R., Targeted delivery of multifunctional magnetic nanoparticles. *Nanomedicine* **2007**, *2*,153–167.
139. Lee, J. H.; Huh, Y. M.; Jun, Y. W.; Seo, J. W.; Jang, J. T.; Song, H. T.; Kim, S.; Cho, E. J.; Yoon, H. G.; Suh, J. S.; Cheon, J., Artificially engineered magnetic nanoparticles for ultra-sensitive molecular imaging. *Nat. Med.* **2007** *13*(1), 95–99.
140. Byrne, J. D.; Betancourt, T.; Brannon-Peppas, L., Active targeting schemes for nanoparticle systems in cancer therapeutics. *Adv. Drug Deliv. Rev.* **2008**, *60*(15), 1615–1626.
141. Lee, L. S.; Conover, C.; Shi, C.; Whitlow, M.; Filpula, D., Prolonged circulating lives of single-chain Fv proteins conjugated with polyethylene glycol: A comparison of conjugation chemistries and compounds. *Bioconjug. Chem.* **1999**, *10*(6), 973–981.
142. Park, J. H.; von Maltzahn, G.; Zhang, L.; Schwartz, M. P.; Ruoslahti, E.; Bhatia, S. N.; Sailor, M. J., Magnetic iron oxide nanoworms for tumor targeting and imaging. *Adv. Mater.* **2008**, *20*(9), 1630–1635.

143. Lin, M. M.; Kim, H. H.; Kim, H.; Dobson, J.; Kim, D. K., Surface activation and targeting strategies of superparamagnetic iron oxide nanoparticles in cancer-oriented diagnosis and therapy. *Nanomedicine* **2010**, *5*(1), 109–133.
144. Li, X.; Qiu, L.; Zhu, P.; Tao, X.; Imanaka, T.; Zhao, J.; Huang, Y.; Tu, Y.; Cao, X., Epidermal growth factor–ferritin H-chain protein nanoparticles for tumor active targeting. *Small* **2012**, *8*(16), 2505–2514.
145. Ji, T.; Zhao, Y.; Wang, J.; Zheng, X.; Tian, Y.; Zhao, Y.; Nie, G., Tumor fibroblast specific activation of a hybrid ferritin nanocage-based optical probe for tumor microenvironment imaging. *Small* **2013**, *9*(14), 2427–2431.
146. Kwon, K. C.; Ko, H. K.; Lee, J.; Lee, E. J.; Kim, K.; Lee, J., Enhanced *in vivo* tumor detection by active tumor cell targeting using multiple tumor receptor-binding peptides presented on genetically engineered human ferritin nanoparticles. *Small* **2016**, *12*(31), 4241–4253.
147. Kim, S.; Jeon, J. O.; Jun, E.; Jee, J.; Jung, H. K.; Lee, B. H.; Kim, I. S.; Kim, S., Designing peptide bunches on nanocage for bispecific or superaffinity targeting. *Biomacromolecules* **2016**, *17*(3), 1150–1159.
148. Zhao, Y.; Liang, M.; Li, X.; Fan, K.; Xiao, J.; Li, Y.; Shi, H.; Wang, F.; Choi, H. S.; Cheng, D.; Yan, X., Bioengineered magnetoferritin nanoprobes for single-dose nuclear-magnetic resonance tumor imaging. *ACS Nano* **2016**, *10*(4), 4184–4191.
149. Alphandéry, E.; Faure, S.; Raison, L.; Duguet, E.; Howse, P. A.; Bazylinski, D. A., Heat production by bacterial magnetosomes exposed to an oscillating magnetic field. *J. Phys. Chem. C.* **2010**, *115*(1), 18–22.
150. Timko, M.; Molcan, M.; Hashim, A.; Skumiel, A.; Muller, M.; Gojzewski, H.; Jozefczak, A.; Kovac, J.; Rajnak, M.; Makowski, M.; Kopcansky, P., Hyperthermic effect in suspension of magnetosomes prepared by various methods. *IEEE Trans. Magn.* **2013**, *49*(1), 250–254.
151. Liu, R.; Liu, J.; Tong, J.; Tang, T.; Kong, W.; Wang, X.; Li, Y.; Tang, J. T. Heating effect and biocompatibility of bacterial magnetosomes as potential materials used in magnetic fluid hyperthermia. *Prog. Nat. Sci. Mater. Int.* **2012**, *22*(1), 31–39.
152. Roman, F., A novel mechanism for ferritin iron oxidation and deposition. *J. Mol. Catal.* **1978**, *4*(1), 75–82.
153. Hao, R.; Xing, R.; Xu, Z.; Hou, Y.; Gao, S.; Sun, S., Synthesis, functionalization, and biomedical applications of multifunctional magnetic nanoparticles. *Adv. Mater.* **2010**, *22*(25), 2729–2742.
154. Xu, C.; Wang, B.; Sun, S., Dumbbell-like Au–Fe_3O_4 nanoparticles for target-specific platin delivery. *J. Am. Chem. Soc.* **2009**, *131*(12), 4216–4217.
155. Uchida, M.; Kang, S.; Reichhardt, C.; Harlen, K.; Douglas, T., The ferritin superfamily: Supramolecular templates for materials synthesis. *Biochim. Biophys. Acta (BBA)-General Subjects* **2010**, *1800*(8), 834–845.
156. Kang, S.; Uchida, M.; O'Neil, A.; Li, R.; Prevelige, P. E.; Douglas, T., Implementation of p22 viral capsids as nanoplatforms. *Biomacromolecules* **2010**, *11*(10), 2804–2809.
157. Yang, C.; Cao, C.; Cai, Y.; Xu, H.; Zhang, T.; Pan, Y., Effects of PEGylation on biomimetic synthesis of magnetoferritin nanoparticles. *J. Nanopart. Res.* **2017**, *19*(3), 101.
158. Zhen, Z.; Tang, W.; Guo, C.; Chen, H.; Lin, X.; Liu, G.; Fei, B.; Chen, X.; Xu, B.; Xie, J., Ferritin nanocages to encapsulate and deliver photosensitizers for efficient photodynamic therapy against cancer. *ACS Nano* **2013**, *7*(8), 6988–6996.
159. Falvo, E.; Tremante, E.; Arcovito, A.; Papi, M.; Elad, N.; Boffi, A.; Morea, V.; Conti, G.; Toffoli, G.; Fracasso, G.; Giacomini, P.; Ceci, P., Improved doxorubicin encapsulation and pharmacokinetics of ferritin-fusion protein nanocarriers bearing proline, serine, and alanine elements. *Biomacromolecules* **2015**, *17*(2), 514–522.
160. Zhang, S.; Zang, J.; Zhang, X.; Chen, H.; Mikami, B.; Zhao, G., "Silent" amino acid residues at key subunit interfaces regulate the geometry of protein nanocages. *ACS Nano* **2016**, *10*(11), 10382–10388.

Changqian Cao (E-mail: changqiancao@mail.iggcas.ac.cn) is an associate professor at the Institute of Geology and Geophysics, Chinese Academy of Sciences. He earned his BSc in Veterinary Medicine from Shenyang Agricultural University and his PhD in Geobiology from the Institute of Geology and Geophysics, Chinese Academy of Sciences. In 2011, he received the honour of Excellent Doctor of Chinese Academy of Sciences. His research focuses on genetic engineering of biomineralization for the biomimetic synthesis of magnetic nanoparticles with highly desirable properties as well as their use in biomedicine and geoscience. Now, he is conducting a multidisciplinary work to explore biological magnetic nanoparticles for imaging of tumours at the earliest stage.

Yongxin Pan (E-mail: yxpan@mail.iggcas.ac.cn) is a full professor at Institute of Geology and Geophysics, Chinese Academy of Sciences, where he is also the director of France-China Bio-Mineralization and Nano-Structures Laboratory. He earned his BSc in Geology from China University of Geosciences and his PhD in geophysics from the Institute of Geophysics, Chinese Academy of Sciences. His research covers many different fields, including biogeomagnetism, mineral and rock magnetism, paleointensity of the Earth's magnetic field, biomineralization of iron minerals, paleomagnetism and their applications to geosciences. He established the innovative lab for biogeomagnetism and biomineralization, a collaborative multidisciplinary research group with biologists and geophysicists to investigate how the geomagnetic field affects living organisms in the present and past.

Section II

Biofunctionalisation of MNPs

5 Main Challenges about Surface Biofunctionalization for the *In Vivo* Targeting of Magnetic Nanoparticles

*Laurent Adumeau, Marie-Hélène Delville and Stéphane Mornet**

CONTENTS

5.1 INTRODUCTION

Magnetic nanoparticles (MNPs) have found applications in various fields of nanomedicine and bioimaging.[1–48] Their magnetic properties make them useful as efficient contrast agents for nuclear magnetic resonance imaging (MRI),[16,35,49] in hyperthermia therapy[16] and also in magnetically guided nanoparticles (NPs).[50] Moreover, MNPs could be functionalized with other components, such as ligands, metal oxides, enzymes, antibodies (Abs) etc. As a result, these biofunctionalized MNPs have shown many applications in various areas, such as molecular MRI, drug and gene delivery, hyperthermia, proteomics and peptidomics analysis in recent years. With the advantages of a large surface area and a possible manipulation by an external magnetic force, MNPs have also been considered as extractive substrates for efficient enzymatic reaction. Among the different varieties of iron oxide NPs, the nontoxic (at low clinical dose for humans between 0.56 to 3 mg Fe/kg of patient body weight[15]), biodegradable and biocompatible γ-Fe_2O_3 and Fe_3O_4 have revealed to be the most commonly used for applications in bio-related fields.[13,14,38,42,51,52]

The design of biofunctionalized MNPs for *in vivo* targeting applications is complex because it must take into account all the specifications required by the biological character of the application. For the MNP design of a given architecture, the chemist must integrate in a conjugation strategy of biomolecules (called thereafter ligand) the choice of the ligand–receptor system and the specifications of the administration route that will contribute to a better targeting efficiency.

* Corresponding author.

This chapter is dedicated to chemists who want to embark on this adventure. The first part consists of (i) the main available bioconjugation routes depending on a selected MNP synthetic method and (ii) the surface modification steps required before conjugation to make the MNP biocompatible. The second part is dedicated to the description of the interactions between the NP surface and its biological environment and a common way to manage them. Finally, in the last part, a description of the current challenges is introduced in terms of benefits and drawbacks identified in the conjugation of MNPs designed for *in vivo* active targeting.

5.2 BASIC PRINCIPLES OF SURFACE BIOCONJUGATION OF MNPs

5.2.1 Different Surface Types of MNPs as a Function of the Synthetic Methods

The design and preparation of functionalized MNPs for their applications in bio-related fields such as drug delivery for example require a strong control of the physicochemical properties of materials in terms of payload drugs and drug carriers, of their behaviours in biological and physiological environments, as well as the targeted functions that will address the relevant medical problems. In order to have efficient drug delivery, the drug carriers based on these MNPs should be able to (i) exhibit as high as possible loading of drug molecules, (ii) protect the drug bioactivity and enhance its biocompatibility and (iii) specifically target the anticipated delivery, keeping the uptake by the normal organs and/or tissue as low as possible. The preparation methods of the magnetic cores with uniform size, controlled shape and desirable compositions[5,42,43,52–54] are also covered in Chapters 1 and 3 of this book. In brief, the most classical ones are presented: coprecipitation,[30,55] thermal decomposition,[5,6,11,14,15,24,30,31,38,41–43,46,49,54,56,57] hydrothermal synthesis,[43,54,58] microemulsion,[57,59] and polyol synthesis,[7,11,14,15,20–21,31,41,42,49,58,60,61] sonochemistry,[5,14,42,43,47,59] microwave-assisted synthesis[42,47,62–63] and reduction–precipitation.[22,32,64]

The available surface of the MNPs will then strongly depend on the given synthetic approach, even if it is generally assumed of hydroxyl groups covering the coordinating sphere of the NPs. These MNPs are therefore amphoteric and may develop surface charges in interaction with their environment depending on the pH of the solution, which is higher or lower than the MNPs point of zero charge. Some of these synthetic methods[52] require the use of surfactants and ligands during the reaction, species which then remain on the particle surface. Table 5.1 summarizes the surface states of MNPs with regards to their synthetic pathways.

TABLE 5.1
Comparison of MNP Surface Features According to Their Synthetic Methods

	Surface Features		
Synthesis Methods (References)	**Dispersion Medium**	**Surface State of Resulting Particles**	**Surface-Capping Agents**
Coprecipitation (30,54,59)	Water	Uncoated NPs, M-OH groups, no molecules, potentially positively or negatively charged	None, added during or after reaction
Hydrothermal/ Solvothermal (22,43,53,54,56,57)	Alcohols/water	Coating needed, added during reaction	None or coated with polyethylene glycol (54)
Sonochemistry (5,14,42,43,57,59)	Water	Uncoated NPs, M-OH groups, no molecules, potentially positively or negatively charged	Needed, added during or after reaction, possible direct grafting of biomolecules on the surface
Polyol (7,11,14,15,20–22,31,41,42,49,52)	Alcohols (DEG, EG, …)	Coating needed, added during or after reaction	Chemisorption of (poly)alcohols
Thermal decomposition (5,6,11,14,15,24,30,31,38,41–43,46,49,53–56,62)	Organic solvent of high boiling T (i.e. octadecene)	Coating needed, added during reaction	Coated with various surfactants carbonyls, fatty acids such as oleic acid, hexadecanediol, oleylamine, hexadecylamine
Microemulsion (31,43,54,56–59)	Nonpolar organic solvent (alkanes)	Coating needed, added during reaction	Coated with various ionic and/or non-ionic surfactants i.e. cetyltrimethlyammonium bromide, sodium dodecylbenzenesulfonate, or AOT, Brij®, NP5,…
Microwave-assisted synthesis (42,47–62)	Organic solvent (i.e. benzyl alcohol)	Coating needed, added during reaction	Coated with fatty acids (oleic acid)
Reduction-Precipitation (22,32,63)	Water	Uncoated NPs, OH groups , no molecules, potentially positively or negatively charged	None, added during or after reaction

5.2.2 Subsequent Relevant Prefunctionalization Steps

Controlling the interactions of nanomaterials with the biological environment is a fundamental challenge. In particular, the NP surface must be covered by a suitable biocompatible coating that must exhibit different functions such as (i) to protect the iron oxide NPs and prevent their degradation when exposed to different environments such as blood or lysosomes, (ii) to prevent the aggregation of the NPs by controlling the attractive van der Waals (VdW) forces and last but not least (iii) to provide the right functional group(s) in the right amount.[2–8,14,15,20,21,24–25,31,37,43,45–47,59,65–68] The resulting inorganic/organic objects then exhibit core–shell architectures, with a shell that may be formed of organic molecules, polymers or inorganic shells and be covalently or adsorbed to the NPs and that may be functionalized by adding various functional groups via classical (bio)-organic routes (see Table 5.2).[69,70]

Most of the time, a prefunctionalization step of the surface is compulsory to provide the attachment of the targeted biomolecules. They are summarized in Figure 5.1, which gathers the different possibilities of functionalization depending on whether the MNP surface is bare and hydrophilic or precoated with the ligand used for their synthesis.

Therefore, the design and preparation of biofunctionalized MNPs require multiple approaches integrating

TABLE 5.2
List of the Different Conjugation Processes of MNPs from Implying Functional Groups Either on the Ligand (Schematically Represented by a Piece of Puzzle) or on the MNPs

Reaction Type	Functionalized NP	Reactant	Final NP
Amide bond formation	NP–NH_2	N-O (NHS ester)	NP–NH (amide)
Amide bond formation	NP–NH_2	HOOC–	NP–NH (amide)
Epoxide opening	NP–NH_2	(epoxide)	NP–NH–, HO
Addition of amine to cyanates	NP–NH_2	XCN–, X = O, S	NP–NH, X, –NH–
Michael addition	NP–SH	(maleimide) N–	NP–S, N–
Amide bond formation	NP–CO_2H	H_2N–	NP, O, NH–
Imine bond formation	NP–CHO	H_2NHN	NP, H, N-NH
Imine bond formation	NP–CHO	H_2NO–	NP, H, N–O
Azide-alkyne cycloaddition	NP–N_3	(alkyne)	NP–N, N=N (triazole)
Ring closing or opening metathesis	NP (alkene)	(alkenes)	NP
Diels-Alder reaction	NP–N (maleimide)	(furan)	NP–N (adduct)

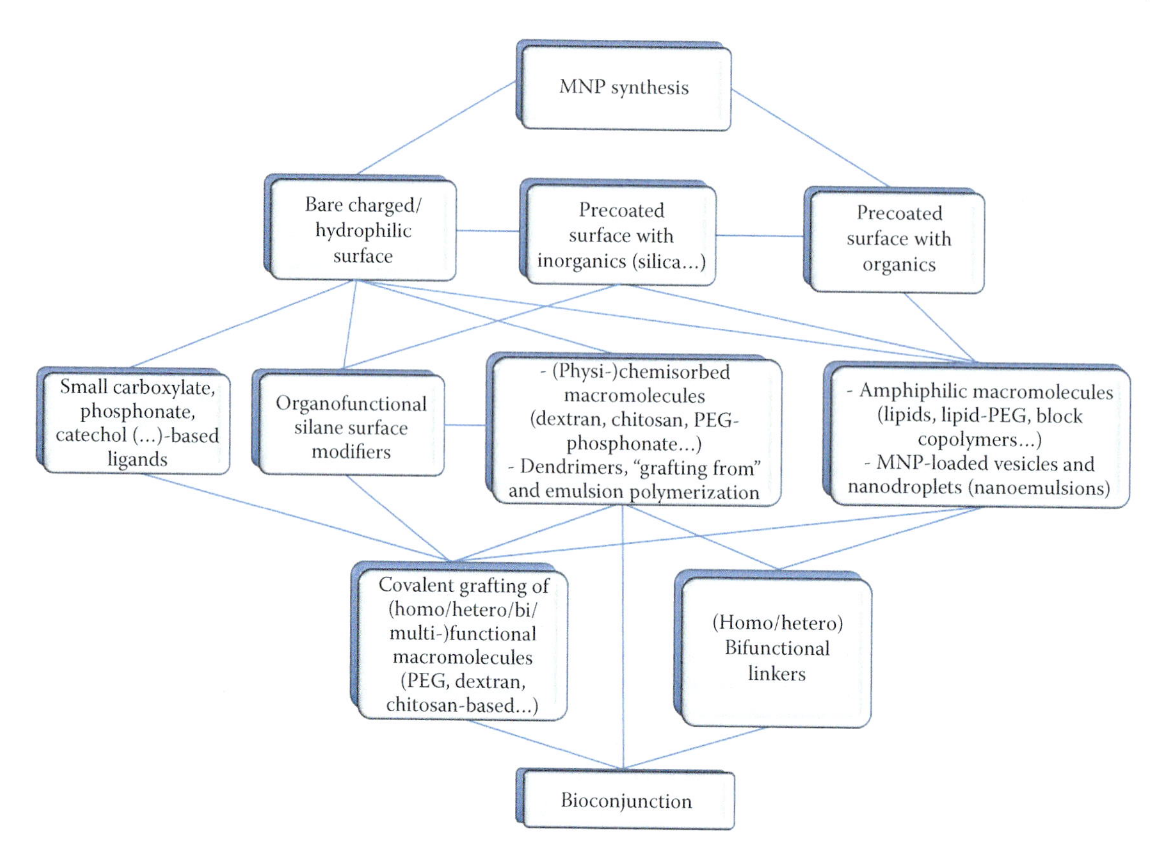

FIGURE 5.1 Chart of different main routes for the development of bioconjugated MNP-based formulation in the context of *in vivo* applications. For better readability, only the boxes showing the main routes used are connected together. Other connections can be added by applying additional treatment for compatibility such as ligand exchange, phase transfer or chemical surface leaching procedures.

multiple criteria according to the given targeted application *in vivo* (drug delivery, MRI, hyperthermia and theranostic approaches). They depend on (i) the physicochemical properties of the nanomaterial itself (degradation, metabolism, potential toxicity and immunogenicity, drug payload and release capacity), (ii) its colloidal stability in physiological media, (iii) its behaviour and responses in the biological or physiological environment (circulation time in blood, ability to cross through physiological barriers etc.) and (iv) the intended functions that address relevant medical problems.

5.2.2.1 Different Types of Anchoring Groups

The preparation of the surface for a further biotargeting depends, as mentioned previously, on its state surface. The bond generated between the surface of iron atoms and the anchoring groups is not so obvious; if it is between a pure ionic nature and a pure covalent one, then it is called iono-covalent bond, with the degree of covalence being strongly dependent on the chemical nature or the anchoring atom (O, N, S, P).[11,15,42,53,71,72]

5.2.2.1.1 Organofunctional Silane Coupling Agents

A large variety of surface functional groups such as amine, carboxyl, hydroxyl, pyridine, amide, aldehyde, epoxy, thiol and others has been described so far (Table 5.2).[14,73] One of the easiest ways to promote the functionalization of MNPs with such organic molecules is to proceed to the surface silanization and the subsequent strengthening of the polysiloxane network. This reaction provides a new surface, which presents satisfying responsivity, low cytotoxicity, high stability under acidic conditions and inertness to redox reactions; it then becomes easy to perform surface chemical modification. Functional alkoxysilane agents include mercaptopropyltriethoxysilane, 3-aminopropyltriethyloxysilane and *p*-aminophenyltrimethoxysilane as the most frequently used to control the surface of the MNP and provide colloidal stability as well as potential interactions with biological molecules such as Abs nucleic acids, enzymes, or proteins.[5–6,74–79]

5.2.2.1.2 Carboxylates and Catechol Derivatives

Other frequently used small stabilizers are based on monomers bearing multiple carboxylate functionalities for the colloidal stabilization of MNPs and at least one carboxylic acid group exposed to the medium. These free carboxylic groups provide negative charges on the surface of particles, making them hydrophilic.[79]

When the MNPs are synthesized in organic solvents, their biocompatibility can be obtained through a ligand exchange of the hydrophobic surfactants with hydrophilic ligands, which consist of two functional groups: one group strongly binds to the NP surface and the other is hydrophilic so that the NPs can be dispersed in an aqueous solution or be further functionalized. Modification of the iron oxide NP surface with a dopamine anchoring group is a nice example of such a ligand exchange with a strong binding of dopamine with the surface Fe(III) via the diol bidentate bonding.[80,81] Other examples of biocompatible ligands[11] with anchoring groups can also be found in the literature, such as water-soluble zwitterionic dopamine sulfonate,[81] 3,4-dihydroxyhydrocinnamic acid[82] and 2,3-dimercaptosuccinic acid.[51,82,83] The ligand shell can further be coupled with other targeted functional molecules via thiol-based linker chemistry.

5.2.2.1.3 Phosphorous Derivatives

However, phosphate- and phosphonate-based ligands reveal to be more efficient due to a much stronger Fe-O-P bond on the oxide surface. These ligands are promising provided they also exhibit other functional groups, as in the case of poly(vinylalcohol phosphate) macromolecules to make the MNPs hydrophilic and biocompatible.[84]

5.2.2.2 Polymers

Coating MNPs with polymer provides an alternative option for the previously mentioned small anchoring agents, leading to particles with the properties of the macromolecular systems grafted to the particle surface. These polymer-functionalized MNPs have been receiving much attention, since polymer coatings not only provide excellent colloidal stability but also increase repulsive forces and balance the magnetic and the VdW attractive forces acting on the MNPs. The main advantages of polymer macromolecules are that they (i) have multiple functional groups on the polymer backbone, which introduce multiple anchoring points to the NP surface, (ii) improve the colloidal stability and (iii) give rise to further reactions with the remaining groups. Both natural[5,6,13,14,20,35,36,39,40,44,47,57,61,64,67,79] (dextran, chitosan, starch, cellulose, gelatin and alginate) and synthetic[6,38,39,52,75,79] (polyethylene glycol [PEG], polyvinyl alcohol [PVA], polylactic acid, polymethylmethacrylate and polyacrylic acid [PAA] for the most important and many others[79]) polymers have been used.

The most common coating materials for MNPs include polystyrene,[85,86] polysaccharides,[13,14,20,37–39,42,46,47,86] polyacids,[87–89] polyols,[47,86,90–92] polypeptides[13,24,31,35,38,42] and polyamines.[86,93–96] Some of these polymers are even able to replace oleate ligands with minor alteration of the hydrodynamic sizes of the now water-dispersible particles.[71]

5.2.2.2.1 Hydrophobic Effects

Monodisperse MNPs are often obtained in organic solvents (Table 5.1) with a hydrophobic capping layer due to the long alkyl chains of the ligands. Therefore, there is a real need to obtain NPs that are water-dispersible, biocompatible and readily surface functionalized for biomedical applications.[11] A smart approach based on the use of amphiphilic molecule encapsulation consists of taking advantage of the hydrophobic effects between the surface ligands and the hydrophobic segments of chosen amphiphilic molecules such as poly(maleic anhydride) derivatives,[97] PEG-derivatized phosphine oxide or PEG–phospholipid. It is also possible to play with hydrophobic effects coupled with coordinating interaction of Tween derivatives.[98] A PEG–phospholipid copolymer has also been used to modify hydrophobic magnetic NPs through a ligand addition strategy.[99] By simply mixing the as-synthesized magnetic NPs with an amphiphilic ligand in a proper solvent, effective ligand addition could be achieved.[99,100]

5.2.2.2.2 Amphiphilic Polymers

Many amphiphilic polymers have also been developed, such as poly(maleic anhydride-alt-1-octadecene), poly-(maleic anhydride alt-1-tetradecene), poly(styrene-block-acrylic acid) copolymers and PEG–block–polylactide.[35,38,42,95,101] An interesting polymer candidate for coating of MNPs, such as PEG-g-poly(ethylene imine), was shown as a good alternative to pure poly(ethylene imine) in terms of colloidal stability and cytotoxicity due to the introduction of the PEG shielding moieties.[95] The carboxylic and amino functional groups present in these polymer backbones are potential sites for further targeted bioconjugation.[102]

5.2.2.3 Inorganic Coating

Besides the polymeric materials described earlier, inorganic materials[46] such as silica,[5–7,13–15,20,35,38–43,52, 65,67,76,79,80,102–107] carbon,[6,41,52] and graphene-based materials have also been used as coating materials. Silica has been so far the most used coating due to its characteristics (easy regulation of the coating process, processability combined with chemical inertness, controlled porosity and optical transparency).[53] Further use of organofunctional coupling agents is once again compulsory (as mentioned previously in Section 5.2.2.1) and generates the formation of an outer layer exhibiting various functional groups such as $-NH_2$, –Br, –CN, –OH, $-CO_2H$, –SH and epoxy, which themselves may undergo transformation to other groups using standard organic methods or methods more dedicated to biofunctionalization.[70] These silica shells may also exhibit interesting properties for biomedical application, such as incorporation of fluorescent dyes into the SiO_2 matrix and controlled mesoporosity useful for the drug loading.[53]

The coating of hydrophobic NPs with inorganic materials was also performed to generate other metal oxides at the surface of MNPs. For example, a sol–gel reaction of tantalum(V) ethoxide in a microemulsion containing Fe_3O_4 NPs provided multifunctional Fe_3O_4@TaOx core–shell NPs with a good biocompatibility and a prolonged circulation time. When intravenously injected, these particles could enhance the contrast of X-ray computer tomography.[108] A TiO_2 coating was also performed on $Fe_3O_4@SiO_2$, and the resulting particles could be modified with dopamine, which attaches onto the surface of the titania substrate. The further immobilization of succinic anhydride onto the surface of the $Fe_3O_4@TiO_2$ NPs via dopamine was then followed by the immobilization of immunoglobulin G (IgG) via amide bonding. The resulting magnetic NPs not only had the capacity to target different pathogenic bacteria but could also effectively inhibit their growth under a low-power ultraviolet (UV) lamp irradiation within a short period.[109]

5.2.3 Common Synthetic Strategies for Bioconjugation

The functional groups grafted on NPs for subsequent conjugation reactions are also found on and complementary to the naturally occurring biomolecules (proteins, DNA, carbohydrates, lipids, amino acids etc.). They are limited to the following: carboxylic acids (R-CO_2H), amines (R-NH_2), hydroxyls (R-OH), ketones (R_1-COR_2), thiols (R-SH), aldehydes (R-CHO) and groups involved in hydrogen bonds (RH interactions),[48,71] which also limits the number and type of reactions that can be used to form a linkage between the MNP and the target biomolecule. The most common chemical reactions involving these groups include *N*-hydroxysuccinimidyl (NHS) ester modification of amines, along with carbodiimide-mediated (1-ethyl-3-(3-dimethylaminopropyl)carbodiimide) condensation of carboxyl groups with amines, maleimide conjugation to thiols and diazonium modification of the phenolic side chain on tyrosine. Table 5.2 illustrates the main conjugation reactions that have already been applied to NPs for their functionalization. It is noteworthy to mention that chemistries can be applied with reactive groups on either the NP or the biological molecule of interest or both.

Many synthetic biomolecules such as peptides and nucleic acids have been revealed to be very useful because they can be modified by any functional group (amines, thiols, carboxyls, biotin, azides and alkynes) and introduced as needed on the NPs. Another commonly used biochemical technique involves the biotin–avidin strong interactions. It has been used for bioconjugation of many NP materials using the grafting of either biotin or avidin on the NP surface.

5.3 NANO–BIO INTERFACE

5.3.1 Specification Analysis for *In Vivo* Applications of Bioconjugated MNPs

MNPs synthesized according to the procedures given in Table 5.1 are not usable directly for *in vivo* applications. A strategy of surface modification, formulation and bioconjugation integrating as much as possible the overall specifications has to be defined depending on the type of use.

After the suspension of NPs in a biological medium, proteins and other biomolecules spontaneously adsorb onto their surface. This leads to the formation of a biomolecular corona. This corona alters the physicochemical interfacial properties of the nanomaterials and provides a new biological identity to the NPs.[110] This biological identity governs the interactions of the NPs with the biological environment and determines the biological functionality of the synthetic objects.

The biomolecular corona formation is driven by the presence of fundamental forces, i.e. electrostatic interactions, VdW interactions and hydrophobic effects.[111] Thus, its composition depends on the surface properties of the nano-objects and on the composition of the medium. The number of parameters influencing the biomolecular corona and its interactions with the biological environment are numerous so that it is rather difficult to predict accurately the interactions (Figure 5.2). Bioinformatics-inspired approaches are developed to help in this prediction.[112]

The biomolecular corona forms almost instantaneously, but the biomolecule adsorption is a dynamic process depending on their affinity for the nanomaterial. During the first stage, the surface is covered with abundant and mobile biomolecules. Then, these biomolecules can be replaced by less abundant and less mobile ones, with a higher affinity for the surface. As a result, the quantitative composition of the corona is not directly correlated to the composition of the medium. The determined dissociation constant (K_d) of complexes between biomolecules and nanomaterials varies from approximately 10^{-4} M to 10^{-9} M,[113] similar to the range of physiological complexes. For example, the K_d of the complex human serum albumin–FePt NPs functionalized with carboxylic acids is of the order of 10^{-6} M.[114] In a dynamic environment, e.g. *in vivo*, the exposure of the NP to a new environment with a different composition, by crossing biological barriers, may lead to only partial displacement of

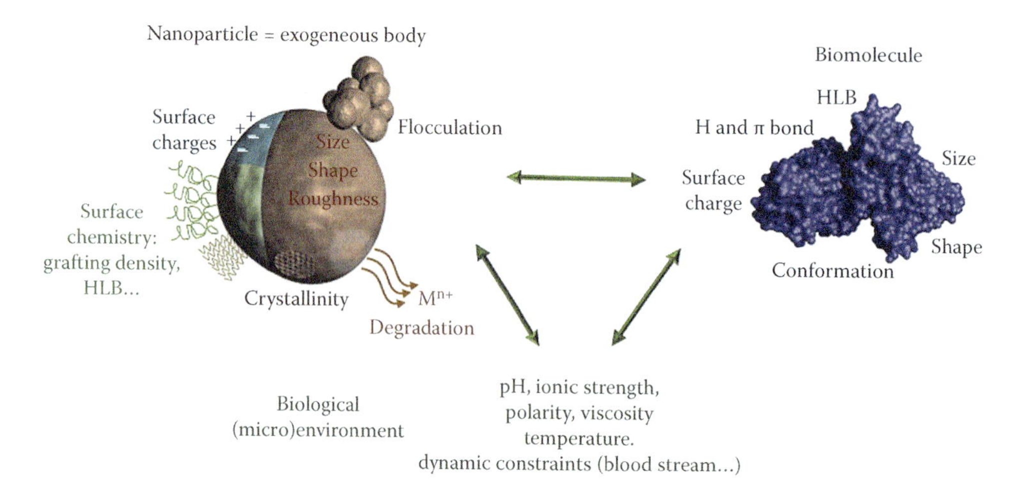

FIGURE 5.2 Illustration of the different physicochemical factors affecting NP–biological environment interactions (HLB, hydrophilic–lipophilic balance).

the original corona. Therefore, the resulting corona would contain the history of all the environments that the NP went through. As a result, the targeting behaviour of bioconjugated NPs optimized *in vitro* is not predictive of the *in vivo* performance.[115] Accordingly, the administration route of MNPs may also drastically affect their biological identity and so, their biodistribution.

For *in vivo* application, noncontrolled adsorption on the NP surface is an important issue. In the blood, administered NPs are exogenous materials liable to be eliminated by the immune defence system, which represents one of the major barriers for systematically administered nanomedicines. Opsonins are plasma biomolecules, in particular immunoglobulin and proteins of the complement, marking antigens and promoting their phagocytosis. Adsorption of opsonins onto the surface of NPs, a process termed as opsonisation, enhances the recognition and uptake of these NPs by the macrophages belonging to the reticuloendothelial system (RES), present in the liver (Kupffer cells), in the spleen and in the bone marrow. This opsonisation leads to the rapid clearance of the nanomaterials from the vascular compartment. Such an effect can be advantageous when these organs are the intended target sites, while in the other case, it may limit the NP delivery to the targeted zone.[116]

Proteins are macromolecules consisting of one or more long chains of amino acid residues. They fold into functional three-dimensional (3D) structures, thanks to weak interactions. During adsorption, proteins may undergo conformational changes. While a preformed albumin corona onto a polymeric NP can extend the blood circulation time by decreasing the complement activation,[117] conformational change undergone by albumin when adsorbed to the silica NP surface increases the NP uptake by human macrophages.[118] Albumin unfolding reveals a cryptic epitope capable of recognition by class A scavenger receptors, a major receptor family associated with the mononuclear phagocyte system (MPS). Protein corona can also interfere with signalling processes. For example, unfolding of adsorbed fibrinogen onto negatively charged PAA conjugated gold NPs leads to the release of inflammatory cytokines by promoting the interactions with the integrin receptor Mac-1.[119] Mahmoudi and coworkers assumed that iron oxide NPs may increase the rate of amyloid β protein fibrillation, depending on the charge of the dextran polymer coating.[120] Protein fibrillation is implicated in amyloidogenic diseases like Parkinson's disease or Alzheimer's disease.

For *in vitro* applications, these interactions may also be harmful, and MNPs can be disrupted. Moreover, it can be easily understood that noncontrolled interactions may drastically reduce the targeting efficiency, e.g. cell labelling, magnetic extraction and purification.

5.3.2 Shielding Approaches

A solution to limit the uncontrolled interactions between NPs and biomolecules consists in modifying their surfaces with charge-neutral, highly hydrophilic polymers to render protein adsorption thermodynamically unfavourable. Macromolecules can be covalently bonded to the surface or just physically adsorbed, but physically adsorbed macromolecules tend to desorb over time in a biological environment. Different polymers have been used to 'passivate' surfaces, such as dextran or PVA, but one of the most used polymers is PEG, which has the chemical formula $HO\text{-}(CH_2\text{-}CH_2\text{-}O)_n\text{-}H$, and its derivatives. This approach is so commonplace that it is known as 'PEGylation.'

The surface density of PEG chains and its molecular weight are the most important parameters determining the interactions of a nanomaterial with proteins. At low density, when macromolecules are too far to interact with each other, the chains adopt a 'mushroom' conformation and do not create an effective barrier against protein adsorption. At high grafting density, macromolecules strongly interact with each other and adopt a 'brush' conformation, creating a large thermodynamic barrier to adsorption. The surface density is crucial to limit biomolecule diffusions to the NP surface, and the shell thickness is important to decrease the intensity of long-distance interactions, i.e. VdW interactions and electrostatic interactions.

The effect of a PEG shell on protein corona formation has been evaluated. Walkey et al.[121] presented that at constant gold NP size, an increase in PEG grafting density resulted in decreased serum protein adsorption, and for the highest PEG density tested (from 0.8 to 1.2 macromolecules/nm², depending on the NP size), 94%–99% of serum protein adsorption was eliminated relative to nongrafted NPs. The study presents also the importance of the NP surface curvature. An inverse correlation has been found between particle size and protein adsorption. For a fixed PEG density, the volume occupied by macromolecules grafted on small NPs is greater than the one of the same macromolecules grafted on a bigger NP. The increase in this volume implies that the conformational freedom of grafted PEG macromolecules increases, weakening the thermodynamic barrier to protein adsorption on smaller NPs. The efficiency of NP uptake by macrophages correlated with variations in serum protein adsorption. At ~0.5 PEG/nm², macrophage uptake was reduced by a factor of 20 to 169 relative to nongrafted NPs.[121] In general, PEGylated NPs in brush conformational regimes display enhanced NP lifetime in the blood stream, but it is not a sufficient condition. In particular, higher grafting densities must compensate the intermolecular spacing generated by high curvature radius.[122] A threshold value R_f/D (with the R_f, Flory radius and D the average distance between neighbouring PEG chains $\left[D = 2(\pi\sigma)^{-\frac{1}{2}}\right]$) can be defined for which an effective stealth behaviour with prolonged blood circulation of NPs can be observed. This threshold was assessed at 2.8 for PEGylated 100 nm polystyrene NPs,[123] whereas it was 4 for 19 nm fluorescent silica NPs.[122] In this last case, an R_f/D of about 4.2 was sufficiently high to increase the blood circulation time up to 24 h and to label xenograft tumours, showing enhanced permeation and retention (EPR) effect.

Although PEGylation is effective for lowering biomolecule adsorption at the surface of NPs, even at high densities, the

PEG corona can never truly eliminate it because of direct interactions between biomolecules and PEG macromolecules.[124] Furthermore, PEGylated NPs may suffer from an 'accelerated blood clearance' (ABC) phenomenon corresponding to the enhanced clearance of second and/or subsequent doses of PEGylated nanomaterials. This phenomenon is coupled with the presence of anti-PEG IgM. However, the immunogenicity of PEG seems to be a function of the nano-object composition.[125]

The limitations of PEG have led to the development of a range of alternative antifouling polymers, such as poly(2-oxazoline)s, peptides and peptoids, and zwitterionic polymers.[126] Zwitterionic polymers are electrically neutral materials, but each monomer is composed of both positively charged functions and negatively charged functions. The high hydration of these polymers due to the presence of electric charges may be responsible for their antifouling properties.[126] One of these promising polymers is poly(carboxybetaine methacrylate) (pCBMA). Recently, the study of Yang et al.[127] on gold NPs coated with PEG or pCBMA has shown that pCBMA coating allowed a longer circulation plasma half-life for the NPs compared to those coated by PEG macromolecules (55.8 h instead of 8.7 h), with no visible ABC phenomenon. Synthetic peptides composed of neutral hydrophilic amino acids may be able to avoid the ABC phenomenon[128] and may overcome the biodegradability issues of PEG and other alternatives.

5.4 CURRENT CHALLENGES IN MNP BIOCONJUGATION FOR THE *IN VIVO* TARGETING

The biological function of the ligand is dictated by specific interactions, e.g. molecular recognition of a specific ligand with a target or interactions between the enzyme and its substrate. Nonspecific interactions can occur directly at the surface of the NP and interfere with the conjugation affecting the orientation or the conformational structure of the ligand. For example, nucleic acids, negatively charged, adsorb onto the surface of positively charged NPs.[129]

For an effective biofunctionalization by conjugation, different criteria have to be taken into account. First, mobility and orientation of the conjugated biomolecule may influence its activity. For example, access to the binding site of a protein, e.g. active site of enzymes, paratope, may be blocked by a bad orientation of the biomolecule. Second, the activity of the biomolecules has to be maintained after conjugation. Biomolecule function is widely dependent on its structure. As mentioned before, adsorption of biomolecules on the NPs may lead to conformational changes and to the loss of their activity. Moreover, adsorption of biomolecules on NPs may be stronger than their affinity for their biological partners, which decreases their bioavailability. For example, transcription by T7 RNA polymerase of DNA adsorbed to cationic gold NPs is completely inhibited *in vitro*.[129] The surface of NPs has to be carefully prepared for the bioconjugation. In the context of the development of theranostic strategies, MNPs are incorporated into a polymer shell, a mesoporous silica matrix or into lipid vesicles constituting the nanocarriers (Figure 5.1).[12,13,33,116,130,131] In this case, the surface chemistry of the nanocarrier must be considered according to the requirements of the theranostic application, notably in regards to the physicochemical and chemical stability of the carrier or also of the drug loading and release.

5.4.1 Categories of Targeting Ligands

Numerous (bio)-molecules possessing molecular recognition properties for tissue-specific targeting can be grafted onto MNPs and then allow molecular imaging as well as therapeutic applications. Targeting ligands can be nucleic acids, monoclonal Abs (MAbs) or their fragments (Figure 5.3), proteins, peptides or smaller molecules such as sugars or vitamins. Here, the most common classes are presented with some benefits and limitations.

5.4.1.1 Nucleic-Acid-Based Ligands

Nucleic acid polymers are well known for their properties of complementary binding. Complementary DNA or RNA strands interact with each other with the formation of hydrogen bonds between pair bases. Adenine interacts with thymine via two hydrogen bonds and cytosine interacts with guanine via three hydrogen bonds. This property has been extensively used for the development of sensitive biosensors.[132] For example, magnetic NPs biofunctionalized with single-stranded DNA have been used for the sensitive detection of bacteria. The detection is based on the aggregation of magnetic NPs hybridized to the bacterial DNA target.[133]

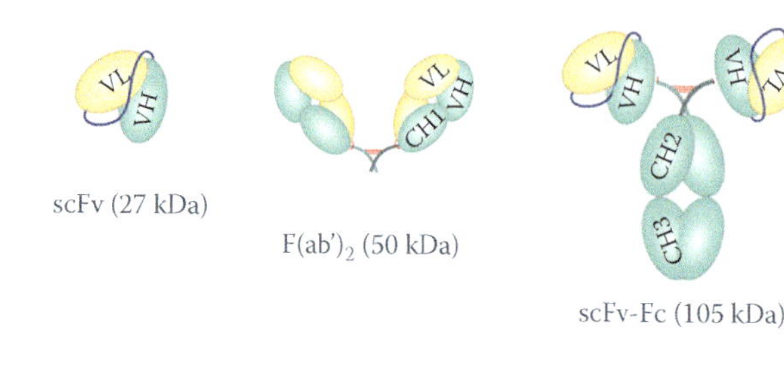

FIGURE 5.3 Examples of Ab fragments (scFv, single-chain variable fragment; Fc, constant fragment) and canonical full length IgG (right).

Aptamers are single-stranded nucleic acids that can fold into specific 3D structures and selectively bind to specific antigen types overexpressed in cancer cells, such as prostate-specific membrane antigen.[134] Most aptamers are obtained through a combinatorial selection process called 'systematic evolution of ligands by exponential enrichment.'[135] Aptamers can be, in principle, selected for any given target. Due to their unique conformational structures, aptamers display high affinity and specificity for their target. They are more stable than proteins and can achieve a higher density for immobilization due to their smaller size. They also have low immunogenicity or toxicity. This advantage makes aptamers excellent sensing elements.[136] The main concern regarding their use is related to their degradation by the nucleases present in high amounts in biological environments.

5.4.1.2 Peptides

Peptides are linear or cyclic sequences composed of two to a few dozen of amino acid residues. Their small size, combined with screening techniques to isolate ligand–substrate combinations, has contributed to an increase in the use of peptides for targeting approaches in the past decade.[137–139] Natural peptides may have the propensity to bind to cell membrane. For example, the tripeptide Arg-Gly-Asp (RGD sequence) is well known to strongly bind $\alpha_v\beta_3$ integrin receptors that are overexpressed on a variety of angiogenic tumour endothelial cells.[140] This property was used to develop RGD-conjugated USPIOs for MRI of tumour angiogenesis targeting[141] and for cancer therapy.[138] The structure–activity relationship of linear and cyclic RGD peptides was compared. The cyclic peptide showed more than 10-fold higher efficiency for tumour targeting compared with its linear counterpart.[139] Cell-penetrating peptides are known to facilitate the cellular uptake of large macromolecules and can be used for the intracellular delivery of NPs.[142] These properties, combined with the EPR effect, have been used in vasculature targeting strategy to favour NP penetration through blood vessels.[143] Other peptides such as the $A\beta_{1-42}$ peptide have been conjugated to MNPs to detect the amyloid deposition in Alzheimer's disease.[144]

Thanks to the phage display technology, new synthetic peptides with a random sequence are selected to bind the desired target.[145] Compared to proteins, peptides can be easily prepared, and their small size enables their bioconjugation with high grafting density on the NPs. However, they are sensitive to degradation by proteases.

5.4.1.3 Proteins

Proteins are also polymers composed of amino acid residues, but they are longer than peptides. These macromolecules fold to adopt a 3D structure, thanks to weak interactions between amino acids residues. Among them, MAbs are one of the most important classes of proteins known for their highly specific binding interaction. The antigen-binding site (paratope) represents a small part of the Ab; the Fc fragment at the base of the Y shape is much less variable and is responsible for the recognition of the protein by the MPS and immune system. MAbs such as IgG are large proteins with a molecular weight of 150 kDa and a hydrodynamic diameter of 10–20 nm that increase the size of the NPs. Smaller Ab formats such as scFv and Fab'2 have been produced by recombinant protein expression to reduce the MAbs to their essential part (Figure 5.3). This limits the increase in the hydrodynamic radius of the final hybrid NPs, their inherent immunogenicity and a rather poor diffusion through physiological barriers.[146]

Other proteins present a high affinity for specific receptors overexpressed on a variety of tumour cells.[147] Among them, transferrin, a glycoprotein of 80 kDa, has been extensively studied as a targeting ligand due to the presence of up-regulated endogenous transferrin receptor 1 (TfR1) on the surface of cancer cells.[148] TfR1s were also targeted by a more exotic MNPs design involving encapsulation of iron oxide NPs within human ferritin protein shells called magnetoferritin.[149] Some proteins are also used for bioconjugation because of their specific biorecognition interactions useful to control the orientation of a protein ligand in order to promote the binding site exposition. Once bound to the NP surface, protein G was used to immobilize in a well-oriented manner anti-horseradish peroxidase IgG via interactions with the Fc region of the Ab.[150] Another representative example of affinity interactions is the well-known streptavidin–biotin binding, which displays the highest affinity constant ($K_d = 4.10^{-14}$ M).[151] Biotinylated herceptins (HER2/neu) were conjugated to commercially available streptavidin-conjugated superparamagnetic iron oxide NPs.[152] HER2/neu receptors are highly expressed in 25% of breast cancers. This bioconjugated MRI contrast agent was used to correlate the different levels of HER2/neu expression on the breast cell membrane. Like the streptavidin–biotin system, the barnase/barstar-based recognition system also possesses a strong binding affinity, K_d, of the order of 10^{-14} M. One of them can be conjugated to an NP while the other can be fused to a protein of interest by genetic engineering, which allows the region-selective grafting of the protein of interest on the NP.[153] However, the barnase is a bacterial protein lethal to the cell when expressed without its inhibitor barstar. Even if such strategies are useful to accurately control the orientation of protein ligands, these supramolecular structures are accompanied by a significant increase in the hydrodynamic diameter of NPs and may be at the origin of nonspecific interactions that can result in an increased uptake from the RES (see also Section 4.4.2). Enzymes are other proteins of interest with their catalytic activity under mild conditions. Immobilization of enzymes on magnetic NPs allows an efficient recovery of the enzyme complex, thereby preventing the enzyme contamination of the final product.[154]

5.4.1.4 Small Molecules

Numerous small biomolecules of interest can be conjugated to NPs. One of the most widely studied small molecules as targeting moiety is folate. Folate is a water-soluble B vitamin B6 essential for cell division and growth. In cancers, folate receptors are overexpressed in a wide variety of tumour cells, including ovarian, brain, breast, colon, renal and lung cancers.[155,156]

The folate ligand has a quite high binding affinity for its receptor (K_d = 10^{-9} M) and thus enables the binding of the conjugated NP.[130] This ligand has been extensively conjugated on MNPs. For instance, MNPs were decorated by PEG and folate to deliver doxorubicin, an anticancer drug, to tumour cells.[157] Carbohydrate moieties such as galactose,[158] glucose[159] and mannose,[160] which are recognized by ubiquitous cellular membrane glycoproteins from lectins family, have also been widely used as targeting ligands.[161] For instance, galactose moieties were conjugated to amino-functionalized (ASPIONs) and lactose-derivatized galactose-terminal ASPION[162] for the specific targeting of asialoglycoprotein receptors overexpressed only on hepatocytes at a high density of 500,000 receptors per cell.[163] These smaller molecular ligands offer the possibility of increasing the affinity of MNPs toward cancer cells through multivalent attachment. In addition, they exhibit a better chemical stability than proteins, peptides or even aptamers or nucleic acids.

5.4.2 Main Critical Parameters Involved in Active Targeting Approaches

The bioconjugation of MNPs changes the physicochemical properties of both targeting molecules and NPs. The size, shape, hydrophilic–lipophilic balance (HLB) and composition are all physicochemical parameters that can affect blood circulation time and the interactions with their targets (Figure 5.4). The immobilized ligands that lose their rotational and translational freedom must be also well oriented in order to promote their binding site exposition. Finally, the density of ligand molecules must be optimized to compensate the rotational motion of MNPs and achieve improved avidity while preserving the selectivity with respect to the targeted zone (see Section 5.4.2.4). The use of too high densities of ligands can also modify the blood circulation and biodistribution kinetics of the biofunctionalized MNPs.

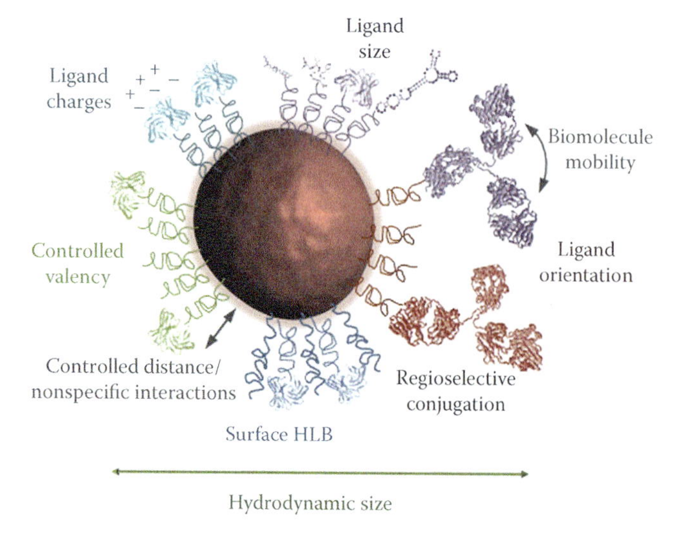

FIGURE 5.4 Schematic representation of the influence of physicochemical properties of the ligand affecting both the pharmacokinetics and the biodistribution of systemic administered bioconjugated MNPs.

5.4.2.1 Effect of the Bioconjugation on the Physicochemical Parameters (Size, Surface Charge and HLB) of the NP Surface

The size of NPs influences their pharmacokinetics and biodistribution. Thus, it must be taken into consideration in the design of bioconjugated NPs. It is also commonly accepted that the charge of NPs affects the systemic circulation times by altering the opsonisation profiles and their recognition by the RES.[164,165] Positively charged particles have been known to form aggregates in the presence of negatively charged serum proteins once intravenously administered.[166] The aggregates are large and often induce transient embolism in the lung capillaries. Negative surface charges can either increase, decrease or have no impact on the NP blood clearance. Parameters like surface charge density in physiological medium and their distribution inside the nanostructure (i.e. core and polymer corona) are at the origin of conflicting findings. Some ligands such as folic acid or proteins can display amphiphilic character, which can also be at the origin of nonspecific interactions of the MNP surface with cells. This effect may be counterbalanced for instance by grafting of long hydrophilic PEG chains.

5.4.2.3 Ligand Orientation

In the case of ligands of great structural complexity, notably when the epitope is localized in a well-specified region such as MAbs and their fragments or aptamers, a regioselective conjugation guaranteeing the correct orientation must be used. The conjugation procedure can be complicated when the ligand does not have an anchoring function sufficiently far from the molecular recognition site. The many reactive functional groups present in protein (mainly amines, carboxylates, cysteine residues and carbohydrate moieties) complicate their oriented immobilization on MNP surface and can lead to cross-linking between NPs. In the case of MAbs, a wrong orientation could further expose the Fc fragment to the outside of the surface, which can result in an increased clearance from the blood. Regioselective strategies have been developed considering the repartition of amino groups in MAbs. At least two types of amine functions are exposed to the medium, the terminal amine that has a pK around 7–8 and the ε-amine moiety of lysine residues that have a pK close to 10.[167] At pH values less than 8.0, the Ab amino terminal groups are the most reactive. At pH values higher than 8.0, the ε-amino groups of Lys residues are more reactive, and as the majority of the lysine residues are located in the Fc portion, the modification should occur preferentially in the Fc portion.[167–169] This property can be exploited by using sulfonated NHS as good leaving groups in order to promote electrostatically the Ab orientation before the formation of peptide bonds[170] or with the same idea, by leaving carboxylate functions bearing negative charges.[171] Postfunctionalization of the Ab by iminothiolane (Traut's reagent) is also performed before coupling on NPs bearing maleimide functional groups.[172,173] Another strategy lies in the presence of a carbohydrate moiety in Fc region to conjugate the

Ab in a proper position on aminated NPs by reductive amination.[167,174] Anti-P selectin half-Abs have been also experimented by reduction of disulphide bonds of the hinge region before conjugation on an MRI contrast agent (versatile USPIOs) for atherosclerosis diseases,[175] but this strategy requires long biochemical studies and the conditions found to enrich the half-Ab fraction can change from one Ab to another. The conjugation of such ligands, in principle, requires the use of a chemical spacer to move the protein away from the NP, thus preventing nonspecific interactions (electrostatic, VdW and hydrophilic–HLB) with the surface. Such interactions can interfere with the orientation and the conformation of the ligand. Long heterobifunctional or homobifunctional PEG macromolecules with different lengths are often used in this aim.[176]

5.4.2.4 Effect of the Multivalence on the Affinity/Avidity and Specificity

One of the most important notions for molecular recognition is the affinity of the NP for its target. In theory, a high affinity increases the efficiency of the targeting because interactions are stronger. It may allow a better *in vivo* delivery to the biological target by compensating the nonspecific interactions described earlier. The affinity of the conjugated NP is dictated by the affinity of the grafted biomolecules, but the strength of interactions between the NP and its target depends also on its valence.

Multivalent interaction mechanisms are complexes, and some of them still need to be finely understood. However, the effect of multivalent interactions can be easily understood with simple notions: the affinity of a couple of biomolecules is defined by the equilibrium dissociation constant K_d, corresponding to the ratio k_{off}/k_{on}, where k_{off} and k_{on} are the kinetic dissociation and association constants, respectively. When low-affinity ligands are used, it is often necessary to increase the valence of MNPs. The Brownian rotational diffusion of the NPs must be also compensated in order to increase the kinetics of binding of the immobilized ligand to its receptor. Multivalent interactions reduce the apparent rate of dissociation due to the clustering and concentration of membrane receptors. Thus, the NPs can leave from the cell membrane only if all the ligand–receptor interactions dissociate at the same time. This clustering is accompanied by the wrapping of the cell membrane, leading to the NP internalisation.[177] The dissociation rate decrease results in the increase of the apparent affinity of the NP for its target, compared to a monovalent NP. The term of 'avidity' is used instead of affinity for multivalent interactions. The avidity of the NP can be increased by up to several orders of magnitude,[178] but it can also decrease the specificity of the targeting. It can be explained by steric hindrance between neighbouring molecules, wrong ligand orientation or competition between ligands for a given receptor. An excess of ligands can also promote nonspecific binding with endothelial or other cells, increase immunogenicity and increase the macrophage uptake of the RES.[179,180]

NP avidity depends also on the surface receptor density.[181] In fact, if the mean distance between the receptors is too large to allow multibinding of the multivalent NP, avidity effect can be discussed. Nevertheless, the control of the valence of the NP is a key parameter for the optimization of the selectivity of the targeting,[182] even if most of the time, the valence corresponds to the average number of biomolecules per NP while the distribution follows the Poisson statistics,[183] leading to heterogeneous NP avidities.[184] Another issue impacting the avidity of multivalent NPs, in correlation with the receptor density, is the decrease in mobility or freedom degree of the immobilized ligands. In this case, the use of longer spacer arms based on heterobifunctional PEG appears once again as a solution, but it has also been demonstrated that they can also prevent the ligands from reaching their receptors.[176,185] A high ligand density is not always the most effective way to improve the targeting since it can result in a decrease in selectivity. For instance, it has been demonstrated that polymer particles coated with lower density of radiolabelled MAb intercellular adhesion molecule 1 (ICAM-1) are more advantageous for positron emission tomography detection of pulmonary inflammation.[182] In the present case, this behaviour can be explained by the receptor mobility at the cell surface, leading to clustering phenomena during the ligand binding.[186] This suggests that the ligand spacing should be considered during the surface bioconjugation step. This trend has also been observed with other NPs conjugated with other targeting agents. Poly(lactic-co-glycolic acid) NPs were conjugated with cyclo-(1,12)-PenITDGEATDSGC (cLABL) peptides at different grafting levels[187] to target ICAM-1 receptors present onto A549 cancer cells. Different grafting conditions were achieved by playing with the ratio of carboxylic and hydroxyl terminations of Pluronic surfactant constituting the surface of these NPs. The conditions corresponding to the highest ligand densities led to the lowest cellular uptake, showing that too high amounts of peptides disturb the receptors clustering.

5.4.3 Preassessment of the Targeting Efficiency

Most of the ligands display proper physicochemical features, and their conjugation to the MNP surface has to be optimized in terms of surface density and orientation to improve as much as possible their avidity for a target molecule. These optimisations can be performed by physical measurement methods such as zeta potential measurements and dynamic light scattering for charge and hydrodynamic diameter assessments. Analytical techniques such as surface plasmon resonance[188] and quartz crystal microbalance[189] are powerful to quantitatively monitor in real-time the interactions between various biomolecules, such as proteins and nucleic acids. These techniques are particularly useful to quantify small aliquots of bioconjugated NPs for desired and undesired molecular interactions. It allows one to study the kinetics of binding and to evaluate their binding affinity/avidity towards targeted molecules immobilized on the sensor surface.[188,189] Immunological methods such as western blot, immunohistology or enzyme-linked immunosorbent assay allow also to reach information about avidity, molecular recognition, targeting efficiency etc. Ligand quantification is an important issue that has to be considered in the development of targeted NPs. At the step of bioconjugation and depending on the nature of the ligand (Ab, DNA, aptamer etc.) and the complexity of the formulation, thermogravimetric analysis is not adapted and requires too much product to estimate the ligand surface density. The amount of ligands is most of the time deduced from gel electrophoresis (polyacrylamide gel electrophoresis for proteins or agarose for nucleic acids), by UV-visible spectroscopy in microvolumes, by liquid chromatography mass spectroscopy or by other spectroscopies such as Fourier-transform infrared spectroscopy and surface-enhanced Raman spectroscopy[189–191] that generally only need small aliquots. Fluorescence and radioisotope labelling can also be useful to control the average amount of ligands for a given developed surface. Even if, as already mentioned previously, *in vitro* studies are not predictive of the *in vivo* results, they seem to remain prerequisites for toxicological data or for the examination of intracellular pathways (i.e. internalization and intracellular release). Finally, the optimisations in terms of surface ligand density have to be performed case by case according to the architecture of the MNP to improve the targeting efficiency/RES clearance balance. For this purpose, screening methods can be helpful to identify the bioconjugated MNPs, that display the highest specificity for a given target or cell line. MNP-based immunoassays-on-chip were developed to screen or detect circulating cancer cells.[46] These screening methods are often coupled with computational approaches. Following the example of combinatorial approaches developed in medicinal chemistry, computational methods involving data analysis methods and statistics are also currently emerging to develop models for the accurate prediction of biological activities or properties of the compounds based on their nanostructure.[192] Predicting methods called QNAR, for 'quantitative nanostructure–activity relationships' integrating diverse NP features (morphological, size, coating) experimentally characterized, combined with surface modifiers characterized by computed chemical descriptors of modified organic molecules, were developed to establish statistically significant relationships between the measured biological effects of NPs and the intrinsic properties of the NPs. For instance, a combinatorial library of 146 NPs designed from cross-linked iron oxide with amine groups (CLIO-NH_2) conjugated with different synthetic molecules was established with the aim to study the effect of multivalency on the specific binding affinity for different cell lines.[193] Different classes of small molecules with amino, sulfhydryl, carboxyl or anhydride functionalities were anchored onto MNPs and stored in multiwell plates for testing. These NPs showed different binding properties toward different targeting proteins using fluorescence detection. This screening enables one to select the NPs displaying the highest specificity for endothelial cells, pancreatic cancer cells and macrophages.

5.5 CONCLUSION AND FUTURE OUTLOOK

This chapter described the main requirements that chemists must consider in bioconjugation of MNPs for a realistic *in vivo* targeting. In particular, the design of bioconjugated MNPs must take into account the changes in physicochemical properties of the particle surface, which is the location of complex interactions with the biological environment. Among all the parameters to be controlled, the density of ligands seems to be the most important one because it affects not only the hydrodynamic diameter but also the surface properties, which themselves influence (i) the stealth required for long blood circulation times and (ii) the avidity and specificity of the targeting.

The multicomponent character of targeted MNP requires preclinical, pharmacokinetics and long-term toxicity studies before use in humans. Even if the impact of each individual component can be distinctively evaluated in preclinical models, the systematic assessment of their synergy in humans become more complex, if not impossible, due to the high cost required and regulatory hurdles. It is therefore necessary to develop preassessment approaches, screening methods and combinatorial approaches that take into account all the physicochemical features required for a particular application in nanomedicine. This type of approach is already being developed for manufactured NPs with the aim of anticipating the toxicological impact on health and the environment during their life cycle ('safe by design' approaches).[192] As it is not possible to accurately predict the MNP *in vivo* fate, these approaches, which could be termed 'targeting-by-design,' would certainly assist in optimizing and selecting the best surface strategies to test for a given pathological situation. To go further, additional information can be found in recent reviews dealing with the use of bioconjugated NPs for bioimaging, drug delivery and theranostics, depending on their administration routes and for different pathological situations.[116,148,192,194,195]

ABBREVIATIONS

ABC accelerated blood clearance
AOT sodium dioctylsulfosuccinate
ASPION amino-functionalized SPIONs
CLIO-NH_2 cross-linked iron oxide with amine groups
CT computed tomography
DEG diethylene glycol
DMSA dimercaptosuccinic acid
DNA deoxyribonucleic acid
EDC 1-ethyl-3-(3-dimethylaminopropyl)carbodiimide
EG ethylene glycol
ELISA enzyme-linked immunosorbent assay
EPR enhanced permeation and retention
FTIR Fourier-transform infrared spectroscopy
HER2/neu human epidermal growth factor receptor-2/
HLB hydrophilic–lipophilic balance
ICAM-1 intercellular adhesion molecule 1
IgG immunoglobulin G
LC-MS liquid chromatography mass spectroscopy
MAbs monoclonal antibodies
Mac-1 macrophage-1 antigen
MNPs magnetic nanoparticles
MPS mononuclear phagocyte system
MRI magnetic resonance imaging
NHS hydroxysuccinimidyl
NPs nanoparticles
NP5 polyoxyethylene (5) nonylphenylether

PAA	poly(acrylic acid)
PAGE	polyacrylamide gel electrophoresis
pCBMA	poly(carboxybetaine methacrylate)
PEG	polyethylene glycol
PEI	poly(ethylene imine)
PET	positron emission tomography
PLA	poly(lactic acid)
PMMA	poly(methylmethacrylate)
PSMA	prostate-specific membrane antigen
PVA	poly(vinyl alcohol)
QCM	quartz crystal microbalance
QNAR	quantitative nanostructure-activity relationships
RES	reticuloendothelial system
RGD	Arg-Gly-Asp tripeptide
RNA	ribonucleic acid
ScFv	single-chain variable fragment
SELEX	systematic evolution of ligands by exponential enrichment
SERS	surface-enhanced Raman spectroscopy
SPIONs	superparamagnetic iron oxide nanoparticles
SPR	surface plasmon resonance
TfR1	transferrin receptor 1
USPIO	ultrasmall superparamagnetic iron oxide
VdW	van der Waals

REFERENCES

1. Ahmad, R., Ali, Z., Mou, X., Wang, J., Yi, H., He, N. 2016. Recent advances in magnetic nanoparticle design for cancer therapy. *J. Nanosci. Nanotechnol.* 16, 9393–9403.
2. Ali, A., Ahmad, M., Akhtar, M. N., Shaukat, S. F., Mustafa, G., Atif, M., Farooq, W. A. 2016. Magnetic nanoparticles (Fe_3O_4 & Co_3O_4) and their applications in urea biosensing. *Russ. J. Appl. Chem.* 89, 517–534.
3. Al-Othman, Z. A. 2012. A review: Fundamental aspects of silicate mesoporous materials. *Materials* 5, 2874–2902.
4. Arnold, R. M., Huddleston, N. E., Locklin, J. 2012. Utilizing click chemistry to design functional interfaces through post-polymerization modification. *J. Mater. Chem.* 22, 19357–19365.
5. Bagheri, S., Julkapli, N. M. 2016. Modified iron oxide nanomaterials, Functionalization and application. *J. Magn. Magn. Mater.* 416, 117–133.
6. Bohara, R. A., Thorat, N. D., Pawar, S. H. 2016. Role of functionalization: Strategies to explore potential nano-bio applications of magnetic nanoparticles. *RSC Adv.* 6, 43989–44012.
7. Brocas, A.-L., Mantzaridis, C., Tunc, D., Carlotti, S. 2013. Polyether synthesis, From activated or metal-free anionic ring-opening polymerization of epoxides to functionalization. *Prog. Polym. Sci.* 38, 845–873.
8. Bruehwiler, D. 2010. Postsynthetic functionalization of mesoporous silica. *Nanoscale* 2, 887–892.
9. Chen, S., Ruenraroengsak, P., Goode, A. E., Ryan, M. P., Porter, A. E., Hu, S., Menzel, R., Smith, E. F., Thorley, A. J., Tetley, T. D., Shaffer, M. S. P. 2014. Aqueous cationic, anionic and non-ionic multi-walled carbon nanotubes, functionalised with minimal framework damage, for biomedical application. *Biomaterials* 35, 4729–4738.
10. Do, T. D., Noh, Y., Kim, M. O., Yoon, J. 2016. In silico magnetic nanocontainers navigation in blood vessels, a feedback control approach. *J. Nanosci. Nanotechnol.* 16, 6368–6373.
11. Gao, Z., Ma, T., Zhao, E., Docter, D., Yang, W., Stauber, R. H., Gao, M. 2016. Small is smarter, nano MRI contrast agents – advantages and recent achievements. *Small* 12, 556–576.
12. Gobbo, O. L., Sjaastad, K., Radomski, M. W., Volkov, Y., Prina-Mello, A. 2015. Magnetic nanoparticles in cancer theranostics. *Theranostics* 5, 1249–1263.
13. Gupta, A. K., Gupta, M. 2005. Synthesis and surface engineering of iron oxide nanoparticles for biomedical applications. *Biomaterials* 26(18), 3995–4021.
14. Hola, K., Markova, Z., Zoppellaro, G., Tucek, J., Zboril, R. 2015. Tailored functionalization of iron oxide nanoparticles for MRI, drug delivery, magnetic separation and immobilization of biosubstances. *Biotechnol. Adv.* 33, 1162–1176.
15. Huang, J., Li, Y., Orza, A., Lu, Q., Guo, P., Wang, L., Yang, L., Mao, H. 2016. Magnetic nanoparticle facilitated drug delivery for cancer therapy with targeted and image-guided approaches. *Adv. Funct. Mater.* 26, 3818–3836.
16. Hudson, R. 2016. Coupling the magnetic and heat dissipative properties of Fe_3O_4 particles to enable applications in catalysis, drug delivery, tissue destruction and remote biological interfacing. *RSC Adv.* 6, 4262–4270.
17. Kanyuk, M. I. 2015. Use of the nanodiamonds in biomedicine. *Biotechnol. Acta* 8, 9-25.
18. Karponis, D., Azzawi, M., Seifalian, A. 2016. An arsenal of magnetic nanoparticles, perspectives in the treatment of cancer. *Nanomedicine* 11, 2215–2232.
19. Khan, S., Danish Rizvi, S. M., Ahmad, V., Baig, M. H., Kamal, M. A., Ahmad, S., Rai, M., Muhammad Zafar Iqbal, A. N., Mushtaq, G., Khan, M. S. 2015. Magnetic nanoparticles, properties, synthesis and biomedical applications. *Curr. Drug Metab.* 16, 685–704.
20. Li, X., Wei, J., Aifantis, K. E., Fan, Y., Feng, Q., Cui, F.-Z., Watari, F. 2016. Current investigations into magnetic nanoparticles for biomedical applications. *J. Biomed. Mater. Res., Part A* 104, 1285–1296.
21. Louie, A. 2010. Multimodality imaging probes, design and challenges. *Chem. Rev.* 110(5), 3146–3195.
22. Ma, X., Gong, A., Chen, B., Zheng, J., Chen, T., Shen, Z., Wu, A. 2015. Exploring a new SPION-based MRI contrast agent with excellent water-dispersibility, high specificity to cancer cells and strong MR imaging efficacy. *Colloids Surf., B* 126, 44–49.
23. Mohammed, L., Ragab, D., Gomaa, H. 2016. Bioactivity of hybrid polymeric magnetic nanoparticles and their applications in drug delivery. *Curr. Pharm. Des.* 22, 3332–3352.
24. Moraes Silva, S., Tavallaie, R., Sandiford, L., Tilley, R. D., Gooding, J. J. 2016. Gold coated magnetic nanoparticles, from preparation to surface modification for analytical and biomedical applications. *Chem. Commun.* 52, 7528–7540.
25. Mrowczynski, R., Markiewicz, R., Liebscher, J. 2016. Chemistry of polydopamine analogues. *Polym. Int.* 65, 1288–1299.
26. Nguyen, D. T., Kim, K.-S. 2014. Functionalization of magnetic nanoparticles for biomedical applications. *Korean J. Chem. Eng.* 31(8), 1289–1305.
27. Nguyen, V. L., Yang, Y., Teranishi, T., Thi, C. M., Cao, Y., Nogami, M. 2015. Biomedical applications of advanced multifunctional magnetic nanoparticles. *J. Nanosci. Nanotechnol.* 15, 10091–10107.
28. Patric, J. P., Veerendra, V., Rajesh, M. 2015. Nanotechnology and its applications in the field of medicine. *World J. Pharm. Res.* 4, 1900–1908.
29. Raj, R., Mongia, P., Sahu, S. K., Ram, A. 2016. Nanocarriers based anticancer drugs, current scenario and future perceptions. *Curr. Drug Targets* 17, 206–228.

30. Ravikumar, C., Bandyopadhyaya, R. 2011. Mechanistic study on magnetite nanoparticle formation by thermal decomposition and coprecipitation routes. *J. Phys. Chem. C* 115(5), 1380–1387.
31. Razzaque, S., Hussain, S. Z., Hussain, I., Tan, B. 2016. Design and utility of metal/metal oxide nanoparticles mediated by thioether end-functionalized polymeric ligands. *Polymers* 8, 156/1–156/26.
32. Shen, Z., Wu, H., Yang, S., Ma, X., Li, Z., Tan, M., Wu, A. 2015. A novel Trojan-horse targeting strategy to reduce the non-specific uptake of nanocarriers by non-cancerous cells. *Biomaterials* 70, 1–11.
33. Shevtsov, M., Multhoff, G. 2016. Recent developments of magnetic nanoparticles for theranostics of brain tumor. *Curr. Drug Metab.* 17, 737–744.
34. Singh, N., Tandan, N., Singh, S. P., Singh, S. P. 2016. Nanoparticles as drug delivery system in modern trends. *World J. Pharm. Pharm. Sci.* 5, 1177–1193.
35. Su, H.-Y., Wu, C.-Q., Li, D.-Y., Hua, A. 2015. Self-assembled superparamagnetic nanoparticles as MRI contrast agents – a review. *Chin. Phys. B* 24, 127506/1–127506/11.
36. Talelli, M., Aires, A., Marciello, M. 2016. Protein-modified magnetic nanoparticles for biomedical applications. *Curr. Org. Chem.* 20, 1252–1261.
37. Tiwari, A. P., Ghosh, S. J., Pawar, S. H. 2015. Biomedical applications based on magnetic nanoparticles, DNA interactions. *Anal. Methods* 7, 10109–10120.
38. Ulbrich, K., Hola, K., Subr, V., Bakandritsos, A., Tucek, J., Zboril, R. 2016. Targeted drug delivery with polymers and magnetic nanoparticles, covalent and noncovalent approaches, release control, and clinical studies. *Chem. Rev.* 116, 5338–5431.
39. van Dongen, S. F. M., de Hoog, H.-P. M., Peters, R. J. R. W., Nallani, M., Nolte, R. J. M., van Hest, J. C. M. 2009. Biohybrid polymer capsules. *Chem. Rev.* 109, 6212–6274.
40. Wibowo, D., Hui, Y., Middelberg, A. P. J., Zhao, C.-X. 2016. Interfacial engineering for silica nanocapsules. *Adv. Colloid Interface Sci.* 236, 83–100.
41. Wu, L., Mendoza-Garcia, A., Li, Q., Sun, S. 2016. Organic phase syntheses of magnetic nanoparticles and their applications. *Chem. Rev.* 116, 10473–10512.
42. Wu, W., Jiang, C. Z., Roy, V. A. L. 2016. Designed synthesis and surface engineering strategies of magnetic iron oxide nanoparticles for biomedical applications. *Nanoscale* 8(47), 19421–19474.
43. Xiao, D., Lu, T., Zeng, R., Bi, Y. 2016. Preparation and highlighted applications of magnetic microparticles and nanoparticles, a review on recent advances. *Microchim. Acta* 183, 2655–2675.
44. Yao, J., Hsu, C.-H., Li, Z., Kim, T. S., Hwang, L.-P., Lin, Y.-C., Lin, Y.-Y. 2015. Magnetic resonance nano-theranostics for glioblastoma multiforme. *Curr. Pharm. Des.* 21, 5256–5266.
45. Yu, Y., Mok, B. Y. L., Loh, X. J., Tan, Y. N. 2016. Rational design of biomolecular templates for synthesizing multifunctional noble metal nanoclusters toward personalized theranostic applications. *Adv. Healthcare Mater.* 5, 1844–1859.
46. Zhu, Y., Kekalo, K., Dong, C. N., Huang, Y.-Y., Shubitidze, F., Griswold, K. E., Baker, I., Zhang, J. X. J. 2016. Magnetic-nanoparticle-based immunoassays-on-chip: Materials synthesis, surface functionalization, and cancer cell screening. *Adv. Funct. Mater.* 26, 3953–3972.
47. Sobczak-Kupiec, A., Venkatesan, J., Alhathal AlAnezi, A., Walczyk, D., Farooqi, A., Malina, D., Hosseini, S. H., Tyliszczak, B. Magnetic nanomaterials and sensors for biological detection. *Nanomedicine* 12(8), 2459–2473.
48. Sapsford, K. E., Algar, W. R., Berti, L., Gemmill, K. B., Casey, B. J., Oh, E., Stewart, M. H., Medintz, I. L. 2013. Functionalizing nanoparticles with biological molecules: Developing chemistries that facilitate nanotechnology. *Chem. Rev.* 113(3), 1904–2074.
49. Shen, Z., Wu, A., Chen, X. 2017. Iron oxide nanoparticle based contrast agents for magnetic resonance imaging. *Mol. Pharmaceutics*, 14(5), 1352–1364.
50. Estelrich, J., Escribano, E., Queralt, J., Busquets, M. A. 2015. Iron oxide nanoparticles for magnetically-guided and magnetically-responsive drug delivery. *Int. J. Mol. Sci.* 16(4), 8070–8101.
51. Maurizi, L., Bisht, H., Bouyer, F., Millot, N. 2009. Easy route to functionalize iron oxide nanoparticles via long-term stable thiol groups. *Langmuir* 25(16), 8857–8859.
52. Wu, W., He, Q., Jiang, C. 2008. Magnetic iron oxide nanoparticles, synthesis and surface functionalization strategies. *Nanoscale Res. Lett.* 3(11), 397–415.
53. Schladt, T. D., Schneider, K., Schild, H., Tremel, W. 2011. Synthesis and bio-functionalization of magnetic nanoparticles for medical diagnosis and treatment. *Dalton Trans.* 40(24), 6315–6343.
54. Lu, A.-H., Salabas, E. L., Schüth, F. 2007. Magnetic nanoparticles: Synthesis, protection, functionalization, and application. *Angew. Chem., Int. Ed.* 46(8), 1222–1244.
55. Kang, Y. S., Risbud, S., Rabolt, J. F., Stroeve, P. 1996. Synthesis and characterization of nanometer-size Fe_3O_4 and γ-Fe_2O_3 particles. *Chem. Mater.* 8(9), 2209–2211.
56. Park, J., An, K., Hwang, Y., Park, J.-G., Noh, H.-J., Kim, J.-Y., Park, J.-H., Hwang, N.-M., Hyeon, T. 2004. Ultra-large-scale syntheses of monodisperse nanocrystals. *Nat. Mater.* 3(12), 891–895.
57. Serge, Y., Tim, L., Perry, E., Frank, G. 2013. Superparamagnetic iron oxide nanoparticles (SPIONs), synthesis and surface modification techniques for use with MRI and other biomedical applications. *Curr. Pharm. Des.* 19(3), 493–509.
58. Daou, T. J., Pourroy, G., Bégin-Colin, S., Grenèche, J. M., Ulhaq-Bouillet, C., Legaré, P., Bernhardt, P., Leuvrey, C., Rogez, G. 2006. Hydrothermal synthesis of monodisperse magnetite nanoparticles. *Chem. Mater.* 18(18), 4399–4404.
59. Santra, S., Tapec, R., Theodoropoulou, N., Dobson, J., Hebard, A., Tan, W. 2001. Synthesis and characterization of silica-coated iron oxide nanoparticles in microemulsion, the effect of nonionic surfactants. *Langmuir* 17(10), 2900–2906.
60. Zeng, L., Luo, L., Pan, Y., Luo, S., Lu, G., Wu, A. 2015. *In vivo* targeted magnetic resonance imaging and visualized photodynamic therapy in deep-tissue cancers using folic acid-functionalized superparamagnetic-upconversion nanocomposites. *Nanoscale* 7(19), 8946–8954.
61. Mahmoudi, M., Laurent, S., Shokrgozar, M. A., Hosseinkhani, M. 2011. Toxicity evaluations of superparamagnetic iron oxide nanoparticles: Cell "vision" versus physicochemical properties of nanoparticles. *ACS Nano* 5(9), 7263–7276.
62. Pascu, O., Carenza, E., Gich, M., Estradé, S., Peiró, F., Herranz, G., Roig, A. 2012. Surface reactivity of iron oxide nanoparticles by microwave-assisted synthesis, comparison with the thermal decomposition route. *J. Phys. Chem. C* 116(28), 15108–15116.
63. Sethi, M., Chakarvarti, S. K. 2015. Hyperthermia techniques for cancer treatment, a review. *Int. J. Pharm. Tech. Res.* 8, 292–299.
64. Qu, S., Yang, H., Ren, D., Kan, S., Zou, G., Li, D., Li, M. 1999. Magnetite nanoparticles prepared by precipitation from partially reduced ferric chloride aqueous solutions. *J. Colloid Interface Sci.* 215(1), 190–192.

65. Niu, M., Pham-Huy, C., He, H. 2016. Core-shell nanoparticles coated with molecularly imprinted polymers, a review. *Microchim. Acta* 183, 2677–2695.
66. Yuan, Y., Wang, W. 2014. Post-synthetic chemical functionalization of oligonucleotides. In *Comprehensive Organic Synthesis II* (Second Edition), edited by P. Knochel. 463–493. Amsterdam: Elsevier.
67. Li, Y., Zhang, X., Deng, C. 2013. Functionalized magnetic nanoparticles for sample preparation in proteomics and peptidomics analysis. *Chem. Soc. Rev.* 42, 8517–8539.
68. Park, J.-W., Park, Y. J., Jun, C.-H. 2011. Post-grafting of silica surfaces with pre-functionalized organosilanes: New synthetic equivalents of conventional trialkoxysilanes. *Chem. Commun.* 47, 4860–4871.
69. Hermanson, G. T. 2008. In *Bioconjugate Techniques* (Second Edition), edited by G. T. Hermanson. 3–212. Academic Press, New York.
70. Thanh, N. T. K., Green, L. A. W. 2010. Functionalisation of nanoparticles for biomedical applications. *Nano Today* 5, 213–230.
71. Erathodiyil, N., Ying, J. Y. 2011. Functionalization of inorganic nanoparticles for bioimaging applications. *Acc. Chem. Res.* 44(10), 925–935.
72. Zeng, J., Jing, L., Hou, Y., Jiao, M., Qiao, R., Jia, Q., Liu, C., Fang, F., Lei, H., Gao, M. 2014. Anchoring group effects of surface ligands on magnetic properties of Fe_3O_4 nanoparticles: Towards high performance MRI contrast agents. *Adv. Mater.* 26(17), 2694–2698.
73. Li, D., Teoh, W. Y., Gooding, J. J., Selomulya, C., Amal, R. 2010. Functionalization strategies for protease immobilization on magnetic nanoparticles. *Adv. Funct. Mater.* 20(11), 1767–1777.
74. Gui, S., Shen, X., Lin, B. 2006. Surface organic modification of Fe_3O_4 nanoparticles by silane-coupling agents. *Rare Met.* 25(6), 426–430.
75. Durdureanu-Angheluta, A., Stoica, I., Pinteala, M., Pricop, L., Doroftei, F., Harabagiu, V., Simionescu, B. C., Chiriac, H. 2009. Glycidoxypropylsilane-functionalized magnetite as precursor for polymer-covered core-shell magnetic particles. *High Perform. Polym.* 21(5), 548–561.
76. Zhang, C., Wängler, B., Morgenstern, B., Zentgraf, H., Eisenhut, M., Untenecker, H., Krüger, R., Huss, R., Seliger, C., Semmler, W., Kiessling, F. 2007. Silica- and alkoxysilane-coated ultrasmall superparamagnetic iron oxide particles: A promising tool to label cells for magnetic resonance imaging. *Langmuir* 23(3), 1427–1434.
77. Drozdov, A. S., Ivanovski, V., Avnir, D., Vinogradov, V. V. 2016. A universal magnetic ferrofluid: nanomagnetite stable hydrosol with no added dispersants and at neutral pH. *J. Colloid Interface Sci.* 468, 307–312.
78. Leena, M., Gomaa, H. G., Ragab, D., Zhu, J. 2017. Magnetic Nanoparticles for environmental and biomedical applications: A review. *Particuology* 30, 1–14.
79. Laurent, S., Forge, D., Port, M., Roch, A., Robic, C., Vander Elst, L., Muller, R. N. 2008. Magnetic iron oxide nanoparticles, synthesis, stabilization, vectorization, physicochemical characterizations, and biological applications. *Chem. Rev.* 108(6), 2064–2110.
80. Xu, C., Xu, K., Gu, H., Zheng, R., Liu, H., Zhang, X., Guo, Z., Xu, B. 2004. Dopamine as a robust anchor to immobilize functional molecules on the iron oxide shell of magnetic nanoparticles. *J. Am. Chem. Soc.* 126(32), 9938–9939.
81. Wei, H., Insin, N., Lee, J., Han, H.-S., Cordero, J. M., Liu, W., Bawendi, M. G. 2012. Compact zwitterion-coated iron oxide nanoparticles for biological applications. *Nano Lett.* 12(1), 22–25.
82. Liu, Y., Chen, T., Wu, C., Qiu, L., Hu, R., Li, J., Cansiz, S., Zhang, L., Cui, C., Zhu, G., You, M., Zhang, T., Tan, W. 2014. Facile surface functionalization of hydrophobic magnetic nanoparticles. *J. Am. Chem. Soc.* 136(36), 12552–12555.
83. Jun, Y., Huh, Y.-M., Choi, J., Lee, J.-H., Song, H.-T., Sungjun, K., Yoon, S., Kim, K.-S., Shin, J.-S., Suh, J.-S., Cheon, J. 2005. Nanoscale size effect of magnetic nanocrystals and their utilization for cancer diagnosis via magnetic resonance imaging. *J. Am. Chem. Soc.* 127(16), 5732–5733.
84. Mohapatra, S., Pramanik, N., Ghosh, S. K., Pramanik, P. 2006. Synthesis and characterization of ultrafine poly(vinylalcohol phosphate) coated magnetite nanoparticles. *J. Nanosci. Nanotechnol.* 6(3), 823–829.
85. Ugelstad, J., Stenstad, P., Kilaas, L., Prestvik, W. S., Herje, R., Berge, A., Hornes, E. 1993. Monodisperse magnetic polymer particles. New biochemical and biomedical applications. *Blood Purif.* 11, 349–69.
86. Chung, H. J., Lee, H., Bae, K. H., Lee, Y., Park, J., Cho, S. W., Hwang, J. Y., Park, H., Langer, R., Anderson, D., Park, T. G. 2011. Facile synthetic route for surface-functionalized magnetic nanoparticles, cell labeling and magnetic resonance imaging studies. *ACS Nano* 5(6), 4329–4336.
87. Wang, G., Zhang, X., Skallberg, A., Liu, Y., Hu, Z., Mei, X., Uvdal, K. 2014. One-step synthesis of water-dispersible ultrasmall Fe_3O_4 nanoparticles as contrast agents for T1 and T2 magnetic resonance imaging. *Nanoscale* 6 5), 2953–2963.
88. Sivakumar, B., Aswathy, R. G., Nagaoka, Y., Suzuki, M., Fukuda, T., Yoshida, Y., Maekawa, T., Sakthikumar, D. N. 2013. Multifunctional carboxymethyl cellulose-based magnetic nanovector as a theragnostic system for folate receptor targeted chemotherapy, imaging, and hyperthermia against cancer. *Langmuir* 29(10), 3453–3466.
89. Hu, F., MacRenaris, K. W., A. Waters, E., Schultz-Sikma, E. A., Eckermann, A. L., Meade, T. J. 2010. Highly dispersible, superparamagnetic magnetite nanoflowers for magnetic resonance imaging. *Chem. Commun.* 46(1), 73–75.
90. Hong, X., Guo, W., Yuan, H., Li, J., Liu, Y., Ma, L., Bai, Y., Li, T. 2004. Periodate oxidation of nanoscaled magnetic dextran composites. *J. Magn. Magn. Mater.* 269(1), 95–100.
91. Mahmoudi, M., Simchi, A., Imani, M., Milani, A. S., Stroeve, P. 2008. Optimal design and characterization of superparamagnetic iron oxide nanoparticles coated with polyvinyl alcohol for targeted delivery and imaging. *J. Phys. Chem. B* 112(46), 14470–14481.
92. Jain, T. K., Foy, S. P., Erokwu, B., Dimitrijevic, S., Flask, C. A., Labhasetwar, V. 2009. Magnetic resonance imaging of multifunctional pluronic stabilized iron-oxide nanoparticles in tumor-bearing mice. *Biomaterials* 30(35), 6748–6756.
93. Arndt, D., Zielasek, V., Dreher, W., Bäumer, M. 2014. Ethylene diamine-assisted synthesis of iron oxide nanoparticles in high-boiling polyol. *J. Colloid Interface Sci.* 417, 188–198.
94. Lee, H. S., Hee Kim, E., Shao, H., Kook Kwak, B. 2005. Synthesis of SPIO-chitosan microspheres for MRI-detectable embolotherapy. *J. Magn. Magn. Mater.* 293(1), 102–105.
95. Schweiger, C., Pietzonka, C., Heverhagen, J., Kissel, T. 2011. Novel magnetic iron oxide nanoparticles coated with poly(ethylene imine)-g-poly(ethylene glycol) for potential biomedical application: Synthesis, stability, cytotoxicity and MR imaging. *Int. J. Pharm.* 408(1–2), 130–137.
96. Babič, M., Horák, D., Trchová, M., Jendelová, P., Glogarová, K., Lesný, P., Herynek, V., Hájek, M., Syková, E. 2008. Poly(L-lysine)-modified iron oxide nanoparticles for stem cell labeling. *Bioconjugate Chem.* 19(3), 740–750.

97. Yu, W. W., Chang, E., Falkner, J. C., Zhang, J., Al-Somali, A. M., Sayes, C. M., Johns, J., Drezek, R., Colvin, V. L. 2007. Forming biocompatible and nonaggregated nanocrystals in water using amphiphilic polymers. *J. Am. Chem. Soc.* 129(10), 2871–2879.
98. Wu, H., Zhu, H., Zhuang, J., Yang, S., Liu, C., Cao, Y. C. 2008. Water-soluble nanocrystals through dual-interaction ligands. *Angew. Chem. Int. Ed.* 47(20), 3730–3734.
99. Na, H. B., Lee, J. H., An, K., Park, Y. I., Park, M., Lee, I. S., Nam, D.-H., Kim, S. T., Kim, S.-H., Kim, S.-W., Lim, K.-H., Kim, K.-S., Kim, S.-O., Hyeon, T. 2007. Development of a T1 contrast agent for magnetic resonance imaging using MnO nanoparticles. *Angew. Chem., Int. Ed. Engl.* 119(28), 5493–5497.
100. Lee, N., Choi, Y., Lee, Y., Park, M., Moon, W. K., Choi, S. H., Hyeon, T. 2012. Water-dispersible ferrimagnetic iron oxide nanocubes with extremely high R2 relaxivity for highly sensitive *in vivo* MRI of tumors. *Nano Lett.* 12(6), 3127–3131.
101. Wang, Y.-X. J., Xuan, S., Port, M., Idee, J.-M. 2013. Recent advances in superparamagnetic iron oxide nanoparticles for cellular imaging and targeted therapy research. *Curr. Pharm. Des.* 19(37), 6575–6593.
102. Yiu, H. H. P. 2011. Engineering the multifunctional surface on magnetic nanoparticles for targeted biomedical applications, a chemical approach. *Nanomedicine* 6(8), 1429–1446.
103. Fang, W., Chen, X., Zheng, N. 2010. Superparamagnetic core-shell polymer particles for efficient purification of his-tagged proteins. *J. Mater. Chem.* 20(39), 8624–8630.
104. Pinho, S. L. C., Laurent, S., Rocha, J., Roch, A., Delville, M.-H., Mornet, S., Carlos, L. D., Vander Elst, L., Muller, R. N., Geraldes, C. F. G. C. 2012. Relaxometric studies of γ-Fe_2O_3@SiO_2 Core shell nanoparticles: When the coating matters. *J. Phys. Chem. C* 116, 2285–2291.
105. Pinho, S. L. C., Pereira, G. A., Voisin, P., Kassem, J., Bouchaud, V., Etienne, L., Peters, J. A., Carlos, L. D., Mornet, S., Geraldes, C. F. G. C., Rocha, J., Delville, M.-H. 2010. Fine tuning of the relaxometry of gamma-Fe_2O_3@SiO_2 nanoparticles by tweaking the silica coating thickness. *ACS Nano* 4(9), 5339–5349.
106. Tang, Y., Zhang, C., Wang, J., Lin, X., Zhang, L., Yang, Y., Wang, Y., Zhang, Z., Bulte, J. W. M., Yang, G.-Y. 2015. MRI/SPECT/fluorescent tri-modal probe for evaluating the homing and therapeutic efficacy of transplanted mesenchymal stem cells in a rat ischemic stroke model. *Adv. Funct. Mater.* 25(7), 1024–1034.
107. Rho, W.-Y., Kim, H.-M., Kyeong, S., Kang, Y.-L., Kim, D.-H., Kang, H., Jeong, C., Kim, D.-E., Lee, Y.-S., Jun, B.-H. 2014. Facile synthesis of monodispersed silica-coated magnetic nanoparticles. *J. Ind. Eng. Chem.* 20(5), 2646–2649.
108. Lee, N., Cho, H. R., Oh, M. H., Lee, S. H., Kim, K., Kim, B. H., Shin, K., Ahn, T.-Y., Choi, J. W., Kim, Y.-W., Choi, S. H., Hyeon, T. 2012. Multifunctional Fe_3O_4/TaO_x Core/shell nanoparticles for simultaneous magnetic resonance imaging and x-ray computed tomography. *J. Am. Chem. Soc.* 134(25), 10309–10312.
109. Chen, W.-J., Tsai, P.-J., Chen, Y.-C. 2008. Functional Fe_3O_4/TiO_2 Core/shell magnetic nanoparticles as photokilling agents for pathogenic bacteria. *Small* 4(4), 485–491.
110. Monopoli, M. P., Åberg, C., Salvati, A., Dawson, K. A. 2012. Biomolecular coronas provide the biological identity of nanosized materials. *Nat. Nanotechnol.* 7(12), 779–786.
111. Nel, A. E., Mädler, L., Velegol, D., Xia T., Hoek, E. M., Somasundaran, P., Klaessig, F., Castranova, V., Thompson, M. 2009. Understanding biophysicochemical interactions at the nano-bio interface. *Nat. Mater.* 8(7), 543–557.
112. Walkey, C. D., Olsen, J. B., Song, F., Liu, R., Guo, H., Olsen, D. W., Cohen, Y., Emili, A., Chan, W. C. 2014. Protein corona fingerprinting predicts the cellular interaction of gold and silver nanoparticles. *ACS Nano* 8(3), 2439–2455.
113. Walkey, C. D., Chan, W. C. W. 2012. Understanding and controlling the interaction of nanomaterials with proteins in a physiological environment. *Chem. Soc. Rev.* 41(7), 2780–2799.
114. Röcker, C., Pötzl, M., Zhang, F., Parak, W. J., Nienhaus, G. U. 2009. A quantitative fluorescence study of protein monolayer formation on colloidal nanoparticles. *Nat. Nanotechnol.* 4(9), 577–580.
115. Hak, S., Helgesen, E., Hektoen, H. H., Huuse, E. M., Jarzyna, P. A., Mulder, W. J. M., Haraldseth, O., Davies, C. d. L. 2012. The effect of nanoparticle polyethylene glycol surface density on ligand-directed tumor targeting studied *in vivo* by dual modality imaging. *ACS Nano* 6, 5648–5658.
116. Reddy, L. H., Arias, J. L., Nicolas, J., Couvreur, P. 2012. Magnetic nanoparticles: Design and characterization, toxicity and biocompatibilityy, pharmaceutical and biomedical applications, *Chem. Rev.* 112, 5818–5978.
117. Peng, Q., Zhang, S., Yang, Q., Zhang, T., Wei, X. Q., Jiang, L., Zhang, C. L., Chen, Q. M., Zhang, Z. R., Lin, Y. F. 2013. Preformed albumin corona, a protective coating for nanoparticles based drug delivery system. *Biomaterials* 34(33), 8521–8530.
118. Mortimer, G. M., Butcher, N. J., Musumeci, A. W., Deng, Z. J., Martin, D. J., Minchin, R. F. 2014. Cryptic epitopes of albumin determine mononuclear phagocyte system clearance of nanomaterials. *ACS Nano* 8(4), 3357–3366.
119. Deng, Z. J., Liang, M., Monteiro, M., Toth, I., Minchin, R. F. 2011. Nanoparticle-induced unfolding of fibrinogen promotes Mac-1 receptor activation and inflammation. *Nat. Nanotechnol.* 6(1), 39–44.
120. Mahmoudi, M., Quinlan-Pluck, F., Monopoli, M. P., Sheibani, S., Vali, H., Dawson, K. A., Lynch, I. 2013. Influence of the physiochemical properties of superparamagnetic iron oxide nanoparticles on amyloid β Protein Fibrillation In Solution. *ACS Chem. Neurosci.* 4(3), 475–485.
121. Walkey, C. D., Olsen, J. B., Guo, H., Emili, A., Chan, W. C. 2012. Nanoparticle size and surface chemistry determine serum protein adsorption and macrophage uptake. *J. Am. Chem. Soc.* 134(4), 2139–2147.
122. Adumeau, L., Genevois, C., Roudier, L., Schatz, C., Couillaud, F., Mornet, S. 2017. Impact of surface grafting density of PEG macromolecules on dually fluorescent silica nanoparticles used for the *in vivo* imaging of subcutaneous tumors. *Biochim. Biophys. Acta, Gen. Subj,* 1861(6), 1581–1595.
123. Yang, Q., Jones, S. W., Parker, C. L., Zamboni, W. C., Bear, J. E., Lai, S. K. 2014. Evading immune cell uptake and clearance requires PEG grafting at densities substantially exceeding the minimum for brush conformation. *Mol. Pharmaceutics* 11, 1250–1258.
124. Halperin, A., Kröger, M. 2009. Ternary protein adsorption onto brushes: Strong versus weak. *Langmuir,* 25(19), 11621–11634.
125. Tagami, T., Nakamura, K., Shimizu, T., Yamazaki, N., Ishida, T., Kiwada, H. 2010. CpG motifs in pDNA-sequences increase anti-PEG IgM production induced by PEG-coated pDNA-lipoplexes. *J. Controlled Release* 142(2), 160–166.
126. Lowe, S., O'Brien-Simpson, N. M., Connal, L. A. 2015. Antibiofouling polymer interfaces, poly(ethylene glycol) and other promising candidates. *Polym. Chem.* 6(2), 198–212.
127. Yang, W., Liu, S., Bai, T., Keefe, A. J., Zhang, L., Ella-Menye, J. R., Li, Y., Jiang, S. 2014. Poly(carboxybetaine) nanomaterials enable long circulation and prevent polymer-specific antibody production. *Nano Today* 9(1), 10–16.
128. Romberg, B., Oussoren, C., Snel, C. J., Carstens, M. G., Hennink, W. E., Storm, G. 2007. Pharmacokinetics of poly (hydroxyethyl-l-asparagine)-coated liposomes is superior over that of PEG-coated liposomes at low lipid dose and upon

repeated administration. *Biochim. Biophys. Acta, Biomembr.* 1768(3), 737–743.

129. McIntosh, C. M., Esposito, E. A., Boal, A. K., Simard, J. M., Martin, C. T., Rotello, V. M. 2001. Inhibition of DNA transcription using cationic mixed monolayer protected gold clusters. *J. Am. Chem. Soc.* 123(31), 7626–7629.
130. Yu, M. K., Park, J., Jon, S. 2012. Targeting strategies for multifunctional nanoparticles in cancer imaging and therapy. *Theranostics* 2(1), 3–44.
131. Lim, E. K., Kim, T., Paik, S. Seungjoo, H., Huh, Y. M., Lee, K. 2015. Nanomaterials for theranostics: Recent advances and future challenges. *Chem. Rev.* 115, 327–394.
132. Abu-Salah, K. M., Zourob, M. M., Mouffouk, F., Alrokayan, S. A., Alaamery, M. A., Ansari, A. A. 2015. DNA-based nanobiosensors as an emerging platform for detection of disease. *Sensors (Basel, Switzerland)* 15(6), 14539–14568.
133. Mezger, A., Fock, J., Antunes, P., Østerberg, F. W., Boisen, A., Nilsson, M., Hansen, M. F., Ahlford, A., Donolato, M. 2015. Scalable DNA-based magnetic nanoparticle agglutination assay for bacterial detection in patient samples. *ACS Nano* 9(7), 7374–7382.
134. Yigit, M. V., Mazumdar, D., Lu, Y. 2008 MRI detection of thrombin with aptamer functionalized superparamagnetic iron oxide nanoparticles. *Bioconjugate Chem.* 19, 412–417.
135. Tuerk, C., Gold, L. 1990. Systematic evolution of ligands by exponential enrichment: RNA ligands to bacteriophage T4 DNA polymerase. *Science* 249, 505–510.
136. Zhou, W., Jimmy Huang, P.-J., Ding, J., Liu, J. 2014. Aptamer-based biosensors for biomedical diagnostics. *Analyst* 139(11), 2627–2640.
137. Kamaly, N., Xiao, Z., Valencia, P. M., Radovic-Moreno, A. F., Farokhzad, O. C. 2012. Targeted polymeric therapeutic nanoparticles: Design, development and clinical translation. *Chem. Soc. Rev.* 41, 2971–3010.
138. Danhier, F., Breton, A. L., Préat, V. 2012. RGD-based strategies to target αvβ3 integrin in cancer therapy and diagnosis. *Mol. Pharmaceutics* 9(11), 2961–2973.
139. Colombo, G., Curnis, F., De Mori, G. M. S., Gasparri, A., Longoni, C., Sacchi, A., Longhi, R., Corti, A. 2002. Structure–activity relationships of linear and cyclic peptides containing the NGR tumor-homing motif. *J. Biol. Chem.* 277, 47891–47897.
140. Varner, J. A., Cheresh, D. A. 1996. Integrins and cancer. *Curr. Opin. Cell Biol.* 8, 724–730.
141. Zhang, C. F., Jugold, M., Woenne, E. C., Lammers, T., Morgenstern, B., Mueller, M. M., Zentgraf, H., Bock, M., Eisenhut, M., Semmler, W., Kiessling, F. 2007. Specific targeting of tumor angiogenesis by RGD-conjugated ultrasmall superparamagnetic iron oxide particles using a clinical 1.5-T magnetic resonance scanner. *Cancer Res.* 67, 1555–1562.
142. Li, H., Tsui, T., Ma, W. 2015. Intracellular delivery of molecular cargo using cell-penetrating peptides and the combination strategies. *Int. J. Mol. Sci.* 16(8), 19518–19536.
143. Roth, L., Agemy, L., Kotamraju, V. R., Braun, G., Teesalu, T., Sugahara, K. N., Hamzah, J., Ruoslahti, E. 2012. Transtumoral targeting enabled by a novel neuropilin-binding peptide. *Oncogene* 31, 3754–3763.
144. Yang, J., Wadghiri, Y. Z., Hoang, D. M., Tsui, W.; Sun, Y., Chung, E., Li, Y., Wang, A., de Leon, M., Wisniewski, T. 2011. Detection of amyloid plaques targeted by USPIO-Aβ1-42 in Alzheimer's disease transgenic mice using magnetic resonance microimaging. *Neuroimage* 55, 1600–1609.
145. Brown, K. C. 2010. Peptidic tumor targeting agents: The road from phage display peptide selections to clinical applications. *Curr. Pharm. Des.* 16(9), 1040–1054.
146. Foon, K. A. 1989. Biological response modifiers. *Cancer Res.* 49, 1621–1639.
147. Vandewalle, B., Granier, A. M., Peyrat, J. P., Bonneterre, J., Lefebvre, J. J. 1985. Transferrin receptors in cultured breast cancer cells. *Cancer Res. Clin. Oncol.* 110, 71–76.
148. Sahoo, S. K., Ma, W., Labhasetwar, V. 2004. Efficacy of transferrin-conjugated paclitaxel loaded nanoparticles in a murine model of prostate cancer, *Int. J. Cancer* 112, 335–340.
149. Fan, K. L., Cao, C. Q., Pan, Y. X., Lu, D., Yang, D. L., Feng, J., Song, L. N., Liang, M. M., Yan, Y. 2012. Magnetoferritin nanoparticles for targeting and visualizing tumour tissues. *Nat. Nanotechnol.* 7, 459–464.
150. Arenal, R., De Matteis, L., Custardoy, L., Mayoral, A., Tence, M., Grazu, V., De La Fuente, J. M., Marquina, C., Ibarra, M. R. 2013. Spatially-resolved EELS analysis of antibody distribution on biofunctionalized magnetic nanoparticles, *ACS Nano* 7(5), 4006–4013.
151. Green, NM. 1990. Avidin and streptavidin. *Methods Enzymol.* 184, 51–67.
152. Artemov, D., Mori, N., Okollie, B., Bhujwalla, Z. M. 2003. MR molecular imaging of the Her-2/neu receptor in breast cancer cells using targeted iron oxide nanoparticles. *Magn. Reson. Med.* 49, 403–408.
153. Nikitin, M. P., Zdobnova, T. A., Lukash, S. V., Oleg A. Stremovskiy, O. A., Deyev, S. M. 2010. Protein-assisted self-assembly of multifunctional nanoparticles. *Proc. Natl. Acad. Sci. U. S. A.* 107(13), 5827–5832.
154. Ansari, S. A., Husain, Q. 2012. Potential applications of enzymes immobilized on/in nano materials: A review. *Biotechnol. Adv.* 30(3), 512–523.
155. Hilgenbrink, A. R., Low, P. S. 2005. Folate receptor-mediated drug targeting: From therapeutics to diagnostics. *J. Pharm. Sci.* 94, 2135–2146.
156. Markert, S., Lassmann, S., Gabriel, B., Klar, M., Werner, M., Gitsch, G., Kratz, F., Hasenburg, A. 2008. Alpha-folate receptor expression in epithelial ovarian carcinoma and non-neoplastic ovarian tissue. *Anticancer Res.* 28, 3567–3572.
157. Zhang, J., Rana, S., Srivastava, R. S., Misra, R. D. K. 2008. On the chemical synthesis and drug delivery response of folate receptor-activated, polyethylene glycol-functionalized magnetite nanoparticles. *Acta Biomater.* 4, 40–48.
158. Bergen, J. M., Von Recum, H. A., Goodman, T. T., Massey, A. P., Pun, S. H. 2006. Gold nanoparticles as a versatile platform for optimizing physicochemical parameters for targeted drug delivery. *Macromol. Biosci.* 6, 506–516.
159. Li, X., Zhou, H., Yang, L., Du, G., Pai-Panandiker, A. S., Huang, X., Yan, B. 2011. Enhancement of cell recognition *in vitro* by dual-ligand cancer targeting gold nanoparticles. *Biomaterials* 32, 2540–2545.
160. Hashida, M., Nishikawa, M., Yamashita, F., Takakura, Y. 2001. Cell-specific delivery of genes with glycosylated carriers. *Adv. Drug Deliv. Rev.* 52 187–196.
161. Zhang, H., Ma, Y., Sun, X.-L. 2010. Recent developments in carbohydrate-decorated targeted drug/gene delivery. *Med. Res. Rev.* 30, 270–289.
162. Huang, G., Diakur, J., Xu, Z., Wiebe, L. I. 2008. Asialoglycoprotein receptor-targeted superparamagnetic iron oxide nanoparticles. *Int. J. Pharm.* 360, 197–203.
163. Shen, Z., Wei, W., Tanaka, H., Kohama, K., Ma, G., Dobashi, T., Maki, Y., Wang, H., Bi, J., Dai, S. 2011, A galactosamine-mediated drug delivery carrier for targeted liver cancer therapy. *Pharmacol. Res.* 64, 410–419.

164. Alexis, F., Pridgen, E., Molnar, L. K., Farokhzad, O. C. 2008. Factors affecting the clearance and biodistribution of polymeric nanoparticles. *Mol. Pharmaceutics* 5, 505–515.
165. Bertrand, N., Leroux, J. C. 2012. The journey of a drug carrier in the body: An anatomo-physiological perspective. *J. Control Rel.* 161, 152–163.
166. Li, S. D., Huang, L. 2008. Pharmacokinetics and biodistribution of nanoparticles. *Mol. Pharmaceutics* 5, 496–504.
167. Puertas, S., Moros, M., Fernández-Pacheco, R., Ibarra, M. R., Grazú, V., de la Fuente, J. M. 2010. Designing novel nano-immunoassays: Antibody orientation versus sensitivity. *J. Phys. D Appl. Phys.* 43, 474012–474019.
168. Hadjidemetriou, M., Al-Ahmady, Z., Mazza, M., Collins, R. F., Dawson, K., Kostarelos, K. 2015. *In vivo* biomolecule corona around blood-circulating, clinically used and antibody-targeted lipid bilayer nanoscale vesicles. *ACS Nano* 9, 8142–8156.
169. Conde, J., Dias, J. T., Grazú, V., Moros, M., Baptista, P. V., de la Fuente, J. M. 2014. Revisiting 30 years of biofunctionalization and surface chemistry of inorganic nanoparticles for nanomedicine. *Front. Chem.* 2(1), 1–27.
170. Parolo, C., de la Escosura-Muñiz, A., Polo, E., Grazú, V., de la Fuente, J. M., Merko, A. 2013. Design, preparation, and evaluation of a fixed-orientation antibody/gold-nanoparticle conjugate as an immunosensing label. *ACS Appl. Mater. Interfaces* 5, 10753–10759.
171. Puertas, S., Batalla, P., Moros, M., Polo, E., Del Pino, P., Guisan, J. M., Grazú, V., de la Fuente, J. M. 2011. Taking advantage of unspecific interactions to produce highly active magnetic nanoparticle-antibody conjugates. *ACS Nano* 5, 4521–4528.
172. Jacobin-Valat, M.-J., Laroche-Traineau, J., Larivière, M., Mornet, S., Sanchez, S., Biran, M., Lebaron, C., Boudon, J., Lacomme, S., Cérutti, M., Clofent-Sanchez, G. 2015. Nanoparticles functionalized with an anti-platelet human antibody for *in vivo* detection of atherosclerotic plaque by Magnetic Resonance Imaging. *Nanomedicine* 11(4), 927–937.
173. Aires, A., Ocampo, S. M., Simões, B. M., Rodríguez, M. J., Cadenas, J. F., Couleaud, P., Spence, K., Latorre, A., Miranda, R., Somoza, A., Clarke, R. B., Carrascosa, J. L., Cortajarena, A. L. 2016. Multifunctionalized iron oxide nanoparticles for selective drug delivery to CD44-positive cancer cells. *Nanotechnology* 27, 065103 (10pp).
174. Ansell, S. M., Tardi, P. G., Buchkowsky, S. S. 1996. 3-(2-Pyridyldithio)propionic acid hydrazide as a cross-linker in the formation of liposome–antibody conjugates. *Bioconjugate Chem.* 7, 490–496.
175. Jacobin-Vallat, M.-J., Deramchia, K., Mornet, S., Hagemeyer, C., Bonetto, S., Robert, R., Biran, M., Massot, P., Miraux, S., Sanchez, S., Bouzier-Sore, A.-K., Franconi, J.-M., Duguet, E., Clofent-Sanchez, G. 2011. Magnetic Resonance Imaging of inducible P-selectin expression in human activated platelets involved in early stages of atherosclerosis. *NMR in Biomed.* 24, 413–424.
176. Stefanick, J. F., Ashley, J. D., Kiziltepe, T., Bilgicer, B. 2013. A systematic analysis of peptide linker length and liposomal polyethylene glycol coating on cellular uptake of peptide-targeted liposomes, *ACS Nano* 7, 2935–2947.
177. Mukherjee, S., Ghosh, R. N., Maxfield, F. R. 1997. Endocytosis, *Physiol. Rev.* 77, 759–803.
178. Hong, S., Leroueil, P. R., Majoros, I. J., Orr, B. G., Baker, J. R., Banaszak Holl, M. M. 2007. The binding avidity of a nanoparticle-based multivalent targeted drug delivery platform. *Chem. Biol.* 14(1), 107–115.
179. Ferrari, M. 2008. Nanogeometry: Beyond drug delivery. *Nat. Nanotechnol.* 3, 131–132.
180. Valencia, P. M., Hanewich-Hollatz, M. H., Gao, W., Karim, F., Langer, R., Karnik, R., Farokhzad, O. C. 2011. Effects of ligands with different water solubilities on self-assembly and properties of targeted nanoparticles. *Biomaterials* 32, 6226–6233.
181. Silpe, J. E., Sumit, M., Thomas, T. P., Huang, B., Kotlyar, A., van Dongen, M. A., Banaszak Holl, M. M., Orr, B. G., Choi, S. K. 2013. Avidity modulation of folate-targeted multivalent dendrimers for evaluating biophysical models of cancer targeting nanoparticles. *ACS Chem. Biol.* 8(9), 2063–2071.
182. Zern, B. J., Chacko, A.-M., Liu, J., Greineder, C. F., Blankemeyer, E. R., Radhakrishnan, R., Muzykantov, V. 2013. Reduction of nanoparticle avidity enhances the selectivity of vascular targeting and PET detection of pulmonary inflammation. *ACS Nano* 7, 2461–2469.
183. Pons, T., Medintz, I. L., Wang, X., English, D. S., Mattoussi, H. 2006. Solution-phase single quantum dot fluorescence resonance energy transfer. *J. Am. Chem. Soc.* 128, 15324–15331.
184. Li, M. H., Choi, S. K., Leroueil, P. R., Baker, J. R. 2014. Evaluating binding avidities of populations of heterogeneous multivalent ligand-functionalized nanoparticles. *ACS Nano* 8, 5600–5609.
185. Mori, A., Klibanov, A. M., Torchilin, V. P., Huang, L. 1991. Influence of the steric barrier activity of amphiphatic poly(ethylene glycol) and ganglioside GM1 on the circulation time of the liposomes and on the target binding of immunoliposomes *in vivo*. *FEBS Lett.* 284, 263–266.
186. Wülfing, C., Sjaastad, M. D., Davis, M. M. 1998. Visualizing the dynamics of T cell activation: Intracellular adhesion molecule 1 migrates rapidly to the T cell/B cell interface and acts to sustain calcium levels. *Proc. Natl. Acad. Sci. U S A.* 95, 6302–6307.
187. Gu, F., Zhang, L., Teply, B. A., Mann, N., Wang, A., Radovic-Moreno, A. F., Langer, R., Farokhzad, O. C. 2008. Precise engineering of targeted nanoparticles by using self-assembled biointegrated block copolymers. *Proc. Natl. Acad. Sci. U S A.* 105, 2586–2591.
188. Schneider, G. S., Bhargav, A. G., Perez, J. G., Wadajkar, A. S., Winkles, J. A., Woodworth, G. F., Kim, A. J. 2015. Surface plasmon resonance as a high throughput method to evaluate specific and non-specific binding of nanotherapeutics. *J. Control. Release* 219, 331–344.
189. Li, L., Mu, Q., Zhang, B. Yan, B. 2010. Analytical strategies for detecting nanoparticle-protein interactions. *Analyst* 135(7), 1519–1530.
190. Zhou, H., Li, X., Lemoff, A., Zhang, B. Yan, B. 2010. Structural confirmation and quantification of individual ligands from the surface of multi-functionalized gold nanoparticles. *Analyst* 135(6), 1210–1213.
191. Zhang, B., Yan, B. 2010. Analytical strategies for characterizing nanoparticle's surface chemistry. *Anal. Bioanal. Chem.* 396, 973–982.
192. Mu, Q., Jiang, G., Chen, L., Zhou, H., Fourches, D., Tropsha, A., Yan, B. 2014. Chemical basis of interactions between engineered nanoparticles and biological systems. *Chem. Rev.* 114, 7740–7781.
193. Weissleder, R., Kelly, K., Sun, E. Y., Shtatland, T., Josephson, L. 2005. Cell-specific targeting of nanoparticles by multivalent attachment of small molecules. *Nat. Nanotechnol.* 23, 1418–1423.
194. Bertrand, N., Wu, J., Xu, X., Kamaly, N., Farokhzad, O. C. 2014. Cancer nanotechnology: The impact of passive and active targeting in the era of modern cancer biology. *Adv. Drug Deliv. Rev.* 66, 2–25.
195. Kamaly, N., He, J. C., Ausiello, D. A., Farokhzad, O. C. 2016. Nanomedicines for renal disease: Current status and future applications. *Nat. Rev. Nephrol.* 12, 738–753.

Stéphane Mornet (E-mail: Stephane.Mornet@icmcb.cnrs.fr) earned his PhD in physical chemistry of condensed matter at the University of Bordeaux in 2002. After 4 years of postdoctoral fellowship at the European Institute of Chemistry and Biology of Bordeaux, he undertook a postdoctoral position from 2006 to 2007 in Italy at the Institute for Health and Consumer Protection (Ispra), where he worked for the European Commission in the field of NP health risk. He is currently a researcher at the French National Centre for Scientific Research (CNRS) at the Institute of Condensed Matter Chemistry of Bordeaux. His research, at the interface of chemistry and biology, focuses on the synthesis of hybrid NPs (magnetic, metallic and luminescent NPs), their surface functionalization (optimized PEGylation routes) and conjugation with biomolecules (antibodies and fragments, proteins, folic acid) for bioimaging (MRI and near-infrared fluorescence imaging) and drug release induced under magnetic field or light (mesoporous NPs).

Laurent Adumeau, after obtaining a research master's degree in structural biochemistry at the University of Bordeaux, he began to work at the interface of chemistry and biology in the area of NPs within the group 'Chemistry of Nanomaterials' at the Institute of Condensed Matter Chemistry of Bordeaux (CNRS). His research focuses on NPs' surface functionalization and on their conjugation with biomolecules for biological applications. He earned his PhD in physical chemistry of condensed matter at the University of Bordeaux in 2015. In 2017, he joined the Centre for BioNano Interactions at the University College Dublin, where he is currently a postdoctoral fellow.

Marie-Hélène Delville is a senior scientist (DR1) at the French National Centre for Scientific Research (CNRS). She earned her PhD in organometallic chemistry at the University of Bordeaux in 1988, a topic on which she worked until 1996. She then moved to the Institute of Condensed Matter of Bordeaux, where she is currently working. She is the head of an International Associated Laboratory (LAFICS/IFLaSSC) between India and France. Her research interests are focused on the fundamental and practical aspects involved in the synthesis of organic–inorganic colloidal nano-objects, with special emphasis on the synthesis, control of shape, surface functionalization of mineral oxide particles and sol–gel chemistry. Her research also includes their use in biomedical applications and their potential toxicity.

6 Experimental Considerations for Scalable Magnetic Nanoparticle Synthesis and Surface Functionalization for Clinical Applications

*Alec P. LaGrow, Maximilian O. Besenhard, Roxanne Hachani and Nguyễn T. K. Thanh**

CONTENTS

6.1 INTRODUCTION

For biomedical applications, magnetic iron oxide nanoparticles (IONPs) (mainly maghemite or magnetite) must be synthesized reproducibly at large enough scales for preclinical trials, clinical trials and ultimately as a medical product. Reproducibility is one of the key difficulties in scalable synthetic procedures and is strongly dependent on methodologies. Chapter 1 of this book has already touched on this issue. Particle formation and stabilization mechanisms differ drastically depending on the precursors and solvents used, *i.e.* polar or nonpolar. Some of the synthetic procedures can be scaled up simply by increasing the size of the reactor and the amounts of reagents. However, this is not an option for most magnetic nanoparticle (MNP) syntheses, which can be notoriously irreproducible due to variations in precursor quality or complicated reaction kinetics.[1,2] The latter are highly sensitive to process conditions, *e.g.* reagent mixing times, spatial homogeneity of temperature and reagent concentrations, heating rates and the order of reactant addition. Continuous methods utilizing flow chemistry, which have been increasingly implemented for nanoparticle (NP) syntheses, have offered solutions to these problems allowing for highly controlled reaction conditions (see Section 6.2.2). Continuous methods can allow laboratory-scale syntheses to be readily scaled up just by running the reactor for longer periods of time, thus producing desirable amounts under identical (well controlled) process conditions.

Once the IONPs are formed, stabilization is essential. For the most common biomedical applications, synthesized NPs have to be stable (no aggregation, agglomeration, dissolution or degradation – see Chapter 1) for long periods of time in aqueous media with a wide range of pH values and salt concentrations, as can be found under physiological conditions.[3] In addition, stabilized particles must be nontoxic and biocompatible. To achieve this, the particles formed must have an adequate coating, typically organic molecules, which will confer stabilization and makes them biocompatible. Such organic stabilizers have already been discussed in detail in Chapter 1 and Chapter 5 in this book on 'Biofunctionalisation of Magnetic Nanoparticles for *In Vivo* Targeting' and are

* Corresponding author.

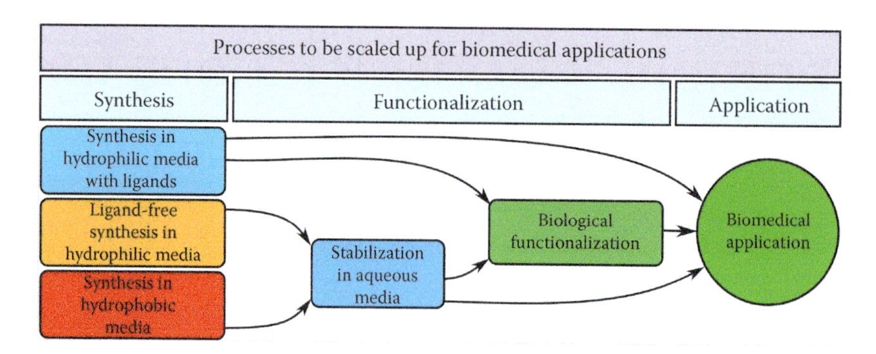

FIGURE 6.1 Schematic of the processes that need to be scaled to produce MNPs for biomedical applications.

generally described as biocompatible molecules with a head group that binds tightly to the NPs' surface and a tail group that is hydrophilic and allows for dispersity in water (where the NP creates a physically and chemically stable colloidal dispersion).[3] To produce water-dispersible and biocompatible NPs, either the particles will be synthesized in the presence of the desired ligand or the ligands will be attached post synthesis via a ligand addition, exchange, cross-linking or assembly step.[4] For applications such as cell targeting etc., further biological functionalization has to be carried out.

Producing larger scales of MNPs for clinical applications involves the implementation of scalable processes starting from the synthetic step through to surface functionalization (see Figure 6.1). Therefore, a robust and well-understood process is required, which allows multiple steps to occur in sequence during the production of the MNPs. Each of these steps will cause unique challenges for the scale-up of production. Therefore, scale-up needs to be considered from the early stages of process development to allow scalable implementation of the multiple operations for production, characterization and quality control, and even cleaning.

In this chapter, we will discuss how to create a multistep synthetic regime to produce MNPs for biological applications. We will review the current scalable synthetic approaches to form NPs and then how to implement stabilization and biological functionalization. Although there are applications in the literature that need hydrophobic particles, such as hyperthermic release of drugs from drug capsules,[5–7] the focus of this chapter will be on aqueous systems as they are by far the most common and challenging.

6.2 SCALE UP OF NP SYNTHESES

Successful scale-up can be understood as the increase in produced nanomaterial maintaining the initial quality attributes achieved at smaller scales. In order to obtain high-quality MNPs (*i.e.* high magnetic moment and crystallinity, having the same shape, monodispersity, desirable size and suitable surface chemistry for intended applications), a systematic development and optimization of the synthesis are required. Hence, small changes in the reaction conditions (*e.g.* the quality of reactants) can have a drastic impact on the synthesized material. This 'sweet spot' of synthetic conditions is what makes scale-up of NP synthesis and functionalization so challenging and for some processes might be near impossible. Every reactor design is characterized by different heat and mass transfer rates, mixing times, shear rates etc. It is simply not possible to keep these characteristics constant while using bigger geometries, although increasing reactor size is frequently the initial strategy attempted to achieve scale-up. For example, the mixing time of the laboratory scale stirred reactor can be adapted in large scale, but usually at the price of faster stirring and, therefore, higher shear rates. In general, larger reactor designs suffer from concentration and temperature gradients, leading to inconsistent reaction conditions, which make it harder to tune the properties of the synthesized nanomaterials. However, to which extent variations in process conditions affect the quality of the synthesized nanomaterial depends on the synthesis, which is why choosing the synthesis is where scale-up starts.

6.2.1 Batch Techniques

Gram scale coprecipitation reactors to form IONPs have been described using a large reactor with mechanical stirring to form IONPs.[8,9] In the work of Kolen'ko et al.,[8] IONPs were precipitated by the addition of a base via a syringe pump during stirring and then a second syringe pump added the stabilizing agent, poly(acrylic acid) (PAA), after the particles had been formed. Large-scale coprecipitation syntheses were also performed under pressure by a hydrothermal approach, where the NPs are synthesized in a pressurized vessel.[8] However, coprecipitation reactions create unique challenges for scale-up, as a slow controlled base addition can create the necessary conditions to form IONPs.[10] Majority of the literature on coprecipitation synthesis reports fast reaction kinetics (initial particle formation within seconds). It was also shown that the particle formation pathway (including phase transformations) can be drastically effected by the mixing conditions[11,12] due to local variations in pH or temperature.[13] Hence, particle size and even phase can change with mixing times and temperature profiles, *i.e.* different results at research laboratories and manufacturing scale. In this way, coprecipitation methods tend to lend themselves better to continuous reactor designs facilitating tight control of mixing conditions and temperature profiles.

High-temperature formation of IONPs of a range of sizes has been synthesized in organic solvents in a scalable fashion.

The work of Park et al.[14] has shown that for the synthesis of IONPs in organic media, large-scale syntheses can be carried out by increasing the reactor size. This has allowed for particles from 3 to 22 nm, with very small size distributions (~5%), to be synthesized in organic solvents, by heating the precursors and surfactants up to ~320°C.[14,15] This approach has been coined as the 'heat up' approach as the reaction creates monodisperse particles by slowly heating the solution to reflux.[16] The reaction kinetics of the heat up system are ideal for a batch technique, where the nucleation occurs via the decomposition of an iron oleate precursor as the solution reaches 320°C.[16] Another batch technique that has shown scalability is the work of Sun et al.[17] with $Fe(acac)_3$ as the precursor that is decomposed at 265°C in the presence of oleylamine, oleic acid and hexadecanediol as ligands. The synthesis also occurred via heating of the precursors up to the reaction temperature.[17]

6.2.2 Continuous Techniques

Reactors for continuous synthesis include continuous-flow stirred-tank reactors and 'millifluidic' (characteristic reactor dimension, *e.g.* capillary diameter, <1 cm) and microfluidic (<1 mm) systems.[18] Microfluidic[19–21] and millifluidic[22–24] systems have been gaining popularity for NP synthesis as the synthesis can be carried out with more control of the reaction time (via residence time),[25,26] mixing times,[27] temperature profiles (including heating stages),[28–30] reaction pressure[31,32] and the separation of nucleation and growth[28,33] with precision that is difficult to achieve with batch techniques. Although microfluidic syntheses are less suitable for large-scale production due to their small channel sizes and thus the small quantities they can produce, they are a promising tool for accelerating the clinical translation of NPs.[34] However, continuous techniques using millifluidic reactors have the capability to produce already considerable amounts, grams per day, of high-quality NPs[26] if operated over long periods (1 mL/min → ~1.5 L/day). Such techniques have also been employed reproducibly to scale up shape-controlled NP systems[31,32] and to trap kinetic products, such as nuclei and seed particles,[25] by precise control over the residence times. Continuous syntheses have also shown a high degree of reproducibility as important parameters such as reductant or coprecipitant and reactant mixing, temperatures and concentrations can be precisely controlled.

Continuous (aqueous) synthesis of IONPs via coprecipitation was demonstrated in millichannels and microchannels using coaxial flow,[35–37] including coaxial turbulent jet mixers[38] and droplets.[39–41] Also, continuously operated spinning disc reactors have been successfully applied for coprecipitation synthesis.[12,42] What these reactors, designs have in common is fast mixing ($t_{mix} \ll 1$ s), which is why they have been considered as beneficial for fast coprecipitation reaction. Lim et al.[38] have shown that the size of synthesized oxide NPs did not change significantly if mixing time was faster than a certain threshold using tetramethylammonium hydroxide as base.

Slightly different reactor designs have been presented for continuous hydrothermal synthesis, usually operating at elevated temperatures and pressures, sometimes even above its critical point, *i.e.* in supercritical water[43–46] (mostly α-Fe_2O_3 was generated initially, which can be transformed to Fe_3O_4 at high temperature under a reducing medium), also including stabilization[47] and biofunctionalization.[48]

Organic synthesis of iron oxide NPs by thermal decomposition utilizing continuous techniques is rare due to the challenging temperature requirements and gas formation of most precursors with decomposition, *e.g.* $Fe(acac)_3$ and $Fe(CO)_5$. Recently, a high-temperature reactor was reported by Jiao et al.[49] for the flow synthesis of PEGylated Fe_3O_4 NPs in a Hastelloy tubes reactor operated at 33 bar to avoid gas formation during $Fe(acac)_3$ decomposition. Glasgow et al.[50] presented a similar reactor design operated at atmospheric pressures (at the outlet) using iron oleate as precursor, in a similar procedure to the work of Park et al.[14]

6.2.2.1 Current Limitations of Continuous Techniques

There are still limitations in the current use of continuous systems, as the reactions are normally carried out in a single step and are not replicating the complexity that is seen in batch systems and carried out by synthetic chemists. For an entirely continuous production of MNPs for clinical applications, future advances in postsynthesis processing steps are needed.

Beyond the NP synthesis, several post-synthesis steps for ligand exchange, ligand coupling or ligand attachment are required.[51–55] Some of the current difficulties to make these post-preparative steps continuous are the process parameters used for the surface modification step. Ligand exchange, addition or shelling steps are usually performed within hours or even days,[56] Frequently, also multiple prepreparative steps are a prerequisite for surface modification.[57]

Purification of MNPs in the laboratory normally uses centrifugation or magnetic separation. Even magnetic separation, which is one of the least intensive methods, often requires multiple cleaning cycles[58] or additional dialysis procedures[56] to remove excess surfactants. For the 'bare' NPs formed via coprecipitation, multiple cleaning steps and pH neutralization are used before the ligands are added to functionalize the particles (see Section 6.3.1). To create a fully continuous process, these steps would also have to be implemented continuously. Magnetic separation has been successfully established in microfluidics for the detection and/or separation of molecules, cells, droplets etc., involving magnetic particles.[59,60] However, reports on the continuous separation of MNPs for manufacturing are rare, but the very recent literature seems to address this gap in continuous processing.[60–64]

Another severe problem, especially for continuous NP synthesis, is fouling. Although flow reactors benefit from their high surface-to-volume ratio as described above, their relatively high surface area poses problems, which can usually be neglected when using batch reactors. Not only that interactions with the reactor walls do affect a larger fraction of the processed material, but also fouling can change the process conditions, *e.g.* reactor volume, pressure gradients, residence times, and heat transfer rates. This is not in keeping with the concept of scale-up via continuous production at constant conditions. Severe fouling, *e.g.* NPs that stick to and grow on the reactor walls during the course of the reaction[1] in the product, limits the ability for flow reactors to be run continuously and leads to contamination, increased size polydispersity, lowered yields and even reactor blockages. However, several reactor designs have been presented that avoid or reduce particle-wall contacts,[65] including the coaxial flow geometry where the NPs are formed at a liquid–liquid interface, away from the capillary walls[35] or in water droplets suspended in an oil.[40]

Another limitation is that while running continuously, flow reactors can exhibit large differences in residence time due to the laminar flow pattern. This can lead to noticeably different growth histories among synthesized particles.[1] One possibility to overcome both obstacles is segmented flow by introducing a secondary carrier phase, such as an immiscible phase that allows the formation of droplets, slugs or plugs and wets the surface of the reactor.[1,26,66] If such a segmented flow cannot be achieved, *e.g.* due to interactions between used liquids or temperature constraints, the usage of gas as the carrier phase can reduce the width of the residence time distribution.[1,67–69] The use of a gaseous carrier phase also minimizes complications that arise for gas release during reduction or precursor decomposition[1,70] and can be utilized as an oxidizing agent[71] or reducing agent.[31,72] Pressurized systems are also often used to keep gaseous products in solution, which are released during a reaction such as the thermal decomposition of $Fe(acac)_3$ and $Fe(CO)_5$.[25,31,32]

Another major challenge for continuous production of MNPs for clinical applications is quality control. It must be guaranteed that the produced NPs are within the strict bounds required for biomedical applications. Since most analyzers that are standardly used for quality control are not designed for continuous sampling procedures, special care has to be taken regarding the control strategy. This involves sampling procedures, as well as methods to discard products that are out of specification for the quality control measurements. Online quality control measurements suitable for continuous production are discussed in Section 6.3.3.2.

The design of a scalable process also includes the selection of the necessary ligands and quantifying the ligands attached to the NP and the stability of the NP dispersion. For continuous production, a biocompatible ligand must be chosen, as well as a chemical that can either be stored in bulk and utilized for the reaction or readily be prepared before the addition of the NP systems. Complex ligands will add additional cost and complexity[56] in any system that will be used industrially. The current stabilization and functionalization processes and characterization methods will be discussed in the next section.

6.3 STABILIZATION AND FUNCTIONALIZATION OF MNPs

IONPs have a wide range of biomedical applications, such as drug delivery, magnetic hyperthermia, immunoassays, diagnostics and environmental testing.[73–80] To efficiently utilize IONPs in biomedical applications, the NP size and shape must be tailored, but so does its surface chemistry. The surface ligands must be chosen carefully as they are critical to ensure the stability and biocompatibility of NPs and can affect their physical properties. The impact that the surface functionalization will have on the nanomaterials and the cell–NP interactions must also be fully understood, thus allowing researchers in this field the ability to critically assess which methods they may consider investigating. To understand surface chemistry and implement it in a scalable way, the interaction between the ligands and the iron oxide must be understood in terms of how readily and tightly they will bind to the NP surface. Once the initial ligand shell has been designed, further functionalization can be carried out to allow the NPs to be a platform to tailor the necessary biofunctionality. The final product must be precisely characterized for the inorganic and the organic components of the functionalized NPs.

Scalability via continuous techniques has been applied in limited examples for surface functionalization and biofunctionality. Continuous techniques offer the same unique advantages detailed previously to the functionalization and biofunctionalization processes, including increased reaction selectivity, shorter reaction times due to faster and more homogeneous mixing rates, better temperature control and online monitoring. An example of surface functionalization of IONPs with microreactors is the work of Abou-Hassan et al.,[81] where they grafted (3-aminopropyl)-triethoxysilane (APTES) to the surface of IONPs continuously. The particles were then further modified in a continuous stream to add a silica shell loaded with fluorescent molecules by reaction with the APTES surface.[81]

6.3.1 Stabilization of MNPs

As mentioned previously, IONPs synthesized via coprecipitation (in aqueous solutions) are stabilized by a negative surface charge in basic solutions. In these cases, ligand addition is needed to form a stable dispersion in a wide range of solvents, salt concentrations and pH values. In the case of ligand addition, the ligand must replace the current counter ion, such as hydroxyl anions or carbonate groups.[57,82]

Alternatively, IONPs are often synthesized in organic media, as this allows for better control of their morphology and often leads to higher crystallinity.[83] In this case, MNPs must be rendered water dispersible through a post-synthesis step: phase transfer or ligand exchange. The chosen ligand should have a higher affinity for the NP surface and confer the desired properties to the NPs. Otherwise, this may lead to a mixed organic shell (mixture of the hydrophobic and hydrophilic ligands), which could impact the colloidal stability of NPs.[84] In most cases, an excess of the new ligand allows for the complete replacement of the original ligand. There are two strategies, which can be explored for the ligand exchange step. The first one involves the removal of the original ligand to obtain bare NPs, which are precipitated before being put in the presence of the new hydrophilic ligand. The main obstacle is the precipitation of bare hydrophobic NPs, which may lead to their irreversible aggregation and impossibility to disperse again. The other strategy is direct exchange of ligands in biphasic dispersions. This relies on choosing a ligand of high enough affinity to replace a labile hydrophobic ligand (for example oleic acid or oleylamine). The method of surface functionalization and the ligand chosen should aim to preserve the morphology of the MNPs synthesized, allow their dispersion and long term stabilization in aqueous solution, while allowing them to preserve their properties for their intended biomedical use. For example, a consequence of ligand exchange of IONPs is their surface oxidation, which may be associated with a decrease in their magnetic properties. Palma et al.[85] observed a core size reduction (4% with citric acid and 10% for 2,3-dimercaptosuccinic acid [DMSA]) of IONPs after ligand exchange of oleic acid coated IONPs as well as a reduction in the saturation magnetization (39% with citric acid and 30% for DMSA). The change in the magnetic properties is related to the removal of surface iron atoms during the ligand exchange process, particle dissolution, oxidation of the particles as well as higher surface magnetic disorder.[86] The strong affinity between certain ligands and the surface of metal oxide NPs, alongside harsh reaction conditions such as long reaction times and high reaction temperatures, may contribute to this surface oxidation of MNPs.

Smolensky and colleagues[87] determined that for an identical polyethylene glycol (PEG), magnetite NPs functionalized with a biphasic protocol possessed better relaxivity values in comparison to those having undergone a complete removal of the initial ligand. In this same study, the authors demonstrated that the magnetic properties of the IONPs were increased when functionalizing the NPs with catechol-derived ligands in comparison to those containing carboxylic groups. More recently, Walter et al.[88] investigated the functionalization of IONPs with dendronized PEGs by a direct ligand exchange reaction (both the hydrophobic MNPs and ligand were dispersed in tetrahydrofuran) or by a ligand exchange and phase transfer reaction (hydrophobic MNPs were dispersed in hexane, whereas the dendrons were dispersed in a mixture of water and methanol). The authors of this study found that while both strategies allowed for the grafting of dendrons to the surface of the IONPs, they did not lead to identical results. Indeed, independently of the ligand, electrostatic interactions during the ligand exchange and phase transfer allowed for stronger anchoring of the ligand, but this method also led to larger NPs, which rapidly precipitated. Ligand exchange remains an intense area of research, and novel methods are being investigated to overcome its limitations. For example, oleic acid functionalized IONPs were transferred into a hydrophilic solution by multistep exchange with commercially available Tiron (4,5-dihydroxy-1,3-benzenedisulfonic acid disodium salt monohydrate), as shown in Figure 6.2.[89] In order to overcome the solvent incompatibility between these two ligands, dopamine was used as an intermediate substitution ligand. While NPs obtained are water dispersible, traces of residual dopamine persisted.[89] A recent study involving the use of milling, a mechanochemical approach, has allowed the authors to overcome the problem of solvent incompatibility

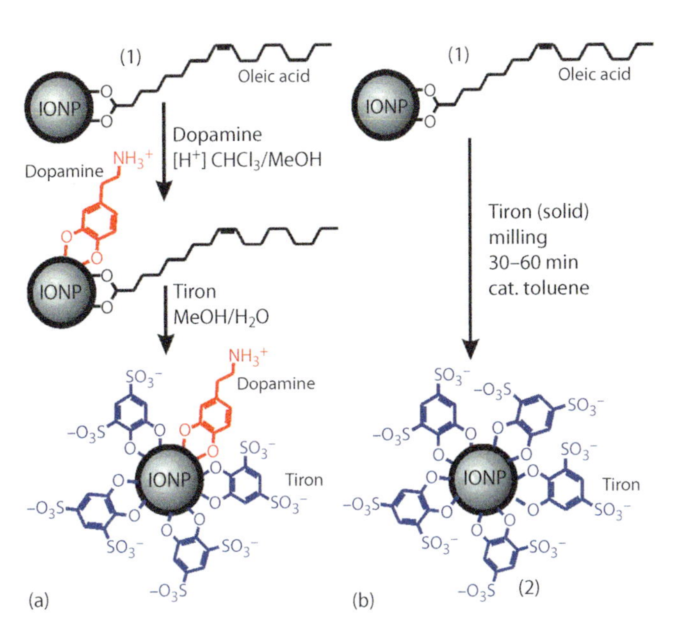

FIGURE 6.2 Synthesis of water-soluble IONPs by (a) two-step ligand exchange in solution, leading to exchange of oleic acid ligands and (b) herein demonstrated one-step mechanochemical solvent-free process.[90] (Korpany, K. V., Mottillo, C., Bachelder, J., Cross, S. N., Dong, P., Trudel, S., Friščić, T., Blum, A. S., 2016. One-step ligand exchange and switching from hydrophobic to water-stable hydrophilic superparamagnetic iron oxide nanoparticles by mechanochemical milling. *Chemical Communications* 52, 3054–3057. Reproduced by permission of The Royal Society of Chemistry.)

during ligand exchange while leading to stable and water dispersible IONPs.[90]

The functional groups with the most affinity towards metal oxide NPs include carboxylic acids, phosphonate or catechol-based ligands.[91–94] Adsorption of carboxylic acids at the surface of metal oxide NPs was determined to occur through coordination of the carboxylate group to the metal atoms at the surface. The carboxylate will act as a ligand for vacant coordination sites of surface metal centres such as Fe^{3+} cations.[95] Deacon and Philips[96] studied the carboxylate–metal coordination modes, and these can be categorized as either monodentate or bidentate. In the case of a bidentate coordination, chelating or bridging modes can be observed. As it can be seen in Figure 6.3, these monodentate and bidentate modes can also be found with phosphonate groups on the surface of IONPs.[47] In addition, phosphonate groups present a significant advantage as they can adsorb in a tridentate mode, which is more stable than the previous two modes mentioned, and this can enhance the colloidal stability of IONPs.[97]

Daou and colleagues[98,99] have compared the surface functionalization of IONPs with phosphonate-derived ligands and carboxylate-derived ligands. These studies demonstrated that the phosphonate ligand had a stronger interaction with the MNP surface than the carboxylate did, but most importantly, it had a higher surface coverage and effectively behaved as an oxidation inhibitor.

Catechol derivatives are excellent ligands for stabilizing iron oxide nanomaterials due to their high affinity to metal

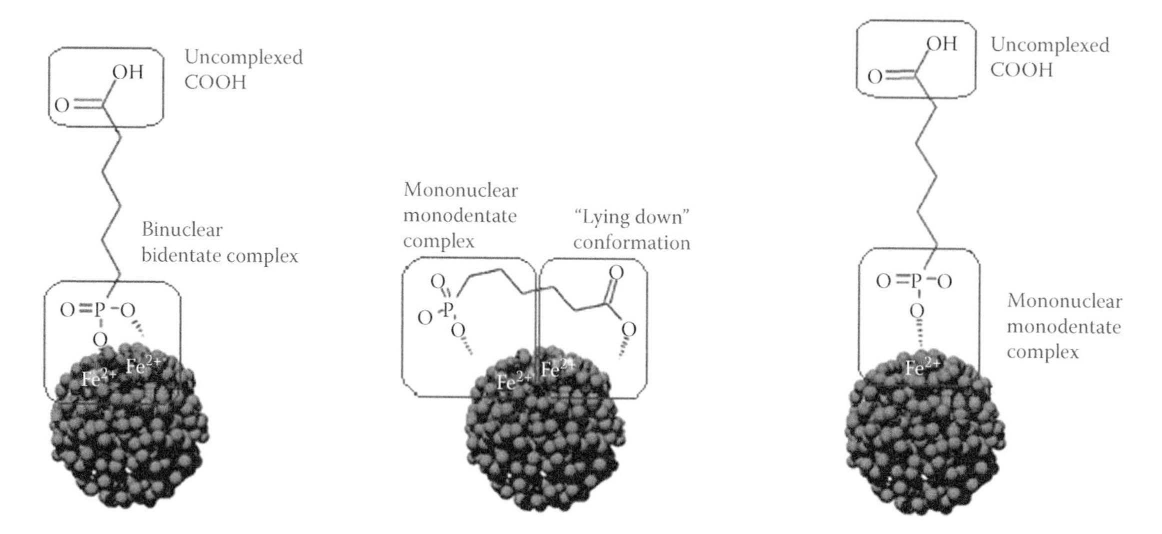

FIGURE 6.3 The most likely conformations of the phosphate-based molecule at the NP surface.[47] (Thomas, G., Demoisson, F., Boudon, J., Millot, N., 2016. Efficient functionalization of magnetite nanoparticles with phosphonate using a one-step continuous hydrothermal process. *Dalton Transactions* 45, 10821–10829. Reproduced by permission of The Royal Society of Chemistry.)

oxide surfaces through the chelation of the two adjacent hydroxyl groups of the enediol group on the benzyl ring.[100,101] However, dopamine has been reported to decompose during functionalization upon reacting with Fe^{3+} through a radical mechanism,[102] which also causes the etching of the IONPs[103] leading to a significant decrease in the core size.[87] A strategy to avoid etching of the IONPs while functionalizing with catechol derivatives is to conjugate the benzyl ring with electron withdrawing substituents, as in the case of nitro-dopamine or caffeic acid.[91] The conjugation of substituents to the catechol derivatives allows the binding affinity of the chelating hydroxyl groups to be optimized to promote colloidal stability.[104]

Polymers such as dextran or other carbohydrates have been used to stabilize and functionalize IONPs approved by the Federal Drug Administration for magnetic resonance imaging (MRI), such as Ferucarbotran or Ferumoxides.[105,106] To increase the stability of IONPs, cross-linkable polymers such as dextran has been extensively used,[107–109] but the use of ammonia during the cross-linking step requires thorough purification to ensure that this will not impact interactions with biological systems.

To date, some of the most common polymers used for the surface modification of IONPs include PEG, PAA or natural polymers such as polysaccharides or polypeptides.[110] The polymer of choice can be functionalized before it is linked to the surface of the IONPs, or it may be linked onto a functional group already present on the surface of the IONP.[111]

One advantage of using a polymer over a lower-molecular-weight monofunctional ligand is their multifunctionality. For example, Huang et al.[112] used a cysteine-terminated PEG polymer, which increased its anchoring onto the surface of IONPs and provided stability through coordination with the carboxylic groups and cross-linking of the polymer by oxidation of the thiol groups into disulfide bridges.

The extensively studied carbodiimide-mediated reaction (Section 5.2.3 of this book) has also been used to graft polymers to the surface of IONPs.[51,113,114] When comparing the carbodiimide reaction *versus* the 'click' alkyne-azide chemistry, recent studies have proven that the latter strategy can lead to a twofold to fivefold increase in the grafting yield of a polymer on the surface of an IONP.[115,116] The average number of cyclic

TABLE 6.1
Average Number of cRGD or cRGD-PEG per NP, Yield Coupling and Colloidal Behaviour (Hydrodynamic Diameter and Surface Charge) Obtained by Click and Carbodiimide Chemistry

	RGD	Grafted Molecule Number	%	D_h (nm)	PDI	Zeta (mV)
Carbodiimide	cRGDfK	6 ± 1	1 ± 0.2	10	0.3	−36
	cRGDfK-PEG-NH_2	12 ± 1	2 ± 0.2	16	0.2	−39
Click	cRGDfK(N_3)	36 ± 1	8 ± 1	15	0.3	−39
	cRGDfK-PEG-N_3	50 ± 1	11 ± 1	19	0.3	−43

Source: Reproduced with permission from Bolley, J., Guenin, E., Lievre, N., Lecouvey, M., Soussan, M., Lalatonne, Y., Motte, L., 2013. Carbodiimide versus click chemistry for nanoparticle surface functionalization: a comparative study for the elaboration of multimodal superparamagnetic nanoparticles targeting αvβ3 integrins. *Langmuir* 29, 14639–14647. Copyright 2013 American Chemical Society.

arginylglycylaspartic acid (cRGD) or cRGD-PEG per NP, yield coupling, colloidal behaviour and polydispersity index (PDI) obtained by both chemical reactions are reported in Table 6.1. Most importantly, this was shown consistently to be specific to the chemical reaction itself independent of the choice of ligand, and in turn this will lead to an increase in binding of IONPs to targeted cells by the ability to graft more targeting peptides on the surface of the NPs.

6.3.2 Biological Functionalization

NPs have been developed to diagnose and treat diseases such as tumours in a personalized manner. This occurs by linking biomolecules to the surface of NPs as targeting moieties to allow the NPs to accumulate in desired sites of the body and/or interact with specific cells. In the case of cancer, two main strategies have been used to specifically address NPs to tumour cells: passive and active targeting.[117] These are illustrated in Figure 6.4.

Passive targeting relies on the enhanced permeability and retention (EPR) effect. In order to proliferate rapidly, tumour cells stimulate the production of novel blood vessels through various biomolecules such as growth factors. These vessels will differ by their anatomy, with large fenestrations between the endothelial cells, allowing the accumulation of various molecules or NPs. Furthermore, tumour tissues usually lack lymphatic drainage; thus, these macromolecules or NPs may remain in these tissues by passive targeting. On the other hand, active targeting of NPs occurs via their surface functionalization with a ligand of choice that will specifically recognize a receptor on a targeted cell. Angiogenesis, the formation of new blood vessels, requires migration and adhesion of endothelial cells on the extracellular matrix (ECM). This will involve adhesion molecules such as integrins, which are heterodimeric transmembrane cell surface receptors, to participate in cell–cell and cell–ECM interactions. They link cells to their surroundings and to proteins of the ECM (collagen, laminin, fibronectin etc.).

The tripeptide RGD (arginin–glycin–aspartate) is present in numerous proteins of the ECM and is a ligand of integrins ($\alpha_v\beta_3$).[118,119] The affinity between these two molecules is related to the protein's conformation and has been the main strategy explored for the active targeting of MNPs to cancer cells. The integrin $\alpha_v\beta_3$ has an extracellular V-type structure in which each subunit has a 'closed headpiece bent' conformation. This structure corresponds to a weak affinity state. The fixation of a ligand, such as the RGD peptide, will lead

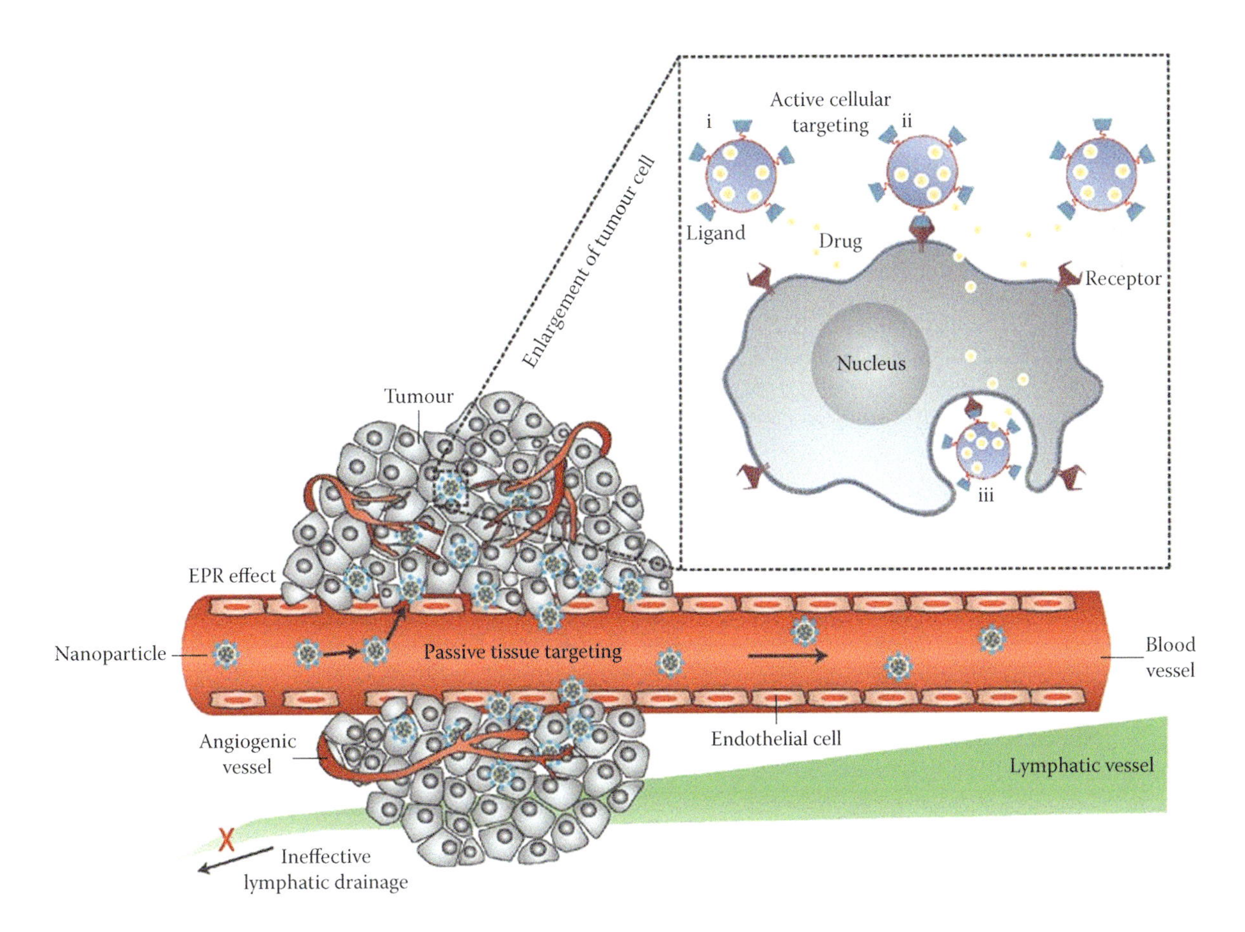

FIGURE 6.4 Polymeric NPs are shown as representative nanocarriers (circles). Passive tissue targeting is achieved by extravasation of NPs through increased permeability of the tumour vasculature and ineffective lymphatic drainage (EPR effect). Active cellular targeting (inset) can be achieved by functionalizing the surface of NPs with ligands that promote cell-specific recognition and binding.[117] (Reprinted by permission from Macmillan Publishers Ltd. *Nature Nanotechnology*. Peer, D., Karp, J. M., Hong, S., Farokhzad, O. C., Margalit, R., Langer, R., 2007. Nanocarriers as an emerging platform for cancer therapy. *Nature Nanotechnology* 2, 751–760. Copyright 2007.)

to a conformational change in which the affinity is increased. Cyclic peptides are more often used than the linear equivalent since the latter was shown, in clinical trials, to accumulate strongly in the liver and not in the desired tumour cells.[120] Also, the linear form of the peptide can have conformations with different affinities for the integrin $\alpha_v\beta_3$ and are more sensitive to proteolysis.

The team of Coll and Dumy was able to target the integrin $\alpha_v\beta_3$ with RAFT(c-[RGDfk])$_4$, the target peptide.[121] The interest of polymers like the RAFT peptide is the possibility to present multiple copies of the RGD peptide. It has a better affinity towards the integrin $\alpha_v\beta_3$ (known to prefer multivalent interactions), improves internalization in cells and confers better signal-to-noise ratio *in vivo*.

In principle, targeting cancer cells by targeting a surface marker they specifically express seems straightforward. However, this strategy must overcome several hurdles before it can be considered within *in vitro* or *in vivo* conditions. Fundamental properties such as the size, shape and surface functionalization of NPs will impact their ability to target specific cells.[122] In a recent study by Ndong et al.,[123] grafting of Herceptin monoclonal antibody on 30 nm and 100 nm aminodextran-coated IONPs was engineered to allow the recognition of the ErbB2 receptor overexpressed on human breast cancer cells. In comparison to small NPs (30 nm), larger NPs (100 nm) allowed for the grafting of multiple ligands to the NP surface and with a higher binding affinity to cells *in vitro*. This study also demonstrated that these results were not predictive of the *in vivo* behaviour of the MNPs. *In vivo*, the smaller 30 nm IONPs showed statistically more significant tumour concentration. This was certainly caused by the accelerated blood clearance of larger IONPs, which did not allow these 100 nm IONPs to reach the tumour tissue.[124] In addition, the shape of the NPs will influence their targeting ability. Several studies have demonstrated that relative to targeted spherical NPs, rod-shaped NPs have a longer blood circulation time and enhanced cell binding properties.[125–128]

Active targeting of MNPs towards specific populations of cells by conjugation of a ligand has been shown to result only in a minor increase in selective uptake.[129] Wilhelm and colleagues[130] recently published a survey which claims that, from the scientific literature published in the last 10 years, only 0.7% (median) of the administered dose of actively targeted NPs is delivered to the solid tumour. A major obstacle to overcome remains the nonspecific uptake by other cell populations such as macrophages of the reticuloendothelial system (RES). For example, several recent studies using IONPs functionalized with the RGD peptide found that 35% of the injected dose accumulated in the liver after *in vivo* administration.[131,132] Schleich et al. demonstrated that RGD functionalized IONPs did accumulate 2.5 times more than nonfunctionalized IONPs in tumour tissues, but this represented only 0.35% of the injected dose. This study concluded that a double targeting strategy using the RGD peptide and magnetic targeting allowed for significant uptake in tumour tissues. Magnetic targeting allows for the accumulation of IONPs in tumour tissues by EPR effect combined with a magnetic force, while the RGD peptide leads to an increase in the uptake of IONPs in cells by receptor-mediated endocytosis. *In vivo*, the EPR effect improves the accumulation of NPs in cancerous tissues; however, this effect is limited when NPs are not actively targeted towards these cells through specific receptor binding.[133,134] Furthermore, MNPs likely accumulate in leaky tumour vasculature but are internalized by the first cells they encounter. This limits their potential penetration depth and their interaction with cancer cells targeted. For example, depending on the nature and stage of the tumour, its vasculature may contain pores that are 50 to 500 nm in size.[135] As mentioned earlier, the size of the NPs therefore also has an impact on their diffusion within tumour tissues and, hence, their ability to target tumour cells. Recently, in order to overcome this lack of penetration depth, Setyawati et al.[136] demonstrated nanodiamond-induced surface-dependent vascular barrier leaking. This effect was mediated by the nanodiamonds causing an increase of intracellular reactive oxygen species and Ca^{2+} ion release. In turn, this allowed for better penetration of doxorubicin (DOX) drug, which could therefore reach subjacent cancer cells.

An additional obstacle towards efficient active targeting of MNPs is the protein corona. Cell–NP interactions are strongly dependent on the nature and amount of serum proteins adsorbed on the surface of the MNPs.[137–139] This was demonstrated by Mirshafiee et al.[140] with fluorescent silica NPs functionalized with bicyclononyne reacting with azide moieties grafted on a silicon substrate. The targeting efficiency dropped by 94% in 10% serum containing media and by 99% in 100% serum containing media, in comparison to pristine NPs. Salvati et al.[141] also found that silica NPs stabilized with PEG and functionalized with transferrin lost their targeting ability when placed in a complex biological environment due to the adsorption of proteins and formation of protein corona. The latter was illustrated by the authors in Figure 6.5.

Dai et al.[142] found that PEG backfilling on gold NPs reduced the number of serum proteins that bind to the surface of NP. In turn, this avoids serum proteins from blocking the binding of the functionalized NPs to the target. A study using polystyrene NPs coated with PEG of different lengths and with different densities demonstrated that these different parameters of the surface ligands influenced the uptake of NPs by human macrophages, as Figure 6.6 shows.[143] In Figure 6.6, R_F/D represents the grafting density, where D is defined as the grafting distance (the distance between two closest neighbouring PEG anchors) and R_F is the Flory radius of the PEG coils in the solution, which is directly dependent on the PEG molecular weight. Optimizing the length, distribution and density of the PEG coating on the NPs helped to minimize their nonspecific uptake by macrophages and also determines the nature of the serum proteins adsorbed and so their potential targeting ability.[144] A recent study illustrates the need to control the parameters of surface ligands for active targeting of NPs. Indeed, dendron micelles functionalized with folic acid (FA) could see their interactions with targeted cells vary from minimal to a 25-fold enhancement depending on the PEG density close

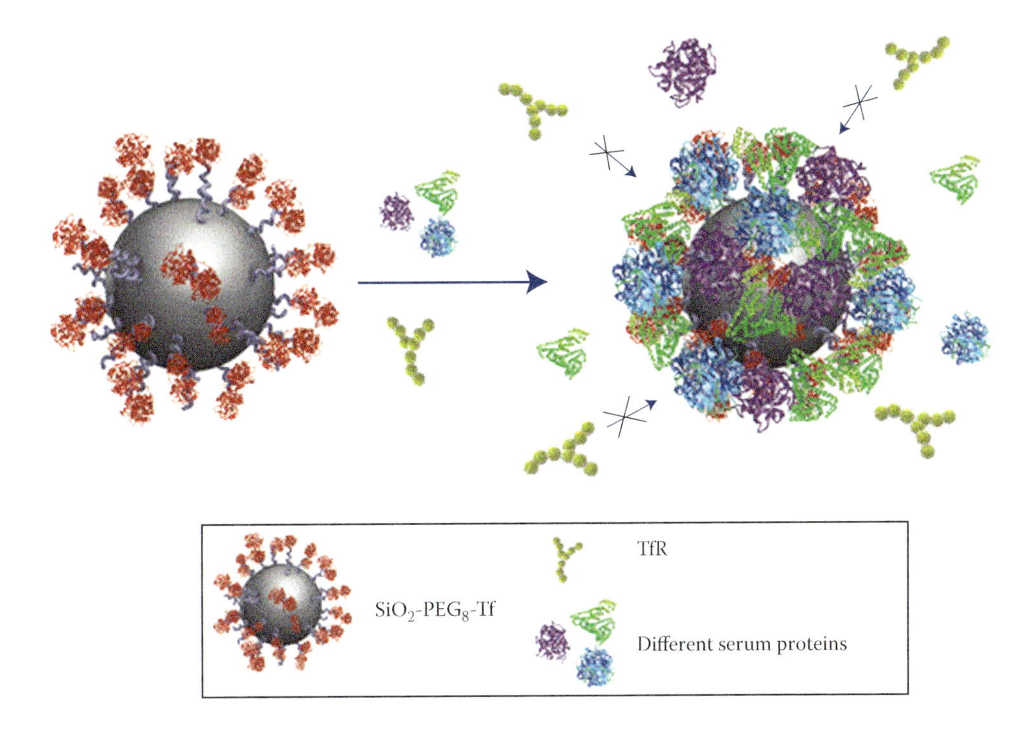

FIGURE 6.5 Schematic representation of blocked Tf–TfR interaction in the presence of foetal bovine serum proteins.[141] (Reprinted by permission from Macmillan Publishers Ltd. *Nature Nanotechnology*. Salvati, A., Pitek, A. S., Monopoli, M. P., Prapainop, K., Bombelli, F. B., Hristov, D. R., Kelly, P. M., Aberg, C., Mahon, E., Dawson, K. A., 2013. Transferrin-functionalized nanoparticles lose their targeting capabilities when a biomolecule corona adsorbs on the surface. *Nature Nanotechnology* 8, 137–143. Copyright 2013.)

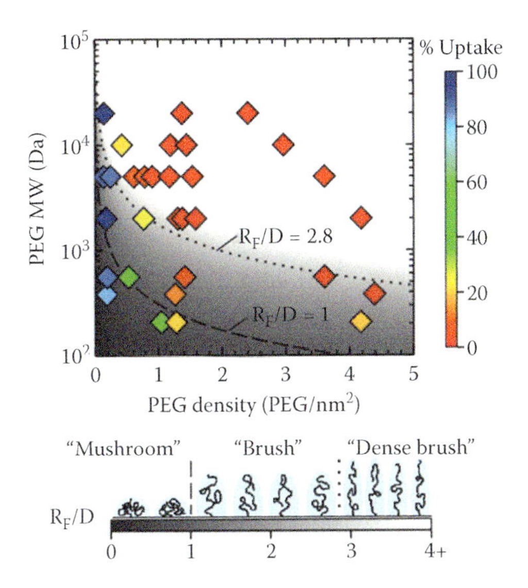

FIGURE 6.6 Phase diagram mapping particle uptake by differentiated THP-1 cells at 4 h as a function of PEG length (MW) and coating density (PEG groups/nm^2). The gray shading represents the various R_F/D values; the transitions between the mushroom–brush and brush–dense brush conformations are indicated by the dashed (R_F/D = 1.0) and dotted (R_F/D = 2.8) lines, respectively. All data represent at least n = 3 independent experiments performed in triplicate.[143] (Reproduced with permission from Yang, Q., Jones, S. W., Parker, C. L., Zamboni, W. C., Bear, J. E., Lai, S. K., 2014. Evading Immune cell uptake and clearance requires PEG grafting at densities substantially exceeding the minimum for brush conformation. *Molecular Pharmaceutics* 11, 1250–1258. Copyright 2014 American Chemical Society.)

to the FA, the surface area of FA accessible to the solvent or the FA mobility.[145]

With folate conjugated IONPs, although their targeting efficiency was not studied, Chen et al.[146] were able to demonstrate that folate binding protein and albumin, which are present in blood serum, caused a cascade of biological events that led to NP agglomeration, thus potentially hindering their targeting ability. Also, despite the adsorption of proteins on the surface of NPs, the latter may retain or even increase their targeting ability if they are large enough and can present multivalent interactions with the targeted receptor.[147]

The effect of the protein corona on the targeting ability of NPs is dependent upon the NP type and the surface functionalization of these NPs. To our knowledge, studies reporting the impact of the protein corona on targeted MNPs remain scarce, but these will be necessary when designing MNPs intended for such applications *in vivo*. Indeed, the physicochemical properties of IONPs will lead to the formation of diverse protein corona, which must be controlled in order to avoid any undesirable uptake into the body (for example in the brain tissue), potentially leading to cytotoxity.[148] This highlights the importance of certain parameters to develop targeted MNPs in biological systems, such as NP morphology, or systemic differences due to diseases, which can affect cell surface markers or cell internalization mechanisms.

Furthermore, the experimental conditions in which the NPs are functionalized or in which they will be used for clinical applications have an impact. The functionalization strategies, which may be used with antibodies or other biomolecules, could lead to nonspecific binding as well as an excess of ligand not bound to the NP in a covalent manner. One can enhance the targeting ability of the nanocarrier by choosing the appropriate grafting method.[149] The experimental conditions influence the NPs and their interactions with cells, as the nature of the protein corona around IONPs has been shown by Liu et al.[150] to be modified in the presence of a static magnetic field during incubation with cells. This had a significant effect on the uptake of IONPs by both normal (3T3 cells) and tumour (HepG2 cells) cell lines, as well as their cytotoxic effects. This is especially relevant when considering potential biomedical applications of IONPs such as MRI or magnetic hyperthermia, which require the application of a magnetic field.

New strategies have been developed to target NPs to cancer cells and to overcome the drawbacks mentioned in this chapter. Recent studies have focused on targeting matrix metalloproteases or various other proteinases that may help with the diagnosis of cancer.[151] Indeed, new nanoparticulate imaging probes, activatable or 'smart' probes, which may be switched from the 'off' to 'on' state upon interaction with a target enzyme, are particularly attractive due to their improved sensitivity and specificity.[152] This principle can also be considered for other biological models where the expression of specific proteins or molecules is triggered in specific conditions. For example, Ye and colleagues[153] developed a redox activated gadolinium-based MRI probe. In the case of malignant and drug-resistant tumours, higher glutathione (GSH) cellular levels would allow for the disulfide reduction to activate these probes leading to their self-assembly through macrocyclization, which in turn leads to an enhanced local r_1 relaxivity by MRI. Gao et al.[154] developed PEG-maleimide coated IONPs labelled with ^{99m}Tc to allow for dual imaging of tumour cells by MRI and single-photon emission computed tomography (SPECT). The maleimide group was linked to a tumour targeting RGD sequence, which in turn was linked through a disulfide bond to a 'self-peptide,' which renders the NP stealth. Their response to the tumour microenvironment is illustrated in Figure 6.7. These NPs are responsive to the high intracellular glutathione (GSH) concentration, within the tumour microenvironment, which induces reduction of the disulfide bonds within their surface coating, leading in turn to high NP uptake through recognition of the RGD sequence by $\alpha_v\beta_3$ integrins. The remaining thiol groups allow for interparticle interactions with maleimide groups present on other IONPs.

The design of actively targeted MNPs is complex and must tackle the biological obstacles that it faces: various cancer types, cancer stage progression, variations of the EPR effect within one tumour and individual expression of the disease, which varies from patient to patient. More recently, research has intensified in the use of NPs to label and/or track specific cell populations in a 'smart' manner and not solely using active targeting of NPs. This can be used to gain a better understanding of the role or localization of a specific cell type in the case of a disease or stem cell (SC) therapy for example.

A relatively simple approach relies on the fact that a nanostructure's morphology may help to address it to a specific cell population. For example, Yi et al.[155] recently found that with identical surface chemistry, polymersomes, in comparison to micelles and filomicelles, could be associated to a higher uptake by dendritic cells and a lower uptake by macrophages of the liver. The benefit of these nanostructures for therapeutic or diagnostic applications in cardiovascular diseases such as

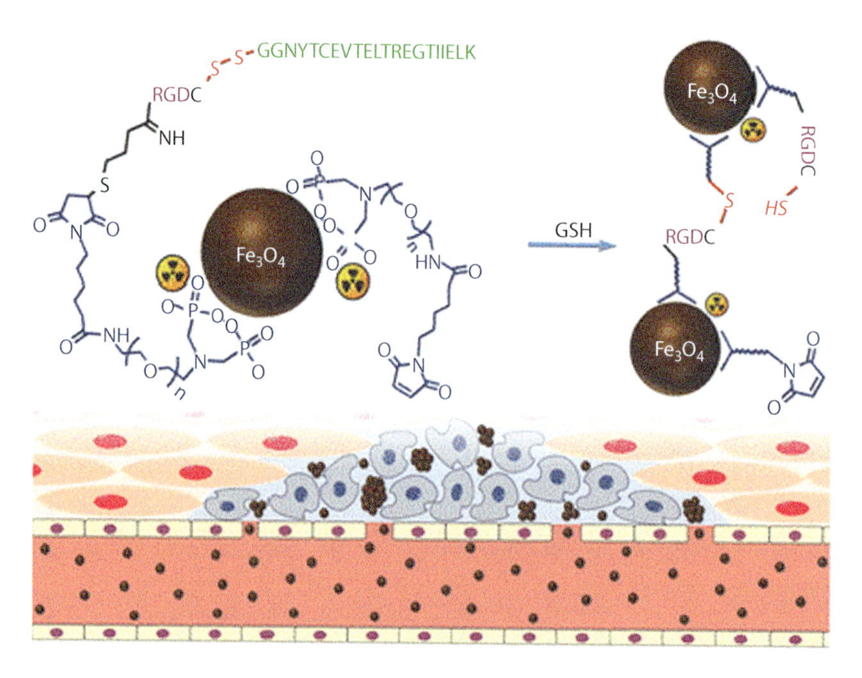

FIGURE 6.7 Schematic drawing of the antiphagocytosis ^{99m}Tc-labeled Fe_3O_4 NPs as bimodal SPECT/MRI imaging probes and their responsiveness to intracellular GSH-triggering within tumour microenvironment for forming particle aggregates through interparticle crosslinking reaction.[154] (Reproduced with permission from Gao, Z., Hou, Y., Zeng, J., Chen, L., Liu, C., Yang, W., Gao, M., 2017. Tumour microenvironment-triggered aggregation of antiphagocytosis 99mTc-labeled Fe_3O_4 nanoprobes for enhanced. *Advanced Materials*. Copyright 2017 John Wiley & Sons.)

atherosclerosis was further confirmed in this study when the authors demonstrated that polymersomes could target dendritic cells both in the spleen and aorta, as well as Ly6C+ monocytes in the blood that accelerate the progression of cardiovascular diseases. Another recent study demonstrated that ferumoxytol, in the case of neuroinflammation, accumulated in astrocyte end-feets surrounding cerebral vessels, astrocyte processes and CD163+/CD68+ macrophages, but not in tumour cells. Alongside MRI, this may reveal useful as a diagnostics tool and a means to guide biopsies in the central nervous system.[156]

While targeting of cancer cells by NPs remains predominant, SCs also represent novel targets of interest, whether it is for labelling purposes during cellular therapy or detection of cancer SCs.[157,158] As discussed earlier, the use of 'smart' NPs may reveal important information regarding SCs used for regenerative medicine purposes. For example, SC apoptosis in a matrix-associated SC implant was detected in a noninvasive manner by MRI using a caspase-3-sensitive nanoaggregation MRI probe (C-SNAM).[159] This could be a potentially useful tool for *in vivo* monitoring of the viability of transplanted SCs and facilitating the development of successful cartilage regeneration techniques or various SC therapies. This platform could represent an innovative tool in the diagnosis of SC failure at an early stage, thus minimizing any health consequences on patients. This approach has also been investigated for monitoring of drug-induced apoptosis using a caspase-3/7 activatable Gd-based MRI probe in mice.[160] Upon reduction and caspase-3/7 activation, the probe undergoes intramolecular cyclization and self-assembly into gadolinium NPs. This will lead to an increase in the r_1 relaxivity in chemotherapy-induced apoptotic cells and tumours that express active caspase-3/7, which can be visualized by MRI. The same principle was also used with IONPs developed to undergo intracellular covalent aggregation in apoptotic cells and tumours, which in turn would allow for enhanced T_2 MRI.[161] As illustrated in Figure 6.8, these MNPs were functionalized with a small peptide [(Ac-Asp-Glu-Val-Asp-Cys(StBu)-Lys-cyanobenzothiazole (CBT)], which could undergo Casp3/7-controlled condensation. Briefly, these MNPs were designed so that after internalization by Casp3/7-activated cells, the disulfide bonds from cysteine were reduced by intracellular GSH, and the peptides were cleaved by the enzymes, thus exposing reactive 1,2-aminothiol groups.[161] Then, the condensation reaction between the free 1,2-aminothiol groups and the cyano groups of the CBT motifs led to cross-linked Fe_3O_4 NP aggregates, which resulted in enhanced T_2 magnetic resonance (MR) signal.[161]

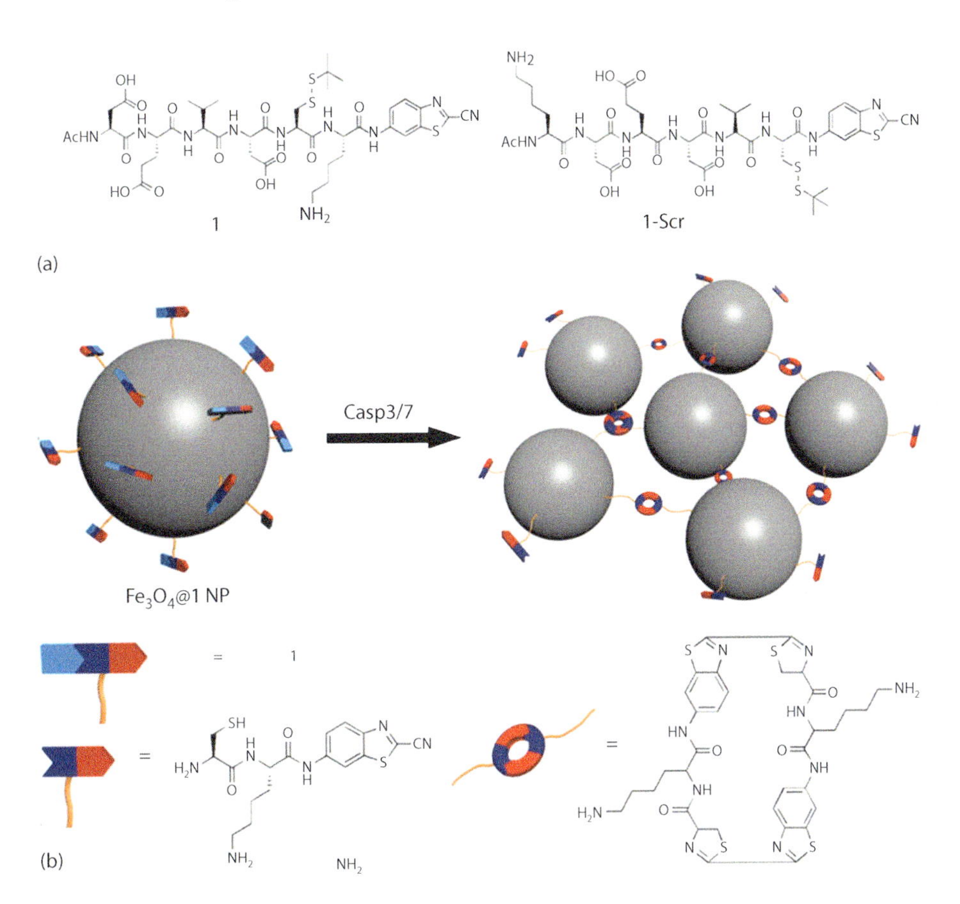

FIGURE 6.8 Chemical structures and schematic illustration of Fe_3O_4 NPs aggregation. (a) Chemical structures of 1 and 1-Scr with (1) Ac-Asp-Glu-Val-Asp-Cys(StBu)-Lys-CBT, which could undergo Casp3/7-controlled condensation, and a control compound (1-Scr) Ac-Lys-Asp-Glu-Asp-Val-Cys(StBu)-CBT (1-Scr) (b) Schematic illustration of intracellular Casp3/7 instructed aggregation of Fe_3O_4@1 NPs.[161] (Reproduced with permission from Yuan, Y., Ding, Z., Qian, J., Zhang, J., Xu, J., Dong, X., Han, T., Ge, S., Luo, Y., Wang, Y., Zhong, K., Liang, G., 2016. Casp3/7-instructed intracellular aggregation of Fe_3O_4 nanoparticles enhances T_2 MR imaging of tumour apoptosis. *Nanoletters*. 16, 2686–2691. Copyright 2016 American Chemical Society. Reprinted with permission.)

The principle of biological targeting of MNPs towards specific cells, especially those expressing disease markers on their surface, is simple. However, research on this matter has intensified but has yet to demonstrate substantial evidence that these targeted MNPs are more efficient in reaching these targeted cells in comparison to nontargeted MNPs. MNPs have been designed to target several diseases, but the most prominent disease focus has been in cancer. Recent developments include the ability to obtain a signal from MNPs after it is induced from specific cellular environments such as the rich GSH intracellular environment in tumour cells or apoptotic cells. From a clinical perspective, the low efficiency of these targeted MNPs implies that the required amount of MNPs to be injected into humans would be high. This would increase the cost and the risk of cytotoxicity due to the larger amounts of the MNP solution administered per dose. We have established that there are several biological barriers that the MNPs will need to face before reaching their targets, and their surface functionalization will play a key role in their interactions with the diverse biological systems they will interact with.

6.3.2.1 Impact of Functionalization on MNPs and Cell–NP Interactions

Determining the effect of NPs on cells is a critical parameter, which determines how the IONPs should be functionalized, as well as whether these will be suitable for *in vivo* or clinical applications. As we have previously mentioned, phosphonates are particularly interesting for the stabilization of IONPs as they allow for enhanced colloidal stability of these MNPs. Several varieties of ligands using this functional group have been studied, such as dendrons or derivatives of PEG for example.[92,94,162–166] Lam et al.[167] demonstrated that the design of these ligands can have an impact on the properties of the MNPs when comparing IONPs obtained after ligand exchange with tetra(ethylene glycol) (TEG)-based phosphonate ligands. At physiological conditions of 0.15 M NaCl, all the conjugated IONPs were found to have good colloidal stability; however, IONPs functionalized with phosphonate–TEG–Me are larger and more influenced by the presence of electrolyte than are those containing phosphonate–TEG–OH. The end group, methyl *versus* hydroxyl, in this case was found to influence the overall colloidal properties of the MNPs. The anchoring of the ligand on the IONP surface also impacts the MNP and the cell–NP interactions. When comparing mono-, bis-, and tri-phosphonate PEG derivatives, Lam et al.[97] found that tri-phosphonate PEG allowed for enhanced magnetic properties of IONPs, enhanced colloidal stability and reduced cellular toxicity through reduced cellular uptake. It was hypothesized that the para-phosphonate arm on the tris-phosphonate ligand chelates differently, perhaps in a tripodal manner, onto the surface of the MNPs, giving rise to increased chelation in comparison to the mono- and bis-functionalized MNPs. When considering IONPs for biomedical applications, such as potential MRI contrast agents, the surface ligand also influences the properties of these IONPs. In the case of potential MRI contrast agents for example, the ability of the ligand to hinder diffusion of water molecules close to the MNP may have an impact on the relaxivity properties.[168] Therefore, carefully choosing which functional groups will anchor directly to the surface of the MNP and also those that will interact with the surrounding environment is critical.

PEG is currently studied extensively to render the surface of the NPs stealth.[169,170] However, whereas previous publications declared PEG to be a nontoxic organic coating for NPs, recent studies demonstrate that the cytotoxicity of PEG-coated MNPs is far more complex and several parameters have an impact. For example, in a recent study by Pisciotti et al.,[171] PEG-coated IONPs were found to have no toxic effects up to concentrations of MNPs of 100 μg/mL, while dextran-coated IONPs presented low cytotoxicity up to 400 μg/mL. The cytotoxicity profile of PEG-coated MNPs will be dependent upon the core size, the cell line considered, PEG coating density, PEG functional groups and PEG molecular weight.[172–175] Overall, a reduced cellular uptake is correlated to the reduced cytotoxic profile of PEG-coated MNPs.

The most important factor to determine the fate of the NPs, once they are in a biological medium, is the formation of the protein corona. The biocompatibility of IONPs relates to the adsorption of proteins on their surface and is influenced by several physicochemical characteristics. For example, a recent study determined that poly(vinyl alcohol)-coated IONPs with negative and neutral surface charge adsorbed more serum proteins, which extended their blood circulation times.[138] In particular, four specific serum proteins were bound with high affinity to negatively charged IONPs. This difference in the composition of the protein corona could decrease the cellular uptake of the IONPs by the RES, which may prolong their blood circulation times.[138] Cedervall et al.[176] were able to quantify the amount of proteins adsorbed on the surface of IONPs as a function of their hydrophobicity. The more hydrophobic a polymer is, the more human serum albumin protein was found to adsorb on the surface of the NPs. More recently, Ashby and his team found that for identical 10 nm IONPs, the degree of hydrophobicity of the NP surface changed the composition of the protein corona.[139] The more hydrophobic the surface of the NP was, the higher quantity of attracted hydrophobic proteins, which formed a more dynamic protein corona with proteins having fast exchange rates. In turn, this had an impact on cell–NP interactions as these quickly exchanging proteins could reach binding equilibrium more rapidly, thus rapidly masking the surface of the NPs upon entry into the biological medium, allowing for their reduced cytotoxic profile and elongated blood circulation times.

An intense area of research includes multimodal applications of polymer-coated NPs coated in areas such as drug delivery, magnetic hyperthermia and MRI. The use of pH- and thermo-responsive polymers can be used for controlled release of a pharmaceutical drug in response to an acidic tumour environment and stimuli by oscillating magnetic field for magnetic hyperthermia applications. A recent study conducted by Hervault et al.[177] showed drug release from IONPs of up to 85.2% of DOX obtained after 48 h at acidic tumour pH under hyperthermia conditions (50°C). The choice of polymer is crucial to optimize the drug loading and release capacity of MNPs. As demonstrated by Quinto et al.,[178] IONPs functionalized with phospholipid–PEG of different lengths led to different DOX loading capacities. Indeed, the drug loading capacity of IONPs with PEG 2000 was significantly higher (30.8% ± 2.2% w/w [DOX/iron]) than those with PEG 5000 (14.5% ± 3.3%). The difference in loading capacity was attributed to the combination of electrostatic and hydrophilic/hydrophobic interactions between the DOX and the IONP coating layer.

6.3.3 Methods of Characterization for Particles and Surface Functionalization

Finally, to understand the ligand addition, exchange and functionalization steps discussed earlier, quantitative and qualitative characterization techniques have to be implemented to characterize the surface chemistry of the NPs. This must be addressed before any clinical application of MNPs may be considered so as to define the new organic coating in terms of the nature and amount of ligands present. We will discuss the techniques that may be used to characterize the NPs before and after surface functionalization and the limitations that these methods may present. We will pay particular attention to methods that can be implemented in line with continuous methods.

6.3.3.1 Particle Size and Hydrodynamic Radius Measurements

Direct measurements of the particle morphology (size and shape) with and without the ligands can be useful for determining ligand coating and ligand exchange. Transmission electron microscopy (TEM) provides information on MNP size, shape, crystallinity and composition through the passing of a focused electron beam through an extremely thin sample. Spectroscopic techniques such as electron energy loss spectroscopy (EELS) and energy dispersive X-ray spectroscopy (EDX) can also give atomic and chemical information. Imaging with electron microscopy techniques like TEM generally gives information on the NP core, as the carbon ligands are much lighter in contrast and often break down in a high-energy environment.[179] Cryo-TEM and liquid cell TEM can be used to image NPs in their liquid environment,[180,181] which is particularly important for studying hydroxide intermediates that can be unstable upon drying.[180,182] Direct measurements of the organic shell can be taken in some cases by proper staining of the organic coating of MNPs, which are otherwise not visible by TEM. Such techniques, also called negative staining, require the use of heavy metal atoms or ions that will provide contrast.[183] These include for example phosphotungstic acid, uranyl acetate or osmium tetroxide.[183]

Small-angle X-ray scattering (SAXS) provides a powerful tool to characterize NPs and even core–shell structures if the difference in electron density between the core, the shell and the solution is sufficient.[184–186] However, the difference in electron density is usually insufficient to study the functionalization of MNPs, which is why SAXS provides mostly information on the particle core geometries and interparticle distances. Still, the structure factor (*i.e.* the contribution of the signal originating from interparticle distance) contains information if the interaction between particles is random (in this case, the structure factor tends towards unity with dilution), attractive or repulsive interaction, as in the case of electrostatic stabilization[187] to gain information on the interaction length between NPs in superlattices.[185,188]

Small-angle neutron scattering (SANS) does allow studying organic coatings, since scattering happens from the nature of the nucleus (*e.g.* number of neutrons, magnetic moments). Therefore, SANS is more sensitive to light elements, providing a higher contrast between the ligands and the solvent. This contrast can be tuned further by using a mixture of heavy and light water as the solvent.[189]

To directly investigate the solvated NP size (hydrodynamic radius), techniques such as dynamic light scattering (DLS), analytical ultracentrifugation (AUC) and differential centrifugal sedimentation (DCS) can be used. These techniques can also be compared to TEM or SAXS to determine the radius of the inorganic part of the NPs.

DLS is used to determine the hydrodynamic diameter of NPs in dispersed state, and the same equipment may be used to determine the sign and amplitude of the surface charges of NPs dispersed in an aqueous or at least polar solution by zeta potential measurements.[190] Fluctuations in the scattered light intensity, due to the Brownian motion of the NPs in suspension, are used to determine the particle diffusion coefficient which is then correlated to the hydrodynamic radius via the Stokes Einstein equation.[191] This method allows for a simple and practical method of characterization of dispersed MNPs across various time points, a wide range of salt or pH conditions or even for MNPs in a cell culture medium. It may provide information on the successful stabilization of MNPs after a ligand exchange reaction or information about the agglomeration of IONPs in high salt conditions.[167] For example, Davis et al.[58] measured the hydrodynamic diameter of oleic-acid-coated IONPs before and after ligand exchange with various small ligands and noted a decrease in the diameter for some of the samples. This seemed to also indicate the removal of excess oleic acid, as later confirmed by thermogravimetric analysis (TGA). Comparative studies of the hydrodynamic radius of MNPs can be carried out with DLS before and after the conjugation of biomolecules on the surface of MNPs; for example Xu et al.[192] measured an increase in the hydrodynamic diameter of 20 nm after bioconjugation of antibodies. However, this method has been shown to be dependent on certain parameters such as the concentration of NP solution, scattering angle and shape anisotropy of NPs.[193] Also, the scattering intensity is proportional to the *sixth power* of the particle diameter, meaning that wide NP–bioconjugate size distributions can mask the presence of smaller conjugates in the sample. The scattering intensity of larger aggregates, particles and coated species can also mask smaller particles in the solution, as well as bimodal populations of NP species in solution.[194] Finally, one may note that DLS can be implemented for *in-situ* analysis during the application of an alternating magnetic field using a VASCO Flex™ remote-head

DLS instrument developed by Cordouan Technologies, as shown by Hemery et al.[195] on MNPs coated by a thermosensitive polymer shell.

To have a fuller view of particles in colloidal solution, separation techniques, such as centrifugation, are coupled to the light scattering technique to differentiate all components in the dispersion. DCS has been used to study the change in the size of gold NPs by studying their sedimentation properties with different ligands and then using the density to determine the change in NP size based on the ligand shell.[196] This technique can study small changes in ligand size with high accuracies.[196,197] AUC also gives the hydrodynamic radius of particles but can be used for much smaller particle or cluster sizes.[25,198]

6.3.3.2 Chemical Measurements

To determine chemical information about the surfactant shell of IONPs, infrared (IR) spectroscopy and TGA is routinely used, while X-ray photo-electron spectroscopy (XPS), EELS, EDX and radiolabeling can also be used with some success. Unfortunately, nuclear magnetic resonance (NMR) spectroscopy cannot be readily used to determine the ligand sphere, due to the magnetic nature of iron atoms,[199] unless using the magic angle spinning technique as for solid-state NMR.[200]

Evaluation of the grafting of ligands on the surface of IONPs can be determined by Fourier transform IR spectroscopy (FTIR), which is routinely used in the mid-IR region (400 to 4000 cm^{-1}), as this provides information on the rotational–vibrational structure of the surface of MNPs. For example, with the regular use of carboxylic-acid-based ligands, by determining the difference between the asymmetric and symmetric stretching modes of the coordinated carboxylate functionality, information can be obtained as to whether or not the COOH group is linked by chemisorption to the NPs.[201,202] This is also applicable to ligands containing phosphonate groups. Indeed, the free ligand will be characterised by the P–O stretching region (1200–900 cm^{-1}), in which two sharp peaks at 1153 and 982 cm^{-1} are assigned to P–O and P–OH, respectively.[94] A broad band around 1073 cm^{-1} was attributed the vibrational mode for the PO_3 group.[203] Upon binding to IONPs, the changes within the P–O stretching region (1200–900 cm^{-1}) demonstrate a strong interaction between the phosphonate head-group and the IONP surface, and these suggest that the Fe atoms within the IONP surface are coordinated by oxygen atoms from the phosphonate groups.[203] In FTIR, the consideration of the Fe-O band is also a good method to evaluate the influence of the surface functionalization reaction on the oxidation of IONPs. Indeed, the core of magnetite will be surrounded by an oxidized layer due to the fact that surface Fe^{2+} cations are very sensitive to oxidation.[204]

Also, when studying iron-oxide-based MNPs, the Fe-O peak of magnetite is found around 570 cm^{-1}, while maghemite will have several peaks between 400 and 800 cm^{-1}, with the number peaks increasing with the structural ordering of its vacancies.[99,205] For example, FTIR was used to characterize IONPs before and after ligand exchange with dendrons.[168] Figure 6.9 compares the Fe-O bands of IONPs before and after dendronization.[168] NS10 is an oleic-acid-coated IONP and displays a single broad band located at 580–590 cm^{-1} and small shoulders between 800 and 600 cm^{-1} attributed to slight surface oxidation. After functionalization, these new peaks tend to demonstrate that the IONPs are characterized by a higher degree of oxidation.

An advantage of FTIR is that it can be utilized in-line in continuous flow reactors to analyze chemical compounds during the reaction.[206] This opens up the ability to use the technique for quality control and continuous monitoring of the ligand addition, exchange or biofunctionalization steps.

Another technique that can be used to determine the elements present on the surface of an NP is XPS. This technique is sensitive to the binding states of the elements, which can be used to study the binding between the ligands and the NP.[25,47,199,207] It has also been employed to study the density of the surfactants bound to an NP surface.[208] Spectroscopic

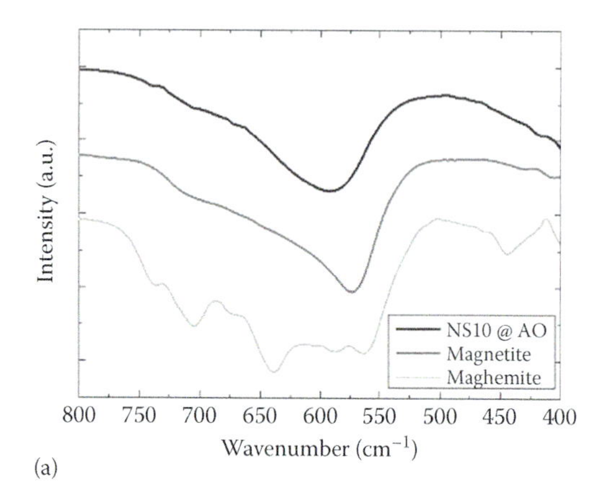

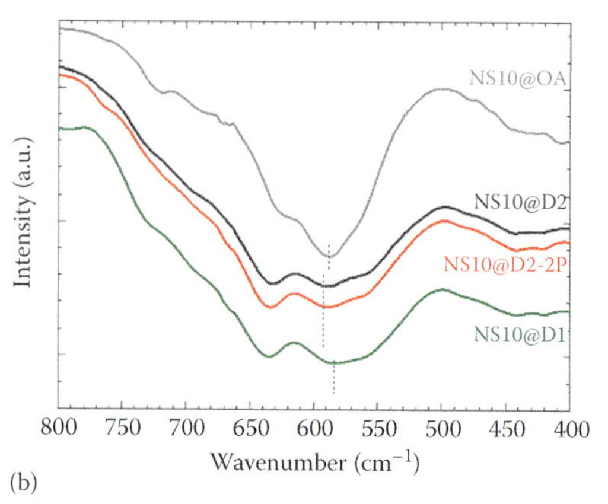

FIGURE 6.9 IR spectra between 800 and 400 cm^{-1} corresponding to the Fe-O bands: (a) NS10@OA before functionalization compared to maghemite and magnetite spectra. (b) NS10 before and after functionalization with dendrons D1, D2, and D2–2P.[168] (Reproduced with permission from Walter, A., Parat, A., Garofalo, A., Laurent, S., Elst, L. V., Muller, R. N., Wu, T., Heuillard, E., Robinet, E., Meyer, F., Felder-Flesch, D., Begin-Colin, S., 2015. Modulation of relaxivity, suspension stability, and biodistribution of dendronized iron oxide nanoparticles as a function of the organic shell design. *Particle & Particle Systems Characterization*, 32, 552–560. Copyright 2015 John Wiley & Sons)

techniques that are attached to an electron microscope such EELS or EDX can also provide information on the elements present with detailed spatial resolution. An example of this is using EDX to study the ratio of iron to the sulphur and phosphorus present in the capping ligands.[209]

For TGA, samples are heated to elevated temperatures under a flow gas (air or nitrogen) while the mass of the sample is monitored, which yields the decomposition curve. This technique is often used to determine the percentage of organic material and gives information on the boiling points or decomposition temperatures of the organic surface ligands.[210–212] For example, in the case of phosphonate-derived PEG functionalization of IONPs, TGA was carried out to assess the ligand exchange reaction.[167] In the TGA measurement, any weight loss prior to 110°C is generally attributed to adsorbed water and tends to be minimal.[167] Functionalized IONPs showed the largest overall weight loss, with a weight decrease of 14% between 200°C and 550°C.[167] This was attributed by Lam et al.[167] to the decomposition of the organic ligand at elevated temperatures. The second weight loss of 11% resulting from rupture between the P–O bonds of phosphonic acid and Fe^{3+} cations on the surface of NPs occurred between 650°C and 850°C. Organic loss at these elevated temperatures has also been observed when peptide dendrimers or other ligands such as polymers were used to functionalize IONPs and is generally correlated to the breaking of bonds between the MNPs and their ligands.[213–215] However, it is difficult to interpret this with certitude as oxidation of nanomaterials may also occur at high temperatures.[216] Also, while quantifying the amount of organic ligand lost, the assumption is that only one type of ligand is present on the surface of the NP, but this has not been proven. Indeed, it is necessary to perform extensive quantitative and qualitative analysis of the functionalized MNPs to determine their surface composition.

To address this problem, in 2014, Davis and colleagues used radiolabeled oleic acid to qualitatively and quantitatively assess the ligand exchange of IONPs by PEG with various terminal groups.[58] This research confirmed hierarchal binding of functional groups on the surface of magnetite NPs, with catechol-derived groups and phosphonate groups having higher binding affinity than amine- and carboxylate-derived PEG ligands, as confirmed by other studies reported in the literature.[217] As Figure 6.10 illustrates, this study was able to demonstrate that even with high-affinity ligands, such as catechol-derived groups, oleic acid still remained on the surface of the NPs after ligand exchange.[58]

In a more recent study by the same team, they confirmed the higher affinity of catechol-derived ligands for IONPs in comparison to other functional groups.[218] They also demonstrated that bifunctional ligands coated the NPs in a more efficient manner than monofunctional ligands did. Also, the conditions in which the ligand exchange reactions are done (*i.e.* pH, temperature, stirring/agitation, time etc.) may influence the yield of this reaction.[167,219]

In order to completely remove any layer of oleic acid on IONPs, Bixner and colleagues[220] found that physically adsorbed species could be removed by repeated extraction in boiling solvents to promote dissociation of carboxylate dimers or by adding a countercharged ligand to prevent reabsorption by ion pair formation. By FTIR and TGA, the authors studied complete removal of oleic acid, which was achieved only by post-coating NPs in a large excess of nitrocatechol-derived ligands.

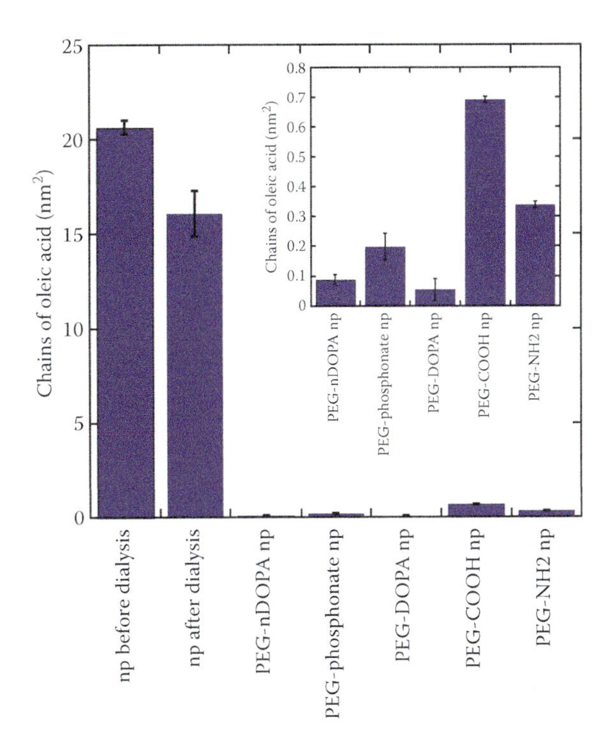

FIGURE 6.10 Chains of oleic acid remaining on the surface of the NPs before and after ligand exchange determined by liquid scintillation counting. The data for the PEG-coated NPs represent the amount of oleic acid remaining after dialysis.[218] (Reproduced with permission from Davis, K., Cole, B., Ghelardini, M., Powell, B. A., Mefford, O. T., 2016. Quantitative measurement of ligand exchange with small-molecule ligands on iron oxide nanoparticles via radioanalytical techniques. *Langmuir* 32, 13716–13727. Copyright 2014 American Chemical Society.)

6.3.3.3 Continuous Characterization Techniques

Continuous characterization of NPs and their surface properties is of great value not only for high throughput analysis, process development and optimization but also for continuous manufacturing with a focus on large-scale production. Since continuous flow reactors process the reactants within small channels (<1 cm) spectroscopic techniques with a relatively low penetration depth (*e.g.* IR, near-IR, Raman or UV/VIS techniques) can be used to provide information from 100% of the produced material if performed at sufficient frequencies. Furthermore, process errors can be detected immediately, allowing a minimum amount of precious material to be discarded. Combining multiple online techniques and statistical analysis allows for further improvements in error detection.[221–223]

NP core properties are most commonly characterized via TEM, X-Ray diffraction, superconducting quantum interference device (SQUID), and SAXS. Unfortunately, most of these techniques are very unlikely to be applied in a production environment. X-ray techniques such as XRD and SAXS can be used *in situ* or continuously at synchrotron facilities.[184] Laboratory SAXS equipment (using line focus for increased signal strength due to simultaneous irradiation of more sample material) has also been applied successfully for online NP characterization.[186,224]

Although DLS was not designed to facilitate continuous measurements of NP size (related to the hydrodynamic diameter), special modifications of the equipment could allow continuous data acquisition.[225,226] Even if not, DLS, and also DCS, can be used for prompt and frequent analysis during continuous production since typical measurement times are less than a minute.

Still, online process characterization of the NP core properties will need to be implemented in the future. Very interesting current attempts address online magnetic measurements of superparamagnetic particles utilizing AC magnetometry/susceptibility[227,228] or the Néel effect.[229,230]

As discussed previously, common characterization strategies for surface functionalization involve spectroscopic techniques such as IR or Raman spectroscopy. Especially the first became standard in recent years for flow chemistry applications due to the rapid development in analytical equipment in the last decade.[206,231–234] IR and Raman were also shown to be able to identify different IONP phases.[11,180] Still, combined functionalization and spectroscopic characterization of MNPs in flow are rare.

6.4 CONCLUSIONS AND OUTLOOK

The use of IONPs in (pre)clinical applications creates a multifaceted problem that must be considered from synthesis to functionalization and biofunctionalization. From a chemistry perspective, these steps need to be understood and controlled while considering the whole process, from synthesis to patient, holistically and increasingly in a scalable and tailorable manner. The synthetic process needs to be able to produce a range of surface functionalities designed for individual applications of the MNPs, such as cell targeting and drug delivery. The most promising method for doing this is using continuous techniques, which can use multiple reactor setups in series, to create a scalable and flexible platform for creating NPs for biomedical applications. These methods are becoming increasingly popular and sophisticated. In the coming years, it is expected that complicated multistep procedures will be automated to create high-value products of any scale needed.

From a developmental perspective, each component needs to be considered due to its utility in the application or its toxicity to the human body. As we have demonstrated throughout this chapter, the surface functionalization of MNPs is a very complex matter. The moment NPs are administered *in vivo*, they are faced with several biological barriers, and their interaction with each of them is controlled by a large number of variables such as the NP core composition and size, the chemical structure of surface ligands and the density of surface ligands for example. The biocompatibility and biodistribution of MNPs are governed by the nature and amount of protein corona formation in biological environments, which in turn depends on the physicochemical properties of MNPs. Using several complementary characterization techniques is required to obtain quantitative and qualitative information on the properties of the functionalized MNPs.

Also, while active targeting has been studied in order to improve the biological distribution of MNPs, these targeted MNPs have been shown to reach their target, but only in an incremental improvement in comparison to nontargeted MNPs.[235] Developing targeted MNPs requires careful consideration regarding the choice of the ligand, the chemical reaction, ensuring a selective display of the ligand on the MNP, studying the behaviour of MNPs *in vitro* and *in vivo* as these differ and determining if the strategy is reproducible and scalable. Nowadays, 'smart' or 'activatable' MNPs are being developed as the information they provide is more precise and dependent upon the biological environment targeted. Future studies will aim to ensure that these 'smart' NPs are reversibly switched from the 'off' to 'on' state upon interaction with a biological target of choice. As suggested by Lazarovits et al.,[236] a database will be needed for NP–serum protein interactions in order to predict their biological function based on nanoparticle design.

ACKNOWLEDGEMENTS

The authors thank EPSRC for funding.

REFERENCES

1. Niu, G., Ruditskiy, A., Vara, M., Xia, Y., 2015. Toward continuous and scalable production of colloidal nanocrystals by switching from batch to droplet reactors. *Chemical Society Reviews* 44, 5806–5820.
2. Zhang, L., Xia, Y., 2014. Scaling up the production of colloidal nanocrystals: Should we increase or decrease the reaction volume? *Advanced Materials* 26, 2600–2606.
3. Thanh, N. T. K., Green, L. A. W., 2010. Functionalisation of nanoparticles for biomedical applications. *Nano Today* 5, 213–230.
4. Boles, M. A., Ling, D., Hyeon, T., Talapin, D. V., 2016. The surface science of nanocrystals. *Nature Materials* 15, 141–153.
5. Bear, J. C., Yu, B., Blanco-Andujar, C., McNaughter, P. D., Southern, P., Mafina, M.-K., Pankhurst, Q. A., Parkin, I. P., 2014. A low cost synthesis method for functionalised iron oxide nanoparticles for magnetic hyperthermia from readily available materials. *Faraday Discussions* 175, 83–95.
6. Che Rose, L., Bear, J. C., McNaughter, P. D., Southern, P., Piggott, R. B., Parkin, I. P., Qi, S., Mayes, A. G., 2016. A SPION-eicosane protective coating for water soluble capsules: Evidence for on-demand drug release triggered by magnetic hyperthermia. *Scientific Reports* 6, 20271.
7. Mertz, D., Sandre, O., Begin-Colin, S., 2017. Drug releasing nanoplatforms activated by alternating magnetic fields. *Biochimica et Biophysica Acta-General Subjects* 1861, 1617–1641.
8. Kolen'ko, Y. V., Bañobre-López, M., Rodríguez-Abreu, C., Carbó-Argibay, E., Sailsman, A., Piñeiro-Redondo, Y., Cerqueira, M. F., Petrovykh, D. Y., Kovnir, K., Lebedev, O. I., Rivas, J., 2014. Large-scale synthesis of colloidal Fe_3O_4 nanoparticles exhibiting high heating efficiency in magnetic hyperthermia. *The Journal of Physical Chemistry C* 118, 8691–8701.
9. Vergés, M. A., Costo, R., Roca, A. G., Marco, J. F., Goya, G. F., Serna, C. J., Morales, M. P., 2008. Uniform and water stable magnetite nanoparticles with diameters around the monodomain–multidomain limit. *Journal of Physics D: Applied Physics* 41, 134003.

10. Blanco-Andujar, C., Ortega, D., Pankhurst, Q. A., Thanh, N. T. K., 2012. Elucidating the morphological and structural evolution of iron oxide nanoparticles formed by sodium carbonate in aqueous medium. *Journal of Materials Chemistry* 22, 12498–12506.
11. Ahn, T., Kim, J. H., Yang, H.-M., Lee, J. W., Kim, J.-D., 2012. Formation pathways of magnetite nanoparticles by coprecipitation method. *The Journal of Physical Chemistry C* 116, 6069–6076.
12. Haseidl, F., Müller, B., Hinrichsen, O., 2016. Continuous-flow synthesis and functionalization of magnetite: Intensified process for tailored nanoparticles. *Chemical Engineering & Technology* 39, 2051–2058.
13. Girod, M., Vogel, S., Szczerba, W., Thünemann, A. F., 2015. How temperature determines formation of maghemite nanoparticles. *Journal of Magnetism and Magnetic Materials* 380, 163–167.
14. Park, J., An, K., Hwang, Y., Park, J.-G., Noh, H.-J., Kim, J.-Y., Park, J.-H., Hwang, N.-M., Hyeon, T., 2004. Ultra-large-scale syntheses of monodisperse nanocrystals. *Nature Materials* 3, 891–895.
15. Kim, B. H., Lee, N., Kim, H., An, K., Park, Y. I., Choi, Y., Shin, K., Lee, Y., Kwon, S. G., Na, H. B., Park, J.-G., Ahn, T.-Y., Kim, Y.-W., Moon, W. K., Choi, S. H., Hyeon, T., 2011. Large-scale synthesis of uniform and extremely small-sized iron oxide nanoparticles for high-resolution T1 magnetic resonance imaging contrast agents. *Journal of the American Chemical Society* 133, 12624–12631.
16. Kwon, S. G., Piao, Y., Park, J., Angappane, S., Jo, Y., Hwang, N.-M., Park, J.-G., Hyeon, T., 2007. Kinetics of Monodisperse iron oxide nanocrystal formation by "heating-up" process. *Journal of the American Chemical Society* 129, 12571–12584.
17. Sun, S., Zeng, H., Robinson, D. B., Raoux, S., Rice, P. M., Wang, S. X., Li, G., 2004. Monodisperse MFe_2O_4 (M = Fe, Co, Mn) nanoparticles. *Journal of the American Chemical Society* 126, 273–279.
18. Zhao, C.-X., He, L., Qiao, S. Z., Middelberg, A. P. J., 2011. Nanoparticle synthesis in microreactors. *Chemical Engineering Science* 66, 1463–1479.
19. Marre, S., Jensen, K. F., 2010. Synthesis of micro and nanostructures in microfluidic systems. *Chemical Society Reviews* 39, 1183–1202.
20. Song, Y. J., Hormes, J., Kumar, C., 2008. Microfluidic synthesis of nanomaterials. *Small* 4, 698–711.
21. Makgwane, P. R., Ray, S. S., 2014. Synthesis of nanomaterials by continuous-flow microfluidics: A review. *Journal of Nanoscience and Nanotechnology* 14, 1338–1363.
22. Phillips, T. W., Lignos, I. G., Maceiczyk, R. M., deMello, A. J., deMello, J. C., 2014. Nanocrystal synthesis in microfluidic reactors: Where next? *Lab on a Chip* 14, 3172–3180.
23. Nightingale, A. M., deMello, J. C., 2013. Segmented flow reactors for nanocrystal synthesis. *Advanced Materials* 25, 1813–1821.
24. Elvira, K. S., i Solvas, X. C., Wootton, R. C. R., deMello, A. J., 2013. The past, present and potential for microfluidic reactor technology in chemical synthesis. *Nat Chem* 5, 905–915.
25. LaGrow, A. P., Besong, T. M. D., AlYami, N. M., Katsiev, K., Anjum, D. H., Abdelkader, A., Costa, P. M. F. J., Burlakov, V. M., Goriely, A., Bakr, O. M., 2017. Trapping shape-controlled nanoparticle nucleation and growth stages *via* continuous-flow chemistry. *Chemical Communications* 53, 2495–2498.
26. Mehenni, H., Sinatra, L., Mahfouz, R., Katsiev, K., Bakr, O. M., 2013. Rapid continuous flow synthesis of high-quality silver nanocubes and nanospheres. *RSC Advances* 3, 22397–22403.
27. Baber, R., Mazzei, L., Thanh, N. T. K., Gavriilidis, A., 2016. Synthesis of silver nanoparticles using a microfluidic impinging jet reactor. *Journal of Flow Chemistry* 6, 268–278.
28. Pan, J., El-Ballouli, A. a. O., Rollny, L., Voznyy, O., Burlakov, V. M., Goriely, A., Sargent, E. H., Bakr, O. M., 2013. Automated Synthesis of photovoltaic-quality colloidal quantum dots using separate nucleation and growth stages. *ACS Nano* 7, 10158–10166.
29. Besenhard, M. O., Neugebauer, P., Ho, C.-D., Khinast, J. G., 2015. Crystal size control in a continuous tubular crystallizer. *Crystal Growth & Design* 15, 1683–1691.
30. Besenhard, M. O., Hohl, R., Hodzic, A., Eder, R. J. P., Khinast, J. G., 2014. Modeling a seeded continuous crystallizer for the production of active pharmaceutical ingredients. *Crystal Research and Technology* 49, 92–108.
31. LaGrow, A. P., Knudsen, K. R., AlYami, N. M., Anjum, D. H., Bakr, O. M., 2015. Effect of precursor ligands and oxidation state in the synthesis of bimetallic nano-alloys. *Chemistry of Materials* 27, 4134–4141.
32. AlYami, N. M., LaGrow, A. P., Joya, K. S., Hwang, J., Katsiev, K., Anjum, D. H., Losovyj, Y., Sinatra, L., Kim, J. Y., Bakr, O. M., 2016. Tailoring ruthenium exposure to enhance the performance of fcc platinum@ruthenium core-shell electrocatalysts in the oxygen evolution reaction. *Physical Chemistry Chemical Physics* 18, 16169–16178.
33. du Toit, H., Macdonald, T. J., Huang, H., Parkin, I. P., Gavriilidis, A., 2017. Continuous flow synthesis of citrate capped gold nanoparticles using UV induced nucleation. *RSC Advances* 7, 9632–9638.
34. Valencia, P. M., Farokhzad, O. C., Karnik, R., Langer, R., 2012. Microfluidic technologies for accelerating the clinical translation of nanoparticles. *Nature Nanotechnology* 7, 623–629.
35. Abou Hassan, A., Sandre, O., Cabuil, V., Tabeling, P., 2008. Synthesis of iron oxide nanoparticles in a microfluidic device: Preliminary results in a coaxial flow millichannel. *Chemical Communications*, 1783–1785.
36. Abou-Hassan, A., Sandre, O., Cabuil, V., Microfluidic Synthesis of iron oxide and oxyhydroxide nanoparticles. In *Microfluidic Devices in Nanotechnology* (ed. C. S. Kumar), John Wiley & Sons, Inc.: 2010; pp. 323–360.
37. Simmons, M., Wiles, C., Rocher, V., Francesconi, M. G., Watts, P., 2013. The preparation of magnetic iron oxide nanoparticles in microreactors. *Journal of Flow Chemistry* 3, 7–10.
38. Lim, J.-M., Swami, A., Gilson, L. M., Chopra, S., Choi, S., Wu, J., Langer, R., Karnik, R., Farokhzad, O. C., 2014. Ultra-high throughput synthesis of nanoparticles with homogeneous size distribution using a coaxial turbulent jet mixer. *ACS Nano* 8, 6056–6065.
39. Chia-Hsien, Y., Yu-Cheng, L., 2013. Use of an adjustable microfluidic droplet generator to produce uniform emulsions with different concentrations. *Journal of Micromechanics and Microengineering* 23, 125025.
40. Frenz, L., El Harrak, A., Pauly, M., Bégin-Colin, S., Griffiths, A. D., Baret, J.-C., 2008. Droplet-based microreactors for the synthesis of magnetic iron oxide nanoparticles. *Angewandte Chemie International Edition* 47, 6817–6820.
41. Kumar, K., Nightingale, A. M., Krishnadasan, S. H., Kamaly, N., Wylenzinska-Arridge, M., Zeissler, K., Branford, W. R., Ware, E., deMello, A. J., deMello, J. C., 2012. Direct synthesis of dextran-coated superparamagnetic iron oxide nanoparticles in a capillary-based droplet reactor. *Journal of Materials Chemistry* 22, 4704–4708.

42. Chin, S. F., Iyer, K. S., Raston, C. L., Saunders, M., 2008. Size selective synthesis of superparamagnetic nanoparticles in thin fluids under continuous flow conditions. *Advanced Functional Materials* 18, 922–927.
43. Hao, Y., Teja, A. S., 2011. Continuous hydrothermal crystallization of α–Fe_2O_3 and Co_3O_4 nanoparticles. *Journal of Materials Research* 18, 415–422.
44. Sue, K., Sato, T., Kawasaki, S.-i., Takebayashi, Y., Yoda, S., Furuya, T., Hiaki, T., 2010. Continuous hydrothermal synthesis of Fe_2O_3 nanoparticles using a central collision-type micromixer for rapid and homogeneous nucleation at 673 K and 30 MPa. *Industrial & Engineering Chemistry Research* 49, 8841–8846.
45. Liang, M.-T., Wang, S.-H., Chang, Y.-L., Hsiang, H.-I., Huang, H.-J., Tsai, M.-H., Juan, W.-C., Lu, S.-F., 2010. Iron oxide synthesis using a continuous hydrothermal and solvothermal system. *Ceramics International* 36, 1131–1135.
46. Xu, C., Teja, A. S., 2008. Continuous hydrothermal synthesis of iron oxide and PVA-protected iron oxide nanoparticles. *The Journal of Supercritical Fluids* 44, 85–91.
47. Thomas, G., Demoisson, F., Boudon, J., Millot, N., 2016. Efficient functionalization of magnetite nanoparticles with phosphonate using a one-step continuous hydrothermal process. *Dalton Transactions* 45, 10821–10829.
48. Thomas, G., Demoisson, F., Chassagnon, R., Popova, E., Millot, N., 2016. One-step continuous synthesis of functionalized magnetite nanoflowers. *Nanotechnology* 27, 135604.
49. Jiao, M., Zeng, J., Jing, L., Liu, C., Gao, M., 2015. Flow synthesis of biocompatible Fe_3O_4 nanoparticles: Insight into the effects of residence time, fluid velocity, and tube reactor dimension on particle size distribution. *Chemistry of Materials* 27, 1299–1305.
50. Glasgow, W., Fellows, B., Qi, B., Darroudi, T., Kitchens, C., Ye, L. F., Crawford, T. M., Mefford, O. T., 2016. Continuous synthesis of iron oxide (Fe_3O_4) nanoparticles *via* thermal decomposition. *Particuology* 26, 47–53.
51. Wu, W., He, Q., Jiang, C., 2008. Magnetic iron oxide nanoparticles: Synthesis and surface functionalization strategies. *Nanoscale Research Letters* 3, 397.
52. Siah, W. R., LaGrow, A. P., Banholzer, M. J., Tilley, R. D., 2013. CdSe quantum dot growth on magnetic nickel nanoparticles. *Crystal Growth & Design* 13, 2486–2492.
53. Ling, D., Hyeon, T., 2013. Chemical design of biocompatible iron oxide nanoparticles for medical applications. *Small* 9, 1450–1466.
54. Lim, J., Majetich, S. A., 2013. Composite magnetic–plasmonic nanoparticles for biomedicine: Manipulation and imaging. *Nano Today* 8, 98–113.
55. McNaughter, P. D., Bear, J. C., Steytler, D. C., Mayes, A. G., Nann, T., 2011. A thin silica–polymer shell for functionalizing colloidal inorganic nanoparticles. *Angewandte Chemie International Edition* 50, 10384–10387.
56. Wei, H., Insin, N., Lee, J., Han, H.-S., Cordero, J. M., Liu, W., Bawendi, M. G., 2012. Compact zwitterion-coated iron oxide nanoparticles for biological applications. *Nano Letters* 12, 22–25.
57. Soares, P. I. P., Alves, A. M. R., Pereira, L. C. J., Coutinho, J. T., Ferreira, I. M. M., Novo, C. M. M., Borges, J., 2014. Effects of surfactants on the magnetic properties of iron oxide colloids. *Journal of Colloid and Interface Science* 419, 46–51.
58. Davis, K., Qi, B., Witmer, M., Kitchens, C. L., Powell, B. A., Mefford, O. T., 2014. Quantitative measurement of ligand exchange on iron oxides *via* radiolabeled oleic acid. *Langmuir* 30, 10918–10925.
59. Xia, N., Hunt, T. P., Mayers, B. T., Alsberg, E., Whitesides, G. M., Westervelt, R. M., Ingber, D. E., 2006. Combined microfluidic-micromagnetic separation of living cells in continuous flow. *Biomedical Microdevices* 8, 299.
60. Pamme, N., Manz, A., 2004. On-chip free-flow magnetophoresis: Continuous flow separation of magnetic particles and agglomerates. *Analytical Chemistry* 76, 7250–7256.
61. Chen, Q., Li, D., Lin, J., Wang, M., Xuan, X., 2017. Simultaneous separation and washing of nonmagnetic particles in an inertial ferrofluid/water coflow. *Analytical Chemistry* 89, 6915–6920.
62. Kumar, V., Rezai, P., 2017. Multiplex inertio-magnetic fractionation (MIMF) of magnetic and non-magnetic microparticles in a microfluidic device. *Microfluidics and Nanofluidics* 21, 83.
63. Zhou, Y., Song, L., Yu, L., Xuan, X., 2017. Inertially focused diamagnetic particle separation in ferrofluids. *Microfluidics and Nanofluidics* 21, 14.
64. Zhou, R., Bai, F., Wang, C., 2017. Magnetic separation of microparticles by shape. *Lab on a Chip* 17, 401–406.
65. Baber, R., Mazzei, L., Thanh, N. T. K., Gavriilidis, A., 2015. Synthesis of silver nanoparticles in a microfluidic coaxial flow reactor. *RSC Advances* 5, 95585–95591.
66. Khan, S. A., Jensen, K. F., 2007. Microfluidic synthesis of titania shells on colloidal silica. *Advanced Materials* 19, 2556–2560.
67. Knossalla, J., Mezzavilla, S., Schuth, F., 2016. Continuous synthesis of nanostructured silica based materials in a gas-liquid segmented flow tubular reactor. *New Journal of Chemistry* 40, 4361–4366.
68. Nightingale, A. M., Phillips, T. W., Bannock, J. H., de Mello, J. C., 2014. Controlled multistep synthesis in a three-phase droplet reactor. *Nature Communications* 5, 3777.
69. Angeli, P., Gavriilidis, A., Taylor flow in microchannels. In *Encyclopedia of Microfluidics and Nanofluidics*, Li, D., Ed. Springer US: Boston, MA, 2008; pp. 1971–1976.
70. Khan, S. A., Duraiswamy, S., 2012. Controlling bubbles using bubbles-microfluidic synthesis of ultra-small gold nanocrystals with gas-evolving reducing agents. *Lab on a Chip* 12, 1807–1812.
71. Sebastian, V., Basak, S., Jensen, K. F., 2016. Continuous synthesis of palladium nanorods in oxidative segmented flow. *Aiche Journal* 62, 373–380.
72. Sebastian, V., Jensen, K. F., 2016. Nanoengineering a library of metallic nanostructures using a single microfluidic reactor. *Nanoscale* 8, 15288–15295.
73. Hervault, A., Thanh, N. T. K., 2014. Magnetic nanoparticle-based therapeutic agents for thermo-chemotherapy treatment of cancer. *Nanoscale* 6, 11553–11573.
74. Hachani, R., Lowdell, M., Birchall, M., Thanh, N. T. K., 2013. Tracking stem cells in tissue-engineered organs using magnetic nanoparticles. *Nanoscale* 5, 11362–11373.
75. Amstad, E., Textor, M., Reimhult, E., 2011. Stabilization and functionalization of iron oxide nanoparticles for biomedical applications. *Nanoscale* 3, 2819–2843.
76. Hola, K., Markova, Z., Zoppellaro, G., Tucek, J., Zboril, R., 2015. Tailored functionalization of iron oxide nanoparticles for MRI, drug delivery, magnetic separation and immobilization of biosubstances. *Biotechnology Advances* 33, 1162–1176.
77. Laurent, S., Saei, A. A., Behzadi, S., Panahifar, A., Mahmoudi, M., 2014. Superparamagnetic iron oxide nanoparticles for delivery of therapeutic agents: Opportunities and challenges. *Expert Opinion on Drug Delivery* 11, 1449–1470.
78. Pankhurst, Q. A., Connolly, J., Jones, S. K., Dobson, J., 2003. Applications of magnetic nanoparticles in biomedicine. *Journal of Physics D: Applied Physics* 36, R167.

79. Niemirowicz, K., Markiewicz, K. H., Wilczewska, A. Z., Car, H., 2012. Magnetic nanoparticles as new diagnostic tools in medicine. *Advances in Medical Sciences* 57, 196–207.
80. Tang, S. C. N., Lo, I. M. C., 2013. Magnetic nanoparticles: Essential factors for sustainable environmental applications. *Water Research* 47, 2613–2632.
81. Abou-Hassan, A., Bazzi, R., Cabuil, V., 2009. Multistep continuous-flow microsynthesis of magnetic and fluorescent gamma-Fe_2O_3@SiO_2 core/shell nanoparticles. *Angewandte Chemie-International Edition* 48, 7180–7183.
82. Blanco-Andujar, C., Ortega, D., Southern, P., Pankhurst, Q. A., Thanh, N. T. K., 2015. High performance multi-core iron oxide nanoparticles for magnetic hyperthermia: Microwave synthesis, and the role of core-to-core interactions. *Nanoscale* 7, 1768–1775.
83. Sun, S., Zeng, H., Robinson, D. B., Raoux, S., Rice, P. M., Wang, S. X., Li, G., 2004. Monodisperse MFe_2O_4 (M= Fe, Co, Mn) nanoparticles. *Journal of the American Chemical Society* 126, 273–279.
84. Sperling, R. A., Parak, W., 2010. Surface modification, functionalization and bioconjugation of colloidal inorganic nanoparticles. *Philosophical Transactions of the Royal Society of London A: Mathematical, Physical and Engineering Sciences* 368, 1333–1383.
85. Palma, S. I., Marciello, M., Carvalho, A., Veintemillas-Verdaguer, S., del Puerto Morales, M., Roque, A. C., 2015. Effects of phase transfer ligands on monodisperse iron oxide magnetic nanoparticles. *Journal of Colloid and Interface Science* 437, 147–155.
86. Rebodos, R. L., Vikesland, P. J., 2010. Effects of oxidation on the magnetization of nanoparticulate magnetite. *Langmuir* 26, 16745–16753.
87. Smolensky, E. D., Park, H.-Y. E., Berquó, T. S., Pierre, V. C., 2011. Surface functionalization of magnetic iron oxide nanoparticles for MRI applications – Effect of anchoring group and ligand exchange protocol. *Contrast Media & Molecular Imaging* 6, 189–199.
88. Walter, A., Garofalo, A., Bonazza, P., Meyer, F., Martinez, H., Fleutot, S., Billotey, C., Taleb, J., Felder-Flesch, D., Begin-Colin, S., 2017. Effect of the functionalization process on the colloidal, magnetic resonance imaging, and bioelimination properties of mono- or bisphosphonate-anchored dendronized iron oxide nanoparticles. *ChemPlusChem* 82, 647–659.
89. Korpany, K. V., Habib, F., Murugesu, M., Blum, A. S., 2013. Stable water-soluble iron oxide nanoparticles using Tiron. *Materials Chemistry and Physics* 138, 29–37.
90. Korpany, K. V., Mottillo, C., Bachelder, J., Cross, S. N., Dong, P., Trudel, S., Friščić, T., Blum, A. S., 2016. One-step ligand exchange and switching from hydrophobic to water-stable hydrophilic superparamagnetic iron oxide nanoparticles by mechanochemical milling. *Chemical Communications* 52, 3054–3057.
91. Ruiz, A., Morais, P., Bentes de Azevedo, R., Lacava, Z. M., Villanueva, A., del Puerto Morales, M., 2014. Magnetic nanoparticles coated with dimercaptosuccinic acid: Development, characterization, and application in biomedicine. *Journal of Nanoparticle Research* 16, 1–20.
92. Das, M., Mishra, D., Dhak, P., Gupta, S., Maiti, T. K., Basak, A., Pramanik, P., 2009. Biofunctionalized, phosphonate-grafted, ultrasmall iron oxide nanoparticles for combined targeted cancer therapy and multimodal imaging. *Small* 5, 2883–2893.
93. Lalatonne, Y., Monteil, M., Jouni, H., Serfaty, J. M., Sainte-Catherine, O., Lievre, N., Kusmia, S., Weinmann, P., Lecouvey, M., Motte, L., 2010. Superparamagnetic bifunctional bisphosphonates nanoparticles: A potential MRI contrast agent for osteoporosis therapy and diagnostic. *Journal of Osteoporosis* 2010, 747852–747852.
94. Lalatonne, Y., Paris, C., Serfaty, J. M., Weinmann, P., Lecouvey, M., Motte, L., 2008. Bis-phosphonates-ultra small superparamagnetic iron oxide nanoparticles: A platform towards diagnosis and therapy. *Chemical Communications*, 2553–2555.
95. Dobson, K. D., McQuillan, A. J., 1999. *In situ* infrared spectroscopic analysis of the adsorption of aliphatic carboxylic acids to TiO_2, ZrO_2, Al_2O_3, and Ta_2O_5 from aqueous solutions. *Spectrochimica Acta Part A: Molecular and Biomolecular Spectroscopy* 55, 1395–1405.
96. Deacon, G., Phillips, R., 1980. Relationships between the carbon-oxygen stretching frequencies of carboxylato complexes and the type of carboxylate coordination. *Coordination Chemistry Reviews* 33, 227–250.
97. Lam, T., Avti, P. K., Pouliot, P., Tardif, J.-C., Rhéaume, É., Lesage, F., Kakkar, A., 2016. Surface engineering of SPIONs: Role of phosphonate ligand multivalency in tailoring their efficacy. *Nanotechnology* 27, 415602.
98. Daou, T. J., Begin-Colin, S., Grenèche, J. M., Thomas, F., Derory, A., Bernhardt, P., Legaré, P., Pourroy, G., 2007. Phosphate adsorption properties of magnetite-based nanoparticles. *Chemistry of Materials* 19, 4494–4505.
99. Daou, T., Greneche, J., Pourroy, G., Buathong, S., Derory, A., Ulhaq-Bouillet, C., Donnio, B., Guillon, D., Begin-Colin, S., 2008. Coupling agent effect on magnetic properties of functionalized magnetite-based nanoparticles. *Chemistry of Materials* 20, 5869–5875.
100. Amstad, E., Gillich, T., Bilecka, I., Textor, M., Reimhult, E., 2009. Ultrastable iron oxide nanoparticle colloidal suspensions using dispersants with catechol-derived anchor groups. *Nano Letters* 9, 4042–4048.
101. Xu, C., Xu, K., Gu, H., Zheng, R., Liu, H., Zhang, X., Guo, Z., Xu, B., 2004. Dopamine as a robust anchor to immobilize functional molecules on the iron oxide shell of magnetic nanoparticles. *Journal of the American Chemical Society* 126, 9938–9939.
102. El-Ayaan, U., Herlinger, E., Jameson, R. F., Linert, W., 1997. Anaerobic oxidation of dopamine by iron (III). *Journal of the Chemical Society, Dalton Transactions*, 2813–2818.
103. Shultz, M. D., Reveles, J. U., Khanna, S. N., Carpenter, E. E., 2007. Reactive nature of dopamine as a surface functionalization agent in iron oxide nanoparticles. *Journal of the American Chemical Society* 129, 2482–2487.
104. Amstad, E., Gehring, A. U., Fischer, H., Nagaiyanallur, V. V., Hähner, G., Textor, M., Reimhult, E., 2011. Influence of electronegative substituents on the binding affinity of catechol-derived anchors to Fe_3O_4 nanoparticles. *The Journal of Physical Chemistry C* 115, 683–691.
105. Tassa, C., Shaw, S. Y., Weissleder, R., 2011. Dextran-coated iron oxide nanoparticles: A versatile platform for targeted molecular imaging, molecular diagnostics and therapy. *Accounts of Chemical Research* 44, 842–852.
106. Anselmo, A. C., Mitragotri, S., 2016. Nanoparticles in the clinic. *Bioengineering & Translational Medicine* 1, 10–29.
107. Osborne, E. A., Atkins, T. M., Gilbert, D. A., Kauzlarich, S. M., Liu, K., Louie, A. Y., 2012. Rapid microwave-assisted synthesis of dextran-coated iron oxide nanoparticles for magnetic resonance imaging. *Nanotechnology* 23, 215602.
108. Shen, W. B., Vaccaro, D. E., Fishman, P. S., Groman, E. V., Yarowsky, P., 2016. SIRB, sans iron oxide rhodamine B, a novel cross-linked dextran nanoparticle, labels human neuroprogenitor and SH-SY5Y neuroblastoma cells and serves as a USPIO cell labeling control. *Contrast Media & Molecular Imaging*.

109. Wasiak, I., Kulikowska, A., Janczewska, M., Michalak, M., Cymerman, I. A., Nagalski, A., Kallinger, P., Szymanski, W. W., Ciach, T., 2016. Dextran nanoparticle synthesis and properties. *PloS One* 11, e0146237.
110. Laurent, S., Forge, D., Port, M., Roch, A., Robic, C., Elst, L. V., Muller, R. N., 2008. Magnetic iron oxide nanoparticles: Synthesis, stabilization, vectorization, physicochemical characterizations, and biological applications. *Chemical Reviews* 108, 2064–2110.
111. Boyer, C., Whittaker, M. R., Bulmus, V., Liu, J., Davis, T. P., 2010. The design and utility of polymer-stabilized iron-oxide nanoparticles for nanomedicine applications. *NPG Asia Mater* 2, 23–30.
112. Huang, G., Zhang, C., Li, S., Khemtong, C., Yang, S.-G., Tian, R., Minna, J. D., Brown, K. C., Gao, J., 2009. A novel strategy for surface modification of superparamagnetic iron oxide nanoparticles for lung cancer imaging. *Journal of Materials Chemistry* 19, 6367–6372.
113. Chen, S. *Polymer-Coated Iron Oxide Nanoparticles for Medical Imaging*. Massachusetts Institute of Technology, Cambridge, MA, 2010.
114. Bloemen, M., Van Stappen, T., Willot, P., Lammertyn, J., Koeckelberghs, G., Geukens, N., Gils, A., Verbiest, T., 2014. Heterobifunctional PEG ligands for bioconjugation reactions on iron oxide nanoparticles. *PloS One* 9, e109475.
115. Thorek, D. L. J., Elias, D. R., Tsourkas, A., 2009. Comparative analysis of nanoparticle-antibody conjugations: Carbodiimide versus click chemistry. *Molecular Imaging* 8, 221–229.
116. Bolley, J., Guenin, E., Lievre, N., Lecouvey, M., Soussan, M., Lalatonne, Y., Motte, L., 2013. Carbodiimide versus click chemistry for nanoparticle surface functionalization: A comparative study for the elaboration of multimodal superparamagnetic nanoparticles targeting αvβ3 integrins. *Langmuir* 29, 14639–14647.
117. Peer, D., Karp, J. M., Hong, S., Farokhzad, O. C., Margalit, R., Langer, R., 2007. Nanocarriers as an emerging platform for cancer therapy. *Nature Nanotechnology* 2, 751–760.
118. Ruoslahti, E., 1996. RGD and other recognition sequences for integrins. *Annual Review of Cell and Developmental Biology* 12, 697–715.
119. Danhier, F., Breton, A. L., Préat, V. r., 2012. RGD-based strategies to target alpha (v) beta (3) integrin in cancer therapy and diagnosis. *Molecular Pharmaceutics* 9, 2961–2973.
120. DeNardo, S. J., Liu, R., Albrecht, H., Natarajan, A., Sutcliffe, J. L., Anderson, C., Peng, L., Ferdani, R., Cherry, S. R., Lam, K. S., 2009. 111In-LLP2A-DOTA polyethylene glycol–targeting α4β1 integrin: Comparative pharmacokinetics for imaging and therapy of lymphoid malignancies. *Journal of Nuclear Medicine* 50, 625–634.
121. Garanger, E., Boturyn, D., Coll, J.-L., Favrot, M.-C., Dumy, P., 2006. Multivalent RGD synthetic peptides as potent αVβ3 integrin ligands. *Organic & Biomolecular Chemistry* 4, 1958–1965.
122. Perrault, S. D., Walkey, C., Jennings, T., Fischer, H. C., Chan, W. C., 2009. Mediating tumor targeting efficiency of nanoparticles through design. *Nano Letters* 9, 1909–1915.
123. Ndong, C., Tate, J. A., Kett, W. C., Batra, J., Demidenko, E., Lewis, L. D., Hoopes, P. J., Gerngross, T. U., Griswold, K. E., 2015. Tumor cell targeting by iron oxide nanoparticles is dominated by different factors *in vitro* versus *in vivo*. *PloS One* 10, e0115636.
124. Arami, H., Khandhar, A., Liggitt, D., Krishnan, K. M., 2015. *In vivo* delivery, pharmacokinetics, biodistribution and toxicity of iron oxide nanoparticles. *Chemical Society Reviews* 44, 8576–8607.
125. Park, J. H., von Maltzahn, G., Zhang, L. L., Derfus, A. M., Simberg, D., Harris, T. J., Ruoslahti, E., Bhatia, S. N., Sailor, M. J., 2009. Systematic surface engineering of magnetic nanoworms for *in vivo* tumor targeting. *Small* 5, 694–700.
126. Park, J. H., von Maltzahn, G., Zhang, L. L., Schwartz, M. P., Ruoslahti, E., Bhatia, S. N., Sailor, M. J., 2008. Magnetic iron oxide nanoworms for tumor targeting and imaging. *Advanced Materials* 20, 1630–1635.
127. Arnida, Janat-Amsbury, M. M., Ray, A., Peterson, C. M., Ghandehari, H., 2011. Geometry and surface characteristics of gold nanoparticles influence their biodistribution and uptake by macrophages. *European Journal of Pharmaceutics and Biopharmaceutics* 77, 417–423.
128. Chithrani, B. D., Ghazani, A. A., Chan, W. C. W., 2006. Determining the size and shape dependence of gold nanoparticle uptake into mammalian cells. *Nano Letters* 6, 662–668.
129. Kunjachan, S., Pola, R., Gremse, F., Theek, B., Ehling, J., Moeckel, D., Hermanns-Sachweh, B., Pechar, M., Ulbrich, K., Hennink, W. E., 2014. Passive vs. active tumor targeting using RGD-and NGR-modified polymeric nanomedicines. *Nano Letters* 14, 972.
130. Wilhelm, S., Tavares, A. J., Dai, Q., Ohta, S., Audet, J., Dvorak, H. F., Chan, W. C., 2016. Analysis of nanoparticle delivery to tumours. *Nature Reviews Materials* 1, 16014.
131. Schleich, N., Po, C., Jacobs, D., Ucakar, B., Gallez, B., Danhier, F., Préat, V., 2014. Comparison of active, passive and magnetic targeting to tumors of multifunctional paclitaxel/SPIO-loaded nanoparticles for tumor imaging and therapy. *Journal of Controlled Release* 194, 82–91.
132. Zhang, F., Huang, X., Zhu, L., Guo, N., Niu, G., Swierczewska, M., Lee, S., Xu, H., Wang, A. Y., Mohamedali, K. A., Rosenblum, M. G., Lu, G., Chen, X., 2012. Noninvasive monitoring of orthotopic glioblastoma therapy response using RGD-conjugated iron oxide nanoparticles. *Biomaterials* 33, 5414–5422.
133. Barua, S., Mitragotri, S., 2014. Challenges associated with penetration of nanoparticles across cell and tissue barriers: A review of current status and future prospects. *Nano Today* 9, 223–243.
134. Maeda, H., 2010. Tumor-selective delivery of macromolecular drugs *via* the EPR effect: Background and future prospects. *Bioconjugate Chemistry* 21, 797–802.
135. Yuan, F., Dellian, M., Fukumura, D., Leunig, M., Berk, D. A., Torchilin, V. P., Jain, R. K., 1995. Vascular permeability in a human tumor xenograft: Molecular size dependence and cut-off size. *Cancer Research* 55, 3752–3756.
136. Setyawati, M. I., Mochalin, V. N., Leong, D. T., 2016. Tuning endothelial permeability with functionalized nanodiamonds. *ACS Nano* 10, 1170–1181.
137. Walkey, C. D., Chan, W. C. W., 2012. Understanding and controlling the interaction of nanomaterials with proteins in a physiological environment. *Chemical Society Reviews* 41, 2780–2799.
138. Sakulkhu, U., Mahmoudi, M., Maurizi, L., Salaklang, J., Hofmann, H., 2014. Protein corona composition of superparamagnetic iron oxide nanoparticles with various physicochemical properties and coatings. *Scientific Reports* 4, 5020.
139. Ashby, J., Pan, S., Zhong, W., 2014. Size and Surface functionalization of iron oxide nanoparticles influence the composition and dynamic nature of their protein corona. *ACS Applied Materials & Interfaces* 6, 15412–15419.
140. Mirshafiee, V., Mahmoudi, M., Lou, K., Cheng, J., Kraft, M. L., 2013. Protein corona significantly reduces active targeting yield. *Chemical Communications* 49, 2557–2559.

141. Salvati, A., Pitek, A. S., Monopoli, M. P., Prapainop, K., Bombelli, F. B., Hristov, D. R., Kelly, P. M., Aberg, C., Mahon, E., Dawson, K. A., 2013. Transferrin-functionalized nanoparticles lose their targeting capabilities when a biomolecule corona adsorbs on the surface. *Nature Nanotechnology* 8, 137–143.
142. Dai, Q., Walkey, C., Chan, W. C. W., 2014. Polyethylene glycol backfilling mitigates the negative impact of the protein corona on nanoparticle cell targeting. *Angewandte Chemie International Edition* 53, 5093–5096.
143. Yang, Q., Jones, S. W., Parker, C. L., Zamboni, W. C., Bear, J. E., Lai, S. K., 2014. Evading immune cell uptake and clearance requires PEG grafting at densities substantially exceeding the minimum for brush conformation. *Molecular Pharmaceutics* 11, 1250–1258.
144. Sanchez, L., Yi, Y., Yu, Y., 2017. Effect of partial PEGylation on particle uptake by macrophages. *Nanoscale* 9, 288–297.
145. Pearson, R. M., Sen, S., Hsu, H.-j., Pasko, M., Gaske, M., Král, P., Hong, S., 2016. Tuning the selectivity of dendron micelles through variations of the poly(ethylene glycol) corona. *ACS Nano* 10, 6905–6914.
146. Chen, J., Klem, S., Jones, A. K., Orr, B., Banaszak Holl, M. M., 2017. Folate-binding protein self-aggregation drives agglomeration of folic acid targeted iron oxide nanoparticles. *Bioconjugate Chemistry* 28, 81–87.
147. Su, G., Zhou, X., Zhou, H., Li, Y., Zhang, X., Liu, Y., Cao, D., Yan, B., 2016. Size-dependent facilitation of cancer cell targeting by proteins adsorbed on nanoparticles. *ACS Applied Materials & Interfaces* 8, 30037–30047.
148. Mahmoudi, M., Sheibani, S., Milani, A. S., Rezaee, F., Gauberti, M., Dinarvand, R., Vali, H., 2015. Crucial role of the protein corona for the specific targeting of nanoparticles. *Nanomedicine* 10, 215–226.
149. Lo Giudice, M. C., Meder, F., Polo, E., Thomas, S. S., Alnahdi, K., Lara, S., Dawson, K. A., 2016. Constructing bifunctional nanoparticles for dual targeting: Improved grafting and surface recognition assessment of multiple ligand nanoparticles. *Nanoscale* 8, 16969–16975.
150. Liu, Z., Zhan, X., Yang, M., Yang, Q., Xu, X., Lan, F., Wu, Y., Gu, Z., 2016. A magnetic-dependent protein corona of tailor-made superparamagnetic iron oxides alters their biological behaviors. *Nanoscale* 8, 7544–7555.
151. Chen, Y.-F., Hong, J., Wu, D.-Y., Zhou, Y.-Y., D'Ortenzio, M., Ding, Y., Xia, X.-H., 2016. *In vivo* mapping and assay of matrix metalloproteases for liver tumor diagnosis. *RSC Advances* 6, 8336–8345.
152. Yan, R., Ye, D., 2016. Molecular imaging of enzyme activity *in vivo* using activatable probes. *Science Bulletin* 61, 1672–1679.
153. Ye, D., Pandit, P., Kempen, P., Lin, J., Xiong, L., Sinclair, R., Rutt, B., Rao, J., 2014. Redox-triggered self-assembly of gadolinium-based mri probes for sensing reducing environment. *Bioconjugate Chemistry* 25, 1526–1536.
154. Gao, Z., Hou, Y., Zeng, J., Chen, L., Liu, C., Yang, W., Gao, M., 2017. Tumor microenvironment-triggered aggregation of antiphagocytosis 99mTc-labeled Fe_3O_4 nanoprobes for enhanced tumor imaging *in vivo*. *Advanced Materials*.
155. Yi, S., Allen, S. D., Liu, Y.-G., Ouyang, B. Z., Li, X., Augsornworawat, P., Thorp, E. B., Scott, E. A., 2016. Tailoring nanostructure morphology for enhanced targeting of dendritic cells in atherosclerosis. *ACS Nano*.
156. McConnell, H. L., Schwartz, D. L., Richardson, B. E., Woltjer, R. L., Muldoon, L. L., Neuwelt, E. A., 2016. Ferumoxytol nanoparticle uptake in brain during acute neuroinflammation is cell-specific. *Nanomedicine: Nanotechnology, Biology and Medicine* 12, 1535–1542.
157. Burke, A. R., Singh, R. N., Carroll, D. L., Torti, F. M., Torti, S. V., 2012. Targeting cancer stem cells with nanoparticle-enabled therapies. *Journal of Molecular Biomarkers & Diagnosis*.
158. Fruscella, M., Ponzetto, A., Crema, A., Carloni, G., 2016. The extraordinary progress in very early cancer diagnosis and personalized therapy: The role of oncomarkers and nanotechnology. *Journal of Nanotechnology* 2016.
159. Nejadnik, H., Ye, D., Lenkov, O. D., Donig, J. S., Martin, J. E., Castillo, R., Derugin, N., Sennino, B., Rao, J., Daldrup-Link, H., 2015. Magnetic resonance imaging of stem cell apoptosis in arthritic joints with a caspase activatable contrast agent. *ACS Nano* 9, 1150–1160.
160. Ye, D., Shuhendler, A. J., Pandit, P., Brewer, K. D., Tee, S. S., Cui, L., Tikhomirov, G., Rutt, B., Rao, J., 2014. Caspase-responsive smart gadolinium-based contrast agent for magnetic resonance imaging of drug-induced apoptosis. *Chemical Science* 5, 3845–3852.
161. Yuan, Y., Ding, Z., Qian, J., Zhang, J., Xu, J., Dong, X., Han, T., Ge, S., Luo, Y., Wang, Y., Zhong, K., Liang, G., 2016. Casp3/7-instructed intracellular aggregation of Fe_3O_4 nanoparticles enhances T2 MR imaging of tumor apoptosis. *Nano Letters* 16, 2686–2691.
162. Garofalo, A., Parat, A., Bordeianu, C., Ghobril, C., Kueny-Stotz, M., Walter, A., Jouhannaud, J., Begin-Colin, S., Felder-Flesch, D., 2014. Efficient synthesis of small-sized phosphonated dendrons: Potential organic coatings of iron oxide nanoparticles. *New Journal of Chemistry* 38, 5226–5239.
163. Salamończyk, G. M., 2015. Efficient synthesis of water-soluble, phosphonate-terminated polyester dendrimers. *Tetrahedron Letters* 56, 7161–7164.
164. Walter, A., Garofalo, A., Parat, A., Jouhannaud, J., Pourroy, G., Voirin, E., Laurent, S., Bonazza, P., Taleb, J., Billotey, C., Vander Elst, L., Muller, R. N., Begin-Colin, S., Felder-Flesch, D., 2015. Validation of a dendron concept to tune colloidal stability, MRI relaxivity and bioelimination of functional nanoparticles. *Journal of Materials Chemistry B* 3, 1484–1494.
165. Sandiford, L., Phinikaridou, A., Protti, A., Meszaros, L. K., Cui, X., Yan, Y., Frodsham, G., Williamson, P. A., Gaddum, N., Botnar, R. M., Blower, P. J., Green, M. A., de Rosales, R. T. M., 2013. Bisphosphonate-anchored PEGylation and radiolabeling of superparamagnetic iron oxide: Long-circulating nanoparticles for *in vivo* multimodal (T1 MRI-SPECT) imaging. *ACS Nano* 7, 500–512.
166. Torrisi, V., Graillot, A., Vitorazi, L., Crouzet, Q., Marletta, G., Loubat, C., Berret, J. F., 2014. Preventing corona effects: Multiphosphonic acid poly(ethylene glycol) copolymers for stable stealth iron oxide nanoparticles. *Biomacromolecules* 15, 3171–3179.
167. Lam, T., Avti, P. K., Pouliot, P., Maafi, F., Tardif, J.-C., Rhéaume, É., Lesage, F., Kakkar, A., 2016. Fabricating water dispersible superparamagnetic iron oxide nanoparticles for biomedical applications through ligand exchange and direct conjugation. *Nanomaterials* 6, 100.
168. Walter, A., Parat, A., Garofalo, A., Laurent, S., Elst, L. V., Muller, R. N., Wu, T., Heuillard, E., Robinet, E., Meyer, F., 2015. Modulation of relaxivity, suspension stability, and biodistribution of dendronized iron oxide nanoparticles as a function of the organic shell design. *Particle & Particle Systems Characterization* 32, 552–560.
169. Alexis, F., Pridgen, E., Molnar, L. K., Farokhzad, O. C., 2008. Factors affecting the clearance and biodistribution of polymeric nanoparticles. *Molecular Pharmaceutics* 5, 505–515.

170. Li, S.-D., Huang, L., 2009. Nanoparticles evading the reticuloendothelial system: Role of the supported bilayer. *Biochimica et Biophysica Acta (BBA)-Biomembranes* 1788, 2259–2266.
171. Mojica Pisciotti, M. L., Lima, E., Vasquez Mansilla, M., Tognoli, V. E., Troiani, H. E., Pasa, A. A., Creczynski-Pasa, T. B., Silva, A. H., Gurman, P., Colombo, L., Goya, G. F., Lamagna, A., Zysler, R. D., 2014. *In vitro* and *in vivo* experiments with iron oxide nanoparticles functionalized with DEXTRAN or polyethylene glycol for medical applications: Magnetic targeting. *Journal of Biomedical Materials Research Part B: Applied Biomaterials* 102, 860–868.
172. Hanot, C. C., Choi, Y. S., Anani, T. B., Soundarrajan, D., David, A. E., 2016. Effects of iron-oxide nanoparticle surface chemistry on uptake kinetics and cytotoxicity in CHO-K1 cells. *International Journal of Molecular Sciences* 17, 54.
173. Silva, A. H., Lima, E., Mansilla, M. V., Zysler, R. D., Troiani, H., Pisciotti, M. L. M., Locatelli, C., Benech, J. C., Oddone, N., Zoldan, V. C., 2016. Superparamagnetic iron-oxide nanoparticles mPEG350- and mPEG2000-coated: Cell uptake and biocompatibility evaluation. *Nanomedicine: Nanotechnology, Biology and Medicine* 12, 909–919.
174. Linot, C., Poly, J., Boucard, J., Pouliquen, D., Nedellec, S., Hulin, P., Marec, N., Arosio, P., Lascialfari, A., Guerrini, A., Sangregorio, C., Lecouvey, M., Lartigue, L., Blanquart, C., Ishow, E., 2017. PEGylated anionic magnetofluorescent nanoassemblies: Impact of their interface structure on magnetic resonance imaging contrast and cellular uptake. *ACS Applied Materials & Interfaces* 9, 14242–14257.
175. Gal, N., Lassenberger, A., Herrero-Nogareda, L., Scheberl, A., Charwat, V., Kasper, C., Reimhult, E., 2017. Interaction of size-tailored PEGylated iron oxide nanoparticles with lipid membranes and cells. *ACS Biomaterials Science & Engineering* 3, 249–259.
176. Cedervall, T., Lynch, I., Lindman, S., Berggård, T., Thulin, E., Nilsson, H., Dawson, K. A., Linse, S., 2007. Understanding the nanoparticle–protein corona using methods to quantify exchange rates and affinities of proteins for nanoparticles. *Proceedings of the National Academy of Sciences* 104, 2050–2055.
177. Hervault, A., Dunn, A. E., Lim, M., Boyer, C., Mott, D., Maenosono, S., Thanh, N. T. K., 2016. Doxorubicin loaded dual pH- and thermo-responsive magnetic nanocarrier for combined magnetic hyperthermia and targeted controlled drug delivery applications. *Nanoscale* 8, 12152–12161.
178. Quinto, C. A., Mohindra, P., Tong, S., Bao, G., 2015. Multifunctional superparamagnetic iron oxide nanoparticles for combined chemotherapy and hyperthermia cancer treatment. *Nanoscale* 7, 12728–12736.
179. Egerton, R. F., Li, P., Malac, M., 2004. Radiation damage in the TEM and SEM. *Micron* 35, 399–409.
180. Baumgartner, J., Dey, A., Bomans, P. H. H., Le Coadou, C., Fratzl, P., Sommerdijk, N. A. J. M., Faivre, D., 2013. Nucleation and growth of magnetite from solution. *Nature Materials* 12, 310–314.
181. Liang, W.-I., Zhang, X., Bustillo, K., Chiu, C.-H., Wu, W.-W., Xu, J., Chu, Y.-H., Zheng, H., 2015. *In situ* study of spinel ferrite nanocrystal growth using liquid cell transmission electron microscopy. *Chemistry of Materials* 27, 8146–8152.
182. LaGrow, A. P., Sinatra, L., Elshewy, A., Huang, K.-W., Katsiev, K., Kirmani, A. R., Amassian, A., Anjum, D. H., Bakr, O. M., 2014. Synthesis of copper hydroxide branched nanocages and their transformation to copper oxide. *The Journal of Physical Chemistry C* 118, 19374–19379.
183. Hurley, K. R., Ring, H. L., Kang, H., Klein, N. D., Haynes, C. L., 2015. Characterization of magnetic nanoparticles in biological matrices. *Analytical Chemistry* 87, 11611–11619.
184. Lassenberger, A., Grünewald, T. A., van Oostrum, P. D. J., Rennhofer, H., Amenitsch, H., Zirbs, R., Lichtenegger, H. C., Reimhult, E., 2017. Monodisperse iron oxide nanoparticles by thermal decomposition: Elucidating particle formation by second-resolved *in situ* small-angle X-ray scattering. *Chemistry of Materials* 29, 4511–4522.
185. LaGrow, A. P., Ingham, B., Toney, M. F., Tilley, R. D., 2013. Effect of surfactant concentration and aggregation on the growth kinetics of nickel nanoparticles. *The Journal of Physical Chemistry C* 117, 16709–16718.
186. Li, T., Senesi, A. J., Lee, B., 2016. Small angle X-ray scattering for nanoparticle research. *Chemical Reviews* 116, 11128–11180.
187. Scheck, J., Wu, B., Drechsler, M., Rosenberg, R., Van Driessche, A. E. S., Stawski, T. M., Gebauer, D., 2016. The molecular mechanism of iron(III) oxide nucleation. *The Journal of Physical Chemistry Letters* 7, 3123–3130.
188. Korgel, B. A., Fullam, S., Connolly, S., Fitzmaurice, D., 1998. Assembly and self-organization of silver nanocrystal superlattices: Ordered "soft spheres". *The Journal of Physical Chemistry B* 102, 8379–8388.
189. Von White, G., Mohammed, F. S., Kitchens, C. L., 2011. Small-angle neutron scattering investigation of gold nanoparticle clustering and ligand structure under antisolvent conditions. *The Journal of Physical Chemistry C* 115, 18397–18405.
190. Lim, J., Yeap, S. P., Che, H. X., Low, S. C., 2013. Characterization of magnetic nanoparticle by dynamic light scattering. *Nanoscale Research Letters* 8, 381–381.
191. Schärtl, W., *Light Scattering from Polymer Solutions and Nanoparticle Dispersions*. Springer: Berlin Heidelberg, 2007.
192. Xu, Y., Baiu, D. C., Sherwood, J. A., McElreath, M. R., Qin, Y., Lackey, K. H., Otto, M., Bao, Y., 2014. Linker-free conjugation and specific cell targeting of antibody functionalized iron-oxide nanoparticles. *Journal of Materials Chemistry B* 2, 6198–6206.
193. Takahashi, K., Kato, H., Saito, T., Matsuyama, S., Kinugasa, S., 2008. Precise measurement of the size of nanoparticles by dynamic light scattering with uncertainty analysis. *Particle & Particle Systems Characterization* 25, 31–38.
194. Fissan, H., Ristig, S., Kaminski, H., Asbach, C., Epple, M., 2014. Comparison of different characterization methods for nanoparticle dispersions before and after aerosolization. *Analytical Methods* 6, 7324–7334.
195. Gauvin, H., Elisabeth, G., Sébastien, L., Andrew, D. W., Elizabeth, R. G., Boris, P., Thomas, B., David, J., Olivier, S., 2015. Thermosensitive polymer-grafted iron oxide nanoparticles studied by *in situ* dynamic light backscattering under magnetic hyperthermia. *Journal of Physics D: Applied Physics* 48, 494001.
196. Krpetić, Ž., Davidson, A. M., Volk, M., Lévy, R., Brust, M., Cooper, D. L., 2013. High-resolution sizing of monolayer-protected gold clusters by differential centrifugal sedimentation. *ACS Nano* 7, 8881–8890.
197. Cioran, A. M., Teixidor, F., Krpetic, Z., Brust, M., Vinas, C., 2014. Preparation and characterization of Au nanoparticles capped with mercaptocarboranyl clusters. *Dalton Transactions* 43, 5054–5061.
198. Carney, R. P., Kim, J. Y., Qian, H., Jin, R., Mehenni, H., Stellacci, F., Bakr, O. M., 2011. Determination of nanoparticle size distribution together with density or molecular weight by 2D analytical ultracentrifugation. *Nature Communications* 2, 335.

199. Willis, A. L., Turro, N. J., O'Brien, S., 2005. Spectroscopic Characterization of the Surface of Iron Oxide Nanocrystals. *Chemistry of Materials* 17, 5970–5975.
200. Polito, L., Colombo, M., Monti, D., Melato, S., Caneva, E., Prosperi, D., 2008. Resolving the structure of ligands bound to the surface of superparamagnetic iron oxide nanoparticles by high-resolution magic-angle spinning NMR spectroscopy. *Journal of the American Chemical Society* 130, 12712–12724.
201. Deacon, G., Huber, F., Phillips, R., 1985. Diagnosis of the nature of carboxylate coordination from the direction of shifts of carbon-oxygen stretching frequencies. *Inorganica Chimica Acta* 104, 41–45.
202. Wu, X., Wang, D., Yang, S., 2000. Preparation and characterization of stearate-capped titanium dioxide nanoparticles. *Journal of Colloid and Interface Science* 222, 37–40.
203. Benyettou, F., Lalatonne, Y., Chebbi, I., Di Benedetto, M., Serfaty, J.-M., Lecouvey, M., Motte, L., 2011. A multimodal magnetic resonance imaging nanoplatform for cancer theranostics. *Physical Chemistry Chemical Physics* 13, 10020–10027.
204. Phuoc, L. T., Jouhannaud, J., Pourroy, G. In *Magnetic Iron Oxide and the Effect of Grafting on the Magnetic Properties.* SPIE Nanosystems in Engineering and Medicine, 2012; pp. 85480M-85480M-9.
205. Baaziz, W., Pichon, B. P., Fleutot, S., Liu, Y., Lefevre, C., Greneche, J.-M., Toumi, M., Mhiri, T., Begin-Colin, S., 2014. Magnetic iron oxide nanoparticles: Reproducible tuning of the size and nanosized-dependent composition, defects, and spin canting. *The Journal of Physical Chemistry C* 118, 3795–3810.
206. Carter, C. F., Lange, H., Ley, S. V., Baxendale, I. R., Wittkamp, B., Goode, J. G., Gaunt, N. L., 2010. ReactIR flow cell: A new analytical tool for continuous flow chemical processing. *Organic Process Research & Development* 14, 393–404.
207. Basly, B., Popa, G., Fleutot, S., Pichon, B. P., Garofalo, A., Ghobril, C., Billotey, C., Berniard, A., Bonazza, P., Martinez, H., Felder-Flesch, D., Begin-Colin, S., 2013. Effect of the nanoparticle synthesis method on dendronized iron oxides as MRI contrast agents. *Dalton Transactions* 42, 2146–2157.
208. Torelli, M. D., Putans, R. A., Tan, Y., Lohse, S. E., Murphy, C. J., Hamers, R. J., 2015. Quantitative determination of ligand densities on nanomaterials by X-ray photoelectron spectroscopy. *ACS Applied Materials & Interfaces* 7, 1720–1725.
209. Demay-Drouhard, P., Nehlig, E., Hardouin, J., Motte, L., Guenin, E., 2013. Nanoparticles under the light: Click functionalization by photochemical thiol-yne reaction, towards double click functionalization. *Chemistry – A European Journal* 19, 8388–8392.
210. Binder, W. H., Weinstabl, H., Sachsenhofer, R., 2008. Superparamagnetic ironoxide nanoparticles *via* ligand exchange reactions: Organic 1,2-diols as versatile building blocks for surface engineering. *Journal of Nanomaterials* 2008, 10.
211. Wydra, R. J., Kruse, A. M., Bae, Y., Anderson, K. W., Hilt, J. Z., 2013. Synthesis and characterization of PEG-iron oxide core-shell composite nanoparticles for thermal therapy. *Materials Science and Engineering: C* 33, 4660–4666.
212. De Palma, R., Peeters, S., Van Bael, M., Van den Rul, H., Bonroy, K., Laureyn, W., Mullens, J., Borghs, G., Maes, G., 2007. Silane ligand exchange to make hydrophobic superparamagnetic nanoparticles water-dispersible. *Chemistry of Materials* 19, 1821–1831.
213. Zhu, Y., Jiang, F. Y., Chen, K., Kang, F., Tang, Z. K., 2011. Size-controlled synthesis of monodisperse superparamagnetic iron oxide nanoparticles. *Journal of Alloys and Compounds* 509, 8549–8553.
214. Yang, G., Zhang, B., Wang, J., Xie, S., Li, X., 2015. Preparation of polylysine-modified superparamagnetic iron oxide nanoparticles. *Journal of Magnetism and Magnetic Materials* 374, 205–208.
215. Zhou, J., Fa, H., Yin, W., Zhang, J., Hou, C., Huo, D., Zhang, D., Zhang, H., 2014. Synthesis of superparamagnetic iron oxide nanoparticles coated with a DDNP-carboxyl derivative for *in vitro* magnetic resonance imaging of Alzheimer's disease. *Materials Science and Engineering: C* 37, 348–355.
216. Mansfield, E., Tyner, K. M., Poling, C. M., Blacklock, J. L., 2014. Determination of nanoparticle surface coatings and nanoparticle purity using microscale thermogravimetric analysis. *Analytical Chemistry* 86, 1478–1484.
217. Guenin, E., Lalatonne, Y., Bolley, J., Milosevic, I., Platas-Iglesias, C., Motte, L., 2014. Catechol versus bisphosphonate ligand exchange at the surface of iron oxide nanoparticles: Towards multi-functionalization. *Journal of Nanoparticle Research* 16.
218. Davis, K., Cole, B., Ghelardini, M., Powell, B. A., Mefford, O. T., 2016. Quantitative measurement of ligand exchange with small-molecule ligands on iron oxide nanoparticles *via* radioanalytical techniques. *Langmuir* 32, 13716–13727.
219. Lassenberger, A., Bixner, O., Gruenewald, T., Lichtenegger, H., Zirbs, R., Reimhult, E., 2016. Evaluation of high-yield purification methods on monodisperse peg-grafted iron oxide nanoparticles. *Langmuir* 32, 4259–4269.
220. Bixner, O., Lassenberger, A., Baurecht, D., Reimhult, E., 2015. Complete exchange of the hydrophobic dispersant shell on monodisperse superparamagnetic iron oxide nanoparticles. *Langmuir* 31, 9198–9204.
221. Kresta, J. V., Macgregor, J. F., Marlin, T. E., 1991. Multivariate statistical monitoring of process operating performance. *The Canadian Journal of Chemical Engineering* 69, 35–47.
222. MacGregor, J. F., Kourti, T., 1995. Statistical process control of multivariate processes. *Control Engineering Practice* 3, 403–414.
223. Besenhard, M. O., Scheibelhofer, O., François, K., Joksch, M., Kavsek, B., 2016. A multivariate process monitoring strategy and control concept for a small-scale fermenter in a PAT environment. *Journal of Intelligent Manufacturing.*
224. Polte, J., Erler, R., Thünemann, A. F., Sokolov, S., Ahner, T. T., Rademann, K., Emmerling, F., Kraehnert, R., 2010. Nucleation and growth of gold nanoparticles studied *via in situ* small angle X-ray scattering at millisecond time resolution. *ACS Nano* 4, 1076–1082.
225. Destremaut, F., Salmon, J.-B., Qi, L., Chapel, J.-P., 2009. Microfluidics with on-line dynamic light scattering for size measurements. *Lab on a Chip* 9, 3289–3296.
226. Wang, X. Z., Liu, L., Li, R. F., Tweedie, R. J., Primrose, K., Corbett, J., McNeil-Watson, F. K., 2009. Online characterisation of nanoparticle suspensions using dynamic light scattering, ultrasound spectroscopy and process tomography. *Chemical Engineering Research and Design* 87, 874–884.
227. Fernandez-Garcia, M. P., Teixeira, J. M., Machado, P., Oliveira, M., Maia, J. M., Pereira, C., Pereira, A. M., Freire, C., Araujo, J. P., 2015. Automatized and desktop AC-susceptometer for the *in situ* and real time monitoring of magnetic nanoparticles' synthesis by coprecipitation. *Review of Scientific Instruments* 86, 043904.
228. Strom, V., Olsson, R. T., Rao, K. V., 2010. Real-time monitoring of the evolution of magnetism during precipitation of superparamagnetic nanoparticles for bioscience applications. *Journal of Materials Chemistry* 20, 4168–4175.

229. Milosevic, I., Warmont, F., Lalatonne, Y., Motte, L., 2014. Magnetic metrology for iron oxide nanoparticle scaled-up synthesis. *RSC Advances* 4, 49086–49089.
230. Wu, K., Wang, Y., Feng, Y., Yu, L., Wang, J.-P., 2015. Colorize magnetic nanoparticles using a search coil based testing method. *Journal of Magnetism and Magnetic Materials* 380, 251–254.
231. McQuade, D. T., Seeberger, P. H., 2013. Applying flow chemistry: Methods, materials, and multistep synthesis. *The Journal of Organic Chemistry* 78, 6384–6389.
232. Leung, S.-A., Winkle, R. F., Wootton, R. C. R., deMello, A. J., 2005. A method for rapid reaction optimisation in continuous-flow microfluidic reactors using online Raman spectroscopic detection. *Analyst* 130, 46–51.
233. Ferstl, W., Klahn, T., Schweikert, W., Billeb, G., Schwarzer, M., Loebbecke, S., 2007. Inline analysis in microreaction technology: A suitable tool for process screening and optimization. *Chemical Engineering & Technology* 30, 370–378.
234. Ashok, P. C., Singh, G. P., Rendall, H. A., Krauss, T. F., Dholakia, K., 2011. Waveguide confined Raman spectroscopy for microfluidic interrogation. *Lab on a Chip* 11, 1262–1270.
235. Falagan-Lotsch, P., Grzincic, E. M., Murphy, C. J., 2016. New advances in nanotechnology-based diagnosis and therapeutics for breast cancer: An assessment of active-targeting inorganic nanoplatforms. *Bioconjugate Chemistry* 28, 135–152.
236. Lazarovits, J., Chen, Y. Y., Sykes, E. A., Chan, W. C. W., 2015. Nanoparticle-blood interactions: The implications on solid tumour targeting. *Chemical Communications* 51, 2756–2767.

Nguyễn T. K. Thanh (E-mail: ntk.thanh@ucl.ac.uk) is the editor of this book. For more information, please see the editor's biography or visit her website: http://www.ntk-thanh.co.uk.

Alec P. LaGrow earned his PhD in chemistry at Victoria University of Wellington in 2012, working on shape controlled syntheses of magnetic nanoparticles. He carried out a postdoctoral fellowship in scalable nanoparticle syntheses at King Abdullah University of Science and Technology and *in situ* nanoparticle characterisation at the University of York. He is currently a postdoctoral research associate at the University College London, working on manufacturing of iron oxide nanoparticles for biomedical applications via continuous flow.

Maximilian O. Besenhard earned his MSc in technical physics where he studied nanoporous metal structures. He then completed his PhD in chemical engineering with a focus on continuous processing via tubular reactors at Graz University of Technology. He joined the University College London in 2016 as a postdoctoral research associate after several years as a scientist at Siemens CT and the Research Center Pharmaceutical Engineering. His research combines process development with computational approaches as well as *in situ* characterization techniques. His current activities address the continuous synthesis and characterization of iron oxide nanoparticles for biomedical applications.

Roxanne Hachani is currently a PhD student at University College London, under the supervision of Professor Nguyễn TK Thanh and Professor Martin Birchall. She earned her bachelor of science in biochemistry at the University Claude Bernard Lyon in 2010 and her master's degree in nanoscale engineering in 2012 at the National Institute of Applied Sciences in Lyon. Her current research projects focus on the synthesis of novel magnetic iron oxide nanoparticles for labelling and tracking stem cells by magnetic resonance imaging. Her research includes developing reproducible iron oxide nanoparticles and their surface functionalization, studying their relaxivity properties and potential cytotoxic effects.

7 Magnetic Polymersomes for MRI and Theranostic Applications

*Adeline Hannecart, Dimitri Stanicki, Luce Vander Elst, Robert N. Muller and Sophie Laurent**

CONTENTS

7.1 POLYMERSOMES FOR BIOMEDICAL APPLICATIONS

7.1.1 General Introduction to Polymersomes

Polymer vesicles or polymersomes (from the Greek '–some' = 'body of') are hollow spheres with a hydrophobic polymeric bilayer membrane and hydrophilic polymeric internal and external coronas, having sizes ranging from tens to thousands of nanometres.[1,2] Polymer vesicles were first reported by Meijer and coworkers in the Netherlands[3] and the group of Eisenberg in Canada[4] in 1995. In 1999, the group of Hammer used the name polymersomes for vesicles made of polymers, in analogy to liposomes.[5]

Polymersomes are formed from the self-assembly of amphiphilic polymers containing a large hydrophobic part.[1] Amphiphilic polymers undergo self-assembly in aqueous solution driven by the hydrophobic effect in order to minimize energetically unfavourable hydrophobic–water interactions. Depending on the ratio between the hydrophobic and the hydrophilic parts of the polymer, diverse morphologies have been found to form spontaneously. The most common morphologies include spherical micelles, rods and vesicles (Figure 7.1).[6] Spherical micelles are composed of a hydrophobic spherical core surrounded by hydrophilic chains, while rods (cylindrical or worm-like micelles) consist of a hydrophobic cylindrical core with a hydrophilic corona surrounding the core. On the other hand, polymersomes are hollow spheres with a hydrophobic bilayer membrane that differs from micelles possessing a solid hydrophobic core, encapsulating only hydrophobic moieties. In contrast, polymersomes can encapsulate both hydrophilic (in their inner aqueous cavity) and hydrophobic (in their hydrophobic membrane) species.[1]

The morphology obtained by the self-assembly of amphiphilic polymers is based on the geometry of the amphiphile, which can vary from conical to cylindrical depending on the hydrophilic-to-hydrophobic block ratio. Israelachvili and coworkers first developed this geometrical approach in 1976 to predict the morphology obtained by the self-assembly of small surfactants.[7] This theory was further transferred to polymeric amphiphiles.[8–10] The morphology of self-assembled nanostructures is determined by a dimensionless 'packing parameter,' P_c, which is defined in Equation 7.1:

$$P_c = v/(a_o l_c), \quad (7.1)$$

where v and l_c are the volume and length of the hydrophobic chains, respectively, and a_o is the area occupied by the

* Corresponding author.

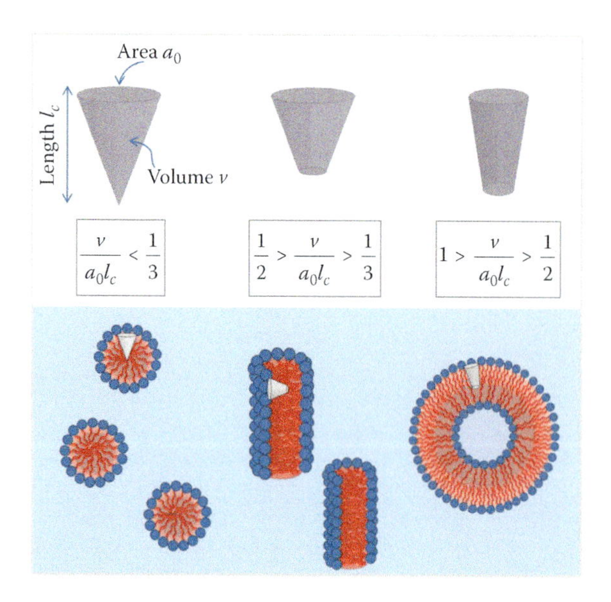

FIGURE 7.1 Packing parameter $Pc = v/(a_0 lc)$ and different morphologies formed from the self-assembly of amphiphilic polymers as determined by the geometry of the amphiphile.[10] (Reprinted with permission from Bleul, R., R. Thiermann, and M. Maskos, Techniques to control polymersome size. *Macromolecules*, 2015. 48(20): p. 7396–7409. Copyright 2015 American Chemical Society.)

hydrophilic headgroup. For values of $P_c < 1/3$, the hydrophilic repulsive forces create highly curved structures (*i.e.* micelles). Increasing the relative hydrophobic fraction results in higher P_c values. When $1/3 < P_c < 1/2$, the production of cylindrical micelles (or rods) is favoured, while for $1/2 < P_c < 1$, vesicular structures are favoured (Figure 7.1).

More recently, Discher and Eisenberg[11] used experimental results obtained from different amphiphilic block copolymers and suggested a rule to predict the self-assembled morphology based on the hydrophilic to total mass fraction, $f_{hydrophilic}$. Copolymers with $0.25 < f_{hydrophilic} < 0.45$ usually form vesicles, while those with $f_{hydrophilic}$ greater than 0.45 will mostly yield micelles.[11] Although the hydrophilic fraction is an important factor, the final morphology is also governed by the chemical composition of the copolymer and the experimental conditions used during the self-assembly process. The self-assembly of poly(ethylene oxide) (PEO)-*b*-poly(ε-caprolactone) (PCL) copolymers illustrates the effect of the chemical composition of the copolymer on the final morphology. In general, polymersomes are formed from copolymers possessing a hydrophilic-to-total mass fraction between 25% and 45%. However, the formation of PEO-*b*-PCL based polymersomes for lower hydrophilic fractions ($f_{hydrophilic}$ between 12% and 28%) has been reported.[12,13] This phenomenon was explained by the lower hydrophobicity and strength of segregation of PCL blocks compared to other hydrophobic blocks like poly(butadiene) (PB).[14] Experimental conditions also play a crucial role on the self-assembly process.[12] Thereby, different morphologies have been formed from the same copolymer (poly(acrylic acid) [PAA]-*b*-poly(styrene) [PS]) in different solvents, with the latter affecting the degree of stretching of the hydrophilic and the hydrophobic blocks and, therefore, the relative volume taken by these blocks.[15]

7.1.2 Polymersomes Preparation Techniques

Several methods for polymer vesicle preparation have been described in the literature. Among them, two methods are broadly applied. The first one is 'nanoprecipitation,' also called the 'solvent displacement' method. That process implies the dissolution of the amphiphilic polymer in a water miscible organic solvent, which is a good solvent for both blocks. An aqueous solution (nonsolvent of the hydrophobic block) is then added, driving the self-assembly of amphiphilic polymers. The organic solvent is finally removed by some means, commonly dialysis or evaporation.[9,12] The formation of vesicle by the nanoprecipitation method depends on the water–organic solvent, water–polymer and organic solvent–polymer interactions, which will change the diffusion process between both solvents and, therefore, the self-assembly conditions. These interactions can be adjusted by the experimental conditions (organic solvent, stirring speed, order of addition, duration of addition, temperature etc.) to control the size and the size distribution of the vesicles. Nanoprecipitation appeared as a reproducible and short time process along with possible scale-up at the industrial level.[16]

The second method is the 'film rehydration' technique. It consists of the prior dissolution of the polymer in a volatile organic solvent such as chloroform, followed by the evaporation of the solvent to form a thin film. Vesicles are then formed by simple rehydration from the lamellar geometry adopted in the concentrated bulk phase. Rehydration proceeds through diffusion of water molecules inside the hydrophilic volume of the lamellae, followed by budding of vesicles into the solution.[13,17,18] Stirring, heating or sonication can influence the swelling process.[10,17,19,20] This technique usually forms vesicles with a rather broad size distribution. Other rehydration techniques like electroformation[5,21] or bulk rehydration[18,22,23] have also been applied to form polymer vesicles.

Other polymersome formation techniques have been more recently reported. Among them, microfluidic approaches have emerged as an interesting methodology allowing the continuous preparation of polymersomes.[10] Water–oil–water double emulsion generated by a microfluidic device has enabled the generation of highly monodisperse polymersomes in the micrometer range.[24] Another continuous approach using micromixer technology and based on the nanoprecipitation process was described as an efficient tool to produce size-controlled polymersomes with sizes below 200 nm.[25,26]

The choice of the preparation method strongly affects the size of the obtained vesicles.[2,9] Postpreparation processes like sonication, freeze–thaw cycles or extrusion through calibrated membranes are frequently applied to obtain an appropriate size distribution.[2,10,19,27] The final vesicle size is known to be a crucial factor for biomedical applications as it affects *in vivo* circulation times, biodistribution and the mechanism of cellular uptake.[9,10] The optimal polymersome diameter for prolonged blood circulation times is reported to be below

150–200 nm due to the uptake of bigger particles by the mononuclear phagocyte system and above 40 nm to inhibit rapid renal clearance.[28]

7.1.3 Polymersomes in Biomedical Research

Polymersomes exhibit great potential for biomedical applications, such as *in vivo* imaging vehicles and/or smart drug delivery applications. Such applications require the use of biodegradable or, at least, biocompatible polymers. To date, polymersomes have been generated from various amphiphilic, copolymers including PEO-*b*-PB, PEO-*b*-poly(ethylethylene), PEO-*b*-PS, PAA-*b*-PS and PEO-*b*-PPS. However, none of these polymersome formulations produces fully biodegradable vesicles.[13]

Among the high variety of chemical structures of polymersomes forming polymers, simple copolymers that have been reported to form biodegradable vesicles are PEO-*b*-poly(lactic acid) (PLA), PEO-*b*-PCL and PEO-*b*-poly(trimethylene carbonate) (PTMC).[29,30] PEO reduces plasma protein adsorption and subsequently decreases uptake by the mononuclear phagocytic system (also called the reticuloendothelial system), prolonging vesicle circulation times in the bloodstream.[17] Polyester-based polymersomes have attracted increased attention in recent years due to their ability of long-time encapsulation of drugs in blood circulation along with their capacity to release them in acidic environments such as intracellular endolysosomes of tumour cells.[29] Ahmed et al.[31] showed that PEO-*b*-PLA-based polymersomes can be dually loaded with hydrophobic anticancer drugs (paclitaxel) and hydrophilic drugs (doxorubicin [DOX]). Hydrolytic degradation of polyester blocks within the endolysosomes induced pH-triggered drug release. *In vivo* studies demonstrated a threefold increase in tumour shrinkage after an intravenous injection of multiple drug-loaded polymersomes compared to injections of the free drugs.[31]

7.1.4 Polymersomes vs. Liposomes

The similarity between liposomes and polymersomes is obvious as they are both constituted of an amphiphilic bilayer enclosing an aqueous compartment. However, major differences discriminate liposomes and polymersomes due to the different building blocks used to obtain vesicles (natural phospholipids for liposomes and amphiphilic polymers for polymersomes). Compared to lipids, polymer chains possess much larger molecular weight, providing thicker, more robust and less permeable membranes.[32]

The membrane thickness depends on the composition of the hydrophobic block and increases with the molecular weight of the hydrophobic block.[33] In particular, polymersomes with membrane thicknesses ranging from 3–5 nm[34] to 200 nm[35] have been reported, whilst liposomes possess a more limited range of membrane thickness (typically 3 to 5 nm).[33] Thicker membranes facilitate the incorporation of hydrophobic compounds.[36,37] For example, the inclusion of hydrophobic nanoparticles into liposome membranes can be difficult because these nanoparticles are usually larger than the hydrophobic bilayer.[37]

The robustness of polymeric membranes was demonstrated by the micropipette aspiration technique. This technique is widely exploited to measure the mechanical properties of cells[19,38] and was further applied to micrometre-size vesicles. Polymersomes were thereby shown to possess higher maximum strain before rupture compared to liposomes due to their thicker membranes.[2,5,8,19,33]

Polymersomes are less permeable to water and small organic molecules compared to liposomes (10 to 20 times lower) as a consequence to their thicker membranes, limiting the passive diffusion.[2,5,19,33] The permeability of polymersome membranes can be modulated by incorporating stimuli responsive hydrophobic blocks for drug delivery purposes.[2,33]

Another advantage of polymersomes comes from the rich diversity of block copolymer chemistries, leading to polymer vesicles with tunable properties. Polymersomes could therefore be designed to achieve given properties, depending on the foreseen applications.[2,5,19] For example, polymersomes exhibiting a wide variety of different drug-controlled release mechanisms in response to specific stimuli such as pH, temperature, ultrasound, alternating magnetic field, oxidation/reduction, light and enzyme degradation have generated broad interest for programmed drug delivery.[39–43]

PEO has been extensively attached to liposomes or other nanoparticle surfaces to extend their blood-circulation half-life ($t_{1/2}$) by reducing nonspecific protein adsorption and delaying mononuclear phagocyte system uptake. This ability, known as 'stealth,' allows nanoparticles to circulate sufficiently long to find their way to target sites such as tumours.[2,44] Upon injection into the circulation, stealth liposomes exhibit a slower clearance from the circulation ($t_{1/2} \approx$ 10–15 h) than conventional liposomes ($t_{1/2} \approx 4$ h). However, liposomes are not able to stably integrate high molar ratios of PEO. In fact, inclusion of PEO increases the hydrophilic mass fraction ($f_{hydrophilic}$), inducing a morphologic transition from vesicles to micelles.[17,45] On the contrary, most polymersomes are intrinsically 'stealth' as they are made from polymers with nonfouling hydrophilic part (such as PEO, dextran and PAA) and do not require post-functionalization with PEO.[46] Photos et al.[17] demonstrated that polymersomes composed of PEO-based copolymers have *in vivo* circulation half-times in rats ($t_{1/2} \approx$ 15–30 h) up to about twofold longer than stealth liposomes, and those containing PEO with higher molecular weight possess longer circulations times ($t_{1/2} \approx 15.8$ h for PEO_{1200}-polymersomes *vs.* $t_{1/2} \approx 28$ h for PEO_{2300}-polymersomes).

These advantages of polymersomes, conferred by higher molecular weight and higher diversity of polymers compared to lipids, overcome some drawbacks encountered with liposomes, as summarized in Figure 7.2. However, one must notice that in contrast to natural phospholipids, polymers are synthetic compounds, limiting their availability. Moreover, polymer syntheses are often performed in the presence of potentially toxic catalysts, which can result in some restrictions for biomedical applications if catalysts are not entirely removed during purification steps.

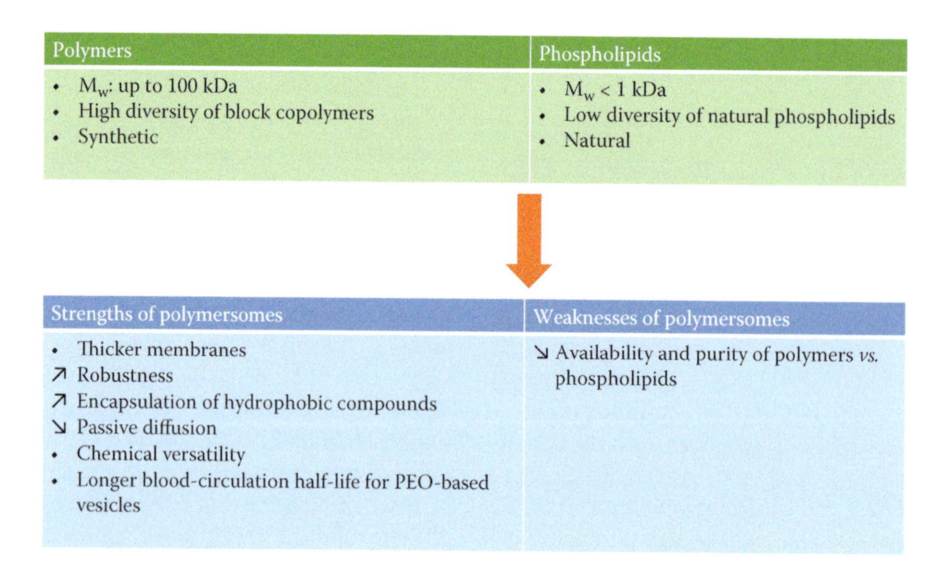

FIGURE 7.2 Summary of the strengths and weaknesses of polymersomes compared to liposomes.

7.2 MAGNETOPOLYMERSOMES

7.2.1 Approaches to Encapsulate Iron Oxide Nanoparticles in Polymersomes

Iron-oxide-based magnetic nanoparticles (magnetite Fe_3O_4 and its oxidized form maghemite γ-Fe_2O_3) are extensively used in nanomedicine due to their unique properties such as high magnetization and biocompatibility. Concerning their *in vivo* applications, they are effective contrast agents for magnetic resonance imaging (MRI) and they can also dissipate heat when they are exposed to an alternating magnetic field (hyperthermia), which can be useful for cancer thermotherapy and also for thermo-triggered drug release.[47–49] Iron oxide nanoparticles can be classified into two main categories. The first class includes small particles of iron oxide (SPIOs), which are particles containing several crystals within the same coating with a hydrodynamic diameter (D_h) usually larger than 40 nm. The second category consists of ultrasmall superparamagnetic iron oxide nanoparticles (USPIO), which are particles containing a single crystal and whose hydrodynamic diameter is smaller than 40 nm.[50]

Combining iron oxide nanoparticles with polymer vesicles is of particular interest to create versatile nanoplatforms for biomedical applications such as imaging and drug delivery. In fact, this approach offers the opportunity to encapsulate several USPIOs inside a vesicle, resulting in an increase in their MRI sensitivity.[26,28,51–55] Moreover, iron oxide nanoparticles and drugs can be dually loaded in a single vesicle, allowing combining diagnostic and therapeutic applications (a concept named thernanostics).[26,54–56] Encapsulation of iron oxide nanoparticles in nanocarriers such as polymersomes allows obtaining stealth vehicles possessing long lifetime in the bloodstream.[55] Polymersome surfaces can be further functionalized by the covalent attachment of targeting peptides or antibodies to allow a specific recognition.[26,55,57]

The first step to develop such nanohybrids (known as magnetopolymersomes) is the selection of the synthetic method and surface coating of iron oxide nanoparticles. To date, numerous syntheses such as coprecipitation, thermal decomposition, hydrothermal synthesis, sol–gel synthesis, microemulsion and sonochemistry methods have been developed to produce iron oxide nanoparticles.[48,58] The commonly used method is the aqueous ferrous and ferric salts coprecipitation in alkaline media as described by Massart. However, it often results in poorly crystalline particles characterized by a quite broad size distribution. A remarkable nonaqueous process yielding highly magnetized, well-crystallized and monodispersed nanoparticles is the thermal decomposition method. The as-obtained nanoparticles possess a hydrophobic surface and are dispersible in an organic solvent.[47,58–60]

Methods for magnetopolymersome preparation are adapted from polymersome preparation techniques. Generally, iron oxide nanoparticles are preformed and then added to the starting organic[26,37,53,54] or aqueous[55] solution. The surface coating of iron oxide nanoparticles determines where they will spatially be located within the polymersome, either in the polymersome lumen (hydrophilic USPIO) or in the polymer membrane (hydrophobic USPIO) (Figure 7.3).

Lecommandoux et al.[61] first described magnetopolymersomes in 2006. They studied the association of vesicle-forming copolymers (PB_{48}-*b*-PGA_{56}) with either aqueous or organic ferrofluids. Hydrophobic USPIOs are confined in the copolymer bilayer, while the association of hydrophilic citrated USPIO gives a mixture of USPIO encapsulated in the lumen and nonencapsulated USPIO.[61]

Another example, showing the importance of the USPIO surface coating on their localization within the polymersomes, was given by Krack and coworkers.[37] They reported the association of vesicle-forming polymers (PI_{53}-*b*-PEO_{28} and $P2VP_{66}$-*b*-PEO_{44}) with trimethylamin-*N*-oxide-coated USPIO. They showed that iron oxide nanoparticles are located

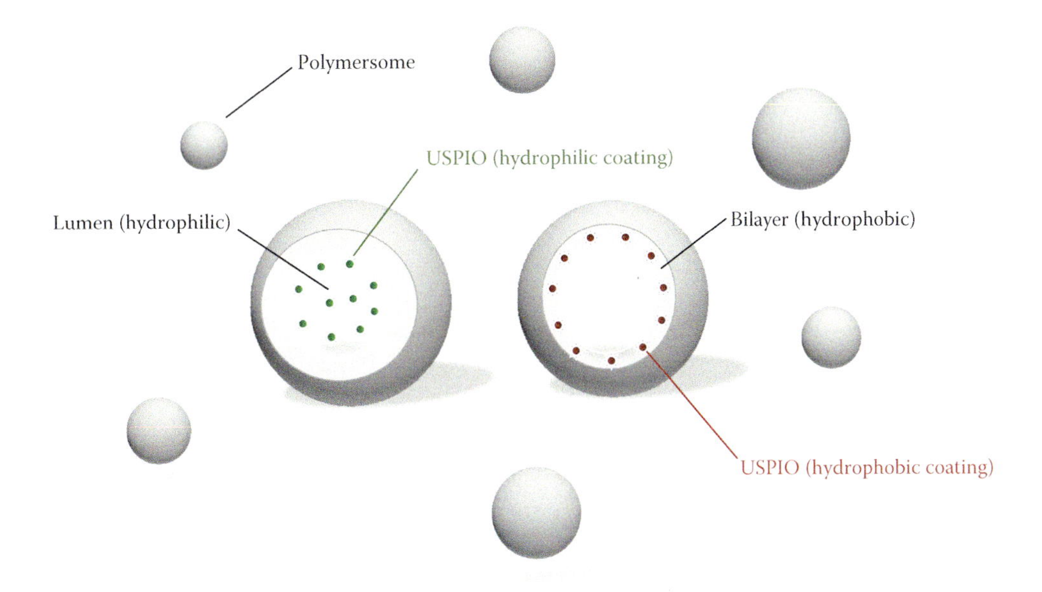

FIGURE 7.3 Encapsulation of iron oxide nanoparticles in polymersomes: either in the vesicle bilayer for hydrophobic USPIO or in the lumen for hydrophilic USPIO.

at the interface between the hydrophobic and the hydrophilic copolymers as trimethylamin-*N*-oxide-coated USPIOs possess a hydrophilic/lipophilic balance between the hydrophobic phase and the hydrophilic phase of the studied copolymers. This leads to a bridging of adjacent bilayers causing an increased tendency to form oligolamellar and multilamellar vesicles. Micrometer-size magnetic vesicles produced by this way are able to migrate under an external magnetic field (mobilities in the range of 12 μm/s for 10-μm-diameter PI-*b*-PEO vesicles and 3 μm/s for 6-μm-diameter P_2VP-*b*-PEO vesicles) and are therefore promising as delivery vehicles for *in vivo* magnetic localization by external magnetic fields.

As highlighted in Table 7.1, most reports on magnetic polymersomes used the nanoprecipitation method with hydrophobic USPIO. Entrapment efficiencies in polymersome membrane close to 100% were reported.[26,28] On the contrary, hydrophilic USPIO should mostly appear in the bulk solution, nonentrapped in the vesicles since the internal volume of vesicles represents only a small fraction of the total aqueous volume as already reported for lipid vesicles.[62] A technique offering the possibility of high encapsulation efficiencies of hydrophilic nanoparticles is based on double W/O/W emulsion techniques. Although this method generally forms micrometer-sized vesicles,[24,63] Yang et al.[64] reported the formation of vesicles having a diameter close to 200 nm and encapsulating water-soluble iron oxide nanoparticles.

Ren et al.[65] reported an original method for loading USPIO in the membrane of PEO-*b*-P(AA-stat-tBA) polymer vesicles. Iron oxide nanoparticles were generated *in situ* within the membrane by adding $FeCl_3$ and $FeCl_2$ solutions: Fe^{3+} and Fe^{2+} cations interact with carboxylate anions in PAA chains and nanoparticles were subsequently formed by adding NaOH.[65] They further used the same strategy to *in situ* deposit iron oxide nanoparticles in biodegradable vesicles made of PEO-*b*-PCL-*b*-PAA copolymers[56] and folic acid-PGA-*b*-PCL copolymers.[66]

A similar strategy based on electrostatic interactions between $AuCl_4^-$ and protonated polymer chains was reported for the incorporation of gold nanoparticles in polymersome membranes. In this study, the protonation of the pH responsive polymer chains localized in the membrane leads to interactions with $AuCl_4^-$. Gold nanoparticles were then *in situ* produced by reduction with $NaBH_4$.[67]

Studies on polymersomes entrapping other nanocrystal structures (gold nanoparticles, quantum dots etc.) have also highlighted a low encapsulation efficiency of water-soluble nanoparticles compared to hydrophobic ones.[68,69] This phenomenon is explained by a significant number of hydrophilic nanocrystals remaining in the aqueous media outside the vesicles. The difficulty in isolating the vesicles from the nonencapsulated water-soluble nanocrystals was also reported.[68]

Other examples highlighting the importance of the nanocrystal surface coating on their localization in polymersomes were reported. For example, by coating nanoparticles (Pb and Au nanoparticles) with amphiphilic copolymers of the same or similar composition as that of the vesicle-forming polymers, a controlled incorporation of nanoparticles in the central portion of vesicle walls was evidenced.[70] Wang et al. showed that PbS, LaOF and TiO_2 nanocrystals localize at both the interior and exterior surfaces of the PS-*b*-PAA vesicles. This was ascribed to a possible binding of the PAA blocks to the surface of the nanocrystals.[68]

Finally, one can notice that an alternative strategy was reported for the formation of tubules and vesicles with Au nanoparticles packed in their membranes.[71] Self-assembly was induced by the

TABLE 7.1
Magnetopolymersomes: USPIO Synthesis and Specifications, Polymer Composition, Self-Assembly Method and Foreseen Applications

Reference	USPIO Synthesis, Surface Coating and Average Diameter	Polymer Composition[a]	Self-Assembly Method	Foreseen Application(s)
Lecommandoux et al.[61]	Coprecipitation H^+ coated (water soluble), citrate coated (water soluble) and Beycostat NE[b] coated (organic solvent soluble) Diameter ≈ 6–7 nm	PBD_{48}-*b*-PGA_{56}	Bulk rehydration	
Sanson et al.[54] and Oliveira et al.[72]	Coprecipitation Beycostat NB09[c] coated (organic solvent soluble) diameter ≈ 6–7 nm	$PTMC_{24}$-*b*-PGA_{19}	Nanoprecipitation	Theranostic nanocarriers (MRI contrast agent and DOX triggered release under radiofrequency oscillating field)
Pourtau et al.[57]	Coprecipitation Beycostat NB09[c] coated (organic solvent soluble)	Mixture of $PTMC_{26}$-*b*-PGA_{20}, $PTMC_{26}$-*b*-PGA_{20}-*b*-PEG_{45}-maleimide and $PTMC_{26}$-*b*-PGA_{20}-FITC	Nanoprecipitation	Targeting and imaging of bone metastases by MRI (trastuzumab targeted magnetic polymersomes)
Arosio et al.[28]	Coprecipitation Beycostat NB09[c] coated (organic solvent soluble) Three different diameters ≈ 6–7 nm, ≈ 8–10 nm, ≈ 10–15 nm	$PTMC_{25}$-*b*-PGA_{12}	Nanoprecipitation	MRI contrast agent
Agut et al.[73]	Coprecipitation Beycostat NB09[c] coated (organic solvent soluble) diameter ≈ 6–7 nm	poly(γ-benzyl-L-glutamate) $(PBLG)_{41}$-*b*-poly(2-dimethylamino)ethyl $methacrylate_{85}$	Nanoprecipitation	Vesicle to micelle transition upon USPIO incorporation
Krack et al.[37]	Thermal decomposition Trimethylamin-*N*-oxide coated (organic solvent soluble) diameters = 8.6 nm and 14.1 nm	PI_{53}-*b*-PEO_{28} and $P2VP_{66}$-*b*-PEO_{44}	Film rehydration	Magnetic localization by external magnetic fields of micrometer-sized magnetic polymersomes
Hickey et al.[52,53]	Thermal decomposition Oleic acid coated (organic solvent soluble) Diameters = 5.6 nm, 5.8 nm, 6.4 nm, 9.9 nm, 10.8 nm, 15.5 nm, 16.3 nm and 19.9 nm	PAA_{38}-*b*-PS_{73}	Nanoprecipitation	MRI contrast agent
Yang et al.[64]	Thermal decomposition followed by ligand addition Tetramethylammonium 11-aminoundecanoate coated (water soluble) Diameter = 6 nm	Mixture of folate-PEG_{45}-*b*-PLA_{194} and NH_2-PEG_{45}-*b*-PLA_{194}	Double emulsion method	Tumour targeting MRI contrast agent
Yang et al.[55]	Thermal decomposition followed by ligand addition Tetramethylammonium 11-aminoundecanoate coated (water soluble) Diameter = 6 nm	Mixture of folate-PEG_{114}-PLA_{293}-PEG_{46}-acrylate and CH_3O-PEG_{114}-PLA_{293}-PEG_{46}-acrylate	Double emulsion method	Tumour targeting theranostic nanocarriers (MRI contrast agent and DOX triggered release under acidic conditions)
Yang et al.[74]	Thermal decomposition followed by ligand addition Tetramethylammonium 11-aminoundecanoate coated (water soluble) Diameter = 6 nm	Mixture folate-PEG_{114}-$PBLG_{157}$-$hydrazone_{14}$-DOX_{13}-PEG_{46}-acrylate and CH_3O-PEG_{114}-$PBLG_{157}$-$hydrazone_{14}$-DOX_{13}-PEG_{46}	Bulk rehydration	Tumour targeting theranostic nanocarriers (MRI contrast agent and DOX triggered release under acidic conditions)
Yang et al.[75]	Thermal decomposition Oleic acid and oleyamine coated USPIO (organic solvent soluble) Diameter = 6 nm	Mixture of folate-PEG_{46}-PLA_{194} and CH_3O-PEG_{46}-PLA_{194}	Double-emulsion method	Tumour targeting theranostic nanocarriers (MRI contrast agent and DOX triggered release under acidic conditions)

(*Continued*)

TABLE 7.1 (CONTINUED)
Magnetopolymersomes: USPIO Synthesis and Specifications, Polymer Composition, Self-Assembly Method and Foreseen Applications

Reference	USPIO Synthesis, Surface Coating and Average Diameter	Polymer Composition[a]	Self-Assembly Method	Foreseen Application(s)
Sun et al.[76]	Thermal decomposition followed by ligand addition Tetramethylammonium 11-aminoundecanoate coated (water soluble) Diameter = 6 nm	PEG_{22}-*b*-PAsp$(DIP)_{58}$	Double emulsion method	Theranostic nanocarriers (MRI contrast agent and DOX triggered release under acidic conditions)
Ren et al.[65]	*In situ* coprecipitation of iron oxide nanoparticles in the vesicle membrane Diameter = 6 nm	PEO_{43}-*b*-P(tBA_{56}-stat-AA_9) and PEO_{43}-*b*-P(tBA_{40}-*stat*-AA_{25})	Nanoprecipitation	MRI contrast agent
Qin et al.[56]	*In situ* coprecipitation of iron oxide nanoparticles in the vesicle membrane Diameter = 1.9 nm	PEO_{43}-*b*-PCL_{98}-*b*-PAA_{25}	Nanoprecipitation	Theranostic nanocarriers (MRI contrast agent and DOX delivery)
Liu et al.[66]	*In situ* coprecipitation of iron oxide nanoparticles in the vesicle membrane Diameter = 7 nm	Folate-PGA_{42}-*b*-PCL_{35}	Nanoprecipitation	Theranostic nanocarriers (MRI contrast agent and DOX delivery)
Bleul et al.[26]	Thermal decomposition oleic acid coated (organic solvent soluble) Diameter = 6 nm	PEO_5-*b*-PPO_{68}-*b*-PEO_5 and COOH-PEO_5-*b*-PPO_{68}-*b*-PEO_5-COOH[d]	Microfluidic (micromixer technology)	Tumour targeting theranostic nanocarriers (MRI contrast agent and camptothecin delivery)
Bixner et al.[77]	Thermal decomposition *N*-palmityl-6-nitrodopamide coated USPIO (organic solvent soluble) Diameter: 3.5 nm	HOOC-PI_{17}-*b*-$PNIPAM_{8.5}$-S-S-CH_3	Nanoprecipitation	Calcein triggered release by an alternating magnetic field

[a] The subscripts indicate the degrees of polymerization.
[b] Beycostat NE is a phosphoric diester type tensioactive.[61]
[c] Beycostat NB09 is an anionic surfactant composed of a mixture of monoesters and diesters of phosphoric acid with alkylphenol chains containing nine ethoxy groups.[54]
[d] The covalent link to Lys3-bombesin was performed post polymersome formation.

film rehydration procedure from Au nanoparticles directly grafted with amphiphilic block copolymers. Morphologies of the as-obtained self-assembly nanostructures depend on the hydrophilic/hydrophobic balance, which can be tuned by varying the diameter of Au nanoparticles and/or the molecular weight of the hydrophilic/hydrophobic blocks of the copolymer.[71]

The incorporation of nanoparticles can remarkably affect the self-assembly structure of block-copolymers. For example, Hickey and coworkers[53] produced magnetopolymersomes by the self-assembly of micelles-forming copolymers with hydrophobic USPIO. This phenomenon was explained by an increase in the relative volume of the hydrophobic block induced by USPIO incorporation, promoting the formation of magnetopolymersomes instead of magnetomicelles (Figure 7.4). Agut et al.[73] showed that self-assembly of vesicles forming poly(2-dimethylamino)ethyl methacrylate-*b*-poly(γ-benzyl-L-glutamate) (PBLG) copolymers with USPIO produced magnetomicelles instead of magnetopolymersomes due to depletion attractions between USPIO spheres in PBLG rods, leading to a vesicle-to-micelle transition.

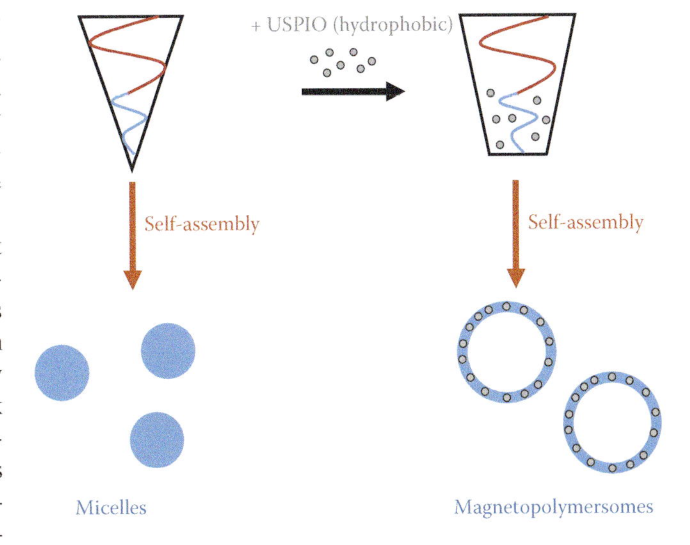

FIGURE 7.4 Scheme of the relative hydrophobic volume ratio change caused by the addition of hydrophobic USPIO inducing the micelle-to-vesicle transition. (Adapted from Yang, X. et al., *Biomaterials*, 31, p. 9065–9073, 2010.)

7.2.2 Characterization of Magnetopolymersomes

In this section, techniques that are most commonly used to characterize the structure of magnetopolymersomes are briefly presented (Table 7.2).

7.2.2.1 Scattering Methods

The dynamic light scattering (DLS) technique gives a mean value of the hydrodynamic radius (R_H) of the vesicles. The mean gyration radius (R_G) is obtained from static light scattering (SLS) measurements, which allows to calculate the ratio $\rho = R_G/R_H$. While vesicles are characterized by ρ values close to 1, ρ values around 0.775 are expected for spherical micelles.[12,54,78] The aggregation number (Z), defined as the number of molecules per aggregate, can be calculated by dividing the weight-average molecular weight of the aggregate ($M_{w,\,agg}$), determined from SLS measurements by the molecular weight of the polymer chain. Vesicular morphologies are characterized by a higher aggregation number than micelles.[78]

The small-angle neutron scattering (SANS) method appears as a reliable technique to characterize vesicular morphologies. The scattering intensity from a suspension of particles is given by the following equation:

$$I(q) = \Phi V_p \Delta\rho^2 P(q) S(q), \quad (7.2)$$

where Φ is the particle volume fraction; V_p is the particle volume; $\Delta\rho$ is the difference of scattering length density between the particles and the solvent; $P(q)$ is the form factor, which gives information about the particle morphology; and $S(q)$ is the structure factor giving information about particle interactions ($S(q)$ is equal to 1 for dilute and noninteracting suspension of particles).[79] As the scattering intensity depends on $\Delta\rho$, different solvent mixtures can be used to match the scattering of one component of the sample. For example, for magnetopolymersomes, pure D_2O is used to almost match the nuclear signal of the USPIO, while the use of a H_2O/D_2O mixture is used to match the copolymer scattering. Scattering intensity in pure D_2O reveals the overall morphology and the membrane thickness. The SANS curve in the H_2O/D_2O mixture matching the copolymer scattering can be divided by the volume fraction and the form factor of USPIO, measured independently. It has been shown that the resulting intra-aggregate structure factor ($S_{intra}(q)$) of the USPIO is associated with the global form factor of the object that they decorated and therefore gives information about their spatial arrangement into aggregates of a given morphology (*i.e.* micellar or vesicular).[54,61,80]

TABLE 7.2
Characterization Techniques for the Structural Analysis of Magnetopolymersomes

Characterization Techniques	Main Characteristics
Light scattering: DLS and SLS	Determination of the hydrodynamic radius (R_H) by DLS
	Determination of the gyration radius (R_G) by SLS
	$\rho = R_G/R_H$ gives an information about the aggregate morphology (micelle, vesicle)
	Determination of the weight-average molecular weight of the aggregate ($M_{w,\,agg}$) and the aggregation number (Z)
SANS	Determination of the aggregate morphology (micelle, vesicle)
	Determination of the membrane thickness
	Determination of the spatial arrangement of USPIO
TEM, cryo-TEM	Size and morphology (2D projections of the structure)
TEM tomography, cryo-TEM tomography	Size and morphology (3D structure)
	Determination of the spatial arrangement of USPIO
AFM	Size and morphology

7.2.2.2 Microscopy Techniques

Microscopy techniques are frequently used to visualize polymersomes. Optical microscopy allows investigating vesicles under physiological conditions but possesses a limited resolution (vesicles with diameters above 1 μm are best suited for optical microscopy), while electron microscopy yields highly resolved images, but specimens need to be dried, stained or frozen.[19] Most reports on magnetopolymersomes used transmission electron microscopy (TEM) to study morphology. As magnetopolymersomes are three-dimensional (3D) structures, it is difficult to elucidate their morphology by simple TEM microscopy using 2D projections of 3D objects. Copolymers have almost no contrast under TEM due to the very low attenuation of the electron beam. Therefore, hybrid polymersomes are largely present as isolated clusters of nanoparticles[26,28,53] (Figure 7.5). TEM tomography is more suitable to confirm

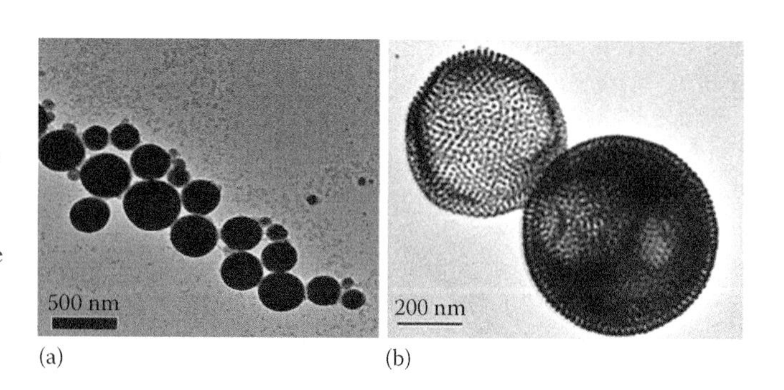

FIGURE 7.5 TEM images of polymersomes loaded with (a) iron oxide nanoparticles[53] and (b) gold nanoparticles.[71] Nanoparticles are seen everywhere in the vesicle in the TEM images. (a, Reprinted with permission from Hickey, R.J. et al., Controlling the self-assembly structure of magnetic nanoparticles and amphiphilic block-copolymers: from micelles to vesicles. *Journal of the American Chemical Society*, 2011. 133(5): p. 1517–1525. Copyright 2011 American Chemical Society. b, Reprinted with permission from He, J. et al., Self-assembly of inorganic nanoparticle vesicles and tubules driven by tethered linear block copolymers. *Journal of the American Chemical Society*, 2012. 134(28): p. 11342–11345. Copyright 2012: American Chemical Society.)

the hollow structure of polymersomes. A series of tilted projection images is collected at different tilt intervals to reconstruct the 3D structure of the sample.[52] However, the morphology could be partly destroyed during the drying step using TEM methods. To overcome this problem, cryo-TEM is used to visualize colloids in the frozen–hydrated state. Therefore, cryo-TEM tomography appears to be the most powerful microscopy technique to visualize 3D structures in a frozen–hydrated state as close to their native state as possible, as already reported in the case of magnetoliposomes.[81]

The atomic force microscopy (AFM) technique has also proved to be helpful to study polymer vesicles. However, one has to be careful when using AFM data as the objects deposit on a mica surface could be different from those in the solution before deposition.[78] For example, a study on magnetopolymersomes has shown a bursting and total spreading of the vesicles onto the mica substrate, resulting in larger diameters appearing on the AFM images compared to the hydrodynamic sizes. The average thicknesses of membranes spread on mica were deduced from the AFM height images.[54]

Besides these structural characterizations, it is also important to investigate the magnetic properties of the resulting magnetopolymersomes. Magnetic behaviour can be measured using a vibrating sample magnetometer or a superconducting quantum interference device magnetometer. Similar magnetization curves for both USPIO and USPIO loaded in polymersomes were reported. The absence of hysteresis from these magnetization curves confirmed the superparamagnetic property of USPIO embedded in polymersomes.[54,64,76] The degree of USPIO encapsulation in polymersomes strongly affects their magnetic properties, which can be measured experimentally by relaxivity measurements,[54] as discussed in more detail in the subsequent section.

7.2.3 Magnetopolymersomes as MRI Contrast Agents

7.2.3.1 Introduction to MRI

7.2.3.1.1 Generalities

Despite its poor sensitivity, MRI is one of the most used medical imaging techniques. This technique allows a precise visualization of organs and tissues by the acquisition of virtual cuts made in all directions of space. Its interest lies in the detection, noninvasively and with an exceptional spatial resolution, of tissue anomalies (malformations or tumours). Based on the principle of nuclear magnetic resonance (NMR), information is obtained by the detection of NMR-sensitive nuclei present in biological tissues. In this short chapter, we will not go into the theoretical aspects of NMR but just give some basics in order to understand the action of contrast agents. Several books describe the technique in more detail.[82–84]

7.2.3.1.2 NMR and Relaxation Phenomenon

The NMR principle applies to atoms possessing a nuclear spin. Typically, MRI focuses on hydrogen atoms (^1H), which are abundantly present in living organisms. As a rotating charge, the proton has a magnetic moment μ, which is represented by a microscopic magnetization vector oriented in the same direction as the rotation axis. Under normal conditions, magnetic moments are randomly oriented, resulting in a net magnetization equal to zero. When the body is placed in an external static magnetic field (B_0), such as in an MRI scanner, all magnetic moments will precess around the magnetic field axis with a specific frequency (Larmor frequency, ω_0) given by the Larmor equation:

$$\omega_0 = (\gamma/2\pi)B_0, \tag{7.3}$$

where γ is the gyromagnetic ratio of the nucleus and B_0 is the magnetic field strength.

A longitudinal magnetization (M_z, magnetization along the z axis, which is parallel to B_0) is produced because a few more spins rotate around B_0 in a parallel rather than in an antiparallel direction. The transverse magnetization (M_{xy}, magnetization along the xy plane, which is perpendicular to B) is equal to zero because each spin rotates with different frequencies. In fact, each proton is subjected to field inhomogeneity and precess with a frequency, which is slightly larger or lower than ω_0.

In a typical NMR experience, a population of nuclei placed in a static external magnetic field (B_0) will be disturbed by the application of a perpendicular oscillating electromagnetic field (B_1), leading to a tilt (generally 90° or 180° angle) of the net magnetization. When the radiofrequency source is switched off, the system returns to its equilibrium state. The return to equilibrium from the excited to the steady state (Figure 7.6) is called relaxation and is characterized by the following:

- T_1, which is the longitudinal relaxation time (spin-lattice). This time defines the growing of the longitudinal magnetization following an exponential law and corresponds to the time taking by M_z to return to 63% of its initial value after a 90° pulse.
- T_2, which is the transverse relaxation time (spin-spin). In this case, this time characterizes the decrease in the transverse magnetization following an exponential function and corresponds to the time required for M_{xy} to decay to 37% of its initial value after a 90° pulse.

Relaxation phenomena are important in the context of MRI as they affect the contrast.[85]

7.2.3.1.3 Magnetic Resonance Imaging

By MRI, it is possible to obtain images displaying the spatial distribution of spin density. The origin of such images is the emission of radiofrequency signals due to nuclei relaxation. Depending on their environment and their dynamics, the relaxation kinetics vary and induce modulations of the intensity of the image depending on the tissue's nature. To code an image, it is essential to precisely discriminate the

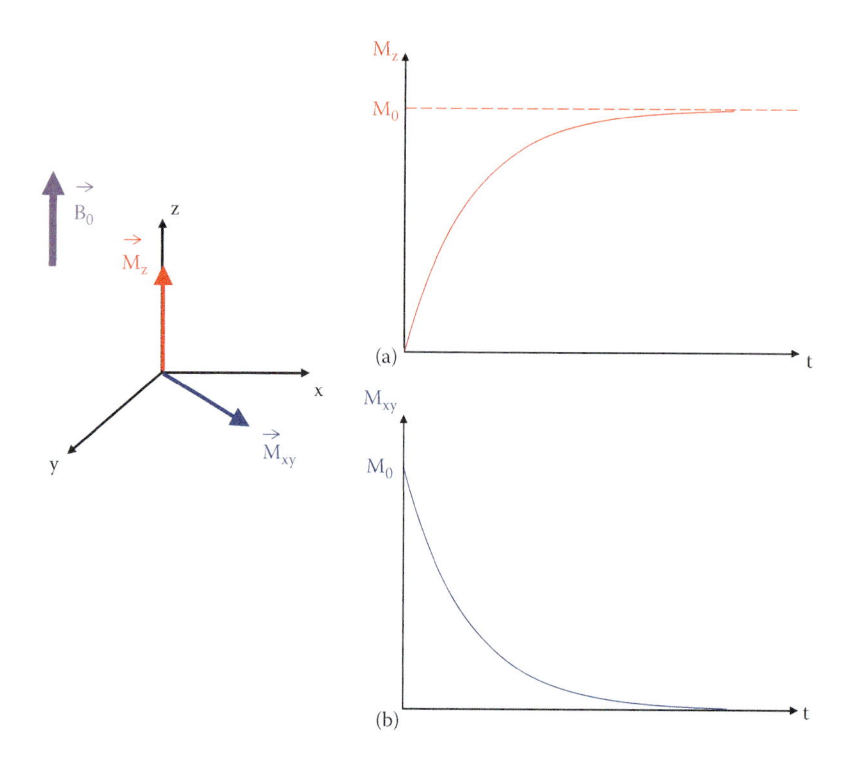

FIGURE 7.6 Return to the equilibrium state: (a) recovery of the magnetization along the z axis after a 90° pulse: M_z (t) = M_0 (1 – exp (–t/T_1)) and (b) decay of the magnetization in the xy plane after a 90° degree pulse: M_{xy} (t) = M_0 exp (–t/T_2).

different elements of volume (*i.e.* the voxels). To proceed, a magnetic field gradient, the intensity of which varies according to a given direction, will be superimposed (by pulses) on the static magnetic field B_0. A gradient $\vec{G}$ is defined as a linear variation in the magnetic field with respect to position. Three gradients (G_x, G_y, G_z) are used to obtain a tridimensional localization. For a proton located in position $\vec{\mu}$ (x, y, z), Larmor frequency will be:

$$\begin{aligned} v &= \frac{\gamma}{2\pi}(B_0 + \vec{G}.\vec{\mu}) \\ v &= \frac{\gamma}{2\pi}(B_0 + G_x x + G_y y + G_z z). \end{aligned} \tag{7.4}$$

These magnetic field gradients will therefore make it possible to differentiate the various proton compositions within the body and enable their mapping.

7.2.3.1.4 Contrast Agents in MRI

The contrast in MRI comes, among other things, from local difference in T_1 (longitudinal relaxation time) and T_2 (transverse relaxation time) and therefore depends on the chemical and physical nature of the tissues and on local variation in proton density. To improve the sensitivity and the specificity of MRI, exogenous substances called MRI contrast agents can be administered.

An MRI contrast agent can alter the signal intensity by shortening T_1 and/or T_2 of water protons located nearby. Reduction of T_1 results in a hypersignal leading to a positive contrast, while shortened T_2 reduces signal giving rise to a negative contrast. The efficiency of an MRI contrast agent is quantified by its longitudinal (r_1) and transverse (r_2) relaxivities, defined as the increase in the proton nuclear relaxation rates R_1 ($1/T_1$) and R_2 ($1/T_2$) of the solvent caused by 1 mM of an active compound per litre (s^{-1} mM^{-1}). In the last decades, the chemistry of contrast agents has continued to evolve with the aim of developing MRI contrast agents with great stability and high efficacy for clinical applications.

Contrast agents for MRI are often classified as positive or T_1-contrast agents and negative or T_2-contrast agents. T_1-contrast agents predominantly reduce the longitudinal relaxation time of protons (T_1) and increase the signal where they are located. Consequently, they appear brighter than pure water. Typically, ions with unpaired electrons such as gadolinium (III), manganese (II) and iron (II) or (III) efficiently reduce T_1. In contrast, iron oxide nanoparticles can predominantly shorten the transverse relaxation time of proton (T_2) by accelerating the dephasing of the proton magnetic moments. Due to their very large magnetic moment, they produce magnetic field inhomogeneities through which water molecules diffuse and consequently speed up the dephasing of the magnetic moments. Iron oxide nanoparticles are therefore mainly used as T_2-contrast agents, which decrease MRI signal intensity and thus darken the regions where they are present. The most effective T_2-weighted contrast agents are those characterized by a higher r_2/r_1 ratio. For example, due to their high r_2/r_1 ratio, SPIO were initially developed as T_2-contrast agents.[86]

7.2.3.2 Use of Magnetopolymersomes in MRI

Recently, magnetopolymersomes were found to induce important T_2-contrast enhancement effect in MRI. Clustering of several USPIOs inside a single nanoplatform is known to further reduce the T_2 relaxation time in comparison to individual USPIO, as already reported for hydrophobic USPIO in micelles[87,88] and hydrophilic USPIO in liposomes.[89,90] Moreover, the encapsulation of USPIO inside the aqueous lumen or in the bilayer of polymer vesicles limits the diffusion of bulk water molecules in the vicinity of USPIO, resulting in an increase in the T_1 relaxation time compared to nonencapsulated USPIO. Magnetopolymersomes with high transverse relaxivities and high r_2/r_1 ratios have been reported, which clearly demonstrates their potential as highly sensitive MRI probes.[26,28,52,54,55,65,66,76] One can quote, for example, the study of Arosio et al.,[28] who compared the relaxivities of Endorem (a commercial superparamagnetic contrast agent composed of several iron oxide nanoparticles coated by dextran) and magnetopolymersomes. They showed that magnetopolymersomes possess r_2/r_1 ratios (at 1.4 T) up to one order of magnitude higher than 10, which was the value measured for Endorem.[28]

The efficiency of magnetopolymersomes as MRI contrast agents will strongly depend on the applied magnetic field. In fact, r_1 decreases rapidly as a function of the applied magnetic field, while r_2 reaches a plateau value,[54] resulting in an increased r_2/r_1 ratio at higher magnetic fields.

Sanson and coworkers measured T_1 and T_2 relaxation times (at 4.7 T) of magnetic polymersomes based on PTMC-*b*-PGA copolymers possessing similar size ($R_H \approx 50$ nm) and different USPIO contents.[54] The transverse relaxivity increased from 81 s^{-1} mM^{-1} to 182 s^{-1} mM^{-1} as the USPIO content increased, while the longitudinal relaxivity remained weak and almost constant (between 2.8 and 3.6 s^{-1} mM^{-1}), hence resulting in a higher r_2/r_1 ratio (up to 52), with a higher USPIO content. The USPIO content was increased up to a threshold value above which hydrophobic USPIO began to aggregate, forming macroscopic clusters. Magnetopolymersomes with a low (approximatively 6.7 μg/mL) MRI detection limit (defined as the copolymer concentration at which the MRI signal intensity decreases to 50% of that of pure water) were therefore obtained. They further functionalized the magnetic polymersome surface with antibodies targeting breast cancer bone metastasis and showed that a higher contrast was observed in comparison to the naked magnetic polymersomes (without antibodies) within the bone metastasis by *in vivo* MRI. Several antibodies can be bound to polymersome surfaces, thus increasing their targeting efficiency.[57] Magnetopolymersomes with low MRI detection sensitivity were also reported in other studies.[64–66,74] Figure 7.7 shows an example of T_2-weighted MRI images obtained with magnetopolymersomes. A very strong negative contrast was produced for an iron concentration in the μM range, confirming that magnetopolymersomes can act as a highly efficient T_2-contrast agent.

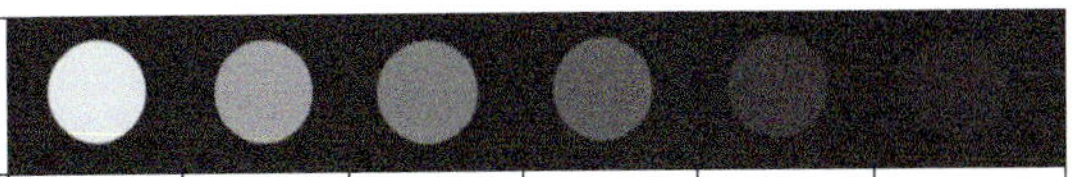

MRI grey level (%)	100	73	54	40	18	2
[Fe] (μM)	0	21	37	49	74	148
C_{weight} (μg/mL)	0	2.9	5	6.7	10	20
$C_{vesicle}$ (nM)	0	0.25	0.42	0.57	0.85	1.7

FIGURE 7.7 T_2-weighted MRI images obtained from magnetic PTMC-*b*-PGA polymersomes at different dilution factors. (Reprinted with permission from Sanson, C. et al., Doxorubicin loaded magnetic polymersomes: theranostic nanocarriers for MR imaging and magneto-chemotherapy. *ACS Nano*, 2011. 5(2): p. 1122–1140. Copyright 2011 American Chemical Society.)

The effect of the overall morphology resulting from the self-assembly of amphiphilic copolymers with USPIO on the magnetic relaxation properties was studied by the group of Hickey et al.[52,53] and Arosio et al.[28] They demonstrated that the individual USPIO size has a tremendous effect on the final transverse relaxivity, with higher r_2 obtained when increasing USPIO diameter. For a given USPIO size, vesicles possessing a larger number of nanoparticles show better transverse relaxivity. As the number of nanoparticles that can be encapsulated per assembly decreases when increasing USPIO size, it appears important to determine which option is better: increasing individual USPIO size or using smaller USPIOs that enable higher loadings. Arosio et al.[28] showed that the individual size of USPIO has the greatest effect on r_2 as better r_2 values were obtained for vesicles loaded with 10–15 nm USPIO at 20% feed weight ratio (FWR; defined as the weight of iron oxide divided by the copolymer weight multiplied by 100) (r_2 at 1.4 T = 280 s^{-1} mM^{-1}) than those loaded with 6–7 nm USPIO at 50% FWR (r_2 at 1.4 T = 114 s^{-1} mM^{-1}). They also demonstrated that, for a given number of encapsulated USPIOs, the cluster structure (micelles or vesicles) has a mild influence on the transverse relaxivity. These results are in accordance with Monte Carlo simulations of spherical micelles and hollow shells of USPIO clusters, leading to similar r_2 values for both geometries.[28] Another parameter affecting transverse relaxivity is polymersome size, as the transverse relaxation rate increases with decreasing polymersome size.[28,53] Finally, for a given USPIO size at a fixed magnetic field, a linear relationship between r_2 and the ratio of the number of encapsulated USPIO (N_{USPIO}) over the hydrodynamic diameter of the clusters (D_H) was obtained at 1.4 T.[28]

$$r_2 \approx N_{USPIO}/D_H \tag{7.5}$$

7.2.4 Magnetopolymersomes as Nanotheranostic Systems

In recent years, several studies have been performed to combine both therapeutic and diagnostic functionalities within a single nanoparticle. These theranostic nanoparticles (nanotheranostic) allow monitoring the efficiency of a proposed treatment in real time. This approach aims to better adapt the treatment for each patient, taking into account the individual variability, and holds great promise in the emerging field of personalized medicine. As exemplified in Figure 7.8, by combining noninvasive imaging and tumour-targeted drug delivery, patients responding more positively to a nanomedicine treatment can be selected. The first step is the selection of patients showing medium to high levels of accumulation of the nanotheranostic formulation within the tumour. For individuals accumulating no or low nanotheranostic agents within the tumour, alternative treatments are proposed. In addition, during the first selection step, patients presenting high levels of nanomedicine accumulation in healthy organs can be excluded in order to attenuate side effects. In the second step, imaging is used to monitor therapeutic efficiency. Although proof of concept of the effectiveness of nanotheranostics has already been demonstrated, clinical trials are still ongoing before its introduction into clinical practice.[91,92]

Given their high efficiency as T_2-contrast agents for MRI, magnetic polymersomes loaded with anticancer drugs have been shown to have high potential for cancer nanotheranostic. Magnetopolymersomes based on PTMC-*b*-PGA copolymers and used as drug carriers for DOX were the first example of their use for cancer theranostics. The release of DOX was faster at higher temperatures due to the semi crystalline nature of the PTMC blocks forming the membrane. An approximatively two-fold enhancement of DOX release rate under local hyperthermia conditions induced by the application of an oscillating magnetic field (2.65 mT at 500 kHz) was subsequently evidenced. Due to their high USPIO loading, magnetopolymersomes are efficient T_2-MRI contrast agents, allowing one to monitor the treatment process by MRI. *In cellulo* studies on HeLa cells further demonstrated that upon cell internalization and exposure to an alternating magnetic field (14 mT at 750 kHz), these magnetic polymersomes boost the release of DOX into the intracellular compartment, resulting in increased cell toxicity.[72]

Other examples using magnetopolymersomes as theranostic nanocarriers were reported by Yang et al.[55,74] They first produced worm-like vesicles composed of triblock copolymers R(methoxy or folate)-PEG_{114}-PLA_x-PEG_{46}-acrylate) dually loaded with USPIO and DOX. The short PEG_{46}-acrylate segments were mostly segregated to the inner hydrophilic vesicle layers and were cross-linked via free radical polymerization for enhanced *in vivo* stability while the long R(methoxy or folate)-PEG_{114} segments were mostly segregated to the outer hydrophilic vesicle layers and provided an active tumour targeting ability. A faster DOX release at pH 5.3 than at pH 7.4 was reported and attributed to a faster degradation of the vesicle membrane at lower pH values. This pH dependent release is of particular interest to improve the efficacy of targeted cancer therapy. Indeed, it is expected that most of the loaded DOX will remain in the vesicle during circulation in the bloodstream (where the pH is approximatively 7.4),

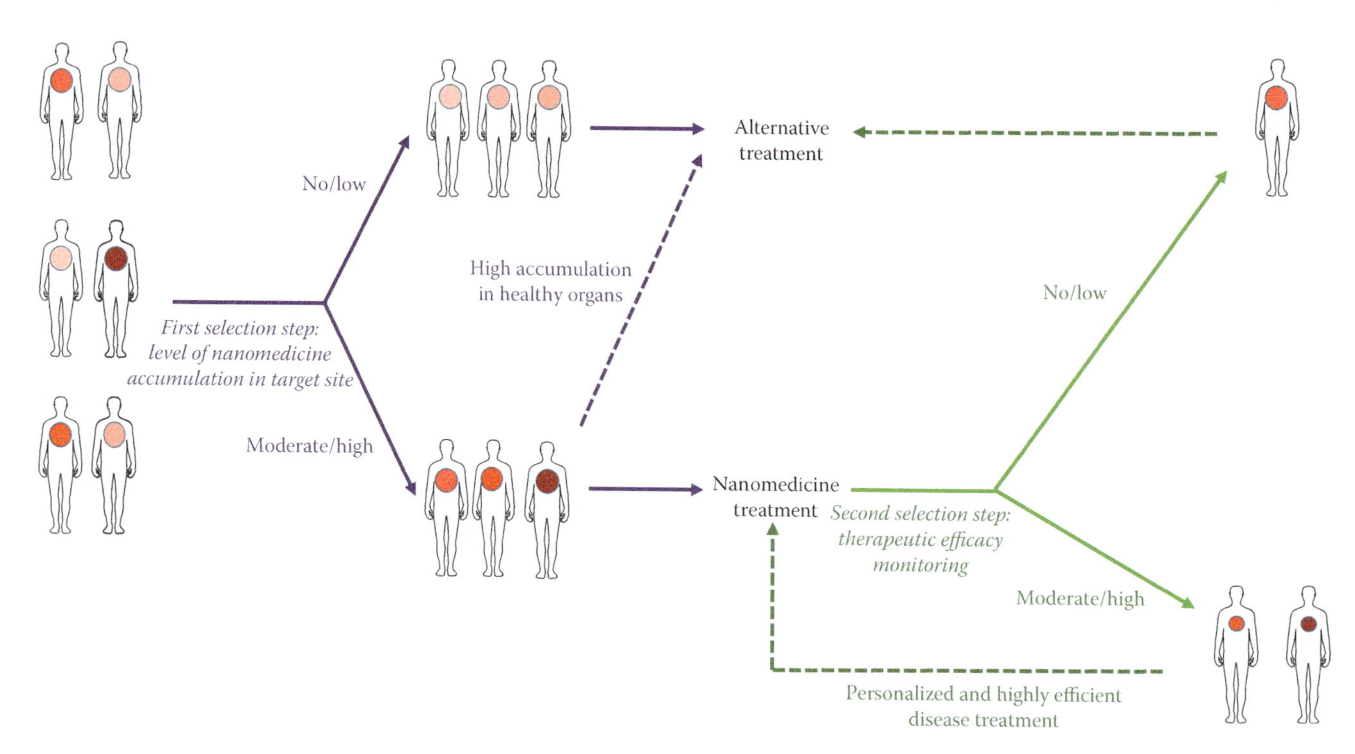

FIGURE 7.8 Scheme of the theranostic approach allowing imaging-guided and personalized nanomedicine: patients are preselected by combining non-invasive information on the target accumulation (first selection step) and therapeutic efficacy (second selection step). (Adapted from Rizzo, L.Y. et al., *Curr. Opin. Biotechnol.*, 24, 1159–1166, 2013.)

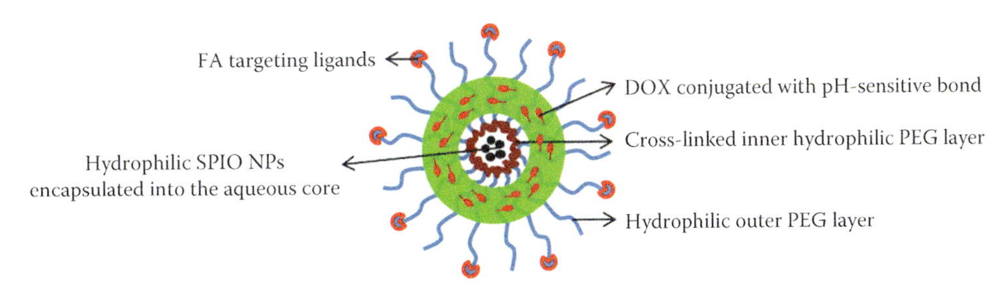

FIGURE 7.9 Scheme of tumour targeting magnetopolymersomes used as theranostic nanocarriers (MRI contrast agent and DOX triggered release under acidic conditions): spherical polymersomes formed from copolymers R(folate or CH_3-O)-poly(ethylene glycol)$_{114}$-poly(glutamate)$_{157}$-hydrozone$_{14}$-doxorubicin$_{13}$-poly(ethylene glycol)$_{46}$-acrylate. (Reproduced with permission from Yang X. et al., *ACS Nano*, 4, 6805–6817, 2010.)

while faster release will occur once the vesicles will be taken up by tumour cells via folate receptor-mediated endocytosis (the pH inside endosomes ranges from 4.5 to 6.5).[55] They further replaced the PLA hydrophobic block by a polyglutamate segment conjugated to DOX through an acid-sensitive hydrazone bond to achieve pH-triggered drug release (Figure 7.9). Higher DOX loading and more pronounced acid-triggered release were achieved via this covalent DOX conjugation.[74]

pH-sensitive theranostic nanocarriers encapsulating both DOX and USPIO were also obtained from vesicles of the diblock copolymer PEG_{22}-*b*-$(PAsp(DIP))_{58}$. DOX release from the vesicles was faster at pH 5 than at pH 7.4 (37°C) due to the protonation of the diisopropylamino groups at pH 5.0, which changed the vesicle membrane from hydrophobic to hydrophilic and induced the disassembly of vesicles. MTT cytotoxicity assay conducted on Bel 7402 cells revealed that DOX loaded magnetopolymersomes showed high cytotoxicity while magnetopolymersomes without DOX did not show significant cell growth inhibition (at a concentration of vesicles up to 0.4 mg/ml). Encapsulation of USPIO in these polymersomes results in an increase in the r_2/r_1 ratio compared to the starting hydrophilic USPIO (100.5 *vs.* 10.5 at 1.4 T). These pH-sensitive magnetic vesicles are therefore promising as an MRI-visible drug delivery system combining rapid intracellular drug release and high MRI detection sensitivity.[76]

7.3 CONCLUSION AND OUTLOOK

Over the past decades, vesicles formed from polymeric amphiphiles have been widely studied and found to have many desirable properties compared to their lipid analogues. Their properties can be tailored to their intended applications. A large variety of polymers has already been designed to form vesicles, but the possibility of polymer chemistry seems to be practically limitless.

Several studies have highlighted the benefit of the inclusion of iron oxide nanoparticles into polymersomes. These so-called magnetopolymersomes possess high transverse-to-longitudinal relaxivity ratios provided by the USPIO clustering effect and have therefore emerged as efficient T_2 MRI contrast agents. Moreover, therapeutic compounds (hydrophilic or hydrophobic) can also be coencaspulated within magnetic polymersomes, extending their applications to the theranostic field.

However, the self-assembly of amphiphilic copolymers to form polymersomes is a rather complex phenomenon depending on several parameters, such as the polymer composition and the experimental conditions used to induce the self-assembly. When designing multifunctional systems such as polymersomes loaded with iron oxide nanoparticles and/or drugs, the number of factors affecting the self-assembly increases. Therefore, the formation of such systems is quite labour-intensive and cost-intensive. The complexity of multifunctional polymersomes could also put some restrictions for their transfer to clinic applications, which need robust and scalable production methods.

Although some challenges need still to be overcome before their introduction to clinic, magnetopolymersomes exhibit improved properties in terms of stability and detection sensitivity, making them interesting candidates for biomedical applications. Future researches are needed to allow reproducible and large-scale syntheses of magnetopolymersomes. The utilization of biocompatible and/or biodegradable polymers in combination with controlled release approaches needs to be privileged.

ABBREVIATIONS

AFM	atomic force microscopy
DLS	dynamic light scattering
DOX	doxorubicin
FITC	fluorescein isothiocyanate
MRI	magnetic resonance imaging
NMR	nuclear magnetic resonance
P2VP	poly(2-vinylpyridine)
PAA	poly(acrylic acid)
P(Asp(DIP))	poly(2-(diisopropylamino)ethyl aspartate)
PB	poly(butadiene)
PBLG	poly(γ-benzyl-l-glutamate)
PCL	poly(ε-caprolactone)
PDMAEMA	poly(2-dimethylamino)ethyl methacrylate
PEE	poly(ethylethylene)
PEG	poly(ethylene glycol)
PEO	poly(ethylene oxide)
PGA	poly(glutamic acid)
PI	poly(isoprene)
PLA	poly(lactic acid)

PNIPAM	poly(N-isopropylacrylamide)
PPS	poly(propylenesulfide)
PS	poly(styrene)
P(tBA)	poly(tert-butyl acrylate)
PTMC	poly(trimethylene carbonate)
RF	radiofrequency
SANS	small-angle neutron scattering
SLS	static light scattering
T_1	longitudinal relaxation time
T_2	transverse relaxation time
TAX	paclitaxel
TEM	transmission electron microscopy
USPIO	ultrasmall superparamagnetic iron oxide nanoparticles

ACKNOWLEDGEMENTS

This work was performed with the financial support of the Fonds National pour la Recherche Scientifique (F.R.S.-FNRS), the FEDER, the Walloon Region, the COST Action TD1402, the Centre for Microscopy and Molecular Imaging (CMMI) supported by the European Regional Development Fund of the Walloon Region, the ARC and UIAP programs. The authors would like to also acknowledge Aurore Van Koninckxloo-Van Bever for her kind help.

REFERENCES

1. Du, J. and R.K. O'Reilly, Advances and challenges in smart and functional polymer vesicles. *Soft Matter*, 2009. 5(19): p. 3544–3561.
2. LoPresti, C. et al., Polymersomes: Nature inspired nanometer sized compartments. *Journal of Materials Chemistry*, 2009. 19(22): p. 3576–3590.
3. van Hest, J.C.M. et al., Polystyrene–dendrimer amphiphilic block copolymers with a generation-dependent aggregation. *Science*, 1995. 268(5217): p. 1592–1595.
4. Zhang, L. and A. Eisenberg, Multiple morphologies of "crew-cut" aggregates of polystyrene-b-Poly(acrylic acid) block copolymers. *Science*, 1995. 268(5218): p. 1728–1731.
5. Discher, B.M. et al., *Polymersomes: Tough vesicles made from diblock copolymers. Science*, 1999. 284(5417): p. 1143–1146.
6. Mai, Y. and A. Eisenberg, Self-assembly of block copolymers. *Chemical Society Reviews*, 2012. 41(18): p. 5969–5985.
7. Israelachvili, J.N., D.J. Mitchell, and B.W. Ninham, Theory of self-assembly of hydrocarbon amphiphiles into micelles and bilayers. *Journal of the Chemical Society, Faraday Transactions 2: Molecular and Chemical Physics*, 1976. 72(0): p. 1525–1568.
8. Blanazs, A., S.P. Armes, and A.J. Ryan, Self-assembled block copolymer aggregates: From micelles to vesicles and their biological applications. *Macromolecular Rapid Communications*, 2009. 30(4–5): p. 267–277.
9. Brinkhuis, R.P., F.P.J.T. Rutjes, and J.C.M. van Hest, Polymeric vesicles in biomedical applications. *Polymer Chemistry*, 2011. 2(7): p. 1449–1462.
10. Bleul, R., R. Thiermann, and M. Maskos, Techniques to control polymersome size. *Macromolecules*, 2015. 48(20): p. 7396–7409.
11. Discher, D.E. and A. Eisenberg, Polymer vesicles. *Science*, 2002. 297(5583): p. 967–973.
12. Adams, D.J. et al., On the mechanism of formation of vesicles from poly(ethylene oxide)-block-poly(caprolactone) copolymers. *Soft Matter*, 2009. 5(16): p. 3086–3096.
13. Qi, W. et al., Aqueous self-assembly of poly(ethylene oxide)-block-poly(?-caprolactone) (PEO-b-PCL) copolymers: Disparate diblock copolymer compositions give rise to nano- and meso-scale bilayered vesicles. *Nanoscale*, 2013. 5(22): p. 10908–10915.
14. Rajagopal, K. et al., Curvature-coupled hydration of semicrystalline polymer amphiphiles yields flexible worm micelles but favors rigid vesicles: Polycaprolactone-based block copolymers. *Macromolecules*, 2010. 43(23): p. 9736–9746.
15. Yu, Y., L. Zhang, and A. Eisenberg, Morphogenic effect of solvent on crew-cut aggregates of amphiphilic diblock copolymers. *Macromolecules*, 1998. 31(4): p. 1144–1154.
16. Sanson, C. et al., Biocompatible and biodegradable poly (trimethylene carbonate)-b-poly(l-glutamic acid) polymersomes: size control and stability. *Langmuir*, 2010. 26(4): p. 2751–2760.
17. Photos, P.J. et al., Polymer vesicles *in vivo*: Correlations with PEG molecular weight. *Journal of Controlled Release*, 2003. 90(3): p. 323–334.
18. Lee, J.C.M. et al., Preparation, stability, and *in vitro* performance of vesicles made with diblock copolymers. *Biotechnology and Bioengineering*, 2001. 73(2): p. 135–145.
19. Kita-Tokarczyk, K. et al., Block copolymer vesicles – Using concepts from polymer chemistry to mimic biomembranes. *Polymer*, 2005. 46(11): p. 3540–3563.
20. Parnell, A.J. et al., The efficiency of encapsulation within surface rehydrated polymersomes. *Faraday Discussions*, 2009. 143(0): p. 29–46.
21. Dimova, R. et al., Hyperviscous diblock copolymer vesicles. *The European Physical Journal E*, 2002. 7(3): p. 241–250.
22. Wittemann, A., T. Azzam, and A. Eisenberg, Biocompatible polymer vesicles from biamphiphilic triblock copolymers and their interaction with bovine serum albumin. *Langmuir*, 2007. 23(4): p. 2224–2230.
23. Ahmed, F. and D.E. Discher, Self-porating polymersomes of PEG–PLA and PEG–PCL: Hydrolysis-triggered controlled release vesicles. *Journal of Controlled Release*, 2004. 96(1): p. 37–53.
24. Lorenceau, E. et al., Generation of polymerosomes from double-emulsions. *Langmuir*, 2005. 21(20): p. 9183–9186.
25. Thiermann, R. et al., Size controlled polymersomes by continuous self-assembly in micromixers. *Polymer*, 2012. 53(11): p. 2205–2210.
26. Bleul, R. et al., Continuously manufactured magnetic polymersomes – A versatile tool (not only) for targeted cancer therapy. *Nanoscale*, 2013. 5(23): p. 11385–11393.
27. Ghoroghchian, P.P. et al., Near-infrared-emissive polymersomes: Self-assembled soft matter for *in vivo* optical imaging. *Proceedings of the National Academy of Sciences of the United States of America*, 2005. 102(8): p. 2922–2927.
28. Arosio, P. et al., Hybrid iron oxide-copolymer micelles and vesicles as contrast agents for MRI: Impact of the nanostructure on the relaxometric properties. *Journal of Materials Chemistry B*, 2013. 1(39): p. 5317–5328.
29. Liu, G.-Y., C.-J. Chen, and J. Ji, Biocompatible and biodegradable polymersomes as delivery vehicles in biomedical applications. *Soft Matter*, 2012. 8(34): p. 8811–8821.
30. Meng, F. et al., Biodegradable polymersomes. *Macromolecules*, 2003. 36(9): p. 3004–3006.

31. Ahmed, F. et al., Shrinkage of a rapidly growing tumor by drug-loaded polymersomes: pH-triggered release through copolymer degradation. *Molecular Pharmaceutics*, 2006. 3(3): p. 340–350.
32. Discher, D.E. and F. Ahmed, Polymersomes. *Annual Review of Biomedical Engineering*, 2006. 8: p. 323–341.
33. Le Meins, J.-F., O. Sandre, and S. Lecommandoux, Recent trends in the tuning of polymersomes' membrane properties. *The European Physical Journal E*, 2011. 34(2): p. 14.
34. Schillén, K., K. Bryskhe, and Y.S. Mel'nikova, Vesicles formed from a poly(ethylene oxide)–poly(propylene oxide)–poly(ethylene oxide) triblock copolymer in dilute aqueous solution. *Macromolecules*, 1999. 32(20): p. 6885–6888.
35. Jenekhe, S.A. and X.L. Chen, Self-assembled aggregates of rod-coil block copolymers and their solubilization and encapsulation of fullerenes. *Science*, 1998. 279(5358): p. 1903–1907.
36. Ahmed, F. et al., Biodegradable polymersomes loaded with both paclitaxel and doxorubicin permeate and shrink tumors, inducing apoptosis in proportion to accumulated drug. *Journal of Controlled Release*, 2006. 116(2): p. 150–158.
37. Krack, M. et al., Nanoparticle-loaded magnetophoretic vesicles. *Journal of the American Chemical Society*, 2008. 130(23): p. 7315–7320.
38. Lee, L.M. and A.P. Liu, The application of micropipette aspiration in molecular mechanics of single cells. *Journal of Nanotechnology in Engineering and Medicine*, 2014. 5(4): p. 0408011–0408016.
39. Li, M.-H. and P. Keller, Stimuli-responsive polymer vesicles. *Soft Matter*, 2009. 5(5): p. 927–937.
40. Onaca, O. et al., Stimuli-responsive polymersomes as nanocarriers for drug and gene delivery. *Macromolecular Bioscience*, 2009. 9(2): p. 129–139.
41. Meng, F., Z. Zhong, and J. Feijen, Stimuli-responsive polymersomes for programmed drug delivery. *Biomacromolecules*, 2009. 10(2): p. 197–209.
42. De Oliveira, H., J. Thevenot, and S. Lecommandoux, Smart polymersomes for therapy and diagnosis: Fast progress toward multifunctional biomimetic nanomedicines. *Wiley Interdisciplinary Reviews: Nanomedicine and Nanobiotechnology*, 2012. 4(5): p. 525–546.
43. Feng, A. and J. Yuan, Smart nanocontainers: Progress on novel stimuli-responsive polymer vesicles. *Macromolecular Rapid Communications*, 2014. 35(8): p. 767–779.
44. Discher, D.E. et al., Emerging applications of polymersomes in delivery: From molecular dynamics to shrinkage of tumors. *Progress in Polymer Science*, 2007. 32(8–9): p. 838–857.
45. Bermúdez, H., D.A. Hammer, and D.E. Discher, Effect of bilayer thickness on membrane bending rigidity. *Langmuir*, 2004. 20(3): p. 540–543.
46. Meng, F. and Z. Zhong, Polymersomes spanning from nano- to microscales: Advanced vehicles for controlled drug delivery and robust vesicles for virus and cell mimicking. *The Journal of Physical Chemistry Letters*, 2011. 2(13): p. 1533–1539.
47. Laurent, S. et al., Magnetic iron oxide nanoparticles for biomedical applications. *Future Medicinal Chemistry*, 2010. 2(3): p. 427–449.
48. Laurent, S. et al., Magnetic iron oxide nanoparticles: Synthesis, stabilization, vectorization, physicochemical characterizations, and biological applications. *Chemical Reviews*, 2008. 108(6): p. 2064–2110.
49. Pankhurst, Q.A. et al., Applications of magnetic nanoparticles in biomedicine. *Journal of Physics D: Applied Physics*, 2003. 36(13): p. R167.
50. Gossuin, Y. et al., Magnetic resonance relaxation properties of superparamagnetic particles. *Wiley Interdisciplinary Reviews: Nanomedicine and Nanobiotechnology*, 2009. 1(3): p. 299–310.
51. Peng, E., F. Wang, and J.M. Xue, Nanostructured magnetic nanocomposites as MRI contrast agents. *Journal of Materials Chemistry B*, 2015. 3(11): p. 2241–2276.
52. Hickey, R.J. et al., Size-controlled self-assembly of superparamagnetic polymersomes. *ACS Nano*, 2014. 8(1): p. 495–502.
53. Hickey, R.J. et al., Controlling the self-assembly structure of magnetic nanoparticles and amphiphilic block-copolymers: From micelles to vesicles. *Journal of the American Chemical Society*, 2011. 133(5): p. 1517–1525.
54. Sanson, C. et al., Doxorubicin loaded magnetic polymersomes: Theranostic nanocarriers for MR imaging and magneto-chemotherapy. *ACS Nano*, 2011. 5(2): p. 1122–1140.
55. Yang, X. et al., Multifunctional SPIO/DOX-loaded wormlike polymer vesicles for cancer therapy and MR imaging. *Biomaterials*, 2010. 31(34): p. 9065–9073.
56. Qin, J. et al., Rationally separating the corona and membrane functions of polymer vesicles for enhanced T2 MRI and drug delivery. *ACS Applied Materials & Interfaces*, 2015. 7(25): p. 14043–14052.
57. Pourtau, L. et al., Antibody-functionalized magnetic polymersomes: *In vivo* targeting and imaging of bone metastases using high resolution MRI. *Advanced Healthcare Materials*, 2013. 2(11): p. 1420–1424.
58. Stanicki, D. et al., Synthesis and processing of magnetic nanoparticles. *Current Opinion in Chemical Engineering*, 2015. 8: p. 7–14.
59. Ling, D. and T. Hyeon, Iron oxide nanoparticles: Chemical design of biocompatible iron oxide nanoparticles for medical applications (small 9–10/2013). *Small*, 2013. 9(9–10): p. 1449–1449.
60. Wei, W. et al., Recent progress on magnetic iron oxide nanoparticles: synthesis, surface functional strategies and biomedical applications. *Science and Technology of Advanced Materials*, 2015. 16(2): p. 023501.
61. Lecommandoux, S. et al., Smart hybrid magnetic self-assembled micelles and hollow capsules. *Progress in Solid State Chemistry*, 2006. 34(2–4): p. 171–179.
62. Garnier, B. et al., Optimized synthesis of 100nm diameter magnetoliposomes with high content of maghemite particles and high MRI effect. *Contrast Media & Molecular Imaging*, 2012. 7(2): p. 231–239.
63. Hayward, R.C. et al., Dewetting instability during the formation of polymersomes from block-copolymer-stabilized double emulsions. *Langmuir*, 2006. 22(10): p. 4457–4461.
64. Yang, X. et al., Tumor-targeting, superparamagnetic polymeric vesicles as highly efficient MRI contrast probes. *Journal of Materials Chemistry*, 2009. 19(32): p. 5812–5817.
65. Ren, T. et al., Multifunctional polymer vesicles for ultrasensitive magnetic resonance imaging and drug delivery. *Journal of Materials Chemistry*, 2012. 22(24): p. 12329–12338.
66. Liu, Q. et al., A superparamagnetic polymersome with extremely high T2 relaxivity for MRI and cancer-targeted drug delivery. *Biomaterials*, 2017. 114: p. 23–33.
67. Du, J. and S.P. Armes, pH-responsive vesicles based on a hydrolytically self-cross-linkable copolymer. *Journal of the American Chemical Society*, 2005. 127(37): p. 12800–12801.
68. Wang, M. et al., Polymer vesicles as robust scaffolds for the directed assembly of highly crystalline nanocrystals. *Langmuir*, 2009. 25(24): p. 13703–13711.

69. Binder, W.H. et al., Guiding the location of nanoparticles into vesicular structures: A morphological study. *Physical Chemistry Chemical Physics*, 2007. 9(48): p. 6435–6441.
70. Mai, Y. and A. Eisenberg, Controlled incorporation of particles into the central portion of vesicle walls. *Journal of the American Chemical Society*, 2010. 132(29): p. 10078–10084.
71. He, J. et al., Self-assembly of inorganic nanoparticle vesicles and tubules driven by tethered linear block copolymers. *Journal of the American Chemical Society*, 2012. 134(28): p. 11342–11345.
72. Oliveira, H. et al., Magnetic field triggered drug release from polymersomes for cancer therapeutics. *Journal of Controlled Release*, 2013. 169(3): p. 165–170.
73. Agut, W. et al., Depletion induced vesicle-to-micelle transition from self-assembled rod-coil diblock copolymers with spherical magnetic nanoparticles. *Soft Matter*, 2011. 7(20): p. 9744–9750.
74. Yang, X. et al., Multifunctional stable and pH-responsive polymer vesicles formed by heterofunctional triblock copolymer for targeted anticancer drug delivery and ultrasensitive MR imaging. *ACS Nano*, 2010. 4(11): p. 6805–6817.
75. Yang, X. et al., Multifunctional polymeric vesicles for targeted drug delivery and imaging. *Biofabrication*, 2010. 2(2): p. 025004.
76. Sun, Q. et al., A pH-sensitive polymeric nanovesicle based on biodegradable poly(ethylene glycol)-*b*-poly(2-(diisopropylamino) ethyl aspartate) as a MRI-visible drug delivery system. *Journal of Materials Chemistry*, 2011. 21(39): p. 15316–15326.
77. Bixner, O. et al., Triggered release from thermoresponsive polymersomes with superparamagnetic membranes. *Materials*, 2016. 9(1): p. 29.
78. Šachl, R. et al., Preparation and characterization of self-assembled nanoparticles formed by poly(ethylene oxide)-block-poly(ε-caprolactone) copolymers with long poly(ε-caprolactone) blocks in aqueous solutions. *Langmuir*, 2007. 23(6): p. 3395–3400.
79. Hocine, S. et al., Polymersomes with PEG corona: Structural changes and controlled release induced by temperature variation. *Langmuir*, 2013. 29(5): p. 1356–1369.
80. Lecommandoux, S. et al., Self-assemblies of magnetic nanoparticles and di-block copolymers: Magnetic micelles and vesicles. *Journal of Magnetism and Magnetic Materials*, 2006. 300(1): p. 71–74.
81. Bonnaud, C. et al., Insertion of nanoparticle clusters into vesicle bilayers. *ACS Nano*, 2014. 8(4): p. 3451–3460.
82. Brian M. Dale, M.A.B., Richard C. Semelka, *MRI: Basic Principles and Applications, 5th Edition*. John Wiley & Sons. 2015.
83. Wagner, B.M., *MRI Made Easy (Imaging Systems, Diagnostic Imaging, and Radiology & Nuclear Medicine Guide): MRI in Practice for MRI Scanner and MRI Safety*. Kindle Edition. 2016.
84. Westbrook, C., *Handbook of MRI Technique, 4th Edition*. J. Wiley & Sons. 2014.
85. Henoumont, C., S. Laurent, and L. Vander Elst, How to perform accurate and reliable measurements of longitudinal and transverse relaxation times of MRI contrast media in aqueous solutions. *Contrast Media & Molecular Imaging*, 2009. 4(6): p. 312–321.
86. Geraldes, C.F.G.C. and S. Laurent, Classification and basic properties of contrast agents for magnetic resonance imaging. *Contrast Media & Molecular Imaging*, 2009. 4(1): p. 1–23.
87. Ai, H. et al., Magnetite-loaded polymeric micelles as ultrasensitive magnetic-resonance probes. *Advanced Materials*, 2005. 17(16): p. 1949–1952.
88. Nasongkla, N. et al., Multifunctional polymeric micelles as cancer-targeted, MRI-ultrasensitive drug delivery systems. *Nano Letters*, 2006. 6(11): p. 2427–2430.
89. Martina, M.-S. et al., Generation of superparamagnetic liposomes revealed as highly efficient MRI contrast agents for *in vivo* imaging. *Journal of the American Chemical Society*, 2005. 127(30): p. 10676–10685.
90. Béalle, G. et al., Ultra magnetic liposomes for MR imaging, targeting, and hyperthermia. *Langmuir*, 2012. 28(32): p. 11834–11842.
91. Mura, S. and P. Couvreur, Nanotheranostics for personalized medicine. *Advanced Drug Delivery Reviews*, 2012. 64(13): p. 1394–1416.
92. Rizzo, L.Y. et al., Recent progress in nanomedicine: Therapeutic, diagnostic and theranostic applications. *Current Opinion in Biotechnology*, 2013. 24(6): p. 1159–1166.

Born in 1967, **Sophie Laurent** (E-mail: Sophie.laurent@umons.ac.be) studied at the University of Mons-Hainaut (UMH), Belgium, from which she graduated with a Lic. Sci. degree in chemistry in 1989. She obtained her PhD in 1993 from the same university, where she was successively appointed assistant, lecturer and associate professor. She joined Prof R.N. Muller's team and was involved in the development (synthesis and physicochemical characterization) of contrast agents (paramagnetic lanthanide complexes and superparamagnetic iron oxide nanoparticles) for MRI. She is coauthor of more than 230 publications in international journals such as *Chemical Reviews* and more than 400 abstracts in international conferences. She collaborates actively with the CMMI in Gosselies, Belgium. Since October 2016, she has been the head of General, Organic and Biomedical Chemistry Unit in the University of Mons and of the UMONS part of the CMMI.

Adeline Hannecart was born in 1989 and earned her master's degree in chemistry at the University of Mons in 2012. At present, she is undertaking a thesis at the laboratory of NMR and molecular imaging with the group of S. Laurent. Her research interests include the synthesis and physicochemical characterization of nanocomposites made of iron oxide nanoparticles and polymers for different biomedical applications, such as their use as MRI contrast agents.

Born in 1984, **Dimitri Stanicki** studied chemistry at the University of Mons-Hainaut. He graduated with a Lic. Sci. degree in 2006 and earned his PhD degree in 2010 at the same institution. He joined Prof. R. N. Muller's team in 2011, where he is now involved in the development (synthesis, surface modification and characterization) of superparamagnetic nanoparticles and the study of their biomedical applications (multimodal and molecular imaging).

Luce Vander Elst was born in 1955. She studied at the University of Mons-Hainaut, where she earned her PhD degree in 1984. She did postdoctoral research on the multinuclear NMR analysis of the metabolism of perfused mammalian hearts at the Medical School of Harvard University in 1986. She works as a professor in the NMR and Molecular Imaging Laboratory of the University of Mons. Her research focuses mainly on high-resolution NMR and the physicochemical characterization of MRI and optical imaging (OI) contrast agents.

Robert N. Muller was born in 1948 and studied at the University of Mons-Hainaut (UMH), Belgium, from which he graduated with a Lic. Sci. degree in chemistry in 1969. He earned his PhD in 1974 at the same university, where he was successively appointed assistant, lecturer and full professor; head of the Department of General, Organic and Biomedical Chemistry from October 2005 until September 2013; dean of the Faculty of Medicine and Pharmacy of the University of Mons from October 2005 until September 2013; and Emeritus Professor of the University of Mons since October 1, 2013. He has been the scientific director of the CMMI since 2011.

He carried out postdoctoral studies in MRI with Paul C. Lauterbur's research group at the State University of New York at Stony Brook in 1981–1982, collaborated with Dr Seymour Koenig in the domain of fast-field cycling relaxometry and was on a sabbatical leave at the Center for Magnetic Resonance (CERM), Florence, Italy, with Professors Ivano Bertini and Claudio Luchinat in 2002–2003.

He is a cofounder of the European Workshop on Nuclear Magnetic Resonance in Medicine; vice chairman of the European Magnetic Resonance Forum Foundation, 1991–present; president of the European Society for Magnetic Resonance in Medicine and Biology, 1987–1988; president of the GRAMM (Groupe de Recherche sur les Applications du Magnétisme en Médecine), 1998–2000; vice chairman of the COST D18 Action, 2003; Founding member of the European Society for Molecular Imaging (ESMI); editor in chief of *Contrast Media and Molecular Imaging* (Wiley); member of the editorial boards of *Magnetic Resonance Materials MAGMA* (Springer); and former member of the editorial board of *Investigative Radiology* (Lippincott).

He has produced more than 270 publications and contributed to six books, mainly in NMR relaxometry, spectroscopy and imaging in the context of the development and applications of contrast agents for Molecular Imaging.

8 Ultrasmall Iron Oxide Nanoparticles Stabilized with Multidentate Polymers for Applications in MRI

*Jung Kwon (John) Oh and Marc-André Fortin**

CONTENTS

8.1 INTRODUCTION

Magnetic resonance imaging (MRI) has developed at an exponential rate over the last decades, and the development of contrast agents to enhance the visualization of organs has followed the same trend. Meanwhile, magnetic nanoparticles that generate either 'positive' or 'negative' contrast for MRI have become one of the most important biomedical applications of nanotechnology. Iron oxide nanoparticles (IONPs) administrated *in vivo* can be used to detect hepatic tumours, for targeting the early signs of several vascular diseases (i.e. atherosclerotic plaque),[1,2] as well as for cell tracking applications[3–7] and imaging-assisted drug delivery.[8–11] All of these applications require IONPs to remain as stable colloids in a variety of harsh physiological environments (blood, lymph, urine, cell culture medium, etc.).

MRI is a high-resolution, noninvasive imaging modality that provides images with excellent anatomical resolution. The technology requires the insertion of the subject in a strong magnetic field, which causes hydrogen proton spins (^{1}H) in the biological tissues to align according to its direction. Then, a precisely tuned radiofrequency (RF) wave is applied through electromagnetic coils and hydrogen protons are excited. Once the RF excitation stops, the energy acquired by the protons is progressively released and it is detectable by 'detection coils.' Codified with the application of gradients, these oscillations are then analysed and interpreted in the form of contrast-modulated images. The time required for these signals to fade out (i.e. the 'relaxation' time) is typically in the order of milliseconds to seconds, depending on acquisition parameters and characteristics intrinsic to each biological tissue (hydration, types of molecules, etc.). Very early in the development of MRI, the use of magnetic substances has been suggested as an efficient way to accelerate the relaxation of ^{1}H protons in several tissues and in particular in the blood. Paramagnetic substances such as gadolinium (Gd) chelates have been widely used until now to increase the contrast in vascularised tissues by using so-called 'positive' or T_1-weighted contrast imaging.[12,13] A stronger presence of Gd in biological tissues containing a moderate amount of this substance leads to signal increase ('positive' contrast). However, concerns related to the toxicity risks associated to this rare earth element justify the development of contrast agents made of more biocompatible substances.[14,15]

In the past three decades, there has been an impressive number of research projects on and product development of IONPs as contrast agents for MRI applications. The iron

* Corresponding author.

oxide nanocrystals making the core of IONPs are usually in the range of 2–15 nm. A significant difference of IONPs compared to Gd-based contrast agents is their superparamagnetic behaviour. In fact, the magnetization of IONPs is much stronger than that of paramagnetic substances, whereas the small size of these crystals impedes the magnetic remanence effect that is characteristic of bigger crystals and bulk ferromagnetic materials. For different reasons all reviewed in the present chapter, IONPs developed for biomedicine until now have been largely applied with so-called T_2/T_2*-weighted contrast imaging acquisition. They have found numerous applications in preclinical and clinical MRI (cell labeling, vascular contrast, lymph node imaging and liver contrast). The presence of IONPs in biological tissues decreases proton excitation and increases transverse relaxation, and T_2/T_2*-weighted contrast parameters emphasize these effects. Although negative contrast agents can be well exploited for the high-sensitivity detection of cells and clusters of materials injected at precise locations in the body, the dark negative contrast provided by T_2/T_2* agents often deliver misleading clinical diagnoses, as they can be confused with signals from bleeding, calcification or metal deposits. This problem is particularly acute in biomedical vascular diagnostics. Therefore, there has been a strong and lasting interest in the past decades to develop IONPs that could be used in T_1-weighted imaging for 'positive' contrast enhancement.

Ultrasmall superparamagnetic ion oxide nanoparticles (USPIOs) with particle diameters < 5 nm have shown promising results as 'positive' contrast agents for T_1-weighted MRI.[16,17] They concentrate a high number of magnetic atoms per unit of contrast agent (several hundreds for nanoparticles with a diameter = 3 nm). In addition, USPIO are considered to be more biocompatible than lanthanides, due to biodegradation and metabolism of iron oxides into iron ions, which can be readily incorporated into the body's natural iron stores. Further, the relaxivity of USPIO can be tuned with varying particle sizes as the magnetic moment of USPIO rapidly decreases with decreasing particle size due to the reduction in volume magnetic anisotropy and surface spin disorders.[18,19]

With an adequate surface coating, USPIO can stay in the blood for relatively long times (minutes up to a few hours), therefore opening possibilities for blood pool imaging and angiography applications. This application, however, requires the IONPs to have a very small magnetic moment and to remain under a certain concentration threshold enabling 'positive' contrast in T_1-weighted MRI. To be used as signal enhancers in T_1-weighted MRI, IONPs should have a small size (usually below 5 nm in core diameter). The short Néel relaxation times and very high specific surface characteristics related to these small-sized crystals attenuate the transverse relaxation associated to strong 'negative' contrast effects. Another important condition is the necessity to achieve excellent and prolonged colloidal stability through the application of nanoparticle coatings enabling the fabrication of highly individual IONPs. Surface coatings impeding the agglomeration of IONPs and leading to highly uniform and small hydrodynamic particle diameter profiles are necessary to preserve the 'positive' contrast enhancement characteristics of IONP preparations. Therefore, control of the surface chemistry of USPIO is essential, not only to prevent individual colloidal USPIO from aggregation but also to impart water-dispersibility, biocompatibility, nonspecific adsorption to cells and colloidal stability.

Several coating processes have been explored, including silica coatings,[20–22] intercalation with amphiphiles,[23,24] ionic interactions[25] and ligand exchange.[26] Among the processes, ligand exchange enables the formation of molecular monolayers at the particle surface. These can be strongly grafted to the crystal surfaces, in order to preserve colloidal stability in a wide range of pH and high electrolyte concentrations.[27] In addition to small water-soluble or hydrophilic molecules having one (monodentate) or two (bidentate) anchoring groups,[28–30] poly(ethylene glycol)-based bidentate ligands or oligomers[16,31–33] have been explored. Polymeric ligands possess multiple anchoring groups in each chain capable of multiple binding interactions with the particle surface (multidentate ligands). As stabilizing ligands at the surface of IONPs is susceptible to undergo a constant equilibrium of adsorption/desorption, the use of multidentate polymeric ligands could endow the particles with enhanced colloidal stability. Several reports have described the synthesis and use of natural polysaccharides,[34,35] homopolymers,[36,37] dendrimers[37] or random copolymers as multidentate ligands.[38,39] However, these polymers suffer from a lack of functionalities (i.e. they are homopolymers), weak affinity to IONPs surfaces (i.e. polysaccharides) or broad molecular weight distribution. The resulting USPIO could offer poor long-term colloidal stability in aqueous dispersion.

One of the most promising strategies developed in the last years, enabling the development of ultrasmall and highly individual IONPs, is the use of multifunctional block copolymers (multidentate block copolymer [MDBC]-USPIO). The strategy focuses on well-defined tunable MDBCs with narrow molecular weight distribution. These MDBCs consist of one anchoring block possessing pendant multidentate anchoring groups such as carboxylates or phosphonates strongly bound to USPIO surfaces, while the other tunable block is varied with components that can promote compatibility with end-use applications. When compared to monodentate, bidentate and even random copolymer ligands, MDBCs exhibit stronger nanocrystal binding characteristics due to more favourable conformational arrangements of multiple anchoring groups tethered from long polymeric chains onto USPIO surfaces.

This chapter aims to describe the fundamental mechanisms of MRI contrast agents based on ultrasmall IONPs. A brief review of the main magnetic characteristics of IONPs is presented, followed by a description of the fundamental concepts behind the relaxation rate of hydrogen protons submitted to magnetic substances. The research community often refers to the concept of 'relaxivity.' The relaxivity of MRI contrast agents depends on their nanoparticulate size and magnetic properties, on the distance between water molecules and their surface and on the exchange between water protons and the surface, among a long list of parameters. These principles will be presented in the context of the development of advanced polymer coatings made of multiple surface binding sites. Multidentate polymers exhibit stronger binding isotherm to the surface of IONPs. These features stemming from MDBC technology are presented and discussed.

8.2 PRINCIPLES OF MRI

MRI is a widely available anatomical and functional imaging modality that provides high-resolution images. In this modality, patients are submitted to strong magnetic fields. Typical magnetic field strengths range from 1 to 3 T, which is 2.10^4 to 6.10^4 times stronger compared to that of the Earth. Figure 8.1 represents each one of the steps leading to the generation of MRI signal. When a patient is in the MRI machine, hydrogen protons (^1H) align their spins (i.e. the magnetic moments associated to each ^1H nucleus) in the direction of the main magnetic field of the scanner (B_0; Figure 8.1a). The sum of each one of the magnetic moments of these spins represents the 'macroscopic magnetization vector' ($\vec{M}$) of the biological tissue. This vector is oriented along the main magnetic field of the scanner. If placed in a magnetic field, the magnetic dipole moments (the spins) will 'precess' around the direction of the magnetic field with an angular frequency (ω_0: the Larmor resonance frequency for hydrogen protons). The Larmor frequency dictates the precession at a given magnetic field strength. Prior to application of the RF, the moments are not in phase with each other.

Then, an RF wavelength is projected to the patient by means of a transmitter coil. Coils are antennae inserted in the MRI system. If this radiation precisely matches the Larmor frequency characteristics of the nucleus at the given magnetic field of the scanner, the hydrogen spins can be excited. As a result, the macroscopic magnetization vector ($\vec{M}$) progressively rotates and develops a vector component in the x–y plane. The experimenter can decide to stop the RF at any time, but preferentially when the vector is aligned in the x–y plane. At this condition, corresponding to the plane perpendicular to the axis of the main magnetic field (Figure 8.1b), the intensity of the oscillating motion detectable by the coils is expected to be maximal.

At this precise moment, $\vec{M}$ rotates within the x–y plane and around the z axis (i.e. the main axis of the scanner). After this operation, commonly called a '90°' pulse, the RF excitation is turned off. Then, within a time span that varies between a few milliseconds to a few seconds, the excited spins 'relax'; i.e. they release energy in their environment. While doing so, the spins go back to their initial energy state. Receiver coils can detect this energy release. The oscillating signals are first codified by the application of magnetic gradients, which enables identification of the spatial origin of the signal. These signals are then digitalized, Fourier-transformed to construct 'k-space' maps, from which the processor reconstructs anatomical maps of the human body. In the last step of the process, the MRI signal is presented in the form of grey-scale voxel matrices. Voxel is the spatial imaging unit in MRI.

8.2.1 Concepts of T_1 and T_2 Relaxation

The coils can detect the oscillation of $\vec{M}$ around 'z.' This provides crucial information about the characteristics of the hydrated biological tissues. In fact, hydrogen spins found in certain tissues relax faster than that found in other environments. Taken macroscopically, these differences make the basis of MRI contrast. With the gradual release of ^1H energy, a progressive recovery of the M_z is observed macroscopically, while the M_{xy} component decreases sharply (Figure 8.1c). Hence, the spins progressively return to their original orientation, and the kinetics of recovery in M_z is referred as the longitudinal relaxation time (T_1). T_1 is an intrinsic

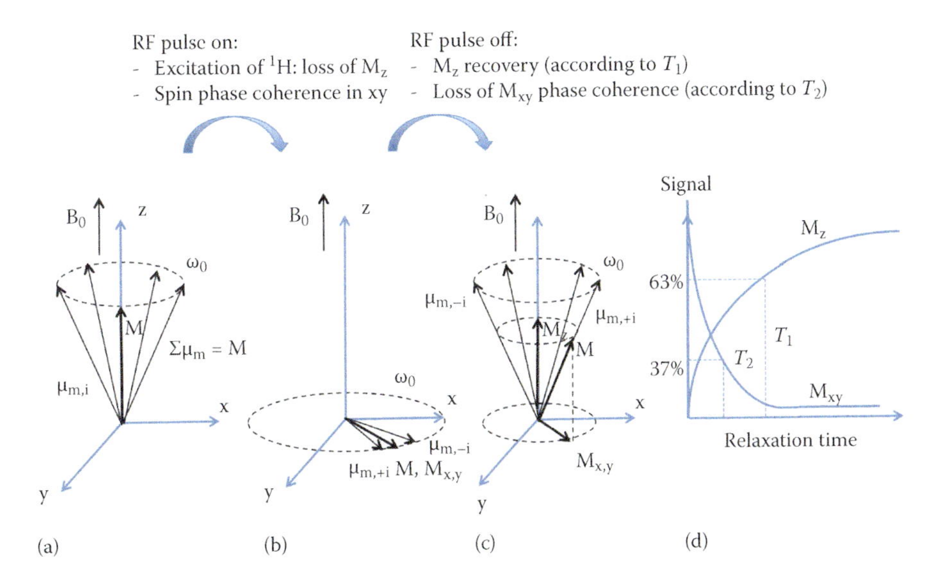

FIGURE 8.1 Schematic representation of the macroscopic magnetization vector generated by MR excitation: ω_0 is the Larmor precession frequency, μ_m refers to the magnetic moments of each spin contributing to M, the macroscopic magnetization vector (with components M_z and $M_{x,y}$).

TABLE 8.1
T_1, T_2 and Proton Densities (ρ) of Brain Tissues at 1.5 T (from List of Common T_1 and T_2 Values in Stark and Bradley[40])

Tissue	T_1 (ms)	T_2 (ms)	ρ (Relative to CSF)
White matter	510	67	0.61
Gray matter	760	77	0.69
Edema	900	126	0.86
Cerebrospinal fluid	2650	180	1.00

characteristic of each one of the biological tissues in the human body (Figure 8.1d). Using so-called T_1-weighted imaging sequences, it is possible to generate contrast between two adjacent tissues showing different longitudinal relaxation times (T_1).

The magnetic moments of neighbouring ^{1}H protons also exert a mutual influence on each other. From the moment the RF frequency is turned off, adjacent ^{1}H spins influence each other and, while doing so, slightly affect their respective Larmor frequency. As a result, some ^{1}H precesses slightly faster, and other slightly slower. The global impact of such mutual influence is a loss of 'x–y' phase coherence. M_{xy} rapidly decreases, in fact within tens of milliseconds in most biological tissues. While doing so, M_z recovery occurs; however, it takes a longer time than the observed M_{xy} decay (T_1 is typically in the order of hundreds of milliseconds). The decay of M_{xy} amplitude (i.e. loss of phase coherence between the hydrogen spins) is described and quantified by T_2: the transverse relaxation time of a given hydrated tissue (Figure 8.1d). As for T_1, T_2 is also an intrinsic characteristic of each biological tissue. Using so-called 'T_2-weighted' imaging sequences, it is possible to generate contrast between two adjacent tissues showing different T_2s. Finally, when using 'gradient echo' sequences, M_{xz} decay is influenced both by the molecular structure of the tissues and by external factors such as the homogeneity of the magnetic field, as well as the local presence of magnetic elements within the tissues (ferromagnetic objects and clusters, superparamagnetic nanoparticles etc.). This has a considerable impact on T_2, and the transverse magnetization decay is then described by the term T_2*, which is not an intrinsic property of the biological tissue. Typical T_1 and T_2 relaxation constants for several tissues at 1.5 T are indicated in Table 8.1.

8.2.2 From Spin Relaxation to MRI Signal

T_1 and T_2, as well as ρ, the density of ^{1}H spins, are the three main characteristics dictating MRI signal and contrast. They are intrinsic to each one of the biological tissues. Signal is also influenced by scanning parameters, such as the echo time (TE) and repetition time (TR), which are basic parameters in spin-echo MRI sequences.[41] The signal (S) detected in a given voxel of the magnetic resonance (MR) image using a basic spin echo sequence, can be calculated by the following equation:

$$S = \rho\left(1 - e^{(-TR/T_1)}\right)\left(e^{-TE/T_2}\right). \tag{8.1}$$

This signal–intensity equation is fundamental in the field of contrast media imaging. A typical signal–intensity curve is provided in Figure 8.2 for cerebrospinal fluid and white matter (data extracted from Table 8.1, for a TE of 10 ms). The dotted line and the arrows appearing at TR = 400 ms indicate the optimal conditions for achieving a 'signal–intensity' contrast in T_1-weighted imaging. Indeed, the difference in signal, or the contrast, between these two tissues is maximal at TR = 650 ms (dotted arrows).

A careful and precise selection of the MRI scanning parameters (i.e. the 'sequence') enables the MR radiologist

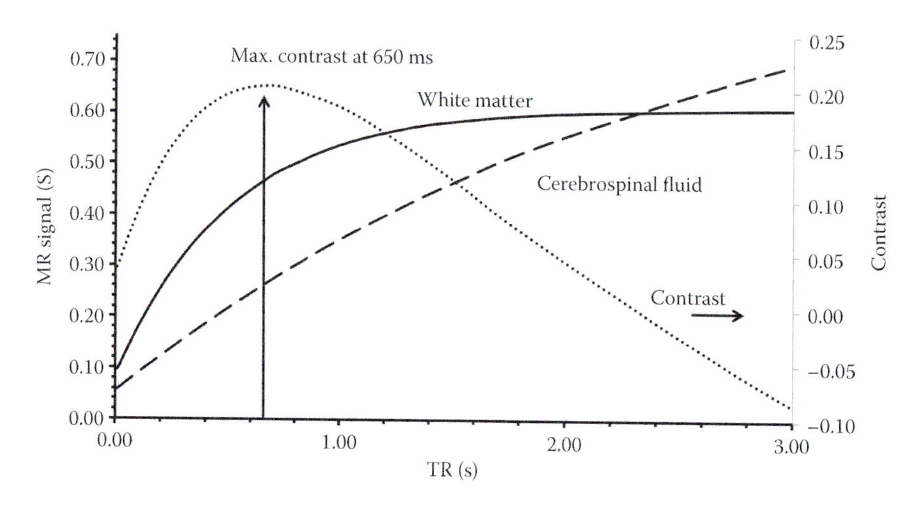

FIGURE 8.2 MR signal depending on longitudinal recovery time (TR) for two typical biological tissues (cerebrospinal fluid and white matter), for a spin echo sequence at TE = 10 ms and at a magnetic field strength of 1.5 T. The contrast (scale at the right) is maximal when using a recovery time (TR) of 650 ms.

TABLE 8.2
Typical Contrast Weighting Parameters (T_1 weighted, T_2 weighted and ρ) in the Spin Echo Pulse Sequence

Contrast Weighting	TE (ms)	TR (ms)
T_1 weighted	5–30 (minimum)	350–750 (intermediate)
T_2 weighted	50–150 (long)	1500–3500 (long)
Proton density	5–30 (minimum)	1500–3500 (long)

to proceed to either T_1-weighted (such as in Figure 8.2) or T_2-weighted images. Usually in T_1-weighted images, tissues showing a short T_1 and a moderate to long T_2 appear the brightest. In T_2-weighted imaging, tissues showing long T_2 are usually brighter, while fast-decaying tissues (i.e. short T_2) appear as dark areas. Table 8.2 lists the general parameters used in clinical MRI to achieve T_1-weighted or T_2-weighted acquisitions.

When two adjacent tissues have similar T_1, T_2 and proton density characteristics, similar signal intensity values are reached, and this impedes the efficient delineation of anatomical information. In such case, it may be necessary to use contrast agents. MRI contrast agents are usually injected in the blood to induce a change in T_1 and T_2 of tissues. The presence of paramagnetic (or superparamagnetic) contrast agents in the blood, in a concentration usually in the range 0.1–2 mM (Gd or Fe, atomic conc.), leads to a significant decrease in T_1 and T_2. This relaxation change can be significantly different depending on the type of contrast media and depending on its concentration.

As mentioned in the introduction, most clinically approved contrast agents are based on Gd^{3+} chelates, which are small molecules that sequestrate the paramagnetic ion Gd^{3+}.[12,13] Paramagnetic contrast agents injected in the blood account for approximately 30% to 40% of all current MRI procedures.[42] However, because each one of the Gd^{3+} ions does not have the capacity to influence more than ~10^9 water molecules per second, molecules bearing only one Gd^{3+} atom are not sufficient to track single molecules or single cells.[43] This lack of sensitivity is one of the main limitations of paramagnetic contrast agents. Adding to this, Gd^{3+} ions are strongly associated with the occurrence of nephrologic systemic fibrosis.[14,15] The development of vascular contrast agents based on less toxic elements than Gd figures among the general objectives of this research field.

Ultrasmall particles of iron oxide (USPIOs) are made of iron oxide cores of mean diameters inferior to 20 nm. As mentioned previously, IONPs, including USPIOs, are generally used as 'negative' contrast agents in T_2/T_2^*-weighted imaging. However, for very small iron oxide cores, and for highly individualized nanoparticle preparations, USPIO/IONP preparations can provide very strong 'positive' contrast enhancement effects. This condition can be achieved only when very small USPIOs are coated with a thin layer of stabilizing ligands, usually polymers. If these nanoparticles do not agglomerate or form clusters *in vivo*, and more precisely in the blood, they can be applied in vascular procedures for T_1-weighted imaging.

Figure 8.3 provides a good illustration of typical differences found in signal intensity between a conventional Gd chelate-based paramagnetic contrast agent and ultrasmall IONPs. The changes in T_1 and T_2 induced by the presence of Gd-based or IONP-based contrast agents are significantly different. For Gd contrast agents, the signal increase in T_1-weighted imaging appears to be more gradual at increasing concentrations. For IONPs, the signal increase is very strong at low concentrations, then reaches a peak, followed by a progressive decrease at high nanoparticle concentrations. The development of USPIOs/IONPs for 'positive' contrast enhancement must take into account this duality of contrast effects in order to reach the right balance, allowing optimal signal enhancement effect.

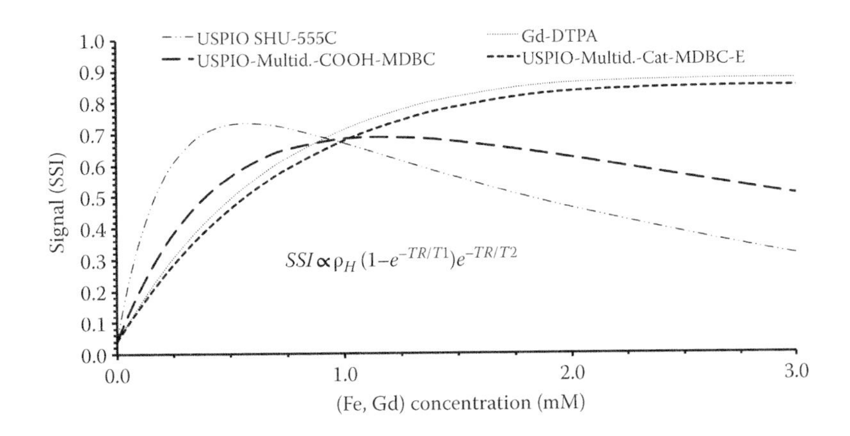

FIGURE 8.3 MR signal enhancement curves for different contrast agents: commercial Gd-DTPA (paramagnetic), commercial SHU-555C (USPIO; superparamagnetic agent) and MDBC-coated USPIOs presented in the last section of this chapter (see Table 8.3 for the list and relaxometric data, as well as Rohrer et al.[42] and Merbach et al.[44]). Magnetic field strength: 1.5 T; spin echo sequence: TE/TR = 10/400 ms. Proton density was normalised to 1.

8.3 STRUCTURE AND MAGNETIC PROPERTIES OF ULTRASMALL IONPs

IONPs made of Fe_2O_3/Fe_3O_4 have core diameters typically in the range 2–15 nm, whereas ultrasmall nanoparticles (USPIOs) usually have a core diameter < 5 nm. Comprehensive reviews have been written on this topic.[45–48] The different synthetic methods available to produce IONPs made of magnetite (Fe_3O_4) and maghemite (γ-Fe_2O_3), have been well described[47,49,50] (see also Chapter 1 of this book). Typical methods to synthesize IONPs include coprecipitation, in constrained environments, thermal decomposition and/or reduction, hydrothermal synthesis and polyol synthesis.[47,51–55] Each one has its own specific advantages. Usually, synthetic techniques based on high-temperature decomposition methods provide good particle size control and the highest degree of particle uniformity, tenability and homogeneity of magnetic properties.[56,57] Such procedures are based first on the synthesis of a metal–oleate complex using metal chloride or nitrates dissolved in organic solvents containing oleic acid (OA). As a second step, the purified oleate complex is dissolved into an organic solvent and heated under inert and anhydrous atmosphere at high temperatures (>200°C). Nanoparticles are formed through decomposition of the metal–oleate precursor and are readily covered by oleate molecules (hydrophobic). A variant of this technique, using high-temperature reaction of iron (III) acetylacetonate ($Fe(acac)_3$) in phenyl ether in the presence of alcohol, OA and oleylamine, can yield ultrasmall monodisperse magnetite nanoparticles with tunable sizes in the range 4–20 nm (Figure 8.4).[58]

However, one of the major prerequisites of magnetic nanoparticles for MRI applications is to achieve a good dispersion in aqueous solvents. Therefore, an additional final step must be introduced in this type of synthetic procedure, to replace the hydrophobic coating with an amphiphilic and biocompatible surfactant. Ligand exchange is a critical step toward the fabrication of monodisperse iron oxide colloids in aqueous conditions, and this topic is addressed in the last section of this chapter.

Once individualized with an appropriate coating of organic or polymeric molecules, the magnetic properties of IONPs can be characterized. The magnetic properties of USPIO are strongly dependent on the diameter of the magnetic crystals. In fact, bulk magnetite/maghemite is usually ferromagnetic, and its magnetic behaviour is guided by the presence of magnetic domains (the 'Weiss domains'). Inside each one of these volumes, the magnetic moments are aligned in different directions. In ultrasmall nanoparticles, however, each particle can accommodate only one of these domains, and this confers different magnetic properties to the nanomaterial. Larger-sized IONPs (7–15 nm) have stronger magnetization and induce 'negative' or T_2/T_2^* contrast more efficiently. On the other hand, ultrasmall nanoparticles (diameter < 5 nm) minimize the Néel relaxation time and thus lead to a decrease in magnetic properties, which are more compatible with the preservation of signal enhancement in T_1-weighted MRI.

Magnetite (Fe_3O_4) and maghemite (γ-Fe_2O_3) are two relatively similar forms of iron oxide (crystal structure and magnetic properties).[46,59] Both are present in superparamagnetic

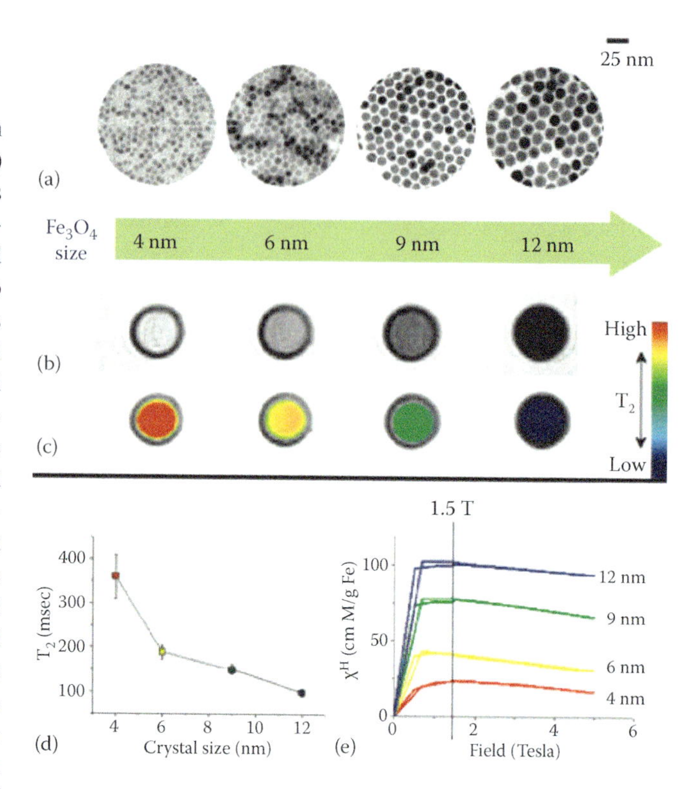

FIGURE 8.4 Size effect of iron oxide nanocrystals synthesized by thermal decomposition on magnetism and MRI contrast (T_2-weighted contrast). (a) Transmission electron microscopy (TEM) images of iron oxide nanocrystals of 4 to 6, 9 and 12 nm in diameter (in nonpolar solvent, i.e. not in water). (b, c) T_2-weighted MR images of nanocrystals in aqueous solutions, showing the impact of size on the negative contrast effect (images taken with a 1.5 T scanner). (d) Graph of T_2 value versus nanocrystal core size. (e) Magnetization values measured by a SQUID magnetometer. (Reproduced with permission from Jun, Y. W., Huh, Y. M., Choi, J. S., Lee, J. H., Song, H. T., Kim, S., Yoon, S., Kim, K. S., Shin, J. S., Suh, J. S., Cheon, J. 2005. Nanoscale size effect of magnetic nanocrystals and their utilization for cancer diagnosis via magnetic resonance imaging. *Journal of the American Chemical Society*. 127 (16), 5732–5733. Copyright 2005 American Chemical Society.)

IONPs. Magnetite is typically preferred due to its superior magnetic properties.[60] Maghemite ($Fe^{3+}[Fe^{2+}, Fe^{3+}]O_4$) often results from the oxidation of magnetite ($Fe^{3+}[Fe^{3+}_{5/3}\ V_{1/3}]O_4$, where V represents a cation vacancy). Bulk magnetite is ferromagnetic. The occurrence of an oxygen-mediated coupling mechanism aligns all the magnetic moments of the iron ions located in the tetrahedral sites of the crystal (8 crystallographic sites per unit structure), whereas all the magnetic moments of the octahedral ions (16 crystallographic sites per unit structure) are aligned in the opposite direction. It is assumed that the magnetic properties of magnetite are provided by uncompensated Fe^{2+} ions, whereas for maghemite, they are provided by that of Fe^{3+} ions.[61]

Paramagnetic nanoparticles are expected to follow Curie's law:

$$M = \chi H = \frac{C}{T} H, \tag{8.2}$$

where χ is the magnetic susceptibility, H is the applied magnetic field (e.g. that of the MRI scanner), T is the absolute temperature and C is the Curie constant, which is specific to each material. Their magnetic moment is not saturated at magnetic field strengths typically used in MRI.[43,62] On the other hand, superparamagnetic IONPs such as USPIOs feature an exceptionally saturation magnetization. They also have a strong Curie constant. As a result, they respond quickly to the application of an external magnetic field, and their magnetization quickly becomes saturated at relatively low magnetic field strengths (Figure 8.5). This difference in magnetic properties between paramagnetic and superparamagnetic nanocrystals underlies one of the fundamental differences between 'positive' and 'negative' contrast agents.[63]

The absence of magnetic remanence is one of the signatures of IONPs (2–15 nm), compared to particles of higher diameters. This characteristic is fundamental to avoid the dramatic consequences that permanent magnetization could have on IONPs injected *in vivo*. Permanent magnetization would lead to strong aggregation of the particles even after these are retrieved from the main magnetic field. The absence of remanence in ultrasmall iron oxide nanocrystals is due to the return to equilibrium of the magnetic moments through Néel relaxation. This phenomenon refers to the relaxation of the global electronic moment of a superparamagnetic crystal.

In fact, superparamagnetism occurs only when nanoparticles are small enough to belong to single magnetic domains.

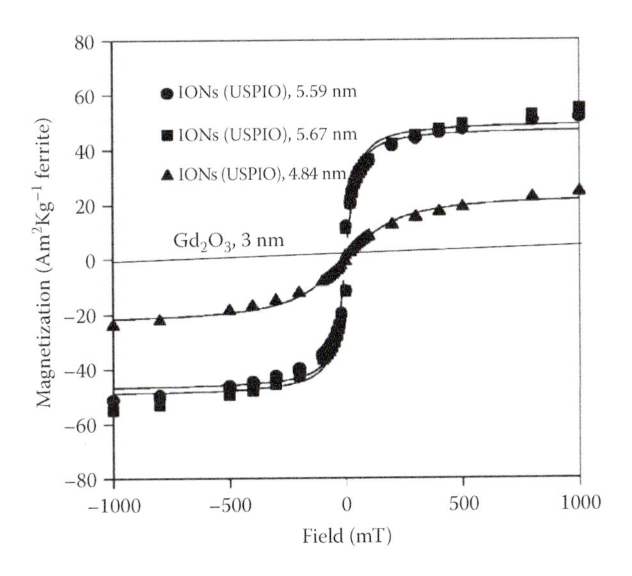

FIGURE 8.5 Magnetometric measurements of USPIOs (superparamagnetic) and Gd_2O_3 (paramagnetic) nanoparticles.[47,63] (Adapted with permission from Laurent, S., Forge, D., Port, M., Roch, A., Robic, C., Elst, L. V., Muller, R. N. 2008. Magnetic iron oxide nanoparticles: synthesis, stabilization, vectorization, physicochemical characterizations, and biological applications. *Chemical Reviews*. 108 (6), 2064–2110. Copyright 2008 American Chemical Society; and Fortin, M. A., Petoral Jr, R. M., Söoderlind, F., Klasson, A., Engströom, M., Veres, T., Käll, P. O., Uvdal, K. 2007. Polyethylene glycol-covered ultra-small Gd_2O_3 nanoparticles for positive contrast at 1.5 T MR clinical scanning. *Nanotechnology*. 18 (39). Copyright 2007 American Chemical Society.)

The magnetic energy of IONPs depends upon the direction of their magnetization vector, and this vector in turn depends on the crystallographic directions (the magneto-crystalline anisotropy field).[46] The directions that minimize the magnetic energy under no external magnetic field are called anisotropy directions, or easy axes (Figure 8.6a). The resulting magnetic moment of a magnetite/maghemite crystal is preferentially aligned along these specific directions. The magnetic energy increases with the tilt angle between the magnetic vectors of the easy directions.[64] The anisotropy of magnetite particles is often assumed to be uniaxial, with a single anisotropy axis. In fact, there are several anisotropy axes dictated by the oxide's crystallographic structure. The anisotropy energy (the amplitude of the curve) is given by the product of the crystal volume (V) times a constant (K_a: the anisotropy constant).

$$E_a = K_a V. \tag{8.3}$$

Large samples of bulk ferromagnetic magnetite/maghemite are divided into Weiss domains (represented in Figure 8.6b). Inside each one of these volumes, the magnetic moments are aligned in different directions. As iron oxide nanocores (such as in USPIOs) are smaller than one of these domains, each nanoparticle is therefore composed of a single domain whose magnetic moment is oriented in a specific direction. In these single domains, the direction of the magnetic moment can flip from one orientation to the other. When the thermal energy, given by kT (k: Boltzman constant; T: absolute temperature), is sufficient to overcome this anisotropy energy barrier, the magnetization fluctuates between the different anisotropy directions, according to the Néel relaxation time (τ_N).[65]

Although τ_N relaxation indirectly influences the hydrogen relaxation times by inducing changes to the magnetic moment of magnetic nanoparticles, τ_N relaxation is a phenomenon entirely distinct from the nuclear relaxation mechanisms of hydrogen protons (^1H) (see next section—the relaxivity).

For dry powders of monodomain IONPs, τ_N indicates the time it takes the magnetization to come back to a state of equilibrium after it is submitted to a strong magnetic field. For highly anisotropic crystals, the crystal magnetization is 'locked' in the easy axes. The Néel relaxation defines the rate of fluctuations that arise from the jumps of the magnetic moment between the different easy axes[66] (Figure 8.6c). In order to flip from one easy direction to the other, the magnetization of a nanoparticle must jump over an anisotropy energy hump. For a superparamagnetic nanoparticle of specific V and K_a, the Néel relaxation time (τ_N) is given by an Arrhenius law that is similar to that describing the activation energy for a chemical reaction[67]:

$$\tau_N = \tau_0 (E_a) e^{\frac{E_a}{kT}}, \tag{8.4}$$

where $\tau_0(E_a)$ is the preexponential factor of the Néel relaxation time expression, which depends on factors such as the volume (V), the specific magnetization of the nanocrystal

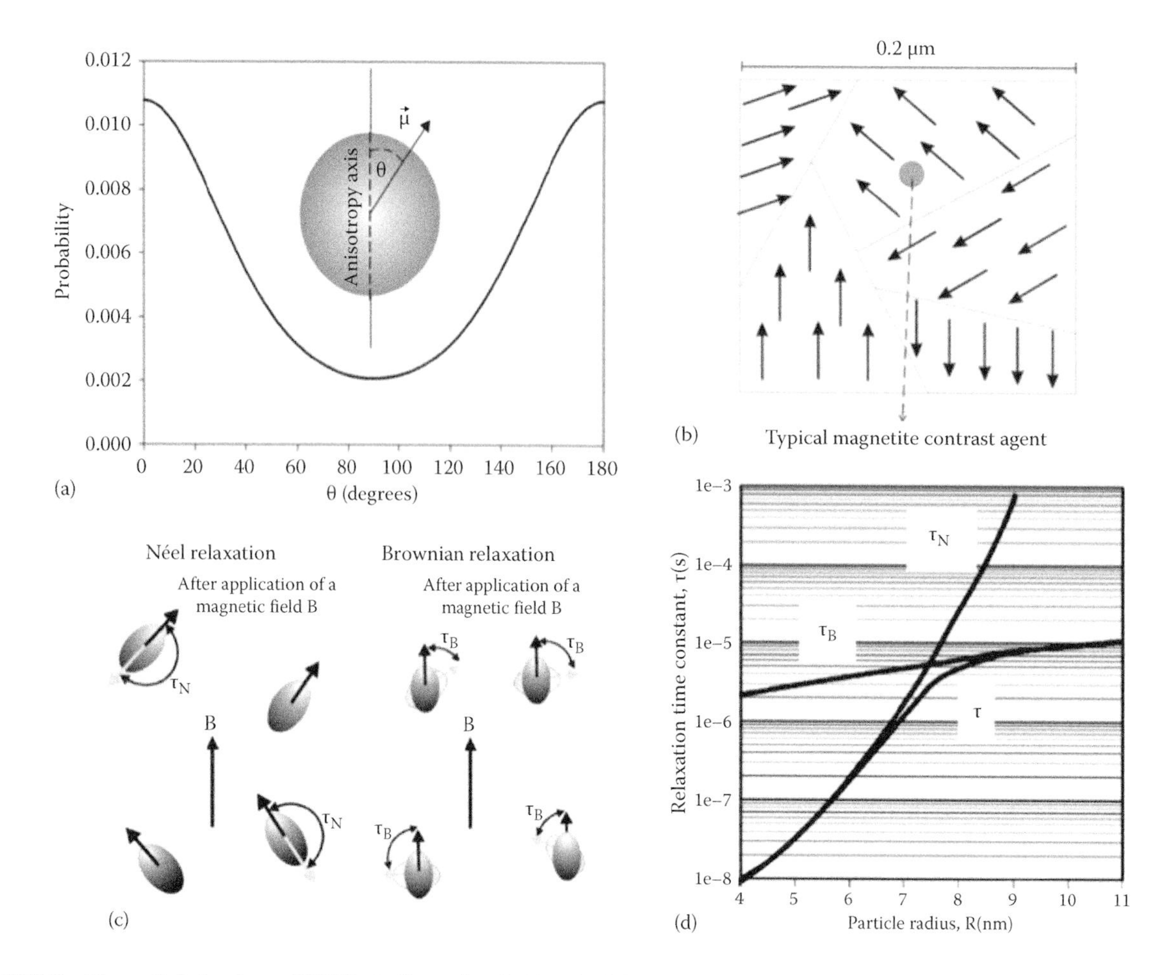

FIGURE 8.6 Magnetic behaviour of IONPs (radius = 5 nm): (a) uniaxial anisotropy for magnetite/maghemite nanoparticles (i.e. the probability of alignment of the magnetic moment in one direction with respect to the angle between this direction and the anisotropy axis); (b) representation of Weiss domains in a large magnetite/maghemite crystal, compared with the dimensions of a typical nanoparticle (the small circle), much smaller than a Weiss domain.[46] (Adapted with permission from Gossuin, Y., Gillis, P., Hocq, A., Vuong, Q. L., Roch, A. 2009. Magnetic resonance relaxation properties of superparamagnetic particles. *Wiley Interdisciplinary Reviews: Nanomedicine and Nanobiotechnology*. 1 (3), 299–310. Copyright 2009 American Chemical Society.) (c) Schematic representation of Néel relaxation and Brownian relaxation.[47] (Laurent, S., Forge, D., Port, M., Roch, A., Robic, C., Elst, L. V., Muller, R. N.: *Chemical Reviews*. 2064–2110. 2008. Copyright Wiley-VCH Verlag GmbH & Co. KGaA. Reproduced with permission.) (d) Relaxation time values plotted as a function of magnetite/maghemite nanoparticle size.[47,66] (Laurent, S., Forge, D., Port, M., Roch, A., Robic, C., Elst, L. V., Muller, R. N.: *Chemical Reviews*. 2064–2110. 2008. Copyright Wiley-VCH Verlag GmbH & Co. KGaA. Reproduced with permission; Reprinted from *Journal of Magnetism and Magnetic Materials*, 252, Rosensweig, R. E., Heating magneticfluid with alternating magnetic field, 370–374, Copyright 2002, with permission from Elsevier.)

and the gyromagnetic ratio of the electron.[60,68,69] Whereas the preexponential factor decreases as the value of anisotropy energy increases, τ_N increases as an exponential function of V because of the second factor of Equation 8.4. For small values of the anisotropy energy and at high temperatures, $E_a \ll kT$ (the exponential term tends to 1) and τ_N is mainly determined by the preexponential term. These conditions are fulfilled, for instance, with individual ultrasmall nanoparticles of iron oxide of $r < 4$ nm. On the other hand, for high anisotropy energies, when $E_a \gg kT$, the evolution of τ_N is mainly dictated by the exponential factor (fast increase with E_a).

According to Equation 8.4, the flipping of the magnetic moment of magnetite/maghemite crystals is observed only for nanoparticles of size $r < 12$ nm. Indeed, for magnetite ($\tau_0 \approx 10^{-9}$ s, $K \approx 13{,}500$ J m^{-3}), τ_N goes from ~500 years for particles of $r = 15$ nm down to the ms for particles of about $r = 10$ nm. Practically, this means that for particles having Néel relaxation times (τ_N) longer than the measurement time, the magnetization curve of the nanoparticle system is irreversible and shows a hysteresis loop. These are referred to as 'frozen single domains.'

For nanoparticles dispersed in a liquid media (a colloid), the return of the magnetization to equilibrium after application of a strong magnetic field is determined by both τ_N and the Brownian relaxation τ_B of the particles. The latter characterizes the rotation of the particle in the fluid and takes into account the viscosity of the solvent (Figure 8.8c). The global magnetic relaxation rate is a sum of two processes:

$$\frac{1}{\tau} = \frac{1}{\tau_N} + \frac{1}{\tau_B}, \tag{8.5}$$

where τ is the global magnetization time and τ_B is the Brownian relaxation time, given by

$$\tau_B = \frac{3V\eta}{kT}, \quad (8.6)$$

where η is the viscosity of the solvent. For large particles, $\tau_B < \tau_N$ because the Brownian component of the magnetic relaxation is proportional to the crystal volume (Equation 8.6) and the Néel relaxation is an exponential function of the volume (Equation 8.4). Inversely, in suspensions of very small nanoparticles ($r < 7$ nm), the magnetic relaxation is mainly driven by the Néel relaxation time, whereas the Brownian contribution tends to be negligible (Figure 8.6d, from Ref. 66).

In summary, superparamagnetism refers to a specific magnetic condition for which ultrasmall particles of a size well inferior to typical Weiss domains of magnetite/maghemite materials can be submitted to high magnetic field strengths, without evidence of magnetic remanence (Figure 8.5). Macroscopically, the magnetization of IONPs suspensions is described by a Langevin function, whose shape depends on the saturation magnetization (M_{sat}) and the size of the magnetite crystals:

$$M(B_0) = M_{sat}L(x), \quad (8.7)$$

where M_{sat} is the magnetization at saturation and L(x) is the Langevin function as:

$$L(x) = \left[\coth(x) - \frac{1}{x}\right] \quad (8.8)$$

with

$$x = \frac{M_s(T)VB_0}{kT}. \quad (8.9)$$

Magnetization curves of ultrasmall IONPs (Figure 8.5) obey these laws. The curves are perfectly reversible because the fast magnetic relaxation allows the system to always stay at thermodynamic equilibrium.[70]

8.4 RELAXOMETRIC PROPERTIES OF ULTRA-SMALL IONPs

Contrast agents interact with the small, hydrogen-containing mobile molecules of biological tissues and fluids. They influence the macroscopic relaxation times (T_1 and T_2) of ^{1}H contained in neighbouring mobile molecules. In turn, this effect modulates the MRI signal (Equation 8.1), which translates into contrast effects in anatomical MR images. The efficiency of MRI contrast agents in decreasing both the T_1 and T_2 of hydrogen protons is referred to as the 'relaxivity.' For IONPs, relaxivity depends on the concentration, as well as on the physicochemical characteristics of the contrast agent: hydrodynamic diameter, diameter of the inner core, magnetization, specific surface, distance of closest approach of hydrogen protons etc. The motion of contrast agents in the fluid is a very important factor influencing the relaxivity, as well as the field-dependence of magnetization. The resulting impact of all of these parameters on the longitudinal and transverse relaxation of water protons is referred to as the 'theory of relaxivity.'

The mathematical theory underlying dipole–dipole (DD) interactions between paramagnetic substances and hydrogen protons was originally developed by Bloembergen, Purcell and Pound in 1948,[71] generalized by Solomon in 1955[72–74] and extended by Bloembergen and Morgan to include electron spin relaxation in 1961.[75] Together, this body of work is often called the Solomon–Bloembergen–Morgan (SBM) theory, and its main principles will be briefly presented here.[12,13,43,46,47]

8.4.1 Introduction to the Theory of Relaxivity and Its Practical Aspects

The effects of contrast agents on the relaxation time of protons (^{1}H protons) are usually measured by relaxometric analysis, which is the technical term that refers to the measurement of T_1 and T_2 of mobile ^{1}H species in aqueous suspensions or in biological tissues. On a practical basis, this consists of preparing fractions of a given contrast agent taken at various concentrations; the T_1 and T_2 of these contrast agents are measured by nuclear MR spectrometry. The measurements are usually performed at 37°C and in a range of magnetic field strength corresponding to the MRI scanners commonly found in clinics (0.5 to 3.0 T). The concentration of magnetic elements (e.g. Fe) is measured by elemental analysis techniques (e.g. atomic absorption spectroscopy, inductively coupled plasma optical emission spectroscopy and inductively coupled plasma mass spectrometry).

Longitudinal and transverse relaxivities (r_1 and r_2) are defined as the increase in water longitudinal or transverse relaxation rates resulting from the presence of 1 mM of a magnetic element (e.g. $Fe^{2+/3+}$) in the solution. In MRI, the impact of contrast agents on the relaxation rate of protons, measured in fixed conditions of magnetic strength and temperature, is described by the following equation:

$$R_i = \frac{1}{T_i} = \left(\frac{1}{T_{i0}}\right) + r_iC \quad (8.10)$$

where $R_{i=1,2}$ is the relaxation rate of the aqueous solution, T_{i0} is the relaxation time of the aqueous media in the absence of the contrast agent, $r_{i=1,2}$ is the relaxivity (usually at 37°C, pH = 7 and B_0 = 0.5, 1.5 or 3.0 T) and C is the contrast agent concentration (in mM of Fe). Therefore, the relaxometric performance of MRI contrast agents is assessed first by measuring their relaxation rates ($1/T_1$ and $1/T_2$), followed by normalizing the data to the elemental concentration of magnetic element. Relaxivity values (r_1 and r_2) are extracted from the slope of the graph given by Equation 8.10. These are often referred to as 'relaxivity curves.' In general, the relaxation rate varies linearly with an increasing concentration of (para)magnetic ions.

The observed relaxation rate ($1/T_i$) is the sum of the intrinsic diamagnetic relaxation rate of water ($1/T_{i,d}$) and the specific contribution of the paramagnetic element ($1/T_{i,p}$), which can also be expressed as follows:

$$\left(\frac{1}{T_{i,obs}}\right)=\left(\frac{1}{T_{i,d}}\right)+\left(\frac{1}{T_{i,p}}\right)=\left(\frac{1}{T_{i,d}}\right)+r_i[\mathrm{Fe}], \quad (8.11)$$

where $i = 1, 2$ and [Fe] is the concentration of iron. The relaxivity (r_1, r_2) is normally expressed in units of $mM^{-1}\,s^{-1}$.

8.4.2 Paramagnetic Contribution to Relaxivity

The relaxation rate due to paramagnetic compounds ($1/T_{i,p}$) is generally divided into two components: the inner-sphere (IS) and the outer-sphere (OS) contributions.[43]

$$\left(\frac{1}{T_{i,p}}\right)=\left(\frac{1}{T_{i,d}}\right)^{IS}+\left(\frac{1}{T_{i,p}}\right)^{OS} \quad (8.12)$$

IS relaxation refers to relaxation enhancement of a solvent molecule directly coordinated to the paramagnetic ion. OS relaxation refers to the relaxation enhancement of solvent molecules in the second coordination sphere and beyond (i.e. bulk solvent; see Figure 8.7). For small paramagnetic contrast agents (e.g. Gd-DTPA), the longitudinal relaxation rate ($1/T_1$) has a strong impact and usually leads to high MRI signal increase.

The IS relaxation relies on the exchange of energy between the spins and the electrons of the paramagnetic elements, which is facilitated when water molecules bind to the magnetic cores. The binding water molecules, denoted 'p' water molecules (red circles in Figure 8.7) rapidly leave the first coordination sphere and are immediately replaced by 'fresh'

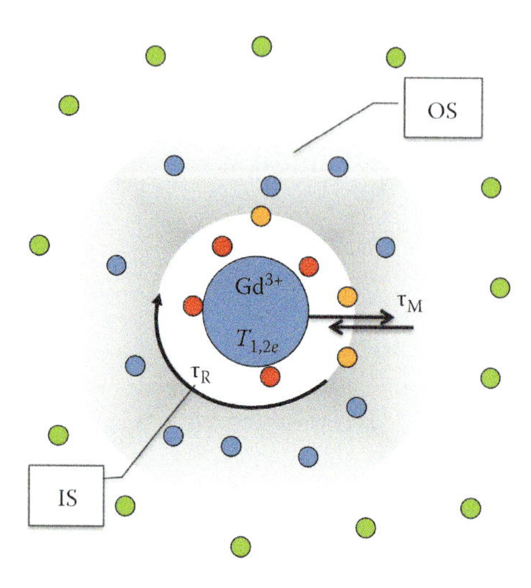

FIGURE 8.7 Schematic representation of the IS, OS, chemical exchange and rotational correlation times guiding the paramagnetic relaxation.

molecules from the matrix ('d' water molecules: orange circles). In close contact with the magnetic ions, hydrogen relaxes faster. The water residence time in the IS (τ_M) is in the order of ~1 ns, and this means the relaxation effect propagates very fast to the rest of the solution (to 'd' protons). $T_{1,2e}$ refers to the electron relaxation time of the paramagnetic elements (see also Section 8.4.3). Each one of the water protons that relax energy participates in the decrease in the overall longitudinal relaxation time of the water solvent. The IS model is at the core of the SBM theory.[72–74]

The IS contributions relate the lifetime, the chemical shift and the relaxation rates of solvent molecules in the IS, to nuclear magnetic relaxation (NMR) observables (T_1 and T_2), as follows[76,77]:

$$\left(\frac{1}{T_1}\right)^{IS}=fq\left(\frac{1}{T_{1M}+\tau_M}\right) \quad (8.13)$$

$$\left(\frac{1}{T_2}\right)^{IS}=fq\frac{1}{\tau_M}\left(\frac{T_{2m}^{-1}\left(\tau_m^{-1}+T_{2m}^{-1}\right)+\Delta\omega_m^2}{\left(\tau_m^{-1}+T_{2m}^{-1}\right)^2+\Delta\omega_m^2}\right) \quad (8.14)$$

$$\Delta\omega_{obs}^{IS}=fq\frac{1}{\tau_M}\left(\frac{\Delta\omega_m}{\left(1+\tau_m T_{2m}^{-1}\right)^2+\tau_m^2\Delta\omega_m^2}\right), \quad (8.15)$$

where the IS superscript refers to inner-sphere, f is the relative concentration of magnetic complexes over water molecules, and q is the number of water molecules in the first coordination sphere of the magnetic element (metal ion), τ_m is the lifetime of the solvent molecule in contact with the magnetic element (note: τ_m is the reciprocal of the solvent exchange rate, k_{ex}). The 'm' subscript refers to the shift or relaxation rate of the solvent molecule in the IS, whereas $\Delta\omega_m$ refers to the chemical shift difference between the magnetic complex and a diamagnetic reference. From Equation 8.13, one sees that if the water exchange rate is fast enough such that $\tau_m \ll T_{1,m}$ then the relaxation rate enhancement experienced by the bulk solvent will depend on the relaxation rate enhancement for the coordinated solvent molecule ($T_{1,m}$).

For ^1H protons, the two relaxation mechanisms operative are the DD mechanism and the scalar (SC) or contact mechanism, such as

$$\frac{1}{T_{im}}=\frac{1}{T_i^{DD}}+\frac{1}{T_i^{SC}}, \quad (8.16)$$

where $i = 1, 2$. The correlation times that define DD and SC relaxation are τ_{ci} and τ_{ei}, respectively. At high field strengths with slowly rotating molecules, the Curie spin relaxation mechanism may become important, but it is considered negligible at the low fields used in MRI (0.5–3 T). Therefore, it is neglected in the current description. The DD component of

$1/T_{1m}$, is different from its SC component, and a comprehensive description of its mechanisms and impact can be found elsewhere.[47]

8.4.3 Main Parameters Affecting the Relaxivity of Contrast Agents

The *rotational correlation time* (τ_R) characterizes the reorientation of the vector between the (para)magnetic ions, or the magnetic cores, and the protons of the water. For a low-molecular-weight complex, τ_R limits the relaxivity of paramagnetic contrast agents at magnetic field strengths used in clinical MRI (0.5–3.0 T). The most immediate and convenient way to estimate the rotational correlation time is through the use of the following equation[13]:

$$\tau_R = \frac{4\pi a^3 \eta}{3kT}, \tag{8.17}$$

where a is the radius of the molecules or the nanoparticles and η is the viscosity of the medium. In general, relaxivities tend to increase at lower temperatures, as a consequence of slower molecular tumbling rates. Comprehensive reviews on the measurement of rotational correlation times can be found elsewhere in the literature.[13,43]

The *hydration number* (q), i.e. the number of water molecules in the IS of the paramagnetic element, also plays an important role in relaxivity. For IONPs, this factor is difficult to measure with precision. It is expected to have only a minor influence on the relaxivity of IONPs, which is mainly driven by OS mechanisms.

Coordinated water residence time (τ_M) describes the rate of water exchange between an IS water molecule and the bulk. The mechanism of IS relaxation is based on an exchange between water molecules surrounding the complex and the water molecules coordinated to the surface of the contrast agent. Consequently, the exchange rate ($k_{ex} = 1/\tau_M$) is an essential parameter for transmitting the relaxation effect to protons in the water matrix.[43]

Finally, *electron relaxation times* (τ_{S1} and τ_{S2}; namely the longitudinal and transverse electron relaxation times) describe the process of return to equilibrium of the magnetization associated to electrons that transit between electronic levels of the magnetic core. These transitions produce fluctuations that allow the relaxation of protons; τ_{S1} and τ_{S2} are magnetic field dependent and they have also a strong influence on r_1 and r_2 curves.

8.4.4 OS Relaxation and the Case of Superparamagnetic Nanoparticles

For superparamagnetic particles, the IS contribution to the relaxation is negligible compared to the dominant OS contribution (the second term of Equation 8.12). It is mainly controlled by the dipolar interaction at long distance between the magnetic moment of the superparamagnetic nanoparticles and the nuclear spin of hydrogen protons. In fact, the magnetic cores influence the local magnetic field around the ^1H protons flowing in their vicinity. Comprehensive equations describing the OS contribution to the longitudinal relaxation rate $\left(\frac{1}{T_1}\right)^{OS}$ can be found in a short list of contributions.[60,78,79] The dipolar intermolecular mechanism in the OS contribution is modulated by the translational correlation time (τ_D), which takes into account the relative diffusion (D) of the paramagnetic centre and the solvent molecule, as well as their distance of closest approach (d), according to[78]

$$\tau_D = \frac{d^2}{D}. \tag{8.18}$$

This expression indicates that viscous solvents and larger particles lead to high translational correlation times, and this of course has a strong influence on the relaxivity of USPIO. Very small USPIOs are characterised by smaller anisotropy energy so that the locking of the particle magnetization onto the anisotropy directions is attenuated. Koenig et al.[79] developed a model discarding anisotropy energy, whereas Roch et al.[80] developed a complete quantum theory introducing E_A, the anisotropy energy, as a quantitative parameter of the problem. The predictions of that model are in good agreement with experimental observations on ultrasmall magnetite particles.

8.4.5 Superparamagnetic Nanoparticles and the Measurement of Relaxivity by Nuclear Magnetic Relaxation Dispersion

Bulte et al.[81] were among the first to measure the influence of the magnetic field on the relaxivities of maghemite nanoparticles. The transverse relaxivity (r_2) of superparamagnetic nanoparticles increases with increasing field and then saturates, in direct relationship with the Langevin function (as in Figure 8.5 and Equation 8.8). For larger superparamagnetic particles, the IS contribution to the relaxation is minor and often completely negligible compared to the dominant OS contribution. The OS contribution is largely dependent on the movement of water molecules near the local magnetic field gradients generated by the superparamagnetic nanoparticles. Overall, the equations describing IS and OS contributions to the relaxivity of magnetic contrast agents involve a relatively large number of parameters (τ_M, q, τ_R, D, r, d, τ_V, τ_{S0} and τ_D). Because of this high number of parameters, it is often difficult to perform an accurate theoretical estimation of the performance of MRI contrast agents at different magnetic field strengths. To characterize contrast agents precisely and reproductively at different magnetic field strengths, nuclear magnetic relaxation dispersion (NMRD) profiling must be used.[82] The field cycling relaxometer has the ability to quickly switch magnetic fields between preset values corresponding to phases of polarization, relaxation at the selected field and detection, respectively.[83] The fitting of NMRD profiles by adequate theories provides information about the average radius (r) of the iron oxide nanocrystals, their specific magnetization (M_s), their anisotropy energy (E_a) as well as their Néel relaxation time (τ_N).[84] Here is a brief list of the main information extracted from NMRD profiles of superparamagnetic nanoparticle suspensions:

- *The average radius (r), or the distance of closest approach:* at high magnetic fields, the relaxation rate depends only on τ_D and the inflection point corresponds to the condition $\omega_I \cdot \tau_D \sim 1$ (Figure 8.8). According to Equation 8.25, the determination of τ_D gives the crystal size r (good complement to high-resolution transmission electron microscopy measurements).
- *The specific magnetization (M_s):* at high magnetic fields, M_s can be obtained from the equation $M_s \approx [(R_{max}/(C \cdot \tau_D)]^{1/2}$, where C is a constant and R_{max} is the maximal relaxation rate.
- *The crystal anisotropy energy (E_a):* the absence or the presence of an inflection point at low magnetic field strengths (10^{-2} – 1) is an indication of the anisotropy energy. For crystals characterized by a high E_a compared to the thermal agitation, the low field dispersion disappears. This was confirmed in a previous work with cobalt ferrites, a high anisotropy energy material.[80]
- *The Néel relaxation time (τ_N):* the relaxation rate at very low fields (R_0) is governed by a 'zero magnetic field' correlation time τ_{c0}, which is equal to τ_N if $\tau_N \ll \tau_D$. However, this situation is often not met, and in this case, τ_N is only reported as qualitative information.

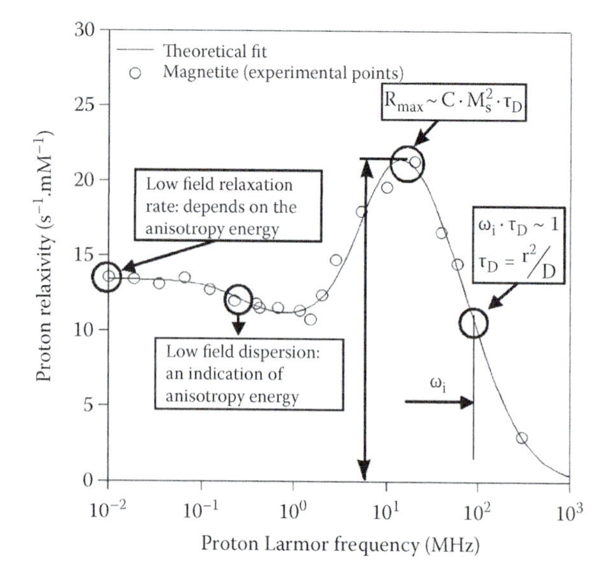

FIGURE 8.8 NMRD profile for magnetite particles in colloidal solution.[85] (With kind permission from Springer Science+Business Media: *Handbook of Experimental Pharmacology*, Contrast agents: magnetic resonance. 2008, 135–165, Burtea, C., Laurent, S., Vander Elst, L., Muller, R. N.)

More information about the interpretation of NMRD curves can be found in a selection of references.[43,47,85] For USPIOs, we invite the reader to refer to the work of Laurent et al., as well as Gossuin et al.[46–47,86]

8.5 MDBC STABILIZATION STRATEGY: NEW PERSPECTIVES IN THE DEVELOPMENT OF AQUEOUS COLLOIDAL IONPs

8.5.1 Ultrasmall IONPs for T_1-Weighted MRI

As presented and described in the previous sections, USPIOs with particle diameters < 5 nm can be used as 'positive' contrast agents for T_1-weighted MRI.[16,17] Their small iron oxide core minimizes the magnetic moment and thereby limits the T_2/T_2* contribution, whereas their relaxometric properties tend to be closer to that of paramagnetic chelates (e.g. low r_2/r_1 ratios, high longitudinal relaxivities – r_1). The relaxivity of USPIOs can be tuned based on particle size, as reflected in Figure 8.4. The magnetic moment of USPIO rapidly decreases with decreasing particle diameter due to the reduction in volume magnetic anisotropy and surface spin disorders.[18,19] In fact, the dimension of the iron oxide cores is a fundamental parameter in the magnetic and relaxometric performance of USPIOs. It must be measured by transmission electron microscopy (TEM) (Figure 8.4a). On the other hand, TEM provides only a two-dimensional projection of the inorganic part of the particle; it does not take into account the presence of the polymer layer or the interaction of the surface with ions and molecules in the fluid in which the particle is dispersed. The hydrodynamic diameter refers to the total volume of the fluid that is affected by the presence of nanoparticles. The hydrodynamic size of particles has a huge incidence on their fate *in vivo*: blood retention, sequestration by immune cells, clearance kinetics etc. It is measured by dynamic light scattering (DLS), which is a volume-rendering analytical technique. Usually, DLS is used on a routine basis to demonstrate the absence of aggregation in a colloid, to assess the quality of the dispersion, and to evaluate the particle size distribution in a liquid. It takes into account not only the particle itself but also its interaction with the neighbouring species in the fluid. For this reason, the hydrodynamic diameter of USPIOs measured by DLS is systematically larger than the size of iron oxide cores measured by TEM. *In vivo*, the colloids must remain stable when submitted to biological fluids, which is not easily achieved considering the high salinity, varying pH conditions and the presence of a large variety of potentially reactive ions. Upon injection in the blood, a corona of proteins usually adsorbs at the surface of magnetic nanoparticles, and this can have a dramatic impact on the hydrodynamic diameter of the contrast agent. Hydrodynamic diameter has a major influence on the tumbling rate of the particles and, as a result, on the relaxometric properties as described in the previous section.

Over the past 20 years, a few USPIO products were developed and commercialised based on their capacity to generate 'positive' MR contrast (see Table 8.3). First, AMI-227 (Sinerem or Combidex) is a 20–40-nm dextran-coated USPIO

TABLE 8.3
Size and Physicochemical Relaxometric Properties of Multidentate-Coated USPIOs, Compared with Other USPIO Systems Coated with Mono/Bidentate Ligands and Small Molecules

Product: Type Of Ligand, Name, *Coating*	NP Core Diameter, TEM (nm)	Hydrodynamic Size, DLS (nm)	r_1 (1.5 T, 37°C)	r_2/r_1	Refs.
Multidentate: COOH-MDBC/USPIO	3.4	34	4.8	4.7	87
Multidentate: Cat-MDBC/USPIO	3.4	19.2	6.8	5.5	88
Multidentate: Cat-MDBC/E-USPIO	1.9	14.8	3.0	1.5	89
Multidentate: OligoPEG-Dopa(Cat)	11	38	0.173 (at 11.75 T)	952	39
Mono/bi-dentate: USPIO, *bis-phosphonate-PEG*	5.5 ± 0.6	24 ± 3	9.5 (at 3 T, RT)	2.97	17
Mono/bi-dentate: ESION (iron oxide), *PO-PEG*	3		4.77 (at 3 T)	6.12	32
Mono/bi-dentate: USPIO, Ferumoxtran-10 (AMI-227), *dextran T10*	4.5	15–30	9.9	6.57	90
Mono/bidentate: USPIO, Ferumoxytol-7228, *carboxlymethldextran*	6.7	30–35	15	5.93	91,92
Mono/bidentate: USPIO, Supravist (SHU-555C), *carboxydextran*	3–5	21	10.7	3.55	44
Mono/bidentate: USPIO, Feruglose NC100150, Clariscan, *PEGylated starch111*	6.43	11.9	~18 (at 0.5 T)	n.a.	93,94
Mono/bidentate: VSOP (iron oxide), C184, *citrate*	7	14	14	2.4	95

with a human blood pool half-life of more than 24 h.[90] It can be used as an MR angiography agent during the early phase of intravenous administration[96–98] and as a MR lymphography agent during the late phase.[99] NC100150 (Clariscan), a PEGylated USPIO, and SHU 555C (Resovist) are bolus-injectable agents developed for MR angiography and perfusion studies.[100] Monocrystalline IONPs and cross-linked IONPs are monocrystalline USPIOs (typically 2–9 nm); the latter is stabilized by a cross-linked aminated dextran coating.[101,102] Finally citrate-coated monocrystalline very-small superparamagnetic iron oxide particles (VSOPs) were also suggested as a new contrast medium for MRI.[95] The values of r_1 and r_2/r_1 in Table 8.3 are valid for homogeneous solutions of the contrast agents. Aggregation occurring in the blood, or inside cells, can lead to a drastic increase in r_2/r_1, with significant impact on the positive contrast enhancement effect.

8.5.2 Molecular Coatings and Ligands Developed for Individualised USPIOs

One of the major conditions to be reached in order to control the magnetometric and relaxometric properties of USPIOs is the development of coatings that isolate each one of the iron oxide cores. Isolating coatings minimizes the formation of clusters, prevents agglomeration and decreases the magnetic interactions between the particles. Several processes have been explored for coating ultrasmall iron oxide cores. Citric acid is a small ligand that binds to the surface of IONPs through carboxylic binding.[103] In particular, this strategy was used to synthesize the commercial product VSOP C184 (4 nm core size).[104] However, citric acid may induce loss of magnetite–maghemite iron oxide crystal structure, with impact on the magnetic properties of the nanoparticles.[105] Its binding to the oxide core is not particularly stable. It is also highly charged, and this may cause the nanoparticles to be very rapidly retrieved from the blood.[106] Other ligand molecules can be used for the stabilization of magnetic nanoparticles in aqueous medium, in particular gluconic acid, dimercaptosuccinic acid, phosphorylcholine as well as phosphate and phosphonates.[47,107] The size, physicochemical and relaxometric characteristics of several of these products can be found in Table 8.3.

In general, small, hydrophilic particles of neutral charge show long plasma half-lives.[7] Among the small, hydrophilic polymer coatings that most efficiently enhance the blood retention of magnetic nanoparticles, figure dextran and (carboxy, carboxymethyl)dextran.[90,108] In particular, the commercial USPIO products ferumoxtran-10 (AMI-227), ferumoxytol, and Supravist (SHU-555C) are all based on iron oxide of core diameters in the range 4–8 nm and of hydrodynamic sizes not larger than 30 nm.[45] In order to improve colloidal stability, biocompatibility and blood retention, PEG is also used at the surface of IONPs.[109–111] Feruglose (Clariscan) was developed using a PEGylated starch coating.[112] The presence of PEG at the surface of magnetic nanoparticles delays the adsorption of proteins at their surface, a process that leads to recognition and elimination of nanoparticles by the macrophage-monocytic system.

As introduced in Section 8.1, most synthetic processes that provide high-quality magnetic nanocrystals are based on high-temperature reaction of organometallic precursors. In fact, uniform IONPs of close to monomodal size distributions are usually synthesized by thermal decomposition procedures leaving layers of hydrophobic molecules such as OA at their surface.[49,57,113] Thus-coated crystals are water-immiscible. The ligands must be removed and replaced by biocompatible,

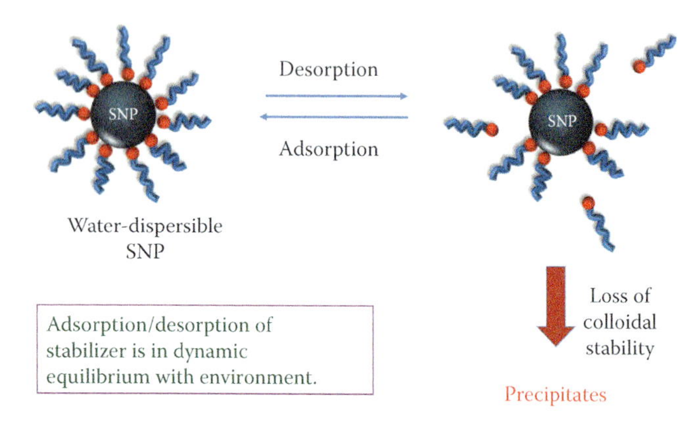

FIGURE 8.9 Schematic representation of the metastable nature of IONPs stabilised with ligands that are susceptible to dynamic adsorption/desorption (e.g. –COOH, catechol).

hydrophilic coatings. The main nanoparticle coating routes that have been explored until now for transferring hydrophobic nanoparticles to aqueous solvents include silanization,[20–22] intercalation with amphiphiles,[23,24] ionic interactions[25] and ligand exchange.[26] Typical examples of each are summarized in a recent review.[114] Among these, ligand exchange is the most facile and versatile process to form USPIOs stabilised with monolayers of ligand molecules. By these procedures, the ligands are grafted at the iron oxide surfaces to provide stable colloids in a wide range of pH and saline contents and in different physiological conditions.[27] However, these strategies often rely on the use of commercially available but ineffective simple ligands or large mass block copolymers. They can provide nanoparticles with limited long-term stability and/or substantially increased hydrodynamic size. In fact, ligands attached at the surface of nanoparticles, if they are not bound through a strong covalent bond, are susceptible to detach through the kinetics of adsorption/desorption mechanisms characteristics of the nature of their bond (see Figure 8.9). The presence of at least one other reactive bond on the same polymer chain greatly limits the risk of agglomeration related to the possible 'bare' nanoparticle surface areas that might occur in time due to the dynamic processes of adsorption/desorption.

8.5.3 Multidentate Polymers for High-Stability USPIO Coatings

Polymers with pendant anchoring groups can bind at several places at the surface of each USPIO, thereby serving as stabilizing multidentate ligands. This approach is increasingly used to transfer iron oxide cores from nonpolar solvents to water, while achieving high-quality molecular monolayers with a good particle separation and colloidal stabilization efficiency. Compared to small molecules and oligomers having one or two anchoring groups (called monodentates and bidentates), multidentate polymeric ligands exhibit stronger binding isotherms due to more favourable conformational arrangements of multiple anchoring groups tethered from long polymeric chains onto USPIO surfaces.[87] As stabilizing ligands undergo a constant equilibrium of adsorption/desorption from the particle surface, the use of multidentate polymeric ligands could endow USPIOs with enhanced colloidal stability. Typical examples of polymeric multidentates include natural polysaccharides,[34,35] homopolymers composed of one monomer units,[36,37] branched dendrimers[37] and random copolymers.[38,39]

The ideal multidentate polymer for ligand exchange should be made of biocompatible units, each one tolerated at very high doses by the human body. They should have narrow molecular weight distribution. Their multiple functional grafting units must allow a strong binding to metal oxide surfaces. Recent advances have focused on block copolymer-based multidentate molecules enabling the development of aqueous colloidal USPIOs stabilized with multifunctional block copolymers.[87,115–119] The strategy centres on the synthesis of well-defined MDBCs consisting of one anchoring block having pendant carboxylates, phosphonates or catechol groups for binding to iron oxide surfaces. The other tuneable block is varied with components that can promote compatibility with end-use applications (Figure 8.10). MDBCs can be designed with a water-soluble block, thus endowing the resultant MDBC/USPIO colloids water dispersible and biocompatible for MRI applications *in vivo*.

Na et al.[39] were among the first to report on multidentate polymers as coatings for IONPs. The researchers synthesized rather large iron oxide cores of 11, 17 and 23 nm TEM diameter using thermal decomposition (Figure 8.11). Then, the surfaces were ligand exchanged with oligo-PEG-COOH and oligo-PEG-dihydroxyphenylalanine (Dopa).

L-Dopa is a catechol derivative precursor to dopamine that is also used as a component of adhesives generated by marine mussels. It exhibits strong affinity to metal oxide nanocrystals. Thus-coated USPIOs had a hydrodynamic diameter of 38 nm (DLS measured for nanoparticles made of 11-nm iron oxide cores). OligoPEG-Dopa ligands provided rapid ligand exchange, and the resulting nanoparticles exhibited greatly enhanced colloidal stability over a broad pH range and in the presence of excess electrolytes. Stability in different aqueous conditions (with immunoglobulin G, in serum and in a broad range of pH and excess salts) was notably improved compared to non-catechol-presenting molecular or oligomer ligands. Relaxometric properties were also measured, although at a nonclinical magnetic field strength (11.75 T; temperature unknown). As expected at such high magnetic field strength, very high transverse relaxivities were achieved, as well as very limited longitudinal ones (listed in Table 8.3). Unfortunately, the products were not measured at lower—clinical—magnetic field strength (e.g. 1.5–3.0 T).

Recently, we reported on the development of a versatile and readily scalable ligand exchange strategy to prepare monodisperse, biocompatible and water-soluble USPIOs stabilized with well-controlled MDBCs having pendant carboxylates.[87] The strategy also extends to pendant catechol groups.[88] The COOH-MDBCs are designed to consist of a poly(methacrylic acid) (PMAA) block having pendant carboxylates as multidentate anchoring groups, binding to USPIO surfaces, and a hydrophilic block composed of pendant oligo(ethylene oxide) chains (PEOMA), endowing biocompatibility and colloidal stability. The MDBC system is designated as

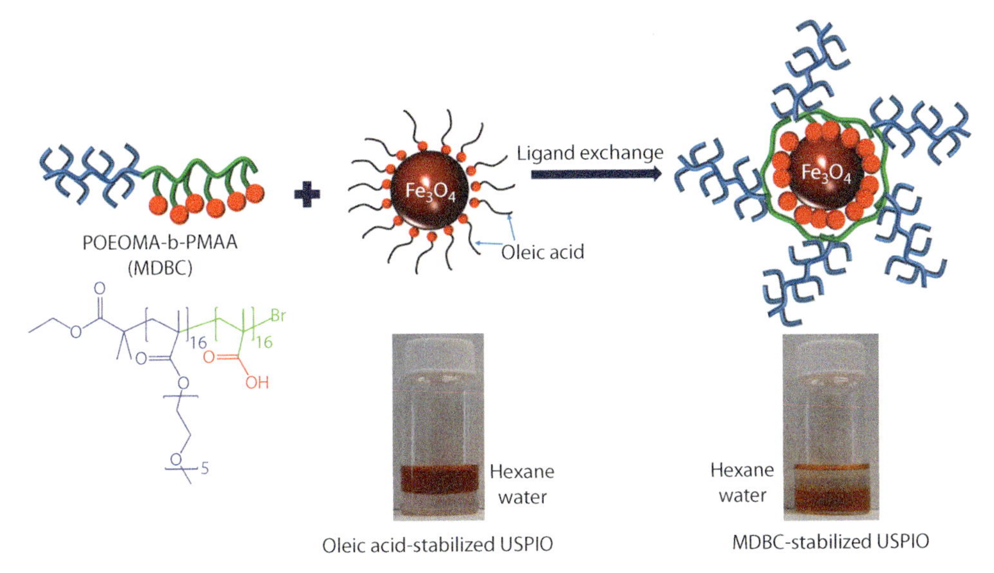

FIGURE 8.10 Schematic illustration of MDBC made of an anchoring polymethacrylate (PMAA) block with pendant carboxylate groups, bound to a hydrophilic polymethacrylate block with pendant oligo(ethylene oxide) groups (POEOMA). The presence of multiple carboxylate groups on the anchoring block secures the attachment of the polymer at the nanoparticle surface and enables efficient and lasting colloidal stability in a large range of pH and saline conditions.[87] (Reproduced with permission from Chan, N., Laprise-Pelletier, M., Chevallier, P., Bianchi, A., Fortin, M. A., Oh, J. K. 2014. Multidentate block-copolymer-stabilized ultrasmall superparamagnetic iron oxide nanoparticles with enhanced colloidal stability for magnetic resonance imaging. *Biomacromolecules*. 15 (6), 2146–2156. Copyright 2014 American Chemical Society.)

PMAA-b-POEOMA (Figure 8.10).[87] The resultant COOH-MDBC allows effective ligand exchange with OA-stabilized iron oxide cores of diameter = 4 nm (TEM data). Thus-formed USPIOs feature monolayers of COOH-MDBCs at the nanoparticle surface. The results from relaxometric properties as well as *in vitro* and *in vivo* MRI demonstrate that aqueous COOH-MDBC/USPIO colloids hold great potential as effective T_1 contrast agent (see Table 8.3 and Figure 8.12).

One concern for carboxylated MDBC-coated USNPs involves their negatively charged surfaces (zeta potential, $\zeta \approx -10$ mV at pH = 7). Further studies suggest that the optimal design of chain lengths of both anchoring and hydrophilic blocks can be required for enhanced colloidal stability in biologically relevant conditions as well as effective T_1-weighted contrast enhancement.[119]

As previously mentioned, catechol groups express a high affinity to metal surfaces.[120,121] Several studies have reported on the fabrication of IONPs with catechol-functionalized polymers: homopolymers,[122] random copolymers[39,123,124] and branched polymers.[125] Recently, the synthesis of well-controlled MDBCs with pendant catechol groups (Cat-MDBCs) by a combination of controlled radical polymerization and postmodification methods was reported.[126] The Cat-MDBC is effective at binding to USPIO surfaces. This approach enables the fabrication of aqueous Cat-MDBC/USPIO colloids at single molecular layers with a very small hydrodynamic diameter (≈20 nm; see Table 8.3). Further, the Cat-MDBC strategy has been utilized to fabricate extremely small USPIOs (E-USPIO—also reported as ESNP; see Figure 8.13) with core diameter ≈2 nm (TEM based), resulting in the formation of aqueous Cat-MDBC/ESNP colloids with a hydrodynamic diameter of ≈16 nm.[89] Promisingly, both aqueous Cat-MDBC/USNP colloids and Cat-MDBC/ESNP colloids exhibited excellent colloidal stability in a broad pH range and in various physiological conditions. No significant protein adsorption was reported. The relaxometric performance of these products also appears in Table 8.3: Cat-MDBC/ESNPs have a longitudinal relaxivity of 3.0 mM^{-1} s^{-1} and an r_2/r_1 = 1.5. These values are close to the relaxivity values of conventional paramagnetic chelates (e.g. Gd-DOTA [1,4,7,10-tetraazacyclododecane-1,4,7,10-tetraacetic acid]) for T_1-weighted contrast enhancement. It suggests that USPIOs could be considered as an alternative to rare-earth chelates in vascular contrast applications.

Finally, multidentate ligands based on poly(isobutylene-alt-maleic anhydride) having dopamine as coordinating moieties and near-infrared fluorophores (indocyanine) were used as efficient coating for IONPs (7 nm core diameter by TEM; ~20 nm hydrodynamic size—DLS). These products were demonstrated as efficient agents for trimodality imaging (near infrared optical, photoacoustic and MRI modalities).[127]

However, certain concerns can be raised about the administration *in vivo* catechol derivatives such as L-Dopa. First, L-Dopa is the precursor to the neurotransmitters dopamine, norepinephrine (noradrenaline) and epinephrine (adrenaline), collectively known as catecholamines. As a drug, it is used in the clinical treatment of Parkinson's disease and dopamine-responsive dystonia. Hypotension, arrhythmia, disturbed respiration, anxiety and psychosis have been reported as side effects of this drug. For these reasons, clinicians try to avoid these side effects by limiting L-Dopa doses as much as possible until absolutely necessary. A second reason for limiting as much as possible the remaining 'free' catechol groups on nanoparticles is the fact that catechol groups can cause toxicity risks to different cells (e.g. endovascular epithelial cells), similar to that of poisonous ivy.[128] Therefore, the consideration of catechol binding

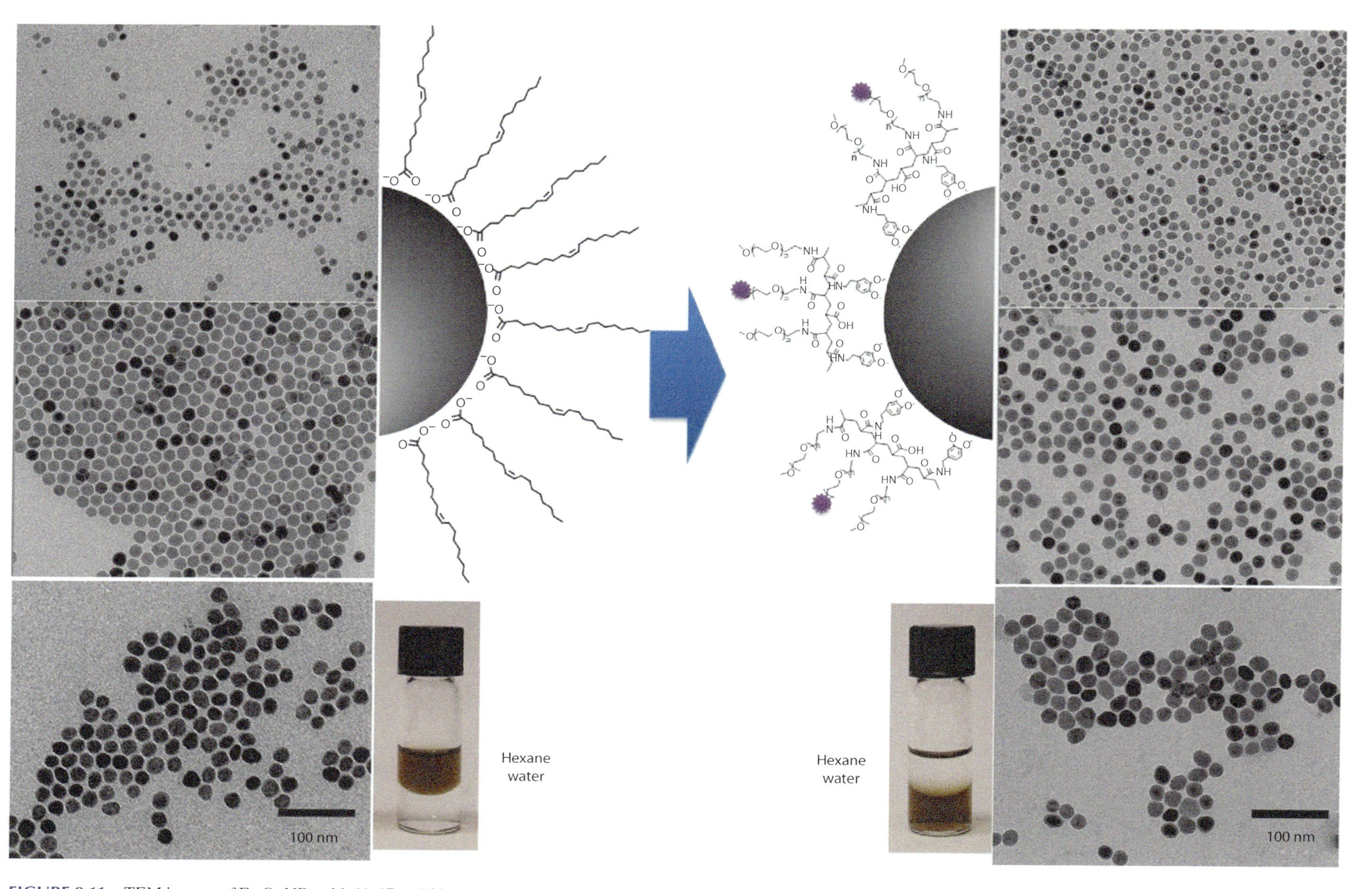

FIGURE 8.11 TEM images of Fe_3O_4 NPs with 11, 17 and 23 nm core size before (left) and after (right) ligand exchange with OligoPEG-Dopa. Schematic representation of the NP with the corresponding surface cap along with images of the organic and aqueous dispersions are shown.[39] (Reproduced with permission from Na, H. B., Palui, G., Rosenberg, J. T., Ji, X., Grant, S. C., Mattoussi, H. 2012. Multidentate catechol-based polyethylene glycol oligomers provide enhanced stability and biocompatibility to iron oxide nanoparticles. *ACS Nano*. 6 (1), 389–399. Copyright 2012 American Chemical Society.)

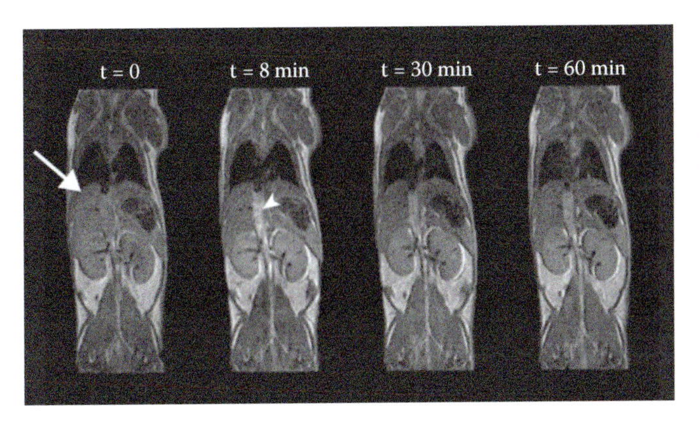

FIGURE 8.12 Positive vascular contrast enhancement in the mouse model. A 30 μL intravenous injection of 7.56 mM Fe of COOH-MDBC/USPIO can clearly enhance the signal in the vasculature (arrowhead) in T_1-weighted imaging (1.0 T, 2D spin-echo, TE/TR: 18/850 ms; see details in Chan et al.[87]). (Reproduced with permission from Chan, N., Laprise-Pelletier, M., Chevallier, P., Bianchi, A., Fortin, M. A., Oh, J. K. 2014. Multidentate block-copolymer-stabilized ultrasmall superparamagnetic iron oxide nanoparticles with enhanced colloidal stability for magnetic resonance imaging. *Biomacromolecules*. 15 (6), 2146–2156. Copyright 2014 American Chemical Society.)

groups for IONP coatings should involve systematic studies aiming at decreasing the amount of pendant 'free' groups and a comprehensive study about the possible side effects that such dopamine derivatives present at the surface of nanoparticles could have on a variety of cells and tissues.

Overall, the recent studies on multidentate ligands confirmed the superiority of this approach for providing enhanced grafting and colloidal stability at the surface of nanoparticles, compared with previous ligand exchange processes based on small molecules and monodentate ligands. Very strong binding affinities were reached by using both carboxylate (–COOH) and catechol (L-Dopa) groups, and the resulting USPIOs were demonstrated as very good potential vascular contrast agents for T_1-weighted MRI.

8.6 CONCLUSION AND PERSPECTIVES

Iron oxide nanocrystals have been integrated in an increasing number of preclinical and clinical MRI procedures. In this chapter, we presented an overview of the properties of USPIOs that makes them efficient contrast agents in MRI. Their magnetic and their relaxometric properties were presented and discussed. The relaxometric ratio at clinical magnetic strengths (typically at 1.5 T) is used to classify the

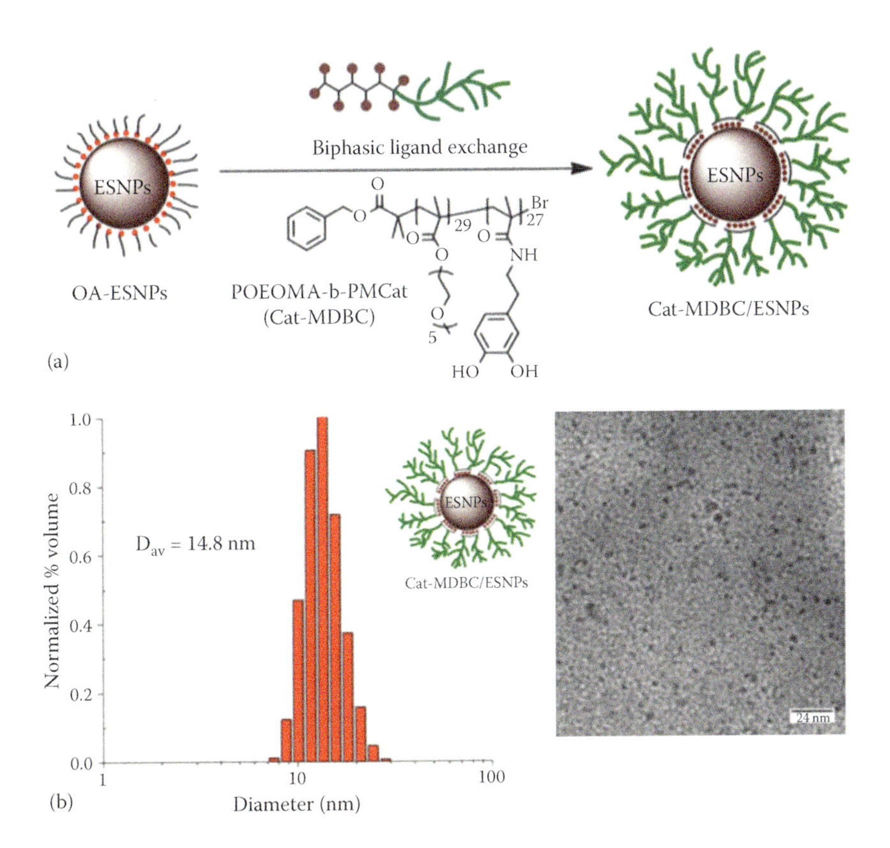

FIGURE 8.13 (a) Illustration of biphasic ligand exchange of Cat-MDBCs on OA-stabilized ESNPs to fabricate colloidally stable Cat-MDBC/ESNP colloids in aqueous dispersion. (b) DLS diagram and TEM image of aqueous Cat-MDBC/ESNP colloids prepared by biphasic ligand exchange of OA-ESNPs in hexane with Cat-MDBCs in water.[89] (Li, P. Z., Xiao, W. C., Chevallier, P., Biswas, D., Ottenwaelder, X., Fortin, M. A., Oh, J. K. 2016. Extremely small iron oxide nanoparticles stabilized with catechol-functionalized multidentate block copolymer for enhanced MRI. *ChemistrySelect*. 1 (13), 4087–4091. Copyright Wiley-VCH Verlag GmbH & Co. KGaA. Reproduced with permission.)

behaviour of contrast agents between 'positive' (i.e. $r_2/r_1 < 5$) and 'negative' ones (i.e. $r_2/r_1 \gg 10$). To be used as efficient 'signal enhancers' in blood pool MRI and for targeted molecular imaging in MRI, USPIOs must have a very small core and a molecular coating well attached at their surface. The main role of surface ligands is to keep nanoparticles as stable colloids in the biological fluids. Without an adequate layer of stabilizing molecules on their surfaces, IONPs usually agglomerate instantly in aqueous conditions. Coating the surface of IONPs with biocompatible polymers such as PEG and related derivatives is essential to the development of more stable colloids. In order to increase the strength of grafting, to minimize the risk of chain detachment due to the breakage of one single bond and to ultimately improve the colloidal stability, new functional polymeric constructs have been developed featuring several functional groups per chain. In this chapter, we presented the latest advances in the field of multidentate polymeric ligands as coatings for IONPs. These new particles of much stronger colloidal stability in biological fluids and producing strong 'positive' contrast enhancement are being tested as blood pool, lymph node and molecular targeting agents.

REFERENCES

1. Chen, W., Cormode, D. P., Fayad, Z. A., Mulder, W. J. M. 2011. Nanoparticles as magnetic resonance imaging contrast agents for vascular and cardiac diseases. *Wiley Interdisciplinary Reviews: Nanomedicine and Nanobiotechnology*. 3 (2), 146–161.
2. Lobatto, M. E., Fuster, V., Fayad, Z. A., Mulder, M. J. M. 2011. Perspectives and opportunities for nanomedicine in the management of atherosclerosis. *Nature Reviews Drug Discovery*. 10 (11), 835–852.
3. Bulte, J. W. M., Douglas, T., Witwer, B., Zhang, S. C., Strable, E., Lewis, B. K., Zywicke, H., Miller, B., van Gelderen, P., Moskowitz, B. M., Duncan, I. D., Frank, J. A. 2001. Magnetodendrimers allow endosomal magnetic labeling and *in vivo* tracking of stem cells. *Nature Biotechnology*. 19 (12), 1141–1147.
4. Cho, M. H., Lee, E. J., Son, M., Lee, J. H., Yoo, D., Kim, J. W., Park, S. W., Shin, J. S., Cheon, J. 2012. A magnetic switch for the control of cell death signalling in *in vitro* and *in vivo* systems. *Nature Materials*. 11 (12), 1038–1043.
5. Lewin, M., Carlesso, N., Tung, C. H., Tang, X. W., Cory, D., Scadden, D. T., Weissleder, R. 2000. Tat peptide-derivatized magnetic nanoparticles allow *in vivo* tracking and recovery of progenitor cells. *Nature Biotechnology*. 18 (4), 410–414.
6. Song, H. T., Choi, J. S., Huh, Y. M., Kim, S., Jun, Y. W., Suh, J. S., Cheon, J. 2005. Surface modulation of magnetic nanocrystals in the development of highly efficient magnetic resonance probes for intracellular labeling. *Journal of the American Chemical Society*. 127 (28), 9992–9993.
7. Mornet, S., Vasseur, S., Grasset, F., Duguet, E. 2004. Magnetic nanoparticle design for medical diagnosis and therapy. *Journal of Materials Chemistry*. 14 (14), 2161–2175.
8. Bergemann, C., Muller-Schulte, D., Oster, J., Brassard, L., Lubbe, A. S. 1999. Magnetic ion-exchange nano- and microparticles for medical, biochemical and molecular biological applications. *Journal of Magnetism and Magnetic Materials*. 194 (1–3), 45–52.
9. McCarthy, J. R., Weissleder, R. 2008. Multifunctional magnetic nanoparticles for targeted imaging and therapy. *Advanced Drug Delivery Reviews*. 60 (11), 1241–1251.
10. Sun, C., Lee, J. S. H., Zhang, M. 2008. Magnetic nanoparticles in MR imaging and drug delivery. *Advanced Drug Delivery Reviews*. 60 (11), 1252–1265.
11. Kelkar, S. S., Reineke, T. M. 2011. Theranostics: Combining imaging and therapy. *Bioconjugate Chemistry*. 22 (10), 1879–1903.
12. Caravan, P. 2006. Strategies for increasing the sensitivity of gadolinium based MRI contrast agents. *Chemical Society Reviews*. 35 (6), 512–523.
13. Caravan, P., Ellison, J. J., McMurry, T. J., Lauffer, R. B. 1999. Gadolinium(III) chelates as MRI contrast agents: Structure, dynamics, and applications. *Chemical Reviews*. 99 (9), 2293–352.
14. Grobner, T., Prischl, F. C. 2007. Gadolinium and nephrogenic systemic fibrosis. *Kidney International*. 72 (3), 260–264.
15. Penfield, J. G., Reilly, R. F., Jr. 2007. What nephrologists need to know about gadolinium. *Nature Clinical Practice. Nephrology*. 3 (12), 654–668.
16. Hu, F. Q., Jia, Q. J., Li, Y. L., Gao, M. Y. 2011. Facile synthesis of ultrasmall PEGylated iron oxide nanoparticles for dual-contrast T-1- and T-2-weighted magnetic resonance imaging. *Nanotechnology*. 22 (24).
17. Sandiford, L., Phinikaridou, A., Protti, A., Meszaros, L. K., Cui, X., Yan, Y., Frodsham, G., Williamson, P. A., Gaddum, N., Botnar, R. M., Blower, P. J., Green, M. A., de Rosales, R. T. 2013. Bisphosphonate-anchored PEGylation and radiolabeling of superparamagnetic iron oxide: Long-circulating nanoparticles for *in vivo* multimodal (T1 MRI-SPECT) imaging. *ACS Nano*. 7 (1), 500–512.
18. Na, H. B., Song, I. C., Hyeon, T. 2009. Inorganic nanoparticles for MRI contrast agents. *Advanced Materials*. 21 (21), 2133–2148.
19. Hu, F., Zhao, Y. S. 2012. Inorganic nanoparticle-based T1 and T1/T2 magnetic resonance contrast probes. *Nanoscale*. 4 (20), 6235–6243.
20. Lee, H., Lee, E., Kim, D. K., Jang, N. K., Jeong, Y. Y., Jon, S. 2006. Antibiofouling polymer-coated superparamagnetic iron oxide nanoparticles as potential magnetic resonance contrast agents for *in vivo* cancer imaging. *Journal of the American Chemical Society*. 128 (22), 7383–7389.
21. El-Boubbou, K., Zhu, D. C., Vasileiou, C., Borhan, B., Prosperi, D., Li, W., Huang, X. 2010. Magnetic Glyco-nanoparticles: A tool to detect, differentiate, and unlock the glyco-codes of cancer via magnetic resonance imaging. *Journal of the American Chemical Society*. 132 (12), 4490–4499.
22. Stanicki, D., Boutry, S., Laurent, S., Wacheul, L., Nicolas, E., Crombez, D., Vander Elst, L., Lafontaine, D. L. J., Muller, R. N. 2014. Carboxy-silane coated iron oxide nanoparticles: A convenient platform for cellular and small animal imaging. *Journal of Materials Chemistry B: Materials for Biology and Medicine*. 2 (4), 387–397.
23. Qin, J., Laurent, S., Jo, Y. S., Roch, A., Mikhaylova, M., Bhujwalla, Z. M., Muller, R. N., Muhammed, M. 2007. A high-performance magnetic resonance imaging T2 contrast agent. *Advanced Materials*. 19 (14), 1874–1878.
24. Foy, S. P., Manthe, R. L., Foy, S. T., Dimitrijevic, S., Krishnamurthy, N., Labhasetwar, V. 2010. Optical imaging and magnetic field targeting of magnetic nanoparticles in tumors. *ACS Nano*. 4 (9), 5217–5224.
25. Babic, M., Horak, D., Trchova, M., Jendelova, P., Glogarova, K., Lesny, P., Herynek, V., Hajek, M., Sykova, E. 2008. Poly(L-lysine)-modified iron oxide nanoparticles for stem cell labeling. *Bioconjugate Chemistry*. 19 (3), 740–750.

26. Chung, H.-J., Lee, H.-S., Bae, K. H., Lee, Y.-H., Park, J.-N., Cho, S.-W., Hwang, J.-Y., Park, H.-W., Langer, R., Anderson, D., Park, T.-G. 2011. Facile synthetic route for surface-functionalized magnetic nanoparticles: Cell labeling and magnetic resonance imaging studies. *ACS Nano*. 5 (6), 4329–4336.
27. Zhang, T., Ge, J., Hu, Y., Yin, Y. 2007. A General approach for transferring hydrophobic nanocrystals into water. *Nano Letters*. 7 (10), 3203–3207.
28. Amstad, E., Textor, M., Reimhult, E. 2011. Stabilization and functionalization of iron oxide nanoparticles for biomedical applications. *Nanoscale*. 3 (7), 2819–2843.
29. Xiao, L.-S., Li, J.-T., Brougham, D. F., Fox, E. K., Feliu, N., Bushmelev, A., Schmidt, A., Mertens, N., Kiessling, F., Valldor, M., Fadeel, B., Mathur, S. 2011. Water-soluble superparamagnetic magnetite nanoparticles with biocompatible coating for enhanced magnetic resonance imaging. *ACS Nano*. 5 (8), 6315–6324.
30. Long, M. J. C., Pan, Y., Lin, H.-C., Hedstrom, L., Xu, B. 2011. Cell compatible trimethoprim-decorated iron oxide nanoparticles bind dihydrofolate reductase for magnetically modulating focal adhesion of mammalian cells. *Journal of the American Chemical Society*. 133 (26), 10006–10009.
31. Hu, F., MacRenaris, K. W., Waters, E. A., Liang, T., Schultz-Sikma, E. A., Eckermann, A. L., Meade, T. J. 2009. Ultrasmall, water-soluble magnetite nanoparticles with high relaxivity for magnetic resonance imaging. *Journal of Physical Chemistry C*. 113 (49), 20855–20860.
32. Kim, B. H., Lee, N., Kim, H., An, K., Park, Y. I., Choi, Y., Shin, K., Lee, Y., Kwon, S. G., Na, H. B., Park, J. G., Ahn, T. Y., Kim, Y. W., Moon, W. K., Choi, S. H., Hyeon, T. 2011. Large-scale synthesis of uniform and extremely small-sized iron oxide nanoparticles for high-resolution T 1 magnetic resonance imaging contrast agents. *Journal of the American Chemical Society*. 133 (32), 12624–12631.
33. Tromsdorf, U. I., Bruns, O. T., Salmen, S. C., Beisiegel, U., Weller, H. 2009. A highly effective, nontoxic t-1 MR contrast agent based on ultrasmall PEGylated iron oxide nanoparticles. *Nano Letters*. 9 (12), 4434–4440.
34. Lee, Y., Lee, H., Kim, Y. B., Kim, J., Hyeon, T., Park, H., Messersmith, P. B., Park, T. G. 2008. Bioinspired surface immobilization of hyaluronic acid on monodisperse magnetitc nanocrystals for targeted cancer imaging. *Advanced Materials*. 20 (21), 4154–4157.
35. Creixell, M., Herrera, A. P., Latorre-Esteves, M., Ayala, V., Torres-Lugo, M., Rinaldi, C. 2010. The effect of grafting method on the colloidal stability and *in vitro* cytotoxicity of carboxymethyl dextran coated magnetic nanoparticles. *Journal of Materials Chemistry*. 20 (39), 8539–8547.
36. Yang, H.-M., Park, C. W., Woo, M.-A., Kim, M. I., Jo, Y. M., Park, H. G., Kim, J.-D. 2010. HER2/neu antibody conjugated poly(amino acid)-coated iron oxide nanoparticles for breast cancer MR imaging. *Biomacromolecules*. 11 (11), 2866–2872.
37. Hofmann, A., Thierbach, S., Semisch, A., Hartwig, A., Taupitz, M., Ruehl, E., Graf, C. 2010. Highly monodisperse water-dispersable iron oxide nanoparticles for biomedical applications. *Journal of Materials Chemistry*. 20 (36), 7842–7853.
38. Lutz, J. F., Stiller, S., Hoth, A., Kaufner, L., Pison, U., Cartier, R. 2006. One-pot synthesis of PEGylated ultrasmall iron-oxide nanoparticles and their *in vivo* evaluation as magnetic resonance imaging contrast agents. *Biomacromolecules*. 7 (11), 3132–3138.
39. Na, H. B., Palui, G., Rosenberg, J. T., Ji, X., Grant, S. C., Mattoussi, H. 2012. Multidentate catechol-based polyethylene glycol oligomers provide enhanced stability and biocompatibility to iron oxide nanoparticles. *ACS Nano*. 6 (1), 389–399.
40. Stark, D. D., Bradley, W. G., *Magnetic Resonance Imaging*. 3rd ed. C.V. Mosby: St-Louis, 1999; Vol. 1–3, p 44.
41. Bushberg, J. T., *The Essential Physics of Medical Imaging*. 3rd ed.; Wolters Kluwer Health/Lippincott Williams & Wilkins: Philadelphia, 2012; p xii.
42. Rohrer, M., Bauer, H., Mintorovitch, J., Requardt, M., Weinmann, H. J. 2005. Comparison of magnetic properties of MRI contrast media solutions at different magnetic field strengths. *Investigative Radiology*. 40 (11), 715–724.
43. Merbach, A. E., Tôth, E., *The Chemistry of Contrast Agents in Medical Magnetic Resonance Imaging*. Wiley: Chichester: New York, 2001; p xii.
44. Simon, G. H., von Vopelius-Feldt, J., Fu, Y., Schlegel, J., Pinotek, G., Wendland, M. F., Chen, M. H., Daldrup-Link, H. E. 2006. Ultrasmall supraparamagnetic iron oxide-enhanced magnetic resonance imaging of antigen-induced arthritis: A comparative study between SHU 555 C, ferumoxtran-10, and ferumoxytol. *Investigative Radiology*. 41 (1), 45–51.
45. Corot, C., Robert, P., Idée, J. M., Port, M. 2006. Recent advances in iron oxide nanocrystal technology for medical imaging. *Advanced Drug Delivery Reviews*. 58 (14), 1471–1504.
46. Gossuin, Y., Gillis, P., Hocq, A., Vuong, Q. L., Roch, A. 2009. Magnetic resonance relaxation properties of superparamagnetic particles. *Wiley Interdisciplinary Reviews: Nanomedicine and Nanobiotechnology*. 1 (3), 299–310.
47. Laurent, S., Forge, D., Port, M., Roch, A., Robic, C., Elst, L. V., Muller, R. N. 2008. Magnetic iron oxide nanoparticles: Synthesis, stabilization, vectorization, physicochemical characterizations, and biological applications. *Chemical Reviews*. 108 (6), 2064–2110.
48. Sosnovik, D. E., Nahrendorf, M., Weissleder, R. 2008. Magnetic nanoparticles for MR imaging: Agents, techniques and cardiovascular applications. *Basic Research in Cardiology*. 103 (2), 122–130.
49. Sun, S., Zeng, H. 2002. Size-controlled synthesis of magnetite nanoparticles. *Journal of the American Chemical Society*. 124 (28), 8204–8205.
50. Di Marco, M., Sadun, C., Port, M., Guilbert, I., Couvreur, P., Dubernet, C. 2007. Physicochemical characterization of ultrasmall superparamagnetic iron oxide particles (USPIO) for biomedical application as MRI contrast agents. *International Journal of Nanomedicine*. 2 (4), 609–622.
51. Stephen, Z. R., Kievit, F. M., Zhang, M. 2011. Magnetite nanoparticles for medical MR imaging. *Materials Today*. 14 (7–8), 330–338.
52. Lu, A. H., Salabas, E. L., Schuth, F. 2007. Magnetic nanoparticles: Synthesis, protection, functionalization, and application. *Angewandte Chemie. International Edition England*. 46 (8), 1222–1244.
53. Rui, H., Xing, R., Xu, Z., Hou, Y., Goo, S., Sun, S. 2010. Synthesis, functionalization, and biomedical applications of multifunctional magnetic nanoparticles. *Advanced Materials*. 22 (25), 2729–2742.
54. Schladt, T. D., Schneider, K., Schild, H., Tremel, W. 2011. Synthesis and bio-functionalization of magnetic nanoparticles for medical diagnosis and treatment. *Dalton Transactions*. 40 (24), 6315–6343.
55. Wu, W., He, Q., Jiang, C. 2008. Magnetic iron oxide nanoparticles: Synthesis and surface functionalization strategies. *Nanoscale Research Letters*. 3 (11), 397–415.
56. Sun, Y. K., Ma, M., Zhang, Y., Gu, N. 2004. Synthesis of nanometer-size maghemite particles from magnetite. *Colloids and Surfaces A: Physicochemical and Engineering Aspects*. 245 (1–3), 15–19.

57. Park, J., An, K. J., Hwang, Y. S., Park, J. G., Noh, H. J., Kim, J. Y., Park, J. H., Hwang, N. M., Hyeon, T. 2004. Ultra-large-scale syntheses of monodisperse nanocrystals. *Nature Materials*. 3 (12), 891–895.
58. Jun, Y. W., Huh, Y. M., Choi, J. S., Lee, J. H., Song, H. T., Kim, S., Yoon, S., Kim, K. S., Shin, J. S., Suh, J. S., Cheon, J. 2005. Nanoscale size effect of magnetic nanocrystals and their utilization for cancer diagnosis via magnetic resonance imaging. *Journal of the American Chemical Society*. 127 (16), 5732–5733.
59. Cornell, R. M., Schwertmann, U., *The Iron Oxides: Structure, Properties, Reactions, Occurrence, and Uses*. VCH: Weinheim; New York, 1996; p xxxi.
60. Laurent, S., Forge, D., Port, M., Roch, A., Robic, C., Vander Elst, L., Muller, R. N. 2008. Magnetic iron oxide nanoparticles: Synthesis, stabilization, vectorization, physicochemical characterizations, and biological applications. *Chemical Reviews*. 108 (6), 2064–110.
61. Neel, L. 1948. Magnetic properties of ferrites: Ferrimagnetism and antiferromagnetism. *Annales de Physique. Paris*. 3, 137–198.
62. Gossuin, Y., Hocq, A., Vuong, Q. L., Disch, S., Hermann, R. P., Gillis, P. 2008. Physico-chemical and NMR relaxometric characterization of gadolinium hydroxide and dysprosium oxide nanoparticles. *Nanotechnology*. 19 (47), 475102.
63. Fortin, M. A., Petoral Jr, R. M., Söoderlind, F., Klasson, A., Engströom, M., Veres, T., Käll, P. O., Uvdal, K. 2007. Polyethylene glycol-covered ultra-small Gd_2O_3 nanoparticles for positive contrast at 1.5 T magnetic resonance clinical scanning. *Nanotechnology*. 18 (39).
64. Crangle, J., *Solid-State Magnetism*. Van Nostrand Reinhold: New York, 1991; p xii.
65. Dormann, J. L. 1981. Superparamagnetism phenomenon. *Revue de Physique Appliquée* 16 (6), 275–301.
66. Rosensweig, R. E. 2002. Heating magnetic fluid with alternating magnetic field. *Journal of Magnetism and Magnetic Materials*. 252, 370–374.
67. Chantrell, R. W., Lyberatos, A., El-Hilo, M., O'Grady, K. 1994. Models of slow relaxation in particulate and thin film materials (invited). *Journal of Applied Physics*. 76 (10), 6407–6412.
68. Dormann, J. L., Spinu, L., Tronc, E., Jolivet, J. P., Lucari, F., D'Orazio, F., Fiorani, D. 1998. Effect of interparticle interactions on the dynamical properties of γ-Fe_2O_3 nanoparticles. *Journal of Magnetism and Magnetic Materials*. 183 (3), L255–L260.
69. Dormann, J. L., D'Orazio, F., Lucari, F., Tronc, E., Prené, P., Jolivet, J. P., Fiorani, D., Cherkaoui, R., Noguès, M. 1996. Thermal variation of the relaxation time of the magnetic moment of γ-Fe_2O_3 nanoparticles with interparticle interactions of various strengths. *Physical Review B – Condensed Matter and Materials Physics*. 53 (21), 14291–14297.
70. Bean, C. P., Livingston, J. D. 1959. Superparamagnetism. *Journal of Applied Physics*. 30, 120S.
71. Bloembergen, N. J., Purcell, E. M., Pound, R. V. 1948. Relaxation effects in nuclear magnetic resonance absorption. *Physical Review*. 73, 679.
72. Solomon, I. 1955. Relaxation processes in a system of two spins. *Physical Review*. 99 (2), 559–565.
73. Solomon, I., Bloembergen, N. J. 1956. Nuclear magnetic interactions in the HF molecule. *Journal of Chemical Physics*. 25, 261.
74. Bloembergen, N. J. 1957. Proton relaxation times in paramagnetic solutions. *The Journal of Chemical Physics*. 27, 573–573.
75. Bloembergen, N. J., Morgan, N. O. 1961. Proton relaxation times in paramagnetic solutions. *Journal of Chemical Physics*. 34, 842.
76. Luz, Z., Meiboom, S. 1964. Proton relaxation in dilute solutions of cobalt (II) and nickel (II) ions in methanol and the rate of methanol exchange of the solvation sphere. *The Journal of Chemical Physics*. 40 (9), 2686–2692.
77. Swift, T. J., Connick, R. E. 1962. NMR-Relaxation mechanisms of O17 in aqueous solutions of paramagnetic cations and the lifetime of water molecules in the first coordination sphere. *The Journal of Chemical Physics*. 37 (2), 307–320.
78. Freed, J. H. 1978. Dynamic effects of pair correlation functions on spin relaxation by translational diffusion in liquids. II. Finite jumps and independent T1 processes. *Journal of the Journal of Chemical Physics*. 68, 4034–4037.
79. Koenig, S. H., Kellar, K. E. 1995. Theory of 1/T1 and 1/T2 NMRD profiles of solutions of magnetic nanoparticles. *Magnetic Resonance in Medicine: Official Journal of the Society of Magnetic Resonance in Medicine/Society of Magnetic Resonance in Medicine*. 34 (2), 227–233.
80. Roch, A., Muller, R. N., Gillis, P. 1999. Theory of proton relaxation induced by superparamagnetic particles. *Journal of Chemical Physics*. 110 (11), 5403–5411.
81. Bulte, J. M., Vymazal, J., Brooks, R. A., Pierpaoli, C., Frank, J. A. 1993. Frequency dependence of MR relaxation times. II. Iron oxides. *Journal of Magnetic Resonance Imaging: JMRI*. 3 (4), 641–648.
82. Muller, R. N., Vallet, P., Maton, F., Roch, A., Goudemant, J. F., Vander Elst, L., Gillis, P., Peto, S., Moiny, F., Van Haverbeke, Y. 1990. Recent developments in design, characterization, and understanding of MRI and MRS contrast media. *Investigative Radiology*. 25 (suppl. 1), S34–S36.
83. Noack, F. 1986. NMR field-cycling spectroscopy: Principles and applications. *Progress in Nuclear Magnetic Resonance Spectroscopy*. 18 (3), 171–276.
84. Muller, R. N., Vander Elst, L., Roch, A., Peters, J. A., Csajbok, E., Gillis, P., Gossuin, Y. 2006. Relaxation by metal-containing nanosystems. *Advances in Inorganic Chemistry*. 57, 239–292.
85. Burtea, C., Laurent, S., Vander Elst, L., Muller, R. N. 2008. Contrast agents: Magnetic resonance. *Handbook of Experimental Pharmacology*. (185 Pt 1), 135–165.
86. Vuong, Q. L., Berret, J. F., Fresnais, J., Gossuin, Y., Sandre, O. 2012. A universal scaling law to predict the efficiency of magnetic nanoparticles as MRI T2-contrast agents. *Advanced Healthcare Materials*. 1 (4), 502–512.
87. Chan, N., Laprise-Pelletier, M., Chevallier, P., Bianchi, A., Fortin, M. A., Oh, J. K. 2014. Multidentate block-copolymer-stabilized ultrasmall superparamagnetic iron oxide nanoparticles with enhanced colloidal stability for magnetic resonance imaging. *Biomacromolecules*. 15 (6), 2146–2156.
88. Li, P. Z., Chevallier, P., Ramrup, P., Biswas, D., Vuckovich, D., Fortin, M. A., Oh, J. K. 2015. Mussel-inspired multidentate block copolymer to stabilize ultrasmall superparamagnetic Fe_3O_4 for magnetic resonance imaging contrast enhancement and excellent colloidal stability. *Chemistry of Materials*. 27 (20), 7100–7109.
89. Li, P. Z., Xiao, W. C., Chevallier, P., Biswas, D., Ottenwaelder, X., Fortin, M. A., Oh, J. K. 2016. Extremely small iron oxide nanoparticles stabilized with catechol-functionalized multidentate block copolymer for enhanced MRI. *ChemistrySelect*. 1 (13), 4087–4091.
90. McLachlan, S. J., Morris, M. R., Lucas, M. A., Fisco, R. A., Eakins, M. N., Fowler, D. R., Scheetz, R. B., Olukotun, A. Y. 1994. Phase I clinical evaluation of a new iron oxide MR contrast agent. *Journal of Magnetic Resonance Imaging*. 4 (3), 301–307.

91. Li, W., Tutton, S., Vu, A. T., Pierchala, L., Li, B. S. Y., Lewis, J. M., Prasad, P. V., Edelman, R. R. 2005. First-pass contrast-enhanced magnetic resonance angiography in humans using ferumoxytol, a novel ultrasmall superparamagnetic iron oxide (USPIO)-based blood pool agent. *Journal of Magnetic Resonance Imaging*. 21 (1), 46–52.
92. Modo, M. M. J. J., Bulte, J. W. M., *Molecular and Cellular MR Imaging*. CRC Press: Boca Raton, 2007.
93. Kellar, K. E., Fujii, D. K., Gunther, W. H. H., Briley-Sæbø, K., Bjørnerud, A., Spiller, M., Koenig, S. H. 2000. NC 100150 injection, a preparation of optimized iron oxide nanoparticles for positive-contrast MR angiography. *Journal of Magnetic Resonance Imaging*. 11 (5), 488–494.
94. Daldrup-Link, H. E., Kaiser, A., Helbich, T., Werner, M., Bjørnerud, A., Link, T. M., Rummeny, E. J. 2003. Macromolecular contrast medium (feruglose) versus small molecular contrast medium (gadopentetate) enhanced magnetic resonance imaging: Differentiation of benign and malignant breast lesions. *Academic Radiology*. 10 (11), 1237–1246.
95. Taupitz, M., Wagner, S., Schnorr, J., Kravec, I., Pilgrimm, H., Bergmann-Fritsch, H., Hamm, B. 2004. Phase I clinical evaluation of citrate-coated monocrystalline very small superparamagnetic iron oxide particles as a new contrast medium for magnetic resonance imaging. *Investigative Radiology*. 39 (7), 394–405.
96. Mayo-Smith, W. W., Saini, S., Slater, G., Kaufman, J. A., Sharma, P., Hahn, P. F. 1996. MR contrast material for vascular enhancement: Value of superparamagnetic iron oxide. *AJR. American Journal of Roentgenology*. 166 (1), 73–77.
97. Stillman, A. E., Wilke, N., Jerosch-Herold, M. 1997. Use of an intravascular T1 contrast agent to improve MR cine myocardial-blood pool definition in man. *Journal of Magnetic Resonance Imaging*. 7 (4), 765–767.
98. Stillman, A. E., Wilke, N., Li, D., Haacke, M., McLachlan, S. 1996. Ultrasmall superparamagnetic iron oxide to enhance MRA of the renal and coronary arteries: Studies in human patients. *Journal of Computed Assisted Tomography*. 20 (1), 51–55.
99. Anzai, Y., Blackwell, K. E., Hirschowitz, S. L., Rogers, J. W., Sato, Y., Yuh, W. T. C., Runge, V. M., Morris, M. R., McLachlan, S. J., Lufkin, R. B. 1994. Initial clinical experience with dextran-coated superparamagnetic iron oxide for detection of lymph node metastases in patients with head and neck cancer. *Radiology*. 192 (3), 709–715.
100. Saeed, M., Wendland, M. F., Engelbrecht, M., Sakuma, H., Higgins, C. B. 1998. Value of blood pool contrast agents in magnetic resonance angiography of the pelvis and lower extremities. *European Radiology*. 8 (6), 1047–1053.
101. Weissleder, R., Lee, A. S., Khaw, B. A., Shen, T., Brady, T. J. 1992. Antimyosin-labeled monocrystalline iron oxide allows detection of myocardial infarct: MR antibody imaging. *Radiology*. 182 (2), 381–385.
102. Weissleder, R., Lee, A. S., Fischman, A. J., Reimer, P., Shen, T., Wilkinson, R., Callahan, R. J., Brady, T. J. 1991. Polyclonal human immunoglobulin G labeled with polymeric iron oxide: Antibody MR imaging. *Radiology*. 181 (1), 245–249.
103. Sahoo, Y., Goodarzi, A., Swihart, M. T., Ohulchanskyy, T. Y., Kaur, N., Furlani, E. P., Prasad, P. N. 2005. Aqueous ferrofluid of magnetite nanoparticles: Fluorescence labeling and magnetophoretic control. *Journal of Physical Chemistry B*. 109 (9), 3879–3885.
104. Wagner, S., Schnorr, J., Pilgrimm, H., Hamm, B., Taupitz, M. 2002. Monomer-coated very small superparamagnetic iron oxide particles as contrast medium for magnetic resonance imaging—Preclinical *in vivo* characterization. *Investigative Radiology*. 37 (4), 167–177.
105. Liu, C., Huang, P. M. 1999. Atomic force microscopy and surface characteristics of iron oxides formed in citrate solutions. *Soil Science Society of America Journal*. 63 (1), 65–72.
106. Naccache, R., Chevallier, P., Lagueux, J., Gossuin, Y., Laurent, S., Vander Elst, L., Chilian, C., Capobianco, J. A., Fortin, M. A. 2013. High relaxivities and strong vascular signal enhancement for NaGdF4 nanoparticles designed for dual MR/optical imaging. *Advanced Healthcare Materials*. 2 (11), 1478–1488.
107. Daou, T. J., Pourroy, G., Greneche, J. M., Bertin, A., Felder-Flesch, D., Begin-Colin, S. 2009. Water soluble dendronized iron oxide nanoparticles. *Dalton Transactions*. 0 (23), 4442–4449.
108. Bulte, J. W. M., Kraitchman, D. L. 2004. Iron oxide MR contrast agents for molecular and cellular imaging. *NMR in Biomedicine*. 17 (7), 484–499.
109. Paul, K. G., Frigo, T. B., Groman, J. Y., Groman, E. V. 2004. Synthesis of ultrasmall superparamagnetic iron oxides using reduced polysaccharides. *Bioconjugate Chemistry*. 15 (2), 394–401.
110. Tiefenauer, L. X., Tschirky, A., Kühne, G., Andres, R. Y. 1996. *In vivo* evaluation of magnetite nanoparticles for use as a tumor contrast agent in MRI. *Magnetic Resonance Imaging*. 14 (4), 391–402.
111. Moghimi, S. M., Hunter, A. C., Murray, J. C. 2001. Long-circulating and target-specific nanoparticles: Theory to practice. *Pharmacological Reviews*. 53 (2), 283–318.
112. Papisov, M. I., Bogdanov Jr, A., Schaffer, B., Nossiff, N., Shen, T., Weissleder, R., Brady, T. J. 1993. Colloidal magnetic resonance contrast agents: Effect of particle surface on biodistribution. *Journal of Magnetism and Magnetic Materials*. 122 (1–3), 383–386.
113. Park, J., Lee, E., Hwang, N. M., Kang, M. S., Kim, S. C., Hwang, Y., Park, J. G., Noh, H. J., Kini, J. Y., Park, J. H., Hyeon, T. 2005. One-nanometer-scale size-controlled synthesis of monodisperse magnetic iron oxide nanoparticles. *Angewandte Chemie. International Edition*. 44 (19), 2872–2877.
114. Oh, J. K., Park, J. M. 2011. Iron oxide-based superparamagnetic polymeric nanomaterials: Design, preparation, and biomedical application. *Progress in Polymer Science*. 36 (1), 168–189.
115. Kumagai, M., Kano, M. R., Morishita, Y., Ota, M., Imai, Y., Nishiyama, N., Sekino, M., Ueno, S., Miyazono, K., Kataoka, K. 2009. Enhanced magnetic resonance imaging of experimental pancreatic tumor *in vivo* by block copolymer-coated magnetite nanoparticles with TGF-beta inhibitor. *Journal of Controlled Release*. 140 (3), 306–311.
116. Torrisi, V., Graillot, A., Vitorazi, L., Crouzet, Q., Marletta, G., Loubat, C., Berret, J. F. 2014. Preventing Corona Effects: Multiphosphonic acid poly(ethylene glycol) copolymers for stable stealth iron oxide nanoparticles. *Biomacromolecules*. 15 (8), 3171–3179.
117. Zhong, Y. Q., Dai, F. Y., Deng, H., Du, M. H., Zhang, X. N., Liu, Q. J., Zhang, X. 2014. A rheumatoid arthritis magnetic resonance imaging contrast agent based on folic acid conjugated PEG-b-PAA@SPION. *J. Mat. Chem. B*. 2 (19), 2938–2946.
118. Basuki, J. S., Jacquemin, A., Esser, L., Li, Y., Boyer, C., Davis, T. P. 2014. A block copolymer-stabilized co-precipitation approach to magnetic iron oxide nanoparticles for potential use as MRI contrast agents. *Polymer Chemistry*. 5 (7), 2611–2620.
119. Chan, N., Li, P. Z., Oh, J. K. 2014. Chain length effect of the multidentate block copolymer strategy to stabilize ultrasmall Fe_3O_4 nanoparticles. *ChemPlusChem*. 79 (9), 1342–1351.

120. Moulay, S. 2014. Dopa/catechol-tethered polymers: Bioadhesives and biomimetic adhesive materials. *Polymer Reviews*. 54 (3), 436–513.
121. Yuen, A. K. L., Hutton, G. A., Masters, A. F., Maschmeyer, T. 2012. The interplay of catechol ligands with nanoparticulate iron oxides. *Dalton Transasctions*. 41 (9), 2545–2559.
122. Martin, M., Salazar, P., Villalonga, R., Campuzano, S., Pingarron, J. M., Gonzalez-Mora, J. L. 2014. Preparation of core-shell Fe_3O_4@poly(dopamine) magnetic nanoparticles for biosensor construction. *Journal of Materials Chemistry. B*. 2 (6), 739–746.
123. Lee, Y. H., Lee, H., Kim, Y. B., Kim, J. Y., Hyeon, T., Park, H., Messersmith, P. B., Park, T. G. 2008. Bioinspired surface immobilization of hyaluronic acid on monodisperse magnetite nanocrystals for targeted cancer imaging. *Advanced Materials*. 20 (21), 4154–4157.
124. Shukoor, M. I., Natalio, F., Therese, H. A., Tahir, M. N., Ksenofontov, V., Panthofer, M., Eberhardt, M., Theato, P., Schroder, H. C., Muller, W. E. G., Tremel, W. 2008. Fabrication of a silica coating on magnetic gamma-Fe_2O_3 nanoparticles by an immobilized enzyme. *Chemistry of Materials*. 20 (11), 3567–3573.
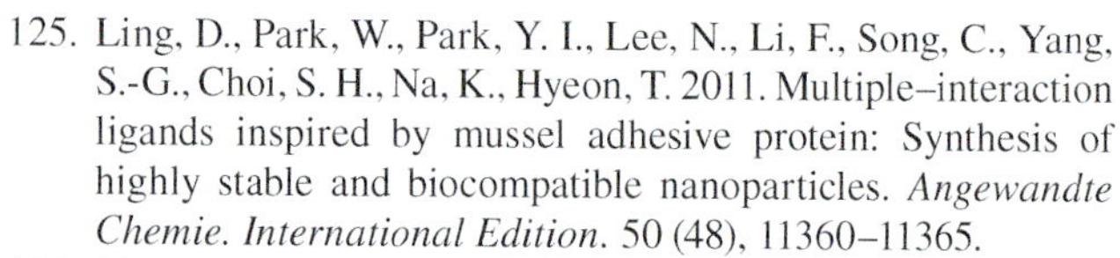
125. Ling, D., Park, W., Park, Y. I., Lee, N., Li, F., Song, C., Yang, S.-G., Choi, S. H., Na, K., Hyeon, T. 2011. Multiple–interaction ligands inspired by mussel adhesive protein: Synthesis of highly stable and biocompatible nanoparticles. *Angewandte Chemie. International Edition*. 50 (48), 11360–11365.
126. Li, P., Chevallier, P., Ramrup, P., Biswas, D., Vuckovich, D., Fortin, M.-A., Oh, J. K. 2015. Mussel-inspired multidentate block copolymer to stabilize ultrasmall superparamagnetic Fe_3O_4 for magnetic resonance imaging contrast enhancement and excellent colloidal stability. *Chem. Mater.* 27 (20), 7100–7109.
127. Wu, Y. Y., Gao, D. Y., Zhang, P. F., Li, C. S., Wan, Q., Chen, C., Gong, P., Gao, G. H., Sheng, Z. H., Cai, L. T. 2016. Iron oxide nanoparticles protected by NIR-active multidentate-polymers as multifunctional nanoprobes for NIRF/PA/MR trimodal imaging. *Nanoscale*. 8 (2), 775–779.
128. Keil, H., Wasserman, D., Dawson, C. R. 1944. The relation of chemical structure in catechol compounds and derivatives to poison ivy hypersensitiveness in man as shown by the patch test. *Journal of Experimental Medicine*. 80 (4), 275–287.

Marc-André Fortin (E-mail: marc-andre.fortin@gmn.ulaval.ca) is a full professor at the department of Mining, Metallurgy and Materials Engineering at Université Laval, Quebec City, Canada. He is the head of the Biomaterials for Imaging Laboratory at the CR-CHUQ, one of the top 8 largest academic hospitals in Canada. Dr Fortin earned his BSc in Materials Engineering at École Polytechnique de Montréal and his PhD degree at the Institut National de la Recherche Scientifique (INRS) in Montreal, in the area of energy and materials sciences. He completed his postdoctoral research training in biomedical imaging and nanosciences at the Universities of Uppsala and Linköping (Sweden; FRQNT, NSERC and Wenner-Gren Fellowships). Since his appointment at Université Laval (2007), Dr Fortin was awarded twice a national research career award from Fonds de la Recherche en Santé du Quebec (FRQS; Junior 1—2008; Junior 2—2012). His laboratory pioneered research on paramagnetic and ultrasmall nanoparticles for molecular and cellular magnetic resonance imaging. At the CR-CHUQ, he has developed a research platform integrating nanoscience tools and small-animal imaging scanners, to facilitate the development and the *in vivo* validation of new nanostructured hybrid materials for biomedical applications. Since 2010, Dr Fortin has published more than 40 articles and holds four patents in this field.

Jung Kwon (John) Oh is a Canada Research Chair Tier II in Nanobioscience (renewed in 2016) and an associate professor in the Department of Chemistry and Biochemistry at Concordia University in Montreal. He earned his PhD degree at the University of Toronto in the area of polymer chemistry and materials science. He then completed his postdoctoral research at Carnegie Mellon University. He has been employed at Korea Chemical Company in Korea and Dow Chemical Company in Michigan for over 10 years. He has authored and coauthored more than 95 publications and holds 18 international patents. His research has been recognized with several prestigious awards, including NSERC Postdoctoral Fellowship of Canada in 2004, PCI Outstanding Paper Award in 2010, Canada Research Chair Award in 2011–2021 and CNC-IUPAC Travel Award in 2013, as well as Dean's award to Excellence in Scholarship-Mid-Career at Concordia University in 2016. His research interests involve the design and processing of macromolecular nanoscale materials for biomedical and industrial applications.

9 Encapsulation and Release of Drugs from Magnetic Silica Nanocomposites

Damien Mertz and Sylvie Bégin-Colin*

CONTENTS

9.1 INTRODUCTION

Functionalized iron oxide nanoparticles (IONPs) are well-known contrast agents for magnetic resonance imaging (MRI).[1–4] They are now designed for magnetic hyperthermia (MH) therapy: when exposed to alternating magnetic fields (AMFs) of appropriate intensity and frequency, these IONPs convert the magnetic wave energy into localized heat, which is released where they are concentrated.[5–7] MH is used to enhance the sensitivity of tumour cells towards chemotherapy or radiotherapy but is more and more developed to trigger thermally induced release of drugs. Indeed, today, the development of blossoming strategies to synthetize polymer-based and/or inorganic nanoplatforms holding magnetic nanoparticles (NPs) paves the way towards the development of drug delivery nanocarriers actuated by external magnetic fields.[5,8–11]

There are various chemical strategies that were developed these last decennia to design drug-loaded magnetic nanocarriers. A first approach is based on the polymer grafting at the surface of magnetic NPs. Such nanocarriers are usually called magnetic core–polymer shell and the drug is bound either by noncovalent loading[12,13] or by covalent coupling[14,15] to the polymer shell. Another approach includes self-assembly methods allowing the design of self-assembled polymer or lipid-based magnetic nanocarriers having either an hydrophobic interior such as nanomicelles[16,17] or a hollow aqueous compartment as in magnetic liposomes,[18,19] polymersomes[20,21] and emulsion capsules.[22,23] In these self-assembled strategies, hydrophobic or hydrophilic drugs are usually encapsulated during the self-assembly procedures. Besides these self-assembled nanocarriers, a third approach is to form covalently or noncovalently cross-linked magnetic polymer-based hydrophilic nanogels[24,25] or hydrophobic polyester-based NPs.[26] A last approach is the use of sol–gel processes to embed the drugs and magnetic NPs in either nonporous or porous magnetic silica composites, which will be detailed in this chapter. For all these systems, the drug loading is usually achieved either by coupling the therapeutic molecules through pH- or thermo-responsive linkers on the IONPs or by encapsulating drugs along with IONPs within the polymer- or inorganic-based nanocarriers. Hence, as described in many reports, pH and temperature are the most reported stimuli used to trigger the drug release.

* Corresponding author.

The drug encapsulation in a nanocarrier can be quantified according to two main parameters: the drug loading efficiency (DLE) and drug loading capacity or content (DLC), whose expressions are respectively:

$$\mathrm{DLE}(\%) = \frac{\text{weight of loaded drugs}}{\text{weight of fed drugs}}$$

$$\mathrm{DLC}(\%) = \frac{\text{weight of loaded drugs}}{\text{weight of the nanocarrier}}$$

In these DLE and DLC expressions, loaded drug is within the nanocarrier, whereas fed drug is the initial amount in contact with the nanocarrier in solution. The DLC has to be as high as possible to achieve enough drug payloads in the nanocarriers while preserving the magnetic NP properties, allowing their *in vivo* uses. The DLE should be maximized to ensure a high efficacy of the drug loading process and to optimize the conditions of the drug loading (determining drug concentrations, mass of the nanocarriers, volume, loading time etc.). These two values depend on various parameters related to the nanoplatform design such as surface area, pore size and surface modification of the nanocomposite. It depends also on the conditions of the drug loading, such as the time, solvent type or drug concentration used during impregnation, etc. It is noteworthy that the amount of the drug released under a given stimulus (pH, T etc.) is also a key parameter to consider regarding the design of such nanocarrier.

Among these encapsulation-based nanosystems briefly overviewed previously, nanocomposites made of nonporous and porous silica are particularly well-suited nanoplatforms to formulate drugs and magnetic NPs because of the huge advantages offered by such inorganic coatings/matrixes, namely

i) The possibility of high drug payloads (≥20%) in comparison with polymer nanovesicles[21,22] or nanospheres[27,28] (DLC in the range 1%–10%);
ii) Food and Drug Administration approval for both IONPs and silica;
iii) A high colloidal stability in aqueous and physiological solutions; and
iv) A high versatility in the surface modification due to facile chemical strategies.

In this book chapter, we present different ways to formulate drug-loaded magnetic silica nanocomposites and the main strategies of surface functionalization used to ensure improved payloads and control of drug release in dosage, location and time. Herein, we will describe the current efficient systems and discuss the main issues associated with drug loading. This chapter will be divided into four main parts:

i) Drug encapsulation in nonporous magnetic silica made by *in situ* sol–gel process (9.2)
ii) Drug loading in magnetic core–mesoporous shell (9.3)
iii) Influence of the surface modification to control drug loading/release (9.4)
iv) Gatekeeping strategies for stimuli responsive drug release (9.5)

9.2 ENCAPSULATION OF DRUGS IN NONPOROUS MAGNETIC SILICA BY *IN SITU* SOL–GEL PROCESS

9.2.1 Drug Sequestration/Coupling: Limited Release in Physiological Conditions

The encapsulation of various molecules (dyes, drugs etc.) within a silica network can be achieved via a straightforward method by reacting the silica source precursors and the drug together during the sol–gel reaction. For instance, in a work by Barbe et al.,[29] Orange II, rhodamine 6G and doxorubicin (DOX) were efficiently encapsulated *in situ* in a (nonmagnetic) silica matrix during the sol–gel process with a high molecule loading efficiency (DLE = 85–98%). In this work, microporous silica particles with tunable size ranging from 50 nm to several tens of microns were formed by a sol–gel procedure combined with water in oil (w/o) emulsion. The size and internal structure of such NPs could be tailored with the sol–gel conditions (such as w/o droplet size, nature of the surfactant or solvent combination) and influenced importantly the drug release rate. Furthermore, DOX was released from 30-nm-size silica NPs (≥80% drug released) by passive diffusion of the drugs out of the silica over 28 d in water, which was explained by a combination of local silica matrix dissolution and drug diffusion.[29]

Alternatively to the noncovalent drug sequestration within a nonporous silica network, the drug can also be covalently coupled to the silica precursor and condensed to magnetic NPs during the sol–gel process. In a work by Li and coworkers,[30] DOX-siloxane-coated IONPs (67 nm) were obtained via a one-pot synthesis by the condensation of isocyanatopropyltriethoxysilane (ICPTES)-DOX at the IONP surface. The ICPTES allowed the formation of a strong urea bond with DOX, which allowed reaching a very high degree of drug encapsulation (DLE = 60%, DLC = 39%). The sustained drug release from this magnetic nanocomposite was investigated in a phosphate-buffered saline (PBS) solution. A very slow and slight release of DOX in PBS (13% in 261 h) was observed, which was attributed to the gradual and limited hydrolysis of the weak bond urea.

Regarding these two latter examples (encapsulation and covalent coupling), although the two strategies were very efficient to bind the drug tightly to the inorganic composite matrix, thereby limiting importantly premature release, a main drawback that can be underlined is the difficulty to trigger the drug release in physiological or biological conditions. Incorporation of efficient magnetic mediators thus appears of high interest to activate the drug release upon an external AMF, which will convert local magnetic energy into local heating.

9.2.2 Nonporous Magnetic Silica Composites with a Drug Release Actuated by Magnetothermal Effects

In this section, we report some examples describing the loading and release of drugs from nonporous magnetic silica nanocomposites that were formed through *in situ* sol–gel approaches. Since 2008, the group of Chen and coworkers have been developing the simultaneous encapsulation of drugs and magnetic NPs in a tight nonporous silica matrix. Hence, ibuprofen (IBU)-loaded magnetic silica composites (DLE = 23%, DLC = 1%), with a size range of 50–60 nm, were obtained by Hu et al.[31] simply by mixing iron salts, silica precursors and the drugs during a one-pot the sol–gel process (Figure 9.1A). The drug release was triggered by the application of an AMF applied on these systems (field strength H = 2.5 kA/m^{-1}, frequency f = 50 kHz), which produced larger porosities and structure enlargements of the silica matrix and, therefore, a rapid drug release. The same authors[32] have also loaded a fluorescent dye in a poly-(*N*-vinyl-2-pyrrolidone) (PVP)-coated silica core (size range = 15–23 nm) and coated it with a single crystal IO shell (4 nm shell) around the dye-loaded PVP–silica core. Here, too, the drug release was triggered by the application of an AMF (magnetic field power P = 15 W, f = 50 kHz), whose interaction with the magnetic shell generated pores/cracks caused by the high local magneto-thermal energy conversion (Figure 9.1B). Recently, in another example, drug-loaded one-pot sol–gel synthesis based on ureasil–poly(ethylene oxide)-Fe_2O_3 nanocomposites[33] allowed the encapsulation of sodium diclofenac. The efficient release of this drug was achieved through the use of an AMF (H = 14.9 kA/m^{-1}, f = 420 kHz), which ensured the drug delivery by the melting of the crystalline poly(ethylene oxide) from temperatures above 38°C, which triggers drug diffusion out of the network.

However, one important issue encountered with such systems is the very high local temperature generated by AMF that may create irreversible structural effects. This may preclude applications, for instance, like pulsatile release or for the encapsulation of biologically active drugs. Since the middle of the 2000s, strategies making use of porous magnetic silica structures appeared as a more suitable solution to load and release drugs in physiological conditions.

9.3 DRUG LOADING IN MAGNETIC CORE–MESOPOROUS SILICA SHELL NANOPARTICLES

9.3.1 Design of Magnetic Core Mesoporous Silica Shell Nanoparticles

9.3.1.1 Conventional Magnetic Core–Mesoporous Silica Shell Nanoparticles

In 2006, Kim et al.[34] reported for the first time encapsulation within mesoporous silica (MS) shells of IO nanocrystals made by thermal decomposition. In this procedure, IONPs capped with oleic acid and dispersed in chloroform were mixed to an aqueous solution of cetyltrimethylammonium bromide (CTAB) surfactants (Figure 9.2A). Using the CTAB surfactant ensured

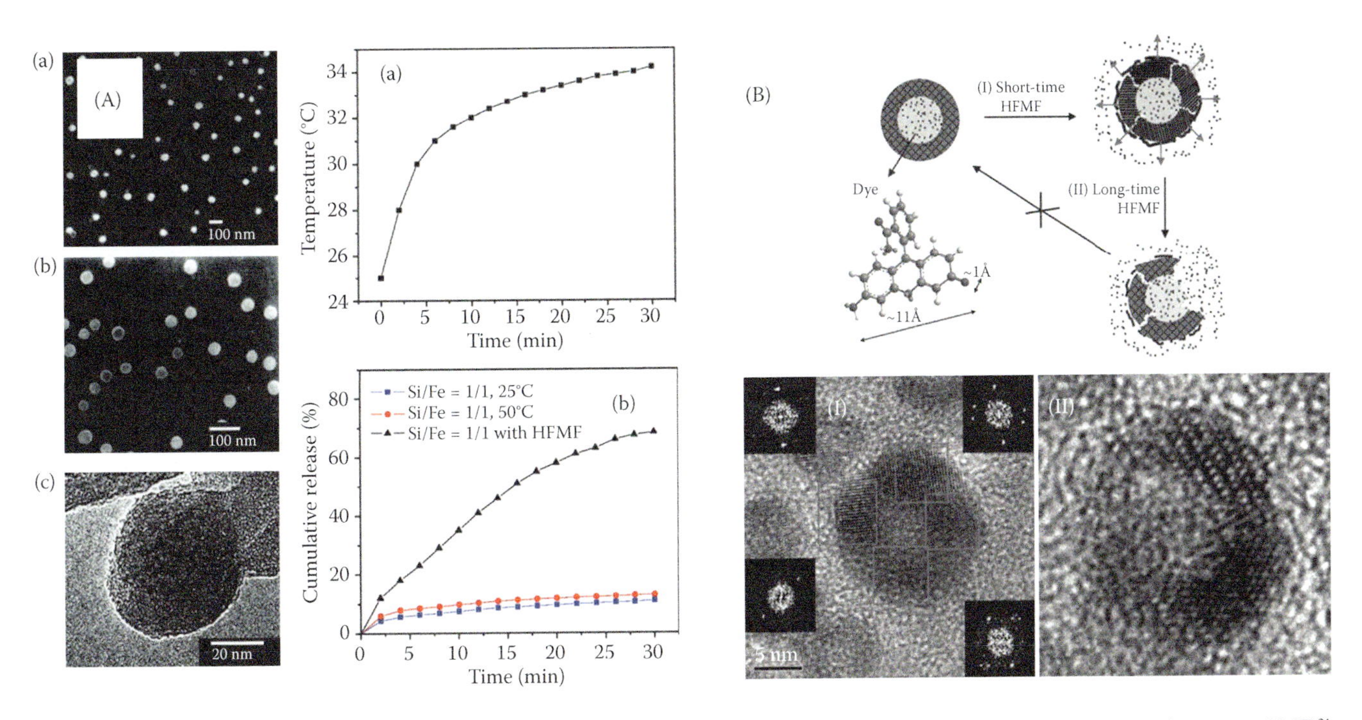

FIGURE 9.1 (A) TEM images, temperature profiles and drug release profiles of IBU-loaded magnetic silica composites upon AMF.[31] (Reproduced with permission from Hu, S.-H.; Liu, T.-Y.; Huang, H.-Y.; Liu, D.-M.; Chen, S.-Y. Magnetic-Sensitive Silica Nanospheres for Controlled Drug Release. *Langmuir* 2008, *24* (1), 239–244. Copyright 2008 American Chemical Society.) (B) Scheme showing the dye release actuated upon AMF from PVP-silica core/IO shell and TEM images of these nanocomposites with a zoomed image showing the crystal lattice of the shell.[32] (Reproduced with permission from Hu, S.-H.; Chen, S.-Y.; Liu, D.-M.; Hsiao, C.-S. Core/Single-Crystal–Shell Nanospheres for Controlled Drug Release via a Magnetically Triggered Rupturing Mechanism. *Adv. Mater.* 2008, *20* (14), 2690–2695. Copyright Wiley-VCH Verlag GmbH & Co. KGaA. Reproduced with permission.)

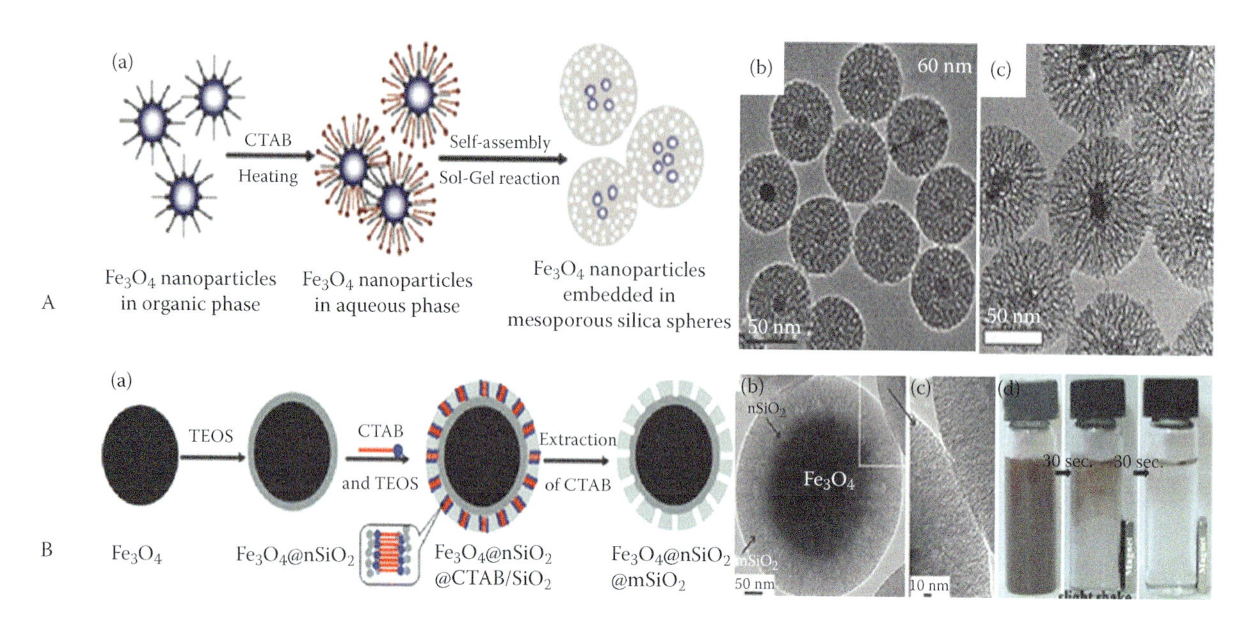

FIGURE 9.2 A—(a) Scheme of the synthesis of individual or multicore IO@MS core–shell magnetic nanocomposites by surfactant-templated formation of porous silica shell.[34] (Reproduced with permission from Kim, J.; Lee, J. E.; Lee, J.; Yu, J. H.; Kim, B. C.; An, K.; Hwang, Y.; Shin, C.-H.; Park, J.-G.; Kim, J.; Hyeon, T. Magnetic Fluorescent Delivery Vehicle Using Uniform Mesoporous Silica Spheres Embedded with Monodisperse Magnetic and Semiconductor Nanocrystals. *J. Am. Chem. Soc.* 2006, *128* (3), 688–689. Copyright 2006 American Chemical Society.) TEM images for small pore (b) or enhanced pore (c) sizes of IO@MS nanocomposites.[35,36] (b, Kim, J.; Kim, H. S.; Lee, N.; Kim, T.; Kim, H.; Yu, T.; Song, I. C.; Moon, W. K.; Hyeon, T.: *Angew. Chem. Int. Ed.* 8438–8441. 2008. Copyright Wiley-VCH Verlag GmbH & Co. KGaA. Reproduced with permission. c, Reprinted from *J. Colloid Interface Sci.*, 362, Zhang, J.; Li, X.; Rosenholm, J. M.; Gu, H. Synthesis and Characterization of Pore Size-Tunable Magnetic Mesoporous Silica Nanoparticles, 16–24, 2011, with permission from Elsevier.) B—(a) Scheme of IONCs coated with a nonporous and then a porous silica shell: IONC@nSiO_2@MS. TEM images of one IONC@ nSiO2@MS particle (b) and a zoomed image between two MS shells (c). (d) Photographs showing their magnetic response with time by using a magnet. (Reproduced with permission from Deng, Y.; Qi, D.; Deng, C.; Zhang, X.; Zhao, D. Superparamagnetic High-Magnetization Microspheres with an Fe_3O_4@ SiO_2 Core and Perpendicularly Aligned Mesoporous SiO_2 Shell for Removal of Microcystins. *J. Am. Chem. Soc.* 2008, *130* (1), 28–29. Copyright 2008 American Chemical Society.[37] For the overall figure: Wang, Y.; Gu, H.: *Adv. Mater.* 576–585. 2015. Copyright Wiley-VCH Verlag GmbH & Co. KGaA. Reproduced with permission.[38])

not only the water-phase transfer of the hydrophobic NPs by interacting with oleic-acid-coated IONPs but also allowed the formation of cetyltrimethylammonium (CTA) micelles in the aqueous solution. The addition of NaOH (or NH_3) and of tetraethyl orthosilicate precursor thus allowed the polymerization of the (negatively charged) silicate monomers around the (positively charged) CTA-coated IONPs and the ordered arrangement of CTA micelles. These micelles acted as templates (or structure directing agents) for the formation of the resulting MS network. A main drawback of this seminal work was the lack of control in the design, as a quite random number of several nanocrystals were loaded within the porous silica matrix. In 2008, a protocol was optimized to afford precisely IO monocore covered with homogenous MS shells.[35] The usual pore size afforded by such an approach is *ca.* 2.5 nm as measured by N_2 isotherm adsorptions. Other approaches using CTAB with variable chain length (from 14 to 18 carbons in the CTAB chain) allowed tuning pore size from 2.4 to 3.4 nm.[39] Later, several groups focused on tuning the pore size especially to allow the embedding of active compounds having bigger dimensions than usual drugs, such as nucleic acids or proteins. For instance, the use of 1,3,5-triisopropyl benzene and decane as pore swelling agents, as reported by Zhang et al.,[36] has allowed reaching pore size up to 6.1 nm in the porous magnetic core/silica shell.

However, regarding the magnetic properties, one main issue associated with such magnetic core–MS shells is the important decrease in the saturation magnetization (M_s measured at 300 K) resulting from the added mass of the silica shell. Indeed, for magnetic silica composites, the measured M_s values are usually referred to the total mass of the composite, and the high majority fraction of silica makes these values particularly low compared to magnetic NPs coated with polymer or organic molecules. For instance, for individual IO@MS core–shell having small pore size[35] (*ca.* 2.5 nm) or enhanced pore size[36] (*ca.* 6.1 nm), M_s values of *ca.* 0.06 emu.g^{-1} and 2.0–2.7 emu.g^{-1}, respectively, were measured. The M_s remained still weak for IONP multicores shelled with small-pore MS (*ca.* 2.5 nm with M_s *ca.* 2.0 emu.g^{-1}).[40] To address this issue, among the various magnetic core-MS shell NPs,[38] this last decade, there was an important development in the design of magnetic MS having a large magnetic core consisting of superparamagnetic controlled aggregates of IO NPs (5–15 nm) called also magnetic iron oxide nanoclusters (IONCs) or raspberry's.

9.3.1.2 Iron Oxide Nanocluster Core @Mesoporous Silica

IO magnetic nanoclusters (IONCs) having a size ranging from 100 to 500 nm are obtained from controlled aggregation of small IONPs (10–15 nm), usually by modified-polyol solvothermal methods. They have recently attracted tremendous interest because of their important gain in terms of saturation magnetization (up *ca.* 85 emu.g^{-1}) combined with their superparamagnetic properties, thus facilitating their magnetic manipulation.[41] In the next section, we focus on magnetic IONCs coated with MS shells and their applications for drug loading and release. It is worthy to note also that there recently have been important works describing the design of magnetic clusters coated with nonporous silica and the possibility to align them in chains by performing the sol–gel process upon a static magnetic field applied.[42–44]

In 2008, Deng et al.[37] described in a seminal work the coating of an MS shell around IONCs synthesized by a solvothermal method in ethylene glycol (300 nm diameter made of 15 nm IONPs) (Figure 9.2B). In a first step, a nonporous silica shell ($nSiO_2$, 20 nm thickness) was deposited through a classical Stöber process. Then they formed an MS shell (70 nm) by applying the sol–gel procedure in the presence of CTAB surfactant as a pore structure agent in a water/ethanol solution (as seen in the previous section). After CTAB extraction in acetone, IONC@$nSiO_2$@MS (500 nm size, 2.3 nm pore size, 365 m^2.g^{-1} surface area and 0.29 cm^3.g^{-1} pore volume) was obtained. A radial (or perpendicular) orientation of the MS mesopores around the IONCs interestingly formed, which was explained by a minimization of surface energy of the CTAB/silicate rod-like complexes. Further, the silica coating was shown to decrease its M_s from 80.7 to 53.3 emu.g^{-1} (with M_s still referred to the total mass of the composite), which nevertheless remained very high to ensure a strong and rapid magnetic response. Later, there were several works that have improved the design of such submicron composites especially by decreasing IONC size (up to 80 nm diameter). Such composites were developed for drug loading and release[45,46] or pollutant removal[47] applications.

9.3.2 Drug Loading in Bare Mesoporous Silica Shell: Issue with Colloidal Stability

There are actually a few works that have reported the loading of hydrophilic and hydrophobic drugs directly within nonmodified (bare) MS (magnetic or not) mainly because of the limited colloidal stability of the resulting drug-loaded nanocomposites. Indeed, the drug loading in bare surface may result in an important loss of the NP colloidal stability (aggregation) and a polymer grafting is required to ensure colloidal stability. Strategies ensuring simultaneous drug loading and improved colloidal stability will be detailed in the next section.

One of the few works that have investigated the interactions of an hydrophilic drug (DOX) directly with nonmodified (bare) MS is reported by Shen et al.[48] In this work, DOX was loaded in nonmagnetic MS NPs (MSNs) with 120, 200 and 360 nm size by simple incubation of DOX in bare MSNs in water at 2 mg.ml^{-1}. They showed very high drug payloads up to DLC = 30% (from ultraviolet [UV]/Vis spectrophotometry method). The same authors have shown the spontaneous DOX release in PBS, at 37°C; however, only 25% of the encapsulated DOX was released in 48 h.

In another example, the encapsulation of not only hydrophobic drugs but also biologically active molecules has been investigated within bare MS capping IONCs. In this work, Zhang et al.[45] investigated the loading of the hydrophobic drug rapamycin and of salmon sperm DNA within the bare MS shell of an IONC core–MS shell (225 nm diameter with 125 nm IONC core size, M_s = 28 emu.g) having mesopores of 4.2 nm. For the loading of rapamycin, the amount of loaded drug was shown to be dependent on the nature of the solvent. The loading capacities of rapamycin in three different solvents, CCl_4, toluene and chloroform, were found respectively at 124.5, 108.5 and 37.8 mg.g^{-1} (measurements by high-performance liquid chromatography methods). Regarding the loading of DNA, salmon sperm DNA (~250 base pairs) could be efficiently loaded by using a 4 M chaotropic salt solution at pH 5.2 with a DLC = 11% (110.7 mg.g, measurements by UV/Vis analysis at 260 nm). In this work, no release of drug or DNA was performed probably because of the difficulty to control the drug release by local/external stimuli or issues to ensure good colloidal stability. As mentioned in the beginning of this section, strategies involving polymer grafting were developed to improve the colloidal stability of drug-loaded composites.

9.3.3 Improving The Colloidal Stability by Polymer Grafting

In several works, grafting of the well-known low fouling polyethylene glycol (PEG) polymer on magnetic or nonmagnetic porous silica (pore size close to 2.5 nm) was performed before the drug loading steps with the aim to ensure a combined high colloidal stability and high drug loading.[35,49,50] However, in these studies, it was not clearly stated if the indicated DLC values, which is in the range [8%–10%], can be attributed to the interaction of the drugs with the pore network or with the polymer chains itself as PEG may perform H-bonding with the drugs.

In other works made by Zink and coworkers, the loading of water-insoluble hydrophobic drugs such as camptothecin (CPT) and paclitaxel (PTX) was achieved in porous (3 nm pore size) nonmagnetic MSNs and magnetic MSNs (denoted MMSNs) after modifying the surface with methyl phosphonate silane groups in dimethyl sulfoxide solvent.[51,52] In the absence of any surface modification, the MSN composites loaded with the hydrophobic drugs were shown to aggregate massively. Conversely, after surface modification with phosphonate groups, a high colloidal stability was obtained for the MSNs. Moreover, for CPT, DLE = 10%, DLC = ~1% and for PTX, DLE = 26%, DLC = 2.6% were obtained respectively. Here, too, the mechanism of drug loading is not fully understood and the drug retention was attributed to a water nonsoluble PTX or CPT drug precipitate within the mesopores. The relatively low DLC values (1%–2%) obtained with this phosphonate grafted MMSNs nevertheless allowed powerful cell apoptosis after incubation with pancreatic cell line (PANC-1 and BxPC-3). This cell death was explained by drug delivery achieved in the cellular hydrophobic regions. Moreover, the NPs displayed the retention of their cytotoxic activity over 2 months' storage at 4°C.

Thus, modifying the surface with a polymer will actually have a (positive) effect on the colloidal stability and also on the interactions with the drugs. However, in these latter examples, we can notice and underline that the interactions between the drugs and the surface were not really well described/understood. The same conclusion can be drawn for the drug loading in bare silica (see Section 9.3.2), as the encapsulation of either hydrophilic or hydrophobic drugs with both high drug loading levels in various solvents could be achieved. This clearly highlights that interactions between the drugs and the silica surface cannot be described with simple assumptions and it may involve various types of intermolecular interactions. Hence, probably a better way to address desired interactions between the drug and the surface is to consider the surface modification with chemical groups of well-defined nature (charged groups, aromatic, H-bond donor receptors). Several groups have recently investigated the influence of the surface modification on the loading and the release of drugs. These works are detailed in Section 9.4.

9.4 INFLUENCE OF CHEMICAL SURFACE MODIFICATION ON DRUG LOADING AND RELEASE

The influence of chemical modification of the MS surface on drug loading and on its release properties was investigated since only a few years ago. The modification of the MS surface by grafting molecule or polymer was shown to be of key importance as it helps to increase the amount of loaded drugs, to facilitate the drug release and also to improve the colloidal stability of the resulting silica NPs. Different methods were investigated by tuning the interactions between drugs and the mesoporous internal surface of the nanocomposite:

i) Electrostatic attractions;
ii) π-π interactions between aromatic cycles; and
iii) H-bonds between H-bond donors/acceptors

9.4.1 Tailoring the Drug Loading/Release by Electrostatic Attractions

In a work by Chang et al.,[40] the surface of MMSNs was tuned with organosilanes having pendant negatively charged groups, including carboxylate ($-COO^-$) and phosphonate ($-PO_3^-$), and compared with nonmodified (silanols functions) surface as a control. The aim was to assess the loading of DOX (pKa = 8.3) within these modified mesopores through electrostatic interactions. The DLE and DLC values associated to the impregnation of DOX in water were determined for these negatively charged surfaces. DLE/DLC values after soaking DOX in water (0.5 mg.mL^{-1}) were found on bare MMSNs at DLE = 12.4%, DLC = 5.8%, on COOH-modified MMSNs at DLE = 23.6%, DLC = 10.5% and on PO_3-modified MMSNs at DLE = 89.8%, DLC = 31%. The increasing values of DLE and DLC were explained by the increasing electrostatic interactions resulting from the increased negative charge occurring from silanol to phosphonate groups as measured by zeta potential. Moreover, the particularly high DLC with phosphonate groups was also explained by a higher pore volume after surface modification compared to the COOH-modified MSNs (1.03 vs. 0.54 cm^3.g^{-1}). DOX release was investigated as a function of pH, which is a key parameter to modulate the charge of both the drug and the MMSN surface and, thus, their resulting electrostatic interactions. With the COOH-modified MMSNs, the release of DOX was shown to be pH dependent in the range of pH = 4.2–7.4, with an increased released amount of drugs obtained by decreasing the pH (from 23% to 56.7% when the pH was lowered from 7.4 to 4.2 over 120 h). This effect was explained by the progressive protonation of the COO^- groups and then weakening of the electrostatic binding, ensuring hydration and release of DOX in slightly acidic aqueous solutions. Conversely, for the phosphonate modification, no clear pH-dependent release of DOX was observed as already 45% of DOX was released at pH 7.4 vs. 57.1% at pH 4.2. Such a similar release behaviour in this pH range was explained by similar electrostatic interactions in this range of pH, between the phosphonate groups and DOX.

9.4.2 Tailoring the Drug Loading/Release by π–π Stacking Interactions

Chang et al.[40] compared also the loading of PTX within bare and phenyl (Ph)-modified MMSNs. The aim of the study was to investigate the effect of the hydrophobic interactions (π–π stacking) between the grafted-Ph groups with the PTX aromatic cycles. MMSNs and Ph-MMSNs were soaked in a PTX solution in dichloromethane. DLE and DLC values of 85.4%/8.8% for bare MMSNs and 71%/7.1% for Ph-modified MMSNs were respectively measured (measurements by UV/Vis spectrophotometry at 227 nm in MeOH). The quite satisfying DLC value of 8.8% obtained with bare silica was explained by the formation of strong H-bonds between the hydroxyl (OH) groups of PTX and the silanols (Si-OH) of the porous silica internal surface. Furthermore, these results indicated that adding a surface modification with Ph groups was not shown to be more relevant for improving the PTX loading. These results were explained by a competition between the hydrophobic and H-bond interactions. Indeed, even if the Ph modification should ensure hydrophobic interactions by π–π stacking, it lowers the H-bond interactions with PTX, and this finally did not provide a gain in the drug payload. Furthermore, the PTX release was investigated in PBS at 37°C over 14 d and the kinetics was performed with both surface modifications. A faster full release of PTX was observed with Ph-MMSNs over 7 d, while a delay of *ca.* 5 d was observed with bare MMSNs (full release only after 12 d). This delay was explained by a stronger retention due to the H-bond interactions between silanols and OH groups of PTX compared to the π–π stacking between PTX and Ph groups.

In another work by Knežević et al.,[53] the loading of two aromatic drugs: 9-aminoacridine (9-AA) (a water soluble positively charged and aromatic drug) and the hydrophobic CPT was investigated in MeOH before and after organic functionalization of the mesopores by phenylethylsilanes (Ph). The surface modification with Ph groups allowed a higher DLC for CPT (from 13.4 to 46.1 μmol.g^{-1} of MMSNs after surface modification with Ph), whereas no notable difference in the DLC values was obtained for 9-AA (from 26.7 to 25 μmol.g^{-1} of MMSNs after surface modification with Ph). For CPT, the DLC increase was explained by the strong π–π stacking between the Ph groups and the aromatic drug. For 9-AA, the Ph group modification did not promote increase of the drug loading. As it was the case with the loading of PTX inside nonmodified MMSNs, as described previously, H-bonds may also occur between 9-AA and nonmodified MMSNs.

Regarding the drug release in PBS, opposite behaviours were observed according to the type of drugs (Figure 9.3). For 9-AA, an efficient drug release was obtained from the bare MMSNs; however, a negligible drug release was observed from the Ph-modified MMSNs (*ca.* 17% vs. 0.75% of 9-AA released over 96 h). These results were explained by the inaccessibility of PBS to enter the mesopores of the Ph-modified MSNs, whereas the bare MMSNs allowed a better solvation of 9-AA and subsequent release in the PBS medium. Conversely, the CPT release was favoured with the Ph-MMSNs and was importantly hindered with bare MMSNs (*ca.* 12.5% vs. 0.8% of CPT released over 96 h). This effect was explained by a progressive hydrolysis of the lactone ring of the CPT into a water-soluble carboxylate group promoting the release of the CPT in PBS solution. The limited release of CPT in bare MMSNs is explained by a strong interaction between the silica surface and the drugs and a lower DLC. Finally, these trends were confirmed with the cell viability studies assessing the release of the drugs from Ph-modified and bare MMSNs in Chinese hamster ovarian (CHO) cells. Indeed, relevant cell toxicity effects towards CHO cells were observed only with Ph-modified MSNs loaded with CPT (57.5% cell viability) and with bare MSNs loaded with 9-AA (55% cell viability), whereas the others cases did not present any cytotoxicity (Figure 9.3, cell viability graphs).

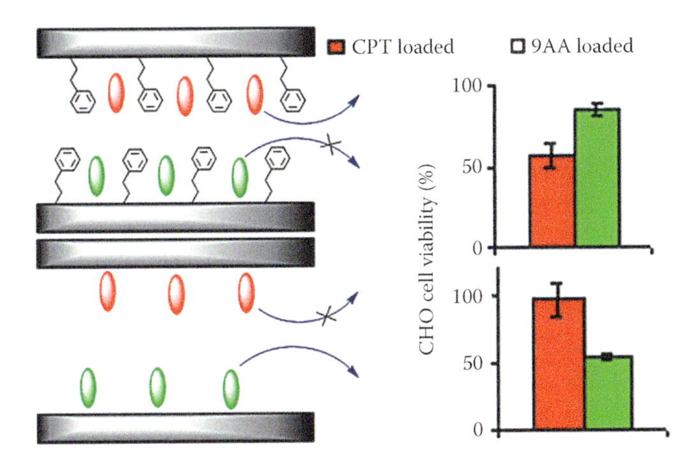

FIGURE 9.3 Schemes illustrating the effects of the surface modification with phenylethyl groups of MMSNs on the loading and delivery of CPT and 9-AA drugs to cancer cells. (Knežević, N. Ž.; Slowing, I. I.; Lin, V. S.-Y.: *ChemPlusChem*. 48–55. 2012. Copyright Wiley-VCH Verlag GmbH & Co. KGaA. Reproduced with permission.)

9.4.3 Tailoring Drug Loading/Release by Tuning H-Bond Interactions

In a work by Gai et al.,[54] IONCs (80 nm size, made of Fe_3O_4) covered with a nonporous thin silica shell ($nSiO_2$) and an MS shell having 2.2-nm-sized radially oriented pores (Fe_3O_4@$nSiO_2$@MS) were functionalized with $NaYF_4$:Yb^{3+}/Er^3 for applications in magnetism and up-conversion emission (green fluorescence for Yb^{3+}/Er^{3+} under 980 nm laser excitation). The effect of the aminopropyltriethoxysilane (APTS) modification ($-NH_2$) of the MS shell on the loading and release of IBU was investigated. The IBU loading within Fe_3O_4@$nSiO_2$@MS@$NaYF_4$:Yb^{3+}/Er^{3+} was achieved in hexane. DLC values of *ca.* 11% and 8% before and after APTS surface modification, respectively, were found. This difference in drug loading was mainly attributed to a decrease in the surface area measured after the APTS modification (from 296 to 193 m^2.g^{-1}). Then, the IBU release was followed in a simulated body fluid (i.e. a mimetic physiological solution) by measuring IBU absorbance in the supernatant at λ = 222 nm. The APTS modification was importantly shown to delay the release of IBU vs. the nonmodification. Indeed, 72 h was needed to release over 87% of IBU for the APTS-modified sample, whereas only 6 h was needed to release 95% of IBU without any surface modification. These results were explained by the assumed strong H-bond interaction between the NH_2 group of APTS and the carboxyl group of IBU, which delayed the release of IBU. Very interestingly, the up-conversion emission intensity of the multifunctional carrier was shown to increase in a correlated way, with the released amount of IBU highlighting the potential of such nanodevices to follow a drug release with the fluorescence emission parameters in real time.

9.5 GATEKEEPING STRATEGIES FOR STIMULI RESPONSIVE DRUG RELEASE

Up to now, all the examples reported in this chapter have described a release that was performed after soaking the drug-loaded composites in an aqueous buffer. In this section, we will detail the strategies that were used to ensure controlled drug release in time and in location upon external or local stimuli. As this book chapter is dedicated to drug loading and release from magnetic nanocomposites, we will focus first in Section 9.5.1 on AMF-triggered release, which is probably the most relevant and promising way to trigger drug release from such nanocomposites. Indeed, as mentioned in the introduction section, an AMF interacting with the magnetic core can convert the magnetic energy into local heating, which can be beneficial for MH therapy and may also activate the drug delivery by magneto thermal action. In Section 9.5.2, some examples of drug release actuated by other stimuli such as light or pH are described as complementary modes to the magneto-thermal release.

9.5.1 Magnetothermal Responsive Drug Release via Thermoresponsive Gatekeepers

The design of magneto-responsive drug-loaded magnetic silica composites requires coating these NPs with thermo-responsive gatekeepers (i.e. a polymer or molecule) that will respond to local heating by a change in its conformation or configuration or even degradation, thus ensuring drug release from the platforms. There are various kinds of thermo-responsive gatekeepers having different thermal responses that will be described in this section.

In 2010, Thomas et al.[55] used a pseudorotaxane macrocycle (cucurbituryl) as a thermo-responsive gatekeeper to ensure the magneto-thermal release of encapsulated DOX (DLC = 4%) from magnetic Zn-doped IO@MSNs modified with aminohexyl-aminomethyltriethoxysilane (AMTS) threads (Figure 9.4A). Hence, upon AMF ($H = 32.4$ kA.m^{-1}, $f = 500$ kHz), which resulted in a local temperature increase, the thermal disassembly of the complex pseudo-rotaxane–AMTS occurred and triggered the drug release. In this work, Thomas et al. showed the possibility to perform continuous or pulsatile release of the drugs from the magnetic nanocomposites while a negligible amount of drugs was released in the absence of AMF. Moreover, they applied this strategy on breast cancer cells (MDA MB 231) and demonstrated a synergic and successful treatment of MH combined with drug delivery to kill cancer cell.

Another way to ensure efficient thermo-responsive gatekeeping is to use the well-established thermal behaviour of a polymer having a lower critical solution temperature (LCST). Indeed, with such polymers, for temperature, $T < \text{LCST}$, the polymer is in an extended configuration and is in a fully hydrated state, whereas for $T > \text{LCST}$, solvation interactions with water are disrupted and the polymer start to aggregate and becomes hydrophobic. Poly(*N*-isopropylacrylamide) (PNIPAM) is probably the most reported LCST polymer and has an LCST of 32°C. In a work by Baeza et al.,[56] the fluorophore fluorescein isothiocyanale (FITC) was loaded in multicore MMSNs (200 nm size), and catalase, an enzyme, was adsorbed on the surface initially coated with a PNIPAM-*b*-poly(ethyleneimine) copolymer shell. Upon AMF ($H = 24$ kA.m^{-1}, $f = 100$ kHz), a simultaneous high release content of both the protein and the FITC dye (higher than 50% in 6 h) was achieved. The release of both components was attributed to an increased local temperature induced by AMF application, which was assumed to be above the copolymer LCST. Then, PNIPAM becoming hydrophobic ensured the protein release from the shell and the opening of the MS pore releasing the embedded drugs by diffusion.

In the same group, Vallet Regi and coworkers investigated also DNA hybridization as a way to ensure drug release by AMF application. Indeed, Ruiz-Hernandez et al.[57] loaded fluorescein in multicore magnetic MS capped with single-strand DNA and closed the pore with its complementary DNA strands to block fluorescein diffusion. Thus, the local heating generated upon AMF application ($H = 24$ kA.m^{-1}, $f = 100$ kHz) ensured the dehybridization of the DNA double strands into single-strand DNA and, thus, the successful release of fluorescein.

Strategies making use of thermo-degradable moieties to open pores were also developed. Recently, Saint-Cricq et al.[58] synthesized a PEG polymer bearing a diazo bond and used it as a gatekeeper for controlling the release of loaded rhodamine upon AMF (Figure 9.4B). Diazo is known as a thermo-degradable bond when the temperature locally reaches 80°C and releases N_2 gas, which allows cleaving the PEG chains and thus opening up the mesopores. Upon AMF application ($H = 375$ kHz, $f = 20$ kA/m), rhodamine was efficiently released through polymer degradation and pore opening.

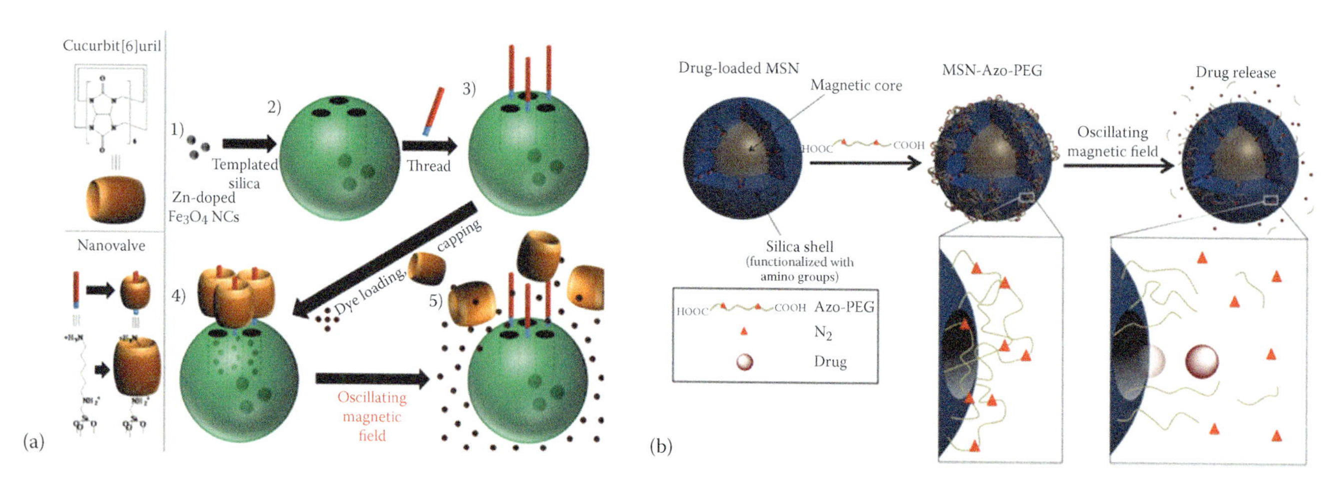

FIGURE 9.4 Schemes illustrating the dye release from MMSNs through magneto-thermal actuation of (a) cucurbituryl capping and of (b) PEG capping bearing a thermodegradable diazo bond. (a, Reproduced with permission from Thomas, C. R.; Ferris, D. P.; Lee, J.-H.; Choi, E.; Cho, M. H.; Kim, E. S.; Stoddart, J. F.; Shin, J.-S.; Cheon, J.; Zink, J. I. Noninvasive Remote-Controlled Release of Drug Molecules in Vitro Using Magnetic Actuation of Mechanized Nanoparticles. *J. Am. Chem. Soc.* 2010, *132* (31), 10623–10625. Copyright 2006 American Chemical Society. b, Saint-Cricq, P.; Deshayes, S.; Zink, J. I.; Kasko, A. M. Magnetic Field Activated Drug Delivery Using Thermodegradable Azo-Functionalised PEG-Coated Core–Shell Mesoporous Silica Nanoparticles. *Nanoscale* 2015, *7* (31), 13168–13172. Reproduced by permission of The Royal Society of Chemistry.)

9.5.2 Other Stimuli-Responsive Release (pH, Light) from Magnetic Silica Composites

In parallel to the magneto-thermal drug release, some works have addressed elegant ways to trigger the drug release from magnetic silica nanocomposites by using other types of stimuli, which may be of interest for biomedical applications such as pH or light.

9.5.2.1 pH-Responsive Release from Drug-Loaded Magnetic Silica

In 2010, Lee et al.[59] designed original magnetic silica nanocomposites by grafting MSNs with IONPs to afford MRI along with pH-responsive drug delivery. Hydrazine bonds were grafted at the surface of the nanocomposites to ensure the coupling of the surface with the ketone group of DOX via the formation of a hydrazone bond. With this covalent coupling, DOX was loaded at 11 μg drug.mg^{-1} MS. Hydrazone bond is pH-sensitive (i.e. it is cleaved in slightly acidic conditions pH ≤ 6), thus ensuring the release of DOX. A very specific pH-responsive release was obtained when the pH decreased from 7.4 to 4, with 4% and 78% of drug released, respectively, over 56 h. Such a strategy was successfully applied to kill breast cancer cell lines (MDA MB 231).

9.5.2.2 Light-Responsive Release from Drug-Loaded Magnetic Silica

In another example by Wang et al.,[46] IONC@nSiO_2@MS (200 nm IONC size, 2.4 nm MS pore size, M_s *ca.* 39.8 emu.g^{-1}) were grafted with trimethoxysilane azobenzene groups. The aim of this work was to use the property of light responsiveness of azo benzene to isomerize (*cis/trans* configuration) at 450 nm and to act as a light-triggered impeller within the mesopores of the nanocomposites. In this work, rhodamine 6G was loaded in ethanol (*ca.* 1% DLC) by opening the azobenzene gates upon light irradiation (azo benzene in configuration *cis*) and then tightly encapsulating rhodamine by switching off the light (azo benzene in configuration *trans*). Then, a periodically controlled dye release was obtained by applying on/off light cycles, demonstrating the controlled pulsatile release of rhodamine upon azobenzene isomerization at 450 nm. Such an elegant approach was also applicable for the loading and the light responsive release of IBU with similar loading and release profiles compared with rhodamine.

9.6 CONCLUSION

The design of theranostic NPs able to combine in one nano-object imaging and targeted therapeutic property is a hot topic in nanomedicine. IONPs are well-known contrast agents for MRI and are currently optimized to provide therapy by MH. However, it is also well known that the combination of MH and chemotherapy is a promising way to treat cancers. Therefore, among the numerous magnetic nanocarriers that can be used for such purposes, magnetic silica nanocomposites are currently strongly investigated as they may allow, in comparison with organic/polymeric magnetic nanocarriers, higher drug payloads.

In this book chapter, we have reviewed the different methods to encapsulate drugs in magnetic silica nanocomposites that are mainly

i) The drug encapsulation directly during the sol–gel process along with the magnetic NPs;
ii) The drug impregnation in porous magnetic core–MS shell and the tuning of the payload by adjusting the surface modification of the porous shells;
iii) The gatekeeping of embedded drugs by thermoresponsive polymers/molecules; and
iv) The drug coupling via a responsive linker (pH, temperature).

For the drugs encapsulated *in situ* during the sol–gel process within the nonporous magnetic silica, a main challenge is the difficulty to release the sequestered drugs, as an important applied magnetic energy is required to ensure the generation of pores or cracks within the silica matrix. For drug impregnation, a wide panel of surface chemical modification allows a high control of the interactions between the internal porous silica surface and the drugs via electrostatic, π–π stacking or H bond interactions. This is a key point to improve the drug payload; however, we can notice that the different studies on drug release reported in this book chapter by modifying the surface of the magnetic composites is mainly performed through a natural leaking in aqueous solutions (or by lowering the pH and thus changing electrostatic or H-bond interactions).

Finally, to address the controlled drug release from magnetic silica composites, there are different strategies, including polymer gatekeepers or host–guest molecules, which can respond to various stimuli (pH, redox, temperature, near-infrared light and magnetic field). Among these stimuli, the activation of the drug release upon an AMF from magnetic nanocomposites is particularly relevant. For this latter point, we anticipate that important challenges will be on the design of the magnetic nanomediators to improve the magneto-thermal transfer and to reduce the dose of magnetic composites administered. Additionally, designing nanocarriers made of biological macromolecules such as proteins based on silica templated approaches will be also of high importance with the aim to design a fully biodegradable or resorbable nanocarrier. Several groups, including our team, have recently developed new strategies for the design of optimized magnetic nanomediators[60,61] or the assembly of MS templated protein-base vesicles.[62–65] We expect that important developments should be also performed towards the possibility of ensuring a pulsatile drug release where a controlled dose of drugs could be released timely and spatially. At least in the near future, it is anticipated that the magnetic nanocomposite releasing drugs upon AMF applied will be developed for preclinical (*in vivo*) and, eventually, clinical applications.

ACKNOWLEDGEMENTS

This article is based upon work from COST Action (RADIOMAG TD1402), supported by COST (European Cooperation in Science and Technology). D.M. thanks the University of Strasbourg for financial support from IDEX-Attractivité framework.

REFERENCES

1. Na, H. B.; Song, I. C.; Hyeon, T. Inorganic Nanoparticles for MRI Contrast Agents. *Adv. Mater.* **2009**, *21* (21), 2133–2148.
2. Walter, A.; Garofalo, A.; Parat, A.; Jouhannaud, J.; Pourroy, G.; Voirin, E.; Laurent, S.; Bonazza, P.; Taleb, J.; Billotey, C.; Elst, L. V.; Muller, R. N.; Begin-Colin, S.; Felder-Flesch, D. Validation of a Dendron Concept to Tune Colloidal Stability, MRI Relaxivity and Bioelimination of Functional Nanoparticles. *J. Mater. Chem. B* **2015**, *3* (8), 1484–1494.
3. Wang, X.-Y.; Mertz, D.; Blanco-Andujar, C.; Bora, A.; Ménard, M.; Meyer, F.; Giraudeau, C.; Bégin-Colin, S. Optimizing the Silanization of Thermally-Decomposed Iron Oxide Nanoparticles for Efficient Aqueous Phase Transfer and MRI Applications. *RSC Adv.* **2016**, *6* (96), 93784–93793.
4. Hachani, R.; Lowdell, M.; Birchall, M.; Hervault, A.; Mertz, D.; Begin-Colin, S.; Thanh, N. T. K. Polyol Synthesis, Functionalisation, and Biocompatibility Studies of Superparamagnetic Iron Oxide Nanoparticles as Potential MRI Contrast Agents. *Nanoscale* **2016**, *8* (6), 3278–3287.
5. Kumar, C. S.; Mohammad, F. Magnetic Nanomaterials for Hyperthermia-Based Therapy and Controlled Drug Delivery. *Adv. Drug Deliv. Rev.* **2011**, *63* (9), 789–808.
6. Périgo, E. A.; Hemery, G.; Sandre, O.; Ortega, D.; Garaio, E.; Plazaola, F.; Teran, F. J. Fundamentals and Advances in Magnetic Hyperthermia. *Appl. Phys. Rev.* **2015**, *2* (4), 41302.
7. Blanco-Andujar, C.; Walter, A.; Cotin, G.; Bordeianu, C.; Mertz, D.; Felder-Flesch, D.; Begin-Colin, S. Design of Iron Oxide-Based Nanoparticles for MRI and Magnetic Hyperthermia. *Nanomed.* **2016**, *11* (14), 1889–1910.
8. Thévenot, J.; Oliveira, H.; Sandre, O.; Lecommandoux, S. Magnetic Responsive Polymer Composite Materials. *Chem. Soc. Rev.* **2013**, *42* (17), 7099–7116.
9. Hervault, A.; Thanh, N. T. K. Magnetic Nanoparticle-Based Therapeutic Agents for Thermo-Chemotherapy Treatment of Cancer. *Nanoscale* **2014**, *6* (20), 11553–11573.
10. Liu, J.; Detrembleur, C.; Mornet, S.; Jérôme, C.; Duguet, E. Design of Hybrid Nanovehicles for Remotely Triggered Drug Release: An Overview. *J. Mater. Chem. B* **2015**, *3* (30), 6117–6147.
11. Mertz, D.; Sandre, O.; Bégin-Colin, S. Drug Releasing Nanoplatforms Activated by Alternating Magnetic Fields. *Biochim. Biophys. Acta* **2017**, *1861* (6), 1617–1641.
12. Kakwere, H.; Leal, M. P.; Materia, M. E.; Curcio, A.; Guardia, P.; Niculaes, D.; Marotta, R.; Falqui, A.; Pellegrino, T. Functionalization of Strongly Interacting Magnetic Nanocubes with (Thermo) Responsive Coating and Their Application in Hyperthermia and Heat-Triggered Drug Delivery. *ACS Appl. Mater. Interfaces* **2015**, *7* (19), 10132–10145.
13. Griffete, N.; Fresnais, J.; Espinosa, A.; Wilhelm, C.; Bée, A.; Ménager, C. Design of Magnetic Molecularly Imprinted Polymer Nanoparticles for Controlled Release of Doxorubicin under an Alternative Magnetic Field in Athermal Conditions. *Nanoscale* **2015**, *7* (45), 18891–18896.
14. Hervault, A.; Dunn, A. E.; Lim, M.; Boyer, C.; Mott, D.; Maenosono, S.; Thanh, N. T. Doxorubicin Loaded Dual pH-and Thermo-Responsive Magnetic Nanocarrier for Combined Magnetic Hyperthermia and Targeted Controlled Drug Delivery Applications. *Nanoscale* **2016**, *8* (24), 12152–12161.
15. Yoo, D.; Jeong, H.; Noh, S.-H.; Lee, J.-H.; Cheon, J. Magnetically Triggered Dual Functional Nanoparticles for Resistance-Free Apoptotic Hyperthermia. *Angew. Chem. Int. Ed.* **2013**, *52* (49), 13047–13051.
16. Kim, H.-C.; Kim, E.; Jeong, S. W.; Ha, T.-L.; Park, S.-I.; Lee, S. G.; Lee, S. J.; Lee, S. W. Magnetic Nanoparticle-Conjugated Polymeric Micelles for Combined Hyperthermia and Chemotherapy. *Nanoscale* **2015**, *7* (39), 16470–16480.
17. Kim, D.-H.; Vitol, E. A.; Liu, J.; Balasubramanian, S.; Gosztola, D. J.; Cohen, E. E.; Novosad, V.; Rozhkova, E. A. Stimuli-Responsive Magnetic Nanomicelles as Multifunctional Heat and Cargo Delivery Vehicles. *Langmuir* **2013**, *29* (24), 7425–7432.
18. Nappini, S.; Bonini, M.; Ridi, F.; Baglioni, P. Structure and Permeability of Magnetoliposomes Loaded with Hydrophobic Magnetic Nanoparticles in the Presence of a Low Frequency Magnetic Field. *Soft Matter* **2011**, *7* (10), 4801–4811.
19. Di Corato, R.; Béalle, G.; Kolosnjaj-Tabi, J.; Espinosa, A.; Clement, O.; Silva, A. K.; Menager, C.; Wilhelm, C. Combining Magnetic Hyperthermia and Photodynamic Therapy for Tumor Ablation with Photoresponsive Magnetic Liposomes. *ACS Nano* **2015**, *9* (3), 2904–2916.
20. Sanson, C.; Diou, O.; Thévenot, J.; Ibarboure, E.; Soum, A.; Brûlet, A.; Miraux, S.; Thiaudière, E.; Tan, S.; Brisson, A.; Dupuis, V.; Sandre, O.; Lecommandoux, S. Doxorubicin Loaded Magnetic Polymersomes: Theranostic Nanocarriers for MR Imaging and Magneto-Chemotherapy. *ACS Nano* **2011**, *5* (2), 1122–1140.
21. Oliveira, H.; Pérez-Andrés, E.; Thevenot, J.; Sandre, O.; Berra, E.; Lecommandoux, S. Magnetic Field Triggered Drug Release from Polymersomes for Cancer Therapeutics. *J. Controlled Release* **2013**, *169* (3), 165–170.
22. Hu, S.-H.; Liao, B.-J.; Chiang, C.-S.; Chen, P.-J.; Chen, I.-W.; Chen, S.-Y. Core–Shell Nanocapsules Stabilized by Single-Component Polymer and Nanoparticles for Magneto-Chemotherapy/Hyperthermia with Multiple Drugs. *Adv. Mater.* **2012**, *24* (27), 3627–3632.
23. Hu, S.-H.; Chen, S.-Y.; Gao, X. Multifunctional Nanocapsules for Simultaneous Encapsulation of Hydrophilic and Hydrophobic Compounds and On-Demand Release. *ACS Nano* **2012**, *6* (3), 2558–2565.
24. Huang, H.-Y.; Hu, S.-H.; Chian, C.-S.; Chen, S.-Y.; Lai, H.-Y.; Chen, Y.-Y. Self-Assembling PVA-F127 Thermosensitive Nanocarriers with Highly Sensitive Magnetically-Triggered Drug Release for Epilepsy Therapy *in Vivo*. *J. Mater. Chem.* **2012**, *22* (17), 8566–8573.
25. Liu, J.; Detrembleur, C.; Debuigne, A.; De Pauw-Gillet, M.-C.; Mornet, S.; Vander Elst, L.; Laurent, S.; Duguet, E.; Jérôme, C. Glucose-, pH-and Thermo-Responsive Nanogels Crosslinked by Functional Superparamagnetic Maghemite Nanoparticles as Innovative Drug Delivery Systems. *J. Mater. Chem. B* **2014**, *2* (8), 1009–1023.
26. Chiang, W.-L.; Ke, C.-J.; Liao, Z.-X.; Chen, S.-Y.; Chen, F.-R.; Tsai, C.-Y.; Xia, Y.; Sung, H.-W. Pulsatile Drug Release from PLGA Hollow Microspheres by Controlling the Permeability of Their Walls with a Magnetic Field. *Small* **2012**, *8* (23), 3584–3588.
27. Kong, S. D.; Sartor, M.; Hu, C.-M. J.; Zhang, W.; Zhang, L.; Jin, S. Magnetic Field Activated Lipid–Polymer Hybrid Nanoparticles for Stimuli-Responsive Drug Release. *Acta Biomater.* **2013**, *9* (3), 5447–5452.

28. Sivakumar Balasubramanian, A. R. G.; Nagaoka, Y.; Iwai, S.; Suzuki, M.; Kizhikkilot, V.; Yoshida, Y.; Maekawa, T.; Nair, S. D. Curcumin and 5-Fluorouracil-Loaded, Folate-and Transferrin-Decorated Polymeric Magnetic Nanoformulation: A Synergistic Cancer Therapeutic Approach, Accelerated by Magnetic Hyperthermia. *Int. J. Nanomedicine* **2014**, *9*, 437.
29. Barbe, C.; Bartlett, J.; Kong, L.; Finnie, K.; Lin, H. Q.; Larkin, M.; Calleja, S.; Bush, A.; Calleja, G. Silica Particles: A Novel Drug-Delivery System. *Adv. Mater.* **2004**, *16* (21), 1959–1966.
30. Li, S.; Ma, Y.; Yue, X.; Cao, Z.; Dai, Z. One-Pot Construction of Doxorubicin Conjugated Magnetic Silica Nanoparticles. *N. J. Chem.* **2009**, *33* (12), 2414–2418.
31. Hu, S.-H.; Liu, T.-Y.; Huang, H.-Y.; Liu, D.-M.; Chen, S.-Y. Magnetic-Sensitive Silica Nanospheres for Controlled Drug Release. *Langmuir* **2008**, *24* (1), 239–244.
32. Hu, S.-H.; Chen, S.-Y.; Liu, D.-M.; Hsiao, C.-S. Core/Single-Crystal–Shell Nanospheres for Controlled Drug Release via a Magnetically Triggered Rupturing Mechanism. *Adv. Mater.* **2008**, *20* (14), 2690–2695.
33. Caetano, B. L.; Guibert, C.; Fini, R.; Fresnais, J.; Pulcinelli, S. H.; Ménager, C.; Santilli, C. V. Magnetic Hyperthermia-Induced Drug Release from Ureasil-PEO-γ-Fe_2O_3 Nanocomposites. *RSC Adv.* **2016**, *6* (68), 63291–63295.
34. Kim, J.; Lee, J. E.; Lee, J.; Yu, J. H.; Kim, B. C.; An, K.; Hwang, Y.; Shin, C.-H.; Park, J.-G.; Kim, J.; Hyeon, T. Magnetic Fluorescent Delivery Vehicle Using Uniform Mesoporous Silica Spheres Embedded with Monodisperse Magnetic and Semiconductor Nanocrystals. *J. Am. Chem. Soc.* **2006**, *128* (3), 688–689.
35. Kim, J.; Kim, H. S.; Lee, N.; Kim, T.; Kim, H.; Yu, T.; Song, I. C.; Moon, W. K.; Hyeon, T. Multifunctional Uniform Nanoparticles Composed of a Magnetite Nanocrystal Core and a Mesoporous Silica Shell for Magnetic Resonance and Fluorescence Imaging and for Drug Delivery. *Angew. Chem. Int. Ed.* **2008**, *47* (44), 8438–8441.
36. Zhang, J.; Li, X.; Rosenholm, J. M.; Gu, H. Synthesis and Characterization of Pore Size-Tunable Magnetic Mesoporous Silica Nanoparticles. *J. Colloid Interface Sci.* **2011**, *361* (1), 16–24.
37. Deng, Y.; Qi, D.; Deng, C.; Zhang, X.; Zhao, D. Superparamagnetic High-Magnetization Microspheres with an Fe3O4@ SiO2 Core and Perpendicularly Aligned Mesoporous SiO2 Shell for Removal of Microcystins. *J. Am. Chem. Soc.* **2008**, *130* (1), 28–29.
38. Wang, Y.; Gu, H. Core–Shell-Type Magnetic Mesoporous Silica Nanocomposites for Bioimaging and Therapeutic Agent Delivery. *Adv. Mater.* **2015**, *27* (3), 576–585.
39. Zhang, L.; Qiao, S.; Jin, Y.; Yang, H.; Budihartono, S.; Stahr, F.; Yan, Z.; Wang, X.; Hao, Z.; Lu, G. Q. Fabrication and Size-Selective Bioseparation of Magnetic Silica Nanospheres with Highly Ordered Periodic Mesostructure. *Adv. Funct. Mater.* **2008**, *18* (20), 3203–3212.
40. Chang, B.; Guo, J.; Liu, C.; Qian, J.; Yang, W. Surface Functionalization of Magnetic Mesoporous Silica Nanoparticles for Controlled Drug Release. *J. Mater. Chem.* **2010**, *20* (44), 9941–9947.
41. Gerber, O.; Pichon, B. P.; Ulhaq, C.; Grenèche, J.-M.; Lefevre, C.; Florea, I.; Ersen, O.; Begin, D.; Lemonnier, S.; Barraud, E.; et al. Low Oxidation State and Enhanced Magnetic Properties Induced by Raspberry Shaped Nanostructures of Iron Oxide. *J. Phys. Chem. C* **2015**, *119* (43), 24665–24673.
42. Kralj, S.; Makovec, D. Magnetic Assembly of Superparamagnetic Iron Oxide Nanoparticle Clusters into Nanochains and Nanobundles. *ACS Nano* **2015**, *9* (10), 9700–9707.
43. Kopanja, L.; Kralj, S.; Zunic, D.; Loncar, B.; Tadic, M. Core–Shell Superparamagnetic Iron Oxide Nanoparticle (SPION) Clusters: TEM Micrograph Analysis, Particle Design and Shape Analysis. *Ceram. Int.* **2016**, *42* (9), 10976–10984.
44. Kralja, S.; Potrčd, T.; Kocbekd, P.; Marchesanb, S.; Makoveca, D. Design and Fabrication of Magnetically Responsive Nanocarriers for Drug Delivery. *Curr. Med. Chem.* **2016**, *23*, 1–16.
45. Zhang, J.; Sun, W.; Bergman, L.; Rosenholm, J. M.; Lindén, M.; Wu, G.; Xu, H.; Gu, H. Magnetic Mesoporous Silica Nanospheres as DNA/drug Carrier. *Mater. Lett.* **2012**, *67* (1), 379–382.
46. Wang, Y.; Li, B.; Zhang, L.; Song, H.; Zhang, L. Targeted Delivery System Based on Magnetic Mesoporous Silica Nanocomposites with Light-Controlled Release Character. *ACS Appl. Mater. Interfaces* **2012**, *5* (1), 11–15.
47. Liu, F.; Tian, H.; He, J. Adsorptive Performance and Catalytic Activity of Superparamagnetic Fe_3O_4@ $nSiO_2$@ $mSiO_2$ Core–Shell Microspheres towards DDT. *J. Colloid Interface Sci.* **2014**, *419*, 68–72.
48. Shen, J.; He, Q.; Gao, Y.; Shi, J.; Li, Y. Mesoporous Silica Nanoparticles Loading Doxorubicin Reverse Multidrug Resistance: Performance and Mechanism. *Nanoscale* **2011**, *3* (10), 4314–4322.
49. Liu, J.; Detrembleur, C.; De Pauw-Gillet, M.-C.; Mornet, S.; Vander Elst, L.; Laurent, S.; Jérôme, C.; Duguet, E. Heat-Triggered Drug Release Systems Based on Mesoporous Silica Nanoparticles Filled with a Maghemite Core and Phase-Change Molecules as Gatekeepers. *J. Mater. Chem. B* **2014**, *2* (1), 59–70.
50. Lee, J. E.; Lee, N.; Kim, H.; Kim, J.; Choi, S. H.; Kim, J. H.; Kim, T.; Song, I. C.; Park, S. P.; Moon, W. K.; Hyeon, T. Uniform Mesoporous Dye-Doped Silica Nanoparticles Decorated with Multiple Magnetite Nanocrystals for Simultaneous Enhanced Magnetic Resonance Imaging, Fluorescence Imaging, and Drug Delivery. *J. Am. Chem. Soc.* **2010**, *132* (2), 552–557.
51. Liong, M.; Lu, J.; Kovochich, M.; Xia, T.; Ruehm, S. G.; Nel, A. E.; Tamanoi, F.; Zink, J. I. Multifunctional Inorganic Nanoparticles for Imaging, Targeting, and Drug Delivery. *ACS Nano* **2008**, *2* (5), 889–896.
52. Lu, J.; Liong, M.; Zink, J. I.; Tamanoi, F. Mesoporous Silica Nanoparticles as a Delivery System for Hydrophobic Anticancer Drugs. *Small* **2007**, *3* (8), 1341–1346.
53. Knežević, N. Ž.; Slowing, I. I.; Lin, V. S.-Y. Tuning the Release of Anticancer Drugs from Magnetic Iron Oxide/mesoporous Silica Core/Shell Nanoparticles. *ChemPlusChem* **2012**, *77* (1), 48–55.
54. Gai, S.; Yang, P.; Li, C.; Wang, W.; Dai, Y.; Niu, N.; Lin, J. Synthesis of Magnetic, Up-Conversion Luminescent, and Mesoporous Core–Shell-Structured Nanocomposites as Drug Carriers. *Adv. Funct. Mater.* **2010**, *20* (7), 1166–1172.
55. Thomas, C. R.; Ferris, D. P.; Lee, J.-H.; Choi, E.; Cho, M. H.; Kim, E. S.; Stoddart, J. F.; Shin, J.-S.; Cheon, J.; Zink, J. I. Noninvasive Remote-Controlled Release of Drug Molecules in Vitro Using Magnetic Actuation of Mechanized Nanoparticles. *J. Am. Chem. Soc.* **2010**, *132* (31), 10623–10625.
56. Baeza, A.; Guisasola, E.; Ruiz-Hernández, E.; Vallet-Regí, M. Magnetically Triggered Multidrug Release by Hybrid Mesoporous Silica Nanoparticles. *Chem. Mater.* **2012**, *24* (3), 517–524.
57. Ruiz-Hernandez, E.; Baeza, A.; Vallet-Regí, M. Smart Drug Delivery through DNA/magnetic Nanoparticle Gates. *ACS Nano* **2011**, *5* (2), 1259–1266.
58. Saint-Cricq, P.; Deshayes, S.; Zink, J. I.; Kasko, A. M. Magnetic Field Activated Drug Delivery Using Thermodegradable Azo-Functionalised PEG-Coated Core–Shell Mesoporous Silica Nanoparticles. *Nanoscale* **2015**, *7* (31), 13168–13172.

59. Lee, J. E.; Lee, D. J.; Lee, N.; Kim, B. H.; Choi, S. H.; Hyeon, T. Multifunctional Mesoporous Silica Nanocomposite Nanoparticles for pH Controlled Drug Release and Dual Modal Imaging. *J. Mater. Chem.* **2011**, *21* (42), 16869–16872.
60. Hugounenq, P.; Levy, M.; Alloyeau, D.; Lartigue, L.; Dubois, E.; Cabuil, V.; Ricolleau, C.; Roux, S.; Wilhelm, C.; Gazeau, F.; et al. Iron Oxide Monocrystalline Nanoflowers for Highly Efficient Magnetic Hyperthermia. *J. Phys. Chem. C* **2012**, *116* (29), 15702–15712.
61. Walter, A.; Billotey, C.; Garofalo, A.; Ulhaq-Bouillet, C.; Lefèvre, C.; Taleb, J.; Laurent, S.; Vander Elst, L.; Muller, R. N.; Lartigue, L.; Gazeau, F.; Felder-Flesch, D.; Begin-Colin, S. Mastering the Shape and Composition of Dendronized Iron Oxide Nanoparticles to Tailor Magnetic Resonance Imaging and Hyperthermia. *Chem. Mater.* **2014**, *26* (18), 5252–5264.
62. Mertz, D.; Wu, H.; Wong, J. S.; Cui, J.; Tan, P.; Alles, R.; Caruso, F. Ultrathin, Bioresponsive and Drug-Functionalized Protein Capsules. *J. Mater. Chem.* **2012**, *22* (40), 21434–21442.
63. Mertz, D.; Cui, J.; Yan, Y.; Devlin, G.; Chaubaroux, C.; Dochter, A.; Alles, R.; Lavalle, P.; Voegel, J. C.; Blencowe, A.; et al. Protein Capsules Assembled via Isobutyramide Grafts: Sequential Growth, Biofunctionalization, and Cellular Uptake. *ACS Nano* **2012**, *6* (9), 7584–7594.
64. Mertz, D.; Tan, P.; Wang, Y.; Goh, T. K.; Blencowe, A.; Caruso, F. Bromoisobutyramide as an Intermolecular Surface Binder for the Preparation of Free-Standing Biopolymer Assemblies. *Adv. Mater.* **2011**, *23* (47), 5668–5673.
65. Mertz, D.; Affolter-Zbaraszczuk, C.; Barthès, J.; Cui, J.; Caruso, F.; Baumert, T. F.; Voegel, J.-C.; Ogier, J.; Meyer, F. Templated Assembly of Albumin-Based Nanoparticles for Simultaneous Gene Silencing and Magnetic Resonance Imaging. *Nanoscale* **2014**, *6* (20), 11676–11680.

Damien Mertz (E-mail: damien.mertz@ipcms.unistra.fr) has been a researcher at the Centre National de la Recherche Scientifique (CNRS) at IPCMS (Institut de Physique et Chimie des Matériaux de Strasbourg) since 2013. He earned his PhD degree in physical chemistry in 2008 at INSERM (French National Health Institute) and the University of Strasbourg. During his PhD, he worked on the design of mechano-bioresponsive polymer films releasing drugs or displaying a biocatalytic action under mechanical stretch. He was a postdoctoral fellow at the University of Melbourne, Australia, in the group of Professor F. Caruso, to develop silica-templated polymer and protein-based micro and nanocapsules for drug delivery and bioimaging applications (2009–2011). His current research activity at IPCMS, CNRS, University of Strasbourg, includes the design and characterization of functional and tunable mesoporous silica to form novel nanocomposites encapsulating drugs, magnetic NPs or carbon-based materials and development also of new templated-polymer and protein-based vesicles or nanocapsules. The main targeted applications of his activities are MRI, magnetic hyperthermia, phototherapy and drug release applications.

Sylvie Bégin-Colin obtained her PhD degree in material chemistry at University of Nancy (France) in 1992. She then entered the CNRS as researcher at the Laboratory of Science and Engineering of Material and of Metallurgy at the Mining Engineering School of Nancy and she has developed researches on physico-chemical modifications induced by ball-milling in oxides. In September 2003, she was appointed professor at the European Engineering School in Chemistry, Material and Polymer (ECPM), of the University of Strasbourg and is currently the director of ECPM. She has also been involved in the Research Commission of the University of Strasbourg since 2013.

She has developed a new research activity at the Institute of Physic and Chemistry of Materials (IPCMS) of Strasbourg on the synthesis, functionalization and organisation of oxide NPs for biomedical, energy and spintronic applications and is head of the team 'functionalized nanoparticles.' A great part of her research activity is devoted to the design of oxide NPs as these nano-objects are highly sought after for their applications in the biomedical field and are also considered as the building blocks of the future nanotechnological devices in fields of spintronic or energy. Most of these studies are made in collaboration with organic chemists, biologists and physicists. Sylvie Bégin has received AOARD, ANR, INCA, Labex, MICA, ARC and Alsace contre le Cancer grants and has participated and participates as partner to different European programs. Her work has been rewarded by the 'Jean-Rist' prize of the French Society for Metallurgy and Materials and Scientific Excellence Award. She holds 147 publications in peer-reviewed journals, two patents and 48 invited conferences.

Section III

In-Vitro Application of MNPs

10 Current Progress in Magnetic Separation-Aided Biomedical Diagnosis Technology

*Sim Siong Leong, Swee Pin Yeap, Siew Chun Low, Rohimah Mohamud and JitKang Lim**

CONTENTS

10.1 INTRODUCTION

In biomedical diagnosis, sample purification is an essential step in which the specific target (such as bacteria, viruses, parasites, cancer cells and molecules of biologically active compounds) is extracted from a complex biological sample containing various components. This sample purification step is crucial to allow more precise disease diagnosis in the subsequent process. In this regard, magnetic separation (MS) is one of the approaches commonly utilized in sample purification for biomedical diagnosis. This approach, which employs magnetophoretic force to execute the separation process, offers several advantages such as (i) high throughput,[1] (ii) low cost,[1] (iii) energy saving (involves minimal usage of additional energy when permanent magnet is used as the magnetic source)[2] and (iv) noninvasive to biological components.[3]

MS of targeted entity from a complex biological sample can be conducted in (a) a positive selection manner or (b) a depletion (or negative selection) manner (Figure 10.1). In the positive selection manner, the targeted entity is magnetically responsive and isolated from other nonmagnetic components upon the application of an external magnetic field.[4] Meanwhile, for MS performed in the depletion (or negative selection) manner, the magnetic responsive species are non-targeted species, which are separated by a magnetic field, leaving behind the desired targets (which are magnetically irresponsive and unseparated) in the suspension.[4]

For certain targets such as deoxygenated red blood cells (RBCs) and magnetotactic bacteria, their isolation from the surrounding medium can be achieved on the basis of their intrinsic magnetic properties through MS.[5] The isolation of RBCs from a whole blood sample by using magnetic field was first demonstrated in 1975 by Melville et al.[6] Later, in an effort to diagnose malaria, it was found that malaria-infected RBCs (i-RBCs) exhibit greater magnetic susceptibility than the healthy RBCs (h-RBCs) and thus can be fractionated out through the MS technique.[7] This separation is made possible

* Corresponding author.

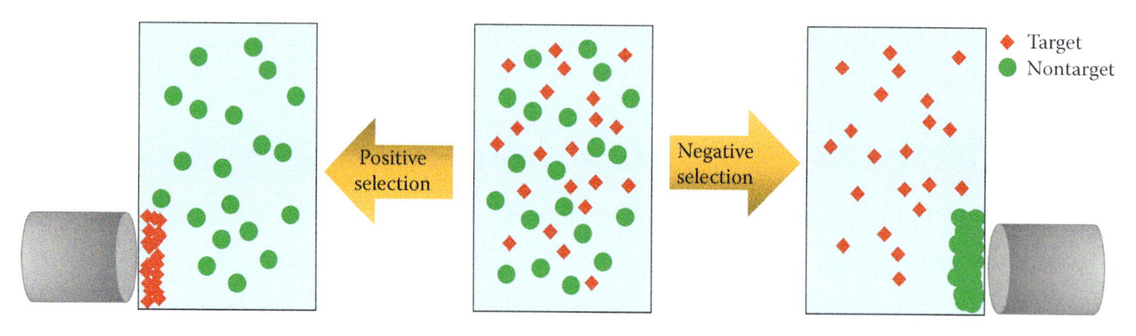

FIGURE 10.1 Illustration of MS in biomedical diagnosis. For positive selection, targeted compounds are magnetically responsive and collected by the magnet (left). On the contrary, nontargeted species are responsive to the externally applied magnetic field and are segregated from the solution, leaving behind the targets in the solution (right).

as malaria parasites tend to convert haemoglobin to haemozoin.[8] Haemozoin contains Fe^{3+} that exhibits greater paramagnetic property in comparison to Fe^{2+}, which is embraced within haemoglobin in h-RBCs. As a result, in comparison to h-RBCs, i-RBCs are preferentially isolated out of the solution by the externally applied magnetic field.[7]

However, not all the targets to be isolated exhibit intrinsic magnetic property and display magnetic responsiveness toward the magnetic field in their natural form. In such a case, targets to be isolated must be artificially labelled with magnetic particles so that they are magnetically responsive towards the externally applied magnetic field throughout MS. In this regard, magnetite (Fe_3O_4), maghemite (γ-Fe_2O_3) and commercially available Dynabeads are examples of the commonly employed magnetic particles for this purpose. The size of the magnetic particles ranges from nanometres (~10 nm) up to micrometres (1–10 μm).[9] As a prerequisite, magnetic particles should be surface modified with specific molecules that can bind with the targeted entity. For instance, in an attempt to capture *Salmonella enterica* (a food pathogen that affects the human gastrointestinal tract) by using the MS technique, Chen et al.[10] employed magnetic particles that had been conjugated with a specific antibody that enables the possible labelling of *S. enterica*. Due to the specific antibody–antigen recognition feature, antibody-conjugated magnetic particles were found to possess higher selectivity towards *S. enterica* over other species of bacteria (such as *Escherichia coli*, *Shigella* spp., *Staphylococcus aureus* and *Spirillum cholera*). Besides labelling the targeted entity with antibody-conjugated magnetic particles, magnetic particles (particularly those with size of nanometer) could be further internalized into the body of the targeted entity. For instance, Pamme and Wilhelm[11] reported that magnetic iron oxide nanoparticles of size <10 nm can be internalized into two types of targeted cells, i.e. mouse macrophages and human ovarian cancer (HeLa) cells. These cells, which had been loaded with magnetic nanoparticles, can then be magnetically segregated by the application of an external magnetic field. Correspondingly, the internalization of magnetic nanoparticles into targeted cells also has been found elsewhere.[12,13] Usually, magnetic nanoparticles penetrate into the targeted cell through endocytotic pathway.[12] As such, the magnetic responsiveness of the targeted cells during the MS process is attributed to the cells' endocytotic capacities to internalize the magnetic nanoparticles.[13] Apart from the specific recognition between antibody and antigen as described previously, magnetic labelling of targeted cells can also be performed by other methods such as direct interaction between targeted entity and surface stabilized magnetic particles in buffer solution, electrostatics interaction between targeted entity and polyelectrolyte stabilized magnetic particles, covalent immobilization etc.[14,15] Additionally, magnetofection can also be employed in magnetic labelling in which magnetic field is manipulated to concentrate magnetic particles with nucleic acid to the targeted cells such that the labelling of magnetic particles on the targeted cells can be performed in a more efficient manner.[15]

Up to now, numerous researches have shown the feasibility of MS technique in the detection of viruses, bacteria, parasites, cancer cells as well as biological molecules.[9,10,12,16–19] Nevertheless, in order to ensure successful implementation of MS technique in biomedical diagnostic applications, it is imperative to allocate substantial effort to design an efficient magnetic separator, which is capable of performing MS in a more effective manner. Thus, it is essential to gain an in-depth understanding on the fundamental mechanism of the MS process, which leads to further development of MS-aided biomedical diagnosis technology. On the basis of the magnitude of the externally applied magnetic field gradient, MS is categorized into high-gradient MS (HGMS) and low-gradient MS (LGMS). A detailed description of the mechanism and working principle of HGMS and LGMS is provided in the next section. Both HGMS and LGMS have been successfully employed to solve various separation and purification problems, ranging from laboratory to industrial scale. However, in contrast to HGMS, design rules for LGMS remain ill-defined.[20] Interestingly, we found that numerous published works relevant to MS-aided biomedical diagnosis were in fact operated under LGMS mode unconsciously. Concurrently, along with recent discoveries in the mechanistic study of magnetophoresis under low magnetic field gradient,[20–24] our understanding towards the underlying working principle of LGMS and driving factors that affect the separation rate has been significantly improved. In particular, those underlying working principles serve as the key factors that determine the effectiveness of MS-aided biomedical diagnosis and should be taken into consideration in the design and optimization of magnetic separator for diagnosis purposes. Hence, it is the main aim of this chapter to discuss the influencing role of several important parameters (such as magnetic particle size, magnetic particle concentration and the spatial distribution of magnetic field [or field gradient]) in the development of MS technique for biomedical diagnostic applications.

10.2 WORKING PRINCIPLES OF MS

10.2.1 High-Gradient MS

As magnetic particle solution is channelled through a separation column (or separation chamber for batch process), magnetic particles will be magnetized by an external magnetic field and acquire net magnetic dipole moment that aligns along the direction of the magnetic field.[25] The response of magnetic dipole moment (possessed by the magnetic particles) towards the external magnetic field gradient will impose a magnetophoretic force $\overrightarrow{F_m}$ onto the magnetic particles. The magnetophoretic force experienced by spherical magnetic particles under an external magnetic field is depicted as follows:[24]

$$\overrightarrow{F_m} = \frac{4}{3}\pi r^3(\vec{M}\cdot\nabla)\vec{B}, \tag{10.1}$$

where $\vec{M}$ is the volumetric magnetization of the magnetic particle, r is the radius of particles, $\vec{B}$ is magnetic field strength and ∇ is the vector differential operator. Under an externally applied inhomogeneous magnetic field, magnetic particles are driven to the region with highest magnetic field strength and are separated from the solution.

Yet, in order to impose deterministic motion on magnetic particles and perform a successful MS, such magnetophoretic force should be strong enough to overcome two opposing forces, namely viscous drag force and Brownian motion.[26] Viscous drag originates from the friction exerted by a viscous fluid on a particle, which performs a relative motion with respect to the given fluid. The viscous drag force $\overrightarrow{F_d}$ experienced by a spherical particle that moves with velocity $\vec{v}$ relative to the surrounding fluid is given by Stokes' law as follows:[27]

$$\overrightarrow{F_d} = -6\pi\eta r_h\vec{v}, \tag{10.2}$$

where η is the viscosity of the surrounding fluid and r_h is the hydrodynamic radius of the moving particle. The negative sign in the equation indicates that the viscous drag force is applied in the opposite direction to the migration velocity of the particle. On the other hand, Brownian motion is the random movement performed by particles in the suspension and it is initiated by the thermal energy possessed by each particle.[28] The intensity of Brownian motion can be reflected from the magnitude of diffusion coefficient D of the given particle suspended in fluid, as given by the Stokes-Einstein equation as demonstrated in the following:[28]

$$D = \frac{kT}{6\pi\eta r_h}, \tag{10.3}$$

where k is Boltzmann constant and T is the absolute temperature of the particle system. The higher diffusion coefficient indicates that the Brownian motion encountered by the particles in the suspension is more remarkable.

According to Equations 10.1 and 10.2, it can be observed that magnetophoretic force and viscous drag force are directly proportional to the r^3 and r of the particle, respectively. Thus, in comparison to viscous drag force, the magnetophoretic force experienced by magnetic particle decays much faster with decreasing particle size. The magnitude of viscous drag force encountered by magnetic particle with extremely tiny dimension (especially magnetic nanoparticle) is almost comparable to the magnetophoretic force exerted on it, rendering the magnetophoretic motion of the particle occurring in an exceptionally slow pace. Furthermore, Brownian motion is also considerably more significant for smaller-sized magnetic particles (as shown in Equation 10.3), and this has caused difficulty in controlling the magnetophoretic motion of the given particles. Consequently, the precise control of nano-sized particle motion by magnetophoretic force is highly challenging due to the interference of viscous drag force and Brownian motion. In this regard, it is essential to introduce an intense magnetophoretic force on the magnetic particles for successful implementation of a separation process.[26] In line with Equation 10.1, the magnitude of magnetophoretic force varies directly with magnetic field gradient. In accordance to this situation, high magnetic field gradient must be introduced in order to achieve successful separation of magnetic particles within a reasonable timescale.

However, such an enormous magnetic field cannot be generated by solely permanent magnet or electromagnet. Specifically, for the magnetic field gradient generated by neodymium ferrum boron (NdFeB) magnet, only a small region adjacent to the edges of the magnet experiences high magnetic field gradient ($\nabla B > 100$ T/m), while most domains within the separator suffer low magnetic field strength and gradient (Figure 10.2a and c). This phenomenon is attributed to the extremely fast decay of magnetic field gradient with the displacement from the magnetic source.[20] It should be emphasized that NdFeB magnet is one of the most powerful rare-earth magnets, which is commercially available, reasonably priced and characterized by relatively high magnetic energy per unit volume (~382 kJ/m^3) in comparison to other types of permanent magnets such as SmCo (~254 kJ/m^3) and Alnico (~59 kJ/m^3) magnets.[29] In a more conventional way, magnetic field gradient throughout the separation chamber can be enhanced by inserting intertwined wires (or any material with relatively high magnetic permeability) inside a column.[30] Such wires have a higher affinity to concentrate a magnetic field line originating from the magnetic source

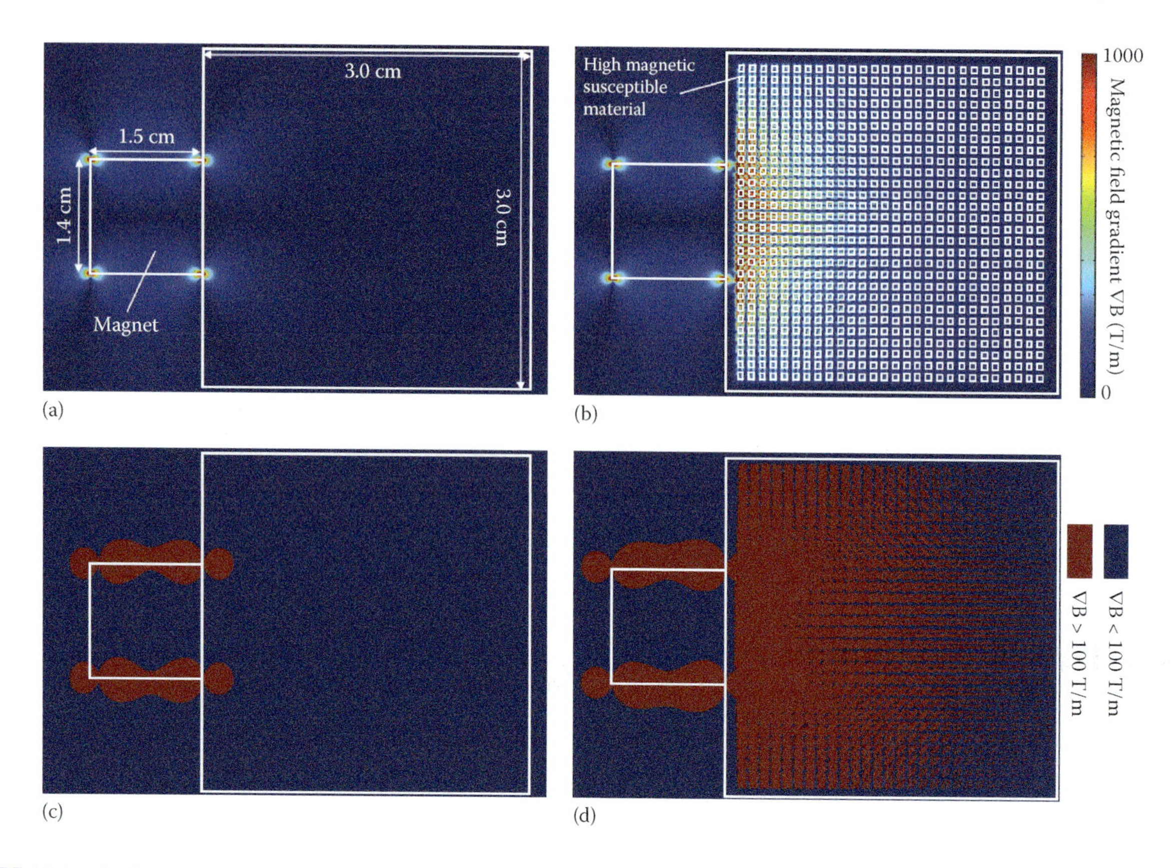

FIGURE 10.2 Surface plot of magnetic field gradient induced by a cylindrical NdFeB magnet (diameter and height are given by 1.4 cm and 1.5 cm, respectively, which have a remanent magnetization of 1.45 T) in a separation chamber which is (a) vacant and (b) filled with relatively high magnetic susceptibility materials with $\chi = 1000$. (c and d) The contour plot of magnetic field gradient for panels a and b, respectively. The magnetic field calculation and generation of surface plot were performed by AC/DC module of COMSOL Multiphysics.

and hence significantly distort the magnetic field within the column.[31] Thereby, an inhomogeneous magnetic field with an extremely high magnetic field gradient is induced in the vicinity of the magnetically susceptible wires (Figure 10.2b and d).[32] Consequently, magnetic particles are subjected to an exceptionally large magnetophoretic attractive force as they approach the magnetizable wires, which in turn is magnetically captured on the surface of the wires. In such a way, magnetic particles are segregated out of the nonmagnetic medium and MS is successfully achieved.[33] Due to the utilization of high magnetic field gradient (∇B exceeds 100 T/m in most regions in the separation column) in this separation scheme, the separation of magnetic particles by using this mechanism is denoted as high-gradient MS.

Nevertheless, the use of magnetic susceptible wires to create high magnetic field gradient in an HGMS column has some drawbacks. First and foremost, the randomly entangled wires have further complicated the magnetic field distribution and flow behaviour of magnetic particle solutions within the column. This scenario has in turn impeded the mathematical analysis as well as theoretical modelling of the HGMS process.[30] Secondly, some of the magnetic particles might be permanently retained on the magnetic susceptible wires after each separation cycle (after the removal of the external magnetic field). The particle retention is mostly due to the fact that magnetic wires might be ferromagnetic in nature and have retained some degree of magnetic moment after the withdrawal of the magnetic field, which induces their magnetization.[29] This situation might render permanent retention of magnetic particles on the wires within the HGMS column, which leads to separation efficiency reduction in the subsequent operation.[34]

10.2.2 Low-Gradient MS

Since most of the disadvantages encountered by an HGMS column arise from its magnetic susceptible wires, it is appealing to remove them from the MS column (or chamber) such that MS can be conducted in a less intricate environment. Removal of the magnetic susceptible wires reduces the nonuniformity of the magnetic field and hence results in a weak magnetic field gradient throughout the MS column (see Figure 10.2 for comparison). Typically, magnetic field gradient suffers substantial decays with increasing distance from the magnet such that the magnitude of magnetic field gradient across the whole separation column/chamber (except the small region adjacent to the edges of magnetic source) is less than 100 T/m (Figure 10.2a and c).[20] Since the MS is carried out under a low magnetic field gradient environment, this MS technique is denoted as low-gradient MS.[24] Owing to the linear relationship between magnetic field gradient and magnetophoretic force, under the LGMS mode, the magnetophoretic attractive force exerted on the magnetic particles is insignificant due to the low magnetic field gradient. As a consequence, LGMS might not guarantee deterministic motion on the magnetic particles along the magnetophoresis migration pathway and gives rise to exceptionally long separation time.

Surprisingly, Yavuz et al.[35] demonstrated the feasibility to capture 12 nm magnetic nanoparticles by a permanent magnet within reasonable timescale. In addition, Cuevas et al. reported that the MS of 200 nm magnetic particles (Estapor M1-020/50)* can be accomplished within 50 seconds when it was subjected to low-field-gradient magnetophoresis (under concentration of 10 g/L).[24] This observed separation is much shorter in comparison to the time predicted by using Equations 10.1 and 10.3, which predicts a separation time of 3337 s.† In addition, MS rate was found to be strongly dependent on the concentration of magnetic particle solution employed, and this phenomenon is contradictory to the prediction of classical magnetophoresis model (see Section 10.4.2.1 for more elaboration). These observations have led us to believe that the accelerated rate of LGMS is triggered by the self-aggregation of magnetic particles under an external magnetic field, which will be described in detail in the following paragraphs.[36]

Owing to the intrinsic magnetic property of magnetic particles, they will be magnetized and acquire some degree of magnetic dipole moment upon exposure to an external magnetic field. In order to minimize the total interacting magnetic energy of the system, magnetic particles (which are acting as 'small magnets') will align themselves spontaneously such that their poles are pointing toward the direction of the magnetic field. When magnetic particles approach one another, they will experience an attractive magnetic dipole–dipole interaction among each other and finally stick together.[37] The particle clusters formed consist of two or more magnetic particles that are bound together by magnetic dipole–dipole force. De Las Cuevas et al.[24] reported the formation of elongated (linear) aggregates which move collectively in the direction of magnetic field gradient as they are exposed to an external magnetic field (Figure 10.3a). Concurrently, Schaller et al.[25] also observed the magnetic-field-induced aggregation of magnetic particles in their attempt to investigate the magnetic particles motion under the influence of a magnetic field (Figure 10.3b). Apart from that, Andreu et al.[38] observed that magnetic particles under a strong magnetic field will assemble into a chain in their simulation according to on-the-fly coarse-grain (CG) model (Figure 10.3c). More interestingly, Erb et al.[39] employed this phenomenon (self-assembly of magnetic particle under magnetic field) to create a complex superstructure from magnetic particles of different sizes (Figure 10.3d).

The aggregation of magnetic particles can be further classified into two categories according to the reversibility of the process, namely reversible aggregation and irreversible aggregation.[40] Reversible aggregation is the aggregation process in which a particle aggregate will disintegrate into individual particles (regains its initial form) upon the removal of the magnetic field, which induces aggregation. On the contrary, irreversible aggregation is the formation of particle aggregate, which is unable to break down into individual particles spontaneously after being magnetized by a magnetic field. The reversibility of particle aggregation can be explained by extended Derjaguin-Landau-Verwey-Overbeek (XDLVO) analysis, which expresses the total interaction energy (electrostatic, van der Waals and magnetic dipole–dipole interaction energies)[41] as the function of displacement between two individual magnetic particles, as demonstrated in Figure 10.4.[29] There are two energy minima appearing in the free energy profile of magnetic particles under the presence of an external magnetic field. The occurrence of primary minimum is a result of the dominance of the strong van der Waals attraction over the electrostatic repulsive force when two magnetic particles are sufficiently close to one another. Magnetic particles must overcome an energy barrier originating from an electrostatic repulsive force in order to come close to each other and get into the primary minimum. Being trapped in this energy well, a magnetic particle pair is unable to disintegrate into individual particles spontaneously. Therefore, this aggregation process is irreversible. In order to regain its initial form (dispersed form), additional energy must be supplied to the irreversible aggregates such that individual particles acquire sufficient energy to overcome the energy barrier from van der Waals attraction. Secondary minimum is created by the magnetic dipole–dipole attractive force between a pair of magnetized particles under the influence of an external magnetic field. The particles will be held together under the presence of magnetic field due to the existence of a secondary minimum. However, upon the removal of the magnetic field, the energy well of the secondary minimum diminishes and those aggregated particles will disintegrate into their individual particles

* Estapor M1-020/50 particles were supplied by Merck Chimie SAS (France) and have diameter of 0.2 μm with 55% mass of ferrite distributed uniformly in a polystyrene matrix. The saturation volumetric magnetization of these particles is about 6×10^4 A/m.

† In order to calculate the separation time of Estapor M1-020/50 magnetic particle (diameter of 200 nm), the following assumptions were made: (i) The flow of magnetic particle is under low Reynold number environment such that the inertial term (acceleration) can be negligible, (ii) the magnetic particle is fully saturated (or fully magnetized) throughout its migration, (iii) a magnetic field gradient ∇B of 30 T/m is assumed and its magnitude remains constant all over the SEPMAG separator used for separation, (iv) the separation time is defined as the time required for a magnetic particle to travel from the centre of the cylindrical cavity to the vessel wall (total distance travelled is 1.5 cm) and (v) essential quantities for the calculation along with their magnitude are particle (or hydrodynamic) radius, r/r_h (100 nm), volumetric magnetization of particle at saturation, M (6×10^4 A/m) as well as viscosity of surrounding fluid, η (0.00089 Pa s). By modifying Equations 10.1 and 10.2,

$$\frac{4}{3}\pi r^3 |\vec{M}||\nabla B| = 6\pi\eta r_h |\vec{v}|.$$

After inserting values stated previously into the equation, magnetophoretic velocity $|\vec{v}|$ is estimated as 4.49 × 10−6 m/s. Therefore, the time required for the particle to travel distance as long as 1.5 cm is

$$\frac{0.015 \text{ m}}{4.49 \times 10^{-6} \text{ m/s}} = 3337 \text{ s}.$$

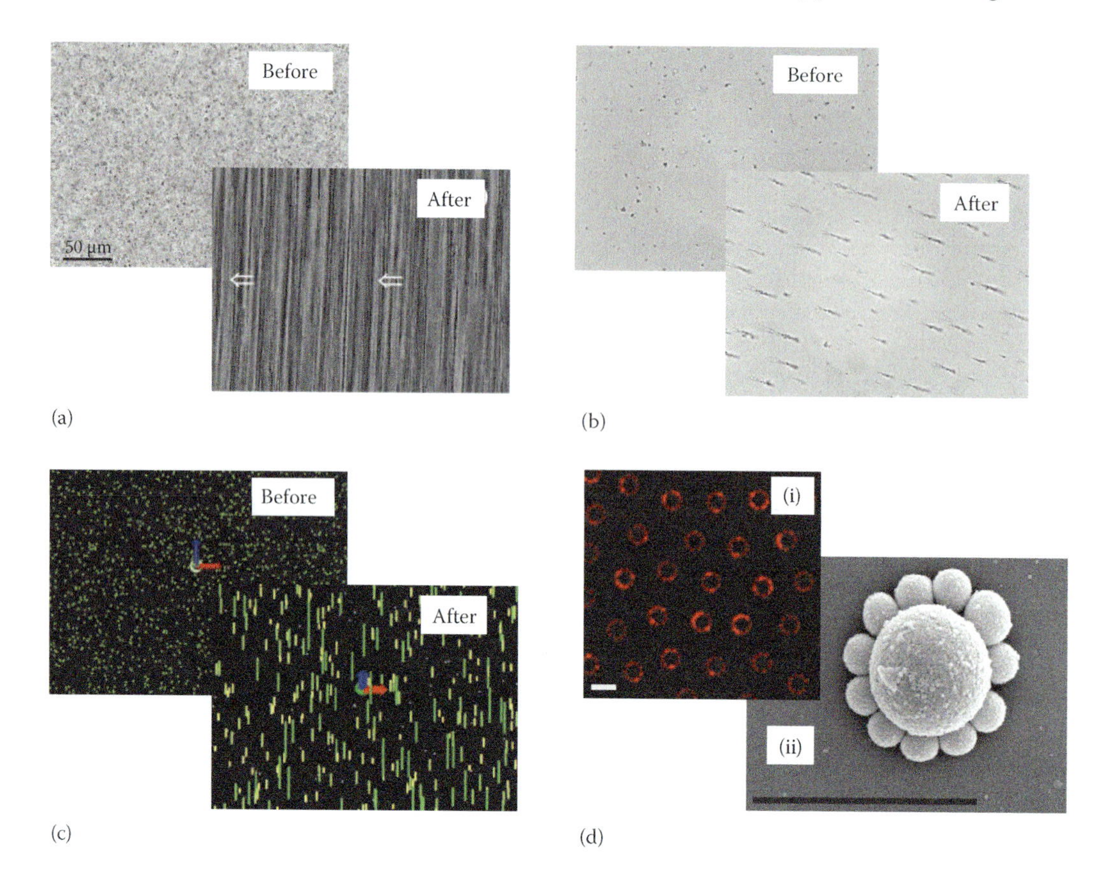

FIGURE 10.3 (a) Optical micrograph showing a magnetic nanoparticle solution composed of 1 g/L Estapor M1-030/40 (superparamagnetic particle with diameter of 0.41 μm) before and after being exposed to magnetic field for 120 s. The white arrows indicate the direction of magnetic particle migration under magnetic field.[24] (Reprinted with permission from De Las Cuevas, G., Faraudo, J., and Camacho, J. 2008. Low-gradient magnetophoresis through field-induced reversible aggregation. *The Journal of Physical Chemistry C* 112:945–950. Copyright 2008 American Chemical Society.) (b) Optical microscopy images of fluidMAG-D nanoparticles from Chemicell GmbH (magnetic particles with diameter of 425 nm) in deionized water before and after exposure to an external magnetic field for few seconds.[25] (Reprinted with permission from Schaller, V., Kräling, U., Rusu, C., Petersson, K., Wipenmyr, J., Krozer, A., Wahnström, G., Sanz-Velasco, A., Enoksson, P., and Johansson, C. 2008. Motion of nanometer sized magnetic particles in a magnetic field gradient. *Journal of Applied Physics* 104:093918. Copyright 2008, American Institute of Physics.) (c) Snapshots of the images for magnetic particles (green dots) before and after exposure to strong magnetic field for 0.28 s. The snapshot was obtained from CG simulation with magnetic coupling parameter $\Gamma = 40$.[38] (Reprinted with permission from Andreu, J. S., Calero, C., Camacho, J., and Faraudo, J., *Physical Review E*, 85, 036709, 2012. Copyright 2012 by the American Physical Society.) (d) (i) Self-assembly of 1.0 μm nonmagnetic particles (red dots) and 2.7 μm paramagnetic particles (the invisible substance in the centre of the red 'ring') into 'Saturn-ring' particles under an externally applied magnetic field. (ii) SEM image of the 'Saturn-ring' superstructure.[39] (Reprinted by permission from Macmillan Publishers Ltd. *Nature*. Erb, R. M., Son, H. S., Samanta, B., Rotello, V. M., and Yellen, B. B. 2009. Magnetic assembly of colloidal superstructures with multipole symmetry. *Nature* 457:999–1002. Copyright 2009.)

without any external driving force. Thus, this phenomenon is known as reversible aggregation.

Owing to its larger size and magnetic volume, particle aggregate experiences a stronger magnetophoretic force, which is able to overcome viscous drag force and Brownian motion. Thus, magnetic particles move collectively towards the magnetic source with much higher magnetophoretic velocity and are separated within a shorter timescale. The process, in which magnetic particles move collectively in a particle aggregate under the influence of magnetic field, is denoted as cooperative magnetophoresis. By the assistance of particle aggregation, MS can be conducted in an accelerated manner. In addition to utilization of larger magnetic particle, particle aggregation serves as an alternative mechanism to enhance the MS rate without the need to sacrifice the large exposure area of the tiny-sized magnetic particles (problem encountered when the size of magnetic particles is enlarged to increase MS rate). In short, LGMS is also a feasible separation scheme to trigger fast magnetophoretic collection.

However, there exist certain criteria for the onset of magnetic particle aggregation under an external magnetic field. In this regard, the concept of magnetic Bjerrum length was introduced to predict the requirements to initiate the aggregation of magnetic particles subjected to magnetophoresis.[24] Magnetic Bjerrum length λ_B is defined as the displacement between two magnetized particles (possess magnetic dipole moments that are pointing to the same direction) in which the

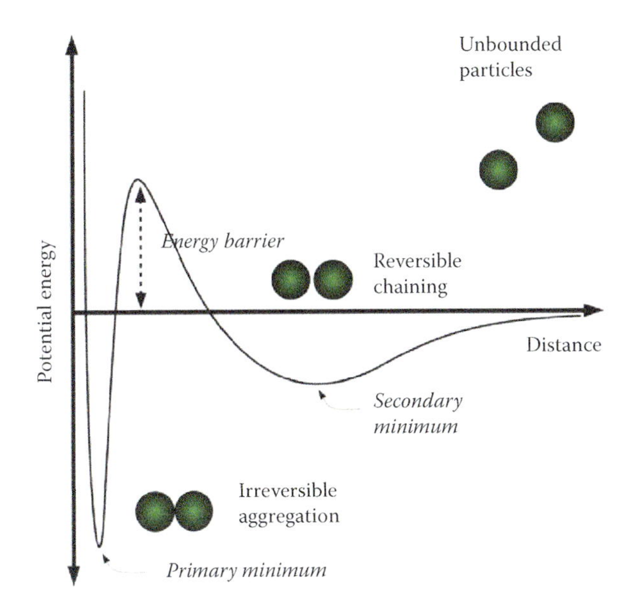

FIGURE 10.4 The free energy profile of two electrostatically stabilized magnetic particles, which is calculated by XDLVO theory with respect to interparticle distance.[40] (Faraudo, J., Andreu, J. S., and Camacho, J. 2013. Understanding diluted dispersions of superparamagnetic particles under strong magnetic fields: a review of concepts, theory and simulations. *Soft Matter* 9:6654–6664. Reproduced by permission of The Royal Society of Chemistry.)

attractive magnetic energy is exactly the same with their thermal energy kT:[24]

$$\lambda_B = \left[\frac{8\pi\mu_0 M^2}{9kT}\right]^{\frac{1}{3}} r^2, \quad (10.4)$$

where μ_0 is the permeability of free space and M is the volumetric magnetization of particles. Aggregation is possible for particle system with $\lambda_B > 2r$. On the contrary, the formation of particle aggregate is beyond the bound of possibilities for the particle system with $\lambda_B < 2r$, in which thermal energy is still overwhelming the magnetic attractive force even two particles are in physical contact with each other (separation distance = $2r$). As the feasibility of LGMS technique is heavily dependent on the emergence of particle aggregation, the magnetic Bjerrum length calculation is very useful in revealing the possibility to execute rapid LGMS for a particular system.

According to the thermodynamic self-assembly theory, Andreu et al.[42] introduced aggregation parameter N^* to characterize the onset of particle aggregation induced by external magnetic field. Interestingly, aggregation parameter N^* is dependent on only two dimensionless numbers, namely volumetric fraction of magnetic particle in the suspension ϕ_0 and magnetic coupling parameter Γ, as given in the following equation:[42]

$$N^* = \sqrt{\phi_0 e^{(\Gamma-1)}}. \quad (10.5)$$

Here, magnetic coupling parameter Γ is the ratio of the maximum magnetic attraction energy between two magnetic particles (it happens when the both fully magnetized particles are in close contact and possess magnetic dipole moments that are pointing towards the same direction) to the thermal energy in equilibrium, as demonstrated in the following equation:

$$\Gamma = \frac{\mu_0 m_s^2}{2\pi d^3 kT}, \quad (10.6)$$

where m_s is the total magnetic dipole moment possessed by each magnetic particle under saturation and d is the diameter of the given magnetic particle. In accordance to this analysis, aggregation of magnetic particles is possible if and only if N^* is greater than unity. Nevertheless, the magnitude of aggregation parameter N^* also reflects the average number of magnetic particles within one particle aggregate in the aggregation dominant regime. Aggregation parameter N^* calculation has provided a quick check on the feasibility to perform LGMS on a particular magnetic particles system.

10.3 APPLICATION OF HGMS AND LGMS IN BIOMEDICAL DIAGNOSIS

Table 10.1 summarizes several selected works that employed HGMS and LGMS in biomedical diagnosis. Magnetic separator was designed to generate high magnetic field gradient in order to capture the targeted cells. For instance, in an attempt to achieve two-dimensional positioning of single bacterial cells, Pivetal et al.[43] fabricated micromagnet arrays with a magnetic field gradient ranging from 2.89×10^4 T/m to 2.5×10^5 T/m. Such a huge local magnetic field gradient is able to trap/immobilize the targeted bacteria cells, which have been prelabelled with magnetic particles. Zimmerman et al.[44] designed a malaria magnetic deposition microscopy (MDM) device to concentrate erythrocytes that have been infected by human malaria species. Their magnetic separator was characterized by magnetic field gradient as high as 804 T/m, which allows the parasite-infected erythrocytes to be magnetically deposited when the cell suspension was channelled through the separator. Meanwhile, Xia et al.[16] developed a microfluidic device equipped with high-gradient magnetic concentrator (HGMC) of comb shape (NiFe microcomb). The HGMC concentrates the magnetic field gradient locally when it was magnetized by NdFeB magnet. Under the influence of the NiFe microcomb structures, the highest magnetic field gradient generated was 290 T/m, as compared to 20 T/m under the absence of NiFe microcomb. It was found that the separator was effective in the separation of *E. coli* cells that have been labelled with magnetic particles of size 130 nm.[16]

Nevertheless, it should be noted that a magnetic field gradient equal to or lower than 100 T/m is also feasible in MS-aided biomedical diagnosis. For example, SuperMag separator (Ocean NanoTech) with magnetic field gradient 100 T/m was successfully employed to isolate magnetic particle-labelled cancer cells (human breast cancer cell line SK-BR3).[12] By using this MS

TABLE 10.1
Selected Reported Works on Magnetic Separator Design and Their Applications in MS-Aided Biomedical Diagnosis

	Magnetic Separator Design				
Ref.	Description on the Magnetic Separator	Batch/ Continuous	Magnetic Field Strength	Magnetic Field Gradient	Application in MS-Aided Biomedical Diagnosis
43	Micromagnet 5-μm-thick hard NdFeB films were sputtered on Si wafers covered by a 100 nm Ta buffer layer. To avoid oxidation of the magnetic thin film, a 100-nm Ta protecting overlayer was placed. Then, micromagnet arrays were formed by using thermomagnetic patterning technique. KrF (248 nm) pulsed excimer laser was irradiated, through a TEM grid, onto the magnetized hard magnetic films. Simultaneously, an external magnetic field was applied in the direction opposite to the initial magnetization direction. This procedure created structures containing arrays of oppositely magnetized micromagnets.	Batch	~0.12 T–~0.29 T	~2.89 × 10^4 T/m–2.5 × 10^5 T/m	Micro-patterning of *E. coli* cells that have been magnetically labelled with magnetic beads.
44	Magnetic field was formed by assembly of permanent magnets with a 1.27 mm interpolar gap. The permanent magnet assembly consists of ferrite magnets (Dexter Magnetic Technologies, Elk Grove Village, Illinois) and a pair of 1016 low-carbon steel pole pieces.	Continuous	1.426 T	804 T/m	Magnetically concentrating parasite-infected erythrocytes for diagnosis of malaria infection.
16	Microfabricated high gradient magnetic field concentrator (HGMC) A NiFe layer was deposited adjacent to a microfluidic channel to create an on-chip HGMC with defined comb-shape. The HGMC locally concentrates the field gradients when magnetized by a neodymium permanent magnet.	Continuous	0.048 T	290 T/m	Separation of *E. coli* cells, which have been labelled with magnetic particles, from a flowing biological fluid.
12	SuperMag separator (Ocean NanoTech)	Batch	No data provided	100 T/m	Magnetic isolation of tumour cells from fresh whole blood. The tumour cells were bound with antibody-conjugated Fe_3O_4 magnetic nanoparticles.
13	NdFeB permanent magnet (Calamit): 5 cm long, 1 cm wide and 4 mm thick. The permanent magnet was positioned on a microchip's surface.	Continuous	0.15 T–0.26 T (from middle of the separation chamber to surface of the magnet)	30 T/m–80 T/m	Sorting of monocytes and macrophages based on the amount of magnetic nanoparticles (γ-Fe_2O_3 of size 8.7 nm) that internalise into the cells.
11	NdFeB magnet (Magnetsales, Swindon, UK) with a thickness of 10 mm and a diameter of 20 mm.	Continuous	0.4 T	<50 T/m	Sorting of mouse macrophages and human ovarian cancer cells (HeLa cells) that have been loaded with magnetic nanoparticles.

approach, the enrichment factor of cancer cell over normal cell was as high as 1:10,000,000. On the other hand, Robert et al.[13] demonstrated that a low magnetic field gradient ranging from 30 to 80 T/m is feasible for the separation of magnetic nanoparticle-internalized monocytes and macrophages. Similarly, in the continuous sorting of macrophages and human ovarian cancer cells (HeLa cells), the cells that were loaded with different amounts of magnetic nanoparticles were channelled through a separation chamber in which the magnetic field was applied in the direction perpendicular to the flow direction.[11]

Even though the magnetic field gradient developed within the separation chamber is less than 50 T/m, the sorting of targeted cells according to the amount of magnetic nanoparticles that were loaded into the cells still could be accomplished.

Nevertheless, in order to optimize MS-aided biomedical diagnosis technology, thorough understanding on the governing parameters that control the performance of MS is inevitable. This topic will be discussed comprehensively in the subsequent sections.

10.4 ROLE OF DIFFERENT CONTROL PARAMETERS IN MS-AIDED BIOMEDICAL DIAGNOSIS

10.4.1 Particle Size

As discussed earlier, the magnetophoretic force F_m and magnetophoretic velocity v experienced by a spherical magnetic particle are directly proportional to r^3 and r^2, respectively. Therefore, magnetic particles of larger size can be magnetophoretically separated at a faster rate in comparison to their smaller counterparts. Accordingly, the size of the magnetic particles plays a crucial role in determining the MS duration and, thus, the overall time consumed to perform biomedical diagnosis.

10.4.1.1 Separation Rate and Selectivity

In consideration of their faster MS rate, larger-sized magnetic particles are recommended for the intention of fast screening during biomedical diagnosis. In this regard, Lin et al.[45] have compared the time taken to magnetically isolate immunomagnetic particles (which are employed in the separation of *E. coli* O157:H7) of difference sizes. They found that the separation efficiency of 30 nm immunomagnetic particle approached 95% only after conducting MS for 60 min. On the contrary, when immunomagnetic particles of size 180 nm were employed, 100% of separation efficiency could be accomplished by performing the MS merely for 1 min. By including the time spent on the immunoreaction (45 min), the total duration for the biomedical diagnosis procedure was approximately 2 h and 1 h when immunomagnetic particles of size 30 nm and 180 nm were used, respectively.

Apart from determining the MS duration, particle size also plays a defining role in influencing the signal readout for target detection during biomedical diagnosis. For instance, Chen et al.[10] reported the critical role of magnetic particle size in dictating the functionality of the MS-based magnetic relaxation switching (MS-MRS) sensor proposed in their recent publication. Here, magnetic particles of two different sizes, i.e. 250 nm (MP_{250}*) and 30 nm (MP_{30}†), were used as magnetic carriers. Upon exposure to an external magnetic field (0.01 T), MP_{250} can be rapidly separated within 1 min due to their high magnetic response (hence experiencing a higher magnetophoretic force, which leads to higher magnetophoretic velocity) toward the magnetic field. On the other hand, owing to their relatively low magnetophoretic velocity under magnetic field, MP_{30} still remains suspended even after 1 h of full exposure to MS (Figure 10.5a). The working principle of MS-MRS sensor is delineated in Figure 10.5b. Briefly, antibody-conjugated MP_{250} and MP_{30} were added into a sample containing the targeted entity, which is *S. enterica* in this case. Owing to the recognition feature between an antigen from targeted entity and an antibody from a functionalized magnetic particle, the targeted entity will bind with magnetic particles present in the solution and embedded in the MP_{30}–target–MP_{250} complex. Once a magnetic field (generated by permanent magnet) is applied across the solution, the larger MP_{250} and MP_{30}–target–MP_{250} complex experiences stronger magnetophoretic force and is segregated from the solution within a shorter timescale. On the contrary, MP_{30} exhibits a much weaker response toward the external magnetic field such that majority of the MP_{30} still remains suspended in the solution after exposure to the external magnetic field for a few minutes. The suspended MP_{30} can significantly alter the transverse relaxation time of the water molecules (T_2 value) of the solution in such a way that a higher concentration of suspended MP_{30} gives rise to lower T_2 value (further details of T_2 value are elaborated in Section 10.4.2.2). The change in T_2 value was measured by a relaxometer and was utilized as the readout to correlate to the concentration of targeted entity in the initial sample (a higher concentration of targeted entity gives rise to larger change in T_2 value). This novel sensing approach is promising and has highlighted the interplay between magnetic particle size and target detection in MS-aided biomedical diagnosis.

10.4.1.2 Magnetic Loading

The size of the magnetic particle also exerts a significant impact on the amount of magnetic particles that can be attached onto a targeted entity. For instance, due to their lower steric exclusion, smaller-sized magnetic particles can be packed onto the surface of a particular cell in much higher quantity.[12] As shown in Figure 10.6a, there are large amounts of 3-mercaptophenylboronic acid/1-decanethiol modified nanometer-sized magnetic particles immobilized onto an *E. coli* cell.[46] On the other hand, in the work reported by Sun et al.,[47] the number of micrometer-sized magnetic particles (Dynabeads M-280 Streptavidin‡ coated with biotinylated antibodies against *E. coli* O157:H7) that can bind onto one unit of *E. coli* O157:H7 reduced significantly (about two magnetic particles per unit *E. coli* O157:H7) in comparison to that of smaller particles. Pamme and Wilhelm[11] anticipated that there would be more than a million of 9-nm magnetic particles that can be attached onto the surface of a 12-μm targeted cell. However, when the magnetic particles were further enlarged to 400 nm and 2.8 μm, the number of magnetic particles attached to one unit of target

* Supplied by Micromod Partikeltechnologie GmbH (Germany), with diameter of 250 nm.

† Purchased from OceanNano Tech (USA), with diameter of 30 nm.

‡ Supplied by Dynal (Lake Success, New York), with a diameter of 2.8 μm. The coefficient of variation (CV) of this particle system is less than 3%. The particles' ferrite content is 12%.

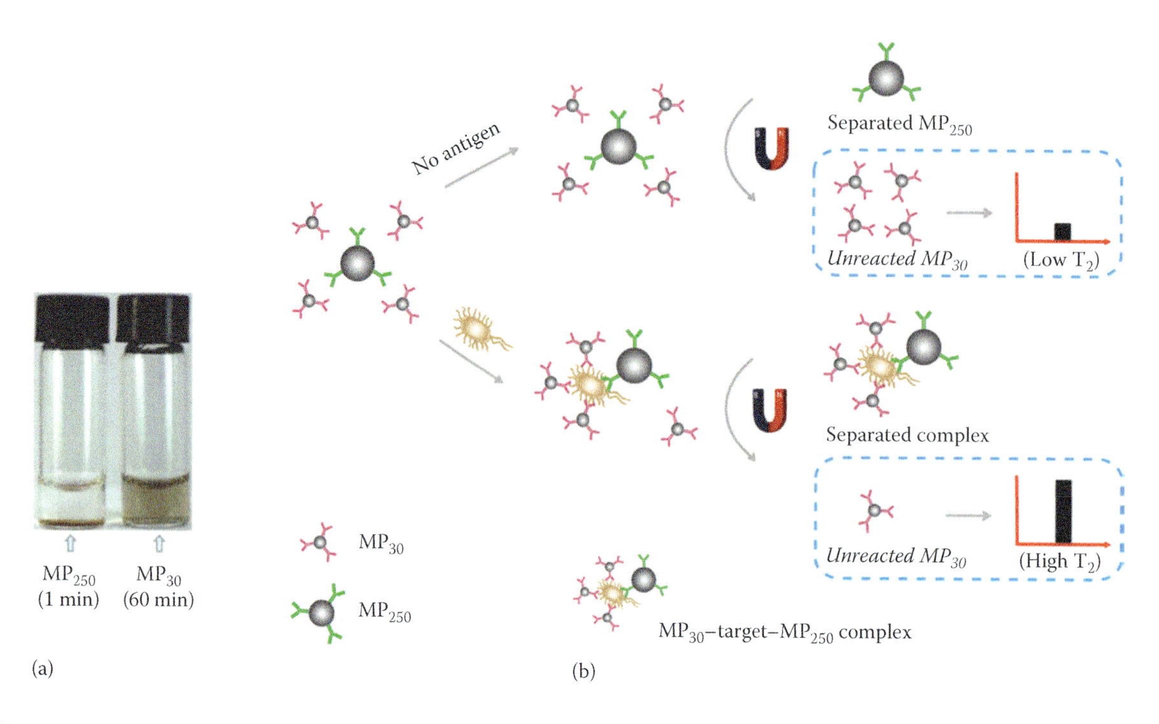

FIGURE 10.5 (a) MS of MP_{250} and MP_{30} under 0.01 T magnetic field. (b) Sensing mechanism of MS-MRS sensor.[10] (Reprinted with permission from Chen, Y., Xianyu, Y., Wang, Y., Zhang, X., Cha, R., Sun, J., and Jiang, X. 2015. One-step detection of pathogens and viruses: combining magnetic relaxation switching and magnetic separation. *ACS Nano* 9:3184–3191. Copyright 2015 American Chemical Society.)

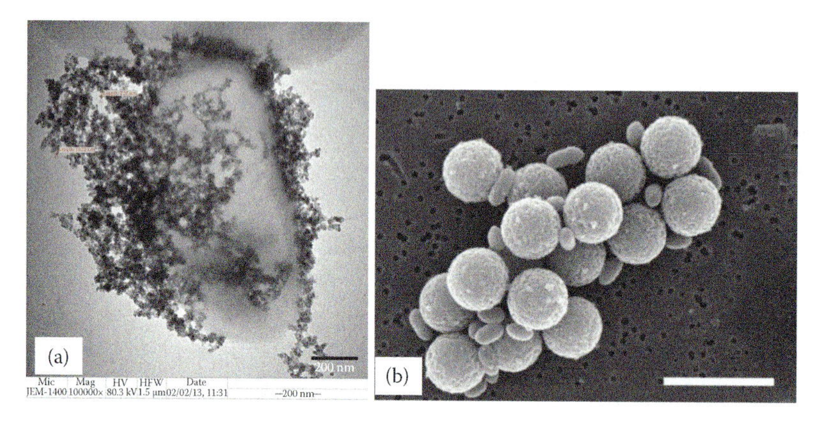

FIGURE 10.6 Electron micrographs show (a) *E. coli* bound with a large number of magnetic nanosized particles (scale bar = 200 nm)[46], and (b) *E. coli* O157:H7 captured by several magnetic microsized particles (Dynabeads M-280).[47] These images clearly illustrate the role of magnetic particle size to the magnetic loading capacity per unit target cell (scale bar = 5.14 µm). (a, Reprinted with permission from Tamer, U., Cetin, D., Suludere, Z., Boyaci, I. H., Temiz, H. T., Yegenoglu, H., Daniel, P., Dinçer, İ., Elerman, Y., *Int. J. Mol. Sci.*, 14, 6223–6240, 2013. b, Reprinted with permission from Sun, W., Khosravi, F., Albrechtsen, H., Brovko, L. Y., and Griffiths, M. W., *J. Appl. Microbiol.*, 92, 1021–1027, 2002. Copyright John Wiley and Sons.)

cell's surface would be significantly reduced to 10^2 and 10^0, respectively. Thus, a smaller particle size is more suitable for use in the analysis of cell surface markers at high density, while larger magnetic particles are more suitable for analysis of rare cell surface markers.

Not to mention, particle size is also one of the key determinants that affect particle endocytosis.[48] Numerous studies have discussed the size-dependent cellular uptake of particles.[49–51] For instance, Chithrani et al.[51] found that HeLa cells are capable of uptaking a larger amount of 50 nm gold nanoparticles in comparison to gold nanoparticles with sizes of 14 nm, 30 nm, 74 nm and 100 nm. Nevertheless, the patterns of cellular uptake could be varied for different material types of particle used.[48] Since the cellular uptake of magnetic particles also exerts significant impact on the efficiency of MS-aided biomedical diagnosis, the effect of particle size on the extent of particle internalization into a given cell should be more intensively studied.

10.4.1.3 Effect of Particle Size on Biotoxicity

Another important concern related to magnetic particle size in biomedical diagnostic application is the biotoxicity effect caused by the direct contact between the magnetic particle and biological cell. This issue is predominantly important for *in vivo* biomedical diagnosis, in which magnetic particles are injected into the human body for target detection. For iron-oxide-based magnetic particles, their detrimental effect is contributed by their small dimension. In particular, nano-sized particles (especially particles of the order of few nanometer length scale) could increase their internalization and interaction with biological tissues.[52] In addition, the physicochemical properties of nano-sized particles vary substantially from their bulk counterparts. Consequently, adverse biological effects can result when nanoparticles are in-used.[52] Apart from imposing harmful effects to the biological system, the nanoparticles behave in a different way throughout their interaction with biological cells in comparison to those of larger-sized particles.[53] In consideration of this phenomenon, size-dependent toxicity seems apparent and therefore must be addressed.

Kunzmann et al.[54] assessed the role of the size of silica-coated iron oxide nanoparticles (CSNPs) towards the viability of human monocyte-derived dendritic cells (MDDCs). The experimental results showed that when the MDDCs were exposed to 100 μg/mL of 30 nm or 50 nm CSNPs, the percentage of cell death increased significantly as compared to the system without CSNPs. Meanwhile, MDDCs exposed to 100 μg/mL of 70 nm CSNPs or 120 nm CSNPs did not show obvious difference in terms of the percentage of cell death as compared to the system without CSNPs.

On the other hand, Raju et al.[55] investigated ocular toxicity induced by 50 nm and 4 μm magnetic particles. Interestingly, they found that there was a significant reduction in the number of corneal endothelial cell in the eyes after injection with 4 μm magnetic particles as compared to the eyes injected with 50 nm magnetic particles or those injected with phosphate-buffered saline (PBS).[55] The authors suggested that the higher toxicity effect induced by the microsized particles could be due to (1) their larger size, which brings about physical trauma, or (2) their prolonged persistence inside the anterior chamber as compared to the nanoparticles.

However, it should be emphasized that not all magnetic particles exhibit a toxicity effect as a consequence of their small dimension. For cobalt and nickel magnetic particles, their toxicity characteristic is a result of the carcinogenic effect imposed by the material itself.[56] Thus, apart from particle size, it is crucial to assess the toxicity of the magnetic particle used in biomedical diagnosis from other aspects such as type of material and particle concentration (see Section 10.4.2.3 for more details).

10.4.2 Particle Concentration

The concentration of magnetic particle solution employed in MS-aided biomedical diagnosis is also an important parameter, which exerts a significant impact on the performance of this technique in detecting the targeted entity in a biological sample. Generally, the concentration of magnetic particle solution affects the efficiency of MS-aided biomedical diagnosis in three aspects, namely (1) MS rate, (2) sensitivity of the detection method and (3) biotoxicity.

10.4.2.1 Separation Rate

The concentration of the magnetic particle solution used greatly influences the magnetophoretic separation rate of the particles, and this influence is more pronounced for the LGMS process.[24] For LGMS to work effectively, a certain level of particle aggregation is needed in order to achieve desirable separation efficiency within a practical timescale. According to Equation 10.5, the aggregation parameter (which represents the number of magnetic particles in one unit of aggregate) varies directly with $\phi_0^{\frac{1}{2}}$.[42] In other words, larger aggregates will be formed within the magnetic particle solution with higher concentration. This phenomenon can be explained microscopically in terms of collision between magnetic particles suspended in the solution. As the concentration of the magnetic particle solution increases, the average displacement between two neighbouring particles is smaller. Thus, while the magnetic particles are driven to move under the presence of magnetic field, the possibility for collision between particles increases, leading to the formation of large aggregate. Particle aggregates have a higher magnetic volume, which subsequently can be magnetically separated within a shorter timescale. According to their magnetophoresis experiments in which Estapor M1-030/40 particles* (410 nm) and Estapor M1-020/50 particles (200 nm) were used, De Las Cuevas et al.[24] reported that the magnetophoresis rate was impressively accelerated if a magnetic particle solution of higher concentration was employed (Figure 10.7a). Similarly, this observation has also been reported by our group for both nanosphere and nanorod undergoing LGMS process promoted by a NdFeB magnet (Figures 10.7b and c).[57]

The nature of cooperative magnetophoresis in LGMS (self-aggregation of magnetic particles) can be fully utilized to enhance the performance of MS-aided biomedical diagnosis. By increasing the concentration of magnetic particle solution subjected to magnetophoresis, the isolation of magnetic particles can be significantly accelerated. Thus, in addition to the utilization of larger-sized magnetic particle or higher magnetic field gradient, increasing the particle concentration employed in MS-aided biomedical diagnosis serves as an alternative to accelerate the MS rate and hence improve the performance of this technology. This alternative route is appealing compared to the employment of a larger-sized particle, which could compromise the total exposure area for magnetic loading (see Section 10.4.1.2 for a more detailed discussion). Additionally, the utilization of cooperative magnetophoresis to boost the MS rate is much more attractive than generating a high magnetic field gradient, which can be costly and energy demanding (superconducting material is required to produce huge electric current).[58] However, up to the present time, there is very limited literature reported on the influences

* Supplied by Merck Chimie SAS (France), with a diameter of 0.41 μm, with 41% mass of ferrite distributed uniformly in a polystyrene matrix. The saturation volumetric magnetization of these particles is about 4.5×10^4 A/m.

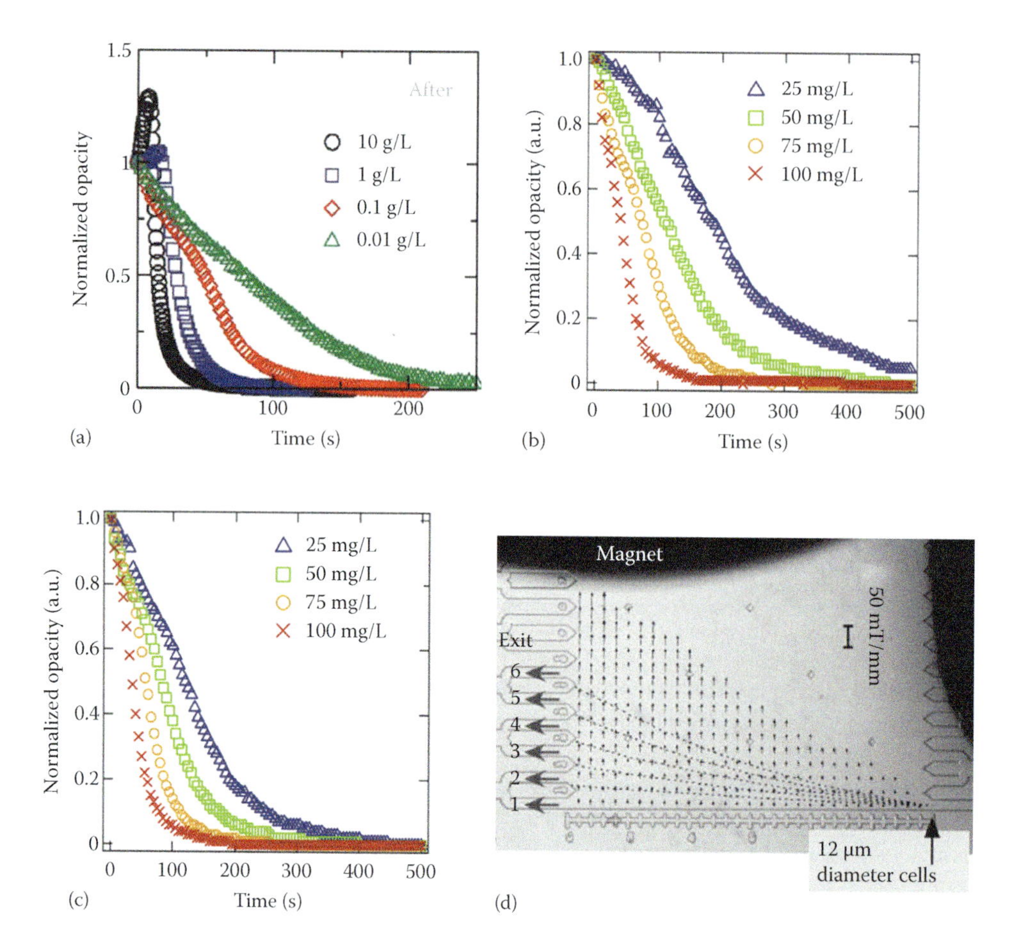

FIGURE 10.7 (a) Magnetophoresis kinetic profiles of Estapor M1-030/40 particles (410 nm) under magnetic field gradient of 30 T/m using a SEPMAG LAB325 2042 separator.[24] (Reprinted with permission from De Las Cuevas, G., Faraudo, J., and Camacho, J. 2008. Low-gradient magnetophoresis through field-induced reversible aggregation. *The Journal of Physical Chemistry C* 112:945–950. Copyright 2008 American Chemical Society.) (b) Magnetophoresis kinetic profiles of PDDA-coated magnetic nanosphere after the exposure to NdFeB magnet. (c) Magnetophoresis kinetic profiles of PDDA-coated magnetic nanorod after exposure to NdFeB magnet.[57] (Reprinted from *Journal of Colloid and Interface Science*, 421, Lim, J., Yeap, S. P., Leow, C. H., Toh, P. Y., and Low, S. C., Magnetophoresis of iron oxide nanoparticles at low field gradient: the role of shape anisotropy. 170–177, 2014, with permission from Elsevier.) (d) The calculated trajectories of 12 μm spherical magnetic cells with different magnetic loadings under a magnetic field gradient of 50 mT/mm. The trajectories (from bottom to top) correspond to magnetic cells with magnetic moments of 4, 8, 12, 16 and 20 × 10^{-13} A/m^2, respectively. The arrows in the diagram represent the vector of grad H imposed by the magnet on the separation chamber, which is calculated by experimentally measuring the trajectories of magnetic beads (based on the average of four series of magnetic beads tracks).[3] (Pamme, N. 2007. Continuous flow separations in microfluidic devices. *Lab on a Chip* 7:1644–1659. Reproduced by permission of The Royal Society of Chemistry.)

of particle concentration on the performance of MS-aided biomedical diagnosis. Henceforth, the adaptation of the cooperative nature of LGMS in MS-aided biomedical diagnosis should be studied more comprehensively.

Furthermore, the magnetic content of a particular cell (or concentration of magnetic particle incorporated in the cell) might also impose a nontrivial effect on the MS rate of the given cell. The magnetic content within a cell can be reflected by its volumetric magnetization (which is magnetic dipole moment per unit volume). Cells loaded with a huge amount of magnetic particles will acquire a larger magnetic dipole moment, display stronger response towards the applied magnetic field and are isolated more rapidly. The most relevant case was reported by Pamme and Wilhelm,[11] in which they showed that targeted cells loaded with a larger amount of magnetic nanoparticles were deflected more significantly to the direction of the applied magnet. In contrast, cells that were loaded with less amount of magnetic nanoparticles exit the separation chamber without showing much deflection by the externally applied magnetic field (Figure 10.7d).[11] Thus, it can be deduced that magnetic particle concentration (within a given cell) exerts a significant influence on the MS rate. In such a way, cells can be fractionated according to their volumetric magnetization (which is determined by the amount of magnetic particles tagged on or internalized into them).

10.4.2.2 Sensitivity

Apart from affecting the rate of MS, the concentration of magnetic particle solution also exerts a notable impact on the sensitivity of MS-aided biomedical diagnosis. This factor is particularly important when magnetic relaxation is coupled with MS to measure the concentration of targeted entity in the biological sample.[59] Particularly, antibody-conjugated magnetic particles are dispersed into the biological sample, which contains the targeted entity (pathogen) to be detected. The interaction between antibody (that is conjugated on magnetic particle) and antigen (from target entity) will induce the aggregation and formation of a much larger particle–target complex.[10] Next, the T_2 value of the biological sample is measured by a relaxometer, and it is used to infer the concentration of the targeted pathogen (or cell) present in the given sample. However, the correlation between the T_2 value and concentration of a targeted entity is significantly influenced by the size[60] and concentration[61] of the magnetic particles being used. For nano-sized magnetic particles, the existence of larger magnetic particle–target aggregate leads to a more significant dehomogenization of a magnetic field around them, which in turn gives rise to a more effective diphase of water protons that diffuse in that region.[62] Thus, in general, the T_2 value of water protons decreases with the particle aggregation (and target concentration). On the contrary, the T_2 value shows an opposite trend with the extent of particle aggregation (and target concentration) when larger-sized (such as micron-sized) magnetic particles are employed. This is because of the fact that fewer water protons are exposed to the magnetic field inhomogeneity as the degree of particle aggregation is more intense because particle aggregate presents in smaller amounts after the formation of particle–pathogen complex, which resulted from antibody–antigen interaction.[62]

Nevertheless, the correlation between T_2 value and target concentration as described earlier might break down in certain situations as it is heavily dependent on the ratio of the magnetic particle to the targeted entity in the biological sample. This phenomenon is observed by Koh et al.[62] in their attempt to quantify bovine serum albumin (BSA; a target entity with the size of 8 nm) by employing nano-sized magnetic particles (70 nm). It was found that the T_2 value decreases with an increment in target (or pathogen) concentration until a minimum is achieved, provided that the given magnetic particle concentration remains constant (Figure 10.8a). This minimum in T_2 value is depicted as the equivalence point. At the equivalence point, all binding sites on the magnetic particles have been occupied by the target (or pathogen) and the formation of the particle–target complex has reached its peak. Beyond the equivalence point, T_2 value shows an opposite trend, in which it somewhat increases with the concentration of the target (or pathogen) present in the given biological sample. This observation is owing to the fact that the target (which is present in excess in comparison to magnetic particles) has saturated the binding sites of the antibody that

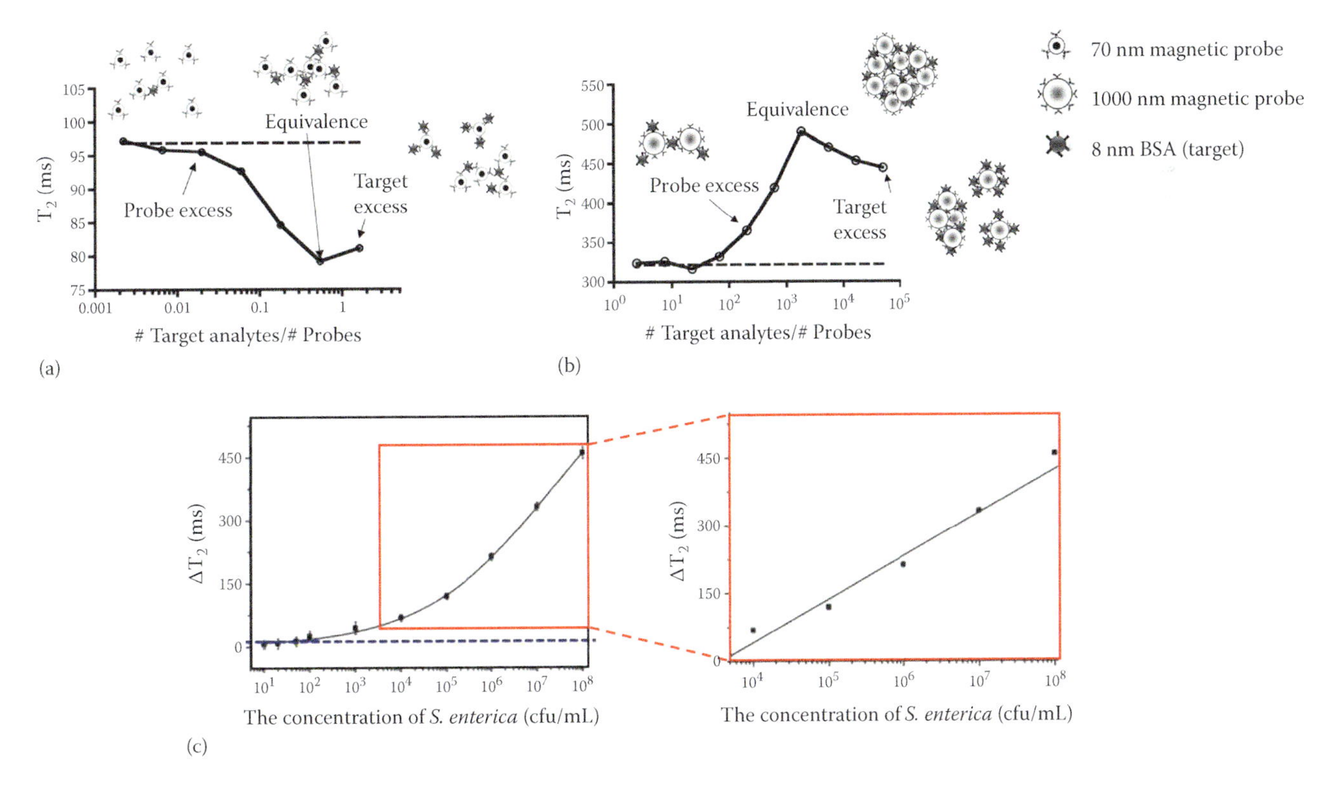

FIGURE 10.8 Plots of T_2 value against the ratio of targets to magnetic probes in the solution that contains (a) 70 nm magnetic probes and 8 nm BSA and (b) 1000 nm magnetic probes and 8 nm BSA.[62] (Reprinted with permission from Koh, I., Hong, R., Weissleder, R., and Josephson, L. 2009. Nanoparticle–target interactions parallel antibody–protein interactions. *Analytical Chemistry* 81:3618–3622. Copyright 2009 American Chemical Society.) (c) Plot showing the change in T_2 value against the concentration of *S. enterica* (target to be detected) in a one-step target detection method proposed by Chen et al. The linear relationship between the two given variables can be found within the *S. enterica* concentration, which ranges from 10^4 to 10^8 cfu/mL.[10] (Reprinted with permission from Chen, Y., Xianyu, Y., Wang, Y., Zhang, X., Cha, R., Sun, J., and Jiang, X. 2015. One-step detection of pathogens and viruses: combining magnetic relaxation switching and magnetic separation. *ACS Nano* 9:3184–3191. Copyright 2015 American Chemical Society.)

are functionalized on the magnetic particles and, hence, the excess target (or pathogen) prevents further aggregation of magnetic particles. The phenomenon in which the excessive target (or pathogen) impedes the aggregation of the fully saturation magnetic particles is known as the hook effect or the prozone effect.[63] As nano-sized and micro-sized magnetic particles give different trends in the relationship between T_2 value and degree of particle aggregation (as described in the previous paragraph), the tabulation of T_2 values against pathogen concentration shows an inverse trend when micron-sized magnetic particles were used (Figure 10.8b) as compared to that when nano-sized magnetic particles were used (Figure 10.8a). In both cases, the intervention of prozone effect can distort the initial correlation between T_2 value and target (or pathogen) concentration, which in turn affects the accuracy of target (or pathogen) detection.[62] Thus, in order to avoid the prozone effect, it is necessary to disperse a sufficient amount of magnetic particles into the given biological sample (the particle concentration should be optimal to target) such that the measurement can be conducted in the sample with target (or pathogen) concentration at the equivalence point.

The impact of magnetic particle concentration on the sensitivity of target detection technology is also observed in the one-step pathogen or virus detection method developed by Chen et al.[10] (see Section 10.4.1.1 and Figure 10.5b for a clearer illustration of this method). The sensitivity of this detection method is highly dependent on the concentration of magnetic particles dispersed in the biological sample. As the concentration of magnetic particles is low, there is an insufficient amount of MP_{30} available in the formation of MP_{30}–target–MP_{250} complex. Thus, it is hard to observe any noticeable change in T_2 value (since all MP_{30} particles has been depleted and involved in the formation of MP_{30}–target–MP_{250} complex) even by using a sample with a greater amount of target (or pathogen). On the contrary, as the magnetic particles have been oversupplied, the fractional depletion of MP_{30} after the MS process (and hence the change in T_2 value) is extremely insignificant to provide reliable and accurate result, which reflects the concentration of target (or pathogen) in the initial biological solution. Furthermore, the linear correlation between the change in T_2 value and the concentration of target (or pathogen) exists only within some particular range of target (or pathogen) concentration (Figure 10.8c). Consequently, it is paramount to tune the concentration of magnetic particle solution to a certain range in order to conduct a precise measurement of the target (or pathogen) concentration by using this technique.

10.4.2.3 Effect of Particle Concentration on Biotoxicity

The biotoxicity of the magnetic particle solution appears as one of the major concerns in MS-aided *in vivo* biomedical diagnosis. Generally, the usage of magnetic particles in biomedical diagnosis requires the particles to be coated with materials (polyelectrolyte such as polydiallyldimethylammonium chloride [PDDA] and poly(sodium(4)styrenesulfonate)), which improves their colloidal stability. Even though most magnetic particles (such as magnetite [Fe_3O_4] and maghemite [γ-Fe_2O_3] particles) are nontoxic from material perspective, the coating material might exert a detrimental effect on the biological tissue and cell. One of the most common mechanisms that underlie the toxicity of magnetic nanoparticles is the generation of reactive oxygen species, which induces oxidative stress and damage on the cells.[64] Therefore, there exists a correlation between the toxicity of the magnetic particle and its particle concentration. In their study on the toxicity of magnetic nanoparticles, Häfeli et al.[65] found that the viability of both human umbilical vein endothelial cells and PC3 prostate cancer cells in magnetic nanoparticle solutions reduces with the concentration increment of magnetite particles coated with triblock copolymers. Concurrently, the decay of cell viability with the increment in the concentration of iron oxide (Fe_2O_3) nanoparticles is also observed in the mixture of PC12 pheochromocytoma clonal cell and (Fe_2O_3) nanoparticles.[66] Furthermore, Kunzmann et al.[54] observed that the percentage death of dendritic cells (MDDC) rises significantly upon the exposure of the core–shell nanoparticles with higher concentration (50–100 μg/mL). Hence, for any *in vivo* biomedical diagnosis, it is particularly important to ensure that the concentration of the magnetic particle solution employed exerts no harmful effect on the biological cells.

10.5 CONSIDERATIONS IN THE DESIGN AND IMPLEMENTATION OF MS-AIDED BIOMEDICAL DIAGNOSIS

10.5.1 Colloidal Stability

The colloidal stability of a magnetic particle is a critical factor that exerts a significant impact on the efficiency of MS-aided biomedical diagnosis. Specifically, magnetic particles should be colloidally stable to ensure that they are well dispersed in the solution throughout the entire labelling process. In contrast, magnetic particles that are colloidally unstable will undergo aggregation and form larger clusters spontaneously. Aggregation reduces the specific surface area of the magnetic particles and thus lessens the available sites for target labelling.

The underlying reason that explains the aggregation and, thus, colloidal instability of magnetic particles is the presence of attractive interaction between particles.[67] These attractive energies are, by nature, van der Waals and/or magnetic dipole–dipole interaction.[67,68] Under the absence of external magnetic field, the existence of magnetic dipole–dipole attraction between the magnetic particles depends on its magnetic properties. After the removal of external magnetic field, which magnetizes the particles, ferrimagnetic and ferromagnetic particles still retain remanent magnetic dipole, whereas superparamagnetic particles will be fully demagnetized (there is no remanent magnetization left). In other words, when ferrimagnetic or ferromagnetic particles are used, magnetic dipole–dipole attraction also contributes to the occurrence of particle aggregation and thus leads to colloidal instability.

However, the colloidal stability of magnetic particles can be strengthened via electrostatic and/or steric repulsion mechanisms. In order to address electrostatic repulsion among particles in the suspension, the particles need to be imparted with either net positive or negative charge. On the other hand, steric repulsion can be imparted by coating magnetic particles with polymer or macromolecules. Magnetic particles can be prevented from aggregation if the total repulsive interaction energy (electrostatic and steric energies) is more intense than the total attractive energy (van der Waals and magnetic dipole–dipole forces). Figure 10.9 shows the interaction energies that exist between two positively charged magnetic particles (Figure 10.9a), two magnetic particles coated with neutrally charged polymers (Figure 10.9b) and two magnetic particles coated with positively charged polymers (Figure 10.9c).

As discussed, magnetic particles that are tailor designed for specific target isolation are required to be functionalized with a specific antibody. In that case, the antibody will be located at the outermost layer of the magnetic particles and thus directly influences the colloidal stability. In biomedical diagnostic application, magnetic particles are dispersed in physiological media with a wide range of working environments. Particularly, when biological media with elevated ionic strength is used, the electrostatic repulsion will be screened and becomes ineffective.[69] Therefore, it is necessary to ensure the colloidal stability of the antibody-conjugated magnetic particles when they are dispersed within a physiological media. For instance, Xu et al.[12] synthesized antibody-conjugated magnetic iron oxide nanoparticles for immunomagnetic separation of tumour cells. They tested the stability of the antibody-conjugated particles in various biological solutions, such as Dulbecco's PBS (DPBS), DPBS + 1% (w/v) BSA, Roswell Park Memorial Institute (RPMI)-1640 medium, RPMI-1640 supplemented with 10% heat-inactivated foetal bovine serum and 1% streptomycin/penicillin, and human plasma. Experimental results showed that the particles were colloidally stable in all the aforementioned biological media. However, particles without antibody conjugation were found to form precipitates in RPMI-1640 medium and lead to the reduction of surface area for specific binding with targeted entity.

It is worth highlighting that even though enhanced colloidal stability can improve the performance of biomedical diagnosis, the MS rate of magnetic particles that are colloidally stable was found to be relatively slower than the MS rate of magnetic particles that are colloidally unstable.[22,57,70] In these works, particle aggregation is less severe for colloidal stable magnetic particles, which subsequently suppresses the cooperative magnetophoresis. The conflicting issue arising here is that colloidally stable magnetic particles solution is required for better target capturing, yet these stabilized magnetic particles could take a longer time to be magnetically separated, which in turn prolongs the diagnosis duration. In view of this conflict, the trade-off concern mentioned here must be assessed in the design of MS technique for biomedical diagnostic application.

10.5.2 Particle Shape

Another criterion worth considering in the selection of magnetic particles for MS-aided biomedical diagnostic application is the shape of the magnetic particles. Such consideration is due to the influential role of particle shape anisotropy towards the MS rate. It was reported that magnetophoresis is preferred in anisometric particles.[71] Lim et al.[57] tested the magnetophoretic separation of sphere-shaped magnetic particles (diameter ~50 nm) and rod-shaped magnetic particles (20 × 300 nm) under low magnetic field gradient generated by a cylindrical NdFeB permanent magnet. The rod-shaped magnetic particles were found to experience faster MS compared to the sphere-shaped magnetic particles. This phenomenon is ascribed to the reason that the alignment of magnetic dipole along the axis of rod-shaped magnetic particles is more stable; thus, the magnetophoresis is less disrupted by thermal randomization (Brownian motion) and viscous drag force. Figure 10.10 schematically illustrates the magnetic dipole alignment induced by an externally applied magnetic field on sphere-shaped and rod-shaped magnetic particles. In fact, it was reported that due to dipole–dipole interaction during magnetophoresis, the sphere-shaped magnetic particles tend to be assembled into linear chain-like structures, which speeds up the magnetic-field-induced transportation.[25,36,37] All these findings indicate the strong impact of particle shape anisotropy on MS rate.

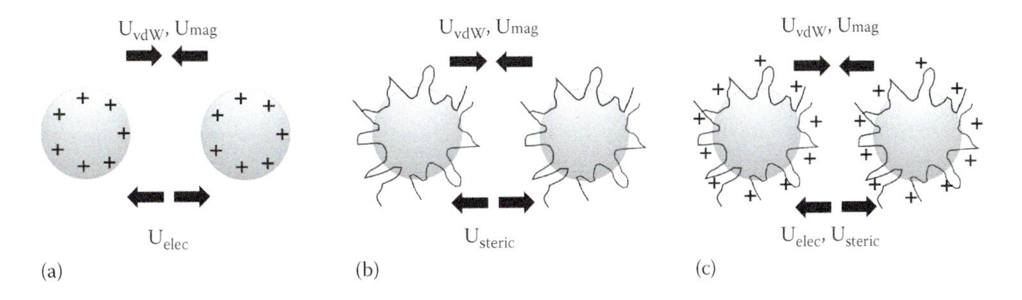

FIGURE 10.9 Possible interaction schemes existing and the interacting energies involved between (a) two positively charged magnetic particles, (b) two magnetic particles coated with neutral-charged polymers and (c) two magnetic particles coated with positively charged (or negatively charged) polymers. (U_{vdW} = van der Waals energy, U_{mag} = magnetic attraction energy, U_{elec} = electrostatic energy, U_{steric} = steric energy).

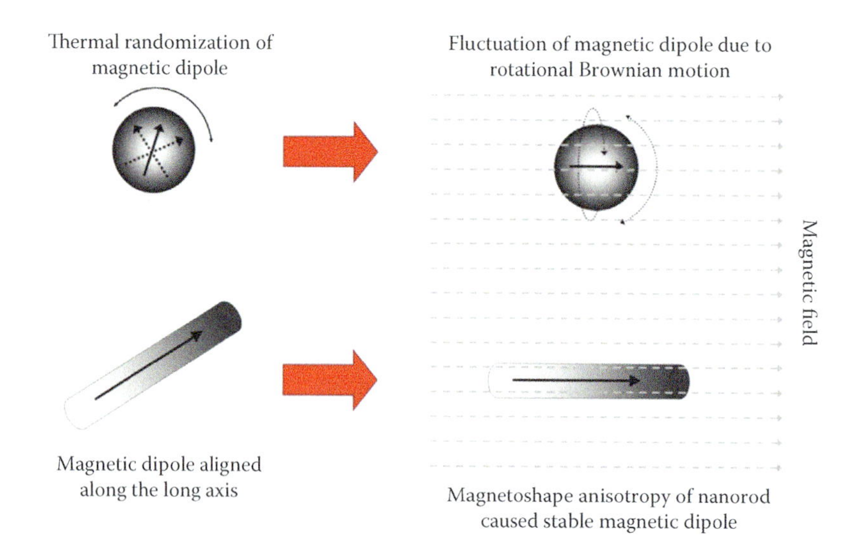

FIGURE 10.10 Schematic diagram showing magnetic dipole alignment of sphere-shaped and rod-shaped magnetic particles under a magnetic field. Note that the rod-shaped particle is less susceptible to fluctuation by rotation Brownian motion as compared to the sphere-shaped counterpart.[57] (Reprinted from *Journal of Colloid and Interface Science*, 421, Lim, J., Yeap, S. P., Leow, C. H., Toh, P. Y., and Low, S. C., Magnetophoresis of iron oxide nanoparticles at low field gradient: the role of shape anisotropy. 170–177, 2014, with permission from Elsevier.)

As mentioned earlier, target cells can be magnetically fractionated based on the amount of magnetic particles that are attached or internalized into the cell. In this regard, particle shape is known to play a significant role in affecting the cellular uptake of particles as well. Kolhar et al.[72] studied the shape-dependent cellular adhesion and internalization of polystyrene (PS) particles in lungs and brain. An *in vivo* experiment showed that anti-Intercellular Adhesion Molecule 1 (ICAM-1) antibody (ICAM-monoclonal antibody [mAb])-coated PS rods (501 ± 43.6 × 123.6 ± 13.3 nm) exhibited higher accumulation in the lungs compared to their spherical (205 ± 0.01 nm, diameter) counterparts. A similar observation was obtained for the accumulation of antitransferrin receptor antibody (TfR-mAb)-coated rod particles in the brain, which is about sevenfold higher than that of TfR-mAb-coated sphere particles. On the other hand, Chithrani et al.[51] investigated the effect of particle aspect ratio on the cellular uptake of gold nanoparticles and found that the number of gold nanoparticles that could be internalized into a HeLa cell reduced with an increment in the aspect ratio of the particles (i.e. 1:1 > 1:3 > 1:5). In other words, sphere-shaped particles (with aspect ratio 1:1) are more likely to internalize into a cell as compared to rod-shaped particles (with aspect ratio 1:3 and 1:5). The surface curvature of the different-shaped particles was speculated as one of the possible reasons contributing to such observation. The conflicting outcomes obtained from both researches imply that more works need to be done and taken into account in order to ensure anisometric magnetic particles can be a useful candidate for MS-aided biomedical diagnostic application.

10.5.3 Specificity

Typically, MS is known as one of the detection techniques that are highly specific in recognizing the targeted entity. For instance, Song et al.[73] developed anti-CD3 (cluster of differentiation 3) mAb-coupled nanobioprobes and anti-prostate-specific membrane antigen mAb-coupled nanobioprobes, which can specifically recognize leukemia (Jurkat T) cells and prostate cancer (LNCaP) cells, respectively, in a sample with targeted cell fraction as low as 0.01%. Additionally, Xu et al.[12] also demonstrated that iron oxide particles, which have been conjugated with antibodies against human epithelial growth factor receptor 2, bind specifically onto the membrane of circulating tumour cells (CTCs), which subsequently lead to the CTC enrichment at a factor of 1:10,000,000 after the exposure to an external magnetic field (with field gradient of 100 T/m).

Even though magnetic cell separation is often highly specific, the specificity of magnetic particles (which acting as the probes) towards the entity to be targeted is one of the most important factors that should be considered in the design of MS-aided biomedical diagnosis. Under this context, the type of magnetic particles and the antigen coating can significantly alter the specificity of the magnetic probes towards an entity to be targeted. Particularly, in their attempt to optimize the detection of *Mycobacterium avium* subsp. Paratuberculosis in milk, Foddai et al.[74] identified that the nonspecific recovery of other *Mycobacterium spp.* is higher when Pathatrix PM-50 beads coated with an affinity-purified polyclonal antibody were used even though these particles give the highest capture efficiency of *M. avium* subsp. Paratuberculosis. On the contrary, AnDiaTec beads coated with a mAb demonstrated not only lower nonspecific recovery of other *Mycobacterium spp.* (about 7% and 4% for *M. avium* and *Mycobacterium bovis* bacillus Calmette-Guérin (BCG), respectively) but also low capture efficiency of *M. avium* subsp. Paratuberculosis. The nonspecific binding between the nontargeted species and magnetic particles was mostly likely contributed by the electrostatic or van der Waals attraction. Therefore, it is essential to evaluate which combination of magnetic carrier and antigen coating is able to give the highest specificity towards the targeted entity. Additionally, it is also equally important to be aware of the tradeoff between specificity and capture efficiency as a magnetic carrier with high specificity might result in poor capture efficiency.

Even though the utilization of larger magnetic particles can lead to more rapid MS, the specificity of the separation process might be adversely affected. As demonstrated by Lin et al.[45] in the separation of *E. coli*, the separation ratio of nontargeted cell (*Salmonella typhimurium* and *Listeria innocua*) increases when magnetic particles with a size of 180 nm were used instead of 30 nm ones. The phenomenon results from the nonspecific binding of the nontargeted cell with the magnetic particles, and larger particles experience much stronger magnetophoretic force, which enables them to be captured more easily. Apart from that, larger magnetic carriers are more likely to entrap nontargeted cells within the magnetic particles and are separated along with the targeted entity during the separation process, rendering the deterioration of the detection specificity.[75]

10.5.4 Hydrodynamic Effect

As the fundamental understanding on the dynamical behaviour of magnetophoresis is inevitable in the design of magnetic separator for MS-aided biomedical diagnosis, hydrodynamic effect appears as a critical factor, which should be taken into consideration in the optimization of MS process. This is owing to the fact that the hydrodynamic effect plays a dominant role in dictating the separation kinetics of low-field-gradient magnetophoresis, as demonstrated in our previous work.[23]

Conventionally, the microscopic picture of LGMS is depicted in such a way: under an externally applied magnetic field, magnetic particles are driven to migrate towards the region where magnetic field strength is the highest by magnetophoretic force, whilst the surrounding fluid remains in stagnant. Under such condition, magnetic particles move and encounter viscous resistance throughout their motion within the fluid, which is unaffected by the particle movement. However, this conventional microscopic picture on LGMS is proven to be inaccurate in some cases. As reported in our previous work,[23] the entire magnetic particle solution (consisting of magnetic particle and surrounding fluid) did not remain in stationary state but was subjected to move collectively throughout the magnetophoresis process under an inhomogeneous magnetic field (magnetic field gradient is extremely strong adjacent to the magnetic source and falls tremendously to virtually zero within the suspension domain located further away from the magnet). This phenomenon arises as a consequence of two-way momentum transfer between magnetic particles and the surrounding fluid. Since the particle motion occurs in fluid, the collision between the moving particles and surrounding fluid is inevitable. Such a collision induces the transfer of momentum from the magnetic particles to the surrounding fluid. Thus, circulating flow is generated within the whole domain of magnetic particle solution, which is subjected to magnetophoresis (Figure 10.11a). This phenomenon, in which convective flow emerges due to the two-way momentum transfer between the magnetic particles and the surrounding fluid, is denoted as hydrodynamic effect of magnetophoresis. However, it should be emphasized that hydrodynamic effect is observed only in the magnetophoresis of magnetic particles under inhomogeneous magnetic field gradient. Under homogeneous magnetic field gradient (for example SEPMAG device), the hydrodynamic effect is not observed and the classical magnetophoresis model is able to accurately predict the dynamical behaviour of the magnetophoresis process.

Owing to the intervention of hydrodynamic effect, the dynamical behaviour of magnetophoresis has been greatly disturbed. Instead of moving along the magnetic field gradient towards the magnetic source, an individual magnetic particle is also subjected to constant displacement by the convective flow. Thus, the trajectory of any magnetic particle subjected to magnetophoresis is not as simple as a smooth curve (or line), which connects its initial and final positions. One of the most noteworthy outcomes of this induced convective flow is the continuous homogenization of the magnetic particle solution during the entire magnetophoresis process (Figure 10.11b). Here, magnetic particles are agitated to the whole domain of the solution by convective flow within a much shorter timescale in comparison to the separation time. Besides, the hydrodynamic effect also plays an indispensable role in accelerating the magnetophoretic separation of magnetic particles from the solution (Figure 10.11c). This induced convective flow sweeps through the region where the magnetic field gradient is weak and particles within are not supposed to experience magnetophoresis. Again, momentum transfer happens, but this time from the fluid back to the particles, and causes their migration along the magnetophoresis pathway. As demonstrated in our previous work on the magnetophoresis of 30 nm magnetic nanoparticles by neodymium iron boron (NdFeB) magnet, theoretical calculation predicted that 1420 h is needed to achieve 99% of magnetic particle removal with no presence of hydrodynamic effect. However, in our previous experiments, hydrodynamic effect is unavoidable and the occurrence of induced convection has shortened the MS time to 51 h (which is approximately 27 times faster in comparison to the case without hydrodynamic effect).

Owing to the accelerated MS rate caused by the hydrodynamic effect, this feature can be exploited to enhance the efficiency of MS-aided biomedical diagnosis. Therefore, it is imperative to ensure the occurrence of magnetophoresis-induced convection within the magnetic separator used in biomedical diagnosis. In order to fulfil this criterion, we defined magnetic Grashof number Gr_m as shown in the following equation:

$$Gr_m = \frac{\nabla B \left(\frac{\partial M}{\partial c} \right)_H (c_s - c_\infty) L_c^3}{\rho v^2}, \tag{10.7}$$

Here, M is the volumetric magnetization of the magnetic particle solution, c_s is magnetic particle concentration on the surface where the magnetic particle is captured and depleted, c_∞ is the magnetic particle concentration of the bulk solution, L_c^3 is the characteristic length, ρ is the mass density of the magnetic particle solution and v is the kinematic viscosity of the

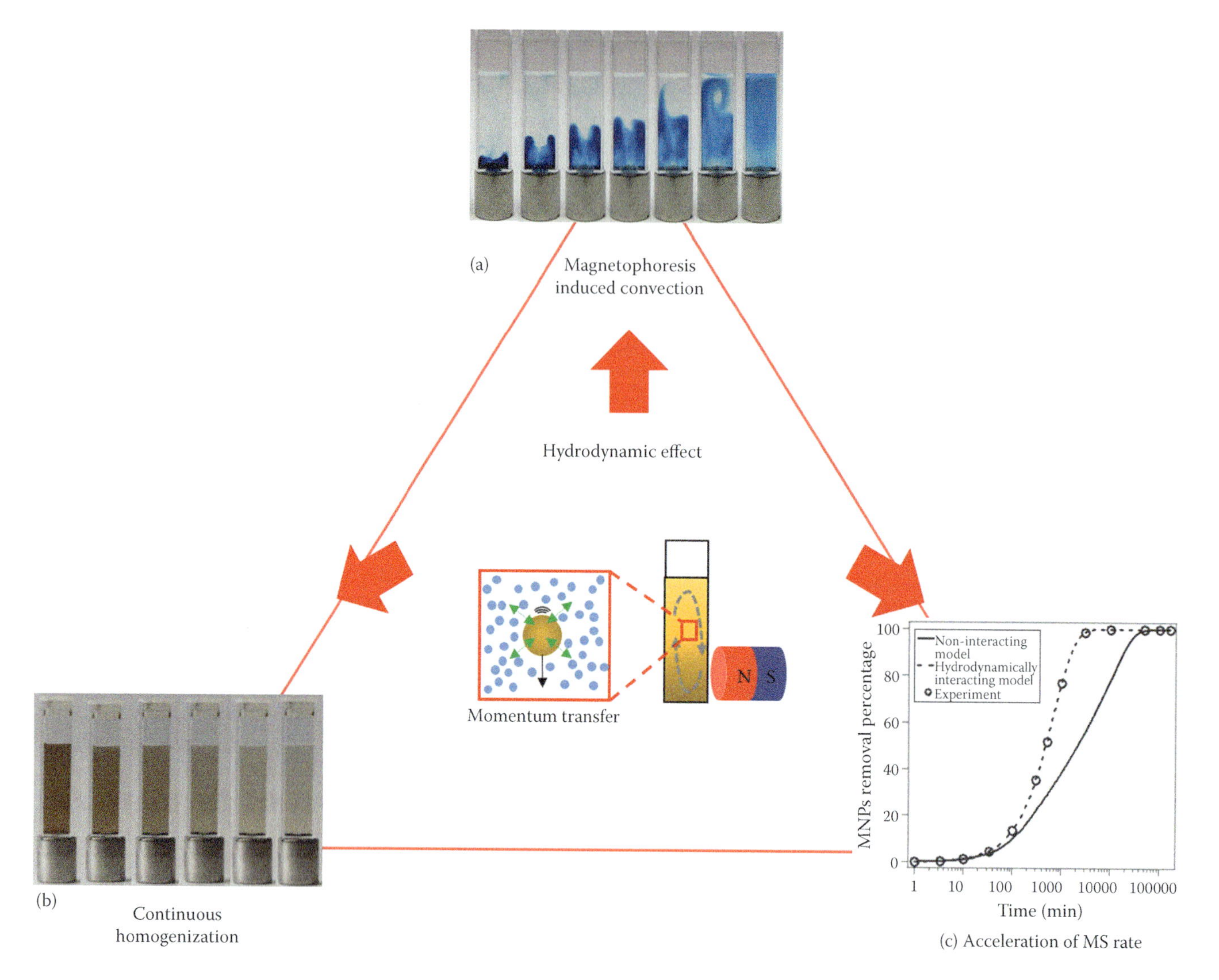

FIGURE 10.11 Consequences of hydrodynamic effect. (a) Time-lapsed photos of 5 mg/L SMG-30 (30 nm magnetite nanoparticle coated with PEG) solution after being subjected to magnetophoresis for 5, 60, 180, 300, 450, 600 and 900 s (from left to right). The function of the injected dye (methylene blue) is to demonstrate the induced convection within the solution throughout the magnetophoresis process. (b) Time-lapsed photos of 100 mg/L SMG-30 solution after being subjected to magnetophoresis for 0, 100, 200, 300, 400 and 500 min (from left to right). The magnetic particle solution remains homogeneous all over the magnetophoresis process as a consequence of continuous agitation performed by the induced convective flow. (c) Comparison between the separation kinetic profiles obtained from the simulation based on conventional model and hydrodynamic model as well as real-time experimental result. The simulation result is performed by COMSOL Multiphysics.[23] (Leong, S. S., Ahmad, Z., and Lim, J. 2015. Magnetophoresis of superparamagnetic nanoparticles at low field gradient: hydrodynamic effect. *Soft Matter* 11:6968–6980. Reproduced by permission of The Royal Society of Chemistry.)

magnetic particle solution. Under this definition, the hydrodynamic effect dominates the dynamical behaviour of MS if magnetic Grashof number Gr_m is greater than unity.

Computational simulation is the most economical way to examine the performance of newly designed magnetic separator for biomedical diagnosis. Nevertheless, the simulation of magnetophoresis-induced convection requires the computational tool to solve Navier-Stokes and drift-diffusion equations, which is fourth order (four variables) and highly nonlinear. For magnetic separator designed in complicated geometry configuration, the simulation might be extremely tedious and require substantial computational power. Besides, we still lack a microscopic view of the hydrodynamic effect, specifically on how it dominates the magnetophoresis process. Despite there still being a lot of challenges to implementing the hydrodynamic effect into MS-aided biomedical diagnosis, it appears as one of the driving factors with huge potential to improve the MS-aided biomedical diagnosis technique in the near future.

10.5.5 Spatial Arrangement of Magnetic Sources

The spatial arrangement of magnetic sources (electromagnet or permanent magnet) is also an important criteria to be discussed in the design and optimization of magnetic separator used for MS-aided biomedical diagnosis. A different arrangement of magnetic sources can generate different spatial magnetic fields that enable magnetophoretic capture of magnetic particles performed in a more efficient manner.

The simplest magnetic source arrangement constitutes of only one permanent or electromagnet, which is placed adjacent to the biological sample (or magnetic particle solution) and acts as the source of magnetic field. Owing to its simplicity, it is widely employed in various applications such as the removal of heavy metal,[76–78] organic dye[79] as well as microalgae[80,81] and denoted as LGMS. Apart from that, the single magnet configuration is also utilized in various biomedical diagnoses and cell separations, which have been reported in various literatures.[10,12,73,82] For instance, in the work reported by Pamme and Wilhelm,[11] single permanent magnet can also be used to sort cells (mouse macrophages and human ovarian cancer cells) loaded with magnetic particles according to their magnetization in a microfluidic device. However, one of the major drawbacks encountered by the utilization of a single magnet in a magnetic separator is the rapid decrease in magnetic field with the displacement from the given magnet.[20] This shortcoming has caused the efficient separation occurring only in an extremely small region in the vicinity of the magnet while most domains within the magnetic separator are characterized by weak magnetic field and gradient (see Figure 10.2c and d).

Alternatively, Zborowski et al.[83] have demonstrated the utilization of quadrupole magnetic separator to isolate human peripheral lymphocytes tagged with magnetic particles. In the quadrupole magnetic separator, magnetic sources are arranged in such a way that four magnetic poles are focusing the magnetic field around a centralized cylindrical area (Figure 10.12a).[84] One of the most important features of quadrupole magnetic field is the linear increase in magnetic field strength B with respect to radial displacement from the axis of the cylindrical tube (Figure 10.12c).[85] Since the magnetophoretic force acting on the magnetic responsive material is directly proportional to the magnetic field gradient (as depicted in Equation 10.1), magnetic particles within the quadrupole magnetic field will experience a constant magnetophoretic force, which points outward along the radial direction. Thus, as magnetic particles are channelled through the tube placed within the quadrupole magnetic field, they will be 'centrifuged' and directed towards the inner wall of the tube (Figure 10.12a and b).[86] The advantages offered by quadrupole magnetic separator are the efficient utilization of magnetic energy and possibility to perform continuous sorting within the given device.[83] Besides, the SEPMAG separator also can generate a magnetic field in which the magnetic particles within its cylindrical cavity will experience the same magnetic field gradient, which is parallel to radial direction.[24,87]

Apart from that, Halbach array is also a possible magnet arrangement that can be implemented in magnetic separator for biomedical diagnosis. Halbach array is a special arrangement of magnets that is able to reinforce the magnetic field strength on one side of the array while weakening the magnetic field on the other side.[85] The Halbach array involves selective arrangement of magnets in which the magnetization of the subsequent magnet is rotated at an angle of 90° in comparison to the previous magnet (Figure 10.13).[88] Thus, by incorporating the Halbach array into the magnetic separator, the magnetic field can be significantly concentrated on the region where MS is intended to be carried out. By intensifying the magnetic field on the separation chamber (the domain where magnetophoresis is conducted), the separated rate can be greatly accelerated and the efficiency of the magnetic separator can be improved. Hoyos et al.[89] have incorporated the Halbach array into a SPLITT magnetic separator, which displays improved efficiency as

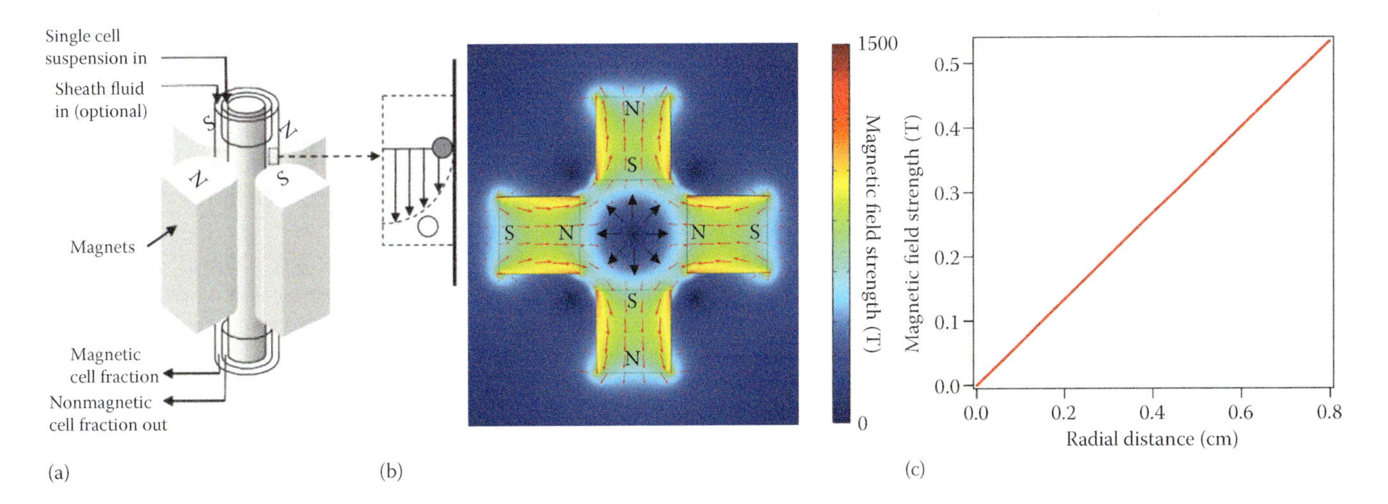

FIGURE 10.12 (a) Schematic diagram of quadrupole magnetic separator with magnet arrangement in N-S-N-S configuration.[86] (Reproduced with permission from Zborowski, M., and Chalmers, J. J. 2011. Rare cell separation and analysis by magnetic sorting. *Analytical Chemistry* 83:8050–8056. Copyright 2011 American Chemical Society.) (b) Surface plot of magnetic field strength created by the quadrupole magnetic system as obtained from simulation performed by COMSOL Multiphysic. The length and width of the magnets are given as 1.5 cm and 1.4 cm, respectively. The remanent magnetization of 1.45 T (NdFeB magnet) is assumed. The red arrows represent the direction of magnetic field lines, while the black arrows point to the direction of the magnetophoretic force acting on magnetic particles. (c) Plot of magnetic field strength against radial distance, which starts at the centre of the quadrupole magnetic system.

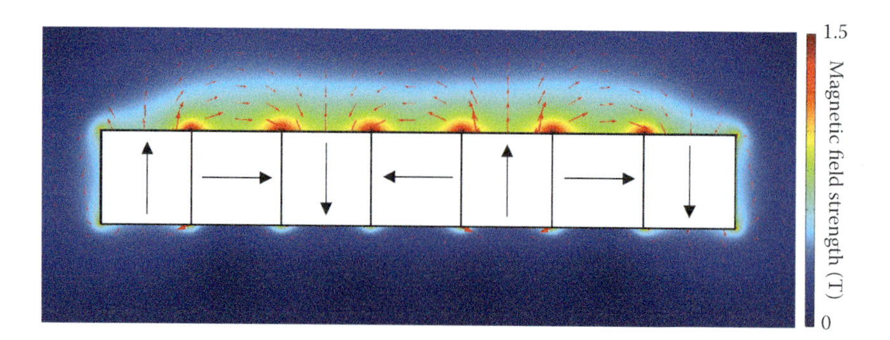

FIGURE 10.13 Surface plot of magnetic field strength generated by Halbach array. The white blocks are 1.5-cm square magnets, with the direction of magnetization denoted by black arrows within the blocks. Red arrows point towards the direction of the magnetic field. This magnet configuration (Halbach array), magnetic field, reinforces each other at the upper part of the ray. On the other hand, the magnetic field cancels each other at the bottom of the array.

compared to conventional SPLITT magnetic separator. In their theoretical study, Babinec et al.[90] also found that the separation time given by magnetic separator incorporated with Halbach array has been shortened by a factor of 25 in comparison to a simple block of permanent magnet. Thus, the Halbach array shows a huge potential to be implemented in magnetic separator for biomedical diagnosis as well as other applications.

The dehomogenization of magnetic field by materials with high magnetic susceptibility also can be implemented in continuous magnetic separator for biomedical diagnosis. This concept is similar to HGMS column in which high magnetic field gradient is created by the distortion of the magnetic field by high magnetic susceptible material. Chen et al.[91,92] have theoretically predicted that the implementation of irregular magnetic susceptible material (ferromagnetic material) in a separator can further boost the separation rate of magnetic particles from human blood. The irregular magnetic susceptible material can be wires (Figure 10.14a) and ferromagnetic prism (Figure 10.14b), which are located near the tubing where MS is conducted.[93] Apart from that, Xia et al.[16] also used NiFe microcomb to generate a higher magnetic field gradient adjacent to the flowing stream of *E. coli*-bound magnetic nanoparticle solution and speed up the isolation of *E. coli* bacteria (Figure 10.14c). This design is more superior in comparison to the conventional HGMS column as the direct contact between biological samples and magnetic susceptible wires can be omitted. This arrangement avoids the permanent retention of magnetic particles on the wires and reduction of magnetic separator efficiency.

In addition to the classical design of magnetic separator (in which the magnetic sources and separation chamber remain stationary throughout the separation process), the actuator-like magnetophoretic device serves as an alternative design, which might further enhance the performance of MS-aided biomedical diagnosis. For instance, Berenguel-Alonso et al.[94] have developed a magnetic actuator that is able to boost the sensitivity of the detection of *E. coli* O157:H7 whole cells. In this particular design, several magnets are placed on the disk, with different radii from the disk centre, and the microfluidic reaction chamber is located above it (Figure 10.15a). Throughout the rotation of the disk, the magnetic particles (which have been tagged with *E. coli* O157:H7 and alkaline phosphatase-labeled anti *E. coli* O157:H7 antibody) in the reaction chamber are driven back and forth by the magnetophoretic actuating force. Such actuation enhances the mixing of the magnetic beads with 4-methylumbelliferyl phosphate (4-MUP) in the reaction chamber and, hence, the dephosphorylation of the 4-MUP to the 4-methylumbelliferone (4-MU), which is catalyzed by the alkaline phosphatase tagged on the magnetic beads. 4-MU is a fluorescent product and the fluorescence emission of the reacted solution is measured in the detection chamber. As illustrated in Figure 10.15b, the actuations of magnetic beads have improved the sensitivity in the detection of *E. coli* O157:H7 in comparison to the case without actuation (movement velocity = 0 mm/s), with the optimal movement velocity at 1.7 mm/s.

In consideration of the critical role played by spatial arrangement of magnetic sources in influencing magnetophoretic separation efficiency, future work on MS-aided biomedical diagnostic application should focus in this area of study.

10.6 COMMERCIALIZED MAGNETIC PARTICLES FOR MAGNETIC CELLS SEPARATION

One of the most popular commercial products used for magnetic cell separation is Invitrogen Dynabeads, which is currently produced by Thermo Fischer Scientific. Dynabeads are superparamagnetic spherical particles with high uniformity of particle size, which are typically of the order of micron size in diameter.[95] Due to their superparamagnetic property, Dynabeads acquire magnetic dipole moment under an externally applied magnetic field and completely lose their magnetism (without any remanent magnetization) upon the removal of the magnetic field. Dynabeads are functionalized with antibodies for specific cell isolation, such as monocytes, granulocytes, endothelial cells, dendritic cells, B cells and T cells. The most significant advantages offered by Dynabeads include insignificant nonspecific binding of nontargeted cells and minimal chemical agglutination. Additionally, owing to their large diameter (of the order of micron size), Dynabeads can be rapidly isolated by applying magnetic field generated by a hand-held magnet. For instance, Helseth and Skodvin[96] reported that the separation of Dynabeads M270 (with radius of 1.4 μm and

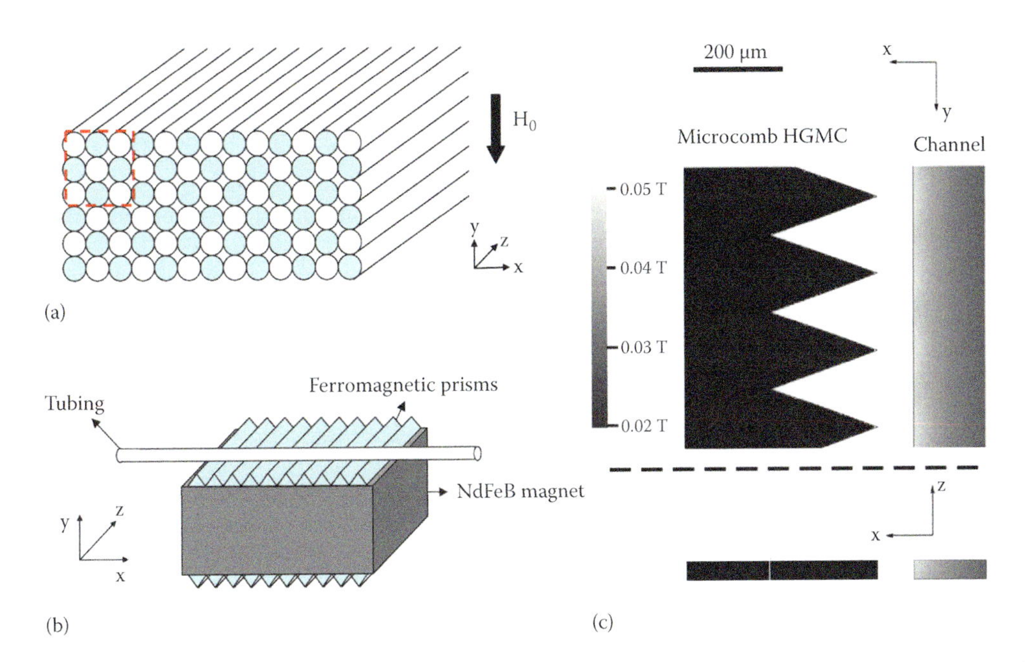

FIGURE 10.14 (a) Arrangement of magnetizable wires (light blue) and tubing where magnetic particle solution are flowing through (white) in an alternate manner.[91] (Reprinted from *Journal of Magnetism and Magnetic Materials*, 320, Chen, H., Bockenfeld, D., Rempfer, D., Kaminski, M. D., Liu, X., and Rosengart, A. J., Preliminary 3-D analysis of a high gradient magnetic separator for biomedical applications, 279–284. Copyright 2008, with permission from Elsevier.) (b) The separator column is placed on the top of an array of ferromagnetic prism.[93] (Reprinted from *Journal of Magnetism and Magnetic Materials*, 313, Chen, H., Kaminski, M. D., Ebner, A. D., Ritter, J. A., and Rosengart, A. J., Theoretical analysis of a simple yet efficient portable magnetic separator design for separation of magnetic nano/micro-carriers from human blood flow, 127–134. Copyright 2007, with permission from Elsevier.) (c) Simulation result of magnetic field distribution in the microfluidic channel where MS is conducted. The magnetic field is generated by a permanent magnet and concentrated by NiFe microcomb high gradient magnetic concentrator (HGMC). The top view (top) and side view (bottom) are demonstrated.[16] (With kind permission from Springer Science+Business Media: *Biomedical Microdevices*, Combined microfluidic-micromagnetic separation of living cells in continuous flow, 8, 2006, 299–308, Xia, N., Hunt, T. P., Mayers, B. T., Alsberg, E., Whitesides, G. M., Westervelt, R. M., and Ingber, D. E.)

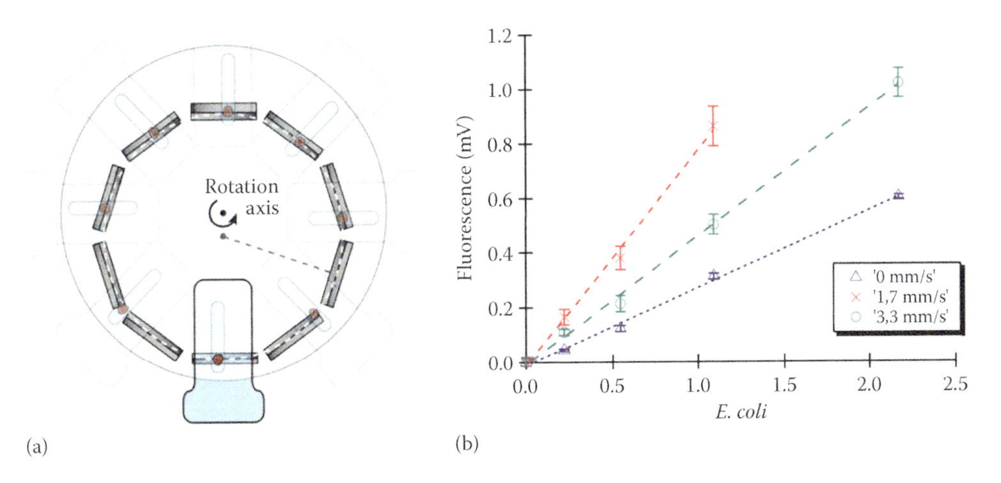

FIGURE 10.15 (a) Top view of the magnetic actuator (disk-shape plate with magnets located on the top) and reaction chamber. As the magnetic actuator is subjected to rotation, the magnetic beads in the reaction chamber will sweep through the whole domain of the chamber as driven by the magnetophoretic force. (b) Calibration curves (graph of fluorescence emission versus *E. coli* concentration) under different moving velocities (or different rotational speed of the magnetic actuator).[94] (With kind permission from Springer Science+Business Media: *Analytical and Bioanalytical Chemistry*, Magnetic actuator for the control and mixing of magnetic bead-based reactions on-chip, 406, 2014, 6607–6616, Berenguel-Alonso, M., Granados, X., Faraudo, J., Alonso-Chamarro, J., and Puyol, M.)

coated with carboxylic group) can be accomplished within a few minutes under the low gradient magnetic field generated by an NdFeB magnet. The fast separation of Dynabeads is contributed by their huge magnetic volume (due to their large particle size) in comparison to their counterparts with nanometre scale. The fast and highly specific cell separation performed by Dynabeads has proven the potential of this commercialized product in clinical application, especially biomedical diagnosis.

Additionally, Miltenyi Biotec has developed MACS MicroBeads, which are coated with antibody that is extremely specific against a particular antigen on the targeted cells.[97] In contrast to Dynabeads, MACS Microbeads have a much smaller dimension (~50 nm) such that they will not activate the cell, do not saturate cell surface epitopes and do not need to be separated for downstream processes. Furthermore, MACS MicroBeads are biodegradable, and hence, they have a very high potential to be implemented in real-time biomedical diagnosis. However, owing to the smaller size and magnetic volume of MACS MicroBeads, high gradient magnetic field must be applied in order to produce effective isolation of magnetically labelled targeted cells. In conjunction with this situation, MACS columns, which consist of the matrix of ferromagnetic spheres with cell-compactible coating, was introduced to amplify the magnetic field gradient by 10,000 times in the domain where MS is conducted.[98] The design of this column has employed the concept of HGMS, which is illustrated in Section 10.2.1.

10.7 CONCLUSION

MS is one of the separation schemes with great potential to be implemented in sample preparation or direct implementation for biomedical diagnosis. In fact, both HGMS and LGMS have been proven to be feasible in the detection of targeted entity, which can be expanded for diagnosis of specific diseases. However, it is crucial to dedicate substantial effort in the design and optimization of magnetic separator for biomedical diagnosis so that this technology can be further advanced. For diagnosis that involves a small-sized sample, the MS system can be miniaturized to improve the detection limit and efficiency of the diagnosis process. In contrast, the design of a more efficient macroscopic-scaled MS system (either as integrated or standalone device) is the major challenge for diagnosis processes that involve a larger amount of samples. Here, a fundamental understanding of MS plays a vital role in designing an optimum magnetic separator for diagnosis purposes. Additionally, we anticipate that the future development of MS-aided biomedical diagnosis technology can rely on the hydrodynamic effect to speed up the MS process. In this regard, the geometry configuration of an MS device (such as sample volume, shape of separation device, magnetic field gradient imposed on particle collection plane as well as its surface area[99]) plays an influential role in dictating the separation rate of magnetically labelled target and, hence, the performance of biomedical diagnosis. Finally, it is also equally important to improve the colloidal stability of the particle system used for magnetic labelling purposes in order to enhance the targeted entity tagging efficiency. Even though the highly stabilized particle system might slow down the MS rate,[70] the occurrence of particle aggregation (cooperative magnetophoresis, which is initiated at higher particle concentration) and induced convection (resulted from hydrodynamic effect) can boost the magnetophoretic collection of magnetically labelled targeted cells without sacrificing the tagging efficiency. In conclusion, the future advancement of MS-aided biomedical diagnosis technology should be focused on the interplay between MS parameters (particle size and concentration), magnetophoresis phenomena (particle aggregation and induced convection) as well as colloidal behaviours of the magnetic particles employed as tagging agents.

REFERENCES

1. Probst, C. E., Zrazhevskiy, P., and Gao, X. 2011. Rapid multitarget immunomagnetic separation through programmable DNA linker displacement. *Journal of the American Chemical Society* 133:17126–17129.
2. Yavuz, C. T., Prakash, A., Mayo, J. T., and Colvin, V. L. 2009. Magnetic separations: From steel plants to biotechnology. *Chemical Engineering Science* 64:2510–2521.
3. Pamme, N. 2007. Continuous flow separations in microfluidic devices. *Lab on a Chip* 7:1644–1659.
4. Zborowski, M. 2007. Commercial magnetic cell separation instruments and reagents. In *Magnetic Cell Separation*, ed. Zborowski, M.; Chalmers, J. J., 265–292. Elsevier, Amsterdam.
5. Gijs, M. A. M., Lacharme, F., and Lehmann, U. 2010. Microfluidic applications of magnetic particles for biological analysis and catalysis. *Chemical Reviews* 110:1518–1563.
6. Melville, D., Paul, F., and Roath, S. 1975. Direct magnetic separation of red cells from whole blood. *Nature* 255:706–706.
7. Nam, J., Huang, H., Lim, H., Lim, C., and Shin, S. 2013. Magnetic separation of malaria-infected red blood cells in various developmental stages. *Analytical Chemistry* 85:7316–7323.
8. Hackett, S., Hamzah, J., Davis, T. M. E., and St Pierre, T. G. 2009. Magnetic susceptibility of iron in malaria-infected red blood cells. *Biochimica et Biophysica Acta* 1792:93–99.
9. Chen, G. D., Alberts, C. J., Rodriguez, W., and Toner, M. 2010. Concentration and purification of human immunodeficiency virus type 1 virions by microfluidic separation of superparamagnetic nanoparticles. *Analytical Chemistry* 82:723–728.
10. Chen, Y., Xianyu, Y., Wang, Y., Zhang, X., Cha, R., Sun, J., and Jiang, X. 2015. One-step detection of pathogens and viruses: Combining magnetic relaxation switching and magnetic separation. *ACS Nano* 9:3184–3191.
11. Pamme, N., and Wilhelm, C. 2006. Continuous sorting of magnetic cells via on-chip free-flow magnetophoresis. *Lab on a Chip* 6:974–980.
12. Xu, H., Aguilar, Z. P., Yang, L., Kuang, M., Duan, H., Xiong, Y., Wei, H., and Wang, A. 2011. Antibody conjugated magnetic iron oxide nanoparticles for cancer cell separation in fresh whole blood. *Biomaterials* 32:9758–9765.
13. Robert, D., Pamme, N., Conjeaud, H., Gazeau, F., Iles, A., and Wilhelm, C. 2011. Cell sorting by endocytotic capacity in a microfluidic magnetophoresis device. *Lab on a Chip* 11:1902–1910.
14. Safarik, I., Maderova, Z., Pospiskova, K., Horska, K., and Safarikova, M. 2014. Chapter 10 Magnetic decoration and labeling of prokaryotic and eukaryotic cells. In *Cell Surface*

Engineering: Fabrication of Functional Nanoshells, ed. Fakhrullin, R., Choi, I., and Lvov, Y., 185–215. The Royal Society of Chemistry, Cambridge.

15. Safarik, I., Pospiskova, K., Baldikova, E., Maderova, Z., and Safarikova, M. 2016. Chapter 5 Magnetic modification of cells. In *Engineering of Nanobiomaterials*, ed. Grumezescu, A., 145–180. William Andrew Publishing, New York.
16. Xia, N., Hunt, T. P., Mayers, B. T., Alsberg, E., Whitesides, G. M., Westervelt, R. M., and Ingber, D. E. 2006. Combined microfluidic-micromagnetic separation of living cells in continuous flow. *Biomedical Microdevices* 8:299–308.
17. Olsvik, O., Popovic, T., Skjerve, E., Cudjoe, K. S., Hornes, E., Ugelstad, J., and Uhlen, M. 1994. Magnetic separation techniques in diagnostic microbiology. *Clinical microbiology reviews* 7:43–54.
18. Lien, K.-Y., Lin, J.-L., Liu, C.-Y., Lei, H.-Y., and Lee, G.-B. 2007. Purification and enrichment of virus samples utilizing magnetic beads on a microfluidic system. *Lab on a Chip* 7, 868–875.
19. Gundersen, S. G., Haagensen, I., Jonassen, T. O., Figenschau, K. J., de Jonge, N., and Deelder, A. M. 1992. Quantitative detection of schistosomal circulating anodic antigen by a magnetic bead antigen capture enzyme-linked immunosorbent assay (MBAC-EIA) before and after mass chemotherapy. *Transactions of the Royal Society of Tropical Medicine and Hygiene* 86:175–178.
20. Lim, J., Yeap, S. P., and Low, S. C. 2014. Challenges associated to magnetic separation of nanomaterials at low field gradient. *Separation and Purification Technology* 123:171–174.
21. Yeap, S. P., Leong, S. S., Ahmad, A. L., Ooi, B. S., and Lim, J. 2014. On size fractionation of iron oxide nanoclusters by low magnetic field gradient. *The Journal of Physical Chemistry C* 118:24042–24054.
22. Yeap, S. P., Toh, P. Y., Ahmad, A. L., Low, S. C., Majetich, S. A., and Lim, J. 2012. Colloidal stability and magnetophoresis of gold-coated iron oxide nanorods in biological media. *The Journal of Physical Chemistry C* 116, 22561–22569.
23. Leong, S. S., Ahmad, Z., and Lim, J. 2015. Magnetophoresis of superparamagnetic nanoparticles at low field gradient: Hydrodynamic effect. *Soft Matter* 11:6968–6980.
24. De Las Cuevas, G., Faraudo, J., and Camacho, J. 2008. Low-gradient magnetophoresis through field-induced reversible aggregation. *The Journal of Physical Chemistry C* 112:945–950.
25. Schaller, V., Kräling, U., Rusu, C., Petersson, K., Wipenmyr, J., Krozer, A., Wahnström, G., Sanz-Velasco, A., Enoksson, P., and Johansson, C. 2008. Motion of nanometer sized magnetic particles in a magnetic field gradient. *Journal of Applied Physics* 104:093918.
26. Lim, J., Lanni, C., Evarts, E. R., Lanni, F., Tilton, R. D., and Majetich, S. A. 2011. Magnetophoresis of nanoparticles. *ACS Nano* 5:217–226.
27. Bird, R. B., Stewart, W. E., and Lightfoot, E. N. 2006. *Transport Phenomena*. John Wiley & Sons, Inc.
28. Berg, H. C. 1993. *Random Walks in Biology*. Princeton University Press, Princeton.
29. Hatch, G. P., and Stelter, R. E. 2001. Magnetic design considerations for devices and particles used for biological high-gradient magnetic separation (HGMS) systems. *Journal of Magnetism and Magnetic Materials* 225:262–276.
30. Moeser, G. D., Roach, K. A., Green, W. H., Alan Hatton, T., and Laibinis, P. E. 2004. High-gradient magnetic separation of coated magnetic nanoparticles. *AIChE Journal* 50: 2835–2848.
31. Ditsch, A., Lindenmann, S., Laibinis, P. E., Wang, D. I. C., and Hatton, T. A. 2005. High-gradient magnetic separation of magnetic nanoclusters. *Industrial & Engineering Chemistry Research* 44:6824–6836.
32. Moeser, G. D., Roach, K. A., Green, W. H., Laibinis, P. E., and Hatton, T. A. 2002. Water-based magnetic fluids as extractants for synthetic organic compounds. *Industrial & Engineering Chemistry Research* 41:4739–4749.
33. Oder, R. 1976. High gradient magnetic separation theory and applications. *IEEE Transactions on Magnetics* 12:428–435.
34. Gómez-Pastora, J., Bringas, E., and Ortiz, I. 2014. Recent progress and future challenges on the use of high performance magnetic nano-adsorbents in environmental applications. *Chemical Engineering Journal* 256:187–204.
35. Yavuz, C. T., Mayo, J. T., Yu, W. W., Prakash, A., Falkner, J. C., Yean, S., Cong, L., Shipley, H. J., Kan, A., Tomson, M., Natelson, D., and Colvin, V. L. 2006. Low-field magnetic separation of monodisperse Fe_3O_4 nanocrystals. *Science* 314:964–967.
36. Faraudo, J., and Camacho, J. 2010. Cooperative magnetophoresis of superparamagnetic colloids: Theoretical aspects. *Colloid and Polymer Science* 288:207–215.
37. Faraudo, J., Andreu, J. S., Calero, C., and Camacho, J. 2016. Predicting the self-assembly of superparamagnetic colloids under magnetic fields. *Advanced Functional Materials* 26: 3837–3858.
38. Andreu, J. S., Calero, C., Camacho, J., and Faraudo, J. 2012. On-the-fly coarse-graining methodology for the simulation of chain formation of superparamagnetic colloids in strong magnetic fields. *Physical Review E* 85:036709.
39. Erb, R. M., Son, H. S., Samanta, B., Rotello, V. M., and Yellen, B. B. 2009. Magnetic assembly of colloidal superstructures with multipole symmetry. *Nature* 457:999–1002.
40. Faraudo, J., Andreu, J. S., and Camacho, J. 2013. Understanding diluted dispersions of superparamagnetic particles under strong magnetic fields: A review of concepts, theory and simulations. *Soft Matter* 9:6654–6664.
41. Toh, P. Y., Ng, B. W., Ahmad, A. L., Chieh, D. C. J., and Lim, J. 2014. The role of particle-to-cell interactions in dictating nanoparticle aided magnetophoretic separation of microalgal cells. *Nanoscale* 6:12838–12848.
42. Andreu, J. S., Camacho, J., and Faraudo, J. 2011. Aggregation of superparamagnetic colloids in magnetic fields: The quest for the equilibrium state. *Soft Matter* 7:2336–2339.
43. Pivetal, J., Royet, D., Ciuta, G., Frenea-Robin, M., Haddour, N., Dempsey, N. M., Dumas-Bouchiat, F., and Simonet, P. 2015. Micro-magnet arrays for specific single bacterial cell positioning. *Journal of Magnetism and Magnetic Materials* 380:72–77.
44. Zimmerman, P. A., Thomson, J. M., Fujioka, H., Collins, W. E., and Zborowski, M. 2006. Diagnosis of malaria by magnetic deposition microscopy. *The American journal of Tropical Medicine and Hygiene* 74:568–572.
45. Lin, J., Li, M., Li, Y., and Chen, Q. 2015. A high gradient and strength bioseparator with nano-sized immunomagnetic particles for specific separation and efficient concentration of E. coli O157:H7. *Journal of Magnetism and Magnetic Materials* 378:206–213.
46. Tamer, U., Cetin, D., Suludere, Z., Boyaci, I. H., Temiz, H. T., Yegenoglu, H., Daniel, P., Dinçer, İ., and Elerman, Y. 2013. Gold-coated iron composite nanospheres targeted the detection of *Escherichia coli*. *International Journal of Molecular Sciences* 14:6223–6240.

47. Sun, W., Khosravi, F., Albrechtsen, H., Brovko, L. Y., and Griffiths, M. W. 2002. Comparison of ATP and *in vivo* bioluminescence for assessing the efficiency of immunomagnetic sorbents for live *Escherichia coli* O157:H7 cells. *Journal of Applied Microbiology* 92:1021–1027.
48. Oh, N., and Park, J.-H. 2014. Endocytosis and exocytosis of nanoparticles in mammalian cells. *International Journal of Nanomedicine* 9:51–63.
49. Rejman, J., Oberle, V., Zuhorn, I. S., and Hoekstra, D. 2004. Size-dependent internalization of particles via the pathways of clathrin-and caveolae-mediated endocytosis. *Biochemical Journal* 377:159–169.
50. Chithrani, B. D., and Chan, W. C. W. 2007. Elucidating the mechanism of cellular uptake and removal of protein-coated gold nanoparticles of different sizes and shapes. *Nano Letters* 7:1542–1550.
51. Chithrani, B. D., Ghazani, A. A., and Chan, W. C. W. 2006. Determining the size and shape dependence of gold nanoparticle uptake into mammalian cells. *Nano Letters* 6:662–668.
52. Nel, A., Xia, T., Mädler, L., and Li, N. 2006. Toxic potential of materials at the nanolevel. *Science* 311:622–627.
53. Oberdörster, G., Oberdörster, E., and Oberdörster, J. 2005. Nanotoxicology: An emerging discipline evolving from studies of ultrafine particles. *Environmental Health Perspectives* 113(7):823–839.
54. Kunzmann, A., Andersson, B., Vogt, C., Feliu, N., Ye, F., Gabrielsson, S., Toprak, M. S., Buerki-Thurnherr, T., Laurent, S., Vahter, M., Krug, H., Muhammed, M., Scheynius, A., and Fadeel, B. 2011. Efficient internalization of silica-coated iron oxide nanoparticles of different sizes by primary human macrophages and dendritic cells. *Toxicology and Applied Pharmacology* 253:81–93.
55. Raju, H. B., Hu, Y., Vedula, A., Dubovy, S. R., and Goldberg, J. L. 2011. Evaluation of magnetic micro-and nanoparticle toxicity to ocular tissues. *PloS One* 6:e17452.
56. Magaye, R., Zhao, J., Bowman, L., and Ding, M. 2012. Genotoxicity and carcinogenicity of cobalt-, nickel-and copper-based nanoparticles (Review). *Experimental and Therapeutic Medicine* 4:551–561.
57. Lim, J., Yeap, S. P., Leow, C. H., Toh, P. Y., and Low, S. C. 2014. Magnetophoresis of iron oxide nanoparticles at low field gradient: The role of shape anisotropy. *Journal of Colloid and Interface Science* 421:170–177.
58. Parker, M. R. 1981. High gradient magnetic separation. *Physics in Technology* 12:263.
59. Kaittanis, C., Naser, S. A., and Perez, J. M. 2007. One-step, nanoparticle-mediated bacterial detection with magnetic relaxation. *Nano Letters* 7:380–383.
60. Hong, R., Cima, M. J., Weissleder, R., and Josephson, L. 2008. Magnetic microparticle aggregation for viscosity determination by MR. *Magnetic Resonance in Medicine* 59:515–520.
61. Tsourkas, A., Hofstetter, O., Hofstetter, H., Weissleder, R., and Josephson, L. 2004. Magnetic relaxation switch immunosensors detect enantiomeric impurities. *Angewandte Chemie International Edition* 43:2395–2399.
62. Koh, I., Hong, R., Weissleder, R., and Josephson, L. 2009. Nanoparticle–target interactions parallel antibody–protein interactions. *Analytical Chemistry* 81:3618–3622.
63. Schiettecatte, J., Anckaert, E., and Smitz, J. 2012. Interferences in immunoassays. In *Advances in Immunoassay Technology*, ed. Chiu, N. H. L., and Christopoulos, T. K., 45–65. InTech.
64. Fu, P. P., Xia, Q., Hwang, H.-M., Ray, P. C., and Yu, H. 2014. Mechanisms of nanotoxicity: Generation of reactive oxygen species. *Journal of Food and Drug Analysis* 22:64–75.
65. Häfeli, U. O., Riffle, J. S., Harris-Shekhawat, L., Carmichael-Baranauskas, A., Mark, F., Dailey, J. P., and Bardenstein, D. 2009. Cell uptake and *in vitro* toxicity of magnetic nanoparticles suitable for drug delivery. *Molecular Pharmaceutics* 6:1417–1428.
66. Pisanic Ii, T. R., Blackwell, J. D., Shubayev, V. I., Fiñones, R. R., and Jin, S. 2007. Nanotoxicity of iron oxide nanoparticle internalization in growing neurons. *Biomaterials* 28:2572–2581.
67. Phenrat, T., Saleh, N., Sirk, K., Kim, H.-J., Tilton, R. D., and Lowry, G. V. 2008. Stabilization of aqueous nanoscale zerovalent iron dispersions by anionic polyelectrolytes: Adsorbed anionic polyelectrolyte layer properties and their effect on aggregation and sedimentation. *Journal of Nanoparticle Research* 10:795–814.
68. Golas, P. L., Louie, S., Lowry, G. V., Matyjaszewski, K., and Tilton, R. D. 2010. Comparative study of polymeric stabilizers for magnetite nanoparticles using ATRP. *Langmuir* 26:16890–16900.
69. Lim, J. K., Majetich, S. A., and Tilton, R. D. 2009. Stabilization of superparamagnetic iron oxide core–gold shell nanoparticles in high ionic strength media. *Langmuir* 25:13384–13393.
70. Yeap, S. P., Ahmad, A. L., Ooi, B. S., and Lim, J. 2012. Electrosteric stabilization and its role in cooperative magnetophoresis of colloidal magnetic nanoparticles. *Langmuir* 28:14878–14891.
71. Roca, A., Costo, R., Rebolledo, A., Veintemillas-Verdaguer, S., Tartaj, P.; Gonzalez-Carreno, T., Morales, M., and Serna, C. 2009. Progress in the preparation of magnetic nanoparticles for applications in biomedicine. *Journal of Physics D: Applied Physics* 42:224002.
72. Kolhar, P., Anselmo, A. C., Gupta, V., Pant, K., Prabhakarpandian, B., Ruoslahti, E., and Mitragotri, S. 2013. Using shape effects to target antibody-coated nanoparticles to lung and brain endothelium. *Proceedings of the National Academy of Sciences* 110:10753–10758.
73. Song, E.-Q., Hu, J., Wen, C.-Y., Tian, Z.-Q., Yu, X., Zhang, Z.-L., Shi, Y.-B., and Pang, D.-W. 2011. Fluorescent-magnetic-biotargeting multifunctional nanobioprobes for detecting and isolating multiple types of tumor cells. *ACS Nano* 5:761–770.
74. Foddai, A., Elliott, C. T., and Grant, I. R. 2010. Maximizing capture efficiency and specificity of magnetic separation for *Mycobacterium avium* subsp. paratuberculosis cells. *Applied and Environmental Microbiology* 76:7550–7558.
75. Chalmers, J. J., Tong, X., Lara, O., and Moore, L. R. 2007. Preparative applications of magnetic separation in biology and medicine. In *Laboratory Techniques in Biochemistry and Molecular Biology*, ed. Zborowski, M., and Chalmers, J. J., 249–264. Elsevier, Amsterdam.
76. Yantasee, W., Warner, C. L., Sangvanich, T., Addleman, R. S., Carter, T. G., Wiacek, R. J., Fryxell, G. E., Timchalk, C., and Warner, M. G. 2007. Removal of heavy metals from aqueous systems with thiol functionalized superparamagnetic nanoparticles. *Environmental Science & Technology* 41:5114–5119.
77. Hu, J., Lo, I. M. C., and Chen, G. 2007. Comparative study of various magnetic nanoparticles for Cr(VI) removal. *Separation and Purification Technology* 56:249–256.
78. Zhou, L., Deng, H., Wan, J., Shi, J., and Su, T. 2013. A solvothermal method to produce RGO-Fe_3O_4 hybrid composite for fast chromium removal from aqueous solution. *Applied Surface Science* 283:1024–1031.
79. Kong, L., Gan, X., Ahmad, A. L. b., Hamed, B. H., Evarts, E. R., Ooi, B., and Lim, J. 2012. Design and synthesis of magnetic nanoparticles augmented microcapsule with catalytic and magnetic bifunctionalities for dye removal. *Chemical Engineering Journal* 197:350–358.

80. Toh, P. Y., Yeap, S. P., Kong, L. P., Ng, B. W., Chan, D. J. C., Ahmad, A. L., and Lim, J. K. 2012. Magnetophoretic removal of microalgae from fishpond water: Feasibility of high gradient and low gradient magnetic separation. *Chemical Engineering Journal* 211–212:22–30.
81. Ge, S., Agbakpe, M., Wu, Z., Kuang, L., Zhang, W., and Wang, X. 2015. Influences of surface coating, UV irradiation and magnetic field on the algae removal using magnetite nanoparticles. *Environmental Science & Technology* 49:1190–1196.
82. Qu, B.-Y., Wu, Z.-Y., Fang, F., Bai, Z.-M., Yang, D.-Z., and Xu, S.-K. 2008. A glass microfluidic chip for continuous blood cell sorting by a magnetic gradient without labeling. *Analytical and Bioanalytical Chemistry* 392:1317–1324.
83. Zborowski, M., Sun, L., Moore, L. R., Stephen Williams, P., and Chalmers, J. J. 1999. Continuous cell separation using novel magnetic quadrupole flow sorter. *Journal of Magnetism and Magnetic Materials* 194:224–230.
84. Moore, L. R., Milliron, S., Williams, P. S., Chalmers, J. J., Margel, S., and Zborowski, M. 2004. Control of magnetophoretic mobility by susceptibility-modified solutions as evaluated by cell tracking velocimetry and continuous magnetic sorting. *Analytical Chemistry* 76:3899–3907.
85. Chalmers, J. J., Zborowski, M., Sun, L., and Moore, L. 1998. Flow through, immunomagnetic cell separation. *Biotechnology Progress* 14:141–148.
86. Zborowski, M., and Chalmers, J. J. 2011. Rare cell separation and analysis by magnetic sorting. *Analytical Chemistry* 83:8050–8056.
87. Andreu, J. S., Camacho, J., Faraudo, J., Benelmekki, M., Rebollo, C., and Martínez, L. M. 2011. Simple analytical model for the magnetophoretic separation of superparamagnetic dispersions in a uniform magnetic gradient. *Physical Review E* 84:021402.
88. Hilton, J. E., and McMurry, S. M. 2012. An adjustable linear Halbach array. *Journal of Magnetism and Magnetic Materials* 324:2051–2056.
89. Hoyos, M., Moore, L., Williams, P. S., and Zborowski, M. 2011. The use of a linear Halbach array combined with a step-SPLITT channel for continuous sorting of magnetic species. *Journal of Magnetism and Magnetic Materials* 323:1384–1388.
90. Babinec, P., Krafčík, A., Babincová, M., and Rosenecker, J. 2010. Dynamics of magnetic particles in cylindrical Halbach array: Implications for magnetic cell separation and drug targeting. *Medical & Biological Engineering & Computing* 48:745–753.
91. Chen, H., Bockenfeld, D., Rempfer, D., Kaminski, M. D., Liu, X., and Rosengart, A. J. 2008. Preliminary 3-D analysis of a high gradient magnetic separator for biomedical applications. *Journal of Magnetism and Magnetic Materials* 320:279–284.
92. Chen, H., Kaminski, M. D., and Rosengart, A. J. 2008. 2D modeling and preliminary *in vitro* investigation of a prototype high gradient magnetic separator for biomedical applications. *Medical Engineering & Physics* 30:1–8.
93. Chen, H., Kaminski, M. D., Ebner, A. D., Ritter, J. A., and Rosengart, A. J. 2007. Theoretical analysis of a simple yet efficient portable magnetic separator design for separation of magnetic nano/micro-carriers from human blood flow. *Journal of Magnetism and Magnetic Materials* 313:127–134.
94. Berenguel-Alonso, M., Granados, X., Faraudo, J., Alonso-Chamarro, J., and Puyol, M. 2014. Magnetic actuator for the control and mixing of magnetic bead-based reactions on-chip. *Analytical and Bioanalytical Chemistry* 406:6607–6616.
95. Thermo Fisher Scientific. Dynabeads® Products & Technology for Magnetic Bead Separation. https://www.thermofisher.com/my/en/home/brands/product-brand/dynal/dynabeads-technology.html
96. Helseth, L. E., and Skodvin, T. 2009. Optical monitoring of low-field magnetophoretic separation of particles. *Measurement Science and Technology* 20:095202.
97. Miltenyi Biotec. MACS Technology MicroBeads—The most trusted and proven MACS MicroBeads. http://www.miltenyibiotec.com/en/products-and-services/macs-cell-separation/macs-technology/microbeads_dp.aspx
98. Miltenyi Biotec. MACS manual Cell Separation Columns. http://www.miltenyibiotec.com/en/products-and-services/macs-cell-separation/manual-cell-separation/columns.aspx
99. Leong, S. S., Ahmad, Z., Camacho, J., Faraudo, J., and Lim, J. 2017. Kinetics of low field gradient magnetophoresis in the presence of magnetically induced convection. *The Journal of Physical Chemistry C* 121:5389–5407.

JitKang Lim (E-mail: chjitkangl@usm.my) earned his PhD degree in chemical engineering at Carnegie Mellon University (CMU) in 2009. He is currently an associate professor of chemical engineering at Universiti Sains Malaysia (USM). Since October 2009, he has been holding a courtesy appointment as the Visiting Research Professor of Physics at CMU. He is actively engaged in international collaboration, working on numerous projects related to magnetic nanoparticles, and has served as visiting scientist/professor of the Korean Advanced Institute of Technology (KAIST), Chinese Academy of Science (CAS) and National University of Singapore (NUS). In addition, he is the founding chair of the American Chemical Society (ACS) Local Chapter in Malaysia. His current research interests focus on both colloidal behaviours and engineering application of magnetic nanoparticles.

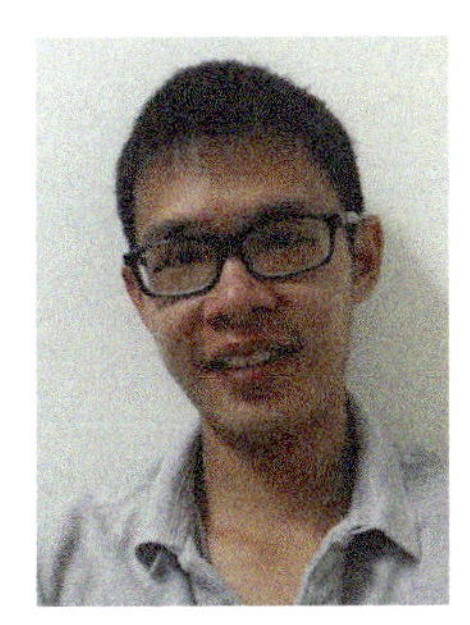

Sim Siong Leong was born in Seremban, Malaysia. He earned his bachelor's degree in chemical engineering at Universiti Sains Malaysia (USM) in 2013. Currently, he is pursuing a PhD degree under the supervision of Assoc. Prof. Dr JitKang

Lim in the School of Chemical Engineering, USM. His area of research is nanoscience and nanotechnology, which includes (i) theoretical modelling of magnetophoresis of magnetic nanoparticles under low magnetic field gradient and (ii) design of high efficient magnetic separator to optimize the recollection of magnetic nanoparticles from its suspension.

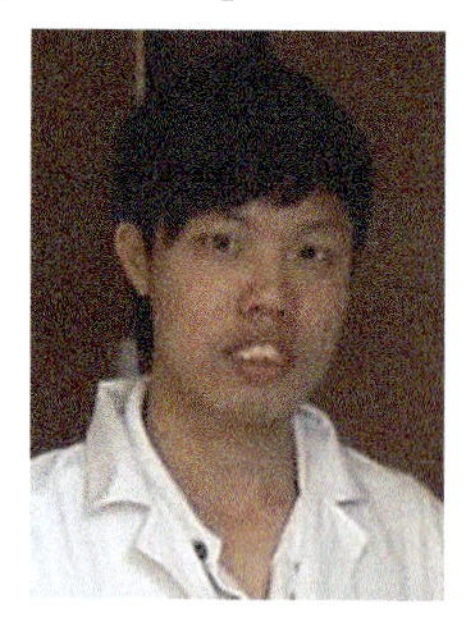

Swee Pin Yeap earned his bachelor's degree in chemical engineering at Universiti Sains Malaysia, USM, in 2011. Later on, he earned his PhD in chemical engineering under the supervision of Assoc. Prof. Dr JitKang Lim at USM, with specialisation in nanoscience and technology. Currently, he holds a lecturer position in the Faculty of Engineering, Technology & Built Environment, UCSI University, Cheras, Malaysia. Over the past 5 years, he has been actively involved in nanoscience research, which includes (i) synthesis and stabilization of magnetic nanoparticles, (ii) fundamental study on magnetophoresis of magnetic nanoparticles, as well as (iii) study on interface interaction between functionalized magnetic nanoparticles with heavy metal ions and natural dissolved organic matters. He is interested to further venture into the research area on colloid and interface science.

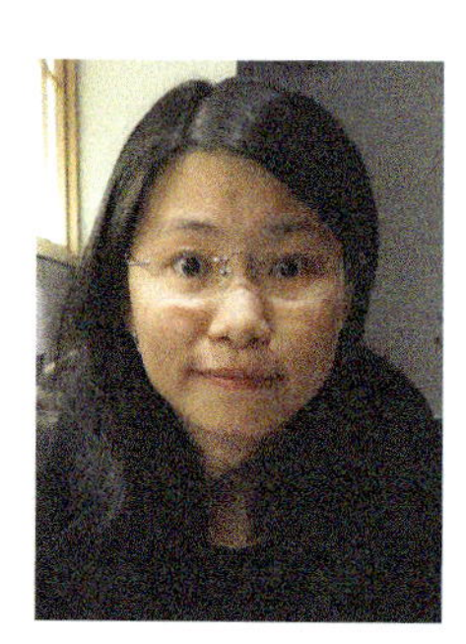

Siew Chun Low earned her PhD in chemical engineering at Universiti Sains Malaysia (USM) in 2010. She is currently a senior lecturer of chemical engineering at USM. Her research is focused on the design and synthesis of polymeric membrane (or advanced materials) tailored towards membrane defouling, biomedical and wastewater applications. Recently, she has been working on the design and synthesis of molecularly imprinting polymer (MIP) for the sensing of creatinine in blood serum, which in turn behaves as a clinical indicator for renal dysfunctions.

Rohimah Mohamud earned her MSc in vaccinology at the School of Health Sciences, University Sains Malaysia, in 2008. Later, in 2014, she graduated from Monash University, Melbourne, Australia, with a PhD in immunology under the supervision of Prof. Magdalena Plebanski, Prof. Robyn O'Hehir, Prof. Jennifer Rolland and Dr Charles Hardy. Her PhD thesis was focused on the effects of nanoparticles on regulatory T cells in the lungs of murine disease model. Currently, she is working as a lecturer at Universiti Sains Malaysia. Her current research interests focus on nanomedicine development and application for allergic diseases.

11 Magnetic Separation in Integrated Micro-Analytical Systems

Kazunori Hoshino

CONTENTS

11.1 INTRODUCTION

Magnetized materials under controlled magnetic fields experience magnetic forces. These magnetic forces have been utilized to separate magnetically labelled biosamples. Immunomagnetic separation is among the most commonly practiced forms of magnetic separation for biomedical applications. Functionalization with antibodies has allowed magnetic carriers to selectively label target objects. Several types of proteins, nucleic acids and cells have been successfully separated through immunomagnetic separation for a variety of applications. Magnetic separation is especially useful when it is combined with the modern microfabrication and nanofabrication technology. Magnetic separation in microfluidic devices has been intensively studied. While a small amount of biosamples flow in a microscale flow channel, magnetic fields from integrated magnets or solenoids can efficiently separate targets and collect them for further analysis. In this chapter, we review integrated micro-analytical systems designed and fabricated for magnetic separation. The chapter is composed of three sections of principles, materials and applications.

In Section 11.2, we discuss the working principles of magnetic separation systems. We describe the magnetization of different materials and forces exerted on magnetized materials. Theoretical consideration on the efficacy of separation systems with different the sizes of samples and magnetic carriers is also presented.

In Section 11.3, we discuss the materials used in micro-analytical systems. The efficacy of a separation system comes down to the (1) choice of magnetic carriers, (2) creation of a controlled magnetic field and (3) design of separation systems. In Section 11.3.1, we discuss the different types of magnetic carriers, including commercially available magnetic microbeads and several magnetic nanoparticles in important studies. In Section 11.3.2, the types of permanent magnets used in several separation systems are discussed. In Section 11.3.3, we describe the designs and methods of separation systems integrated with actively controlled magnets. Microfluidic separation systems from various groups are also described here.

In Section 11.4, practical applications of magnetic separation in biomedical applications are described. Focus will be on immunomagnetic separation of cells. In particular, techniques to separate circulating tumour cells (CTCs), including the author's recent studies, are reviewed to discuss the challenges and promises of magnetic separation in engineered micro total analytical systems.

11.2 PRINCIPLES

11.2.1 Magnetization

Magnetization is the vector field induced in a material when it is exposed to a magnetic field. When a magnetic field H is imposed to a sample, its response in magnetization M is characterized by the magnetization hysteresis loop, as shown in Figure 11.1. As H increases, an initially unmagnetized material reaches the saturated state ($M = M_s$), and remains magnetized ($M = M_r$) when H is reduced to 0. M_s and M_r are called

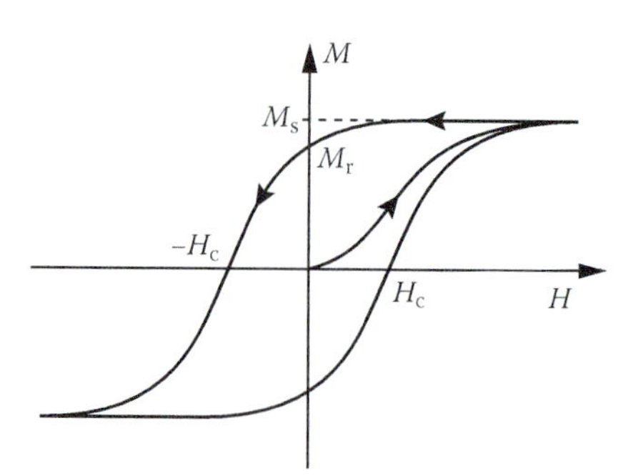

FIGURE 11.1 Hysteresis loop characterizes magnetic materials.

the saturated magnetization and the remanence. The coercive field $H = -H_c$ is needed to change the sign of M. Based on M_s, M_r, and H_c in the hysteresis curve, magnetic materials are classified into soft or hard magnetic materials. *Hard magnetic materials*, or hard magnets, have broad magnetization loops and retain their magnetization in zero magnetic field. *Soft magnetic materials*, or soft magnets, have narrow loops and lose their magnetization when the magnetic field is removed. Soft magnetic materials are sometimes called *electromagnets* because they are often used in electrically induced magnets whose magnetization can be actively switched or flipped.

In this chapter, we describe mainly three types of magnets in terms of functionality: (1) magnetic carriers, (2) permanent magnets and (3) actively controlled magnets.

11.2.1.1 Magnetic Carriers

Magnetic carriers attach to target objects and carry them under an applied magnetic field. In magnetic separation, carriers should not retain magnetism when no magnetic field is applied. Otherwise, carriers will form aggregation or clumping as a result of attractive forces between carriers and lose proper functionality. In this chapter, we model magnetic carriers as a spherical material that is magnetized with a linear magnetization function, namely

$$\mathbf{H} = \Delta\chi\mathbf{M}, \tag{11.1}$$

where $\Delta\chi$ is the carrier's effective susceptibility relative to the medium (see Figure 11.2).

When the size of a magnetic material is as small as a few to few tens of nanometres, it may be composed of a single domain in which magnetization occurs in a uniform direction. Such a nanomaterial shows a characteristic called *superparamagnetism*, which is well modelled with the linear function shown in Figure 11.2. Superparamagnetic materials do not

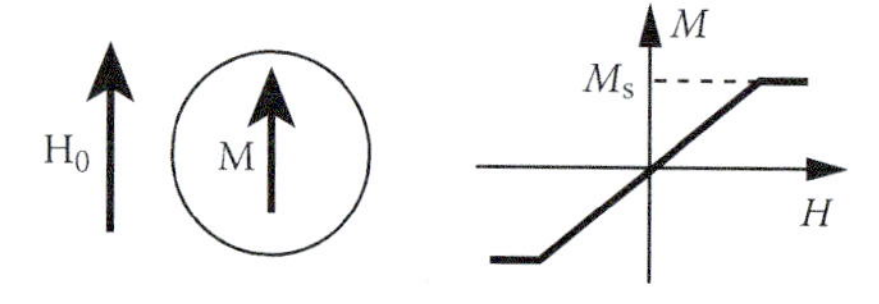

FIGURE 11.2 Magnetic carrier modelled with a linear magnetization function.

retain magnetism when the magnetic field is removed and are preferred for magnetic carriers. Materials that have been used as magnetic carriers include metal nanoparticles, metal oxide nanoparticles, and polymer microbeads which bind magnetic materials inside.

11.2.1.2 Permanent Magnets

Fixed hard magnets are used as permanent magnets that provide constant magnetic fields. They are useful because once magnetized, they work alone without any power supply. In the design of magnetic separation systems, permanent magnets are often simply modelled as fixed magnetic dipoles.

11.2.1.3 Actively Controlled Magnets

Actively controlled magnets are used in analytical systems in which magnetization needs to be changed during operation. They are often micromagnets integrated in a microfluidic system. Use of soft magnets is preferred when time-controlled direction or manipulation of target objects is needed. Magnetization of soft materials is modelled considering the magnetic anisotropy and the demagnetizing field. Unless the material is spherical or the applied magnetic field is aligned to the 'easy axis' of the material, the magnetization vector is not parallel to the applied field. It depends on the shape of the material.

The magnetization $\mathbf{M}$ under the magnetic field $\mathbf{H}_0$ will experience a torque, $\mathbf{T}_{\text{field}}$, given as follows:

$$\mathbf{T}_{\text{field}} = \mathbf{M} \times \mathbf{H}_0. \tag{11.2}$$

When a magnetic material is magnetized, the magnetization generates the demagnetization field $\mathbf{H}_\text{d}$ in the magnet. Because of the magnetic anisotropy, the magnetization and the demagnetization occur in different orientations. This induces anisotropic force $\mathbf{T}_\text{d}$ by the demagnetization $\mathbf{H}_\text{d}$.

$$\mathbf{T}_\text{d} = \mathbf{M} \times \mathbf{H}_\text{d}. \tag{11.3}$$

The orientation of magnetization satisfies that these torques are equal and opposite to each other.

$$\mathbf{T}_{\text{field}} = -\mathbf{T}_\text{d} \tag{11.4}$$

The magnetization vector $\mathbf{M}$ is generated somewhere between that of the external magnetic field and the easy axis of the material (see Figure 11.3a). If the material is unrestrained, the overall torque on the material will rotate it until it is eventually aligned with the magnetic field. If the material is restricted, or mechanically fixed in a system, the magnetization stays 'off-axis' of the magnetic field and the easy axis. Magnetization becomes stronger when the direction of the easy axis is aligned to the applied magnetic field. This characteristic is used to 'focus' the magnetic field in integrated separation systems. As we will see in many examples, when there is a magnetic object with sharp edges, edges aligned to the magnetic field create larger field to attract more magnetic carriers.

It is difficult to find an analytical solution for the magnetization of a body with irregular shapes. Ellipsoid geometries that have semi axes a, b and c for the three principle axes x, y and z, respectively, are often used as soft magnet models because magnetization can be analytically solved.[1] It is possible to consider anisotropy induced by shapes.

When an ellipsoid is uniformly magnetized in a magnetic field $\mathbf{H}_0$, the relationship between the component of the internal magnetic field and the magnetization in a principle axis i ($I = x, y, z$) is given as

$$H_i = (H_0)_i - \frac{\gamma_i}{4\pi} M_i \quad i = x, y, z. \tag{11.5}$$

Here, γ_i is the demagnetization factor and can be analytically solved and described.[1] For a sphere, the internal magnetic field is

$$\gamma_x = \gamma_y = \gamma_z = 4\pi/3. \tag{11.6}$$

When an analytical solution for other shapes is needed, a commonly used simplification is to assume an infinite cylinder ($a \gg b$ and $a \gg c$) or an infinite plate ($a \gg c$ and $b \gg c$), as shown in Figure 11.3. In the case of an infinite cylinder,

$$\gamma_x = 0, \quad \gamma_y = \gamma_z = 2\pi. \tag{11.7}$$

And for the case of an infinite thin plate,

$$\gamma_x = 4\pi, \quad \gamma_y = \gamma_z = 0. \tag{11.8}$$

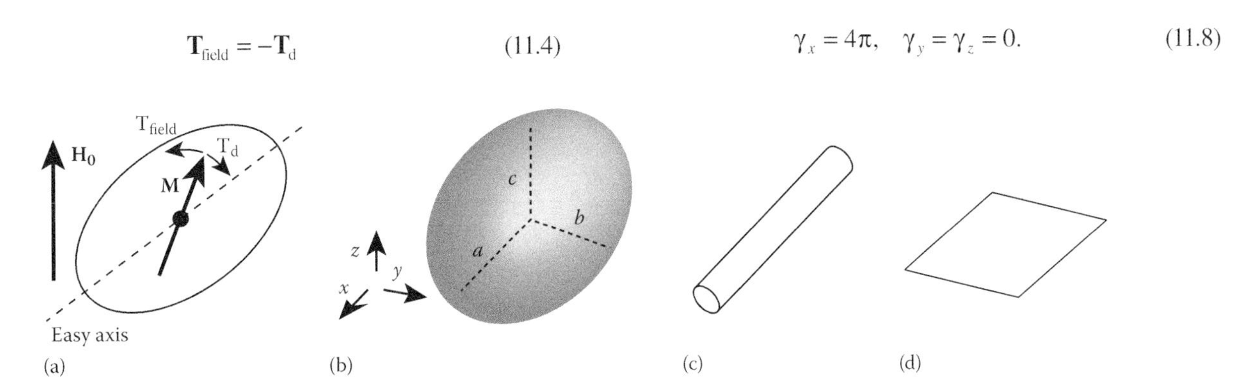

FIGURE 11.3 (a) Magnetization of an anisotropic soft material comes between H_0 and the easy axis. (b) Ellipsoid is frequently used as a model to study magnetization of soft materials. (c) Infinite cylinder and (d) infinite thin plate are modelled as an ellipsoid.

With these assumptions, the anisotropic torque can be easily calculated and the magnetization can be found.

11.2.2 Magnetic Force

We model a magnetic carrier as a magnetic dipole to calculate the force used in magnetic separation. The magnetic force that acts on a carrier with a magnetic dipole moment (**m**) under a magnetic field (**B**) is expressed in the following equation:

$$\mathbf{F}_m = (\mathbf{m} \cdot \nabla)\mathbf{B}. \tag{11.9}$$

The dipole moment **m** is given as the product of the carrier magnetization **M** and the volume V. Using a linear magnetization model (1), the dipole moment is written as:

$$\mathbf{m} = V\mathbf{M} = V\Delta\chi\mathbf{H}. \tag{11.10}$$

The magnetic field in a dilute suspension of particles in water can be approximated as

$$\mathbf{B} = \mu_0\mathbf{H}, \tag{11.11}$$

where $\mu_0 = 4\pi \times 10^{-7}\ \mathrm{T \cdot m \cdot A^{-1}}$ is the magnetic permeability of a vacuum. From Equations 11.9 through 11.11, $\mathbf{F}_m$ is given as

$$\mathbf{F}_m = \frac{V\Delta\chi}{\mu_0}(\mathbf{B} \cdot \nabla)\mathbf{B}. \tag{11.12}$$

When there is no electric current in the system, Ampère's law gives $\nabla \times \mathbf{B} = 0$, and Equation 11.12 can be simplified into

$$(\mathbf{B} \cdot \nabla)\mathbf{B} = \frac{1}{2}\nabla(\mathbf{B} \cdot \mathbf{B}) = \frac{1}{2}\nabla B^2, \tag{11.13}$$

where B is the magnitude of the magnetic field **B**, or it can be simply referred to as the magnetic field intensity. From Equations 11.12 and 11.13, the force $\mathbf{F}_m$ is found as

$$\mathbf{F}_m = \frac{V\Delta\chi}{2\mu_0}\nabla B^2. \tag{11.14}$$

In biomedical applications, carriers are usually suspended in an aqueous solution. Drag force works against the magnetic force described earlier, defining the velocity of magnetic carriers in the process of separation. When we assume a spherical carrier, the drag force that applies on the carrier in the liquid medium follows the Stokes law and is given by

$$\mathbf{F}_d = 3\pi \cdot \eta \cdot D \cdot \Delta\mathbf{v}, \tag{11.15}$$

where η is the medium viscosity, $\Delta\mathbf{v}$ is the carrier velocity relative to the medium and D is the carrier diameter. It is important to note that $\Delta\mathbf{v}$ is the 'relative' velocity. In many cases, especially within micro-analytical systems, magnetic carriers travel along the medium flow in a microfluidic channel. Only the velocity component relative to the medium flow works for the drag force, while the total velocity has to be considered for carrier transportation. The relative velocity is found by assuming quasi-static motion, where the two forces $\mathbf{F}_d$ and $\mathbf{F}_m$ equal each other:

$$\mathbf{F}_d = \mathbf{F}_m. \tag{11.16}$$

Equations 14 through 16 and $V = (1/6) \cdot \pi D^3$ give the carrier instant velocity:

$$\Delta v = \frac{D^2\Delta\chi}{36\mu_0\eta}\nabla B^2. \tag{11.17}$$

When multiple carriers are attached to a much larger object such as a cell (*ca.* 5–15 μm in diameter), D and $\Delta\chi$ need to be modified accordingly. Details are discussed in Section 11.2.3. Again, $\Delta\mathbf{v}$ is a relative velocity to the medium. When the medium flows in a fluidic device at a velocity of $\mathbf{v}_m$, the total velocity $\mathbf{v}_p$ of the moving nanoparticle becomes

$$\mathbf{v}_p = \mathbf{v}_m + \Delta\mathbf{v}. \tag{11.18}$$

11.2.3 Magnetic Separation

11.2.3.1 Magnetic Labelling

In magnetic separation, target objects (molecules, cells, other particles) need to be magnetically labelled. In a typical assay, magnetic carriers are functionalized with affinity ligands that attach to the targets to be separated. Antibodies, which recognize specific proteins, are commonly used as a ligand for biomedical applications. Affinity ligands used for magnetic separation other than antibodies include nitrilotriacetic acid, glutathione, trypsin, trypsin inhibitor and gelatine.[2] We will review different types of functionalization and assay to bind magnetic carriers in Section 11.4.

11.2.3.2 Size of Magnetic Carriers

The size of magnetic carriers is an important parameter in designing a magnetic separation system. Here, we discuss the effect of particle sizes on magnetic carrier-based separation. One important factor in a magnetic separation assay is the surface-to-volume ratio of the magnetic carriers. When carriers are surface-functionalized, the surface area on each carrier defines the amount of capturing molecules or affinity ligand. On the other hand, the magnetic force used in the separation step is proportional to the carrier volume. Surface-to-volume ratio is thus a parameter that can be used to compare the efficacy of labelling and the intensity of magnetic forces.

Figure 11.4a and b illustrates different scenarios with respect to the sizes of magnetic carriers and targets. Figure 11.4a shows cases where analyte molecules are smaller than

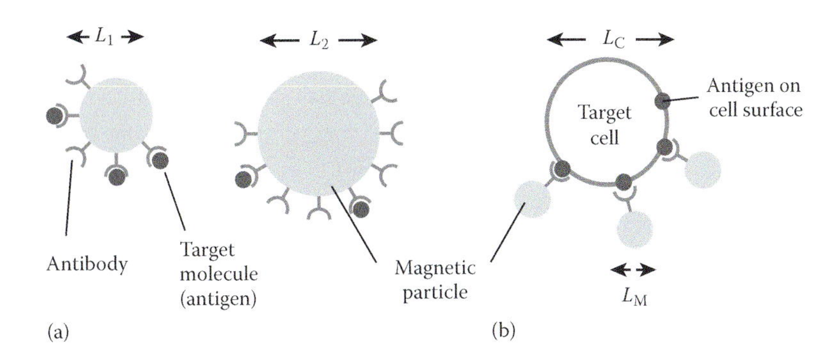

FIGURE 11.4 Size effect of magnetic carriers. (a) Magnetic particles capture smaller targets such as proteins or DNA pieces. (b) Magnetic particles label larger targets such as cells.

the carriers. Separation of bacteria,[3,4] viruses[5] or macromolecules, including DNAs,[6,7] proteins[8,9] and exosomes,[10] is categorized in this group. Two carriers with different diameters of L_1 and L_2 are compared in the figure. In order to consider the size effect, we assume that carriers are spheres, and we use the diameter of the carrier to be the characteristic length L.

The amount of capturing molecules that can be immobilized on the particle surface is proportional to L^2. If we simply consider labelling efficiency E_{label} as the number of capturing molecules per particle volume,

$$E_{\text{label}} \propto L^2/L^3 = L^{-1}. \tag{11.19}$$

This means that smaller carriers are more efficient in terms of labelling. For this reason, magnetic nanoscale particles are intensively studied as carriers for microbiological assays.

However, we need to consider the other factor, which is the forces that act on the magnetic carriers. The velocity of the carrier is found by solving the equilibrium equation for the drag force and the magnetic force as we have already seen in Equation 11.16. Stokes law in Equation 11.15 describes that the viscous force that acts on a small sphere in a liquid medium is proportional to the diameter L (note: for a larger fluidic system like a wing of an airplane, drag is proportional to the surface area, i.e. the square of the characteristic length). The magnetic force acting on the carrier is defined by the total volume V, which is proportional to L^3. The equilibrium equation for the viscous force and the magnetic force expressed in Equation 11.17 is thus proportional to L^2.

$$\Delta \mathbf{v} = \frac{R^2 \Delta\chi}{9\mu_0 \eta} \nabla B^2 \propto L^2 \tag{11.20}$$

As one can see from this equation, larger particles have larger separation speed. For this consideration, larger carriers are faster.

Figure 11.4b illustrates the separation of a target larger than the carriers. This is usually the case with magnetic cell separation. Antibodies attached on the carriers bind to antigen expressed on the cell surface. In this case, the viscous force is defined by the diameter of the cell, not by the diameter of carriers.

When we consider the magnetic force acting on the cell, we model a labelled cell as a weakly magnetized sphere with average magnetic volumetric susceptibility $\Delta\chi_C$. Here, $\Delta\chi_C$ is found by the number of magnetic particles N attached and the cell volume.

$$\Delta\chi_c = N \frac{L_M^3}{L_C^3} \Delta\chi M, \tag{11.21}$$

where $\Delta\chi_F$ is the volumetric susceptibility of the magnetic carrier and L_M and L_C are diameters of the magnetic particle and the cell, respectively. As one can see in Equation 11.21, larger particles can create larger magnetic forces with smaller number of particles; however, as we discussed earlier, larger particles have smaller chance of labelling, because the number of capturing N molecules per magnetic volume is smaller. Smaller particles have more chances of binding to target cells and larger can be expected. However, since L_M is smaller, smaller particles are not necessarily more efficient. In either case of Figure 11.4a or b, the optimal size of the carrier depends on the balance between labelling efficiency and the separation speed, and it changes depending on applications. For example, if the target molecules or objects are very rare (which is the case with CTCs, discussed in Section 11.4), smaller particles may be beneficial because the labelling step is the critical step in such a system. On the other hand, if the target is abundant and simple and quick separation is desired, larger particles or beads may be chosen. In practice, micrometre-sized beads as Dynabeads (diameter: 1 μm–4.5 μm) rather than nanometre-sized particles are often used for cell separation. We will see several types magnetic carriers with different materials, sizes and functionalization in the following section.

11.3 MATERIALS

In this section, we review the materials that compose magnetic separation systems: (1) magnetic carriers, (2) permanent magnets and (3) integrated actively controlled magnets.

11.3.1 Magnetic Carriers Used in Magnetic Separation Systems

Table 11.1 summarizes the types of magnetic carriers and separation methods described in this chapter.

TABLE 11.1
Magnetic Carriers and Separation Methods Reviewed in This Chapter

Size	Magnetic Material	Coating	Target	Separation Method	Product Names, References
3–5 μm	Fe_3O_4	Polymer (with porous silica and quantum dots)	Separation among beads	Small volume on glass coverslip	Sathe et al.[11]
1–4.5 μm	Fe_2O_3 Fe_3O_4	Polymer	Cells, bacteria, proteins, DNA, mRNA	Test tubes, microfluidics, tweezers	Dynabeads[3,12–15]
200 nm–8 μm	–	Polystyrene (fluorescent)	Cells, proteins, DNA	Microfluidics, test tubes	SPHERO magnetic particles[16]
200–500 nm	Iron oxide		*Escherichia coli*, proteins, DNA	Microfluidics, test tube	Bio-Adembeads[17]
250 nm	Fe_3O_4	Graphite oxide	Tumour cells (immunoassay)	Microfluidics	Yu et al.[18]
100–200 nm	Fe_3O_4	Polymer	Tumour cells (immunoassay)	Robotic system (test tube)	CellSearch ferrofluid[19]
50 nm	Iron oxide	Dextran	Cells (immunoassay)	MACS columns, robotic system	MACS microbead[20]
45 nm	Iron oxide	Dextran	Tumour cells (uptake)	MACS columns	Lewin et al.[21]
3–15 nm	Fe_2O_3	Dextran	Tumour cells (uptake)	Flow cell	Wilhelm et al.[22]
5–10 nm	Fe_3O_4	Gold	Tumour cells (immunoassay)	Microfluidics	Wu et al.[23]

The diameters of magnetic carriers range from ~5 nm to ~5 μm. Smaller carriers are typically metal or metal oxide nanoparticles, while larger carriers are commercially available polymer microbeads, which contain magnetic nanoparticles inside. Most magnetic carriers comprise of iron oxide as the magnetic material. Maghemite (γ-Fe_2O_3) and magnetite (Fe_3O_4) are very commonly used to construct carriers. They are easily available and found to be biocompatible.[24] The particle sizes are typically about 5–20 nm in diameter.

In many cases, the carrier's surface must be modified or added with functional ligands. Carriers are typically coated with a few atomic layers of stable materials, including organic polymers, oxides, gold and platinum to facilitate surface functionalization. Proper surface coating also allows magnetic nanoparticles to be suspended in suitable solvents, forming homogeneous solutions.

11.3.1.1 Commercially Available Magnetic Carriers

The most commonly utilized types of magnetic carriers are commercially available polystyrene beads, which encapsulate magnetic materials for magnetization. Dynabeads (ThermoFisher Scientific) is based on the invention of Ugelstad[3,15] and is now available with different sizes (4.5 μm, 2.8 μm and 1 μm) and different types of functionalization. According to the company's product information, Dynabeads contain 17%–37% magnetic particles, which are a mixture of maghemite (γ-Fe_2O_3) and magnetite (Fe_3O_4). Other types of polystyrene and silica-based magnetic particles are commercially available, as shown in the table.

With the advent of nanotechnology-based particle synthesis, different sizes and shapes of magnetic nanoparticles have been studied. CellSearch, the first CTC screening method approved by the Food and Drug Administration (FDA), is based on immunomagnetic separation. CellSearch utilizes Fe_3O_4 magnetic nanoparticles encapsulated in polymers. As shown in the transmission electron microscopy (TEM) photograph in Figure 11.5, the diameter of polymer bound particle is approximately ~100 nm. The polymer binding helps increase the average diameter of the carriers. As we discussed in Section 11.2.3, if the particles are too small, they will not create a sufficient magnetic force to move labelled tumour cells.

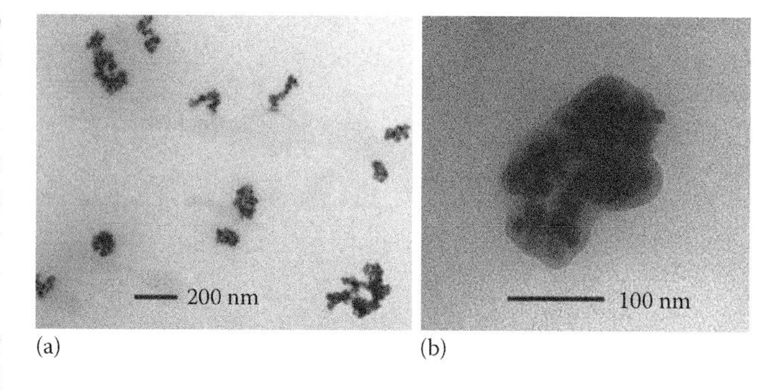

FIGURE 11.5 Ferrofluid particles used in the CellSearch method.[19] (a) TEM image. Average particle size is ~100 nm. (b) Zoom up TEM image of a single particle. (K. Hoshino, Y. Huang, N. Lane, M. Huebschman, J.W. Uhr, E.P. Frenkel, X. Zhang, Microchip-based immunomagnetic detection of circulating tumour cells, *Lab on a Chip*. 11 (2011) 3449–2457. Reproduced by permission of The Royal Society of Chemistry.)

11.3.1.2 Multifunctional Carriers

There have been approaches to integrate multiple functions of magnetic labelling, cellular uptake and optical labelling into a single nanoparticle. When fluorescent molecules are incorporated in a particle, it can be used as a fluorescent marker for immunofluorescence imaging. When magnetic particles are coated with gold, optical scattering can be used for spectroscopy imaging. Here, we introduce studies of multifunctional nanoparticle development.

Gao et al.[25] reviewed the synthesis and applications of multifunctional magnetic nanoparticles for biomedical applications. They discuss strategies used in synthesis of multifunctional magnetic nanoparticles in two steps, as shown in Figure 11.6. The first strategy is surface functionalization with biomolecules, including antibodies, ligands and receptors. Biofunctionalization of magnetic nanoparticles provides highly selective binding capabilities needed for biomedical applications, including drug delivery, bacterial detection, protein purification and toxin decorporation. The second strategy utilizes sequential growth or coating to effectively integrate magnetic particles with other functional nanostructures, such as quantum dots, nanodrugs, and metals. Through the integration, a single particle may demonstrate multiple functions in nanoscale to then be used in dual-functional molecular imaging such as combined magnetic resonance imaging (MRI) and fluorescence imaging.

Sathe et al.[11] developed silica microbead that contained colloidal quantum dots and Fe_3O_4 crystals for fluorescence imaging and magnetic separation. Mesoporous silica beads were sequentially or simultaneously mixed with a suspension of 32 nM iron oxide nanocrystals and a suspension of 100 nM quantum dots (QDs). This mixture resulted in the doping of silica beads with QDs and magnetic crystals. The silica microbeads were obtained from Alltech Inc. IL. The diameters of the beads were 3–5 μm, and the average pore diameter was 30 nm. The beads were coated with polymers, one of which is octylamine poly(acrylic acid). Magnetized beads were separated by placing suspension in a tube on a 1.5 T NdFeB permanent magnet.

Noble metal nanoparticles are known to demonstrate a size- and shape-dependent absorption spectrum. This absorption spectrum results from localized surface plasmon resonance, which is oscillation of electrons at the metal–air interface induced by light. Different colours of metal nanoparticles have long been used to colour-stained glass. Localized surface plasmon resonance has been studied for applications in optical markers for biomedical imaging and sensing. They have been used for labelling of cancer cells. El-Sayed et al.[26] used gold nanoparticles (average size: 35 nm) functionalized with antiepidermal growth factor receptor (EGFR) antibodies to label nonmalignant (HaCaT) and malignant (HOC 313 and HSC 3) epithelial cells. It is possible to coat the surface of magnetic nanoparticles with gold to add the similar optical characteristics.

Wu et al.[23] reported on a multi-antibody assay consisting of multifunctional nanoparticles. There have been several studies to develop magnetic nanoparticles suitable for CTC separation. Many CTCs express epithelial cell adhesion molecule (EpCAM), since carcinomas are tumours developed from epithelial cells. However, it is known that cells growing in a tumour may undergo epithelial–mesenchymal transition (EMT). Due to tumour heterogeneity and EMT, metastatic tumour cells often lose their expression of the epithelial cell-specific antigen. The absence of this antigen limits the efficacy of immunomagnetic assay relying on EpCAM. EpCAM-based assays may show low capturing efficiency for CTCs from mesenchymal-like cancer.

Use of multiple antibodies is one approach to capture cells that are not expressing EpCAM. Wu et al.[23] used Fe_3O_4/Au core/shell nanostructures to synthesize multifunctional nanoparticles to address the previously mentioned problems associated with EpCAM-based assays. The gold shell enables novel imaging schemes based on plasmonic scattering. Additionally, the gold shell allows for conjugation of functional ligands on nanoparticle surface for specific labelling. Antibodies are conjugated directionally, leaving the antigen binding sites available for cell array. This approach improves the specificity of the assay (Figure 11.7). The diameter of the nanoparticles used in this study is less than 10 nm, which is smaller than nanoparticles used in other immunomagnetic separation studies. The advantages of using smaller nanoparticles include the possible reduction in steric hindrance and increased permeability. The gold shell, which can be as thin as 1 nm, permits easy surface functionalization while simultaneously keeping the magnetic core large enough for efficient magnetic separation.

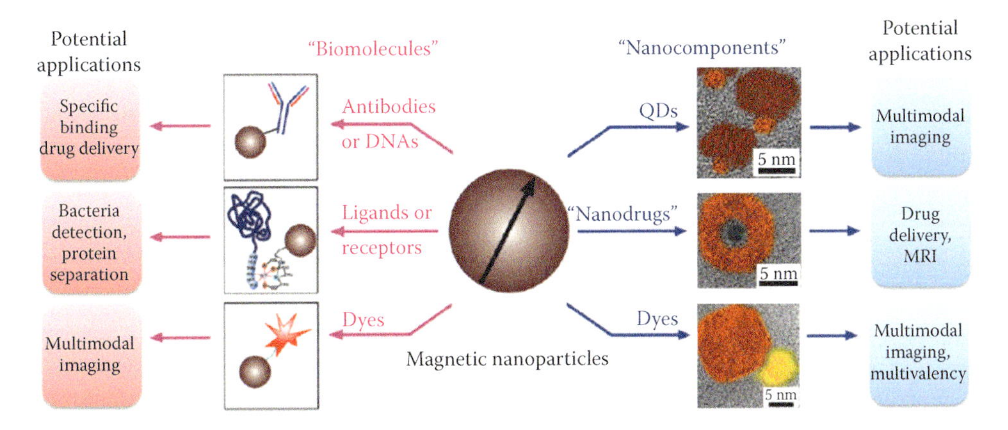

FIGURE 11.6 Strategies for multifunctional magnetic nanoparticles synthesis and their applications.[25] (Reprinted with permission from J. Gao, H. Gu, B. Xu, Multifunctional magnetic nanoparticles: design, synthesis, and biomedical applications, *Accounts of Chemical Research*. 42 (2009) 1097–1107. Copyright 2009 American Chemical Society.)

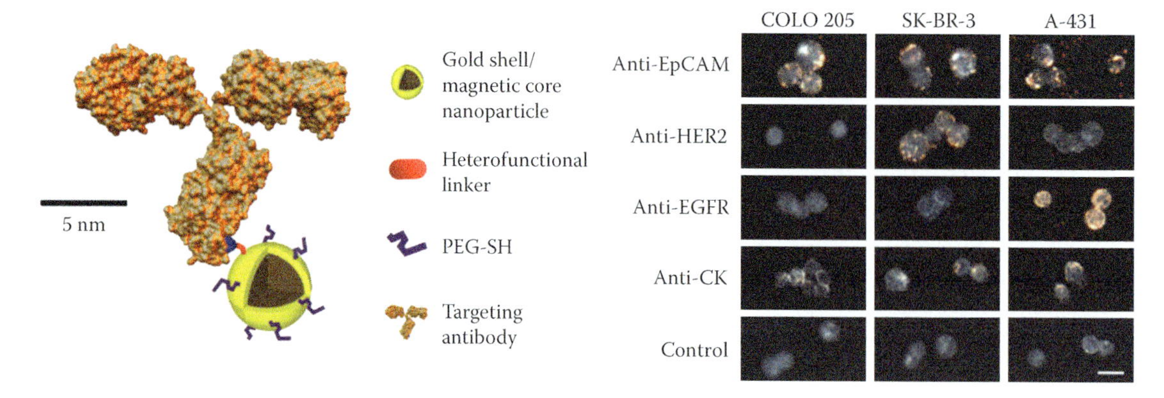

FIGURE 11.7 Multifunctional Fe_3O_4/Au core/shell nanostructures tested with multiple antibodies and cell lines.[23] (Reprinted with permission from C. Wu, Y. Huang, P. Chen, K. Hoshino, H. Liu, E.P. Frenkel, J.X. Zhang, K.V. Sokolov, Versatile immunomagnetic nanocarrier platform for capturing cancer cells, *ACS Nano.* 7 (2013) 8816–8823. Copyright 2011 American Chemical Society.)

The Fe_3O_4/Au core/shell nanoparticles were synthesized through thermal decomposition of iron(III) acetylacetonate in a mixture of oleylamine and oleic acid, with the following reduction step of gold acetate on the iron oxide seeds. The nanoparticles displayed a narrow size distribution in the water phase. TEM measurements of more than 200 particles showed the mean diameter to be 6.2 ± 0.8 nm. TEM images showed a successful gold coating that appears darker than Fe_3O_4 precursors (Figure 11.8b).

11.3.2 Permanent Magnets Used in Separation Systems

11.3.2.1 Test-Tube-Based Separation Systems

The simplest way to conduct magnetic separation is to use a test tube and a permanent magnet, as shown in Figure 11.9a. In this case, the magnitude of the magnetic field simply decays as the distance from the magnet increases. This method is often used to separate proteins or nucleic acids

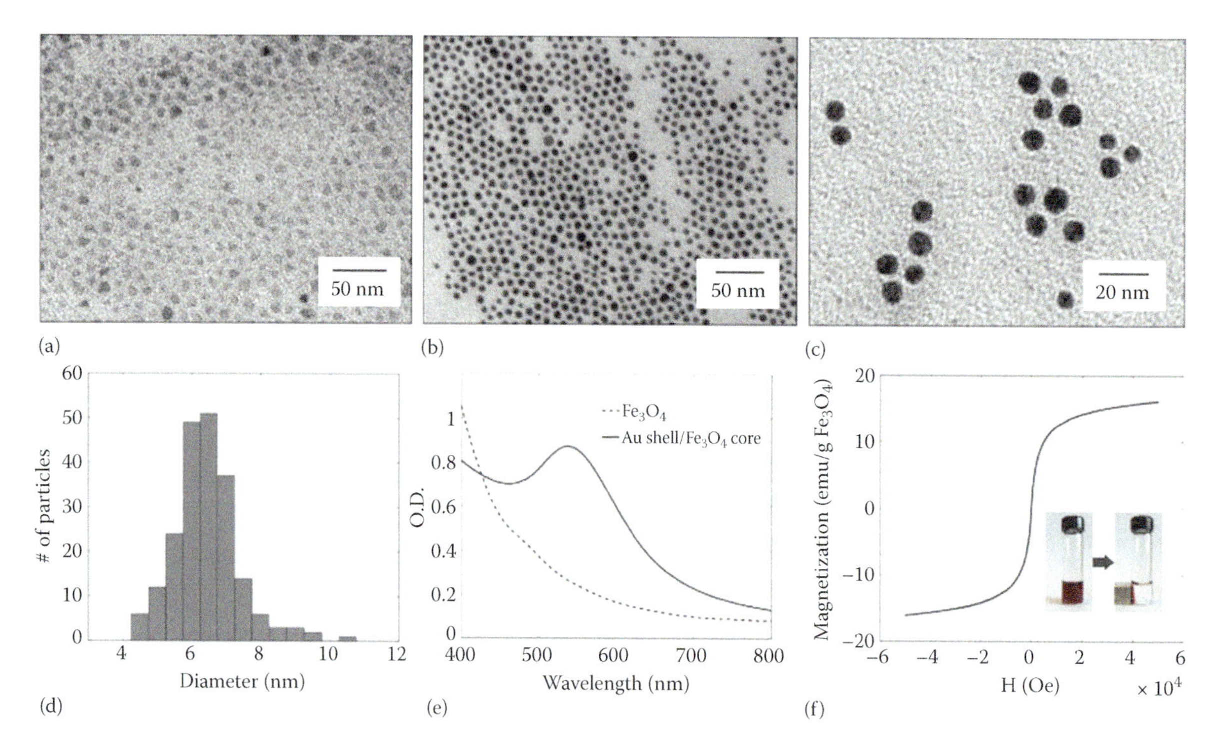

FIGURE 11.8 TEM images and basic characterization of multifunctional Fe_3O_4/Au core/shell nanostructures. (a) Fe_3O_4 nanoparticles in hexane before gold coating. (b) Gold shell-coated nanoparticles in hexane. (c) Gold shell-coated nanoparticles in the aqueous phase. (d) Nanoparticle size distribution determined from TEM images. (e) UV-vis spectra of Fe_3O_4 nanoparticles (dashed) and gold shell-coated nanoparticles (solid) in hexane. (f) Magnetization hysteresis of gold shell-coated nanoparticles. The inset shows separation of nanoparticles from a colloidal suspension.[23] (Reprinted with permission from C. Wu, Y. Huang, P. Chen, K. Hoshino, H. Liu, E.P. Frenkel, J.X. Zhang, K.V. Sokolov, Versatile immunomagnetic nanocarrier platform for capturing cancer cells, *ACS Nano.* 7 (2013) 8816–8823. Copyright 2011 American Chemical Society.)

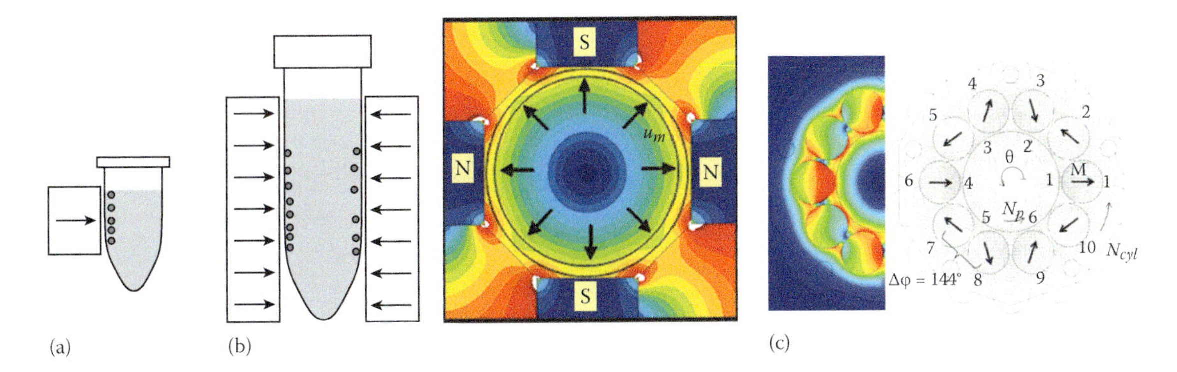

FIGURE 11.9 Typical tube-based separation. (a) Small tube with a single permanent magnet. (b) Magnetic field distribution induced by the arranged permanent magnets.[27] (The right panel in b: Reprinted with permission from M. Zborowski, J.J. Chalmers, Rare cell separation and analysis by magnetic sorting, *Analytical Chemistry*. 83 (2011) 8050–8056. Copyright 2011 American Chemical Society.) (c) Circular arrangement of 10 magnets to create an axisymmetric field.[28] (Reprinted with permission from P. Joshi, P.S. Williams, L.R. Moore, T. Caralla, C. Boehm, G. Muschler, M. Zborowski, Circular Halbach array for fast magnetic separation of hyaluronan-expressing tissue progenitors, *Analytical Chemistry*. 87 (2015) 9908–9915. Copyright 2015 American Chemical Society.)

in a small test tube (typically polymerase chain reaction [PCR] tube). When using a small test tube, a single permanent magnet will be sufficient to create magnetic field gradient needed to separate magnetic carriers. In addition, if the target molecules are small, carriers can move freely in the test tube and can be easily separated, as discussed in Section 11.2.3.

However, when the container or the tube is larger, it is necessary to create a magnetic field gradient that covers the entire medium and can efficiently guide magnetically labelled samples. Commercially available devices utilize an arrangement of multiple magnets to create large spatial changes to the magnetic field along the entire container. Figure 11.9b is an example of the magnet arrangement used to create a gradient in a standard test tube.[27] This type of arrangement is used for tubes typically sized 5–10 mL. The induced field in this type of magnet arrangement is also shown. Note that the north poles of the two opposing magnets (and the two south poles of the other pair) are facing each other and the magnetic field intensity is zero at the centre of the tube. This results in the increasing magnetic gradient going outward from the centre, moving magnetic carriers on the inside wall of the tube. The magnets used in the CellSearch system are arranged in this form. The CellSearch system includes robotic arms and pipetting systems. The process is practically identical to that of which a person would manually do with test tubes. In another study, 10 cylindrical magnets with rotating polarities, instead of rectangular magnets, were used to generate a higher and uniform field gradient.[28]

11.3.2.2 Permanent Magnets with Microfluidic Systems

Microfluidic separation holds benefits in the design of the magnetic field, that is unless large volumes of biosample (typically >~10 mL) need to be processed. The distance from the magnet to the sample can be small, and the distance that labelled cells need to travel can also be small. Figure 11.10 is an example of the magnetic arrangement designed for a microfluidic channel.[19] Different from a test tube, microfluidic channels tend to have a thin and flat structure; this is because most microfluidic channels are fabricated based on the photolithography process, as we described in Section 11.3.3. In this example, magnets are arrayed with alternate polarities to create large spatial changes of the magnetic field. The magnets used are grade N42 NdFeB blocks with a maximum energy product of 42 MG Oe. The largest gradient value ∇B^2 is found at the edges where two magnets meet. The panel showing the zoom-up view illustrates the magnitude and the directions of magnetic field at the edge location.

Figure 11.11 shows an interesting method of creating an array of permanent magnets in a lab-on-a-chip (LOC) microfluidic system.[29,30] Dumas-Bouchiat et al. introduced the technique of thermomagnetic patterning to create a two-dimensional (2D) array of micromagnets with alternating

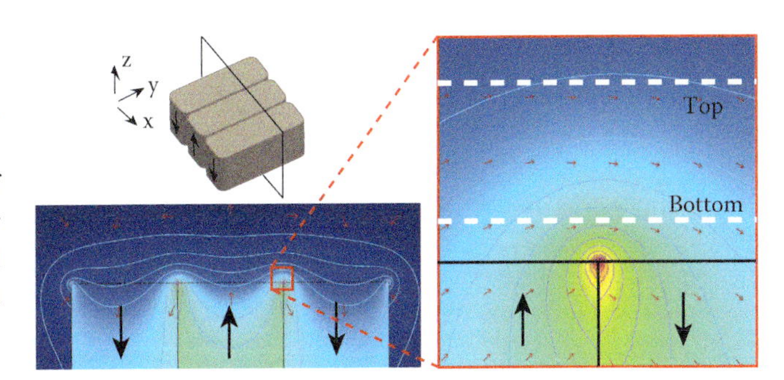

FIGURE 11.10 Arrays of permanent magnets with alternating polarities are used to create larger magnetic field gradient in a microfluidic chip.[19] (K. Hoshino, Y. Huang, N. Lane, M. Huebschman, J.W. Uhr, E.P. Frenkel, X. Zhang, Microchip-based immunomagnetic detection of circulating tumour cells, *Lab on a Chip*. 11 (2011) 3449–2457. Reproduced by permission of The Royal Society of Chemistry.)

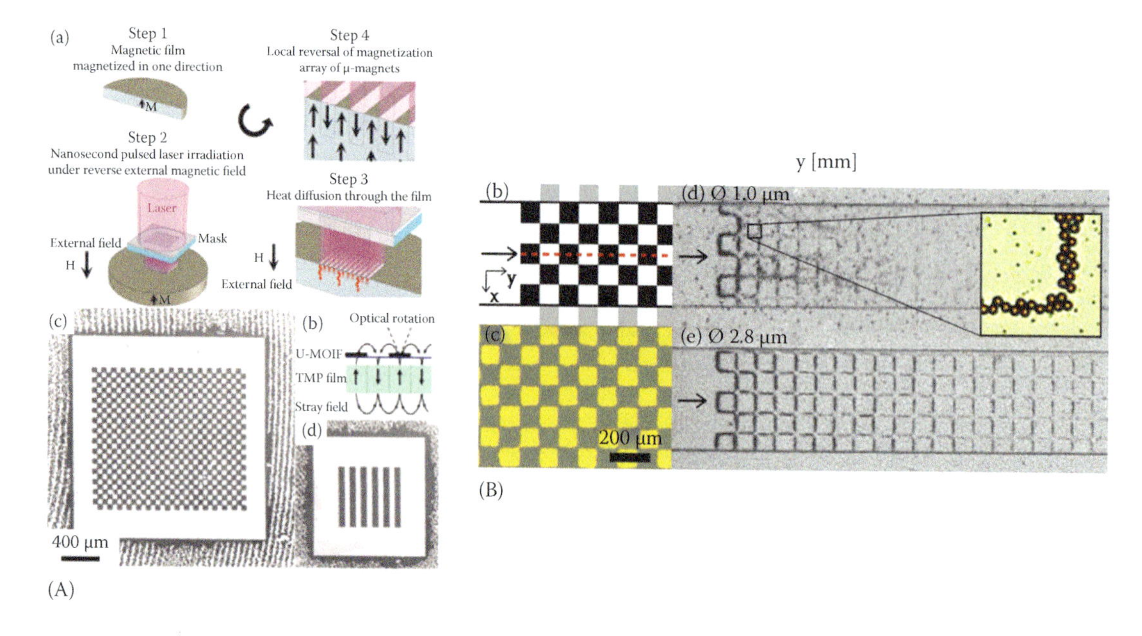

FIGURE 11.11 (A) Microarray of permanent magnets fabricated through selective laser heating.[30] (Reprinted with permission from F. Dumas-Bouchiat, L. Zanini, M. Kustov, N. Dempsey, R. Grechishkin, K. Hasselbach, J. Orlianges, C. Champeaux, A. Catherinot, D. Givord, Thermomagnetically patterned micromagnets, *Applied Physics Letters*. 96 (2010) 102511. Copyright 2010, American Institute of Physics.) (B) Application of microbead separation in a microfluidic channel.[29] (Reprinted with permission from L. Zanini, N. Dempsey, D. Givord, G. Reyne, F. Dumas-Bouchiat, Autonomous micro-magnet based systems for highly efficient magnetic separation, *Applied Physics Letters*. 99 (2011) 232504. Copyright 2011, American Institute of Physics.)

polarities. In their fabrication of the magnets, the hard magnetic film of NdFeB is initially magnetized in one direction. The magnets with the other polarity were made through partial magnetization, where a patterned pulsed laser irradiation locally heated the magnet film through an optical mask and the opposite magnetic field was applied for magnetization. This method is based on the characteristic that the magnetic materials are at a lower magnetic field when the temperature is higher. At the irradiated areas, the summation of the applied field and the demagnetizing field overcomes the coercive field, which is the field needed to demagnetize the areas. The depth of the reversed magnetization was estimated to be in the range of 1.1–1.2 μm.

Many magnetic separation systems use the arrangement of magnet with alternating or rotating polarities. One notable characteristic in such arrays is that the effective magnetic field is generated only in areas close to the magnets. The spatially alternating magnetic fields cancel each other and the magnetic field is weaker when the distance from the magnets becomes larger than the size of magnets. Examples in Figures 11.10 and 11.11 are efficient because they are used in thinner microfluidic channels.

11.3.3 Actively Controlled Magnets Used in Separation Systems

Often integrated in a magnetic separation system are small pieces of soft magnets that function as local magnets. Such additional magnets make the distance between targets and the source of magnetic field even closer to make the attraction force larger.

11.3.3.1 MACS System

In the early study of magnetic separation, steel wool was inserted in a small column[31] to locally enhance the magnetic field. This idea served as the basis of the commercially available MACS system.[32] Figure 11.12 is the diagram showing the working principle of the MACS system. The currently available MACS column uses ferromagnetic spheres, instead of steel wool, as the local enhancement of the magnetic field. The spheres are coated to prevent direct contact with samples. Spheres are arranged with spaces several times larger than target cells to allow them to go through freely while preventing creation of large cell aggregates. Unlike the magnetic sorter we saw in Figure 11.9, the magnets used in the MACS are aligned in the same polarity (see Figure 11.12). If there were no inserted magnetic spheres to create a large magnetic field within the column (steel wool or ferromagnetic spheres), this would not be a good design because the magnetic field in the column is almost uniform and the magnetic carriers would not experience any magnetic force. However, with the insertion of magnetic field, it is more beneficial to apply a larger magnetic field with magnets of the same polarity rather than using the designs we discussed in Section 11.2.

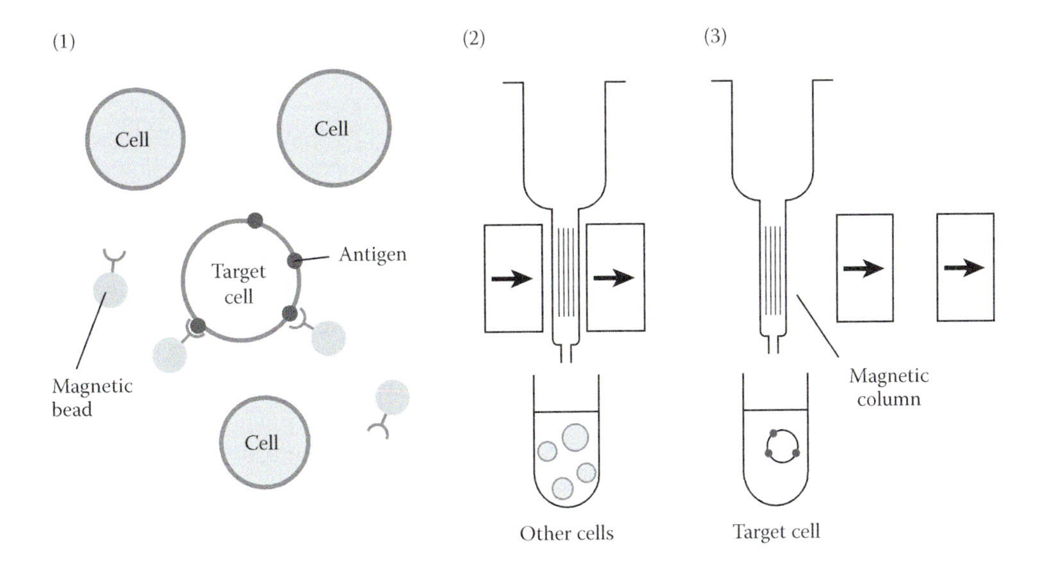

FIGURE 11.12 Separation procedure with the MACS system.

11.3.3.2 Microfluidic Systems with Integrated Micromagnets

Microfluidic systems are miniaturized, integrated microscale fluidic systems often based on lithographic fabrication processes. It has been a topic of interest in biomedical engineering since the simple replica moulding technique was introduced to create polydimethylsiloxane (PDMS)-based microfluidic systems.[33,34] Microfluidic systems are the key technology for studies in LOC or micro total analysis systems (μTAS). The main advantage of using LOC systems or μTAS is that miniaturized systems make it possible to integrate multiple functions within a very small experimental setup. Microfluidic channels require less volume in the process, and materials can be easily made disposable. Microfluidic systems can also potentially make the system more cost-effective. In terms of throughput, the total sample volume that can be processed in a certain time, the fact that the channel volume is small may not necessarily be a benefit for magnetic separation. However, downsizing of systems brings target cells or molecules closer to magnets, allowing for more efficient and rapid reactions. Miniaturized microfluidic channels may result in faster response times. Miniaturization allows for the development of compact yet fully functional microfluidic separation devices. Here, we first overview technologies used in the fabrication of microfluidic systems and describe microfluidic-system-based magnetic separation systems, followed by a discussion on the benefit and drawbacks of microfluidic magnetic separation.

The use of microfluidic systems has become a very popular approach because the replica moulding technique,[33,34] also known as the 'soft lithography' technique, can be easily used in laboratory settings without introducing usually expensive semiconductor fabrication tools. Figure 11.13 shows an example of PDMS-based microfluidic system used for magnetic separation.[4] The microfluidic device was designed to remove magnetically labelled *Escherichia coli* bacteria from a sample containing a mixture of red blood cells (RBCs) and bacteria. The device shows the typical design used in magnetic separation with a microsystem. The channel has multiple inlets and outlets. As the sample flows in the main channel, labelled targets are separated toward the integrated magnet and come

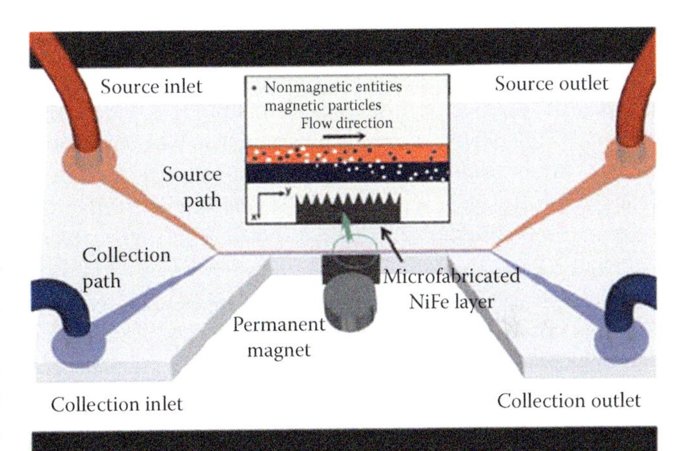

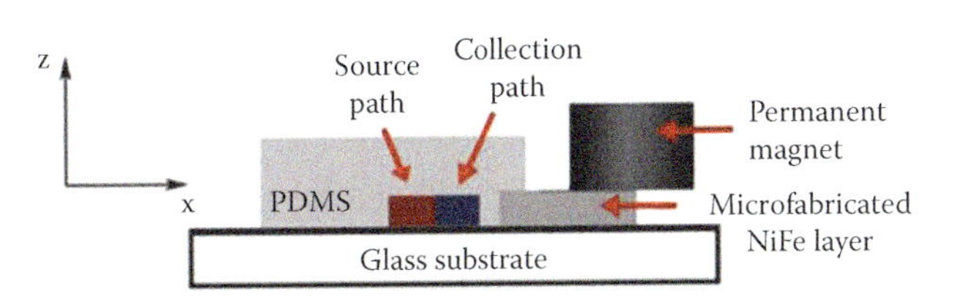

FIGURE 11.13 Microfluidic magnetic separation.[4] (With kind permission from Springer Science+Business Media: *Biomedical Microdevices*, Combined microfluidic–micromagnetic separation of living cells in continuous flow, 8, 2006, 299–308, N. Xia, T.P. Hunt, B.T. Mayers, E. Alsberg, G.M. Whitesides, R.M. Westervelt, D.E. Ingber.)

out from one of the outlets. The magnet is an electroplated permalloy (80% Ni, 20% Fe) with size $l \times h \times w$ = 12 mm × 50 µm × 3.8 mm. It is shaped like a comb or a saw, which contains sharp edges (300 µm deep, 200 µm spacing) to induce large magnetic field gradient.

There are several microfabrication techniques to integrate microscale magnetic materials within the microfluidic channel. Magnetic materials including Ni or Fe-Ni can be deposited through typical processes like sputtering, evaporation or electroplating. Metal films can be patterned through a chemical etching process similar to the methods used for fabrication of integrated circuits.

When evaporation or sputtering is used, the thickness of the deposited metal film is typically a few hundred nanometres or about a micrometre at most. Because these deposition processes are usually conducted at higher temperature, thicker films will experience thermally induced internal stress at a room temperature and become very unstable. As we see in Figure 11.3d, thin soft magnets can be modelled as an infinite thin film, where magnetization occurs mostly in the x or y direction (parallel to the substrate), while magnetization in the z direction (direction normal to the substrate) is small.

For thinner metal films up to ~200 nm, the technique called 'lift-off' is typically used. In this process, the part of the metal film deposited on the photoresist will be 'lifted off' along with the photoresist, leaving the metal patterns deposited though the open windows. Relatively thicker films may be patterned through wet-etching. Nitric acid, ferric chloride and potassium hydroxide are among the commonly used etchants for metal wet etching. Metal films as thick as a few hundred micrometres or about a micrometre may be patterned through wet-etching. However, in wet-etching processes, the etchant usually goes under the photoresist film and overetches the metal film. Because of this 'overetch' or 'undercut,' the minimum feature size that can be patterned through etching is related to the thickness of the film to be etched. Dry-etching using a plasma induced in a vacuum chamber is also used for metal etching and usually displays smaller effects of overetching, rather than wet-etching. However, etch rates are usually very low and not used for thick films over a micrometre (Figure 11.14).

When metal films thicker than a few micrometres need to be deposited, electroplating is often used. Because electroplating can be made at a temperature close to a room temperature, films thicker than 10 µm can be deposited without causing problems with thermally induced internal stress. In electroplating, mechanically patterned grooves or guides are used to define the metal pattern. The grooves are often patterned thick photoresist such as AZ4000 series and, in other cases silicon wafers that are deeply etched by deep reactive ion etching. The surface of an electroplated metal film tends to be coarse and uneven because deposition is governed by current flow on the surface of substrate. Mechanical and chemical polishing processes are often used with electroplating.

Smistrup et al.[16] used permalloy micromagnets and an external magnetic field of 50 mT to create a magnetic bead separator. The micromagnets are patterned in stripes and measured $l \times h \times w$ = 4400 µm × 50 µm × 150 µm in a microfluidic channel sized $l \times h \times w$ = 13,500 µm × 80 µm × 200 µm. Particles separated are 1 µm fluorescent magnetic beads Spherotec FCM-1052-2. Electroplating was used to fabricate thick micromagnets. They further modified the system with a 2D composed of small alternately arranged Nd–Fe–B permanent magnets, each of which is sized 2 mm × 2 mm × 2 mm to characterize the separation efficiency for 250 nm Nanomag-D plain silica beads.[35] Inglis et al.[20] used 2-µm-thick nickel stripes to separate continuously flowing Leukocytes (white blood cells). Anti-CD45 Microbeads from Miltenyi Biotech were

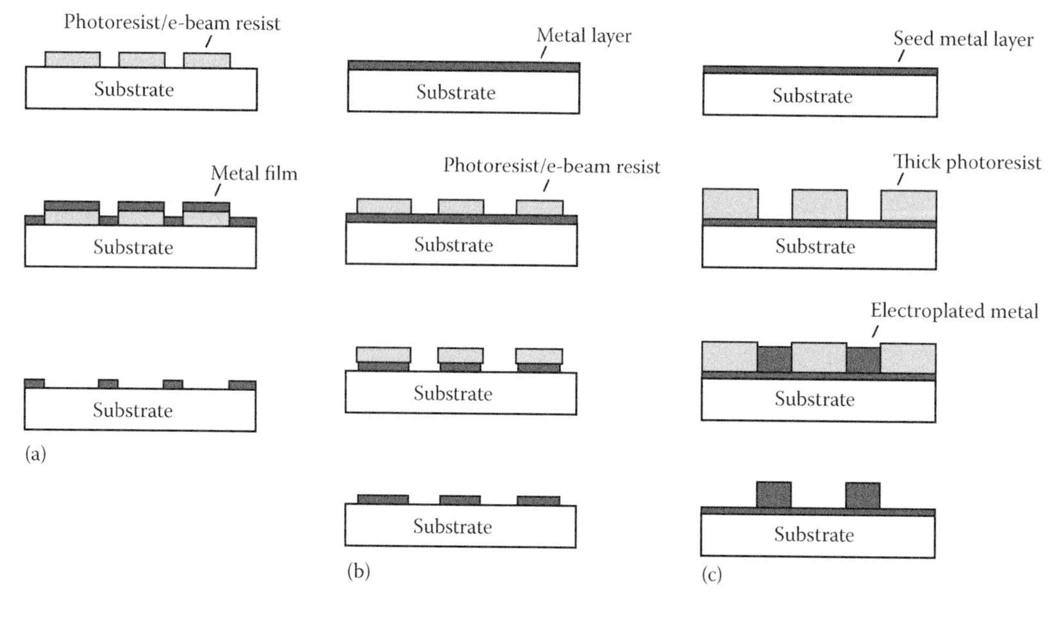

FIGURE 11.14 Typical metal patterning processes. (a) Lift off is used for deposition of thin metal films (up to a few hundred nanometres). (b) Etching can be used for relatively thicker metal films. Metal film thickness affects the minimum feature size. (c) Electroplating is used for deposition of thick metal films up to about a hundred micrometres.

used. CD45 is a type of protein expressed in all leukocytes. The diameters of the beads were in the range of 20–100 nm, and an average magnetic moment of 1.8×10^5 μB was estimated. In order to create a 2-μm-thick nickel pattern, they first created stripe-shaped grooves on a silicon substrate, and a nickel film deposited on the grooved substrate was chemically and mechanically polished to obtain a flat substrate with patterned magnets. A similar approach of using a stripe patterned magnets has been used by Kim et al.[36] for separation of CTCs. Lou et al.[37] used a similar stripe-patterned device as Inglis to separate aptamers through systematic evolution of ligands by exponential enrichment (SELEX) process. As will be discussed in Section 11.4, magnetic separation plays an important role in the SELEX process for aptamer selection.

Figure 11.15 shows an example of integrated micromagnets made from a 200-nm-thick nickel film reported by Chen et al.[38] The lift-off technique was used to pattern the magnets from a nickel film. External magnets were arranged in the same way as used in Hoshino et al.[19] (see Figure 11.10). The system was used for separation of cancer cells from blood. Without the micromagnet array, cells and free particles tend to aggregate in small packed areas because magnetized particles are easily attracted to a few points that have a strong magnetic field. Once particles start to aggregate in a certain area, the aggregation increases the magnetic field and further attract nanoparticles. As discussed in Section 11.3.2, alternating arrays have a very intense field along the sides of the magnets. Without micromagnets, cells and particles were attracted around the edges of the external magnets (Figure 11.15c, right, and d, bottom). When captured cells are hidden behind aggregation of nanoparticles, the following immunofluorescence analysis becomes difficult, weakening the efficacy of the entire system. When micromagnets were introduced, they create local sharp peaks of magnetic field and serve as an array of attraction sites. Cells and particles were separated in larger areas (Figure 11.15c, left, and d, top) and captured cells were easily visible.

Yu et al.[17] fabricated 9-μm-thick nickel patterns through electroplating. They were magnetized with two permanent magnets placed in parallel with opposite NS poles to create a uniform magnetic field. The pattern has sharp edges to control local magnetic field distribution. Different sizes of microbeads, including 500-nm-diameter beads (02150 AdemTech, Pessac) and 1.05-μm-diameter beads (MyOne carboxylic acid magnetic beads, Dynal), were tested. Lung cancer cells (A549) were labelled with wheat germ agglutinin functionalized magnetic beads and captured by micromagnets. Deng et al.[39] used electroplated nickel posts (7 μm in height and 15 μm in diameter, and 40 μm spacing) in a microfluidic channel. A film of UV-curable polyurethane transferred onto a substrate using a PDMS mould was used as a mask of electroplating. Tested in the microchannel were 4.5-μm-diameter magnetic beads (M-450 Dynal). Liu et al.[40] deposited a 6-μm-thick permalloy wave form pattern on the substrate through electroplating and encapsulated it with polystyrene. Live Jurkat cells, which were 10 μm in average diameter, labelled with StemCell Technologies EasySep Human CD3 positive selection cocktail, were captured in the microchannel. Permanent magnets were used to magnetize the pattern.

Once an aggregation of nanoparticles or microbeads is formed, it works as a small magnet that attracts more particles and grows in a positive feedback loop. When this local aggregation is controlled, it can be also used in a similar way as the photolithographically patterned magnets. Saliba et al.[41] patterned magnetic ink (ferrofluidMJ300, Liquids Research,

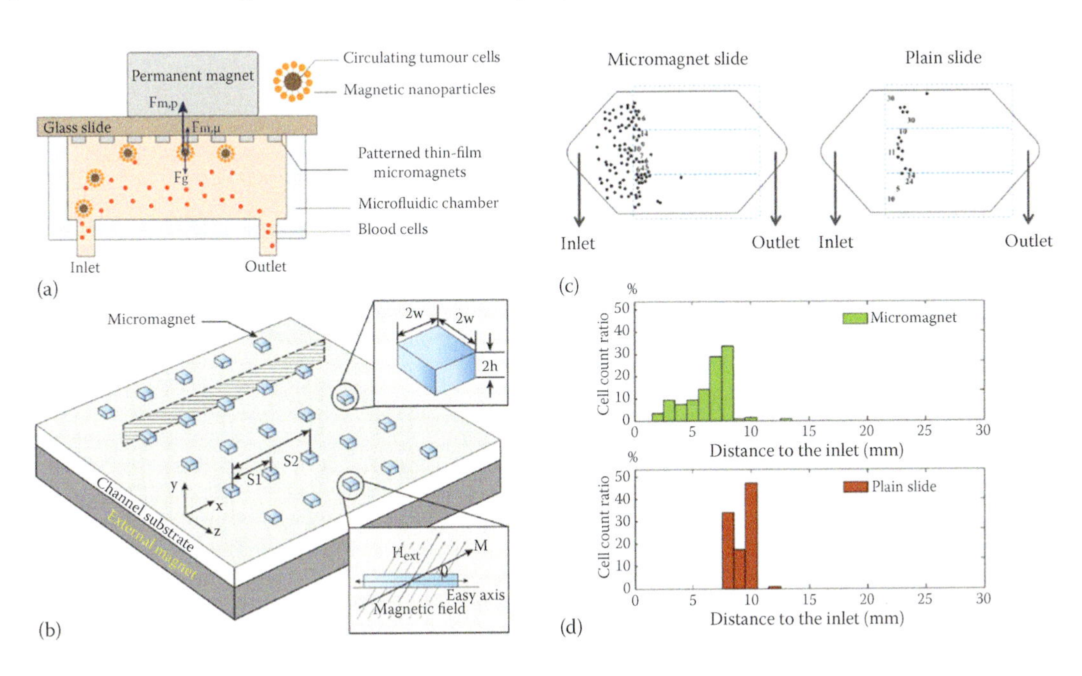

FIGURE 11.15 Patterned micromagnets for separation of CTCs. Arrayed magnets prevent aggregation of nanoparticles in a small spot. (a) Schematic of the microchip. (b) Micromagnet design. (c) Locations of the captured COLO205 cells. (d) Distribution histogram of the captured cells.[38] (From Chen, P., Huang, Y., Hoshino, K., Zhang, J.X., *Sci. Rep.* 5, 2015. With permission.)

United Kingdom) through micro-contact printing technique to create a micromagnet array. Magnetic beads with a diameter of 4.5 μm (Dynal) introduced in the microchannel self-aligned to form pillars that capture target cells. The 4.5-μm beads functionalized with anti-CD19 monoclonal antibody were used to capture mixtures of T (Jurkat cell line) and B (Raji cell line) lymphoid cells, and anti-EpCAM Dynal beads were used to capture MCF7 breast cancer cells.

Chen et al.[42] used inkjet printing technique to pattern magnetic nanoparticles (fluidMAG-ARA chemicell) with a hydrodynamic diameter of 100 nm to form a micromagnet array. A commercially available printer (Fuji DMP-2800 Dimatrix Materials Printer, FUJIFILM) was used to pattern the particles on a standard glass slide, which was then heated at 100°C to evaporate the liquid. A 10-nm-thick SiO_2 film was deposited onto the pattern through plasma-enhanced chemical vapour deposition to coat the array. Figure 11.16a, b and c shows scanning electron microscopy (SEM) images, an atomic force microscopy (AFM) image and an AFM profile of the micromagnet pattern, respectively. Whole blood spiked with COLO205 colon cancer cells was incubated with EpCAM functionalized magnetic nanoparticles (Ferrofluid, Janssen Diagnostic, J&J; average diameter ~100 nm, see also Figure 11.5 for TEM) and tested in the microfluidic channel with the printed micromagnet array. The experiment with the micromagnets showed a better capture efficiency by 26% compared to the rate with a plain glass slide substrate.

Yu et al.[18] used graphite oxide (GO)-coated magnetic nanoparticles to provide magnetic pillars with antibodies. They synthesized pristine water-soluble Fe_3O_4 nanoparticles through a hydrothermal method. The approximate diameter of the particles was 250 nm. The particles were mixed with GO aqueous solution through a magnetic stir and sonication to coat the surface with GO. The GO-coated nanoparticles were then functionalized with EpCAM. The carboxylic groups on GO was first chemically activated and conjugated with streptavidin. The particles were then incubated with biotinylated anti-EpCAM, allowing streptavidin-biotin binding to occur. The nanoparticles were tested in a microfluidic channel integrated with micropillars, which sized 50 μm in diameter and 15 μm in height. Under a magnetic field, nanoparticles were attracted to the pillars to decorate their surface. When HCT116 colorectal cancer cells were introduced in the microchannel, they were trapped on the surface of pillars. They demonstrated capture efficiencies better than 70% from culture medium and 40% from blood. It was also possible to release captured HCT116 cells with an efficiency of 92.9% when the magnetic field is removed.

11.3.3.3 Active Control Using Solenoids

A magnetic field can be electrically induced using solenoids. Solenoids are advantageous when it is necessary to change the magnetic field actively. By changing the intensity of applied magnetic field, the magnetization of the integrated magnet can be switched. Switching control can be used to capture and release target molecules or cells. It is also possible to manipulate magnetic carriers within a fluidic chamber by using multiple solenoids and controlling field intensities.

Earhart et al.[43] developed a separation device, named magnetic sifter, to separate cancer cells with anti-EpCAM magnetic nanoparticles MAG999 (R&D Systems). The magnetic sifter is a silicon wafer with etches through square holes sized 40 × 40 μm made through a silicon nitride membrane. The density of pores is ~200 pores/mm^2. A 12-μm-thick soft magnet film of permalloy is deposited on the silicon nitride membrane to function as an active magnet. The device is capable of releasing captured cells when the external magnetic field applied from a solenoid is released.

There have been attempts to create microcoils within microfluidic systems. Choi et al.[44] fabricated a planar copper wire in a serpentine pattern (10 μm width and spacing, 7 μm thick) through electroplating and covered it with Ni 81% Fe 19% permalloy. A polyimide film is spin-coated between the

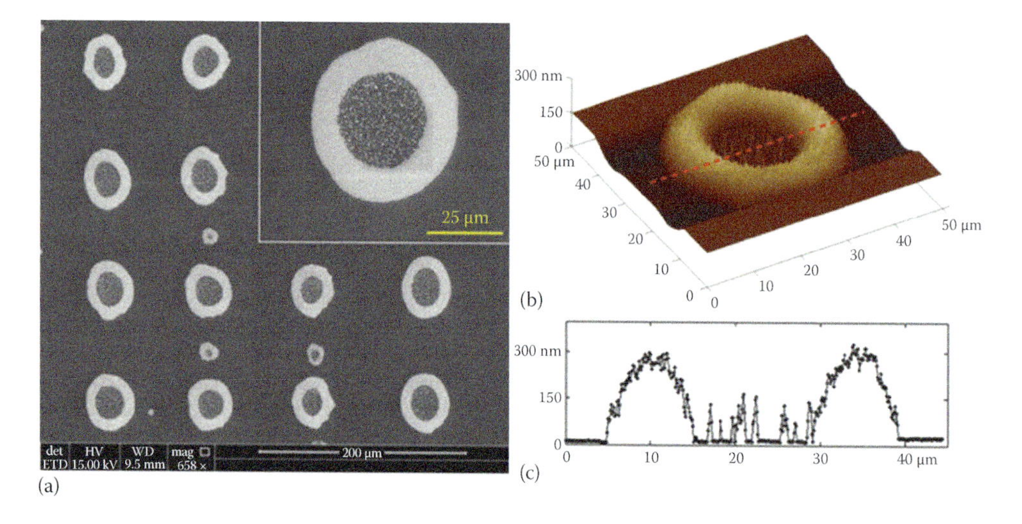

FIGURE 11.16 Inkjet printed micromagnet array. (a) SEM photograph of printed micromagnets. (b) AFM topographic measurement of a micromagnet. (c) Cross-sectional profile of the AFM measurement.[42] (With kind permission from Springer Science+Business Media: *Annals of Biomedical Engineering*, Inkjet-print micromagnet array on glass slides for immunomagnetic enrichment of circulating tumour cells, 44, 2016, 1710–1720, P. Chen, Y. Huang, G. Bhave, K. Hoshino, X. Zhang.)

copper wire and the permalloy layer for insulation. Magnetic beads with diameters of 0.8–1.3 μm with a 60% solid content of magnetite core were used in the study. Smistrup et al.[14] fabricated microcoils on the backside of the silicon substrate. A 60-μm-wide, 25-μm-high copper wire is wound by 12 turns and covered with a 25-μm-thick nickel yoke. Dynal MyOne (diameter 1.05 μm, density 1.8×10^3 kg/m^3, permeability $\mu = 2.5$) has been tested in a simulation and an experiment. Chiou et al.[13] made an array of 3D coils that are integrated to manipulate a microbead attached with a DNA piece. A copper wire ($w \times t = 80 \times 25$ μm, spacing 100 μm) was wound 30 turns around a $Ni_{80}Fe_{20}$ permalloy core ($w \times t = 300$ μm × 15 μm). Microbeads (M280, Dynal, 2.8 μm) were used in the study. The largest drawback of using the integrated coils is the limit of current that can be applied. Even though all reviewed studies utilize electroplating techniques to create thicker (typically ~20 μm) copper wires, it is difficult to apply a large current, which will damage the wiring due to heat accumulation.

When soft magnetic materials are used as inserted local magnets, it is easy to control magnetization through an external magnetic field rather than solenoids directly integrated in a microsystem. Utilization of external solenoids and soft magnetic materials is an effective approach. By changing the orientations of the applied magnetic field, it is also possible to control the magnetization direction of the inserted local magnets. With this approach, larger solenoids placed outside of the microfluidic system can be used. de Vries et al.[12] demonstrated magnetic tweezers that rely on micron scale multiple magnetic poles. Figure 11.15a shows the design of the multipole magnetic tweezers. Magnetic fields are created by external solenoids and the integrated poles focus the magnetic field in the centre. The poles used in their system are 8-μm-thick electroplated cobalt micropatterns that can be positioned close to the target. Cobalt was used for the high saturation magnetization (1.8 T) and its durability in aqueous environments. Figure 11.17b shows the fabricated cobalt poles. The field gradient created in the system was measured by the force exerted on a magnetic bead (Dynabead M280 from Dynal). The bead

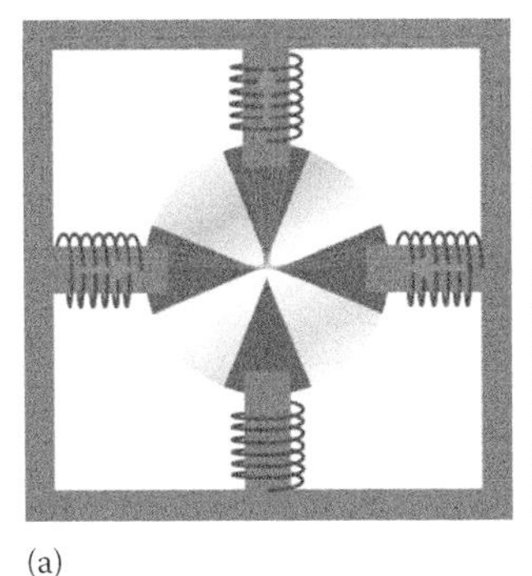

(a)

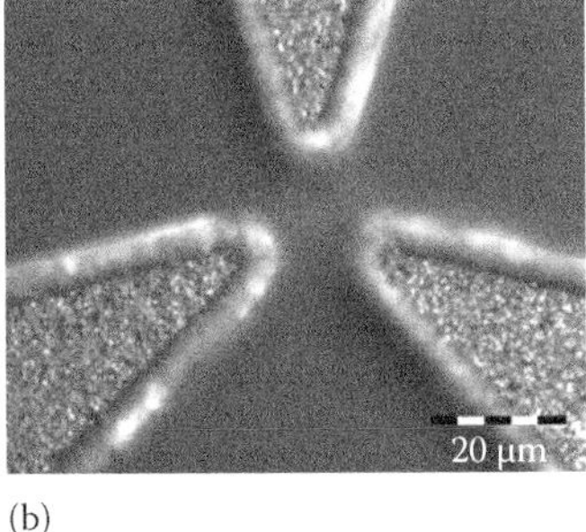

(b)

FIGURE 11.17 (a) Illustration of a four-pole magnetic tweezers and (b) microscopic image of a three-pole magnetic tweezers.[12] (Reprinted from *Biophysical Journal*, 88, A.H. de Vries, B.E. Krenn, R. van Driel, J.S. Kanger, Micro magnetic tweezers for nanomanipulation inside live cells, 2137–2144, Copyright 2005, with permission from Elsevier.)

is attached to a pipette tip, which has a known spring constant. Forces up to 1000 pN were measured. For cell manipulation, Dynal 'MyOne' (diameter 1.05 μm) and 47% γ-Fe_3O_4 beads (diameter 0.35 μm) from Bangs Laboratory (Fischers, Indiana) were used. Forces up to 120 pN and 12 pN were exerted for the 1.05 μm and 0.35 μm beads, respectively. Magnetic beads were introduced in live cells by cellular uptake (phagocytosis). Granulocytes isolated from fresh blood through centrifuge are mixed with a suspension of magnetic beads and incubated for 30 min. The magnetized cells were prepared on glass slides treated with poly-L-lysine to allow for adhesion of the cells on the substrate. Cell experiments with both 1.05 μm and 0.35 μm beads demonstrated easy manipulation of interior of a cell. A similar approach was used by Donolato et al.,[45] where zigzag-shaped permalloy ($Ni_{80}Fe_{20}$) wires are integrated in a microfluidic channel to allow for manipulation of yeast cells labelled with magnetic beads. The 30-nm-thick permalloy pattern was patterned through e-beam lithography and lift-off. The yeast cells are labelled by MyOne carboxylic acid beads from Dynal (diameter 1.05 μm) functionalized with Concanavalin-A, which has a high binding force with the cell surface. Individual yeast cells are manipulated and released through the use of five external solenoids.

Figures 11.18–11.20 show our recent effort of creating micromagnets that can capture and release magnetized cancer cells.[46] A photolithographically patterned 200–250-nm thick nickel film is deposited on the bottom surface of the microchannel. Chemical wet etching was used for patterning. When an external magnetic field is applied, the nickel patterns are magnetized and work as micromagnets. As we discussed earlier, when the magnet is thinner, magnetization in the z direction is very weak and decays very quickly. These thin magnets patterned on the bottom are not suitable to separate cells floating higher in the microchannel, which has a thickness of 800 μm. In order to cover the entire microchannel with an effective magnetic field, this method captures cells in two steps. The external magnetic field is applied with a solenoid with a yoke with tapered ends. Since the distance between the ends is closest at the bottom, the magnetic field is also strongest at the bottom. Magnetized cells are attracted to the area with strongest magnetic field, i.e. the bottom substrate. The patterned micromagnets capture cells that come closer to the substrate. Magnetic field gradient, ∇B^2 (see Equation 11.17), was calculated to estimate the force acting on cells. Figure 11.18b shows the z-axis magnetic force calculated for different edge angles of the yoke ($\alpha = 25°$, 45° and 60°). The channel substrate was located at $z = 1$ mm to move cells downward with the negative z-directional force. Figure 11.18c shows the magnetic field intensity between the two edges of the yoke. The microchannel is located in the middle (-0.5 mm $< x <$ 0.5 mm) of the two poles, where the magnetic gradient in the x direction is small.

Magnetic forces from the integrated micromagnets are also calculated using values of B_x, B_y and B_z at $z = 0$, 10, 25 and 40 μm above the micromagnet. Figure 11.18b shows the graph of z-directional force versus the position in the y axis. Larger magnetic force is obtained when cells are closer to the

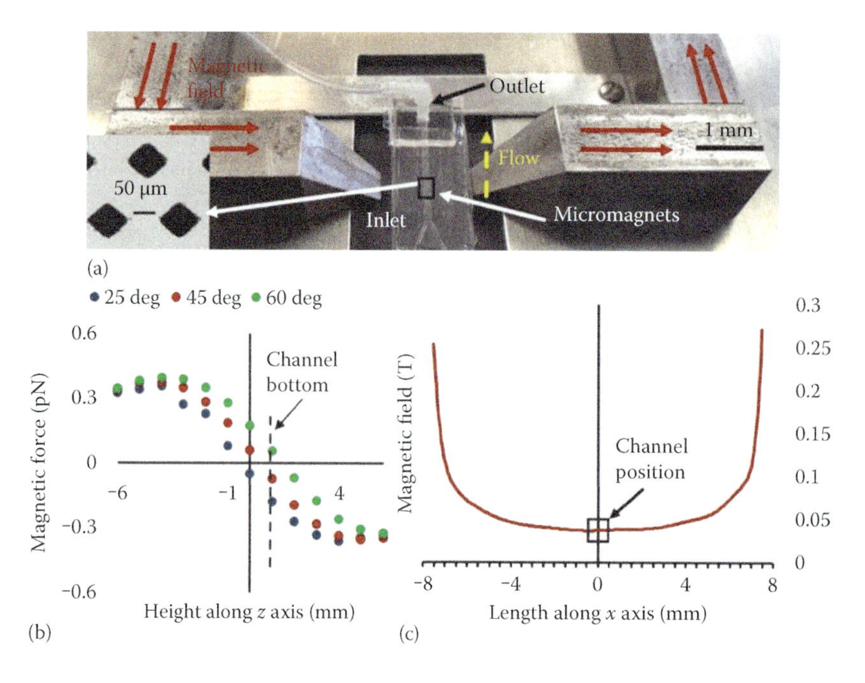

FIGURE 11.18 Experimental setup of the cell capturing system for real-time single-cell analysis. (a) Microfluidic device located between the solenoid. (b) Calculated magnetic force a cell experienced in the z-axis. (c) The intensity of the magnetic field between the two poles.[46] (Reprinted from *Journal of Magnetism and Magnetic Materials*, D. Jaiswal, A.T. Rad, M. Nieh, K.P. Claffey, K. Hoshino, Micromagnetic cancer cell immobilization and release for real-time single cell analysis, Copyright 2017, with permission from Elsevier.)

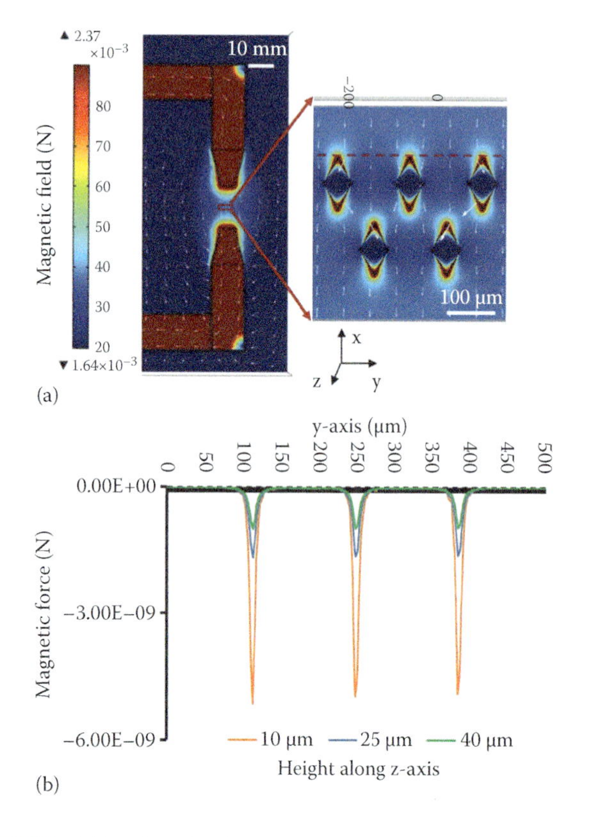

FIGURE 11.19 Calculated magnetic field and magnetic force induced by the active micromagnets. (a) COMSOL simulation of magnetic field on the edges of micromagnets. (b) Analysis of forces experienced by cells.[46] (Reprinted from *Journal of Magnetism and Magnetic Materials*, D. Jaiswal, A.T. Rad, M. Nieh, K.P. Claffey, K. Hoshino, Micromagnetic cancer cell immobilization and release for real-time single cell analysis, Copyright 2017, with permission from Elsevier.)

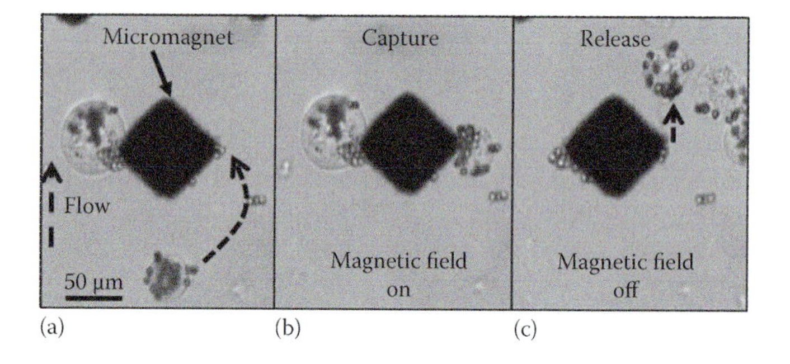

FIGURE 11.20 Cell capture and release by the active micromagnets. (a) Cells were attracted to a higher magnetic field at the edges of micromagnets. (b) Cells were captured by the magnet. (c) When magnetic field was turned off, the cells were released.[46] (Reprinted from *Journal of Magnetism and Magnetic Materials*, D. Jaiswal, A.T. Rad, M. Nieh, K.P. Claffey, K. Hoshino, Micromagnetic cancer cell immobilization and release for real-time single cell analysis, Copyright 2017, with permission from Elsevier.)

micromagnets. The magnetic force calculated at $z = 10$ μm is 5 times larger than the force at $z = 40$ μm.

When the current going through the external solenoid is turned off, the magnetic field is reduced and the paramagnetic micromagnets release the captured cells. Figure 11.20 shows the sequence of cell capture and release. This method is useful when different cells are sequentially studied for real-time responses. The cells shown in the figure are MCF7 (breast cancer cell line) labelled with 4.5 μm polystyrene beads functionalized with EpCAM.

Figure 11.21 shows an example of active control of magnetization in a microfluidic system.[47] A 100-nm-thick soft permalloy

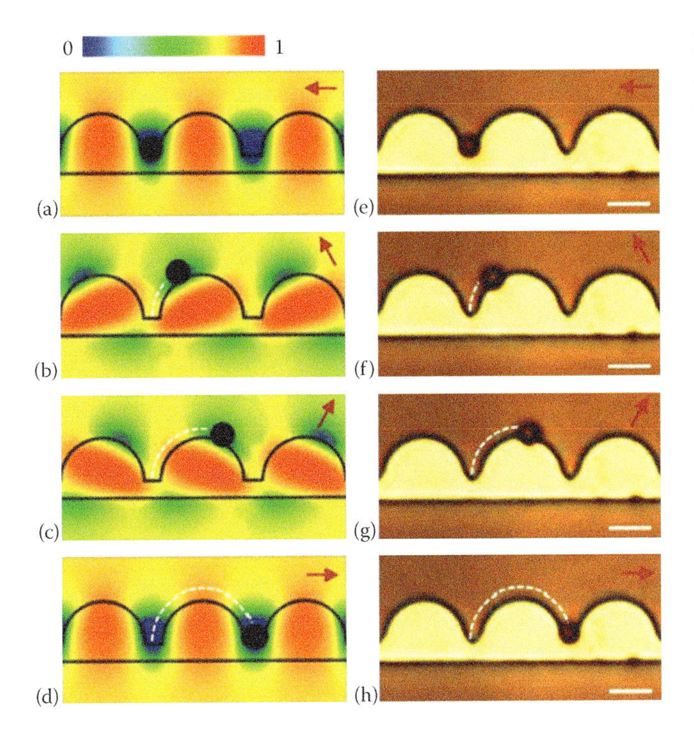

FIGURE 11.21 Active control of magnetic beads in a microfluidic channel. Scale bar 5 μm. (a–d) Rotating magnetic field. (e–h) Corresponding magnetic bead trajectory.[47] (From Lim, B., Reddy, V., Hu, X., Kim, K., Jadhav, M., Abedini-Nassab, R., Noh, Y., Lim, Y.T., Yellen, B.B., Kim, C., *Nat. Commun.*, 5, 2014. With permission.)

($Ni_{82.6}Fe_{17.4}$) film was lithographically patterned to form active magnets. Because of the semicircular design, the magnet has similar magnetization characteristics for different directions and it is possible to magnetize it with different orientations. Pairs of external solenoids are used to create a magnetic field with different angles. As we have seen in Figures 11.19 and 11.20, magnetization is largest at the end of a magnet. When magnetization direction was rotated as shown in the figure, a magnetic bead moved, following the direction. Beads used in the experiments are 2.8-μm and 4.5-μm-diameter Dynabeads. In this study, they have overlaid a gold pattern on top of the magnets and applied current to induce additional magnetic field that actively controlled paths of moving particles (Figure 11.21).

11.4 APPLICATIONS

11.4.1 Applications of Magnetic Separation

11.4.1.1 Proteins

Magnetic carriers functionalized with affinity or hydrophobic ligands can be used to separate proteins.[2] Affinity ligands including streptavidin, antibodies, protein A and protein G are commonly used for selective protein separation. Immunomagnetic separation is a method where antibodies are used as capturing molecules to selectively bind magnetic carriers and specific proteins or other molecules. An antibody is a protein that is 10–20 nm in length and 2–3 nm in height.[48] They show high affinity or selectivity to targets, including pathogens,[49,50] proteins expressed in malignant cells,[51] food allergens,[52] toxins[53] etc.; thus, they can be used for separation of such materials. Most importantly, immunomagnetic cell separation relies on binding of functionalized particles to antigen proteins expressed on the surface of target cells. Particles used for immunomagnetic separation selectively target proteins expressed in the cells to be sorted. Details of cell separation methods are described in the following section.

In the practice of immunomagnetic assay, an aqueous sample containing target materials is first added with a suspension of functionalized magnetic particles and is incubated for a certain time to allow magnetic particles to bind to the target. The test tube containing the sample may be gently rotated or shaken during incubation to prevent sedimentation and promote binding of particles. However, vortex or sonication is not usually used to avoid damages to the target objects. There are two types of magnetic separation, namely positive selection and negative selection. Positive selection is a method in which desired molecules or cells are magnetized, separated and collected from the sample. On the other hand, negative selection labels and separates undesired materials to be removed from the sample.

11.4.1.2 DNA/RNA

Separation of DNA/RNA is the first step in many types of biochemical and biomedical genomic analysis. Purification of samples is a crucial step in applications, including amplification, detection, hybridization and sequencing. Magnetic carriers functionalized with complementary DNA or RNA sequences or DNA binding proteins are used to capture RNA and DNA.[54] Since magnetic beads would separate both total DNA and RNA, either DNA or RNA need to be destroyed to obtain only DNA or RNA. An RNAse or a DNAse can be used for such processes.[54]

Nucleic acids can be also used as an affinity ligand to separate molecules other than DNA/RNA pieces. An aptamer is a nucleic acid that can specifically bind to target molecules, including proteins, peptides, amino acids and whole cells, with the binding affinities comparable to those of antibodies.[55,56] Aptamers are typically oligonucleotides such as RNA and DNA or small peptide molecules. They are smaller than antibodies and are able to reach concealed protein epitopes.[55] It is possible to chemically modify aptamers chemically to add reporters and other functional groups. Mass production of aptamers can be achieved through *in vitro* processes.

Aptamers are nucleotides that are typically sized 30–40 pairs in length and function as a primary binding site. They are separated from a process called systematic evolution of ligands by exponential enrichment (SELEX)[57–59] from a library of random collection (10^{12}–10^{14}) of DNA or RNA sequences. The library is incubated with the target, and sequences that bind to the target are separated and others are washed. Selected samples are amplified through the PCR, and binding and separation are repeated multiple rounds.

The SELEX process usually relies on magnetic separation. For example, if the target molecule is a protein, the target proteins are first immobilized on magnetic carriers. Free proteins that do not attach to the carriers are washed away. A library of nucleic acids sized 30–40 pairs in length is mixed with the carrier and incubated for about 30 min. The carriers are magnetically separated and unbounded or weakly bounded sequences are washed. Contaminants in the separated samples are removed using sodium dodecyl sulfate polyacrylamide gel electrophoresis. The bounded sequences are amplified through PCR. The same process is repeated for about six rounds, and the binding affinity is characterized through a ligand-biding assay. Selected aptamers are sequenced to visualize and record the structure.

11.4.1.3 Bacteria and Viruses

Immunoassay or DNA/RNA assay-based magnetic separation can be used to separate bacteria and viruses. Magnetic separation used for separation of bacteria and viruses in microbiology was summarized by Olsvik et al.[3] Bacteria studied with magnetic separation include *E. coli*, *Salmonella* and *Shigella* species. Viruses studied include enteroviruses, cytomegalovirus and HIV-1. We have reviewed microfluidic magnetic separation of *E. coli* bacteria[4] in the previous section. Gao et al.[25] reported on a method of bacteria separation with FePt nanoparticles, which are functionalized with vancomycin, an antibiotic used to treat bacterial infections, to detect vancomycin-resistant pathogens. In this experiment, vancomycin-functionalized particles that are in a separation system using a small magnet are compared to particles in nonspecific capping (FePt-NH_2), as illustrated in Figure 11.22a and b. While the functionalized particles successfully enriched vancomycin-resistant bacteria, nonspecific particles failed to capture them.

11.4.2 Cell Separation: Alternative Methods

Cell separation has been one of the most important applications for magnetic separation. In the following sections, we review the methods and applications of different cell separation types and discuss application-specific issues. First, we review alternative methods that do not rely on magnetic carriers and describe the advantages and disadvantages, which will be followed by the discussion on magnetic carrier based cell separation.

11.4.2.1 Physical Separation

Separation of cells can be categorized into two types depending on whether the method uses biochemical assay, which relies on biochemical binding, including antigen–antibody binding (immunoassay), biotin–avidin binding and DNA and RNA hybridization. Physical separation does not depend on biochemical binding. The most commonly utilized method is centrifugation to separate cells based on their density. After centrifugation of blood, cells are sorted in different layers. Leukocytes (white blood cells), thrombocytes (platelets) and possible cancer cells are separated into a thin layer called a buffy coat between the layers containing erythrocytes (red cells) and plasma. Commonly utilized types of separation media include Ficoll and OncoQuick.[60]

In other cases, physical filters, i.e. membrane with pores, are used to separate cells. This relies on the size difference among cells. For example, cancer cells tend to be larger (typical average diameter ~15 μm) than normal white blood cells (typically smaller than 10 μm), and they may be separated with a physical filter. The advantage of physical separation is the throughput. Since the process is usually a simple filtration, a large amount of sample blood may be processed in a short time. Problems are relatively low purity and capture rates.

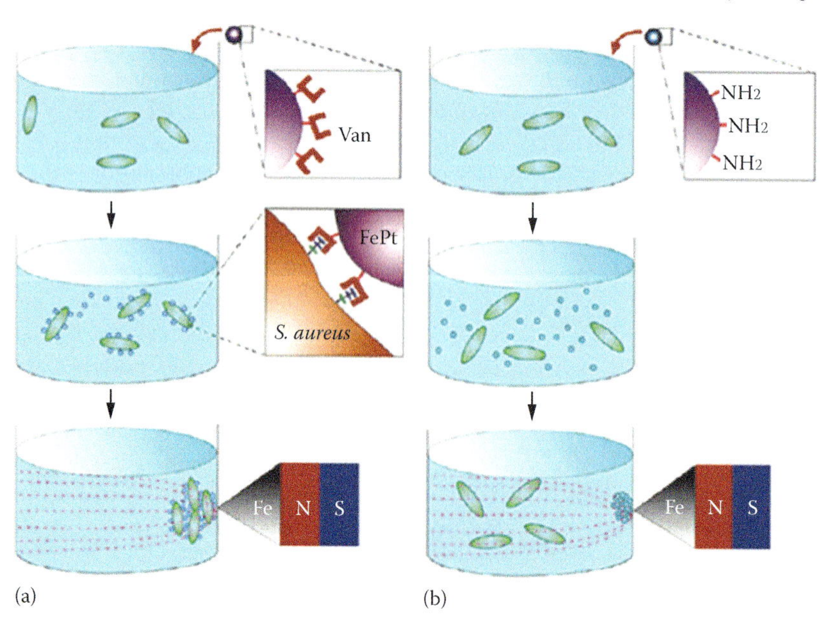

FIGURE 11.22 (a) Separation of bacteria with vancomycin-conjugated FePt nanoparticles. (b) Control with FePt-NH_2 nanoparticles.[25] (Reprinted with permission from J. Gao, H. Gu, B. Xu, Multifunctional magnetic nanoparticles: design, synthesis, and biomedical applications, *Accounts of Chemical Research*. 42 (2009) 1097–1107. Copyright 2009 American Chemical Society.)

The purity may be affected by some types of blood cells, which are larger than typical cancer cells. The capture rate may be affected by losing cancer cells that are smaller than typical blood cells. Physical screening may be beneficial if it is used as a prescreening step before the main separation step.

11.4.2.2 Fluorescence-Activated Cell Sorting

Flow cytometry, or more specifically fluorescence-activated cell sorting (FACS), is the other very commonly used cell sorting method other than magnetic separation. We will describe the principle and characteristics of FACS and compare it with magnetic separation.

Flow cytometry is a method used to count and/or sort cells based on fluorescence intensity. Usually, it is combined with immunofluorescence.[61–63] Flow cytometry is as routinely used in clinical practice as magnetic separation. It is composed of a flow chamber, laser excitation sources, a microscopic detector and a sorting mechanism. The laser-based fluorescence microscope allows for the fluorescence intensity analysis of single cells flowing in the capillary, which is capable of flowing cells in the exact centre of the channel. By flowing a faster sheath flow that surrounds the main flow containing cells, flowing cells are focused in the middle of the flow chamber. The method is called hydrodynamic focusing, allowing cells to be in the focal point of the laser-based fluorescence microscopic detector. The microscopic detector, in principle, is an integrated fluorescence microscope. It is very common that multiple biomarkers are used to study complex marker expression levels. The sorting system is similar to an inkjet printer head. The nozzle is vibrated by an actuator and flowing fluid is ejected as droplets. An individual droplet is electrically charged positive or negative, depending on the direction in which the droplet should be ejected. Based on the fluorescence intensity data, flowing cells can be sorted into multiple tubes or wells. Companies providing flow cytometry systems include Becton, Dickinson and Company (BD), Beckman Coulter, Sony and Millipore. Fluorescence-activated cell sorting is the name of the system commercialized by BD and is now used as a general name for flow cytometry. As described here, the most significant feature of FACS is the capability to count and sort cells one by one. This is the largest advantage of FACS over magnetic separation, but it has issues as well. With a most advanced flow cytometer, cells passing through the capillary can be processed at a rate of more than several thousand per second. However, it is still not fast enough to process millions of cells prepared in multiple tubes. Many studies use magnetic separation for an initial enrichment process before flow cytometry analysis for individual cell analysis. A good example is the report on stem cell separation by Spangrude et al.,[64] which will be described in a later section.

11.4.2.3 Microfluidic Separation

Other types of immunoassay include microfluidic affinity-based immunoassay, where sidewalls in a channel are functionalized with antibodies and target cells are captured on the walls.[65] It is very typical that many columns[66] or grooves[67] are integrated in the microchannel to promote cell–wall interaction.

11.4.3 Cell Separation: Methods of Magnetic Labelling

11.4.3.1 Magnetic Susceptibility of Cells

Although not as significant as common magnetic materials, any material, including a cell, shows magnetization $\chi\mathbf{M}$ in response to applied magnetic field $\mathbf{H}$. When χ is in the range of 10^{-6} to 10^{-1}, the material is classified as paramagnetic. The material is diamagnetic when χ is negative and in the range of -10^{-6} to -10^{-3}. It has been reported[68] that deoxy and methemoglobin in erythrocytes (RBCs) show paramagnetic properties with the susceptibility of 0.265×10^{-6} and 0.301×10^{-6}, respectively, while oxyhaemoglobin shows diamagnetic properties with the susceptibility of -0.0147×10^{-6}. The change in magnetic characteristics is a result of different forms of oxygen binding. Bonds in deoxy and methemoglobin are ionic, and their unpaired electrons make them paramagnetic. Oxyhaemoglobin has covalent bonds, and the lack of unpaired electrons results in the diamagnetic properties. Zborowski et al.[68] prepared erythrocytes with deoxyhaemoglobin, methemoglobin and oxyhaemoglobin and measured their mobilities under an applied magnetic field.

Disease-infected blood cells may show unique magnetic characteristics. Some cells are magnetized by themselves when infected and can be separated without using magnetic particles for labelling. Malaria parasites use haemoglobin in the host cell as a source of amino acids. They convert free hemes released from digested haemoglobin to particulate hemozoin. Hemozoin is a distinctive, clearly visible crystal within infected erythrocytes. St Pierre et al.[69] used a laboratory strain of parasites to magnetically enrich cells and used a superconducting quantum interference device magnetometer to find volumetric susceptibility of $(1.88 \pm 0.60) \times 10^{-6}$. Zborowski et al. developed a fluidic separation device composed of a rubber spacer and polyester sheets. The lumen of the fluidic channel is 6.4 mm × 0.25 mm. The device was placed on a pair of permanent magnets arranged with an interpolar gap of 1.27 mm, which showed a magnetic field intensity of 1.426 T at the midline.[70] They tested blood samples collected from 55 plasmodium-infected individuals, using a conventional blood smear and the fluidic separation device. The detection with the fluidic device showed the better rates for the three different parasite development stages.[71]

11.4.3.2 Cellular Uptake

Other methods utilize cellular uptake of nanoparticles. In this scenario, cells uptake magnetic nanoparticles within cellular membranes. For this purpose, the particles have to be as small as or smaller than the order of 100 nm to allow for cellular uptake. Internalization of magnetic nanoparticles are typically studied as contrast enhancement in MRI but also utilized for cell separation.

In an early study, Lewin et al.[21] developed superparamagnetic nanoparticles derivatized by short HIV-tat peptides to track the distribution and differentiation of stem cells and progenitor cells. HIV-tat is a regulatory protein encoded by the tat gene in HIV-1. Peptide sequences derived from tat proteins have been used to promote cellular internalization of several marker proteins. They utilized the same method for efficient internalization of magnetic nanoparticles that are modified with tat-sequences. The core of the particle is 5 nm monocrystalline iron oxide, which was stabilized by a dextran coating. The overall size of the particle was ~45 nm. The particles were internalized by hematopoietic and neural progenitor cells with quantities of up to 10–30 pg of magnetic ion per cell. Intravenous injection into mice enabled detection of labelled single cells through MRI. They did not observe the effect of iron incorporation on cell viability. They further demonstrated magnetic separation of cells that had homed to bone marrow using a separation column (Miltenyi Biotec, columns type LS+). The separation experiment is described in Figure 11.23B.

A typical technique for nanoparticle surface modification for cellular uptake is the use of poly(ethylene glycol) (PEG) and folic acid.[72] Zhang et al.[72] immobilized PEG and folic acid on the surface of magnetic nanoparticles and reported *in vitro* cellular internalization and targeted uptake to specific cells. They cultured mouse microphage cells (RAW 264.7) and human breast cancer cells (BT20) for 48 h in a medium containing the particles and quantified the amount of internalization by inductively coupled plasma emission spectroscopy. They found that the amount of PEG particles internalized by macrophage cells was much lower than the unmodified particles. On the other hand, breast cancer cells showed more internalization for both PEG and folic acid particles, suggesting the use of PEG and folic acid functionalization for cancer therapy and diagnosis. Wilhelm et al.[22] discussed the effect of nanoparticle coating with dextran, bovine serum albumin and immunoglobulin G for intercellular uptake in mouse microphage cells (RAW 264.7) and human ovarian tumour cells (HeLa). They used maghemite (γ-Fe_2O_3) particles with TEM observed diameters ranging from 3 nm to 15 nm. Cells incubated with magnetic particles were suspended under a magnetic field and their migration was quantified through the experiment shown in Figure 11.23B. Silica (SiO_2) coating is another approach to improve biocompatibility. Kim et al.[73] coated cobalt ferrite magnetic nanoparticles (average diameter ~9 nm) with silica to form 50-nm-diameter nanoparticles and observed endocytosis in lung cancer cells (A549 cells). The use of silica allowed them to include organic fluorescence dye (rhodamine B) within the particles.

11.4.3.3 Immunomagnetic Assay

In an early study by Molday and Molday[74] reported in 1977, fluorescent magnetic microspheres containing Fe_3O_4 and fluorescein isothiocyanate (FITC) for magnetization and fluorescence, respectively, were used for magnetic separation of RBC and lymphoid cell. Now, immunomagnetic separation is widely used for the separation of cells. By choosing proper biomarkers, cells expressing specific proteins can be labelled and separated. Separation kits for cancer cells, stem cells and blood cell have been commercialized and widely used among researchers. Antigens used for immunomagnetic separation should be strongly expressed in most of target cells, while nontargeted cells should not express the same antigen. In some cases, it is difficult to find a single type of antigen that satisfies such a requirement. Combinations of multiple-target antigens may be used to overcome the problem. Cancer cells are heterogeneous and diverse and are sometimes difficult to label with one type of antibody. The use of 'cocktail' of antibodies can be effective, as reported by Wu et al.[23]

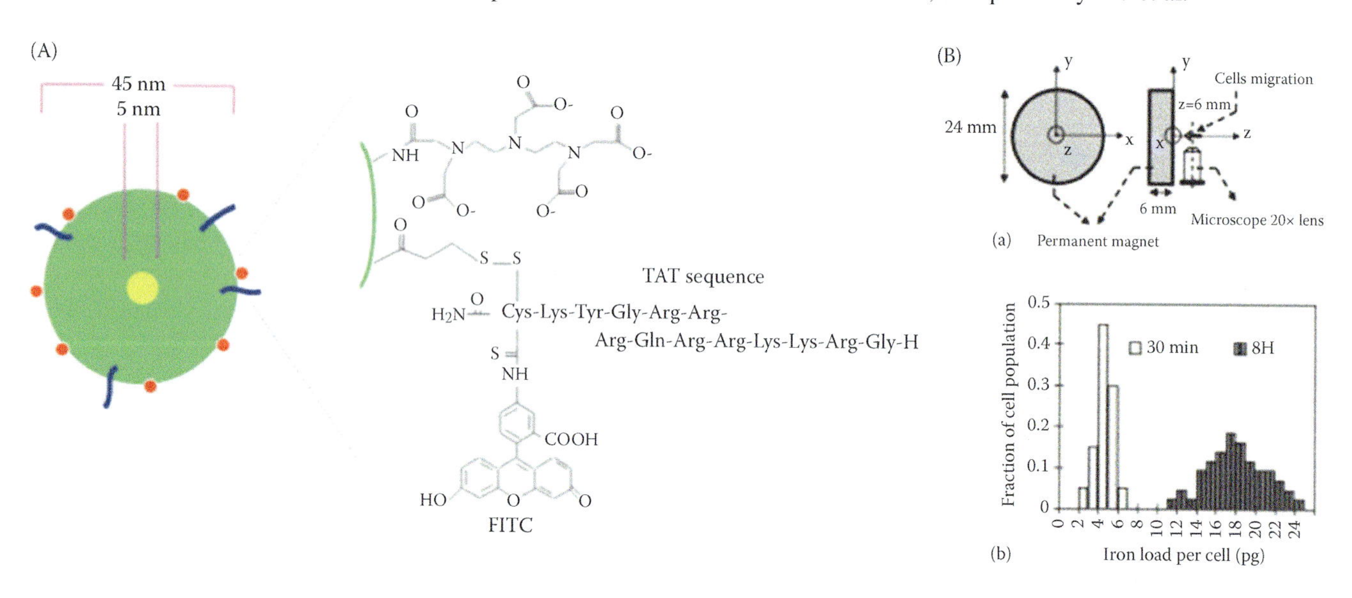

FIGURE 11.23 Nanoparticle internalization study for magnetic separation. (A) Nanoparticles functionalized with short HIV-tat peptides.[21] (Reprinted by permission from Macmillan Publishers Ltd. *Nature Biotechnology,* M. Lewin, N. Carlesso, C. Tung, X. Tang, D. Cory, D.T. Scadden, R. Weissleder, Tat peptide-derivatized magnetic nanoparticles allow *in vivo* tracking and recovery of progenitor cells, *Nature Biotechnology*. 18 (2000) 410–414. Copyright 2000.) (B) Microscopic observation of cell migration under a magnetic field was used to quantify the amount of internalized particles.[22] (Reprinted from *Biomaterials*, 24, C. Wilhelm, C. Billotey, J. Roger, J. Pons, J. Bacri, F. Gazeau, Intracellular uptake of anionic superparamagnetic nanoparticles as a function of their surface coating, 1001–1011, Copyright 2003, with permission from Elsevier.)

Parameters used to assess the efficacy of immunomagnetic assay include capture rate, throughput, purity and cell integrity. Capture rate is the rate of captured cells per all the cells that exist in the sample liquid. In many cases of disease detection, separation of rare infected or tumourous cells are required, and thus, having a higher capture rate is a crucial factor. On the other hand, for some cases where target cells are abundant and only a part of them need to be retrieved, capture rate is not an important parameter.

Throughput is the rate at which the sample is processed, or the amount of sample that can be processed in a certain time. For some cases, the amount of processed liquid volume may be referred to as throughput. In other cases, number of tests rather than the volume may be the throughput. Cases where larger volumes of samples need to be processed include blood cleansing, where immunomagnetic separation is used to remove infected cells, and detection of very rare cells, as we will discuss later for CTC analysis. In other cases such as DNA separation, the PCR steps can efficiently amplify target genes from a very small amount of samples. In this case, number of tests, or number of target genes, that can be performed in a certain time rather than a sample volume may be more important.

Purity refers to the rate of real positive cells among the total number of cells that are separated. Lower purity means cells that are separated may contain a considerable amount of false-positive cells. Purity is crucial in detection of diseases. Analysis that shows frequent false-positive detection does not make sense for clinical disease detection. In some cases, immunomagnetic separation is followed by additional immunofluorescence screening or gene amplification steps through PCR to cover possible false-positive detection.

Cell integrity is related to the reliability of the separation method. Separated cells usually go through additional analysis steps after separation. It is desirable that the separated cells are not damaged or changed by the separation step. In some cases, separated cells will go through live cell analysis or a cell culture step.

11.4.4 Cell Separation: Cells Sorted in Immunomagnetic Separation

11.4.4.1 Stem Cells

Stem cells are cells that have the ability to differentiate into multiple types of specialized cells and to produce more stem cells through cell division. The term stem cell usually refers to either totipotent or pluripotent stem cell. Totipotent stem cells are stem cells that are able to divide from a single cell to produce all differentiated cells in the body. Pluripotent stem cells can differentiate into any of the three germ layers, namely endoderm (interior stomach lining, gastrointestinal tract and lungs), mesoderm (muscle, bone, blood and urogenital) or ectoderm (epidermal tissues and nervous system). Multipotent cells that are able to differentiate into multiple types of cells are also sometimes called stem cells. Tissues contain populations of stem cells that can produce cellular progeny, which differentiate into mature tissue phenotypes. In certain tissues, there are reservoirs of stem cells called stem-cell niches to repair damage when needed. For example, bone marrow contains a small number of pluripotent stem cells and produces approximately 400 billion myeloid cells every day. Separation of stem cells from normal cells is a very important step in bone marrow transplant. Stem cells are an attractive source of cells for medical and engineering applications. Magnetic separation of stem cells has been actively studied in the field of regenerative medicine. Magnetic particles functionalized with stem cell markers including CD33, CD34, CD90 and CD133 are commercially available.

In the early study reported by Spangrude et al.,[64] a combination of negative and positive separation was used to separate pluripotent stem cells from mouse bone marrow cells. Firstly, T cells are removed using anti-CD4 and anti-CD8 as the markers for negative selection, followed by positive selection based on anti-Thy1.1 (CD90.1). Approximately 2.0% of the original cells were recovered after these steps and further studied through FACS. Kato and Radbruch[75] used CD34 functionalized magnetic beads in the MACS separation system and the frequency of CD34+ cells was enriched from 0.18% ± 0.052% among leukocytes to 38.6%–87.1%. In the report, the sample was further purified through multiparameter FACS.

11.4.4.2 Blood Cells

Separation of leukocytes from blood samples is one of the very important practical applications of magnetic separation. Immune cells including T cells or B cells have been separated from blood samples. Furdui et al.[76] used a silicon-based microfluidic system to demonstrate separation of Jurkat cells from human blood. In their experiment; the Jurkat cells were mixed with blood cells at a ratio of 1:10,000.

11.4.4.3 Circulating Tumour Cells

CTCs are tumour cells that have detached from a primary tumour and circulate in a blood stream. CTCs are believed to seed subsequent metastatic tumours in distant organs. Detection of CTCs from blood has been drawing attention because a simple blood testing may provide information needed for early cancer detection or diagnosis of disease activity. Immunomagnetic separation of CTCs can be categorized into conventional manual separation, robotic separation and micro fluidic separation.

Here, we discuss the unique engineering issues associated with immunomagnetic detection of CTCs. The capture rate, which indicates the sensitivity of the system, is one of the most crucial factors because most significant characteristic of CTC detection is the extreme rarity of tumour cells in blood. The number of CTCs per leukocytes is thought to be 10^{-6} to 10^{-7}. If the amount of blood that can be processed is doubled, the chance of finding tumour cells is also doubled. In clinical blood testing, it is common to draw patient blood samples with multiple 10-mL tubes. The system has to be able to process that amount of blood in a reasonable amount of time. In this point of view, microfluidic systems that process samples at relatively small flow rates are not suitable. Purity is the rate of real tumour cells among cells that are detected in the analysis. Because of the rarity of CTCs, immunomagnetic separation is almost always followed by additional screening steps.

Many types of cancer are categorized as a carcinoma, where tumours develop from epithelial cells. EpCAM is a type of molecule that mediates cell–cell adhesion in epithelia. Because EpCAM is exclusively found in epithelial cells and not in blood cells, it is very commonly used to separate circulating epithelial cells, which are most likely tumour cells. Currently, EpCAM may be the only successful biomarkers that can be used to separate many types of tumour cells. A problem associated with the use of EpCAM is EMT, which is a cellular process where epithelial cells lose cell–cell adhesion and polarity and gain properties similar to mesenchymal cells. Carcinoma cells often lose cell–cell adhesion, meaning they lose EpCAM expression. Some types of tumour cells that have gone through EMT are known to escape EpCAM-based screening. Other types of biomarkers used for CTC separation include human epidermal growth factor receptor 2 (HER2) for breast cancer cells and prostate-specific antigen for prostate cancer cells. Because they are also found in normal cells, they are not as an efficient marker as EpCAM in terms of the purity of selection. In CTC detection, they are often used as an additional antibody in combination with the use of EpCAM.

Wu et al.[23] used a combination of multiple antibodies with magnetic carriers and studied the capture yield for different cell lines.

The magnetic carriers used are gold shell/magnetic core nanoparticles (see Section 11.3.1). Three cell lines, COLO 205 (colorectal cancer, EpCAM+, HER2−, EGFR− and CK+), SK-BR-3 (a breast cancer, EpCAM+, HER2+, EGFR− and CK+) and A-431 (skin cancer, EpCAM+, HER2-, EGFR+ and CK+), were tested. Nanoparticles are functionalized with either EpCAM, HER2, EGFR or CK antibodies. The targeting specificity was characterized by assessing the binding of nanoparticles through dark-field microscopy. Nanoparticles attached on cells show a yellow-orange colour in dark-field images, while a gray-bluish colour indicates the endogenous scattering on unlabelled cells.

All cell lines were labelled well with anti-EpCAM nanoparticles. Anti-HER2 and anti-EGFR particles label only SK-BR-3 and A-431 cells, respectively. Mixture of functionalized particles that target various cancer antigens can be used to improve efficacy of CTC detection. A mixture of anti-EGFR and anti-EpCAM nanoparticles showed an increased capture yield of 93% for A-431 cells, while anti EGFR alone showed a yield of 79%. Combination of anti-HER2 and anti-EpCAM for SK-BR-3 showed a yield of 93% compared to 69% alone with HER2 particles (Table 11.2). Wu et al.[23] also demonstrated capture of cancer cells with low EpCAM expression and compared it with manually reproduced CellSearch assay. Blood samples are spiked with basal like breast cancer cell line BT20, which shows low EpCAM expression. A mixture of anti-Mucl nanoparticles and anti-EpCAM showed a capture rate of 78%, while that with EpCAM-based CellSearch assay was 44%.

TABLE 11.2
Capture Efficiency in Spike Experiments in Whole Blood Samples from a Normal Volunteer

Nanocarriers against Antibody	Cell Line	Capture Yield (%)
EGFR[a]	A-431 (skin)	79 ±8
EGFR + EpCAM[b]	A-431	93 ±2
HER2[c]	SK-BR-3 (breast)	69 ±7
HER2 + EpCAM	SK-BR-3	93 ±10
EpCAM	COLO 205 (colon)	75 ±9
CK[d]	COLO 205	79 ±6
EpCAM	BT-20 (breast)	45 ±8
EpCAM + MUC1[e]	BT-20	78 ±10

Source: Reprinted with permission from C. Wu, Y. Huang, P. Chen, K. Hoshino, H. Liu, E.P. Frenkel, J.X. Zhang, K.V. Sokolov, Versatile immunomagnetic nanocarrier platform for capturing cancer cells, *ACS Nano*. 7 (2013) 8816–8823. (Copyright 2013 American Chemical Society.)

[a] Antiepidermal growth factor receptor 1.
[b] Antiepithelial cell adhesion molecule.
[c] Antiepidermal growth factor receptor 2.
[d] Anticytokeratin.
[e] Antimucin 1.

11.4.5 Cell Separation: Analysis Beyond Magnetic Separation

Chen et al.[38] reported on the introduction of patterned micromagnets, which made localized magnetic fields up to eightfold stronger than the system without them. The system has been tested with four cancer cell lines, namely SK-BR-3 (breast), MCF7 (breast), COLO 205 (colorectal) and PC3 (prostate), demonstrating capture rates over 97%,[77] followed by testing of clinical samples. Table 11.3 summarizes the successful separation of CTCs from patient blood samples.

Since the immunomagnetic separation process does not completely eliminate nonspecifically bound white blood cells, the sample needs to be further studied through immunofluorescence. Especially in the case of CTC detection, most of the cells fixed on a slide are leukocytes, and cancer cells are only a small part of cells on the slide. Huang et al.[77] stained cells for DNA, Cytokeratin (CK) and CD45 using three different markers: DAPI (blue), FITC (green) and Alexa Fluor 594 (red), respectively. Cancer cells originated from epithelial cells exhibit DAPI+, CK+, and CD45−. On the other hand, leukocytes show DAPI+, CK− and CD45+. Figure 11.24 shows tumour cells captured from patient blood samples and observed under immunofluorescence. In addition to fluorescent signals, morphological characteristics are important information in cancer cell identification. These cells are separated and identified by a trained observer and further confirmed by physicians. For CTC studies, this multistep manual identification process still plays an important role to demonstrate clinical significance.

TABLE 11.3

Capture Yields Measured for Multiple Antibodies and Cell Lines

Sample Number	Cancer Type	Gender	Ferrofluid Blood (μL/ml)	Flow Rate (μl/hr)	Screening Volume (mL)	Number of CTCs	Number of CTCs/7.5 mL of Blood Found from *CellSearch*™
1	Colon cancer	M	7.5	2.5	5.0	1	N/A
2	Lung cancer	F	7.5	2.5	10.0	1	N/A
3	Prostate cancer	M	7.5	2.5	7.5	13	0
4	Breast cancer	F	7.5	2.5	7.5	6	0
5	Breast cancer	F	7.5	2.5	7.5	3	N/A
6	Breast cancer	F	7.5	2.5	5.0	10	N/A
7	Breast cancer	F	7.5	2.5	7.5	2	N/A
8	Breast cancer	F	7.5	2.5	7.5	1	N/A
9	Breast cancer	F	7.5	2.5	10.0	22	N/A
10	Breast cancer	F	7.5	2.5	5.0	215	N/A
11	Breast cancer	F	7.5	2.5	7.5	2	N/A
12	Breast cancer	F	7.5	2.5	7.5	6	N/A
13	Breast cancer	F	7.5	2.5	5.0	7	N/A

Source: From Huang, Y.Y., Chen, P., Wu, C.H., Hoshino, K., Sokolov, K., Lane, N., Liu, H., Huebschman, M., Frenkel, E., Zhang, J.X., *Sci. Rep.*, 5, 16047, 2015. With permission.

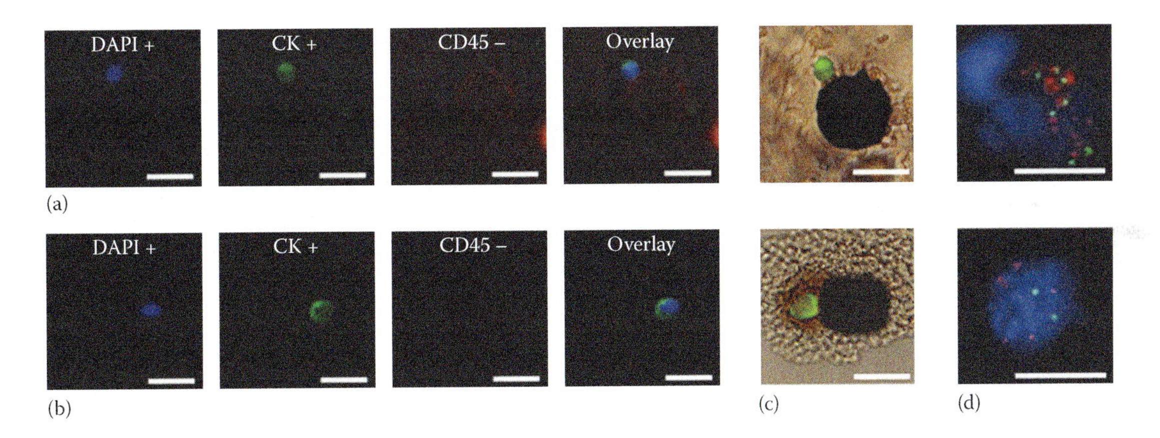

FIGURE 11.24 (a, b) Immunofluorescence images, (c) bright field images and (d) FISH analysis of cells captured from patient samples.[76] (From Huang, Y.Y., Chen, P., Wu, C.H., Hoshino, K., Sokolov, K., Lane, N., Liu, H., Huebschman, M., Frenkel, E., Zhang, J.X., *Sci. Rep.*, 5, 16047, 2015. With permission.)

Once CTCs are identified, detailed genomic analysis may be conducted for further characterization of captured cells. Figure 11.24d shows the result of fluorescence *in situ* hybridization (FISH) performed for cancer cells from patient samples 9 and 10 in Table 11.3. Samples 9 and 10 showed more than 16 copies and more than 6 copies of HER2, respectively. Amplified HER2 with ratio numbers larger than 2.2 indicates clear characteristics seen in progressed HER2+ breast cancer. The copy number of HER2 is associated with shorter survival and increased risk of reoccurrence. This study demonstrated the use of captured CTCs as biomarkers for genomic disease analysis.

For cancer genomic analysis, PCR is one of the most widely utilized techniques. The promising advantage of PCR analysis of CTC[78–83] is that the genomic information of the disease may be obtained through a simple blood testing rather than invasive tumour biopsy. As indicated in Table 11.3, the number of tumour cells captured from 5–10 mL of blood samples is often fewer than 10. A PCR analysis combined with detailed fluorescence analysis is preferred. Conventional fluorescence-activated cell sorting[84–86] (FACS) tools are more suitable for sorting larger numbers of cells in a simple intensity-based analysis. Detailed immunofluorescence to study cell morphology and protein expressions should be also made along with PCR analysis to correctly identify the capture CTCs. As we discussed earlier, sample purity is a critical parameter for PCR analysis. Blood cells should be eliminated from PCR samples to improve the purity of captured cancer cells.

Hoshino et al.[87] proposed a method of single-cell PCR analysis using the microfluidic immunomagnetic separation assay.

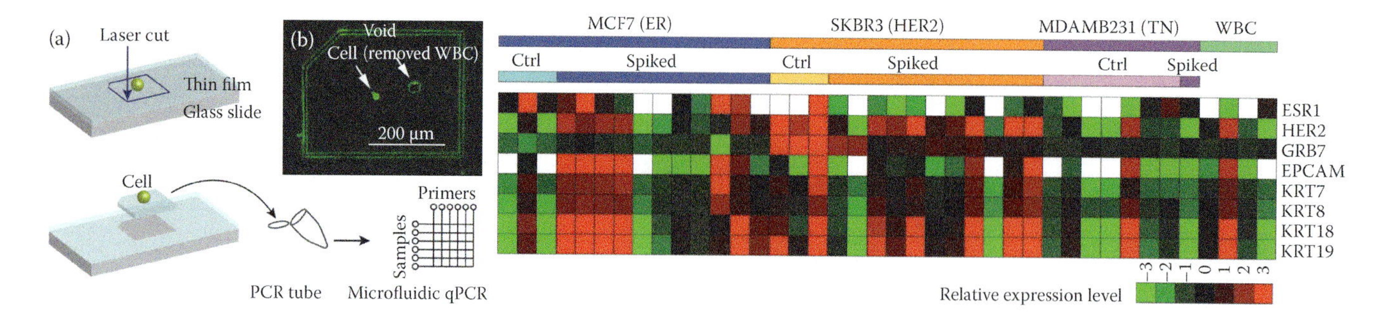

FIGURE 11.25 Single cell PCR analysis of captured cells. (Left panel) (a) Cells were captured on a thin polymer film. (b) Cells were picked up through laser microdissection. (Right panel) Heat map of the result of single cell PCR analysis.[86] (From Hoshino, K., Chung, H., Wu, C., Rajendran, K., Huang, Y., Chen, P., Sokolov, K., Kim, J., Zhang, J., *J. Circ. Biomark.*, 4, 11, 2015. With permission.)

The separation assay is similar to the one described by Huang *el al.*[77] The difference is that the bottom substrate is coated with a thin polymer film that can be laser-cut in small pieces. Using the laser microdissection technique,[88] a small piece of polymer film that contains fluorescently observed cancer cells can be cut and picked up for the following single-cell PCT analysis. Figure 11.25a shows the experimental procedure. A laser-cut small piece of the polymer film is shown in Figure 11.25b. A small void in the piece was also made by laser cutting to eliminate white blood cells from the piece. FITC cytokelatin and red (Alexa Fluor 594) CD45 were used to identify single cancer cells and white blood cells, respectively. The Zeiss PALM MicroBeam Laser Microdissection was used in the measurement.

Breast cancer cell lines were used to assess the efficacy of this method. Three cell lines, namely MCF7, SKBR3 and MDAMB231, were used to represent the three common types of breast cancer, namely ER/PR+, HER2 and triple negative. In the PCR analysis, primers for 10 different genes, all of which are commonly used for breast cancer diagnosis, were tested. A single-cell RNA extraction kit (RNAGEM tissue, ZyGEM Corp. Ltd.) was used to extract mRNA from the cells captured on a laser cut polymer piece. Complementary DNA (cDNA) is obtained through reverse transcription using SuperMix (Quanta Biosciences). A mixture of primers for target gene sequences were then added to the cDNA sample and was preamplified by 20–23 cycles before the analysis in the microfluidic qPCR system. The BioMark HD System,[89] from Fluidigm Corporation, was used for the qPCR analysis. The result of single-cell analysis is shown in Figure 11.25. A heat map showing the result of single-cell analysis matched with the reference analysis made with a few thousand cells. Most primers used in this analysis showed the trend expected for ER/PR+, HER2 and triple negative cells, although the difference is less significant in single-cell analysis. Some of the markers often used for cancer identification (e.g. ER and HER2) showed relatively large deviations. Other markers (e.g. GRB7) showed small deviations, and they may be used to supplement disease profiling based on single-cell analysis.

11.5 CONCLUSION

We have shown theories, materials, designs and instrumentation of integrated microsystem specifically designed for the purpose of magnetic separation. Examples from successful commercial devices and important studies reported by several groups have been reviewed to discuss practical aspects of magnetic separation-based analytical systems. The magnetic field induced by a magnetic material is highly dependent on its size and shape. Microfabrication of magnetic materials has a strong advantage in realizing highly controlled manipulation of biomaterials. Use of integrated device-based magnetic separation will be an important trend in designing systems for biomedical detection, analysis and diagnosis.

In the later sections, immunomagnetic cell sorting and analysis were discussed as one of the promising applications of magnetic separation systems and compared with other techniques, including FACS. Integration of microscale magnets permits separation, manipulation and analysis of single cells from samples. Among cell separation applications, stem cells and CTCs are the most important subjects that have been intensively studied, and commercial applications are expected. Integrated total analytical systems further permit downstream cell analysis following magnetic separation.

11.6 FUTURE PERSPECTIVES

The two most important advantages of integrated microanalytical systems are the capabilities to process a small amount of sample volumes and a large number of different samples in parallel, which ensures the potential of future low-cost, high-throughput analyses in clinical applications. Several promising approaches have been reviewed in this chapter.

Currently the most commonly utilized carriers for magnetic separation are commercially available magnetic beads. However, the efficacy of separation is greatly influenced by the size and magnetization of the carriers. Optimal choice of magnetic carriers, which matches the design of integrated micromagnets in the microfluidic system, will improve the applicability of the method in clinical applications. We have seen a number of excellent studies in the development of magnetic particles designed for separation. Further studies will allow for effective analytical systems that take full advantage of micro integrated systems to bring benefits in clinical applications.

REFERENCES

1. J. Osborn, Demagnetizing factors of the general ellipsoid, *Physical Review.* 67 (1945) 351.
2. I. Safarik, M. Safarikova, Magnetic techniques for the isolation and purification of proteins and peptides, *BioMagnetic Research and Technology.* 2 (2004) 1.
3. O. Olsvik, T. Popovic, E. Skjerve, K.S. Cudjoe, E. Hornes, J. Ugelstad, M. Uhlen, Magnetic separation techniques in diagnostic microbiology, *Clinical Microbiology Reviews.* 7 (1994) 43–54.
4. N. Xia, T.P. Hunt, B.T. Mayers, E. Alsberg, G.M. Whitesides, R.M. Westervelt, D.E. Ingber, Combined microfluidic–micromagnetic separation of living cells in continuous flow, *Biomedical Microdevices.* 8 (2006) 299–308.
5. W. Chang, H. Shang, R.M. Perera, S. Lok, D. Sedlak, R.J. Kuhn, G.U. Lee, Rapid detection of dengue virus in serum using magnetic separation and fluorescence detection, *Analyst.* 133 (2008) 233–240.
6. M. Uhlen, Magnetic separation of DNA, *Nature.* 340 (1989) 733–734.
7. P.S. Doyle, J. Bibette, A. Bancaud, J.L. Viovy, Self-assembled magnetic matrices for DNA separation chips, *Science.* 295 (2002) 2237.
8. H. Gu, K. Xu, C. Xu, B. Xu, Biofunctional magnetic nanoparticles for protein separation and pathogen detection, *Chemical Communications.* (2006) 941–949.
9. J. Kim, Y. Piao, N. Lee, Y.I. Park, I. Lee, J. Lee, S.R. Paik, T. Hyeon, Magnetic nanocomposite spheres decorated with NiO nanoparticles for a magnetically recyclable protein separation system, *Advanced Materials.* 22 (2010) 57–60.
10. C. Théry, S. Amigorena, G. Raposo, A. Clayton, Isolation and characterization of exosomes from cell culture supernatants and biological fluids, *Current Protocols in Cell Biology.* (2006) 3.22.1–3.22.29.
11. T.R. Sathe, A. Agrawal, S. Nie, Mesoporous silica beads embedded with semiconductor quantum dots and iron oxide nanocrystals: Dual-function microcarriers for optical encoding and magnetic separation, *Analytical Chemistry.* 78 (2006) 5627–5632.
12. A.H. de Vries, B.E. Krenn, R. van Driel, J.S. Kanger, Micro magnetic tweezers for nanomanipulation inside live cells, *Biophysical Journal.* 88 (2005) 2137–2144.
13. C. Chiou, Y. Huang, M. Chiang, H. Lee, G. Lee, New magnetic tweezers for investigation of the mechanical properties of single DNA molecules, *Nanotechnology.* 17 (2006) 1217.
14. K. Smistrup, O. Hansen, H. Bruus, M.F. Hansen, Magnetic separation in microfluidic systems using microfabricated electromagnets – Experiments and simulations, *Journal of Magnetism and Magnetic Materials.* 293 (2005) 597–604.
15. J.W. Vanderhoff, M.S. El-Aasser, J. Ugelstad, US Patent 4,177,177, Polymer emulsification process. (1979).
16. K. Smistrup, T. Lund-Olesen, M.F. Hansen, P.T. Tang, Microfluidic magnetic separator using an array of soft magnetic elements, *Journal of Applied Physics.* 99 (2006) 08P102.
17. X. Yu, X. Feng, J. Hu, Z. Zhang, D. Pang, Controlling the magnetic field distribution on the micrometer scale and generation of magnetic bead patterns for microfluidic applications, *Langmuir.* 27 (2011) 5147–5156.
18. X. Yu, R. He, S. Li, B. Cai, L. Zhao, L. Liao, W. Liu, Q. Zeng, H. Wang, S. Guo, Magneto-controllable capture and release of cancer cells by using a micropillar device decorated with graphite oxide-coated magnetic nanoparticles, *Small.* 9 (2013) 3895–2901.
19. K. Hoshino, Y. Huang, N. Lane, M. Huebschman, J.W. Uhr, E.P. Frenkel, X. Zhang, Microchip-based immunomagnetic detection of circulating tumor cells, *Lab on a Chip.* 11 (2011) 3449–2457.
20. D.W. Inglis, R. Riehn, R. Austin, J. Sturm, Continuous microfluidic immunomagnetic cell separation, *Applied Physics Letters.* 85 (2004) 5093–5095.
21. M. Lewin, N. Carlesso, C. Tung, X. Tang, D. Cory, D.T. Scadden, R. Weissleder, Tat peptide-derivatized magnetic nanoparticles allow *in vivo* tracking and recovery of progenitor cells, *Nature Biotechnology.* 18 (2000) 410–414.
22. C. Wilhelm, C. Billotey, J. Roger, J. Pons, J. Bacri, F. Gazeau, Intracellular uptake of anionic superparamagnetic nanoparticles as a function of their surface coating, *Biomaterials.* 24 (2003) 1001–1011.
23. C. Wu, Y. Huang, P. Chen, K. Hoshino, H. Liu, E.P. Frenkel, J.X. Zhang, K.V. Sokolov, Versatile immunomagnetic nanocarrier platform for capturing cancer cells, *ACS Nano.* 7 (2013) 8816–8823.
24. A.K. Gupta, M. Gupta, Synthesis and surface engineering of iron oxide nanoparticles for biomedical applications, *Biomaterials.* 26 (2005) 3995–4021.
25. J. Gao, H. Gu, B. Xu, Multifunctional magnetic nanoparticles: Design, synthesis, and biomedical applications, *Accounts of Chemical Research.* 42 (2009) 1097–1107.
26. I.H. El-Sayed, X. Huang, M.A. El-Sayed, Surface plasmon resonance scattering and absorption of anti-EGFR antibody conjugated gold nanoparticles in cancer diagnostics: Applications in oral cancer, *Nano Letters.* 5 (2005) 829–834.
27. M. Zborowski, J.J. Chalmers, Rare cell separation and analysis by magnetic sorting, *Analytical Chemistry.* 83 (2011) 8050–8056.
28. P. Joshi, P.S. Williams, L.R. Moore, T. Caralla, C. Boehm, G. Muschler, M. Zborowski, Circular Halbach array for fast magnetic separation of hyaluronan-expressing tissue progenitors, *Analytical Chemistry.* 87 (2015) 9908–9915.
29. L. Zanini, N. Dempsey, D. Givord, G. Reyne, F. Dumas-Bouchiat, Autonomous micro-magnet based systems for highly efficient magnetic separation, *Applied Physics Letters.* 99 (2011) 232504.
30. F. Dumas-Bouchiat, L. Zanini, M. Kustov, N. Dempsey, R. Grechishkin, K. Hasselbach, J. Orlianges, C. Champeaux, A. Catherinot, D. Givord, Thermomagnetically patterned micromagnets, *Applied Physics Letters.* 96 (2010) 102511.
31. R. Molday, L. Molday, Separation of cells labeled with immunospecific iron dextran microspheres using high gradient magnetic chromatography, *FEBS Letters.* 170 (1984) 232–238.
32. S. Miltenyi, W. Müller, W. Weichel, A. Radbruch, High gradient magnetic cell separation with MACS, *Cytometry.* 11 (1990) 231–238.
33. G.M. Whitesides, The origins and the future of microfluidics, *Nature.* 442 (2006) 368–273.
34. J.R. Anderson, D.T. Chiu, H. Wu, O.J. Schueller, G.M. Whitesides, Fabrication of microfluidic systems in poly (dimethylsiloxane), *Electrophoresis.* 21 (2000) 27–40.
35. M. Bu, T.B. Christensen, K. Smistrup, A. Wolff, M.F. Hansen, Characterization of a microfluidic magnetic bead separator for high-throughput applications, *Sensors and Actuators A: Physical.* 145 (2008) 430–436.
36. S. Kim, S. Han, M. Park, C. Jeon, Y. Joo, I. Choi, K. Han, Circulating tumor cell microseparator based on lateral magnetophoresis and immunomagnetic nanobeads, *Analytical Chemistry.* 85 (2013) 2779–2786.

37. X. Lou, J. Qian, Y. Xiao, L. Viel, A.E. Gerdon, E.T. Lagally, P. Atzberger, T.M. Tarasow, A.J. Heeger, H.T. Soh, Micromagnetic selection of aptamers in microfluidic channels, *Proceedings of the National Academies of Science U. S. A.* 106 (2009) 2989–2994.
38. P. Chen, Y. Huang, K. Hoshino, J.X. Zhang, Microscale magnetic field modulation for enhanced capture and distribution of rare circulating tumor cells, *Scientific Reports.* 5 (2015) 8745.
39. T. Deng, M. Prentiss, G.M. Whitesides, Fabrication of magnetic microfiltration systems using soft lithography, *Applied Physics Letters.* 80 (2002) 461–463.
40. W. Liu, N. Dechev, I.G. Foulds, R. Burke, A. Parameswaran, E.J. Park, A novel permalloy based magnetic single cell micro array, *Lab on a Chip.* 9 (2009) 2381–2390.
41. A.E. Saliba, L. Saias, E. Psychari, N. Minc, D. Simon, F.C. Bidard, C. Mathiot, J.Y. Pierga, V. Fraisier, J. Salamero, V. Saada, F. Farace, P. Vielh, L. Malaquin, J.L. Viovy, Microfluidic sorting and multimodal typing of cancer cells in self-assembled magnetic arrays, *Proceedings of the National Academies of Science U. S. A.* 107 (2010) 14524–14529.
42. P. Chen, Y. Huang, G. Bhave, K. Hoshino, X. Zhang, Inkjet-print micromagnet array on glass slides for immunomagnetic enrichment of circulating tumor cells, *Annals of Biomedical Engineering.* 44 (2016) 1710–1720.
43. C.M. Earhart, C.E. Hughes, R.S. Gaster, C.C. Ooi, R.J. Wilson, L.Y. Zhou, E.W. Humke, L. Xu, D.J. Wong, S.B. Willingham, Isolation and mutational analysis of circulating tumor cells from lung cancer patients with magnetic sifters and biochips, *Lab on a Chip.* 14 (2014) 78–88.
44. J. Choi, C.H. Ahn, S. Bhansali, H.T. Henderson, A new magnetic bead-based, filterless bio-separator with planar electromagnet surfaces for integrated bio-detection systems, *Sensors and Actuators B: Chemical.* 68 (2000) 34–29.
45. M. Donolato, A. Torti, N. Kostesha, M. Deryabina, E. Sogne, P. Vavassori, M.F. Hansen, R. Bertacco, Magnetic domain wall conduits for single cell applications, *Lab on a Chip.* 11 (2011) 2976–2983.
46. D. Jaiswal, A.T. Rad, M. Nieh, K.P. Claffey, K. Hoshino, Micromagnetic cancer cell immobilization and release for real-time single cell analysis, *Journal of Magnetism and Magnetic Materials.* 427 (2017) 7–13.
47. B. Lim, V. Reddy, X. Hu, K. Kim, M. Jadhav, R. Abedini-Nassab, Y. Noh, Y.T. Lim, B.B. Yellen, C. Kim, Magnetophoretic circuits for digital control of single particles and cells, *Nature Communications.* 5 (2014) 3846.
48. D.R. Davies, S. Chacko, Antibody structure, *Accounts of Chemical Research.* 26 (1993) 421–427.
49. P.D. Skottrup, M. Nicolaisen, A.F. Justesen, Towards on-site pathogen detection using antibody-based sensors, *Biosensors and Bioelectronics.* 24 (2008) 339–248.
50. V. Velusamy, K. Arshak, O. Korostynska, K. Oliwa, C. Adley, An overview of foodborne pathogen detection: In the perspective of biosensors, *Biotechnology Advances.* 28 (2010) 232–254.
51. D.M. Goldenberg, E.E. Kim, F.H. DeLand, J.v. Nagell, N. Javadpour, Clinical radioimmunodetection of cancer with radioactive antibodies to human chorionic gonadotropin, *Science.* 208 (1980) 1284–1286.
52. I. Mohammed, W.M. Mullett, E.P. Lai, J.M. Yeung, Is biosensor a viable method for food allergen detection? *Analytica Chimica Acta.* 444 (2001) 97–102.
53. D.A. Blake, R.M. Jones, R.C. Blake II, A.R. Pavlov, I.A. Darwish, H. Yu, Antibody-based sensors for heavy metal ions, *Biosensors and Bioelectronics.* 16 (2001) 799–809.
54. S. Berensmeier, Magnetic particles for the separation and purification of nucleic acids, *Applied Microbiology and Biotechnology.* 73 (2006) 495–504.
55. K. Song, S. Lee, C. Ban, Aptamers and their biological applications, *Sensors.* 12 (2012) 612–631.
56. S. Song, L. Wang, J. Li, C. Fan, J. Zhao, Aptamer-based biosensors, *TraC: Trends in Analytical Chemistry.* 27 (2008) 108–117.
57. A.D. Ellington, J.W. Szostak, *In vitro* selection of RNA molecules that bind specific ligands, *Nature.* 346 (1990) 818–822.
58. A.D. Keefe, S.T. Cload, SELEX with modified nucleotides, *Current Opinion in Chemical Biology.* 12 (2008) 448–456.
59. C. Tuerk, L. Gold, Systematic evolution of ligands by exponential enrichment: RNA ligands to bacteriophage T4 DNA polymerase, *Science.* 249 (1990) 505–510.
60. R. Rosenberg, R. Gertler, J. Friederichs, K. Fuehrer, M. Dahm, R. Phelps, S. Thorban, H. Nekarda, J. Siewert, Comparison of two density gradient centrifugation systems for the enrichment of disseminated tumor cells in blood, *Cytometry.* 49 (2002) 150–158.
61. H.M. Shapiro, *Practical Flow Cytometry*, Wiley-Liss, Hoboken, NJ, 2005.
62. H.M. Davey, D.B. Kell, Flow cytometry and cell sorting of heterogeneous microbial populations: The importance of single-cell analyses, *Microbiological Reviews.* 60 (1996) 641–696.
63. A. Krishan, H. Krishnamurthy, S. Totey, *Applications of Flow Cytometry in Stem Cell Research and Tissue Regeneration*, Wiley-Blackwell, Hoboken, NJ, 2011.
64. G.J. Spangrude, S. Heimfeld, I.L. Weissman, Purification and characterization of mouse hematopoietic stem cells, *Science.* 241 (1988) 58–62.
65. U.A. Gurkan, T. Anand, H. Tas, D. Elkan, A. Akay, H.O. Keles, U. Demirci, Controlled viable release of selectively captured label-free cells in microchannels, *Lab on a Chip.* 11 (2011) 3979–2989.
66. S. Nagrath, L.V. Sequist, S. Maheswaran, D.W. Bell, D. Irimia, L. Ulkus, M.R. Smith, E.L. Kwak, S. Digumarthy, A. Muzikansky, Isolation of rare circulating tumour cells in cancer patients by microchip technology, *Nature.* 450 (2007) 1235–1239.
67. S.L. Stott, C.H. Hsu, D.I. Tsukrov, M. Yu, D.T. Miyamoto, B.A. Waltman, S.M. Rothenberg, A.M. Shah, M.E. Smas, G.K. Korir, F.P. Floyd Jr, A.J. Gilman, J.B. Lord, D. Winokur, S. Springer, D. Irimia, S. Nagrath, L.V. Sequist, R.J. Lee, K.J. Isselbacher, S. Maheswaran, D.A. Haber, M. Toner, Isolation of circulating tumor cells using a microvortex-generating herringbone-chip, *Proceedings of the National Academies of Science U. S. A.* 107 (2010) 18392–18397.
68. M. Zborowski, G.R. Ostera, L.R. Moore, S. Milliron, J.J. Chalmers, A.N. Schechter, Red blood cell magnetophoresis, *Biophysical Journal.* 84 (2003) 2638–2645.
69. S. Hackett, J. Hamzah, T. Davis, T. St Pierre, Magnetic susceptibility of iron in malaria-infected red blood cells, *Biochimica et Biophysica Acta (BBA)-Molecular Basis of Disease.* 1792 (2009) 93–99.
70. P.A. Zimmerman, J.M. Thomson, H. Fujioka, W.E. Collins, M. Zborowski, Diagnosis of malaria by magnetic deposition microscopy, *American Journal of Tropical Medicine and Hygiene.* 74 (2006) 568–572.
71. S. Karl, M. David, L. Moore, B.T. Grimberg, P. Michon, I. Mueller, M. Zborowski, P.A. Zimmerman, Enhanced detection of gametocytes by magnetic deposition microscopy predicts higher potential for *Plasmodium falciparum* transmission, *Malaria Journal.* 7 (2008) 66.
72. Y. Zhang, N. Kohler, M. Zhang, Surface modification of superparamagnetic magnetite nanoparticles and their intracellular uptake, *Biomaterials.* 23 (2002) 1553–1561.

73. J. Kim, T. Yoon, K. Yu, M.S. Noh, M. Woo, B. Kim, K. Lee, B. Sohn, S. Park, J. Lee, Cellular uptake of magnetic nanoparticle is mediated through energy-dependent endocytosis in A549 cells, *Journal of Veterinary Science*. 7 (2006) 321–226.
74. R. Molday, S. Yen, A. Rembaum, Application of magnetic microspheres in labelling and separation of cells, *Nature*. *268* (1977) 437–438.
75. K. Kato, A. Radbruch, Isolation and characterization of CD34 hematopoietic stem cells from human peripheral blood by high-gradient magnetic cell sorting, *Cytometry*. 14 (1993) 384–292.
76. V.I. Furdui, J.K. Kariuki, D.J. Harrison, Microfabricated electrolysis pump system for isolating rare cells in blood, *Journal of Micromechanics and Microengineering*. 13 (2003) S164.
77. Y.Y. Huang, P. Chen, C.H. Wu, K. Hoshino, K. Sokolov, N. Lane, H. Liu, M. Huebschman, E. Frenkel, J.X. Zhang, Screening and molecular analysis of single circulating tumor cells using micromagnet array, *Scientific Reports*. 5 (2015) 16047.
78. A. Strati, A. Markou, C. Parisi, E. Politaki, D. Mavroudis, V. Georgoulias, E. Lianidou, Gene expression profile of circulating tumor cells in breast cancer by RT-qPCR, *BMC Cancer*. 11 (2011) 422.
79. E.A. Punnoose, S.K. Atwal, J.M. Spoerke, H. Savage, A. Pandita, R. Yeh, A. Pirzkall, B.M. Fine, L.C. Amler, D.S. Chen, Molecular biomarker analyses using circulating tumor cells, *PloS One*. 5 (2010) e12517.
80. A.A. Powell, A.H. Talasaz, H. Zhang, M.A. Coram, A. Reddy, G. Deng, M.L. Telli, R.H. Advani, R.W. Carlson, J.A. Mollick, Single cell profiling of circulating tumor cells: Transcriptional heterogeneity and diversity from breast cancer cell lines, *PloS One*. 7 (2012) e33788.
81. K. Sakaizawa, Y. Goto, Y. Kiniwa, A. Uchiyama, K. Harada, S. Shimada, T. Saida, S. Ferrone, M. Takata, H. Uhara, Mutation analysis of BRAF and KIT in circulating melanoma cells at the single cell level, *British Journal of Cancer*. 106 (2012) 939–946.
82. D. Ramsköld, S. Luo, Y. Wang, R. Li, Q. Deng, O.R. Faridani, G.A. Daniels, I. Khrebtukova, J.F. Loring, L.C. Laurent, Full-length mRNA-Seq from single-cell levels of RNA and individual circulating tumor cells, *Nature Biotechnology*. 30 (2012) 777–782.
83. D.T. Ting, B.S. Wittner, M. Ligorio, N.V. Jordan, A.M. Shah, D.T. Miyamoto, N. Aceto, F. Bersani, B.W. Brannigan, K. Xega, Single-cell RNA sequencing identifies extracellular matrix gene expression by pancreatic circulating tumor cells, *Cell Reports*. 8 (2014) 1905–1918.
84. L.A. Doyle, W. Yang, L.V. Abruzzo, T. Krogmann, Y. Gao, A.K. Rishi, D.D. Ross, A multidrug resistance transporter from human MCF-7 breast cancer cells, *Proceedings of the National Academies of Science U. S. A*. 95 (1998) 15665–15670.
85. A. Müller, B. Homey, H. Soto, N. Ge, D. Catron, M.E. Buchanan, T. McClanahan, E. Murphy, W. Yuan, S.N. Wagner, Involvement of chemokine receptors in breast cancer metastasis, *Nature*. 410 (2001) 50–56.
86. M.E. Prince, R. Sivanandan, A. Kaczorowski, G.T. Wolf, M.J. Kaplan, P. Dalerba, I.L. Weissman, M.F. Clarke, L.E. Ailles, Identification of a subpopulation of cells with cancer stem cell properties in head and neck squamous cell carcinoma, *Proceedings of the National Academies of Science U. S. A*. 104 (2007) 973–978.
87. K. Hoshino, H. Chung, C. Wu, K. Rajendran, Y. Huang, P. Chen, K. Sokolov, J. Kim, J. Zhang, An immunofluorescence-assisted microfluidic single cell quantitative reverse transcription polymerase chain reaction analysis of tumour cells separated from blood, *Journal of Circulating Biomarkers*. 4 (2015) 11.
88. M.R. Emmert-Buck, R.F. Bonner, P.D. Smith, R.F. Chuaqui, Z. Zhuang, S.R. Goldstein, R.A. Weiss, L.A. Liotta, Laser capture microdissection, *Science*. 274 (1996) 998–1001.
89. S.L. Spurgeon, R.C. Jones, R. Ramakrishnan, High throughput gene expression measurement with real time PCR in a microfluidic dynamic array, *PloS One*. 3 (2008) e1662.

Kazunori Hoshino (E-mail: hoshino@engr.uconn.edu) earned a PhD at the University of Tokyo, Tokyo, Japan, in 2000. He worked for the University of Tokyo from 2003 to 2006 as a lecturer in the Department of Mechano-Informatics, School of Information Science and Technology, where he conducted several government-funded projects as the principal investigator. From 2006 to 2013, he worked for the University of Texas at Austin as a senior research associate in the Department of Biomedical Engineering. In 2014, he joined the University of Connecticut, where he currently works as an assistant professor of biomedical engineering. His research interests include (1) nano/micro-electro-mechanical systems (NEMS/MEMS)-based detection and analysis of cancer cells and (2) nanoscale and microscale mechanical sensing and optical imaging. He has more than 100 peer-reviewed publications and is the inventor of six US patents and 12 Japanese patents.

12 Magnetic Nanoparticles for Organelle Separation

*Mari Takahashi and Shinya Maenosono**

CONTENTS

12.1 INTRODUCTION: COMMON MAGNETIC SEPARATION IN BIOMEDICAL FIELDS

Magnetic nanoparticles have attracted much attention from the biomedical field because of their characteristic sizes and physical properties. Taking advantage of the features of magnetic nanoparticles, there have been several studies on magnetic separation of cells,[1–4] bacteria[5–7] and proteins[8–10] for a broad range of biological purposes since early times. Furthermore, magnetic separation has come into wider use for the isolation of cellular organelles in recent years. In this chapter, we focus on use of the magnetic separation technique for cellular organelles. To begin with, we briefly introduce several studies regarding conventional magnetic separation of cells, bacteria and proteins. We then focus on magnetic separation of cellular organelles, such as endosomes, exosomes and mitochondria, and explain the biological importance of the technique. Finally, we touch upon how to improve magnetic nanoparticles to implement the magnetic separation technique for a wide variety of cellular organelles and make the technique more versatile.

* Corresponding author.

12.1.1 Magnetic Separation of Cells and Bacteria

Magnetic separation of cells has been performed for a long time for various purposes. Here, several recent studies are introduced. Fan et al.[1] synthesized iron oxide nanoparticles that were covered with a gold shell (Fe_3O_4@Au core@shell nanoparticles) and had a mean size of 55 nm for cancer cell separation (Figure 12.1a). The surface of these Fe_3O_4@Au nanoparticles was modified with a specific aptamer (S6 aptamer) that can conjugate with the HER2 receptor, a type of epidermal growth factor receptor, on the cellular surface. The expression level of HER2 receptors is cell type dependent. For example, SKBR-3 cells (human breast cancer cell line) express HER2 at the rate of 6.3×10^6/cell, while LNCaP cells (human prostate cancer cell line) do not express HER2. They added Cy3 fluorescent dye and S6 aptamer-modified Fe_3O_4@Au nanoparticles to a mixture of SKBR-3 (10^3 cells/mL) and LNCaP (10^7 cells/mL) cells. After 2 h of incubation, the suspension was subjected to magnetic separation. Next, the magnetically separated substances and the supernatant were analyzed using several microscopic techniques. As expected, SKBR-3 cells were selectively separated, while LNCaP cells remained in the supernatant (Figure 12.1b). In addition, they demonstrated that SKBR-3 cells bound to the S6 aptamer-modified Fe_3O_4@Au nanoparticles via HER2 receptors were selectively killed when the mixture of SKBR-3 and

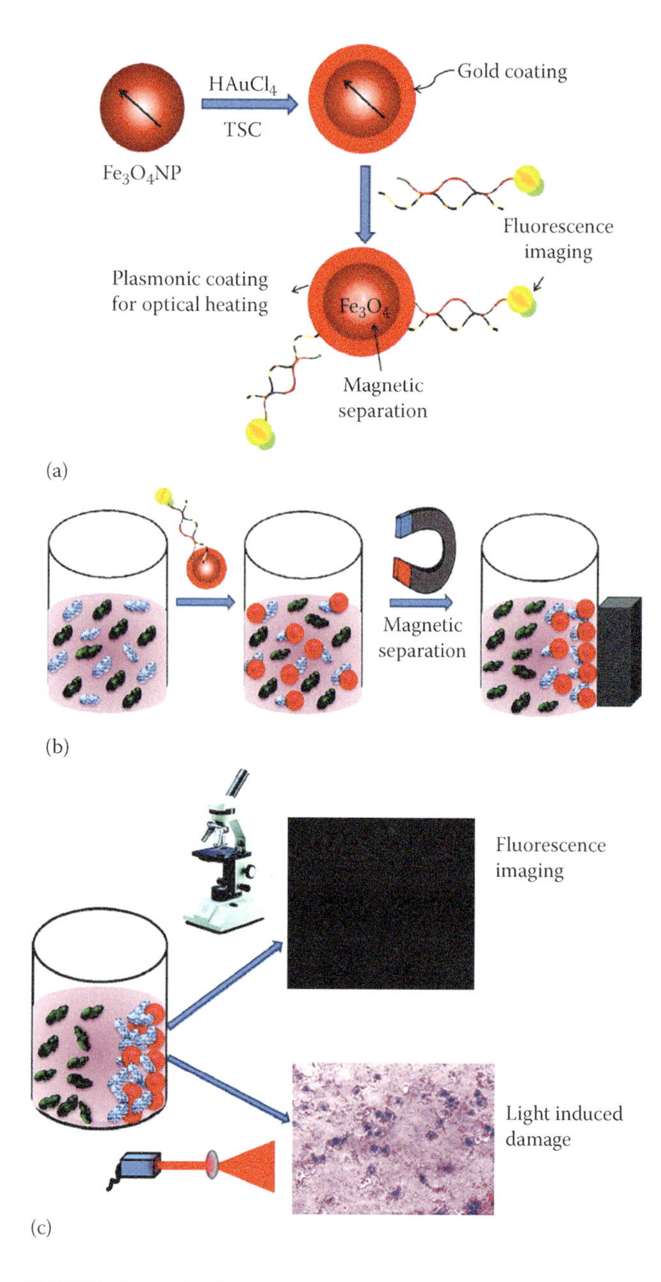

FIGURE 12.1 (a) Schematic representation showing the synthesis of S6 aptamer-conjugated multifunctional magnetic core–gold shell nanoparticles. (b) Schematic representation showing the separation of specific cancer cells using S6 aptamer-conjugated plasmonic/magnetic nanoparticles. (c) Schematic representation showing the selective fluorescence imaging and targeted photothermal destruction of specific cancer cells.[1] (Reprinted with permission from Fan, Z., M. Shelton, A. K. Singh, D. Senapati, S. A. Khan, and P. C. Ray. 2012. Multifunctional plasmonic shell–magnetic core nanoparticles for targeted diagnostics, isolation, and photothermal destruction of tumour cells. *ACS Nano* 6:1065–73. Copyright 2012 American Chemical Society.)

LNCaP cells was irradiated with a near-infrared laser (Figure 12.1c; blue shows dead cells stained with trypan blue).

In another case, enrichment of stem cells was achieved by magnetic separation for the purpose of regenerative medicine and cell therapy.[4] Endothelial progenitor cells (EPCs) are stem cells that differentiate into endothelial cells and thus contribute to the formation of blood vessels and healing of tissues. However, the proportion of EPCs relative to all cells in adult blood is only 0.01%–0.0001%. Therefore, it is necessary to enrich EPCs using techniques such as Ficoll-Paque gradient centrifugation followed by cell culture to obtain sufficient numbers of EPCs for cell therapy, and this process usually requires several days. Wadajkar et al.[2] prepared iron oxide particles that had sizes ranging from 50 to 100 μm and were covered with functional multilayers. These iron oxide particles have four components. The outer layer contains EPC-specific anti-CD34 antibodies. The second layer below the outer surface is made of thermoresponsive poly(*N*-isopropylacrylamide-co-allylamine) to provide a surface for cell attachment and detachment depending on changes in temperature. This polymer has a hydrophobic nature at temperatures above the critical temperature (32°C–34°C) to support cell adhesion but becomes hydrophilic at temperatures below the critical temperature and enhances cell detachment from the particle surface. The polymer layer also contains vascular endothelial growth factors that can be rapidly released for EPC proliferation. The central material consists of a biodegradable core of poly(lactide-co-glycolic acid) (PLGA) microparticles containing basic fibroblast growth factor that are released when PLGA degrades for EPC differentiation. The iron oxide particles are conjugated with PLGA. By adding these magnetic beads to a blood sample and applying an external magnetic field, they succeeded in separating and enriching EPCs from blood samples. This multifunctionality of particles makes magnetic cell separation promising for cell therapy.

Infectious diseases caused by multidrug-resistant bacteria (MDRB) are one of the most serious problems in medicine worldwide. Because MDRB are highly resistant to most commonly used antibiotics, there is an urgent demand for new medical treatments for MDRB-related infectious diseases. Fan et al.[5] demonstrated that magnetic iron core–plasmonic gold shell nanoparticles can be utilized for removal and killing of MDR *Salmonella* by magnetic separation and photothermal destruction. M3038 antibodies, which are specific for *Salmonella* DT104, were conjugated onto the core–shell nanoparticles. *Salmonella* DT104 was incubated with the M3038-conjugated nanoparticles, and the *Salmonella*-attached nanoparticles were magnetically separated using a bar magnet. They further showed that detection of *Salmonella* became possible by surface-enhanced Raman scattering (SERS) through the formation of a huge number of 'hot spots' by aggregation of magnetic–plasmonic core@shell nanoparticles.

12.1.2 Magnetic Separation of Proteins

Regarding separation of proteins, one of the main purposes is the collection and purification of immunoglobulins. Hu et al.[9] showed that immunoglobulin G (IgG) could be effectively collected by magnetic separation using iron oxide nanoparticles covered with a mesoporous silica shell (Fe_3O_4@SiO_2 core@shell nanoparticles) that ranged in size from 50 to 100 nm, after modification of the nanoparticle surface with protein G that can bind with IgG. They further demonstrated that the magnetic separation efficiency for IgG was significantly enhanced using Fe_3O_4@SiO_2 nanoparticles, compared with magnetic nanoparticles without the mesoporous silica shell. This effect would arise from the highly specific surface area provided by the mesoporous silica shell. In another case, Masthoff et al.[10] fabricated chelating agent-modified iron oxide nanoparticles with a mean diameter of 8 nm. After coordinating Ni^{2+} cations with the chelating agent-modified iron oxide nanoparticles, a histidine (His)-tagged antibody was magnetically purified utilizing the His-Ni^{2+} coordinated interaction. These magnetic separation techniques are promising for purification of antibodies and antibody fragments necessary for diagnosis and treatment.

Magnetic separation of proteins can also be applied for diagnosis of diseases. For example, Tang et al.[11] proposed a sensitive detection method for protein biomarkers combining localized surface plasmon resonance (LSPR) biosensing and magnetic separation. They separately prepared Fe_3O_4 nanoparticles and Au nanorods and modified the surfaces of both probes with an anticardiac troponin I (cTnI) antibody. It should be noted that cTnI is a cardiac marker for myocardial infarction diagnosis. When they mixed the anti-cTnI antibody-modified Au nanorods with blood plasma, an LSPR peak redshift of 6 nm was observed through selective binding of cTnI protein to the Au nanorods, because the surface of metal nanoparticles is highly sensitive to changes in the local refractive index, and thus various types of LSPR-based colorimetric biosensors have been proposed.[12,13] Meanwhile, when they first mixed the anti-cTnI antibody-modified Fe_3O_4 nanoparticles with blood plasma to magnetically separate cTnI, followed by mixing with the Au nanorods in an aqueous phase, as shown in Figure 12.2, an LSPR peak redshift of 13 nm was observed.[11] The LSPR peak shift and the cTnI concentration showed a linear relationship. They further reported that the sensitivity for cTnI biomarker detection was amplified by up to sixfold when they combined magnetic separation and LSPR biosensing, compared with LSPR biosensing alone.

12.2 IMPORTANCE OF MAGNETIC SEPARATION OF CELLULAR ORGANELLES

A cell is composed of different compartments, so-called cellular organelles, as illustrated in Figure 12.3. Cellular organelles play critical roles in metabolism, transport, immunity and homeostasis, which are indispensable for maintenance of life. There are various proteins on/in cellular organelles that work cooperatively to result in biological actions. However, in many organelles, the proteins present have not been completely identified. The identification and characterization of proteins on/in cellular organelles are quite important not only for cell biology but also for drug discovery and clinical diagnosis.

To identify proteins on/in cellular organelles of interest, there are several conventional techniques. Gene manipulation, including gene knockdown and gene knockout, is a powerful tool that is frequently used in the field of biology. If a phenotype is changed after disruption or suppression of a specific gene, the change can be attributed to the protein encoded by the gene. In this manner, the function of the protein can be revealed. Unfortunately, it is not always easy to identify the proteins that are actually active on/in cellular organelles solely by gene manipulation because a gene and its protein do not always have a one-to-one relationship and because gene manipulation cannot provide information about posttranslational modifications.

Isolation of target cellular organelles has attracted attention as a complementary technique because it enables

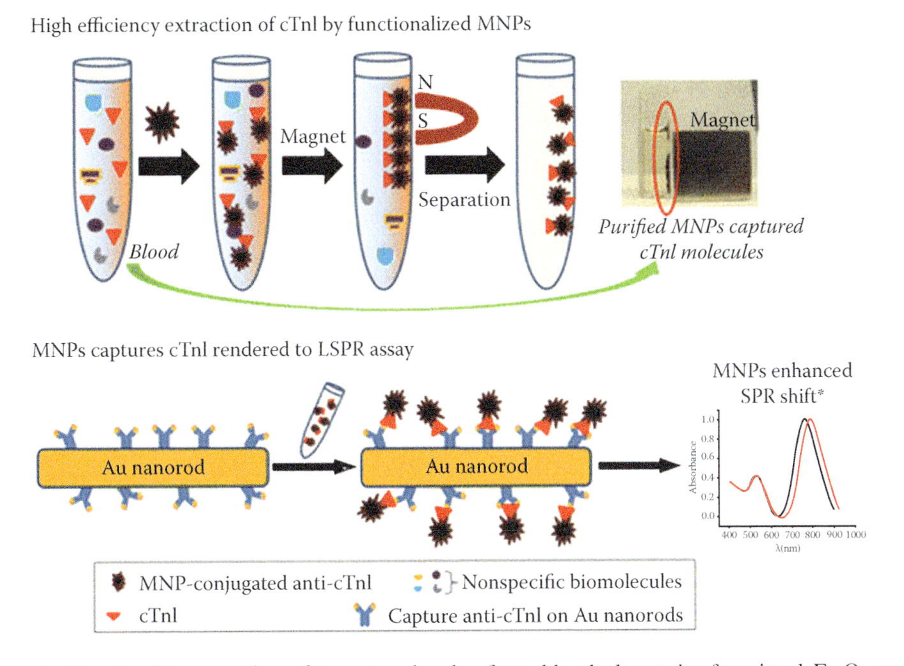

FIGURE 12.2 Schematic showing bioseparation of target molecules from blood plasma by functional Fe_3O_4 magnetic nanoparticles (MNPs), followed by the MNP-mediated nanoSPR assay. The application of MNP results in an enhancement of the LSPR shift at peak absorption wavelength.[11] (Reprinted with permission from Tang, L., J. Casas, and M. Venkataramasubramani. 2013. Magnetic nanoparticle mediated enhancement of localized surface plasmon resonance for ultrasensitive bioanalytical assay in human blood plasma. *Anal. Chem.* 85:1431–9. Copyright 2013 American Chemical Society.)

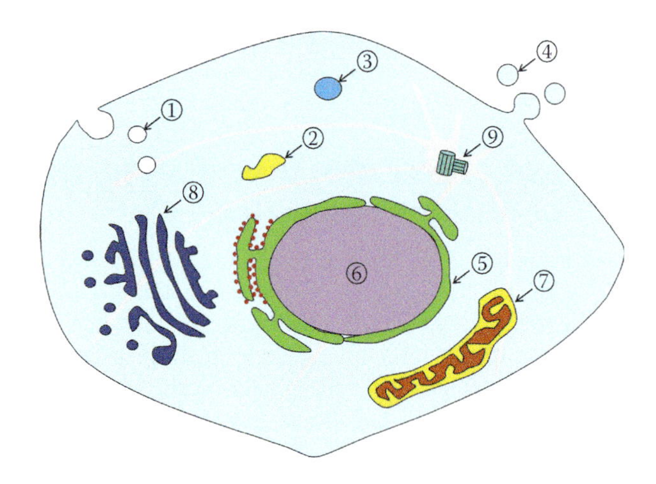

FIGURE 12.3 Illustration of an animal cell: ① endosome, ② lysosome, ③ peroxisome, ④ exosome, ⑤ endoplasmic reticulum, ⑥ nucleus, ⑦ mitochondria, ⑧ Golgi body and ⑨ centrioles.

direct analysis of intact proteins on/in the isolated organelles. Several commonly used centrifugation techniques for organelle isolation are briefly described here. To isolate cellular organelles by centrifugation, the cell membrane must be first disrupted by one of the following methods: (i) osmotic shock method, (ii) sonication method, (iii) mild detergent method, (iv) French press method or (v) tissue grinder method (Figure 12.4).[14] All of these methods have advantages and disadvantages. For example, physical disruption methods often cause deformation and/or aggregation of proteins owing to heat produced by the equipment and thus need to be performed at low temperature. In addition, these methods are poorly reproducible in terms of cell membrane disruption efficiency. Meanwhile, chemical disruption methods require strict experimental procedures under appropriate conditions; otherwise, the proteins on the cellular organelles can become dissolved. The solution obtained after gentle disruption of the cell membrane, designated the cell homogenate, contains intact cellular organelles. Differential centrifugation is a widely used technique to isolate certain cellular organelles. First, the homogenate is centrifuged at low speed to separate intact cells, nuclei and cytoskeletons from other cellular components. Second, the supernatant is centrifuged at medium speed to precipitate mitochondria, lysosomes and peroxisomes. Third, the supernatant is centrifuged at high speed to obtain microsomes and other small vesicles. Finally, the supernatant is centrifuged at very high speed to obtain ribosomes, viruses and large macromolecules, as shown in Figure 12.5a.

Density gradient centrifugation can be used to separate cellular organelles based on their physical properties such as size, mass and density. There are two typical methods for density gradient centrifugation: rate zonal centrifugation, also called velocity sedimentation (Figure 12.5b), and isopycnic centrifugation, also called sedimentation equilibrium (Figure 12.5c).[14] In the rate zonal centrifugation method, a sample solution is gently layered on a sucrose gradient solution (typically 5–20 wt%) preliminarily prepared in a centrifuge tube and then subjected to ultracentrifugation. After the centrifugation, several bands are formed within the tube depending on the buoyant density of each cellular component. In this way, different organelles of similar size and mass can be separated. The fractionated component is then subjected to column chromatography or gel electrophoresis to purify the proteins present in the fraction. In the isopycnic centrifugation method, a sample solution is mixed with a sucrose or cesium chloride solution and then subjected to ultracentrifugation. During the centrifugation, a density gradient is spontaneously formed in the solution, and thus different organelles of similar size and mass can be separated, similar to the case for the rate zonal centrifugation method. Unfortunately, heterologous organelles that are barely different in size and mass cannot be effectively separated by these centrifugation techniques. Moreover, superficial proteins on the organelles can become detached during the lengthy and harsh centrifugation processes. Therefore, the demand for novel techniques for isolation of cellular organelles with maintenance of their intracellular state has increased to identify proteins and/or lipids on/in these organelles.

In response to this situation, magnetic separation of cellular organelles has become important in recent years because it is capable of selectively isolating target cellular organelles rapidly and mildly irrespective of size and mass. As described previously, magnetic separation of cells and/or proteins has been widely used for medical diagnosis and therapy. In contrast, magnetic separation of cellular organelles is much less common than magnetic cell/bacteria/protein separation and often used only in the field of basic biology. As described in the following sections, endosomes,[15–19] exosomes[20–25] and mitochondria[26–29] have been successfully isolated using the magnetic separation technique. There have also been some reports on magnetic separation of other organelles such as peroxisomes,[30–32] lysosomes,[33,34] Golgi vesicles[35] and macropinosomes.[36]

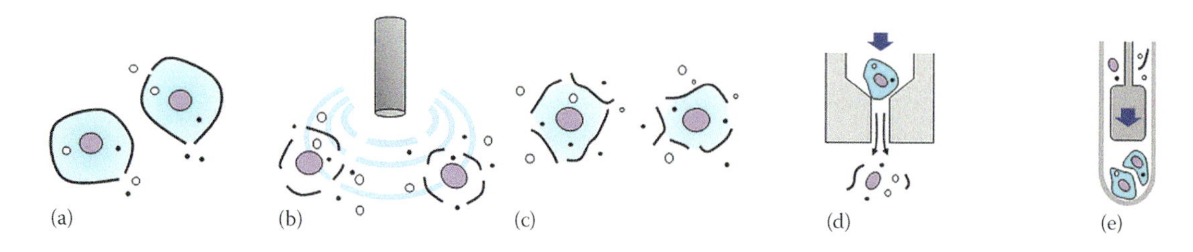

FIGURE 12.4 Schematic illustrations of several methods for cell membrane disruption: (a) osmotic shock method, (b) sonication method, (c) mild detergent method, (d) French press method and (e) tissue grinder method.

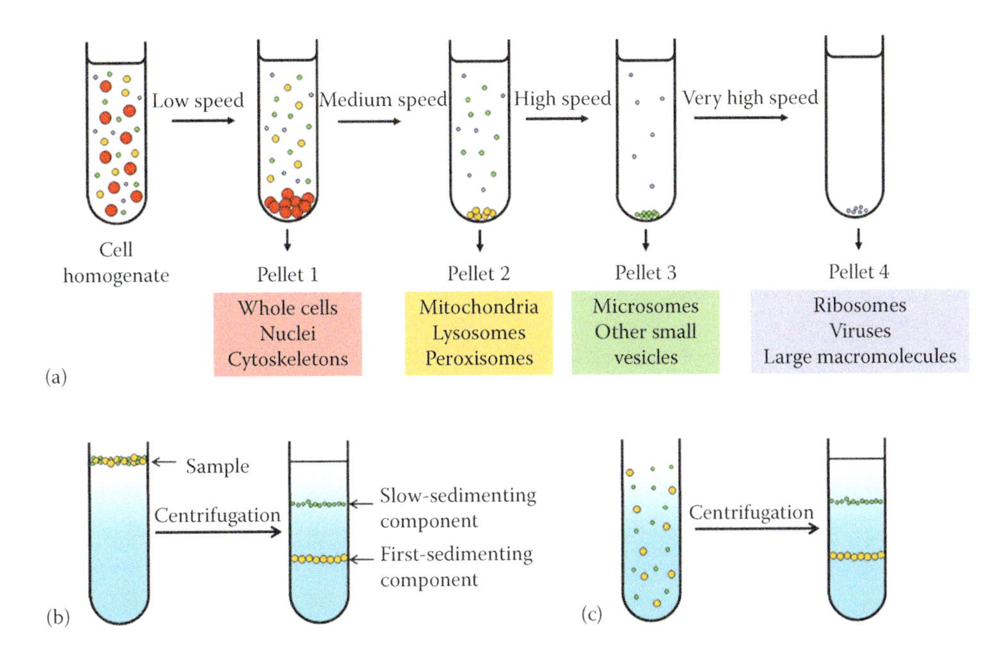

FIGURE 12.5 Schematic illustrations of several commonly used centrifugation techniques for organelle isolation: (a) Differential centrifugation, (b) rate zonal centrifugation and (c) isopycnic centrifugation.

12.3 MAGNETIC SEPARATION OF ENDOSOMES

Endosomes are small membrane vesicles formed by an endocytic process in a cell, during which extracellular materials or receptors on the cell membrane are incorporated. There are two general types of endocytosis: phagocytosis and pinocytosis.[37] Phagocytosis takes place actively in phagocytes such as macrophages and monocytes. Pinocytosis is divided into four categories: macropinocytosis, clathrin-mediated endocytosis, caveolae-mediated endocytosis and clathrin- and caveolae-independent endocytosis. Figure 12.6 shows several endocytic entry ways into a cell.[37] Extracellular materials incorporated via endocytosis are transported to lysosomes, resulting in degradation of the materials, or returned to the cellular membrane for recycling of receptors. During this material transport process, endosomes change their state (early endosomes, late endosomes and recycling endosomes). Early endosomes undergo differentiation into late endosomes or recycling endosomes depending on the situation. In the case of receptor-mediated endocytosis, for example, both ligand and receptor are simultaneously incorporated. The ligand is finally transferred to a lysosome, followed by degradation, while the receptor is returned to the cell membrane and reused several hundreds of times by recycling endosomes.[38] Endocytosis regulates various cellular processes, including cell motility, cell determination, nutrient uptake and microbial invasion.[16] Therefore, functional disorders of endocytosis can cause many diseases such as familial hypercholesterolemia[39] and cancer.[40] Low-density lipoprotein (LDL) is incorporated into cells via receptor-mediated endocytosis. Patients with familial hypercholesterolemia have dysfunction of the LDL receptor and thus cannot uptake LDL appropriately.[39] For this reason, their blood LDL concentration increases, thereby enhancing the potential for arteriosclerosis, which may lead to serious diseases such as myocardial infarction.

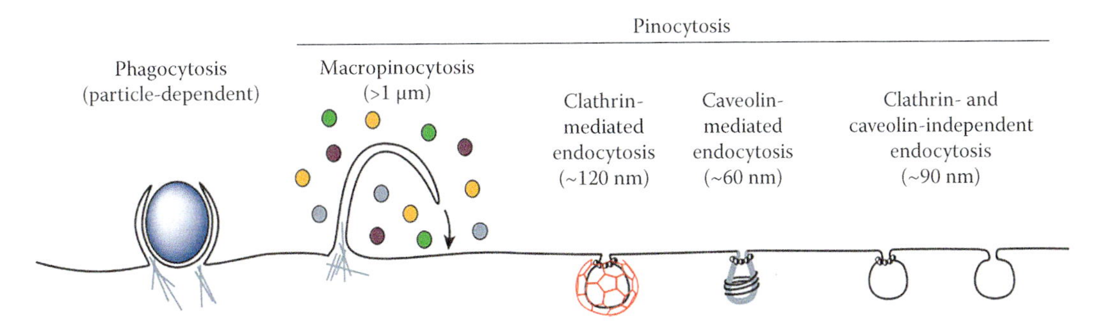

FIGURE 12.6 Multiple portals of entry into the mammalian cell. The endocytic pathways differ with regard to the size of the endocytic vesicle, the nature of the cargo (ligands, receptors and lipids) and the mechanism of vesicle formation.[37] (Reprinted by permission from Macmillan Publishers Ltd. *Nature*. Conner, S. D., and S. L. Schmid. 2003. Regulated portals of entry into the cell. *Nature* 422:37–44. Copyright 2003.)

12.3.1 Magnetic Separation of Different States of Endosomes

Niemann-Pick type C (NPC) disease belongs to a group of lysosomal storage diseases, and affected patients cannot properly metabolize LDL-derived cholesterol. Most patients with NPC disease (~95%) have mutations in the *NPC1* gene, while the remaining patients have mutations in the *NPC2* gene. Although the role of NPC2 protein in the pathogenesis of NPC disease is unclear, deficiency in either the *NPC1* gene or the *NPC2* gene produces indistinguishable cellular phenotypes. Based on these observations, it can be hypothesized that a lack of NPC1 protein affects the function and/or location of NPC2 protein. To test this hypothesis, Chen et al.[15] isolated late endosomes from wild-type (Wt) mice and mice with mutations in NPC1 protein (designated NPC1 mice) by magnetic separation using dextran-coated superparamagnetic iron oxide nanoparticles (SPIONs), which were synthesized by coprecipitation of $FeCl_2$ and $FeCl_3$.[33] They intravenously injected the dextran-coated SPIONs into a mouse tail vein. Since SPIONs cannot be metabolized, they accumulate in the liver and are endocytosed by liver cells. Mice were euthanized at 0, 0.5, 1, 2 or 4 h after the SPION injection, and their livers were excised, homogenized and centrifuged. The obtained postmitochondrial supernatant was subjected to magnetic separation. The magnetically separated fraction was analyzed by transmission electron microscopy (TEM) and western blotting (WB). Figure 12.7a shows a TEM image of endocytic vesicles. The arrows indicate SPIONs. Since SPIONs are thought to migrate through the endosomal–lysosomal pathway in a time-dependent manner, an early endosomal marker (Rab5), late endosomal marker (Rab7) and late endosomal and lysosomal marker (α-GalNAcase) were investigated by WB.

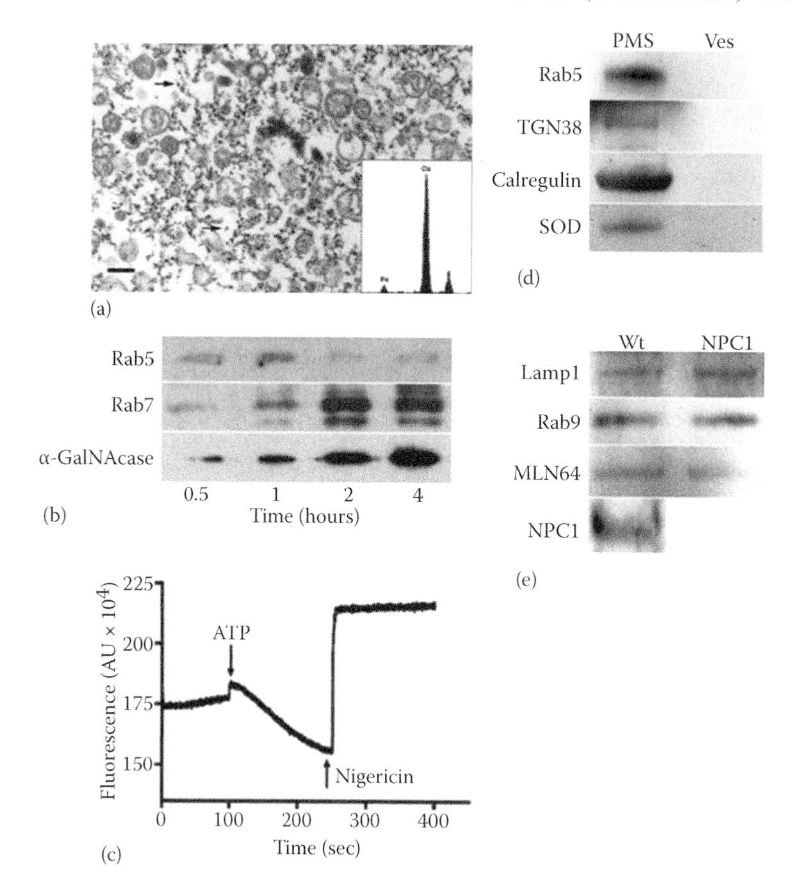

FIGURE 12.7 (a) Electron micrograph of a typical vesicle preparation showing purified endocytic vesicles; arrows indicate particles composed of colloidal iron. Inset, elemental composition of electron-dense spots determined by EDS. The copper peaks result from the copper grids on which the specimen was mounted. Scale bar, 0.5 μm. (b) Vesicles isolated at 0.5, 1, 2 and 4 h after particle injection were analyzed for marker proteins associated with specific vesicle populations: Rab5 for early endosomes, Rab7 for late endosomes and α-*N*-acetylgalactosaminidase (α-GalNAcase) for late endosomes/lysosomes. Vesicles isolated at 1, 2 and 4 h after particle injection were enriched for early endosomal, late endosomal and E/L markers, respectively. (c) The functional integrity of purified vesicles was routinely determined by an AO fluorescence-quenching assay. Addition of ATP initiates vesicle acidification via the action of the proton ATPase-driven quenching of AO fluorescence. This effect is reversed by the addition of the proton ionophore nigericin. (d) Crude PMS and purified late endosomes (Ves) from Wt mice were characterized for their content of nonlate endosomal proteins: early endosomes (Rab5), TGN (TGN38), endoplasmic reticulum (calregulin) and mitochondria (SOD). (e) Purified vesicles from Wt and NPC1 mice contain equivalent levels of late endosomal marker proteins, indicating similar uptake and endocytosis kinetics in Wt and NPC1 mouse livers. Furthermore, vesicles isolated from Wt mice contained significant amounts of NPC1 protein, indicating that these vesicles are NPC1-containing late endosomes.[15] (From Chen, F.W., Gordon, R.E., Ioannou, Y.A. *Biochem. J.*, 390, 549–561, 2005. With permission.)

As shown in Figure 12.7b, maximal association of Rab5 was observed between 0.5 and 1 h after the injection, while maximal association of Rab7 was observed at 2 h after the injection. Meanwhile, α-GalNAcase showed maximal association at 4 h after the injection. These findings indicate that specific organelles (early endosomes, late endosomes and lysosomes) can be enriched by controlling the time after injection of SPIONs. They harvested late endosomes from Wt and NPC1 mice by magnetic separation and then analyzed the lipid profiles of the late endosomes by high-performance thin-layer chromatography. Thus, a significant difference in the lipid profiles of late endosomes was observed between Wt and NPC1 mice.

Chen et al.[15] also investigated endosomal transport. Rab proteins regulate many steps of membrane trafficking, and two Rab proteins are important for transport by late endosomes. Rab9 regulates membrane trafficking from late endosomes to the *trans*-Golgi network (TGN), while Rab7 regulates membrane trafficking from late endosomes to lysosomes. It was confirmed that Rab7 association with late endosomes was increased by threefold in NPC1 mice compared with Wt mice, while no difference was observed in Rab9 association. These results indicate that the association–dissociation equilibrium of Rab7 is disturbed in NPC1 mouse late endosomes. In addition, the band of NPC2 protein observed for NPC1 mouse late endosomes was much broader than that for Wt mouse late endosomes. They attributed this broadening of the NPC2 band to variations in NPC2 glycosylation. Based on these results, it is thought that NPC2 protein is also compromised in NPC1 mice, presumably because of an interaction between NPC1 protein and NPC2 protein.

12.3.2 Magnetic Separation of Receptor-Mediated Endosomes

Another example of magnetic separation of endosomes was reported by Wittrup et al.[16,17] Macromolecular ligands such as polylysine/cationic lipid–DNA complexes, cationic polymers and antimicrobial peptide–DNA complexes are internalized by cell-surface heparin sulfate proteoglycan (HSPG)-mediated endocytosis.[41] As some viruses also utilize HSPG-mediated internalization for entry into host cells, HSPGs have become an important target molecule for drug delivery.[42] However, the exact function of HSPGs in the endocytotic process has remained unclear, such as whether HSPGs are true internalizing receptors or just receptors for initial cell-surface attachment. Wittrup et al.[16] prepared AO4B08 antibody (HSPG ligand)-conjugated superparamagnetic nanoparticles and added them to HeLa cells together with fluorophore-conjugated AO4B08 antibody. After incubation for a certain duration, the cells were washed and treated with trypsin. The postnuclear supernatant (PNS) was then harvested by centrifugation. Subsequently, the PNS was subjected to magnetic separation using a magnetic separation tool (PickPen; Bio-Nobile). Figure 12.8a shows epifluorescence microscopic and TEM images of the PNS and magnetically isolated fraction. Localization of the superparamagnetic nanoparticles in the endocytic vesicles was clearly observed. By monitoring the fluorescence before and after the magnetic isolation, it was possible to roughly estimate the yield and purity of the isolated vesicles. The results indicated that the magnetic isolation successfully enriched the AO4B08 antibody-conjugated SPION-containing endocytic vesicles with high yield. They further determined the HSPG concentration in the magnetically isolated vesicles using metabolically labeled ^{35}S-sulfate. Thus, it was observed that HSPGs were enriched by approximately 20-fold in the magnetically isolated fraction compared with the PNS and nonmagnetic fraction. These results confirmed that HSPGs were internalized into the cell by endocytosis, indicating that HSPGs are true internalizing receptors for macromolecular cargos.[16]

In another study, Li et al.[18] separated insulin receptor-mediated endosomes. Cell signaling and endocytosis have traditionally been regarded as distinct processes. However, it has recently been recognized that these processes are closely linked, and endocytosis has been found to regulate receptor signaling cascades.[43,44] Therefore, elucidation of the relationships between signaling and endocytosis has become an important topic in the field of cell biology. The Ras/Raf-1/MEK1/Erk1-2 cascade regulates cell proliferation and differentiation, even though the precise role of this signaling cascade in endocytosis remains controversial.[45] When insulin binds to an insulin receptor expressed on a cell surface, the receptor initiates a phosphorylation cascade.[46] Li et al.[18] prepared insulin-conjugated ferrofluid nanoparticles (Fe-INS) with a diameter of 10 nm simply by mixing ferrofluid (FerroTec) with insulin molecules. They successfully internalized Fe-INS into HIRcB cells through clathrin-coated vesicles by incubating Fe-INS with HIRcB cells (Figure 12.9). It was found that the size of the vesicles containing Fe-INS became larger with increasing incubation time from 5 min to 60 min. Specifically, the vesicle size was about 100–200 nm after a 5-min incubation but increased to around 1–2 μm after a 60-min incubation, as shown in Figure 12.10. The HIRcB cells containing Fe-INS were homogenized and allowed to flow through a magnetic column placed on a magnetized column holder. The magnetically attracted fraction was then eluted from the column after its removal from the holder. The eluted fractions with different incubation times were analyzed by WB, and the results revealed that the localized Fe-INS underwent a transition from early endosomes to lysosomes with increasing incubation time without contamination by endoplasmic reticulum or Golgi. Interestingly, when they used bovine serum albumin-conjugated ferrofluid nanoparticles (Fe-BSA), these nanoparticles were not internalized into the cell.

Li et al.[18] also investigated the activation of proteins involved in the Ras/Raf-1/MEK/Erk cascade in a time-dependent manner. Their results indicated that the signaling cascade takes place on endocytic vesicles that are trafficked along with insulin and insulin receptor.

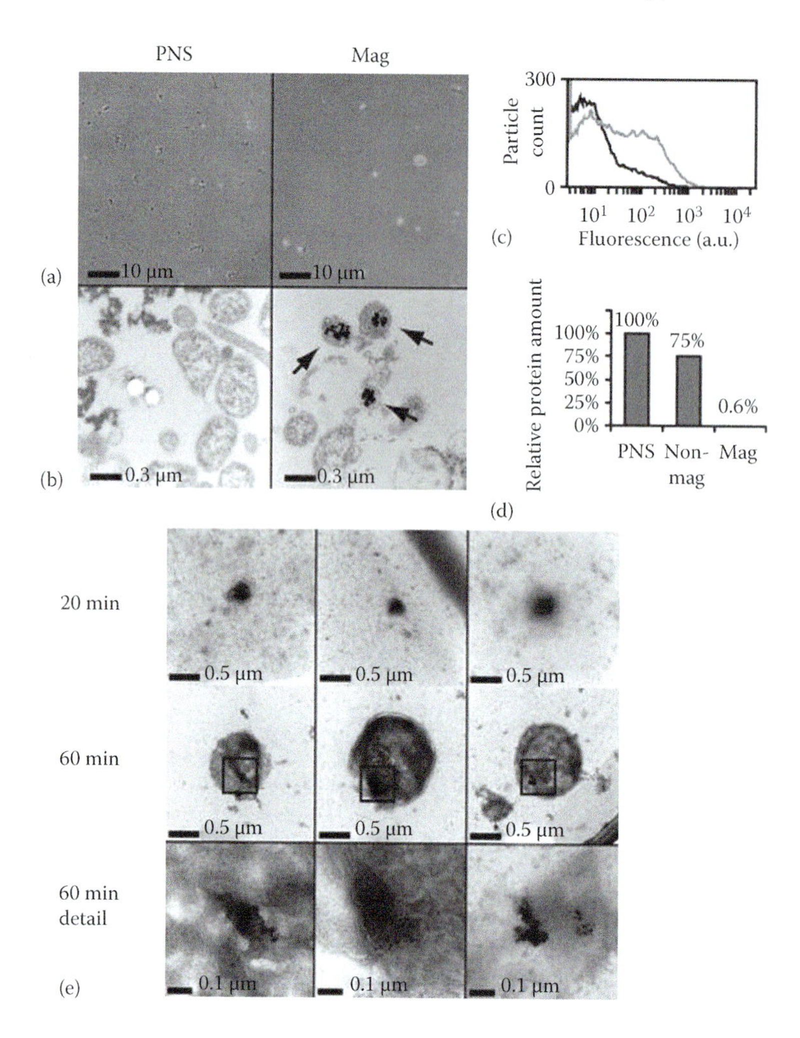

FIGURE 12.8 Isolation of endocytic vesicles containing magnetic nanoparticle-conjugated AO4B08 antibody complexes. HeLa cells were incubated with AO4B08-F and AO4B08-M for 1 h, extensively washed and trypsinized, and mechanically disrupted. The resulting subcellular particles were visualized using epifluorescence microscopy and differential interference contrast (a) or by electron microscopy (b) prior to (*PNS*) and after (*Mag*) magnetic separation. The arrows in (b) indicate vesicles containing magnetic particles. (c) Flow cytometry analysis of PNS (black) and magnetic (green) fractions shows the enrichment of fluorescent particles in the magnetic fraction. (d) Relative protein amounts in the respective fraction from a representative experiment. (e) Subcellular structures of the magnetic fraction were negatively stained and analyzed by electron microscopy, showing intact vesicles containing magnetic nanoparticles. *Lower panel*, high magnification images of the indicated areas show the *dotted structure* of magnetic particles.[16] (From Wittrup, A., Zhang, S.-H., Ten Dam, G.B. et al., *J. Biol. Chem.*, 284, 32959–32967, 2009. With permission.)

12.4 MAGNETIC SEPARATION OF EXOSOMES

Cells release different types of extracellular membrane vesicles, such as exosomes (~40–100 nm), microvesicles (~50 nm to several micrometres) and apoptotic bodies (~50–5000 nm).[47] Exosomes are specific subtypes of extracellular membrane vesicles[48] and have been widely studied at both the biochemical and functional levels. Exosomes are secreted upon fusion of multivesicular endosomes with the cell surface. In general, exosomes are classified according to their size, density and expression of certain markers such as CD63, CD81 and CD82.[49] They have pleiotropic biological roles, including antigen presentation and intercellular transfer of protein cargo, mRNA, microRNA, lipids and oncogenic potential.[23] Exosomes are also involved in cancer, neurodegenerative disease progression, cardiovascular disease and infectious disease.[50,51] Differential centrifugation and ultracentrifugation are conventional methods for isolation of exosomes.[50] Several exosome detection kits are commercially available. The main advantages of magnetic separation over the conventional separation techniques are higher accuracy, shorter processing time and separation capability of certain exosomes from other types of exosomes, as described in the following.

12.4.1 Magnetic Separation of Exosomes Derived from Cancer Cells

Tauro et al.[22,23] isolated exosomes by magnetic separation. First, they compared the exosome isolation efficiencies among different separation techniques, including ultracentrifugation,

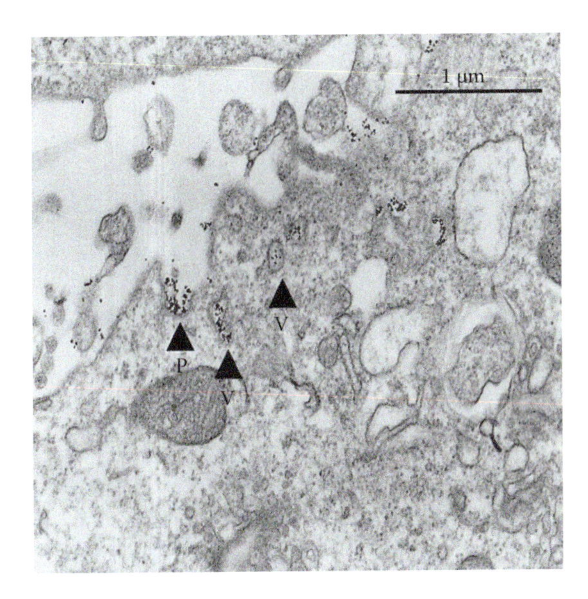

FIGURE 12.9 Electron micrograph of endocytosed iron-insulin (Fe-INS) beads. HIRcB cells were incubated with Fe-INS for 5 min. At the end of the incubation, the cells were chilled, harvested, fixed with glutaraldehyde and examined by electron microscopy. Clathrin-coated pits (P) and vesicles (V) are highlighted with arrows.[18] (Reproduced with permission from Li, H.-S., Stolz, D.B., Romero, G. *Traffic*, 6, 324–334, 2005. Copyright 2005 John Wiley and Sons.)

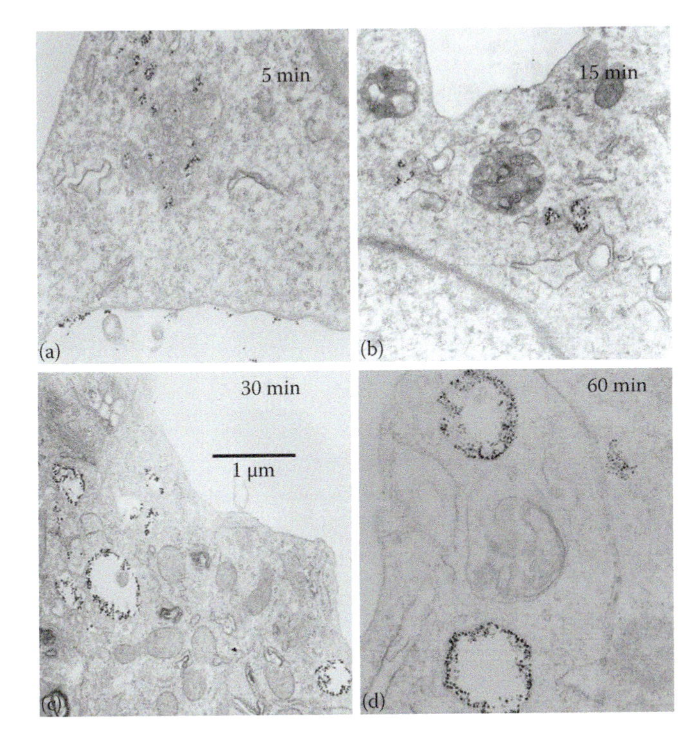

FIGURE 12.10 Electron microscopic study of the traffic of Fe-INS beads in HIRcB cells. The cells were incubated with Fe-INS for the specified times (5, 15, 30 and 60 min), harvested, fixed and examined with an electron microscope.[18] (Reproduced with permission from Li, H.-S., Stolz, D.B., Romero, G. *Traffic*, 6, 324–334, 2005. Copyright 2005 John Wiley and Sons.)

differential centrifugation and immunomagnetic separation.[22] As a result, immunomagnetic separation was found to enrich exosomes more effectively than the other methods. Next, they investigated the physiological role of exosomes in colorectal cancer biology.[23] Because the tumour microenvironment plays critical roles in tumour initiation, progression and metastasis, detailed characterization of exosomes, including proteomics analyses, is important to understand their possible roles in tumour microenvironment regulation. They cultured human colon carcinoma LIM1863 cells to grow organoids. Two distinct populations of exosomes exist in culture medium of LIM1863 cells: A33-exosomes and EpCAM-exosomes. These two types of exosomes are indistinguishable from one another by TEM observation and both contain general exosome markers such as TSG101, Alix and HSP70. For isolation of the two types of exosomes, the culture medium was centrifuged at low speed to remove intact cells, cell detritus and large membrane particles. The supernatant was then centrifuged to remove shed microvesicles. Subsequently, the supernatant was filtered and concentrated to obtain the concentrated culture medium (CCM). Anti-A33-coupled magnetic beads composed of SPIONs (Dynabeads; Invitrogen) were mixed with the CCM, followed by magnetic separation. After separation of the A33-exsosomes, the residual CCM was treated with anti-EpCAM-coupled SPION-based magnetic beads (EpCAM MicroBeads; Miltenyi Biotec) to isolate EpCAM-exosomes. Figure 12.11a shows TEM images of the magnetically isolated A33-exsosomes and EpCAM-exosomes. Two common exosome markers, Alix and TSG101, were detected in both A33-exosomes and EpCAM-exosomes by WB, as shown in Figure 12.11b. To clearly distinguish A33-exosomes from EpCAM-exosomes, a proteomics analysis was performed. Thus, 1024 and 898 proteins were identified in A33-exosomes and EpCAM-exosomes, respectively. Of these, 340 and 214 proteins were found to be unique to A33-exosomes and EpCAM-exosomes, respectively, while the remaining 684 proteins were common to both types of exosomes (Figure 12.11c). Interestingly, several MHC class I molecules were exclusively observed in A33-exosomes, while neither MHC class I nor MHC class II molecules were observed in EpCAM-exosomes. Colocalization of EpCAM, claudin-7 and CD44 was observed in EpCAM-exosomes. These proteins interact with one another to promote tumour progression.

12.4.2 Magnetic Separation of Exosomes Derived from Different Immune Cells

Clayton et al.[20] proposed a rapid and versatile technique for analysis of membrane proteins on exosomes based on immunomagnetic separation and flow cytometry. They used five cell lines: B-lymphocytes (B cells) derived from two donors [B-LCL(DA) and B-LCL(MA)], human Burkitt lymphoma line (Daudi) cells deficient in MHC class I, human chronic myelogenous leukemia line (K562) cells deficient in MHC class II and monocyte-derived dendritic cells. It should be noted that the B-lymphocytes were immortalized with

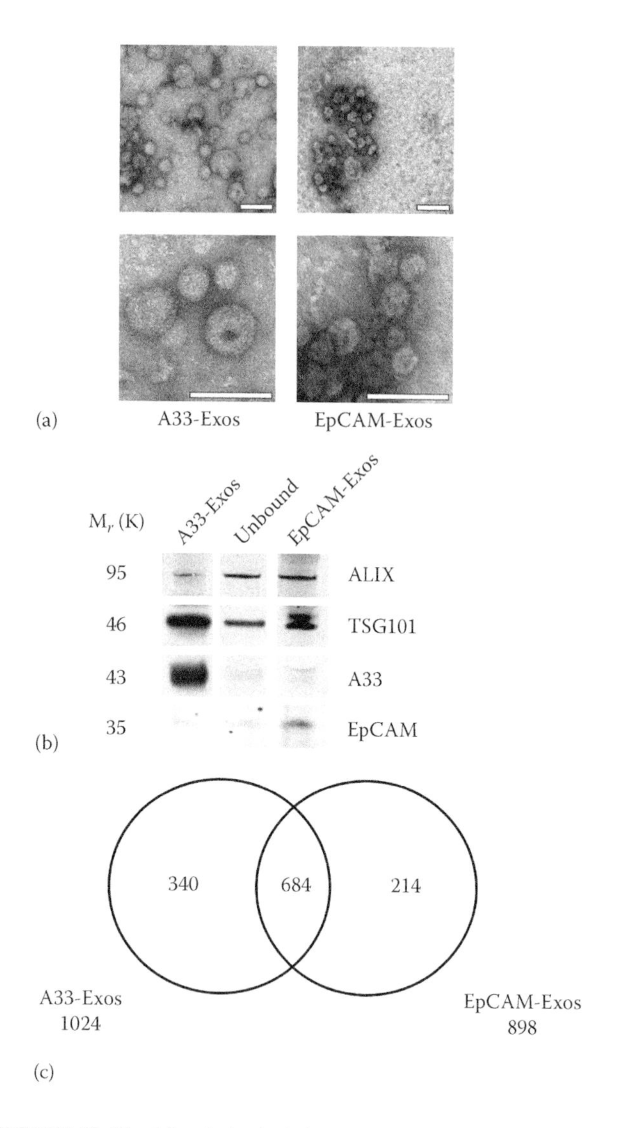

FIGURE 12.11 Morphological characterization and proteome analysis of LIM1863 cell-derived A33- and EpCAM-Exos. (a) Electron micrographs of A33- and EpCAM-Exos negatively stained with uranyl acetate and examined at 200 kV; scale bar = 100 nm. (b) Western blot analysis of A33-Exos, unbound material (flow-through of anti-A33 antibody capture beads) and EpCAM-Exos (10 μg per lane) for Alix (PDCD6IP), TSG101, A33 and EpCAM. (c) Two-way Venn diagram depicting the overlap of exosomal proteins derived from A33- and EpCAM-Exos. A total of 684 proteins were common to both exosomal datasets, and 340 and 214 proteins were unique to A33- and EpCAM-Exos, respectively.[23] (From Tauro, B.J., Greening, D.W., Mathias, R.A., Mathivanan, S., Ji, H., Simpson, R.J. *Mol. Cell. Proteomics*, 12, 587–598, 2013. With permission.)

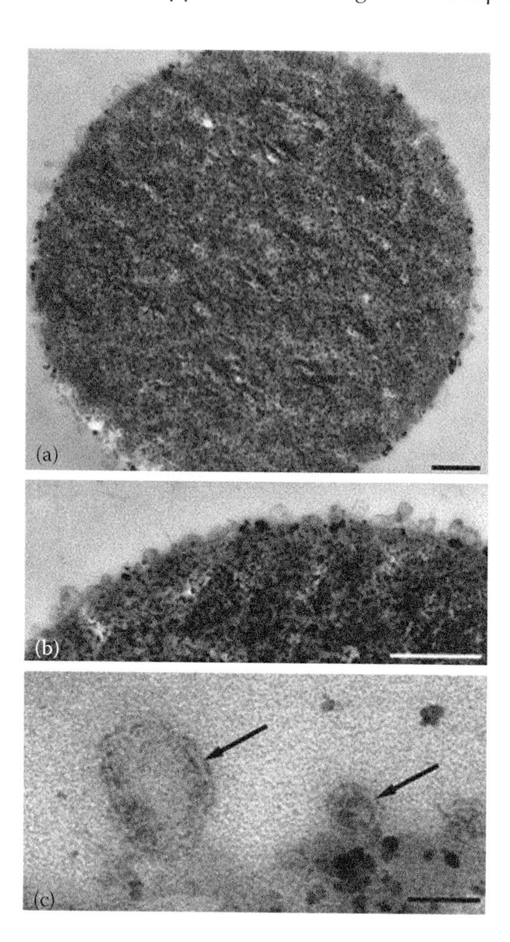

FIGURE 12.12 Ultrathin sections of exosome-coated beads. Beads were coated under saturating conditions, with exosomes derived from B-LCL(DA) cells and ultrathin sections viewed under transmission electron microscope. The bead surface is well coated with vesicles: (a) bar = 0.5 μm, which are seen at higher magnification; (b) bar = 0.5 μm. At higher magnification, membrane structures on two differently sized vesicles are visible (arrows); (c) bar = 50 nm.[20] (Reprinted from *J. Immunol. Methods*, 247, Analysis of antigen presenting cell derived exosomes, based on immuno-magnetic isolation and flow cytometry, 247, 163–174, Clayton, A., J. Court, H. Navabi et al., Copyright 2001, with permission from Elsevier.)

Epstein-Barr virus in advance. The culture medium was centrifuged from low to high speed to remove intact cells and cell debris. Magnetic particles with a mean diameter of 4.5 μm conjugated with anti-human leukocyte antigen antibodies were then added to the cell-free supernatants, followed by magnetic separation. Exosomes were magnetically separated from the B-LCL(DA), B-LCL(MA), Daudi, K562 and dendritic cell supernatants. Figure 12.12 shows TEM images of a single magnetic particle coated with extracellular vesicles derived from B-LCL(DA) cells. Based on a size analysis, ~70% of the vesicles were thought to be exosomes. Next, FITC-conjugated IgG antibodies, which were conjugated to magnetic particles, were selectively bound to specific proteins expressed on exosomes for flow cytometric analysis. The results clearly showed that B-cell-derived exosomes expressed MHC class I and MHC class II molecules. Daudi-cell-derived exosomes were most strongly stained for MHC class II molecules and there was no staining for MHC class I molecules, as expected. Meanwhile, the K562 cell line, which is deficient in MHC class II molecules, did not secrete material that became bound to the magnetic particles. The dendritic-cell-derived exosomes expressed abundant MHC class I and II molecules. The merit of their technique is that it requires a short time (approximately half a day) for qualitative and semiquantitative analysis of proteins on exosomes.[20]

12.4.3 Magnetic Separation and Simultaneous Detection of Exosomes

Zong et al.[25] combined immunomagnetic separation and SERS-based sensing techniques to detect tumour-cell-derived exosomes with enhanced sensitivity. Tumour-cell-derived exosomes are deeply involved in modulation of the tumour microenvironment. Therefore, tumour-cell-derived exosomes are considered an important cancer biomarker, and analysis of these exosomes can provide insightful information on the pathological stages of cancerous diseases. Zong et al. synthesized two nanoparticles: Fe_3O_4@SiO_2 core@shell nanoparticles (size of approximately 120–220 nm) and Au@Ag@SiO_2 core@shell@shell nanorods containing Raman reporter molecules in their structure. Anti-CD63 and anti-HER2 antibodies were conjugated to the surface of the Fe_3O_4@SiO_2 nanoparticles (magnetic probes) and Au@Ag@SiO_2 nanorods (SERS probes), respectively. In their study, SKBR-3 and normal human embryonic lung fibroblasts (MRC-5) were used as model cells for cancerous and normal cells, respectively. An exosome stock solution was prepared from cell culture medium using an exosome precipitation kit (Exo-Quick-TC; System Biosciences). Next, the SERS and magnetic probes were simultaneously added to the exosome stock solution to form sandwich-type immunocomplexes (i.e. Fe_3O_4@SiO_2/exosome/Au@Ag@SiO_2). After incubation, the immunocomplexes were collected with a magnet and subjected to SERS measurements. Based on a quantitative analysis, the SERS-based method showed high sensitivity, with a detection limit of 1200 exosomes. The SERS signal of SKBR-3 cell-derived exosomes was 3.8 times higher than that of MRC-5 cell-derived exosomes. These findings are consistent with the fact that tumour cells usually secrete more exosomes than normal cells do.[52] This SERS-based method requires only 2 h from collection to identification of exosomes, being much shorter than conventional methods that normally require about 12 h in total.[25]

12.5 MAGNETIC SEPARATION OF MITOCHONDRIA

Mitochondria participate in various biological processes, including energy generation, production of oxygen free radicals, maintenance of intracellular calcium homeostasis, signal transduction and apoptosis.[27] Therefore, mitochondrial dysfunction often leads to the development of various diseases.[53] It has been reported that mitochondrial dysfunction is related to obesity,[54] cancer,[55] neurodegenerative disease[56] and cardiac disease,[57] among others. For precise investigation of the role and performance of functional mitochondria, isolation of mitochondria is crucially important, and thus, several methods, such as density gradient centrifugation and differential centrifugation, have been used for this purpose. Although differential centrifugation, which was developed during the 1950s[58] and modified in many ways, is the most frequently used method, it has limitations, such as its time-consuming process and lack of purity. Meanwhile, density gradient centrifugation can isolate mitochondria with high purity but also requires a long time and shows contamination depending on the sample type.[28] Furthermore, the bilayer structure of mitochondria can be destroyed under certain physiological conditions. As these conventional separation techniques have problems in terms of purity, reproducibility and ease of process, many researchers have become interested in magnetic isolation of mitochondria.

12.5.1 Magnetic Separation of Mitochondria Derived from Mouse Tissues

Franko et al.[28] proposed a standardized isolation method for mitochondria by demonstrating their isolation from various mouse tissues. They utilized anti-TOM22 (translocase of outer mitochondria membrane 22 homolog) antibody-coupled magnetic beads (MicroBeads; Miltenyi Biotec) for magnetic separation of mitochondria. To magnetically separate mitochondria, tissue lysates were prepared using a Mitochondria Extraction Kit (Miltenyi Biotec). The tissue lysates were treated with the anti-TOM22 MicroBeads, followed by magnetic separation using a MACS separator (Miltenyi Biotec). After the magnetic separation, the magnetically labeled eluted fraction and nonlabelled (wash) fraction were analyzed by TEM and WB. Figure 12.13 shows TEM images of magnetically isolated mitochondria from the heart (Figure 12.13a), muscle (Figure 12.13b), brain (Figure 12.13c) and liver (Figure 12.13d). The small black dots observed outside the membrane in these images correspond to the anti-TOM22 MicroBeads. By carefully examining the morphology of the isolated mitochondria, it was revealed that ~90% of the mitochondria were intact. WB analysis was performed to detect mitochondrial markers (TOM20 and cytochrome c) and confirm the presence of mitochondrial proteins in the eluted fractions. Thus, both TOM20 and cytochrome c were detected in the eluted fractions from the heart, brain and liver tissues, while none of these markers were detected in the wash fractions. On the contrary, PDIA2 (endoplasmic reticulum marker) and PEX1 (peroxisomal protein marker) were rarely detected in the eluted fractions, while they were detected in the wash fractions. These results suggested that mitochondria were successfully enriched. To assess the function of the magnetically separated mitochondria, the oxygen consumption rates of mitochondria derived from the skeletal muscle, liver and brain were analyzed by the Clark electrode method. The results revealed that the magnetically separated mitochondria from these tissues maintained their function.

Finally, Franko et al.[28] compared their magnetic separation technique with differential centrifugation in terms of purity and quality of the isolated mitochondria. In the case of differential centrifugation, both TOM20 and cytochrome c were detected in the precipitate. Unfortunately, PDIA2 was also detected in the precipitate. These results suggested that the separation of mitochondria and endoplasmic reticulum was not successful using differential centrifugation. They further confirmed that the purity of their magnetically separated mitochondria was higher than that of mitochondria separated

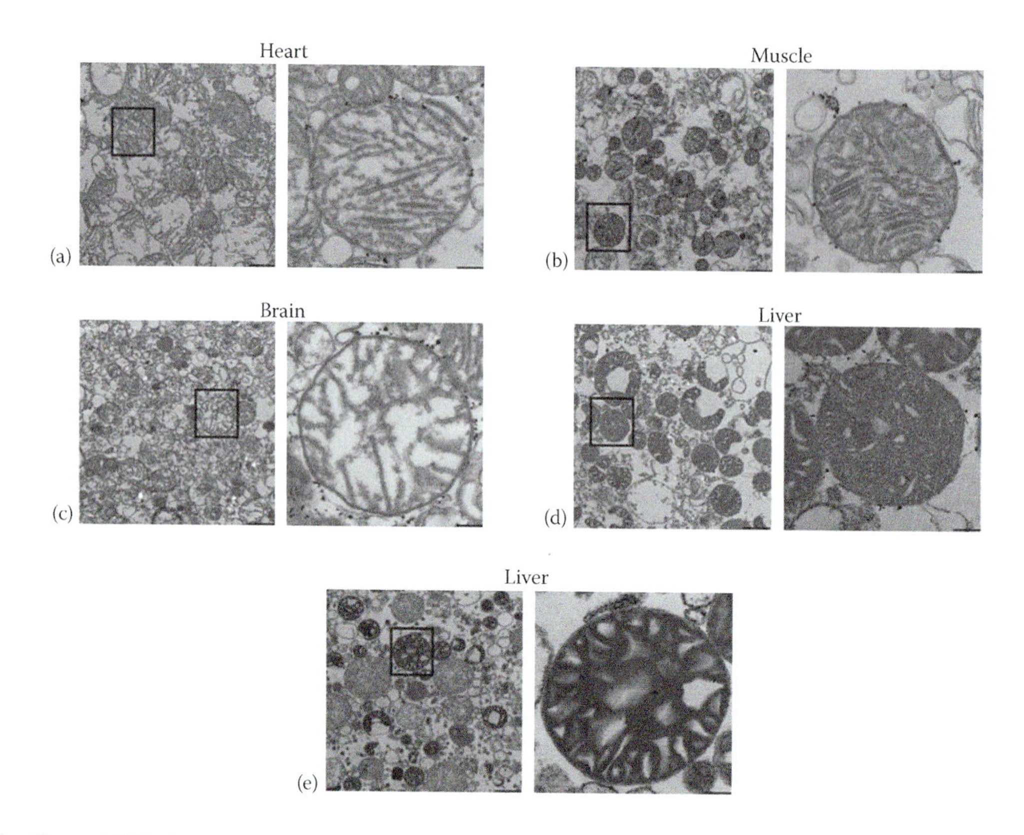

FIGURE 12.13 The anti-TOM22 magnetic beads method yields intact mitochondria. Anti-TOM22 magnetic bead method was applied to purify mitochondria from (a) heart, (b) skeletal muscle, (c) brain and (d) liver tissues. Mitochondria were isolated by differential centrifugation technique from (e) liver tissue. Mitochondrial morphology was investigated by TEM from four biological samples per group and representative pictures are shown. Pictures were taken at 4,000× (left panel) and 20,000× (right panel) magnification and black bars display 1 mm (left panel) and 200 nm (right panel), respectively. Black dots at the outer membrane in the pictures in a, b, c and d represent anti-TOM22 magnetic beads.[28] (From Franko, A., Baris, O.R., Bergschneider, E. et al., *PLoS One* 8, e82392, 2013. With permission.)

by differential centrifugation using liquid chromatography–tandem mass spectrometry.

12.5.2 Magnetic Separation of Mitochondria from Cells

Tang et al.[27] also compared a magnetic separation technique for mitochondria with other conventional separation methods such as differential centrifugation and density gradient centrifugation. They conjugated an anti-monoamine oxidase-A (MAO-A; mitochondrial marker) antibody to Fe_3O_4@SiO_2 core@shell nanoparticles with a mean diameter of approximately 220 nm. Meanwhile, HepG2 and HeLa cells were cultured and then homogenized. Anti-MAO-A-conjugated Fe_3O_4@SiO_2 nanoparticles were added to the cell lysates, followed by incubation for 20 min at 4°C. Subsequently, a magnet was placed on the tube wall to collect the magnetic nanoparticles. The supernatant was discarded and the magnetically attracted substances on the tube wall were washed. After the washing, the magnetically attracted fraction was collected. The authors described that the purity of their isolated mitochondria was more than 90%, while the mitochondria isolated by differential centrifugation were contaminated by other subcellular components. In addition, in the case of HepG2 cells, the magnetic separation method gave a twofold higher yield than differential centrifugation and a threefold higher yield than density gradient centrifugation. The respiratory functions of the mitochondria isolated by the different methods were compared, and the magnetically isolated mitochondria were found to exhibit the highest activity. It was also confirmed that the purity of the magnetically separated mitochondria was higher than that of the mitochondria separated by either differential centrifugation or density gradient centrifugation. These results showed that magnetic separation was a superior isolation technique with higher yield and greater mitochondrial purity and activity compared with conventional centrifugation techniques. In addition, the magnetic separation requires only about 1 h from cell harvesting to isolation of mitochondria.

12.6 MULTIFUNCTIONAL NANOPARTICLES FOR VERSATILE ISOLATION OF CELLULAR ORGANELLES

12.6.1 Requirements for Magnetic Probes for Versatile Isolation of Cellular Organelles

Several examples of magnetic separation of cellular organelles have been introduced in the previous sections. Although some types of cellular organelles could be successfully isolated/enriched, there remain various limitations to the magnetic isolation of cellular organelles in terms of general versatility. The most important step in the magnetic separation of cellular organelles is the conjugation of magnetic probes to the target organelle. In the cases of magnetic separation of exosomes and mitochondria, the magnetic probes were conjugated to the target organelles in a culture medium or lysate.[20,22,23,25,27,28] Meanwhile, in the case of magnetic separation of endosomes, the magnetic probes were internalized into endosomes by endocytosis.[15–18] However, because the characteristic size of commercially available magnetic probes is typically more than 50 nm, they are too large for general use in the isolation of various small membrane vesicles, such as clathrin-coated vesicles, because of large steric hindrance.[18] Therefore, smaller magnetic probes are desired for more efficient and versatile isolation of cellular organelles. Troublingly, however, when the volume of magnetic probes becomes too small, the saturation magnetization decreases, mainly through thermal fluctuation of spins and surface disorders. To solve this problem, a magnetic material that intrinsically exhibits high saturation magnetization is desired for use as a probe material. Currently, most of the common magnetic probes are SPION-based probes because they are chemically stable and exhibit low cytotoxicity. However, the saturation magnetization of Fe_3O_4 (approximately 446 emu/cm^3)[59] is not very high compared with those of other soft magnetic materials such as Fe (1745.9 emu/cm^3),[60] $L1_0$-type ordered FeNi (1270 emu/cm^3)[61] and FeCo (1790 emu/cm^3),[59] and thus it would be better to replace SPIONs with nanoparticles consisting of another soft magnetic material with higher saturation magnetization for isolation of cellular organelles.

In addition, since the key step in successful isolation of target organelles is the conjugation/incorporation of magnetic probes to/into target organelles, it is imperative to confirm the colocalization of the probes with the target organelles, especially for the isolation of organelles that differentiate relatively quickly, such as certain organelles in the endocytic and autophagic pathways. TEM and immuno-TEM techniques are most often used to confirm the colocalization of probes with target organelles, as presented earlier (see Figures 12.7a, 12.8b and e, 12.10a–d, 12.11a, 12.12a–c and 12.13a–d). Unfortunately, TEM observation requires much more effort and time than do optical microscopic imaging techniques such as fluorescence microscopy and confocal laser scanning microscopy. Furthermore, in the case of TEM, it is impossible to perform live imaging. For utilization of optical microscopic techniques to confirm the colocalization of probes with target organelles, the probes must have visualization capabilities as well.

12.6.2 Magnetic–Plasmonic Hybrid Nanoparticles

Magnetic–plasmonic hybrid nanoparticles are one of the promising candidates for the versatile isolation of cellular organelles because they simultaneously hold magnetic separation and visualization capabilities. Although various types of magnetic–plasmonic hybrid nanoparticles have been reported, such as Au-Fe_3O_4 hybrid nanoparticles (Figures 12.14 and 12.15),[5,62–64] Co@Au core@shell nanoparticles,[65] FePt-Au hybrid nanoparticles,[66] FePt-Ag hybrid nanoparticles[67] and Fe_3O_4@C-Ag nanoparticles (Figure 12.16),[68] few nanoparticles can satisfy the requirements for the versatile isolation of cellular organelles described previously. To fulfil these requirements, Ag@FeCo@Ag core@shell@shell nanoparticles have recently been created.[69–71] In these double-shell structured nanoparticles, the Ag core is responsible for providing the LSPR properties that enable the visualization of the nanoparticles by optical microscopy. In addition, the Ag core suppresses the oxidation of the FeCo shell by electron transfer from the Ag core to the FeCo shell. The outer Ag shell has other roles in further suppressing oxidation of the FeCo shell and providing a platform for surface modification of the nanoparticles using thiol–metal interactions. The Ag@FeCo@Ag nanoparticles were synthesized by a combination of polyol and hot injection methods.[69–71] Briefly, the Ag precursor, polyol and capping ligands were placed into a flask

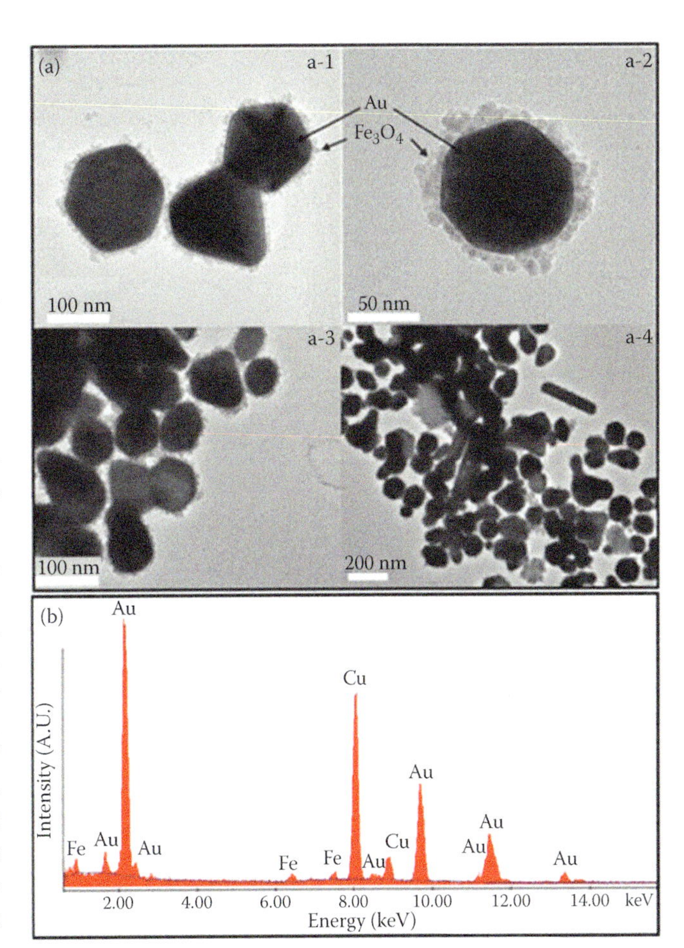

FIGURE 12.14 Characterization of Fe_3O_4-decorated Au NPs using TEM (a) and EDS (b). For the majority of the NPs, the Au core is quasi-spherical with an average size of 90 nm (a-1, a-2); however, hexagonal, triangular and elongated Au cores ranging from 80 to 150 nm are also found (a-3, a-4). The shell surrounding the Au cores is formed by a compact arrangement of iron oxide NPs with a size around 7 nm. Fe, Au and Cu are present in the EDS spectrum (b). The presence of Cu is due to the copper grid used for the TEM/EDS experiments.[63] (Reprinted with permission from Mezni, A., I. Balti, A. Mlayah, N. Jouini, and L. S. Smiri. 2013. Hybrid Au-Fe_3O_4 nanoparticles: plasmonic, surface enhanced Raman scattering, and phase transition properties. *J. Phys. Chem. C* 117:16166–74. Copyright 2013 American Chemical Society.)

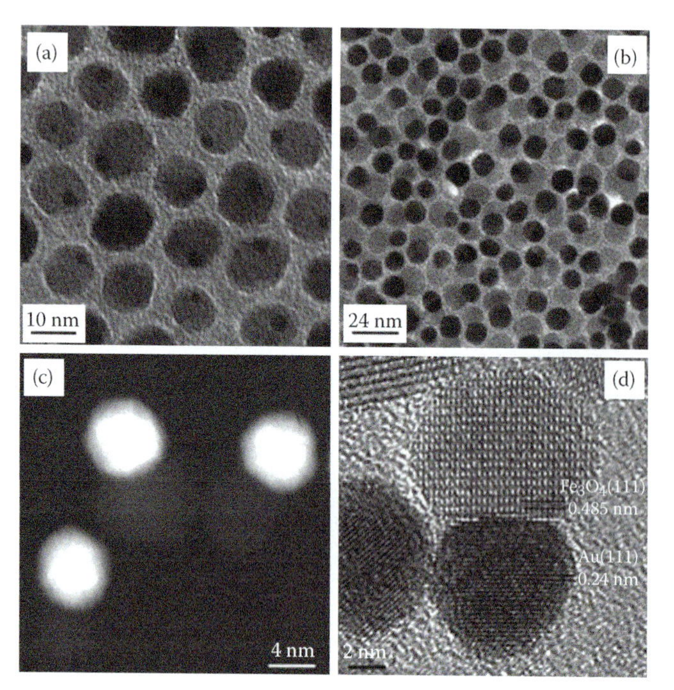

FIGURE 12.15 TEM and STEM images of the dumbbell-like Au-Fe_3O_4 NPs: (a) TEM image of the 3–14 nm Au-Fe_3O_4 NPs, (b) TEM image of the 8–14 nm Au-Fe_3O_4 NPs, (c) HAADF-STEM image of the 8–9 nm Au-Fe_3O_4 NPs and (d) HRTEM image of one 8–12 nm Au-Fe_3O_4 NP.[64] (Reprinted with permission from Yu, H., M. Chen, P. M. Rice, S. X. Wang, R. L. White, and S. Sun. 2005. Dumbbell-like bifunctional Au-Fe_3O_4 nanoparticles. *Nano Lett.* 5:379–82. Copyright 2005 American Chemical Society.)

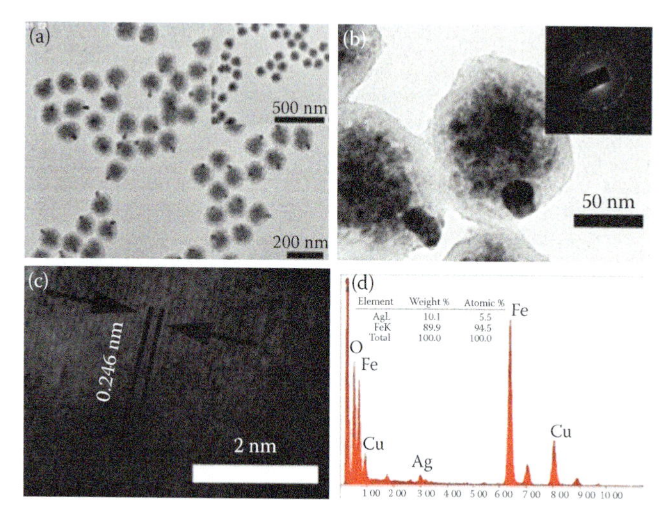

FIGURE 12.16 (a) Low and (b) high magnified TEM images of the dumbbell-like Fe_3O_4@C-Ag hybrid NPs synthesized at $[Ag^+]$ = 0.25 mM and 30°C for 15 min. The inset in (a) is a TEM image of the Fe_3O_4@C NPs; the inset in (b) is the electron diffraction pattern of the dumbbell-like hybrid NPs. (c) The lattice fringe of Ag nanocrystals. (d) EDS and elemental ratio of a single dumbbell-like hybrid NP.[68] (Reprinted with permission from Wang, H., J. Shen, Y. Li et al. 2013. Porous carbon protected magnetite and silver hybrid nanoparticles: morphological control, recyclable catalysts, and multicolor cell imaging. *ACS Appl. Mater. Interfaces* 5:9446–53. Copyright 2013 American Chemical Society.)

at room temperature, and the temperature was increased up to 250°C. During the increase in temperature, two stock solutions were sequentially injected into the reaction solution. The first stock solution containing Fe and Co precursors was injected into the reaction solution at 170°C. When the temperature reached 250°C, the second stock solution containing the Ag precursor was injected. Following the injection of the second stock solution, the reaction was continued for 15 min at 250°C. After the reaction, the nanoparticles were washed and dried. A TEM image of the resulting Ag@FeCo@Ag nanoparticles is shown in Figure 12.17. The size of the nanoparticles was calculated to be around 15 nm. An energy dispersive X-ray spectroscopy (EDS) line profile analysis confirmed the double shell structure. A distinct LSPR peak was observed at 409 nm in the ultraviolet-visible spectrum of the nanoparticles, and the saturation magnetization of the nanoparticles was measured to be 36.4 emu/g at 300 K.[69]

To assess the feasibility of magnetic separation of cellular organelles using the Ag@FeCo@Ag nanoparticles, the nanoparticles were electrostatically attached to liposomes and the magnetic migration behaviour of the liposomes was examined under a confocal laser scanning microscope. In particular, the hydrophobic surface capping ligands of the as-synthesized nanoparticles were exchanged with poly-*L*-lysine-containing thiol groups to render the nanoparticles water-dispersible. The nanoparticles were then nonspecifically adsorbed on the surface of DPPC (1,2-dipalmitoyl-*sn*-glycero-3-phosphocholine) liposomes in aqueous solution. The migration behaviour of the nanoparticle-loaded liposomes under an external magnetic field was monitored by plasmon scattering of the nanoparticles using a confocal laser scanning microscope, as shown in Figure 12.18. The results clearly indicated that the Ag@FeCo@Ag nanoparticles, which have both magnetic separation and imaging capabilities, can be promising probes for the versatile isolation of various cellular organelles

12.7 CONCLUSIONS AND FUTURE OUTLOOK

The magnetic separation technique has recently been applied not only for isolation of cells, bacteria and proteins but also for isolation of cellular organelles to characterize the proteins on/in the target organelles. In this chapter, several examples of magnetic separation of cellular organelles such as endosomes, exosomes and mitochondria were reviewed. It has been clearly demonstrated that the magnetic separation technique has advantages over conventional separation techniques such as differential centrifugation and density gradient centrifugation in terms of purity and ease of use. However, because the magnetic probes used in almost all of the studies were SPION-based, single-function and relatively large probes, the magnetic separation technique lacks versatility. To expand the magnetic separation technique to other cellular organelles, ultrasmall probes with magnetic separation and imaging capabilities are required. By enabling the on-demand isolation of various cellular organelles, developing and utilizing multifunctional probes to identify proteins on/in the organelles and analyzing the function of these proteins, it will become possible to offer new insights not only into basic biology but also into clinical medicine.

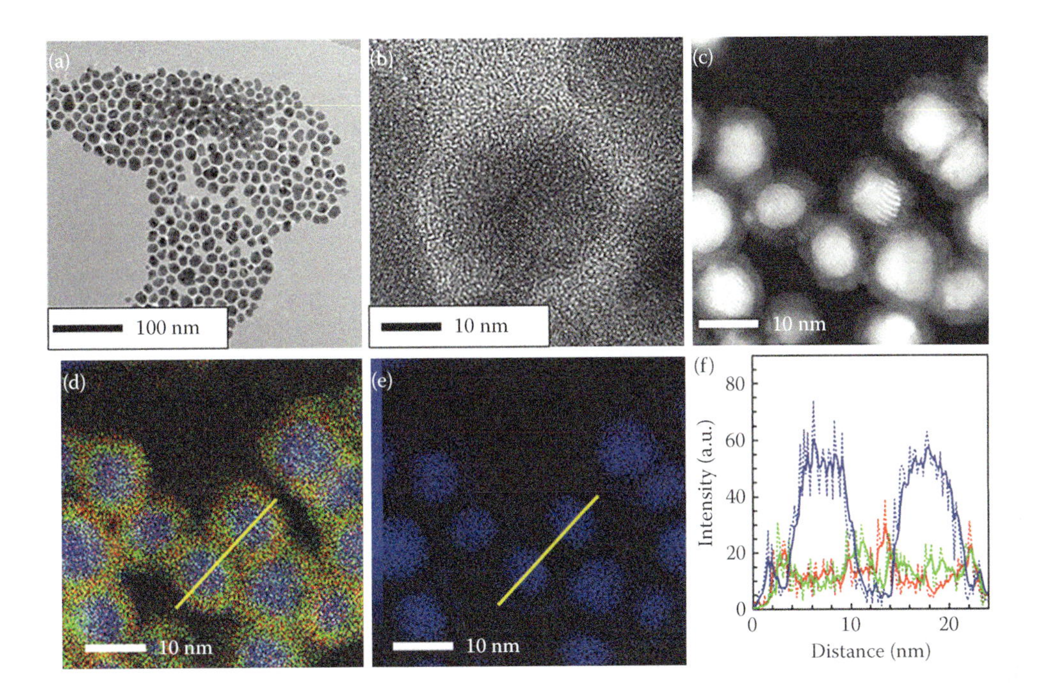

FIGURE 12.17 (a) TEM, (b) HRTEM, (c) STEM-HAADF and (d) and (e) EDS elemental mapping images (merged and Ag L-edge images, respectively) of Ag-core NPs. (f) EDS line profile at the centre of the NPs indicated by the yellow line in (d) and (e). The dashed and solid lines are the raw and low-pass-filtered profiles, respectively. The blue, green and red in (d)–(f) correspond to Ag L-, Co K- and Fe K-edge intensities, respectively.[71] (Reprinted with permission from Takahashi, M., P. Mohan, K. Higashimine, D. M. Mott, and S. Maenosono. 2016. Transition of exchange bias from the linear to oscillatory regime with the progression of surface oxidation of Ag@FeCo@Ag Core@Shell@Shell nanoparticles. *J. Appl. Phys.* 120:134301. Copyright 2016, American Institute of Physics.)

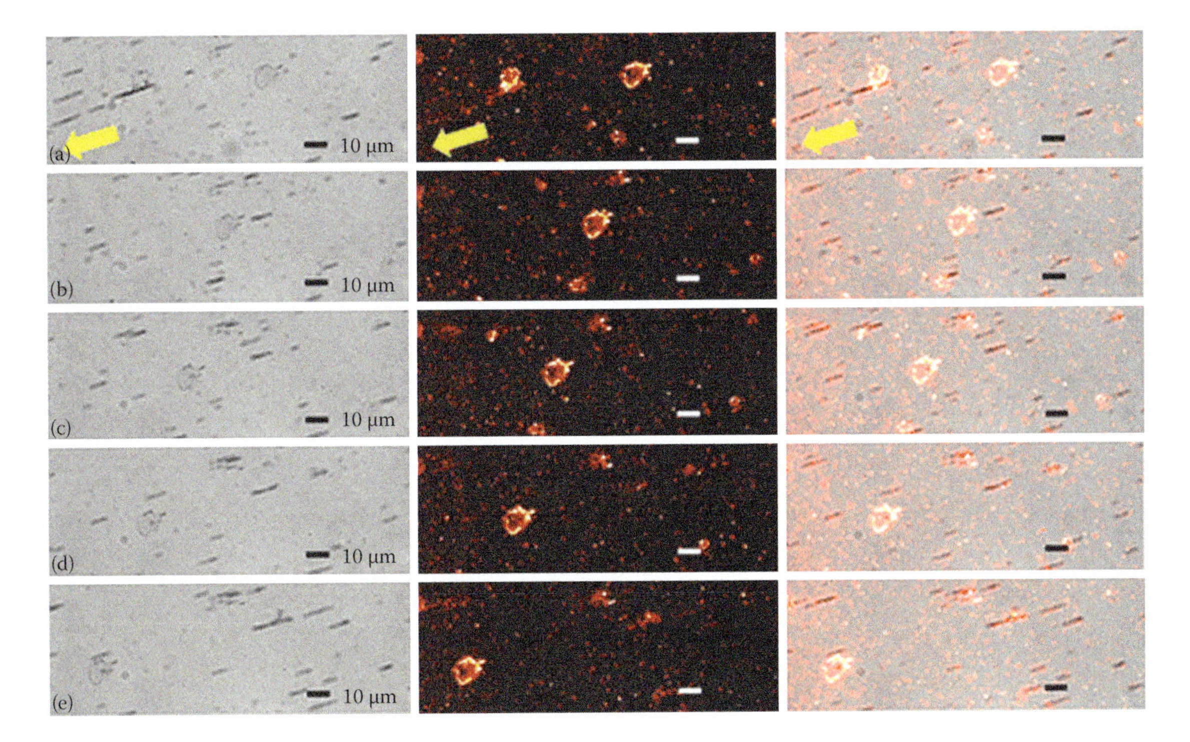

FIGURE 12.18 Snapshots of CLSM images: transmission (left), plasmon scattering (middle) and merged (right) images. The images were obtained at (a) $t = 53$ s, (b) $t = 62$ s, (c) $t = 72$ s, (d) $t = 82$ s and (e) $t = 92$ s. Yellow arrows indicate the direction of magnetic force.[69] (Reprinted with permission from Takahashi, M., P. Mohan, A. Nakade et al. 2015. Ag/FeCo/Ag core/shell/shell magnetic nanoparticles with plasmonic imaging capability. *Langmuir* 31:2228–36. Copyright 2015 American Chemical Society.)

REFERENCES

1. Fan, Z., M. Shelton, A. K. Singh, D. Senapati, S. A. Khan, and P. C. Ray. 2012. Multifunctional plasmonic shell–magnetic core nanoparticles for targeted diagnostics, isolation, and photothermal destruction of tumor cells. *ACS Nano* 6:1065–73.
2. Wadajkar, A. S., S. Santimano, L. Tang, and K. T. Nguyen. 2014. Magnetic-based multi-layer microparticles for endothelial progenitor cell isolation, enrichment, and detachment. *Biomaterials* 35:654–63.
3. Sestier, C., and D. Sabolovic. 1998. Particle electrophoresis of micrometric-sized superparamagnetic particles designed for magnetic purification of cells. *Electrophoresis* 19:2485–90.
4. Chen, H., Y. Zeng, W. Liu, S. Zhao, J. Wu, and Y. Du. 2013. Multifaceted applications of nanomaterials in cell engineering and therapy. *Biotechnol. Adv.* 31:638–53.
5. Fan, Z., D. Senapati, S. A. Khan et al. 2013. Popcorn-shaped magnetic core-plasmonic shell multifunctional nanoparticles for the targeted magnetic separation and enrichment, label-free SERS imaging, and photothermal destruction of multidrug-resistant bacteria. *Chem. Eur. J.* 19:2839–47.
6. Verbarg, J., W. D. Plath, L. C. Shriver-Lake et al. 2013. Catch and release: Integrated system for multiplex detection of bacteria. *Anal. Chem.* 85:4944–50.
7. Behra, M., N. Azzouz, S. Schmidt et al. 2013. Magnetic porous sugar-functionalized PEG microgels for efficient isolation and removal of bacteria from solution. *Biomacromolecules* 14:1927–35.
8. Wang, Y., Z. Ye, J. Ping, S. Jing, Y. Ying. 2014. Development of an aptamer-based impedimetric bioassay using microfluidic system and magnetic separation for protein detection. *Biosens. Bioelectron.* 59:106–11.
9. Hu, J., S. Huang, X. Huang, Z. Kang, and N. Gan. 2014. Superficially mesoporous $Fe_3O_4@SiO_2$ core shell microspheres: Controlled syntheses and attempts in protein separations. *Micropor. Mesopor. Mater.* 197:180–4.
10. Masthoff, I. C., F. David, C. Wittmann, and G. Garnweitner. 2014. Functionalization of magnetic nanoparticles with high-binding capacity for affinity separation of therapeutic proteins. *J. Nanopart. Res.* 16:2164.
11. Tang, L., J. Casas, and M. Venkataramasubramani. 2013. Magnetic nanoparticle mediated enhancement of localized surface plasmon resonance for ultrasensitive bioanalytical assay in human blood plasma. *Anal. Chem.* 85:1431–9.
12. Sepúlveda, B., P. C. Angelomé, L. M. Lechuga, and L. M. Liz-Marzán. 2009. LSPR-based nanobiosensors. *Nano Today* 4:244–51.
13. Haes, A. J., and R. P. Van Duyne. 2002. A nanoscale optical biosensor: Sensitivity and selectivity of an approach based on the localized surface plasmon resonance spectroscopy of triangular silver nanoparticles. *J. Am. Chem. Soc.* 124:10596–604.
14. Alberts, B., D. Bray, K. Hopkin et al. 2004. *Essential cell biology, second edition.* Garland Science Taylor and Francis Group.
15. Chen, F. W., R. E. Gordon, and Y. A. Ioannou. 2005. NPC1 late endosomes contain elevated levels of non-esterified ('free') fatty acids and an abnormally glycosylated form of the NPC2 protein. *Biochem. J.* 390:549–61.
16. Wittrup, A., S.-H. Zhang, G. B. Ten Dam et al. 2009. ScFv antibody-induced translocation of cell-surface heparan sulfate proteoglycan to endocytic vesicles. *J. Biol. Chem.* 284:32959–67.
17. Wittrup, A., S. H. Zhang, K. J. Svensson et al. 2010. Magnetic nanoparticle-based isolation of endocytic vesicles reveals a role of the heat shock protein GRP75 in macromolecular delivery. *Proc. Natl. Acad. Sci. U.S.A.* 107:13342–7.
18. Li, H.-S., D. B. Stolz, and G. Romero. 2005. Characterization of endocytic vesicles using magnetic microbeads coated with signalling ligands. *Traffic* 6:324–34.
19. Nakamura, N., J. R. Lill, Q. Phung et al. 2014. Endosomes are specialized platforms for bacterial sensing and NOD2 signalling. *Nature* 509:240–4.
20. Clayton, A., J. Court, H. Navabi et al. 2001. Analysis of antigen presenting cell derived exosomes, based on immuno-magnetic isolation and flow cytometry. *J. Immunol. Methods* 247:163–74.
21. Ji, H., M. Chen, D. W. Greening et al. 2014. Deep sequencing of RNA from three different extracellular vesicle (EV) subtypes released from the human LIM1863 colon cancer cell line uncovers distinct mirna-enrichment signatures. *PLoS One* 9:e110314.
22. Tauro, B. J., D. W. Greening, R. A. Mathias et al. 2012. Comparison of ultracentrifugation, density gradient separation, and immunoaffinity capture methods for isolating human colon cancer cell line LIM1863-derived exosomes. *Methods* 56:293–304.
23. Tauro, B. J., D. W. Greening, R. A. Mathias, S. Mathivanan, H. Ji, and R. J. Simpson. 2013. Two distinct populations of exosomes are released from LIM1863 colon carcinoma cell-derived organoids. *Mol. Cell. Proteomics* 12:587–98.
24. Hong, C. S., L. Muller, M. Boyiadzis, and T. L. Whiteside. 2014. Isolation and characterization of CD34+ blast-derived exosomes in acute myeloid leukemia. *PLoS One* 9:e103310.
25. Zong, S., L. Wang, C. Chen et al. 2016. Facile detection of tumor-derived exosomes using magnetic nanobeads and SERS nanoprobes. *Anal. Methods* 8:5001–8.
26. Hornig-Do, H.-T., G. Günther, M. Bust, P. Lehnartz, A. Bosio, and R. J. Wiesner. 2009. Isolation of functional pure mitochondria by superparamagnetic microbeads. *Anal. Biochem.* 389:1–5.
27. Tang, B., L. Zhao, R. Liang, Y. Zhang, and L. Wang. 2012. Magnetic nanoparticles: An improved method for mitochondrial isolation. *Mol. Med. Rep.* 5:1271–6.
28. Franko, A., O. R. Baris, E. Bergschneider et al. 2013. Efficient isolation of pure and functional mitochondria from mouse tissues using automated tissue disruption and enrichment with anti-TOM22 magnetic beads. *PLoS One* 8:e82392.
29. Kappler, L., J. Li, H.-U. Häring et al. 2016. Purity matters: A workflow for the valid high-resolution lipid profiling of mitochondria from cell culture samples. *Sci. Rep.* 6:21107.
30. Lüers, G. H., R. Hartig, H. Mohr et al. 1998. Immuno-isolation of highly purified peroxisomes using magnetic beads and continuous immunomagnetic sorting. *Electrophoresis* 19:1205–10.
31. Wang, Y., T. H. Taylor, and E. A. Arriaga. 2012. Analysis of the bioactivity of magnetically immunoisolated peroxisomes. *Anal. Bioanal. Chem.* 402:41–9.
32. Kikuchi, M., N. Hatano, S. Yokota, N. Shimozawa, T. Imanaka, and H. Taniguchi. 2004. Proteomic analysis of rat liver peroxisome: Presence of peroxisome-specific isozyme of Lon protease. *J. Biol. Chem.* 279:421–8.
33. Rodriguez-Paris, J. M., K. V. Nolta, and T. L. Steck. 1993. Characterization of lysosomes isolated from *Dictyostelium discoideum* by magnetic fractionation. *J. Biol. Chem.* 268: 9110–6.

34. Diettrich, O., K. Mills, A. W. Johnson, A. Hasilik, and B. G. Winchester. 1998. Application of magnetic chromatography to the isolation of lysosomes from fibroblasts of patients with lysosomal storage disorders. *FEBS Lett.* 441:369–72.
35. Mura, C. V., M. I. Becker, A. Orellana, and D. Wolff. 2002. Immunopurification of Golgi vesicles by magnetic sorting. *J. Immunol. Methods* 260:263–71.
36. Journet, A., G. Klein, S. Brugière et al. 2012. Investigating the macropinocytic proteome of *Dictyostelium* amoebae by high-resolution mass spectrometry. *Proteomics* 12:241–5.
37. Conner, S. D., and S. L. Schmid. 2003. Regulated portals of entry into the cell. *Nature* 422:37–44.
38. Maxfield, F. R., and T. E. McGraw. 2004. Endocytic recycling. *Nat. Rev. Mol. Cell Biol.* 5:121–32.
39. Garcia, C. K., K. Wilund, M. Arca et al. 2001. Autosomal recessive hypercholesterolemia caused by mutations in a putative LDL receptor adaptor protein. *Science* 292:1394–8.
40. Mellman, I., and Y. Yarden. 2013. Endocytosis and cancer. *Cold Spring Harb. Perspect. Biol.* 5:a016949.
41. Sandgren, S., F. Cheng, and M. Belting. 2002. Nuclear targeting of macromolecular polyanions by an HIV-Tat derived peptide: Role for cell-surface proteoglycans. *J. Biol. Chem.* 277:38877–83.
42. Spillmann, D. 2001. Heparan sulfate: Anchor for viral intruders? *Biochimie* 83:811–7.
43. Platta, H. W., and H. Stenmark. 2011. Endocytosis and signaling. *Curr. Opin. Cell Biol.* 23:393–403.
44. Di Fiore, P. P., and P. De Camilli. 2001. Endocytosis and signaling: An inseparable partnership. *Cell* 106:1–4.
45. Andresen, B. T., M. A. Rizzo, K. Shome, and G. Romero. 2002. The role of phosphatidic acid in the regulation of the Ras/MEK/Erk signaling cascade. *FEBS Lett.* 531:65–8.
46. Boucher, J., A. Kleinridders, and C. R. Kahn. 2014. Insulin receptor signaling in normal and insulin-resistant states. *Cold Spring Harb. Perspect. Biol.* 6:a009191.
47. Van der Meel, R., M. K. Durka, W. W. Van Solinge, and R. M. Schiffelers. 2014. Toward routine detection of extracellular vesicles in clinical samples. *Int. J. Lab. Hematol.* 36:244–53.
48. Mathivanan, S., H. Ji, and R. J. Simpson. 2010. Exosomes: Extracellular organelles important in intercellular communication. *J. Proteomics* 73:1907–20.
49. Simons, M., and G. Raposo. 2009. Exosomes – vesicular carriers for intercellular communication. *Curr. Opin. Cell Biol.* 21:575–81.
50. De Toro, J., L. Herschlik, C. Waldner, and C. Mongini. 2015. Emerging roles of exosomes in normal and pathological conditions: New insights for diagnosis and therapeutic applications. *Front. Immunol.* 6:203.
51. Kalani, A., A. Tyagi, and N. Tyagi. 2014. Exosomes: Mediators of neurodegeneration, neuroprotection and therapeutics. *Mol. Neurobiol.* 49:590–600.
52. Antonyak, M. A., and R. A. Cerione. 2015. Emerging picture of the distinct traits and functions of microvesicles and exosomes. *Proc. Natl. Acad. Sci. U.S.A.* 112:3589–90.
53. Vafai, S. B., and V. K. Mootha. 2012. Mitochondrial disorders as windows into an ancient organelle. *Nature* 491:374–83.
54. Bournat, J. C., and C. W. Brown. 2010. Mitochondrial dysfunction in obesity. *Curr. Opin. Endocrinol. Diabetes Obes.* 17:446–52.
55. Kiebish, M. A., X. Han, H. Cheng, J. H. Chuang, and T. N. Seyfried. 2008. Cardiolipin and electron transport chain abnormalities in mouse brain tumor mitochondria: Lipidomic evidence supporting the Warburg theory of cancer. *J. Lipid Res.* 49:2545–56.
56. Johri, A., and M. F. Beal. 2012. Mitochondrial dysfunction in neurodegenerative diseases. *J. Pharmacol. Exp. Ther.* 342:619–30.
57. Wallace, K. B. 2007. Adriamycin-induced interference with cardiac mitochondrial calcium homeostasis. *Cardiovasc. Toxicol.* 7:101–7.
58. De Duve, C., B. C. Pressman, R. Gianetto, R. Wattiaux, and F. Appelmans. 1955. Tissue fractionation studies. 6. Intracellular distribution patterns of enzymes in rat-liver tissue. *Biochem. J.* 60:604–17.
59. Maenosono, S., and S. Saita. 2006. Theoretical assessment of FePt nanoparticles as heating elements for magnetic hyperthermia. *IEEE Trans. Magn.* 42:1638–42.
60. Thanh, N. T. K. 2011. *Magnetic nanoparticles: From fabrication to clinical applications.* Boca Raton: CRC Press.
61. Kojima, T., M. Mizuguchi, and K. Takanashi. 2016. Growth of $L1_0$-FeNi thin films on Cu(001) single crystal substrates using oxygen and gold surfactants. *Thin Solid Films* 603:348–52.
62. Lee, Y., M. A. Garcia, N. A. F. Huls, and S. Sun. 2010. Synthetic tuning of the catalytic properties of Au-Fe_3O_4 nanoparticles. *Angew. Chem. Int. Ed.* 49:1271–4.
63. Mezni, A., I. Balti, A. Mlayah, N. Jouini, and L. S. Smiri. 2013. Hybrid Au-Fe_3O_4 nanoparticles: Plasmonic, surface enhanced Raman scattering, and phase transition properties. *J. Phys. Chem. C* 117:16166–74.
64. Yu, H., M. Chen, P. M. Rice, S. X. Wang, R. L. White, and S. Sun. 2005. Dumbbell-like bifunctional Au-Fe_3O_4 nanoparticles. *Nano Lett.* 5:379–82.
65. Xu, Y. H., J. Bai, and J.-P. Wang. 2007. High-magnetic-moment multifunctional nanoparticles for nanomedicine applications. *J. Magn. Magn. Mater.* 311:131–4.
66. Choi, J.-S., Y.-W. Jun, S.-I. Yeon, H. C. Kim, J.-S. Shin, and J. Cheon. 2006. Biocompatible heterostructured nanoparticles for multimodal biological detection. *J. Am. Chem. Soc.* 128:15982–3.
67. Trang, N. T. T., T. T. Thuy, K. Higashimine, D. M. Mott, and S. Maenosono. 2013. Magnetic-plasmonic FePt@Ag core–shell nanoparticles and their magnetic and SERS properties. *Plasmonics* 8:1177–84.
68. Wang, H., J. Shen, Y. Li et al. 2013. Porous carbon protected magnetite and silver hybrid nanoparticles: Morphological control, recyclable catalysts, and multicolor cell imaging. *ACS Appl. Mater. Interfaces* 5:9446–53.
69. Takahashi, M., P. Mohan, A. Nakade et al. 2015. Ag/FeCo/Ag core/shell/shell magnetic nanoparticles with plasmonic imaging capability. *Langmuir* 31:2228–36.
70. Takahashi, M., K. Higashimine, P. Mohan, D. M. Mott, and S. Maenosono. 2015. Formation mechanism of magnetic-plasmonic Ag@FeCo@Ag core–shell–shell nanoparticles: Fact is more interesting than fiction. *CrystEngComm.* 17:6923–9.
71. Takahashi, M., P. Mohan, K. Higashimine, D. M. Mott, and S. Maenosono. 2016. Transition of exchange bias from the linear to oscillatory regime with the progression of surface oxidation of Ag@FeCo@Ag Core@Shell@Shell nanoparticles. *J. Appl. Phys.* 120:134301.

Shinya Maenosono (E-mail: shinya@jaist.ac.jp) leads his research group at the Japan Advanced Institute of Science and Technology (JAIST). His research in JAIST has focused on two main areas of interest in the field of materials chemistry and nanotechnology. The first area involved wet chemical synthesis of semiconductor nanoparticles with controlled size, shape and composition for energy conversion device applications. The second area has focused on the synthesis and bioapplication development of monometallic and alloyed multimetallic nanoparticles.

Mari Takahashi is a PhD student at the Japan Advanced Institute of Science and Technology (JAIST). In April 2013, she started her master's studies in JAIST under the supervision of Prof. Shinya Maenosono, working on the synthesis of magnetic–plasmonic hybrid nanoparticles for cellular organelles. In April 2015, she started her PhD studies in the same group to expand her master's work into the magnetic isolation of autophagosomes using magnetic–plasmonic hybrid nanoparticles.

13 Magnetic Nanoparticle-Based Biosensing

*Kai Wu, Diqing Su, Yinglong Feng and Jian-Ping Wang**

CONTENTS

13.1 INTRODUCTION

In recent years, MNPs have been widely used in clinical and medical applications. MNPs are used as contrast agents in magnetic particle imaging (MPI)[1–6] and magnetic resonance imaging (MRI),[7,8] carriers for drug delivery,[9–12] heating sources for hyperthermia cancer treatments,[13–15] labels for magnetic biosensing,[16–23] etc. MNPs-based biosensing strategies have received considerable attention because of their unique advantages over other techniques. For example, iron oxide MNPs are biocompatible, nontoxic, inexpensive to produce, environmentally safe, and physically and chemically stable. In addition, most biological samples exhibit diamagnetic or paramagnetic properties, and thus highly sensitive measurements can be performed in visually obscured samples without further processing.[24]

MNPs are generating much enthusiasm in biotechnology and biomedicine, with particular emphasis in clinical diagnostics and therapy.[25,26] Nowadays, MNPs can be easily conjugated with biologically important constituents such as nucleic acids (i.e., DNA and RNA),[27,28] peptides,[29,30] antibodies,[31] polymers (i.e., PEGs),[32] drugs,[33] tumour markers,[34–36] enzymes[32,37] and other proteins for biological diagnostics and therapy. Figure 13.1 gives a schematic view of a multifunctional MNP.

This chapter will focus on the use of MNPs for the detection of target ligands based on two detection platforms: search coil-based biosensors (volumetric-based biosensing platform, discussed in Section 13.3) and giant magnetoresistor (GMR) biosensor (surface-based biosensing platform, discussed in Section 13.4).

The volumetric-based biosensing platforms measure analytical signals that come from the entire detection volume, which makes assays simple and fast.[38] Representative examples of volumetric sensors include nuclear magnetic resonance (NMR) devices,[39,40] magnetic susceptometers,[38,41] and conventional superconducting quantum interference devices (SQUIDs).[42,43] The volumetric-based sensor has a drawback: the acquired signal is the average of the whole volume.

* Corresponding author.

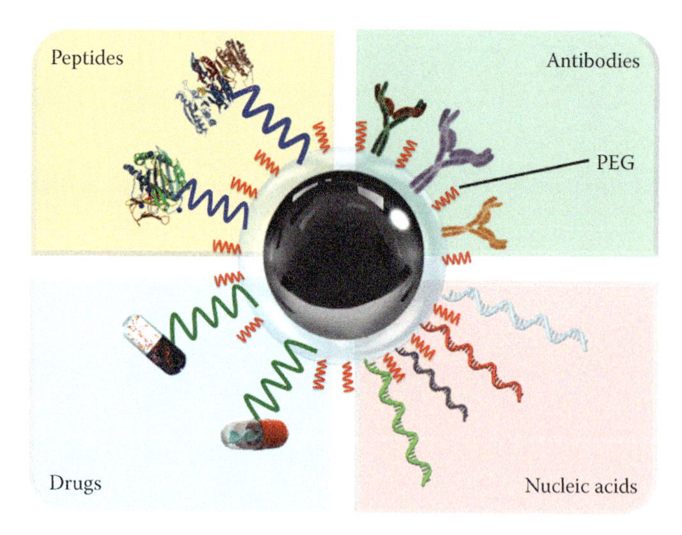

FIGURE 13.1 Schematic representation of multifunctional MNP, polyethylene glycol (PEG) coatings, used to improve solubility and decrease immunogenicity.

Surface-based biosensing platforms directly detect individual magnetic objects near the sensing elements. These sensors generally achieve higher sensitivity and finer resolution than volumetric ones; however, they require target samples to be placed in close proximity to the sensor surface. Such an arrangement limits the assay configuration, and typically causes the assays to be more time-consuming (usually one test takes 15–30 min).[16,44] To date, many different types of magnetometers (i.e., GMR sensors, Hall effect sensors) have been developed as surface-based biosensors.

13.2 BACKGROUND

13.2.1 Search Coil-Based Biosensor

13.2.1.1 Introduction to Search Coil Biosensor

MNP detection for biological and medical applications has been achieved by a variety of sensing schemes. The search coil-based sensing scheme is one of the best candidates among them for future point-of-care (POC) devices and systems because of its unique integrated features:[45] relatively high sensitivity at room temperature, dynamic volume detection (nonsurface binding), intrinsic superiority in measuring alternating current (AC) magnetic field, functionality as an antenna for wireless information transmission, application driven properties such as low cost, portability and ease of use.[19] There are two main streams of detecting schemes based on search coils: one is frequency mixing method,[19,46–55] and the other is mono-frequency method.[56–60] Both methods have demonstrated their feasibilities of using MNPs for immunoassay applications as well as monitoring the *in vivo* binding process.

Dual-frequency search coil (frequency mixing method). In frequency mixing-based search coil detection scheme, a low frequency field is applied to drive the MNPs into the nonlinear region up to saturation[46,47] and a high frequency field is applied to modulate the nonlinearity of the magnetization in the high frequency region where the noise floor is lower. The response signals at combinatorial frequencies are collected. This method will be discussed in detail in the following sections. The Krause, Nikitin and Wang groups have successfully used this dual-frequency search coil scheme for the immunoassay related applications.[46–48]

Mono-frequency search coil. In this method, only one sinusoidal magnetic field is applied to MNPs.[59–61] The induced magnetization is slightly distorted due to Brownian relaxation of the MNPs, which results in harmonics that can be measured. The main difference from the aforementioned mixing frequency search coil system is the use of only one sinusoidal magnetic field, the relaxation process being studied using the phases of the 3rd (3f) and the 5th (5f) harmonics and amplitude ratio of the 5th over the 3rd harmonics. Usually, this kind of apparatus consists of one sinusoidal current driven coil and one differentially rounded pickup coil. The frequency for the sinusoidal magnetic field is generated by a phase lock amplifier which is also used to amplify and record the harmonics.[59–64] The Weaver group has successfully used this mono-frequency search coil detection scheme for immunoassays and for measuring viscosity and temperature.[58–60,65]

13.2.1.2 Superparamagnetism

For sufficiently small MNPs, the magnetic moment can randomly flip direction under the influence of thermal fluctuation, this behaviour being named superparamagnetism. Superparamagnetism occurs for single domain MNPs, which implies average diameters around a few nanometres to tens of nanometres, depending on the material.[66] As the size of the particles increases, single domain develops into a multidomain magnetic structure consisting of different magnetic domains, each one with different orientation of the magnetization separated by domain walls. Figure 13.2 shows the qualitative evolution from superparamagnetic to multidomain along with NP size.

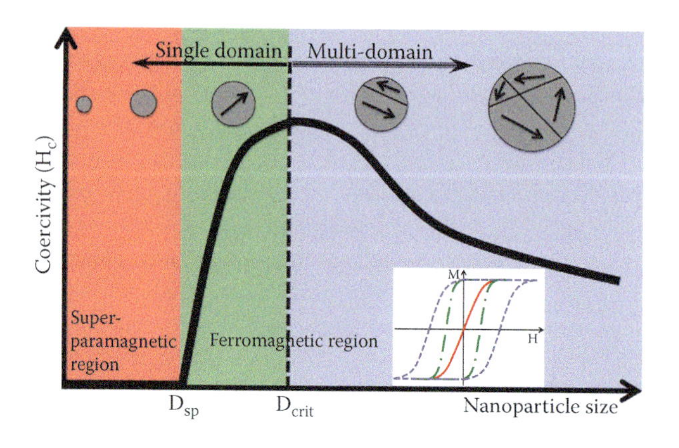

FIGURE 13.2 Transition from superparamagnetic to multidomain region. The inset figure shows qualitative behaviours of the size-dependent coercivity of MNPs. D_{sp} and D_{crit} are, respectively, the transition sizes from superparamagnetic to a blocked state, and from single to multidomain regions.

The sum of all the magnetic moments of individual atoms in a superparamagnetic nanoparticle (SPION) is treated as a single giant magnetic moment (macrospin). For single-domain MNPs with uniaxial anisotropies, the magnetic energy has two minima that are separated by the energy barrier E_B. The orientations of macrospins at the two energy minima are antiparallel to each other and both are aligned in a preferred axis called the easy axis. At finite temperature, thermal fluctuations cause fast flipping of macrospin between two energy minima. The mean time between flips obeys the Néel–Arrhenius law: $\tau_N = \tau_0 \exp(E_B/k_BT)$, where τ_0 is a material-specific relaxation time, which is of the order of 10^{-9}s, k_B is Boltzmann constant and T is temperature in Kelvin. This characteristic time τ_N should be compared with measurement time τ_m. When $\tau_m \gg \tau_N$, the macrospin will flip several times during the measurement and the time-average magnetic moment is zero. In contrast, if $\tau_m \ll \tau_N$, the macrospin of the NP does not change orientation during the measurement and is considered to be blocked. The measured magnetic moment will exhibit apparent ferromagnetic behaviour with characteristic hysteresis in the magnetization curve upon the externally applied magnetic field.[67] In the former case, the NP is superparamagnetic whereas in the latter case it is blocked. A transition between superparamagnetic and blocked states occurs when $\tau_m = \tau_N$. Therefore, this transition depends on the nature of the material, the size and shape of NPs, and thus can be described in terms of the smallest size of NP blocked at a defined temperature or in terms of the temperature where the superparamagnetic transition occurs for a given size of NP.

The transition size D_{sp} from superparamagnetic to blocked states for several ferromagnetic materials, and the critical size D_{crit} between single- and multidomain states can be found in Krishnan.[68] For the SPIONs with blocking temperature much below room temperature, their individual magnetic dipoles are randomly oriented due to thermal fluctuations and they do not display a net magnetic moment unless an external magnetic field is applied.

When a MNP suspension (no interparticle interactions) is exposed to an external magnetic field, their magnetic moments tend to align with the field, leading to a net magnetization. The induced magnetization of the whole system is a function of the applied field. The exact function is complicated but with the assumption that all the particles are identical and all orientations of easy axes are possible, it can be approximated by the Langevin function. The magnetization of the MNP suspension is expressed as the Langevin function of the ratio between the magnetic energy and the thermal energy as:

$$M(t) = m_s cL\left(\frac{m_s H(t)}{k_B T}\right), \tag{13.1}$$

where

m_s is the magnetic moment of each NP
c is concentration of NPs in the sample
L is the Langevin function
H(t) is the externally applied magnetic field

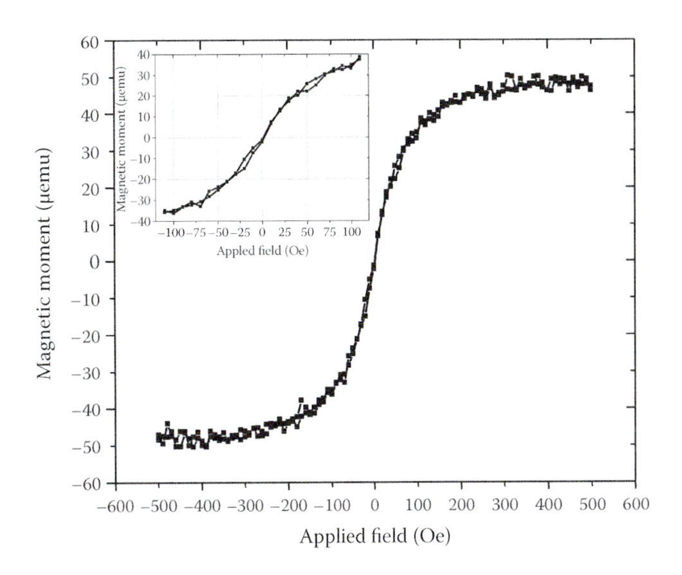

FIGURE 13.3 Magnetization curve of SPIONs suspension measured at room temperature by vibrating sample magnetometer. (Reprinted with permission from Wu, K.; Liu, J.; Wang, Y.; Ye, C.; Feng, Y.; Wang, J.-P., *Applied Physics Letters*, *107* (5), 053701. Copyright 2015, American Institute of Physics.)

Figure 13.3 shows the magnetic response curve of superparamagnetic iron oxide NPs (SPIONs) under an externally applied magnetic field. These SPIONs (Ocean NanoTech Inc.) with an average magnetic core diameter of 25 nm and coated with approximately 4 nm of oleic acid and amphiphilic polymer shells, are dispersed in 0.02% sodium azide in deionized (DI) water. The hysteresis curve of SPIONs shows that their magnetic response to the external magnetic field is linear for small fields (less than 50 Oe) and nonlinear as the magnitude of driving field reaches 100 Oe.

13.2.1.3 Néel and Brownian Relaxation Mechanisms

Due to the magnetic anisotropy in each MNP, the magnetic moment has usually only two stable orientations antiparallel to each other, separated by an energy barrier.[69] At finite temperature, there is a finite probability for the magnetic moment to flip and reverse its direction due to thermal fluctuation. The flip process is called relaxation and the mean time between two flips is called relaxation time. There exist two relaxation mechanisms for MNPs under AC magnetic fields: Néel and Brownian relaxations. The Néel relaxation process is the rotation of macrospin inside a stationary particle, whereas the Brownian process is the rotation of the entire particle along with its macrospin. The magnetization dynamics of MNPs are usually characterized by effective relaxation time τ_{eff}, which is dependent on both Brownian relaxation time τ_B and Néel relaxation time τ_N.[70–72] Both relaxation processes are dependent on the frequency and amplitude of the applied magnetic fields.[63,73,74] The effective relaxation time τ_{eff} of a NP governs its ability to follow the external fields, and it is related to the

Brownian and Néel relaxation time as follows, considering that the mechanisms are independent (this mathematical model is a rough approximation, more precise models can be found from the literature; see Usov and Liubimov):[75]

$$\frac{1}{\tau_{eff}} = \frac{1}{\tau_B} + \frac{1}{\tau_N}. \tag{13.2}$$

Zero-field Néel relaxation time is expressed as:

$$\tau_N = \tau_0 \exp\left(\frac{KV_M}{k_B T}\right), \tag{13.3}$$

where

K is the effective anisotropy constant of the NP

$V_M = \pi D^3/6$ is the magnetic core volume of particle, D is the diameter of the magnetic core

Zero-field Brownian relaxation time is expressed as:

$$\tau_B = \frac{3\eta V_H}{k_B T}, \tag{13.4}$$

where

η is the viscosity of the MNP suspension

$V_H = \pi(D + d)^3/6$ is the hydrodynamic volume of the particle, d is twice the thickness of the coating layer

Although Equations 13.3 and 13.4 are zero-field relaxation time models, they also apply for low field approximation (≤100 Oe). Note that $\tau_{eff} \approx \tau_B$ for larger MNPs due to the fact that τ_N increases more rapidly than τ_B as the particle size increases. The simulated Néel, Brownian, and effective relaxation time as functions of magnetic core diameter D are plotted in Figure 13.4 for an assembly of SPIONs dispersed in a solution with viscosity η = 1 cP – tested at temperature T = 300 K – with a coating layer thickness of d = 4 nm and an anisotropy constant K = 90000 ergs/cm^3. For these SPIONs with magnetic core diameter smaller than 20 nm, the overall relaxation process is dominated by the Néel process, whereas, the Brownian relaxation mechanism dominates if the core diameter exceeds 20 nm.

Brownian relaxation of SPIONs has been proposed and investigated for a variety of biosensing applications.[48,49,58,65,76–78] Due to their sensibility to the characteristics of the suspension environment such as temperature[58,77] and viscosity of solution,[55] and hydrodynamic volume of particle.[76]

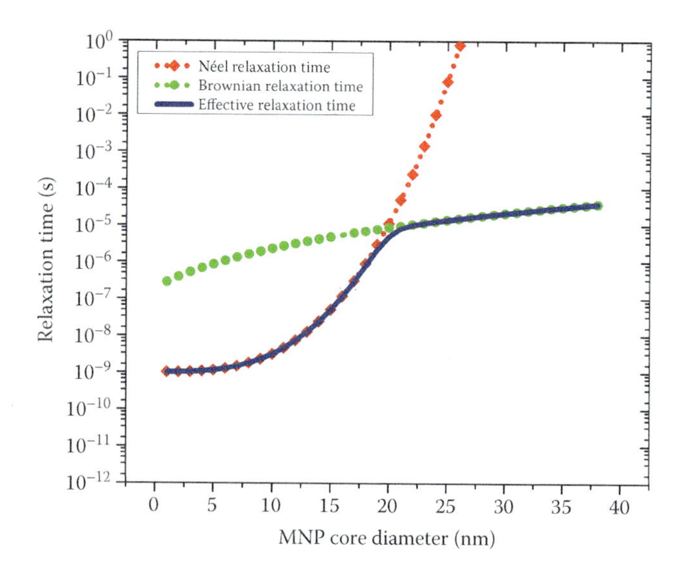

FIGURE 13.4 Néel, Brownian, and effective relaxation time as function of magnetic core diameters for SPIONs dispersed in a solution with viscosity η = 1 cP and tested at temperature T = 300 K. (Reprinted with permission from Wu, K.; Liu, J.; Wang, Y.; Ye, C.; Feng, Y.; Wang, J.-P., *Applied Physics Letters*, *107* (5), 053701. Copyright 2015, American Institute of Physics.)

13.2.1.4 Nonlinear Magnetic Response

Let us assume a MNP suspension is exposed to a mixture of high frequency and low frequency AC magnetic fields expressed as:[46,47]

$$H(t) = A_L \cos(2\pi f_L t) + A_H \cos(2\pi f_H t), \tag{13.5}$$

where

A_L and A_H are the amplitude of low and high frequency magnetic fields

f_L and f_H are the frequency of low and high frequency magnetic fields

The amplitude of the low frequency magnetic field is high enough to drive SPIONs into the nonlinear region periodically. The high frequency magnetic field is responsible for driving the induced harmonics to the high frequency region where a lower pink noise (1/f noise) is expected.[46] The magnetic responses of MNPs exposed to magnetic field H at frequencies f_L and f_H is schematically depicted in Krause et al.[46] The Fourier transform of the response signal exhibits not only the excitation components f_L and f_H but the additionally odd harmonics $3f_L$, $5f_L$, …$3f_H$, $5f_H$, and odd frequency mixing harmonics $f_H \pm 2f_L$ (the 3rd harmonics), $f_H \pm 4f_L$ (the 5th harmonics), etc.[46,47,50,54,69,79]

Herein, we formulate a mathematical description of these manifest relationships. Taylor expansion of the Langevin function in Equation 13.1 is shown below:[19,49,52]

$$\begin{aligned}\frac{M(t)}{m_s c} &= L\left(\frac{m_s H(t)}{k_B T}\right)\\ &= \frac{1}{3}\left(\frac{m_s}{k_B T}\right)H(t) - \frac{1}{45}\left(\frac{m_s}{k_B T}\right)^3 H(t)^3\\ &\quad + \frac{2}{945}\left(\frac{m_s}{k_B T}\right)^5 H(t)^5 + \cdots\\ &= \cdots + \left[-\frac{1}{60}A_H A_L^2\left(\frac{m_s}{k_B T}\right)^3 + \frac{1}{252}A_H^3 A_L^2\left(\frac{m_s}{k_B T}\right)^5\right.\\ &\quad \left. + \frac{1}{378}A_H A_L^4\left(\frac{m_s}{k_B T}\right)^5 + \cdots\right]\\ &\quad \bullet \cos\left[2\pi\left(f_H \pm 2f_L\right)t\right]\\ &\quad + \left[\frac{1}{1512}A_H A_L^4\left(\frac{m_s}{k_B T}\right)^5 + \cdots\right]\\ &\quad \bullet \cos\left[2\pi\left(f_H \pm 4f_L\right)t\right] + \cdots\end{aligned} \tag{13.6}$$

In Equation 13.6, the odd harmonics at combinatorial frequencies are derived. However, the phase delay of magnetization M(t) to the low frequency applied driving field H(t) is not included. If the relaxation time rises to the point that it cannot be ignored, the phase delay of the magnetization has to be considered. Let us assume that the particle's magnetization has a phase delay ϕ_H to the high frequency field and a phase delay ϕ_L to the low frequency field, and assume ϕ_H is independent of the low frequency field. The major mixing frequency components are revised as follows:

$$\begin{aligned}\frac{M(t)\big|_{f_H+2f_L}}{m_s c} &= \left[\frac{1}{60}A_H A_L^2\left(\frac{m_s}{k_B T}\right)^3 + \cdots\right]\\ &\quad \bullet \cos\left[2\pi\left(f_H + 2f_L\right)t - \phi_H - 2\phi_L + \pi\right],\end{aligned} \tag{13.7}$$

$$\begin{aligned}\frac{M(t)\big|_{f_H-2f_L}}{m_s c} &= \left[\frac{1}{60}A_H A_L^2\left(\frac{m_s}{k_B T}\right)^3 + \cdots\right]\\ &\quad \bullet \cos\left[2\pi\left(f_H - 2f_L\right)t - \phi_H + 2\phi_L + \pi\right].\end{aligned} \tag{13.8}$$

According to Faraday's law of induction, collected voltage signal from the pick-up coil has a 90° phase shift from the magnetization of particles. The phases of the mixing frequencies at $f_H \pm 2f_L$ from the voltage signal are:

$$\phi_{f_H \pm 2f_L} = -\left(\phi_H \pm 2\phi_L - \frac{3\pi}{2}\right). \tag{13.9}$$

The relaxation phase delay ϕ_H and ϕ_L can therefore be calculated by:

$$\phi_H = -\frac{\phi_{f_H+2f_L} + \phi_{f_H-2f_L} + \pi}{2}, \tag{13.10}$$

$$\phi_L = -\frac{\phi_{f_H+2f_L} - \phi_{f_H-2f_L}}{4}, \tag{13.11}$$

When f_L is so low that $\phi_L \approx 0$, the phase at the 3rd harmonic will have a simple relationship with the phase at the high frequency:

$$\phi_H = -\phi_{f_H \pm 2f_L} - \frac{\pi}{2}. \tag{13.12}$$

The above derivations are based on the assumption that ϕ_H is a constant number independent of the low frequency field. However, as is aforementioned, the phase delay, or the relaxation time process characteristic of the SPION, is modulated by the offset field, thus ϕ_H cannot be approximated as a static value. To better understand the phase modulation, the low frequency field can be considered as a quasistatic offset field. The overall magnetization is a combination of the quasistatic low-frequency magnetization and the high frequency magnetization.[69] The phase delay ϕ_H to the high-frequency field depends on the absolute amplitude of the offset field. When this offset field is swept with a frequency of f_L, the time-varying phase delay $\phi_H(t)$ will be modulated with a base frequency of $2f_L$, and with a modulation factor α:[19]

$$\phi_H(t) = \phi_{H0} + \alpha \times \cos(4\pi f_L t) + \cdots, \tag{13.13}$$

where

ϕ_{H0} is the high frequency phase delay at zero offset field

Because of the complexity of the magnetization under two AC magnetic fields, when the modulation factor α is large, ϕ_H does not hold the physical meaning of high frequency magnetization phase delay any longer. In this chapter, ϕ_{H0} is used as a rough estimation of the phase delay to the high frequency field.

It is worthwhile to mention the Debye model which is valid for small amplitude and low frequency AC field.[80] When the frequency of the AC magnetic field is low, the particle's magnetization can follow the excitation field tightly, and the susceptibility χ is a real number. As the excitation frequency increases, the particle's magnetization cannot follow the excitation field, and the relaxation process introduces a phase in the complex AC susceptibility. The relationship between relaxation time τ_{eff} and phase delay ϕ of AC susceptibility can be calculated using Debye model:[81]

$$\chi(\omega) = \frac{\chi_0}{1 + j\omega\tau_{eff}} = \frac{\chi_0}{\sqrt{1 + (\omega\tau_{eff})^2}} e^{-j\tan^{-1}(\omega\tau_{eff})} = |\chi| e^{-j\phi}, \quad (13.14)$$

$$\phi = \tan^{-1}(\omega\tau_{eff}), \quad (13.15)$$

where
χ_0 is the static susceptibility
ω is the angular frequency

13.2.2 Magnetic Field Sensors for Biosensing

This section will mainly review chip-based magnetic biosensors for POC sensing in medical application. These sensors use MNPs as labels to correlate electrical signal to the concentration of the analytes. The magnetic biosensor has advantages of high sensitivity, wide dynamic range, biocompatibility between magnetic particles and human cells[82,83] and most importantly, it is matrix insensitive, which means minimal interference from magnetic background in biological samples. Herein, magnetoresistance-based (MR) biosensors and Hall sensors are introduced, and a brief comparison between these magnetic biosensors is shown in Table 13.1.

GMR sensor. After the discovery of the giant magnetoresistance effect in 1988, the GMR sensor has been well developed and successfully used in hard disk drive heads since late 1990s. GMR sensors, also known as spin-valves, have demonstrated the advantage of their high sensitivity to satisfy the challenge of increasing data densities as well as data rates. Almost at the same time, in 1998, researchers in the Naval Research Laboratory first demonstrated the use of GMR sensors as biosensors to detect bindings in bimolecular level, such as bindings between antigen–antibody, DNA–DNA, etc. They also demonstrated the systematic view for such biosensors. Magnetic particles are used as tags to infer the number of captured analytes, and the Wheatstone bridge is incorporated in the biosensing system to readout and amplify the electrical signal. Researchers kept most of these features later in their research.

The GMR effect is a quantum mechanism effect caused by spin-dependent scattering of conduction electrons with different spin relative to that of the ferromagnetic layers. A GMR sensor consists of a multilayer structure of alternating ferromagnetic and nonmagnetic layers (see Figure 13.5).

The magnetization of the fixed layer is defined by thermal magnetic field annealing via an antiferromagnetic exchange coupling between the fixed layer and the antiferromagnetic layer under it. However, the magnetization of the free layer can rotate freely in response to the externally applied field.

A sandwich structure-based immunoassay is the typical approach in GMR biosensor applications. The capture antibody is firstly immobilized on the sensor surface, then the analyte of interest is added into the solution to specifically bind to the capture antibody, magnetic labelled detection antibodies are

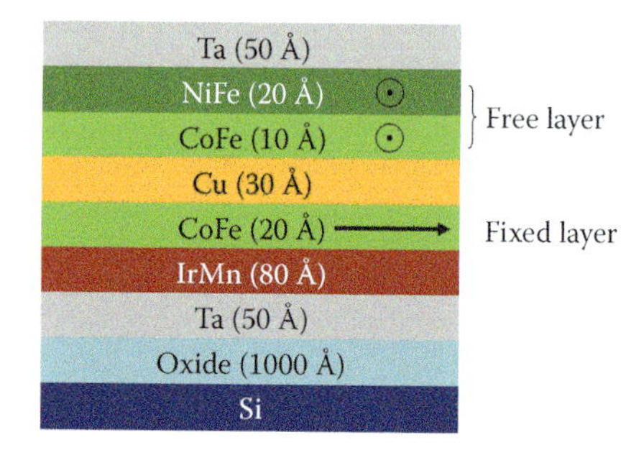

FIGURE 13.5 Typical GMR stack with a multilayer structure consisting of alternative ferromagnetic layers and nonmagnetic layers. The magnetic configuration of GMR sensors is typically defined in 90 degrees in order to have a linear response region of resistance to external field.

TABLE 13.1
Comparison between GMR and Hall Biosensors

	Magnetic Particles	Detection Limit	Notes	References
GMR sensor	250 nm Nanomag®-D particles	1 fM	DNA assay	84
GMR sensor	12.8 nm FeCo particles	600 copies of streptavidin; ~zM	High moment FeCo MNPs, Protein assay	20
GMR sensor	50 nm Fe_2O_3 MACS particles	5 pM	Protein assay	85
GMR sensor	50 nm Fe_2O_3 MACS particles	0.04 (ng/mL)	Protein assay	16
GMR sensor	50 nm Fe_2O_3 MACS particles	~50 aM	Protein assay	86
GMR sensor	500 nm particles	0.8 pM	Protein assay	87
MTJ sensor	16 nm Fe_3O_4	100 nM	DNA assay	18
Hall sensor	~200 nm magnetic beads	1.8 nT/(Hz)$^{1/2}$	DNA assay; field sensitivity	23

applied into the solution to form the complete sandwich structure form the complete sandwich structure (see Section 13.4.2).

MTJ sensor. Magnetic tunnel junction (MTJ) has a similar stack structure compared to a GMR sensor, except that the nonmagnetic metal layer has been replaced with an ultrathin insulation oxide barrier layer composed of Al_2O_3 and/or MgO, which are the most common barrier layers. Similar to the GMR effect, electrons will experience spin-dependent tunnelling, and the junction will show different resistance according to the different configurations of the magnetizations in the top and bottom magnetic layers.

MTJ sensors have a much higher MR ratio compared to that of GMR sensors (the highest MR ratio for MTJ is reported to be 604% at room temperature by S. Ikeda and H. Ohno's group of Tohoku Univeristy[88]; the highest MR ratio for GMR is reported to be 110% at room temperature[89]), which results in a higher sensitivity especially at smaller field. However, the requirements for biological applications such as a larger biological functionalization surface of the sensor and low drift will not favour MTJ. Large area MTJ is not easy to fabricate because of the high quality requirement of the oxide-insulating layer; a single pinhole defect will burn the device. These are the main reasons why people rarely use MTJ sensors for immunoassay related biological applications.

Hall sensor. The classic Hall effect describes the phenomenon of generation of a voltage difference across a current-carrying electrical conductor placed in a magnetic field. The Hall sensor has been extensively used as a magnetic field sensor in many applications. The Hall sensor has demonstrated its feasibility of detecting magnetic beads and biological molecules.[22,23,90,91] A typical method of magnetic bead detection is the AC phase tracking method as shown in Figure 13.6. A constant DC current is applied to drive the Hall sensor. A DC magnetic field H_0 (100–200 Oe) is applied to the sensor perpendicularly. An AC magnetic field (several hundred Hz) is added in either z-direction or x-direction. The harmonic term of the Hall voltage will be measured with a lock-in amplifier, which is proportional to the magnetic induction produced by the bead.

The Hall sensor is further developed by optimizing semiconductor materials to improve field sensitivity.[23] The Hall sensor array also demonstrated the feasibility of multiplexing testing,[90] combined with advantages such as no requirements for washing and purification steps, which makes the Hall sensor an attractive candidate for biological detection.

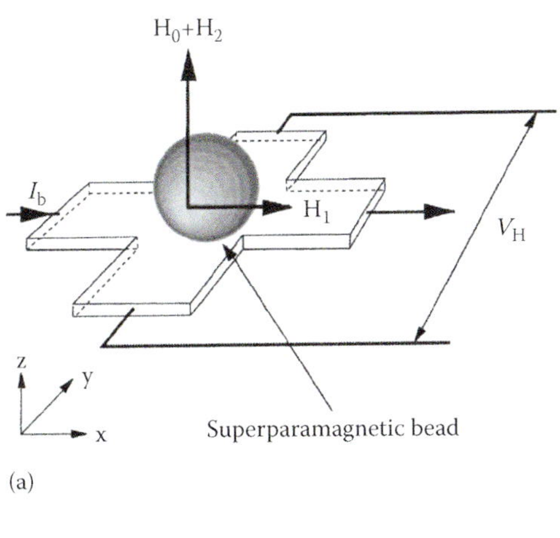

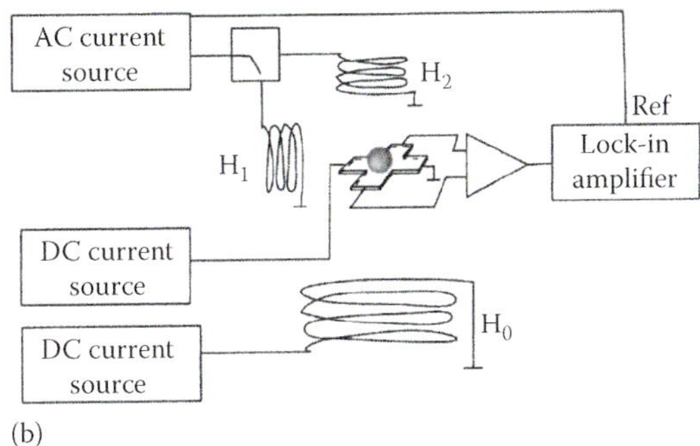

FIGURE 13.6 Testing setup of detection of a single bead using a Hall sensor. (a) A single superparamagnetic nanoparticle is centered on a Hall sensor, biased with the current I_b. A DC magnetic field H_0 is applied perpendicular to the sensor. An AC field is added either in the z direction (H_2) or in the x direction (H_1). (b) Schematic view of the measurement setup. The Hall voltage is measured with a lock-in amplifier.[92] (Reprinted with permission from Besse, P.-A.; Boero, G.; Demierre, M.; Pott, V.; Popovic, R., *Applied Physics Letters*, *80* (22), 4199–4201. Copyright 2002, American Institute of Physics.)

13.3 SEARCH COIL BIOSENSORS

The search coil biosensor is one type of volumetric-based biosensing platform. In this immunoassay system, the surfaces of MNPs dominated by the Brownian relaxation process are coated with probes (such as detection antibody, protein, DNA, or any other types of ligands) that can specifically bind with the target ligands. The specific binding between probes and targets increases the hydrodynamic sizes of MNPs, and as a result the phase delay of magnetization increases and the frequency spectrum of odd harmonics decrease. The number/concentration of target ligands from the fluidic sample is proportional to the difference in phase delay and frequency spectrum before and after adding the samples.

13.3.1 Search Coil-Based Immunoassays

13.3.1.1 3D Immunoassays

Figure 13.7 shows the schematic system setup and signal chain of a typical search coil-based biosensing system.[19,48–50,52–55] A PC with data acquisition card (DAQ) generates two sinusoidal waves, which are amplified by two instrument amplifiers (IAs). Band-pass-filters (BPFs) are used to suppress higher harmonics introduced by IAs. Then filtered waves drive excitatory coils to generate oscillating fields: a high frequency field f_H and a low frequency field f_L. The response signals at combinatorial frequencies $f_H + 2f_L$ (the 3rd harmonic) and $f_H + 4f_L$ (the 5th harmonic) are collected. One pair of differentially wound pick-up coils collects induced voltage and phase signals from MNPs and sends them back to a band-stop filter (BSF) before being digitalized on the DAQ. The entire setup is controlled via a PC using the LABVIEW program to control the instrument and the MATLAB® program for signal processing.

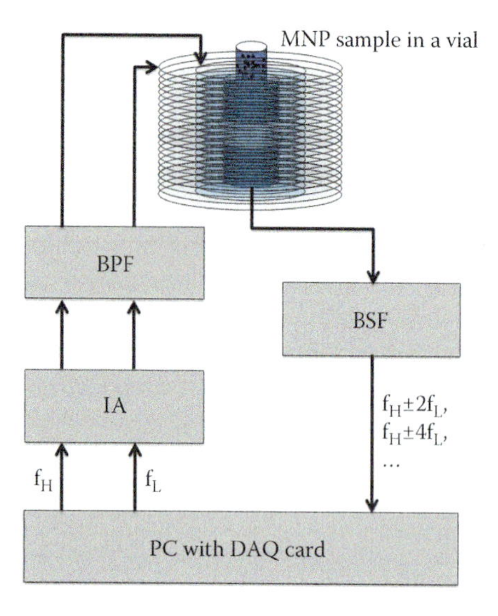

FIGURE 13.7 General schematic setup of a search coil detection system. The entire system used here records the nonlinear magnetic response of MNPs under two different AC magnetic fields.

Three commercially available SPION samples (Ocean NanoTech Inc.) are used for Brownian relaxation-based immunoassay studies: SHP35 (SPIONs with 35 nm magnetic core diameter, 4 nm oleic acid, and amphiphilic polymer coatings, 0.1 mL, 5 mg/mL in H_2O carboxylic acid solution), IPG35 (SHP35 conjugated with around 10 nm protein G layer, 0.1 mL, 1 mg/mL), and IPG35-Ab (IPG35 conjugated with around 10 nm goat antihuman IgG-HRP with ratio 1:100, 0.1 mL, 1 mg/mL). The phase delays from these three SPIONs are measured under AC magnetic fields. Phase delays of the frequency mixing components are measured by scanning the frequency of f_H up to 10 kHz. The results are averaged over 10 independent experiments and plotted in Figure 13.8a. Experimental results are fitted by the Debye model.

The hydrodynamic size distributions of these three SPION samples are measured by a dynamic light scattering (DLS) instrument (Brookhaven Instruments Corp., Holtsville, NY) and plotted in Figure 13.8b. The wide hydrodynamic size distribution of IPG35-Ab indicates that IPG35 binds with antibodies in varying degrees of binding affinities.

The binding process of the antibodies to protein G coated IPG35 SPIONs is detected by monitoring real-time phase delay at the 3rd harmonic (shown in Figure 13.8c). Instead of the time-consuming whole frequency scan, the frequency f_H is fixed at 4 kHz. At 50 seconds, 0.05 mL antibody (goat antihuman IgG HRP conjugated, 5 mg/mL) is added into 0.1 mL of IPG35 sample. As is shown in Figure 13.8d, the antibodies gradually bind to protein G on the surface of IPG35 SPIONs until all the binding sites are occupied by antibodies. The hydrodynamic size gradually increases due to the conjugation of antibodies onto IPG35 SPIONs, and as a result the phase

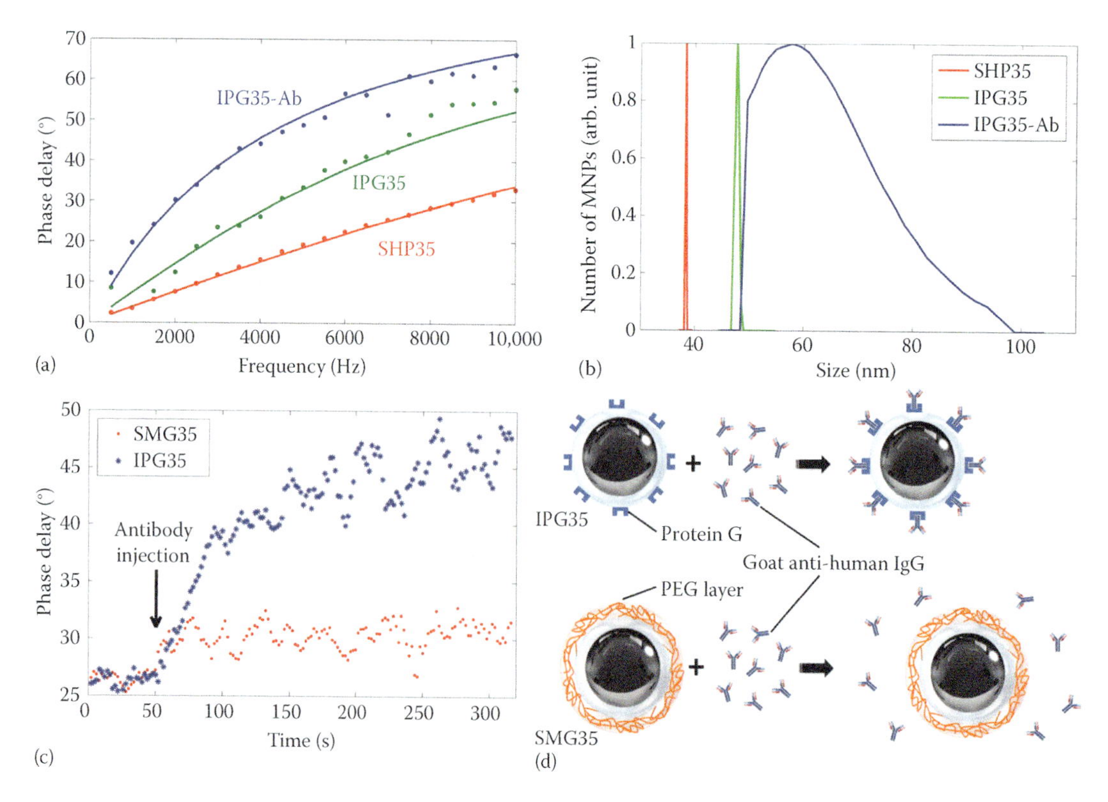

FIGURE 13.8 (a) Experimental plots (dots) of phase delay of the 3rd harmonic along scanning frequency f_H for three SPION samples in water solution. The Debye model (solid lines) is plotted to compare with the experimental data. (b) Size distribution of SHP35, IPG35, and IPG35-Ab samples measured by DLS. (c) Real-time measurement of phase delay of IPG35 and control sample SMG35, with antibody injection at 50 seconds. (d) Schematic view of the binding process between goat antihuman IgG and protein G. PEG layer on SMG35 effectively blocks the binding of IgG. (Reprinted with permission from Tu, L.; Jing, Y.; Li, Y.; Wang, J.-P., *Applied Physics Letters*, *98* (21), 213702. Copyright 2011, American Institute of Physics.)

delay of magnetization increases and gradually reaches a plateau, which is very close to that of the standard IPG35-Ab. The negative control experiment is conducted by adding 0.05 mL antibodies to 0.1 mL of SMG35 sample (SHP35 conjugated with around 10 nm PEG layer, 1 mg/mL). The phase delay measurement in Figure 13.8c shows very little increase because the PEG coating layer on SMG35 SPIONs effectively blocks the nonspecific binding with antibodies. The slight increase is possibly due to the higher viscosity from the abundant antibodies in the solution.

The experimental results show that this search coil system accurately detects the binding between the protein G and antibodies in real time. This study shows the potential capability of this biosensing platform for fundamental biological research, disease diagnostic and drug discovery.

13.3.1.2 Quasi-3D Immunoassays

Nikitin's group[93] have proposed and demonstrated a 3D fibre filter-based magnetic immunoassay device in which the fibre serves as a solid phase to provide large reaction surface, quick reagent mixing, and antigen immunofiltration directly during the assay. By introducing a nontransparent 3D fibre structure, the reaction surface is tremendously increased compared to the surface-based biosensing platform we have introduced in Section 13.2.2. However, there is still a gap between this design and the volumetric-based biosensing in which NPs serve as reaction surfaces (see Section 13.2.1). Herein, in this section, we name this experimental design as Quasi-3D Immunoassay.

By employing a sandwich-type immunoassay structure and nonlinear magnetic response of MNPs, they tested out staphylococcal enterotoxin A (SEA) and toxic shock syndrome toxin (TSST) from neat milk without sample preparation. The limits of detections (LODs) are 4 and 10 pg/mL for TSST and SEA obtained from 30 mL samples, respectively. TSST and SEA are widely present in the environment and they are frequently responsible for diverse serious and potentially fatal illnesses such as severe gastrointestinal diseases and toxic shock. One of the most common ways to contract staphylococcal enterotoxins (SEs) is through contaminated foods and water. Thermal processing such as pasteurization or heating kills only bacteria and does not affect SEs. Thus, the detection of SEs in food should be done by immunological methods.

The 3D filters are put into tips compatible with standard automatic pipets. There are two testing formats in regards to how the fluidic sample is dispensed. Figure 13.9a shows the design of express magnetic immunoassay (Express MIA) which is intended for analysis of smaller sample volumes; the sample is dispensed simultaneously through all tips in cycles by a modified 12-channel electronic pipet and each cycle lasts for 30 s. Figure 13.9b shows the design of high-volume magnetic immunoassay (HV MIA) in which a larger sample volume is pumped through the solid phase by a peristaltic pump. Sample volume is determined by the time and rate of pumping and, in practice, can be virtually unlimited.

In this design, cylindrical porous 3D filters (see Figure 13.9c, and surface morphology of fibre filters in Figure 13.9e)

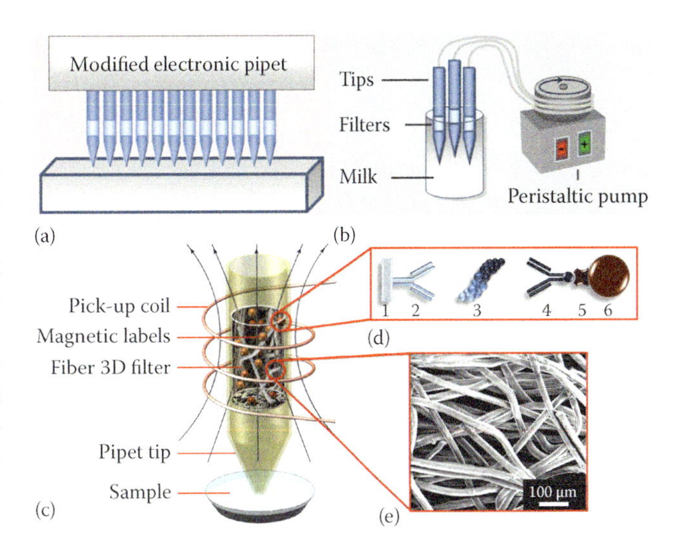

FIGURE 13.9 (a) Express MIA setup. (b) High volume MIA setup. (c) The 3D solid phase located inside a pipet tip. (d) Magnetic immunosandwich on 3D fibre filter used as a solid phase: 1, filter surface; 2, capture antibody; 3, antigen; 4, biotinylated tracer antibody; 5, streptavidin; 6, SPIONs. (e) Surface morphology of cylindrical 3D fibre filters obtained with scanning electron microscope (SEM).[94] (Reprinted with permission from Orlov, A. V.; Khodakova, J. A.; Nikitin, M. P.; Shepelyakovskaya, A. O.; Brovko, F. A.; Laman, A. G.; Grishin, E. V.; Nikitin, P. I., *Analytical Chemistry*, *85* (2), 1154–1163. Copyright 2012 American Chemical Society.)

are made of twisted polyethylene fibres and coated by polypropylene (PP). Each 3D filter is a cylinder with external diameter of 3 mm and height of 5 mm. The sandwich immunoassay structure is schematically shown in Figure 13.9d and described as follows: (1), the filter surfaces are sequentially put into 96% ethanol for 10 min, 50% ethanol for 10 min, washed in 0.1 M bicarbonate buffer pH 9.4 twice, incubated in a vortex overnight in 0.1 mL per filter of capture antibody (25 μg/mL in bicarbonate buffer), and washed with 0.1% solution of Tween-20 in a phosphate-buffered saline (PBS) for 2 min. Then the filter surfaces are blocked by 5% bovine serum albumin (BSA) in PBS for 1 h followed by washing in 0.1% solution of Tween-20 in PBS for 2 min. (2), monoclonal antibodies to TSST and SEA are produced in ascites fluid from inbred specific pathogen-free (SPF) mice BALB/c. These antibodies are further purified by combination of ammonium sulphate fractionation and ion-exchange chromatography on a mono Q column.[95] (3), either SEA or TSST are spiked into milk samples. The spiked toxin concentrations varied from 1 pg/mL to 100 ng/mL. (4), the monoclonal antibodies are biotinylated using NHS-LC-Biotin obtained from Pierce. (5) and (6) are the streptavidin coated SPIONs (50 nm) obtained from Magnisense SE, France.

In the testing process, after all the samples have passed through the cylindrical porous 3D filters, each filter is washed by 400 μL of 0.1% solution of Tween-20 in PBS for 1 min. Then followed by 7 min dispensing of biotinylated tracer antibody (15 μg/mL in PBS, 150 μL per tip). Finally, 0.1 mg/mL solution of streptavidin-coated SPIONs of 100 μL per tip are dispensed for 5 min with a further washing procedure.

The bounded SPIONs will be sensed by the search coil system. The detection principle is based on recording the nonlinear response of SPIONs in a combined magnetic field generated at two frequencies f_H and f_L. The amplitude of the low frequency field of f_L = 20 Hz is up to 70 Oe, whereas the amplitude of the high frequency field $f_H \approx$ 100 kHz does not exceed 6 Oe. The magnetic response is measured at the 3rd harmonics ($f_H + 2f_L$). The output signals are proportional to the quantity of NPs inside the pick-up coils. Each measurement cycle lasts for 4 s and the values are averaged over 12 measurement cycles.

The calibration curves of SEA and TSST from Express MIA and HV MIA are plotted in Figure 13.10. Calibration of Express MIA is carried out with sample volume of 150 μL and total assay time is 25 min. For HV MIA, a total sample volume of 30 mL is pumped through each tip at a flow rate of 0.33 mL/min, the assay time is 2 h. From calibration curves shown in Figure 13.10, the LOD is determined by 2σ criterion where the specific signal exceeded the double standard deviation of the signal from negative control samples (milk without toxin).

Table 13.2 gives a brief comparison between several immunoassay methods for SE detections. The improved assay performance is due to the highly sensitive search coil-based SPION detection and 3D fibre filters employed as a solid phase. This design is simple, rapid, and affordable and can be applied for food safety control, *in vitro* diagnostics, and veterinary applications, etc.

13.3.2 Viscosity Measurements

Recall that in Section 13.2.1.3, we have introduced two relaxation mechanisms in SPIONs. For those SPIONs with core diameters larger than 20 nm, Brownian relaxation mechanism will dominate. Here, we rewrite the effective relaxation time τ_{eff} for Brownian relaxation process-dominated SPIONs as follows:

$$\tau_{eff} \approx \tau_B = \frac{3\eta V_H}{k_B T}. \tag{13.16}$$

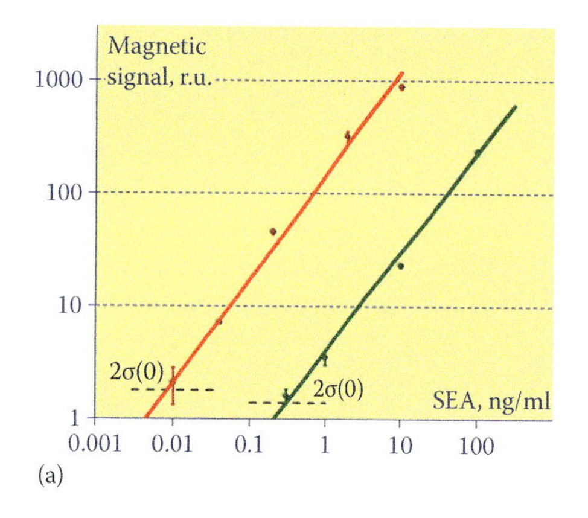

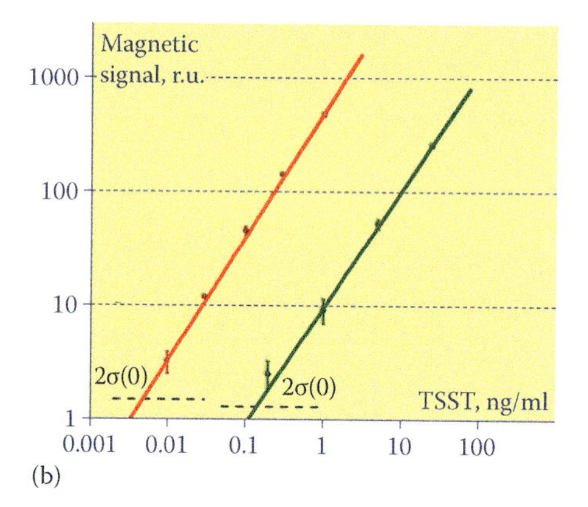

FIGURE 13.10 Calibration curves of (a) SEA and (b) TSST. Red line: HV MIA. Green line: Express MIA.[94] (Reprinted with permission from Orlov, A. V.; Khodakova, J. A.; Nikitin, M. P.; Shepelyakovskaya, A. O.; Brovko, F. A.; Laman, A. G.; Grishin, E. V.; Nikitin, P. I., *Analytical Chemistry*, *85* (2), 1154–1163. Copyright 2013 Copyright 2012 American Chemical Society.)

TABLE 13.2
Comparison of Several Immunoassay Methods for SE Detection

Assay Types	LOD	Assay Time[a]	Note	Reference
Enzyme-Linked Immunosorbent Assay (ELISA)	28.2 pg/mL for SEA	2 h	Commercialized.	96,97
VIDAS™ SET2 (bioMérieus, Marcy-l'Etoile, France)	<0.5 ng/g for SEA and SEB. <1 ng/g for SEC_2. ~1 ng/g for SED and SEE.	80 min	Multiplexing. Commercialized.	98,99
RIDASCREEN SET kit (R-Biopharm GmbH, Darmstadt, Germany)	0.20–0.75 ng/g for SEs	<3 h	Multiplexing. Commercialized.	100
Fluorescent hydrogel biochip	0.1–0.5 ng/mL for SEs	2 h	Multiplexing.	95
Cantilever sensors	50 pg/mL for SEB	N/A		101
Evanescent wave fluorescence array biosensors	0.5 ng/mL	30 min	Multiplexing.	102
Search coil-based sensors	10 pg/mL for SEA	2 h		93

[a] Assay time does not include sample preparation required for all these methods.

According to Faraday's law, the induced voltage in a pair of pick-up coils is expressed as:

$$u(t) = -S_0 v \frac{d}{dt} M(t) \quad (13.17)$$

where

v is the total volume of MNP solution

S_0 is the sensitivity of pick-up coils

A sampling rate of F_s = 500 kHz is performed to collect discrete voltage signal according to the Nyquist criterion. By performing discrete-time Fourier transform (DFT) on the discrete-time voltage signal, the spectra density *vs.* frequency is collected.[50,51] In this chapter, we use 'amplitude' to represent the spectral density, which along with the phase delay (or phase lag) information are used to describe the harmonic signals generated from MNPs.

For Brownian relaxation process-dominated SPIONs, as the viscosity of SPION solution increases, the effective relaxation time τ_{eff} increases, phase delay at the 3rd harmonic increases, and the amplitude detected at the 3rd harmonic decreases. It is worth mentioning that this relationship chain works well when SPIONs are Brownian relaxation process-dominated. Herein, we repeat the simulation described in Figure 13.4 and vary the viscosity η from 1 to 1000 cP. The simulation result from Figure 13.11 shows that SHP-25 are Brownian relaxation dominated when η < 100 cP, as the viscosity exceeds 100 cP, the magnetic response of these SPIONs gradually changes to Néel relaxation dominated process, which is no longer susceptible to the viscosity of solution.

In this section, SHP-25 is used for the measurement of liquid viscosity. Upon the stimulation of AC fields, these MNPs will realign the magnetic moments to the fields through Brownian relaxation process. Thus, by keeping the hydrodynamic volumes of MNPs and the temperature of the solution unchanged, we are able to estimate viscosity of any liquid by analysing the harmonic signals generated from MNPs that are evenly mixed in that liquid.[52,54,55]

To plot a calibration curve, nine liquid solutions are prepared by mixing glycerol (Sigma–Aldrich, concentration ≥99%, density 1.25 g/mL) and DI water in different volume ratios (listed in Table 13.3). G1-4W means the volume ratio of glycerol to DI water is 1 to 4. The rest are named in the same manner. These mixtures are sonicated for 2 h before being mixed evenly and then viscosities are pretested by the AR-G2 rheometer (TA instruments) as standard values. The system setup is given in Figure 13.7. A high frequency field of 15 kHz and amplitude of 10 Oe along with a low frequency field of 50 Hz and amplitude of 100 Oe are applied in this experiment. By adding DI water and glycerol mixtures, MNPs will experience a higher viscosity environment and the effective relaxation time increases, hence the phase delay increases and the amplitudes at induced harmonics decrease as a result.

Herein, the amplitude at the 3rd harmonic is collected in real time before and after adding the mixtures so as to cancel out the noise induced by the unbalanced pick-up coil, white noise, and 1/f noise. During the testing process, the background noise floor signal is collected for 20 s, after which a plastic vial containing 50 μL of SHP-25 is inserted into the pick-up coil. The induced harmonic signal is collected for 20–30 s. Then 200 μL of glycerol-DI water mixture is added to the MNP sample and the signal is collected for another 30–50 s. Three independent experiments are carried out for each mixture. The median is chosen from three independent tests to plot the real-time amplitude trend graph of the 3rd harmonic in Figure 13.12.

As is clearly shown in Figure 13.12a, the amplitude signal at the 3rd harmonic drops to varying degrees for different mixtures. The 3rd harmonic amplitude-drops in percentage ΔA% for different viscosities at 300 K are averaged over three independent experiments and plotted in Figure 13.12b. The calibration curve plotted in Figure 13.12b can be used as a standard diagram for determining viscosities of any unknown liquid. By comparing the harmonic signal before and after adding an unknown liquid, the corresponding ΔA%

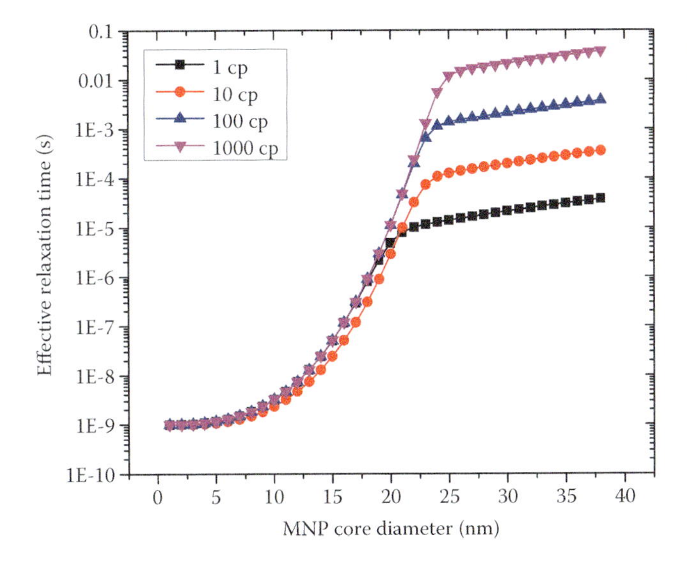

FIGURE 13.11 Simulated effective relaxation time as a function of viscosity.

TABLE 13.3
Viscosities of DI Water and Glycerol Mixtures at 300 K

Mixture	Viscosity (cP)
DI water	0.99
G1-4W	1.95
G1-3W	2.29
G1-2W	3.36
G1-1W	7.94
G2-1W	22.99
G3-1W	44.78
G4-1W	136.58
Glycerol	1087.43

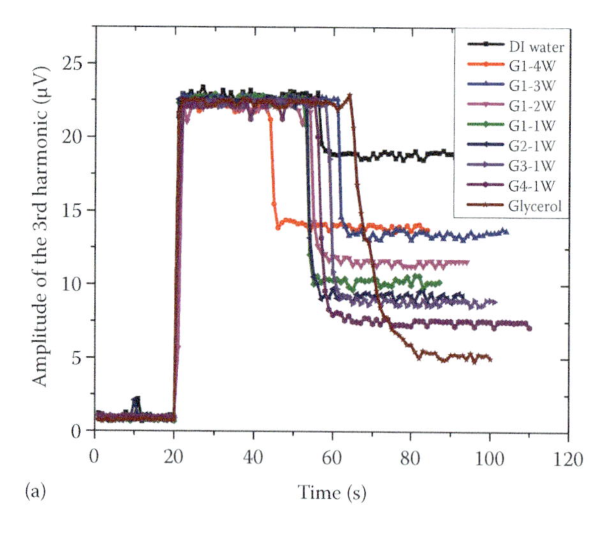

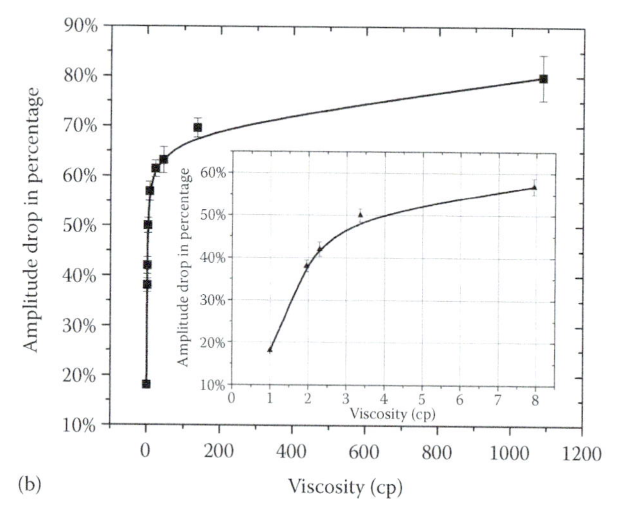

FIGURE 13.12 (a) Real-time amplitude signal of the 3rd harmonic for SHP25 in 9 mixtures tested at 300 K. (b) Amplitude drop in percentage *vs.* viscosity for SHP25 in 9 mixtures at 300 K.[52] (Reprinted with permission from Wu, K.; Liu, J.; Wang, Y.; Ye, C.; Feng, Y.; Wang, J.-P., *Applied Physics Letters*, *107* (5), 053701. Copyright 2015 American Institute of Physics.)

is calculated, and then by comparing this data with our calibration curve we can find out the corresponding viscosity.

A serum sample (male human serum type AB, Sigma–Aldrich) is tested. This product consists of haemoglobin ≤ 20 mg/dL and endotoxin ≤ 10 EU/mL. 50 μL of SPION solution and 200 μL of serum are tested at 300 K. The $\Delta A\%$ of serum sample is calculated to be 36.9%. By fitting this data into the calibration curve, the viscosity is estimated to be 1.8 cP at 300 K. While the standard viscosity value of this serum sample is 1.74 cP tested by AR-G2 rheometer. The whole *in vitro* viscosity measuring process takes less than 1.5 min and achieves an error rate lower than 5%.

This experimental design features kits for determining the viscosity of any liquids. The kits include: (i) Brownian relaxation process-dominated MNPs; (ii) a standard diagram which is based on a plot of the amplitude-drop with viscosity in percentage of this type of MNPs; (iii) a search coil system capable of picking up the harmonic signals; and optionally (iv) disposable plastic vials for holding the MNPs and liquid.

13.4 GMR BIOSENSORS

The GMR biosensor is one type of surface-based biosensing platform. Let us take a sandwich structure-based GMR biosensing system as an example. The GMR sensor surface is biofunctionalized with capture antibodies, which will specifically recognize and bind with target ligands from fluidic samples, then MNP tagged with detection antibodies are added. The detection antibodies will specifically bind to target ligands thus forming a sandwich structure and immobilizing MNPs to the near surface of GMR sensors. By applying an external magnetic field to the system, the stray fields from those MNPs alter the magnetization of the free layer of GMR stacks, and as a result an electrical signal is read out as an indication of the abundance of target ligands from the sample.

13.4.1 Fabrication of GMR Sensors

In general, the fabrication process of GMR sensors can be divided into two parts, namely, the deposition of GMR multilayers and the patterning of the deposited GMR films. In this section, a commonly used process flow will be introduced to give readers a comprehensive understanding of how to make GMR sensors that can be integrated with the biological processes discussed in the next section.

Magnetron sputtering technique is often used to deposit GMR stacks. In this technique, a large electric field (a few hundred to a few thousand electron volts) is employed to generate a plasma between two electrodes. The cathode surface is immersed in a magnetic field such that the electrons around it can be trapped and keep interacting with the gas molecules, which not only speeds up the ionization process, but also reduces the pressure needed to generate the plasma.[103] Then, the positively charged ions in the plasma are accelerated by the electric field and bombard the target on the cathode with energy high enough to kick out the target atoms. Some of the atomized metal targets will land on the substrate and form the desired film. The GMR stacks are often grown on Si/SiO_2 wafers and typically contain several metallic layers.[21] Other depositing techniques, such as ion beam deposition and electrodeposition have also been reported.[104]

Once we have a silicon wafer with GMR stacks on it, the next step is the patterning process, which basically includes definition of the shape of the sensor, deposition of electrodes along with the circuits, and deposition of the passivation layers. Some techniques that are employed in the patterning process will be introduced first, followed by a schematic illustration of the whole process flow.

Photolithography is a technique that is used several times in the patterning process. During the photolithography process, the sample is coated with a light-sensitive polymer called photoresist. The solubility of photoresist will either increase or decrease when exposed to light, which is known as positive or negative photoresist, respectively. When light is shining through a mask onto the photoresist, only the solubility of the parts exposed to light will change. Consequently, by immersing the exposed photoresist into a certain developer, the pattern on the mask will be transferred to the photoresist. In the GMR sensor fabrication process, two types of photolithography processes are commonly used, namely, contact lithography and projection lithography. In contact lithography, the resist is in direct contact with the mask, while in projection lithography, the image is projected at a distance from the mask, resulting in a reduced pattern on the sample. In general, projection lithography has a higher resolution, but is also more expensive than contact lithography.

In order to etch the GMR film into a desired shape, an anisotropic etching process is needed to remove the area that is not covered by the photoresist. Since a wet etching process usually involves corrosive chemicals and is isotropic, it is not an ideal choice for GMR fabrication. To minimize the damage and optimize the precision of the geometry, dry etching, especially ion milling, is often used. In ion milling, a plasma is created. The ions can kick out the target atoms under a large acceleration electric field. By adjusting the angle between the injected ions and the sample, one can easily control the magnetic property of the sensors.[105]

In most cases, a chip sensor may contain several or even tens of sensors under serial and parallel connections. Therefore, wiring processes are needed. The wires are made of metals with high conductivity such as gold and copper, which, just like GMR stacks can be deposited by magnetron sputtering. Other physical vapour deposition techniques, such as e-beam evaporation, which uses a bunch of electrons to vaporize the metal target, are also appropriate candidates.

To prevent current leakage and other environmental damage to the device, passivation layers are needed. Oxides such as Al_2O_3 and SiO_2 are often deposited on the sensor after the wiring process. Chemical vapour deposition (CVD) is mostly used for the deposition of SiO_2 due to its high deposition rate and moderate cost. In this technique, gases like SiH_4 and $SiCl_2H_2$ can react with O_2 and N_2O at high temperature to produce SiO_2. Since this process is isotropic, the resulting film can have excellent step coverage, which is good for the passivation of the device. However, the quality of the film deposited by CVD is low due to the large deposition rate compared to that deposited by e-beam evaporation.

Atomic layer deposition (ALD) is the most common way for the deposition of Al_2O_3. Reactive gases are also used in ALD, but unlike those in CVD, they are inserted as several pulses. In the case of Al_2O_3, the film surface will first be functionalized with a hydroxyl group. Then a pulse of trimethylaluminium (TMA) is induced to react with these sites and create Al-O bonding. After all the sites are consumed, the remaining TMA will be pumped out of the chamber and a single layer of Al_2O_3 is deposited on the substrate. The above process will be repeated multiple times to get the desired thickness of Al_2O_3 layer.

A typical process flow of GMR sensor fabrication is shown in Figure 13.13.

13.4.2 Surface Biofunctionalization

A sandwich structure consisting of capture antibody, target antigen, biotinylated detection antibody and streptavidin labelled MNPs is employed in the surface biofunctionalization process.[21] Sandwich structure is the most commonly used detection scheme for immunoassay application.

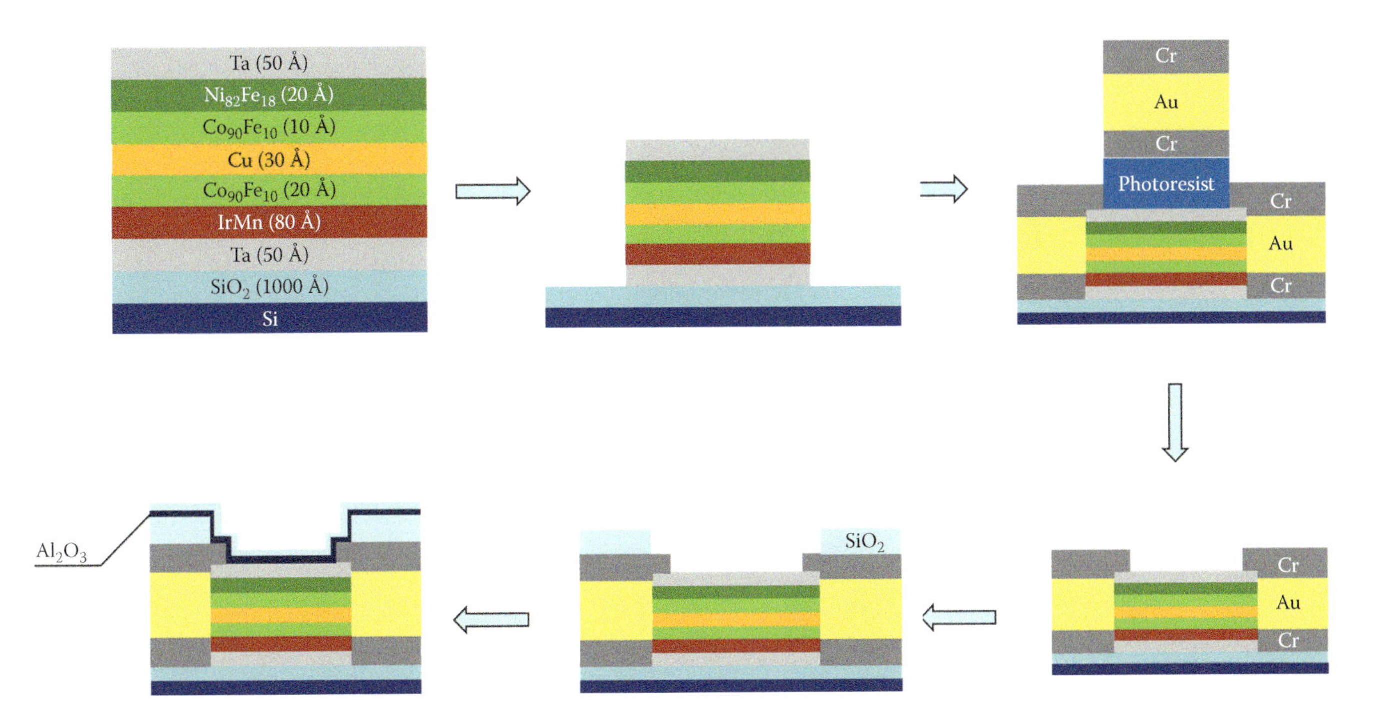

FIGURE 13.13 Process flow of GMR sensor fabrication.

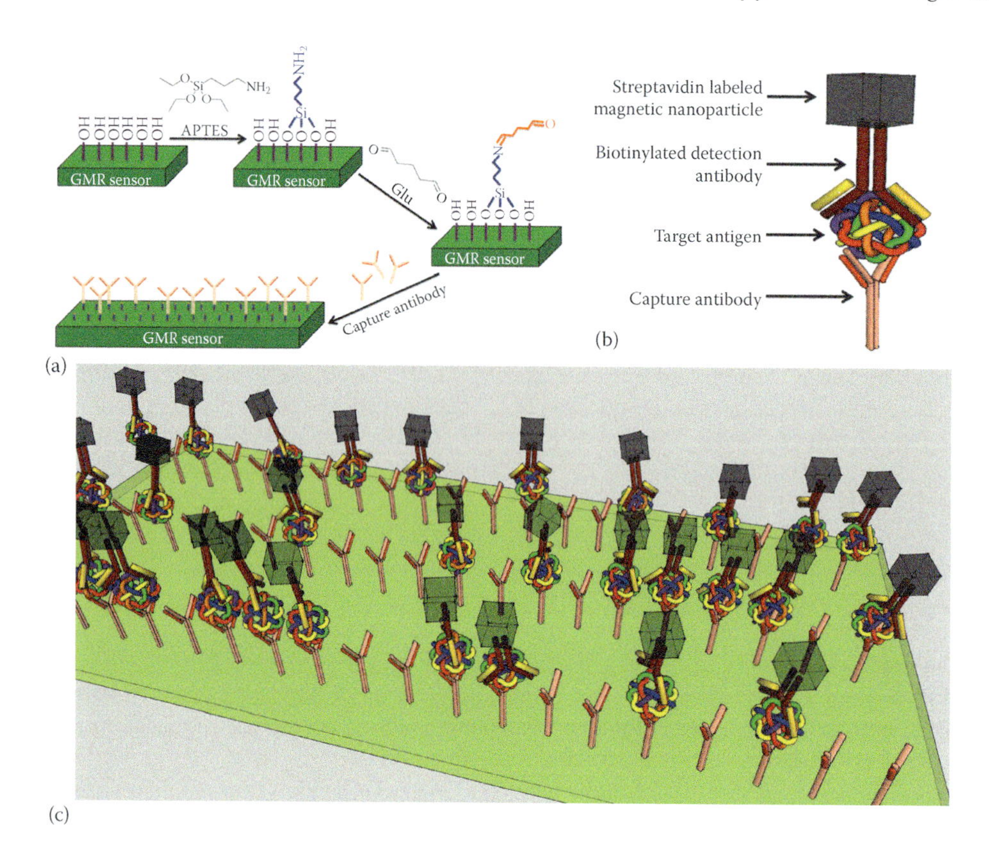

FIGURE 13.14 (a) Schematic illustration of surface biofunctionalization process. (b) Schematic drawing of typical sandwich structure. (c) Schematic illustration of sandwich structure-based immunoassay.[21] (From Krishna, V. D.; Wu, K.; Perez, A. M.; Wang, J.-P., *Frontiers in Microbiology*, *7*, 400, 2016. Copyright 2016 Frontiers. Reprinted with permission.)

As demonstrated in Figure 13.14a, the chip is first immersed in 1% 3-aminopropyltriethoxysilane (APTES) solution, allowing for the APTES to bind with the hydroxyl group on the SiO_2 surface. The sensor area is then covered with glutaraldehyde and incubated for several hours to introduce aldehyde groups to the system.[106] The terminal aldehyde groups generated on the sensor surface allow subsequent covalent bonding of biomolecules containing amino groups onto the GMR sensor.[16,107]

Next, the capture antibody is robotically printed and immobilized on the sensor surface due to the covalent bonding between the amino groups on the protein and the aldehyde groups induced by previous functionalization. In a multiplex sensing system, some sensors are printed with bovine serum albumin (BSA) instead of capture antibody, serving as the negative control group. Then, a reaction well made of polymethyl methacrylate (PMMA) is attached to the chip and loaded with BSA to reduce nonspecific surface binding sites. After incubating, BSA is removed, followed by the addition of fluidic samples (containing one or more types of antigens, macromolecular proteins, peptides, etc.) and detection antibodies. Finally, the streptavidin labelled MNPs (Miltenyi Biotec, Inc., Auburn, CA, USA; Catalog No. 130-048-101) are added and bond with the detection antibody through the biotin-streptavidin interaction (see Figure 13.14b and c). During the real-time measurement process, the reaction well attached on the GMR sensor is preloaded with 30 μL of PBS solution to wet the sensor surface. After 15 min of stabilization, 30 μL of streptavidin labelled MNP solution is added into the reaction well and signals are collected for another 35 min.

13.4.3 Detection Principle

A GMR effect is observed in structures with several ferromagnetic layers separated by nonmagnetic layers. When the ferromagnetic layers are magnetized in opposite direction, the electrical resistance will be much higher compared with those magnetized in the same direction. This effect can be explained by the two-channel model. In this model, the conductivity in metallic layers is assumed to be the result of up-spin and down-spin conducting channels in parallel configuration.

The most commonly used GMR structure in biomedical detection is the spin valve (SV) structure due to its high field sensitivity and linear response to applied field. In SV structure, GMR sensors are often patterned into rectangular bars such that the magnetization of the free layer lies along the longitudinal direction because of the shape anisotropy. Meanwhile, the magnetization of the pinned layer is defined along the transverse direction by applying a field during post-annealing process.

In biomarker detection, only the sites with target antigens are tagged with MNPs. Then, these NPs are excited by an external magnetic field, and the resulting fringe field is detected by the GMR sensors. Thus, the signal generated by GMR sensors is directly related to the number of the target antigen on the sensor surface, which leads to quantitative concentration analysis of the biomarkers. As for the choice of NPs, ferromagnetic NPs tend to agglomerate during the detection process due to the magnetostatic interaction; SPIONs, with zero net moment in the absence of field, are ideal candidate for biodetection process. During a real detection process, a bias field is often needed to bias the sensor to its most sensitive region.

13.4.4 GMR-Based Immunoassays

13.4.4.1 Detection System Setup and Signal Measurement

In this work, each GMR chip has a size of 16 mm × 16 mm with 8 × 8 sensor array in its centre (see Figure 13.15a and b). Each sensor has a size of 120 μm × 120 μm containing five GMR strip groups connected in series and each group contains 10 GMR strips connected in parallel (see Figure 13.15c). Each strip with a size of 120 μm × 750 nm is separated by a 2 μm gap (see Figure 13.15d).

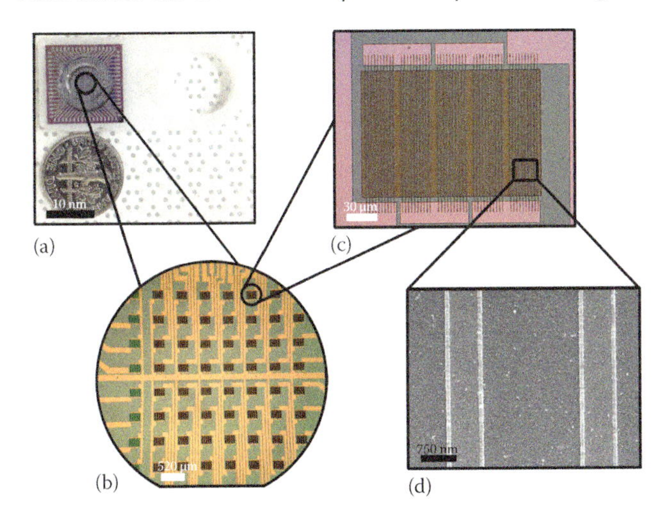

FIGURE 13.15 (a) Fabricated GMR chip. (b) 8 × 8 sensor array. (c) One GMR biosensor structure. (d) GMR strip structure.[21] (From Krishna, V. D.; Wu, K.; Perez, A. M.; Wang, J.-P., *Frontiers in Microbiology*, 7, 400, 2016. Copyright 2016 Frontiers. Reprinted with permission.)

A well-developed lab-based bench top system is introduced here. In this system (see Figure 13.16), a probe station with a 17 × 4 pin array is connected to the pads of a GMR chip.[16] An alternating current with frequency of 1 kHz flows through the main bus. An in-plane sinusoidal magnetic field with amplitude of 30 Oe and frequency of 50 Hz is applied along the short axis direction.

13.4.4.2 Multiplex Immunoassay

Usually, a single protein biomarker is not qualified to specifically and sensitively identify the disease at an early stage. Therefore, the use of a panel of biomarkers to identify at-risk individuals with adequate confidence is needed for early diagnosis of disease.

In this section, a multiplex immunoassay method using GMR biosensors is introduced. Three promising candidate biomarkers for early diagnostic of cardiovascular disease are investigated: pregnancy-associated plasma protein A (PAPP-A),[108,109] proprotein convertase subtilisin/kexin type 9 (PCSK9),[110,111] and suppression of tumorigenicity 2 (ST2).[112]

For multiplex experiments with three different proteins on one GMR chip, BSA (10 mg/mL, serve as negative control group) and three capture antibodies (PAPP-A, PCSK9, and ST2 at 500 μg/mL) are robotically spotted onto individual sensors. Each experiment sensor is covered with 1.2 nL protein, and the negative control group sensor is covered with 1.2 nL BSA. The detailed process has been discussed in Section 13.4.2. Each fluidic sample with the volume of 100 μL containing multiple target proteins is pipetted into the reaction well and incubated for 1 h at room temperature. All of the sensors on the chip are exposed to the same multiple proteins

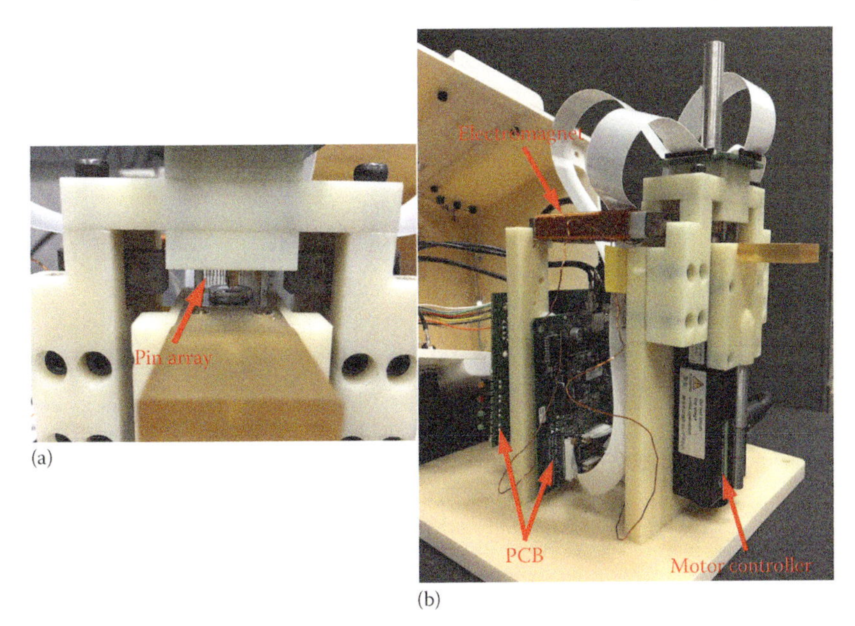

FIGURE 13.16 (a) Photograph of pin array aligning and moving close to the electrode pads of a GMR chip. (b) Schematic design of the bench top station.

solution and different capture antibodies on different sensors can bind to their corresponding antigens.

Three protein biomarkers, PAPP-A, PCSK9, and ST2, are chosen as model protein analytes, which are simultaneously and sensitively analyzed by a lab-based GMR biosensing platform. The limit of detection (LOD) of this assay[16] for PAPP-A is 1 ng/mL with a dynamic detection range up to 4 orders (1 ng/mL–10 μg/mL). The dynamic ranges for PCSK9 and ST2 assays are also up to 4 orders of magnitude, and their LODs are 433.4 pg/mL and 40 pg/mL, respectively. It is also noteworthy that this system can be used for other types of multiplexed immunoassay panels such as cancer and infectious disease detection. Furthermore, the possibility of analyzing biomarkers in blood serum is also demonstrated, which shows a promising application prospect in real clinical settings.

13.4.5 Magnetic Detection of Virus

Influenza A viruses (IAVs) are common respiratory pathogen infecting many hosts including humans, pigs (swine influenza virus or SIV) and birds (avian influenza virus or AIV). Monitoring of swine and avian influenza viruses in the wild, in farms, and in live bird markets is critical for detection of newly emerging influenza viruses with significant potential impacts on human and veterinary public health. Various technologies have been developed for rapid, sensitive, and specific detection of viruses using nanotechnology-based approaches.[113,114] These technologies use NPs in combination with electrical or electrochemical detection.[115–119] This section will introduce a sensitive detection scheme for influenza virus using GMR biosensors. Using SIV H3N2v as representative virus, the LOD of GMR biosensor array is 150 $TCID_{50}$/mL viruses.

Swine IAV strain H3N2v and negative control (mock) are treated with 1% octylphenoxypolyethoxyethanol (IGEPAL CA-630) to disrupt virus particles before detection by GMR biosensor and enzyme linked immunosorbent assay (ELISA). Influenza A capture antibody (MAB8800; EMD Millipore Corporation, Temecula, CA, USA), which is specific to IAV nucleoprotein, is robotically printed in a volume of 1.2 nL to each sensor using the sci-FLEXARRAYER S5 (Scienion AG, Germany). A mouse anti-influenza A monoclonal antibody specific to IAV NP is chosen as capture antibody (MAB8257B; EMD Millipore Corporation, Temecula, CA, USA).

Results are compared with the ELISA method. The real-time binding curves for influenza virus are shown in Figure 13.17a, and averaged signals in Figure 13.17b. The signal from negative control group does not show any obvious rise, which indicates that the signals are specific to IAV. The LOD of GMR sensor for detection of IAV is around 150 $TCID_{50}$/mL compared to 250 $TCID_{50}$/mL by ELISA.

To further confirm the binding of MNPs to the GMR sensor, these GMR chips are investigated by field-emission gun

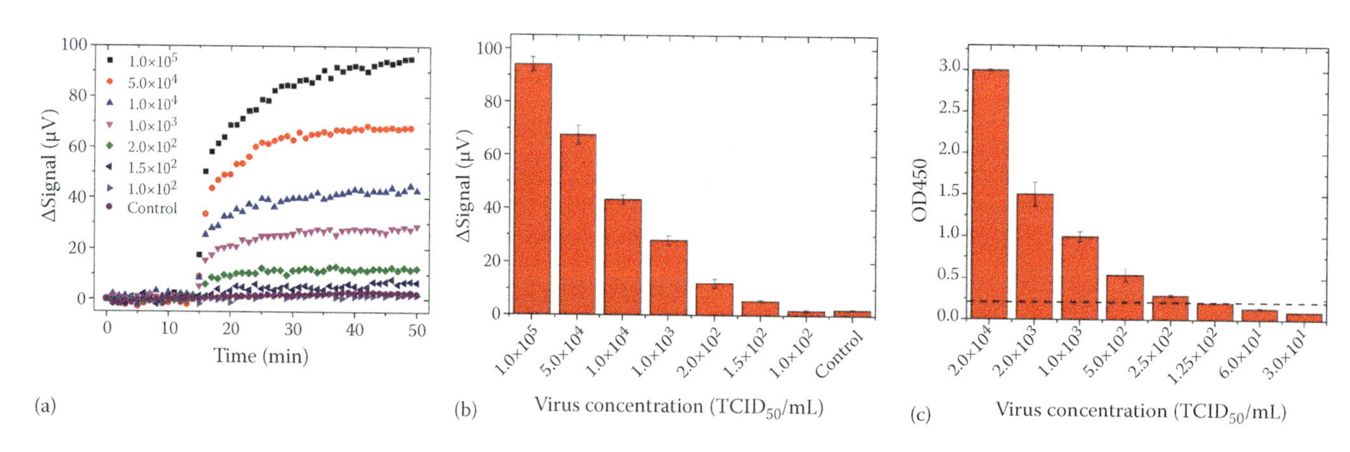

FIGURE 13.17 (a) Binding curves in real time on GMR biosensor. (b) Signals averaged over the last 10 data points from different concentrations of IAV and mock in GMR biosensor. (c) OD450 signal from ELISA with different concentration of IAV. Dotted line indicates the cut-off value.[21] (From Krishna, V. D.; Wu, K.; Perez, A. M.; Wang, J.-P., *Frontiers in Microbiology*, 7, 400, 2016. Copyright 2016 Frontiers. Reprinted with permission.)

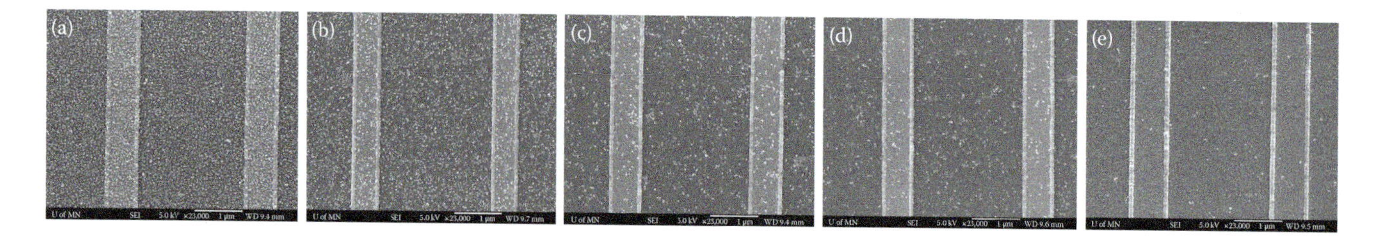

FIGURE 13.18 FEG-SEM images of MNPs bound onto GMR sensors after immunoassays. (a) Biotin-BSA positive control. (b) 1.0×10^4 $TCID_{50}$/mL virus. (c) 1.0×10^3 $TCID_{50}$/mL virus. (d) 2.0×10^2 $TCID_{50}$/mL virus. (e) BSA negative control.[21] (From Krishna, V. D.; Wu, K.; Perez, A. M.; Wang, J.-P., *Frontiers in Microbiology*, 7, 400, 2016. Copyright 2016 Frontiers. Reprinted with permission.)

scanning electron microscopy (FEG-SEM). GMR chips are rinsed with DI water to wash away unbound MNPs and dried by nitrogen gas. Then, these chips are coated with 5 nm of platinum (Pt) and observed by FEG-SEM. As shown in Figure 13.18, the number of bound MNPs per unit area increases as the concentration of influenza virus increases.

In summary, this work extends the application of GMR-based assay for virus detection. As nucleoprotein is localized within the virus particle, a non-ionic, nondenaturing detergent IGEPAL CA-630 is used to disrupt the virus particles in the sample. The results demonstrated that the GMR biosensor is able to detect viral concentration with dynamic range of 3 orders of magnitude (1.5×10^2 to 1.0×10^5 $TCID_{50}$/mL). This is relevant to nasal samples of infected swine, which has been reported to contain 10^3 to 10^5 $TCID_{50}$/mL viral particles.[120]

13.4.6 Magnetic Detection of Mercury Ions

GMR sensing technology has also been introduced into the pollutant monitoring area. Wang et al. has reported a sensing strategy employing GMR biosensor and DNA chemistry for the detection of mercury ion (Hg^{2+}).[121] Mercury ion can specifically bind between two DNA thymine bases and lead to the formation of a thymine- Hg^{2+}-thymine (T-Hg^{2+}-T) pair.[122] Herein, this T-Hg^{2+}-T complex chemistry and complementary DNA with deliberately designed T-T mismatches are introduced and combined with a GMR biosensing system for sensitive and selective Hg^{2+} detection. The detection architecture is similar to the sandwich DNA hybridization assay[84,123–125] and is briefly illustrated in Figure 13.19.

The capture DNA oligomers are immobilized onto the GMR sensor surface followed by biotin-labelled DNA (biotin-DNA) with T-T mismatches to capture DNA and Hg^{2+}. In the absence of Hg^{2+}, biotin-DNA would rarely be hybridized to the immobilized capture DNA because of their mismatched base pairs. In contrast, the biotin-DNA can be bound and hybridized to the GMR sensor surface with the presence of Hg^{2+} due to the T-Hg^{2+}-T complex. The amount of bound biotin-DNA is expected to increase as the amount of Hg^{2+} ions increases, and finally lead to an increased number of MNPs after streptavidin-labelled MNPs are bound to the GMR sensor surface via the biotin-streptavidin interaction.

This assay takes advantage of the high selectivity of thymine-thymine (T-T) pair for Hg^{2+} and achieves a LOD of 10 nM in both buffer and natural water, which is the maximum mercury level in drinking water regulated by U.S. Environmental Protection Agency (EPA). The dynamic range for Hg^{2+} detection is up to 3 orders of magnitude (10 nM to 10 μM). Furthermore, as a versatile and strong contender in molecular diagnostics, GMR bioassay can be applied not only in Hg^{2+} detection, but also has great potential for monitoring other pollutants in environmental and food samples.

13.4.7 Competition-Based Magnetic Bioassays

Based on the detection principle of GMR sensors, the sensitivity of a GMR sensor strongly depends on the distance (d) between the centre of the MNP and the free layer of the sensor[20,124]: the magnetic signal is proportional to $1/d^3$. Most GMR detections of protein-based biomolecules employ a three-layer (sandwich structure) approach, as shown in Figure 13.14b. In the foregoing sections, a sandwich structure-based detection scheme has been reported and applied for different kinds of assays. This detection scheme increases the distance between the MNP and sensor by incorporating a detection antibody, compared to a two-layer approach (see Figure 13.20b and c), thus compromising the detection sensitivity. Theoretically, this two-layer approach could enhance the detection sensitivity by 3.4–39 times,[124] depending on the size and orientation of the antibodies and biomolecules.

Li et al.[124] compared the LOD of these two detection schemes by quantifying interleukin-6 (IL-6). By using the same device, the LOD of IL-6 is 20.76 pM using the three-layer approach and it is 373 fM in the two-layer approach, which is 55 times more sensitive than the three-layer approach. These results confirmed the superior detecting sensitivity of the two-layer approach over the traditional three-layer approach. Furthermore, this competition-based two-layer approach has

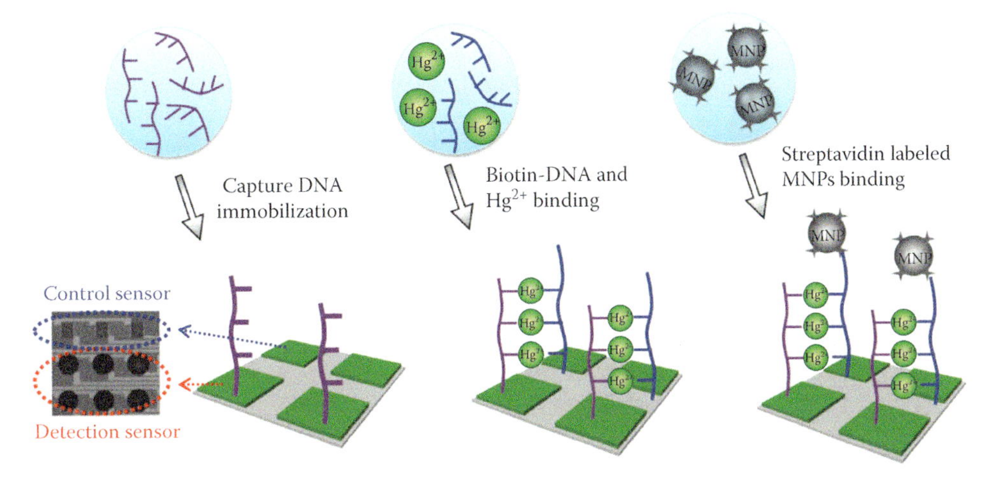

FIGURE 13.19 Schematic illustration of Hg^{2+} detection using GMR biosensor.[121] (Reprinted with permission from Wang, W.; Wang, Y.; Tu, L.; Klein, T.; Feng, Y.; Li, Q.; Wang, J.-P., *Analytical Chemistry*, *86* (8), 3712–3716. Copyright 2014 American Chemical Society.)

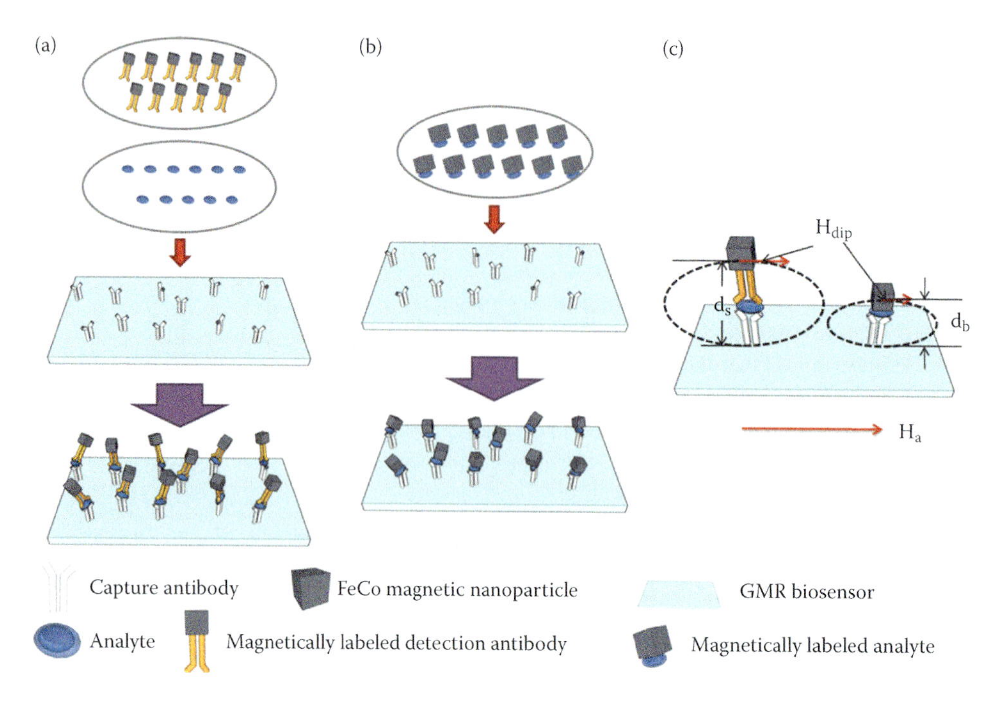

FIGURE 13.20 (a) Three-layer (sandwich structure) approach: capture antibody-analyte-detection antibody-MNP. (b) Two-layer approach: capture antibody-analyte-MNP. (c) GMR biosensor working principle.[124] (Reprinted with permission from Li, Y.; Srinivasan, B.; Jing, Y.; Yao, X.; Hugger, M. A.; Wang, J.-P.; Xing, C., *Journal of the American Chemical Society*, *132* (12), 4388–4392. Copyright 2010 American Chemical Society.)

also demonstrated the feasibility of directly quantifying IL-6 in unprocessed human serum samples.

13.4.8 Wash-Free Magnetic Bioassay

During the GMR-based biomarker detection, MNPs can not only attach to the sensor surface through the specific interaction between antigen and antibody, but also through nonspecific adhesion induced by magnetic interactions between particles and sensor surface, of which the latter one will introduce large background signals. As a result, a washing step is commonly employed before the detection to aspirate the unbound particles. However, the need for washing also eliminates the possibility of real-time measurement of the binding process. To solve these problems, a wash-free method was developed by Wang et al.[126] In their experiments, the target antigen and detection antibody were first introduced to the system by a microfluidic channel. The MNPs were then flowed through the sensor surface (Figure 13.21). It was shown that the lateral fluidic force acting on MNPs could overcome the magnetic trapping force on the sensor surface, resulting in negligible nonspecific adhesion.

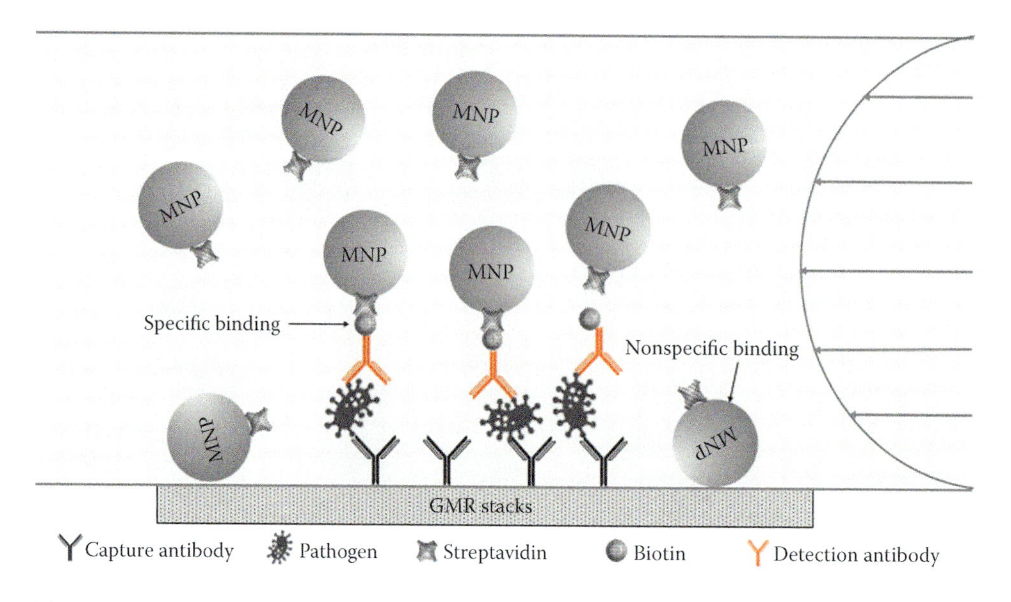

FIGURE 13.21 Schematic illustration of the wash-free method.

A ten-fold improvement in signal magnitude was realized by the employment of MNPs with larger size (the size of MNPs increased from 50 nm to 150 nm) whose application had been limited by their severe nonspecific adhesion problem. The detection limit for C-reactive protein (CRP) was also reduced from 225 to 23.7 pg/mL.

13.5 CONCLUSION AND PERSPECTIVES

To date, chip-based GMR SVs along with MNPs have become a powerful tool for high sensitivity, real-time electrical readout, and rapid biomolecule detection method.[17,86,127–132] The fabrication and integration of GMR biosensors are compatible with the large multiplex technology and the current very large-scale integration (VLSI) technology, therefore it is possible to lower down the cost if mass production is carried out. Moreover, GMR chips can be integrated with not only electronics but also microfluidics for immunoassay applications.[125,133] In addition, GMR biosensors are matrix-insensitive and therefore their performance is very robust and unaffected by environmental factors such as temperature and pH.[134] The high sensitivity of this detecting system opens new avenues for the detection of biomolecules involved in the aetiology of various diseases, especially chronic ones, such as cancer. It can also be used for monitoring pollutants in the environment and food.

The search coil biosensor along with MNPs has emerged over the last several years and is compatible with different immunoassay formats. The detection method is based on the nonlinear magnetization of SPIONs, the remarkable feature that allows reliable and easy discrimination of these particles from paramagnetic materials, which allows very robust analysis of nontransparent liquids of suspensions. This detection platform can be used for medical diagnostics, POC devices, food pathogen detection, water analysis, etc.

Logically, biosensing strategies based on MNPs have gained considerable attention during the last decade. However, despite this progress, a gap remains between lab trials and clinical diagnosis. It can be anticipated that in the near future, immunoassays will be performed on progressively smaller biofluidic samples to enable multiplexing. Furthermore, an inexpensive, easy-to-use, high sensitivity, and reliable POC device will enable long-term monitoring so that treatment of chronic diseases can be tailored to individual patients.

ACKNOWLEDGEMENTS

The authors thank Dr Venkatramana D. Krishna from the College of Veterinary Medicine, University of Minnesota, for the fruitful discussions.

REFERENCES

1. Gleich, B.; Weizenecker, J., Tomographic imaging using the nonlinear response of magnetic particles. *Nature* **2005,** *435* (7046), 1214–1217.
2. Ferguson, R. M.; Khandhar, A. P.; Kemp, S. J.; Arami, H.; Saritas, E. U.; Croft, L. R.; Konkle, J.; Goodwill, P. W.; Halkola, A.; Rahmer, J., Magnetic particle imaging with tailored iron oxide nanoparticle tracers. *IEEE Transactions on Medical Imaging* **2015,** *34* (5), 1077–1084.
3. Arami, H.; Khandhar, A. P.; Tomitaka, A.; Yu, E.; Goodwill, P. W.; Conolly, S. M.; Krishnan, K. M., In vivo multimodal magnetic particle imaging (MPI) with tailored magneto/optical contrast agents. *Biomaterials* **2015,** *52*, 251–261.
4. Rahmer, J.; Halkola, A.; Gleich, B.; Schmale, I.; Borgert, J., First experimental evidence of the feasibility of multi-color magnetic particle imaging. *Physics in Medicine and Biology* **2015,** *60* (5), 1775.
5. Bauer, L. M.; Situ, S. F.; Griswold, M. A.; Samia, A. C. S., Magnetic particle imaging tracers: State-of-the-art and future directions. *Journal of Physical Chemistry Letters* **2015,** *6* (13), 2509–2517.
6. Murase, K.; Aoki, M.; Banura, N.; Nishimoto, K.; Mimura, A.; Kuboyabu, T.; Yabata, I., Usefulness of magnetic particle imaging for predicting the therapeutic effect of magnetic hyperthermia. *Open Journal of Medical Imaging* **2015,** *5* (02), 85.
7. Merbach, A. S.; Helm, L.; Tóth, É., *The Chemistry of Contrast Agents in Medical Magnetic Resonance Imaging*. John Wiley & Sons, Chichester, United Kingdom: 2013.
8. Vlaardingerbroek, M. T.; Boer, J. A., *Magnetic Resonance Imaging: Theory and Practice*. Springer Science & Business Media, Berlin/Heidelberg, Germany: 2013.
9. Sun, C.; Lee, J. S.; Zhang, M., Magnetic nanoparticles in MR imaging and drug delivery. *Advanced Drug Delivery Reviews* **2008,** *60* (11), 1252–1265.
10. Hola, K.; Markova, Z.; Zoppellaro, G.; Tucek, J.; Zboril, R., Tailored functionalization of iron oxide nanoparticles for MRI, drug delivery, magnetic separation and immobilization of biosubstances. *Biotechnology Advances* **2015,** *33* (6), 1162–1176.
11. Oka, C.; Ushimaru, K.; Horiishi, N.; Tsuge, T.; Kitamoto, Y., Core–shell composite particles composed of biodegradable polymer particles and magnetic iron oxide nanoparticles for targeted drug delivery. *Journal of Magnetism and Magnetic Materials* **2015,** *381*, 278–284.
12. Kakwere, H.; Leal, M. P.; Materia, M. E.; Curcio, A.; Guardia, P.; Niculaes, D.; Marotta, R.; Falqui, A.; Pellegrino, T., Functionalization of strongly interacting magnetic nanocubes with (thermo) responsive coating and their application in hyperthermia and heat-triggered drug delivery. *ACS Applied Materials & Interfaces* **2015,** *7* (19), 10132–10145.
13. Yu, L.; Liu, J.; Wu, K.; Klein, T.; Jiang, Y.; Wang, J.-P., Evaluation of hyperthermia of magnetic nanoparticles by dehydrating DNA. *Scientific Reports* **2014,** *4*, 7216.
14. Sakellari, D.; Brintakis, K.; Kostopoulou, A.; Myrovali, E.; Simeonidis, K.; Lappas, A.; Angelakeris, M., Ferrimagnetic nanocrystal assemblies as versatile magnetic particle hyperthermia mediators. *Materials Science and Engineering: C* **2016,** *58*, 187–193.
15. Shah, R. R.; Davis, T. P.; Glover, A. L.; Nikles, D. E.; Brazel, C. S., Impact of magnetic field parameters and iron oxide nanoparticle properties on heat generation for use in magnetic hyperthermia. *Journal of Magnetism and Magnetic Materials* **2015,** *387*, 96–106.
16. Wang, Y.; Wang, W.; Yu, L.; Tu, L.; Feng, Y.; Klein, T.; Wang, J.-P., Giant magnetoresistive-based bio-sensing probe station system for multiplex protein assays. *Biosensors and Bioelectronics* **2015,** *70*, 61–68.

17. Baselt, D. R.; Lee, G. U.; Natesan, M.; Metzger, S. W.; Sheehan, P. E.; Colton, R. J., A biosensor based on magnetoresistance technology. *Biosensors and Bioelectronics* **1998,** *13* (7), 731–739.
18. Shen, W.; Schrag, B. D.; Carter, M. J.; Xie, J.; Xu, C.; Sun, S.; Xiao, G., Detection of DNA labeled with magnetic nanoparticles using MgO-based magnetic tunnel junction sensors. *Journal of Applied Physics* **2008,** *103* (7), 7A306.
19. Tu, L. Detection of magnetic nanoparticles for bio-sensing applications. University of Minnesota, 2013.
20. Srinivasan, B.; Li, Y.; Jing, Y.; Xu, Y.; Yao, X.; Xing, C.; Wang, J. P., A detection system based on giant magnetoresistive sensors and high-moment magnetic nanoparticles demonstrates zeptomole sensitivity: Potential for personalized medicine. *Angewandte Chemie International Edition* **2009,** *48* (15), 2764–2767.
21. Krishna, V. D.; Wu, K.; Perez, A. M.; Wang, J.-P., Giant magnetoresistance-based biosensor for detection of influenza A virus. *Frontiers in Microbiology* **2016,** *7*, 400. doi:10.3389/fmicb.2016.00400.
22. Sandhu, A.; Kumagai, Y.; Lapicki, A.; Sakamoto, S.; Abe, M.; Handa, H., High efficiency Hall effect micro-biosensor platform for detection of magnetically labeled biomolecules. *Biosensors and Bioelectronics* **2007,** *22* (9), 2115–2120.
23. Togawa, K.; Sanbonsugi, H.; Sandhu, A.; Abe, M.; Narimatsu, H.; Nishio, K.; Handa, H., High sensitivity InSb Hall effect biosensor platform for DNA detection and biomolecular recognition using functionalized magnetic nanobeads. *Japanese Journal of Applied Physics* **2005,** *44* (11L), L1494.
24. Haun, J. B.; Yoon, T. J.; Lee, H.; Weissleder, R., Magnetic nanoparticle biosensors. *Wiley Interdisciplinary Reviews: Nanomedicine and Nanobiotechnology* **2010,** *2* (3), 291–304.
25. Conde, J.; Dias, J. T.; Grazú, V.; Moros, M.; Baptista, P. V.; de la Fuente, J. M., Revisiting 30 years of biofunctionalization and surface chemistry of inorganic nanoparticles for nanomedicine. *Frontiers in Chemistry* **2014,** *2*, 48.
26. Kango, S.; Kalia, S.; Celli, A.; Njuguna, J.; Habibi, Y.; Kumar, R., Surface modification of inorganic nanoparticles for development of organic-inorganic nanocomposites – A review. *Progress in Polymer Science* **2013,** *38* (8), 1232–1261.
27. Jiang, S.; Eltoukhy, A. A.; Love, K. T.; Langer, R.; Anderson, D. G., Lipidoid-coated iron oxide nanoparticles for efficient DNA and siRNA delivery. *Nano Letters* **2013,** *13* (3), 1059–1064.
28. Mout, R.; Moyano, D. F.; Rana, S.; Rotello, V. M., Surface functionalization of nanoparticles for nanomedicine. *Chemical Society Reviews* **2012,** *41* (7), 2539–2544.
29. Nitin, N.; LaConte, L.; Zurkiya, O.; Hu, X.; Bao, G., Functionalization and peptide-based delivery of magnetic nanoparticles as an intracellular MRI contrast agent. *JBIC Journal of Biological Inorganic Chemistry* **2004,** *9* (6), 706–712.
30. Scarberry, K. E.; Dickerson, E. B.; McDonald, J. F.; Zhang, Z. J., Magnetic nanoparticle–peptide conjugates for in vitro and in vivo targeting and extraction of cancer cells. *Journal of the American Chemical Society* **2008,** *130* (31), 10258–10262.
31. Varshney, M.; Li, Y., Interdigitated array microelectrode based impedance biosensor coupled with magnetic nanoparticle–antibody conjugates for detection of *Escherichia coli* O157: H7 in food samples. *Biosensors and Bioelectronics* **2007,** *22* (11), 2408–2414.
32. Virkutyte, J.; Varma, R. S., Green synthesis of metal nanoparticles: Biodegradable polymers and enzymes in stabilization and surface functionalization. *Chemical Science* **2011,** *2* (5), 837–846.
33. Bao, G.; Mitragotri, S.; Tong, S., Multifunctional nanoparticles for drug delivery and molecular imaging. *Annual Review of Biomedical Engineering* **2013,** *15*, 253–282.
34. Gong, J.-L.; Liang, Y.; Huang, Y.; Chen, J.-W.; Jiang, J.-H.; Shen, G.-L.; Yu, R.-Q., Ag/SiO 2 core-shell nanoparticle-based surface-enhanced Raman probes for immunoassay of cancer marker using silica-coated magnetic nanoparticles as separation tools. *Biosensors and Bioelectronics* **2007,** *22* (7), 1501–1507.
35. Chon, H.; Lee, S.; Son, S. W.; Oh, C. H.; Choo, J., Highly sensitive immunoassay of lung cancer marker carcinoembryonic antigen using surface-enhanced Raman scattering of hollow gold nanospheres. *Analytical Chemistry* **2009,** *81* (8), 3029–3034.
36. Stoeva, S. I.; Lee, J.-S.; Smith, J. E.; Rosen, S. T.; Mirkin, C. A., Multiplexed detection of protein cancer markers with biobarcoded nanoparticle probes. *Journal of the American Chemical Society* **2006,** *128* (26), 8378–8379.
37. Ren, Y.; Rivera, J. G.; He, L.; Kulkarni, H.; Lee, D.-K.; Messersmith, P. B., Facile, high efficiency immobilization of lipase enzyme on magnetic iron oxide nanoparticles via a biomimetic coating. *BMC Biotechnology* **2011,** *11* (1), 1.
38. Lee, H.; Shin, T.-H.; Cheon, J.; Weissleder, R., Recent developments in magnetic diagnostic systems. *Chemical Reviews* **2015,** *115* (19), 10690–10724.
39. Hore, P. J., *Nuclear Magnetic Resonance*. Oxford University Press, Oxford, United Kingdom: 2015.
40. Jackman, L. M.; Sternhell, S., *Application of Nuclear Magnetic Resonance Spectroscopy in Organic Chemistry: International Series in Organic Chemistry*. Elsevier, Amsterdam, Netherlands: 2013.
41. Trisnanto, S. B.; Kitamoto, Y., Optimizing coil system for magnetic susceptometer with widely-adjustable field-strength and frequency. *Japanese Journal of Applied Physics* **2016,** *55* (2S), 02BD02.
42. Adolphi, N. L.; Butler, K. S.; Lovato, D. M.; Tessier, T.; Trujillo, J. E.; Hathaway, H. J.; Fegan, D. L.; Monson, T. C.; Stevens, T. E.; Huber, D. L., Imaging of Her2-targeted magnetic nanoparticles for breast cancer detection: Comparison of SQUID-detected magnetic relaxometry and MRI. *Contrast Media & Molecular Imaging* **2012,** *7* (3), 308–319.
43. Chieh, J.-J.; Huang, K.-W.; Lee, Y.-Y.; Wei, W.-C., Dual-imaging model of SQUID biosusceptometry for locating tumors targeted using magnetic nanoparticles. *Journal of Nanobiotechnology* **2015,** *13* (1), 1.
44. Choi, J.; Gani, A. W.; Bechstein, D. J.; Lee, J.-R.; Utz, P. J.; Wang, S. X., Portable, one-step, and rapid GMR biosensor platform with smartphone interface. *Biosensors and Bioelectronics* **2016,** *85*, 1–7.
45. Caruso, M. J.; Bratland, T.; Smith, C. H.; Schneider, R., A new perspective on magnetic field sensing. *Sensors-Peterborough* **1998,** *15*, 34–47.
46. Krause, H.-J.; Wolters, N.; Zhang, Y.; Offenhäusser, A.; Miethe, P.; Meyer, M. H.; Hartmann, M.; Keusgen, M., Magnetic particle detection by frequency mixing for immunoassay applications. *Journal of Magnetism and Magnetic Materials* **2007,** *311* (1), 436–444.
47. Nikitin, P. I.; Vetoshko, P. M.; Ksenevich, T. I., New type of biosensor based on magnetic nanoparticle detection. *Journal of Magnetism and Magnetic Materials* **2007,** *311* (1), 445–449.
48. Tu, L.; Jing, Y.; Li, Y.; Wang, J.-P., Real-time measurement of Brownian relaxation of magnetic nanoparticles by a mixing-frequency method. *Applied Physics Letters* **2011,** *98* (21), 213702.

49. Tu, L.; Wu, K.; Klein, T.; Wang, J.-P., Magnetic nanoparticles colourization by a mixing-frequency method. *Journal of Physics D: Applied Physics* **2014,** *47* (15), 155001.
50. Wu, K.; Batra, A.; Jain, S.; Wang, J.-P., Magnetization response spectroscopy of superparamagnetic nanoparticles under mixing frequency fields. *IEEE Transactions on Magnetics* **2016,** *52* (7). doi: 10.1109/TMAG.2015.2513746.
51. Wu, K.; Batra, A.; Jain, S.; Ye, C.; Liu, J.; Wang, J.-P., A simulation study on superparamagnetic nanoparticle based multitracer tracking. *Applied Physics Letters* **2015,** *107* (17), 173701.
52. Wu, K.; Liu, J.; Wang, Y.; Ye, C.; Feng, Y.; Wang, J.-P., Superparamagnetic nanoparticle-based viscosity test. *Applied Physics Letters* **2015,** *107* (5), 053701.
53. Wu, K.; Wang, Y.; Feng, Y.; Yu, L.; Wang, J.-P., Colorize magnetic nanoparticles using a search coil based testing method. *Journal of Magnetism and Magnetic Materials* **2015,** *380*, 251–254.
54. Wu, K.; Ye, C.; Liu, J.; Wang, Y.; Feng, Y.; Wang, J.-P., In vitro viscosity measurement on superparamagnetic nanoparticle suspensions. *IEEE Transactions on Magnetics* **2016,** *52* (7). doi: 10.1109/TMAG.2016.2529426.
55. Wu, K.; Yu, L.; Zheng, X.; Wang, Y.; Feng, Y.; Tu, L.; Wang, J.-P. In *Viscosity Effect on the Brownian Relaxation Based Detection for Immunoassay Applications*, 2014 36th Annual International Conference of the IEEE Engineering in Medicine and Biology Society, IEEE: 2014; pp 2769–2772.
56. Ludwig, F.; Heim, E.; Schilling, M., Characterization of superparamagnetic nanoparticles by analyzing the magnetization and relaxation dynamics using fluxgate magnetometers. *Journal of Applied Physics* **2007,** *101* (11), 113909.
57. Ludwig, F.; Kazakova, O.; Barquín, L. F.; Fornara, A.; Trahms, L.; Steinhoff, U.; Svedlindh, P.; Wetterskog, E.; Pankhurst, Q. A.; Southern, P., Magnetic, structural, and particle size analysis of single-and multi-core magnetic nanoparticles. *IEEE Transactions on Magnetics* **2014,** *50* (11), 1–4.
58. Rauwerdink, A. M.; Hansen, E. W.; Weaver, J. B., Nanoparticle temperature estimation in combined ac and dc magnetic fields. *Physics in Medicine and Biology* **2009,** *54* (19), L51.
59. Rauwerdink, A. M.; Weaver, J. B., Harmonic phase angle as a concentration-independent measure of nanoparticle dynamics. *Medical Physics* **2010,** *37* (6), 2587–2592.
60. Rauwerdink, A. M.; Weaver, J. B., Measurement of molecular binding using the Brownian motion of magnetic nanoparticle probes. *Applied Physics Letters* **2010,** *96* (3), 033702.
61. Rauwerdink, A. M.; Giustini, A. J.; Weaver, J. B., Simultaneous quantification of multiple magnetic nanoparticles. *Nanotechnology* **2010,** *21* (45), 455101.
62. Rauwerdink, A. M.; Weaver, J. B., Concurrent quantification of multiple nanoparticle bound states. *Medical Physics* **2011,** *38* (3), 1136–1140.
63. Weaver, J. B.; Rauwerdink, A. M.; Sullivan, C. R.; Baker, I., Frequency distribution of the nanoparticle magnetization in the presence of a static as well as a harmonic magnetic field. *Medical Physics* **2008,** *35* (5), 1988–1994.
64. Zhang, X.; Reeves, D. B.; Perreard, I. M.; Kett, W. C.; Griswold, K. E.; Gimi, B.; Weaver, J. B., Molecular sensing with magnetic nanoparticles using magnetic spectroscopy of nanoparticle Brownian motion. *Biosensors and Bioelectronics* **2013,** *50*, 441–446.
65. Weaver, J. B.; Harding, M.; Rauwerdink, A. M.; Hansen, E. W. In *The Effect of Viscosity on the Phase of the Nanoparticle Magnetization Induced by a Harmonic Applied Field*, SPIE Medical Imaging, International Society for Optics and Photonics: 2010; pp 762627–762627-8.
66. Krishnan, K. M.; Pakhomov, A. B.; Bao, Y.; Blomqvist, P.; Chun, Y.; Gonzales, M.; Griffin, K.; Ji, X.; Roberts, B. K., Nanomagnetism and spin electronics: Materials, microstructure and novel properties. *Journal of Materials Science* **2006,** *41* (3), 793–815.
67. Singamaneni, S.; Bliznyuk, V. N.; Binek, C.; Tsymbal, E. Y., Magnetic nanoparticles: Recent advances in synthesis, self-assembly and applications. *Journal of Materials Chemistry* **2011,** *21* (42), 16819–16845.
68. Krishnan, K. M., Biomedical nanomagnetics: A spin through possibilities in imaging, diagnostics, and therapy. *IEEE Transactions on Magnetics* **2010,** *46* (7), 2523–2558.
69. Wu, K.; Tu, L.; Su, D.; Wang, J.-P., Magnetic dynamics of ferrofluids: Mathematical models and experimental investigations. *Journal of Physics D: Applied Physics* **2017,** *50* (8), 085005.
70. Ceccon, A.; Tugarinov, V.; Bax, A.; Clore, G. M., Global dynamics and exchange kinetics of a protein on the surface of nanoparticles revealed by relaxation-based solution NMR spectroscopy. *Journal of the American Chemical Society* **2016,** *138* (18), 5789–5792.
71. Soukup, D.; Moise, S.; Céspedes, E.; Dobson, J.; Telling, N. D., In situ measurement of magnetization relaxation of internalized nanoparticles in live cells. *ACS Nano* **2015,** *9* (1), 231–240.
72. Chen, Y.; Xianyu, Y.; Wang, Y.; Zhang, X.; Cha, R.; Sun, J.; Jiang, X., One-step detection of pathogens and viruses: Combining magnetic relaxation switching and magnetic separation. *ACS Nano* **2015,** *9* (3), 3184–3191.
73. Dieckhoff, J.; Eberbeck, D.; Schilling, M.; Ludwig, F., Magnetic-field dependence of Brownian and Néel relaxation times. *Journal of Applied Physics* **2016,** *119* (4), 043903.
74. Deissler, R. J.; Wu, Y.; Martens, M. A., Dependence of Brownian and Néel relaxation times on magnetic field strength. *Medical Physics* **2014,** *41* (1), 012301.
75. Usov, N.; Liubimov, B. Y., Dynamics of magnetic nanoparticle in a viscous liquid: Application to magnetic nanoparticle hyperthermia. *Journal of Applied Physics* **2012,** *112* (2), 023901.
76. Chung, S.-H.; Hoffmann, A.; Guslienko, K.; Bader, S.; Liu, C.; Kay, B.; Makowski, L.; Chen, L., Biological sensing with magnetic nanoparticles using Brownian relaxation. *Journal of Applied Physics* **2005,** *97* (10), 10R101.
77. Weaver, J. B.; Rauwerdink, A. M.; Hansen, E. W., Magnetic nanoparticle temperature estimation. *Medical Physics* **2009,** *36* (5), 1822–1829.
78. Tu, L.; Klein, T.; Wang, W.; Feng, Y.; Wang, Y.; Wang, J.-P., Measurement of Brownian and Néel relaxation of magnetic nanoparticles by a mixing-frequency method. *IEEE Transactions on Magnetics* **2013,** *49* (1), 227–230.
79. Wu, K.; Schliep, K.; Zhang, X.; Liu, J.; Ma, B.; Wang, J. P., Characterizing physical properties of superparamagnetic nanoparticles in liquid phase using Brownian relaxation. *Small* **2017**. doi: 10.1002/smll.201604135.
80. Mac Oireachtaigh, C.; Fannin, P., Investigation of the nonlinear loss properties of magnetic fluids subject to large alternating fields. *Journal of Magnetism and Magnetic Materials* **2008,** *320* (6), 871–880.
81. Shliomis, M.; Raikher, Y., Experimental investigations of magnetic fluids. *IEEE Transactions on Magnetics* **1980,** *16* (2), 237–250.
82. Harisinghani, M. G.; Barentsz, J.; Hahn, P. F.; Deserno, W. M.; Tabatabaei, S.; van de Kaa, C. H.; de la Rosette, J.; Weissleder, R., Noninvasive detection of clinically occult lymph-node metastases in prostate cancer. *New England Journal of Medicine* **2003,** *348* (25), 2491–2499.

83. Weissleder, R.; Reimer, P.; Lee, A.; Wittenberg, J.; Brady, T., MR receptor imaging: Ultrasmall iron oxide particles targeted to asialoglycoprotein receptors. *AJR. American Journal of Roentgenology* **1990,** *155* (6), 1161–1167.
84. Martins, V.; Cardoso, F.; Germano, J.; Cardoso, S.; Sousa, L.; Piedade, M.; Freitas, P.; Fonseca, L., Femtomolar limit of detection with a magnetoresistive biochip. *Biosensors and Bioelectronics* **2009,** *24* (8), 2690–2695.
85. Hall, D.; Gaster, R.; Lin, T.; Osterfeld, S.; Han, S.; Murmann, B.; Wang, S., GMR biosensor arrays: A system perspective. *Biosensors and Bioelectronics* **2010,** *25* (9), 2051–2057.
86. Gaster, R. S.; Xu, L.; Han, S.-J.; Wilson, R. J.; Hall, D. A.; Osterfeld, S. J.; Yu, H.; Wang, S. X., Quantification of protein interactions and solution transport using high-density GMR sensor arrays. *Nature Nanotechnology* **2011,** *6* (5), 314–320.
87. Dittmer, W.; De Kievit, P.; Prins, M.; Vissers, J.; Mersch, M.; Martens, M., Sensitive and rapid immunoassay for parathyroid hormone using magnetic particle labels and magnetic actuation. *Journal of Immunological Methods* **2008,** *338* (1), 40–46.
88. Ikeda, S.; Hayakawa, J.; Ashizawa, Y.; Lee, Y.; Miura, K.; Hasegawa, H.; Tsunoda, M.; Matsukura, F.; Ohno, H., Tunnel magnetoresistance of 604% at 300 K by suppression of Ta diffusion in Co Fe B/Mg O/Co Fe B pseudo-spin-valves annealed at high temperature. *Applied Physics Letters* **2008,** *93* (8), 082508.
89. Coehoorn, R., Giant magnetoresistance and magnetic interactions in exchange-biased spin-valves. *Handbook of Magnetic Materials* **2003,** *15*, 1–197.
90. Sandhu, A.; Handa, H., Practical Hall sensors for biomedical instrumentation. *IEEE Transactions on Magnetics* **2005,** *41* (10), 4123–4127.
91. Aytur, T.; Foley, J.; Anwar, M.; Boser, B.; Harris, E.; Beatty, P. R., A novel magnetic bead bioassay platform using a microchip-based sensor for infectious disease diagnosis. *Journal of Immunological Methods* **2006,** *314* (1), 21–29.
92. Besse, P.-A.; Boero, G.; Demierre, M.; Pott, V.; Popovic, R., Detection of a single magnetic microbead using a miniaturized silicon Hall sensor. *Applied Physics Letters* **2002,** *80* (22), 4199–4201.
93. Orlov, A. V.; Khodakova, J. A.; Nikitin, M. P.; Shepelyakovskaya, A. O.; Brovko, F. A.; Laman, A. G.; Grishin, E. V.; Nikitin, P. I., Magnetic immunoassay for detection of staphylococcal toxins in complex media. *Analytical Chemistry* **2012,** *85* (2), 1154–1163.
94. Orlov, A. V.; Khodakova, J. A.; Nikitin, M. P.; Shepelyakovskaya, A. O.; Brovko, F. A.; Laman, A. G.; Grishin, E. V.; Nikitin, P. I., Magnetic immunoassay for detection of staphylococcal toxins in complex media. *Analytical Chemistry* **2013,** *85* (2), 1154–1163.
95. Rubina, A. Y.; Filippova, M.; Feizkhanova, G.; Shepeliakovskaya, A.; Sidina, E.; Boziev, K. M.; Laman, A.; Brovko, F.; Vertiev, Y. V.; Zasedatelev, A., Simultaneous detection of seven staphylococcal enterotoxins: Development of hydrogel biochips for analytical and practical application. *Analytical Chemistry* **2010,** *82* (21), 8881–8889.
96. Kuang, H.; Wang, W.; Xu, L.; Ma, W.; Liu, L.; Wang, L.; Xu, C., Monoclonal antibody-based sandwich ELISA for the detection of staphylococcal enterotoxin A. *International Journal of Environmental Research and Public Health* **2013,** *10* (4), 1598–1608.
97. Clarisse, T.; Michèle, S.; Olivier, T.; Valérie, E.; Jacques-Antoine, H.; Michel, G.; Florence, V., Detection and quantification of staphylococcal enterotoxin A in foods with specific and sensitive polyclonal antibodies. *Food Control* **2013,** *32* (1), 255–261.
98. Vernozy-Rozand, C.; Mazuy-Cruchaudet, C.; Bavai, C.; Richard, Y., Comparison of three immunological methods for detecting staphylococcal enterotoxins from food. *Letters in Applied Microbiology* **2004,** *39* (6), 490–494.
99. Bennett, R. W., Staphylococcal enterotoxin and its rapid identification in foods by enzyme-linked immunosorbent assay-based methodology. *Journal of Food Protection®* **2005,** *68* (6), 1264–1270.
100. Park, C.; Akhtar, M.; Rayman, M., Evaluation of a commercial enzyme immunoassay kit (RIDASCREEN) for detection of staphylococcal enterotoxins A, B, C, D, and E in foods. *Applied and Environmental Microbiology* **1994,** *60* (2), 677–681.
101. Campbell, G. A.; Medina, M. B.; Mutharasan, R., Detection of Staphylococcus enterotoxin B at picogram levels using piezoelectric-excited millimeter-sized cantilever sensors. *Sensors and Actuators B: Chemical* **2007,** *126* (2), 354–360.
102. Ligler, F. S.; Taitt, C. R.; Shriver-Lake, L. C.; Sapsford, K. E.; Shubin, Y.; Golden, J. P., Array biosensor for detection of toxins. *Analytical and Bioanalytical Chemistry* **2003,** *377* (3), 469–477.
103. Swann, S., Magnetron sputtering. *Physics in Technology* **1988,** *19* (2), 67.
104. Heremans, J., Solid state magnetic field sensors and applications. *Journal of Physics D: Applied Physics* **1993,** *26* (8), 1149.
105. Kustov, M.; Laczkowski, P.; Hykel, D.; Hasselbach, K.; Dumas-Bouchiat, F.; O'Brien, D.; Kauffmann, P.; Grechishkin, R.; Givord, D.; Reyne, G., Magnetic characterization of micropatterned Nd–Fe–B hard magnetic films using scanning Hall probe microscopy. *Journal of Applied Physics* **2010,** *108* (6), 063914.
106. Wang, W.; Wang, Y.; Tu, L.; Klein, T.; Feng, Y.; Wang, J.-P., Surface modification for Protein and DNA immobilization onto GMR biosensor. *IEEE Transactions on Magnetics* **2013,** *49* (1), 296–299.
107. Wang, W.; Wang, Y.; Tu, L.; Feng, Y.; Klein, T.; Wang, J.-P., Magnetoresistive performance and comparison of supermagnetic nanoparticles on giant magnetoresistive sensor-based detection system. *Scientific Reports* **2014,** *4*, 5716.
108. Bonaca, M. P.; Scirica, B. M.; Sabatine, M. S.; Jarolim, P.; Murphy, S. A.; Chamberlin, J. S.; Rhodes, D. W.; Southwick, P. C.; Braunwald, E.; Morrow, D. A., Prospective evaluation of pregnancy-associated plasma protein-a and outcomes in patients with acute coronary syndromes. *Journal of the American College of Cardiology* **2012,** *60* (4), 332–338.
109. Li, Y.; Zhou, C.; Zhou, X.; Song, L.; Hui, R., PAPP-A in cardiac and non-cardiac conditions. *Clinica Chimica Acta* **2013,** *417*, 67–72.
110. Cui, Q.; Ju, X.; Yang, T.; Zhang, M.; Tang, W.; Chen, Q.; Hu, Y.; Haas, J. V.; Troutt, J. S.; Pickard, R. T., Serum PCSK9 is associated with multiple metabolic factors in a large Han Chinese population. *Atherosclerosis* **2010,** *213* (2), 632–636.
111. Huijgen, R.; Boekholdt, S. M.; Arsenault, B. J.; Bao, W.; Davaine, J.-M.; Tabet, F.; Petrides, F.; Rye, K.-A.; DeMicco, D. A.; Barter, P. J., RETRACTED: Plasma PCSK9 levels and clinical outcomes in the TNT (Treating to New Targets) Trial: A nested case-control study. *Journal of the American College of Cardiology* **2012,** *59* (20), 1778–1784.
112. Rehman, S. U.; Mueller, T.; Januzzi, J. L., Characteristics of the novel interleukin family biomarker ST2 in patients with acute heart failure. *Journal of the American College of Cardiology* **2008,** *52* (18), 1458–1465.

113. Lee, D.; Chander, Y.; Goyal, S. M.; Cui, T., Carbon nanotube electric immunoassay for the detection of swine influenza virus H1N1. *Biosensors and Bioelectronics* **2011,** *26* (8), 3482–3487.
114. Nidzworski, D.; Pranszke, P.; Grudniewska, M.; Król, E.; Gromadzka, B., Universal biosensor for detection of influenza virus. *Biosensors and Bioelectronics* **2014,** *59*, 239–242.
115. Patolsky, F.; Zheng, G.; Hayden, O.; Lakadamyali, M.; Zhuang, X.; Lieber, C. M., Electrical detection of single viruses. *Proceedings of the National Academy of Sciences of the United States of America* **2004,** *101* (39), 14017–14022.
116. Tam, P. D.; Van Hieu, N.; Chien, N. D.; Le, A.-T.; Tuan, M. A., DNA sensor development based on multi-wall carbon nanotubes for label-free influenza virus (type A) detection. *Journal of Immunological Methods* **2009,** *350* (1), 118–124.
117. Shirale, D. J.; Bangar, M. A.; Park, M.; Yates, M. V.; Chen, W.; Myung, N. V.; Mulchandani, A., Label-free chemiresistive immunosensors for viruses. *Environmental Science & Technology* **2010,** *44* (23), 9030–9035.
118. Driskell, J. D.; Jones, C. A.; Tompkins, S. M.; Tripp, R. A., One-step assay for detecting influenza virus using dynamic light scattering and gold nanoparticles. *Analyst* **2011,** *136* (15), 3083–3090.
119. Singh, R.; Sharma, A.; Hong, S.; Jang, J., Electrical immunosensor based on dielectrophoretically-deposited carbon nanotubes for detection of influenza virus H1N1. *Analyst* **2014,** *139* (21), 5415–5421.
120. Lekcharoensuk, P.; Lager, K. M.; Vemulapalli, R.; Woodruff, M.; Vincent, A. L.; Richt, J. A., Novel swine influenza virus subtype H3N1, United States. *Emerging Infectious Diseases* **2006,** *12* (5), 787.
121. Wang, W.; Wang, Y.; Tu, L.; Klein, T.; Feng, Y.; Li, Q.; Wang, J.-P., Magnetic detection of mercuric ion using giant magnetoresistance-based biosensing system. *Analytical Chemistry* **2014,** *86* (8), 3712–3716.
122. Ono, A.; Togashi, H., Highly selective oligonucleotide-based sensor for mercury (II) in aqueous solutions. *Angewandte Chemie International Edition* **2004,** *43* (33), 4300–4302.
123. Mulvaney, S.; Cole, C.; Kniller, M.; Malito, M.; Tamanaha, C.; Rife, J.; Stanton, M.; Whitman, L., Rapid, femtomolar bioassays in complex matrices combining microfluidics and magnetoelectronics. *Biosensors and Bioelectronics* **2007,** *23* (2), 191–200.
124. Li, Y.; Srinivasan, B.; Jing, Y.; Yao, X.; Hugger, M. A.; Wang, J.-P.; Xing, C., Nanomagnetic competition assay for low-abundance protein biomarker quantification in unprocessed human sera. *Journal of the American Chemical Society* **2010,** *132* (12), 4388–4392.
125. Zhi, X.; Liu, Q.; Zhang, X.; Zhang, Y.; Feng, J.; Cui, D., Quick genotyping detection of HBV by giant magnetoresistive biochip combined with PCR and line probe assay. *Lab on a Chip* **2012,** *12* (4), 741–745.
126. Bechstein, D. J.; Lee, J.-R.; Ooi, C. C.; Gani, A. W.; Kim, K.; Wilson, R. J.; Wang, S. X., High performance wash-free magnetic bioassays through microfluidically enhanced particle specificity. *Scientific Reports* **2015,** *5*, 11693.
127. Rife, J.; Miller, M.; Sheehan, P.; Tamanaha, C.; Tondra, M.; Whitman, L., Design and performance of GMR sensors for the detection of magnetic microbeads in biosensors. *Sensors and Actuators A: Physical* **2003,** *107* (3), 209–218.
128. Graham, D. L.; Ferreira, H. A.; Freitas, P. P., Magnetoresistive-based biosensors and biochips. *TRENDS in Biotechnology* **2004,** *22* (9), 455–462.
129. Schotter, J.; Kamp, P.-B.; Becker, A.; Pühler, A.; Reiss, G.; Brückl, H., Comparison of a prototype magnetoresistive biosensor to standard fluorescent DNA detection. *Biosensors and Bioelectronics* **2004,** *19* (10), 1149–1156.
130. Millen, R. L.; Kawaguchi, T.; Granger, M. C.; Porter, M. D.; Tondra, M., Giant magnetoresistive sensors and superparamagnetic nanoparticles: a chip-scale detection strategy for immunosorbent assays. *Analytical Chemistry.* **2005,** *77* (20), 6581–6587.
131. Loureiro, J.; Ferreira, R.; Cardoso, S.; Freitas, P.; Germano, J.; Fermon, C.; Arrias, G.; Pannetier-Lecoeur, M.; Rivadulla, F.; Rivas, J., Toward a magnetoresistive chip cytometer: Integrated detection of magnetic beads flowing at cm/s velocities in microfluidic channels. *Applied Physics Letters* **2009,** *95* (3), 034104.
132. Loureiro, J.; Andrade, P.; Cardoso, S.; Da Silva, C.; Cabral, J.; Freitas, P., Magnetoresistive chip cytometer. *Lab on a Chip* **2011,** *11* (13), 2255–2261.
133. Xu, L.; Yu, H.; Akhras, M. S.; Han, S.-J.; Osterfeld, S.; White, R. L.; Pourmand, N.; Wang, S. X., Giant magnetoresistive biochip for DNA detection and HPV genotyping. *Biosensors and Bioelectronics* **2008,** *24* (1), 99–103.
134. Zhang, Y.; Zhou, D., Magnetic particle-based ultrasensitive biosensors for diagnostics. *Expert Review of Molecular Diagnostics* **2012,** *12* (6), 565–571.

Dr Jian-Ping Wang (E-mail: jpwang@umn.edu) is a Distinguished McKnight University professor of Electrical and Computer Engineering, and a member of the graduate faculty in Physics, Chemical Engineering and Materials Science and Biomedical Engineering at the University of Minnesota. He is the associate director of the Center for Micromagnetics and Information Technologies (MINT) at the University of Minnesota. He also leads the Center for Spintronic Materials, Interfaces and Novel Architectures (C-SPIN), a research collaboration of 33 experts from 19 universities that investigates groundbreaking technologies for the next generation of microelectronics. The centre is positioned to address the key challenges of next generation computing and memory. He received the information storage industry consortium (INSIC) technical award in 2006 for his pioneering work in exchange coupled composite magnetic media. His research has resulted in more than 300 scientific publications, more than 40 patents, three start-up companies, and breakthroughs that could revolutionize medical and environmental testing, including biosensors that can detect disease from a single drop of body fluid. He earned global attention for his research on $Fe_{16}N_2$, a potential powerful rare-earth-free magnet that could replace expensive and less environmentally friendly rare earth magnets in wind turbines, motors and generators.

Kai Wu is currently a PhD candidate in the Department of Electrical and Computer Engineering at the University of Minnesota. In 2013, he earned his BS degree in Electrical Engineering with distinction from Northwestern Polytechnical University in China. His research is multidisciplinary and focuses on fabricating GMR nanosensors; biofunctionalizing sensor surface for biomarker detection; characterizing magnetization dynamics of SPIONs; developing a handheld platform based on GMR biosensors for on-site bioassays; and developing a search coil system for multiplexed immunoassays. He has published 15 journal articles and submitted 1 patent in these fields. He is a recipient of the Interdisciplinary Doctoral Fellowship (2016), the National Scholarship from the Chinese Ministry of Education (2010 and 2012), the Yajun Wu Fellowship (2011), the travel award from IEEE Magnetic Summer School (2016), awards from the IEEE Engineering in Medicine and Biology Society (2014) and the Council of Graduate Students from University of Minnesota (2016), as well as the best poster award from IEEE Magnetics Society (2016).

Diqing Su has been a Materials Science PhD student at University of Minnesota since 2015. She received her BS in Materials Physics at Fudan University, China, in 2015. Her research interest lies in the detection of biomarkers based on GMR sensors. The objective of her work is to accomplish early diagnosis of diseases and the fabrication of flexible biosensors which can conformally situate on biological tissues.

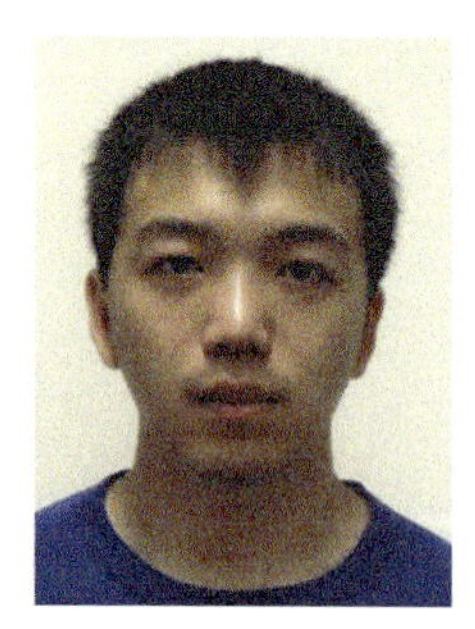

Yinglong Feng is currently a PhD candidate in the Department of Electrical and Computer Engineering at the University of Minnesota. He earned his BS degree in Control Science and Technology from Xi'an Jiaotong University in 2010. His research interests are primarily in the development of GMR and MTJ magnetic biosensors.

Section IV

In-Vivo Application of MNPs

14 Immunotoxicity and Safety Considerations for Iron Oxide Nanoparticles

*Gary Hannon, Melissa Anne Tutty and Adriele Prina-Mello**

CONTENTS

14.1 INTRODUCTION: CLINICAL APPLICATION OF IRON OXIDE NPs

Iron oxide nanoparticles (IONPs) possess unique characteristics that make them an attractive materials for a range of clinical applications. Many of these are discussed throughout the chapters in this book and include drug/gene delivery,[1,2] diagnostic imaging,[3] tissue engineering[4] and hyperthermia therapy.[5] Interestingly, in the literature, the range of diseases associated with these magnetic nanoparticles (NPs) is broad and includes cancer,[6] anaemia,[7] multiple sclerosis[8] and microbial infection.[9] Currently, IONP are the most clinically tested inorganic NPs.[10,11] However, despite this huge interest, and high biocompatibility, the therapeutic potential of IONP has yet to be successfully met in the clinic and many FDA (U.S. Food and Drug Administration) and EMA (European Medicines Agency) approved diagnostic IONP-formulations have been discontinued. There are many proposed reasons for this poor translation to the clinic, nonetheless oxidative stress and immunotoxicity caused by IONP are certainly contributing factors that must be considered and overcome in future translational research.

Feridex®, (Endorem®), Cliavist, Gastromark™ (Lumirem®) and Ferumoxtran-10 are IONP-containing MRI contrast agents that were approved for clinical use but that have subsequently been discontinued by their respective manufacturers.[11] The reasons for these failures are difficult to pinpoint in each individual case, and the literature exploring these reasons is scarce.[12] Likewise, intravenous IONP formulations have been shown to carry a risk of systemic inflammation and oxidative stress,[13,14] and although IONP dextran formulations (low and high molecular weight) reduced the incidence of adverse reactions, hypersensitive reactions are still reported.[14,15] Reactive oxygen species (ROS) have long been an outstanding toxicity factor associated with these nanoformulations which relates to the majority of their side-effects,[16,17] including immunotoxicity.[18] Importantly, a dextran coating is common to Feridex® (Endorem®), Cliavist and Ferumoxtran-10. Dextran–IONP formulations have been widely reported in the literature, proving to be a hugely versatile system for drug delivery, MRI, positron emission tomography (PET) and photodynamic therapy.[19] However, cytotoxic and immunotoxic effects have also been reported with these coatings.[20,21] Oxidative and immunostimulatory effects related to these formulations do occur, and may be holding potentially clinically relevant IONPs back.[22–24] Critically, the majority of IONP under preclinical and clinical development currently incorporate a dextran coating on their surface. Furthermore, one must consider that IONP can interact with cells and proteins of the immune

* Corresponding author.

TABLE 14.1
Clinical Trials Ongoing with Ferumoxytol (Trade Name Feraheme® in the United States and Rienso® in the European Union)

Application	Disease	Clinicaltrials. Gov Identifier
Imaging	Multiple sclerosis and demyelinating diseases	NCT01973517
	Brain tumours and central nervous system lymphoma	NCT00978562, NCT02359097, NCT00103038, NCT00659126, NCT02452216, NCT02466828, NCT00660543
	Bone sarcomas, osteomyelitis and osteonecrosis	NCT01336803, NCT02893293
	Triple negative breast cancer	NCT01770353
	Oesophageal neoplasms	NCT02253602, NCT02857218
	Lymphoma and sarcoma	NCT01542879
	Neuroendocrine neoplasm, cervical, breast, gastric, nonsmall cell lung cancer, small cell lung cancer and ovarian cancer	NCT02631733
	Colorectal cancer	NCT01983371
	Pancreatic cancer	NCT00920023
	Head and neck cancer	NCT01895829
	Lymph node cancer	NCT01815333
	Prostate, kidney and bladder cancer	NCT02141490
	Thyroid cancer	NCT01927887
	Neuroinflammation in epilepsy	NCT02084303
	Type-1 diabetes progression	NCT01521520
	Peripheral arterial disease	NCT00707876
	Myocardial inflammation/infarction	NCT02319278, NCT01995799
	Atherosclerosis and stroke	NCT01674257
	Whole body imaging: Cancer staging	NCT01542879
	Kidney transplant rejection	NCT02006108
	Hereditary haemorrhagic telangiectasia	NCT02977637
	Uteroplacental flow, perfusion, oxygenation and inflammation	NCT02791568
	Chronic kidney disease	NCT02997046
	Paediatric congenital heart disease	NCT02752191
	Solid tumours and breast cancer: Measuring tumour associated macrophages and predict treatment response	NCT01770353
	Inflammation during migraines	NCT02549898
	HIV infections: Inflammation in the brain	NCT02678767
Anaemia	Peritoneal dialysis patients	NCT01942460
	Preoperative management for cardiac surgery	NCT02189889
	Chronic kidney disease	NCT01227616
	Restless leg syndrome occurring with anaemia	NCT02499354

system when introduced into the blood circulation. IONP have been shown to interact with leukocytes and red blood cells, as well as thrombin, complement and many other plasma proteins depending on the IONPs physiochemical characteristics (PCC), including size, charge and surface chemistry.[25–27] Thus, avoiding or preventing harmful levels of oxidative stress and unintentional interactions with the immune system may prove clinically significant for the potential use of IONP.

With this issue in mind, it is worth mentioning the most successful IONP clinically approved to date, Ferumoxytol. It is an IONP with a hydrodynamic diameter of ~30 nm, coated with the modified dextran layer, polyglucose sorbitol carboxymethylether. Originally approved as a contrast agent for MRI, Ferumoxytol has since been tested extensively in a variety of clinical applications (Table 14.1), while also showing promising magnetic hyperthermia properties after a preclinical validation step.[28] This is one of the few IONP that has not been discontinued, showing a high level of flexibility in the clinic with a favourable safety profile. Therefore, IONP with a high level of efficacy and tolerability *in vivo* may have multiple therapeutic and diagnostic applications.

Limiting or preventing the immunotoxic effects of IONP requires early consideration of potential undesired interactions with the immune system, as well as enabling efficient design and specific immunotoxicity evaluation, thus saving time and financial resources. Therefore, the overall goal is to improve the carrier/particle design and preclinical assessment with a view to further enhance their translational value in the clinical setting as diagnostic, treatment or theranostic products.

14.2 IMMUNOTOXICITY AND SAFETY ISSUES ASSOCIATED WITH IONP

ROS have a major role in the immunotoxicity associated with IONP. Upon administration, IONP internalize into cells where they can induce ROS upregulation. If this process occurs within cells of the immune system, this may cause undesired immune dysfunctions, reducing the safety and efficacy of the drug. Hypersensitivity and other immune modulating effects with IONP are commonly observed during clinical application.[29,30] Physicochemical characteristics are often responsible for these effects, nonetheless if properly controlled they can also help predicting them, thus facilitating safer designs. Capitalizing on the past decade's extensive physicochemical and biological characterization knowledge, and expertise developed internationally, this chapter will describe the key aspects in immunotoxicity and safety associated with the translational process of IONP.

14.2.1 IONP Immunotoxic Profile

14.2.1.1 Coagulation System

The coagulation system deals with the haemostatic balance of procoagulant and anticoagulant factors in the blood. These factors are tightly regulated by a panel of blood cells (e.g. platelets and leukocytes) and endothelial cells.[31] IONP have been shown to interact with these cells and factors involved in this delicate balance. Evidence in the literature suggests both a procoagulant and anticoagulant effect on rodents upon exposure to these nanoformulations. An imbalance in this network can result in thrombotic or haemorrhagic issues. Zhu M-T et al.[32] evaluated the pulmonary responses of two different maghemite NPs (22 and 280 nm in size). In particular, a significantly prolonged prothrombin and activated partial thromboplastin time (time taken for blood to clot) was observed in rats intratracheally administered with low doses of 22 nm IONP (0.8 mg/kg/body weight) in comparison to the saline-treated controls.[32] This prolonged coagulation time has been attributed to oxidation and inflammation derived from these IONP. A prolonged coagulation time is a known risk factor in disseminated intravascular coagulation (DIC),[33] a life-threatening condition related to the depletion of coagulation factors and abnormal haemorrhaging.[34]

In contrast, Nemmar et al. found a conflicting result with magnetite (Fe_3O_4) with a size distribution between 4 and 6 nm.[35] After intravenous administration at doses of 2.0 and 10.0 μg/kg of body weight into mice, a statistically significant ($p < 0.05$ and $p < 0.001$, respectively) shortening of prothrombin and activated partial thromboplastin time occurred. A significant shortening of thrombotic occlusion time for both doses was also observed. Therefore, a dose-dependent procoagulant effect was observed in this case. Moreover, serum concentrations of PAI-1, a factor of the coagulation system involved in fibrinolysis, showed a significant, dose-dependent increase upon IONP exposure. Finally, platelet aggregation and cardiac oxidative stress (significant increases in ROS, lipid peroxidation and superoxide dismutase) were also shown to be dose-dependent. Therefore, thrombotic events and oxidative damage to the cardiovascular system are potential toxic effects with this IONP. A similar result was observed using polyampholyte-coated magnetite NPs (4.5 ± 1.4 nm).[36] This study reported a reduced prothrombin time for these particles in rabbit plasma at concentrations above 7.8 mM, concluding that safe administration was below this concentration. Conflicting findings such as these suggest that disorders in the coagulation system caused by IONP may be size-dependent. More studies are needed, however, to decipher the true extent of the role that PCC may play on this delicate system. Clearly, coagulation assessments are necessary in the early immunotoxic evaluation of an IONP.

14.2.1.2 Haemolysis

Haemolysis is the damage to red blood cells (RBC) in the form of cell membrane rupture which induces the release of intracellular proteins such as haemoglobin into the blood.[37] Haemolysis carries an array of pathological conditions such as anaemia, jaundice and even thrombotic events, as recent literature suggests.[38,39] Literature is scarce on the interactions with IONP and RBC; however, a study by Gaharwar and Paulraj identified changes to the RBC of Wistar rats administered with IONP.[25] IONP with size ranging between 30 and 35 nm were intravenously administered at 7.5, 15.0 and 30.0 mg/kg of bodyweight once a week for 28 days. Blood samples were obtained weekly, and haematological and oxidative parameters were measured. Significant decreases in RBC count percentages were observed at weeks 2 and 3 in rats administered with 15 and 30 mg/kg of body weight as compared to untreated controls. Haemoglobin levels were shown to decrease significantly in weeks one and two for all doses of the NP. Further results indicated an induction of lipid peroxidation in a time- and dose-dependent manner, where a 30 mg/kg of bodyweight IONP dose significantly reduced the antioxidants glutathione, catalase and superoxide dismutase in a time-dependent manner. Lesser doses showed varied significance over each time point. It was concluded that the IONP were able to reduce RBC in a dose- and time-dependent manner. In 2015, Ran et al. evaluated the dose- and time-dependant haemolytic activity of uncoated magnetite at a size distribution of 60–90 nm.[40] Doses from 25 μg/ml caused significant increases of phosphatidylserine exposure on the surface of RBC (human blood). This is a marker for RBC damage and eventually leading to eryptosis. A similar response was observed *in vivo* with a 12 mg/kg dose proving sufficient to induce significant levels of erythrocyte cell death in circulation. Significant dose- and time-dependant increases in ROS levels were also noted in this study. It appears IONP have a significant haemolytic role that may be time- and dose-dependent. It is therefore necessary to carefully select time points when evaluating this immunotoxicity. ROS appears to have a role in this effect and its upregulation could induce a systemic inflammation followed by a large-scale immunotoxic response.

14.2.1.3 Opsonization and Monocyte–Phagocytic System

When administered intravenously, NPs are instantly covered in blood-derived proteins. Blood plasma contains thousands of proteins that can bind to the surface of NPs, changing their PCC and altering their *in vivo* function.[41,42] Opsonin proteins are a subset of these and refer to any protein involved in phagocytic uptake of pathogens in the body, and include complement, nonspecific immunoglobulins, fibronectin and type-1 collagen. Once bound, they act as signals for phagocytic uptake and can result in off-target accumulation and immunotoxic effects.[43]

There is ample literature discussing the beneficial effect of IONP interactions with the monocyte-phagocytic system (MPS) in terms of site-specific targeting,[8] but immunotoxic effects related to key components of this process may hamper the beneficial effects in clinical setting, if not carefully considered.[44,45] Complement proteins make up 5% of the total globulins in serum and play a fundamental role in the uptake of NPs. Moreover, activation of this system can result in immune stimulation, inflammation and potentially anaphylaxis.[46] Complement activation-related pseudoallergy (CARPA) is an acute immune toxicity disease initiated through a complex network of immune cell signalling leading to hypersensitive reactions.[47] CARPA has been described as one of the fundamental reasons why IONP fail, with hypersensitivity in particular highlighted as a contributing factor to the discontinuation of dextran IONP (Feridex® and Ferumoxtran-10), two promising MRI contrast agents.[44] Dextran coatings have been noted as considerable factors in the onset of complement activation, but iron formulations with modifications to this polysaccharide could reestablish their clinical relevance.[48] Inturi et al. evaluated dextran-coated iron oxide nanoworms for their complement adsorption and subsequent uptake to leukocytes and platelets in normal and tumour-bearing mice, as well as in healthy and cancer patient blood samples.[26] The study found differences in leukocyte internalization via complement-dependent pathways between rodents and humans. Nanoworms showed complement-dependent uptake into monocytes, lymphocytes and neutrophils in both normal and tumour-bearing mice. In healthy humans and cancer patients, eosinophils, monocytes, neutrophils and lymphocytes internalized the nanoworms. Interestingly, this internalization was prevented in both mice and human blood when coadministered with complement inhibitors. This suggests a potential for combination therapies utilizing these two agents in the future to enhance IONP efficacy *in vivo*. Moreover, cross-linking and hydrogelation of the dextran coat with epichlorohydrin on these nanoworms was able to reduce opsonisation by over 70% in comparison to the noncrosslinked nanoworms. In humans, opsonisation was reduced by an average of 60%. A more recent study by Chen et al. recognized that the dextran shell on SPION may not directly induce opsonisation but the initial adsorption of serum and plasma proteins form a more attractive scaffold for complement 3 (C3) to bind.[49] This was confirmed when dextran-SPION were precoated with human serum/plasma, washed and then recoated and washed again. The recoated SPION underwent a significant increase in white blood cell uptake *in vitro* against both the precoated and uncoated SPION. The experiment was repeated *in vivo* and showed a faster C3-dependant opsonisation with precoated SPION. Inclusion of antiproperdin, an inhibitor for the alternative pathway of complement activation, or ethylenediaminetetraacetic acid (EDTA), an inhibitor of all complement pathways, could block opsonisation. Evaluating the interaction of IONP with opsonins may allow for potential design adjustments or combination therapies to be considered providing a better chance of clinical translation in the future.

14.2.1.4 Protein Corona: A Potential Influence?

The protein corona refers to the entire protein layer adsorbed onto a NP's surface, which forms their biological identity in the blood. This is a dynamic coating that can alter the PCC of the NP (e.g. size, surface charge and colloidal stability), effecting its *in vivo* function and distribution.[50] There is limited literature on the characterization of the protein corona with IONP that provides details of its overall *in vivo* safety and efficacy. Importantly, human serum proteins have been shown to reduce the magnetization saturation, heating capabilities and cellular uptake of magnetic NPs,[51] while foetal bovine serum (FBS) can increase relaxivity of negatively charged SPION and decrease relaxivity in positively charged SPION.[52] Furthermore, some literature has reviewed the potential of the protein corona to elicit immunotoxic effects. The protein corona may indirectly induce immune suppression, stimulation or avoidance.[53,54] Proteins derived from coronas of a variety of NPs were found to possess immunotoxic potential when bound to the NP surface. Interestingly, bovine serum albumin has been shown to inhibit monocyte uptake but induce macrophage uptake when bound to polymer NPs coated with poly(methacrylic acid), suggesting alternative immunological responses depending on the local environment.[55] Generation of ROS has been also linked to corona formation. *In vitro* studies show different levels of ROS are generated depending on the charge and media in which IONP are incubated. Bare and NH_2-IONP treated with cells in culture media supplemented with FBS had no change in ROS levels in relation to untreated control whereas slight dose-dependent increases could be observed with cells treated in the absence of serum.[56] Suggesting a potential protective role for the corona in this case.

In addition to ROS levels and immunotoxicity, corona studies can provide greater knowledge on the distribution of the NP *in vivo*. SPIONs of a size range between 3.5 and 15 nm with negatively, neutrally and positively charged surfaces form a protein corona after incubation in FBS. It was found that the smaller SPION bound to the low-molecular weight proteins whereas the larger SPION bound to high-molecular weight proteins, but concentration and composition of these coronas appeared random. *In vivo* studies showed the 3.5 nm neutral and negative SPION to be localized in the brain 5 min after intravenous administration, whereas positively charged ones did not.[57] Therefore size and charge may significantly affect the protein corona formation and distribution of SPIONs. A further point to consider is that the protein corona formation has been shown to vary between individuals[58] and diseases,[59] making protein corona-specific design for iron oxide nanomedicines difficult. However, there is literature to suggest that the formation of a protein corona may be used as a diagnostic tool for cancer,[60] therefore, a level of similarity may indeed exist within a disease. This may be an important design factor to consider as clinically successful IO-formulations may not be immediately effective for alternative diseases or orphan applications. Considerations about the specific corona for diseases may be an effective means of efficient transfer.

Similarly, Sakulkhu et al. evaluated the corona formation on SPION coated with either dextran or polyvinyl alcohol polymer (PVA) with a neutral, negative or positive charge (14.2).[27] The dextran-coated SPIONs represent a relevant clinical model for these particles, as highlighted in Section 14.1. Specific protein adsorption with FBS proteins was characterized and described as a percentage of the total protein corona for each SPION. Moreover, each SPION blood circulation time was evaluated in female Lewis rats (Table 14.2).

Results reported in Table 14.3 revealed that about 20% of proteins on the surface of PVA-SPION and dextran-SPION were shared, indicating large protein adsorption differences between coatings. More proteins were found to bind to PVA than dextran-SPION, resulting in longer blood circulation time for PVA-SPION. Charge also seemed to have a role in the components and affinities of the corona in both SPION, with different proteins showing a preference and binding strength

TABLE 14.2
Characteristics of SPION Tested for Their Protein Corona

SPION	Functional Group	Hydrodynamic Diameter (nm)	Zeta Potential (mV)
PVA coated	NH_2	37.9 ± 3.0	+16.7 ± 1.5
	OH	28.3 ± 2.1	+1.5 ± 2.2
	COOH	38.1 ± 4.2	−5.9 ± 0.9
Dextran coated	NH_2	18.8 ± 0.2	+36.5 ± 1.8
	OH	19.7 ± 0.4	−17.9 ± 1.1
	COOH	18.3 ± 0.2	−26.7 ± 0.5

Note: PVA: polyvinyl alcohol polymer.

for different surface charges. Interestingly, there are no proteins which bind predominantly to neutrally charged dextran-SPION, whereas all other SPION have a unique protein cloak when in circulation (Table 14.3).

Such studies point to the complexity underlying the interactions between NPs and proteins of the blood. Importantly, magnetic fields can also alter the corona formed onto magnetic particles, changing their cellular uptake.[61] Corona characterization alone is not enough and must be coupled with uptake, distribution, toxicity and functional assays to build an in-depth picture of *in vivo* efficacy. A study emphasizing the power of the corona,[62] noted that the presence of serum proteins was vital for folic acid-functionalized IONP internalization into ovarian cancer cells. Vogt et al. concluded the presence of a protein corona promoted silica coated-SPION internalization into primary human macrophages and increased r_1/r_2 values while no changes were observed with dextran-SPION.[63] Therefore, efficient evaluation of the protein corona early in the design and testing of an IONP may provide a means for predicting initial cytotoxicity, immunotoxicity, distribution, uptake and functional characteristics of the NP *in vivo*, before it enters further animal and clinical studies. Furthermore, considerations in patient-to-patient variability and increasing the power of these studies is a must for reliable interpretation in the future.

TABLE 14.3
Protein Corona Specific to Dextran and PVA-SPION over Various Surface Charges

A. Positively Charged PVA and Dextran Coated SPION: Protein Quantification. Note Last Column Details Common Protein Identified on Both SPION.

(+) PVA-SPION 69%	(+) Dextran-SPION 10%	(+) Common 21%
Hemicentin-1	Complement factor B	Complement C3
α2-macroglobulin	α1-antitrypsin-like protein	Collagen α1(I) chain
Complement factor H	Haemoglobin subunit α-I/II	Serotransferrin
Fibrinogen α-chain	Apolipoprotein A-II	Prothrombin
Cytochrome P450 2C5		Serum albumin
α2-Antiplasmin		α-fetoprotein
Vitamin D-binding protein		Kiniogen-1
α1B-glycoprotein		Fetuin-B

B. Negatively Charged PVA and Dextran Coated SPION: Protein Quantification. Note Last Column Details Common Protein Identified on Both SPION.

(−) PVA-SPION 64%	(−) Dextran-SPION 15%	(−) Common 21%
Plasminogen	α1-Antiproteinase	Serrotransferrin
Complement factor B	Thyroxine-binding globulin	Prothrombin
Fibulin-1	Endopin-1	Serum albumin
Testis-specific Y-encoded-like protein 2	Fetuin-B	α-Fetoprotein
Mutated melanoma-associated antigen 1	Transthyretin	Kininogen-1
Vitronectin	Haemoglobin subunit α	Fibrinogen α-chain
α2-antiplasmin	Apolipoprotein A-II	Cytochrome P450 2C5
Vitamin D-binding protein		Antithrombin-III
α1B-glycoprotein		Factor XIIa inhibitor
Serine protease inhibitor A3F		Angiotensinogen

C. Neutrally Charged PVA and Dextran Coated SPION: Protein Quantification. Note Dextran Column with No Protein Binding. Common Proteins Identified on Both SPION Are of Low Predominance.

(0) PVA-SPION 83%	(0) Dextran-SPION 0%	(0) Common 17%
Ecoto-ADP-ribosyltransferase 4		Serotransferrin
Apolipoprotein A-I		Prothrombin
α1-Acid glycogoprotein		Serum albumin
Tetracan		α-Fetoprotein
Complement C3		Kininogen-1
α2-Macroglobulin		Fibrinogen α-chain

14.2.1.5 Immune System

IONPs can be engineered to stimulate, suppress or avoid the immune system, depending on the desired *in vivo* function. Intentional immune stimulation related to IONP can be seen in their potential role as adjuvants in nanovaccines,[64] while unintentional stimulation may result in immunotoxic effects related to hypersensitivity.[29] By contrast, their immune suppression properties include reducing inflammation and hypersensitivity.[65,66] However, these mechanisms are yet to be fully elucidated with IONP and could therefore be harmful if they occur unintentionally.[67]

The major regulators of the immune system are T lymphocytes (T cells), the main effector cells in immune responses. Cytokines are required for activation, differentiation and T cell function.[68] Many papers have investigated the effects of IONP on regulation of T cells and cytokines. T cells have long been known to regulate the levels of iron in the blood, with low levels of these cells giving poorer prognosis for individuals with haemochromatosis.[69] Likewise, iron deficiency anaemia has also shown to impact the levels of T cells.[70] It is therefore suggested that IONP may have a major role in the regulation of these cells. A study revealed that Cliavist administered into mice at doses of 0.6 and 10.0 mg Fe/kg deregulated pro-inflammatory cytokine production (IL-6, TNF-α and IFN- γ) through interactions with T-helper 1 (Th1) cells and macrophages to significantly suppress inflammatory effects associated with delayed-type hypersensitivity.[66] Fe_3O_4 NPs ranging from 20.3 to 95.5 nm, with a surface charge of approximately −21 mV, did not influence the levels of cytokines derived from Th1 or T-helper 2 cells (IL-2, IL-4, IL-10 and IFN- γ) in mice who were orally administered doses up to 1.2 g/kg of NP.[71] The levels of T lymphocytes ($CD3^+CD4^+$ and $CD3^+CD8^+$), however, were significantly higher in mice treated at this high dose, compared to untreated controls. Elsewhere, <25.0 nm dextran coated IONP produced no immunological or oxidative response (up to 1 mg/ml at 24 h), while 9 nm dextran-coated IONP were shown to significantly increase T and B-lymphocyte proliferation and IL-1β levels in rats 7 days after exposure with 10.0 mg/kg of the NPs.[21,72] This immune modulation was short lived and levels normalized after 14 days. Time points for T lymphocyte modulation *in vivo* is therefore critical to these studies in the future.

Monocytes are bone marrow-derived progenitors with primary roles in phagocytosis and inflammation. Upon stimulation, these cells migrate to sites of inflammation where they differentiate into dendritic cells or macrophages to clear potential infections.[73] IONP have been shown to interact with a variety of monocytes inducing potential immunotoxicities. Xu et al. evaluated the response of monocytes to IONP *in vitro*.[74] These NPs (with hydrodynamic sizes between 26.0 and 30.0 nm) were coated with either poly(acrylic acid) (PAA), polyethylenimine (PEI) or poly(ethylglycol) (PEG) leading to a negative (−50.0 mV), positive (51.0 mV) or neutral charge (−3.0 mV) respectively. With concentrations of up to 50.0 μg/ml, no significant changes in TNF-α and TLR-2 were established. Furthermore, cell viability was only slightly affected with each of these NPs, thus suggesting no major innate immune response and minimum toxicity upon treatment with each IONP.[74]

Couto et al. investigated the interaction with PAA-coated and uncoated IONP with neutrophils.[75] This *in vitro* test, with both particles close to 10 nm, revealed both could significantly induce the activation of neutrophils, signified by the level of oxidative bursts within the neutrophils (100.0 μg/ml for PAA-IONP and 20.0 μg/ml for bare-IONP). Furthermore, at 100.0 μg/ml, PAA-IONP induced a significant increase in apoptosis in neutrophils while the percentage of apoptotic cells decreased significantly upon exposure to naked IONP. This suggests that a PAA coating can be toxic to these cells while the uncoated IONP are protective.

Strehl et al. provided a comprehensive study of the interaction between amino-polyvinyl alcohol coated-SPION (a-PVA-SPION) and the immune system.[76] These SPION, characterized at 31 ± 10 nm and 22 ± 6 mV, were assessed in human whole blood (19 rheumatoid arthritis and 18 healthy individuals) to closely resemble the *in vivo* scenario. Furthermore, individual cells were isolated ($CD4^+$ T cells, $CD14^+$ monocytes and $CD15^+$ granulocytes) to evaluate specific SPION–cell interactions. Results showed no change in survival to the human immune cells in whole blood experiments but a dose-dependent increase of cytokines was noted (Table 14.4). IL-1β upregulation was shown to be primarily derived from granulocytes and monocytes. Moreover, the survival of monocyte-derived macrophages was significantly increased when exposed to SPION (100–1000 μg/ml) early in its differentiating process. This paper suggested a 'potentially clinically relevant interaction of a-PVA-SPION with immune cells.'

TABLE 14.4
Significant Change in Levels of Cytokines in Healthy and Rheumatoid Arthritis Patients after Treatment with 100 μg/ml of a-PVA-SPION

Rheumatoid Arthritis Patients	Healthy Individuals
IL-1β	IL-1β
IL-4	IL-4
IL-6	IL-6
IL-8	IL-8
IL-9	MIP-1b
IFN- γ	RANTES
MCP-1	
MIP-1a	
MIP-2b	
PDGF	

Note: Exposure of 20 hours can significantly increase cytokine expression in blood. This response may vary between diseases.

There is a lot of literature on the role of IONP interactions with immune cells. Slight changes in PCC can have a significant effect on the immunological role of the formulation so it is essential that this be extensively researched during the design and testing process. Moreover, time-points are essential to establish the long-term effects of these drugs for which evidence is largely lacking.

14.3 MAIN ELEMENTS OF IMMUNOTOXICITY ASSESSMENT

Immunotoxicity still represents a huge concern for clinically approved IONP and those in the pipeline, as presented above. It is therefore essential that immunotoxicity is evaluated early and efficiently during the development and testing of these NPs.

14.3.1 Endotoxin Contamination and Sterility

One well-established issue with regards to NP immunotoxicity and safety is the obstacle of endotoxin contamination.

Endotoxin (or lipopolysaccharide, LPS) is a large hydrophobic molecule that forms part of the cell wall of gram-negative bacteria. It plays a large role in the failure of many nanodrugs in their early stage and, as discussed by Dobrovolskaia and McNeil in 2012, endotoxin contributes to the early failure of more than 30% of nanoformulations during their preclinical studies.[77] This can be seen specifically with regard to IONP, with the vast majority of them either failing before they reach clinical trials, or being discontinued once on the market.

In mammalian cells, endotoxin binds to the TLR-4 receptor and causes upregulation, secretion and synthesis of a variety of proinflammatory factors, such as IL-6, TNF-α and IL-1β. This, in turn, causes both inflammation and immune activation, with as little as 0.1 EU/ml (endotoxin is quantified in the regulatory setting as endotoxin units, EU) upregulating inflammatory gene expression significantly.[78,79] Since endotoxin is responsible for a variety of serious health implications in humans, ranging from respiratory problems such as asthma and pulmonary inflammation,[80,81] to fever, shock and impaired organ function,[82] there is the need for reliable detection and quantification methods for assessing endotoxin contamination.

Endotoxin is commonly found in glassware and laboratory reagents and chemicals, and complete sterilization of this contaminant from materials is rendered difficult due to its ubiquitous and heat-resistant nature. Unless pyrogen-free materials are used during the synthetic stage, NPs will likely become contaminated with endotoxin early on, an issue that causes significant impact on both *in vivo* and *in vitro* experiments. TiO_2 NPs have shown high affinity for endotoxin resulting in a synergistic increase in IL-6 and TNF-α secretion followed by macrophage activation *in vitro*.[83] This can explain the inflammatory responses experienced during *in vivo* studies.[84] By contrast, a similar study by Grosse et al.[85] revealed endotoxin could readily adsorb to 10 and 30 nm IONP coated with oleic acid and amphiphilic polymer, which reduced TLR-4 activation of endotoxin and its CD14/TLR-4 dependent cellular internalization. Moreover, after treating primary human monocytes with these NPs, TNF-α produced by endotoxin was reduced.[85] Differences in affinities for endotoxin may alter the downstream immune response differently. With regards to preventative strategies, a review by Li and Boraschi[86] discussed in depth the various detection and elimination methods for endotoxin-contaminated nanoformulations. Here, an emphasis was placed on sterility during the synthesis phase, with an overall goal of reducing the level of endotoxin early on, so as to not rely on testing and elimination processes further down the line.[86]

The original test used for determining the presence of endotoxin in nanoformulations was the rabbit pyrogen test (RPT); however, in 1973 this was somewhat replaced by the quantitative, time-effective and ethically sound limulus amoebocyte lysate (LAL) assay. The RPT test is undertaken by measuring the increase in a rabbit's temperature following the administration of a product containing endotoxin. As endotoxin is a pyrogen (fever causing substance), an increase in temperature is observed in the rabbits. This is undertaken to mimic the effect endotoxin would have in humans, with the subsequent temperature increase measured in order to determine the level of endotoxin present in the administered sample. As well as the lack of practicality associated with the RPT in terms of time, labour (each animal needs to be monitored for 3 h postinjection at 30 min intervals) and ethical issues, it has also been noted that false positives are common, and there is difficulty in determining whether the results yielded are due to endotoxin presence or other endogenous or exogenous factors. It is also difficult to manage the test animals, as all animals must remain free from temperature increase for minimum of 2 weeks prior to experiments, and also must be untouched for 2 days after the rest period before more tests. Furthermore, large quantities of animals may be needed, making this test laborious; consequently, in recent years, there has been a great move towards using the LAL assay for endotoxin testing.[87] Despite this move towards LAL testing, there are still times when the traditional method is preferred. The issue of interference and the difficulty in detecting endotoxin contamination *in vitro* may be the reason why in certain cases researchers revert back to the more complicated *in vivo* pyrogen test, as illustrated by Vermeij et al.[88] This study reported the effects of bare 14.0 ± 2 nm SPION and 45.0 ± 3 nm PVA-coated SPION on experimental arthritis *in vivo*. This study utilized TLR-4 -/- mice and exploited the mechanism by which LPS binds to TLR-4. When this occurs, the proinflammatory cytokines TNFα and IL-1β are released. When both SPIONs were injected into the mice at high doses (locally: 24.0 μg Fe in 6 μl, and intravenously: 33.3 mg Fe/kg in 200 μl), inflammatory responses similar to those in C57BL/6 mice were observed, illustrating that signalling was not mediated by TLR-4, and disqualifying the likelihood of endotoxin contamination. This can be seen as an example of a situation where returning to the traditional testing method proved more functional.[88]

The LAL assay is much less time consuming and can be routinely undertaken in research laboratories, making it the current assay of choice for easy determination of the presence of endotoxin in biological products, medicines and nanomaterials. This is the current assay requested by regulatory authorities as a means of assessing endotoxin contamination in wide arrays of nanomaterials, including IONPs. The presence of endotoxin can be detected and quantified using the LAL assay down to levels as low as 0.01 EU/ml, with certain assays having an even greater sensitivity of 0.005 EU/ml. Generally, three forms of LAL assays are used for endotoxin detection and quantification: (1) gel-clot LAL, (2) chromogenic LAL and (3) turbidimetric LAL. Each of these assays has their own characteristics, format and mechanism of action (as shown in Figure 14.1). The chromogenic LAL is fully quantitative with high sensitivity, while the gel-clot LAL is low cost, is simpler and needs no special equipment.[86] A summary of the various LAL formats, their advantages and disadvantages can be found in the 1995 review by Hurley J.C., *Endotoxemia: methods of detection and clinical correlates*.[89]

When dealing with IONP, the LAL format most suitable is the chromogenic LAL, as it provides the least interference and the most accurate results, provided suitable controls are

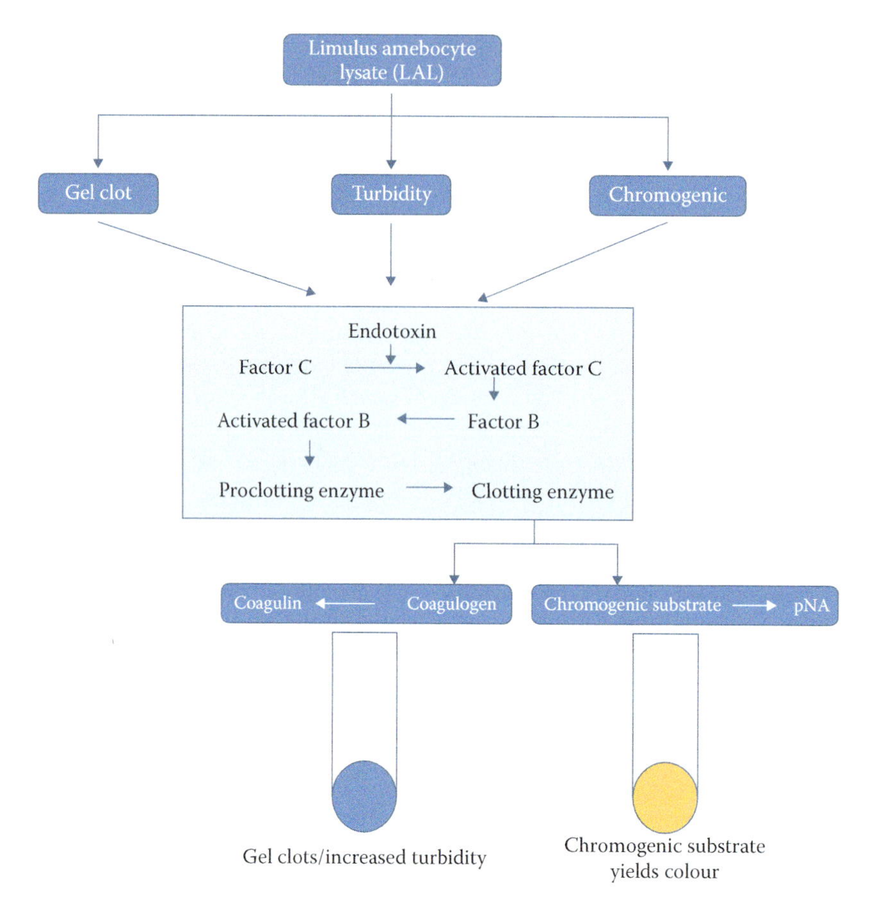

FIGURE 14.1 Three LAL formats and their mechanisms of action. Endotoxin induces a cascade of events associated with different measurable endpoints.

used. In a recent study undertaken by Dobrovolskaia et al., Ferumoxytol was tested using three different LAL formats (gel-clot, chromogenic endpoint and kinetic turbidimetric) and yielded consistent, low endotoxin levels for all three.[90] This reassuring result may be among the reasons why Ferumoxytol has been so successful in the clinic to date.

It is also worth noting that LAL assays are not independent of challenges, such as assay interference when used to test NPs. In fact, NP chemistry, concentration and coating can all interfere with the results of LAL assays, and it is vital to choose the right LAL format for the NP being tested. While it is possible to overcome issues with interference via sample dilution, care must be taken not to dilute the sample to the point that the endotoxin is undetectable.[91] Careful consideration to interference/enhancement controls is also necessary for successfully endotoxin evaluations using these assays.

Despite being cost effective, the gel-clot LAL suffers from result variability, is only semiquantitative and has much lower sensitivity than its counterparts. High concentrations of metal NPs have been shown to interfere with the clotting cascade mechanisms assessed within the assay.[92] Iron–silica NPs of size 3–5 nm interfered with this assay, preventing the gel to clot, while larger iron–silica NPs did not. Moreover, Endorem®, the dextran-coated IONP with a smaller size than the smaller iron–silica NPs was also observed not to interfere with this assay at concentrations 10-fold that of the small iron–silica concentrations used in the assay.[77,93] It is also worth noting the results are subjective and based on observations. As detailed by the U. S. pharmacopoeial (USP) in Chapter 89 of USP 29, the gel-clot LAL is also commonly used as a deciding factor in situations where there is ambiguity or doubt over results of the other LAL formats.[94] Despite the USP recommendation, this does not seem to be the case, and the gel-clot LAL assay is not an appropriate format to resolve discrepancies.[90]

Both the turbidimetric and chromogenic LAL assays have low specificity and are expensive, requiring special incubating plates or tube readers. The kinetic forms of both assays are time consuming and require special equipment, and the endpoint assays are less sensitive, with the endpoint turbidimetric assay not being commercially available. The main issue with the kinetic and chromogenic LAL assays lies in the interference seen with both. NPs at high optical densities interfere with the turbidimetric assay, with absorbencies close to 405 nm or 540 nm interfering with the chromogenic assay (Table 14.5). Despite the understanding that the chromogenic LAL is the most appropriate format for IONP, there is still data to suggest that these NPs will interfere with the chromogenic assay. As detailed by Grosse et al.,[85] where there is ambiguity over the accuracy of the LAL format used, a more sensitive cell-based reporter assay may be utilized. And so in this study, after discovering interference with the chromogenic

TABLE 14.5
Mechanisms of Interference Summarized with the Gel Clot, Chromogenic and Turbidimetric Assays

Assay	Gel Clot	Chromogenic	Turbidimetric
Mechanisms of interference	Solutions outside pH range 6 to 8	Solutions outside pH range 6 to 8 NPs with high absorbance in the 405–410 nm or 540–550 nm region of light	Solutions outside pH range 6 to 8 NPs with intrinsically high optical density

Note: The gel-clot assay has the broadest specificity of the three assays. NPs that interfere with the clotting cascade through mechanisms such as protein denaturing or chelating ions can interfere with each of these assays. These interferences can manifest as either inhibitions or enhancements within the assay. Lowering the concentration of these nanoformulations may remove this interference. Using more than one LAL format is advised for confirming results.

assay, the luciferase reporter cell assay was used as an appropriate replacement.[85] Consideration can be therefore made in a summary format where not all NPs can be tested using the same LAL format, as reported in Table 14.5.

As endotoxin testing can quickly become a costly endeavour if multiple variations of NP are studied, this may introduce an additional cost for the smaller laboratories. It has also been shown that no LAL format is optimal for multiple types of clinical-grade nanoformulations, including the SPION Ferumoxytol, and other NP formulations such as Abraxane® (nanoalbumin-bound paclitaxel) and Depocyt® (lipid-based particles encapsulating cytarabine).[90]

When discussing IONP sterility assessment *in vitro*, it is also important to discuss the issue of mycoplasma. Mycoplasma and its detection is determined using the recommendations from the EU-NCL (European Nanomedicine Characterisation Laboratory) in their standardized set of guidelines for NP characterization, the assay cascade. Various methods for detection of mycoplasma may be employed, but the most common utilizes PCR (polymerase chain reaction). Following incubation with an appropriate indicating cell line (to allow for amplification of low-grade mycoplasma growth), a PCR system may be used to detect mycoplasma DNA that encodes specific 16S rRNAs. This PCR system, as described by Gopalkrishna et al. is undertaken in combination with RFLP (restriction fragment length polymorphism) to detect and identify species-specific mycoplasma that are involved in contamination.[95] Optical biosensors based on fluorescence resonance energy transfer (FRET) may also be used for detection, as can fluorescence microscopy employing suitable fluorescent dyes.[96]

Ultimately, advancements in the area of nanomaterial sterility determination will be the key for the advancement of nanomaterials. In order to preserve their future in the clinic, there is great need to focus attention on contamination early on in the process to maximize success.[97]

14.3.2 Blood Compatibility

In addition to sterility assessments, it is vital to have a standardized set of tests or assays to evaluate potential interactions that IONP may have with the blood components. To fully quantify the interactions with blood components, various areas need to be systematically studied, including coagulation, haemolysis, platelet aggregation, lymphocyte proliferation and activation of the complement system.[98]

In relation to immunotoxicity, there are no *in vitro* tests that are validated as a suitable replacement for *in vivo* testing. A combination of *in vitro* and *in vivo* is recommended to fully assess NP formulations for their blood compatibility. Various test methods are described in ISO Standard 10993-4, a recognized suitable document that discusses tests for blood compatibility with medical devices.[99] Blood compatibility tests are not required for conventional pharmaceuticals, so these are the only toxicity guidelines currently available for the assessment of NP formulations.[99] The haemolysis assay, a simple test to assess blood compatibility, is a particularly significant test to perform under the blanket of haematology, since it measures how RBC membranes behave when they come into contact with nanomaterials. RBC are isolated by centrifugation and washed with saline, then are suspended in phosphate-buffered saline (PBS). A RBC suspension is then prepared and incubated at 37°C for 1 hour, before being centrifuged again. The absorbance of the resulting supernatant is then read at 541 nm, and quantity of released haemoglobin that has been destroyed by NPs is measured. For this assay, a dose–effect assay should be observed, using a realistic range of concentrations, as well as a suitable positive and negative control. For NPs who show satisfactory blood compatibility, i.e. no disturbance, they are suitable for IV administration. White blood cell count or leukocyte count may also be studied to observe the impact of immunotoxicities associated with these components.[100]

When considering blood compatibility, coagulation or clotting must also be considered. Clotting must be observed in a concentration range of NP. For IONP, it is expected that an increase in concentration will have a greater impact on coagulation that a lower concentration.[101] A similar assay may also be undertaken to determine anticoagulant activity and plasma clotting time.

Thrombosis must also be taken into consideration when examining NP blood compatibility to assess the effect of IONP exposure on platelet count and aggregation.[100]

Finally, complement activation following exposure to IONP should be evaluated. Tests relating to the complement system fall under the category of immunological analysis and include complement activation assays for a variety of complement complexes, including iC3b, SC5b–9, C4D and iC3b.[102]

Owing to the variability in IONP composition, inconsistencies between data sets and comparability among experiments are reported. For example, the interference of IONP on blood coagulation, in a similar way to how they affect platelet aggregation, is altered depending on particle size, concentration and coating.[77] Considering their uses as contrast agents in diagnostics, IONP may have future applications as drug delivery agents through various administration routes, therefore it is vitally important to expand the variety of tests available to assess the interaction these particles will have with blood components.

14.4 IMPROVING EARLY DESIGN, ASSESSMENT AND SAFETY CONSIDERATIONS

Avoiding adverse effects on the immune system is the next step to implement and maintain IONPs in the clinic. A broad knowledge of potential blood component interactions and careful consideration to sterile synthesis can improve the safety and efficacy of IONP. To achieve this point, novel design for immune avoidance, paired with efficient immunotoxic assessment and controlled sterility throughout these stages is critical (Figure 14.2).

14.4.1 Efficient Design of IO Formulations

IONP are hugely flexible in design. Avoiding dose-dependent adverse effects could be as simple as changing the physiochemical characteristics of the IONP. Size varies hugely between IONP for clinical applications; for instance, Ferumoxytol® has an average hydrodynamic size of 31.0 nm, whereas Cliavist is about 60 nm, and Feridex® (Endorem®) between 120 and 180 nm.[28,103] A small change in size can have a fundamental effect on the biodistribution of NPs.[104] Moreover, polymer coatings (e.g. poly(ethyl glycol)) aid in avoidance of the immune system.[105,106] Among the IONP approved for clinical use, iron–dextran formulations have dominated this setting, although they have been discontinued, whereas coatings such as sucrose, gluconate and maltose are seemingly safer.[10] Recent research has revealed that such coatings can improve the circulation time and immunotoxic profile of IONP. Naked 5 and 30 nm IONP had a sixfold cytotoxic effect on porcine aortic endothelial cells exposed to 0.5 mg/ml of drug over IONP coated with PEG, dextran and polysaccharide.[20] Coated IONP did not affect cell viability and elongation, whereas naked IONP significantly increased ROS levels in these cells as compared to the former. Despite these improvements, PEGylated IONP are yet to advance to the clinic and dextran coatings appear insufficient at covering the entire iron oxide core.[107] Other notable biocompatible coatings reported in the literature, and in use with IONP, include polyvinyl alcohol, dimercaptosuccinic acid, chitosan and alginate.[5,108,109] It is worth extensively searching the literature to identify the most suitable coating for the desired *in vivo* application. Coatings have been shown to aid or hinder specific functions of IONP. For instance, a comprehensive review examined the current trends in polymer coatings for IONP in the field of stem cell labelling.[110] It was discussed that

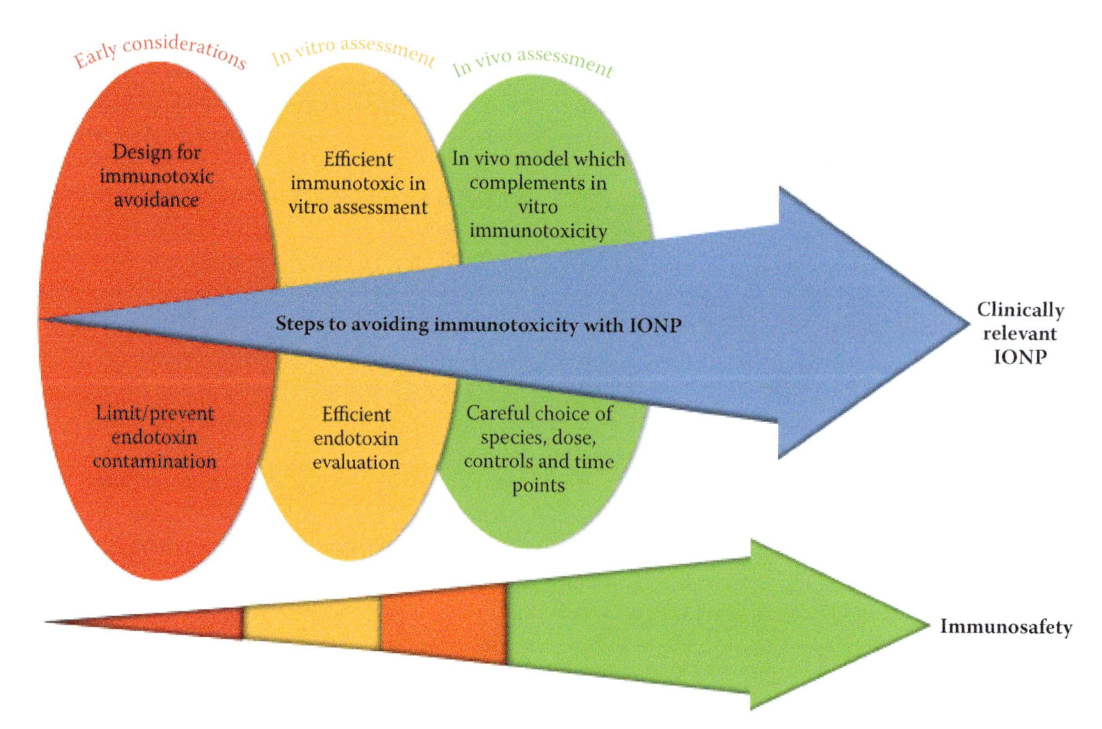

FIGURE 14.2 The way forward: improving the path to the clinic for IONP. Immunotoxic effects are major hurdles for IONP translation into the clinic. Efficient considerations to these at the development and *in vitro/in vivo* stages may provide better success for these promising nanoformulations in the future.

polymer coatings can have unique characteristics that may be favourable depending on the mechanisms for cell-specific uptake. Coating thickness has been shown to influence r_1 and r_2 relaxivity and hyperthermia potential of IONP.[111,112]

Therefore, biocompatible coatings appear to be a double-edged sword in the structure of IONP. Although providing a means for significant improvement in circulation time and reducing immunotoxic effects, they can also provide a false sense of security with regards to safety and efficacy that can only be fully investigated during *in vivo* or clinical studies. Moreover, coatings can be insufficient at completely covering the iron oxide core, exposing sites for involuntary interactions. Research into potential coatings for novel IONP is necessary and must be considered along with the desired clinical application of the technology.

14.5 CONCLUSION AND FUTURE PERSPECTIVES

Iron oxide NPs are a hugely promising nanomaterial designed and already used in a wide array of clinical applications. Nonetheless, several formulations have been withdrawn or discontinued in the clinic because of mild to overt adverse effects, including immunotoxicity, in spite of their proven technical performance. However, more recent developments in the literature have reconsidered such interactions with the immune system, increasing knowledge in this field. Improvements to early full cascade characterization, focused on efficient and sterile design, as well as immunotoxic assessment may enable this innovative technology to reach its full potential to revolutionize the diagnosis and treatment of many diseases in the near future.

ACKNOWLEDGEMENTS

The authors would like to thank the following funding agencies: Irish Research Council (IRC) postgraduate fellowship for G.H. financial support. NOCANTHER project (European Commission H2020, grant ref# 685795) for M.A.T. financial support. EU-NCL project (EC, H2020, grant ref# 654190) for the critical aspects associated with the characterization of IONP.

REFERENCES

1. Li, D.; Tang, X.; Pulli, B.; Lin, C.; Zhao, P.; Cheng, J.; Lv, Z.; Yuan, X.; Luo, Q.; Cai, H.; Ye, M., Theranostic nanoparticles based on bioreducible polyethylenimine-coated iron oxide for reduction-responsive gene delivery and magnetic resonance imaging. *International Journal of Nanomedicine* **2014,** *9*, 3347–3361.
2. Unterweger, H.; Tietze, R.; Janko, C.; Zaloga, J.; Lyer, S.; Dürr, S.; Taccardi, N.; Goudouri, O.-M.; Hoppe, A.; Eberbeck, D.; Schubert, D. W.; Boccaccini, A. R.; Alexiou, C., Development and characterization of magnetic iron oxide nanoparticles with a cisplatin-bearing polymer coating for targeted drug delivery. *International Journal of Nanomedicine* **2014,** *9*, 3659–3676.
3. Briley-Saebo, K. C.; Mani, V.; Hyafil, F.; Cornily, J.-C.; Fayad, Z. A., Fractionated feridex and positive contrast: In vivo MR imaging of atherosclerosis. *Magnetic Resonance in Medicine* **2008,** *59* (4), 721–730.
4. Ziv-Polat, O.; Margel, S.; Shahar, A., Application of iron oxide nanoparticles in neuronal tissue engineering. *Neural Regeneration Research* **2015,** *10* (2), 189–191.
5. Kossatz, S.; Grandke, J.; Couleaud, P.; Latorre, A.; Aires, A.; Crosbie-Staunton, K.; Ludwig, R.; Dähring, H.; Ettelt, V.; Lazaro-Carrillo, A.; Calero, M.; Sader, M.; Courty, J.; Volkov, Y.; Prina-Mello, A.; Villanueva, A.; Somoza, Á.; Cortajarena, A. L.; Miranda, R.; Hilger, I., Efficient treatment of breast cancer xenografts with multifunctionalized iron oxide nanoparticles combining magnetic hyperthermia and anti-cancer drug delivery. *Breast Cancer Research: BCR* **2015,** *17* (1), 66.
6. Revia, R. A.; Zhang, M., Magnetite nanoparticles for cancer diagnosis, treatment, and treatment monitoring: Recent advances. *Materials Today* **2016,** *19* (3), 157–168.
7. Spinowitz, B. S.; Kausz, A. T.; Baptista, J.; Noble, S. D.; Sothinathan, R.; Bernardo, M. V.; Brenner, L.; Pereira, B. J. G., Ferumoxytol for treating iron deficiency anemia in CKD. *Journal of the American Society of Nephrology: JASN* **2008,** *19* (8), 1599–1605.
8. Mahmoudi, M.; Sahraian, M. A.; Shokrgozar, M. A.; Laurent, S., Superparamagnetic iron oxide nanoparticles: Promises for diagnosis and treatment of multiple sclerosis. *ACS Chemical Neuroscience* **2011,** *2* (3), 118–140.
9. Arakha, M.; Pal, S.; Samantarrai, D.; Panigrahi, T. K.; Mallick, B. C.; Pramanik, K.; Mallick, B.; Jha, S., Antimicrobial activity of iron oxide nanoparticle upon modulation of nanoparticle-bacteria interface. *Scientific Reports* **2015,** *5*, 14813.
10. Anselmo, A. C.; Mitragotri, S., Nanoparticles in the clinic. *Bioengineering & Translational Medicine* **2016,** *1* (1), 10–29.
11. Anselmo, A. C.; Mitragotri, S., A review of clinical translation of inorganic nanoparticles. *The AAPS Journal* **2015,** *17* (5), 1041–1054.
12. Wang, Y.-X. J., Current status of superparamagnetic iron oxide contrast agents for liver magnetic resonance imaging. *World Journal of Gastroenterology* **2015,** *21* (47), 13400–13402.
13. Garneata, L., Intravenous iron, inflammation, and oxidative stress: Is iron a friend or an enemy of uremic patients? *Journal of Renal Nutrition* **2008,** *18* (1), 40–45.
14. Agarwal, R.; Vasavada, N.; Sachs, N. G.; Chase, S., Oxidative stress and renal injury with intravenous iron in patients with chronic kidney disease. *Kidney International* **2004,** *65* (6), 2279–2289.
15. Auerbach, M.; Ballard, H., Clinical use of intravenous iron: Administration, efficacy, and safety. *ASH Education Program Book* **2010,** *2010* (1), 338–347.
16. Luo, C.; Li, Y.; Yang, L.; Wang, X.; Long, J.; Liu, J., Superparamagnetic iron oxide nanoparticles exacerbate the risks of reactive oxygen species-mediated external stresses. *Archives of Toxicology* **2015,** *89* (3), 357–369.
17. Maqusood, A.; Hisham, A. A.; Javed, A.; Khan, M. A. M.; Daoud, A.; Saud, A., Iron oxide nanoparticle-induced oxidative stress and genotoxicity in human skin epithelial and lung epithelial cell lines. *Current Pharmaceutical Design* **2013,** *19* (37), 6681–6690.
18. Park, E.-J.; Oh, S. Y.; Kim, Y.; Yoon, C.; Lee, B.-S.; Kim, S. D.; Kim, J. S., Distribution and immunotoxicity by intravenous injection of iron nanoparticles in a murine model. *Journal of Applied Toxicology* **2016,** *36* (3), 414–423.
19. Tassa, C.; Shaw, S. Y.; Weissleder, R., Dextran-coated iron oxide nanoparticles: A versatile platform for targeted molecular imaging, molecular diagnostics, and therapy. *Accounts of Chemical Research* **2011,** *44* (10), 842–852.
20. Yu, M.; Huang, S.; Yu, K. J.; Clyne, A. M., Dextran and polymer polyethylene glycol (peg) coating reduce both 5 and 30 nm iron oxide nanoparticle cytotoxicity in 2d and 3d cell culture. *International Journal of Molecular Sciences* **2012,** *13* (5), 5554.

21. Syama, S.; Gayathri, V.; Mohanan, P. V., Assessment of immunotoxicity of dextran coated ferrite nanoparticles in albino mice. *Molecular Biology International* **2015,** *2015*, 10.
22. Chun Soo Lim; Vaziri, N. D., The effects of iron dextran on the oxidative stress in cardiovascular tissues of rats with chronic renal failure. *Kidney International* **2004,** *65* (5), 1802–1809.
23. Maísa Silva; Joyce Ferreira da Costa Guerra; Ana Flávia Santos Sampaio; Wanderson Geraldo de Lima; Marcelo Eustáquio Silva; A. M. L. Pedrosa, Iron dextran increases hepatic oxidative stress and alters expression of genes related to lipid metabolism contributing to hyperlipidaemia in murine model. *BioMed Research International* **2015,** *2015*, 9.
24. Neiser, S.; Koskenkorva, T. S.; Schwarz, K.; Wilhelm, M.; Burckhardt, S., Assessment of dextran antigenicity of intravenous iron preparations with enzyme-linked immunosorbent assay (ELISA). *International Journal of Molecular Sciences* **2016,** *17* (7), 1185.
25. Gaharwar, U. S.; Paulraj, R., Iron oxide nanoparticles induced oxidative damage in peripheral blood cells of rat. *Journal of Biomedical Science and Engineering* **2015,** *8* (4), 13.
26. Inturi, S.; Wang, G.; Chen, F.; Banda, N. K.; Holers, V. M.; Wu, L.; Moghimi, S. M.; Simberg, D., Modulatory role of surface coating of superparamagnetic iron oxide nanoworms in complement opsonization and leukocyte uptake. *ACS Nano* **2015,** *9* (11), 10758–10768.
27. Sakulkhu, U.; Mahmoudi, M.; Maurizi, L.; Salaklang, J.; Hofmann, H., Protein corona composition of superparamagnetic iron oxide nanoparticles with various physicochemical properties and coatings. *Scientific Reports* **2014,** *4*, 5020.
28. Bullivant, J. P.; Zhao, S.; Willenberg, B. J.; Kozissnik, B.; Batich, C. D.; Dobson, J., Materials characterization of feraheme /ferumoxytol and preliminary evaluation of its potential for magnetic fluid hyperthermia. *International Journal of Molecular Sciences* **2013,** *14* (9), 17501–17510.
29. Szebeni, J.; Fishbane, S.; Hedenus, M.; Howaldt, S.; Locatelli, F.; Patni, S.; Rampton, D.; Weiss, G.; Folkersen, J., Hypersensitivity to intravenous iron: Classification, terminology, mechanisms and management. *British Journal of Pharmacology* **2015,** *172* (21), 5025–5036.
30. Bircher, A. J.; Auerbach, M., Hypersensitivity from intravenous iron products. *Immunology and Allergy Clinics of North America* **2014,** *34* (3), 707–723.
31. Palta, S.; Saroa, R.; Palta, A., Overview of the coagulation system. *Indian Journal of Anaesthesia* **2014,** *58* (5), 515–523.
32. Zhu, M.-T.; Feng, W.-Y.; Wang, B.; Wang, T.-C.; Gu, Y.-Q.; Wang, M.; Wang, Y.; Ouyang, H.; Zhao, Y.-L.; Chai, Z.-F., Comparative study of pulmonary responses to nano- and submicron-sized ferric oxide in rats. *Toxicology* **2008,** *247* (2–3), 102–111.
33. Venugopal, A., Disseminated intravascular coagulation. *Indian Journal of Anaesthesia* **2014,** *58* (5), 603–608.
34. Ilinskaya, A. N.; Dobrovolskaia, M. A., Nanoparticles and the blood coagulation system. Part II: safety concerns. *Nanomedicine (London, England)* **2013,** *8* (6), 969–981.
35. Nemmar, A.; Beegam, S.; Yuvaraju, P.; Yasin, J.; Tariq, S.; Attoub, S.; Ali, B. H., Ultrasmall superparamagnetic iron oxide nanoparticles acutely promote thrombosis and cardiac oxidative stress and DNA damage in mice. *Particle and Fibre Toxicology* **2015,** *13*, 22.
36. Wang, Q.; Shen, M.; Zhao, T.; Xu, Y.; Lin, J.; Duan, Y.; Gu, H., Low toxicity and long circulation time of Polyampholyte-coated magnetic nanoparticles for blood pool contrast agents. *Scientific Reports* **2015,** *5*, 7774.
37. Alfano, K. M.; Tarasev, M.; Meines, S.; Parunak, G., An approach to measuring RBC haemolysis and profiling RBC mechanical fragility. *Journal of Medical Engineering & Technology* **2016,** *40* (4), 162–171.
38. L'Acqua, C.; Hod, A. E., New perspectives on the thrombotic complications of haemolysis. *British Journal of Haematology* **2014,** *168*, 175–185.
39. Thompson, W. P., Hemolytic jaundice: Its diagnosis, behavior and treatment. *Bulletin of the New York Academy of Medicine* **1939,** *15* (3), 177–187.
40. Ran, Q.; Xiang, Y.; Liu, Y.; Xiang, L.; Li, F.; Deng, X.; Xiao, Y.; Chen, L.; Chen, L.; Li, Z., Eryptosis Indices as a novel predictive parameter for biocompatibility of Fe(3)O(4) magnetic nanoparticles on erythrocytes. *Scientific Reports* **2015,** *5*, 16209.
41. Owens I, D. E.; Peppas, N. A., Opsonization, biodistribution, and pharmacokinetics of polymeric nanoparticles. *International Journal of Pharmaceutics* **2006,** *307* (1), 93–102.
42. Pieper, R.; Gatlin, C. L.; Makusky, A. J.; Russo, P. S.; Schatz, C. R.; Miller, S. S.; Su, Q.; McGrath, A. M.; Estock, M. A.; Parmar, P. P.; Zhao, M.; Huang, S.-T.; Zhou, J.; Wang, F.; Esquer-Blasco, R.; Anderson, N. L.; Taylor, J.; Steiner, S., The human serum proteome: Display of nearly 3700 chromatographically separated protein spots on two-dimensional electrophoresis gels and identification of 325 distinct proteins. *Proteomics* **2003,** *3* (7), 1345–1364.
43. Sobot, D.; Mura, S.; Couvreur, P., Nanoparticles: Blood Components Interactions. In *Encyclopedia of Polymeric Nanomaterials*, Kobayashi, S.; Müllen, K., Eds. Springer Berlin Heidelberg: Berlin, Heidelberg, 2014; pp 1–10.
44. Banda, N. K.; Mehta, G.; Chao, Y.; Wang, G.; Inturi, S.; Fossati-Jimack, L.; Botto, M.; Wu, L.; Moghimi, S. M.; Simberg, D., Mechanisms of complement activation by dextran-coated superparamagnetic iron oxide (SPIO) nanoworms in mouse versus human serum. *Particle and Fibre Toxicology* **2014,** *11*, 64.
45. Moghimi, S. M.; Andersen, A. J.; Ahmadvand, D.; Wibroe, P. P.; Andresen, T. L.; Hunter, A. C., Material properties in complement activation. *Advanced Drug Delivery Reviews* **2011,** *63* (12), 1000–1007.
46. Simberg, D., Iron oxide nanoparticles and the mechanisms of immune recognition of nanomedicines. *Nanomedicine* **2016,** *11* (7), 741–743.
47. Szebeni, J., Complement activation-related pseudoallergy: A new class of drug-induced acute immune toxicity. *Toxicology* **2005,** *216* (2–3), 106–121.
48. Wang, G.; Inturi, S.; Serkova, N. J.; Merkulov, S.; McCrae, K.; Russek, S. E.; Banda, N. K.; Simberg, D., High-relaxivity superparamagnetic iron oxide nanoworms with decreased immune recognition and long-circulating properties. *ACS Nano* **2014,** *8* (12), 12437–12449.
49. Chen, F.; Wang, G.; Griffin, J. I.; Brenneman, B.; Banda, N. K.; Holers, V. M.; Backos, D. S.; Wu, L.; Moghimi, S. M.; Simberg, D., Complement proteins bind to nanoparticle protein corona and undergo dynamic exchange in vivo. *Nat Nano* **2016,** advance online publication.
50. Pino, P. d.; Pelaz, B.; Zhang, Q.; Maffre, P.; Nienhaus, G. U.; Parak, W. J., Protein corona formation around nanoparticles – From the past to the future. *Materials Horizons* **2014,** *1* (3), 301–313.
51. Yallapu, M. M.; Chauhan, N.; Othman, S. F.; Khalilzad-Sharghi, V.; Ebeling, M. C.; Khan, S.; Jaggi, M.; Chauhan, S. C., Implications of protein corona on physicochemical and biological properties of magnetic nanoparticles. *Biomaterials* **2015,** *46*, 1–12.

52. Amiri, H.; Bordonali, L.; Lascialfari, A.; Wan, S.; Monopoli, M. P.; Lynch, I.; Laurent, S.; Mahmoudi, M., Protein corona affects the relaxivity and MRI contrast efficiency of magnetic nanoparticles. *Nanoscale* **2013,** *5* (18), 8656–8665.
53. Lee, Y. K.; Choi, E.-J.; Webster, T. J.; Kim, S.-H.; Khang, D., Effect of the protein corona on nanoparticles for modulating cytotoxicity and immunotoxicity. *International Journal of Nanomedicine* **2015,** *10*, 97–113.
54. Corbo, C.; Molinaro, R.; Parodi, A.; Toledano Furman, N. E.; Salvatore, F.; Tasciotti, E., The impact of nanoparticle protein corona on cytotoxicity, immunotoxicity and target drug delivery. *Nanomedicine* **2016,** *11* (1), 81–100.
55. Yan, Y.; Gause, K. T.; Kamphuis, M. M. J.; Ang, C.-S.; O'Brien-Simpson, N. M.; Lenzo, J. C.; Reynolds, E. C.; Nice, E. C.; Caruso, F., Differential roles of the protein corona in the cellular uptake of nanoporous polymer particles by monocyte and macrophage cell lines. *ACS Nano* **2013,** *7* (12), 10960–10970.
56. Mbeh, D. A.; Mireles, L. K.; Stanicki, D.; Tabet, L.; Maghni, K.; Laurent, S.; Sacher, E.; Yahia, L. H., Human alveolar epithelial cell responses to core–shell superparamagnetic iron oxide nanoparticles (SPIONs). *Langmuir* **2015,** *31* (13), 3829–3839.
57. Mahmoudi, M.; Sheibani, S.; Milani, A. S.; Rezaee, F.; Gauberti, M.; Dinarvand, R.; Vali, H., Crucial role of the protein corona for the specific targeting of nanoparticles. *Nanomedicine* **2015,** *10* (2), 215–226.
58. Hajipour, M. J.; Laurent, S.; Aghaie, A.; Rezaee, F.; Mahmoudi, M., Personalized protein coronas: A "key" factor at the nanobiointerface. *Biomaterials Science* **2014,** *2* (9), 1210–1221.
59. Hajipour, M. J.; Raheb, J.; Akhavan, O.; Arjmand, S.; Mashinchian, O.; Rahman, M.; Abdolahad, M.; Serpooshan, V.; Laurent, S.; Mahmoudi, M., Personalized disease-specific protein corona influences the therapeutic impact of graphene oxide. *Nanoscale* **2015,** *7* (19), 8978–8994.
60. Caputo, D.; Papi, M.; Coppola, R.; Palchetti, S.; Digiacomo, L.; Caracciolo, G.; Pozzi, D., A protein corona-enabled blood test for early cancer detection. *Nanoscale* **2017,** *9* (1), 349–354.
61. Liu, Z.; Zhan, X.; Yang, M.; Yang, Q.; Xu, X.; Lan, F.; Wu, Y.; Gu, Z., A magnetic-dependent protein corona of tailor-made superparamagnetic iron oxides alters their biological behaviors. *Nanoscale* **2016,** *8* (14), 7544–7555.
62. Krais, A.; Wortmann, L.; Hermanns, L.; Feliu, N.; Vahter, M.; Stucky, S.; Mathur, S.; Fadeel, B., Targeted uptake of folic acid-functionalized iron oxide nanoparticles by ovarian cancer cells in the presence but not in the absence of serum. *Nanomedicine: Nanotechnology, Biology and Medicine* **2014,** *10* (7), 1421–1431.
63. Vogt, C.; Pernemalm, M.; Kohonen, P.; Laurent, S.; Hultenby, K.; Vahter, M.; Lehtiö, J.; Toprak, M. S.; Fadeel, B., Proteomics analysis reveals distinct corona composition on magnetic nanoparticles with different surface coatings: Implications for interactions with primary human macrophages. *PLOS ONE* **2015,** *10* (10), e0129008.
64. Clauson, R. M.; Scheetz, L.; Berg, B.; Chertok, B., Immunoactive dna-tethered nanocomplexes of antigen-cpg and iron-oxide nanoparticles as potential cancer nanovaccines. *Frontiers in Bioengineering and Biotechnology.* doi:10.3389/conf.FBIOE.2016.01.02025.
65. Hwang, J.; Lee, E.; Kim, J.; Seo, Y.; Lee, K. H.; Hong, J. W.; Gilad, A. A.; Park, H.; Choi, J., Effective delivery of immunosuppressive drug molecules by silica coated iron oxide nanoparticles. *Colloids and Surfaces B: Biointerfaces* **2016,** *142*, 290–296.
66. Shen, C.-C.; Liang, H.-J.; Wang, C.-C.; Liao, M.-H.; Jan, T.-R., Iron oxide nanoparticles suppressed T helper 1 cell-mediated immunity in a murine model of delayed-type hypersensitivity. *International Journal of Nanomedicine* **2012,** *7*, 2729–2737.
67. Ngobili, T. A.; Daniele, M. A., Nanoparticles and direct immunosuppression. *Experimental Biology and Medicine* **2016,** *241* (10), 1064–1073.
68. Broere, F.; Apasov, S. G.; Sitkovsky, M. V.; van Eden, W., A2 T cell subsets and T cell-mediated immunity. In *Principles of Immunopharmacology*, 3rd revised and extended edition, Nijkamp, F. P.; Parnham, M. J., Eds. Birkhäuser Basel: Basel, 2011; pp 15–27.
69. Sousa, M. d., T lymphocytes and iron overload: Novel correlations of possible significance to the biology of the immunological system. *Memórias do Instituto Oswaldo Cruz* **1992,** *87*, 23–29.
70. Attia, M. A.; Essa, S. A.; Nosair, N. A.; Amin, A. M.; El-Agamy, O. A., Effect of iron deficiency anemia and its treatment on cell mediated immunity. *Indian Journal of Hematology and Blood Transfusion* **2009,** *25* (2), 70–77.
71. Wang, J.; Chen, B.; Jin, N.; Xia, G.; Chen, Y.; Zhou, Y.; Cai, X.; Ding, J.; Li, X.; Wang, X., The changes of T lymphocytes and cytokines in ICR mice fed with Fe(3)O(4) magnetic nanoparticles. *International Journal of Nanomedicine* **2011,** *6*, 605–610.
72. Easo, S. L.; Mohanan, P. V., In vitro hematological and in vivo immunotoxicity assessment of dextran stabilized iron oxide nanoparticles. *Colloids and Surfaces B: Biointerfaces* **2015,** *134*, 122–130.
73. Shi, C.; Pamer, E. G., Monocyte recruitment during infection and inflammation. *Nature Reviews Immunology* **2011,** *11* (11), 762–774.
74. Xu, Y.; Sherwood, J. A.; Lackey, K. H.; Qin, Y.; Bao, Y., The responses of immune cells to iron oxide nanoparticles. *Journal of Applied Toxicology* **2016,** *36* (4), 543–553.
75. Couto, D.; Freitas, M.; Vilas-Boas, V.; Dias, I.; Porto, G.; Lopez-Quintela, M. A.; Rivas, J.; Freitas, P.; Carvalho, F.; Fernandes, E., Interaction of polyacrylic acid coated and non-coated iron oxide nanoparticles with human neutrophils. *Toxicology Letters* **2014,** *225* (1), 57–65.
76. Strehl, C.; Gaber, T.; Maurizi, L.; Hahne, M.; Rauch, R.; Hoff, P.; Häupl, T.; Hofmann-Amtenbrink, M.; Poole, A. R.; Hofmann, H.; Buttgereit, F., Effects of PVA coated nanoparticles on human immune cells. *International Journal of Nanomedicine* **2015,** *10*, 3429–3445.
77. Dobrovolskaia, M. A.; McNeil, S. A., *Handbook of Immunological Properties of Engineered Nanomaterials (Frontiers in Nanobiomedical Research)* 1st ed.; World Scientific Publishing Company, Singapore; 1 edition (February 21, 2013): 2012; p 720.
78. Elsabahy, M.; Wooley, K. L., Cytokines as biomarkers of nanoparticle immunotoxicity. *Chemical Society Reviews* **2013,** *42* (12), 5552–5576.
79. Oostingh, G. J.; Casals, E.; Italiani, P.; Colognato, R.; Stritzinger, R.; Ponti, J.; Pfaller, T.; Kohl, Y.; Ooms, D.; Favilli, F.; Leppens, H.; Lucchesi, D.; Rossi, F.; Nelissen, I.; Thielecke, H.; Puntes, V. F.; Duschl, A.; Boraschi, D., Problems and challenges in the development and validation of human cell-based assays to determine nanoparticle-induced immunomodulatory effects. *Particle and Fibre Toxicology* **2011,** *8* (1), 8.
80. Liu, A. H., Endotoxin exposure in allergy and asthma: Reconciling a paradox. *The Journal of Allergy and Clinical Immunology* **2002,** *109* (3), 379–392.

81. Inoue, K.; Takano, H.; Yanagisawa, R.; Hirano, S.; Sakurai, M.; Shimada, A.; Yoshikawa, T., Effects of airway exposure to nanoparticles on lung inflammation induced by bacterial endotoxin in mice. *Environmental Health Perspectives* **2006,** *114* (9), 1325–1330.
82. Castegren, M.; Skorup, P.; Lipcsey, M.; Larsson, A.; Sjölin, J., Endotoxin tolerance variation over 24 h during porcine endotoxemia: Association with changes in circulation and organ dysfunction. *PLoS ONE* **2013,** *8* (1), e53221.
83. Bianchi, M. G.; Allegri, M.; Costa, A. L.; Blosi, M.; Gardini, D.; Del Pivo, C.; Prina-Mello, A.; Di Cristo, L.; Bussolati, O.; Bergamaschi, E., Titanium dioxide nanoparticles enhance macrophage activation by LPS through a TLR4-dependent intracellular pathway. *Toxicology Research* **2015,** *4* (2), 385–398.
84. Park, E.-J.; Yoon, J.; Choi, K.; Yi, J.; Park, K., Induction of chronic inflammation in mice treated with titanium dioxide nanoparticles by intratracheal instillation. *Toxicology* **2009,** *260* (1–3), 37–46.
85. Grosse, S.; Stenvik, J.; Nilsen, A. M., Iron oxide nanoparticles modulate lipopolysaccharide-induced inflammatory responses in primary human monocytes. *International Journal of Nanomedicine* **2016,** *11*, 4625–4642.
86. Li, Y.; Italiani, P.; Casals, E.; Tran, N.; Puntes, V. F.; Boraschi, D., Optimising the use of commercial LAL assays for the analysis of endotoxin contamination in metal colloids and metal oxide nanoparticles. *Nanotoxicology* **2015,** *9* (4), 462–473.
87. Booth, C., The limulus amoebocyte lysate (LAL) assay – a replacement for the rabbit pyrogen test. *Developments in Biological Standardization* **1986,** *64*, 271–275.
88. Vermeij, E. A.; Koenders, M. I.; Bennink, M. B.; Crowe, L. A.; Maurizi, L.; Vallee, J. P.; Hofmann, H.; van den Berg, W. B.; van Lent, P. L.; van de Loo, F. A., The in-vivo use of superparamagnetic iron oxide nanoparticles to detect inflammation elicits a cytokine response but does not aggravate experimental arthritis. *PLoS One* **2015,** *10* (5), e0126687.
89. Hurley, J. C., Endotoxemia: Methods of detection and clinical correlates. *Clinical Microbiology Reviews* **1995,** *8* (2), 268–292.
90. Dobrovolskaia, M. A.; Neun, B. W.; Clogston, J. D.; Grossman, J. H.; McNeil, S. E., Choice of method for endotoxin detection depends on nanoformulation. *Nanomedicine (London)* **2014,** *9* (12), 1847–1856.
91. US Department of Health and Human Services, *Guideline on Validation of the Limulus Amebocyte Lysate Test as an End-Product Endotoxin Test for Human and Animal Parenteral Drugs, Biological Products, and Medical Devices*; US Department of Health and Human Services: 1987.
92. Smulders, S.; Kaiser, J. P.; Zuin, S.; Van Landuyt, K. L.; Golanski, L.; Vanoirbeek, J.; Wick, P.; Hoet, P. H. M., Contamination of nanoparticles by endotoxin: Evaluation of different test methods. *Particle and Fibre Toxicology* **2012,** *9*, 41.
93. Kucki, M.; Cavelius, C.; Kraegeloh, A., Interference of silica nanoparticles with the traditional Limulus amebocyte lysate gel clot assay. *Innate Immunity* **2014** *20* (3), 327–336.
94. United States Pharmacopeial Convention, USP 29, NF 24: The United States Pharmacopeia, the National Formulary. United States Pharmacopeial Convention: 2005.
95. Gopalkrishna, V.; Verma, H.; Kumbhar, N. S.; Tomar, R. S.; Patil, P. R., Detection of Mycoplasma species in cell culture by PCR and RFLP based method: Effect of BM-cyclin to cure infections. *Indian Journal of Medical Microbiology* **2007,** *25* (4), 364–368.
96. Darlington, G. J., Detection of Mycoplasma in Mammalian cell cultures using fluorescence microscopy. *CSH Protocols* **2006,** *2006* (1). doi:10.1101/pdb.prot4351.
97. Desai, N., Challenges in development of nanoparticle-based therapeutics. *AAPS Journal* **2012,** *14* (2), 282–295.
98. Huang, H.; Lai, W.; Cui, M.; Liang, L.; Lin, Y.; Fang, Q.; Liu, Y.; Xie, L., An evaluation of blood compatibility of silver nanoparticles. *Scientific Reports* **2016,** *6*, 25518.
99. Dobrovolskaia, M. A.; Germolec, D. R.; Weaver, J. L., Evaluation of nanoparticle immunotoxicity. *Nature Nanotechnology* **2009,** *4* (7), 411–414.
100. ISO, Biological evaluation of medical devices – Part 4: Selection of tests for interactions with blood. *10993–10994(en).*
101. Liao, S. H.; Liu, C. H.; Bastakoti, B. P.; Suzuki, N.; Chang, Y.; Yamauchi, Y.; Lin, F. H.; Wu, K. C. W., Functionalized magnetic iron oxide/alginate core-shell nanoparticles for targeting hyperthermia. *International Journal of Nanomedicine* **2015,** *10*, 3315–3328.
102. Salvador-Morales, C.; Sim, R. B., Complement Activation. In *Handbook of Immunological Properties of Engineered Nanomaterials (Frontiers in Nanobiomedical Research)* Dobrovolskaia, M. A.; McNeil, S. A., eds. World Scientific Publishing Company, Singapore; 2012; p. 720.
103. Wang, Y.-X. J., Superparamagnetic iron oxide based MRI contrast agents: Current status of clinical application. *Quantitative Imaging in Medicine and Surgery* **2011,** *1* (1), 35–40.
104. Blanco, E.; Shen, H.; Ferrari, M., Principles of nanoparticle design for overcoming biological barriers to drug delivery. *Nature Biotechnology* **2015,** *33* (9), 941–951.
105. García-Jimeno, S.; Estelrich, J., Ferrofluid based on polyethylene glycol-coated iron oxide nanoparticles: Characterization and properties. *Colloids and Surfaces A: Physicochemical and Engineering Aspects* **2013,** *420*, 74–81.
106. Jokerst, J. V.; Lobovkina, T.; Zare, R. N.; Gambhir, S. S., Nanoparticle PEGylation for imaging and therapy. *Nanomedicine (London, England)* **2011,** *6* (4), 715–728.
107. Simberg, D.; Park, J.-H.; Karmali, P. P.; Zhang, W.-M.; Merkulov, S.; McCrae, K.; Bhatia, S.; Sailor, M.; Ruoslahti, E., Differential proteomics analysis of the surface heterogeneity of dextran iron oxide nanoparticles and the implications for their in vivo clearance. *Biomaterials* **2009,** *30* (23–24), 3926–3933.
108. Laurent, S.; Forge, D.; Port, M.; Roch, A.; Robic, C.; Vander Elst, L.; Muller, R. N., Magnetic Iron oxide nanoparticles: Synthesis, stabilization, vectorization, physicochemical characterizations, and biological applications. *Chemical Reviews* **2008,** *108* (6), 2064–2110.
109. Calero, M.; Chiappi, M.; Lazaro-Carrillo, A.; Rodríguez, M. J.; Chichón, F. J.; Crosbie-Staunton, K.; Prina-Mello, A.; Volkov, Y.; Villanueva, A.; Carrascosa, J. L., Characterization of interaction of magnetic nanoparticles with breast cancer cells. *Journal of Nanobiotechnology* **2015,** *13*, 16.
110. Barrow, M.; Taylor, A.; Murray, P.; Rosseinsky, M. J.; Adams, D. J., Design considerations for the synthesis of polymer coated iron oxide nanoparticles for stem cell labelling and tracking using MRI. *Chemical Society Reviews* **2015,** *44* (19), 6733–6748.
111. Hajesmaeelzadeh, F.; Shanehsazzadeh, S.; Grüttner, C.; Daha, F. J.; Oghabian, M. A., Effect of coating thickness of iron oxide nanoparticles on their relaxivity in the MRI. *Iranian Journal of Basic Medical Sciences* **2016,** *19* (2), 166–171.
112. Kolhatkar, A. G.; Jamison, A. C.; Litvinov, D.; Willson, R. C.; Lee, T. R., Tuning the magnetic properties of nanoparticles. *International Journal of Molecular Sciences* **2013,** *14* (8), 15977–16009.

Adriele Prina-Mello (E-mail: prinamea@tcd.ie) is an Ussher Professor in Translational Nanomedicine, the director of the LBCAM laboratory and AMBER and CRANN Principal Investigator.

Prof. Prina-Mello's scientific interests are focused on advanced translation research in NanoMedicine (in vitro/in vivo diagnostic, imaging and therapeutics), microfluidic, biomedical devices and tissue engineering applications of nanotechnology and nanomaterials.

The continuous exploration of the dynamic interaction between nanodeveloped products and biologically relevant models constitute the basic ground for Prof. Prina-Mello's multidisciplinary scientific work within and outside the Trinity Translational Medicine Institute (TTMI), the AMBER (Advanced Materials and BioEngineering Research as Science Foundation Ireland funded centre), and the CRANN Nanoscience Institute and the School of Medicine.

Prof. Prina-Mello is part of the Executive Board of the European Technology Platform of Nanomedicine, as Chair of the Characterization and Toxicology working group. At the European level, he has been involved in several EC-H2020 and FP7 projects such as the EU-NCL, NoCanTher, AMCARE, MULTIFUN, NAMDIATREAM and others.

Among these, he is the principal investigator behind the TCD participation to the European Nanomedicine Characterization Laboratory infrastructure project (H2020-Infra-1). Prof. Prina-Mello is also a lecturer in Nanomedicine and Translational Nanomedicine at the TTMI.

Publications and Project details available at:

Google Scholars – search Prina-Mello A
Research Gate – search Prina-Mello A
Website info: http://www.tcd.ie/IMM/lbcam/
Website info: http://ambercentre.ie/people/dr-adriele-prina-mello

Gary Hannon studied biology and chemistry in Maynooth University. He then completed a master in translational oncology at Trinity College, Dublin (TCD) where he joined Prof. Prina-Mello's group at the LBCAM as part of the Nanomedicine theme at the Trinity Translational Medicine Institute (TCD). After completing this project, and the summer school of molecular medicine from Jena University Hospital, he returned to begin a PhD with the Nanomedicine group. His main interests lie in nanomaterials for the treatment and diagnosis of cancer. His current research focuses on magnetic hyperthermia for the treatment of cancer.

In October 2016 he was awarded an Irish Research Council (IRC) funded postgraduate studentship to pursue his PhD degree.

Melissa Anne Tutty is a PhD candidate student in the Nanomedicine and Molecular Imaging Group, at the Trinity Translational Medicine Institute, Trinity College Dublin. She earned a BSc Biochemistry degree from University College Dublin in 2013, followed by an MSc in Pharmaceutical Analysis, with Distinction, from Trinity College Dublin in 2014. In the past, she has worked in chemical synthesis and characterization of large molecules. Her current research interests include characterization of novel nanomedicinal products using cell-based and in vivo assays.

15 Impact of Core and Functionalized Magnetic Nanoparticles on Human Health

Bella B. Manshian, Uwe Himmelreich and Stefaan J. Soenen*

CONTENTS

Increasingly more new formulations of superparamagnetic iron oxide NPs (IONPs) are being produced as magnetic resonance imaging (MRI) contrast agents due to their high relaxivity, which can significantly increase the accuracy of disease diagnosis. Furthermore, magnetic hyperthermia using IONPs to treat cancer is the highlight of the current decade. While the race is on to manufacture the optimal NP for diagnosis and treatment of disease, the question remains as to how toxic are these formulations to human health and what are the main factors affecting their potential toxicity?

IONPs are one of the very few NPs that are currently in clinical trials and several formulations (e.g. Endorem® [known as Feridex® in United States], Resovist®) already have a long-standing history of clinical use as they are considered relatively safe to human health. However, despite their clinical use, the production of the clinically approved formulations has been halted due to a lack of sales.[1] If so, then why is the concern about their toxicity still the subject of many publications?[2] The answer lies in the characteristic details of the different IONPs synthesized and studied. In addition to Chapter 14 on 'Immunotoxicity and Safety Considerations for IONPs', in this chapter we will discuss the mechanisms that underlie potential toxicity of IONPs and the influence of the core, the different surface coatings used and the size of IONPs in relation to their potential for generating toxic effects. Finally, we will look into factors affecting their cellular uptake levels and thus their toxicity keeping in mind the role played by different routes of administration and the various concentrations tested.

15.1 IONP TOXICITY OVERVIEW

As mentioned above, the clinical history of dextran- or carboxydextran-coated IONPs as liver contrast agents for MRI in combination with the low level of toxicity for Fe^{2+} (2 mM) has contributed to the initial belief that IONPs were safe.[3,4] However, from the year 2000 onwards, the field of nanotechnology gained significant interest and the interactions of nanomaterials with their biological environment also became the subject of more rigorous safety testing. Initial studies performed by Brunner et al.[5] revealed that toxic effects of IONPs occurred at concentrations approximately 40 times lower than the chemical toxicity of iron ions. These initial studies revealed the power of *in vitro* toxicity testing as a method to more rapidly evaluate NP toxicity compared to more laborious animal studies. In this study, IONPs, without any specific chemical coating, were compared to two so-called control particles, being amorphous silica (regarded as low cytotoxic) and crocidolite asbestos (regarded as highly toxic). This study already demonstrated that NP toxicity was largely influenced by the solubility of the NP and as toxicity levels were rather high, it revealed that IONPs by themselves could elicit toxic effects that could not be fully ascribed to their iron ions, but were specifically NP-associated.

The use of the dextran-coated IONPs for off-license applications, such as the direct labelling of stem or immune cells to enable noninvasive monitoring of stem-cell therapy was also met with several complications.[6] The dextran-coated NPs themselves had a low intrinsic endocytosis capacity, and cellular uptake levels therefore remained fairly low,[7] impeding

* Corresponding author.

efficient *in vivo* detection by MRI. To overcome this problem, the IONPs were complexed with polycationic transfection agents that would improve cellular uptake levels of the IONPs.[8] The IONP-transfection agent combination leads to complexes with noncontrollable surface features which have been reported to affect cellular functionality, e.g. impairing chondrogenic or osteogenic differentiation of mesenchymal stem cells.[7,9,10] To date, it remains unclear whether this issue is due to the IONPs, the transfection agent or particular cell labelling protocols used since several other studies did not observe any effect on mesenchymal stem cell differentiation.[11] The generation of such disparate data has generated concern regarding the true safety of IONPs for biomedical use, and the subject is still actively studied. Based on the available information in literature, some main mechanisms that have been linked to IONP toxicity can be defined and will be discussed in the following sections.

15.1.1 Induction of Reactive Oxygen Species

One of the main causes of NP-mediated toxicity in general, and for IONPs in particular is the generation of reactive oxygen species (ROS), which can have dramatic effects on cellular well-being by affecting the intrinsic cellular redox capabilities of the cells. Much effort has been put into understanding the effect of ROS on cellular well-being, where a three-step tier approach has been suggested.[12] In an initial phase, cells can be exposed to low levels of ROS, which can easily be dealt with by the intrinsic defensive capabilities of the cells against oxidative damage. The level of these defences are cell type-dependent, which explains the wide variability in the level of cellular damage due to oxidative stress.[13] In a second step, ROS levels can further increase and overpower the cellular defence levels, resulting in oxidative stress. In a third step, persisting oxidative stress can result in oxidative damage, which typically result in DNA damage, or peroxidation of lipids or proteins. These cellular effects are also observed in animal models, where antioxidant imbalance and lipid peroxidation were observed in all major organs in mice after intravenous administration of IONPs.[14] This is in part due to the nature of the cellular internalization of nanomaterials, which are typically internalized by means of active endocytosis,[15] resulting in the presence of the nanomaterials in endosomes and lysosomes. The IONPs will thus be exposed to the acidic lumen of these cellular organelles. For IONPs reactive oxygen intermediates can then be produced by the Fenton or Haber–Weiss reactions.[16,17] The generation of ROS by IONPs has been described in a wide range of studies for different formulations, where changes in NP size or surface coating will mainly impact the kinetics and amplitude of ROS generation, but cannot really overcome it (see Sections 15.3 and 15.4). Also here, disparate data have been generated, where in several studies, IONPs have been described to possess intrinsic peroxidase-like activities that can scavenge any reactive oxygen intermediates.[18] In follow-up studies, it was shown that carboxydextran-coated IONPs possessed intrinsic peroxidase-like activity which diminished intracellular ROS. In combination with the presence of free iron ions, in part generated by the intracellular degradation of the IONPs, this was found to stimulate mesenchymal stem cell proliferation, rather than inhibit it.[19]

15.1.2 Degradation of IONPs

As mentioned above, IONPs will be exposed to a more acidic microenvironment following cellular endocytosis. IONPs have been shown to be prone to acidic degradation, resulting in the leaching of iron ions from the actual NPs.[20] Upon the gradual degradation of the iron oxide core, free iron ions can be shuttled out of the endocytic compartment into the normal cellular iron pool. This in turn will affect the cytoplasmic iron concentrations, which can then affect cellular functionality by, for example, altering the level of transferrin receptor expression, and can affect cellular proliferation capacity by altering the expression of cyclins and cyclin-dependent kinases.[19,21] As mentioned above, free iron ions that are generated in the acidic microenvironment of the cellular organelles can then also contribute to the generation of ROS and hereby directly result in cell damage. In particular for cells of the central nervous system, this is an important parameter as excessive IONP accumulation can lead to a disruption of normal iron metabolism, which is a characteristic hallmark resembling that of several neurodegenerative disorders.[22] In particular pathophysiological conditions, such as cirrhosis, excess iron might be a risk factor because of impacts of elevated lipid metabolism, disruption of iron homeostasis and the aggravated loss of liver functions.[23] Release of iron ions can lead to both short-term and long-term reactions, where initially mitochondrial damage is observed, whereas longer exposures can result in a multitude of effects.[22] Studying the long-term degradation of IONPs has been proven to be rather difficult due to the lack of suitable models. Recently, stem-cell spheroids have been used as a tissue model to track intracellular magnetic nanoparticle (MNP) transformations during long-term tissue maturation. In this model, massive degradation of iron oxide nanocubes was found, resulting in near-complete NP degradation over a month of tissue maturation (Figure 15.1).[24]

However, the rather rapid NP degradation only had minimal effects on iron homeostasis. The same group also studied the biotransformation of IONPs injected into the bloodstream of mice and found that superparamagnetic maghemite NPs were transformed into poorly-magnetic iron species over a three-month period.[25] Using artificial fluids mimicking the endosomal lumen with varying pH levels, the pH-dependent degradation of different types of IONPs could also be confirmed.[20,26] More in-depth studies revealed that regardless of the type of IONP, degradation resulted into two distinct groups, one with rapidly dissolved nanocrystals and a second group that appeared to be more resistant.[27]

15.1.3 Intracellular IONP Levels and Cellular Responses

The cellular internalization of IONPs can also affect cellular homeostasis by the mere presence of the NPs in the cellular organelles. As endosomes and lysosomes are filled

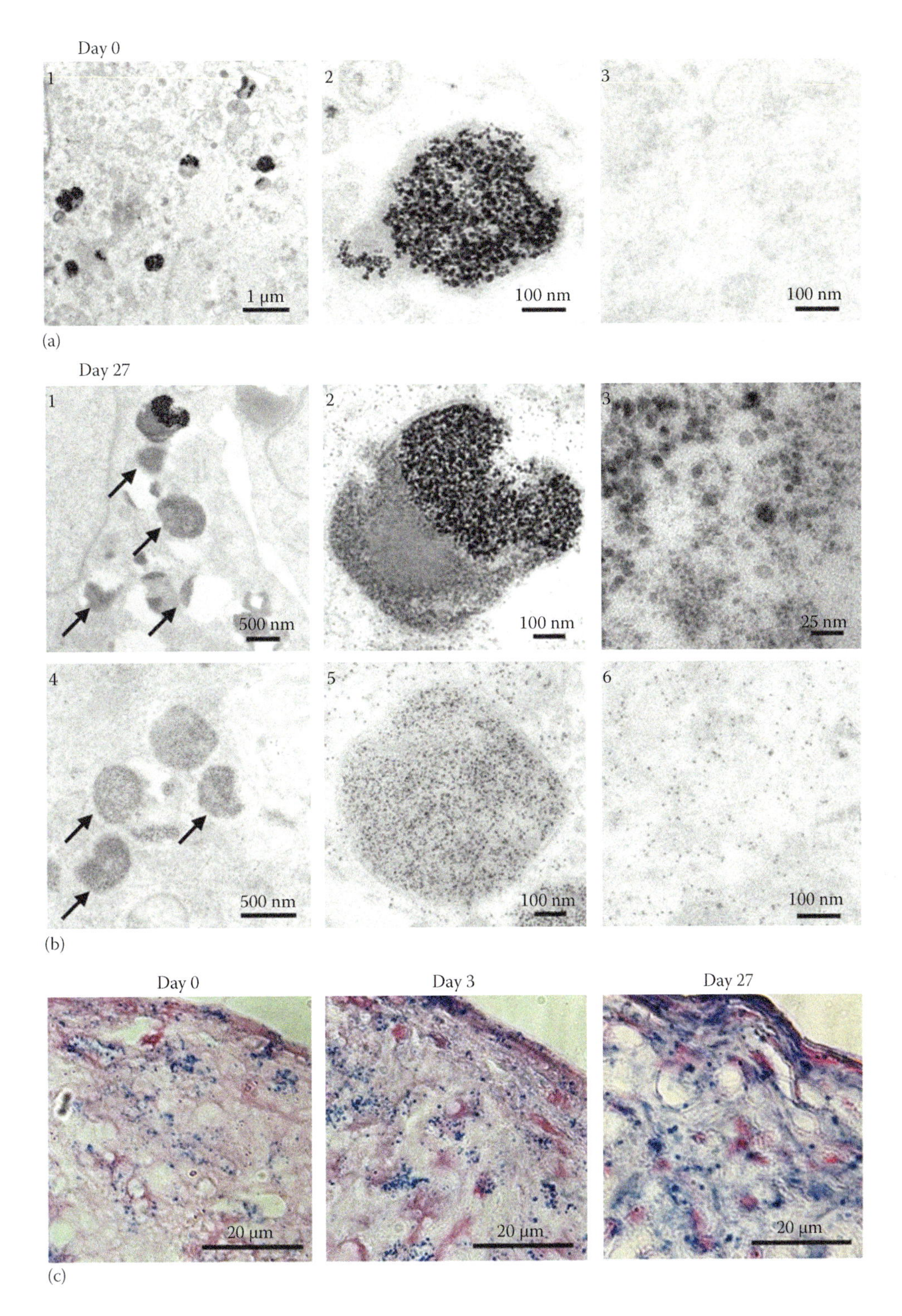

FIGURE 15.1 Nanoparticle imaging within the tissue: Evidence of nanoscale degradation. (a, b) Transmission electron microscopy of spheroid tissues containing NPs on the day of spheroid formation (day 0) (a); and after 27 d of maturation (day 27) (b). At day 0, all NPs are confined inside endosomes (lysosomes) (a1 and a2), and no dark nanospots can be seen in the cytoplasm (a3). At day 27, the endosomes are still filled with NPs, but most of them (arrows) contain less dark nanospots identified as ferritin. In the enlarged zones b2, b3 and b4, the endosome is a hybrid, filled with both ferritin and intact NPs. In b5, only ferritin spots remain in the endosome. Ferritin spots are also detected outside the endosomes, throughout the cytoplasm (see enlarged b6). (c) Sections of spheroids at different maturation times and stained with Pearls' reagent that colours iron blue. At day 0 and day 3, numerous small blue spots are detected throughout the cells. At day 3, a diffuse blue colour starts to appear, reflecting the appearance of ferritin loaded with iron ions in the cytoplasm. At day 27, fewer blue spots are detected, while the entire cell cytoplasm is blue.[24] (Reprinted with permission from F. Mazuel, A. Espinosa, N. Luciani, M. Reffay, R. Le Borgne, L. Motte et al., *ACS Nano*, 10, 7627–7638. Copyright 2016 American Chemical Society.)

with non- or slowly-degrading NPs, they are often not fully functional, resulting in a reduced degradative capacity of the cells.[28] In order to compensate for this, the cell can activate degradative pathways, which are often accompanied by the induction of autophagy. As autophagy is also a cellular defence mechanism to clear damaged organelles such as mitochondria, autophagy can also be induced by oxidative stress.[29] As a result, autophagy has been frequently linked with IONP toxicity both *in vitro* and *in vivo*.[30–33] The induction of autophagy can either have prosurvival effects[34,35] or result in cellular damage.[36] This apparent discrepancy can be attributed to the functional role of autophagy in cellular

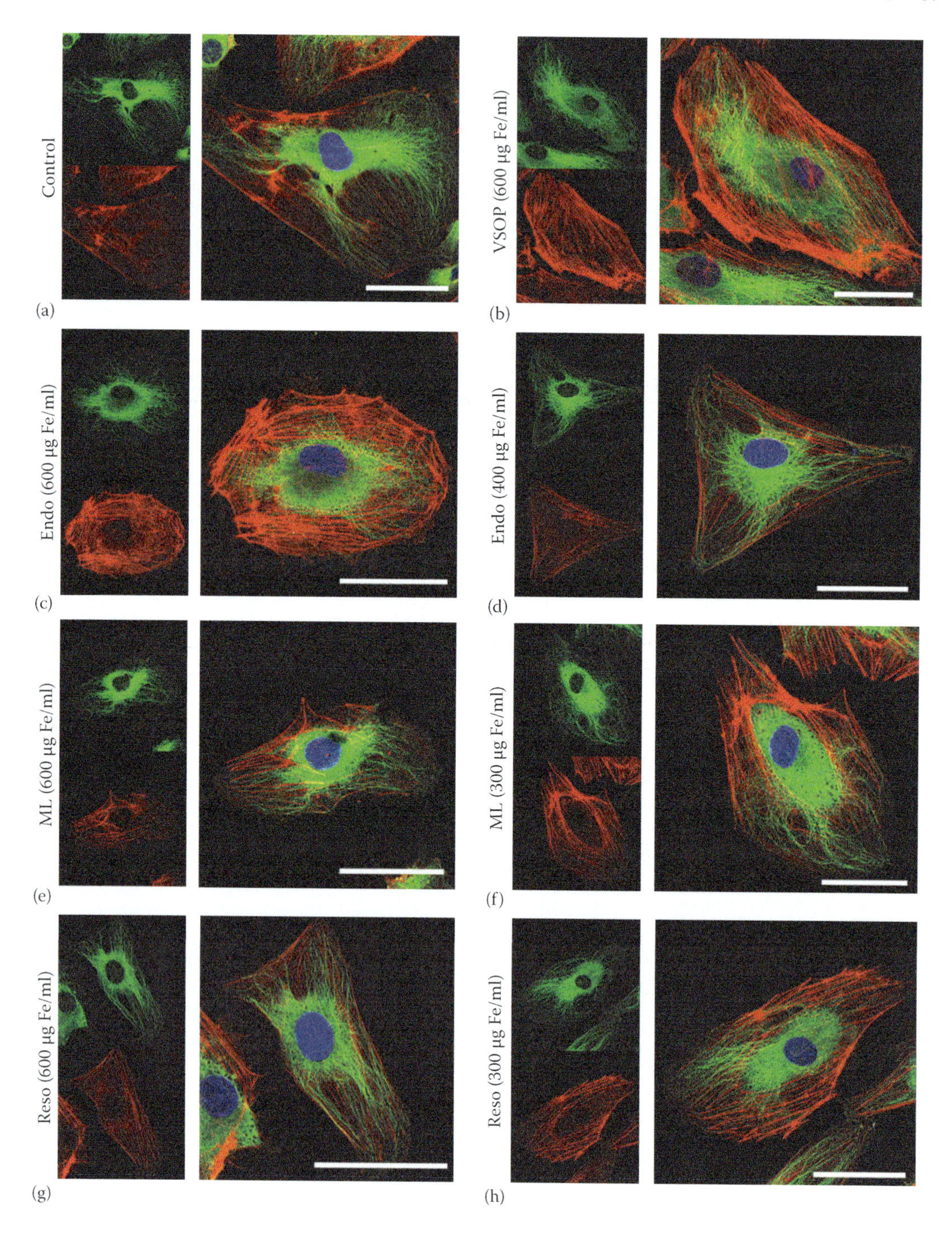

FIGURE 15.2 High intracellular NP concentrations affect cell spreading. Confocal micrographs of hBOECs incubated for 24 h either untreated (a) or incubated with (b) citrate-coated IONPs (VSOP), (c) dextran-coated IONPs (Endorem), (e) lipid-coated IONPs (MLs) or (g) carboxydextran-coated IONPs (Resovist) at 600 mg Fe/ml for 24 h, (d) dextran-coated (Endorem) at 400 mg Fe/ml, (f) lipid-coated (MLs) or (h) carboxydextran-coated (Resovist) at 300 mg Fe/ml for 24 h. Cells were kept in culture for an additional 3 days. F-actin is coloured red, the nucleus is coloured blue by 4',6-diamidino-2-phenylindole (DAPI) staining and α-tubulin is indicated in green. Green and red channels are shown separately for every image. Scale bars: 50 µm.[26] (Reprinted from *Biomaterials*, 32, S.J.H. Soenen, U. Himmelreich, N. Nuytten and M. De Cuyper, 195–205, Copyright 2011, with permission from Elsevier.)

signalling. As mentioned above, autophagy is typically seen as a cellular protective mechanism, where it will be induced to aid the cell in efficiently clearing away damaged organelles and help in recycling components and nutrients that can be further used for cellular growth. As such, cellular damage will often be joined by elevated levels of autophagy[37]; this in itself does not always result in cell death, but rather pro-survival. Often, autophagy induction has been defined as the main reason for nanomaterial-mediated cell death, but this was mainly a misnomer due to the observed elevated levels of autophagy in stressed and dying cells as the NP-induced damage was too severe for the cell to recover from by means of autophagy induction.[38] Autophagy signalling typically inhibits apoptosis, resulting in a sensitive balance between both signalling mechanisms; the NP-mediated damage can favour either mechanisms.[39] Persistent and elevated levels of autophagy have been found to result in direct cell death, as the cell almost 'eats itself'.[40]

Apart from autophagy, high levels of IONPs and the intracellular presence of high levels of slowly degrading NPs have been suggested to impose physical stress on the cells by steric hindrance, affecting the architecture of the cellular cytoskeleton (Figure 15.2).[41,42] As the cellular cytoskeleton is involved in a wide variety of different signalling cascades, this can result in a wide variety of effects, including reduced cell proliferation or cell migration,[43–45] which then impedes the use of these NPs for labelling of stem cells.

15.2 INFLUENCE OF THE MAGNETIC CORE

In terms of MNPs, IONPs have most frequently been used, but other formulations also exist, such as cobalt ferrite, iron platinum, gadolinium or manganese NPs, which have been described to possess better magnetic properties; however, the latter two have been shown to cause highly significant toxicity,[46,47] while there is a lack of toxicity information for FeCo and FePt particles. Thus, to date, the bulk of toxicity studies have also been performed on IONPs. Based on these studies, the main mechanisms underlying MNP toxicity as described above have been selected. More detailed descriptions of the toxicity of IONPs can be found elsewhere.[4] Studies on iron platinum or cobalt ferrite NPs are rarer, and comparative studies between the different types of core materials are lacking, hampering any attempt to define the degree of toxicity of the different NP types. Similar to IONPs, iron platinum NPs display degradation behaviour that results in the leaching of iron ions.[48] The degradation rate of the iron platinum NPs was much higher than that of pure iron, but no comparison is made with iron oxide. The released iron ions were found to inhibit cell proliferation,[48] and by linking these NPs to a cancer cell-specific peptide, the iron release could be harnessed as a potential antitumour therapy.[49] A preliminary study further found a clear indication of genotoxicity by iron platinum NPs,[50] but considering their degradation properties and associated induction of oxidative stress, this needs to be studied more in-depth.

Cobalt ferrite NPs have been found to induce high levels of oxidative stress, which in turn can result in cell death through activation of pro-apoptotic signalling pathways.[51,52] The particles have displayed clear dose–response behaviour in view of their toxicity.[52,53] Upon comparing the toxicological responses of cobalt ferrite NPs across different cell types, clear cell type-dependent effects have been observed, which is typical for nanotoxicity studies. This can be due, in part, to differences in the intrinsic endocytic uptake capacity of the cell types or differences in their antioxidant properties. In cancer cells, cobalt ferrite NPs were found to affect cancer stemness markers in one type of cancer cell line, but not in the other.[54] Cobalt ferrite NPs were also found to induce DNA damage and induced either apoptosis or necrosis, depending on the cell type studied.[55] These studies do however need to be considered with some precautions, as the concentrations at which cytotoxicity was observed between the different studies varied by approximately a 100-fold (0.95–100 μg/ml). Additionally, the assays performed may have suffered from interference by the NPs, explaining why cell viability was reduced at concentrations lower than 100 μg/ml, but viability was increased at concentrations exceeding 100 μg/ml.[55] Such interferences are quite common for optically dense NPs and fluorescence or absorbance measurements[56] and should be checked by means of appropriate controls.[56–58] Cobalt ferrite NPs have also been found to result in toxicity on different tissue levels, being mainly the liver, where they primarily end up following intravenous administration, but also in the lung and kidneys.[59,60] The high levels of oxidative stress, in part due to degradation of the NPs, has also been linked to toxicity on embryonic development.[46,61]

15.3 INFLUENCE OF DIFFERENT SURFACE COATINGS

Various types of coating agents have been used for the generation of 'biocompatible' IONPs, including synthetic or natural polymers, polysaccharides such as dextran, lipids, poly(ethylene glycol) (PEG), silica, organic (carbon) or inorganic (gold) shells.[62] The nature of the surface coating will contribute to the toxicity of the IONPs, by means of providing colloidal stability, determining the extent of interaction with the cells and protecting the NPs from the (acidic) environment and thereby impeding degradation. It is important to note that the coating in itself can sometimes be toxic such as, for instance, for cationic lipids.[63] Any observed toxicity can therefore be caused by either the iron oxide core itself or the coating. The coating together with the iron oxide cores also form an entirely novel entity that can result in toxic effects not observed by either of the components on their own. As such, the coating plays an important role in the final safety profile of the NPs. Table 15.1 presents an overview of toxicological data generated on bare, dimercaptosuccinic acid (DMSA)-coated, dextran-coated, PEGylated- and silica-coated IONPs, revealing the importance of the coating, but also highlighting the disparate nature of the data, making it impossible to draw any

TABLE 15.1
Overview of Several Historically Important and Some Recent Toxicology-Related Findings for Bare, Dimercaptosuccinic Acid (DMSA-) Coated, (Carboxy-) Dextran-Coated, Pegylated and Silica-Coated IONPs

Cell Type	Important Findings	Ref.
	Bare NP	
Human dermal fibroblasts	Cell morphology affected at 48 h post NP uptake	64
Human hepatoma and lung adenocarcinoma cell lines	No effects on cell viability could be noted up to 100 μg Fe/ml injected suspension?	65
A549 human alveolar cancer cell line	Only very low toxic effects were noted in terms of cell viability, DNA or mitochondrial damage. No differences between nano- or micrometre sized particles	66
Human mesothelioma, rat fibroblasts	Rodent fibroblasts were relatively insensitive to IONPs whereas human mesothelioma showed greatly reduced viability	5
Murine alveolar macrophage, human macrophage and epithelial cell lines	Significant reductions in cell viability for all three cell types tested. Results were comparable with those observed when exposing cells to asbestos	67
Human aortic endothelial cells	No induction of inflammatory responses at concentrations up to 50 μg Fe/ml injected suspension	68
Human umbilical vein endothelial cells	No cell death observed up to 400 μg/ml injected suspension, but significant induction of autophagy resulting in inflammation and endothelial dysfunction	33
Mouse peritoneal macrophages (RAW 264.7)	Oxidative stress induces organelle damage which results in autophagy induction and finally cell death	69
A549 human alveolar cancer cell line, IMR-90 human lung fibroblast cells	Bare IONPs resulted in selective toxicity in cancer cells by the induction of oxidative stress-mediated organelle damage resulting in autophagy	70
	Dimercaptosucccinic Acid (DMSA)-Coated	
14 different cell lines, including adult, progenitor, immune and tumour cells	High uptake of DMSA-coated particles without any effects on cell viability and functionality. Endothelial progenitor cell migration and tube formation is unaffected upon labelling	7
Human fibroblasts	Cell viability and mitochondrial activity were decreased at higher concentrations	71
Human melanoma	No effects on cell viability and morphology, in contrast to citrate or lauric acid-coated particles	72
Human cervical carcinoma cell line	Low uptake but no cytotoxic effects in contrast to heparin-coated particles, which induced abnormal mitotic spindle formation	73
Rat pheochromocytoma cell line	Dose-dependent reduction in cell viability, cell adhesion and nerve growth factor-induced neurite outgrowth	74
Mouse hepatoma (Hepa1-6), human monocyte (THP-1), human hepatoma (HepG2), mouse macrophages (264.7)	Gene expression profiling revealed that the DMSA coated IONPs resulted in a significant upregulation of cysteine-rich proteins in all 4 cell types	75
NCTC 1469 nonparenchymal hepatocytes and C57/Bl6 mice	Little cyto- or genotoxicity was observed on the NCTC cells; *in vivo*, the particles mainly accumulated in liver, spleen and lungs after which they are biotransformed into nonsuperparamagnetic forms of iron without clear toxicity	76
Mouse macrophages (264.7)	Gene expression profiling revealed clear changes in genes involved in intracellular iron homeostasis, indicating cellular adaptation to iron overload	77
	(Carboxy-)dextran-Coated Particles	
Human dermal fibroblasts	48 h post IONP uptake, cell morphology, viability, proliferation and migration were decreased to a greater extent than uncoated IONPs	64
Rat skeletal myoblasts	Induction of free radicals, decreased cell proliferation and cell viability	78
Human mesenchymal stem cells	Endorem with poly-L-lysine did not affect cell viability, proliferation, osteogenic or adipogenic differentiation but impeded chondrogenic differentiation	9
Swine endothelial progenitor cells	Viability and proliferation of Ferucarbotran-labelled cells was not affected, but adhesion capacity was increased and migration impeded in a dose-dependent manner	45
Human blood outgrowth endothelial cells, murine C17.2 neural progenitor cells	High intracellular concentrations of ferucarbotran, ferumoxides or lipid-coated IONPs impeded cell proliferation, affected cell spreading, focal adhesion maturation and focal adhesion kinase-signalling	41

(*Continued*)

TABLE 15.1 (CONTINUED)

Overview of Several Historically Important and Some Recent Toxicology-Related Findings for Bare, Dimercaptosuccinic Acid (DMSA-) Coated, (Carboxy-) Dextran-Coated, Pegylated and Silica-Coated IONPs

Cell Type	Important Findings	Ref.
Murine C17.2 neural progenitor cells, rat pheochromocytoma cells	Ferucarbotran and ferumoxide-labelled cells displayed intracellular degradation of the iron oxide core, resulting in induction of free radicals, increased expression of transferrin receptor 1 and impeded nerve growth factor-induced neurite outgrowth Effects were most outspoken for citrate-coated particles	70
Human mesenchymal stem cells	Ferucarbotran did not affect cell viability, induce free radicals, nor affect mitochondrial membrane potential, and did not impede differentiation. At later time points, lysosomal degradation was evident	79
Rat mesenchymal stem cells	Ferucarbotran without transfection agent results in an increased expression of transferrin receptor 1	21
Human mesenchymal stem cells	Ferucarbotran possesses intrinsic peroxidase-like activity and diminishes intracellular reactive oxygen species. Upon degradation, free iron stimulates cell growth and accelerates cell cycle progression	19
Human mesenchymal stem cells	Ferucarbotran inhibits osteogenic differentiation in a dose-dependent manner and activates cellular signalling molecules. All effects were the result of free iron as they could be impeded by the use of the iron chelator desferrioxamine	10
Human peripheral blood samples, male Wistar rats	Up to 1 mg/ml injected suspension, dextran-coated IONPs did not induce haemolysis, but showed variable immune responses based on the model used	80
Head and neck squamous cancer cells	Carboxydextran-coated IONPs resulted in concentration-dependent apoptosis and resulted in secretion of pro-inflammatory cytokines at high concentrations	81
Adult zebrafish	Acute toxicity of cross-linked aminated dextran-coated NPs on zebrafish brain by increased iron levels, causing reduced acetylcholinesterase activity and induction of caspase 8 and 9 activity	82
	PEGylated Particles	
Chinese hamster ovary cells	PEGylation of IONPs reduced cytotoxicity compared to aminated or dextran-coated IONPs. PEG chain length revealed that shorter PEG chains resulted in lowest toxicity levels due to lowest cellular uptake	83
C6 rat glioma cells, human cervical cancer (HeLa)	PEGylated IONPs displayed no toxicity at concentrations up to a dose of 1 mg/ml injected suspension	84
Mouse NIH 3T3 fibroblasts and Balb/c mice	PEGylated IONPs displayed no significant toxicity up to 192 μM Fe^{3+} and were found to be efficiently cleared from the body by hepatobiliary route within 14 d without toxicity	85
Vascular smooth muscle cells	Apart from one condition, no differences in toxicity could be observed between short and long PEG chains or 2-methoxyethyleneamine-coated IONPs	86
Bone marrow derived macrophages	Moderate toxicity observed for PEGylated IONPs even at the lowest dose (0.3 μg Fe/ml), but no dose–response relationship. No release of inflammation or anti-inflammatory cytokines up to 30 μg Fe/ml	87
Human cervical cancer cells (HeLa), Balb/c mice	No toxicity of differently sized (5, 15, 30 nm diameter) IONPs coated with PEG up to 200 μg Fe/ml. Persistent presence of the differently sized PEGylated IONPs in liver and spleen with minor increases in liver enzyme activity up to 1 month post IONP administration	77
	Silica-Coated Particles	
Human neuroblastoma SHSY5Y, human glioblastoma A172 cells	Silica-coated IONPs were more cytotoxic than oleic acid-coated ones. Toxicity of both IONP types could be improved by presence of serum components. Clear interference of both IONP types with absorbance readouts were observed and appropriate controls were included	88
Human cervical cancer (HeLa), human lung adenocarcinoma (A549) cells	Passivation of the silica coating significantly reduces IONP cytotoxicity by reducing degradation and oxidative stress levels	67
Mouse osteoblast precursor cells MC3T3	Silica-coating of IONPs reduces cytotoxicity and oxidative stress compared to bare IONPs	89
Human bronchoalveolar cells (BEAS2B)	Silica coating of IONPs reduces acidic degradation and improves cell viability	90

Source: S.J.H. Soenen and M. De Cuyper, *Nanomedicine (Lond)*, 5, 1261–75, 2010. Copyright 2010 Future Science Group Ltd. Adapted with permission.

conclusions as to which surface coating is optimally suited for biomedical applications.

The coating of the IONPs will also contribute to the surface charge, which is an important determinant in the degree of interaction of the IONPs with biological molecules and cells. For example, dextran-coated (neutral) IONPs were found to exhibit low cytotoxicity compared to anionic DMSA-coated IONPs, which was mainly attributed to the higher uptake levels of the latter.[91] A positive surface charge can also result in a high level of interaction with the negatively charged cell membrane and polycationic coatings (e.g. polydimethylamine) that have been found to induce cell membrane damage.[92] Minor differences in the nature of the coating can already have far-ranging impacts on the toxicity profile of IONPs. For example, IONPs coated with PEG chains of different molecular weight revealed that longer PEG chains resulted in lower toxicity on cultured cells, while resulting in higher liver and kidney toxicity after intravenous administration.[93] Surface passivation also plays an important role, where silica-coated IONPs either left bare or further passivated with amino- or sulphonate silanes revealed that the surface passivated IONPs exhibit far less cytotoxicity compared to their bare silica-coated counterparts (Figure 15.3)[94] despite similar cellular uptake levels.

This was mainly attributed to differences in the degradation properties of the IONPs, where surface passivation was

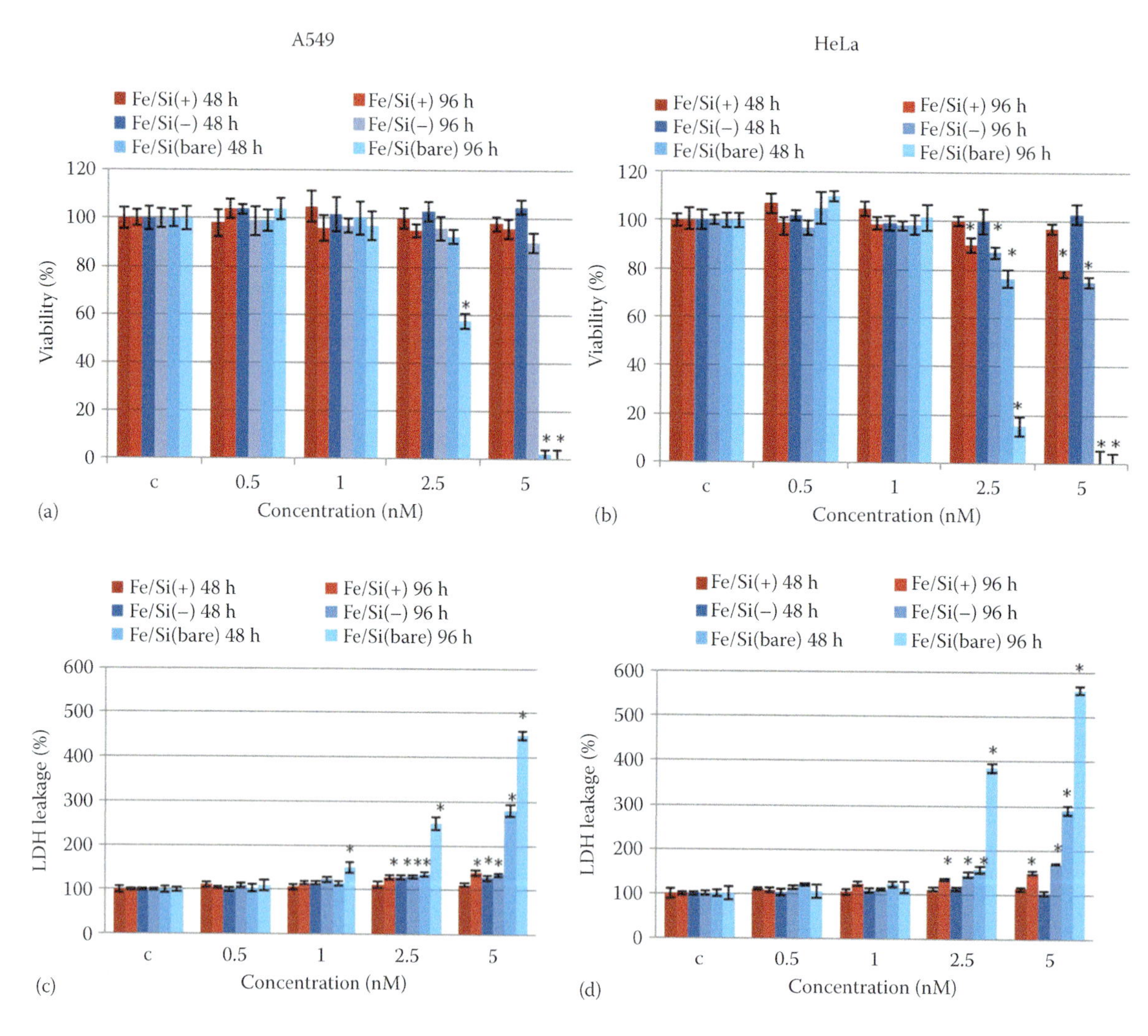

FIGURE 15.3 Effect of bare and passivated Fe_3O_4/SiO_2 NPs on the viability and membrane damage in two cell lines (A549 and HeLa). (a, b) WST-8 proliferation assay and (c, d) Lactase dehydrogenase (LDH) assay on A549 and HeLa cells incubated with increasing concentrations (0.5, 1, 2.5, 5 nM) of bare and passivated Fe_3O_4/SiO_2 NPs at different times (48 and 96 h). C identifies the negative control in the absence of NPs. Viability of NPs-treated cells is expressed relative to nontreated control cells. As positive control (P), cells were incubated with 5% dimethylsulphoxide (DMSO) in WST-8 assay and 0.9% Triton X-100 in LDH assay (not shown). Data are reported as mean ± SD from three independent experiments; $^*P < 0.05$ compared with control ($n = 8$).[94] (From M.A. Malvindi et al., *PLoS One*, 9, e85835. Copyright 2014 Plos One. Reprinted with permission, through open access publishing.)

found to reduce IONP degradation and thereby also reduced the levels of oxidative stress induced by the IONPs. Apart from a chemically applied coating, cellular exposure to IONPs in serum-containing medium will also result in the formation of a protein corona surrounding the NPs, which in most cases reduces the toxicity level of IONPs compared to that of particles without any surrounding proteins.[95] As previously mentioned, cytotoxicity profiles will also be cell type specific, where different cell types can activate different detoxification pathways for the IONPs.[96]

As mentioned above, the coating will determine the degree of exposure of the iron oxide core to the surrounding environment and the associated degradation. In initial studies, intracellular degradation of IONPs was found to be one of the main causes of toxicity in various cell types, and the rate of degradation was determined by the nature of the coating.[97] In later studies, it has been shown that a polymer coating controls surface reactivity and that availability and access of chelating agents to the crystal surface govern the degradation rate of IONPs.[98] The extensive biodegradation reported for many types of IONPs can be efficiently inhibited by using inert molecules, such as gold shells as an inorganic coating.[99]

15.4 INFLUENCE OF THE SIZE AND SHAPE OF IONPs

Apart from the chemical nature of the core and the available surface chemistry, the size and shape of MNPs is another important determinant in their toxicity profile. For most types of NPs, smaller sized spherical NPs are typically considered to be more cytotoxic, which is in part due to the elevated ratio of surface area over volume. For a higher total surface area, the interface between the NPs and their surrounding environment also increases, elevating the chance for NP-mediated cellular damage. To date, the effect of size and shape of IONPs has not received much attention. This is explained in part by the fact that the size of the total particles preferably remains below a certain size that renders the NPs superparamagnetic rather than ferromagnetic.[100] However, multidomain particles that are superparamagnetic in total while individual domains can be ferromagnetic are equally interesting for biomedical use, enabling a wider range of NP diameter.

In this small size range, it has been observed that IONPs ranging from 9 to 20 nm diameter displayed no differences in their toxicity level, while their cellular uptake levels were affected.[101] Studies on IONPs of 5 and 30 nm diameter revealed that highest levels of ROS induction were found with the 30 nm diameter IONPs.[102] For IONPs of 15, 20 and 50 nm diameter, the highest levels of toxicity were observed for 20 nm diameter IONPs.[103] PEGylated IONPs of 5, 15 and 30 nm diameter were also found not to affect cell viability at concentrations up to 200 μg Fe/ml. Interestingly, following 1 month of intravenous administration in mice, IONPs were found to persist in the liver and spleen of the mice, displaying low levels of liver damage up to 1 month after IONP administration (Figure 15.4).[104] The shape of IONPs has also been shown to affect their toxicity levels, where rod-shaped IONPs were found to be more cytotoxic than spherical ones.[105]

Together, these data suggest a size-dependent effect on IONP toxicity, but only for IONPs of widely differing sizes. Despite the increase of surface area over volume ratio for smaller IONPs, higher levels of toxicity are typically observed for particles of 20–30 nm diameter. This may be explained by the intrinsic cellular uptake efficiency of the NPs which has also been shown to be size-dependent and is maximal for particles of approximately 40 nm diameter.[106] The intrinsic toxicity of smaller sized IONPs may therefore be higher, but the lower level of cellular internalization will mask this effect. It is therefore of great interest to study the toxicity of different NPs with respect to their cellular uptake levels.[39,97,107]

15.5 INFLUENCE OF THE EXPOSURE CONDITIONS USED

The toxicity of MNPs remains unresolved to date. This is mainly due to the wide variability in different types of MNPs, varying in size, shape, surface coating or core chemistry. While all aspects mentioned above have been defined as single parameters, all data mentioned must always be interpreted with care as a change in a single parameter can automatically affect others.[108] For instance, applying a novel coating will affect the surface charge of the NPs, but also the colloidal stability, hydrodynamic diameter and susceptibility to acidic degradation. Therefore, a novel coating will affect NP toxicity by altering the degradation rate, the degree of cellular internalization and the extent of interaction of the NPs with their biological surroundings. Assigning any observed toxic effect to a particular physicochemical parameter therefore remains highly difficult and the ability to synthesize IONPs based on a safe-by-design procedure remains elusive to date.

Apart from the wide differences in NP specifics, further difficulties in data interpretation also lie in the exact exposure conditions used and the methods used to analyze cell well-being. As previously mentioned, optically dense NPs such as IONPs have been described to interfere with a wide range of common biochemical toxicity assays such as the MTT assay, which can result in artificially high levels of cell viability (Figure 15.5).[42,55,109–111]

Appropriate controls must be included to minimize the chance of any interferences occurring, but typically, one can obtain the best results by including multiple assays for every

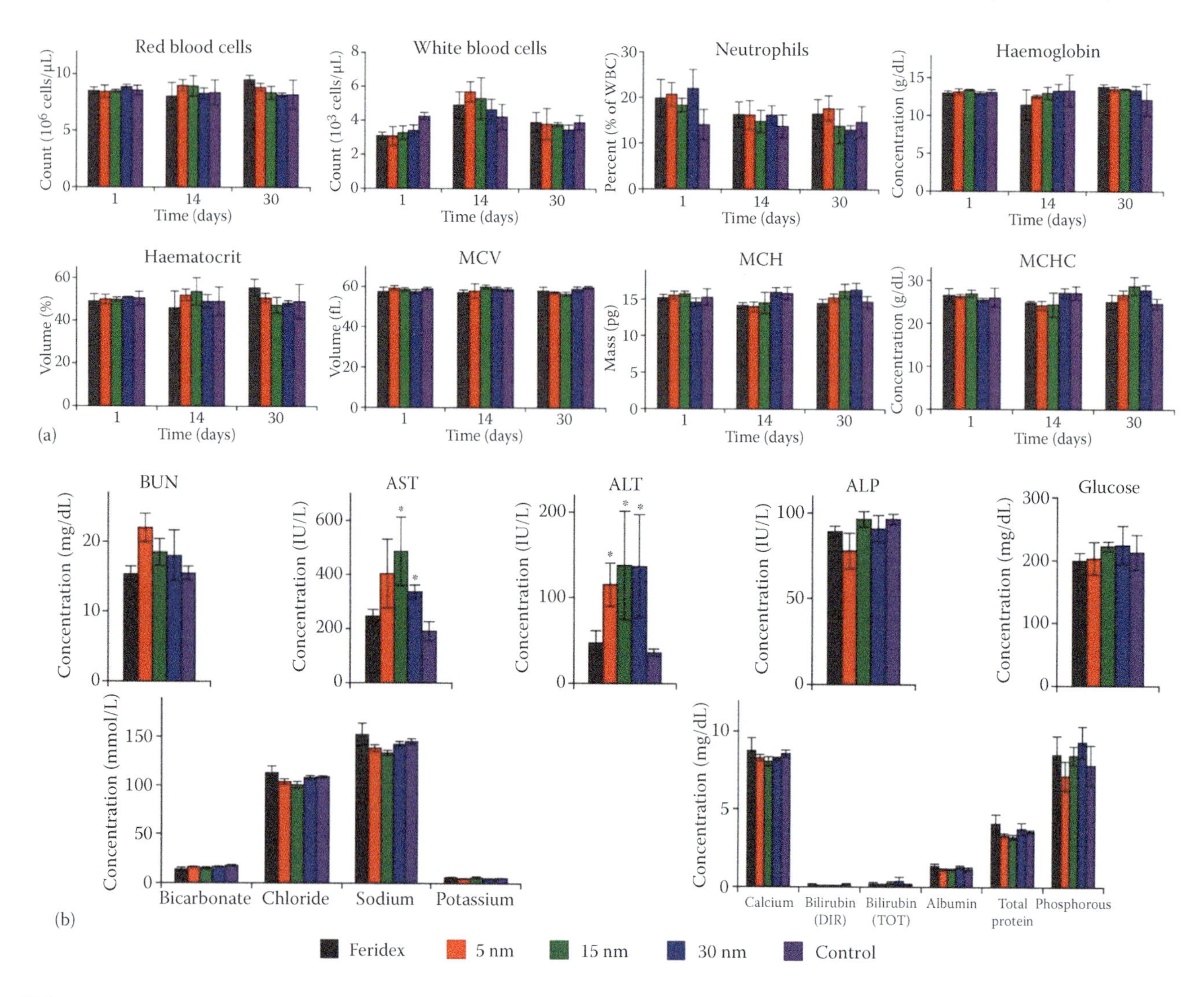

FIGURE 15.4 Haematology and blood chemistry of mice following injection of IONPs, prepared via the organometallic route with sizes of ~5, ~15 or ~30 nm and coated with PEG-phospholipid, or commercially obtained Feridex (as indicated). (a) Red blood cell number, white blood cell number, percentage of neutrophils among white blood cells, haemoglobin concentration, haematocrit, mean corpuscular volume (MCV), mean corpuscular haemoglobin (MCH) and mean corpuscular haemoglobin concentration (MCHC) of mice 1, 14 and 30 days after intravenous injection with IONPs (5 mg Fe/kg). (b) Concentration of blood urea nitrogen (BUN), aspartate transaminase (AST), alanine transaminase (ALT), alkaline phosphatase (ALP), glucose, bicarbonate, chloride, sodium, potassium, calcium, direct bilirubin (DIR), total bilirubin (TOT), albumin, total protein and phosphorus of mice 30 days after intravenous injection with IONPs (5 mg Fe/kg). Statistical analyses were performed with Student's t test (*$p < 0.05$ for the difference between IONPs and phosphate buffered saline (PBS), two-tailed, unpaired, $n = 4–6$, error bars = standard deviation).[104] (Reprinted with permission L. Gu, R.H. Fang, M.J. Sailor and J.-H. Park, *ACS Nano*, 6, 4947–4954. Copyright 2012 American Chemical Society.)

single parameter tested, preferably using assays that are based on different readout mechanisms, ranging from typical biochemical absorbance assays to microscopic evaluations of fluorescently stained cells.[56,57,62] Other difficulties in interpreting toxicity data lie in the differences in the exact exposure conditions used, where exposure times can vary from 1 hour to multiple days and NP concentrations can vary widely.[109] The use of different cell types, each with their own specific properties and intrinsic defence levels against oxidative stress, can also influence the outcome of any toxicity studies performed.[96] To overcome this issue, toxicity studies are preferably performed on multiple cell types at different time points for a wide range of concentrations. Although requiring more work, this can be facilitated by using more high-throughput methods such as kinetic analysis of live cells and automated microscopy.[110,112,113]

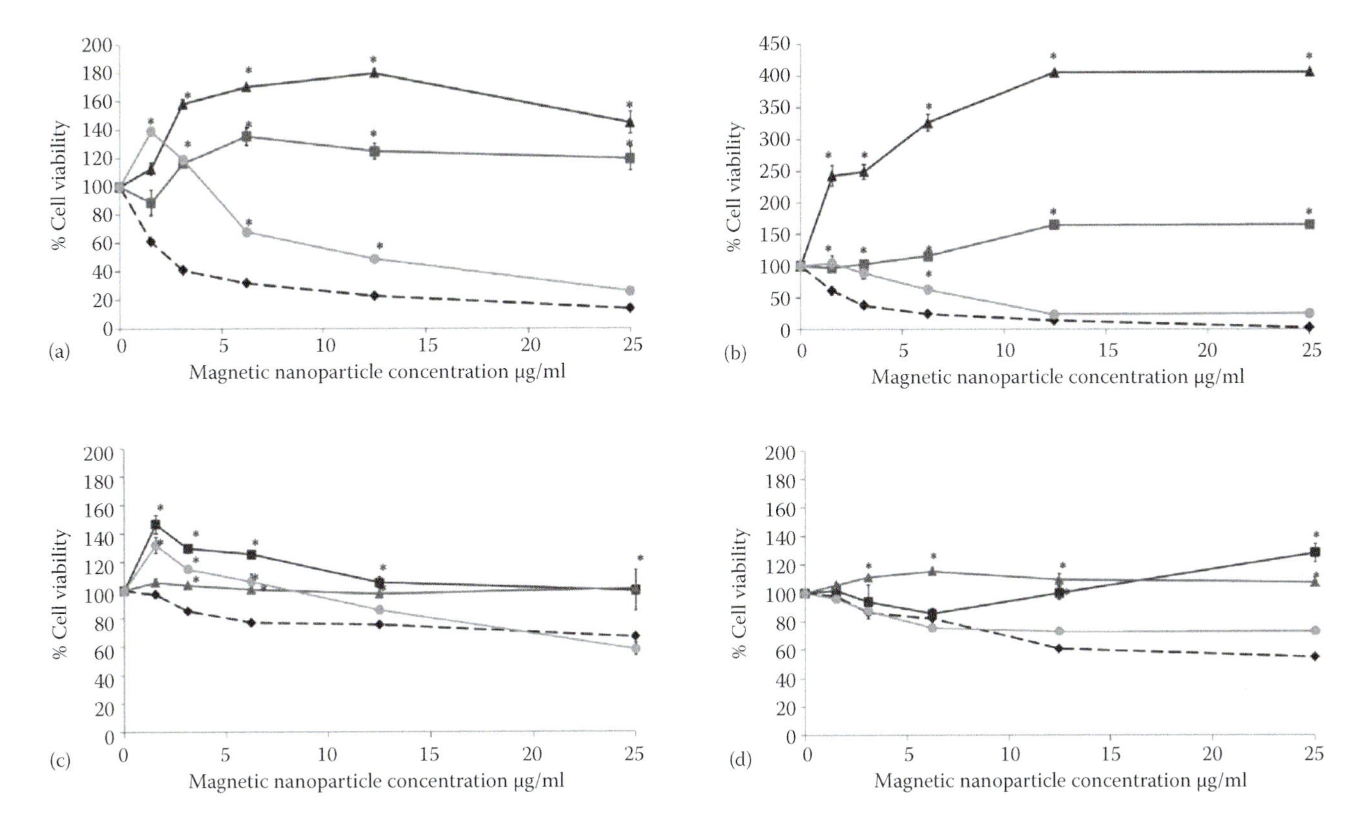

FIGURE 15.5 Cell viability of SH-SY5Y and RAW 264.7 cells. (a) SH-SY5Y cells incubated with poly(ethylene imine)-coated MNPs (MNP-PEI); (b) RAW 264.7 cells incubated with MNP-PEI NPs; (c) SH-SY5Y cells incubated with MNP-PEI-PEG NPs; and (d) RAW 264.7 cells incubated with MNP-PEI-PEG NPs. The cells were incubated at different concentrations as indicated over a 72 h incubation. Cell viability was determined using common assays including MTS assay (square), CellTiter-Blue assay (triangle), CellTiter-Glo assay (circle) and trypan blue counting (diamond). ($n = 3 \pm$ SE). Asterisk denotes significantly increased level of cell viability compared with trypan blue measurement ($p < 0.05$).[111] (With kind permission from Springer Science+Business Media: *Nanoscale Res. Lett.*, 7, 2012, 77, C. Hoskins, L. Wang, W. Cheng and A. Cuschieri.)

15.6 CONCLUSION AND FUTURE OUTLOOK

The increased use of MNPs for biomedical applications is resulting in a higher level of exposure, which in turn raises questions regarding the safety and possible shortcomings regarding the use of these nanomaterials. While a significant data have been generated on the cytotoxicity of MNPs, detailed studies that provide long-term exposure data in animals or patients are scarce. Initial assessments of the cytotoxicity of MNPs were quite difficult to carry out due to the lack of appropriate standards, controls or procedures. While this has been largely addressed since, the reliability of any data can sometimes be questionable, as it is often unclear whether appropriate controls were taken into consideration to rule out any interference of the NPs with the assays performed. The advent of new technologies and methods, including high-throughput analysis, will aid in the further development of this field, where large datasets can be generated on multiple parameters involved in the cytotoxicity of large sets of different MNPs, covering both a wide concentration range and multiple time points. These datasets can be used for *in silico* analysis, where NP-associated parameters can be linked to particular effects of their cytotoxicity profile. Additionally, meta-analysis studies which compare the results obtained for all published studies thus far will further help to filter out any data that may have suffered from NP interference, and try as well to appropriately define the most important biological and physicochemical parameters involved in NP cytotoxicity. Together, the use of high-throughput assays, along with *in silico* models, can pave the way for predictive toxicity studies, where computational models can be generated to predict NP toxicity levels without having to perform actual wet-lab experiments. These models and datasets will then also have to be further validated with clinical data or by preclinical animal studies. These efforts will provide a strong background to predict the toxicity profiles of generated MNPs and to describe the conditions under which they can be safely used.

ACKNOWLEDGEMENTS

This work was supported by the FWO-Vlaanderen (S.J.S., KAN 1514716N to B.B.M.), the KU Leuven program financing IMIR (PF 2010/017), the Flemish agency for Innovation through Science and Technology (IWT SBO MIRIAD, IWT SBO NanoComit).

REFERENCES

1. S.J. Soenen, S.C. de Smedt and K. Braeckmans, Limitations and caveats of magnetic cell labeling using transfection agent complexed iron oxide nanoparticles, *Contrast Media Mol. Imaging* 7 (2012), pp. 140–152.
2. S. Laurent, A.A. Saei, S. Behzadi, A. Panahifar and M. Mahmoudi, Superparamagnetic iron oxide nanoparticles for delivery of therapeutic agents: Opportunities and challenges, *Expert Opin. Drug Deliv.* 11 (2014), pp. 1449–1470.
3. R. Weissleder, D. Stark, B. Engelstad, B. Bacon, C. Compton, D. White et al., Superparamagnetic iron oxide: Pharmacokinetics and toxicity, *Am. J. Roentgenol.* 152 (1989), pp. 167–173.
4. V.I. Shubayev, T.R. Pisanic and S. Jin, Magnetic nanoparticles for theragnostics, *Adv. Drug Deliv. Rev.* 61 (2009), pp. 467–477.
5. T.J. Brunner, P. Wick, P. Manser, P. Spohn, R.N. Grass, L.K. Limbach et al., In vitro cytotoxicity of oxide nanoparticles: Comparison to asbestos, silica, and the effect of particle solubility, *Environ. Sci. Technol.* 40 (2006), pp. 4374–4381.
6. S.M. Cromer Berman, Kshitiz, C.J. Wang, I. Orukari, A. Levchenko, J.W.M. Bulte et al., Cell motility of neural stem cells is reduced after SPIO-labeling, which is mitigated after exocytosis, *Magn. Reson. Med.* 69 (2013), pp. 255–262.
7. C. Wilhelm and F. Gazeau, Universal cell labelling with anionic magnetic nanoparticles, *Biomaterials* 29 (2008), pp. 3161–3174.
8. H.S. Kim, Y. Choi, I.C. Song and W.K. Moon, Magnetic resonance imaging and biological properties of pancreatic islets labeled with iron oxide nanoparticles, *NMR Biomed.* 22 (2009), pp. 852–856.
9. L. Kostura, D.L. Kraitchman, A.M. Mackay, M.F. Pittenger and J.W.M. Bulte, Feridex labeling of mesenchymal stem cells inhibits chondrogenesis but not adipogenesis or osteogenesis, *NMR Biomed.* 17 (2004), pp. 513–517.
10. Y.-C. Chen, J.-K. Hsiao, H.-M. Liu, I.-Y. Lai, M. Yao, S.-C. Hsu et al., The inhibitory effect of superparamagnetic iron oxide nanoparticle (Ferucarbotran) on osteogenic differentiation and its signaling mechanism in human mesenchymal stem cells, *Toxicol. Appl. Pharmacol.* 245 (2010), pp. 272–279.
11. A.S. Arbab, G.T. Yocum, A.M. Rad, A.Y. Khakoo, V. Fellowes, E.J. Read et al., Labeling of cells with ferumoxides-protamine sulfate complexes does not inhibit function or differentiation capacity of hematopoietic or mesenchymal stem cells, *NMR Biomed.* 18 (2005), pp. 553–559.
12. A. Nel, T. Xia, L. Mädler and N. Li, Toxic potential of materials at the nanolevel, *Science* 311 (2006), pp. 622–627.
13. B. Díaz, C. Sánchez-Espinel, M. Arruebo, J. Faro, E. de Miguel, S. Magadán et al., Assessing methods for blood cell cytotoxic responses to inorganic nanoparticles and nanoparticle aggregates, *Small* 4 (2008), pp. 2025–2034.
14. A. Sabareeswaran, E.B. Ansar, P.R.V. Harikrishna Varma, P.V. Mohanan and T.V. Kumary, Effect of surface-modified superparamagnetic iron oxide nanoparticles (SPIONS) on mast cell infiltration: An acute in vivo study, *Nanomed. Nanotechnol. Biol. Med.* 12 (2016), pp. 1523–1533.
15. N. Bohmer and A. Jordan, Caveolin-1 and CDC42 mediated endocytosis of silica-coated iron oxide nanoparticles in HeLa cells, *Beilstein J. Nanotechnol.* 6 (2015), pp. 167–176.
16. J.-M. Idee, M. Port, I. Raynal, M. Schaefer, B. Bonnemain, P. Prigent et al., *Superparamagnetic Nanoparticles of Iron Oxides for Magnetic Resonance Imaging Applications, in Nanotechnologies for the Life Sciences*, Wiley-VCH Verlag GmbH & Co. KGaA, Weinheim, Germany, 2007.
17. A.S. Arbab, L.A. Bashaw, B.R. Miller, E.K. Jordan, B.K. Lewis, H. Kalish et al., characterization of biophysical and metabolic properties of cells labeled with superparamagnetic iron oxide nanoparticles and transfection agent for cellular MR imaging, *Radiology* 229 (2003), pp. 838–846.
18. L. Gao, J. Zhuang, L. Nie, J. Zhang, Y. Zhang, N. Gu et al., Intrinsic peroxidase-like activity of ferromagnetic nanoparticles, *Nat. Nanotechnol.* 2 (2007), pp. 577–583.
19. D.-M. Huang, J.-K. Hsiao, Y.-C. Chen, L.-Y. Chien, M. Yao, Y.-K. Chen et al., The promotion of human mesenchymal stem cell proliferation by superparamagnetic iron oxide nanoparticles, *Biomaterials* 30 (2009), pp. 3645–3651.
20. S.J.H. Soenen, U. Himmelreich, N. Nuytten, T.R. Pisanic, A. Ferrari and M. De Cuyper, Intracellular nanoparticle coating stability determines nanoparticle diagnostics efficacy and cell functionality, *Small* 6 (2010), pp. 2136–2145.
21. R. Schäfer, R. Kehlbach, J. Wiskirchen, R. Bantleon, J. Pintaske, B.R. Brehm et al., Transferrin receptor upregulation: In vitro labeling of rat mesenchymal stem cells with superparamagnetic iron oxide, *Radiology* 244 (2007), pp. 514–523.
22. T. Coccini, F. Caloni, L.J. Ramírez Cando and U. De Simone, Cytotoxicity and proliferative capacity impairment induced on human brain cell cultures after short- and long-term exposure to magnetite nanoparticles, *J. Appl. Toxicol.* 37 (2016), pp. 361–373.
23. Y. Wei, M. Zhao, F. Yang, Y. Mao, H. Xie and Q. Zhou, Iron overload by superparamagnetic iron oxide nanoparticles is a high risk factor in cirrhosis by a systems toxicology assessment, *Sci. Rep.* 6 (2016), pp. 29110.
24. F. Mazuel, A. Espinosa, N. Luciani, M. Reffay, R. Le Borgne, L. Motte et al., Massive intracellular biodegradation of iron oxide nanoparticles evidenced magnetically at single-endosome and tissue levels, *ACS Nano* 10 (2016), pp. 7627–7638.
25. M. Levy, N. Luciani, D. Alloyeau, D. Elgrabli, V. Deveaux, C. Pechoux et al., Long term in vivo biotransformation of iron oxide nanoparticles, *Biomaterials* 32 (2011), pp. 3988–3999.
26. S.J.H. Soenen, U. Himmelreich, N. Nuytten and M. De Cuyper, Cytotoxic effects of iron oxide nanoparticles and implications for safety in cell labelling, *Biomaterials* 32 (2011), pp. 195–205.

27. M. Lévy, F. Lagarde, V.-A. Maraloiu, M.-G. Blanchin, F. Gendron, C. Wilhelm et al., Degradability of superparamagnetic nanoparticles in a model of intracellular environment: Follow-up of magnetic, structural and chemical properties, *Nanotechnology* 21 (2010), pp. 395103.
28. S.T. Stern, P.P. Adiseshaiah and R.M. Crist, Autophagy and lysosomal dysfunction as emerging mechanisms of nanomaterial toxicity, *Part. Fibre Toxicol.* 9 (2012), p. 20.
29. B. Halamoda Kenzaoui, C. Chapuis Bernasconi, S. Guney-Ayra and L. Juillerat-Jeanneret, Induction of oxidative stress, lysosome activation and autophagy by nanoparticles in human brain-derived endothelial cells., *Biochem. J.* 441 (2012), pp. 813–821.
30. J. Du, W. Zhu, L. Yang, C. Wu, B. Lin, J. Wu et al., Reduction of polyethylenimine-coated iron oxide nanoparticles induced autophagy and cytotoxicity by lactosylation, *Regen. Biomater.* 3 (2016), pp. 223–229.
31. X. Zhang, H. Zhang, X. Liang, J. Zhang, W. Tao, X. Zhu et al., Iron oxide nanoparticles induce autophagosome accumulation through multiple mechanisms: Lysosome impairment, mitochondrial damage, and er stress, *Mol. Pharm.* 13 (2016), pp. 2578–2587.
32. X. Li, J. Feng, R. Zhang, J. Wang, T. Su, Z. Tian et al., Quaternized chitosan/alginate-Fe_3O_4 magnetic nanoparticles enhance the chemosensitization of multidrug-resistant gastric carcinoma by regulating cell autophagy activity in mice, *J. Biomed. Nanotechnol.* 12 (2016), pp. 948–961.
33. L. Zhang, X. Wang, Y. Miao, Z. Chen, P. Qiang, L. Cui et al., Magnetic ferroferric oxide nanoparticles induce vascular endothelial cell dysfunction and inflammation by disturbing autophagy, *J. Hazard. Mater.* 304 (2016), pp. 186–195.
34. T.-C. Tseng, F.-Y. Hsieh and S. Hsu, Increased cell survival of cells exposed to superparamagnetic iron oxide nanoparticles through biomaterial substrate-induced autophagy, *Biomater. Sci.* 4 (2016), pp. 670–677.
35. X. Mao, M. Shi, L. Cheng, Z. Zhang and Z. Liu, Ferroferric oxide nanoparticles induce prosurvival autophagy in human blood cells by modulating the Beclin 1/Bcl-2/VPS34 complex, *Int. J. Nanomed.* 10 (2014), pp. 207–216.
36. J.-H. Kim, J. Sanetuntikul, S. Shanmugam and E. Kim, Necrotic cell death caused by exposure to graphitic carbon-coated magnetic nanoparticles, *J. Biomed. Mater. Res. Part A* 103 (2015), pp. 2875–2887.
37. E.-J. Park, D.-H. Choi, Y. Kim, E.-W. Lee, J. Song, M.-H. Cho et al., Magnetic iron oxide nanoparticles induce autophagy preceding apoptosis through mitochondrial damage and ER stress in RAW264.7 cells, *Toxicol. Vitr.* 28 (2014), pp. 1402–1412.
38. K. Peynshaert, B.B. Manshian, F. Joris, K. Braeckmans, S.C. De Smedt, J. Demeester et al., Exploiting intrinsic nanoparticle toxicity: The pros and cons of nanoparticle-induced autophagy in biomedical research, *Chem. Rev.* 114 (2014), pp. 7581–7609.
39. B.B. Manshian, S. Munck, P. Agostinis, U. Himmelreich and S.J. Soenen, High content analysis at single cell level identifies different cellular responses dependent on nanomaterial concentrations, *Sci. Rep.* 5 (2015), p. 13890.
40. Y. Liu and B. Levine, Autosis and autophagic cell death: The dark side of autophagy, *Cell Death Differ.* 22 (2015), pp. 367–376.
41. S.J.H. Soenen, N. Nuytten, S.F. De Meyer, S.C. De Smedt and M. De Cuyper, High intracellular iron oxide nanoparticle concentrations affect cellular cytoskeleton and focal adhesion kinase-mediated signaling, *Small* 6 (2010), pp. 832–842.
42. S.J.H. Soenen, E. Illyes, D. Vercauteren, K. Braeckmans, Z. Majer, S.C. De Smedt et al., The role of nanoparticle concentration-dependent induction of cellular stress in the internalization of non-toxic cationic magnetoliposomes, *Biomaterials* 30 (2009), pp. 6803–6813.
43. S.J. Soenen, B. Manshian, J.M. Montenegro, F. Amin, B. Meermann, T. Thiron et al., Cytotoxic effects of gold nanoparticles: A multiparametric study, *ACS Nano* 6 (2012), pp. 5767–5783.
44. C.Y. Tay, P. Cai, M.I. Setyawati, W. Fang, L.P. Tan, C.H.L. Hong et al., Nanoparticles strengthen intracellular tension and retard cellular migration, *Nano Lett.* 14 (2014), pp. 83–88.
45. J.-X. Yang, W.-L. Tang and X.-X. Wang, Superparamagnetic iron oxide nanoparticles may affect endothelial progenitor cell migration ability and adhesion capacity, *Cytotherapy* 12 (2010), pp. 251–259.
46. F. Ahmad, X. Liu, Y. Zhou and H. Yao, An in vivo evaluation of acute toxicity of cobalt ferrite ($CoFe_2O_4$) nanoparticles in larval-embryo zebrafish (*Danio rerio*), *Aquat. Toxicol.* 166 (2015), pp. 21–28.
47. D. Pan, A.H. Schmieder, S.A. Wickline and G.M. Lanza, Manganese-based MRI contrast agents: Past, present, and future, *Tetrahedron* 67 (2011), pp. 8431–8444.
48. T. Huang, J. Cheng and Y.F. Zheng, In vitro degradation and biocompatibility of Fe–Pd and Fe–Pt composites fabricated by spark plasma sintering, *Mater. Sci. Eng.* C 35 (2014), pp. 43–53.
49. C. Xu, Z. Yuan, N. Kohler, J. Kim, M.A. Chung and S. Sun, FePt Nanoparticles as an Fe reservoir for controlled Fe release and tumor inhibition, *J. Am. Chem. Soc.* 131 (2009), pp. 15346–15351.
50. S. Maenosono, R. Yoshida and S. Saita, Evaluation of genotoxicity of amine-terminated water-dispersible FePt nanoparticles in the Ames test and in vitro chromosomal aberration test, *J. Toxicol. Sci.* 34 (2009), pp. 349–354.
51. M. Ahamed, M.J. Akhtar, M.A.M. Khan, H.A. Alhadlaq and A. Alshamsan, Cobalt iron oxide nanoparticles induce cytotoxicity and regulate the apoptotic genes through ROS in human liver cells (HepG2), *Colloids Surf. B Biointerf.* 148 (2016), pp. 665–673.
52. L. Horev-Azaria, G. Baldi, D. Beno, D. Bonacchi, U. Golla-Schindler, J.C. Kirkpatrick et al., Predictive toxicology of cobalt ferrite nanoparticles: Comparative in-vitro study of different cellular models using methods of knowledge discovery from data, *Part. Fibre Toxicol.* 10 (2013), p. 32.
53. V. Mariani, J. Ponti, G. Giudetti, F. Broggi, P. Marmorato, S. Gioria et al., Online monitoring of cell metabolism to assess the toxicity of nanoparticles: The case of cobalt ferrite, *Nanotoxicology* 6 (2012), pp. 272–287.
54. V. Pašukonienė, A. Mlynska, S. Steponkienė, V. Poderys, M. Matulionytė, V. Karabanovas et al., Accumulation and biological effects of cobalt ferrite nanoparticles in human pancreatic and ovarian cancer cells, *Medicina (B. Aires).* 50 (2014), pp. 237–244.

55. M. Abudayyak, T. Altincekic Gurkaynak and G. Özhan, In vitro toxicological assessment of cobalt ferrite nanoparticles in several mammalian cell types, *Biol. Trace Elem. Res.* 175 (2016), pp. 458–465.
56. K.J. Ong, T.J. Maccormack, R.J. Clark, J.D. Ede, V.A. Ortega, L.C. Felix et al., Widespread nanoparticle-assay interference: Implications for nanotoxicity testing, *PLoS One* 9 (2014), p. e90650.
57. N.A. Monteiro-Riviere, A.O. Inman and L.W. Zhang, Limitations and relative utility of screening assays to assess engineered nanoparticle toxicity in a human cell line, *Toxicol. Appl. Pharmacol.* 234 (2009), pp. 222–235.
58. S.J. Soenen and M. De Cuyper, Assessing iron oxide nanoparticle toxicity in vitro: Current status and future prospects, *Nanomedicine (Lond)* 5 (2010), pp. 1261–1275.
59. A. Hanini, M. El Massoudi, J. Gavard, K. Kacem, S. Ammar and O. Souilem, Nanotoxicological study of polyol-made cobalt-zinc ferrite nanoparticles in rabbit, *Environ. Toxicol. Pharmacol.* 45 (2016), pp. 321–327.
60. D.W. Hwang, D.S. Lee and S. Kim, Gene expression profiles for genotoxic effects of silica-free and silica-coated cobalt ferrite nanoparticles, *J. Nucl. Med.* 53 (2012), pp. 106–112.
61. C. Di Guglielmo, D.R. López, J. De Lapuente, J.M.L. Mallafre and M.B. Suàrez, Embryotoxicity of cobalt ferrite and gold nanoparticles: A first in vitro approach, *Reprod. Toxicol.* 30 (2010), pp. 271–276.
62. S.J.H. Soenen and M. De Cuyper, Assessing iron oxide nanoparticle toxicity in vitro: Current status and future prospects, *Nanomedicine (Lond).* 5 (2010), pp. 1261–1275.
63. S.J.H. Soenen, A.R. Brisson and M. De Cuyper, Addressing the problem of cationic lipid-mediated toxicity: The magnetoliposome model, *Biomaterials* 30 (2009), pp. 3691–3701.
64. C.C. Berry, S. Wells, S. Charles, G. Aitchison and A.S.G. Curtis, Cell response to dextran-derivatised iron oxide nanoparticles post internalisation, *Biomaterials* 25 (2004), pp. 5405–5413.
65. S. Liu, L. Long, Z. Yuan, L. Yin and R. Liu, Effect and intracellular uptake of pure magnetic Fe3O4 nanoparticles in the cells and organs of lung and liver., *Chin. Med. J. (Engl).* 122 (2009), pp. 1821–1825.
66. H.L. Karlsson, J. Gustafsson, P. Cronholm and L. Möller, Size-dependent toxicity of metal oxide particles – A comparison between nano- and micrometer size, *Toxicol. Lett.* 188 (2009), pp. 112–118.
67. K. Soto, K.M. Garza and L.E. Murr, Cytotoxic effects of aggregated nanomaterials, *Acta Biomater.* 3 (2007), pp. 351–358.
68. I.M. Kennedy, D. Wilson, A.I. Barakat and HEI Health Review Committee, Uptake and inflammatory effects of nanoparticles in a human vascular endothelial cell line, *Res. Rep. Health. Eff. Inst.* 136 (2009), pp. 3–32.
69. E.-J. Park, H.N. Umh, S.-W. Kim, M.-H. Cho, J.-H. Kim and Y. Kim, ERK pathway is activated in bare-FeNPs-induced autophagy, *Arch. Toxicol.* 88 (2014), pp. 323–336.
70. M.I. Khan, A. Mohammad, G. Patil, S.A.H. Naqvi, L.K.S. Chauhan and I. Ahmad, Induction of ROS, mitochondrial damage and autophagy in lung epithelial cancer cells by iron oxide nanoparticles, *Biomaterials* 33 (2012), pp. 1477–1488.
71. M. Auffan, L. Decome, J. Rose, T. Orsiere, M. De Meo, V. Briois et al., In vitro interactions between DMSA-coated maghemite nanoparticles and human fibroblasts: A physicochemical and cyto-genotoxical study, *Environ. Sci. Technol.* 40 (2006), pp. 4367–4373.
72. E.R.L. de Freitas, P.R.O. Soares, R. de P. Santos, R.L. dos Santos, J.R. da Silva, E.P. Porfirio et al., In vitro biological activities of anionic gamma-Fe2O3 nanoparticles on human melanoma cells, *J. Nanosci. Nanotechnol.* 8 (2008), pp. 2385–2391.
73. A. Villanueva, M. Cañete, A.G. Roca, M. Calero, S. Veintemillas-Verdaguer, C.J. Serna et al., The influence of surface functionalization on the enhanced internalization of magnetic nanoparticles in cancer cells, *Nanotechnology* 20 (2009), p. 115103.
74. T.R. Pisanic, J.D. Blackwell, V.I. Shubayev, R.R. Fiñones and S. Jin, Nanotoxicity of iron oxide nanoparticle internalization in growing neurons, *Biomaterials* 28 (2007), pp. 2572–2581.
75. L. Zhang, X. Wang, J. Zou, Y. Liu and J. Wang, DMSA-coated iron oxide nanoparticles greatly affect the expression of genes coding cysteine-rich proteins by their DMSA coating, *Chem. Res. Toxicol.* 28 (2015), pp. 1961–1974.
76. R. Mejías, L. Gutiérrez, G. Salas, S. Pérez-Yagüe, T.M. Zotes, F.J. Lázaro et al., Long term biotransformation and toxicity of dimercaptosuccinic acid-coated magnetic nanoparticles support their use in biomedical applications, *J. Control. Release* 171 (2013), pp. 225–233.
77. Y. Liu and J. Wang, Effects of DMSA-coated Fe3O4 nanoparticles on the transcription of genes related to iron and osmosis homeostasis, *Toxicol. Sci.* 131 (2013), pp. 521–536.
78. E.J. van den Bos, A. Wagner, H. Mahrholdt, R.B. Thompson, Y. Morimoto, B.S. Sutton et al., Improved efficacy of stem cell labeling for magnetic resonance imaging studies by the use of cationic liposomes, *Cell Transplant.* 12 (2003), pp. 743–756.
79. C.-Y. Yang, J.-K. Hsiao, M.-F. Tai, S.-T. Chen, H.-Y. Cheng, J.-L. Wang et al., Direct labeling of hMSC with SPIO: The long-term influence on toxicity, chondrogenic differentiation capacity, and intracellular distribution, *Mol. Imaging Biol.* 13 (2011), pp. 443–451.
80. S.L. Easo and P.V. Mohanan, In vitro hematological and in vivo immunotoxicity assessment of dextran stabilized iron oxide nanoparticles, *Colloids Surf. B Biointerf.* 134 (2015), pp. 122–130.
81. A. Lindemann, B.M. Fraederich, R. Pries, B. Wollenberg, K. Lüdtke-Buzug and K. Graefe, Biological impact of superparamagnetic iron oxide nanoparticles for magnetic particle imaging of head and neck cancer cells, *Int. J. Nanomed.* (2014), p. 5025.
82. G.M.T. de Oliveira, L.W. Kist, T.C.B. Pereira, J.W. Bortolotto, F.L. Paquete, E.M.N. de Oliveira et al., Transient modulation of acetylcholinesterase activity caused by exposure to dextran-coated iron oxide nanoparticles in brain of adult zebrafish, *Comp. Biochem. Physiol. Part C Toxicol. Pharmacol.* 162 (2014), pp. 77–84.
83. C. Hanot, Y. Choi, T. Anani, D. Soundarrajan and A. David, Effects of iron-oxide nanoparticle surface chemistry on uptake kinetics and cytotoxicity in CHO-K1 cells, *Int. J. Mol. Sci.* 17 (2015), p. 54.
84. G. Yuan, Y. Yuan, K. Xu and Q. Luo, Biocompatible PEGylated Fe3O4 nanoparticles as photothermal agents for near-infrared light modulated cancer therapy, *Int. J. Mol. Sci.* 15 (2014), pp. 18776–18788.

85. L. Dai, Y. Liu, Z. Wang, F. Guo, D. Shi and B. Zhang, One-pot facile synthesis of PEGylated superparamagnetic iron oxide nanoparticles for MRI contrast enhancement, *Mater. Sci. Eng. C* 41 (2014), pp. 161–167.
86. Y.C. Park, J.B. Smith, T. Pham, R.D. Whitaker, C.A. Sucato, J.A. Hamilton et al., Effect of PEG molecular weight on stability, T2 contrast, cytotoxicity, and cellular uptake of superparamagnetic iron oxide nanoparticles (SPIONs), *Colloids Surf. B Biointerf.* 119 (2014), pp. 106–114.
87. A. Lak, J. Dieckhoff, F. Ludwig, J.M. Scholtyssek, O. Goldmann, H. Lünsdorf et al., Highly stable monodisperse PEGylated iron oxide nanoparticle aqueous suspensions: A nontoxic tracer for homogeneous magnetic bioassays, *Nanoscale* 5 (2013), p. 11447.
88. C. Costa, F. Brandão, M. João, S. Costa, V. Valdiglesias, G. Kiliç et al., In vitro cytotoxicity of superparamagnetic iron oxide nanoparticles on neuronal and glial cells. Evaluation of nanoparticle interference with viability tests, *J. Appl. Toxicol.* 36 (2016), pp. 361–372.
89. R.K. Singh, T.-H. Kim, K.D. Patel, J.C. Knowles and H.-W. Kim, Biocompatible magnetite nanoparticles with varying silica-coating layer for use in biomedicine: Physicochemical and magnetic properties, and cellular compatibility, *J. Biomed. Mater. Res. Part A* 100A (2012), pp. 1734–1742.
90. O. Baber, M. Jang, D. Barber and K. Powers, Amorphous silica coatings on magnetic nanoparticles enhance stability and reduce toxicity to in vitro BEAS-2B cells, *Inhal. Toxicol.* 23 (2011), pp. 532–543.
91. Y. Luengo, S. Nardecchia, M.P. Morales and M.C. Serrano, Different cell responses induced by exposure to maghemite nanoparticles, *Nanoscale* 5 (2013), p. 11428.
92. C.J. Rivet, Y. Yuan, D.-A. Borca-Tasciuc and R.J. Gilbert, altering iron oxide nanoparticle surface properties induce cortical neuron cytotoxicity, *Chem. Res. Toxicol.* 25 (2012), pp. 153–161.
93. A.H. Silva, E. Lima, M.V. Mansilla, R.D. Zysler, H. Troiani, M.L.M. Pisciotti et al., Superparamagnetic iron-oxide nanoparticles mPEG350– and mPEG2000-coated: Cell uptake and biocompatibility evaluation, *Nanomedicine Nanotechnology, Biol. Med.* 12 (2016), pp. 909–919.
94. M.A. Malvindi, V. De Matteis, A. Galeone, V. Brunetti, G.C. Anyfantis, A. Athanassiou et al., Toxicity assessment of silica coated iron oxide nanoparticles and biocompatibility improvement by surface engineering, *PLoS One* 9 (2014), p. e85835.
95. V. Escamilla-Rivera, M. Uribe-Ramírez, S. González-Pozos, O. Lozano, S. Lucas and A. De Vizcaya-Ruiz, Protein corona acts as a protective shield against Fe3O4-PEG inflammation and ROS-induced toxicity in human macrophages, *Toxicol. Lett.* 240 (2016), pp. 172–184.
96. M. Mahmoudi, S. Laurent, M.A. Shokrgozar and M. Hosseinkhani, Toxicity evaluations of superparamagnetic iron oxide nanoparticles: Cell "vision" versus physicochemical properties of nanoparticles, *ACS Nano* 5 (2011), pp. 7263–7276.
97. S.J.H. Soenen, U. Himmelreich, N. Nuytten, T.R. Pisanic, A. Ferrari and M. De Cuyper, Intracellular nanoparticle coating stability determines nanoparticle diagnostics efficacy and cell functionality, *Small* 6 (2010), pp. 2136–2145.
98. L. Lartigue, D. Alloyeau, J. Kolosnjaj-Tabi, Y. Javed, P. Guardia, A. Riedinger et al., Biodegradation of iron oxide nanocubes: High-resolution in situ monitoring, *ACS Nano* 7 (2013), pp. 3939–3952.
99. Y. Javed, L. Lartigue, P. Hugounenq, Q.L. Vuong, Y. Gossuin, R. Bazzi et al., Biodegradation mechanisms of iron oxide monocrystalline nanoflowers and tunable shield effect of gold coating, *Small* 10 (2014), pp. 3325–3337.
100. S. Mornet, S. Vasseur, F. Grasset and E. Duguet, Magnetic nanoparticle design for medical diagnosis and therapy, *J. Mater. Chem.* 14 (2004), p. 2161.
101. R.G. Mendes, B. Koch, A. Bachmatiuk, A.A. El-Gendy, Y. Krupskaya, A. Springer et al., Synthesis and toxicity characterization of carbon coated iron oxide nanoparticles with highly defined size distributions, *Biochim. Biophys. Acta – Gen. Subj.* 1840 (2014), pp. 160–169.
102. M. Yu, S. Huang, K.J. Yu and A.M. Clyne, Dextran and polymer polyethylene glycol (peg) coating reduce both 5 and 30 nm iron oxide nanoparticle cytotoxicity in 2D and 3D cell culture, *Int. J. Mol. Sci.* 13 (2012), pp. 5554–5570.
103. A.R. Murray, E. Kisin, A. Inman, S.-H. Young, M. Muhammed, T. Burks et al., Oxidative stress and dermal toxicity of iron oxide nanoparticles in vitro, *Cell Biochem. Biophys.* 67 (2013), pp. 461–476.
104. L. Gu, R.H. Fang, M.J. Sailor and J.-H. Park, In vivo clearance and toxicity of monodisperse iron oxide nanocrystals, *ACS Nano* 6 (2012), pp. 4947–4954.
105. J.H. Lee, J.E. Ju, B. Il Kim, P.J. Pak, E.-K. Choi, H.-S. Lee et al., Rod-shaped iron oxide nanoparticles are more toxic than sphere-shaped nanoparticles to murine macrophage cells, *Environ. Toxicol. Chem.* 33 (2014), pp. 2759–2766.
106. W. Jiang, B.Y.S. Kim, J.T. Rutka and W.C.W. Chan, Nanoparticle-mediated cellular response is size-dependent, *Nat. Nanotechnol.* 3 (2008), pp. 145–150.
107. B.B. Manshian, A.M. Abdelmonem, K. Kantner, B. Pelaz, M. Klapper, C. Nardi Tironi et al., Evaluation of quantum dot cytotoxicity: Interpretation of nanoparticle concentrations versus intracellular nanoparticle numbers, *Nanotoxicology* 10 (2016), pp. 1318–1328.
108. D. Hühn, K. Kantner, C. Geidel, S. Brandholt, I. De Cock, S.J.H. Soenen et al., Polymer-coated nanoparticles interacting with proteins and cells: Focusing on the sign of the net charge, *ACS Nano* 7 (2013), pp. 3253–3263.
109. S.J. Soenen, M. De Cuyper, Assessing cytotoxicity of (iron oxide-based) nanoparticles: An overview of different methods exemplified with cationic magnetoliposomes, *Contrast Media Mol Imaging* 4 (2009), pp. 207–219.
110. R. Hachani, M.A. Birchall, M.W. Lowdall, K. Georgios, L.D. Tung, B.B. Manshian et al., Assessing cell-nanoparticle interactions by high content imaging of biocompatible iron oxide nanoparticles as potential contrast agents for magnetic resonance imaging, *Sci. Rep.* 7 (2017), p. 7850.
111. C. Hoskins, L. Wang, W. Cheng and A. Cuschieri, Dilemmas in the reliable estimation of the in-vitro cell viability in magnetic nanoparticle engineering: Which tests and what protocols?, *Nanoscale Res. Lett.* 7 (2012), p. 77.
112. B.B. Manshian, D.F. Moyano, N. Corthout, S. Munck, U. Himmelreich, V.M. Rotello et al., High-content imaging and gene expression analysis to study cell-nanomaterial interactions: The effect of surface hydrophobicity, *Biomaterials* 35 (2014), pp. 9941–9950.
113. B.B. Manshian, C. Pfeiffer, B. Pelaz, T. Heimerl, M. Gallego, M. Möller et al., High-content imaging and gene expression approaches to unravel the effect of surface functionality on cellular interactions of silver nanoparticles, *ACS Nano* 9 (2015), pp. 10431–10444.

Bella B. Manshian (E-mail: bella.manshian@kuleuven.be) earned her PhD in human genetics and molecular biology at the College of Medicine at Swansea University. During her PhD, Dr Manshian gained valuable experience in the field of child developmental medicine and stem cell research while working as a research assistant in the Developmental Medicine Group at Swansea University. She was one of the few researchers who developed and optimized the first successful method for the perfusion of progenitor stem cells for therapeutic use during a collaborative work with Celgene, Ltd. She then worked as a postdoctoral research officer in the Nanotoxicology and DNA Damage Group where she also worked with industrial partner Royal Mint in determining toxicity of new products used for the production of money coins. Since 2014, Dr Manshian has joined the KU Leuven, where the focus is on studying the interaction of various nanomaterials and biological systems in view of biomedical applications of these materials. Dr Manshian has authored more than 50 publications in international peer-reviewed journals and 2 book chapters.

Uwe Himmelreich has studied physical chemistry. He has more than 25 years of research experience in magnetic resonance imaging and spectroscopy. He is the coordinator of the KU Leuven core facility Molecular Small Imaging Center (MoSAIC) with approximately 150 users. He is currently the advisor of 4 MSc students and 8 PhD students. In addition to the development of novel, multimodal imaging approaches, his main interest is also the development and validation of new contrast agents for in vivo cell imaging and potential theranostic applications. This includes iron oxide-based nanoparticles for the validation of novel approaches in cancer treatment monitoring. A number of cancer models are available in his laboratory, ranging from xenograft models to orthotopic models of glioma, head, neck, and pancreas tumours. He has published more than 200 papers in peer-reviewed journals.

Dr Stefaan J. Soenen has more than 8 years of experience on bio–nano interactions and biomedical use of nanomaterials. His research focuses on setting up novel methods for understanding how nanomaterials interact in a biological environment and to translate this knowledge into biomedical applications. His main interests lie in the development, validation, and use of advanced optical imaging methods for rapid, in-depth analysis of bio–nano interactions and employing nanomaterials for anticancer purposes. He has published more than 70 papers in peer-reviewed journals.

16 Magnetic Nanoparticles for Cancer Treatment Using Magnetic Hyperthermia

*Laura Asín**, *Grazyna Stepien**, *María Moros*†,
Raluca Maria Fratila†, *and Jesús Martínez de la Fuente*

CONTENTS

16.1 INTRODUCTION

The antitumour effect of hyperthermia treatment is known and has been practiced for a long time using not very sophisticated, yet effective technologies such as saunas and hot water baths.[1] The demonstration that cancer cells are more sensitive to heat than normal ones[2,3] and the huge advances in the field of nanotechnology have given researchers new tools to redesign heat treatments, making them more specific and therefore reducing the unpleasant side effects of treatments currently applied. That is why in recent years many studies have focused on the development of new materials as heating agents. Their antitumour efficacy is being probed either alone or in combination with other antitumour treatments such as radiotherapy or chemotherapy in order to achieve a synergistic effect among them, as will be discussed in the last section of this chapter.

Despite the fact that other nanomaterials are being tested as heating sources, such as gold nanoparticles (NPs), magnetic nanoparticles (MNPs) are the 'stars' of these new heating agents. MNP-based hyperthermia, usually known as magnetic hyperthermia (MHT), is one of the first applications of nanotechnology in biomedicine that has been already introduced in the clinic for particular cases of cancer treatment.

In this chapter, we provide an overview of recent advances in the field of cancer therapy using MHT. We first introduce the main concepts of MHT, followed by a discussion regarding the challenges associated with ensuring an adequate heating efficiency of MNPs in the intracellular environment. The last part of the chapter is focused on applications of MHT in combination with other cancer therapeutic modalities.

16.2 INTERACTIONS BETWEEN MAGNETIC NANOPARTICLES AND ALTERNATING MAGNETIC FIELDS: AN OVERVIEW OF THE MECHANISMS INVOLVED IN MAGNETIC HYPERTHERMIA

Magnetic hyperthermia is based on the exposure of MNPs to external electromagnetic fields in the range of radiofrequency (RF), which is situated in the region of the electromagnetic wave spectrum that corresponds to low and medium frequencies (30–30000 kHz) (Figure 16.1). These electromagnetic fields can enter into the human body in order to achieve an internal and localized heating when MNPs are present. Electromagnetic waves are composed of an electric and a magnetic component. At a suitable frequency, magnetic moments from MNPs couple to the magnetic component of the radiation. At this state, MNPs are able to absorb energy from the magnetic field and release it to the environment as heat. This is the main concept of heating generation by MNPs in MHT. It is important to note that to date there is no evidence about health and environmental problems caused by these nonionizing magnetic fields.[4] Due to possible heating of biological tissues caused by eddy currents and other phenomena that have been described and registered in the guidelines from the International Commission on Non-Ionizing Radiation Protection (ICNIRP),[5] the limits for magnetic field amplitudes and frequencies for *in vivo* experiments, have to be carefully selected.

* Contributed equally to the present work.
† Corresponding author.

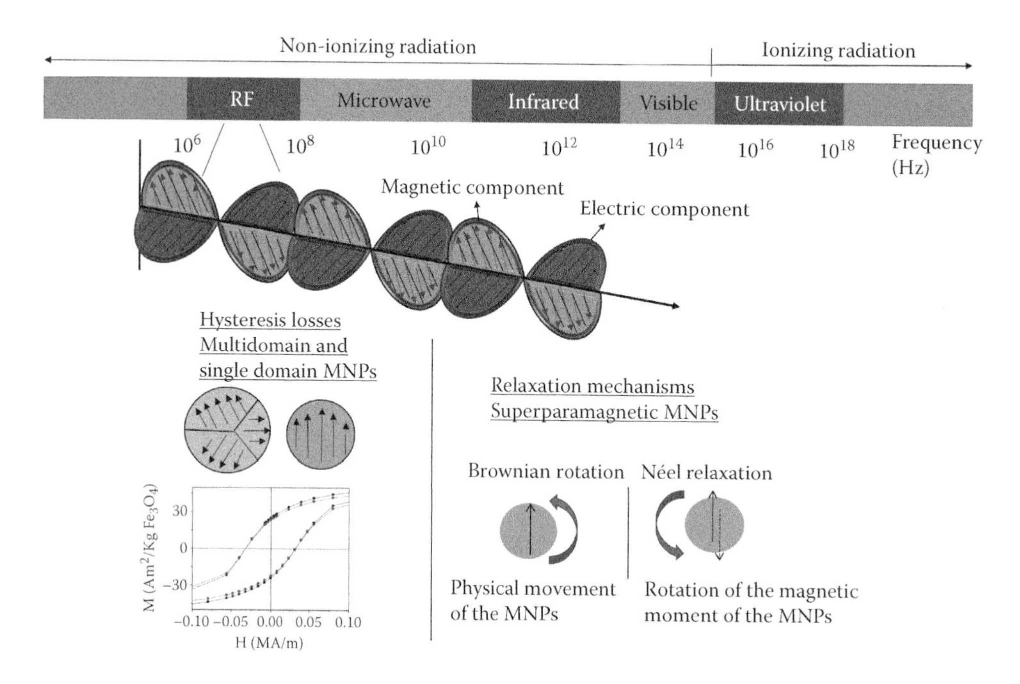

FIGURE 16.1 Upper panel: region of the electromagnetic wave spectrum where an electromagnetic field in the range of the RF is shown. Lower panel: heat generation mechanisms in magnetic hyperthermia, including hysteresis losses in multidomain and single domain MNPs, and Brownian and Néel relaxation in superparamagnetic MNPs.

There are several mechanisms involved in the heat dissipation stage (see Figure 16.1).

Magnetic hysteresis losses are described in multidomain particles.[6] However, the use of these particles in biomedicine is limited, because when the external magnetic field is removed the magnetization of the sample does not go back to zero. Therefore, this remanent magnetization could cause NP aggregation, triggering serious problems *in vivo*. This is why one of the main requisites for the use of MNPs in biomedical applications, among others that will be discussed in the next Section (16.3), is the capability of not staying magnetized when the external magnetic field is turned off; in other words, the MNPs need to be superparamagnetic. In superparamagnetic single-domain particles once the external magnetic field is removed, the macroscopic magnetization tends to vanish with a typical relaxation time, converting this magnetic energy into thermal energy. There are two mechanisms involved in the magnetic relaxation of the MNPs (Figure 16.1). The first one is the external rotation of the particles within the carrier fluid, which occurs when MNPs respond to the alternating current (AC) magnetic field through physical rotation, which is called Brownian relaxation. Brownian relaxation time (τ_B) is given by the following expression (Equation 16.1):

$$\tau_B = \frac{3\eta V_H}{k_B T} \tag{16.1}$$

where η is the viscosity coefficient of the matrix fluid, k_B is the Boltzmann constant (1.38×10^{-23} JK^{-1}), T is the temperature, and V_H is the hydrodynamic volume of the particles.

On the other hand, superparamagnetic MNPs that are physically fixed or in a viscous medium can relax by other mechanism based on thermal processes, called Néel relaxation. The reason for the reversal of the magnetic moments after removing the external magnetic field is the thermal fluctuation $k_B T$. The energy barrier that has to be overcome to reverse the magnetization is given by KV, where K in the magnetic anisotropy and V the volume of the particle. The Néel relaxation time (τ_N) is given by the following expression (Equation 16.2):

$$\tau_N = \tau_0 e^{\frac{KV}{k_B T}} \tag{16.2}$$

where τ_0 is approximately 10^{-9} s. Both expressions are valid for small applied magnetic fields and considering negligible particle–particle interactions. Both Brownian and Néel processes may be present in a typical ferrofluid. The effective relaxation time, τ, considering the two mechanisms as individual ones that can occur in parallel is given by Equation 16.3[7]:

$$\frac{1}{\tau} = \frac{1}{\tau_B} + \frac{1}{\tau_N}. \tag{16.3}$$

Two different terms for both Néel and Brown mechanism, depending on the energy source that provokes the flipping or rotation, are usually used.[8] The first one refers to the relaxation that occurs in the zero-field state when the thermal energy from the surrounding medium is responsible for the flipping or rotation. These are the so-called *zero-field Brownian relaxation* and *zero-field Néel relaxation*. The second one refers to externally field forced NPs, so-called *Brownian forcing* and *Néel forcing*. Here, it is claimed that the use of the terms Brownian and Néel relaxation is inadequate, considering externally forced NPs whose state is far away from equilibrium.[9] In a recent work, new names have been proposed for these four magnetic hyperthermia mechanisms: Brownian and Néel relaxation, and Brownian and Néel forcing for the externally field forced MNPs. In the same paper, the authors show a magnetic hyperthermia strategy where the heat is released by the friction of magnetically rotating nanowires.[10]

Many recent works focus on understanding better the contributions of these magnetic relaxation mechanisms in order to be able to tune and to adapt the design of the MNPs to obtain optimal heat production. A recent study has demonstrated, using AC hysteresis loops, that both relaxation mechanisms coexist and that Brownian relaxation, depending on MNPs size and fluid viscosity, is much faster than Néel relaxation and can occur after Néel relaxation.[11] This is in contradiction with one of the basic theories regarding magnetic relaxation, which claims that the faster relaxation dominates the entire process. An example of this is the proposed theory that Néel is the main mechanism in small MNPs, especially superparamagnetic ones, and at very high frequencies of AC magnetic fields. This is because it is much faster than Brownian mechanism and thus dominates the entire process. On the other hand, Brownian mechanisms for MNPs having larger sizes are favoured.[12] A recent publication estimates the relative contributions of these two mechanisms by performing dynamic magnetic susceptibility (DMS) measurements, and, although the method has some limitations and cannot be generally applied, it is a simple way to determine the relative fractions of both Brownian and Néel particles in a fluid.[13]

Moving to *in vitro* applications, MNPs under the influence of an external magnetic field can trigger cell death without increasing the macroscopic temperature. Some authors claim that the localized heat generated is enough to provoke cellular destruction but not to raise the macroscopic temperature.[14,15] In these cases, cellular effects are triggered by heat generated by the mechanisms described above (see Section 16.3 for a more in-depth discussion). However, other authors defend the idea that MNPs in the presence of the magnetic field have a mechanical effect due to the forces that the agglomerates of MNPs 'feel' inside the cells, therefore MNPs under saturated magnetic fields cause cell death by a mechanism that is not based on heat generation.[16] The physical parameter used to measure the heating efficiency of MNPs can be found in the literature with different names: specific power absorption (SPA), specific absorption rate (SAR) or specific power loss (SPL), among others. All of them are almost synonymous and can be defined as the energy dissipated per unit of mass of MNPs (W/g) (Equation 16.4):

$$\mathrm{SAR} = \frac{c_{\mathrm{LIQ}}\delta_{\mathrm{LIQ}}}{\phi}\left(\frac{\Delta T}{\Delta t}\right) = \frac{m_{\mathrm{LIQ}}c_{\mathrm{LIQ}}}{m_{\mathrm{NP}}}\left(\frac{\Delta T}{\Delta t}\right) \quad (16.4)$$

where δ_{LIQ} and $\phi = m_{NP}/V_{LIQ}$ are the liquid density and the concentration in terms of weight of the NPs in the colloid, respectively. C_{LIQ} is the specific heat of the solvent and m_{NP} is the mass of MNPs. These are the most used parameters to compare the heating efficiency of MNPs in the literature, but all of them present the inconvenience of the dependency on the frequency (f) and the magnetic field amplitude (H) used. To overcome this drawback, another magnitude has been defined (Equation 16.5): intrinsic loss power (ILP) given by[17]

$$\mathrm{ILP} = \mathrm{SAR,\ SAP,\ SPL}/(\mathrm{H}^2 f) \quad (16.5)$$

Although the validity of this parameter depends on the values of frequency and magnetic field amplitudes applied, it is an appropriate tool to normalize and to successfully compare the heating efficiency of the MNPs used for magnetic hyperthermia between different researchers.

It is noteworthy to mention that for *in vivo* experiments there is a limitation of frequency and magnetic field amplitude in order not to damage the tissue (e.g. by eddy currents heating). The limit of the product of these two parameters is $H * f < 5 \times 10^8\ \mathrm{Am^{-1}s^{-1}}$. Several *in vivo* studies have confirmed that this limitation has to be taken into account when characterizing MNPs, as frequently researchers use values of these parameters that would not be accepted for use in clinical applications.[18,19]

16.3 REQUISITES FOR EFFICIENT MAGNETIC HYPERTHERMIA IN BIOLOGICAL CONTEXTS

Ideally, MHT therapeutic procedure should have high efficiency while requiring the minimum amount of the heating agent. Thus, on the one hand, a great deal of fundamental research in MHT is focused on obtaining MNPs with high values of SAR. On the other hand, MNPs intended for biomedical applications, including MHT, must fulfil stringent requirements of biocompatibility, colloidal stability, noncytotoxicity, and ability to avoid clearance by the mononuclear phagocyte system (MPS). Significant advances have been made so far in the synthesis of materials for MHT, as well as their physicochemical and toxicological characterization.[6,20–22] In this section, we will focus on another important aspect that cannot be neglected when designing MNPs for MHT applications, in order to prevent excellent candidates from *in vivo* failure: the impact that the interaction between MNPs and the cellular environment can have on their heating efficiency. As described in Section 16.2, heat generation by MNPs in solution is based on magnetic hysteresis losses or on magnetic relaxation processes; the latter can occur through changes in the direction of magnetic moments within each particle (Néel relaxation) and rotation of the whole MNP throughout the medium (Brownian relaxation). It should be noted that SAR values depend not only on the *intrinsic properties* of the MNPs (size, crystallinity, anisotropy) but also on the *parameters* (field amplitude, frequency) *of the applied magnetic field* and on their *environment* (viscosity and dielectric constant of the medium, dipolar interactions between MNPs, possibility to rotate freely, etc.). Therefore, the heating efficiency of MNPs in an intracellular environment can change drastically

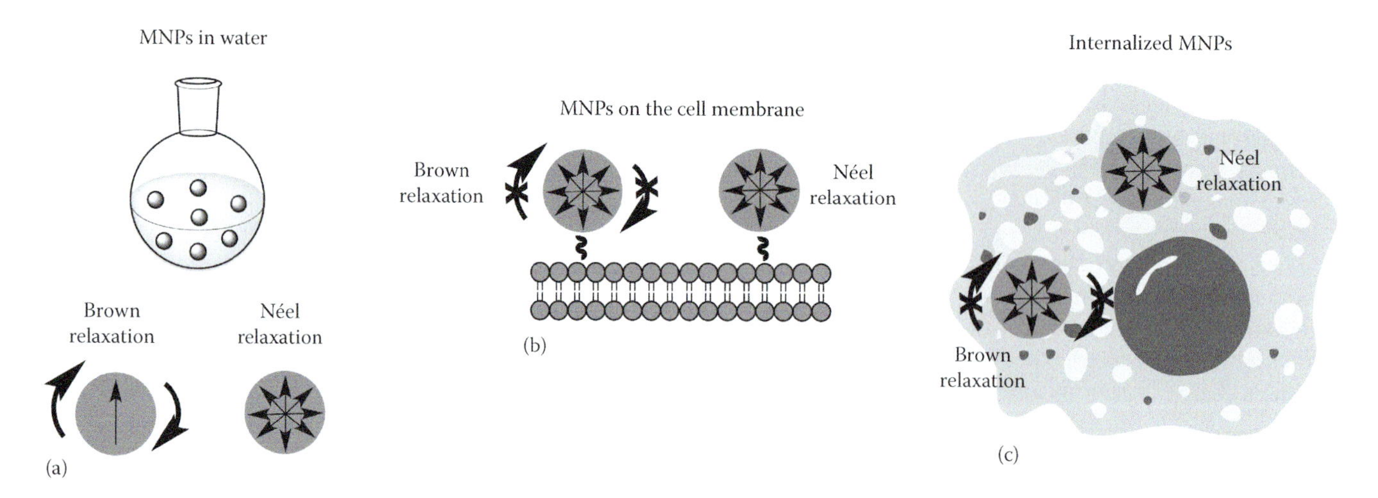

FIGURE 16.2 Contribution of relaxation mechanisms to heat generation by MNPs. In solution, both Néel relaxation and Brownian rotation can occur (a). When particles are attached to the cell membrane (b) or internalized (c), only Néel relaxation is expected to be responsible for heat generation.

when compared to the as-synthesized material. A schematic representation of the effect of the subcellular localization on the heating generation mechanisms is provided in Figure 16.2.

16.3.1 Assessment of Heating Efficiency of Magnetic Nanoparticles *In Vitro*

During the last decade, more and more attention has been paid to the investigation of the heating capability and/or heating mechanisms of MNPs under conditions closely resembling the *in vitro* and *in vivo* scenarios, both experimentally (by measuring the heat generated by MNPs in solvents with high viscosity such as glycerol and/or at different subcellular localizations) and by means of theoretical simulations. In one of the pioneering studies addressing this aspect, Fortin et al. established that internalized MNPs generated heat almost exclusively by Néel relaxation, regardless of their composition (maghemite, γ–Fe_2O_3 or cobalt ferrite, $CoFe_2O_4$), although in water one of the two mechanisms was predominant for each type of NP (Néel for maghemite and Brownian for cobalt ferrite).[23] Interestingly, despite having much higher SAR values in water than maghemite MNPs, cobalt ferrite MNPs showed a much more dramatic reduction of their heating efficiency when internalized (almost by a factor of 6) than maghemite MNPs (SAR value fell by about one-half). Following up, in one of the most comprehensive *in vitro* studies to date, C. Wilhelm and collaborators demonstrated that different nanomaterials with high SAR values in aqueous and glycerol solutions displayed a systematic decrease in the heating efficiency in the cellular environment.[24] The authors investigated a broad range of MHT candidates (having different sizes, compositions, and morphologies) both in living cells mimicking a tumour environment and in fixed cells incorporating the MNPs at different subcellular localizations: attached to the cell membrane or confined in intracellular compartments (Figure 16.3). Maghemite MNPs (Figure 16.3a, 10 nm) and iron oxide nanocubes (Figure 16.3e, 18 nm) showed a rapid decrease in their SAR value upon dispersion in 'minitumour' samples, with values reaching half of the initial ones after 90 min of interaction with cells. Similarly, in fixed samples, all nanomaterials experienced systematically lower SAR values, and the process was independent of the subcellular localization of the particles, although significant differences were observed between the different types of materials. As in the previous example, the fall in the SAR value was more pronounced for MNPs generating heat mainly through Brownian relaxation, although these materials had by far the highest SAR values in solution. This observation highlights the importance of a thorough characterization of the heating behaviour of the MNPs, going well beyond mere SAR measurements of MNP suspensions in water.

In a recent report, Soukup et al. reported the analysis of intracellular heating of MNPs using AC magnetic susceptibility measurements.[25] This approach, used for the first time to probe the microenvironment of MNPs inside live cells, revealed that for blocked fixed magnetite MNPs (24 nm diameter) a clear reduction in Brownian relaxation occurred both in glycerol and inside cells. This observation was in agreement with the previous findings of Dutz et al., who reported similar magnetization behaviour for MNPs immobilized in gelatin and injected into the tumour tissue (see Section 16.3.2).[26] In contrast, AC susceptibility curves for superparamagnetic biogenic magnetite NPs (11 nm diameter) demonstrated that Néel relaxation was not sensitive to the intracellular environment. Noteworthy, the Brownian relaxation of blocked MNPs could be restored after lysing the cells and the recovered particles maintained the integrity of the magnetic core and coating. Although this particular type of MNPs was not deemed suitable for MHT *in vitro* and *in vivo*, the possibility of 'recycling' the NPs after the destruction of cancer cells by hyperthermia treatment *in vivo* seems very appealing as it would significantly reduce the dosage needed for therapeutic effect (provided that the particles have the required heating efficiency in the intracellular environment). This study also drew attention to another important aspect to keep in mind when applying MHT in the cellular environment: the increase

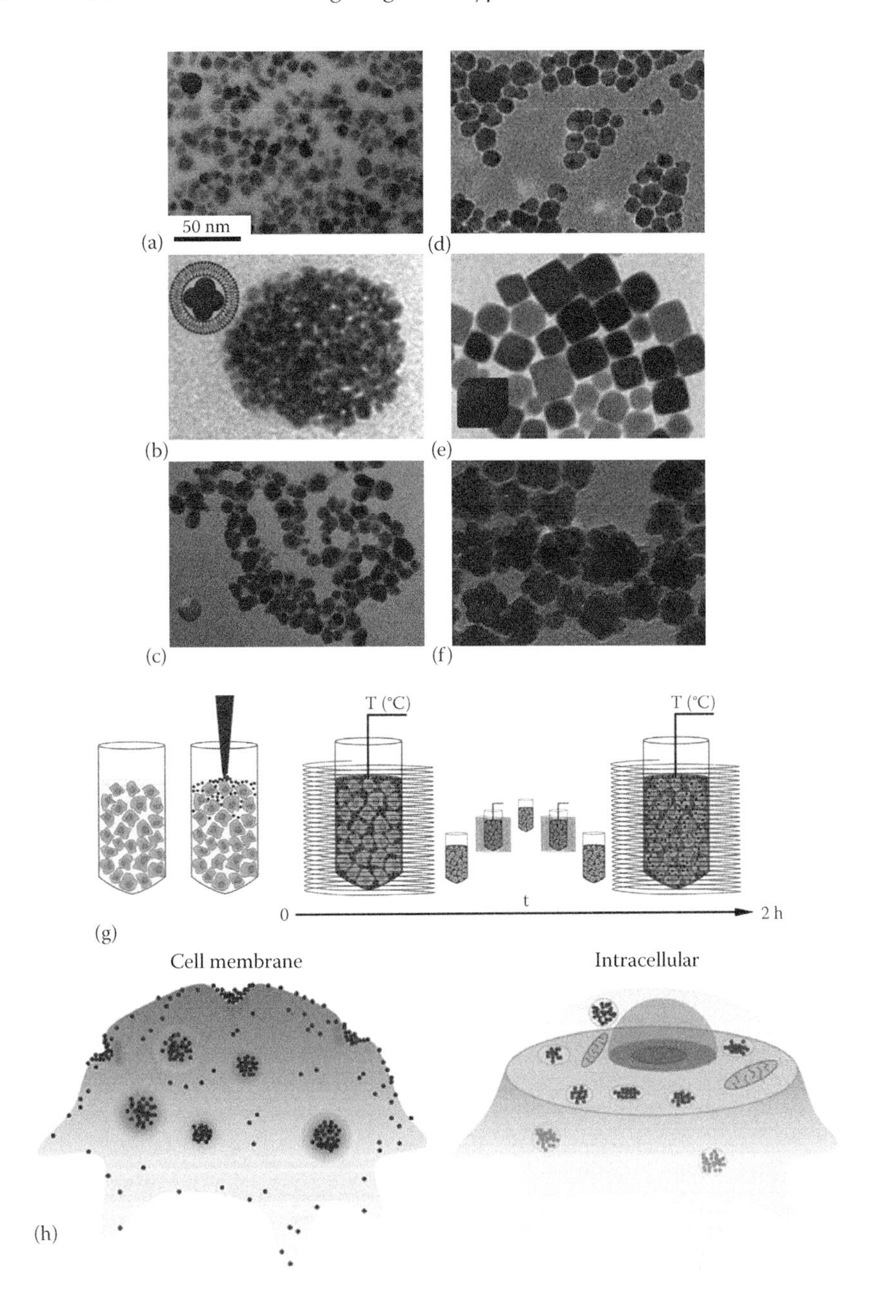

FIGURE 16.3 Evaluation of the heating efficiency of different nanomaterials (a–f) in living cells (microtumours, g) and in fixed cells at distinct subcellular localizations (h). (a) 10 nm maghemite MNPs; (b) 250 nm assemblies of the same MNPs in liposomes; (c) 10 nm cobalt ferrite MNPs; (d) 15 nm iron oxide/gold dimers; (e) 18 nm iron oxide nanocubes; (f) 25 nm iron oxide nanoflowers.[24] (Reprinted from *Biomaterials,* 35, (24), Di Corato, R., Espinosa, A., Lartigue, L., Tharaud, M., Chat, S., Pellegrino, T., Ménager, C., Gazeau, F., Wilhelm, C., 6400–6411, Copyright 2015, with permission from Elsevier.)

in the hydrodynamic diameter of the particles due to agglomeration and/or interaction with proteins (formation of the so-called protein corona, PC). In cell culture medium, transmission electron microscopy (TEM) micrographs showed a PC formed around both types of MNPs; interestingly, the PC was detached upon lysis, although no further studies were conducted to elucidate the mechanism behind the PC release.

Taking into account all the factors that can influence the heating behaviour of MNPs *in vitro* and *in vivo*, predicting the impact of intracellular interactions on the heating efficiency of internalized MNPs by means of numerical simulations becomes vital, as demonstrated by Goya and coworkers.[27] Simulations of the intracellular SAR values for two types of MNPs with similar core size and composition, but different surface coating, were carried out considering three important parameters: viscosity of the medium, magnetic anisotropy, and dipolar interactions between MNPs. The obtained values were found to closely resemble the experimental ones and represented the highest *in vitro* SAR values reported to date.

Another interesting observation has been recently reported by Ammar and coworkers, who showed that the intracellular heating efficiency can also depend on the type of cellular line.[28] Although both healthy (human umbilical vein endothelial cells, HUVEC) and malignant cells (human glioblastoma astrocytoma, U87-MG) reached the same final temperature upon incubation with maghemite MNPs and MHT treatment under similar conditions, the SAR values were found to be higher for MNPs internalized by glioma cells when compared to endothelial ones. Accordingly, the percentage of cellular death after MHT was higher for U87-MG than for HUVEC cells (56% vs. 20%). This effect was attributed to the already reported fact that cancer cells are more sensitive to up-shifts in temperature than healthy ones, although another explanation for the increased cell death percentage in malignant cells could be the difference in the cellular uptake of the MNPs: U87-MG cells internalized twice as many NPs than HUVEC cells, as established from X-ray fluorescence spectroscopy and electrophoresis measurements.

In view of the difficulties that must be overcome for reaching high heating efficiency in the cellular environment, intracellular MHT-assisted drug delivery could be a more promising therapeutic application than MHT alone. Several examples of combined chemo-magnetotherapy are discussed in Section 16.4.

16.3.2 Heating Efficiency of MNPs Immobilized in Tumours

From a practical point of view of MHT therapy, direct injection of MNPs into the tumour is preferred for easily accessible (e.g. superficial) tumours, since it ensures a high local concentration of particles. However, in this case, the heating efficiency of the MNPs is very likely to be also affected by their immobilization in the tumour tissue. Dutz et al. observed a reduction of approximately 35% in the SAR value of magnetic multicore nanoparticles (40–60 nm-cluster size as determined by TEM) when the measurement was carried out with particles immobilized in gelatin instead of suspended in water.[26] When administered by intratumoural injection to mice, the MNPs showed a similar behaviour as when immobilized in gelatin, with a degree of immobilization in the tumour tissue of more than 80%, as determined from magnetic measurements *ex vivo*. However, since the heating efficiency of the immobilized MNPs was still reasonably high, a low concentration (0.9% by mass) was enough to induce a tumour-localized temperature increase of 25 K after 140 s of application of the magnetic field (25 kA m^{-1}, 400 kHz). Gazeau and coworkers monitored the heating properties of iron oxide nanocubes, which are one of the most efficient nanoheaters in solution, after direct injection into carcinoma xenografts in mice.[29] The nanocubes preserved their heating properties during the first days post-injection, while still located in the extratumoural matrix; this indicates that the particles were still mobile and able to generate heat through Néel and Brownian mechanisms. However, MNPs removed from the tumour by macrophages and found in the spleen showed markedly deteriorated magnetic and heating properties, due to intracellular confinement and partial degradation.

16.4 MAGNETIC HYPERTHERMIA USED IN COMBINATION WITH OTHER APPROACHES FOR CANCER TREATMENT

To date, it is broadly documented that standard hyperthermia (HT) is being applied in combination with various established cancer treatments.[30] Indeed, a synergistic interaction between heat and radiation therapy and/or chemotherapy has been confirmed in clinical trials. For instance, by studying the clinical outcomes from 38 clinical trials where 1717 patients were treated with radiotherapy alone and 1761 with radiotherapy and HT, it was shown that 39.8% of the patients treated with radiotherapy alone responded completely, while 54.9% of those treated with the combination of treatments responded completely.[31] Similarly, a long-term study on nonmuscle-invasive bladder cancer (NMIBC) treated with either intravesical thermochemotherapy or intravesical chemotherapy alone using mitomycin-C (MMC), found that the 10-year disease-free survival rate for thermochemotherapy and chemotherapy alone were 53% and 15%, respectively ($P < 0.001$); this confirmed the efficacy of this adjuvant approach for NMIBC at long-term follow-up, even in patients with multiple tumours.[32]

This synergic interaction has its origin in the fact that HT application can cause a wide range of changes, altering tumour physiology, biology, and immunology, finally leading to the loss of the cellular homeostasis and the increase of the effect caused by radiotherapy and/or chemotherapy. Also, HT treatment may induce antitumoural immunity,[33] as discussed further (Figure 16.4).[34]

The effects of the application of a HT treatment in cells are diverse. For instance, it is well known that tumours can develop radioresistance, which arises from the leaky and disorganized vasculature, the hypoxic regions of the tumour, and the acidosis among others.[35] This can be overcome through the mechanism of vasodilation and subsequent increased blood (oxygen) supply to the hypoxic regions of the tumour induced by the HT.[36] Song et al. reported that the tumour oxygenation status in Fischer rats after heating for 30 min at 42.5°C was about three-fold greater than that in the control tumours.[37] The increase in blood flow provoked by a HT treatment was also found to increase drug concentration in the tumour area, resulting in an increased intracellular drug uptake and enhanced DNA damage.[38] Moreover, changes that occur in the cell membrane upon heat application (increase of the cell membrane permeability) can lead not only to a drop in cytosolic pH and change ion homeostasis (e.g. Na^+, K^+ or Ca^{2+}),[39,40] but can also affect the transport of molecules such as polyamines, glucose, and anticancer drugs[41–43] leading to a higher accumulation in the tumour of the latter. Besides, even a small increase in temperature can trigger protein unfolding, entanglement, unspecific aggregation, and as a result, inactivation of their synthesis.[44] Therefore, denaturation of the proteins in charge of DNA repair can cause a rapid inhibition of DNA replication, augmenting the proportion of cellular lethal damage induced by radiotherapy or chemotherapy.[45–47] Together, all these effects may cause the arrest of the cell cycle and a stagnation of growth and proliferation.

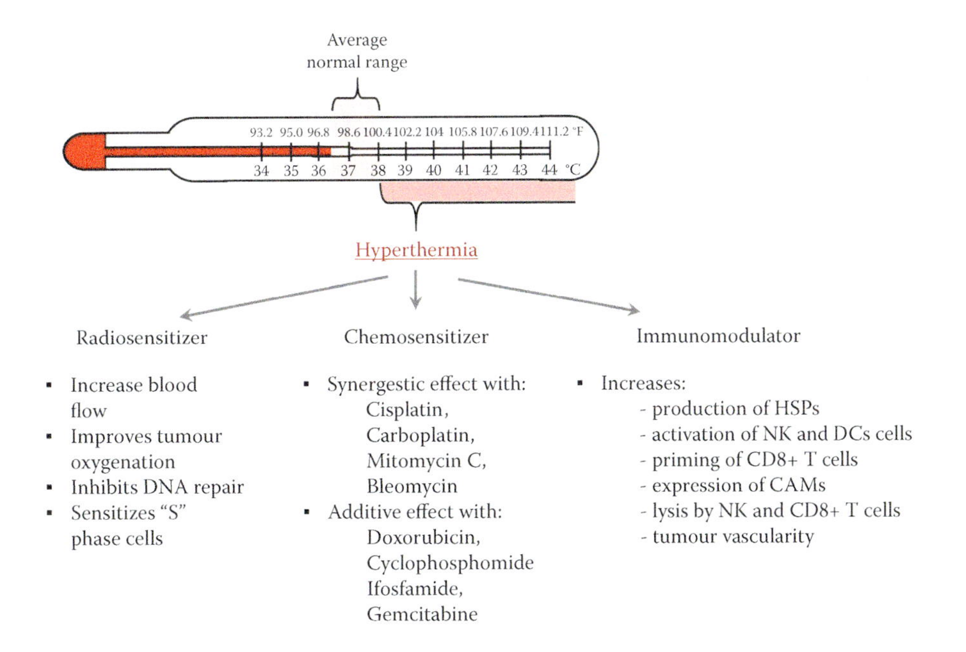

FIGURE 16.4 Multiform action of HT and its effect as a radiosensitizer, chemosensitizer, and immunomodulator. HSPs: heat shock proteins; NK: natural killer; DCs: dendritic cells; CAM: cell adhesive molecule.

Traditional HT (e.g. intraperitoneal hyperthermic chemoperfusion, metal antennas) methods are often very invasive; they provide heating that is nonuniform or nonspecific to the tumour and can cause severe side effects.[48,49] In this context, MNPs' ability to produce local hyperthermia upon the application of an alternating magnetic field present several advantages (Figure 16.5).

First of all, MHT can improve the selectivity of the treatment. This can happen as even deeply situated tumours can be potentially reached by MNPs functionalized with molecules that specifically target the tumour, the vasculature, and/or hypoxic tumour cells that enable crossing the blood–brain barrier.[50–52] However, often it is not possible to direct the MNPs only to the tumours, even when functionalized with targeting moieties. For this reason, direct delivery of the MNPs to the tumoural cells increases the effectiveness of the MHT and decreases the secondary effects by generating the heat only in tissues containing MNPs. Moreover, the heating is more efficient and homogeneous than for macroscopic implants, and it can be repeated several times just after a single administration of MNPs.[53] In fact, *in vitro* it has been demonstrated that the MHT requires a temperature of 6°C lower than that required with external HT to produce a similar cytotoxic effect in human neuroblastoma SH-SY5Y cells.[54] Using other cell type (i.e. melanoma B16 cells) and smaller MNPs, this difference in heating was estimated to be of 12°C.[55] In this case, it has to be noted that the difference in heating was analyzed by using cellular transcripts and not cellular death, and therefore the

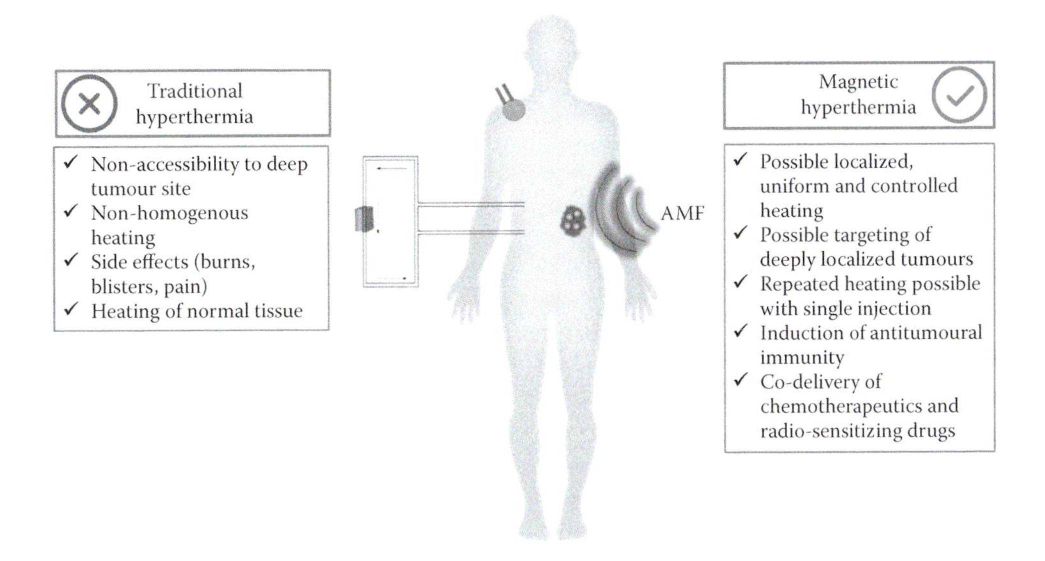

FIGURE 16.5 Comparison of traditional and magnetic hyperthermia. AMF: alternating magnetic field.

effects could be studied at earlier time points. Importantly, the same difference in temperature sensing was found when using an invertebrate model animal, suggesting a general mechanism underlying the response of eukaryotic cells to the delivery of sublethal thermal doses. Last, MHT can also be utilized for a simultaneous and controlled delivery of drugs.[56]

16.4.1 MHT Alone and in Combination with Radiotherapy

Clinical trials with MHT were initiated in 2001, at the time when Jordan et al. presented the prototype of a whole body magnetic field applicator MFH®300F.[57] HT using MNPs was found to be feasible and well tolerated in a pilot clinical study performed by Jordan and colleagues, on previously irradiated and locally recurrent prostate carcinoma.[58] In this study 15 nm MNPs coated with an aminosilane-type shell (MagForce® MFL AS, MagForce® Nanotechnologies GmbH, Berlin, Germany; 120 mg/mL of Fe) were injected transperineally into the prostate under general anaesthesia and transrectal ultrasound/fluoroscopy guidance, and a magnetic field of 5 kA/m strength was maintained during 60 min. During the first treatment, the maximum and the minimum intraprostatic temperatures measured by four thermometry probes reached 44.7°C and 40.65°C, respectively. The same group, two years later, reconfirmed the feasibility of thermotherapy on prostate cancer using the same MNPs and magnetic applicator systems; additionally, they established a method for noninvasive assessment of temperature by correlating computed tomography-derived thermal data with direct temperature mappings in spatially defined measurement points.[59]

Other clinical trial enrolled 22 patients that suffered from recurrent tumours of different nature, i.e. rectal cancer, ovarian cancer, or sarcoma.[60] In this case, the magnetic field strength varied from 3.0–6.0 kA/m in the pelvis, up to 7.5 kA/m in the thoracic and neck region and >10.0 kA/m for the head. Although it was possible to reach 40°C in 86% of the target volume, the coverage with ≥ 42°C was unsatisfactory. It was therefore concluded that an improvement of the temperature distribution was required, either by increasing the amount of injected MNPs or by elevating the field strength.

The feasibility and tolerability of the thermotherapy presented by Jordan et al. were also tested on recurrent glioblastoma multiforme in a phase I clinical trial study.[61] In this case, MagForce® NPs (112 mg/mL of Fe) were injected into nonresected recurrent or primary tumours and a magnetic field with a frequency of 100 kHz and variable strength of 2.5–18 kA/m was applied for 60 min. The MHT was combined with radiotherapy, whereas during each week of irradiation two heat treatments were applied. The MHT caused the median maximum intratumoural temperatures to rise up to 44.6°C and showed signs of local tumour control. The clear therapeutic benefits to prolongation of survival was demonstrated for recurrent glioblastoma patients submitted to a treatment similar as described above in a phase II clinical trial study (12 nm MNPs NanoTherm® AS1, 112 mg/mL of Fe; magnetic field 100 kHz adjusted from 2 to 15 kA/m; a median dose of 30 Gy).[62] The combined treatment of MHT and radiotherapy led to longer OS-2 (overall survival) compared to conventional therapies[63,64] (a median overall survival of 13.4 months in a cohort of 59 glioblastoma patients). Postmortem brain autopsies revealed that the MNPs were dispersed within the tumour necrosis, mainly distributed to the sites of injection.[65] In addition, they were found in macrophages, rather than in glioblastoma cells.

Although these results are promising, it is important to note that different types of cells can show diverse sensitivity to MHT treatment alone or in combination with radiotherapy. For instance, Attaluri et al. reported that human prostate cancer cell lines PC3 or LAPC-4 showed differences in response to heat and ionizing radiation *in vitro*.[66] In fact, LAPC-4 cells are more susceptible to the effects of heat, measured by the clonogenic assay, i.e. the ability of damaged cells to replicate after the treatment. The combination of 60 min of HT with 5 Gy radiation leads to a smaller fraction of survival cells in both types of cells, demonstrating that radiation in combination with hyperthermia is more effective than hyperthermia alone. Again, LAPC-4 cells were more sensitive to the treatment than PC3. This effect could be due to the fact that LAPC-4 cells possess mutated BRCA2 that increases the sensitivity to radiation damage, while PC3 cells possess wild-type BRCA2, displaying less radiation sensitivity. Further, when MHT was combined with radiotherapy, response measured for PC3 tumours *in vivo* was modest when compared to that obtained for LAPC-4 tumours.

16.4.2 MHT and Chemotherapy

Some of the commonly known limitations of chemotherapy, which is another key cancer treatment together with radiotherapy, can be overcome with the employment of NPs and MHT.[67] For example, MNPs can be designed to act as effective drug carriers circumventing systemic toxicity and low drug concentrations in the tumour resulting from traditional chemotherapy.[68] At the same time, a controlled release of the drug activated by the heat can be obtained. Although up to date clinical trials are focused on low temperature thermosensitive liposomal doxorubicin (LTLD) for local release of the drug by conventional HT,[69] efforts are also ongoing to develop MNPs for a triggered drug release.[70] Importantly, MNPs are also being explored as possible drug carriers that would enable thermal synergism with chemotherapeutic agents. This is because hyperthermia can enhance the cytotoxic effect of some of the chemotherapeutic drugs. For instance, the cytotoxic effect of alkylating agents such as cyclophosphamide, ifosfamide, or platinum-based chemical agents/drugs was found to increase several times when the temperature is raised from normal body temperatures to 41°C.[71] However, not all the drugs used for chemotherapy have an increased effect when used in combination with heat, and in other cases it is not possible to establish a general trend due to the diversity of effects between cellular lines.[71] This can be due to the fact that the major effects of hyperthermia in cells are caused in the phase S of the cell cycle, while some chemotherapeutic drugs exert their action in other phases of the cell cycle.[53]

16.4.3 MHT and Immunomodulation

Importantly, it has been demonstrated that MHT can also induce antitumour immune responses, by the production of heat shock proteins (HSP) or activation of antigen-presenting cells among others.[72] Since the pioneering work of Janase et al. in 1998, many other works have been devoted to describing this immunomodulation, where regression of solid tumours not exposed to heat can be observed.[73]

For instance, Toraya-Brown et al. induced dermal tumours in mice on both flanks.[74] 100 nm MNPs were intradermally injected only into left-flank tumours, where the alternating magnetic field (AMF) was applied. The AMF field strength was adjusted so that the tumour temperature was maintained at 42.5–43°C for 30 min, while the rest of the body was maintained at the normal temperature. After 5 days, heated tumours disappeared, whereas those nonheated grew slower, probably due to an immune response. To investigate if an immune response was implied, the intratumoural cytokine and chemokine concentrations 5 days after MHT were analyzed, finding that 25 out of 32 cytokines and chemokines had increased levels in the heated tumours. Interestingly, there were chemokines known to attract natural killer (NK) and dendritic cells (DCs), which are critical in antitumour immunity. Subsequently, $CD8^+$ T cells were activated in the draining lymph node, showing this animal's greater resistance to secondary tumours.

Takada et al. used MNPs bound to a melanogenesis substrate, N-propionyl-cysteaminylphenol (NPrCAP), as the basis for developing melanoma immunotherapy.[75,76] The rationale for using this substrate is that NPrCAP can be selectively incorporated into melanoma cells, produces free radicals, and results in apoptotic cell death. To synthesize the complex, 10 nm MNPs were coated with aminosilane and conjugated with NPrCAP via maleimide cross-linkers (NPrCAP/M). The NPs were injected into the centre of B16 mouse melanoma tumour nodules and AMF was applied at 43°C for 30 min. Although the growth of primary transplants was the same using NPrCAP/M with or without heat, there was a significant difference in the tumour growth inhibition of the secondary distant transplants (which were not treated with NPrCAP/M) when using heat. Expression of HSP70 in the primary tumours and of neutrophilic leukocytes, macrophages, and $CD4^+$ and $CD8^+$ T cells was observed around and within the rechallenge tumours, suggesting the involvement of a melanoma-specific T cell immunity. Although immunomodulation has been widely described for HT, almost no information is given for MHT clinical trials in this respect.

16.5 CONCLUSIONS AND FUTURE PERSPECTIVES

MHT research is a complex, multi- and interdisciplinary field involving material scientists, chemists, physicist, biologists, and medical doctors. Sometimes, this complexity inevitably generates some 'heated' debates among specialists in different fields, but it is safe to say that important advances have been made during the past two decades, especially regarding the overall understanding of the mechanisms of heat generation by MNPs. While MNP design and synthesis must still be focused on obtaining nanomaterials with high heating capability, researchers must always keep in mind all the aspects that can impact negatively on the heating efficiency of the material under *in vitro* and *in vivo* conditions. In fact, MHT in clinics is still in its infancy although HT treatment applied by conventional methods *in vivo* is regarded as a useful treatment in combination with radiotherapy and chemotherapy. Taking into account all the advantages that MHT could provide over other HT techniques, greater efforts should be devoted to pushing this technology forward. The use of MHT in the future could be greatly improved by the application of noninvasive thermometry, the development of MHT equipment specifically designed for diverse body regions, the purposeful tailoring of MNPs that are delivered to the tumour without the need of intratumoural injection, and the design of larger clinical trials.

REFERENCES

1. Glazer, E. S., Curley, S. A. 2011. The ongoing history of thermal therapy for cancer. *Surgical Oncology Clinics of North America* 20 (2), 229–235.
2. Woodhall, B., Pickrell, K. L., Georgiade, G. N., Mahaley, M. S., Dukes, H. T. 1960. Effect of hyperthermia upon cancer chemotherapy – Application to external cancers of head and face structures. *Annals of Surgery* 151 (5), 750–759.
3. Hahn, G. M. 1974. Metabolic aspects of the role of hyperthermia in mammalian cell inactivation and their possible relevance to cancer treatment metabolic aspects of the role of hyperthermia in mammalian cell inactivation and their possible relevance to cancer treatment. *Cancer Research* 34 (11), 3117–3123.
4. Goya, G. F., Asín, L., Ibarra, M. R. 2013. Cell death induced by ac magnetic fields and magnetic nanoparticles: Current state and perspectives. *International Journal of Hyperthermia* 29 (8), 810–818.
5. Vecchia, P., Matthes, R. 2009. *Exposure to High Frequency Electromagnetic Fields, Biological Effects and Health Consequences (100 kHz-300 GHz)*; Paolo Vecchia, Rüdiger Matthes, Gunde Ziegelberger, James Lin, Richard Saunders, A. S., Eds.; ICNIRP 16/2009; Munich, 2009.
6. Dutz, S., Hergt, R. 2014. Magnetic particle hyperthermia – A promising tumour therapy? *Nanotechnology* 25 (45), 452001.
7. Rosensweig, R. E. E. 2002. Heating magnetic fluid with alternating magnetic field. *Journal of Magnetism and Magnetic Materials* 252, 370–374.
8. Deissler, R. J., Wu, Y., Martens, M. A. 2013. Dependence of Brownian and Néel relaxation times on magnetic field strength. *Medical Physics* 41 (1), 12301.
9. Reeves, D. B., Weaver, J. B. 2014. Approaches for modeling magnetic nanoparticle dynamics. *Critical Reviews in Biomedical Engineering* 42 (1), 85–93.
10. Egolf, P. W., Shamsudhin, N., Pané, S., Vuarnoz, D., Pokki, J., Pawlowski, A. G., Tsague, P. et al. 2016. Hyperthermia with rotating magnetic nanowires inducing heat into tumour by fluid friction. *Journal of Applied Physics* 120 (6), 64304.

11. Ota, S., Kitaguchi, R., Takeda, R., Yamada, T., Takemura, Y. 2016. Rotation of magnetization derived from Brownian relaxation in magnetic fluids of different viscosity evaluated by dynamic hysteresis measurements over a wide frequency range. *Nanomaterials* 6 (9), 1–11.
12. Hergt, R., Dutz, S., Zeisberger, M. 2010. Validity limits of the Néel relaxation model of magnetic nanoparticles for hyperthermia. *Nanotechnology* 21 (1), 15706.
13. Maldonado-Camargo, L., Torres-Díaz, I., Chiu-Lam, A., Hernández, M., Rinaldi, C. 2016. Estimating the contribution of Brownian and Néel relaxation in a magnetic fluid through dynamic magnetic susceptibility measurements. *Journal of Magnetism and Magnetic Materials* 412, 223–233.
14. Asín, L., Goya, G. F., Tres, A., Ibarra, M. R. 2013. Induced cell toxicity originates dendritic cell death following magnetic hyperthermia treatment. *Cell Death and Disease* 4 (4), e596.
15. Creixell, M., Bohórquez, A. C., Torres-Lugo, M., Rinaldi, C. 2011. EGFR-targeted magnetic nanoparticle heaters kill cancer cells without a perceptible temperature rise. *ACS Nano* 5 (9), 7124–7129.
16. Hapuarachchige, S., Kato, Y., Ngen, E. J., Smith, B., Delannoy, M., Artemov, D. 2016. Non-temperature induced effects of magnetized iron oxide nanoparticles in alternating magnetic field in cancer cells. *PLoS One* 11 (5), 1–12.
17. Kallumadil, M., Tada, M., Nakagawa, T., Abe, M., Southern, P., Pankhurst, Q. A. 2009. Suitability of commercial colloids for magnetic hyperthermia. *Journal of Magnetism and Magnetic Materials* 321 (10), 1509–1513.
18. Atkinson, W. J., Brezovich, I. A., Chakraborty, D. P. 1984. Usable frequencies in hyperthermia with thermal seeds. *IEEE Transactions on Biomedical Engineering* BME-31 (1), 70–75.
19. Gneveckow, U., Jordan, A., Scholz, R., Brüß, V., Waldöfner, N., Ricke, J., Feussner, A., Hildebrandt, B., Rau, B., Wust, P. 2004. Description and characterization of the novel hyperthermia- and thermoablation-system MFH®300F for clinical magnetic fluid hyperthermia. *Medical Physics* 31 (6), 1444–1451.
20. Laurent, S., Dutz, S., Häfeli, U. O., Mahmoudi, M. 2011. Magnetic fluid hyperthermia: Focus on superparamagnetic iron oxide nanoparticles. *Advances in Colloid and Interface Science* 166 (1–2), 8–23.
21. Kozissnik, B., Bohorquez, A. C., Dobson, J., Rinaldi, C. 2013. Magnetic fluid hyperthermia: Advances, challenges, and opportunity. *International Journal of Hyperthermia* 29 (8), 706–714.
22. Blanco-Andujar, C., Walter, A., Cotin, G., Bordeianu, C., Mertz, D., Felder-Flesch, D., Begin-Colin, S. 2016. Design of iron oxide-based nanoparticles for MRI and magnetic hyperthermia. *Nanomedicine* 11 (14), 1889–1910.
23. Fortin, J. P., Gazeau, F., Wilhelm, C. 2008. Intracellular heating of living cells through Néel relaxation of magnetic nanoparticles. *European Biophysics Journal* 37 (2), 223–228.
24. Di Corato, R., Espinosa, A., Lartigue, L., Tharaud, M., Chat, S., Pellegrino, T., Ménager, C., Gazeau, F., Wilhelm, C. 2014. Magnetic hyperthermia efficiency in the cellular environment for different nanoparticle designs. *Biomaterials* 35 (24), 6400–6411.
25. Soukup, D., Moise, S., Céspedes, E., Dobson, J., Telling, N. D. 2015. In Situ Measurement of magnetization relaxation of internalized nanoparticles in live cells. *ACS Nano* 9 (1), 231–240.
26. Dutz, S., Kettering, M., Hilger, I., Müller, R., Zeisberger, M. 2011. Magnetic multicore nanoparticles for hyperthermia-influence of particle immobilization in tumour tissue on magnetic properties. *Nanotechnology* 22, 265102.
27. Sanz, B., Calatayud, M. P., De Biasi, E., Lima, E., Mansilla, M. V., Zysler, R. D., Ibarra, M. R., Goya, G. F. 2016. In silico before in vivo: How to predict the heating efficiency of magnetic nanoparticles within the intracellular space. *Scientific Reports* 6 (December), 38733.
28. Hanini, A., Lartigue, L., Gavard, J., Schmitt, A., Kacem, K., Wilhelm, C., Gazeau, F., Chau, F., Ammar, S. 2016. Thermosensitivity profile of malignant glioma u87-mg cells and human endothelial cells following γ-Fe_2O_3 NPs internalization and magnetic field application. *RSC Advances* 6 (19), 15415–15423.
29. Kolosnjaj-Tabi, J., Di Corato, R., Lartigue, L., Marangon, I., Guardia, P., Silva, A. K. A., Luciani, N. et al. 2014. Heat-generating iron oxide nanocubes: Subtle "destructurators" of the tumoral microenvironment. *ACS Nano* 8 (5), 4268–4283.
30. Wust, P., Hildebrandt, B., Sreenivasa, G., Rau, B., Gellermann, J., Riess, H., Felix, R., Schlag, P. 2002. Hyperthermia in combined treatment of cancer. *The Lancet Oncology* 3 (8), 487–497.
31. Datta, N. R., Ordonyez, S. G., Gaipl, U. S., Paulides, M. M., Crezee, H., Gellermann, J., Marder, D., Puric, E., Bodis, S. 2015. Local hyperthermia combined with radiotherapy and-/or chemotherapy: Recent advances and promises for the future. *Cancer Treatment Reviews* 41 (9), 742–753.
32. Colombo, R., Salonia, A., Leib, Z., Pavone-Macaluso, M., Engelstein, D. 2011. Long-term outcomes of a randomized controlled trial comparing thermochemotherapy with mitomycin-c alone as adjuvant treatment for non-muscle-invasive bladder cancer (NMIBC). *BJU International* 107 (6), 912–918.
33. Kobayashi, T., Kakimi, K., Nakayama, E., Jimbow, K. 2014. Antitumour immunity by magnetic nanoparticle-mediated hyperthermia. *Nanomedicine* 9 (11), 1715–1726.
34. Hildebrandt, B., Wust, P., Ahlers, O., Dieing, A., Sreenivasa, G., Kerner, T., Felix, R., Riess, H. 2002. The cellular and molecular basis of hyperthermia. *Critical Reviews in Oncology/Hematology* 43 (1), 33–56.
35. Yoshimura, M., Itasaka, S., Harada, H., Hiraoka, M. 2013. Microenvironment and radiation therapy. *BioMed Research International*, Article ID 685308.
36. Shetake, N. G., Balla, M. M. S., Kumar, A., Pandey, B. N. 2016. Magnetic hyperthermia therapy: An emerging modality of cancer treatment in combination with radiotherapy. *Journal of Radiation and Cancer Research* 7 (1), 13–17.
37. Song, C. W., Shakil, A., Osborn, J. L., Iwata, K. 2009. Tumour oxygenation is increased by hyperthermia at mild temperatures. *International Journal of Hyperthermia* 25 (2), 91–95.
38. van der Zee, J. 2002. Heating the patient: A promising approach? *Annals of Oncology* 13 (8), 1173–1184.
39. Bates, D. A., Le Grimellec, C., Bates, J. H., Loutfi, A., Mackillop, W. J. 1985. Effects of thermal adaptation at 40 degrees c on membrane viscosity and the sodium-potassium pump in Chinese hamster ovary cells. *Cancer Research* 45 (10), 4895–4899.
40. Ruifrok, A. C. C., Kanon, B., Konings, A. W. T. 1987. Heat-induced k + loss, trypan blue uptake, and cell lysis in different cell lines: Effect of serum. *Radiation Research* 109 (2), 303–309.
41. Gerner, E. W., Cress, A. E., Stickney, D. G., Holmes, D. K., Culver, P. S. 1980. Factors regulating membrane permeability alter thermal resistance. *Annals of the New York Academy of Sciences* 335 (1), 215–233.
42. Lecavalier, D., Mackillop, W. J. 1985. The effect of hyperthermia on glucose transport in normal and thermal-tolerant Chinese hamster ovary cells. *Cancer Letters* 29 (2), 223–231.

43. Bates, D. A., Mackillop, W. J. 1986. Hyperthermia, adriamycin transport, and cytotoxicity in drug-sensitive and -resistant Chinese hamster ovary cells hyperthermia, adriamycin transport, and cytotoxicity in drug-sensitive and -resistant Chinese hamster ovary cells. *Cancer Research* 46 (11), 5477–5481.
44. Lepock, J. R. 2005. How do cells respond to their thermal environment? *International Journal of Hyperthermia* 21 (8), 681–687.
45. Warters, R. L., Roti, J. L. R. 1982. Hyperthermia and the cell nucleus. *Radiation Research* 92 (3), 458–462.
46. Lui, J. C. K., Kong, S. K. 2007. Heat shock protein 70 inhibits the nuclear import of apoptosis-inducing factor to avoid DNA fragmentation in TF-1 cells during erythropoiesis. *FEBS Letters* 581 (1), 109–117.
47. Kühl, N. M., Kunz, J., Rensing, L. 2000. Heat shock-induced arrests in different cell cycle phases of rat c6-glioma cells are attenuated in heat shock-primed thermotolerant cells. *Cell Proliferation* 33 (3), 147–166.
48. van der Zee, J., Vujaskovic, Z., Kondo, M., Sugahara, T. 2008. The Kadota fund international forum 2004–clinical group consensus. *International Journal of Hyperthermia* 24 (2), 111–122.
49. Cihoric, N., Tsikkinis, A., van Rhoon, G., Crezee, H., Aebersold, D. M., Bodis, S., Beck, M. et al. 2015. Hyperthermia-related clinical trials on cancer treatment within the clinicaltrials.gov registry. *International Journal of Hyperthermia* 31 (6), 609–614.
50. Reddy, G. R., Bhojani, M. S., McConville, P., Moody, J., Moffat, B. A., Hall, D. E., Kim, G. et al. 2006. Vascular targeted nanoparticles for imaging and treatment of brain tumors. *Clinical Cancer Research* 12 (22), 6677–6686.
51. Karim, R., Palazzo, C., Evrard, B., Piel, G. 2016. Nanocarriers for the treatment of glioblastoma multiforme: Current state-of-the-art. *Journal of Controlled Release* 227, 23–37.
52. Moros, M., Delhaes, F., Puertas, S., Saez, B., de la Fuente, J. M., Grazú, V., Feracci, H. 2016. Surface engineered magnetic nanoparticles for specific immunotargeting of cadherin expressing cells. *Journal of Physics D: Applied Physics* 49 (5), 54003.
53. Chatterjee, D. K., Diagaradjane, P., Krishnan, S. 2011. Nanoparticle-mediated hyperthermia in cancer therapy. *Therapeutic Delivery* 2 (8), 1001–1014.
54. Sanz, B., Calatayud, M. P., Torres, T. E., Fanarraga, M. L., Ibarra, M. R., Goya, G. F. 2017. Magnetic hyperthermia enhances cell toxicity with respect to exogenous heating. *Biomaterials* 114, 62–70.
55. Moros, M., Ambrosone, A., Stepien, G., Fabozzi, F., Marchesano, V., Castaldi, A., Tino, A., de la Fuente, J. M., Tortiglione, C. 2015. Deciphering intracellular events triggered by mild magnetic hyperthermia in vitro and in vivo. *Nanomedicine (London)* 10 (14), 2167–2183.
56. Kumar, C. S. S. R., Mohammad, F. 2011. Magnetic nanomaterials for hyperthermia-based therapy and controlled drug delivery. *Advanced Drug Delivery Reviews* 63 (9), 789–808.
57. Jordan, A., Scholz, R., Maier-Hauff, K., Johannsen, M., Wust, P., Nadobny, J., Schirra, H. et al. 2001. Presentation of a new magnetic field therapy system for the treatment of human solid tumors with magnetic fluid hyperthermia. *Journal of Magnetism and Magnetic Materials* 225 (1–2), 118–126.
58. Johannsen, M., Gneveckow, U., Eckelt, L., Feussner, A., Waldöfner, N., Scholz, R., Deger, S., Wust, P., Loening, S. A., Jordan, A. 2005. Clinical hyperthermia of prostate cancer using magnetic nanoparticles: Presentation of a new interstitial technique. *International Journal of Hyperthermia* 21, 637–647.
59. Johannsen, M., Gneveckow, U., Thiesen, B., Taymoorian, K., Cho, C. H., Waldöfner, N., Scholz, R., Jordan, A., Loening, S. A., Wust, P. 2007. Thermotherapy of prostate cancer using magnetic nanoparticles: Feasibility, imaging, and three-dimensional temperature distribution. *European Urology* 52 (6), 1653–1662.
60. Wust, P., Gneveckow, U., Johannsen, M., Böhmer, D., Henkel, T., Kahmann, F., Sehouli, J., Felix, R., Ricke, J., Jordan, A. 2006. Magnetic nanoparticles for interstitial thermotherapy – Feasibility, tolerance and achieved temperatures. *International Journal of Hyperthermia* 22 (8), 673–685.
61. Maier-Hauff, K., Rothe, R., Scholz, R., Gneveckow, U., Wust, P., Thiesen, B., Feussner, A. et al. 2007. Intracranial thermotherapy using magnetic nanoparticles combined with external beam radiotherapy: Results of a feasibility study on patients with glioblastoma multiforme. *Journal of Neuro-Oncology* 81 (1), 53–60.
62. Maier-Hauff, K., Ulrich, F., Nestler, D., Niehoff, H., Wust, P., Thiesen, B., Orawa, H., Budach, V., Jordan, A. 2011. Efficacy and safety of intratumoral thermotherapy using magnetic iron-oxide nanoparticles combined with external beam radiotherapy on patients with recurrent glioblastoma multiforme. *Journal of Neuro-Oncology* 103 (2), 317–324.
63. Stupp, R., Mason, W. P., van den Bent, M. J., Weller, M., Fisher, B., Taphoorn, M. J. B., Belanger, K. et al. 2005. Radiotherapy plus concomitant and adjuvant temozolomide for glioblastoma. *New England Journal of Medicine* 352 (10), 987–996.
64. Stupp, R., Hegi, M. E., Mason, W. P., van den Bent, M. J., Taphoorn, M. J., Janzer, R. C., Ludwin, S. K. et al. 2009. Effects of radiotherapy with concomitant and adjuvant temozolomide versus radiotherapy alone on survival in glioblastoma in a randomised Phase III study: 5-year analysis of the EORTC-NCIC Trial. *The Lancet Oncology* 10 (5), 459–466.
65. van Landeghem, F. K. H., Maier-Hauff, K., Jordan, A., Hoffmann, K. T., Gneveckow, U., Scholz, R., Thiesen, B., Brück, W., von Deimling, A. 2009. Post-mortem studies in glioblastoma patients treated with thermotherapy using magnetic nanoparticles. *Biomaterials* 30 (1), 52–57.
66. Attaluri, A., Kandala, S. K., Wabler, M., Zhou, H., Cornejo, C., Armour, M., Hedayati, M. et al. 2015. Magnetic nanoparticle hyperthermia enhances radiation therapy: A study in mouse models of human prostate cancer. *International Journal of Hyperthermia* 31 (4), 359–374.
67. Hervault, A., Thanh, N. T. K. 2014. Magnetic nanoparticle-based therapeutic agents for thermo-chemotherapy treatment of cancer. *Nanoscale* 6 (20), 11553–11573.
68. Fanciullino, R., Ciccolini, J., Milano, G. 2013. Challenges, expectations and limits for nanoparticles-based therapeutics in cancer: A focus on nano-albumin-bound drugs. *Critical Reviews in Oncology/Hematology* 88 (3), 504–513.
69. Dou, Y., Hynynen, K., Allen, C. 2017. To heat or not to heat: Challenges with clinical translation of thermosensitive liposomes. *Journal of Controlled Release* 249, 63–73.
70. Kakwere, H., Leal, M. P., Materia, M. E., Curcio, A., Guardia, P., Niculaes, D., Marotta, R., Falqui, A., Pellegrino, T. 2015. Functionalization of strongly interacting magnetic nanocubes with (thermo)responsive coating and their application in hyperthermia and heat-triggered drug delivery. *ACS Applied Materials and Interfaces* 7 (19), 10132–10145.
71. Torres-Lugo, M., Rinaldi, C. 2013. Thermal potentiation of chemotherapy by magnetic nanoparticles. *Nanomedicine (London)* 8 (10), 1689–1707.

72. Skitzki, J. J., Repasky, E. A., Evans, S. S. 2009. Hyperthermia as an immunotherapy strategy for cancer. *Current Opinion in Investigational Drugs* 10 (6), 550–558.
73. Yanase, M., Shinkai, M., Honda, H., Wakabayashi, T., Yoshida, J., Kobayashi, T. 1998. Antitumour immunity induction by intracellular hyperthermia using magnetite cationic liposomes. *Japanese Journal of Cancer Research* 89 (7), 775–782.
74. Toraya-Brown, S., Sheen, M. R., Zhang, P., Chen, L., Baird, J. R., Demidenko, E., Turk, M. J., Hoopes, P. J., Conejo-Garcia, J. R., Fiering, S. 2014. Local hyperthermia treatment of tumors induces CD8+ T cell-mediated resistance against distal and secondary tumors. *Nanomedicine: Nanotechnology, Biology and Medicine* 10 (6), 1273–1285.
75. Jimbow, K., Ishii-Osai, Y., Ito, S., Tamura, Y., Ito, A., Yoneta, A., Kamiya, T. et al. 2013. Melanoma-targeted chemothermotherapy and in situ peptide immunotherapy through HSP production by using melanogenesis substrate, NPrCAP, and magnetite nanoparticles. *Journal of Skin Cancer* 2013, Article ID 742925.
76. Takada, T., Yamashita, T., Sato, M., Sato, A., Ono, I., Tamura, Y., Sato, N. et al. 2009. Growth inhibition of re-challenge B16 melanoma transplant by conjugates of melanogenesis substrate and magnetite nanoparticles as the basis for developing melanoma-targeted chemo-thermo-immunotherapy. *Journal of Biomedicine and Biotechnology* 2009, Article ID 457936.

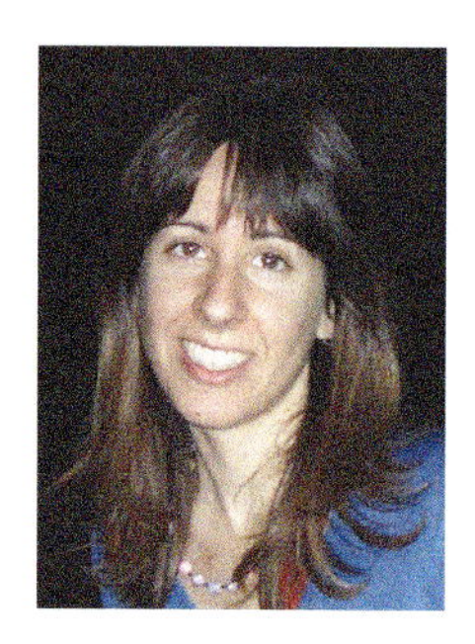

Dr María Moros (E-mail: m.moros@isasi.cnr.it) (Zaragoza, Spain) graduated as a pharmacist in 2003 at the University of Navarra and earned her PhD in 2012 at the Instituto de Nanociencia de Aragón (INA, University of Zaragoza). She carried out her first postdoctoral research (2013–2015) at the INA with the group led by Dr Jesús Martínez de la Fuente, working for a Starting Grant European Project (Nanopuzzle) whose objective was to set up the facilities for performing *in vivo* magnetic hyperthermia experiments with nanoparticles. In 2015, she was awarded the Marie Sklodowska-Curie Fellowship at the Istituto di Scienze Applicate e Sistemi Intelligenti – CNR (Naples, Italy) to start a research project based on the use of invertebrate animals to screen the heating capabilities of gold nanoparticles for optical hyperthermia.

Dr Raluca Maria Fratila (E-mail: rmfratila@gmail.com) (Petrosani, Romania) earned her PhD in Chemistry from the University Politehnica Bucharest (Romania) in 2005. She performed postdoctoral work at the University of Basque Country, San Sebastian, Spain (2006–2008), and at the University of Twente, Enschede, The Netherlands (2009–2013). In November 2013, she became a Marie Curie COFUND-ARAID researcher at the Institute of Nanoscience of Aragón (INA), University of Zaragoza, Spain. Currently, she is a Marie Sklodowska-Curie researcher at the Aragon Materials Science Institute (University of Zaragoza, Spain), and in 2016 she was awarded a prestigious Ramón y Cajal grant (Spanish Government, 2017–2022). Her research interests include bio-organic and bio-orthogonal chemistry, magnetic resonance imaging (MRI), magnetic hyperthermia, and biofunctionalization of magnetic nanoparticles for biomedical applications.

Dr Laura Asín (Zaragoza, Spain) graduated as biochemist in 2006 in the University of Zaragoza and earned her PhD in 2012 in the Instituto de Nanociencia de Aragón (INA-University of Zaragoza). After that, she worked in the group 'Nanotechnology and Apoptosis (NAP)' at the Instituto de Ciencia de los Materiales de Aragón (ICMA) led by Dr Jesús Martínez de la Fuente with the objective to develop a nanosensor to detect analytes using gold nanoparticles and transfer the knowledge to the market in collaboration with Nanoimmunotech S. L. In October 2015, she received a postdoctoral fellowship, 'Juan de la Cierva formación' from MINECO, and has since been carrying out her postdoctoral research in the NAP group from ICMA-CSIC. Her research focuses on the use of magnetic nanoparticles *in vitro* for biological applications and *in vivo* magnetic hyperthermia studies.

Dr Grazyna Stepien (Myszkow, Poland) graduated in biology and biotechnology from the University of Silcsia, Poland in 2011/2012. In 2017, she earned her PhD at the Institute of Nanoscience of Aragón (University of Zaragoza), where she was working in the Nanotherapy and Nanodiagnostics Group under the supervision of Dr Jesús Martinez de la Fuente and Dr Maria Moros. Her PhD study was focused on biological effects caused by magnetic nanoparticles designed for magnetic hyperthermia.

Dr Jesús Martínez de la Fuente (Barakaldo, Spain) (E-mail: jmfuente@unizar.es) earned his PhD in 2003 at the University of Seville, Spain. In 2007, he received a permanent ARAID position at the Institute of Nanoscience of Aragón (University of Zaragoza), where he established the Nanotherapy and Nanodiagnostics Group. His main research interests include the development of general and simple strategies for the biofunctionalization of nanoparticles and surfaces. As principal investigator, Dr de la Fuente has been the recipient of several prestigious grants, including a European Research Council-Starting Grant (NANOPUZZLE, 2010–2015), a European Research Council-Proof of Concept Grant (HOTFLOW, 2017–2018), an ERANET project (NANOTRUCK, 2009–2012), and a FP7-NMP (NAREB, 2014–2018). To date, he has published more than 150 papers, with more than 5000 citations and an h-factor of 38. Since 2003, he is associate member of the Centre for Cell Engineering at the University of Glasgow (UK), and since 2013, he is a visitor professor at Jiao Tong University (Shanghai, China) in the context of the prestigious 'Shanghai-1000 People Plan'. Since 2014, he is a permanent researcher at the Spanish National Research Council, Aragon Materials Science Institute (Zaragoza, Spain).

17 Nanoparticles for Nanorobotic Agents Dedicated to Cancer Therapy

*Mahmood Mohammadi, Charles Tremblay, Ning Li, Kévin Gagné, Maxime Latulippe, Maryam S. Tabatabaei and Sylvain Martel**

CONTENTS

17.1 INTRODUCTION

The growing interest in magnetic nanoparticles (MNPs) is due to their use for fighting cancers in a targeted manner. Indeed, despite the majority of cancers (> 85%) being initially localized, modern treatments often resort to chemotherapy, with the systemic injection of toxic drugs affecting healthy organs and tissues. Not only does this approach result in a poor therapeutic index (1–2%), but the severe secondary effects on the patients, often affecting vital functions, greatly limit the efficacy of such treatments by constraining their frequency and the amount of administered drugs. Even the most advanced chemotherapeutics that are conjugated with special molecules destined to increase specificity to cancer cells, result in a very low therapeutic index (< 2%). That is mainly because with systemic injection they get eliminated by the body before they reach the diseased region. As such, the concept of direct drug targeting (DDT) has been proposed where therapeutic agents are guided from the injection point through the shortest physiological route leading to the target, which in cancer therapy would typically be a solid tumour. This capability would yield a much higher therapeutic index while decreasing the toxicity and secondary effects to a negligible level. In the patient's point of view, this translates to a better quality of life and increased chance of a cure. On the economic side, there is a potential for much more cost-efficient therapy with reduced hospital stays and a lower number of treatments.

Achieving DDT requires a propulsion mechanism capable of steering and navigating therapeutic agents in the desired bifurcations after their injection in the vascular network. Due to current limitations of modern technologies in miniaturization of electronics, thrusters, implementing some level of intelligence, and other functionalities enabling efficient navigation in the vascular network, microrobots cannot be fabricated small enough to reach the smallest arterioles (< 50 μm) and beyond in order to reach the active cancer cells. DDT methods typically rely on the remote magnetic actuation of magnetized microscale agents in large blood vessels, which combine drugs with MNPs. The big advantages of this type of actuation which make it particularly suitable for the task are its biocompatibility and full penetration of the human body. This allows for the noninvasive remote control of the agents by applying magnetic fields and gradients to generate forces and torques.

Put simply, the agents are pulled in the direction where the magnetic field's strength increases and tends to align with the field. Higher the gradient, higher the force's strength will be. Similarly, higher will be the field strength and higher the torque. Since in DDT the magnetic volume of the agents is very small and flows are very fast, high magnetic gradient strengths (ideally greater than 200 mT/m) are required to achieve sufficient steering forces. If NPs are magnetized by the B-field, such as in the case of superparamagnetic NPs, there is an additional need for a high field strength that is sufficient to bring the particles at saturation magnetization in order to maximize the actuation force induced. In medical nanorobotics, superparamagnetic NPs are favoured due to the fact that they lose their magnetization once no longer exposed

* Corresponding author.

to a magnetic field, which is essential to avoid the formation of large clusters once the DDT treatment has been completed.

Beside the induction of a directional propelling force, superparamagnetic NPs are also used in nanorobotic agents or nanorobots for real-time tracking purpose or to assess the level of targeting following navigation of the nanorobotic agents. Indeed, the ability to locate nanorobots is essential to provide a feedback on the treatment's evolution as improper targeting is not suitable. The localization is either realized during the NP steering (real time) or after targeting. The lack of sensitivity or resolution to magnetic particles of imaging modalities such as positron emission tomography (PET), computed tomography (CT) scans, and others, make them unsuited for the tracking of the NPs in the context of DDT. On the other hand, magnetic-based imaging modalities such as magnetic resonance imaging (MRI) and magnetic particle imaging (MPI) provide relatively good spatial and temporal resolution as well as high sensitivity to magnetic particles.

For cancer therapy applications when complementing the drug delivery, the NPs embedded or attached to the nanorobotic agents can also be used as biological excitation agents as well as therapeutic agents. As described in the section discussing hyperthermia, superparamagnetic NPs are used as a direct means to provide medical treatments. They are magnetically heated in order to transfer thermal energy to the surrounding physiological environment. This results in localized tissue necrosis. In the particular case of brain tumour, hyperthermia finds another use as an unlocking process to allow tumour targeting by opening the blood–brain barrier (BBB). The BBB is an additional and major obstacle for therapeutic agents to act on brain tumours. In this temperature sensitive membrane, localized heating of the injected nanorobots is used to increase its permeability in the brain thus allowing an effective drug delivery in the heated region. The interesting fact being that the same superparamagnetic NPs used to propel and track the nanorobots can also be used to temporarily and reversibly open the BBB.

As described, robotic functionalities have been developed for nanorobotic agents, allowing them to increase the therapeutic effects while minimizing systemic toxicity. A more detailed discussion concerning the functionalities and characteristics of these nanorobots is provided in the following sections.

17.2 MAIN TYPES OF MAGNETIC NPs USED BY NANOROBOTIC AGENTS

The characteristics of NPs can be exploited to implement new versions of nanorobotic agents with different functionalities. For instance, at such a small scale, NPs often exhibit quantum effect. Electrons are not distributed anymore and are confined in a finite 3-D nanospace, which is a 'particle in the box' quantum mechanical problem, that makes their electronic wave function having discrete energy levels. For example, gold NPs have energy-size dependence to their electronic potential making them colour-dependent based on their size.[1] In the case of MNPs, the same apply to spin quantification where the exchange energy distance between spins, i.e. the domain wall reversal thickness, is larger than the size of the particles making them magnetic monodomain where the value of magnetization can only take two discrete value of +m and −m even if the particle itself isn't a monocrystal. These particles are referred to be in a superparamagnetic state.[2] The most widely spread type of medical MNP and the one most often used in nanorobotic agents are the superparamagnetic iron oxide (SPION). Their commercial availability in various forms, sizes, surface molecules, and their low toxicity make them ideal candidates for their fast acceptation and intense development.[3] The counterpart is that their spinel ferrimagnetic state does not allow high mass magnetization compared to soft ferromagnetic materials like iron or nickel. Recent works tried to address this limitation through the synthesis of NPs in iron alloys or with pure iron in graphene matrix.[4]

Apart from the quantization effects, the high surface to volume ratio is another critical and interesting aspect of NPs in the context of medical nanorobotics. This feature makes them very reactive to their environment, which makes them great candidates to be integrated in matrix as long as good mixing can occur. NPs can also be dispersed in liquid leading to a quasi-solution type of liquid if sedimentation is prevented or if it takes place after a very long time. This liquid mobility is of great interest for surface functionalization that can make the NPs very complex component for nanorobotic devices. Nanocapsules can be filled with some liquid or solid and have surface molecules that give them suspension properties or biomobility/affinity properties. For example, a nanoliposome can be filled with a drug and have a specific antibody at its surface to bind to or to be integrated in a specific cell.[5] The mixing with a solid-phase material is also of great interest as the properties of the material can retain some properties of the NPs or even benefit from an improvement of the bulk properties. It is out of the scope of this chapter to discuss them as they have so many applications for mechanical, optical, thermal, electrical, and chemical properties modifications. Our interest applies more specifically to NPs mixing with nanocomponents for nanorobots. For example, superparamagnetic NPs can be integrated in a larger object transferring the superparamagnetic property to an object that would usually be too large to demonstrate superparamagnetic behaviour.[6]

17.3 AGGREGATION OF NANOROBOTIC AGENTS

Remote navigation of nanorobotic agents by inducing a directional force on superparamagnetic NPs embedded in each nanorobot using magnetic gradients is a challenging task, especially at the human scale. This is mainly due to the fact that the magnetic field and gradient decays as $1/d^3$ and $1/d^4$ from the magnet, respectively (d being the separating distance), and because of the small volume of the NPs compared to the high viscous forces that prevent them from moving freely.[7,8] The magnetic force scales as $1/r^3$ the size of the particle (r is the radius of the spherical particle) while viscous forces scale as $1/r^2$ while being proportional to particle velocity. For

NPs, this means that due to their relatively large surface and low volume, very high gradients need to be used to induce a suitable acceleration. This acceleration leads to an increase in velocity and a proportional increase in viscous force giving a very low speed limit to a single NP. A simple but elegant way to address this problem is to diminish the surface to volume ratio by combining many NPs together. This can be done through inclusion of NPs in a larger matrix or more simply by aggregating them in a cluster. Both methods have been used to implement nanorobotic agents, each with their advantages and drawbacks. Natural aggregation of MNPs can be achieved by applying a sufficiently high magnetic field. In this case, all the NPs will be magnetized in the same direction and when two particles get sufficiently close, they will attract each other and form an aggregate. This process can lead to very large number of NPs aggregated in a needle shape if placed inside a clinical MRI scanner (Figure 17.1). Indeed, in medical nanorobotics, MRI scanner provides the high field strength required to bring the MNPs at saturation magnetization, while the imaging gradients can be used to induce a 3-D directional propelling force on the MNPs embedded in the nanorobotic agents. Furthermore, MRI is ideal to detect MNPs that also act as MRI contrast agents for tracking and targeting assessment purposes. But the direction of the high uniform magnetic field along the long axis of the tunnel of the MRI scanner forces the NPs to form needle-like aggregates due to dipolar interactions of the MNPs along the easy axis. In a static flow, there is no limit to the length of the aggregate but their width will be limited by the surface adhesion force that counterbalances the repulsion force of the magnetic field on the sides. In a flow with shearing forces such as in the blood stream, these aggregates will have a maximum length defined by their magnetic moment, surface properties, the external magnetic field strength, the vessel size, and the flow velocity. Interestingly, the surface properties of the NPs can be tuned to tailor the aggregation as needed. For example, some ferrofluidic colloidal dispersions are designed to make sure there will be no interaction between the NPs, hence preventing an aggregation even when subjected to a very strong magnetic field. While at the opposite, NPs with different surface properties can be tailored for a specific aggregation size in a range of fluid flows and magnetic strengths. The same approaches can also be used during the synthesis of nanorobotic agents.

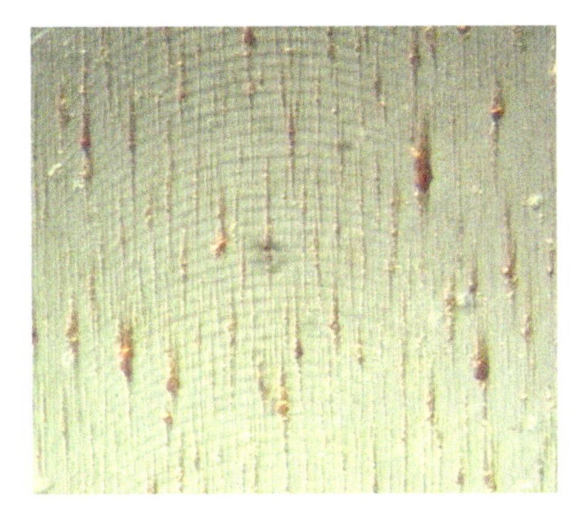

FIGURE 17.1 Aggregation of bare Fe_3O_4 NPs in 1.5 T MRI bore. (Courtesy of Dr Mathieu.)

17.4 NAVIGATION AND TARGETING METHODS FOR MNPs

The simplest approach for drug targeting consists of using a permanent magnet placed near the diseased region in order to attract MNPs towards the targeted site.[9] Once the MNPs circulate in the vicinity of the magnet, NPs coated with drug molecules can diffuse through the tissues and towards the targeted area. This approach is however highly limited in terms of control (attraction only) and distances of the intervention. Indeed, since the magnetic field and gradient strengths decrease exponentially with distance from the magnetic source, this method can only provide sufficient attraction forces near surface tissues without the possibility for targeting deeper in the tissues, unless the magnet is inserted inside the body which then translates into a much more invasive intervention. Other approaches relying on a mobile magnet attached to a robotic arm,[10] or multiple magnet systems allowing a pushing force on magnetic particles,[11] have been proposed to achieve better control. Nevertheless, these approaches still suffer from the lack of magnetic field and gradient strengths in deep tissues, and do not provide the control capabilities required for navigating agents through multiple bifurcations in the vascular network.

To achieve better control in magnetic actuation, several systems based on different configurations of electromagnetic coils, known as electromagnetic actuation (EMA) systems, have been proposed.[12,13] These systems are able to control small magnetic devices in up to six degrees of freedom (DOF), although only 3 DOF are required for DDT. While these platforms can generate strong gradients over a relatively long distance from the coils, they cannot provide the field strength required to magnetize the particles at saturation when scaled for interventions on humans. A recent example is based on a MPI system. Indeed, although MPI is an interesting technology that can be used for the magnetic actuation of NPs[14] with real-time tracking, this platform is still limited at the human scale in term of sufficient field strength to bring MNPs at saturation magnetization.

A more suitable approach at present for the navigation of magnetic particles in the vascular network is known as magnetic resonance navigation (MRN).[15,16] By placing the patient inside an MRI scanner such as the ones widely available in hospitals, MRN benefits from the strong static magnetic field of the scanner (typically 1.5 T or 3 T) that is sufficient to bring MNPs at saturation magnetization. The imaging gradient coils of the scanner are then used to control in real time the required directional forces during the transit of the MNPs typically embedded in a microscale therapeutic carrier along a planned trajectory in the vasculature. The conventional clinical MRI scanners being typically limited to gradient strengths around 40–80 mT/m, the efficiency of MRN for navigating in larger arteries depends on an adequate control of the blood flow. The latter is typically achieved by using a balloon catheter, which when inflated will slow down the transit velocity of the magnetic particles and thus allow more time for the forces to act on them between each bifurcation. Additional gradient coils providing higher gradient strengths (> 300 mT/m) can be added inside the scanner[7] but at the cost of preventing whole-body interventions due to the resulting reduced diameter of the tunnel.

More recently, dipole field navigation (DFN)[17] has been proposed, where ferromagnetic cores are inserted inside an MRI scanner and positioned around the patient in order to generate much stronger magnetic gradients (> 300 mT/m) in the whole body. By distorting the scanner's homogeneous field, the cores can shape a magnetic path that ensures the magnetic therapeutic agents follow a predefined route towards a target. But this capability to provide both a high field and high gradient strengths is obtained at the cost of more complex models and algorithms required for properly positioning the cores as a function of the vasculature, as well as limiting the imaging capabilities for tracking the particles in real time. With statically positioned cores, DFN also has the advantage of allowing the constant injection of agents, which would result in a faster intervention compared with MRN where a number of agent boluses have to be individually and sequentially navigated until the desired dose is reached. There are, however, theoretical limitations in the steering resolutions achievable with DFN since high and abrupt changes in the gradient between two neighboured points require potentially unobtainable distortions of the field, whereas in MRN the gradients can be switched in a fraction of seconds. To cope with this limitation, DFN could also be implemented with mobile cores, whose positions could be dynamically controlled using the scanner gradient coils, or even a combination of static and mobile cores. The mobile cores could be moved according to a bolus of agents to apply all the desired forces during transit toward the target.

There is probably no perfect magnetic actuation method adapted for all situations for the direct targeting of superparamagnetic NPs embedded in therapeutic agents. In a clinical setting, the aforementioned approaches, and probably others to come, would have to be used in a complementary fashion, especially for more complex pathways and deeply located target sites.

17.5 LOCALIZATION OF MNPs

The choice of the MNPs used in navigable nanorobotic agents for cancer therapy is based on the importance of their interaction with the same physical phenomenon as the one exploited by magnetic-based imaging modalities, i.e., magnetism. As a result, the impact of the nanorobotic agents on their neighbourhood in the resulting image is direct and unavoidable, resulting in a high sensibility of the magnetic-based imaging modalities to the same MNPs used for propelling the agents during navigation in the vascular network. The imaging platform can therefore be used not only for steering purposes, as mentioned previously, but also for tracking these NPs and to visualize them once they have reached their targets, hence opening the door to theranostic methods relying on nanorobotic agents. These capabilities also allow physicians to realize real-time tracking as well as planning the targeting trajectory by means of pre-interventional acquisitions. Because MNPs distort the magnetic field, they can be easily detected using magnetic-based imaging modalities such as MRI and MPI, two modalities that can also be used to actuate MNPs. The following sections explain the concepts of these imaging technologies and the impact of the MNPs' presence inside the imaging field on the resulting image.

17.6 MAGNETIC RESONANCE IMAGING

Magnetic resonance imaging technology is based on the pioneering work of E. L. Hahn in 1950[18] who developed the spin echo experiment on which was subsequently built this imaging modality. Since then, more than sixty years of research has led to highly specialized material resources and imaging sequences allowing high resolution 3-D image acquisition of soft tissues, real-time imaging, etc. This widespread technology is already used as a source of information for cancer diagnosis as well as for angiography; it is an essential tool for targeting path planning.

The fact that all living cells contain large numbers of hydrogen atoms is the cornerstone of MRI. Such atoms have spin-½ nucleus. Using the classical depiction of a spin's evolution when immersed in a magnetic field B_0, the proton's spin precesses around the magnetic field's direction at the Larmor frequency. As each spin has a slightly higher chance to be in the lower energy state (parallel to the direction of the MRI homogeneous field B_0) noted $E_{+½}$, a macroscopic amount of spin displays an excess of +½-spins of about one in 10^6 at room temperature. A resulting net magnetization M_0 in the field direction therefore emerges from the latter excess. M_0 can then be tilted away from the main field direction by using a weaker radio frequency (RF) oscillating magnetic field with an oscillation frequency $f = \omega_0/2\pi$. Once in the transverse plane, M_0 slowly realigns with B_0 while precessing around the direction of B_0. According to Faraday's law, such oscillating magnetic field would produce an oscillating signal in a conductive loop if it is positioned near the source of these oscillations.[19] The latter oscillation constitutes the signal used to obtain the image in MRI. Local magnetic environment encountered in the human body results in spin dephasing and additional signal loss. The rate at which the decays are happening hides the mean to discriminate different magnetic environments; i.e., different tissues. Spatial encoding is typically realized through the application of magnetic field gradients and clever combinations of sequence events (RF pulses and others, see Ref. 20) in order to fill the Fourier k-space and finally obtain an image by applying an inverse Fourier transform on the collected data.

The highly inhomogeneous local magnetic field generated by MNPs results in a fast spin dephasing in their vicinity. This important signal loss translates itself to a low intensity dot in the resulting image; i.e., the particle is localized. As discussed previously, different factors affect the behaviour and the magnetization of the NPs. The optimization of the nanorobots' steering or hyperthermia characteristics hence yields the additional advantage to generate greater field disturbances and thus easier tracking. Recent work[21] has shown that the localization of a single particle is possible as the low intensity dot can be approximately 50 times larger than the magnetic particle itself. Generally, smaller particles are selected to act as contrast agents requiring numerous NPs embedded in each nanorobotic agent to achieve detectability.[22] Similarly, the accumulation of a sufficient number of MNPs in the nanorobotic agent increases the induced propulsive force during navigation.

By choosing the timely optimized sequence[23] and by reducing the field of view of a MRI acquisition, one can obtain a high-speed image of regions such as the targeting path presumably followed by the NPs. This important feature allows real-time tracking of the magnetic particles and, therefore, the nanorobotic agents as well. Furthermore, using a MRI machine for multiple aspects of a cancer therapy intervention decreases the cost of a fully adapted intervention room. The main drawbacks of this technology concern its incompatibility with almost any standard engineered devices since no relatively large ferromagnetic components can get anywhere close to the MRI tunnel's entrance without risking potentially dramatic events for the MRI machine and/or the device in question.

17.7 MAGNETIC PARTICLE IMAGING

Magnetic particle imaging is a relatively new imaging modality introduced by Gleich and Weizenecker in 2005.[24] The method takes advantage of the magnetic hysteresis phenomenon of SPIONs, a class of biocompatible contrast agents used in MRI.[25] As the surrounding magnetic field increases, such material's magnetization moves nonlinearly from a linear regime at low fields to another linear regime at higher fields. A weak oscillating magnetic field is therefore superimposed on a strong static magnetic field featuring a narrow field free point (FFP). The oscillating field is produced by three orthogonal pairs of auxiliary coils that are used as a mean to move the FFP through the sample. Magnetic materials outside the FFP do not generate much signal. On the other hand, the magnetization's variations of such material when crossing the FFP generates an important oscillating signal; in that case the material is exclusively subjected to the oscillating field. The nonlinearity of the material's magnetization induces auxiliary odd harmonics in the data collected which become the exploited signal (see Figure 17.2).[26] After acquiring signals from the whole sample, the spatial distribution of the magnetic material is obtained using adapted signal processing.[26–28]

The MPI technology has demonstrated to be highly sensitive to magnetic particles and at the same time being a relatively high resolution modality both temporarily and spatially.[29] These reasons explain the interest the method generated when proposed in 2005. The technology is also seen as an alternative imaging modality, since patients with fragile

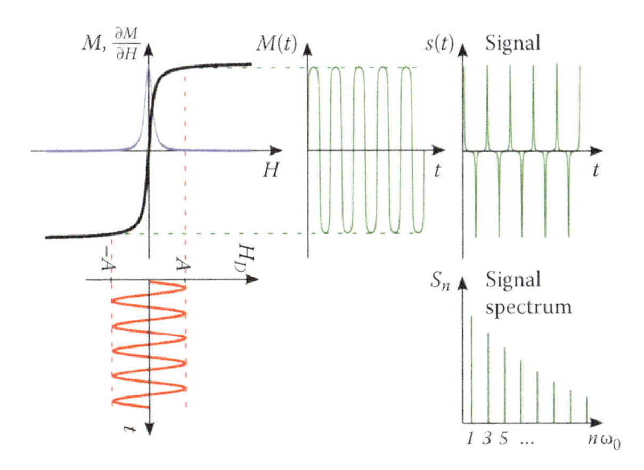

FIGURE 17.2 Principles of signal generation within magnetic particle imaging (MPI).[28] (Reproduced from Rahmer, J. et al., *BMC Med. Imaging*, 9, 4, 2009.)

livers do not tolerate repeated exposure to iodinated contrast agent typically used in CT scans.[27,30] The aim of the method is to use SPIONs as contrast agents to produce anatomical images, thus opening the door to theranostic use of the MPI technology. For navigable nanorobotic agents, MPI is also interesting since as mentioned earlier, the gradient field of MPI can also be used to directionally move magnetic objects. However, MPI is not yet ready for clinical use. The challenges encountered while scaling up the small proof-of-concept setups are still not completely overcome. The problem is similar to the one of scaling up nuclear magnetic resonance (NMR) spectrometers to human size in order to perform MRI; it is not conceivable to obtain the desired FFP homogeneity and field gradients required at such scale by simply using the same design for the whole apparatus. The geometry relative to the field strength, coil design, type of magnetic particle[31] and auxiliary field oscillation frequency are factors to properly combine in order to move FFP through a sufficiently large volume while being competitive with other imaging modalities in terms of resolution. Submillimetre spatial resolution would require magnetic field gradients of approximately 10 T/m,[27] which is quite significant and extremely challenging technologically. To put that in perspective, the magnetic field gradient at the tunnel's entrance of a 3 T MRI machine is around 5 T/m. The technological challenges to be overcome are therefore important for MPI researchers.

17.8 HYPERTHERMIA PRODUCED BY MNPs IN NANOROBOTIC AGENTS

Medication can be encapsulated and overlaid on the nanorobotic agents and actively released by means such as local temperature change, chemical reaction, antibody binding, ligands mediated interactions, radiation, or vibration. Encapsulation of the drug molecules shields healthy cells from further toxic exposure to chemicals. This type of release formulation reduces unintended chemical side effects and reduces unnecessary high dosage administration.

Hyperthermia or an elevation of the temperature at the site of treatment can be combined with the release of drug molecules to enhance the therapeutic effects.[32,33] As such, the same superparamagnetic NPs used to propel and track the nanorobotic agents can also be used for hyperthermia. As local hyperthermia created by the superparamagnetic NPs of the nanorobotic agents can also temporarily open the blood–brain barrier (BBB) to allow larger therapeutic molecules to pass from the bloodstream into the brain tissue.[34]

Fever is a common sign for an infection or other viral diseases. It is accompanied by an immune system response that recognizes the potential harmful foreign antigen and reacts to destroy it. Inspired by this natural thermotherapy, 'hyperthermia', the artificial elevation of body temperature has become an interesting topic in oncology.

Studies suggest that hyperthermia is associated with immunological reactions and heat above 42–43°C can be cytotoxic for tumour environments that have a low level of oxygen and a low pH. It deforms their protein structure and cell morphology, leading to cell necrosis and therefore shrinking the overall tumour size.[35–37] It can also be used as a complementary therapy to chemotherapy, radiotherapy, and surgical resection to yield the maximum benefit while minimizing their applied drug dosage and toxic side effects.[38] It is also reported that hyperthermia can substantially enhance drug uptake above specific temperatures.[39] All of which make this technique a potential candidate for cancer therapy.

Hyperthermia setup consists of an inductive coil with high alternating current passing through it. According to Lenz law, the rapid alternating movement of electron charges in the coil creates an alternating magnetic field. Materials under such an alternating electromagnetic field exhibit heat through three processes: (1) Dielectric loss for nonconductive material, (2) Eddy current for conductive material, and (3) Magnetic loss for magnetic material.

Human tissue can be considered as a heterogeneous mixture of conductive and nonconductive materials. Dielectric and Eddy current convection profiles depend on the applied alternating electromagnetic field frequency. Therefore, depending on the tissue type and its proportion in the treatment area, one effect can overpower the other in elevating body temperature. In oncology, it is desired to minimize dielectric heating of the target tissue and focus more on controlled heating. This means delivering heat from an injected conductive material to the vicinity of the targeted tissue. In other words, the goal is to gain more spatial precision and depth in treating cancerous tumours without stimulating or damaging nearby healthy cells. This can be done by loading conductive materials on a nanorobotic agent as a means for delivering localized heat to the desired site following the navigation phase, since Eddy current loss will transfer the energy much faster than dielectric loss. Eddy current heat dissipation is highly proportional to the size of the conductive materials and at the nanometer scale, becomes negligible.[40] Therefore, MNPs under hyperthermia become perfect candidates for generating heat directly to the surface of where they are attached.

The MNPs used for propulsion and tracking of the nanorobotic agents dissipate heat by hysteresis or relaxation losses. These two processes dominate depending on magnetic domain size. For multidomain MNPs, hysteresis loss is the main source of heat. Within an applied external magnetic field, magnetic moments of the single domain NPs align with the new field and after a short time, they relax back to their original orientation. Heat is therefore generated in exchange for realigning the magnetic moment to its original orientation (Néel relaxation) or from thermal agitation of particles rotating against the viscous drag of the medium where they are embedded in. Via these MNPs, delivered to a desired site by means of any delivery agents such as nanorobots, hyperthermia could be applied effectively to treat cancerous cells.

A novel application of hyperthermia was recently published regarding drug delivery to the brain. The importance of this application comes from the fact that the outer surface of brain arterioles is sheathed with structures that makes them impermeable to 98% of the chemotherapy drugs. Only molecules with molecular weight less than 400 kDa can pass through.[41] Hyperthermia of MNPs transfers heat to the endothelial cells in brain capillaries, leading to temporal disruption of the BBB.[34] Nanorobot agents can either be fabricated so that they can become magnetically excited by an alternating field, or loaded with appropriate particles to perform hyperthermia after they reached their final destination. A more complex nanorobot agent would have drugs encapsulated as well such that when the BBB is opened, the drug molecules could be released, penetrating into the brain tissue.

As mentioned above, an alternating magnetic field can make MNPs yield much heat. This phenomena, known as magnetic hyperthermia, has also been used for thermal ablation of cancerous cells. In order to effectively kill the cells, the temperature beside them needs to be maintained at a desired value of 42°C. Usually, 10–100 mg of particles per cm^3 in the tumour tissue is required based on their thermal properties.[42] The high concentration of MNPs becomes a challenge due to difficulties in controlling the NPs inside the body. Under this circumstance, the microencapsulation of MNPs in nanorobotic agents provides the possibility of increasing the targeting efficiency of NPs, and preventing side effects while increasing the effectiveness of targeted therapies based on heat delivery. NPs can be encapsulated into a biodegradable polymeric matrix, which increases the controllability of the NPs during magnetic navigation due to an increase of the effective volume of magnetic material per nanorobotic agent. After reaching the tumour, these NPs can be concentrated in the targeted area and the heat could also be used to initiate the degradation process of polymeric materials. Therapeutic drug molecules can also be encapsulated into microspheres such as liposomes or micelles. The encapsulation enables drug molecules to have a prolonged time for drug release[43] which is desirable in many instances.

17.9 MICROENCAPSULATION

Polymers are usually used as the base material of microencapsulation in nanorobotic agents. Compared with synthetic ones, the natural ones generally exhibit good biocompatibility. Among the types of encapsulation materials, alginate and poly lactic-coglycolic acid (PLGA) are the two most widely used polymers. Alginate is an anionic polysaccharide which is widely distributed in the cell walls of brown algae. When the alginate molecules are in contact with divalent cations, the molecules can link together. After that, the NPs in alginate solution are locked into the alginate hydrogel capsules.

PLGA is a type of copolymer that, due to its biodegradability and biocompatibility, has been widely used in therapeutic agents approved by the Food and Drug Administration (FDA). The first navigable therapeutic nanorobotic agents were made of drug molecules and MNPs encased in a PLGA matrix. During the encapsulation process, usually methylene chloride is used to dissolve PLGA, followed by the method of interfacial precipitation used for the encapsulation.

Many techniques have been investigated and applied as encapsulation approaches, including mainly electrostatic spray, air jet, and microfluidic channel.[44] The electrostatic spray has significant merits such as the possibility of sterile operation, small size particles, and ease of manipulation. Electrostatic force, also known as the Coulomb interaction or Coulomb force, is defined as the attraction or repulsion between two charged objects. Figure 17.3 shows a simple encapsulation setup by using the electrostatic force. When a high voltage is applied, the charge can be distributed on the meniscus surface of a liquid during droplets forming, which enables them to drop down from a needle tip with the help of electrostatic force.[45] Therefore, the size of droplets can be decreased. A higher voltage means a higher electrostatic force and a smaller particle size. Using this method, the diameter of the particles can be scaled down to 200 μm. However, the liquid droplets can easily become a continuous flow jet if the surface tension is close to 0.[46] In such a case, the droplets may break into small and uniform drops. As such, the voltage needs to be maintained at a proper scale depending on the distance of the needle tip and the liquid surface below.

The coaxial air jet around the needle can also generate a force that enables the droplets to drop from the needle tip. This method can produce microparticles ranging from a few micrometres to several millimetres, which is the same as when using the electrostatic force. By increasing the airflow velocity, the particles' diameter decreases but the size dispersion increases. Figure 17.4 shows the image of the microparticles formed after MNPs were encapsulated in alginate by air-jet technique.

Emulsification-based techniques are the most commonly used method for NPs encapsulation.[6,47] After mixing two kinds of immiscible fluids, polymer-containing droplets can be formed and broken into smaller ones if shear stress is applied on the fluids. Vibration and stirring are often used to cause shear stress. The limitation of these two methods is that the power is not evenly distributed in fluids, resulting in a large size distribution and difficulty in morphology control. Recently, promising applications of microfluidic channel-based emulsification have been demonstrated in the encapsulation of biosamples and NPs. The techniques are known as shear-focusing methods, which relies on the shear effect under externally applied shear forces. Many structures of microfluidic channel have been proposed for particles encapsulation. One of the simplest designs is to use two liquid phases: a dispersal phase and continuous phase.[42,48,49] After injection into a microfluidic chip, the liquid of the dispersal phase, which consists of dissolved polymer and encapsulated materials such as MNPs, is divided into small droplets at the flow-focusing junction. An oil-in-water (O/W) emulsion is formed due to the shear force induced by the liquid flow of the continuous phase, as shown in Figure 17.5. A water-in-oil (W/O) emulsion can also be made by interchanging the two phases. The last step is to create microspheres through solvent extraction and/or evaporation from the droplets. The emulsion, after the first junction, can be sheared again if a second junction is designed in the chip. The oil-in-water-in-oil (O/W/O) or water-in-oil-in-water (W/O/W) structures have been widely used in many research areas such as bacteriology and pharmacology.

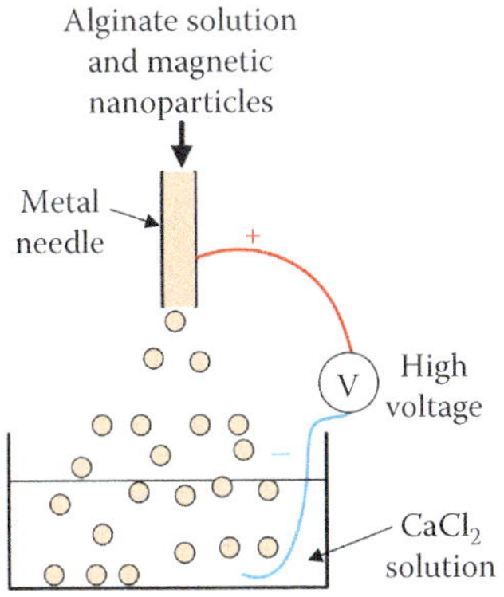

Alginate solution and magnetic nanoparticles
Metal needle
High voltage
$CaCl_2$ solution

FIGURE 17.3 Micro encapsulation with electrostatic force technique in alginate setup. (Courtesy of Dr Li.)

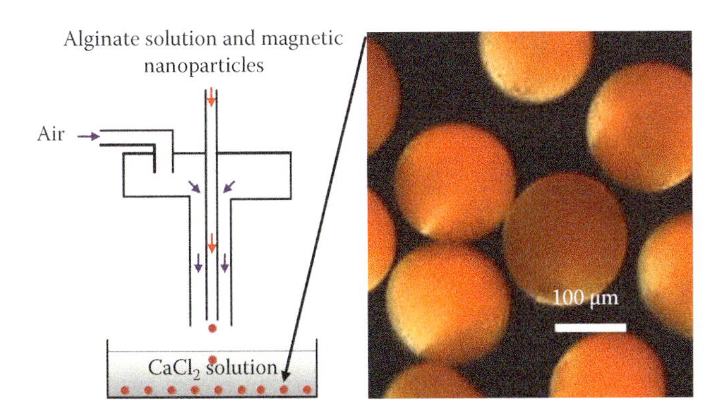

FIGURE 17.4 MNPs encapsulated in alginate by air-jet technique. (Courtesy of Dr Li.)

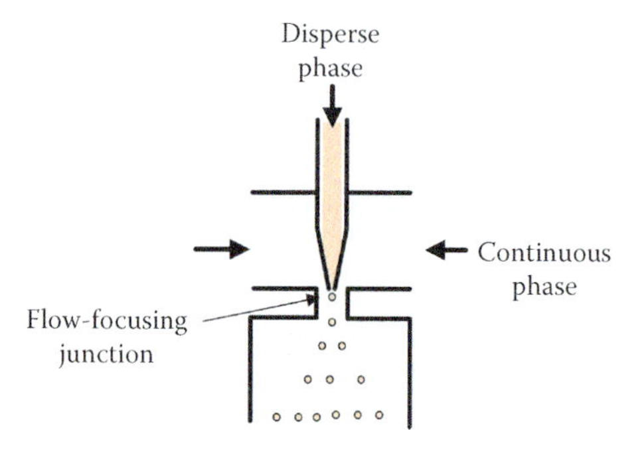

FIGURE 17.5 Microfluidic flow-focusing encapsulation technique. (Courtesy of Dr Li.)

Different core environments in microparticles greatly increase the diversity of the encapsulation materials.

17.10 DIAGNOSTICS

Cancer microenvironment plays an important role in treatment options. Depending on the type of tumour, rigidity, metastasis level, and hypoxic regions, the drug dosage and treatment choice can vary, leading to more or less aggressive chemotherapy. As NPs are routinely used in medical research and clinical trials, their role is also rapidly evolving in imaging and early cancer diagnosis detection. When embedded in nanorobotic agents NPs would implement image-guided drug carriers that not only deliver the therapeutics, but also provide vital information about the pathological site. Current imaging modalities such as MRI, computerized tomography (CT), positron emission tomography (PET), ultrasound (US), photo-acoustic imaging (PAI), and various optical imaging techniques use NPs with appropriate coatings as contrast agents. Depending on the type of examination required, each imaging modality can provide the necessary information for further diagnosis about the stage of the disease. It is important to meet pharmacokinetic demands when guiding image-therapeutic NPs. Due to their small size, if they extravasate out of the vasculature, they will leave compartments that cause unspecific background noise for imaging.[50] For therapeutics, it is sufficient to penetrate the interstitial space of the tumour; however, for diagnosis, binding with tumour cells and reaching out to hypoxic regions may also prove to be important. For that matter, magnetotactic bacteria acting as nanorobotic agents could adequately address this issue.

17.11 MAGNETOTACTIC BACTERIA

Scientists working in the field of nanotechnology are in search of autonomous micro/nanorobots that are equipped with a steering control system, sensors, and some level of autonomy for various applications. Presently due to technological constraints these robots cannot be made artificially at the size required to reach deep regions of tumours.

The use of magnetotactic bacteria (MTB) strain MC-1[51] as micro/nanorobots with a directional autonomous propulsion system has been presented and proposed for the first time by our group. In the robotic system relying on MTB, the flagellum provides the autonomous propulsion system. The orientation of the bacterium is controlled by applying a magnetic field on a chain of nanoscale particles known as magnetosomes (Figure 17.6). These particles act as a navigational compass embedded in each bacterial cell and allow very accurate control over the displacement of the bacterium, in a fashion similar to a steering control system. The proposed approach bypasses current technological constraints, not only by integrating and exploiting the molecular motor of the bacteria as a means of autonomous propulsion, but more importantly, it proposes and validates a method to control their swimming path using computer software. Recent progress in this field has been stated in several papers,[52–56] with experimental results showing the feasibility of such a method and its potential in many applications. These include but are not limited to the fast detection of pathogenic bacteria, biosensors, and biocarriers in microfluidic systems. The MTB could also be used in a bacterial micro-factory concept or as an autonomous bio-sensing micro-robots in aqueous environments. Among all potential applications, the most considerable is the exploitation of MTB for targeted drug transport in cancer therapy. Recent works[52] show that these bacteria can be controlled by magnetic field and target active cancer cells in tumours.

MC-1 strain is a microaerophilic bacterium found in marine environments. This organism uses the Earth's magnetic field to navigate towards a depth corresponding to its optimal oxygen concentration for growth. This phenomenon, called magneto-aerotaxis, reduces tridimensional aerotaxis searches to a one-dimensional search in which bacterial cells swim to reach optimum oxygen concentrations. Magnetotaxis combined with aerotaxis requires two bacterial organelles: (1) magnetosomes which are membrane-based nanometer-scaled ferromagnetic particles that orient the bacterial cells along the Earth's magnetic field, and are typically oriented perpendicular to a line drawn between the bases of two flagella bundles; and (2) two flagella bundles that propel each bacterium like a biological motor along the magnetic field.[57] The latter is influenced by the level of oxygen when the magnetic field strength is sufficiently low.

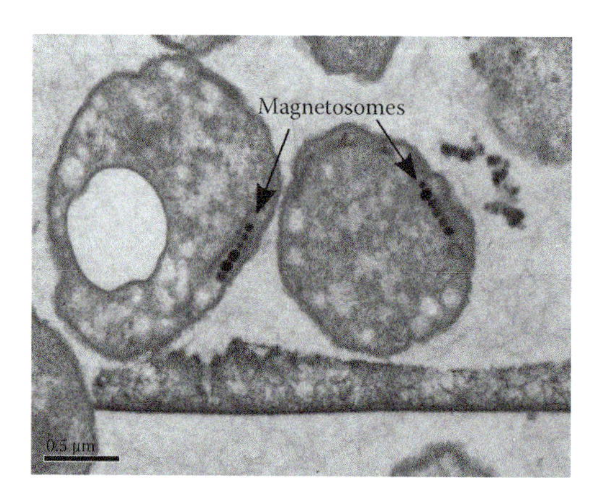

FIGURE 17.6 Magnetosomes in MC-1 bacteria. (Courtesy of Dr Mohammadi.)

Based on previous investigations and similar to observations made with other magnetotactic bacteria, the swimming direction of MC-1 cells is influenced by magnetotaxis combined with aerotaxis and controlled by a two-component aerotactic sensory mechanism. Depending on the oxygen concentration, MC-1 cells swim parallel or antiparallel to the Earth's magnetic field. The direction in which the bacteria navigate is determined by the sense of flagellar rotation which is controlled by the oxygen concentration.[58] Under oxic conditions the bacterium swims parallel to Earth's magnetic field in order to reach a zone with optimum oxygen concentration. The anoxic zone reverses the sense of flagellar rotation, causing the bacterium to swim antiparallel to the Earth's magnetic field in order to reach an optimum oxygen zone which in tumour targeting corresponds to regions of active and fast duplicating cancer cells. In a magnetic field sufficiently higher than the Earth's magnetic field, the oxygen concentration does not influence the swimming direction. Under this condition, the bacterium lines up and swims parallel to B (magnetic field direction) and when the magnetic field is reversed, the bacterium changes its direction, performs a U-turn, lines up once again in the magnetic field and swims parallel to B due to the directional torque induced on the chain of MNPs synthesized in the cell. However, our recent experimental results show that the oxygen concentration still has an influence on the behaviour of bacteria subjected to a high magnetic field. In a magnetic field greater than the Earth's magnetic field, the oxygen concentration controls the swimming speed of the bacteria.[53] In anoxic conditions, the swimming speed is significantly reduced compared with the speed observed in oxic zones in which the bacteria reach their maximum velocity. Therefore, when the MC-1 cells are subjected to a magnetic field slightly higher that 0.5 Gauss (the Earth's magnetic field) the directional motions of these cells becomes exclusively influenced by magnetotaxis and completely controlled by electronics and computers with the swimming speed being influenced by the oxygen concentration.

How MTB exactly produce their magnetosomes is still quite unknown. Several research laboratories are presently working on these MTB. Their findings helped us acquire some knowledge on the genetics and cell biology of magnetosome formation which we can summarize here.[59–61] Most magnetosome-specific proteins are encoded by genes organized within a discrete genomic region termed the magnetosome island (MAI).[59] The magnetosome-related genes are very well conserved across different classes of bacteria. Some of the genes involved in the magnetosomes synthesis have been identified[61,62]; these genes have been implicated in the formation of the magnetosome membrane, organization of the magnetosome chain, and the biomineralization of magnetite. Most of these genes involved in magnetosome formation are generally called mam (for magnetosome membrane).

The magnetosomes are made of nanometer-sized magnetic crystals. These particles are very useful in biotechnology and can be used in enzyme immobilization, formation of magnetic antibodies for various immunofluorescence, gene targeting or drug delivery, and magnetic resonance imaging.[63,64] The shape and composition of magnetosomes are species- and strain-specific. This feature indicates that magnetosome morphology is controlled by the genome of the bacterium. The magnetosomes morphology is an important factor for different applications in bacterial robotics. Identifying all the genes controlling magnetosome formation would be an important asset for controlling the synthesis of magnetosomes and their morphology.

Identifying the minimum of genes required for magnetosome synthesis would also make it possible to create a gene cassette, allowing, through genetic engineering, the creation of new generations of magnetotactic bacterial robots exploitable in various biomedical applications. As an example, cancer therapy based on facultative or obligate anaerobic bacteria acting as 'Trojan horses' invading tumour tissues would be facilitated by engineering bacteria to make them become magnetic.[65]

17.12 CONCLUSION

MNPs play an important and critical role in the development and synthesis of nanorobotic agents dedicated to medical applications, especially for the delivery of therapeutics to spatially defined regions in the body. The characteristics of these NPs can be tuned to implement functionalities that otherwise would not be possible to implement at such small scales. The type of material, the geometry including the anisotropy constant, the overall size, the number of NPs, and the configuration of the cluster are just a few examples of factors that must be considered when implementing MNPs in nanorobotic agents dedicated to specific tasks.

REFERENCES

1. Link, S. & El-Sayed, M. A. Shape and size dependence of radiative, non-radiative and photothermal properties of gold nanocrystals. *Int. Rev. Phys. Chem.* **19,** 409–453 (2000).
2. Mørup, S. & Hansen, M. F. in *Handbook of Magnetism and Advanced Magnetic Materials* (John Wiley & Sons, Ltd, Chichester, England, 2007). Editors: Helmut Kronmueller and Stuart Parkin. doi:10.1002/9780470022184.hmm409
3. Revia, R. A. & Zhang, M. Magnetite nanoparticles for cancer diagnosis, treatment, and treatment monitoring: Recent advances. *Mater. Today (Kidlington).* **19,** 157–168 (2016).
4. Tuček, J. et al. Air-stable superparamagnetic metal nanoparticles entrapped in graphene oxide matrix. *Nat. Commun.* **7,** 12879 (2016).
5. Torchilin, V. P. Recent advances with liposomes as pharmaceutical carriers. *Nat. Rev. Drug Discov.* **4,** 145–160 (2005).
6. Pouponneau, P., Leroux, J.-C., Soulez, G., Gaboury, L. & Martel, S. Co-encapsulation of magnetic nanoparticles and doxorubicin into biodegradable microcarriers for deep tissue targeting by vascular MRI navigation. *Biomaterials* **32,** 3481–3486 (2011).
7. Mathieu, J.-B. & Martel, S. Steering of aggregating magnetic microparticles using propulsion gradients coils in an MRI Scanner. *Magn. Reson. Med.* **63,** 1336–1345 (2010).
8. Eberbeck, D., Wiekhorst, F., Steinhoff, U. & Trahms, L. Aggregation behaviour of magnetic nanoparticle suspensions investigated by magnetorelaxometry. *J. Phys. Condens. Matter* **18,** S2829 (2006).

9. Amirfazli, A. Nanomedicine: Magnetic nanoparticles hit the target. *Nat. Nano.* **2,** 467–468 (2007).
10. Mahoney, A. W. & Abbott, J. J. Generating rotating magnetic fields with a single permanent magnet for propulsion of untethered magnetic devices in a lumen. *IEEE Trans. Robot.* **30,** 411–420 (2014).
11. Shapiro, B., Dormer, K. & Rutel, I. B. A two-magnet system to push therapeutic nanoparticles. *AIP Conf. Proc.* **1311,** 77–88 (2010).
12. Kummer, M. P. et al. OctoMag: An electromagnetic system for 5-dof wireless micromanipulation. *IEEE Trans Robot.* **26,** 1006–1017 (2010).
13. Jeong, S. et al. Penetration of an artificial arterial thromboembolism in a live animal using an intravascular therapeutic microrobot system. *Med. Eng. Phys.* **38,** 403–410 (2016).
14. Mahmood, A., Dadkhah, M., Kim, M. O. & Yoon, J. A novel design of an mpi-based guidance system for simultaneous actuation and monitoring of magnetic nanoparticles. *IEEE Trans. Magn.* **51,** 1–5 (2015).
15. Martel, S. et al. Automatic navigation of an untethered device in the artery of a living animal using a conventional clinical magnetic resonance imaging system. *Appl. Phys. Lett.* **90,** 114105 (2007).
16. Belharet, K., Folio, D. & Ferreira, A. MRI-based microrobotic system for the propulsion and navigation of ferromagnetic microcapsules. *Minim. Invasive Ther. Allied Technol.* **19,** 157–169 (2010).
17. Latulippe, M. & Martel, S. Dipole field navigation: Theory and proof of concept. *IEEE Trans. Robot.* **31,** 1353–1363 (2015).
18. Hahn, E. L. Spin echoes. *Phys. Rev.* **80,** 580–594 (1950).
19. Griffiths, D. J. *Introduction to Electrodynamics* (Addison-Wesley, Prentice Hall, Upper Saddle River, NJ, USA, 1998).
20. Brown, R. W., Cheng, Y.-C. N., Haacke, E. M., Thompson, M. R. & Venkatesan, R. *Magnetic Resonance Imaging. Magnetic Resonance Imaging* (John Wiley & Sons Ltd, Hoboken, NJ, USA, 2014). doi:10.1002/9781118633953.ch1
21. Shapiro, E. M. et al. MRI detection of single particles for cellular imaging. *Proc. Natl. Acad. Sci. U. S. A.* **101,** 10901–10906 (2004).
22. Mikawa, M. et al. Paramagnetic water-soluble metallofullerenes having the highest relaxivity for MRI contrast agents. *Bioconjug. Chem.* **12,** 510–514 (2001).
23. Uecker, M. et al. Real-time MRI at a resolution of 20 ms. *NMR Biomed.* **23,** 986–994 (2010).
24. Gleich, B. & Weizenecker, J. Tomographic imaging using the nonlinear response of magnetic particles. *Nature* **435,** 1214–1217 (2005).
25. Kim, J. et al. Multifunctional uniform nanoparticles composed of a magnetite nanocrystal core and a mesoporous silica shell for magnetic resonance and fluorescence imaging and for drug delivery. *Angew. Chem.* **120,** 8566–8569 (2008).
26. Borgert, J. et al. Fundamentals and applications of magnetic particle imaging. *J. Cardiovasc. Comput. Tomogr.* **6,** 149–153 (2012).
27. Goodwill, P. W. et al. X-space MPI: Magnetic nanoparticles for safe medical imaging. *Adv. Mater.* **24,** 3870–3877 (2012).
28. Rahmer, J., Weizenecker, J., Gleich, B. & Borgert, J. Signal encoding in magnetic particle imaging: Properties of the system function. *BMC Med. Imaging* **9,** 4 (2009).
29. Krishnan, K. M. Biomedical nanomagnetics: A spin through possibilities in imaging, diagnostics, and therapy. *IEEE Trans. Magn.* **46,** 2523–2558 (2010).
30. Goodwill, P., Krishnan, K. & Conolly, S. in *Magnetic Nanoparticles* 523–540 (CRC Press, Boca Raton, FL, 2012). Editor: Nguyen TK Thanh.
31. Ferguson, R. M., Minard, K. R., Khandhar, A. P. & Krishnan, K. M. Optimizing magnetite nanoparticles for mass sensitivity in magnetic particle imaging. *Med. Phys.* **38,** 1619–1626 (2011).
32. Hervault, A. & Thanh, N. T. K. Magnetic nanoparticle-based therapeutic agents for thermo-chemotherapy treatment of cancer. *Nanoscale* **6,** 11553–11573 (2014).
33. Hervault, A. et al. Doxorubicin loaded dual pH- and thermo-responsive magnetic nanocarrier for combined magnetic hyperthermia and targeted controlled drug delivery applications. *Nanoscale* **8,** 12152–12161 (2016).
34. Tabatabaei, S. N., Girouard, H., Carret, A.-S. & Martel, S. Remote control of the permeability of the blood–brain barrier by magnetic heating of nanoparticles: A proof of concept for brain drug delivery. *J. Control. Release* **206,** 49–57 (2015).
35. Cavaliere, R. et al. Selective heat sensitivity of cancer cells. Biochemical and clinical studies. *Cancer* **20,** 1351–1381 (1967).
36. Hildebrandt, B. et al. The cellular and molecular basis of hyperthermia. *Crit. Rev. Oncol. Hematol.* **43,** 33–56 (2002).
37. Dieing, A. et al. The effect of induced hyperthermia on the immune system. *Prog. Brain Res.* **162,** 137–152 (2007).
38. Wust, P. et al. Hyperthermia in combined treatment of cancer. *Lancet. Oncol.* **3,** 487–497 (2002).
39. Purushotham, S. et al. Thermoresponsive core-shell magnetic nanoparticles for combined modalities of cancer therapy. *Nanotechnology* **20,** 305101 (2009).
40. Habib, A. H. et al. The role of eddy currents and nanoparticle size on AC magnetic field–induced reflow in solder/magnetic nanocomposites. *J. Appl. Phys.* **111,** 07B305 (2012).
41. Pardridge, W. M. Blood–brain barrier drug targeting: The future of brain drug development. *Mol. Interv.* **3,** 51,90–105 (2003).
42. Bokharaei, M. et al. Production of monodispersed magnetic polymeric microspheres in a microfluidic chip and 3D simulation. *Microfluid. Nanofluidics* **20,** 6 (2016).
43. S. S. Bansode, S. K. Banarjee, D. D. Gaikwad, S. L. Jadhav, R. M. T. Microencapsulation: A review. *Int. J. Pharm. Sci. Rev. Res.* **1,** (2010).
44. Zhang, W. & He, X. Encapsulation of living cells in small (approximately 100 microm) alginate microcapsules by electrostatic spraying: A parametric study. *J. Biomech. Eng.* **131,** 74515 (2009).
45. Berkland, C., Pack, D. W. & Kim, K. K. Controlling surface nano-structure using flow-limited field-injection electrostatic spraying (FFESS) of poly(D,L-lactide-co-glycolide). *Biomaterials* **25,** 5649–5658 (2004).
46. Kühtreiber, W., Lanza, R. P. & Chick, W. L. *Cell Encapsulation Technology and Therapeutics.* (Birkhäuser Basel, Boston, USA, 1999).
47. Pouponneau, P., Leroux, J.-C. & Martel, S. Magnetic nanoparticles encapsulated into biodegradable microparticles steered with an upgraded magnetic resonance imaging system for tumour chemoembolization. *Biomaterials* **30,** 6327–6332 (2009).
48. Teh, S.-Y., Khnouf, R., Fan, H. & Lee, A. P. Stable, biocompatible lipid vesicle generation by solvent extraction-based droplet microfluidics. *Biomicrofluidics* **5,** 44112–44113 (2011).
49. Tan, Y.-C., Hettiarachchi, K., Siu, M., Pan, Y.-R. & Lee, A. P. Controlled microfluidic encapsulation of cells, proteins, and microbeads in lipid vesicles. *J. Am. Chem. Soc.* **128,** 5656–5658 (2006).

50. Kiessling, F., Mertens, M. E., Grimm, J. & Lammers, T. Nanoparticles for imaging: Top or flop? *Radiology* **273,** 10–28 (2014).
51. Meldrum, F. C., Mann, S., Heywood, B. R., Frankel, R. B. & Bazylinski, D. A. Electron microscopy study of magnetosomes in a cultured coccoid magnetotactic bacterium. *Proc. R. Soc. London. Ser. B Biol. Sci.* **251,** 231 LP-236 (1993).
52. Felfoul, O. et al. Magneto-aerotactic bacteria deliver drug-containing nanoliposomes to tumour hypoxic regions. *Nat. Nanotechnol.* (2016). doi:10.1038/nnano.2016.137
53. Martel, S. & Mohammadi, M. Switching between magnetotactic and aerotactic displacement controls to enhance the efficacy of MC-1 magneto-aerotactic bacteria as cancer-fighting nanorobots. *Micromachines* **7,** (2016).
54. Taherkhani, S., Mohammadi, M., Daoud, J., Martel, S. & Tabrizian, M. Covalent binding of nanoliposomes to the surface of magnetotactic bacteria for the synthesis of self propelled therapeutic agents. *ACS Nano* 5049–5060 (2014). doi:10.1021/nn5011304
55. de Lanauze, D., Felfoul, O., Turcot, J.-P., Mohammadi, M. & Martel, S. Three-dimensional remote aggregation and steering of magnetotactic bacteria microrobots for drug delivery applications. *Int. J. Rob. Res.* **33,** 359–374 (2014).
56. Martel, S., Mohammadi, M., Felfoul, O., Zhao, L. & Pouponneau, P. Flagellated magnetotactic bacteria as controlled MRI-trackable propulsion and steering systems for medical nanorobots operating in the human microvasculature. *Int. J. Rob. Res.* **28,** (2009).
57. Frankel, R. B., Bazylinski, D. A., Johnson, M. S. & Taylor, B. L. Magneto-aerotaxis in marine coccoid bacteria. *Biophys. J.* **73,** 994–1000 (1997).
58. Frankel, R. B., Williams, T. J. & Bazylinski, D. A. Magnetoreception and Magnetosomes in Bacteria. in *Magneto-Aerotaxis.* (ed. Schüler, D.), 1–24 (Springer, Berlin, 2007). doi:10.1007/7171_2006_036
59. Komeili, A. Molecular mechanisms of magnetosome formation. *Annu. Rev. Biochem.* **76,** 351–366 (2007).
60. Jogler, C. & Schuler, D. Genomics, genetics, and cell biology of magnetosome formation. *Annu. Rev. Microbiol.* **63,** 501–521 (2009).
61. Uebe, R. & Schuler, D. Magnetosome biogenesis in magnetotactic bacteria. *Nat Rev Micro* **14,** 621–637 (2016).
62. Schuler, D. Genetics and cell biology of magnetosome formation in magnetotactic bacteria. *FEMS Microbiol. Rev.* **32,** 654–672 (2008).
63. Arakaki, A., Nakazawa, H., Nemoto, M., Mori, T. & Matsunaga, T. Formation of magnetite by bacteria and its application. *J. R. Soc. Interface* **5,** 977 LP-999 (2008).
64. Yan, L. et al. Magnetotactic bacteria, magnetosomes and their application. *Microbiol. Res.* **167,** 507–519 (2012).
65. Ben-Jacob, E. Engineering Trojan-horse bacteria to fight cancer. *Blood* **122,** 619–620 (2013).

Sylvain Martel (E-mail: sylvain.martel@polymtl.ca), Fellow of the Canadian Academy of Engineering, is Director of the NanoRobotics Laboratory at the Ecole Polytechnique Montréal, Campus of the University of Montréal. He pioneered several biomedical technologies including platforms for remote surgeries and cardiac mapping systems when at McGill University, and new types of brain implants for decoding neuronal activities in the motor cortex when at the Massachusetts Institute of Technology (MIT). Presently, he is leading an interdisciplinary team involved in the development of navigable therapeutic agents and interventional platforms for cancer therapy. Prof. Martel is internationally recognized as the pioneer in a new paradigm in drug delivery known as direct targeting where therapeutics are navigated in the vascular network towards solid tumours using the most direct route.

Mahmood Mohammadi has been a scientist in NanoRobotics Laboratory at the Ecole Polytechnique Montréal since 2006. He earned a PhD in microbiology and molecular genetics from Laval University (Quebec, Canada) in 2003. His research activities are focused to the development of new hybrid engineered micro- or nanorobotics systems, i.e., systems made of synthetic and biological components. More specifically, he has concentrated his efforts on the integration of special bacteria as bio-actuators and controllable micro- or nano-self-propulsion systems for critical applications such as tumour targeting for enhanced therapeutic effect in cancer therapy.

Charles Tremblay earned his bachelor in engineering physics from the Ecole Polytechnique Montreal in 2003. Since 2005, he has worked there as a research associate with the Laboratoire de NanoRobotique where he is currently a PhD student. His work is devoted to the integration of multiscale bio/physics technicalities in interventional *in vivo* robotics. He also supports the design of robotic medical theaters for actuated tools and drugs.

Ning Li has been a PhD candidate from the Ecole Polytechnique de Montréal, Canada since 2014. He has earned a BS degree in mechanical engineering from the Naval Aeronautical Engineering Academy, Qingdao, China, in 2009, and an MS degree in mechanical engineering in 2012 from Nanjing University of Aeronautics and Astronautics, Nanjing, China. His work has been focused on the manipulation of micro/nano-scale entities by acoustic, electrical, and electromagnetic methods. His current research aims at developing methods and MR-safe devices which can be used to facilitate magnetic resonance navigation of untethered magnetic microparticles for targeted intra-arterial therapies

Kévin Gagné has been a PhD student in Biomedical Engineering at the Nanorobotics Laboratory of the Ecole Polytechnique de Montréal since 2016. He got his bidisciplinary degree in Mathematics and Physics at the University of Montreal, Canada, in 2015. He started his Master degree at the Nanorobotics Laboratory of the Ecole Polytechnique de Montréal where he now continues his project. His research interests are mainly related to magnetic resonance imaging in nonuniform magnetic environments, more precisely in the context of the detection of micro and nanoparticles using magnetic-based modalities. His thesis aims to use, during treatment, both magnetic resonance imaging and magnetic particle imaging to track recently used magneto-aerotactic bacteria's strain MC-a for cancer therapy.

Maxime Latulippe earned both a BE degree in Computer Engineering in 2011, and an MS degree in Computer Science in 2013 from Laval University in Quebec, Canada. He is presently a PhD candidate with the NanoRobotics Laboratory in the Department of Computer and Software Engineering and the Institute of Biomedical Engineering at the Ecole Polytechnique Montréal, Canada. His current research interests are in medical robotics and in the magnetic navigation of untethered agents in the human body, especially for the targeted delivery of drugs.

Maryam S. Tabatabaei has been a biomedical engineer PhD student at Polytechnique Montreal, Canada, since 2015. She earned her BSc degree in physics from McGill University, Canada, in 2013. Following an internship in the department of pharmacology, faculty of medicine, University of Montreal, Canada, she began her MSc degree in biomedical engineering at Polytechnique Montreal in 2014, then fast-tracking to PhD a year after. Her field of research is to conduct multidisciplinary studies between pharmacology and biomedical engineering, focusing on hyperthermia of magnetic nanoparticles and drug delivery to the brain. The objective of her thesis is to overcome the challenge of the blood–brain barrier with the aid of magnetotactic bacteria as thermal ablation agent and drug carriers under hyperthermia to increase the treatment efficiency of the brain tumours, investigated under *in vivo* conditions.

18 Smart Nanoparticles and the Effects in Magnetic Hyperthermia *In Vivo*

Ingrid Hilger

CONTENTS

18.1 INTRODUCTION

Hyperthermia (also called thermal therapy or thermotherapy) is a type of cancer treatment in which body tissue is exposed to temperatures greater than the physiological ones. Several decades ago, it was discovered that high temperatures could damage and kill cancer cells. In consequence, several clinical trials have been performed since focusing on the treatment of many types of cancer, including sarcoma, melanoma, cancers of the head and neck, brain, lung, oesophagus, breast, bladder, rectum, appendix, cervix and peritoneal lining (mesothelioma).[1–11] Hereto, heat has been applied to the human body via regional or whole-body heat delivery systems using ultrasound, radiofrequency, microwaves, infrared radiation, or hot water. Many of these studies, but not all, have shown a significant reduction in tumour size when hyperthermia is combined with other treatments.[1] The main challenges for traditional hyperthermia are the occurrence of unavoidable heating of healthy tissues and the limited penetration of heat into body tissues by microwave, laser and ultrasound energy. Not all of these studies have shown increased survival in patients receiving the combined treatments.[2] One of the main reasons is inadequate positioning of the temperature spot right to the tumour area and insufficient methods to monitor the temperature around the tumour area.

In this context, the means of exploitation of iron oxide (IO) nanoparticles (NPs) as a source to generate heat have been considered and intensively studied. Hereto, different NP sizes, shapes (e.g. spheroids, rods), as well as doping them with further metals of unknown biocompatibility and biodegradability in the human body have been assessed (e.g. Refs. 3–8). IONPs are smart vehicles to provide heating from inside the body. Provided a selective deposition in the area of interest (tumour), they could selectively heat the target tissue and exhibit unlimited tissue penetration of heating when exposing the target organ to an alternating magnetic field. The procedure of inducing heating with magnetic NPs (MNPs) has been termed as 'magnetic hyperthermia' or 'magnetic thermoablation' (Figure 18.1). The procedure is minimally invasive, since it only requires the application of the magnetic material to the tumour region.

18.2 NP SPECIFICATIONS FOR MAGNETIC HYPERTHERMIA

Not every IONP formulation is suitable for hyperthermia. Among the different physicochemical requirements related to the NPs, the heating potential is very important with respect to the dosages that have to be applied to the tumour region in order to be able to inactivate target tumour cells. In this context, the heating potential is defined by the amount of heating delivered per unit mass and time as a consequence of the exposure of the NPs to an alternating magnetic field.[9]

Depending on the morphological features of the magnetic material (NP size, shape and microstructure), different mechanisms are responsible for the delivery of heating. With regard to multidomain MNPs (sizes larger than approximately 100 nm depending on the magnetic field parameters) heating is delivered by displacements of the domain wall (hysteresis losses). NPs with small diameters (e.g. lower than 100 nm)

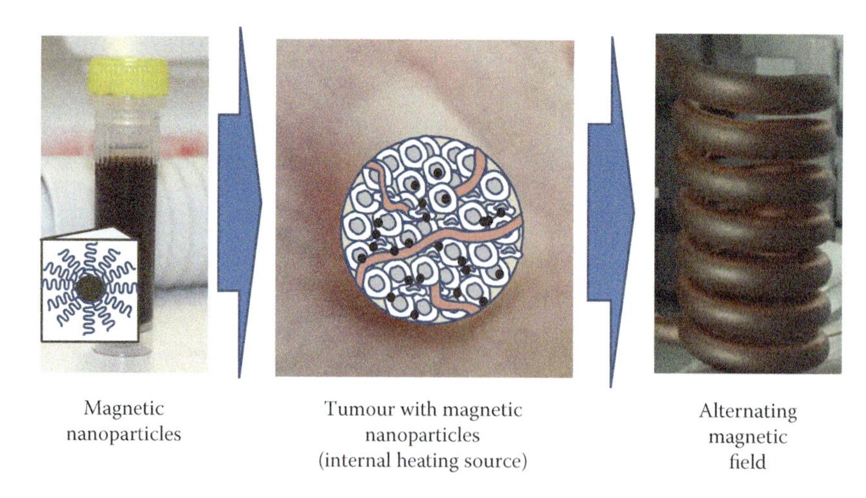

FIGURE 18.1 Principle of magnetic hyperthermia.

are 'superparamagnetic'. The critical upper size limit for the observation of superparamagnetism is approximately 25 nm for magnetite and 30 nm for maghemite.[10] They exhibit single magnetic domains and induce heating as result of loss processes during the reorientation of the magnetization in the magnetic field or frictional losses in the case the NP is able to rotate in the surrounding medium.[11] Due to their suspension features and FDA-approval for MRI applications, superparamagnetic NPs have been increasingly investigated in the last decade for hyperthermic purposes.[12]

Improved aqueous precipitation procedures for magnetic IONP synthesis have been developed in order to further unveil NP features with optimal heating potential. One procedure focused on the cyclic growth of MNPs without nucleation of new particle cores during precipitation. Further, nanoparticle formulations are made up of magnetic multicore NPs, which consist of single cores of approximately 10 nm forming dense clusters in the size range from 40 to 80 nm. The very high heating potential makes these multicore particles highly suitable for hyperthermia application. Through *in vivo* experiments, therapeutically suitable temperatures were reached after 20 s of heating for a particle concentration in the tumour of 1% (w/w) and field parameters of H = 24 kA/m and f = 410 kHz.[13] Despite this, the literature is rich in studies related to the determination of the heating potential of iron oxide MNPs, but comparability is limited due to the different magnetic field parameters used. For this reason, the concept intrinsic loss power (ILP) was introduced.[14] It is a physical magnitude correction to compare heating efficiencies obtained under different magnetic field parameters.[15]

Beyond the determined values of the heating potential *in vitro*, it has been shown that the specific absorption rate (SAR) is distinctly affected by the degree of MNP immobilization. For example, when MNPs are injected in the tumour they might adhere to connective tissue and cells. In this context, the highest SAR values were observed with MNP suspensions in water, where the SAR was strongly reduced after the immobilization (e.g. in polyvinyl alcohol). This aspect should be considered when determining the SAR of MNP for magnetic hyperthermia, since the structural properties of MNPs determine their heating behavior.[9] All in all, to get the optimal heating potential of MNPs for magnetic hyperthermia, we have to consider their size, shape, microstructure and their immobilization degree in tumour microstructures on one hand, and the frequency and the amplitude of the alternating magnetic field they are being exposed to on the other.

18.3 MAGNETIC FIELD APPLICATORS FOR MAGNETIC HYPERTHERMIA

In principle, magnetic fields for hyperthermia are produced by so-called magnetic field applicators. For preclinical research, they are basically made up of a coil or an electromagnet.[16] The configuration of the magnetic field inductor may be specified according to the magnetic field parameters required in relation to the specific iron oxide formulation used. One particular challenge is related to the occurrence of eddy currents, since the magnetic field strength and frequency cannot be increased unlimitedly. Eddy currents might affect temperature sensitive organs. The tolerated maximum value of eddy current heating is known to be of approximately 25 mW/mL tissue. Also, the heart and peripheral nerves might be affected due to the presence of magnetic fields. For example, muscle tissue might respond to defined frequencies of an alternating magnetic field. In particular, the threshold amplitude for cardiac muscle stimulation is higher than that of muscle stimulation and frequencies lower than 5.5 kHz might stimulate the heart and peripheral nerves.[17,18] To avoid excessive heating of the human body, other researchers have postulated that the product between frequency and amplitude of the magnetic field should not exceed 4.85×10^8 A/(m × s). This critical value was very well tolerated by a volunteer exposed to a heating power of 200 W.[19,20] Furthermore, in the clinical situation, 18 kA/m and 100 kHz were well tolerated.[21] It was postulated that 4.7 kA/m[22] is the threshold of the B field value at which stimulation of nerves and muscles does not occur.

18.4 IMPACT OF HEATING ON TARGET TUMOUR CELLS

Intensive studies have been performed to elucidate how heating can kill tumour cells (Figure 18.2). When tumour cells are exposed to temperatures above 45°C, increased motion of the molecules will occur (Brownian motion). Heating can also raise increased metabolism and transition of cellular structures, as well as DNA, RNA and proteins to a disordered state. Accordingly, protein aggregation, insolubilization, increased fluidity of cellular membranes, disruption in ion permeability through cellular membranes, inhibition of amino acid transport, morphological changes as consequence of damaged cytoskeletal system, and inhibition of DNA repair are expected to occur.[23] These are factors which can increase the sensitivity of tumour cells to radio- and chemotherapy. Moreover, transcriptional activation of the heat shock protein 70 has been proposed to be used as molecular thermometer to sense cells' response to magnetic hyperthermia.[24] Also, the alteration of the tumour microenvironment,[25,26] blood perfusion,[27] and immunological function were reported (e.g. see Ref. [28]). When combining hyperthermia with radiotherapy, one can foster hypoxic regions in the tumour by heating, which are normally insensitive to radiotherapy. Similarly, hyperthermia can enhance the action of alkylating chemotherapeutics.[29,30] Cancer cells can be more sensitive to hyperthermia than normal cells as a result of the specific environment[1] and this effect is not related to the cell per se as was postulated in the 1970s.[31] The apparent higher sensitivity of tumour cells against heat is due to physiological aspects. Namely, since the architecture of the tumour vasculature is chaotic, regions with hypoxia and low pH levels are found in tumours in contrast to normal tissues. Therefore, cells in low perfused tumour areas are particularly more sensitive to hyperthermia.[32]

In particular, temperatures above 50°C can induce tumour cell necrosis and coagulation.[23] The critical temperature for inducing cell death via heating as sole therapeutic modality is between 51°C and 55°C for an exposure time of 4 min.[33] The corresponding deposited heating dose would be between 47 and 61°C × min.[34]

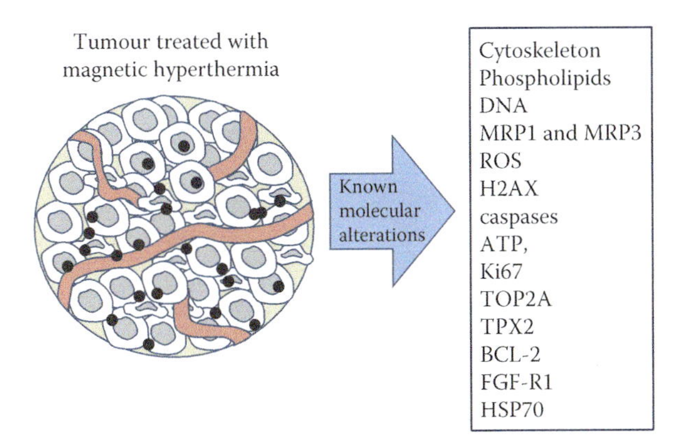

FIGURE 18.2 Impact of (magnetically induced) heating on molecular structures in the tumour.

Beyond this, if one combines the NP exposure with hyperthermia and mitomycin C treatments, a distinct modification of the multidrug resistance protein (MRP) 1 and 3 expression levels take place, which are not associated to *de novo* mRNA expression, but rather to an altered translocation of MRP 1 and 3 to the cell membrane. It is the result of the shifting of intracellular MRP storage pools at the protein level due to the nanoparticle and hyperthermia based production of reactive oxygen species (ROS).[35] The application of heating at temperatures above 55°C for 5 min, leads to distinct DNA damages, which cannot be repaired by endogeneous DNA repair systems,[33] e.g. by mediation of gamma H2AX foci formation,[36] a decrease of cellular ATP,[37] decreased number of vital cells, altered expression of procaspases and altered mRNA expression of Ki-67, TOP2A and TPX2.[38,39] It was also shown to induce host-immune response in addition to local tumour cell killing.[40]

The intratumoural application of the magnetic material, as analyzed in several preclinical studies using laboratory animals, led to pycnotic cell nuclei, an early sign of apoptosis,[41] and reduced tumour volumes with increasing time after treatment.[42] Since no alteration of the intracellular MNP accumulation took place after therapy, multiple treatment sessions are possible.[43,44] Similar results were observed after implanting stick type carboxymethyl-cellulose (CMC)-magnetite particles into rat tumours.[38,45–47]

The application of thermoablative temperatures (2 min) led to a down regulation of BCL2 and FGF-R1 on the protein level while the HSP70 level remained unchanged. Coincidently, the tumour tissue was damaged by heat, resulting in large apoptotic and necrotic areas in regions with high MNP concentration. This is an indication that high temperature magnetically induced heating (> 45°C) reduces the expression of tumour-related proteins.[48]

18.5 TEMPERATURE DISTRIBUTION IN THE TUMOUR REGION

Based on the different experiments on cell in culture conducted in the 1970s, it was found that for *in vivo* treatments, the whole tumour volume should be heated at 43°C for 90 min. From this the so-called CEM43T90 was derived, i.e. the cumulative equivalent time at 43°C in almost 90% of the tumour.[49] Thus, the CEM43T90 was introduced to monitor the temperature distribution in the tumour site, particularly when heating sources were placed outside the human body. Overall, it became obvious that temperatures higher than 43°C homogeneously covering the tumour area were rarely achievable in clinical hyperthermia using external heating sources.

Interestingly, magnetic particle induced hyperthermia significantly reduced or inhibited tumour growth compared to untreated tumours, even though the CEM43T90 temperatures were lower than reported in the literature.[50] This was the result of inducing intracellular hyperthermic spots, which could be identified by tumour surface temperature analysis by infrared thermography. Although no direct correlation between size of these high temperature spots and tumour volume at the end of the experiment was possible, a high temperature area was necessary to induce therapeutic effects. Emanating from these spots, intratumoural temperature gradients might have led to cell death up to a certain distance from the magnetic material. Due to the localized appearance of MNP in tumour tissue after intratumoural injection and the induction of intracellular hyperthermia, we finally hypothesized that lower CEM43T90 temperatures were needed to achieve efficient hyperthermia treatment *in vivo* when induced by MNPs (interstitial heating) compared to other methods, where the heat is delivered from the outside of tumours.[46]

Beyond the advances of interstitial hyperthermia, a crucial parameter for the success of magnetic hyperthermia is temperature distribution.[51] It is of crucial importance to pay attention to the intratumoral NP distribution in order to avoid the occurrence of regions of temperature under dosage within the tumour which might cause insufficient cell death (Figure 18.3).[52] The use of MRI is not favourable for monitoring the intratumoural MNP distribution, since the local NP accumulation required for heating purposes produces large susceptibility artefacts, which impair image quality. In contrast, CT-imaging is capable of depicting intratumourally applied MNP amounts and consequently allows a modulation of the generated temperatures within the tumour tissue. A distinct tumour reduction and a higher therapeutic success of magnetic hyperthermia were achieved with knowledge of the intratumoural MNP distribution. Furthermore, the surrounding nontumour tissue can be preserved while effectively treating breast cancer.[53]

Temperature monitoring during magnetic hyperthermia is still a challenging issue. Mostly, the use of catheters, which enclose a temperature sensor, have been proposed (e.g. see Ref. [39]). The majority of clinical studies on hyperthermia measured temperature on only one to four points of the tumour surface, not allowing for correct temperature distribution analysis, or the achieved thermal dose was not reported at all. In contrast for thermoablation of tumours (e.g. high-frequency ultrasound) magnetic resonance thermometry via the proton resonance frequency shift technique [54] seems to be useful, but is not suitable for magnetic hyperthermia, since the presence of MNP disturbs proton magnetic susceptibility.

18.6 THERAPEUTIC STRATEGIES OF HYPERTHERMIA *IN VIVO*

One widely investigated technique to deposit magnetic material into a tumour region is the intratumoural application of the magnetic material which allows for 'interstitial' heating of tumour cells. It has the advantage that the amounts deposited can be easily controlled and that comparatively high dosages can be achieved compared to other NP application modalities.[33,55] In consequence, even magnetic materials with comparatively low SAR could be suitable for magnetic hyperthermia as long as the required volumes of the NP suspensions are still several times lower (e.g. by a factor of 10 or more) than that of the tumour.

Interstitial heating can also be related to lower temperatures than expected in particular if high blood flow rates are in the proximity of the tumour (heat convection), the concentration of MNPs is too low, or the exerted heating potential is not optimal. Moreover, the intratumoural NP application seldom induces homogeneous NP distributions within the tumour, since the high interstitial pressures at the tumour area often lead to irregular distribution patterns even at slow infiltration rates of the magnetic material. This effect, in fact, produces heterogeneous temperature dosages and areas which escape from exposure to lethal temperatures.[41] For this reason a selective uptake of NPs by the tumour target cells has been desired. In this context, intracellular heating spots can take place after exposure of labeled cells to an alternating magnetic field, where lower CEM43T90 might be effective.[47] Extensive *in vitro* investigations performed so far have shown that the cellular uptake of IONPs is a very complex procedure. Namely, it does not only depend on the cell type, but also on the size, shape, surface charge and chemistry of the materials. In general, the intracellular accumulation into phagocytizing cells (macrophages) is higher compared to those cells that do not (e.g. differentiated somatic and tumour cells). Within the cells, most of the NPs (bare or functionalized ones) have been shown to be accumulated into endo-lysosomes.[56,57]

Interstitial heating of tumours has been already tested in clinical situations, where MNPs have been applied by the use

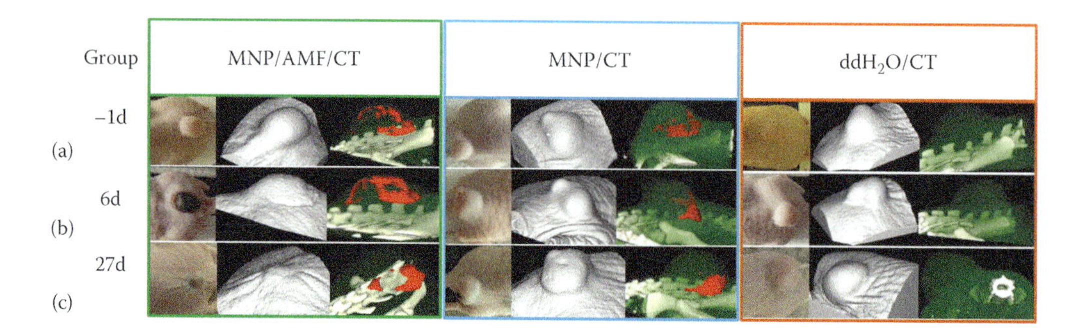

FIGURE 18.3 Assessment of the MNP distribution and localization in human breast cancer xenografts (MDA-MB-231) by micro-CT imaging over time. (a) MNP/AMF/CT represents tumours with loaded MNPs, exposed to an AMF and revealed by CT imaging; (b) MNP/CT; and (c) ddH$_2$O/CT. From left to right: tumour morphology; the morphology as CT image; and segmented CT image. (Dahring, H, Grandke, J, Teichgraber, U, Hilger, I.: *Mol Imaging Biol*. 2015. 17(6). 763–9. Copyright Wiley-VCH Verlag GmbH & Co. KGaA. Reproduced with permission.)

of specific stereotactic devices in combination with CT imaging. In this context, glioblastoma multiforme[58] and recurrent prostate cancer[59] were treated after intratumoural application of aminosilane coated NPs. Provided that NPs are retained in the tumour tissue, several consecutive thermal therapies are possible.[60] Brain tumour autopsies of treated patients presented dispersed or aggregated NPs. In particular, glioblastoma tumours are good targets for magnetic hyperthermia since NPs were phagocytosed by macrophages.[61] This minimally invasive therapy performed in the clinical setting was shown to be well tolerated.

In general, treating tumours via intratumourally administered NPs is effective as shown by preclinical and first clinical studies. Particular challenges are the requirement of stereotactic methods of application of the magnetic material, the implementation of noninvasive thermometry, as well as the homogeneous distribution of the magnetic material.

18.6.1 Passive Targeting of NPs for Magnetic Hyperthermia

In order to obtain a more homogeneous distribution of MNPs in the tumour region, people have been considering passive targeting after intravenous application (Figure 18.4). In this case, the accessibility of the NPs to the tumour is allowed by their neovascularization and by the faculty of the NPs to extravasate to the tumour interstitium as a consequence of the specific tumour vessel architecture. Namely, tumour vessels are irregular in shape, leaky, defective (lack of basal membrane, endothelial cells poorly aligned) and dilated[62] in contrast to normal ones. An increased retention of the NPs is attributed to the lack of lymphatic drainage in tumours.[63] To facilitate extravasation NPs should be small in size, e.g. lower than the diameter of vascular leakages of around 200 and 400 nm.[64] Importantly, this requirement counteracts with a greater heating potential, for which larger NP core sizes are required.[65]

Despite the fact that the magnetic material is administered intravasally, and that there is no need to access the tumour by means of stereotactic methods, the accumulation of NPs at the tumour site cannot be controlled as well as when intratumoural application is performed. Nowadays, it is known that a NP size of 10–100 nm is optimal for tumour infiltration via the vascular system. Small NP sizes foster the accumulation, and consequently, the penetration into the tumour tissue. Moreover, monodispersed NPs, in particular, are expected to display the same biological half-life, biodistribution and target affinity *in vivo*.[66]

In contrast, and in order to achieve high NP dosages in the tumour, one has to postulate that, in principle, almost all injected NPs should reach the tumour region. This aim has been rarely achieved so far. Importantly, in almost all cases the accumulated amounts in the tumour were of only 1 to 5% of the injected MNP amount.[67,68] The majority of NPs are taken up by the mononuclear phagocyte system (MPS) due to the high content of macrophages present in the liver, spleen and lymphatic tissues. Uptake takes place due to the opsonin adsorption (particularly immunoglobulins, albumin and components of the complement system) on the NPs which occurs previously in the blood system.[69,12] Even the use of 'stealth' MNPs with increased circulation time, and a reduced uptake by the MPS,[69] has not led to a breakthrough so far.

The use of external magnets capable of retaining the MNPs passing through the tumour might be a promising alternative in the future,[70] even though quite large NP core diameters are needed for it to work.[71]

18.6.2 Actively Targeted NPs in the Tumour Site for Magnetic Hyperthermia

Active targeting of MNPs for hyperthermia has often been expected to improve the selectivity for tumour cells.[72,73] This strategy implies the incorporation of functional groups and

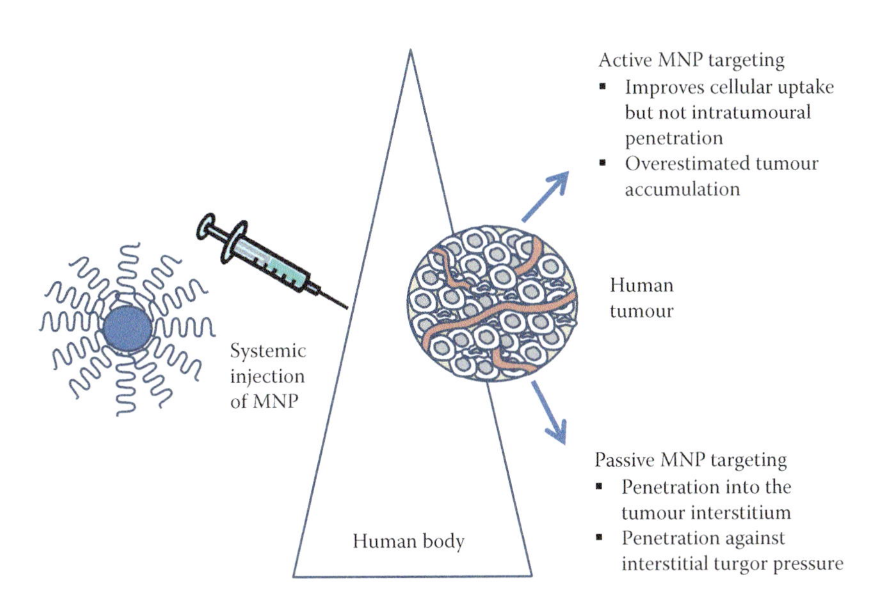

FIGURE 18.4 Challenges related to the accumulation of MNPs after systemic (intravenous) injection in humans.

affinity ligands to the NP surface coating. In analogy to passive targeting hyperthermia, the magnetic material is sought to accumulate particularly at the tumour interstitium surrounding the hyper-vascularized areas of the tumour. Here, NP retention at the tumour site is not only modulated by the lack of lymphatic drainage but also by target-affinity binding and internalization in specific cells.

When considering among all potential targets, those localized on endothelial cells of tumours are the most easily accessible. In contrast, molecular structures in the tumour interstitium (e.g. tumour cells) might be more challenging to address, since the introduction of functionalized MNPs could easily exceed in size the diameter of fenestrae of leaky vasculature, and therefore hinder NP extravasation and diffusion into the interstitium (hindered EPR-effect[63]). Interestingly, the migration features of the NPs into the interstitium are governed by the extent of target-affinity and avidity of ligands to molecular structures in the tumour region.[74] Thus, targeting endothelial cells of the tumour region might be a feasible alternative to a targeted interstitial accumulation of NPs.

Even though intensive efforts have been made so far[75,76] in relation to 'actively targeted hyperthermia,' it has not reached the clinical setting so far. Different reasons are responsible for it: (a) large NP diameters favour heating capabilities but not NP targeting,[12] (b) the vascularization degree of the tumour determines the targeting efficacy, (c) the SARs of all NPs are still too low to reach therapeutic temperatures when only 1 to 5% of the injected NP dose reaches the tumour, (d) active targeting efficacy of NPs has been overestimated from experimentation in isolated cell systems,[77] € the targeting moieties on the NP surface foster NP cell internalization but not its permeability into the tumour region and (f) the introduction of new components to the MNP formulation which implicates costly and long-lasting procedures for authorization by the FDA.

18.7 IDEA OF COMBINING MAGNETIC HYPERTHERMIA WITH MRI

IONPs were originally implemented for tumour diagnosis in clinical situations.[78] Namely, they have been used as T_2 contrast agents for approximately 30 years. As a result, there is significant experience and knowledge on iron-oxide-based NP formulations in MRI applications in radiology. Recently, it has been repeatedly postulated that the utilization of NPs in MRI and magnetic hyperthermia could reveal several synergies concerning morphological NP features and biological parameters. Nevertheless, from the present knowledge, one can derive that larger amounts are needed for magnetic hyperthermia purposes as compared to diagnostic oncology applications,[12] particularly after intravenous application if we consider that approximately less than 1% of the injected MNP dose will be accumulated in the tumour.[67] On the other hand, heating effects are only desired in the case of magnetic hyperthermia and not in MRI applications. For this reason, the corresponding combinations between magnetic hyperthermia and MRI will prospectively be restricted to separate devices with different magnetic field parameters (Figure 18.5).[12]

18.8 COMBINATION OF HYPERTHERMIA WITH CHEMOTHERAPY AND RADIOTHERAPY

Magnetic hyperthermia is an important adjuvant for chemotherapy, since it promotes the chemosensitivity in tumour cells. Chemosensitivity of tumours has been recognized to be a complex pathobiological phenomenon and although new antineoplastic agents have become available in recent decades, they have not necessarily increased the cure rate of a significant portion of cancers when applied alone. Intensive studies on the effects of drugs on the cellular metabolism have shown that drug combinations addressing different cellular pathways

Magnetic hyperthermia

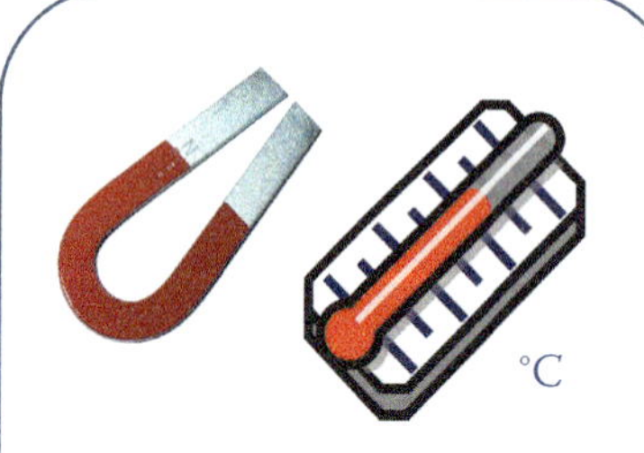

Ideal requirements

- Selective and local heating
- Large amounts of MNPs at the tumour area
- No MNP sequestration by the MPS
- Magnetic field parameters: 20–40 mT, 100–400 kHz

MRI

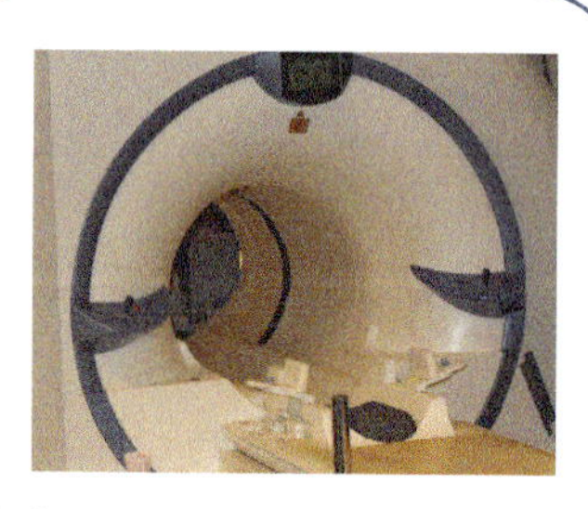

Ideal requirements

- No heating
- Amounts of MNPs systemically distributed
- MNP Sequestration by the MPS
- Magnetic field parameters: 1.5 or 3 T; 64 or 123 MHz

FIGURE 18.5 Challenges of combining magnetic hyperthermia with MRI. MPS: mononuclear phagocyte system.

provide better tumour therapy outcomes. In this regard, the combination of chemo- and radiotherapy with heat has been suggested and tested, where heating is considered to produce thermo- and radio-chemosensitization. The synergy is complex and depends upon many variables including temperature, exposure time, exposure sequence and drug concentration.[79] In particular, the term thermochemo-sensitization has been defined as the dose required for inducing cytotoxicity without heat, compared with the dose necessary to provoke cytotoxicity in the presence of heat.[80] Interestingly, this ratio is not necessarily linear in nature.[80]

In clinical studies, such combinations have been studied in relation to hyperthermia applied by external heating sources (e.g. see Ref. [81]). Concerning magnetic hyperthermia, the combination of magnetically induced heating and cisplatin,[82] mitomycin C,[35] doxorubicin DOX,[83] and methotrexate (MTX)[84] have been studied in preclinical settings.

In particular, when superparamagnetic IONPs were electrostatically functionalized with either the Nucant (N6L) multivalent pseudopeptide, doxorubicin, or both, a gradual inter- and intracellular release of the ligands was attained and NP uptake in cells was increased by the N6L functionalization. In combination with hyperthermia, the functionalized NPs were more cytotoxic to breast cancer cells than the respective free ligands. *In vivo* a substantial tumour growth inhibition was achieved. The proliferative activity of the remaining tumour tissue was distinctly reduced (Figure 18.6).[46]

In the context of radiation therapy, ionizing radiation is cytotoxic to tumour cells since they generally show an increased metabolism, higher rates of glycolysis and enhanced radiosensitivity as compared to normal cells, making them more vulnerable to radiation. Ionizing radiation induces irreversible damage to the DNA of injured cells. Radiotherapy is challenging as far as avoiding radiation exposure to the surrounding healthy tissues. Therefore, there are several attempts to reduce the radiation dose affecting healthy organs. This means that it is important to delimit as precisely as possible the volume of the injured region to be irradiated. This area delimitation also includes some normal tissue that could be somehow affected by tumour cells and for which removal is recommended in order to achieve a better control of the tumour. The combination of radiotherapy–magnetic hyperthermia with radiation will prospectively be of great importance in this context.[85,86]

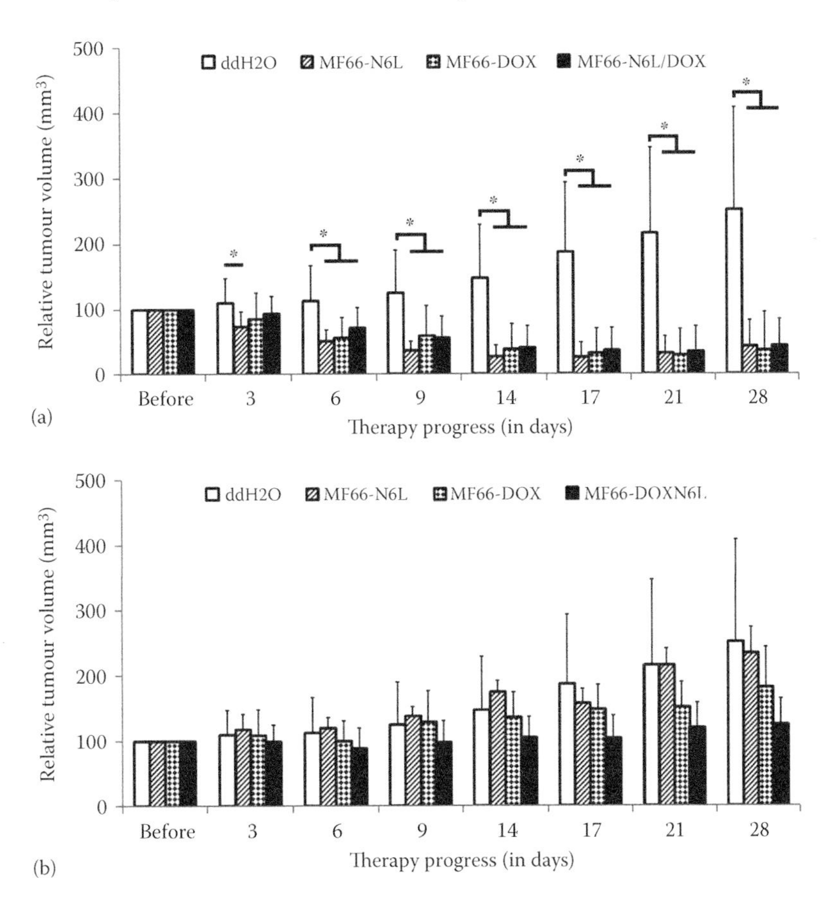

FIGURE 18.6 Magnetic hyperthermia treatment with functionalized MNP of human breast cancer xenografts (MDA-MB-231) in mice led to a strongly significant reduction of tumour volume compared to untreated animals. (a) Tumour volume development after magnetic hyperthermia (60 min at H = 15.4 kA/m, f = 435 kHz) with one of the three functionalized MNP formulations compared to untreated control that received ddH20 intratumourally. (b) Effect of intratumoural presence of functionalized MNP without hyperthermia treatment on tumour volume. (* $p \leq 0.05$ [Mann-Whitney-U-Test: treated vs. untreated]). AMF: alternating magnetic field; fMNP: functionalized MNPs; MH: magnetic hyperthermia; DOX: doxorubicin; M66: used MNP formulation; N6L: pseudopeptide Nucant. (Kossatz S, Grandke J, Couleaud P, Latorre A, Aires A, Crosbie-Staunton K et al.: *Breast Cancer Research*. 2015. 17.66. Copyright Wiley-VCH Verlag GmbH & Co. KGaA. Reproduced with permission.)

18.9 CONCLUSIONS AND FUTURE OUTLOOK

Current knowledge shows that IONPs are smart vehicles for magnetic hyperthermia application in oncological tumour therapy. They have been shown to fulfil specific physicochemical requirements (e.g. narrow size distribution, high stability, NP diameters lower than 200 nm, high SAR) which distinctly impact tumour cellular function upon exposure to an alternating magnetic field. Among the different strategies to deliver the magnetic material, the intratumoural application is favoured, since it allows a distinct control of the amounts to be deposited. In contrast, the often claimed feasibility of targeting IONPs via the vascular system and the local tumour vascularization (active NP targeting) has not been met so far, due to the presence of diverse physiological aspects reviewed in this chapter. An ideal technique to monitor the presence of iron oxide inside the tumours might be CT imaging. IONPs are also ideal vehicles for the attachment of themotherapeutic agents, making it possible to combine magnetic hyperthermia with chemotherapy and radiotherapy.

ACKNOWLEDGEMENTS

Part of my investigations cited in this book chapter were funded by German Research Foundation and the European Commission (7th Framework Program [MuliFUN] and Horizon-2020 [NoCanTher]).

REFERENCES

1. Wust P, Hildebrandt B, Sreenivasa G, Rau B, Gellermann J, Riess H et al. Hyperthermia in combined treatment of cancer. *Lancet Oncology*. 2002;3(8):487–97.
2. Falk MH, Issels RD. Hyperthermia in oncology. *International Journal of Hyperthermia*. 2001;17(1):1–18.
3. Hammad M, Hempelmann R. Enhanced specific absorption rate of bi-magnetic nanoparticles for heating applications. *Materials Chemistry and Physics*. 2017;188:30–8.
4. Kafrouni L, Savadogo O. Recent progress on magnetic nanoparticles for magnetic hyperthermia. *Progress in Biomaterials*. 2016;5(3–4):147–60.
5. Ling Y, Tang XZ, Wang FJ, Zhou XH, Wang RH, Deng LM et al. Highly efficient magnetic hyperthermia ablation of tumors using injectable polymethylmethacrylate-Fe_3O_4. *Rsc Advances*. 2017;7(5):2913–8.
6. Palihawadana-Arachchige M, Nemala H, Naik VM, Naik R. Effect of magnetic dipolar interactions on temperature dependent magnetic hyperthermia in ferrofluids. *Journal of Applied Physics*. 2017;121(2).
7. Simeonidis K, Morales MP, Marciello M, Angelakeris M, de la Presa P, Lazaro-Carrillo A et al. In-situ particles reorientation during magnetic hyperthermia application: Shape matters twice. *Scientific Reports*. 2016;6.
8. Yang TI, Chang SH. Controlled synthesis of metallic iron nanoparticles and their magnetic hyperthermia performance in polyaniline composite nanofibers. *Nanotechnology*. 2017;28(5).
9. Ludwig R, Stapf M, Dutz S, Muller R, Teichgraber U, Hilger I. Structural properties of magnetic nanoparticles determine their heating behavior – An estimation of the in vivo heating potential. *Nanoscale Research Letters*. 2014;9.
10. Krishnan KM. Biomedical Nanomagnetics: A spin through possibilities in imaging, diagnostics, and therapy. *IEEE Transactions on Magnetics*. 2010;46(7):2523–58.
11. Hergt R, Andra W, d'Ambly CG, Hilger I, Kaiser WA, Richter U et al. Physical limits of hyperthermia using magnetite fine particles. *IEEE Transactions on Magnetics*. 1998;34(5):3745–54.
12. Hilger I, Kaiser WA. Iron oxide-based nanostructures for MRI and magnetic hyperthermia. *Nanomedicine*. 2012;7(9):1443–59.
13. Dutz S, Muller R, Eberbeck D, Hilger I, Zeisberger M. Magnetic nanoparticles adapted for specific biomedical applications. *Biomedical Engineering / Biomedizinische Technik*. 2015;60(5):405–16.
14. Kallumadil M, Tada M, Nakagawa T, Abe M, Southern P, Pankhurst QA. Suitability of commercial colloids for magnetic hyperthermia (vol 321, pg 1509, 2009). *Journal of Magnetism and Magnetic Materials*. 2009;321(21):3650–1.
15. Carrey J, Mehdaoui B, Respaud M. Simple models for dynamic hysteresis loop calculations of magnetic single-domain nanoparticles: Application to magnetic hyperthermia optimization. *Journal of Applied Physics*. 2011;109(8).
16. Lacroix LM, Carrey J, Respaud M. A frequency-adjustable electromagnet for hyperthermia measurements on magnetic nanoparticles. *The Review of Scientific Instruments*. 2008;79(9):093909.
17. Reilly JP. *Applied Bioelectricity: From Electrical Stimulation to Electropathology*. Berlin: Springer, 1998.
18. Pankhurst QA, Thanh NTK, Jones SK, Dobson J. Progress in applications of magnetic nanoparticles in biomedicine. *Journal of Physics D—Applied Physics*. 2009;42(22).
19. Atkinson W, Brezovich I, Chakraborty DP. Usable frequencies in hyperthermia with thermal seeds. *IEE—Transactions on Biomedical Engineering* 1984;BME-31:70–5.
20. Staufer A, Cetas TC, Fletcher AM, Deyoung DW, Dewhirst MW, Oleson JR et al. Observations on the use of ferromagnetic implants for inducing hyperthermia. *IEE—Transactions on Biomedical Engineering*. 1984;BME-31(76–90).
21. Johannsen M, Gneveckow U, Eckelt L, Feussner A, Waldofner N, Scholz R et al. Clinical hyperthermia of prostate cancer using magnetic nanoparticles: Presentation of a new interstitial technique. *International Journal of Hyperthermia*. 2005;21(7):637–47.
22. Harvey PR, Katznelson E. Modular gradient coil: A new concept in high-performance whole-body gradient coil design. *Magnetic Resonance in Medicine*. 1999;42(3):561–70.
23. Hilger I, Andra W, Hergt R, Hiergeist R, Schubert H, Kaiser WA. Electromagnetic heating of breast tumors in interventional radiology: In vitro and in vivo studies in human cadavers and mice. *Radiology*. 2001;218(2):570–5.
24. Moros M, Ambrosone A, Stepien G, Fabozzi F, Marchesano V, Castaldi A et al. Deciphering intracellular events triggered by mild magnetic hyperthermia in vitro and in vivo. *Nanomedicine*. 2015;10(14):2167–83.
25. Thrall DE, Larue SM, Pruitt AF, Case B, DeWhirst MW. Changes in tumour oxygenation during fractionated hyperthermia and radiation therapy in spontaneous canine sarcomas. *International Journal of Hyperthermia*. 2006;22(5):365–73.
26. Vujaskovic Z, Poulson JM, Gaskin AA, Thrall DE, Page RL, Charles HC et al. Temperature-dependent changes in physiologic parameters of spontaneous canine soft tissue sarcomas after combined radiotherapy and hyperthermia treatment. *International Journal of Radiation Oncology Biology Physics*. 2000;46(1):179–85.
27. Song CW, Park H, Griffin RJ. Improvement of tumor oxygenation by mild hyperthermia. *Radiation Research*. 2001;155(4):515–28.

28. Multhoff G, Gaipl U. Molekulare und immunologische Effekte der Hyperthermie auf Tumorprogression und Metastasierung. *Onkologe*. 2010;11:1043–50.
29. Hildebrandt B, Wust P, Ahlers O, Dieing A, Sreenivasa G, Kerner T et al. The cellular and molecular basis of hyperthermia. *Critical Reviews in Oncology/Hematology*. 2002; 43(1):33–56.
30. Schildkopf P, Ott OJ, Frey B, Wadepohl M, Sauer R, Fietkau R et al. Biological rationales and clinical applications of temperature controlled hyperthermia – Implications for multimodal cancer treatments. *Current Medicinal Chemistry*. 2010;17(27): 3045–57.
31. Overgaard K, Overgaard J. Investigations on possibility of a thermic tumor therapy – I. Short-wave treatment of a transplanted isologous mouse mammary-carcinoma. *European Journal of Cancer*. 1972;8(1):65–78.
32. Vaupel P, Kallinowski FK, Okunieff P. Blood flow, oxygen and nutrient supply, and metabolic microenvironment of human tumors: A review. *Cancer Research*. 1989;49:6449–65.
33. Hilger I, Rapp A, Greulich KO, Kaiser WA. Assessment of DNA damage in target tumor cells after thermoablation in mice. *Radiology*. 2005;237(2):500–6.
34. Hilger I, Frühauf S, Andrä W, Hiergeist R, Hergt R, Kaiser WA. Magnetic heating as a therapeutic tool. *Thermology International*. 2001;11:130–6.
35. Franke K, Kettering M, Lange K, Kaiser WA, Hilger I. The exposure of cancer cells to hyperthermia, iron oxide nanoparticles, and mitomycin C influences membrane multidrug resistance protein (MRP) expression levels. *International Journal of Nanomedicine*. 2013;8 351–63.
36. Hori T, Kondo T, Lee H, Song CW, Park HJ. Hyperthermia enhances the effect of beta beta-lapachone to cause gamma gamma H2AX formations and cell death in human osteosarcoma cells. *International Journal of Hyperthermia*. 2011;27(1):53–62.
37. Krupka TM, Dremann D, Exner AA. Time and dose dependence of pluronic bioactivity in hyperthermia-induced tumor cell death. *Experimental Biology and Medicine*. 2009;234(1):95–104.
38. Ludwig R, Teran FJ, Teichgraeber U, Hilger I. Nanoparticle-based hyperthermia distinctly impacts ROS production, Ki-67, TOP2A, TPX2-expression and apoptosis induction in pancreatic cancer. *International Journal of Hyperthermia*. 2017;12:1009–1018.
39. Ludwig R, Teran FJ, Teichgraeber U, Hilger I. Nanoparticle-based hyperthermia distinctly impacts production of ROS, expression of Ki-67, TOP2A, and TPX2, and induction of apoptosis in pancreatic cancer. *International Journal of Nanomedicine*. 2017;12:1009–18.
40. Yanase M, Shinkai M, Honda H, Wakabayashi T, Yoshida J, Kobayashi T. Antitumor immunity induction by intracellular hyperthermia using magnetite cationic liposomes. *Japanese Journal of Cancer Research*. 1998;89(7):775–82.
41. Hilger I, Hiergeist R, Hergt R, Winnefeld K, Schubert H, Kaiser WA. Thermal ablation of tumors using magnetic nanoparticles: An *in vivo* feasibility study. *Investigative Radiology*. 2002;37(10):580–6.
42. Jordan A, Wust P, Fähling H. Inductive heating of ferrimagnetic particles and magnetic fluids: Physical evaluation of their potential for hyperthermia. *International Journal of Hyperthermia*. 1993;9:51–68.
43. Kettering M, Richter H, Wiekhorst F, Bremer-Streck S, Trahms L, Kaiser WA et al. Minimal-invasive magnetic heating of tumors does not alter intra-tumoral nanoparticle accumulation, allowing for repeated therapy sessions: An in vivo study in mice. *Nanotechnology*. 2011;22(50):505102.
44. Ito A, Tanaka K, Honda H, Abe S, Yamaguchi H, Kobayashi T. Complete regression of mouse mammary carcinoma with a size greater than 15 mm by frequent repeated hyperthermia using magnetite nanoparticles. *Journal of Bioscience and Bioengineering*. 2003;96(4):364–9.
45. Ohno T, Wakabayashi T, Takemura A, Yoshida J, Ito A, Shinkai M et al. Effective solitary hyperthermia treatment of malignant glioma using stick type CMC-magnetite. In vivo study. *Journal of Neuro-Oncology*. 2002;56(3):233–9.
46. Kossatz S, Grandke J, Couleaud P, Latorre A, Aires A, Crosbie-Staunton K et al. Efficient treatment of breast cancer xenografts with multifunctionalized iron oxide nanoparticles combining magnetic hyperthermia and anti-cancer drug delivery. *Breast Cancer Research*. 2015;17:66.
47. Kossatz S, Ludwig R, Dahring H, Ettelt V, Rimkus G, Marciello M et al. High therapeutic efficiency of magnetic hyperthermia in xenograft models achieved with moderate temperature dosages in the tumor area. *Pharmaceutical Research*. 2014;31(12):3274–88.
48. Stapf M, Pompner N, Kettering M, Hilger I. Magnetic thermoablation stimuli alter BCL2 and FGF-R1 but not HSP70 expression profiles in BT474 breast tumors. *International Journal of Nanomedicine*. 2015;10:1931–9.
49. Franckena M, Fatehi D, de Bruijne M, Canters RAM, van Norden Y, Mens JW et al. Hyperthermia dose-effect relationship in 420 patients with cervical cancer treated with combined radiotherapy and hyperthermia. *European Journal of Cancer*. 2009;45(11):1969–78.
50. Kossatz S, Ludwig R, Dähring H, Ettelt V, Rimkus G, Marciello M et al. High therapeutic efficiency of magnetic hyperthermia in xenograft models achieved with moderate temperature dosages in the tumor area. *Pharmaceutical Research—Dordr*. 2014;31(12):3274–88.
51. Perez CA, Sapareto SA. Thermal dose expression in clinical hyperthermia and correlation with tumor response/control. *Cancer Research*. 1984;44:4818s–25s.
52. Van Vulpen M, De Leeuw AA, Raaymakers BW, Van Moorselaar RJ, Hofman P, Lagendijk JJ et al. Radiotherapy and hyperthermia in the treatment of patients with locally advanced prostate cancer: Preliminary results. *BJU International*. 2004;93(1):36–41.
53. Dahring H, Grandke J, Teichgraber U, Hilger I. Improved hyperthermia treatment of tumors under consideration of magnetic nanoparticle distribution using micro-CT imaging. *Molecular Imaging and Biology*. 2015;17(6):763–9.
54. Streicher MN, Schafer A, Ivanov D, Muller DK, Amadon A, Reimer E et al. Fast accurate MR thermometry using phase referenced asymmetric spin-echo EPI at high field. *Magnetic Resonance in Medicine*. 2014;71(2):524–33.
55. Dutz S, Kettering M, Hilger I, Muller R, Zeisberger M. Magnetic multicore nanoparticles for hyperthermia – Influence of particle immobilization in tumour tissue on magnetic properties. *Nanotechnology*. 2011;22(26):265102.
56. Villanueva A, Canete M, Roca AG, Calero M, Veintemillas-Verdaguer S, Serna CJ et al. The influence of surface functionalization on the enhanced internalization of magnetic nanoparticles in cancer cells. *Nanotechnology*. 2009;20(11).
57. Kettering M, Richter H, Wiekhorst F, Bremer-Streck S, Trahms L, Kaiser WA et al. Minimal-invasive magnetic heating of tumors does not alter intra-tumoral nanoparticle accumulation, allowing for repeated therapy sessions: An in vivo study in mice. *Nanotechnology*. 2011;22(50).
58. Maier-Hauff K, Rothe R, Scholz R, Gneveckow U, Wust P, Thiesen B et al. Intracranial thermotherapy using magnetic

nanoparticles combined with external beam radiotherapy: Results of a feasibility study on patients with glioblastoma multiforme. *Journal of Neuro-Oncology*. 2007;81(1):53–60.
59. Johannsen M, Gneueckow U, Thiesen B, Taymoorian K, Cho CH, Waldofner N et al. Thermotherapy of prostate cancer using magnetic nanoparticles: Feasibility, imaging, and three-dimensional temperature distribution. *European Urology*. 2007;52(6):1653–62.
60. Kettering K, Dornberger V, Lang R, Vonthein R, Suchalla R, Bosch RF et al. Enhanced detection criteria in implantable cardioverter defibrillators: Sensitivity and specificity of the stability algorithm at different heart rates. *Pacing and Clinical Electrophysiology*. 2001;24(9 Pt 1):1325–33.
61. van Landeghem FKH, Maier-Hauff K, Jordan A, Hoffmann KT, Gneveckow U, Scholz R et al. Post-mortem studies in glioblastoma patients treated with thermotherapy using magnetic nanoparticles. *Biomaterials*. 2009;30(1):52–7.
62. Folkman J. Angiogenesis in cancer, vascular, rheumatoid and other disease. *Nature Medicine*. 1995;1(1):27–31.
63. Maeda H, Wu J, Sawa T, Matsumura Y, Hori K. Tumor vascular permeability and the EPR effect in macromolecular therapeutics: A review. *Journal of Controlled Release*. 2000;65(1–2):271–84.
64. Yuan F, Dellian M, Fukumura D, Leunig M, Berk DA, Torchilin VP et al. Vascular-permeability in a human tumor xenograft – Molecular-size dependence and cutoff size. *Cancer Research*. 1995;55(17):3752–6.
65. Hergt R, Hiergeist R, Hilger I, Kaiser WA, Lapatnikov Y, Margel S et al. Maghemite nanoparticles with very high AC-losses for application in RF-magnetic hyperthermia. *Journal of Magnetism and Magnetic Materials*. 2004;270(3):345–57.
66. Svenson S. Theranostics: Are we there yet? *Molecular Pharmacology*. 2013;10(3):848–56.
67. Lammers T, Kiessling F, Hennink WE, Storm G. Drug targeting to tumors: Principles, pitfalls and (pre-) clinical progress. *Journal of Controlled Release*. 2012;161(2):175–87.
68. Svenson S. Theranostics: Are we there yet? *Molecular Pharmaceutics*. 2013;10 848–56.
69. Moghimi SM, Hunter AC, Murray JC. Long-circulating and target-specific nanoparticles: Theory to practice. *Pharmacological Reviews*. 2001;53(2):283–318.
70. Alexiou C, Arnold W, Klein RJ, Parak FG, Hulin P, Bergemann C et al. Locoregional cancer treatment with magnetic drug targeting. *Cancer Research*. 2000;60(23):6641–8.
71. Hergt R, Hiergeist R, Zeisberger M, Schuler D, Heyen U, Hilger I et al. Magnetic properties of bacterial magnetosomes as potential diagnostic and therapeutic tools. *Journal of Magnetism and Magnetic Materials*. 2005;293(1):80–6.
72. Quarta A, Bernareggi D, Benigni F, Luison E, Nano G, Nitti S et al. Targeting FR-expressing cells in ovarian cancer with Fab-functionalized nanoparticles: A full study to provide the proof of principle from in vitro to in vivo. *Nanoscale*. 2015;7(6):2336–51.
73. Yin PT, Shah S, Pasquale NJ, Garbuzenko OB, Minko T, Lee KB. Stem cell-based gene therapy activated using magnetic hyperthermia to enhance the treatment of cancer. *Biomaterials*. 2016;81:46–57.
74. Rudnick SI, Adams GP. Affinity and avidity in antibody-based tumor targeting. *Cancer Biotherapy and Radiopharmaceuticals*. 2009;24(2):155–61.
75. DeNardo SJ, DeNardo GL, Miers LA, Natarajan A, Foreman AR, Gruettner C et al. Development of tumor targeting bioprobes (In-111-chimeric L6 monoclonal antibody nanoparticles) for alternating magnetic field cancer therapy. *Clinical Cancer Research*. 2005;11(19):7087s–92s.
76. DeNardo SJ, DeNardo GL, Natarajan A, Miers LA, Foreman AR, Gruettner C et al. Thermal dosimetry predictive of efficacy of In-111-ChL6 nanoparticle AMF-induced thermoablative therapy for human breast cancer in mice. *Journal of Nuclear Medicine*. 2007;48(3):437–44.
77. Hilger I, Leistner Y, Berndt A, Fritsche C, Haas KM, Kosmehl H et al. Near-infrared fluorescence imaging of HER-2 protein over-expression in tumour cells. *European Radiology*. 2004;14:1124–9.
78. Mclachlan SJ, Morris MR, Lucas MA, Fisco RA, Eakins MN, Fowler DR et al. Phase-I clinical-evaluation of a new iron-oxide Mr contrast agent. *Journal of Magnetic Resonance Imaging*. 1994;4(3):301–7.
79. Issels RD. Hyperthermia adds to chemotherapy. *European Journal of Cancer*. 2008;44(17):2546–54.
80. Petin VG, Kim JK, Zhurakovskaya GP, Dergacheva IP. Some general regularities of synergistic interaction of hyperthermia with various physical and chemical inactivating agents. *International Journal of Hyperthermia*. 2002;18(1):40–9.
81. Issels RD, Lindner LH, Ghadjar P, Reichardt P, Hohenberger P, Verweij J et al. Improved overall survival by adding regional hyperthermia to neo-adjuvant chemotherapy in patients with localized high-risk soft tissue sarcoma (HR-STS): Long-term outcomes of the EORTC 62961/ESHO randomized phase III study. *European Journal of Cancer*. 2015;51:S716-S.
82. Kettering M, Zorn H, Bremer-Streck S, Oehring H, Zeisberger M, Bergemann C et al. Characterization of iron oxide nanoparticles adsorbed with cisplatin for biomedical applications. *Physics in Medicine and Biology*. 2009;54(17): 5109–21.
83. Kossatz S, Ludwig R, Daehring H, Ettelt V, Rimkus G, Marciello M et al. High therapeutic efficiency of magnetic hyperthermia in xenograft models achieved with moderate temperature increases in the tumor area. *Pharmaceutical Research*. 2014;DOI: 10.1007/s11095-014-1417-0.
84. Stapf M, Pompner N, Teichgraber U, Hilger I. Heterogeneous response of different tumor cell lines to methotrexate-coupled nanoparticles in presence of hyperthermia. *International Journal of Nanomedicine*. 2016;11:485–500.
85. Borasi G, Nahum A, Paulides MM, Powathil G, Russo G, Fariselli L et al. Fast and high temperature hyperthermia coupled with radiotherapy as a possible new treatment for glioblastoma. *Journal of Therapeutic Ultrasound*. 2016;4.
86. Crezee J, van Leeuwen CM, Oei AL, van Heerden LE, Bel A, Stalpers LJA et al. Biological modelling of the radiation dose escalation effect of regional hyperthermia in cervical cancer. *Radiation Oncology*. 2016;11.

Ingrid Hilger (E-mail: Ingrid.Hilger@med.uni-jena.de) is head of the Department 'Experimental Radiology' at the University Hospital Jena, Germany. Born in Argentina, she studied biology at the Christian-Albrechts-University in Kiel, Germany, and earned her diploma in 1990. She has performed several studies in biology in South America and Asia. After becoming interested in human biology and biochemistry, she earned her PhD at the University Hospital of Hannover, Germany, in 1996. Since then, she has focused her research activities on the areas of therapeutic nanotechnology and *in vivo* meso/macroscopic molecular preclinical imaging. She was awarded the Walter Friedrich Prize in 2003, and since 2008, she is a full professor at the University Hospital Jena, Germany. Presently, she is the spokeswoman of the German molecular imaging network and the German representative of the European Society of Molecular Imaging.

19 Noninvasive Guidance Scheme of Magnetic Nanoparticles for Drug Delivery in Alzheimer's Disease

*Ali Kafash Hoshiar, Tuan-Anh Le, Faiz Ul Amin, Xingming Zhang, Myeong Ok Kim and Jungwon Yoon**

CONTENTS

19.1 INTRODUCTION

Alzheimer's disease (AD) is a progressive neurodegenerative disease affecting a large proportion of the ageing population; it is predicted that it will affect 1 in 85 people globally by 2050.[1,2] As cognitive function and synaptic integrity in AD patients are gradually lost, selective neurons die and abnormal plaques form in the brains of AD patients.[3] Given that there is no effective cure for the disease to date, once the disease has progressed, the treatment strategies currently available for AD become useless.[4] Despite recent developments in AD treatment, targeted drug delivery (TDD) to the brain remains an open challenge.

With 'traditional' therapeutics, the whole body is exposed to high concentrations of potentially toxic drugs. Although the drugs reach their target locations, they also affect healthy tissues, causing undesirable side effects. Thus, to enhance treatment efficacy and maintain an optimal dose at a desired location, TDD systems have been developed.[5] Nanoparticle (NP)-based molecular transport has been the subject of recent strategies to enhance delivery and reduce toxicity[6–10] in diverse brain-related diseases, including AD. However, noninvasive delivery of therapeutic or diagnostic agents across an intact blood–brain barrier (BBB) remains a major challenge.

The BBB constitutes an important physiological barrier that prevents effective targeting of the brain with most diagnostic and therapeutic agents. Targeting and imaging therapeutic agents to the brain is a clinically important problem, but technologies to alter their pharmaco-distribution remain limited. Thus, developing strategies for the specific delivery of reagents across the intact BBB with minimal toxicity remains a challenge. The ideal strategy for the BBB crossing should not damage the intact BBB, the carrier should be nontoxic in nature, the drug transport should be selective and the drug load should be sufficient. The main strategies for the BBB crossing can be classified into four types as follows: (1) the molecular modifications, (2) the BBB bypassing, (3) the BBB disruption and (4) the noninvasive nanorobotic based solutions.[11] The molecular modifications elevate the drugs uptake across intact BBB. In this method, drug carriers are designed to act like a Trojan horse to cross the BBB. Although many researches aim to improve the practicality of this method, the molecular modification alone fails to deliver drugs with high targeting performance in a sufficient dose to a targeted brain region.[12] Using intranasal pathways to bypass the BBB has been also studied as an alternative. However, since the whole brain is not

* Corresponding author.

accessible using this method, the selective targeting in the brain remains a major challenge in this approach. Besides, other barriers should be explored for use as alternative pathways.[13] The temporary disruption of the BBB, such as with high-intensity focused ultrasound[14] or controlled hyperthermia, has been suggested as an alternative to selectively provide the BBB crossing.[15,16] Despite the effectiveness of the method in selective BBB crossing, the disruption is considered to be a biologically hazardous act. The magnetic-based nanorobotic system for BBB crossing and TDD has been emerging as an alternative solution. The magnetic force noninvasively guides NPs to the desired region with a monitoring technique. Moreover, the magnetic actuation can improve the drug update.

Magnetic nanoparticles (MNPs), in particular, are being developed based on their unique properties in response to magnetic fields. MNPs can cross the intact BBB noninvasively with no apparent toxicity, which may be beneficial in the treatment of central nervous system (CNS) diseases. Unlike other delivery methods (such as that mediated by antibodies to specific cell-surface receptors), magnetically-mediated translocation and transportation of MNPs across the intact BBB does not appear to induce deleterious signal transduction events and shows minimal accumulation in other organs.[17] Micro-nanorobotic systems for TDD, based on noninvasive magnetic force actuation, have been developed for the guidance of MNPs within the vasculature. Magnetic resonance imaging (MRI)-based systems to steer microparticles in the vasculature have also been described.[18,19] The requirement for more precise particle steering resulted in the development of magnetic-based microrobotic platforms.[20–23] Despite these benefits, however, the size factor prevents full interactions between particles and cells, and prevents passage of the former through the BBB for TDD within the brain.

Therefore, we developed a novel nanorobotic platform with magnetic field function (FF) actuation to deliver therapeutic agents to the brain. The FF, a positive–negative pulsed magnetic field generated by electromagnetic coils, is an open loop actuation, developed based on the simulation of vascular systems to optimize the number of particles reaching the desired target.[24,25] *In vivo* experiments demonstrated that nanocarriers successfully crossed the BBB and reached deep regions of the brain with an optimized magnetic FF.[26] To improve the performance of drug delivery to the brain, a FF was designed for steering aggregated particles.[27,28] Moreover, to establish a feedback control loop scheme,[29] a magnetic particle imaging (MPI)-based 1D nanorobotic platform was developed for manipulating MNPs.[30] The suggested approach has been primarily designed for the Alzheimer disease as it covers a broad section of the brain and the primary goal here is the BBB crossing. However, further improvement in the feedback scheme will lead to better steering performance and the proposed scheme has the potential for tumour treatment.

To provide a fully functional nanorobotic platform, a real-time monitoring scheme is needed. Thus, an MPI approach is proposed to address this. MPI is an imaging technique based on the nonlinear magnetization characteristics of superparamagnetic iron oxide nanoparticles. The MPI concept involves two major features: the particles should emit some characteristic signal and this signal should provide information about the amount of magnetic material. The method was first proposed by Gleich and Weizenecker in 2005[31] and has been developed rapidly since then. Today, it is used in human imaging (angiography, cell tracking and cancer imaging).[32,33] MPI not only provides fast and high-sensitivity data with millimetre-scale resolution, but also has the potential to revolutionize the biomedical imaging field.[34–36]

Section 19.2 of this chapter describes the proposed magnetic drug delivery scheme with a special electromagnetic actuator, composed of a coil and core for a high gradient field, and a magnetic FF to prevent sticking. Section 19.3 explains the imaging capability of MNPs by adding additional electromagnetic systems to the proposed actuator, which can lead to a feedback control scheme for the proposed drug delivery system. Section 19.4 describes the proposed drug delivery scheme for crossing the BBB effectively in AD, using open loop actuation. Section 19.5 concludes this chapter.

19.1.1 AD and Its Treatment

Several hypotheses have been proposed to explain AD pathology; the Aβ cascade hypothesis is the major one considered here.[37,38] According to this hypothesis, Aβ peptides are generated from the catalytic cleavage of the transmembrane glycoprotein amyloid precursor protein (APP) through enzymes: β-secretase at the APP N terminal and γ-secretase at the APP C terminus.[39] Currently, only five drugs are approved by the U.S. Food and Drug Administration (FDA) for the treatment of AD: (1) the AChEIs, (2) rivastigmine (Exelon), galantamine (Razadyne, Reminyl), (3) tacrine (Cognex), (4) donepezil (Aricept) and (5) the NMDA receptor antagonist memantine (Namenda).[40,41] Recently, the neuroprotective effects of different therapeutic agents against neurological disorders, like AD and other CNS conditions, have been reported by our group in different animal models, including mice and rats.[26,42–54]

In an AD model, we studied the expression of activated AMPK (p-AMPKα [Thr172]) and SIRT1 because they are both cellular energy-sensing enzymes. We demonstrated that the neuroprotective agent osmotin induced the activation of AMPK and SIRT1 both *in vitro* (APP-transfected SH-SY5Y cells) and *in vivo* (APP/PS1 mice, at 5-, 9- and 12-month-old) AD models (Figure 19.1).

A variety of chemical drugs have been discovered and developed over the past several decades, but some problems like fast elimination/denaturation and degradation are still the main obstacles. Large numbers of attempts to solve these problems have been made by using high dose or multitreatment of the drugs. However, it is very dangerous for efficient therapy, because if overdoses (out of the range of therapeutic windows) are used, nonspecific toxicity of drugs could ensue. One approach to solve these problems was the packaging of the drugs into a particulate carrier system like MNPs and using the electromagnetic actuation for the targeted delivery. We utilize electromagnetic actuation to improve the targeting performance and elevate BBB crossing.

Next, we used different techniques to determine whether osmotin affected the amyloidogenic pathway of Aβ production

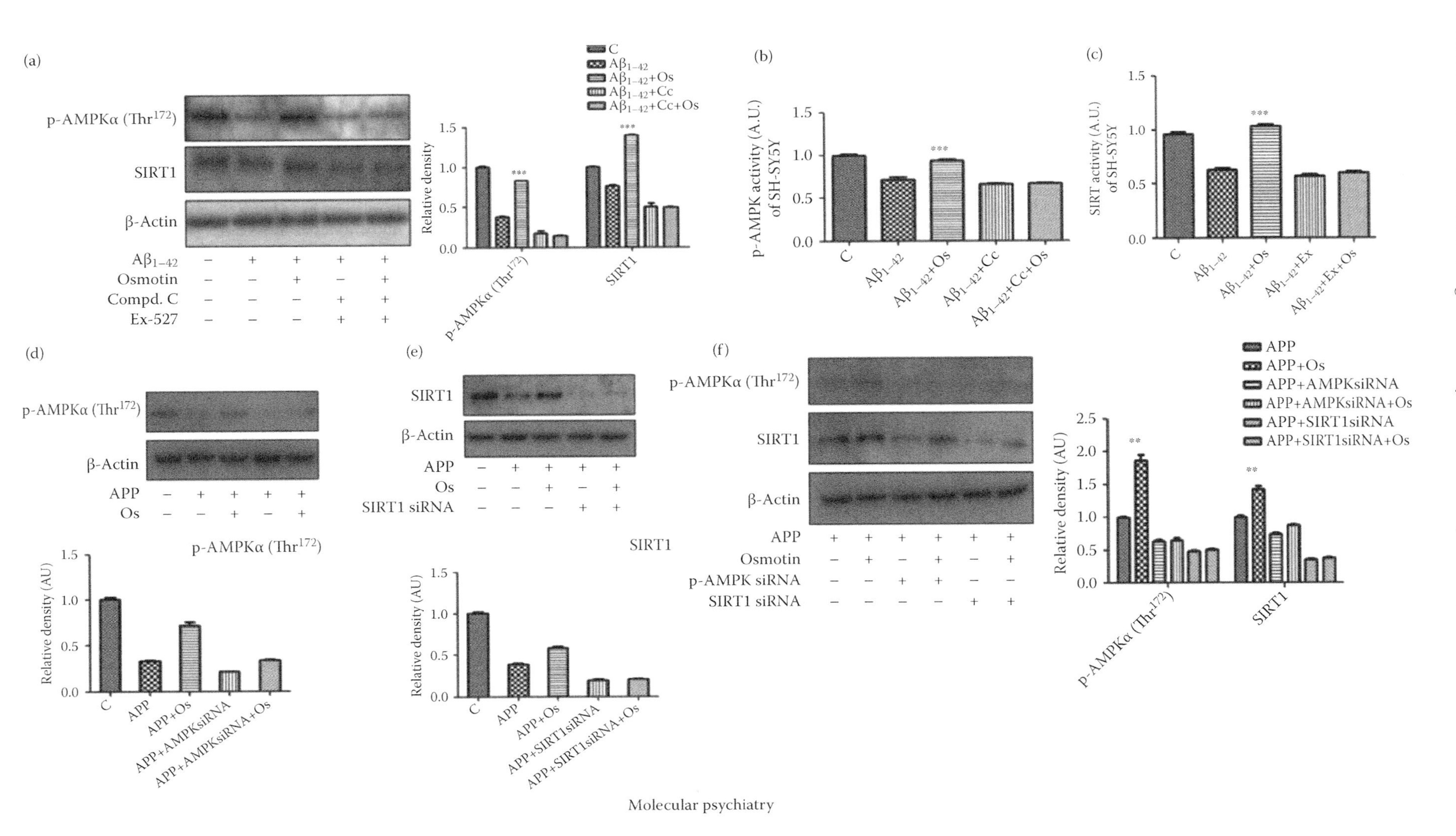

FIGURE 19.1 Osmotin increased phosphorylated AMP-activated protein kinase (p-AMPK) and sirtuin 1 (SIRT1) levels and their activities *in vitro* and *in vivo*. (a) Western blot analysis of p-AMPK and SIRT1 in lysates of SH-SY5Y cells after treatment with $A\beta_{1-d42}$ (5 μM), osmotin (Os, 0.2 μM), compound C (C, 20 μM) and EX527 (80 μM) for 24 h and (b, c) their activity histograms. (d–f) Western blot analysis with the respective density histograms of p-AMPK and SIRT1 in lysates of APPswe/ind-transfected (for 72 h) SH-SY5Y cells subjected to small interfering RNA (siRNA)-induced silencing of AMPK and SIRT1 for 48 h after 24 h of treatment with Os (0.2 μM). (*Continued*)

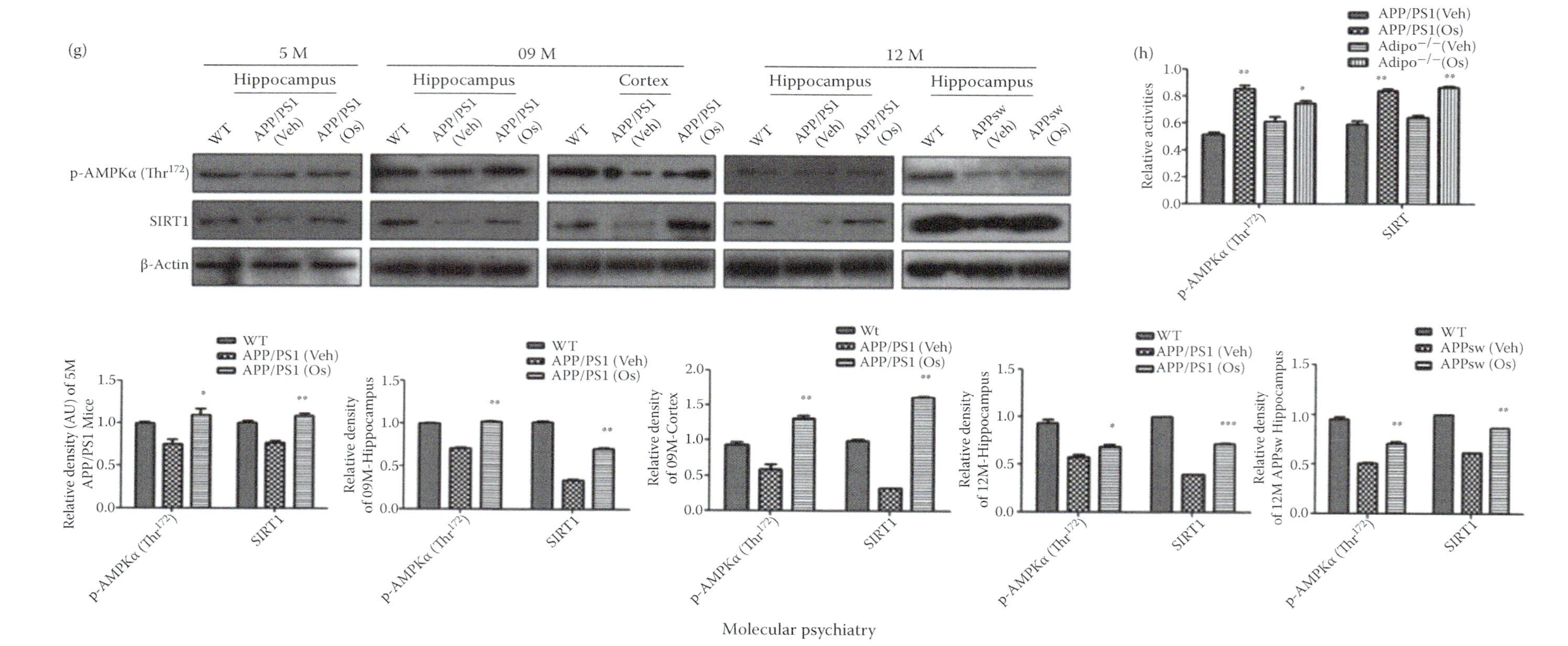

FIGURE 19.1 (CONTINUED) Osmotin increased phosphorylated AMP-activated protein kinase (p-AMPK) and sirtuin 1 (SIRT1) levels and their activities *in vitro* and *in vivo*. (g) Western blot analyses of p-AMPK and SIRT1 levels in hippocampal extracts from 5-, 9- and 12-month-old wild-type (WT) and vehicle (Veh)- or Os-treated (for a short time) APP/presenilin 1 (APP/PS1) and APPSW mice and (h) AMPK and SIRT activity histograms in brain homogenates of Adipo$^{-/-}$ and APP/PS1 mice. The assays were performed three times with similar results, and representative data from one experiment with triplicate samples are shown. The bands were quantified using the Sigma Gel system (SPSS, Chicago, IL, USA), and differences are presented as histograms derived from a one-way analysis of variance (ANOVA) followed by a t-test and as the mean ± SEM for the indicated proteins (n = 5/group). After developing the respective antibodies, membranes were reprobed for the β-actin signal, shown as a loading control. Histograms depict analyses of band density relative to the density of WT on the same blot. Significance: *$P < 0.05$, **$P < 0.01$ and ***$P < 0.001$.[54] (Reprinted by permission from Macmillan Publishers Ltd. *Molecular psychiatry*, Shah, SA, Yoon, GH, Chung, SS, Abid, MN, Kim, TH, Lee, HY, Kim, MO. 22, 407–416, copyright 2017.)

and deposition in APP/PS1 mice of different ages. Western blot analysis of protein extracts showed that the levels of APP, the amyloidogenic β-secretase BACE1 and Aβ were higher in the brain tissues of vehicle-treated APP/PS1 mice of different ages, including 5-, 9- and 12-month-old mice. Conversely, osmotin administration significantly lowered the levels of these proteins, as shown in the results obtained using brain tissues from APP/PS1 mice (Figure 19.2a). The levels of soluble Aβ1-42 and Aβ1-40 were measured using enzyme-linked immunosorbent assays (ELISAs) with brain homogenates of transgenic mice after osmotin treatment. The results indicated that osmotin treatment reduced the levels of soluble Aβ1-42 and Aβ1-40 significantly in the hippocampi and cortices of APP/PS1 mice (Figure 19.2b,c). We also performed thioflavin S and Aβ antibody (6E10) immunostaining to examine both 12- and 16-month-old APP/PS1 mouse brains after osmotin administration for 4 wk. The results indicated that osmotin reduced the amounts of Aβ aggregation and deposition in the cortices and hippocampi of APP/PS1 mice significantly after osmotin administration for 4 weeks (Figure 19.2d,e).

19.1.2 BBB Crossing with Magnetic Force

Currently, there is no cure for neurological diseases such as AD and even the treatment options are extremely limited.[55] Primary challenges in the diagnosis (and treatment) of AD are to overcome the restrictive mechanism of the BBB and deliver adequate levels of drugs to the hippocampus region, located deep within the brain. Several studies have shown that the hippocampus plays an important role in learning and memory.[56]

Based on the pioneering idea proposed by Freeman et al.,[57] that fine iron particles could be transported through the vascular system and concentrated at a particular region of the body using a magnetic field, the use of magnetic particles for the delivery of drugs or antibodies to organs or tissues altered by disease has become an active field of research.[58] Magnetically-guided delivery strategies have the potential to enhance the therapeutic profile of a broad range of pharmaceuticals by increasing their levels at a desired site of action while reducing off-target effects.[59] The feasibility of using MNPs for the targeted delivery of small molecule pharmaceuticals and gene vectors has been assessed in several animal model studies.[60–64] MNPs have also attracted attention due to their relatively low toxicity profile. Their superparamagnetic properties ensure particle stability during storage and use, and their responsiveness to applied magnetic fields can be exploited for magnetically-guided particle targeting[65] or imaging.[66]

19.2 MAGNETIC DRUG DELIVERY TO THE BRAIN

19.2.1 Overview of the Proposed Drug Delivery Scheme

MNPs constitute one of the major particle types used to deliver drugs with noninvasive magnetic forces. However, Cherry et al.[67] demonstrated that the magnetic force acting on MNPs could not overcome the drag force due to blood flow in vessels, and that it is nearly impossible to deliver MNPs to brain cells using standard magnetic actuation systems. In our guiding scheme, the electromagnets are used only for steering, because the blood flow propels the particles. Altering the radial position of the particles moving with the blood inside the vessels can allow them to be directed to the desired vessel at a bifurcation. If the particles are positioned within the 'safe zone' (Figure 19.7) of the vessel before arriving at the bifurcation, they will reach the desired vessel branch. The effects of high blood velocities have been studied to design the FF performance in the bifurcations.[25,27] It also has been suggested that in practice there are several ways for blood flow control like reducing the blood velocity and the modulated flow control which can be achieved with a special tool referred to as a flow control release catheter.[68]

The proposed drug delivery scheme was aimed at improving NP delivery to the brain. To achieve this, we first imaged the patient's brain using MRI and created a 3D model of the brain vessels.[69] Then, we assessed all of the possible paths between the injection point and destination and selected the best conditions for actuation, considering the limitations of the actuation and control system. Finally, the particle trajectories inside the vessels were calculated by simulating the magnetic field of the electromagnet actuation system and blood flow. The introduced platform can be used to design magnetic field variation during all phases of particle delivery to improve the particle delivery rate to the target. Using this approach, offline navigation of the particles inside the vascular system would be possible (Figure 19.3).

19.2.2 Electromagnetic Actuator for Guidance of NPs

As the first step, we developed a special electromagnetic actuation system and evaluated its ability to steer NPs within vessels. The electromagnetic actuator consisted of two sets of coils (7000 turns, wire diameter: 1.0mm) with a core to increase the magnetic field density (cobalt–iron alloy, Vacoflux 50). Currents of up to 17 A (gradient field strength 7.9 T/m) can be applied with the system. The core diameter is 60 mm, the coil length is 110 mm and the diameter of the coil is 180 mm as illustrated in Figure 19.4. The performance time for the system is considered to be 10 min to avoid overheating. For continuous use, a cooling system should be included. The proposed system is optimized for *in vitro* (Y-shape channel in Figure 19.4) and *in vivo* (mice) experiments. The coils should be reconfigured for higher scale testing.

The proposed design has a differential current coil (DCC) arrangement,[70] combining the properties of Helmholtz and Maxwell coil configurations in one set of coils. With the DCC approach, two coils are applied in each direction, and the proposed coil configuration generates a gradient field by increasing the current density in one coil, while decreasing it in the other, attracting particles towards the coil with the higher current density. Because the currents in both coils are in the same direction, the field intensity can be kept higher than the value

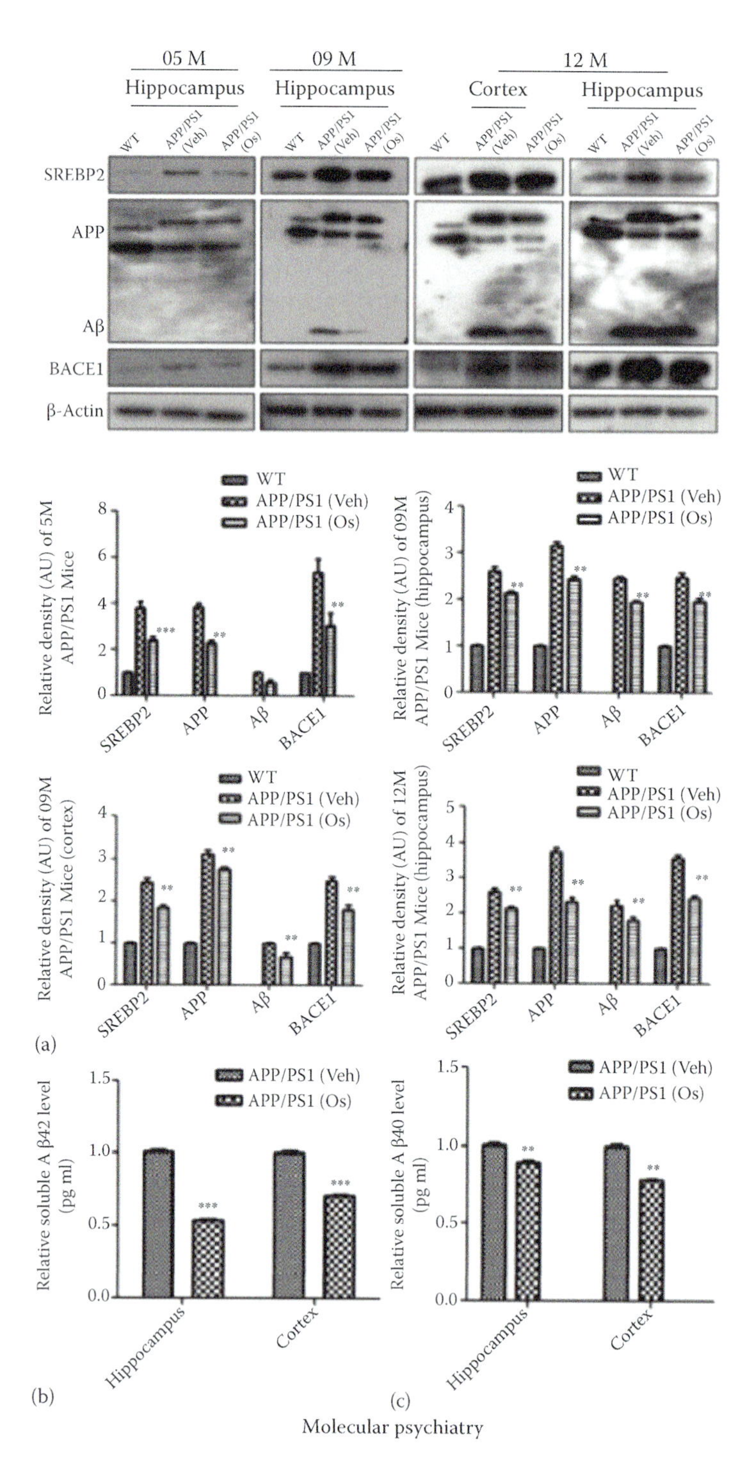

FIGURE 19.2 Osmotin reduced the amyloid burden by inhibiting sterol regulatory element-binding protein 2 (SREBP2) expression in APP/presenilin 1 (APP/PS1) mice. (a) Western blot analysis of SREBP2, BACE1, APP and amyloid-β (Aβ) levels in hippocampal and cortex extracts from 5-, 9- and 12-month-old wild-type (WT) and vehicle (Veh)- or osmotin (Os)-treated (for a short time) APP/PS1 mice. Representative blots and histograms depicting analyses of band density relative to WT bands are shown. The membranes were reprobed for β-actin as a housekeeping control. The bars indicate means ± SEMs (n = 5/group). (b, c) ELISA histograms (relative) of soluble (b) Aβ1-42 and (c) Aβ1-40 in cortex and hippocampal brain homogenates of APP/PS1 mice with or without osmotin treatment. The methods and procedures recommended by the manufacturer were followed and experiments were performed in triplicate. *(Continued)*

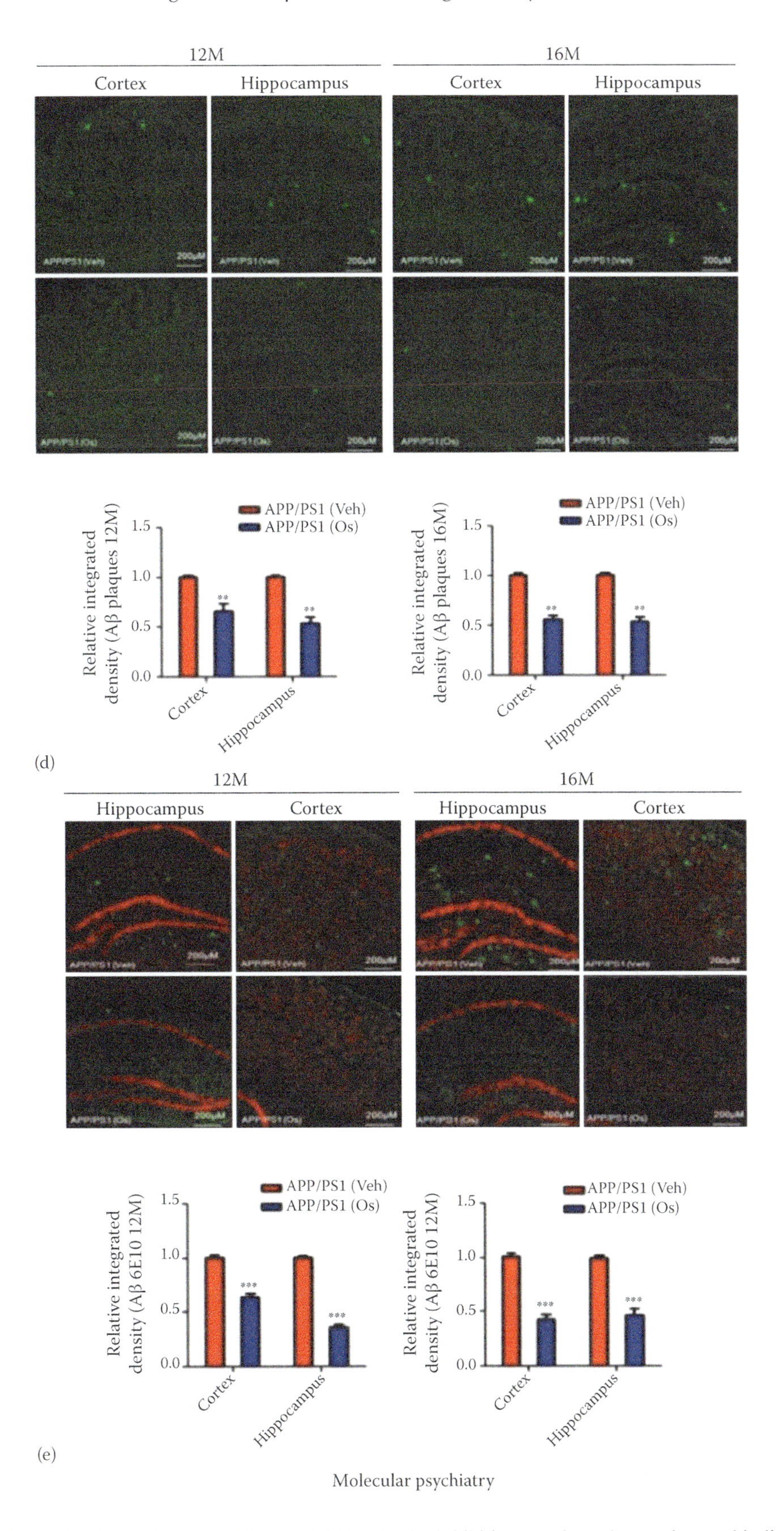

FIGURE 19.2 (CONTINUED) Osmotin reduced the amyloid burden by inhibiting sterol regulatory element-binding protein 2 (SREBP2) expression in APP/presenilin 1 (APP/PS1) mice. (d, e) Fluorescence images with relative density histograms of cortex and hippocampal regions of 12- and 16-month-old Veh- and Os-treated (4 weeks) APP/PS1 mice, indicating the localization of (d) Aβ plaques and (e) Aβ (6E10) aggregates. Significance: $*P < 0.05$, $**P < 0.01$ and $***P < 0.001$.

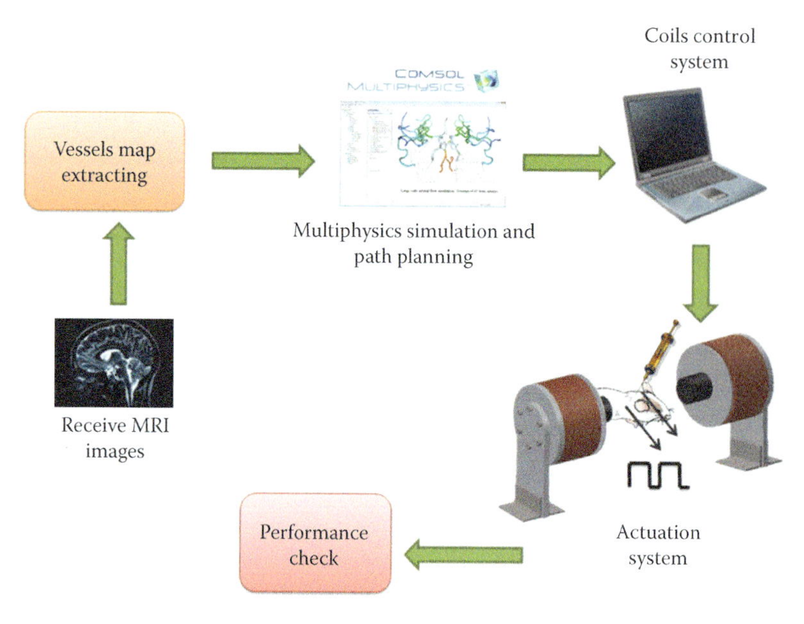

FIGURE 19.3 Nanorobotic drug delivery scheme using an electromagnetic actuator and an offline navigation simulator with particle trajectories inside the vessels.

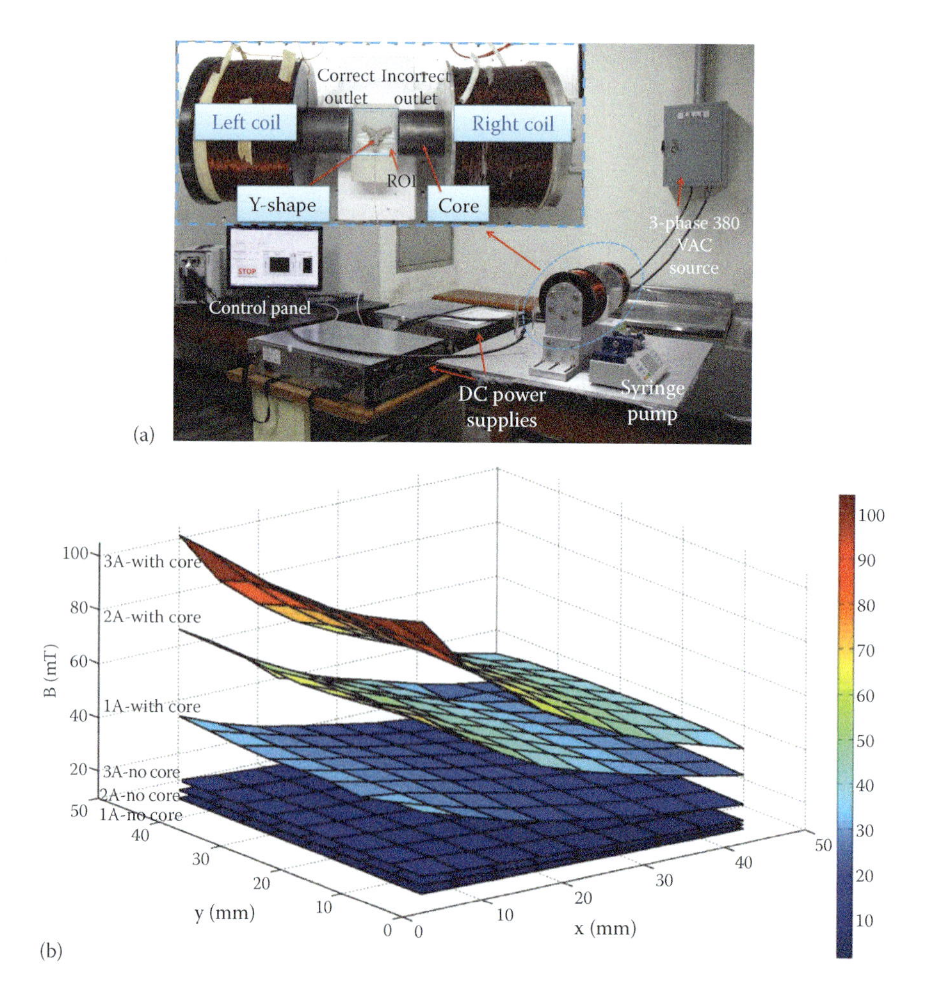

FIGURE 19.4 Electromagnetic actuation system: (a) Experimental setup.[27] (Reprinted from *Journal of Magnetism and Magnetic Materials*, 427, Hoshiar AK, Le T-A, Amin FU, Kim MO, Yoon J, Studies of aggregated nanoparticles steering during magnetic-guided drug delivery in the blood vessels, 181–187, Copyright 2017, with permission from Elsevier.) (b) Magnetic field surface of the left coil in the ROI with and without a core.[71] (Do TD, Amin FU, Noh Y, Kim MO, Yoon J. Functionalized magnetic force enhances magnetic nanoparticle guidance: from simulation to crossing of the blood–brain barrier in vivo. *IEEE Transactions on Magnetics*. 52:1–4 © 2016 IEEE.)

TABLE 19.1
Experimental Conditions

MNP material	Iron powder (Fe)	Flow velocity	13.3 *mm/s*
MNPs diameter	800 nm	Diameter of the Y-shaped channel	1.0 *mm*
MNPs purity	>99.9%	Fluid environment	Water

needed for magnetization. Using DCC, it is possible to make the proposed system more compact, reducing costs and power consumption, as well as using the capability of coils with a core to improve the magnetophoretic force. To improve the efficiency of the electromagnets for MNP guidance, the structural parameters of the cores and coils were chosen, based on simulation results, to provide the highest magnetic force in the region of interest, which was set as the size of a mouse brain.[24]

We found that increasing the width or height of the coil increased the magnetic field intensity and gradient. It was also found that the wire diameter affected the optimum design. Investigations of core parameters demonstrated that the distance between cores influenced both the magnetic field intensity and gradient, whereas the minimum radius of the cores affected only the gradient, and its impact on the magnetic intensity was minor. Moreover, the effect of tip length on both the magnetic field intensity and gradient was slight. Certain constraints should also be considered in the system design. The output power supply and inner and outer coil diameters were considered to be constraints. Based on these and the simulation results, the 'best' configuration[24] was chosen to satisfy the largest minimum force factor in the region of interest (ROI), as shown in Figure 19.4. Figure 19.4a illustrates the experimental setup and dimensions and Figure 19.4b shows the magnetic field surface in the ROI of a coil by comparing a core.[71] The core can significantly strengthen the magnetic field and its gradient at the ROI (about tenfold).

To confirm the effectiveness of the proposed electromagnetic guidance system, we performed experiments with different magnetic fields. Table 19.1 lists the specifications of the MNPs and the experimental conditions.[72] The experiments were carried out with a constant magnetic field and only one coil-core was used to generate a unidirectional force. The exit rates for the particles in the correct outlet with respect to the particles in the incorrect outlet increases compared to the increase in particle radius and in case of the use of the iron core (Figure 19.4). It also confirms that the developed electromagnetic actuator can generate a force adequate for NP steering within the vasculature. (The correct and incorrect outlets are user-defined terms.)

Figure 19.5 shows snapshots of the location of the MNPs in the channel at different times using the electromagnetic actuator at a current of 3 A.[65] It also illustrates the rate of MNPs exiting in the desired outlet for different input currents. As the current in the coil increases, the rate of MNPs reaching the correct outlet rises. The relationship between the percentage of the MNPs that reached the correct outlet and the magnitude of current was approximately linear in both simulations and experiments. The rate of MNPs reaching the correct outlet increases with respect to the rise in magnetic force (which is a function of coil current). This effect demonstrates the ability of the proposed actuator in guiding the MNPs in the desired direction. However, we were not able to apply currents larger than 3 A, due to the MNPs' sticking inside the Y-channel and the syringe. The experimental results showed that the electromagnetic actuator was able to steer 800 nm MNPs effectively in a Y-shaped channel. The similar results for effective MNPs steering with smaller particles are illustrated in Figure 19.6.

19.2.3 Functionalized Magnetic Field for Sticking Prevention and Efficient Guidance

The particle-vessel sticking issue under a constant magnetic field hindered successful MNP guidance. Particles inside the channels were attracted to regions near the walls by the constant magnetic field that was intended to guide the particles to the desired outlet of the channel. The velocity of the fluid flow close to the walls is lower than at other positions inside the vessel and, consequently, the drag force on the particles in these regions is weaker. As a result, the particles move very slowly or the friction force between the channel inner surface and the particles may overcome the flow drag force and, thus, they stick to the wall and remain in the channel.[25]

The sticking and concentration of particles near a Y-shaped channel has been reported before.[73] Likewise,

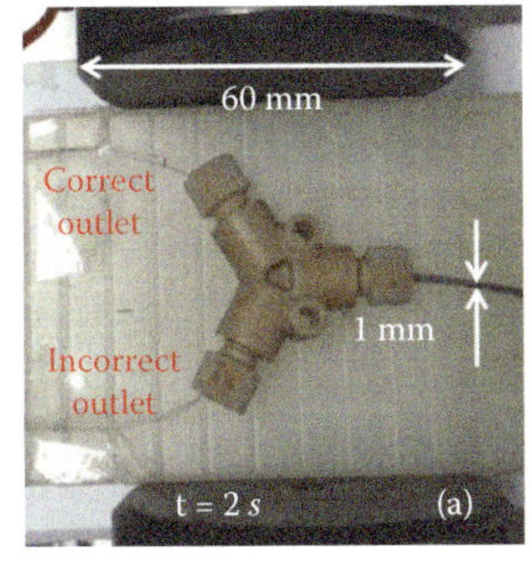

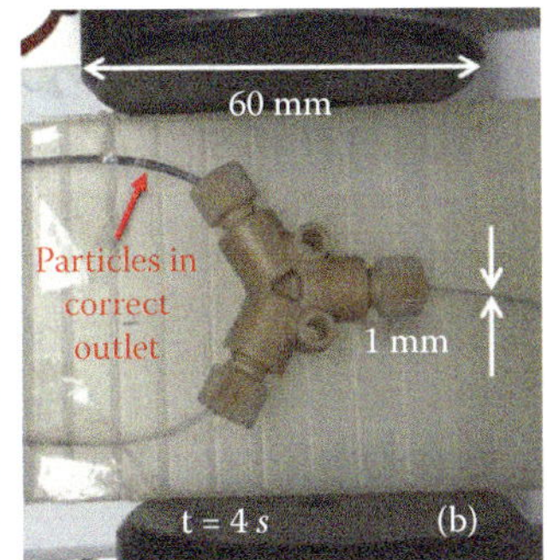

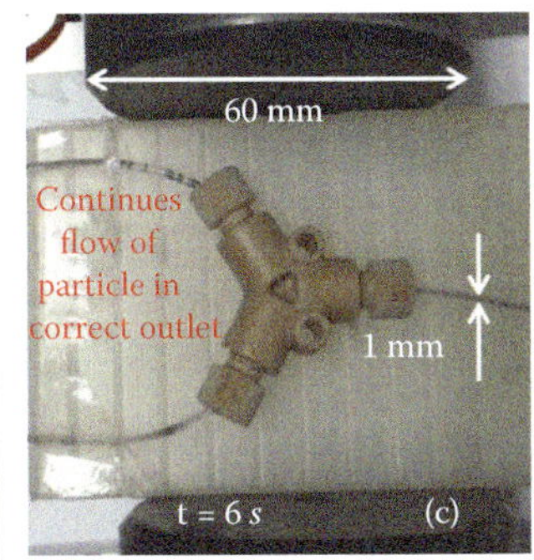

FIGURE 19.5 Snapshots of the locations of MNPs in the channel at various times using the electromagnetic actuator at a current of 3 A (top), and experimental results of MNP guidance using different currents in a Y-shaped channel.

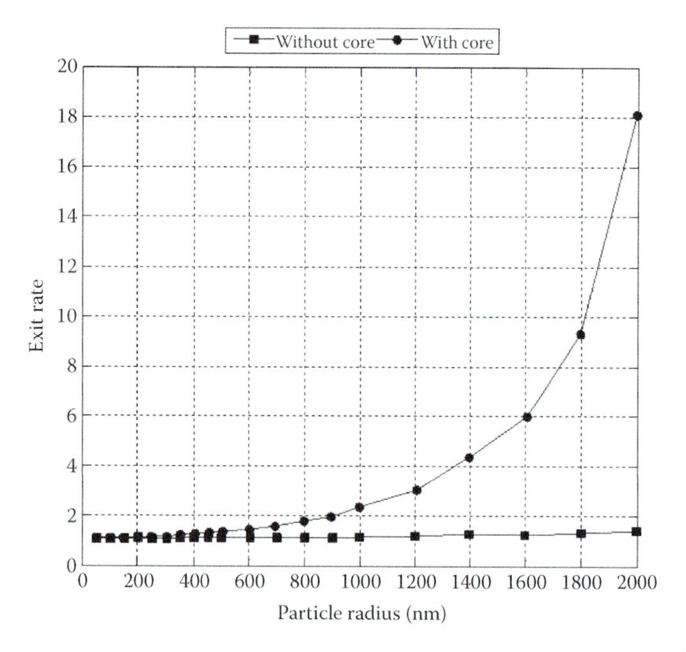

FIGURE 19.6 Exit rate variation with respect to particle diameter.

the adherence of NPs to vessels with a constant magnetic field was examined experimentally and through simulation, respectively.[74,75] Although sticking was reduced slightly by changing the shape of the magnet, this is not a general solution because the study focused only on the special case of particles sticking at the injection site.[75] In Chertok et al.,[32] the steering efficiency, in the special case of a two-dimensional (2D) Y-shaped channel, was increased using the aggregation properties of the particles and a special magnetic field. However, the aggregations orthogonal to the flow are not desirable in drug delivery to the brain due to the clogging of branches.

A magnetic FF is a unitless multiplier function that varies with time, which defines how the magnetic field gradient changes over time to maximize the number of particles reaching the desired outlet by keeping them within the safe zone. This concept is demonstrated in Figure 19.7.[25,27,76]

The FF can guide the MNPs by changing the magnetic field gradient:

$$\nabla H_f = FF(t).\nabla H \tag{19.1}$$

where ∇H is the magnetic field gradient, and $FF(t)$ is the designed FF. The actuation force for steering MNPs is introduced using the ∇H_f. The safe zone concept (Figure 19.7) is initially considered for an open loop design for $FF(t)$ based on simulation. Furthermore, the MPI is proposed for a real-time visual feedback (described in Section 19.3). The MPI is a precise (<1 mm) and real-time (<0.3 s) imaging technique which is developed alongside the actuation scheme to enable real-time visual feedback.

For the open loop scheme, the FF is designed through simulation of MNPs' movement in a vascular system. Optimum conditions for FF depends on the blood vessel size, blood flow velocity, particle size and the magnetic field applied.[25]

For wide ranges of these parameters, three coefficient numbers are generalized: vessel elongation, normal exit time and force rate.[25] As a general trend with a Y-shaped vessel channel (Figure 19.7) with a low exit normal time, low numbers of cycles are more efficient; however, for high normal exit times higher numbers of cycles are better. Furthermore, the best number of cycles for FFs is related directly to the vessel elongation ratio. An increase in force rate also leads to an increase in the number of cycles in most cases. Numerous sticking and guidance simulation results for the FF in Y-shaped or more realistic 3D vessels can be found in Tehrani et al.[25]

For 3D vessel simulations, we first selected a vessel located in the human brain. Next, we extracted its geometry using MRI images via a special procedure.[69] Based on the extracted information, we created a 3D model of the vessel. Finally, we imported this model into the COMSOL software (COMSOL Inc., Palo Alto, CA, USA) prior to the simulation studies. The final 3D model of the channel, with some of the geometric characteristics used for modelling, is shown in Figure 19.8.

We chose the outlet with the longest route and the most bifurcations (Out 6). Blood flow was at a steady, creeping flow, which flowed into the channel from the inlet and exited the channel from the outlets. We selected fluid modelling parameters based on their similarity to blood behavior.[24] In this simulation, for simplicity, the duty ratio was fixed at 0.7. Blood flow inside the channels was simulated as a steady-state laminar flow and its velocity profile was calculated using the computational fluid dynamics module of COMSOL.

Table 19.2 shows the percentages of particles at the desired outlet during simulations using FF schemes with varying frequencies. The data in Table 19.2 clearly indicate that in cases

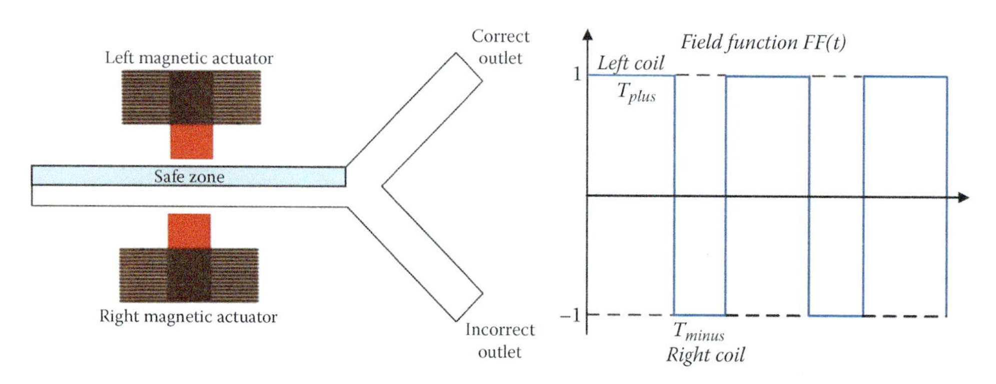

FIGURE 19.7 Mechanism of NP steering within blood vessels with field function (FF[t]) guidance.

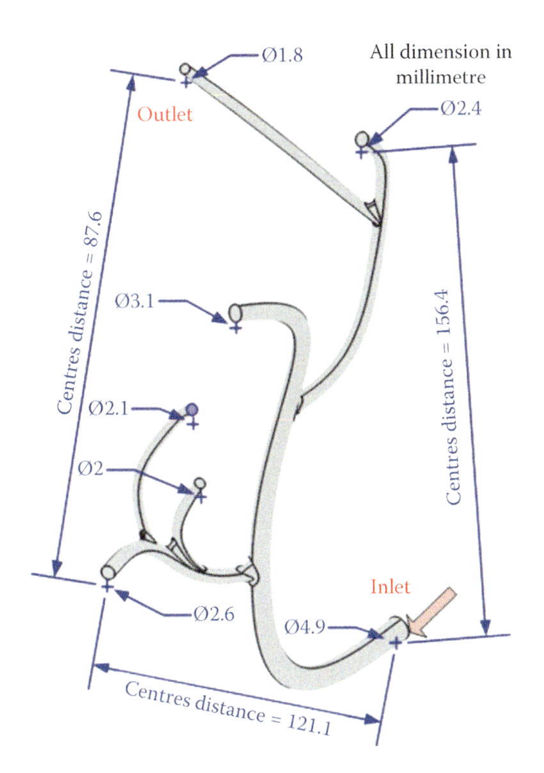

FIGURE 19.8 3D vessel model in simulation studies.[71] (Do, TD, Amin, FU, Noh, Y, Kim, MO, Yoon, J., *IEEE Transactions on Magnetics*, 52, 1–4, 2016. © 2016 IEEE.)

where the magnetic field was kept constant (0 Hz), particles were unable to reach the correct outlet, regardless of the input current. This result occurred because all of the particles became stuck to vessel walls before they could reach the correct outlet. For the various FF schemes, we observed that the MNPs did reach the correct outlet, depending on the frequency and input current values. We observed optimal results at 0.5 Hz. As the frequency increased, fewer particles were stuck on the vessel walls, because the positive time, T_{plus}, was decreased. The negative time, T_{minus}, was long enough to release the particles from the walls (see Figure 19.7 for T_{plus} and T_{minus}). However, when the frequency was too high (> 0.5 Hz), the T_{minus} became too short, and the particles did not have enough time to release from the walls. The results of these simulations suggested that the FF could effectively enhance MNP guidance by preventing sticking at vessel walls. Furthermore, we observed optimal results when the frequency of the FF was 0.5 Hz. Further *in vivo* experimental results for this open loop design are discussed in Section 19.4.

19.2.4 Simulations of Aggregated MNP Steering in Blood Vessels

The particle-tracing module of COMSOL Multiphysics is used for the simulation of the particles steering. The Newtonian dynamic model is considered as:

$$m_i \frac{dv_i}{dt} = F_{MF} + F_{drag} + F_m \quad (19.2)$$

where the index i indicates particle i. v_i is the particle velocity, F_{MF} is the magnetic force, F_{drag} is the hydrodynamic drag force and F_m is the gravitational force. To use Newtonian mechanics, radius of the particles is considered to be greater than 500 nm to exclude the Brownian effect. These forces have been defined in the COMSOL platform to study the steering performance.[67]

Despite the improvement in the BBB crossing using the proposed electromagnetic guidance scheme, we observed that particles aggregated in the brain after crossing the BBB.[26] In real-world use, aggregation phenomena may reduce system performance and should be considered during the design stage of a magnetic guidance system. In previous analyses, particles were considered to be single and aggregation effects were not included, therefore the magnetic force, in the presence of a high blood velocity, was ineffective because it did not have enough power to influence their movement. However, aggregated particles show a better response with respect to magnetic force.

A Y-shaped channel was considered for FF simulations with aggregations of MNPs. The maximum value of the magnetic field gradient applied to the channel, H_g, was chosen based on the average magnitude that the proposed actuation system could generate; thus, the values are realistic and reachable. The proposed magnetic actuators work practically under a 1–6 A current, which generates a magnetic field gradient (H_g) between 0.3 and 2.3 A/m^2. From an extensive study of geometry, environmental and process parameters, it was found that a FF with a duty cycle of 3 to 1 (the ratio of T_{plus} to T_{minus}) could improve the steering performance of MNPs greatly.[25] The FF design was improved further using simulations.[71,76] However, the dipole and contact forces result in particle aggregation, which was not considered in the initial design. Thus, a computational platform was developed to study the aggregation effects.[27]

TABLE 19.2
Percentage of Particles at the Correct Outlet Simulations

Current Frequency	1A	2A	3A	4A	5A	6A
0 Hz	0%	0%	0%	0%	0%	0%
0.25 Hz	2.66%	2.13%	1.96%	1.93%	1.86%	1.86%
0.5 Hz	6.51%	6.15%	10.0%	6.01%	5.98%	5.71%
1 Hz	0%	0%	2.15%	2.82%	1.13%	2.43%

Note: Results are for the outlet ϕ1.8 (shown in Figure 19.8).

To simulate the aggregated MNPs under a FF, first, velocity profile data were extracted using the COMSOL software. Then, the forces were computed based on the modelling. The contact condition in each time step determined particle integration, and the magnetic and drag forces governed particle movements. Based on the FF properties (T_{plus} and T_{minus}) the direction of the magnetic field changed, causing particle movement within the vessel. Finally, the number of particles reaching the desired outlet was determined.

The particles move inside the vessel under the guidance of the applied FF and reached the correct outlet. Figure 19.9 shows the aggregation simulation within a Y-shaped channel. Initially, single particles integrate and generate aggregates (t_0). Then, the column-shaped aggregates move based on the designed FF at different velocities and reach the bifurcation. Finally, as shown in Figure 19.9, particles reach the desired outlet at t_{Final}. Using this simulation model, the performance of the FF can be demonstrated by computing the number of particles that reach the correct outlet.

Figure 19.10 shows the particle distribution pattern. As expected, under the magnetic force, the number of particles reaching the correct outlet was higher than that in the previous estimation.[25] Figure 19.10c and d shows the particle distribution in low-flow velocities. At a low velocity, the sticking of NPs occurs and the FF concept was introduced to address this issue. We found that the sticking issue was solved and evidence of particles has been found in the mouse brain (details in Hoshier et al.)[27] which confirms the results presented in Figure 19.10c and d.

Moreover, the growth in the number of particles reaching the incorrect outlet was also evident. The main reason for this is due to the increase in the magnetic force as a result of aggregation. The force factor clearly demonstrates this trend, because growth in the force factor results in an increased number of particles reaching the incorrect outlet. This analysis is consistent with our recent particle distribution analysis experiments.[26]

The aggregation phenomenon could also explain the reason for the remaining particles' increase in Figure 19.10d. Aggregation leads to growth in the magnetic force, which also leads to an increase in the vertical velocity of the particles. Rapid vertical movement of the particles prevents particle movement with the flow (horizontal movement) and results in an increase in the remaining particles. Consequently, although the sticking issue was solved, particles remained in the vessel because the magnetic force was not adjusted appropriately.

Although aggregation effectively improved the actuation force, it also influenced TDD performance markedly. Thus, an investigation of aggregation can facilitate improvements in targeting efficiency. The proposed aggregation model simulates steering performance in a Y-shaped vessel by means of a FF, and the results agree with those obtained experimentally. Moreover, velocity, magnetic intensity and particle size were found to influence steering efficiency.[27] These results will facilitate a better design for a FF to enable precise guidance of NPs within the vasculature.

19.3 REAL-TIME NAVIGATION OF MNPS WITH MPI

In this section, we describe an MPI-based real time navigation system that sequentially actuates and monitors MNPs in a noninvasive manner. In practical use, the NPs are in aggregated form and these aggregates, which are millimetre and submillimetre in size, are detectable by the MPI. The MPI is capable of illustrating the particles' concentration.

An MPI-based monitoring scheme was designed based on an excitation field at a high frequency of 40 kHz and a low amplitude of 0.2 mT/μ_0, which provided a safe scheme due to a reduced probability of unpleasant peripheral nerve stimulation. The proposed MPI scheme is also feasible for enlarging the field of view (FOV; the range of observation in each monitoring period).[30] Based on this imaging concept and our electromagnetic actuator,[24] a one-dimensional navigation system for a mouse-size was built and validated with both 90 nm and 5 nm particles.[29] The results suggest that the MPI-based

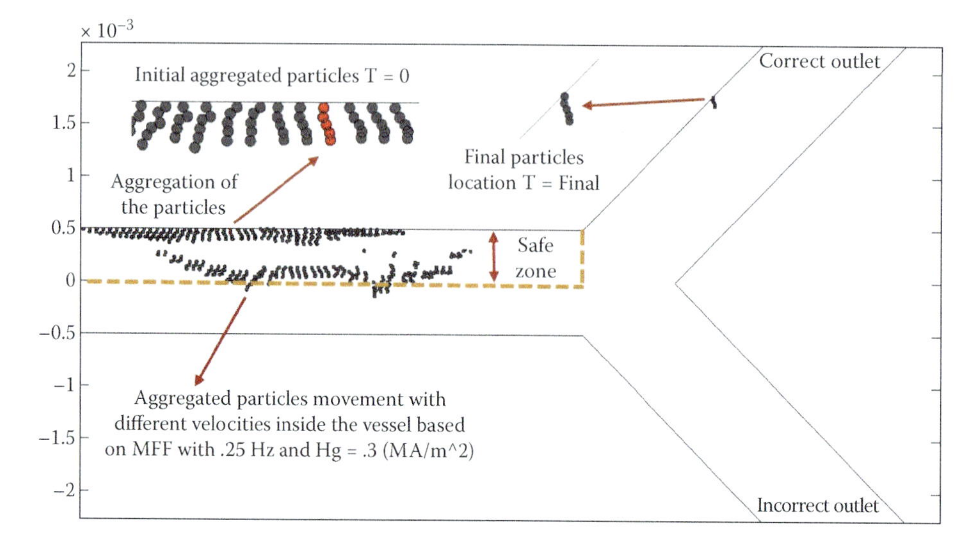

FIGURE 19.9 Steering of the aggregated particles in a Y-shaped channel with the FF(t).[27] (Reprinted from *Journal of Magnetism and Magnetic Materials*, 427, Hoshiar, AK, Le, T-A, Amin, FU, Kim, MO, Yoon, J., 181–187, Copyright 2017, with permission from Elsevier.)

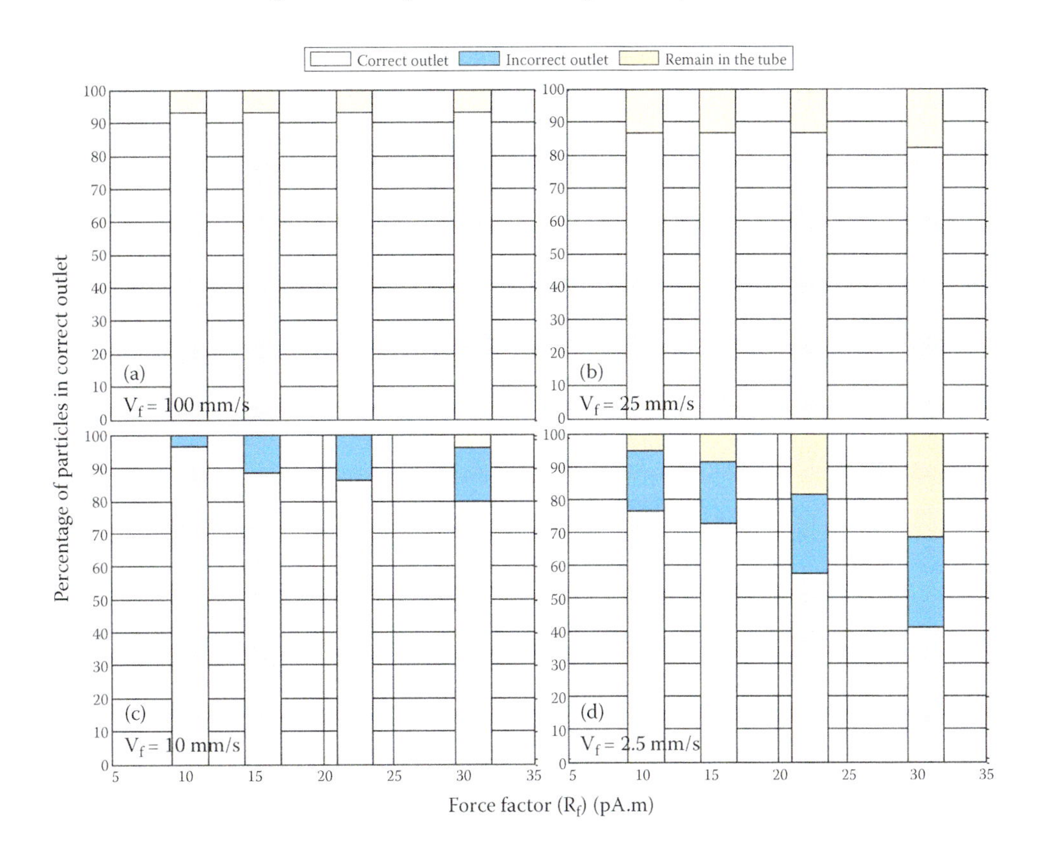

FIGURE 19.10 Results of steering performance simulation for a Y-shaped vessel with dv = 1 mm, L_v = 10 mm and FF duty cycle of 0:25 Hz with equal T_{plus} and T_{minus} and R_f = [10:5; 15:7; 22:3; 30:7] pA:m and four velocity conditions, (a) V_b = 100 mm/s, (b) V_b = 25 mm/s, (c) V_b = 10 mm/s, and (d) V_b = 2:5 mm/s.[27] (Reprinted from *Journal of Magnetism and Magnetic Materials*, 427, Hoshiar, AK, Le, T-A, Amin, FU, Kim, MO, Yoon, J., 181–187, Copyright 2017, with permission from Elsevier.)

navigation system was capable of controlling NPs in real-time with a closed-loop control scheme.

Although the MPI systems initially were designed for small animal experiments, the concept of a human size field free point (FFP) was introduced in 2013.[77] The concept was further studied in a mini-pig size.[78] Due to the implementation limitation, the development of a rabbit-sized field free line scanner was started with a fast 2D imaging. Since a large FOV and fast acquisition are required, the upscaling of MPI for human size is still challenging. The possible approaches for extending FOV to human size is studied by Bringout et al.[78,79]

19.3.1 The Schematic of MPI-Based Monitoring

The MPI-based monitoring system consists of four sets of coils: excitation, receive, selection-driver and cancellation coils. To generate the desired signal characteristics, a high frequency excitation field is applied to the MNPs by the excitation coil, and the MNPs 'respond' by inducing a voltage signal in the receive coil. For spatial encoding, a gradient magnetic field, called the selection field, is applied. To detect the particles' position, a FFP needs to be generated inside, with which the change in particle magnetization can be detected due to the excitation field. The selection-driver coils are designed to create and move the FFP within the FOV. The trajectory of the FFP in the FOV depends on the current signal's amplitude and frequency in the selection-drive coils.

The excitation field also induces signals from the receive coil, which do not have particle information. In the absence of particles in the FFP, however, the receive coil will induce a high voltage signal as a result of excitation coil effects. Thus, the cancellation coil is used to eliminate this undesired signal.[31] Figure 19.11 shows the coil configuration for the MPI monitoring system.

By controlling the trajectories of FFP over FOV and monitoring the particle magnetization responses, the MNPs'

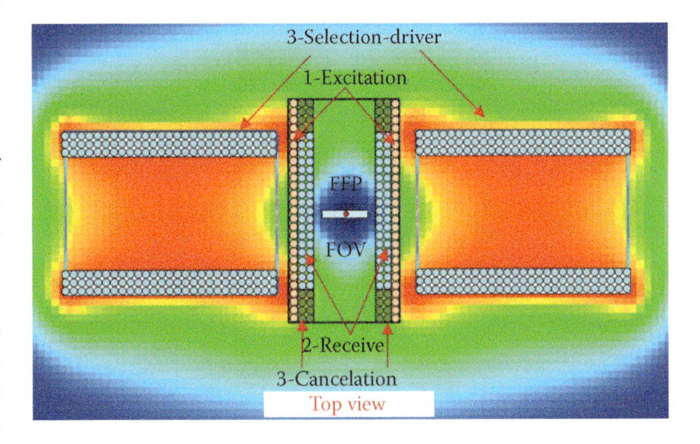

FIGURE 19.11 Schematic of MPI monitoring system. The left and right coils are responsible for generating and moving the FFP and the middle coil, which consists of three different layers, generating excitation, receiving a signal and filtering undesired signals.

concentration can be mapped in an image. The FFP can be moved in 3D using three orthogonal homogenous magnetic fields provided by three drive coil sets.

In the proposed low-amplitude electromagnetic field (LAEF) MPI scheme, we used a drive field with a low frequency to move the FFP, while an excitation field with low amplitude and high frequency was used to generate oscillating magnetization. The cancellation coil was used to filter the empty coil signal.[80] We can measure the concentration of particles in FFP from the amplitude of the signal received by applying the amplitude modulation of the excitation field. Because a precise single excitation frequency is important, a band-pass filter can diminish harmonic interference induced by the power unit for the excitation field. The signal detection and processing procedure for the LAEF MPI are shown in Zhang et al.[30] Then, the MPI image can be acquired in real time using the X-space reconstruction method.[81]

19.3.2 Integration of MPI and Magnetic Actuation System

Several methods have been suggested to develop TDD with feedback control, such as by using ultrasound for locating solid microsize particles,[82] or using a microscope to track visible particles.[83] Magnetic resonance navigation (MRN) is the most effective method for feedback control of TDD. MRN is based primarily on MNPs embedded in microcarriers, which are controlled and tracked using MRI systems.[84,85] However, major challenges in a TDD system include generating sufficiently high gradient fields to steer the MNPs and tracking MNPs in real time, to enable precise targeting.[86] The dipole field navigation has been suggested to provide at the same time a higher magnetic field strength and magnitude (gradient of 400 mT per meter and 3T field strength at human scale).[87]

To address the real-time monitoring issue, MPI has emerged as a revolutionary imaging technique that enables real-time monitoring of MNPs. The proposed MPI monitoring system was associated with our electromagnetic actuator to form a hybrid navigation system. The MPI-based navigation system is capable of steering the MNPs in real-time with a closed-loop control scheme.

The system schematic shows four tasks: (1) the electromagnetic actuation task, for changing the particle's position (performed with the selection-driver coils), which lasts 0.1 s; (2) the relaxation task, which lasts 0.05 s; (3) the MPI acquisition task, which lasts 0.3 s; and (4) the reconstruction and control task, which is responsible for the MPI signal processing, image reconstruction and controller routine. The first three tasks are performed sequentially; however, the reconstruction and control task is performed in parallel to tasks 2 and 3. (See Figure 19.12.)

During the MPI task, a 0.7-A AC current was used to create and control the FFP. A 2-A direct current was used to offset each driver-cancellation coils with the DCC structure, leading to a 4-cm FFP scanning range. The FFP scanning frequency of 0.1 s was determined by the frequency of the AC current component, which was 5 Hz. The magnetic field gradient used to steer the MNPs was generated by a 5-A driver-selection coil. The field gradient was 3.5 T/m during the MPI and 8.75T/m during the electromagnetic actuation.

19.3.3 MPI-Based Real Time Navigation System

To implement the MPI-based navigation system, two types of MNPs were used: (1) MNPs of 90 nm in diameter with a core size of 60–70 nm (MF-COO-0090, MagQu Co., Ltd.), and (2) MNPs of 45–65 nm in diameter with a 5–6-nm core size (Resovist, Meito Sangyo Co., Ltd.). The MNPs information on particle concentration indicates that the MagQu particle sample has a concentration of 8 mg of Fe/ml and the Resovist particles has a concentration of 55 mg of Fe/ml. In this step of the work, the particles are only for monitoring purposes and the BBB crossing is not targeted in this phase. Figure 19.13 shows a control block diagram for the proposed navigation scheme of MNPs using a hybrid of the MPI and the electromagnetic actuator. The results from experiments with the Resovist MNPs are shown in Figure 19.14. The grayscale images were reconstructed from MPI images, which show the concentrations of the MNPs. A pure black colour indicates an MNP concentration of zero and a pure white colour indicates the highest concentration of MNPs in the suspension.

The experiments showed that the navigation system could successfully monitor the positions of MNPs and provide information for the control of actuation. Our 1D MNP navigation system had a 2-Hz system refresh frequency (repetition frequency), meaning that the MPI image frame and control field were updated every 0.5 s. This is an appropriate frequency for navigating capillaries, where the velocity of blood is ~1 mm/s.[88]

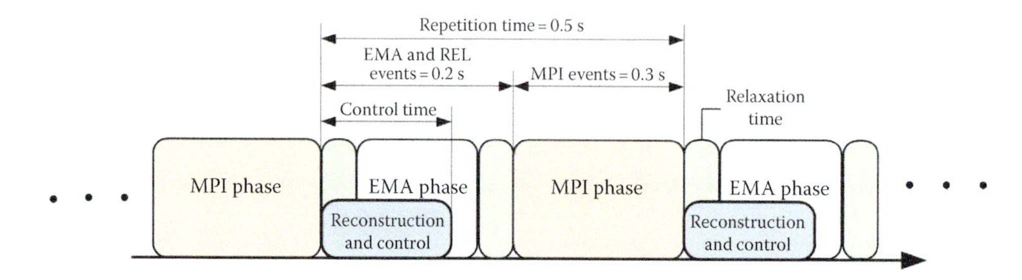

FIGURE 19.12 Schematic diagram of the hybrid MPI navigation system. The monitoring and control scheme consists of three sequential phases of EMA, MPI, relaxation and a parallel reconstruction phase. (Reprinted from *Journal of Magnetism and Magnetic Materials*, 427, Zhang, X, Le, T-A, Yoon, J., 345–351, Copyright 2017, with permission from Elsevier.)

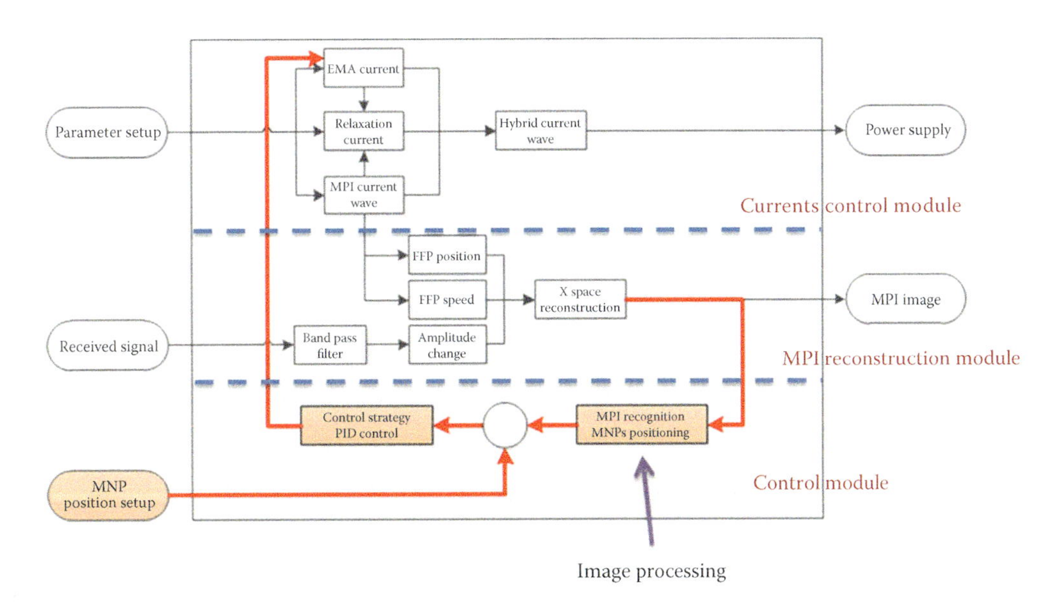

FIGURE 19.13 Control block diagram for navigation of MNPs.

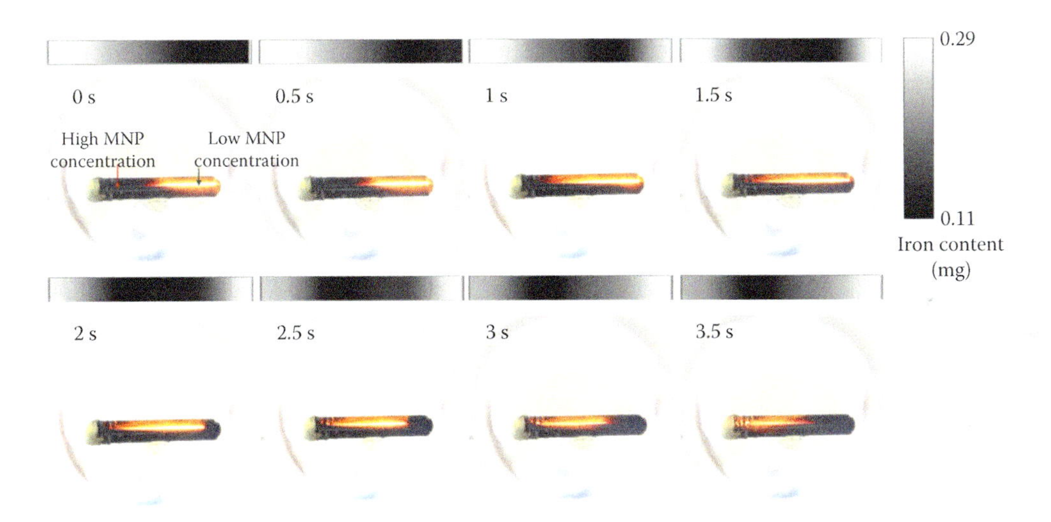

FIGURE 19.14 Steering of Resovist MNPs with the MPI-based navigation system. The black and white toolbar is used to demonstrate the particles' location. The white colour indicates a high concentration of the particles and the black colour, lower particle concentration. (Reprinted from *Journal of Magnetism and Magnetic Materials*, 427, Zhang, X, Le, T-A, Yoon, J., 345–351, Copyright 2017, with permission from Elsevier.)

A navigation system for the real-time actuation and monitoring of NPs is essential for precise targeting and diagnosis with spatial information about the NPs. It can also be cost-efficient, compact and optimized for precise targeting of the NPs. The current design is suitable for a mouse brain, but it is not directly scalable to a human brain. The results of our experiments and the methods described provide valuable data for the potential future enhancement of minimally invasive surgeries, intervention systems and procedures. Our new techniques can be adapted to a variety of applications, particularly the targeted delivery of nanomedicines.

19.4 AD MAGNETIC DRUG TARGETING

The hippocampal region of the brain should be targeted for AD treatments. Because the hippocampus is located in a deep site within the brain, targeting the region is difficult to achieve. More importantly, the BBB prevents particles > 200 nm from reaching the region of interest.[89,90] It is known that an external magnetic field can improve the transport of MNPs through cell barriers and the BBB by endocytotic processes.[91,92]

The MNPs, as drug carriers, transport the drug to the hippocampus and enhance treatment performance markedly. Drug delivery, enhanced by the proposed magnetic actuation system, overcomes the restrictions of the BBB and delivers adequate drug levels to the hippocampus. An optimized scheme for electromagnetic guidance in AD treatment has been suggested.[26]

19.4.1 Best Conditions for Crossing the BBB

MDT refers to adding drugs to magnetizable particles and then applying magnetic fields to concentrate them at disease locations, such as solid tumours, regions of infection or blood clots.[93,94] The idea of using permanent magnets for drug delivery has been investigated for the past 30 years.[95] This approach, however, fails to deliver micro–nano agents to deep tissues or provide quick controlled responses. Using a constant magnetic force or permanent magnets may cause the particles to aggregate or stick to vessel walls, which can lead to blockages. To address this concern, we previously suggested the use of a functionalized magnetic field (FMF), or a FF, to replace the constant magnetic force using an electromagnetic actuator with a high-gradient magnetic field.[24,25] Sticking and aggregation were prevented by intentionally changing the direction of the magnetic field. We also observed that, relative to a constant magnetic field, the rate of MNP uptake and transport across the normal, intact BBB was enhanced by a positive/negative pulsed magnetic field through *in vivo* experiments in mice.[26] Previously, Min et al. used a permanent magnet with a pulsed magnetic field in experiments with cells crossing barriers *in vitro*.[92]

We introduced a novel electromagnetic TDD actuator for guiding MNPs to a region of disease inside the brain. We studied the delivery of MNPs in the brains of normal mice with an intact BBB. We demonstrated the delivery of MNPs to the brain controlled by an external electromagnetic field. Mice were injected with fluorophore-labeled MNPs intravenously and the applied electromagnetic field was kept constant or pulsed on and off. Our study established the ability of an external magnetic field to regulate the MNPs' distribution into the CNS, demonstrating particles crossing the BBB and accumulating within the brain, which can be beneficial for treating CNS diseases. The results suggested that the actuator could be used to enable 770-nm nanocarriers to cross the BBB. The experimental studies confirm that aggregated particles experience a higher magnetic field, which can elevate the BBB crossing performance.

Fluorescent carboxyl magnetic particles (yellow, 1% w/v) used were 700–900 mm (mean: 770 nm) in diameter. According to Debbage et al., the one-pass circulation time of a mouse is ~15 s. Thus, after injecting mice with the nanocarriers, the magnetic field was turned on for 1 min to enable sufficient uptake of the nanocarriers through the BBB.[96]

Figure 19.15 shows confocal micrographs of brains of mice that received an intravenous injection of fluorescent MNPs (0.4 mL) with no magnetic field (Figure 19.15a and b), a constant magnetic field (Figure 19.15c and d) and a pulsed magnetic field (Figure 19.15e–g). There was no significant accumulation of MNPs within the brain in the absence of an applied magnetic field (Figure 19.15a and b). Uptake of fluorescent MNPs into the brain was increased when animals were exposed to a magnetic field (using the FF concept)

FIGURE 19.15 Confocal microscope images showing the uptake and transport of MNPs to the brain of mice. 0.4 mL MNPs was injected intravenously to all of the tested mice except the control group to whom we injected the same amount of saline. Scale bar: 50 μm (a–g).

(Figure 19.15c–g). Figure 19.15(c and d) shows the results of applying a constant magnetic field using input currents of 1 and 3 A, respectively, while Figure 19.15 (e–g) shows the results with a positive–negative-pulsed magnetic field and a 3-A input current at frequencies of 0.25, 0.5 and 1 Hz, respectively. Under constant magnetic field conditions (Figure 19.15c,d), as the input current increased from 1 to 3 A, the magnetic field and magnetic field gradient became higher, resulting in higher magnetic force generation. Figure 19.15d shows more MNP uptake into the brain versus Figure 19.15c. However, a higher magnetic force with a constant magnetic field increases the aggregation of particles, which has a negative effect in particles crossing the BBB.[75] This also occurred for particles that had crossed the BBB, as can be seen in Figure 19.15d. Thus, to eliminate the aggregation phenomenon and increase the uptake of particles into the brain, we propose the concept of the FF (previously described in Section 19.2).[25]

With the FF concept (Figure 19.15e–g), the rate of particle transport across the BBB was increased significantly compared to that in the presence of a constant magnetic field with the same current (Figure 19.15d). Additionally, the apparent intracellular mass of MNPs internalized in the brain tissues under various magnetic field conditions (Figure 19.15c–g) was considerably higher under the FF function than a constant magnetic field, indicating that pulsing (based on the FF) of the magnetic field promoted both the uptake and transport of MNPs across the BBB. It seemed that by pulsing the electromagnetic field, the aggregation of particles could be reduced[25] and this may have caused more particles to cross the BBB. By applying the FF, the rate of particle transport across the BBB increased significantly compared to the transport under a CMF with the same current ($P < 0.0001$)[73] (Figure 19.16).

Therefore, the FF can be used to enhance the transport of MNPs across the BBB. In contrast, when the magnetic field was constant, large magnetized aggregates formed, resulting in a markedly smaller fraction of particles being transported across the BBB. These results suggest that a constant magnetic field is less effective at promoting MNP transport because aggregated MNP clusters cannot penetrate the BBB. Consistent with our findings, continuous exposure to high magnetic fields did not necessarily lead to a greater targeting performance compared to a weaker magnetic field.[68] More importantly, our results indicate the importance of exploring the effects of variation in the magnetic field in the context of *in vivo* drug/gene delivery applications. Our results suggest that the interaction of a pulsed magnetic field (base on the FF) with particle dosing, the magnetic strength and overall duration of applying the magnetic field may also be exploited to obtain the most effective, selective and reliable effects.

Moreover, we demonstrated the delivery effects of osmotin-loaded magnetic nanoparticles (OMNPs) in the brains of $A\beta_{1-42}$-treated AD mice using an electromagnetic function (FF; 6 A, 0.5 Hz) to guide the magnetic particles into brain regions, especially to the hippocampus, which significantly recovered and reversed $A\beta_{1-42}$-induced neurotoxicity. We showed for the first time that, compared with osmotin alone, electromagnetic actuator-guided OMNPs reduced Aβ accumulation, BACE-1 expression, synaptotoxicity, memory impairment and tau hyperphosphorylation in an $A\beta_{1-42}$-injected mouse model.

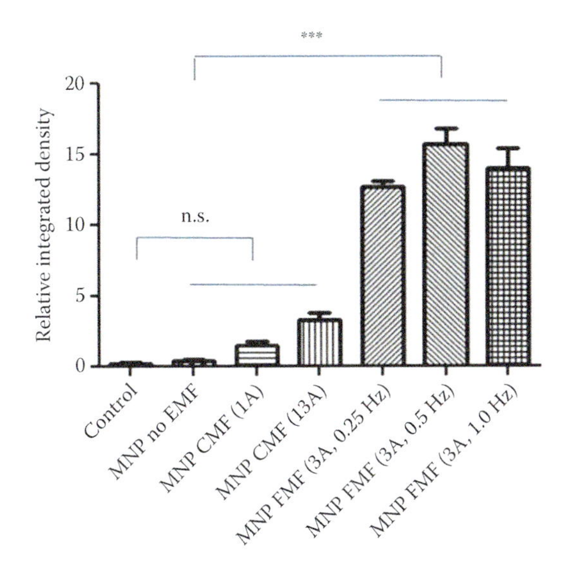

FIGURE 19.16 Histogram showing results of confocal microscopy analysis using the ImageJ software. Data are expressed as the mean ± SEM of experiments performed in triplicate (n = 3). Significant differences were evaluated using a one-way ANOVA followed by Student's t-test. *** indicates a significant difference vs. the control group and CMF (*** $P < 0.0001$). EMF: electromagnetic field; CMF: constant magnetic field; and FMF: functionalized magnetic field, or FF. (Do, TD, Amin, FU, Noh, Y, Kim, MO, Yoon, J., *IEEE Transactions on Magnetics*, 52, 1–4, © 2016 IEEE.)

19.5 CONCLUSIONS AND FUTURE OUTLOOK

Although the synthesis and fabrication of NPs for TDD have improved, important issues remain in the development for a propulsion system within blood vessels. The main obstacle in successful MNPs steering is the size of the NPs, which requires high power to generate a sufficient actuation force. First, we described a new actuation system that can be used in a compact experimental setup for studying MNP drug delivery systems. The configuration and dimensions of this system were chosen to cover the size of the mouse brain, as an initial goal. The results demonstrated that this system could generate the magnetic field necessary for NP steering within the vascular system. Although the current design is adequate for the mouse brain, the design parameters of the proposed system are not directly scalable for human brain use and further studies are needed to extend its applicability to human subjects.

It was observed experimentally that the FF concept solved the sticking issue and facilitated MNP movement. The results when using a FF showed a significant improvement in drug uptake in comparison with a pulsed or constant magnetic field. The 'best' conditions for *in vivo* drug uptake were studied. Promising experimental results encouraged us to further develop the nanorobotic approach for in-brain drug delivery. A simulation platform for TDD was developed by considering aggregation effects to simulate aggregated NPs' steering under the designed FF. As long as they were retained within the safe zone by the FF, the particles moved rapidly towards the desired outlet. However, the FF must be modified to enable precise guidance with respect to different force factors (R_f particle size and gradient of magnetic intensity). Moreover, the results of aggregated particle steering demonstrated the particle distributions in mouse experiments. Thus, the design of the FF should be further improved to elevate the targeting performance in consideration of particles' aggregations.

The MPI-based system enabled real-time monitoring of MNPs. A navigation system for real-time actuation and monitoring of the NPs is essential for precise targeting and diagnosis with NPs based on spatial information. It can also be cost-efficient, compact and optimized for the precise targeting of NPs. The results of our experiments and the methods described here provide valuable data for the potential future enhancement of minimally invasive surgeries, intervention systems and procedures. Our new techniques can be adapted to various applications, particularly the direct delivery of nanomedicines. The current system is suitable for a mouse brain-sized workspace. Future work will focus on 3D system extension and human size upscaling.

The low-amplitude-excitation-field MPI, with the advantage of very low amplitude for a high frequency excitation field, is feasible for enlarging the size of the FOV, because we can combine a core with a coil for a high magnetic gradient field with an enlarged workspace. The disproportionate scaling of magnetic fields and drag forces over human-sized workspaces will require higher gradient fields and magnetic field strengths. For *in vivo* imaging at a human scale, several design improvements and innovations must be introduced to the present system. Real-time implementation of a mouse-sized system and guidance experiments in a vascular network with many bifurcations are also under development. Moreover, the promising results with MPI-based real-time monitoring encouraged us to create a virtual reality environment for MNP steering. Subsequently, the guidance schemes at different branches of blood vessels as well as *in vivo* experiments with the proposed real-time MNP monitoring system should be studied. We also continue to study the use of the Osmotin-loaded MNP carriers since Osmotin has been studied extensively for the treatment of AD, and the transport of Osmotin-loaded MNPs to the mouse brain using the proposed electromagnetic drug delivery system yields promising results.[97]

The methods and experiments described in this chapter provide valuable data for the potential future enhancement of drug delivery systems with MPI feedback. Our novel electromagnetic actuation scheme can be adapted to the drug delivery platform for brain drug targeting. Future works should cover the design and implementation of the 2D real-time navigation system. Also, scaling-up of the system should be considered and *in vivo* real-time imaging (with MPI) and navigation of MNPs should be further studied.

REFERENCES

1. 2014 Alzheimer's disease facts and figures. *Alzheimer's & Dementia: The Journal of the Alzheimer's Association.* 2014;10:e47–92. Epub 2014/05/13.
2. Brookmeyer R, Johnson E, Ziegler-Graham K, Arrighi HM. Forecasting the global burden of Alzheimer's disease. *Alzheimer's & Dementia: The Journal of the Alzheimer's Association.* 2007;3:186–91. Epub 2007/07/01.
3. Hardy J, Selkoe DJ. The amyloid hypothesis of Alzheimer's disease: Progress and problems on the road to therapeutics. *Science (New York, NY).* 2002;297:353–6. Epub 2002/07/20.
4. Scahill RI, Schott JM, Stevens JM, Rossor MN, Fox NC. Mapping the evolution of regional atrophy in Alzheimer's disease: Unbiased analysis of fluid-registered serial MRI. *Proceedings of the National Academy of Sciences of the United States of America.* 2002;99:4703–7. Epub 2002/04/04.
5. Dilnawaz F, Sahoo SK. Therapeutic approaches of magnetic nanoparticles for the central nervous system. *Drug Discovery Today.* 2015;20:1256–64. Epub 2015/06/24.
6. Kreuter J. Nanoparticulate systems for brain delivery of drugs. *Advanced Drug Delivery Reviews.* 2001;47:65–81. Epub 2001/03/17.
7. Kreuter J, Shamenkov D, Petrov V, Ramge P, Cychutek K, Koch-Brandt C, Alyautdin R. Apolipoprotein-mediated transport of nanoparticle-bound drugs across the blood–brain barrier. *Journal of Drug Targeting.* 2002;10:317–25. Epub 2002/08/08.
8. Olivier JC. Drug transport to brain with targeted nanoparticles. *NeuroRx.* 2005;2:108–19.
9. Shubayev VI, Pisanic TR, 2nd, Jin S. Magnetic nanoparticles for theragnostics. *Advanced Drug Delivery Reviews.* 2009;61:467–77. Epub 2009/04/25.
10. Vauthier C, Labarre D, Ponchel G. Design aspects of poly(alkylcyanoacrylate) nanoparticles for drug delivery. *Journal of Drug Targeting.* 2007;15:641–63. Epub 2007/11/29.
11. Martel S. Learning from our failures in blood–brain permeability: What can be done for new drug discovery? *Expert Opinion on Drug Discovery.* 2015;10:207–11. Epub 2015/02/07.
12. Pardridge WM. The blood–brain barrier: Bottleneck in brain drug development. *NeuroRx.* 2005;2:3–14. Epub 2005/02/18.
13. Wu H, Hu K, Jiang X. From nose to brain: Understanding transport capacity and transport rate of drugs. *Expert Opinion on Drug Delivery.* 2008;5:1159–68. Epub 2008/09/27.
14. Tabatabaei SN, Girouard H, Carret AS, Martel S. Remote control of the permeability of the blood–brain barrier by magnetic heating of nanoparticles: A proof of concept for brain drug delivery. *Journal of Controlled Release.* 2015;206:49–57. Epub 2015/03/01.
15. Tabatabaei SN, Girouard H, Carret A-S, Martel S. Remote control of the permeability of the blood–brain barrier by magnetic heating of nanoparticles: A proof of concept for brain drug delivery. *Journal of Controlled Release.* 2015;206:49–57.
16. Mesiwala AH, Farrell L, Wenzel HJ, Silbergeld DL, Crum LA, Winn HR, Mourad PD. High-intensity focused ultrasound selectively disrupts the blood–brain barrier in vivo. *Ultrasound in Medicine & Biology.* 2002;28:389–400. Epub 2002/04/30.
17. Tosi G, Vergoni AV, Ruozi B, Bondioli L, Badiali L, Rivasi F, Costantino L, Forni F, Vandelli MA. Sialic acid and glycopeptides conjugated PLGA nanoparticles for central nervous system targeting: In vivo pharmacological evidence and biodistribution. *Journal of Controlled Release.* 2010;145:49–57. Epub 2010/03/27.
18. Martel S, Mathieu JB, Felfoul O, Chanu A, Aboussouan E, Tamaz S, Pouponneau P, Yahia L, Beaudoin G, Soulez G, Mankiewicz M. A computer-assisted protocol for endovascular target interventions using a clinical MRI system for controlling untethered microdevices and future nanorobots. *Computer Aided Surgery.* 2008;13:340–52. Epub 2008/11/26.

19. Mathieu JB, Martel S. Magnetic microparticle steering within the constraints of an MRI system: Proof of concept of a novel targeting approach. *Biomedical Microdevices*. 2007;9:801–8. Epub 2007/06/15.
20. Choi H, Cha K, Choi J, Jeong S, Jeon S, Jang G, Park J-o, Park S. EMA system with gradient and uniform saddle coils for 3D locomotion of microrobot. *Sensors and Actuators A: Physical*. 2010;163:410–7.
21. Jeong S, Choi H, Choi J, Yu C, Park JO, Park S. Novel electromagnetic actuation (EMA) method for 3-dimensional locomotion of intravascular microrobot. *Sensors and Actuators A: Physical*. 2010;157:118–25.
22. Kummer MP, Abbott JJ, Kratochvil BE, Borer R, Sengul A, Nelson BJ, OctoMag: An electromagnetic system for 5-DOF wireless micromanipulation. 2010 3–7 May 2010.
23. Kummer MP, Abbott JJ, Kratochvil BE, Borer R, Sengul A, Nelson BJ. OctoMag: An Electromagnetic System for 5-DOF Wireless Micromanipulation. *IEEE Transactions on Robotics*. 2010;26:1006–17.
24. Tehrani MD, Kim MO, Yoon J. A Novel Electromagnetic actuation system for magnetic nanoparticle guidance in blood vessels. *IEEE Transactions on Magnetics*. 2014;50:1–12.
25. Tehrani MD, Yoon JH, Kim MO, Yoon J. A novel scheme for nanoparticle steering in blood vessels using a functionalized magnetic field. *IEEE Transactions on Biomedical Engineering*. 2015;62:303–13.
26. Do TD, Ul Amin F, Noh Y, Kim MO, Yoon J. Guidance of magnetic nanocontainers for treating Alzheimer's disease using an electromagnetic, targeted drug-delivery actuator. *Journal of biomedical nanotechnology*. 2016;12:569–74. Epub 2016/06/10.
27. Hoshiar AK, Le T-A, Amin FU, Kim MO, Yoon J. Studies of aggregated nanoparticles steering during magnetic-guided drug delivery in the blood vessels. *Journal of Magnetism and Magnetic Materials*. 2017;427:181–187. Epub 2016/11/06.
28. Le TA, Hoshiar AK, Do TD, Yoon J. A modified functionalized magnetic field for nanoparticle guidance in magnetic drug targeting. *13th International Conference on Ubiquitous Robots and Ambient Intelligence (URAI)*, 19–22 August 2016, pp. 493–496: IEEE.
29. Mahmood A, Dadkhah M, Kim MO, Yoon J. A novel design of an MPI-based guidance system for simultaneous actuation and monitoring of magnetic nanoparticles. *IEEE Transactions on Magnetics*. 2015;51:1–5.
30. Zhang X, Le T-A, Yoon J. Development of a real time imaging-based guidance system of magnetic nanoparticles for targeted drug delivery. *Journal of Magnetism and Magnetic Materials*. 2017;427:345–351. Epub 2016/10/13.
31. Gleich B, Weizenecker J. Tomographic imaging using the nonlinear response of magnetic particles. *Nature*. 2005;435:1214–7.
32. Rahmer J, Antonelli A, Sfara C, Tiemann B, Gleich B, Magnani M, Weizenecker J, Borgert J. Nanoparticle encapsulation in red blood cells enables blood-pool magnetic particle imaging hours after injection. *Physics in Medicine and Biology*. 2013;58:3965–77. Epub 2013/05/21.
33. Saritas EU, Goodwill PW, Croft LR, Konkle JJ, Lu K, Zheng B, Conolly SM. Magnetic particle imaging (MPI) for NMR and MRI researchers. *Journal of Magnetic Resonance*. 1997; 2013;229:116–26. Epub 2013/01/12.
34. Knopp T, Buzug TM. Introduction. Magnetic Particle Imaging: An Introduction to Imaging Principles and Scanner Instrumentation. Berlin: Springer; 2012. pp. 1–9.
35. Knopp T, Sattel TF, Biederer S, Rahmer J, Weizenecker J, Gleich B, Borgert J, Buzug TM. Model-based reconstruction for magnetic particle imaging. *IEEE Transactions on Medical Imaging*. 2010;29:12–8.
36. Weizenecker J, Gleich B, Rahmer J, Dahnke H, Borgert J. Three-dimensional real-time in vivo magnetic particle imaging. *Physics in Medicine and Biology*. 2009;54:L1-110. Epub 2009/02/11.
37. Frautschy SA, Cole GM. Why pleiotropic interventions are needed for Alzheimer's disease. *Molecular Neurobiology*. 2010;41:392–409. Epub 2010/05/04.
38. Sultana R, Perluigi M, Butterfield DA. Oxidatively modified proteins in Alzheimer's disease (AD), mild cognitive impairment and animal models of AD: Role of Abeta in pathogenesis. *Acta Neuropathologica*. 2009;118:131–50. Epub 2009/03/17.
39. Selkoe DJ. The molecular pathology of Alzheimer's disease. Neuron. 1991;6:487–98. Epub 1991/04/01.
40. Auld DS, Kornecook TJ, Bastianetto S, Quirion R. Alzheimer's disease and the basal forebrain cholinergic system: Relations to beta-amyloid peptides, cognition, and treatment strategies. *Progress in Neurobiology*. 2002;68:209–45. Epub 2002/11/27.
41. Farlow MR, Miller ML, Pejovic V. Treatment options in Alzheimer's disease: Maximizing benefit, managing expectations. *Dementia and Geriatric Cognitive Disorders*. 2008; 25:408–22. Epub 2008/04/09.
42. Ahmad A, Ali T, Park HY, Badshah H, Rehman SU, Kim MO. Neuroprotective effect of fisetin against amyloid-beta-induced cognitive/synaptic dysfunction, neuroinflammation, and neurodegeneration in adult mice. *Molecular Neurobiology*. 2016. Epub 2016/03/06.
43. Ali Shah S, Ullah I, Lee HY, Kim MO. Anthocyanins protect against ethanol-induced neuronal apoptosis via GABAB1 receptors intracellular signaling in prenatal rat hippocampal neurons. *Molecular Neurobiology*. 2013;48:257–69. Epub 2013/05/07.
44. Ali T, Badshah H, Kim TH, Kim MO. Melatonin attenuates D-galactose-induced memory impairment, neuroinflammation and neurodegeneration via RAGE/NF-K B/JNK signaling pathway in aging mouse model. *Journal of Pineal Research*. 2015;58:71–85. Epub 2014/11/18.
45. Ali T, Kim MJ, Rehman SU, Ahmad A, Kim MO. Anthocyanin-loaded PEG-gold nanoparticles enhanced the neuroprotection of anthocyanins in an Abeta1-42 mouse model of Alzheimer's disease. *Molecular Neurobiology*. 2016. Epub 2016/10/13.
46. Ali T, Kim MO. Melatonin ameliorates amyloid beta-induced memory deficits, tau hyperphosphorylation and neurodegeneration via PI3/Akt/GSk3beta pathway in the mouse hippocampus. *Journal of Pineal Research*. 2015;59:47–59. Epub 2015/04/11.
47. Ali T, Yoon GH, Shah SA, Lee HY, Kim MO. Osmotin attenuates amyloid beta-induced memory impairment, tau phosphorylation and neurodegeneration in the mouse hippocampus. *Scientific Reports*. 2015;5:11708. Epub 2015/06/30.
48. Amin FU, Shah SA, Kim MO. Glycine inhibits ethanol-induced oxidative stress, neuroinflammation and apoptotic neurodegeneration in postnatal rat brain. *Neurochemistry International*. 2016;96:1–12. Epub 2016/04/09.
49. Badshah H, Ali T, Kim MO. Osmotin attenuates LPS-induced neuroinflammation and memory impairments via the TLR4/NFkappaB signaling pathway. *Scientific Reports*. 2016;6:24493. Epub 2016/04/21.
50. Naseer MI, Ullah I, Narasimhan ML, Lee HY, Bressan RA, Yoon GH, Yun DJ, Kim MO. Neuroprotective effect of osmotin against ethanol-induced apoptotic neurodegeneration in the developing rat brain. *Cell Death & Disease*. 2014;5:e1150. Epub 2014/03/29.

51. Rehman SU, Shah SA, Ali T, Chung JI, Kim MO. Anthocyanins reversed D-galactose-induced oxidative stress and neuroinflammation mediated cognitive impairment in adult rats. *Molecular Neurobiology*. 2016. Epub 2016/01/08.
52. Shah SA, Lee HY, Bressan RA, Yun DJ, Kim MO. Novel osmotin attenuates glutamate-induced synaptic dysfunction and neurodegeneration via the JNK/PI3K/Akt pathway in postnatal rat brain. *Cell Death & Disease*. 2014;5:e1026. Epub 2014/02/01.
53. Shah SA, Yoon GH, Ahmad A, Ullah F, Ul Amin F, Kim MO. Nanoscale-alumina induces oxidative stress and accelerates amyloid beta (Abeta) production in ICR female mice. *Nanoscale*. 2015;7:15225–37. Epub 2015/09/01.
54. Shah SA, Yoon GH, Chung SS, Abid MN, Kim TH, Lee HY, Kim MO. Novel osmotin inhibits SREBP2 via the AdipoR1/AMPK/SIRT1 pathway to improve Alzheimer's disease neuropathological deficits. *Molecular Psychiatry*. 2017;22:407–416. Epub 2016/03/24.
55. Karran E, Mercken M, De Strooper B. The amyloid cascade hypothesis for Alzheimer's disease: An appraisal for the development of therapeutics. *Nature Reviews Drug Discovery*. 2011;10:698–712. Epub 2011/08/20.
56. Squire LR. Memory and the hippocampus: A synthesis from findings with rats, monkeys, and humans. *Psychological Review*. 1992;99:195–231. Epub 1992/04/01.
57. Freeman MW, Arrott A, Watson JHL. Magnetism in medicine. *Journal of Applied Physics*. 1960;31:S404–S5.
58. Goodwin S, Peterson C, Hoh C, Bittner C. Targeting and retention of magnetic targeted carriers (MTCs) enhancing intra-arterial chemotherapy. *Journal of Magnetism and Magnetic Materials*. 1999;194:132–9.
59. Laurent S, Forge D, Port M, Roch A, Robic C, Vander Elst L, Muller RN. Magnetic iron oxide nanoparticles: Synthesis, stabilization, vectorization, physicochemical characterizations, and biological applications. *Chemical Reviews*. 2008;108: 2064–110. Epub 2008/06/12.
60. Chorny M, Polyak B, Alferiev IS, Walsh K, Friedman G, Levy RJ. Magnetically driven plasmid DNA delivery with biodegradable polymeric nanoparticles. *FASEB Journal*. 2007;21:2510–9. Epub 2007/04/04.
61. Lewin M, Carlesso N, Tung CH, Tang XW, Cory D, Scadden DT, Weissleder R. Tat peptide-derivatized magnetic nanoparticles allow in vivo tracking and recovery of progenitor cells. *Nature Biotechnology*. 2000;18:410–4. Epub 2000/04/05.
62. Ma HL, Qi XR, Ding WX, Maitani Y, Nagai T. Magnetic targeting after femoral artery administration and biocompatibility assessment of superparamagnetic iron oxide nanoparticles. *Journal of Biomedical Materials Research Part A*. 2008;84:598–606. Epub 2007/07/10.
63. Namdeo M, Saxena S, Tankhiwale R, Bajpai M, Mohan YM, Bajpai SK. Magnetic nanoparticles for drug delivery applications. *Journal of Nanoscience and Nanotechnology*. 2008;8:3247–71. Epub 2008/12/05.
64. Veiseh O, Gunn JW, Kievit FM, Sun C, Fang C, Lee JS, Zhang M. Inhibition of tumor-cell invasion with chlorotoxin-bound superparamagnetic nanoparticles. *Small*. 2009;5:256–64. Epub 2008/12/18.
65. Sonvico F, Mornet S, Vasseur S, Dubernet C, Jaillard D, Degrouard J, Hoebeke J, Duguet E, Colombo P, Couvreur P. Folate-conjugated iron oxide nanoparticles for solid tumor targeting as potential specific magnetic hyperthermia mediators: Synthesis, physicochemical characterization, and in vitro experiments. *Bioconjugate Chemistry*. 2005;16:1181–8. Epub 2005/09/22.
66. Janib SM, Moses AS, MacKay JA. Imaging and drug delivery using theranostic nanoparticles. *Advanced Drug Delivery Reviews*. 2010;62:1052–63. Epub 2010/08/17.
67. Cherry EM, Maxim PG, Eaton JK. Particle size, magnetic field, and blood velocity effects on particle retention in magnetic drug targeting. *Medical Physics*. 2010;37:175–82. Epub 2010/02/24.
68. Martel S. Magnetic navigation control of microagents in the vascular network: Challenges and strategies for endovascular magnetic navigation control of microscale drug delivery carriers. *IEEE Control Systems*. 2013;33:119–34.
69. Hassan S, Yoon J. A hybrid approach for vessel enhancement and fast level set segmentation based 3d blood vessel extraction using MR brain image. *IEEE 7th International Conference on Nano/Molecular Medicine and Engineering (NANOMED)*, 10–13 November 2013, pp. 77–82: IEEE.
70. Dadkhah M, Kumar N, Yoon J. Design and simulation of a 3D actuation system for magnetic nano-particles delivery system. *International Conference on Intelligent Robotics and Applications*, 2013, pp. 177–187: Springer.
71. Do TD, Amin FU, Noh Y, Kim MO, Yoon J. Functionalized magnetic force enhances magnetic nanoparticle guidance: From simulation to crossing of the blood–brain barrier in vivo. *IEEE Transactions on Magnetics*. 2016;52:1–4.
72. Yoon J, Noh Y, Tehrani M, Do TD, Kim M. *An Electromagnetic Guidance System of Magnetic Nanoparticles for Targeted Drug Delivery*. 59th Annual Conference on Magnetism and Magnetic Materials; 2014.
73. Larimi MM, Ramiar A, Ranjbar AA. Numerical simulation of magnetic nanoparticles targeting in a bifurcation vessel. *Journal of Magnetism and Magnetic Materials*. 2014;362:58–71.
74. Vartholomeos P, Mavroidis C. In silico studies of magnetic microparticle aggregations in fluid environments for MRI-guided drug delivery. *IEEE Transactions on Biomedical Engineering*. 2012;59:3028–38.
75. Chertok B, David AE, Yang VC. Brain tumor targeting of magnetic nanoparticles for potential drug delivery: Effect of administration route and magnetic field topography. *Journal of Controlled Release*. 2011;155:393–9. Epub 2011/07/19.
76. Do TD, Noh Y, Kim MO, Yoon J. An optimized field function scheme for nanoparticle guidance in magnetic drug targeting systems. 2015 Sept. 28 2015–Oct. 2 2015: Publisher.
77. Borgert J, Schmidt JD, Schmale I, Bontus C, Gleich B, David B, Weizenecker J, Jockram J, Lauruschkat C, Mende O, Heinrich M, Halkola A, Bergmann J, Woywode O, Rahmer J. Perspectives on clinical magnetic particle imaging. *Biomedizinische Technik Biomedical Engineering*. 2013;58:551–6. Epub 2013/09/13.
78. Bringout G, Wojtczyk H, Tenner W, Graeser M, Grüttner M, Haegele J, Duschka R, Panagiotopoulos N, Vogt FM, Barkhausen J, Buzug TM. A high power driving and selection field coil for an open MPI scanner. *3rd International Workshop on Magnetic Particle Imaging (IWMPI)*, 23–24 March 2013, p. 1: IEEE.
79. Bringout G, Gräfe K, Buzug TM. Performance and safety evaluation of a human sized FFL imager concept. *5th International Workshop on Magnetic Particle Imaging (IWMPI)*, 26–28 March 2015, p. 1: IEEE.
80. Schulz V, Straub M, Mahlke M, Hubertus S, Lammers T, Kiessling F. A field cancellation signal extraction method for magnetic particle imaging. *IEEE Transactions on Magnetics*. 2015;51:1–4.
81. Goodwill PW, Conolly SM. The X-Space formulation of the magnetic particle imaging process: 1-D signal, resolution, bandwidth, SNR, SAR, and magnetostimulation. IEEE Transactions on Medical Imaging. 2010;29:1851–9.

82. Khalil ISM, Abelmann L, Misra S. Magnetic-based motion control of paramagnetic microparticles with disturbance compensation. *IEEE Transactions on Magnetics*. 2014;50:1–10.
83. Khalil ISM, Ferreira P, Eleut R, x00E, rio, Korte CLd, Misra S, Magnetic-based closed-loop control of paramagnetic microparticles using ultrasound feedback. *IEEE International Conference on Robotics and Automation (ICRA)*, 31 May 2014–7 June 2014, pp. 3807–3812: IEEE.
84. Martel S, Mathieu J-B, Felfoul O, Chanu A, Aboussouan E, Tamaz S, Pouponneau P, Yahia LH, Beaudoin G, Soulez G, Mankiewicz M. Automatic navigation of an untethered device in the artery of a living animal using a conventional clinical magnetic resonance imaging system. *Applied Physics Letters*. 2007;90:114105.
85. Mathieu JB, Martel S. Steering of aggregating magnetic microparticles using propulsion gradients coils in an MRI Scanner. *Magnetic Resonance in Medicine*. 2010;63:1336–45. Epub 2010/05/01.
86. Shapiro B, Kulkarni S, Nacev A, Muro S, Stepanov PY, Weinberg IN. Open challenges in magnetic drug targeting. *Wiley Interdisciplinary Reviews Nanomedicine and Nanobiotechnology*. 2015;7:446–57. Epub 2014/11/08.
87. Latulippe M, Martel S. Dipole field navigation: Theory and proof of concept. *IEEE Transactions on Robotics*. 2015;31:1353–63.
88. Ivanov KP, Kalinina MK, Levkovich YI. Blood flow velocity in capillaries of brain and muscles and its physiological significance. *Microvascular Research*. 1981;22:143–55.
89. Georgieva JV, Kalicharan D, Couraud PO, Romero IA, Weksler B, Hoekstra D, Zuhorn IS. Surface characteristics of nanoparticles determine their intracellular fate in and processing by human blood–brain barrier endothelial cells in vitro. *Molecular Therapy*. 2011;19:318–25. Epub 2010/11/04.
90. Pardeshi C, Rajput P, Belgamwar V, Tekade A, Patil G, Chaudhary K, Sonje A. Solid lipid based nanocarriers: An overview. *Acta Pharmaceutica* (Zagreb, Croatia). 2012;62:433–72. Epub 2013/01/22.
91. Kong SD, Lee J, Ramachandran S, Eliceiri BP, Shubayev VI, Lal R, Jin S. Magnetic targeting of nanoparticles across the intact blood–brain barrier. *Journal of Controlled Release*. 2012;164:49–57. Epub 2012/10/16.
92. Min KA, Shin MC, Yu F, Yang M, David AE, Yang VC, Rosania GR. Pulsed magnetic field improves the transport of iron oxide nanoparticles through cell barriers. *ACS Nano*. 2013;7:2161–71.
93. Bar J, Herbst RS, Onn A. Targeted drug delivery strategies to treat lung metastasis. *Expert Opinion on Drug Delivery*. 2009;6:1003–16. Epub 2009/08/12.
94. Torchilin VP. Passive and active drug targeting: Drug delivery to tumors as an example. *Handbook of Experimental Pharmacology*. 2010:3–53. Epub 2010/03/11.
95. Mosbach K, Schroder U. Preparation and application of magnetic polymers for targeting of drugs. *FEBS Letters*. 1979;102:112–6. Epub 1979/06/01.
96. Debbage PL, Griebel J, Ried M, Gneiting T, DeVries A, Hutzler P. Lectin intravital perfusion studies in tumor-bearing mice: Micrometer-resolution, wide-area mapping of microvascular labeling, distinguishing efficiently and inefficiently perfused microregions in the tumor. *The Journal of Histochemistry and Cytochemistry*. 1998;46:627–39. Epub 1998/05/30.
97. Amin FU, Hoshiar AK, Do TD, Noh Y, Shah SA, Khan MS, Yoon J, Kim MO. Osmotin-loaded magnetic nanoparticles with electromagnetic guidance for the treatment of Alzheimer's disease. *Nanoscale*. 2017;9:10619–10632. Epub 2017/05/03.

Jungwon Yoon (E-mail: jyoon@gist.ac.kr) earned his PhD degree at the Department of Mechatronics, Gwangju Institute of Science and Technology (GIST), Gwangju, Korea, in 2005. He was a senior researcher at the Electronics Telecommunication Research Institute (ETRI), Daejeon, Korea. He was a visiting fellow at the Department of Rehabilitation Medicine, Clinical Center, National Institutes of Health, Bethesda, from 2010 to 2011. He was a professor at the School of Mechanical and Aerospace Engineering, Gyeongsang National University, from 2005 to 2017. In 2017, he joined the School of Integrated Technology, Gwangju Institute of Science and Technology, Gwangju, Korea, where he is currently an associate professor. He serves as a technical editor for *IEEE/ASME Transactions on Mechatronics* and associate editor for *Frontiers in Robotics* and *AI*. His current research interests include bio-nano robot control, virtual reality haptic devices and rehabilitation robots. He has authored and coauthored more than 90 peer-reviewed journal articles and patents.

Ali Kafash Hoshiar is a senior researcher in Gyeongsang National University (GNU) and an assistant professor in Azad university of Qazvin, Iran. He earned his PhD in mechanical engineering from Azad University of Science and Research Branch, Tehran in 2014. His research interests are mainly focused on the micro/nano robotics systems. In the past, he conducted research on the AFM based manipulation techniques. His current research centres on modelling, simulation and implementation of the micro and nano robotic systems for targeted drug delivery. The magnetic actuation based system with MPI monitoring is a revolutionary idea that will provide doctors with noninvasive tools to treat central nervous system diseases such as Alzheimer's.

Tuan Anh Lea earned both BSc and MSc degrees in electrical engineering from Hanoi University of Science and Technology, Hanoi, Viet Nam, in 2009 and 2013, respectively. In 2014, he joined as a researcher the School of Mechanical and Aerospace Engineering, Gyeongsang National University, Jinju, Korea, where he is currently a PhD candidate. His current research interests include electromagnetics, actuators, magnetic particle imaging and magnetic drug targeting devices.

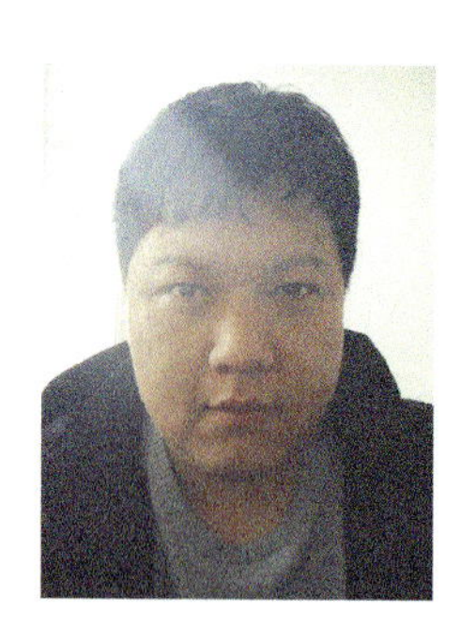

Xingming Zhang was born in Harbin, Republic of China, in 1985. He earned BE, MEng and PhD degrees in marine engineering from Dalian Maritime University, Dalian, People's Republic of China, in 2004, 2010 and 2014, respectively. He was a senior researcher in Gyeongsang National University, Jinju, Republic of Korea. He is currently a lecturer at the Harbin Institute of Technology, Weihai, China. His current research interests include MPI, MPI-based nano-robot control and microparticle sensors.

Faiz Ul Amin was born in Mardan, within the Khyber Pakhtunkhwa Province of Pakistan on February 15, 1982. He earned a Bachelor of Biotechnology degree in 2006 from the Center of Biotechnology at the University of Peshawar, Pakistan. In 2009, he earned an MS degree from the Quaid-i-Azam University, in Islamabad in biotechnology. Faiz Ul Amin took up a lecturership in 2010 at Institute of Molecular Biology and Biotechnology, University of Lahore, Pakistan. He earned his PhD in applied life sciences from Gyeongsang National University, Republic of Korea in 2017. Faiz Ul Amin has published more than a dozen research articles in refereed international journals such as *Nanoscale*, *Scientific Reports*, *Journal of Nanobiotechnology*, *Journal of Biomedical Nanotechnology*, *Neurochemistry International*, etc.

Myeong Ok Kim earned a PhD degree from the Department of Biology, Gyeongsang National University (GNU), Gyeongnam, Korea, in 1994. She was a postdoctoral researcher at the Seoul National University, Seoul, Korea, from 1995 to 1996. She was a visiting researcher at the Neurological Primate Research Center, Oregon University, from 1999 to 2000. In 1998, she joined the School of Natural Science, Gyeongsang National University, Jinju, Korea, where she is currently a full professor of Biology and Applied Life Science. Her current research interests include screening compounds which have neuroprotective effects on neurodegenerative disorders such as Alzheimer's disease. She has authored or coauthored more than 100 peer-reviewed journal articles and patents.

20 Design, Fabrication and Characterization of Magnetic Porous PDMS as an On-Demand Drug Delivery Device

*Ali Shademani, Hongbin Zhang and Mu Chiao**

CONTENTS

* Corresponding author.

20.1 INTRODUCTION

20.1.1 Localized Drug Delivery

Many diseases such as cancer and arthritis are localized and might be better treated by delivering drugs to the specific disease region. However, most drugs are given by oral or intravenous injection methods so that the drug is delivered to the target area by the blood stream. This requires administration of high doses of drugs in order that dilution still permits a local therapeutic concentration. With drug clearance from the bloodstream beginning as soon as the drug is administered, these routes of administration require frequent repeat dosing to maintain efficacious local concentrations. These drug regimes expose all body compartments to drugs that may be toxic. Therefore, there is increasing interest in developing better drug delivery systems rather than trying to modify drug characteristics to reduce toxicity.[1]

Numerous intravenous drug delivery methods using encapsulated drug in a nanosystem have been studied. However, these methods are still many years away from being clinically relevant or effective. Other methods aim to localize a controlled-release deposit of a drug at the disease site so that a continuous stream of drug is released there. These systems may be injectable polymeric-based systems or mechanical devices usually requiring surgical deployment such as MEMS-based delivery systems. This chapter will first review state-of-the-art drug delivery devices and methods, followed by discussion of a new magnetic porous polymer drug delivery system.

20.1.2 Controlled Drug Release System

The important characteristics of oral or intravenous drug delivery systems are the maintenance of effective local concentrations in the blood stream while avoiding unacceptable general toxicity issues. As illustrated in Figure 20.1, the

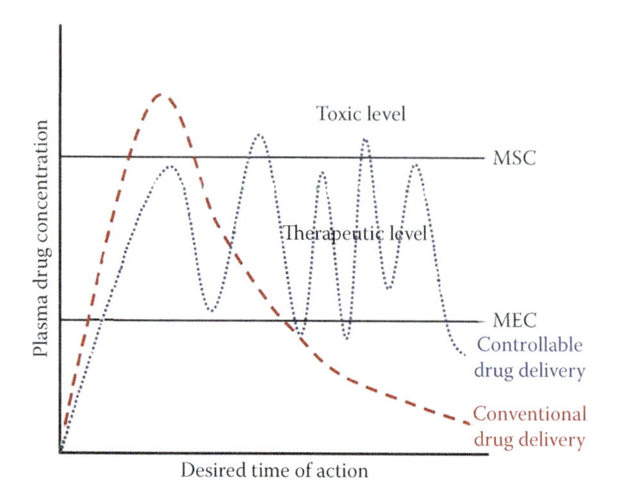

FIGURE 20.1 Representation of different types of drug delivery according to therapeutic window.[1] (Reprinted with permission from D. Das and S. Pal, *RSC Adv.*, vol. 5, no. 32, pp. 25014–25050, 2015. Copyright 2010 American Chemical Society.)

permitted drug concentration in blood is restricted by two factors. The upper limit is defined as the maximum safe concentration (MSC) above which the drug becomes toxic, and the lower limit is defined as the minimum effective concentration (MEC) below which the drug is ineffective. The red-dotted curve corresponds to conventional drug delivery, which is carried out either by direct administration using a syringe or taking tablets. As shown in Figure 20.1 this method has little control over the drug dosage, and higher drug doses must be injected in order to compensate for the dissipation and clearance of the therapeutic agent. However, such regimes result in an initial dose burst in the bloodstream, which may exceed the MSC for some period. After some time, the blood concentration of the drug will drop due to clearance so that eventually it falls below the MEC. When systemic delivery of a drug is needed, a controlled release of the drug into the blood stream is preferred, which might be capable of maintaining the drug concentration within the therapeutic window, as shown in the blue dotted lines in Figure 20.1 This can be achieved conventionally by a passive form of drug delivery, which will be described in Section 20.1.3.

Localized drug delivery using an implantable system close to the disease site can prevent the drug concentration from exceeding the MSC. For all systems the adjustability of drug dose (i.e. release) is important. Although some of the proposed drug delivery devices have shown successful controllable release, they cannot provide that much flexibility in altering the release rate or in changing the drug dose. The condition of the patient may change and drug dose may need to be increased or decreased based on toxicity and efficacy considerations. Thus, a drug delivery device that can be 'actively' controlled will be able to adapt to new conditions.

20.1.3 Passive Controlled Delivery of Drugs

In passive forms of drug delivery, the medication release begins as the delivery agent is situated in the body (either by implanting or swallowing, etc.). Although the drug release rate cannot be actively controlled externally with this approach, some methods have used the local tissue environment (such as pH responsive systems) to partially control the amount of discharged drug.[2] This section will review two most common passive control forms of drug delivery that aim to maintain the drug within the therapeutic window.

20.1.3.1 Polymeric Drug Delivery

Polymeric drug delivery systems have been developed over the last 30 years. The drug is embedded within a polymer matrix and implanted in the body so the drug is released over time. This approach has the advantages of maintaining drug concentrations in the blood stream for systemic applications like releasing hormone-based contraceptives,[3] or in local disease areas. The systems may also promote the solubility of drugs,[4] and enhance the pharmacokinetic behaviour of drugs.[5] Although polymer–drug conjugates are mostly associated with constant release rates, there have been efforts to improve the controllability and make them active drug delivery agents by incorporating techniques such as using pH sensitive polymers[6] or temperature triggered polymer.[7] These techniques, in combination with other stimulatory methods such as ultrasound, electromagnetic radiation, heating, light and so on, have led to active controlled release.[8] However, injecting on-demand exact doses of drugs or altering the release rate of drugs in real time remains a challenge in this field.

20.1.3.2 Osmosis-Based Methods

Osmosis is a phenomenon that relies on the generation of a chemical concentration gradient and provides a way of actuation that can push the drug out of an aqueous reservoir. This approach has been implemented both in oral drug deliveries and in implantable micropumps.[9] In oral delivery, tablets with semipermeable membranes are usually employed and contain a hard core where the drug is regularly situated, and an aperture for the drug ejection.[10,11] In addition, implantable micropump devices have been manufactured using similar concepts.[12,13] In both cases, water diffusion through the tablet or micropump causes the membrane to deform and push the drug out from the allocated repository. This method requires no power but a robust design of the pump[14] and is favoured by inexpensive and large-scale production factors.[15,16] However, this approach is a passive delivery of drug; a dynamic precise dosage control and a pulsatile release rate has not been achieved yet with such systems.

20.1.4 Active Controlled Delivery of Drugs

In contrast to passive controlled release, active controlled release may be able to provide on-demand, or temporal control of drug release profile. External stimuli such as electrical, magnetic or ultrasonic, have been used to achieve active controlled drug delivery. As mentioned in Section 20.1.3.1, the ultrasound may be incorporated with some passive approach, such as a polymeric-based system, to enhance the release protocol. However, the purpose of this section is to introduce and survey another active drug delivery systems, known as reservoir-based drug delivery devices. In these systems, a reservoir is designated for the drug payload, which is further released by employing a method of stimuli. Below, two most common types of actuation have been studied; however, a general summary of these devices can be found in Table 20.1.

20.1.4.1 Electrical Stimuli

Electrical stimulus offers a way of triggering drug delivery from an implantable device. Several microfabricated devices have been demonstrated where a microreservoir holding the drug or drug solution is sealed with a thin metallic electrode.[17,18] An applied electrical current leads to the dissolution of the top electrode, leading to eventual exposure of the reservoir's content. A modified version of this method utilized another electrode at the bottom of the well to produce bubbles to push the drug out of the reservoir.[19] Other methods have been proposed based on drug loaded-hydrogel systems, where drug delivery occurs by applying an electrical field through the hydrogel matrix.[20] The electrical field either forces the entrapped ionized drug particles to expel[21] or leads the hydrogel to deswell and liberate the drug.[22] Even though pulsatile on-demand release has been demonstrated, power sources and wiring connections can still be limiting factors, which restrict their usage, especially when it comes to implantation.

20.1.4.2 Magnetic Stimuli

Among various triggering methods, magnetic actuation has shown potential in providing remote-controlled drug delivery that requires no power source directly on the device. Furthermore, the systems may employ a simple permanent magnet for actuation of drug release on demand. In addition to broad investigations conducted on magnetic nanoparticles used for targeted drug delivery where magnetic fields halt the movement of circulating drug loaded nanoparticles (NPs) at the target site,[23,24] recently implantable devices incorporated with magnetic elements have attracted attentions.

Cai et al. used Fe_3O_4 particles inside the drug reservoir to obstruct the pores of the permeable membrane covering the reservoir, controlling the drug diffusion.[25] However, this device requires a reversible magnetic field for both on/off modes of the drug diffusion. Rahimi et al. employed a thermosensitive hydrogel triggered by a resonant heater which was remotely actuated by a frequency of the applied magnetic field.[26] This hydrogel served as a valve, so that the loaded drug was exposed and diffused into the local environment when it shrank. However, local temperature increases may be an issue for both patient comfort and the device operation.

Ferrogels are another promising substance suitable for drug delivery purposes. Studies on the fundamental components employed in ferrogel formation have established the capability of this material to be used as a controlled drug delivery agent.[27] Zhao et al. reported a highly permeable ferrogel fabricated for pulsatile on-demand cell and drug delivery.[28] However, apart from the difficulties in fabrication process, such as sensitive temperature conditions, incessant drug diffusion from ferrogel has remained an issue.

Pirmoradi et al. created a magnetic membrane with a micro-aperture that covered a drug reservoir and was deflected when a magnetic field was applied.[29,30] Inward deflection of the magnetic membrane caused the drug solution to be pumped out of the reservoir through a micro-aperture in the membrane. This concept has recently been improved and proposed for the treatment of prostate cancer by shaping the device into a cylinder, which may be implanted through a gauge 10 needle system.[31] Instead of having a magnetic particle encapsulated in a membrane, a small magnetic block attached to the membrane is responsible for the membrane deflection when an external magnetic field is applied. Two different versions of this device have been fabricated: a PDMS device, which is the larger device, and the 3D-printed one, which is small enough to be implanted by a needle.

Porous structures made by Fe_3O_4 have shown promising adsorption and release of methyl blue as a model drug.[32] Although this magnetic porous surface could not release the methyl blue in a DC magnetic field, oscillating magnetic fields allowed for considerable release. Nevertheless,

TABLE 20.1
Summarized Reservoir-Based Drug Delivery Methods

Application	Method of Actuation	Device Shape	Delivered Agent	Reference
Prostate cancer	Osmosis	Cylinder; 4 mm dia., 45 mm length	Leuprolide	12
General	Electrochemical	Pyramid shape reservoirs (Larger area: 800×800 μm², Smaller area: 50×50 μm², Total volume: 120 nL)	^{14}C-labeled mannitol	18
Tissue regeneration	Osmosis	Consists of a reservoir (area of 2 cm², depth of 50 μm), where several channels are connected (50×50 μm²). Various channels lengths were studied (2 mm, 4 mm, 8 mm)	Basic fibroblast growth factor (bFGF)	13
Ocular MEMS Drug delivery device	Manually apply pressure	Consists of a square shape reservoir (roughly 8×8 mm²), and a connected cannula (10×1×1 mm³)	Dyed deionized water Phenylephrine	34
General	Magnetic	Overall 12×12×5 mm³, 9×9 arrays of microchannels (0.5×0.5×0.5 mm³)	DNA and vitamin B2	25
General	Electrochemical	Overall 4.4×2.3×22 mm³	Vasopressin	19
General	Magnetic	Membrane thickness varies between 0.15–0.3 mm	Sodium fluorescein	33
General	Electrochemical (Electrolysis)	Consists of: 1. Reservoir (18 mm dia., 3.5 mm height, volume = 560 μL). 2. Cannula (length = cut on site, ID = 0.305 mm, OD = 0.610 mm). 3. Pump electrodes (width = 20 μm, gap = 100 μm)	siRNA	35
Transdermal (Microneedle)	Thermal	Overall 14×14×8 mm³	Dyed DI-water Evans Blue dye	36
Drug/Cell delivery	Magnetic	Cylinder with a height of about 15 mm and diameter of about 8 mm. However, size can be changed depending on the application	Mitoxantrone Plasmid DNA Chemokine SDF-1α	28
Proof of concept	Radiofrequency magnetic fields	Overall 9.5×8.3×1 mm³	Food-colour dye Fluorescein	26
General	Magnetic	Size of the reservoir was 6 mm in diameter and 550 μm in depth	Docetaxel Trypan blue	29
General	Electrical (thermosensitive)	Overall size of 20×20×3.8 mm³	Rhodamine B	7
Eye	Magnetic	Overall device seize: about 3 mm thick and 8 mm dia.	Docetaxel	37
General (Diabetic rat)	NIR	Nano composite membrane with a thickness of 134 ± 14 μm. The diameter of final device was about 13 mm	Insulin	38
Prostate cancer	Magnetic	Larger device overall dimension: (OD = 5 mm, ID = 3 mm, length = 12 mm) Smaller device overall dimensions: (OD = 2 mm, ID = 1 mm, length = 12 mm)	Docetaxel Methylene blue	31
General	Electrical	Device has a reservoir with 3 mm diameter and 1.5 mm length, and an aperture with 0.5 mm diameter and 0.2 mm height	FITC-dextran Methylene blue	21

combining DC and AC fields resulted in better release since it is believed that magnetic particles are aligned in a DC field while an AC field leads to oscillation, thus drug discharge occurs. The Fe_3O_4 structure has been reported to be biocompatible, but this porous material has limited capacity in absorbing the drug.

Hoare et al. used a thermosensitive composite membrane covering a drug reservoir.[33] This membrane was fabricated by embedding engineered superparamagnetic magnetite NPs along with thermosensitive nanogels into ethyl cellulose, utilized as the base of the membrane. Upon exposing this membrane to an oscillating magnetic field, magnetite NPs generated heat and caused the entrapped nanogels to shrink. This will cause the drug to diffuse through the temporary created cavity. By removing the magnetic field, the device cools down leading the nanogels to reswell and retain their initial condition. Therefore, the cavity will be closed and drug diffusion terminates. Table 20.1 summarizes various proposed active controlled drug delivery methods and highlights their features.

20.1.5 Current Challenges

As discussed in previous sections, various drug delivery systems have been proposed based on different applications and methods of release. However, many challenging issues remain to be tackled when it comes to implantation. Biocompatibility, size limitation, long-term functionality and dosage control are common practical issues. In active controlled drug delivery, the type of stimuli is a crucial factor affecting device performance. Electrical stimulus requires power and wire connections, limiting their size and therefore its application in implantable devices. To obviate the need for power, other methods using smart materials such as near infrared devices,[38] temperature-based systems,[7] or ultrasound triggered approaches[39] have been proposed. However, when such devices are implanted, they are no longer accessible to external stimuli. Furthermore, once implanted, disturbances such as body temperature, biofilm formation and local hormone concentration variations may interfere with device performance. Magnetic actuation requires no power, but when patients are in the proximity of large magnetic fields (such as MRI machines), the device function might be compromised.

20.1.6 Magnetic Sponge as an On-Demand Drug Delivery Device

In the following sections, a drug delivery device has been developed based on a magnetic sponge, which is shrunk when a magnetic field is applied.[40] The sponge contraction leads to the inward deflection of the attached membrane resulting in the release of part of the drug payload. The amount of shrinkage is dependent on the magnetic properties of the sponge and is proportional to the strength of the applied magnetic field. The base material used in the fabrication of the sponge is PDMS, which is mixed with carbonyl iron microparticles (4–7 μm) to obtain the magnetic PDMS. This magnetic PDMS is further rendered porous by utilizing a solvent casting and particulate leaching technique. Although this sponge can solely be used as a drug delivery agent, a separated reservoir has been designed and fabricated in order to enhance controllability and decrease drug leakage.

Overall, this device is 6 mm in diameter and less than 2 mm in thickness. Device release performance has been tested using MB and docetaxel as surrogate and real drugs, respectively. Figure 20.2 shows a schematic exploded view of the device components including, reservoir, magnetic sponge and membrane.

This device requires no power, can release drugs on-demand and can adjust the amount of ejected drug by either tuning the magnetic field or increasing the number of actuations. The release may be initiated by a strong magnet (e.g. N52 neodymium magnet). The operational magnetic field strength is much higher than the magnetic field caused by common electrical devices such as a cell phone or laptop ensuring that no background interference may occur.

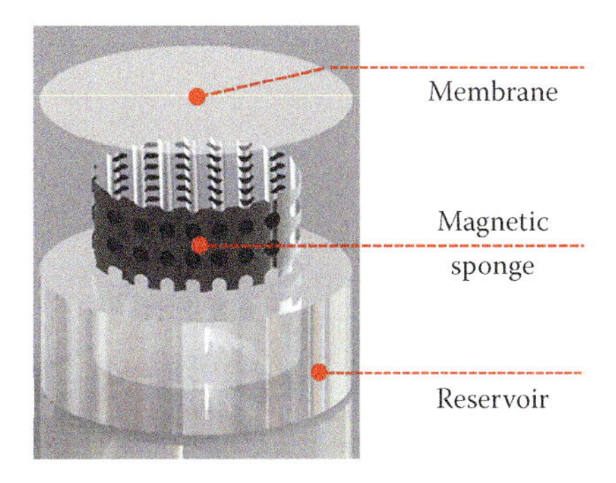

FIGURE 20.2 Schematic view of the device components.[40] (Reprinted with permission from A. Shademani, H. Zhang, J. K. Jackson, and M. Chiao, *Adv. Funct. Mater.*, p. 1604558, 2016. Copyright 2010 American Chemical Society.)

20.2 MATERIALS AND METHODS

20.2.1 Porous PDMS

Different methods have been reported for porous PDMS fabrication. However, most of them are based on a solvent casting and particulate leaching (SCPL) technique.[41–44] The aim of this technique is to introduce solid particles within the structure of PDMS and extract them after curing the PDMS. Vacant spaces are created in the polymer matrix, which makes it porous. As described, this technique has two main parameters: particle selection and method of extraction. Although, some suggested NaCl as the particulate agent,[43,44] with later removal by dissolving the salt, a much easier approach is proposed herein (based on sugar) where similar results are obtained in a fast and convenient way.

20.2.2 Magnetic Porous PDMS

To manufacture a magnetic porous PDMS, two separated tasks should be carried out. One task focuses on the porous scaffold and the other one concerns magnetization of the PDMS. Magnetic porous PDMS is created by merging these two aspects.

20.2.2.1 Task 1: Preparation of a Porous Scaffold

As mentioned earlier, sugar is used in our method as the soluble particle. The following steps indicate the fabrication process of a porous scaffold:

- Step 1: 20 ml of household sugar was mixed with about 0.5 ml of water, and then poured into a small size Petri dish.
- Step 2: By pressing the sugar lump firmly and uniformly, sugar particle contact was ensured. Controlling the amount of sugar and the final volume of the lump plays a crucial role in determining the mechanical properties of the subsequent magnetic porous PDMS.
- Step 3: The existing moisture was evaporated by placing the Petri dish in a convection oven for about 15 min.

20.2.2.2 Task 2: Magnetic PDMS

- Step 1: First of all, PDMS (Sylgard 182 Silicone Elastomer, Dow Corning Corporation) was made with a 1:30 cross-linker to base ratio.
- Step 2: In order to induce the magnetic properties, CI ferromagnetic microparticles were added into the PDMS. The concentration of the CI particles dictates the quality of magnetic properties of the PDMS. Thus, three different CI to PDMS wt% ratios were used (50%, 100% and 150%) and the best ratio was chosen (the criteria used to find the best concentration is explained later on). In order to achieve better dispersion of CI microparticles in the PDMS matrix, about 1 ml of isopropanol was dispensed during the mixing process, which later evaporated. Isopropanol also makes the mixture less viscous, facilitating stirring and dispersion. The mixing process takes about 10 min.

20.2.2.3 Magnetic Porous PDMS

Once the porous scaffold and magnetic PDMS were prepared, the magnetic sponge was made by adding magnetic PDMS to the sugar lump, which occurs as follows:

- Step 1: The magnetic PDMS was poured on the porous scaffold made out of sugar. The magnetic PDMS penetrated through the pores of the sugar lump until it encompassed the whole structure and reached the bottom of the Petri dish
- Step 2: The magnetic PDMS was cured in an oven at 70°C for 3 h to form the magnetic porous PDMS scaffold.
- Step 3: Eventually, several samples were cut from the scaffold and immersed in water for 2–4 h to dissolve the sugar. Warm water and stirring may be utilized to accelerate this step. Dissolved sugar particles allowed for vacant spaces, which form the pores of the sponge and the chain connection of sugar particles results in interconnected pores. Figure 20.3 shows the magnetic PDMS scaffold out of which one sample was cut, with the final magnetic porous PDMS structure having dimensions of 6 mm in diameter and 5 mm in height.

20.2.3 Characterization of Magnetic Porous PDMS

The magnetic sponge has two crucial elements: magnetic properties and porosity. Both of these directly impact the mechanical properties and shrinkage. As mentioned previously, magnetic porous PDMS has been made with three different carbonyl iron concentrations. In this section, magnetic sponge behaviour in various magnetic fields is studied with the intent of finding the best CI to PDMS wt% ratio.

20.2.3.1 Porosity

Porosity is defined as the ratio of the vacant space volume to the total volume of material. Although, the pores' configuration is generated through a random process, porosity is the main parameter in determining the mechanical properties of a porous substance. The porosity of the sponge is dependent on the sugar structure and can be altered either by changing the amount of the sugar and/or the sugar lump compression.

To evaluate the porosity of the sponge, three different samples were selected randomly from three scaffolds fabricated separately and independently. Each sample was then immersed in water and placed in a vacuum so they were filled with water. As the density of the sponge is less than water, a magnet may be used to submerge the samples deep into water. The shrinkage caused by the magnet assists in venting the air inside the sponge. After about 20 min, samples were removed and weighed. By subtracting the weight of sponge before and after submerging, the amount of absorbed water is calculated. The weight of the water entrapped in pores of the sponge is equivalent to the volume of the total pores of the sponge. Hence, the obtained porosity was 0.63 ± 0.02 with the low deviation (less than 3.5%) verifying the reproducibility of sponge fabrication.

20.2.3.2 Carbonyl Iron Concentration

Cross-sectional views of the magnetic sponge including SEM images from samples with different concentrations are provided in Figure 20.4. White dots represent the CI microparticles, which are more evident as the concentration rises.

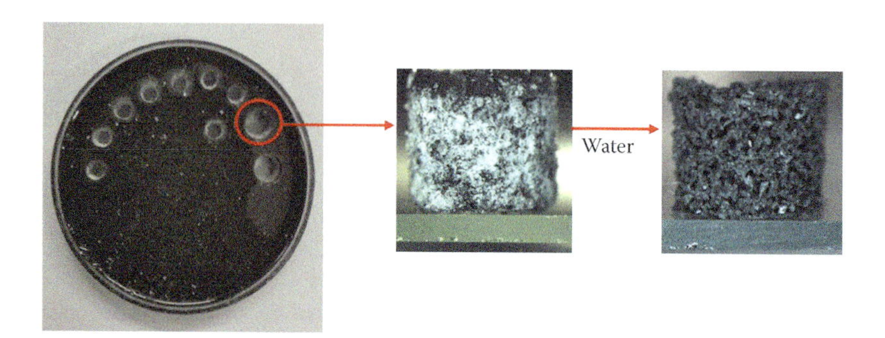

FIGURE 20.3 From left to right; magnetic porous PDMS scaffold, sample punched from the scaffold, and the consequent magnetic sponge after submerging the sample in water.

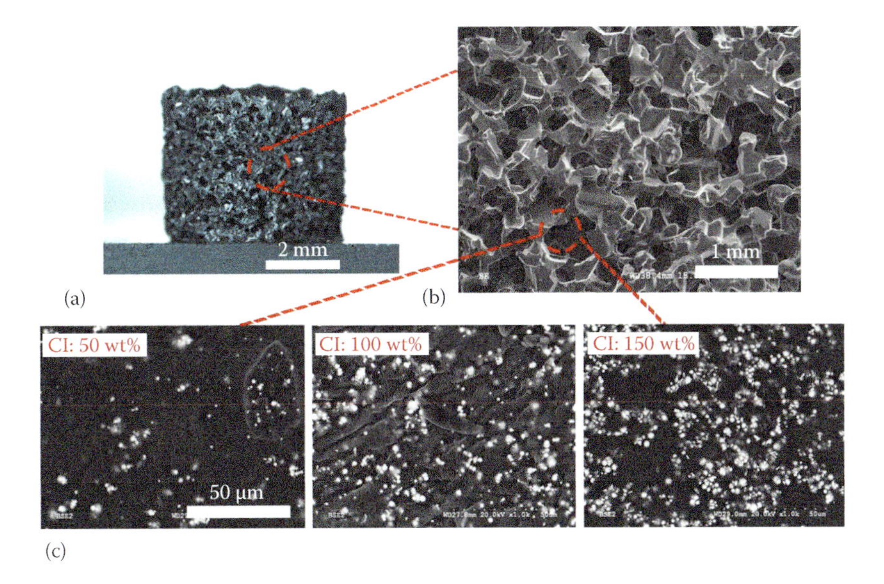

FIGURE 20.4 (a) Sponge, (b) Cross section of the magnetic porous PDMS, (c) Different CI to PDMS wt% ratios, from left to right respectively, 50%, 100%, and 150%. The scale bar is the same for all three states.[40] (Reprinted with permission from A. Shademani, H. Zhang, J. K. Jackson, and M. Chiao, *Adv. Funct. Mater.*, p. 1604558, 2016. Copyright 2010. American Chemical Society.)

Increasing the concentration of CI ferromagnetic particles improves the magnetic properties of the sponge and consequently raises the magnetic force exerted on the sponge in a magnetic field. However, the main objective is to achieve as much shrinkage as possible in various magnetic fields since it corresponds to more drug release. Accordingly, in order to characterize the influence of CI concentration on the sponge deformation, a test was conducted in which the amount of sponge displacement was measured as a function of different magnetic field strengths. Samples with different concentrations were chosen with roughly the same size (5 mm in diameter and about 6.5 mm in length), placed on a glass slide and exposed to different magnetic field strengths using a magnet at various distances from the sponge. Furthermore, the sponges' shrinkage was recorded and the displacements were quantified by image processing.

As shown in Figure 20.5, the sample with the ratio of 100% was squeezed more than 50% as anticipated, but the 150% ratio had less displacement than 100% as well. In fact, another facet of introducing microparticles into the PDMS matrix is the elevation of the elastic modulus. In this regard, adding carbonyl iron particles will increase the elasticity modulus of PDMS considerably, especially at high concentrations (more than 100%).[45] Thus, this parameter acts as a resistive factor opposing the magnetic force applied to the sponge. In order to investigate this possibility, the elasticity modulus of all samples with different ratios, including a porous PDMS, were measured using a thermo-mechanical analyzer (TMA 2940-Q series, TA Instruments, DE, USA).

The result has been plotted in Figure 20.6, which shows the increase in elasticity modulus as the concentration of the CI particles increases. However, the elevated modulus that occurs at a 150% ratio is much greater than the increase between the 50% and 100%, or 0% to 50% ratios. Similar trends were observed and reported by Li et al.[45] Hence, for the

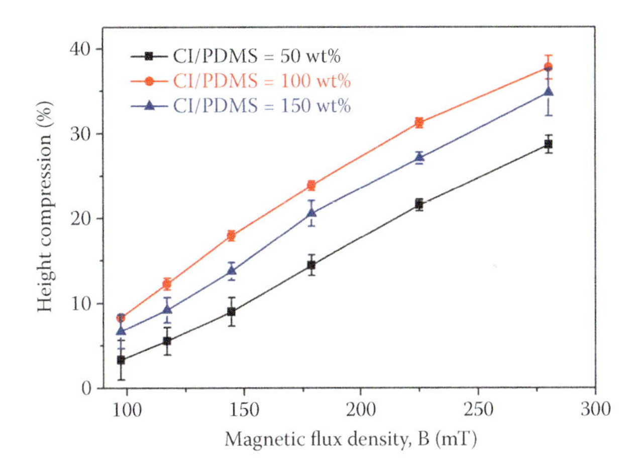

FIGURE 20.5 Displacement of sponges with different CI concentrations in various magnetic fields. (Reprinted with permission from A. Shademani, H. Zhang, J. K. Jackson, and M. Chiao, *Adv. Funct. Mater.*, p. 1604558, 2016. Copyright 2010. American Chemical Society.)

drug delivery device purpose the 100% ratio was selected. The acquired elasticity modulus are 4.56, 4.76, 5.03 and 6.43 kPa for porous PDMS, and magnetic sponges with CI to PDMS wt% ratio of 50%, 100% and 150%, respectively.

20.2.4 Drug Delivery Device Fabrication

As noted earlier, this magnetic sponge could be used solely as a drug delivery agent. In this perspective, one approach might be to immerse the magnetic sponge in a drug solution and let it absorb the drug. Shrinkage and relaxation of the sponge in an on/off magnetic field may then provide an active pumping out of the drug solution. However, the controllability of such a system is relatively weak and the background continuous release from the sponge is a major imperfection. Thus,

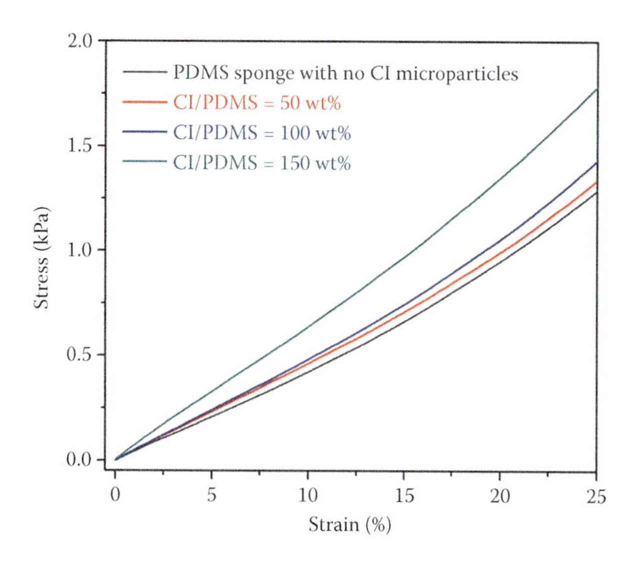

FIGURE 20.6 Stress versus strain for different sponges with 0, 50%, 100%, and 150% CI/PDMS wt% ratios.[40] (Reprinted with permission from A. Shademani, H. Zhang, J. K. Jackson, and M. Chiao, *Adv. Funct. Mater.*, p. 1604558, 2016. Copyright 2010. American Chemical Society.)

to enhance the controllability and improve the delivery performance, a device has been designed and fabricated which frames the sponge. This device is composed of three main components, which are as follows:

1. Reservoir: The drug is loaded inside a reservoir, where the magnetic sponge is placed. This reservoir not only contains and protects the sponge, but it also can hold a high drug payload. This reservoir has a depth of about 1.5 mm and an outer diameter of 6 mm.
2. Magnetic sponge: According to the previous section, a magnetic porous PDMS cylinder with 100% CI/PDMS wt% is made and further cut to fit the reservoir. Therefore, the overall dimensions of sponge are approximately 4 mm in diameter and 1.5 mm in length.
3. Membrane: A thin PDMS membrane (10 μm) is manufactured to seal the reservoir and decrease device leakage.

Once all of these three main elements are prepared, the final device is constructed by assembling these parts. The membrane and reservoir are made from PDMS (Sylgard 184 Silicone Elastomer, Dow Corning Corporation). In the following sections, the design perspective and fabrication process of the aforementioned components are reviewed in detail.

20.2.4.1 Reservoir

To fabricate the reservoir, a positive mould was designed in Solidworks® and built using a 3D printer (Asigo Pico, CA, USA) using Plas White, a UV curable polymer. This mould, shown in Figure 20.7, consists of pillars with a height of 1.5 mm and diameter of 4 mm.

PDMS is made with a 1:10 cross-linker to base ratio, and poured into the mould. The PDMS was then desiccated in a vacuum chamber for 30 min and cured in an oven at 70°C for 3 h. Even though PDMS could be cured faster at higher temperatures (e.g. 200°C), the glass transition temperature of the 3D printed mould is about 83°C, which restricts the curing temperature. The PDMS layer was later unmoulded and reservoirs were punched with an outer diameter of 6 mm. The final thickness of the reservoir is less than 2 mm (approximately 1.5 mm depth of the reservoir and about 0.5 mm base thickness of the reservoir). However, it can be adjusted precisely by controlling the exact volume of the PDMS disposed into the mould and the height of the pillar. Although smaller reservoirs have also been made with a different mould, this particular size has been chosen based on the ocular drug delivery device dimension previously described by our group.[37]

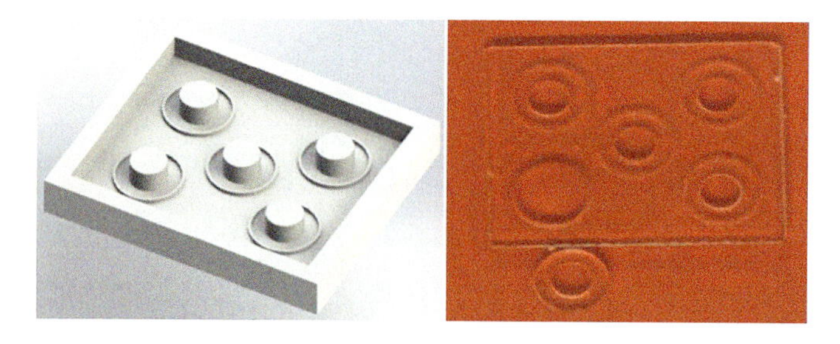

FIGURE 20.7 Mold on the right side and the demolded PDMS out of which one sample reservoir was punched.[40] (Reprinted with permission from A. Shademani, H. Zhang, J. K. Jackson, and M. Chiao, *Adv. Funct. Mater.*, p. 1604558, 2016. Copyright 2010. American Chemical Society.)

20.2.4.2 Membrane

As noted above, the membrane will eventually cover the top of the reservoir after the sponge is placed. The fabrication process of the membrane is as follows:

- Step 1: A polyacrylic acid (PAA) solution was prepared by mixing PAA powder (M_w = 1800, Sigma-Aldrich, ON, Canada) with distilled water to achieve a 25% w/v concentration. This step requires about 10 min vortex mixing of the solution followed by filtering, using a sterile 0.45 μm PVDF syringe filter (Millipore Corporation, Ma, USA).
- Step 2: A precleaned glass slide surface was treated by plasma to enhance surface wettability.
- Step 3: The filtered PAA was spun on the glass slide in two steps (10 s at 500 rpm, 30 s at 800 rpm). The glass slide was later placed on a hot plate at 150°C for 5 min to evaporate water and cooled down afterwards over 5 min.
- Step 4: PDMS was made with a 1:10 curing ratio of agent to prepolymer ratio and degassed in a vacuum chamber for 45 min. The prepared PDMS was further poured on the PAA-coated glass slide and spun in two steps (20 s at 500 rpm, 3 min. at 3500 rpm). The PDMS membrane was then cured on hot plate at 150°C in 5 min.

The final thickness of the membrane was measured to be roughly 10 μm using a surface profilometer (Wyko, VEECO Metrology Group, AZ, USA).

20.2.4.3 Assembling

Initially, the desired amount of the drug was deposited into the reservoir. In the Results and Discussions section (Section 20.3), we illustrated why there is no concern regarding the limit of loaded drug into the reservoir. The magnetic sponge was cut to size and snugly fitted into the reservoir. Plasma surface treatment was applied to both the reservoir (including the sponge) and the membrane, which enabled covalent and irreversible binding of the membrane to the reservoir. At this stage, the device was completely sealed. By immersing this set into water, the sacrificial layer made by PAA was dissolved and the device is freed. Eventually, an aperture of size of 90×90 μm^2 was created at the centre of the membrane by laser ablation using Nd:YAG laser (Quicklaze, New Wave Research, Sunnyvale, CA). The green laser is utilized with the wavelength of 523 nm, 20 pulses per inch (PPI System Inc) and about 10 μm/s scanning speed. Figure 20.8 shows the actual size of the device relative to one Canadian dollar.

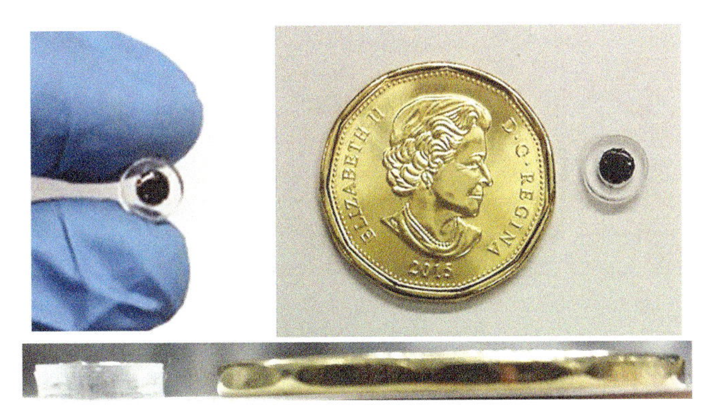

FIGURE 20.8 Real device, in comparison with one Canadian dollar. This device is also roughly as thick as a Canadian dollar.[40] (Reprinted with permission from A. Shademani, H. Zhang, J. K. Jackson, and M. Chiao, *Adv. Funct. Mater.*, p. 1604558, 2016. Copyright 2010. American Chemical Society.)

SEM images from cross sections of the device are also provided in Figure 20.9, where the covalent bonding among reservoir, sponge and the membrane is evident in insets (i) and (ii).

20.2.4.4 Plasma Surface Treatment

In all the aforementioned fabrication processes, air plasma treatment has been performed for 30 seconds, at approximately 700 mtorr pressure using a Harrick plasma chamber (Ithaca, NY, USA). These parameters are determined to provide the strongest bonding and the best surface wettability for PDMS.[46]

20.2.4.5 Device Activation

As noted earlier, the drug was loaded during the assembly process and prior to placement of the magnetic sponge. The reservoir was subsequently sealed by a membrane. In this state, if the device was immersed in a solution or positioned in a biological environment (e.g. tissue), no release would occur. Since the deposited drug is in solid form and has not been dissolved yet, the device is inactive. Device activation takes place when the reservoir is filled with a solvent.

To activate the device, it was submerged in a designated media (water or phosphate buffer saline [PBS] including 1% w/v bovine serum albumin [BSA]) and placed in a vacuum chamber for about 10 minutes to evacuate any air from the reservoir. By venting the chamber, the aqueous media entered

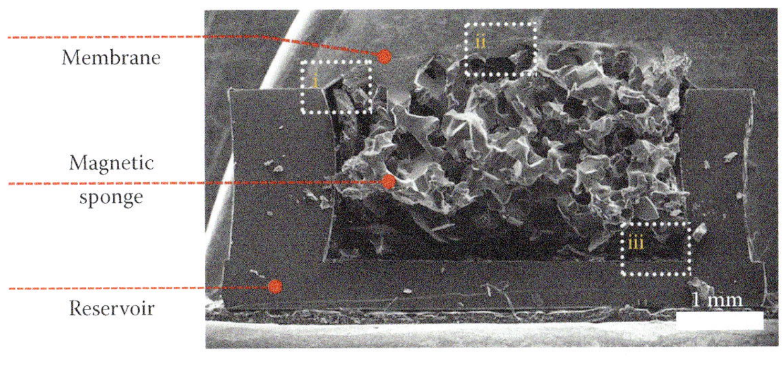

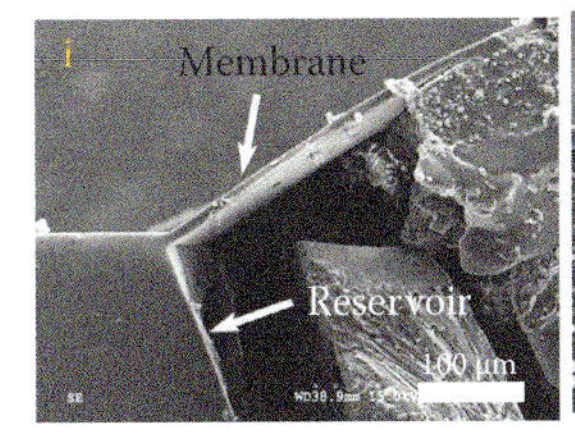

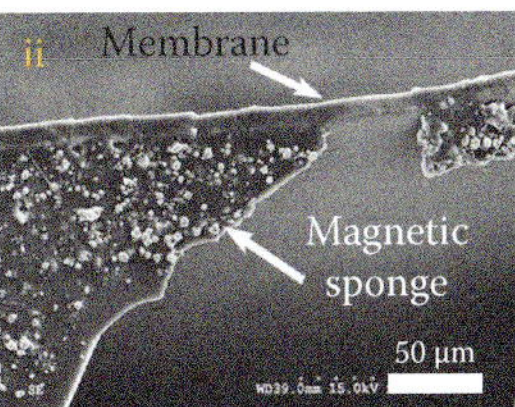

FIGURE 20.9 SEM image from cross section of the proposed drug delivery device.[40] (Reprinted with permission from A. Shademani, H. Zhang, J. K. Jackson, and M. Chiao, *Adv. Funct. Mater.*, p. 1604558, 2016. Copyright 2010. American Chemical Society.)

the device through the aperture and dissolved some of the drug loaded into the reservoir.

20.2.5 Device Characterization

To assess the performance of the drug delivery device, two release tests were performed using a model drug (MB) and an anticancer drug (Docetaxel). In this section, each experiment's conditions are described.

20.2.5.1 Methylene Blue Release

The first step involved using a coloured (visible) model drug to test ejection behaviour. MB was selected as the surrogate drug due to its strong spectrometry properties and high solubility. A solution of this substance was made with the concentration of 10 mg/ml in water. Each device was then filled with about 20 µl of this solution, resulting in 200 µg of MB in the reservoir.

The device was placed at the bottom of a 20 ml vial and submerged in 5 ml of distilled water. A 120 mT magnetic field was applied to actuate the device only once for 5 s, followed by a 10 min dormant phase. This operation was repeated for 90 min to evaluate cumulative MB ejection and to test device functionality over several release cycles. Two 1 ml samples were taken out from the solution before and after release to measure the amount of MB released in each actuation. The amount of released MB was quantified by reading the absorbance of the samples at a wavelength of 662 nm using a UV-Vis spectrophotometer (50 BIO, Varian Medical Systems Inc., Palo Alto, CA, USA). The obtained absorbance was converted to the concentration and the amount of MB that existed in the solution. In order to be able to relate the absorbance to the concentration, a standard curve was first obtained by measuring the absorbance of solutions of MB with known concentrations. Figure 20.10 shows the discharged MB out of the device, which travelled about 7 mm into the medium.

20.2.5.2 Docetaxel Release

To assess device functionality to release a real drug, docetaxel (DTX) was selected as a well-established anticancer medication, which is used to treat several cancer diseases such as prostate, breast, gastric, head and neck and nonsmall cell lung cancer, according to the National Cancer Institute (NCI). Docetaxel has antiproliferative characteristics which inhibit cell replication leading to the death of the cancerous cells in the tumour.[47] Numerous studies have demonstrated docetaxel treatment efficacy based on different treatment protocols.[48–50] Inevitable side effects such as myelosuppression, neutropenia and nail toxicity have been reported.[51–53]

To prepare the drug solution, 50 µl of tritium labelled DTX (50 µCi/200 µl) in ethanol (Moravek Biochemicals Inc., Brea, CA, USA) was mixed with 4 mg of unlabelled DTX in 50 µl of dichloromethane resulting in a 40 mg/ml of DTX solution. Each device was then loaded with 400 µg DTX and 50 µg MB, which was used as a visual aid to detect injection. Since DTX solubility has been reported to be very low in water (around 5 µg/ml),[54] PBS (pH 7.4) including 1% BSA (subsequently referred to as PBS/BSA solution) was chosen as the media for docetaxel release test as DTX has shown better solubility in this solution which is more physiological in nature.[31] The device was situated at the bottom of a 20 ml vial in 4 ml of the PBS/BSA solution and actuated five times, each of which consisted of 7 s of a magnetic field followed by 5 s of relaxation when the device was not actuated. This actuation interval, which takes about 1 min, was repeated for 2 h with 20 min layoff. DTX release over three consecutive days was also studied and the results are provided in the next section. The general low solubility of DTX results in low ejected concentrations, which makes detection difficult. Thus, longer and more actuation protocols were chosen for increasing the amount of released DTX to improve detection.

Two samples of 500 µL were extracted before each actuation interval to measure the background DTX concentration of solution. After the actuation interval, three samples of 500 µL were pipetted into the scintillation vials. Each sample took one scintillation vial, which was filled earlier with about 5 ml of Cytoscint liquid scintillation fluid (Fisher Scientific, Fair Lawn, NJ, USA). The scintillation vials were then stirred using a vortex mixer and the disintegrations per minute (DPM) read using a liquid scintillation counter (Tri-Carb, PerkinElmer, Waltham, MA, USA). The standard curve had previously been provided from specified specimens with predetermined concentrations, enabling us to convert the raw DPM data to the weight of DTX released into the solution in terms of nanograms. Three scintillation vials with 5 ml of scintillation fluid, which were free of DTX, were used to measure the background radiation.

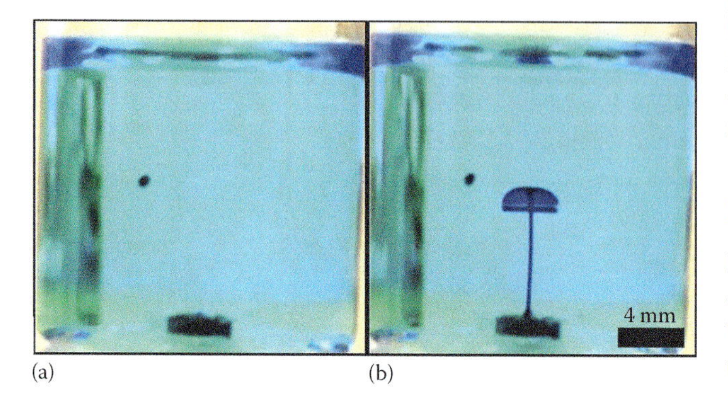

FIGURE 20.10 Methylene blue release, (a) dormant phase, (b) actuation.[40] (Reprinted with permission from A. Shademani, H. Zhang, J. K. Jackson, and M. Chiao, *Adv. Funct. Mater.*, p. 1604558, 2016. Copyright 2010. American Chemical Society.)

20.3 RESULTS AND DISCUSSIONS

In the previous section, device fabrication and subsequent experiments carried out to qualify device performance have been described. This section is dedicated to the data obtained from those tests and a discussion of the results.

20.3.1 Methylene Blue Release

MB is a highly soluble substance in water. Its solubility has been reported around 40 mg/ml in water by the

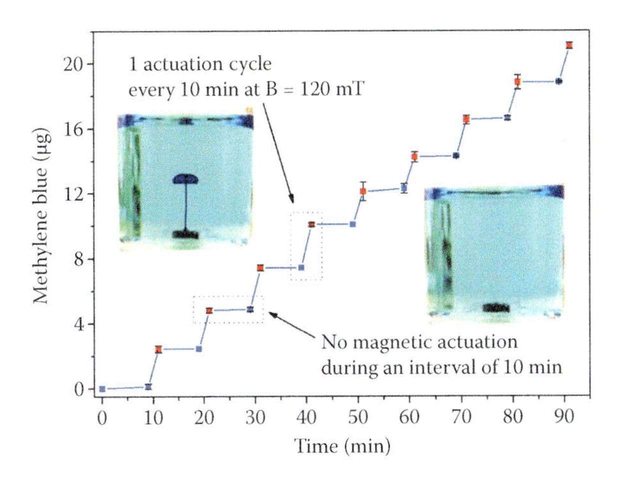

FIGURE 20.11 Cumulative methylene blue release over time.[40] (Reprinted with permission from A. Shademani, H. Zhang, J. K. Jackson, and M. Chiao, *Adv. Funct. Mater.*, p. 1604558, 2016. Copyright 2010. American Chemical Society.)

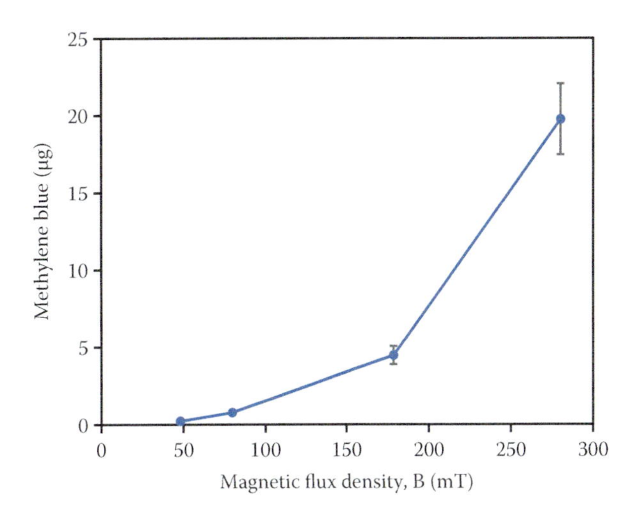

FIGURE 20.12 MB release as a function of magnetic field strength.

manufacturer. Hence, it is expected that the amount of MB in the reservoir decreases as the number of actuations increases or following consecutive actuation. However, ease of use and excellent spectrometry characteristics make it a favourable candidate as a surrogate drug as the release may be visualized in real time. Furthermore, unlike the docetaxel, it does not have any radioactive properties, making its handling much simpler.

As depicted in Figure 20.11, this device released 2.28 ± 0.23 μg MB per actuation (n=36) in a 120 mT magnetic field. As mentioned in the previous section, each device was filled with 200 μg MB. Considering the released amount, this implies that the device will be depleted after about 87 actuations in this magnetic field, which was enough to demonstrate use, and measure repeated MB weights.

20.3.1.1 Influence of the Magnetic Field

Another factor affecting the amount of released MB is the strength of the magnetic field. Stronger magnetic fields exert more force on the magnetic sponge. Although the magnitude of force is proportional to the magnetic properties of the sponge (i.e. the concentration of the CI magnetic microparticles), the ultimate sponge shrinkage is dictated by its elasticity modulus, as illustrated in Section 20.2.3.2. However, for one particular magnetic sponge, stronger magnetic fields lead to more shrinkage which results in more drug ejection.

In order to evaluate the effect of the applied magnetic field, the device was exposed to different magnetic field strengths and the consequent release was quantified. However, other conditions such as the media (water) remained unaltered as described for MB release in Section 20.2.5.1.

The results shown in Figure 20.12 indicate that more MB was released as the magnetic field strength increased. Thus, the magnetic field strength can potentially be used as a critical parameter to adjust the amount of introduced drug according to the treatment protocol.

20.3.2 Controlled Docetaxel Release

Since docetaxel is colourless, MB was mainly used to visually confirm the ejection characteristics. As mentioned in Section 20.2.5.2, the radioactivity of released DTX was measured and later converted to the weight of DTX. The device was actuated often to raise the amount of DTX in the solution and make the samples more radioactive for improved detection. Due to the low solubility of DTX, the selected dormant phase was also longer (20 min) to ensure that the DTX was dissolved thoroughly in the reservoir for the next actuation.

20.3.2.1 Experimental Results

Cumulative DTX release is presented in Figure 20.13, which shows 40.34 ± 4.23 ng DTX released in each actuation interval. To evaluate device release consistency, this

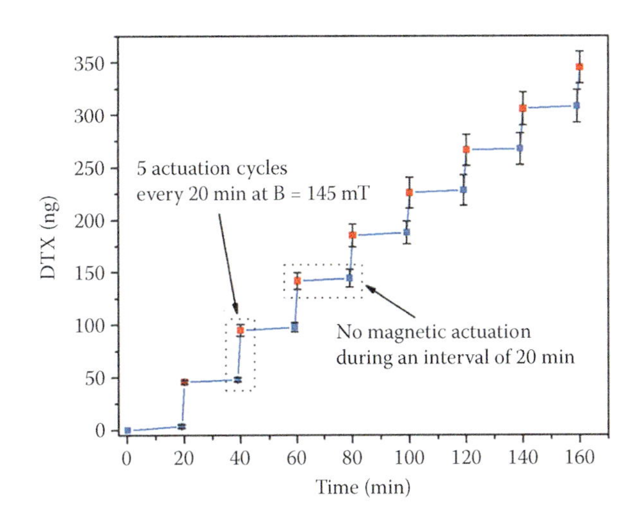

FIGURE 20.13 Cumulative docetaxel release in a 180 MT magnetic field strength.[40] (Reprinted with permission from A. Shademani, H. Zhang, J. K. Jackson, and M. Chiao, *Adv. Funct. Mater.*, p. 1604558, 2016. Copyright 2010. American Chemical Society.)

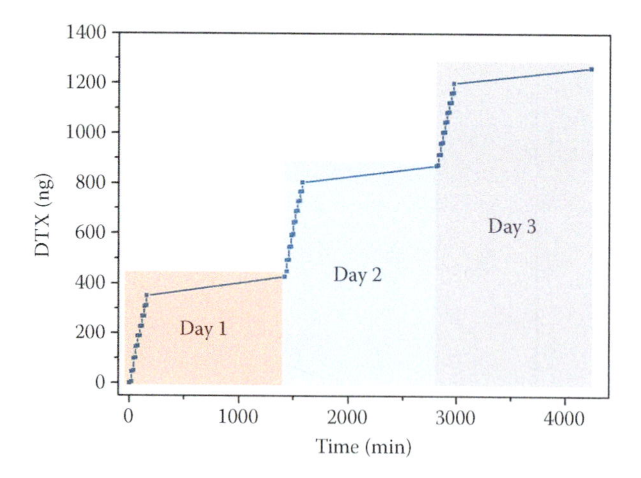

FIGURE 20.14 Three-consecutive-day docetaxel release.[40] (Reprinted with permission from A. Shademani, H. Zhang, J. K. Jackson, and M. Chiao, *Adv. Funct. Mater.*, p. 1604558, 2016. Copyright 2010. American Chemical Society.)

determination was repeated several times. The device was also actuated for about 3 h over three consecutive days. After each actuation period, the device sat in PBS/BSA solution for the rest of day until the next actuation (i.e. next day). The results are plotted in Figure 20.14, which shows roughly 350 ng DTX released each day. As explained before, each device contained about 400 μg DTX including 50 μg MB. Therefore, this device released only about 0.01% and 0.088% of the drug supply in each actuation interval and each day, respectively. In other words, the device might be depleted after 10,000 actuation intervals. Nonetheless, it is believed that the designated reservoir is capable of storing even more drug, up to 1 mg. Thus, there is no concern regarding depletion of the drug in the short term. Generally, this drug works as an antiproliferative agent at 10 ng/ml so the amounts of drug released here are suitable for therapeutic needs. However, the optimal amount of drug released might be determined according to the treatment protocol and the course of therapy.

20.3.2.2 Background Leakage

During the device fabrication process, an aperture is created by laser ablation, which means that there is a permanent opening in the device. Although the main objective of this aperture is to redirect the drug out of the reservoir once the device is actuated, the drug can potentially leave the device by diffusion while no actuation occurs. This phenomenon is due to the much higher drug dosage inside the device, which leads to drug diffusion through the aperture into the surrounding media with a much lower concentration of drug. This is one of the negative features of the reservoir-based drug delivery devices, which incorporate apertures without valves. In this device, because of the interconnected pores of sponge, the drug solution is somehow entrapped within the sponge structure, reducing fluid movement and restricting such leakage. However, due to the high concentration gradient, this diffusion might never be fully stopped unless a valve is provided for the aperture.

In order to find out the rate of background leakage, the device was left in PBS/BSA solution for more than 20 h. However, to compute the leakage rate, the initial and final concentration of the solution were determined by taking out two and three 500 μl samples at the beginning and at the end of the test, respectively. This device had a background leakage of 0.060 ± 0.004 ng/min (n=4), which can also be seen in Figure 20.14 as the line connecting two consecutive actuation periods. In localized drug delivery, this leakage probably occurs in the target zone (e.g. tumour) and may be acceptable.

20.3.2.3 In Vitro Cell Study

Docetaxel was used in a different drug delivery device and the corresponding DTX activity on the PC3 prostate cancer cells was determined.[29] As revealed in Figure 20.15, cell death commenced at around a concentration of 10 ng/ml, which drastically decreased cell proliferation by 60%. Thus, ng/ml concentrations of this drug are adequate for an anticancer effect. Furthermore, since the amount of released DTX in the previous section was computed for a 3 ml medium, the concentration of released DTX is about 13.45 ng/ml, which is within the active range of the DTX (more than 10 ng/ml). The amount of released drug can be adjusted to attain the favourable dosage.

Various dosages of DTX in the range of 30–75 mg/m^2 body-surface area have been systemically administrated[48–50,55,56] for tumour treatment. Body-surface area is a unit defined to allot a suitable dose of the drug to each individual with respect to their own weight and height. In this regard, different formulae have been proposed so far to calculate the value.[57] As a rough estimation, if the prescribed dose and the body surface area of one typical patient are considered to be 30 mg/m^2 and 2 m^2, respectively, the amount of DTX introduced to the body in

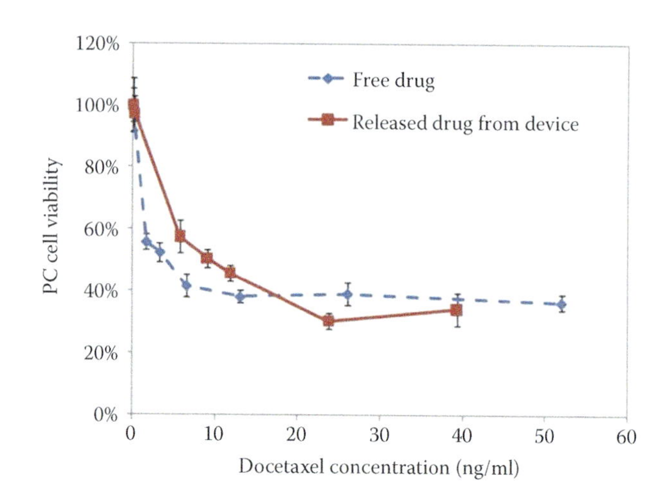

FIGURE 20.15 Inhibition of PC3 cell proliferation using docetaxel.[30] (Reprinted with permission from F.N. Pirmoradi, J.K. Jackson, H.M. Burt, and M. Chiao, *Lab Chip*, pp. 3072–3080, 2011. Copyright 2010. American Chemical Society.)

direct administration would be 60 mg. As mentioned earlier, this amount of DTX will spread through the whole body and may be toxic to other healthy organs. Localized delivery of DTX will require less dosage.[58] According to the experimental data obtained in the DTX release test and cell study, only 2 ng DTX is sufficient to inhibit cell proliferation to roughly 40% (200 μl of each concentration was used).[29]

Localized drug delivery is therefore able to treat local disease sites effectively while releasing only tiny amounts of residual drug into the bloodstream,

20.4 CONCLUSIONS

We have introduced a drug delivery device whose function is based on a magnetic sponge actuation mechanism. The drug is added to a magnetic sponge constructed from polydimethylsiloxane. Then it is deformed when applying an external magnetic field that leads to discharge its content. Magnetic sponge fabrication starts with mixing PDMS with carbonyl iron (CI) to achieve magnetic PDMS, which subsequently becomes porous by pouring it onto a sugar lump, followed by curing and dissolution of the sacrificial sugar. The magnetic sponge shrinkage has been determined as the main criteria to optimize the appropriate CI/PDMS wt% ratio. The shrinkage was found to be dependent on two competing factors: magnetic properties and the elastic modulus of the sponge. However, both of these factors rely on the CI concentration. Hence, the 100% CI/PDMS wt% ratio has shown more contraction in various magnetic fields than the two other ratios of 50% and 150%.

This sponge is able to return to its initial condition once the magnetic field is removed, and applying on/off magnetic fields results in effective pumping out of the drug solution. Furthermore, sponge shrinkage can be controlled by tuning the magnetic field strength. To achieve the most drug release, the magnetic field orientation should be perpendicular to the device. However, angled actuation would lead to less release, where the transverse magnetic field application may not be able to actuate the device at all. The magnetic sponge may absorb the therapeutic compound by submerging it in a drug solution. Nonetheless, a device has been fabricated composed of a reservoir, magnetic sponge and sealing membrane. This device obviates the need for power, and is able to release the drug on-demand at the prescribed dosage by applying a specified magnetic field.

REFERENCES

1. Das, D. and Pal, S., 2015. Modified biopolymer-dextrin based crosslinked hydrogels: Application in controlled drug delivery. *RSC Advances, 5*(32), pp. 25014–25050.
2. Schmaljohann, D., 2006. Thermo-and pH-responsive polymers in drug delivery. *Advanced Drug Delivery Reviews, 58*(15), pp. 1655–1670.
3. Pang, X., Du, H.L., Zhang, H.Q., Zhai, Y.J. and Zhai, G.X., 2013. Polymer–drug conjugates: Present state of play and future perspectives. *Drug Discovery Today, 18*(23), pp. 1316–1322.
4. Terwogt, J.M.M., ten Bokkel Huinink, W.W., Schellens, J.H., Schot, M., Mandjes, I.A., Zurlo, M.G., Rocchetti, M., Rosing, H., Koopman, F.J. and Beijnen, J.H., 2001. Phase I clinical and pharmacokinetic study of PNU166945, a novel water-soluble polymer-conjugated prodrug of paclitaxel. *Anti-Cancer Drugs, 12*(4), pp. 315–323.
5. Vasey, P.A., Kaye, S.B., Morrison, R., Twelves, C., Wilson, P., Duncan, R., Thomson, A.H., Murray, L.S., Hilditch, T.E., Murray, T. and Burtles, S., 1999. Phase I clinical and pharmacokinetic study of PK1 N-(2-hydroxypropyl) methacrylamide copolymer doxorubicin: First member of a new class of chemotherapeutic agents—drug-polymer conjugates. *Clinical Cancer Research, 5*(1), pp. 83–94.
6. Wu, H., Zhu, L. and Torchilin, V.P., 2013. pH-sensitive poly (histidine)-PEG/DSPE-PEG co-polymer micelles for cytosolic drug delivery. *Biomaterials, 34*(4), pp. 1213–1222.
7. Yang, R., Gorelov, A.V., Aldabbagh, F., Carroll, W.M. and Rochev, Y., 2013. An implantable thermoresponsive drug delivery system based on Peltier device. *International Journal of Pharmaceutics, 447*(1), pp. 109–114.
8. Vilar, G., Tulla-Puche, J. and Albericio, F., 2012. Polymers and drug delivery systems. *Current Drug Delivery, 9*(4), pp. 367–394.
9. Yadav, S.M., Pareek, A.K., Kumar, M., Gupta, S. and Garg, S., 2015. Osmotic drug delivery system: A new approach. *Inter Jour Pharma Tech and Biotech, 2*(1), pp. 11–25.
10. Verma, R.K. and Garg, S., 2001. Drug delivery technologies and future directions. *Pharmaceutical Technology, 25*(2), pp. 1–14.
11. Rosen, H. and Abribat, T., 2005. The rise and rise of drug delivery. *Nature Reviews Drug Discovery, 4*(5), pp. 381–385.
12. Wright, J.C., Leonard, S.T., Stevenson, C.L., Beck, J.C., Chen, G., Jao, R.M., Johnson, P.A., Leonard, J. and Skowronski, R.J., 2001. An in vivo/in vitro comparison with a leuprolide osmotic implant for the treatment of prostate cancer. *Journal of Controlled Release, 75*(1), pp. 1–10.
13. Ryu, W., Huang, Z., Prinz, F.B., Goodman, S.B. and Fasching, R., 2007. Biodegradable micro-osmotic pump for long-term and controlled release of basic fibroblast growth factor. *Journal of Controlled Release, 124*(1), pp. 98–105.
14. Herrlich, S., Spieth, S., Messner, S. and Zengerle, R., 2012. Osmotic micropumps for drug delivery. *Advanced Drug Delivery Reviews, 64*(14), pp. 1617–1627.
15. Verma, R.K. and Garg, S., 2004. Development and evaluation of osmotically controlled oral drug delivery system of glipizide. *European Journal of Pharmaceutics and Biopharmaceutics, 57*(3), pp. 513–525.
16. Santus, G. and Baker, R.W., 1995. Osmotic drug delivery: A review of the patent literature. *Journal of Controlled Release, 35*(1), pp. 1–21.
17. Santini, J.T., Cima, M.J. and Langer, R., 1999. A controlled-release microchip. *Nature, 397*(6717), pp. 335–338.
18. Maloney, J.M., Uhland, S.A., Polito, B.F., Sheppard, N.F., Pelta, C.M. and Santini, J.T., 2005. Electrothermally activated microchips for implantable drug delivery and biosensing. *Journal of Controlled Release, 109*(1), pp. 244–255.
19. Chung, A.J., Huh, Y.S. and Erickson, D., 2009. A robust, electrochemically driven microwell drug delivery system for controlled vasopressin release. *Biomedical Microdevices, 11*(4), pp. 861–867.
20. Murdan, S., 2003. Electro-responsive drug delivery from hydrogels. *Journal of Controlled Release, 92*(1), pp. 1–17.
21. Yi, Y.T., Sun, J.Y., Lu, Y.W. and Liao, Y.C., 2015. Programmable and on-demand drug release using electrical stimulation. *Biomicrofluidics, 9*(2), p. 022401.

22. Ramanathan, S. and Block, L.H., 2001. The use of chitosan gels as matrices for electrically-modulated drug delivery. *Journal of Controlled Release, 70*(1), pp. 109–123.
23. Veiseh, O., Gunn, J.W. and Zhang, M., 2010. Design and fabrication of magnetic nanoparticles for targeted drug delivery and imaging. *Advanced Drug Delivery Reviews, 62*(3), pp. 284–304.
24. Mody, V.V., Cox, A., Shah, S., Singh, A., Bevins, W. and Parihar, H., 2014. Magnetic nanoparticle drug delivery systems for targeting tumor. *Applied Nanoscience, 4*(4), pp. 385–392.
25. Cai, K., Luo, Z., Hu, Y., Chen, X., Liao, Y., Yang, L. and Deng, L., 2009. Magnetically triggered reversible controlled drug delivery from microfabricated polymeric multireservoir devices. *Advanced Materials, 21*(40), pp. 4045–4049.
26. Rahimi, S., Sarraf, E.H., Wong, G.K. and Takahata, K., 2011. Implantable drug delivery device using frequency-controlled wireless hydrogel microvalves. *Biomedical Microdevices, 13*(2), pp. 267–277.
27. Liu, T.Y., Hu, S.H., Liu, K.H., Liu, D.M. and Chen, S.Y., 2008. Study on controlled drug permeation of magnetic-sensitive ferrogels: Effect of Fe3O4 and PVA. *Journal of Controlled Release, 126*(3), pp. 228–236.
28. Zhao, X., Kim, J., Cezar, C.A., Huebsch, N., Lee, K., Bouhadir, K. and Mooney, D.J., 2011. Active scaffolds for on-demand drug and cell delivery. *Proceedings of the National Academy of Sciences, 108*(1), pp. 67–72.
29. Pirmoradi, F.N., Jackson, J.K., Burt, H.M. and Chiao, M., 2011. On-demand controlled release of docetaxel from a battery-less MEMS drug delivery device. *Lab on a Chip, 11*(16), pp. 2744–2752.
30. Pirmoradi, F.N., Jackson, J.K., Burt, H.M. and Chiao, M., 2011. A magnetically controlled MEMS device for drug delivery: Design, fabrication, and testing. *Lab on a Chip, 11*(18), pp. 3072–3080.
31. Zachkani, P., Jackson, J.K., Pirmoradi, F.N. and Chiao, M., 2015. A cylindrical magnetically-actuated drug delivery device proposed for minimally invasive treatment of prostate cancer. *RSC Advances, 5*(119), pp. 98087–98096.
32. Mustapić, M., Al Hossain, M.S., Horvat, J., Wagner, P., Mitchell, D.R., Kim, J.H., Alici, G., Nakayama, Y. and Martinac, B., 2016. Controlled delivery of drugs adsorbed onto porous Fe3O4 structures by application of AC/DC magnetic fields. *Microporous and Mesoporous Materials, 226*, pp. 243–250.
33. Hoare, T., Santamaria, J., Goya, G.F., Irusta, S., Lin, D., Lau, S., Padera, R., Langer, R. and Kohane, D.S., 2009. A magnetically triggered composite membrane for on-demand drug delivery. *Nano Letters, 9*(10), pp. 3651–3657.
34. Lo, R., Li, P.Y., Saati, S., Agrawal, R.N., Humayun, M.S. and Meng, E., 2009. A passive MEMS drug delivery pump for treatment of ocular diseases. *Biomedical Microdevices, 11*(5), p. 959.
35. Gensler, H., Sheybani, R., Li, P.Y., Lo, R., Zhu, S., Yong, K.T., Roy, I., Prasad, P.N., Masood, R., Sinha, U.K. and Meng, E., 2010, January. Implantable MEMS drug delivery device for cancer radiation reduction. In *Micro Electro Mechanical Systems (MEMS), 2010 IEEE 23rd International Conference on* (pp. 23–26). IEEE.
36. Mousoulis, C., Ochoa, M., Papageorgiou, D. and Ziaie, B., 2011. A skin-contact-actuated micropump for transdermal drug delivery. *IEEE Transactions on Biomedical Engineering, 58*(5), pp. 1492–1498.
37. Pirmoradi, F.N., Ou, K., Jackson, J.K., Letchford, K., Cui, J., Wolf, K.T., Gräber, F., Zhao, T., Matsubara, J.A., Burt, H. and Chiao, M., 2013, January. Controlled delivery of antiangiogenic drug to human eye tissue using a MEMS device. In *Micro Electro Mechanical Systems (MEMS), 2013 IEEE 26th International Conference on* (pp. 1–4). IEEE.
38. Timko, B.P., Arruebo, M., Shankarappa, S.A., McAlvin, J.B., Okonkwo, O.S., Mizrahi, B., Stefanescu, C.F., Gomez, L., Zhu, J., Zhu, A. and Santamaria, J., 2014. Near-infrared–actuated devices for remotely controlled drug delivery. *Proceedings of the National Academy of Sciences, 111*(4), pp. 1349–1354.
39. Sirsi, S.R. and Borden, M.A., 2014. State-of-the-art materials for ultrasound-triggered drug delivery. *Advanced Drug Delivery Reviews, 72*, pp. 3–14.
40. Shademani, A., Zhang, H., Jackson, J.K. and Chiao, M., 2017. Active regulation of On-demand drug delivery by magnetically triggerable microspouters. *Advanced Functional Materials, 27*(6).
41. Cha, K.J. and Kim, D.S., 2011. A portable pressure pump for microfluidic lab-on-a-chip systems using a porous polydimethylsiloxane (PDMS) sponge. *Biomedical Microdevices, 13*(5), p. 877.
42. King, M.G., Baragwanath, A.J., Rosamond, M.C., Wood, D. and Gallant, A.J., 2009. Porous PDMS force sensitive resistors. *Procedia Chemistry, 1*(1), pp. 568–571.
43. Pedraza, E., Brady, A.C., Fraker, C.A. and Stabler, C.L., 2013. Synthesis of macroporous poly (dimethylsiloxane) scaffolds for tissue engineering applications. *Journal of Biomaterials Science, Polymer Edition, 24*(9), pp. 1041–1056.
44. Pedraza, E., Brady, A.C., Fraker, C.A., Molano, R.D., Sukert, S., Berman, D.M., Kenyon, N.S., Pileggi, A., Ricordi, C. and Stabler, C.L., 2013. Macroporous three-dimensional PDMS scaffolds for extrahepatic islet transplantation. *Cell Transplantation, 22*(7), pp. 1123–1135.
45. Li, J., Zhang, M., Wang, L., Li, W., Sheng, P. and Wen, W., 2011. Design and fabrication of microfluidic mixer from carbonyl iron–PDMS composite membrane. *Microfluidics and Nanofluidics, 10*(4), pp. 919–925.
46. Bhattacharya, S., Datta, A., Berg, J.M. and Gangopadhyay, S., 2005. Studies on surface wettability of poly (dimethyl) siloxane (PDMS) and glass under oxygen-plasma treatment and correlation with bond strength. *Journal of Microelectromechanical Systems, 14*(3), pp. 590–597.
47. Ferlini, C., Scambia, G., Distefano, M., Filippini, P., Isola, G., Riva, A., Bombardelli, E., Fattorossi, A., Panici, P.B. and Mancuso, S., 1997. Synergistic antiproliferative activity of tamoxifen and docetaxel on three oestrogen receptor-negative cancer cell lines is mediated by the induction of apoptosis. *British Journal of Cancer, 75*(6), pp. 884–891.
48. Beer, T.M., Pierce, W.C., Lowe, B.A. and Henner, W.D., 2001. Phase II study of weekly docetaxel in symptomatic androgen-independent prostate cancer. *Annals of Oncology, 12*(9), pp. 1273–1279.
49. Schuette, W., Nagel, S., Blankenburg, T., Lautenschlaeger, C., Hans, K., Schmidt, E.W., Dittrich, I., Schweisfurth, H., von Weikersthal, L.F., Raghavachar, A. and Reißig, A., 2005. Phase III study of second-line chemotherapy for advanced non–small-cell lung cancer with weekly compared with 3-weekly docetaxel. *Journal of Clinical Oncology, 23*(33), pp. 8389–8395.
50. Sinibaldi, V.J., Carducci, M.A., Moore-Cooper, S., Laufer, M., Zahurak, M. and Eisenberger, M.A., 2002. Phase II evaluation of docetaxel plus one-day oral estramustine phosphate in the treatment of patients with androgen independent prostate carcinoma. *Cancer, 94*(5), pp. 1457–1465.

51. Hong, J., Park, S.H., Choi, S.J., Lee, S.H., Lee, K.C., Lee, J.I., Kyung, S.Y., An, C.H., Lee, S.P., Park, J.W. and Jeong, S.H., 2007. Nail toxicity after treatment with docetaxel: A prospective analysis in patients with advanced non-small cell lung cancer. *Japanese Journal of Clinical Oncology, 37*(6), pp. 424–428.
52. Shiota, M., Yokomizo, A., Takeuchi, A., Kiyoshima, K., Inokuchi, J., Tatsugami, K. and Naito, S., 2014. Risk factors for febrile neutropenia in patients receiving docetaxel chemotherapy for castration-resistant prostate cancer. *Supportive Care in Cancer, 22*(12), pp. 3219–3226.
53. Winther, D., Saunte, D.M., Knap, M., Haahr, V. and Jensen, A.B., 2007. Nail changes due to docetaxel—a neglected side effect and nuisance for the patient. *Supportive Care in Cancer, 15*(10), p. 1191.
54. Yin, Y., Cui, F., Mu, C., Chung, S.J., Shim, C. and Kim, D.D., 2009. Improved solubility of docetaxel using a microemulsion delivery system: Formulation optimization and evaluation. *Asian J Pharm Sci, 4*(6), pp. 331–339.
55. Petrylak, D.P., Tangen, C.M., Hussain, M.H., Lara Jr, P.N., Jones, J.A., Taplin, M.E., Burch, P.A., Berry, D., Moinpour, C., Kohli, M. and Benson, M.C., 2004. Docetaxel and estramustine compared with mitoxantrone and prednisone for advanced refractory prostate cancer. *New England Journal of Medicine, 351*(15), pp. 1513–1520.
56. Tannock, I.F., de Wit, R., Berry, W.R., Horti, J., Pluzanska, A., Chi, K.N., Oudard, S., Théodore, C., James, N.D., Turesson, I. and Rosenthal, M.A., 2004. Docetaxel plus prednisone or mitoxantrone plus prednisone for advanced prostate cancer. *New England Journal of Medicine, 351*(15), pp. 1502–1512.
57. Verbraecken, J., Van de Heyning, P., De Backer, W. and Van Gaal, L., 2006. Body surface area in normal-weight, overweight, and obese adults. A comparison study. *Metabolism, 55*(4), pp. 515–524.
58. Zahedi, P., De Souza, R., Piquette-Miller, M. and Allen, C., 2009. Chitosan–phospholipid blend for sustained and localized delivery of docetaxel to the peritoneal cavity. *International Journal of Pharmaceutics, 377*(1), pp. 76–84.

Mu Chiao (E-mail: muchiao@mech.ubc.ca) is a professor in the Department of Mechanical Engineering at University of British Columbia (Canada). He earned his BS degree from National Taiwan University (Taiwan) and a PhD from the University of California-Berkeley (USA). His research interests include MEMS, BioMEMS/drug delivery, MEMS/protein interaction, and micro optical scanner for biological applications.

Ali Shademani is a PhD student in the Department of Biomedical Engineering at the University of British Columbia (Canada). He earned his Master of Applied Science degree in Mechanical Engineering from the University of British Columbia and his BSc from the University of Tehran (Iran). His research is mainly focused on drug delivery devices, microfluidics and BioMEMS.

Hongbin Zhang is currently a postdoctoral research fellow in the Department of Mechanical Engineering, University of British Columbia (Canada). He earned a BS degree from Sichuan University (China) and a PhD from the University of Science and Technology Beijing (China). His current research interests focus on soft materials, controlled drug delivery systems, surface modification and antifouling and tissue engineering.

21 Magnetic Particle Transport in Complex Media

Lamar O. Mair, Aleksandar N. Nacev, Sagar Chowdhury, Pavel Stepanov, Ryan Hilaman, Sahar Jafari, Benjamin Shapiro and Irving N. Weinberg*

CONTENTS

21.1 INTRODUCTION

Since the late 1970s and early 1980s, researchers have proposed using magnetic fields to guide and concentrate particles to specific regions of the body.[1–3] By doing so, specific payloads could be more effectively concentrated at targeted locations *in vivo*. Payloads may consist of molecules, drugs, proteins, nucleic acids or even other particles. Since the initial proposal of magnetic drug targeting, attempts to concentrate particles[4] and treat human cancers[5] have shown some promise. However, many challenges remain.[6] One such challenge is understanding and optimizing magnetic particle transport through the dense, tortuous, viscoelastic environments of tissues, biofluids and biopolymers.[7] Magnetically-targeted particles inherently come into contact with a range of biopolymer environments, and these environments contain a broad spectrum of proteins in numerous and varied organizational structures. Some biopolymer environments, such as blood, consist of freely floating, loosely interacting proteins and cells. Others, such as extracellular matrices, consist of mesh-like networks of fibrous proteins. Here we review a range of concepts and findings in the field of magnetically guided micro- and nanoparticle (NP) transport in biological materials. Comprehensive overviews of drug transport across biological barriers can be found elsewhere.[8,9]

In this chapter, we discuss why magnetic particle transport through biopolymers is scientifically interesting, and introduce specific challenges that may be considered in designing magnetic particles for transport through biopolymers. Section 21.2 introduces concepts in engineering the particle–biopolymer interface and focuses specifically on particle coatings that actively alter the surrounding proteins in order to enhance particle transport. Section 21.3 discusses recent advances in understanding magnetically-directed particle motion through extracellular matrix, mucus and skin. Section 21.4 reviews novel magnetic field arrangements for moving particles using combined static and alternating magnetic fields. Section 21.5 reviews transport of new particle shapes through viscoelastic biopolymers, including established geometries such as rods, as well as more recently deployed magnetic helices and rolled up sheets. Section 21.6 offers an outlook and perspective on future directions.

21.2 ENGINEERING THE INTERFACE

The biological environment presents micro- and NPs with a broad spectrum of pore sizes, protein conformations and electrostatic charges. As such, particle transport can be dictated by particle–biopolymer interactions. Of the particle interactions that take place *in vivo*, protein–particle interactions have remained a primary focus of the literature due their significance in determining both how the biological environment responds to the particle, as well as how particles transport through the biopolymer. Once particles enter a biological environment (*in vitro* or *in vivo*), a protein corona forms. The protein corona is composed of proteins bound to or transiently associated with a particle. Protein coronas can form in less

* Corresponding author.

than 30 s, and are dynamic in terms of the quantity of proteins associated with particles.[10] As proteins interact with a particle, they induce a cascade of cellular signaling responses, which ultimately dictate the particle's fate. Some researchers have suggested that the significance of the protein–NP complex warrants it being considered a separate biological entity.[11] The protein corona is composed of suspended proteins that interact with the particle with a spectrum of binding kinetics. These interactions are based on particle composition, shape, angle of curvature, surface functionalization, roughness, porosity and hydrophilicity or hydrophobicity.[12]

In addition to suspended proteins and biomolecules, magnetic particles moving through tissue also interact sterically with structural, fibre-forming proteins, such as collagen and laminin. The dense, mesh-like environment of elastic filaments created by such proteins serves as a mechanical barrier to long-range transport. Magnetically guided transport necessitates engineering particle surfaces for overcoming both steric and electrostatic obstacles to transport. There are several methods of classifying NP coatings. One useful categorization scheme divides coatings into either passive or active surfaces. Passive coatings, such as polyethylene glycol (PEG), minimize protein adhesion to particles, but leave the surrounding protein(s) unaltered after the particle–protein interaction. PEG is perhaps the most commonly used particle coating, and numerous studies and reviews have demonstrated PEG's ability to minimize particle–protein interactions and elucidated its mode of action.[13–20] PEG surface coatings neutralize particle zeta potentials, minimizing particle entanglement in the electrostatically and ionically diverse environments of biopolymers.[21,22] As PEG and other passive coatings have been extensively discussed in the literature, we focus here on active coatings. Active coatings enhance particle transport by altering the proteins in the vicinity of the particle. Typically, active techniques involve enzymes capable of actively degrading protein organization, thereby diminishing the effects of steric barriers.

Early work by Kuhn et al. compared particle velocities as they moved through a mixture of extracellular matrix and collagen under magnetic guidance.[23] The authors studied transport velocities for ~300 nm diameter paramagnetic particles (Micromod GmbH) with PEG, bovine serum albumin (BSA) or PEG-collagenase coatings. While PEG- and BSA-coated particles showed minimal translational velocity (< 2 μm/h) over the course of 8 h, PEG-collagenase-coated particles moved at an average of 90 μm/h over an initial 8 h period. Over the course of five days, PEG-collagenase-coated particles slowed to 20 μm/h due to the slow decrease in collagenase enzyme activity (Figure 21.1). However, demonstration of enhanced mobility over the course of 5 days indicated that enhanced particle transport can be maintained for clinically relevant periods. Motivation for disrupting collagen networks lies in the fact that collagen is a major barrier to delivery in most tumours.[24–26]

Collagen, however, is not the only transport inhibiting protein in the tumour microenvironment. Hyaluronic acid (HA) is present in the extracellular matrix, and also plays a significant

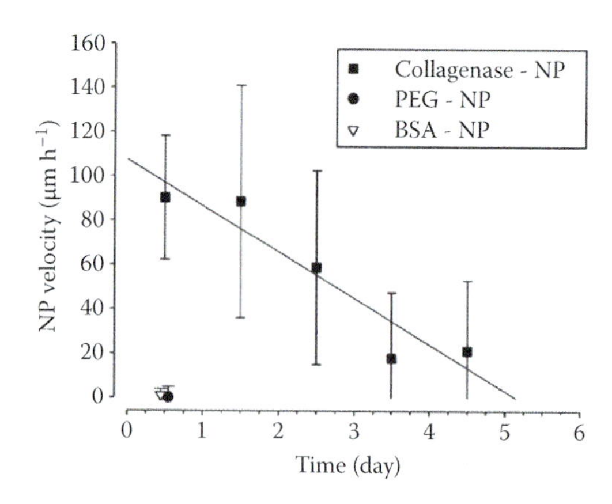

FIGURE 21.1 MNPs coated with collagenase (square data points) move significantly more quickly through collagen-rich extracellular matrix as compared with PEG- or BSA-coated MNPs (circle and triangle data points, respectively).[23] (Reprinted with permission from S.J. Kuhn, S.K. Finch, D.E. Hallahan, T.D. Giorgio, "Proteolytic surface functionalization enhances in vitro magnetic nanoparticle mobility through extracellular matrix," *Nano Lett.* 6 (2006) 306–312. Copyright 2006 American Chemical Society.)

role in suppressing both diffusive and driven particle transport. Hyaluronidase is an enzyme that catalyzes the degradation of hyaluronic acid. Recently, Zhou et al. embedded recombinant human hyaluronidase PH20 within the PEG shell of PLGA-PEG particles, demonstrating increased particle penetration into solid tumours for particles containing hyaluronidase in their PEG shells.[27] Recombinant human hyaluronidase PH20 has garnered significant attention for its ability to increase the distribution of drug in a tumour volume, and the application is currently undergoing clinical testing. In a human clinical trial (Phase IB) study, PEG-PH20 particle treatments were well tolerated by patients with advanced pancreatic cancer, and the PEG-PH20 particle treatments demonstrated the potential for providing a therapeutic benefit for patients with high HA content tumours.[28]

While passively transporting FDA-approved therapeutic particles may operate sufficiently with a PEG coating, magnetically targeted particles will likely need onboard surface chemistry capable of actively disrupting protein networks in order to move through such networks over long distances. Efforts to combine protein disruptors and PEG onto the surfaces of particles significantly increases the particles' chances of success for *in vivo* transport and targeting.

21.3 PARTICLE MOTION IN THREE MODEL BIOLOGICAL POLYMERS

Viscoelastic environments are characterized by nonlinear relationships between the viscoelastic material's modulus and the rate of an applied strain. In biomaterials, nonlinearity arises from two primary phenomena. First, constituent biomaterial molecules demonstrate inherent viscoelasticity: proteins are elastic and demonstrate strain-rate dependent mechanics. Fibre-like protein strands maintain that elasticity, to varying degrees. Second, interactions among biomolecules and protein strands contribute to physical entanglement of the various molecular and supramolecular constituents of a given biopolymer, biofluid or tissue. In this section, we discuss transport through extracellular matrix, mucus and skin. However, readers interested in transport through liver, kidney and brain should see the study comparing transport of various particles through excised liver, kidney and brain.[29]

21.3.1 Salient Features of Viscoelastic Environments: The 'Particle's Eye View'

To a translating particle, viscoelastic biomaterials such as mucus, tissues and extracellular matrices, appear as biphasic materials. Here, the solid phase is composed of fibrillar proteins, with each fibre or fibre type having a characteristic dimension. Some biomaterials may have both fibrillar and globular proteins which, from a 'particle's eye view', both contribute to the elastic component of the environment. The liquid phase generally consists of salts, proteins and other floating components. Within this biphasic structure, there are pores of varying shapes and sizes. Tissues often contain pores whose sizes span orders of magnitude, and typical pore sizes may range from 10 nm to 10 μm. Various models of transport of solid particles through porous media have been applied to particle transport in biopolymers. Three commonly used models are the Renkin model,[30] the Ogston model,[31,32] and the Phillips model.[33,34] These models deal with diffusive transport of particles through porous media and form an excellent basis for understanding particulate transport through mesh-like biopolymers. Readers interested in acquiring a more detailed understanding of diffusive transport are directed to excellent reviews on the topic by Amsden,[35] as well as Cu and Saltzman.[36]

Magnetically-guided particle motion through such porous environments is a function of particle size, particle shape, particle magnetization, particle surface coating, pore dimensions, fibre dimensions, matrix rigidity, applied magnetic force and viscosity of the fluid phase. The fluid phases of biopolymers often have viscosities ranging from 1 to 1000 mPa s (1× to 1000× the viscosity of water). As magnetic particles are driven under a magnetic gradient, they move through the fluid phase of the biopolymer and interact with components in the fibrous solid phase of the biopolymer. As they move, magnetically driven particles may slide past fibres of the solid phase and continue moving, become entangled in the fibrous solid phase of the biopolymer and stop moving, experience stick-slip motion as they are intermittently entangled then freed from steric barriers, or may become entangled in the fibrous phase and induce matrix deformation as they pull the proteins of the matrix along with them. Depending on particle size and available space for particle transport through the fluid phase (porosity), a combination of the previously listed transport modes may take place for a given applied force, particle geometry and surface coating. Significantly, the heterogeneity of biopolymer environments means that magnetically driven particle transport through these biopolymers is, likewise, heterogeneous. Clinicians, scientists and engineers working in the field of magnetophoretic transport and drug delivery may move closer to their intended goals by considering the heterogeneity of transport through tissue, and engineering particles and magnetic drive systems tuned for driving particles through the biopolymer of interest.

21.3.2 Transport through the Extracellular Matrix

One biopolymer of interest is the extracellular matrix (ECM) of the tumour microenvironment. Magnetic drug targeting has aimed to improve the delivery of cancer therapies by concentrating drugs at the tumour site. Ever since the first clinical trial on head and neck cancer patients,[4] research groups have used magnetic fields to guide drug-loaded magnetic particles towards solid tumours and retain them there, in the hopes of increasing the local dose of drug in the tumour while decreasing the systemic dose. The tumour environment, however, presents numerous biological barriers to long-range motion,[25,26] the ECM being a primary barrier. Understanding the barriers will assist in designing particles capable of effectively moving through such barriers. Significant work has been done in elucidating the transport physics in the tumour volume, including the extracellular matrix, and readers specifically interested in particle transport in tumour volumes are directed to insightful literature on the topic.[37–40]

Particles circulating through the tumour vasculature are preferentially deposited into the extracellular matrix of the tumour via the enhanced permeability and retention (EPR) effect.[41–44] The EPR effect is typically attributed to the leaky nature of tumour vasculature: the pores of tumour vasculature are commonly more porous than the vasculature in healthy tissues, resulting in particles between 30 and 200 nm in diameter being readily extravasated from the vasculature into the extracellular matrix.[45] The ECM presents a sterically and electrostatically complex environment[21] consisting of interconnected collagen bound together by connective proteins (such as laminin and entactin), proteoglycans and glycosaminoglycans. The environment contains a wide range of pore dimensions and severely limits particle and molecule motion in the ECM.[37] Electron microscopy of basement membranes has revealed pores ranging from 5 μm to 5 nm.[46] As passage through the ECM is essential for long-range distribution of particles across entire tumour volumes, understanding transport through this matrix may enable the engineering of particles adept at move through the ECM.

Early studies on magnetic particle transport in ECM-like materials were performed in collagen gels and tested the transport efficiency of 290 nm PEG coated particles, 270 nm silica coated particles and 800 nm PEG coated particles under quantitatively assessed magnetic fields.[47] While 290 nm PEG coated particles moved 1.5 ± 0.71 mm/h, 270 nm silica coated particles and 800 nm PEG coated particles moved at only 0.21 ± 0.07 mm/h and less than 0.01 mm/h, respectively. In this work, the authors note that the magnetic nanoparticles (MNPs) move as a bolus, and speculate whether pores in the particle bolus could allow the passage of ECM components through pores in the bolus. However, in assessing the sizes of ECM components such as collagen, laminin and proteoglycans, whose sizes range from ~40 to 280 nm, the authors conclude that these proteins would not have moved through crevices in the bolus structure, instead positing that the bolus moved as a solid entity.

Building on the work by Kuhn et al., Child et al. studied magnetic field-induced transport of NPs in 3D cell culture containing collagen gels. In their study, Child et al. delivered 100 and 200 nm diameter fluorescent iron oxide NPs to cells and found that, by applying a magnetic field, they achieved a fivefold increase in MNP penetration depth (as compared with no magnetic field application).[48] Here, penetratin, a cell-penetrating peptide, was bound to the particle surface and used to increase cellular uptake of particles. The authors posit that combined functionalizations of penetratin and collagenase may further increase transport through 3D cell cultures while simultaneously increasing cellular uptake.

21.3.3 Transport through Mucus

Mucus is one of many critical barriers that must often be traversed in order to effectively deliver drugs to organs in the respiratory, gastrointestinal and reproductive systems. Mucus is a polymer-based hydrogel composed primarily of glycoproteins and water. Mucins, a category of high molecular weight glycoproteins, give mucus its characteristic sticky, viscoelastic properties, and make particle transport through mucus difficult. The sinuses, throat, lungs, stomach, intestines and cervix are lined with mucus, and particles delivering medicines to these tissues must first traverse the mucosal lining before reaching their targets. Magnetic particle transport through mucus has been particularly motivated by the desire to target therapies in the lung and upper respiratory tract.[49,50] Here, mucus rheology often means the difference between normal and healthy respiratory functioning. Diseases such as cystic fibrosis and chronic obstructive pulmonary disease are accompanied by abnormally thick or excessive levels of mucus. Magnetically guided particles capable of crossing mucosal barriers may be able to increase the amount of drug reaching underlying respiratory tract tissues.

As is the case with many other biofluids and tissues, particle transport through mucus is limited by filtration effects of mucus pore size,[51] as well as electrostatic effects of particle–mucus interactions.[52] In assessing magnetically driven particle transport through mucus, pore size and electrostatic interactions have been the primary matrix properties under consideration. However, the interlinked network of mucins generates mucus with a broad range of mechanical properties. By combining magnetophoresis studies with optical tweezer experiments, Kirch et al. found that mucus scaffold rigidity played a significant role in limiting particle transport over length scales of tens of micrometres. Comparing particle motion in mucus and hydroxyethylcellulose gels, Kirch et al. elucidated the role of micro-, meso- and macroscopic phenomena in determining particle mobility. In pulling particles (~200 nm diameter) through mucus and hydroxyethylcellulose gels the authors discovered that, while hydroxyethylcellulose deformed readily, mucus scaffolds were unyielding. Thus, while particles moving through hydroxyethylcellulose could traverse the pores of the matrix or proceed by bending and stretching of the fibres, particles in mucus were caged by less flexible fibres and, under the applied force, could not deform the mucus scaffold enough to achieve significant translation distances.[53] Kirch et al.'s findings make it clear that a third component, network rigidity, plays an important role in determining how driven particles move through biopolymers such as mucus. While much work has focused on matrix fibre density, pore size and viscosity of the fluid phase of the biopolymer network, the work by Kirch et al. suggests that network rigidity plays a determining role for particles of a certain size, and suggests that analysis of pore size or volume fraction may be insufficient for predicting transport properties of MNPs in mucus.

Previous work by Ally et al. successfully applied magnetic fields to retain particles (iron spheres, 1 to 3 μm in diameter) in the mucus layer above ciliated cells of an excised frog palate. However, the authors did not report significant particle transport through the mucus layer, suggesting that the 116 mT field and 2.5 T/m field gradient were insufficient to induce particle transport.[54] However, magnetically guiding particles through mucus has been demonstrated. Using larger field strengths (250 mT) and gradients (60 T/m), Economou et al. successfully moved ~35 nm diameter particles through a ~100 μm thick layer of mucus generated by primary normal human tracheobronchial epithelial (NHTE) cells.[55] It is possible that the small diameter of these particles allowed the particles used by Economou et al. to move through the mucus matrix with minimal interactions with the rigidly structured mucus components.

21.3.4 Transport through the Skin

The notion of using magnetic fields to move materials through the skin dates back to the Sushruta Samhita, a Sanskrit medical text written circa 800 BCE, which reported the surgical procedures of removing a metallic splinter via the application of a magnetic force.[56] Modern applications of magnetophoresis target moving therapeutics *into* the body by passage through the skin.

Skin is a multilayered organ, with each layer offering its own form of protection from the passage of toxins and pathogens. These protection mechanisms also restrict MNP motion. Particles seeking entrance to the body via the skin must first sequentially penetrate the stratum corneum (SC), the viable epidermis and the dermis. The SC is the first layer of the skin, and is the most difficult layer to cross. Particles may move through the SC via intracellular, intercellular or appendage pathways (via follicles, sebaceous glands or sweat glands; Figure 21.2).[57]

The SC is primarily composed of dead, flattened cells, whose cytoplasms are filled with the filamentous protein keratin. This organization severely restricts particle transport, and passage via inter- and intracellular pathways are

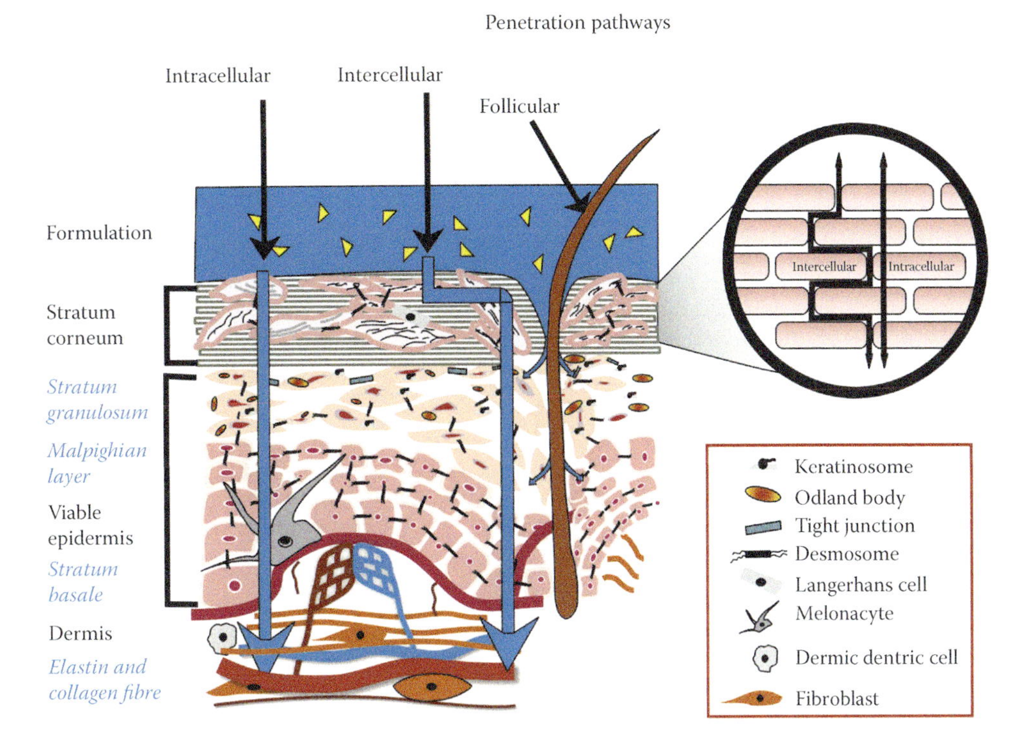

FIGURE 21.2 Several penetration pathways exist through the skin: intracellular, intercellular, sweat glands, sebaceous glands, and follicular. However, follicular glands are the primary passageways accessible by MNPs over 10 nm.[57] (Reprinted from *Journal of Controlled Release*, 164, A. Alexander, S. Dwivedi, Ajazuddin, T.K. Giri, S. Saraf, S. Saraf, D.K. Tripathi, "Approaches for breaking the barriers of drug permeation through transdermal drug delivery," 26–40, Copyright 2012, with permission from Elsevier.)

minimal.[58] Theoretical and experimental assessments have suggested that skin organization does not allow passage of particles greater than ~40 nm via intra- or intercellular pathways.[58,59] However, appendage pathways such as sweat glands and follicles contain sufficiently large pore space for passage through the SC. Thus, MNPs aimed at transdermal drug delivery tend to penetrate skin via follicular pathways.

Recently, Rao et al. demonstrated the follicular transport effect by magnetically driving epirubicin-loaded superparamagnetic iron oxide nanoparticles (SPIONs) through the SC and into deeper layers of excised human cadaver skin.[60] TEM analysis of skin cross sections demonstrated significant particle accumulation within follicular pathways and no particle accumulation far from follicular pathways. Within the follicular pathway through the skin resides sebum, an oily substance that serves to waterproof and lubricate the skin and hair. At the microscopic level, transdermal passage through follicular pathways relies on magnetic guidance through sebum. To date, no detailed reports on magnetic guidance through this biological material exist, likely due to the incipient stage of magnetically guided transdermal drug delivery research. Future research elucidating how MNPs move through sebum, and how this transport depends on particle and magnetic driving field parameters may generate information useful for the field.

NP permeation through human skin is a matter of ongoing debate, and research suggests that only particles below 10 nm permeate healthy skin via nonfollicular routes.[61] As research into iron oxide NPs for skin penetration has recently suggested that penetration is possible, even without the application of driving magnetic fields,[62] it seems that the field of transdermal drug delivery with magnetic particles is a topic with considerable space for innovative methods of moving and concentrating particles.

21.4 COMBINED FIELDS TO ENHANCE TRANSPORT

Impressive levels of magnetic particle accumulation in various targeted organs has been achieved *in vivo*, and recent efforts have extrapolated models to human dimensions.[63] Still, significant hurdles exist[6] and methods for increasing particle transport and accumulation are needed in order to make the process of drug targeting more efficient. Recently, sources of time-varying, uniform magnetic fields ('AC fields') and temporally constant magnetic field gradients ('DC gradients') have been combined to guide magnetic particles. In these experiments, DC magnetic gradients pull particles in one direction, while AC fields are applied perpendicular to the direction of particle motion. These methods increase particle velocities as compared with DC gradient pulling only. Various mechanisms have been proposed to explain the phenomenon.

By combining an AC field (50 or 100 Hz) and a DC gradient, MacDonald et al. demonstrated increased magnetophoresis of 239 nm diameter magnetite particles in a viscous gel (Surgilube).[65] Combining AC fields and DC gradients generated NP velocities that were consistently twice as fast as DC gradients alone. The twofold increase in particle velocity was observed for particle loadings ranging from 10% w/w to 50% w/w. The authors suggest that increased velocities are due to local reductions in the viscosity of the gel surrounding the particles, positing that AC fields induce NP dynamics that result in local shear thinning. Such enhanced local shear thinning would allowing particles to move more easily through the gel, as compared with DC only forces. Due to the low oscillation frequency of the AC fields, hyperthermic effects were not suspected (hyperthermia induced heating generally occurs in the range 0.1 to 260 MHz).

In similarly arranged experiments, Soheilian et al. also demonstrated increased transport velocities of MNPs under dynamic magnetic fields[65] (Figure 21.3). Experiments pulling 150 nm paramagnetic particles through Teflon membranes with 2 μm diameter pores were performed under DC and combined DC and AC (2.5 Hz) magnetophoresis. The authors observed significant chaining and agglomeration of particles under DC magnetophoresis, and noted that agglomerates often grew larger than 2 μm in size, thus excluding them from passage through the pores of the Teflon membranes. Application of a time varying magnetic field induced de-aggregation, and enhanced particle transport through the pores of the membrane by a factor of ~2.

These experiments demonstrate that there are substantial gains in particle transport speed when particles experience AC magnetic fields combined with DC magnetic gradients. They suggest that further exploration in the space of actuation frequency and direction, in combination with particles designed to operate and respond to more complex magnetic fields, may result in significant advances in MNP guidance.

21.5 NONSPHERICAL PARTICLES IN VISCOELASTIC BIOMATERIALS

While most magnetic drug targeting experiments use spherical particles, researchers have devised several new techniques for engineering the shapes of magnetic particles.[66–71] However, few experiments have focused on how shape impacts particle motion through viscoelastic media. In this section, we review experiments manipulating magnetic rods, helices and rolled up sheets in various biopolymer environments.

21.5.1 Rods

The role of geometry: Transport through biopolymers is a size-dependent phenomenon: large particles are sterically hindered by the biomaterial protein structure, while small particles can navigate the open, low viscosity spaces between proteins. However, magnetic guidance is also a force-dependent phenomenon: for a given magnetic material and a given magnetic field and gradient, larger particles will experience greater magnetic force due to their larger overall magnetization. In biomaterials, larger force does not always result in larger velocity. Here we discuss the motion of magnetic rods whose magnetizations are primarily along the long axis of the rod. While particles whose magnetizations are along the short axis of the rod have been generated and actuated,[72–75] such rods have not yet been used in magnetophoretic transport experiments through biomaterials.

Cylinders are of interest for transport through biomaterials because they present a small diameter in their direction of magnetophretic transport, but experience larger forces than similar diameter spheres due to their lengths (given aspect ratios greater than 1). It is helpful to consider a cylinder and a sphere of equivalent volume, the sphere having a radius 10× that of the cylinder. Two such particles will experience similar

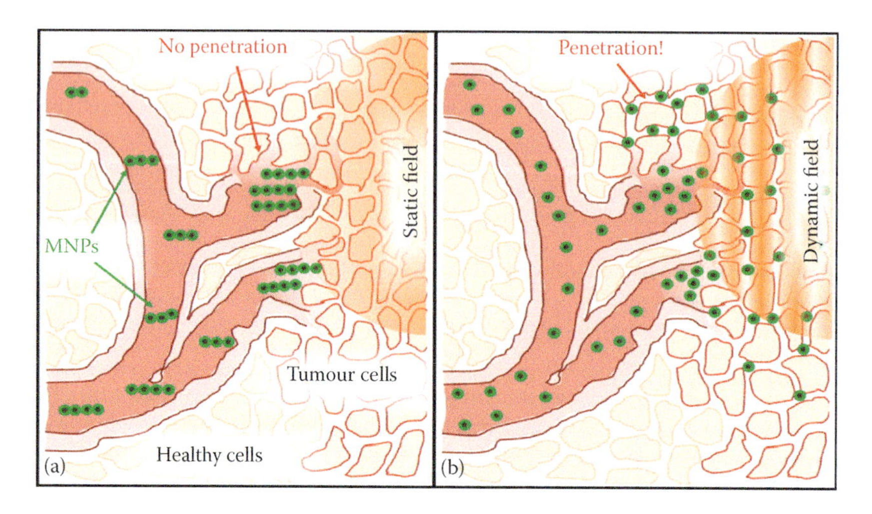

FIGURE 21.3 Using combined AC fields and DC gradients, MNP agglomerates were successfully dispersed, which improved MNP penetration through a synthetic matrix.[64] (Reprinted with permission from R. Soheilian, Y.S. Choi, A.E. David, H. Abdi, C.E. Maloney, R.M. Erb, "Toward accumulation of magnetic nanoparticles into tissues of small porosity," *Langmuir.* 31 (2015) 8267–8274. Copyright 2015 American Chemical Society.)

forces in a field and field gradient. However, in a biopolymer matrix containing pores of various sizes, the cylinder is able to move through a significantly larger percentage of pores due to its narrower cross section. The cylinder, however, in presenting a larger cross section in the direction of transit, is more susceptible to being trapped by steric barriers of the matrix.

Nanorod magnetophoresis in Matrigel: Previously, nickel rods were used to show that nanorod diameter and overall force on the particle determine if particles move with constant velocities, or via stick-slip motion, through collagen-rich, reconstituted basement membranes (Matrigel).[46] Using time invariant magnetic field gradients, nickel nanorods were pulled through Matrigel. Nanorods were coated with poly(ethylene glycol) prior to being mixed with Matrigel, so as to minimize nonspecific protein–particle interactions. Nanorods with diameters of 250 and 18 nm whose lengths varied between 1 and 6 μm were tested. In these studies, 250 nm diameter nanorods moved through the matrix at constant velocities, suggesting that the larger diameter rods were perpetually entangled in the matrix and had diameters too large for moving through the pores of the matrix. SEM imaging of Matrigel indicated a wide range of pore sizes, with some pores as large as 5 μm (Figure 21.4a). For 250 nm diameter rods, the magnetophoretic force induced matrix deformation: instead of the rods moving through the matrix, they dragged the matrix along with them under a constant magnetic force. While pores large enough for the 250 nm diameter rods existed in the matrix, such pores were not sufficiently connected to one another to allow unhindered long-range motion through the large pores. The motion of 250 nm diameter rods was drastically different from that observed for small diameter nanorods. In the same gel and under the same magnetic field and gradient conditions, 18 nm diameter rods experienced stick-slip motion (Figure 21.4b–d). They alternated between moving quickly through pores of the gel, and being fully stuck in the gel. Unlike 250 nm diameter rods, 18 nm diameter rods were observed moving against the magnetic gradient as well as laterally with respect to the field gradient direction.[46,76] For small particles moving through dense matrices, the observation suggests there may be some advantage to using smaller forces to pull on magnetic particles, as these forces would allow for diffusive forces to reorient the particles to paths of less resistance. Intermittent application of strong and weak/no forces may act to momentarily supply maximum force to move particles, while also allowing particle diffusion to enhance the possibility the particle finds a path of lower resistance, or is allowed to diffusively disentangle itself from a particularly dense region of the matrix.

Magnetic particle chaining may achieve a similar transport enhancement. As spherical SPIONs and SPION-based microparticles align in a magnetic field, such particles chain together and, in doing so, may move through the matrix with motion akin to nanorods. As demonstrated by Soheilian, disrupting such chains with transverse AC magnetic fields may further improve transport through porous media.[65]

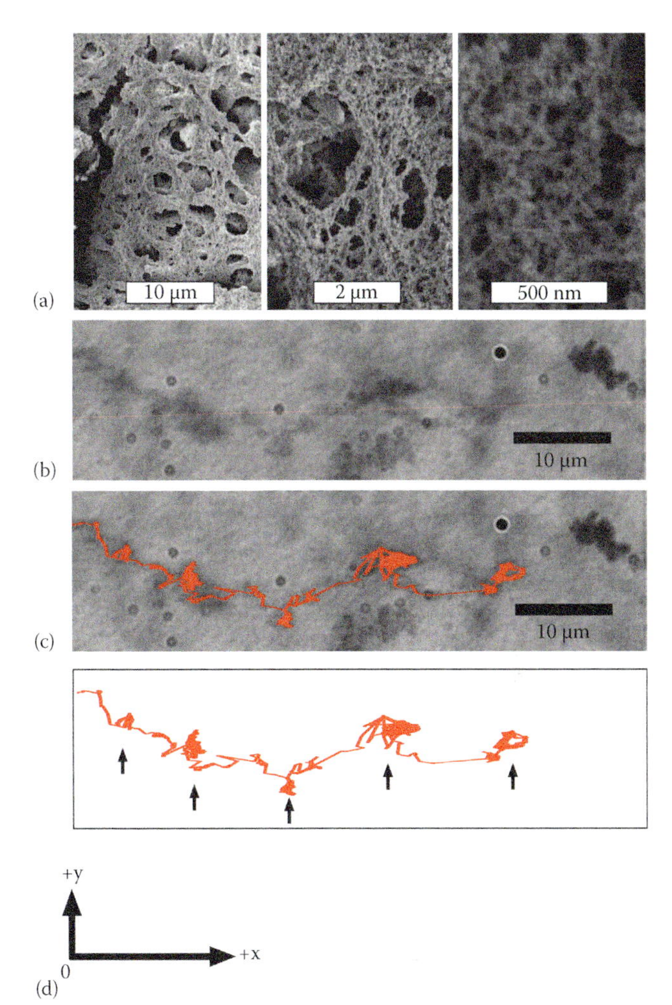

FIGURE 21.4 (a) Scanning electron micrographs show Matrigel, a collagen-rich basement membrane. (b) A rod is pulled from left to right under constant magnetic field and gradient. Image shows a minimum intensity projection of the rod as it moves over time. (c) Tracking reveals that rods move via stick-slip motion, as they are intermittently trapped by the fibres of the polymer. (d) Arrows point to locations at which the rod is trapped.[46] (From L.O. Mair, R. Superfine, "Single particle tracking reveals biphasic transport during nanorod magnetophoresis through extracellular matrix," *Soft Matter.* 10 (2014) 4118–4125. Reproduced by permission of The Royal Society of Chemistry.)

21.5.2 Helices

Recently, several groups have manufactured and guided helical particles for on-chip manipulation of particles, cells, bacteria, as well as generating flows.[69,77,78] These helical particles are biomimetic in their shape and mode of swimming: like the rotating flagella of many bacteria, helical particles generate propulsion by rotating asymmetric coils under magnetic guidance. Several techniques exist for synthesizing helical magnetic particles. In one method, spherical microspheres or nanospheres are deposited onto a surface, followed by evaporation of nonmagnetic via glancing angle deposition onto a rotating substrate. The process generates helical propellers on each micro- or nanosphere. After synthesis of the helical propeller, particles are removed from the substrate and redeposited onto a separate substrate for deposition of a magnetic layer.[68] The magnetic layer coats one side of the helix and is generally magnetized perpendicular to the long axis of the helix. The glancing angle deposition technique has also been used to deposit magnetic helical propellers directly onto the spherical particles.[79–81] In a second method for making magnetic helices, thin layers of differing materials are deposited as a long, thin strip on a supporting substrate. Etching the supporting substrate releases the thin strip, and strain associated with lattice mismatch between the thin layers of different materials results in coiling of the released thin strip.[82–84] A third method for making helices uses template-guided electroplating to grow palladium nanohelices.[85] After growth, the nanohelices were removed from their templates, deposited onto substrates and coated with a magnetic layer.[86] This class of helical particles composes a class of highly capable micromachines.

Magnetically actuated particles in low Reynolds number environments may be magnetically translated using only two modes. One mode is the commonly used magnetic gradient pulling. Another mode allows for translation via the breaking of some symmetry of motion. Symmetry may be broken by the proximity of a solid–liquid boundary,[87–91] by deformations of the particle shape,[92,93] or by built-in asymmetries in the particles.[68,79,84] The nanohelices discussed here do not move via magnetic gradient pulling, but rather rotate around their long axes in step with a rotating, uniform magnetic field. The rotating, uniform magnetic field generates a torque around the long axis of the helical particle, inducing rotation.[93] For magnetic helices, the asymmetry of the helix breaks the symmetry of motion during rotation around the long axis, generating a propulsive force parallel to the helix long axis. Helices rotate around their long axes, with this rotation taking place in the plane of the applied rotating magnetic field. They translate perpendicular to that field and are steered by reorienting the rotating magnetic field. Below the critical frequency, the rotation of the helices moves in-step with the magnetic field and increasing rotational magnetic field frequency increases translation velocity. As rotating magnetic field frequency increases at a given field strength, drag forces increase proportionally. The step-out frequency (also called the critical frequency) is the rotational frequency above which a helix does not rotate in-step with the applied magnetic field.[69,78,94] Above the step-out frequency, translational motion is slow and erratic, as the particle intermittently couples with the applied magnetic field. Initial demonstrations of helical particles performed in Newtonian environments demonstrated precise steering and velocity control.[68]

Recently, these helical particles were manipulated in diluted blood,[95] hyaluronan gels,[79] and gastrointestinal mucus.[81] These first experiments of nanohelices in viscoelastic environments have elucidated important concepts in the physics of helical transport through biofluids. In the first experimental evidence of nanoscale helices moving through viscoelastic media, Venugopalan et al. moved 1 μm wide, ~5 μm long helices through human blood.[95] In blood, Venugopalan et al. observed that helices demonstrated large deviations in translational velocity as a function of applied magnetic field frequency. Deviations in velocity are due to helices colliding with cells and protein barriers at varying rates as they move through blood. Additionally, Venugopalan et al. observed distinct stick-slip behaviour of helix transport. Over the course of 4 s, helices were observed to move at ~6 μm/s for 1.5 s, cease motion for 1.5 s, then resume motion at ~6 μm/s for 1 s. Authors attribute sticking motion to colloidal jamming of blood cells, and note that fully hindered motion typically persisted for 1–2 s.

In testing motion through hyaluronan (HA) gels, Schamel et al. rotated microhelices (450 nm wide, 2.5 μm long) and nanohelices (120 nm wide, 400 nm long) at varying rotational frequencies, observing the resulting translational velocities for motors in 3 mg/ml and 5 mg/ml HA78 (Figure 21.5). At these concentrations, HA forms extended networks of overlapping polymer chains, the gel having a mesh size on the order of tens of nanometres. The authors acknowledge that, at the HA concentrations studied, the gels were only weakly viscoelastic. However, even for weakly viscoelastic HA solutions, helix size plays a significant role in motion. Microhelices experience 10 Hz step-out frequency at HA concentrations of 5 mg/ml. Much smaller nanohelices successfully rotated at 80 Hz in the same 5 mg/ml HA solutions. The significant increase in the step-out frequency experience by nanohelices is presumably due to the fact that nanohelices were able to move through the pores of the HA, interacting with the viscoelastic proteins components only intermittently. Perhaps most interesting,

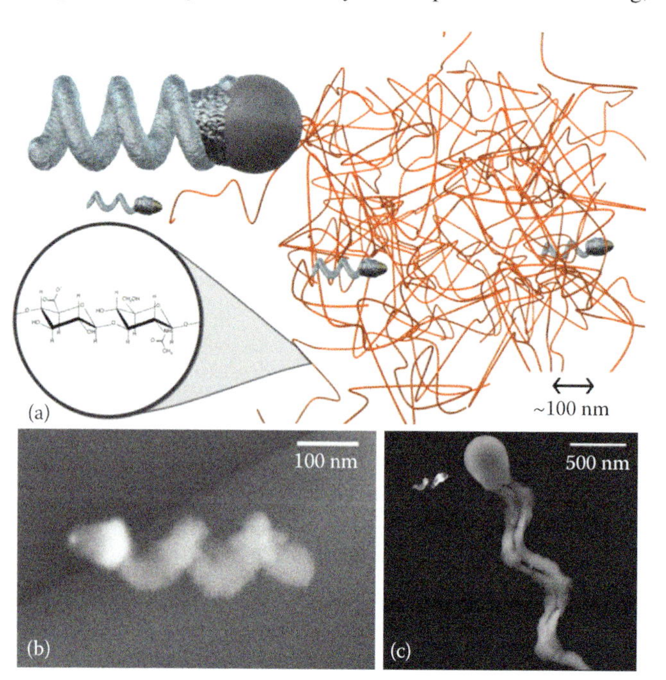

FIGURE 21.5 Micro- and nanoscale helices in hyaluronan gels. (a) HA els contain pores on the order of tens of nanometers. (b) Nanoscale helices with diameters ~120 nm were able to move through 3 mg/ml and 5 mg/ml gels, while larger microscale helices (c) were sterically trapped by HA gels, unable to rotate above 10 Hz.[79] (Reprinted with permission from D. Schamel, A.G. Mark, J.G. Gibbs, C. Miksch, K.I. Morozov, A.M. Leshansky, P. Fischer, "Nanopropellers and their actuation in complex viscoelastic media," *ACS Nano*. 8, 8794–8801. doi:10.1021/nn502360t. Copyright 2014 American Chemical Society.)

25 Hz rotational manipulation of nanohelices enhanced translation velocities in 5 mg/ml HA, as compared with 3 mg/ml. At 50 and 80 Hz, translational velocities were unaffected by increasing HA concentration from 3 to 5 mg/ml, suggesting that nanohelices are able to move through 3 and 5 mg/ml HA gels with minimal inhibitory interactions with the viscoelastic mesh phase of the gel.

Walker et al. used porcine gastric mucins (PGMs) as a model system for studying the transport of magnetically rotated microhelices.[81] As discussed previously, the mucin network creates a dense and sticky network, which significantly hinders both diffusive and driven particle motion. While unmodified synthetic particles experience significant motion suppression, some microorganisms have developed tools for efficiently moving through mucus. The microorganism *Helicobacter pylori* secretes large local quantities of urease. Urease catalyzes the hydrolysis of urea, releasing ammonia. This chemical cascade serves to increase the pH of the surrounding mucus, thereby inducing a sol-gel transition capable of liquefying the mucus in the vicinity of the *H. pylori* cell. Taking a cue from nature, Walker et al. modified the surfaces of nanohelices with urease and demonstrated that, by doing so, nanohelices could be efficiently moved through porcine gastric mucin solutions.

Due to their steerability and their ability to move through viscoelastic biofluids without large magnetic gradients, the future of helical particles seems promising. Helical particles are unique in that their motion inherently induces a local shear in the surrounding media. For many biological fluids that experience shear thinning, such motion may significantly enhance transport by decreasing the effective viscosity in the vicinity of the particle.[96]

21.5.3 Rolled Up Sheets for Drilling

Future applications of micro- and nanoscale particles for surgery-like tasks will require particles that not only transport through, but also modify tissues. Future applications of particles may include tasks such as removing regions of tissue for biopsy or drilling microscale holes in specific tissues. To that end, Xi et al. used rotationally manipulated ferromagnetic microtubes to drill 4 µm wide holes in excised liver tissue.[97] These microscale particles consisting of rolled up tubes of photoresist–Ti–Fe–Cr layers, with metal layers only 5 nm thick. Due to the thinness of the tube wall and a built-in angular cutting angle at the tube edges, the tube edges have geometries similar to the bevelled edge of a hypodermic needle. When rotated atop a piece of pig liver tissue for tens of minutes to several hours, the microtubes were able to penetrate the tissue. It is important to note that the solution for drilling experiments contained both microtubes for drilling, as well as a common soap (50% v/v). This soap served to increase the viscosity of the solution, demonstrating the potential for the microtubes to operate in environments similar to physiological environments. However, from the perspective of tissue transport for magnetic particles, this soap may have also served to lubricate the particle–tissue interface. These experiments demonstrate the ability to drill through tissues using rotationally manipulated microscale particles and form a promising basis for other applications in which microscale particles drill through tissues.

21.6 OUTLOOK AND FUTURE DIRECTIONS

The field of magnetic particle transport through complex environments has only recently begun to explore the enormous parameter space of complex biomaterials, composite magnetic fields and uniquely shaped particles. Nature has evolved objects with meticulously controlled surface chemistries, textures, shapes and dimensions for specific purposes. It is encouraging to see scientists and engineers developing magnetic particles that borrow from nature's toolbox in order to enhance transport. As we learn further details about the biological barriers that must be traversed in order to deliver drugs to their intended targets, new particles and actuation mechanisms will be devised. Promising future directions include combining magnetic particles with microbes,[98] as well as shape- and modulus-tuned particles capable of partially deforming so as to more efficiently move through biological barriers.

REFERENCES

1. K. Widder, A. Senyel, G. Scarpelli, "Magnetic microspheres: A model system of site specific drug delivery in vivo," *Proc Soc Exp Biol Med.* 158 (1978) 141–146.
2. K. Widder, A. Senyei, "Drug targeting: Magnetically responsive albumin microspheres - A review of the system to date," *Gynecol Oncol.* 12 (1981) 1–13.
3. K. Widder, R. Morris, G. Poore, D. Howard, A. Senyei, "Selective targeting of magnetic albumin microspheres containing low-dose doxorubicin: Total remission in Yoshida sarcoma-bearing rats," *Eur J Cancer Clin Oncol.* 19 (1983) 135–139.
4. A.S. Lübbe, C. Bergemann, H. Riess, F. Schriever, P. Reichardt, K. Possinger, M. Matthias, B. Dörken, F. Herrmann, R. Gürtler, P. Hohenberger, N. Haas, R. Sohr, B. Sander, A.J. Lemke, D. Ohlendorf, W. Huhnt, D. Huhn, "Clinical experiences with magnetic drug targeting: A phase I study with 4'-epidoxorubicin in 14 patients with advanced solid tumors," *Cancer Res.* 56 (1996) 4686–4693.
5. C. Alexiou, W. Arnold, R.J. Klein, F.G. Parak, P. Hulin, C. Bergemann, W. Erhardt, S. Wagenpfeil, A.S. Lubbe, "Locoregional cancer treatment with magnetic drug targeting," *Cancer Res.* 60 (2000) 6641–6648.
6. B. Shapiro, S. Kulkarni, A. Nacev, S. Muro, P.Y. Stepanov, I.N. Weinberg, "Open challenges in magnetic drug targeting," *WIRES Nanomed Nanobiotechnol.* 7 (2015) 446–457.
7. P. Ruenraroengsak, J.M. Cook, A.T. Florence, "Nanosystem drug targeting: Facing up to complex realities," *J Control Release.* 141 (2010) 265–276. doi:10.1016/j.jconrel.2009.10.032
8. R.E. Serda, *Mass Transport of Nanocarriers*, CRC Press, Taylor & Francis Group, New York, 476 pages, ISBN 9789814364416. 2012.
9. S. Muro (Ed.), *Drug Delivery across Physiological Barriers*, CRC Press, Taylor & Francis Group, LLC, Boca Raton, FL, 2016.

10. S. Tenzer, D. Docter, J. Kuharev, A. Musyanovych, V. Fetz, R. Hecht, F. Schlenk, D. Fischer, K. Kiouptsi, C. Reinhardt, K. Landfester, H. Schild, M. Maskos, S.K. Knauer, R.H. Stauber, "Rapid formation of plasma protein corona critically affects nanoparticle pathophysiology," *Nat Nanotechnol.* 8 (2013) 772–781. doi:10.1038/nnano.2013.181
11. I. Lynch, T. Cedervall, M. Lundqvist, C. Cabaleiro-Lago, S. Linse, K.A. Dawson, "The nanoparticle-protein complex as a biological entity; a complex fluids and surface science challenge for the 21st century," *Adv Colloid Interface Sci.* 134–135 (2007) 167–174. doi:10.1016/j.cis.2007.04.021
12. A.E. Nel, L. Mädler, D. Velegol, T. Xia, E.M. V Hoek, P. Somasundaran, F. Klaessig, V. Castranova, M. Thompson, "Understanding biophysicochemical interactions at the nano-bio interface," *Nat Mater.* 8 (2009) 543–557. doi:10.1038/nmat2442
13. N.D. Winblade, I.D. Nikolic, A.S. Hoffman, J.A. Hubbell, "Blocking adhesion to cell and tissue surfaces by the chemisorption of a poly-L-lysine-graft-(poly(ethylene glycol); phenylboronic acid) copolymer," *Biomacromolecules.* 1 (2000) 523–533.
14. N.D. Winblade, H. Schmo, M. Baumann, A.S. Hoffman, J.A. Hubbell, "Sterically blocking adhesion of cells to biological surfaces with a surface-active copolymer containing poly(ethylene glycol) and phenylboronic acid," *J Biomed Mater Res A.* 59 (2001) 618–631. doi:10.1002/jbm.1273
15. Y. Zhang, N. Kohler, M. Zhang, "Surface modification of superparamagnetic magnetite nanoparticles and their intracellular uptake," *Biomaterials.* 23 (2002) 1553–1561.
16. S.K. Lai, D.E. O'Hanlon, S. Harrold, S.T. Man, Y.-Y. Wang, R. Cone, J. Hanes, "Rapid transport of large polymeric nanoparticles in fresh undiluted human mucus," *Proc Natl Acad Sci USA.* 104 (2007) 1482–1487. doi:10.1073/pnas.0608611104
17. Y. Wang, S.K. Lai, A. Pace, R. Cone, "Addressing the PEG mucoadhesivity paradox to engineer nanoparticles that 'slip' through the human mucus barrier," *Angew Chem.* 47 (2009) 9726–9729. doi:10.1002/anie.200803526.Addressing
18. S.K. Lai, Y.-Y. Wang, J. Hanes, "Mucus-penetrating nanoparticles for drug and gene delivery to mucosal tissues," *Adv Drug Deliv Rev.* 61 (2009) 158–171. doi:10.1016/j.addr.2008.11.002
19. J.S. Suk, Q. Xu, N. Kim, J. Hanes, L.M. Ensign, "PEGylation as a strategy to improve nanoparticle-based drug and gene delivery," *Adv Drug Deliv Rev.* 99 (2015) 28–51. doi:10.1016/j.addr.2015.09.012
20 Q. Xu, L.M. Ensign, N.J. Boylan, A. Schön, X. Gong, J.-C. Yang, N.W. Lamb, S. Cai, T. Yu, E. Freire, J. Hanes, "Impact of surface polyethylene glycol (PEG) density on biodegradable nanoparticle transport in mucus ex vivo and distribution in vivo," *ACS Nano.* 9 (2015) 150824093910007. doi:10.1021/acsnano.5b03876
21 O. Lieleg, R.M. Baumgärtel, A.R. Bausch, "Selective filtering of particles by the extracellular matrix: An electrostatic bandpass," *Biophys J.* 97 (2009) 1569–1577. doi:10.1016/j.bpj.2009.07.009
22. O. Lieleg, K. Ribbeck, "Biological hydrogels as selective diffusion barriers," *Trends Cell Biol.* 21 (2011) 543–551. doi:10.1016/j.tcb.2011.06.002
23. S.J. Kuhn, S.K. Finch, D.E. Hallahan, T.D. Giorgio, "Proteolytic surface functionalization enhances in vitro magnetic nanoparticle mobility through extracellular matrix," *Nano Lett.* 6 (2006) 306–312.
24. G. Alexandrakis, E.B. Brown, R.T. Tong, T.D. McKee, R.B. Campbell, Y. Boucher, R.K. Jain, "Two-photon fluorescence correlation microscopy reveals the two-phase nature of transport in tumors," *Nat Med.* 10 (2004) 203–207. doi:10.1038/nm981
25. R.K. Jain, T. Stylianopoulos, "Delivering nanomedicine to solid tumors," *Nat Rev Clin Oncol.* 7 (2010) 653–664. doi:10.1038/nrclinonc.2010.139
26. V.P. Chauhan, T. Stylianopoulos, Y. Boucher, R.K. Jain, "Delivery of molecular and nanoscale medicine to tumors: Transport barriers and strategies," *Ann Rev Chem Biomolec Eng.* 2 (2011) 281–298. doi:10.1146/annurev-chembioeng-061010-114300
27. H. Zhou, Z. Fan, J. Deng, P.K. Lemons, D.C. Arhontoulis, W.B. Bowne, H. Cheng, "Hyaluronidase embedded in nanocarrier PEG shell for enhanced tumor penetration and highly efficient antitumor efficacy," *Nano Lett.* 16 (2016) 3268–3277. doi:10.1021/acs.nanolett.6b00820
28. S.R. Hingorani, W.P. Harris, J.T. Beck, B.A. Berdov, S.A. Wagner, E.M. Pshevlotsky, S.A. Tjulandin, O.A. Gladkov, R.F. Holcombe, R. Korn, N. Raghunand, S. Dychter, P. Jiang, H.M. Shepard, C.E. Devoe, "Phase Ib study of PEGylated recombinant human hyaluronidase and gemcitabine in patients with advanced pancreatic cancer," *Clin Cancer Res.* 22 (2016) 2848–2854. doi:10.1158/1078-0432.CCR-15-2010
29. S. Kulkarni, B. Ramaswamy, E. Horton, S. Gangapuram, A. Nacev, D. Depireux, M. Shimoji, B. Shapiro, "Quantifying the motion of magnetic particles in excised tissue: Effect of particle properties and applied magnetic field.," *J Magn Magn Mater.* 393 (2015) 243–252. doi:10.1016/j.jmmm.2015.05.069
30. E.M. Renkin, "Filtration, diffusion, and molecular sieving through porous cellulose membranes," *J Gen Physiol.* (1954) 225–243.
31. A. Ogston, "The spaces in a uniform random suspension of fibres," *Trans Faraday Soc.* 54 (1958) 1754–1757.
32. A.G. Ogston, B.N. Preston, J.D. Wells, "On the transport of compact particles through solutions of chain-polymers," *Proc R Soc Lond. A Math Phys Sci.* 333 (1973) 297–316.
33. D.S. Clague, R.J. Phillips, "A numerical calculation of the hydraulic permeability of three-dimensional disordered fibrous media," *Phys Fluids.* 9 (1997) 1562. doi:10.1063/1.869278
34. R.J. Phillips, "A hydrodynamic model for hindered diffusion of proteins and micelles in hydrogels," *Biophys J.* 79 (2000) 3350–3. doi:10.1016/S0006-3495(00)76566-0
35. B. Amsden, "Solute diffusion within hydrogels. Mechanisms and models," *Macromolecules.* 31 (1998) 8382–8395.
36. Y. Cu, W.M. Saltzman, "Mathematical modeling of molecular diffusion through mucus," *Adv Drug Deliv Rev.* 61 (2009) 101–114. doi:10.1016/j.addr.2008.09.006
37. R.K. Jain, "Transport of molecules in the tumor interstitium: A review," *Cancer Res.* 57 (1987) 3039–3051.
38. T. Stylianopoulos, C. Wong, M.G. Bawendi, R.K. Jain, D. Fukumura, "Multistage nanoparticles for improved delivery into tumor tissue," *Methods Enzymol.* 508 (2012) 109–130. doi:10.1016/B978-0-12-391860-4.00006-9
39. H. Holback, Y. Yeo, "Intratumoral drug delivery with nanoparticulate carriers," *Pharm Res.* 28 (2011) 1819–1830. doi:10.1007/s11095-010-0360-y
40. S.A. Abouelmagd, H. Hyun, Y. Yeo, "Extracellularly activatable nanocarriers for drug delivery to tumors," *Exp Opin Drug Deliv.* 11 (2014) 1601–1618. doi:10.1016/j.micinf.2011.07.011.Innate

41 H. Maeda, J. Wu, T. Sawa, Y. Matsumura, K. Hori, "Tumor vascular permeability and the EPR effect in macromolecular therapeutics: A review," *J Control Rel.* 65 (2000) 271–284. doi:10.1016/S0168-3659(99)00248-5
42. H. Maeda, G.Y. Bharate, J. Daruwalla, "Polymeric drugs for efficient tumor-targeted drug delivery based on EPR-effect," *Eur J Pharm Biopharm.* 71 (2009) 409–419. doi:10.1016/j.ejpb.2008.11.010
43. J. Fang, H. Nakamura, H. Maeda, "The EPR effect: Unique features of tumor blood vessels for drug delivery, factors involved, and limitations and augmentation of the effect," *Adv Drug Deliv Rev.* 63 (2011) 136–151. doi:10.1016/j.addr.2010.04.009
44. V. Torchilin, "Tumor delivery of macromolecular drugs based on the EPR effect," *Adv Drug Deliv Rev.* 63 (2011) 131–135. doi:10.1016/j.addr.2010.03.011
45. F.M. Kievit, M. Zhang, "Cancer nanotheranostics: Improving imaging and therapy by targeted delivery across biological barriers," *Adv Mater.* 23 (2011) H217–H247. doi:10.1002/adma.201102313
46. L.O. Mair, R. Superfine, "Single particle tracking reveals biphasic transport during nanorod magnetophoresis through extracellular matrix," *Soft Matter.* 10 (2014) 4118–4125. doi:10.1039/x0xx00000x
47. S.J. Kuhn, D.E. Hallahan, T.D. Giorgio, "Characterization of superparamagnetic nanoparticle interactions with extracellular matrix in an in vitro system," *Ann Biomed Eng.* 34 (2006) 51–58. doi:10.1007/s10439-005-9004-5
48. H.W. Child, P. a Del Pino, J.M. De La Fuente, A.S. Hursthouse, D. Stirling, M. Mullen, G.M. McPhee, C. Nixon, V. Jayawarna, C.C. Berry, "Working together: The combined application of a magnetic field and penetratin for the delivery of magnetic nanoparticles to cells in 3D," *ACS Nano.* 5 (2011) 7910–7919. doi:10.1021/nn202163v
49. P. Dames, B. Gleich, A. Flemmer, K. Hajek, N. Seidl, F. Wiekhorst, D. Eberbeck, I. Bittmann, C. Bergemann, T. Weyh, L. Trahms, J. Rosenecker, C. Rudolph, "Targeted delivery of magnetic aerosol droplets to the lung," *Nat Nanotechnol.* 2 (2007) 495–499. doi:10.1038/nnano.2007.217
50. C. Ruge, J. Kirch, C.-M. Lehr, "Pulmonary drug delivery: From generating aerosols to overcoming biological barriers – Therapeutic possibilities and technological challenges." *Respir Med.* 2600 (2013) 1–12. doi:10.1016/S2213-2600(13)70072-9
51. N.N. Sanders, S.C. De Smedt, E. Van Rompaey, P. Simoens, F. De Baets, J. Demeester, "Cystic fibrosis sputum: A barrier to the transport of nanospheres," *Am J Respir Crit Care Med.* 162 (2000) 1905–1911. doi:10.1164/ajrccm.162.5.9909009
52. S. Mura, H. Hillaireau, J. Nicolas, S. Kerdine-ro, B. Le Droumaguet, C. Delome, "Biodegradable nanoparticles meet the bronchial airway barrier: How surface properties affect their interaction with mucus and epithelial cells," *Biomacromolecules.* 12 (11) (2011), 4136–4143.
53. J. Kirch, A. Schneider, B. Abou, A. Hopf, U.F. Schaefer, M. Schneider, C. Schall, C. Wagner, C.-M. Lehr, "Optical tweezers reveal relationship between microstructure and nanoparticle penetration of pulmonary mucus," *Proc Natl Acad Sci USA.* 109 (2012) 18355–18360. doi:10.1073/pnas.1214066109
54. J. Ally, W. Roa, A. Amirfazli, "Use of mucolytics to enhance magnetic particle retention at a model airway surface," *J Magn Magn Mater.* 320 (2008) 1834–1843. doi:10.1016/j.jmmm.2008.02.162
55. E.C. Economou, S. Marinelli, M.C. Smith, A.A. Routt, V.V. Kravets, H.W. Chu, K. Spendier, Z.J. Celinski, "Magnetic nanodrug delivery through the mucus layer of air–liquid interface cultured primary normal human tracheobronchial epithelial cells," *BioNanoScience.* 6 (2016) 235–242. doi:10.1007/s12668-016-0216-y
56. G. Sarton, *Introduction to the History of Science. Vol. I: From Homer to Omar Khayyám*, Williams & Wilkins Company, Philadelpia, 1927.
57. A. Alexander, S. Dwivedi, Ajazuddin, T.K. Giri, S. Saraf, S. Saraf, D.K. Tripathi, "Approaches for breaking the barriers of drug permeation through transdermal drug delivery," *J Control Rel.* 164 (2012) 26–40. doi:10.1016/j.jconrel.2012.09.017
58. A. Vogt, B. Combadiere, S. Hadam, K.M. Stieler, J. Lademann, H. Schaefer, B. Autran, W. Sterry, U. Blume-Peytavi, "40nm, but not 750 or 1,500 nm, nanoparticles enter epidermal CD1a+ cells after transcutaneous application on human skin.," *J Investig Dermatol.* 126 (2006) 1316–1322. doi:10.1038/sj.jid.5700226
59. A. Gautam, D. Singh, R. Andvijayaraghavan, "Dermal exposure of nanoparticles: An understanding," *J Cell Tissue Res.* 11 (2011) 2703–2708.
60. Y.F. Rao, W. Chen, X.G. Liang, Y.Z. Huang, J. Miao, L. Liu, Y. Lou, X.G. Zhang, B. Wang, R.K. Tang, Z. Chen, X.Y. Lu, "Epirubicin-loaded superparamagnetic iron-oxide nanoparticles for transdermal delivery: Cancer therapy by circumventing the skin barrier," *Small.* 11 (2015) 239–247. doi:10.1002/smll.201400775
61. B. Baroli, "Penetration of nanoparticles and nanomaterials in the skin: Fiction or reality?," *J Pharm Sci.* 99 (2010) 21–50.
62. U.M. Musazzi, B. Santini, F. Selmin, V. Marini, F. Corsi, R. Allevi, A.M. Ferretti, D. Prosperi, F. Cilurzo, M. Colombo, P. Minghetti, "Impact of semi-solid formulations on skin penetration of iron oxide nanoparticles," *J Nanobiotechnol.* 15 (2017) 14. doi:10.1186/s12951-017-0249-6
63. K.T. Al-Jamal, J. Bai, J.T.W. Wang, A. Protti, P. Southern, L. Bogart, H. Heidari, X. Li, A. Cakebread, D. Asker, W.T. Al-Jamal, A. Shah, S. Bals, J. Sosabowski, Q.A. Pankhurst, "Magnetic drug targeting: Preclinical in vivo studies, mathematical modeling, and extrapolation to humans.," *Nano Lett.* 16 (2016) 5652–5660. doi:10.1021/acs.nanolett.6b02261
64. R. Soheilian, Y.S. Choi, A.E. David, H. Abdi, C.E. Maloney, R.M. Erb, "Toward accumulation of magnetic nanoparticles into tissues of small porosity," *Langmuir.* 31 (2015) 8267–8274. doi:10.1021/acs.langmuir.5b01458
65. C. MacDonald, G. Friedman, J. Alamia, K. Barbee, B. Polyak, "Time-varied magnetic field enhances transport of magnetic nanoparticles in viscous gel," *Nanomedicine (Lond.).* 5 (2010) 65–76. doi:10.2217/nnm.09.97
66. L.E. Euliss, J.A. DuPont, S. Gratton, J. DeSimone, "Imparting size, shape, and composition control of materials for nanomedicine," *Chem Soc Rev.* 35 (2006) 1095–1104. doi:10.1039/b600913c
67. J.A. Champion, Y.K. Katare, S. Mitragotri, "Making polymeric micro- and nanoparticles of complex shapes," *Proc. Natl. Acad. Sci. USA.* 104 (2007) 11901–11904. doi:10.1073/pnas.0705326104
68. A. Ghosh, P. Fischer, "Controlled propulsion of artificial magnetic nanostructured propellers," *Nano Lett.* 9 (2009) 2243–5. doi:10.1021/nl900186w
69. L. Zhang, K.E. Peyer, B.J. Nelson, "Artificial bacterial flagella for micromanipulation," *Lab Chip.* 10 (2010) 2203–2215. doi:10.1039/c004450b

70. J. Nunes, K.P. Herlihy, L. Mair, R. Superfine, J.M. DeSimone, "Multifunctional shape and size specific magneto-polymer composite particles," *Nano Lett.* 10 (2010) 1113–1119. doi:10.1021/nl904152e
71. G. Zabow, S.J. Dodd, A.P. Koretsky, "Shape-changing magnetic assemblies as high-sensitivity NMR-readable nanoprobes," *Nature.* 520 (2015) 73–77. doi:10.1038/nature14294
72. J.C. Love, A.R. Urbach, M.G. Prentiss, G.M. Whitesides, "Three-dimensional self-assembly of metallic rods with submicron diameters using magnetic interactions," *J Am Chem Soc.* 125 (2003) 12696–12697.
73. V. Garcia-Gradilla, J. Orozco, S. Sattayasamitsathit, F. Soto, F. Kuralay, A. Pourazary, A. Katzenberg, W. Gao, Y. Shen, J. Wang, "Functionalized ultrasound-propelled magnetically guided nanomotors: Toward practical biomedical applications," *ACS Nano.* 7 (10) (2013) 9232–9240. doi:10.1021/nn403851v
74. S. Ahmed, W. Wang, L.O. Mair, R.D. Fraleigh, S. Li, L.A. Castro, M. Hoyos, T.J. Huang, T.E. Mallouk, "Steering acoustically propelled nanowire motors toward cells in a biologically compatible environment using magnetic fields," *Langmuir.* 29 (2013) 16113–16118. doi:10.1021/la403946j
75. L.O. Mair, B.A. Evans, A. Nacev, P.Y. Stepanov, R. Hilaman, S. Chowdhury, S. Jafari, W. Wang, B. Shapiro, I.N. Weinberg, "Magnetic microkayaks: Propulsion of microrods precessing near a surface by kilohertz frequency, rotating magnetic fields," *Nanoscale.* 9 (2017) 3375–3381. doi:10.1039/x0xx00000x
76. L.O. Mair, I.N. Weinberg, A. Nacev, M.G. Urdaneta, P. Stepanov, R. Hilaman, S. Himelfarb, R. Superfine, "Analysis of driven nanorod transport through a biopolymer matrix," *J Magn Magn Mater.* 380 (2015) 295–298.
77. L. Zhang, J.J. Abbott, L. Dong, K.E. Peyer, B.E. Kratochvil, H. Zhang, C. Bergeles, B.J. Nelson, "Characterizing the swimming properties of artificial bacterial flagella." *Nano Lett.* 9 (2009) 3663–3667. doi:10.1021/nl901869j
78. K.E. Peyer, L. Zhang, B.J. Nelson, "Localized non-contact manipulation using artificial bacterial flagella," *Appl Phys Lett.* 99 (2011) 174101. doi:10.1063/1.3655904
79. D. Schamel, A.G. Mark, J.G. Gibbs, C. Miksch, K.I. Morozov, A.M. Leshansky, P. Fischer, "Nanopropellers and their actuation in complex viscoelastic media," *ACS Nano.* 8 (2014) 8794–8801. doi:10.1021/nn502360t
80. J.G. Gibbs, A.G. Mark, T.-C. Lee, S. Eslami, D. Schamel, P. Fischer, "Nanohelices by shadow growth," *Nanoscale.* 6 (2014) 9457–9466. doi:10.1039/c4nr00403e
81. D. Walker, B.T. Käsdorf, H. Jeong, O. Lieleg, P. Fischer, "Enzymatically active biomimetic micropropellers for the penetration of mucin gels," *Sci Adv.* 1 (2015) E1500501, pgs. 1–7. doi:10.1126/sciadv.1500501
82. L. Zhang, E. Deckhardt, A. Weber, C. Schönenberger, D. Grützmacher, "Controllable fabrication of SiGe/Si and SiGe/Si/Cr helical nanobelts," *Nanotechnology.* 16 (2005) 655–663. doi:10.1088/0957-4484/16/6/006
83. L. Zhang, E. Ruh, D. Grutzmacher, L. Dong, D.J. Bell, B.J. Nelson, C. Schonenberger, "Anomalous coiling of SiGe/Si and SiGe/Si/Cr helical nanobelts," *Nano Lett.* 6 (2006) 1311–1317.
84. L. Zhang, J.J. Abbott, L. Dong, B.E. Kratochvil, D. Bell, B.J. Nelson, "Artificial bacterial flagella: Fabrication and magnetic control," *Appl Phys Lett.* 94 (2009) 64107. doi:10.1063/1.3079655
85. L. Liu, S.H. Yoo, S.A. Lee, S. Park, "Wet-chemical synthesis of palladium nanosprings," *Nano Lett.* 11 (2011) 3979–3982. doi:10.1021/nl202332x
86. J. Li, S. Sattayasamitsathit, R. Dong, W. Gao, R. Tam, X. Feng, S. Ai, J. Wang, "Template electrosynthesis of tailored-made helical nanoswimmers," *Nanoscale.* 6 (2013) 9415–9420. doi:10.1039/c3nr04760a
87. P. Tierno, R. Golestanian, I. Pagonabarraga, F. Sagués, "Controlled swimming in confined fluids of magnetically actuated colloidal rotors," *Phys Rev Lett.* 101 (2008) 218304. doi:10.1103/PhysRevLett.101.218304
88. P. Tierno, F. Sagués, "Steering trajectories in magnetically actuated colloidal propellers," *Eur Phys J E Soft Matter.* 35 (2012) 9748. doi:10.1140/epje/i2012-12071-4
89. L. Zhang, T. Petit, Y. Lu, B.E. Kratochvil, K.E. Peyer, R. Pei, J. Lou, B.J. Nelson, "Controlled propulsion and cargo transport of rotating nickel nanowires near a patterned solid surface," *ACS Nano.* 4 (2010) 6228–34. doi:10.1021/nn101861n
90. L.O. Mair, B. Evans, A.R. Hall, J. Carpenter, A. Shields, K. Ford, M. Millard, R. Superfine, "Highly controllable near-surface swimming of magnetic Janus nanorods: Application to payload capture and manipulation," *J Phys D Appl Phys.* 44 (2011) 125001. doi:10.1088/0022-3727/44/12/125001
91. H.W. Tung, D.F. Sargent, B.J. Nelson, "Protein crystal harvesting using the RodBot: A wireless mobile microrobot," *J Appl Crystallogr.* 47 (2014) 692–700. doi:10.1107/S1600576714004403
92. E.M. Purcell, "Life at low Reynolds number," *Am J Phys.* 45 (1977) 3–11. doi:10.1119/1.10903
93. J.J. Abbott, K.E. Peyer, M.C. Lagomarsino, L. Zhang, L. Dong, I.K. Kaliakatsos, B.J. Nelson, "How should microrobots swim?," *Int J Robot Res.* 28 (2009) 1434–1447. doi:10.1177/0278364909341658
94. A. Ghosh, P. Mandal, S. Karmakar, A. Ghosh, "Analytical theory and stability analysis of an elongated nanoscale object under external torque," *Phys Chem Chem Phys.* 15 (2013) 10817–10823. doi:10.1039/c3cp50701g
95. P.L. Venugopalan, R. Sai, Y. Chandorkar, B. Basu, S. Shivashankar, A. Ghosh, "Conformal cytocompatible ferrite coatings facilitate the realization of a nanovoyager in human blood," *Nano Lett.* 14 (2014) 1968–1975. doi:10.1021/nl404815q
96. B.J. Nelson, K.E. Peyer, "Micro- and nanorobots swimming in heterogeneous liquids," *ACS Nano.* 8 (2014) 8718. doi:10.1021/nn504295z
97. W. Xi, A. a Solovev, A.N. Ananth, D.H. Gracias, S. Sanchez, O.G. Schmidt, "Rolled-up magnetic microdrillers: Towards remotely controlled minimally invasive surgery," *Nanoscale.* 5 (2013) 1–4. doi:10.1039/c2nr32798h
98. O. Felfoul, M. Mohammadi, S. Taherkhani, D. de Lanauze, Y. Zhong Xu, D. Loghin, S. Essa, S. Jancik, D. Houle, M. Lafleur, L. Gaboury, M. Tabrizian, N. Kaou, M. Atkin, T. Vuong, G. Batist, N. Beauchemin, D. Radzioch, S. Martel, "Magneto-aerotactic bacteria deliver drug-containing nanoliposomes to tumour hypoxic regions," *Nat Nanotechnol.* 11 (2016) 941–947. doi:10.1038/nnano.2016.137

Lamar O. Mair (E-mail: Lamar.Mair@gmail.com) is a materials engineer at Weinberg Medical Physics, Inc. He earned a BS in materials science and engineering from the University of Florida and a PhD in materials science from the University of North Carolina at Chapel Hill. Prior to joining Weinberg Medical Physics, Lamar was a postdoctoral researcher at the National Institute of Standards and Technology's Center for Nanoscale Science and Technology. His research activities are in the areas of magnetic particle synthesis and manipulation.

Aleksandar N. Nacev is a senior engineer at Weinberg Medical Physics, Inc. He earned a BS in aerospace engineering and a PhD in bioengineering from the University of Maryland, College Park. Prior to joining Weinberg Medical Physics, he was a postdoctoral researcher at the University of Maryland, College Park. His research activities are in the areas of magnetic particle manipulation, fast MRI and electropermanent magnetics.

Sagar Chowdhury is a NSF/ASEE Research Fellow at Weinberg Medical Physics (WMP). His current research interest is image guided drug therapy. Before joining WMP, he was a postdoctoral research associate at Purdue University, focusing on autonomous navigation, robot motion planning and surgical robotics.

Dr Chowdhury received his PhD in mechanical engineering from the University of Maryland. He worked on automated optical micromanipulation of biological cells. His research at the University of Maryland was recognized with several awards including: ASME CIE best dissertation award, George Harhalakis Outstanding Systems Engineering Graduate Student Award, Dean's Doctoral Research Award and an ASME CIE best paper award.

Prior to that Dr Chowdhury obtained his MSc in mechanical engineering from the University of Oklahoma. During his masters, he worked on studying the shape similarity to develop common platform for a family of products. His research at OU was recognized with the 2009 Outstanding Research Performance Award by the Aerospace and Mechanical Engineering Department. He also received the best paper award in the 6th International Conference on Design Education (DEC) in 2009.

Dr Chowdhury earned his bachelor's degree in Mechanical Engineering from Bangladesh University of Engineering and Technology in 2007.

Pavel Stepanov is a cofounder of Brain Bio and is a photonics expert working as a senior imaging physicist at Weinberg Medical Physics. Prior to Weinberg Medical Physics, Pavel served as vice-president of Product Development at Naviscan PET systems. Previously, Pavel was a scientist at Novosibirsk Nuclear Physics Institute (Russia) and Boston University. He holds a master's degree in physics and bachelor's degree in business administration from Novosibirsk University.

Ryan Hilaman is a staff engineer at Weinberg Medical Physics, Inc. He earned a BS in bioengineering from the University of Maryland, College Park. Prior to joining Weinberg Medical Physics he was a biomedical engineer at PCTest Engineering Laboratory. Ryan's research interests are in MRI and electropermanent magnetics.

Sahar Jafari earned her BSc in materials science and engineering with a focus on ceramics from International University of Imam Khomeini, Iran in 2003. She earned her MS and PhD in materials science and engineering with a magnetic materials concentration from Iran University of Science and Technology, Tehran, Iran in 2006 and 2014, respectively. She has held appointments with the National Institute for Materials Science in Japan (May 2009–March 2012) as a researcher, and as departmental chair and assistant professor at Ahvaz Azad University, Iran (September 2012–March 2015). She currently works with Dr Irving Weinberg at Weinberg Medical Physics, Inc. on additive manufacturing of electropermanent magnetic materials in addition to drug delivery projects using magnetic nanoparticles. In general, Dr Jafari's research interests are in the areas of biomaterials, bioceramics and magnetic materials.

Benjamin Shapiro is a professor in the Fischell Department of Bioengineering at the University of Maryland, College Park. He earned a BS in aerospace engineering from the Georgia Institute of Technology and a PhD in control and dynamical systems from the California Institute of Technology. His research interests include magnetic particle manipulation, drug delivery and control of microbiological systems. His lab webpage is www.controlofmems.umd.edu.

Dr Irving N. Weinberg has had the good fortune of observing successful medical innovators at the PET lab in UCLA and the NIH Clinical Center's Laboratory of Diagnostic Radiology Research. He earned a PhD in experimental plasma physics and then underwent clinical training at University of Miami and Johns Hopkins Hospital. He continues part-time practice as a radiologist. Dr Weinberg has had a part in launching four FDA-approved products that have been used by over one million Americans with suspected or confirmed cancer. Having succeeded in actualizing these innovations, and having turned over their marketing to venture-backed teams, ten years ago he transitioned to the young field of image-guided therapy and diagnosis with pulsed magnetic fields and magnetic particles. Several of these patents have been licensed or sold to companies active in the medical imaging field, including Philips and Promaxo. He now enjoys mentoring young investigators and working with strategic partners to commercialize medical innovations.

22 Magnetic Nanoparticles for Neural Engineering

Gerardo F. Goya and Vittoria Raffa*

CONTENTS

22.1 HISTORICAL SUMMARY AND STATE OF THE ART

Nerve damage and neurological pathologies are two problems of significant medical and economic impact because of the hurdles of losing nerve functionality as a consequence of nerve injury or degenerative diseases (ND). Nerve regeneration is a complex biological phenomenon.[1] In the peripheral nervous system (PNS), nerves regenerate spontaneously only when injuries are minor. Short gaps can be repaired directly by mobilization of the proximal and distal stumps with end-to-end coaptation and epineural suturing. Long nerve gaps greater than 2 cm require additional material to bridge the defect. The current repair method is the use of autologous nerve grafts (autografts), which provide the regenerating axons with a natural guidance channel populated with functioning Schwann cells surrounded by their basal lamina.[1] Nerve autografting, however, is far from an optimal treatment, and there is suboptimal functional recovery despite technical excellence. These grafts are taken primarily from the sural nerve of the patient. Surveys of the clinical literature show that approximately half of patients with median and ulnar nerve repairs experience satisfactory motor and sensory recovery.[2] The main reasons for the poor functional recovery rates associated with autografts are unavailability of motor nerves (these grafts are primarily sensory) and mismatch in axonal size.[3] The use of autograft has also the disadvantages associated to the requirement for a second surgical site (donor site morbidity, donor site mismatch and the possibility of painful neuroma formation and scarring).[4] The use of nerve guidance conduits (NGCs) is the only clinically approved alternative to the autograft for the treatment of large peripheral nerve injuries. They provide a conduit during the nerve regeneration process for the diffusion of growth factors secreted by the injured nerve ends and to limit the injury site infiltration by scar tissue.[5] However, commercially available devices, based on biodegradable polymer or collagen-based hollow tubes, do not match the regenerative levels of autografts, providing good performances only for short defects (<2 cm) but poor functional recovery for longer nerve gaps.[6] Current knowledge suggests combining the use of NGCs with strategies of molecular or cellular therapies. Molecular therapies deals with the delivery of molecules such as guidance cues (netrins, ephrins, semaphorins and other molecules capable of orientating migrating and growing cells) and factors influencing neuronal growth (e.g. growth factors, neurotransmitters, extracellular matrix proteins).[7] Cell therapies involve cell transplantation to reduce tissue loss, promote axonal regeneration, facilitate myelination of axons or promote the secretion of factors sustaining the regeneration process. The nanotechnology applied to neural development, repair and

* Corresponding author.

protection aspire to implement these approaches for optimal regeneration and recovery of function and to solve those drawbacks arising from the invasiveness of macroscopic implants, including the dependence of the overall performance of such implants to physiological reactions (e.g. fibrosis). Low invasiveness and high selectivity of the growth stimulation are usually conflicting requirements and thus new approaches must be pursued in order to overcome such limitations.

Repairing therapies for those injuries of the central nervous system (CNS) of either the brain or spinal cord are much more challenging. Because the brain coordinates all higher-level functions and communicates with the PNS through the spinal cord, the cellular responses to a mechanical insult and posttrauma situation are numerous, and they are not well understood. Anyhow, once the CNS injury is produced, it initiates a cascade of deleterious events that can affect both cell body and axonal function, resulting in continued dysfunction and prolonged degeneration. For this type of CNS damage, cell therapy is the only therapeutic strategy that proved to work so far. Injured central axons do not spontaneously regenerate. However, the knowledge accumulated during the last two decades challenged the notion that neurons of CNS lack regeneration ability. Although in the 19th century Santiago Ramon y Cajal first suggested the idea that central axons could regenerate but the CNS does not offer a permissive environment, the extended concept of neuroregeneration, including the possibility of neurogenesis and neuroplasticity is a relatively recent one. The notion that neurogenesis is possible led to the idea of implanting viable cells as a therapeutically sound approach in neuroscience. Experimentally, this was demonstrated by transplanting a sciatic nerve explant into optic nerve lesions: the optical nerve regenerated across the graft but growth ceased as soon as axons had crossed the graft and reached the interface with the CNS.[8] It is likely that the regenerative potential of central axons is expressed when the CNS glial environment is changed to that of the PNS.[9] It was proposed to bypass the problem by transplanting some specific cell type, which could provide a permissive environment for elongated axon growth, similarly to the Schwann cells of peripheral nerves.[10] Thirty years later this hypothesis was tested for the first time in a 38-year-old male with a complete chronic thoracic spinal cord injury (SCI). This patient received an autologous sural nerve graft to bridge an 8-mm gap and the transplantation of glia olfactory ensheathing cells (OECs) in the proximal and distal nerve stumps; he experienced functional regeneration of supraspinal connections.[11] Some of clinical studies using cell therapy have been or are being conducted for the treatment of chronic SCI,* traumatic SCI,[†] amyotrophic lateral sclerosis,[‡] Parkinson's Disease,[§] cervical and thoracic SCI,[¶] age-related macular degeneration,** etc. These are early stage trials (phase I/II) to assess the safety of the treatment, which is essentially unknown despite a large amount of data available from preclinical experimentation.

Nanotechnology comes into neuroscience to provide additional ways to tackle the above-mentioned problems. Since nanoparticles (NPs), and more generally nanostructures, can be made small enough to interact with subcellular structures, the possibilities of intracellular targeting and actuation on damaged neural cells are countless. Inorganic NPs can be engineered as drug carriers alone for releasing neuroregenerative drugs, as reported using hollow silica NPs with porous walls to control the drug release kinetics.[12] Also, different physical properties of the NP's core or coating can be used to trigger the release, providing spatial and temporal control of the dose. The possibility of surface functionalization of NPs adds potential increase of specificity and/or hydrophobicity solutions for already existing therapeutic drugs. Among these strategies, the use of noncontact forces such as magnetic fields provides alternatives for remote NP actuation and activation. This chapter will focus on the new solutions nanotechnology can provide for neurological diseases, through engineered MNPs applied to neuroprotection and neuroregeneration. Also, the application of MNPs as magnetic actuators to position or guide neural cells by an external magnetic field will be described and discussed. In the first part we have included a description of the magnetism related to MNPs, as well as the theoretical framework for magnetic field interactions with biological systems. In the second part of the chapter, we offered an outline of the different strategies based on the use of MNPs and magnetic fields, applied to (a) neuroprotection in neurodegenerative diseases and (b) nerve regeneration following injury. We also describe and discuss those relevant MNP-based strategies successfully employed to remotely guide neuronal growth under the action of magnetic fields.

22.2 MAGNETISM OF SINGLE-DOMAIN NANOPARTICLES

The possibility of remote actuation on a nanoscale object has been understood since long ago as a way to manipulate biological systems.[13] From the physical point of view, the action-at-a-distance is a consequence of the interaction among any magnetic dipole (the basic entity in magnetostatics) having magnetic moment $\mathbf{m}$ and the magnetic flux density $\mathbf{B}$, also known as magnetic induction.[14] In the general case, the force on a magnetic moment $\mathbf{m}$ exerted under a magnetic induction $\mathbf{B}$ is given by the expression

$$\mathbf{F} = \nabla(\mathbf{m} \cdot \mathbf{B}), \tag{22.1}$$

where the spatial derivative implies that a nonuniform field is required to apply forces. In addition, the $\mathbf{B}$ field exerts a torque $\mathbf{N} = \mathbf{m} \times \mathbf{B}$ on the magnetic moment $\mathbf{m}$ that will align the dipole parallel to $\mathbf{B}$. Therefore, for those applications that require maximizing the magnetic forces between the external field $\mathbf{B}$ and the magnetic moment $\boldsymbol{\mu}$ of the MNPs, the usual strategies rely on (a) the design of optimized magnetic field

* www.clinicaltrials.gov. Studies ref. NCT01772810, NCT02688049

† www.clinicaltrials.gov. Study ref. NCT02326662

‡ www.clinicaltrials.gov. Studies refs. NCT01730716, NCT01640067, NCT01348451

§ www.clinicaltrials.gov. Study ref. NCT02452723

¶ www.clinicaltrials.gov. Studies refs. NCT02163876, NCT01321333

** www.clinicaltrials.gov. Study ref. NCT01632527

profiles, and (b) the synthesis of MNPs with large magnetic moments. The former choice is rather old and there is an extensive bibliography on numerical methods and magnetic field configurations.[15–17] For a comprehensive review of nanomagnetism and magnetic properties of MNPs the reader is referred to the comprehensive work of D. Ortega (Chapter 1: Structure and Magnetism in Magnetic Nanoparticles in the book *Magnetic Nanoparticle: From Fabrication to Clinical Applications*).[18]

The choice of the material for the magnetic core of MNPs is related to the physical and magnetic properties of the corresponding bulk phase. However, below a given critical particle diameter $d < d_{crit}$ (with $30 \leq d_{crit} \leq 100\ nm$, depending on the material's nature) the magnetic structure of the particle's core is different than the bulk material in the sense that domain walls collapse into a single magnetic domain. A deeper analysis of the concepts of magnetic domains, magnetic order in small particles and superparamagnetism is beyond the scope of this chapter, and the reader is referred to Ortega[15] and the classic book by B.D. Cullity[6] (Chapter 8). The value of d_{crit} is determined by the magnetic anisotropy (K) and the exchange stiffness coefficient (A) of the bulk material, and $d < d_{crit}$ defines a size regime below which the magnetic cores are magnetically ordered in a single direction. Therefore, this spin alignment results in a net magnetic moment of several hundreds of Bohr magnetons (Bohr magneton is the elementary unit of magnetic moment, defined in SI units in terms of the electron charge e, and mass me, and the reduced Planck constant $\hbar$, by $\mu_B = e\hbar/2m_e$). The magnetostatic energy of a single-domain MNP (i.e. the magnetic energy in the presence of an externally applied magnetic field) is proportional to its volume V, and this energy competes with the thermal energy to keep the magnetic moment spatially fixed.[16] Around room temperature (i.e. within the 25–45°C range), where most biomedical uses occur, the thermal energy can be of the same order than the magnetostatic energy for small applied fields. Therefore, the thermally induced magnetic relaxation impairs the magnetic alignment of **m** and **B** diminishing the magnetization at low fields. For MNPs with average size <30 nm thermal relaxation is predominant and thus affects the efficacy of those biomedical applications that require full magnetic saturation at room temperature. For these applications, the design of MNPs must consider average particle size and/or magnetic anisotropy large enough to prevent thermal relaxation.

22.2.1 Magnetic Field–Magnetic Nanoparticle Interactions

The strategy of using MNPs to actuate cells mechanically can be traced back to the year 1920, when W. Seifriz[17] proposed the use of 'minute particles of magnetic material' to measure the elasticity of the cell cytoplasm. Since then, a large amount of theoretical and experimental work on magnetically loaded cells has been reported.[18] The physical concept behind this approach is based on the interaction between the magnetic (dipole) moment **m** of MNPs and a spatially inhomogeneous magnetic field **B**,[19] as described by Equation 22.1. For a spherical MNP composed of magnetite, Fe_3O_4, with diameter d = 50 nm, a magnetic moment of $\boldsymbol{m} \approx 7 \times 10^{-17}\ Am^2$ can be estimated.[20] Assuming a commercially available NdFeB magnet (e.g. type N50) of cubic shape with dimensions 1 × 1 × 1 cm³, a single MNP located at a distance of 5 mm from the surface will experience an average force $F \approx 2.1 \times 10^{-15}\ N$. This force is larger than the gravitational force ($\sim 10^{-18}\ N$).[21] In addition, biomedical applications imply that the MNPs are immersed in a fluid and therefore the Stokes law predicts that any particle moving with velocity $\vec{v}$ will experience a drag force $\vec{F}_D$ given by $\vec{F}_D = 6\pi\eta r\vec{v}$, where η is the viscosity of the medium and r is the radius of the MNP. This force is size-dependent, but for the applications in quasi-stationary conditions such as those existing in a cell culture, the velocity factor makes this force small enough to discard it.[22] On the other hand, diffusional forces due to Brownian motion are also size dependent and cannot be neglected in colloidal systems at room temperature. A complete analysis of the influence of Brownian forces of single-domain MNPs under external magnetic field requires the use of stochastic approaches, such as the stochastic Eulerian–Lagrangian method, which is beyond the scope of the present description. Summarizing, to produce a measurable pulling-magnetic force, the MNPs have to be designed to maximize their magnetic moment **m** under the **B** values applied, and the magnetic field profile must be planned to produce enough field gradients for the experimental conditions required.[23]

22.2.2 Physical Features of Magnetic Nanoparticles

Two iron oxides, namely magnetite (Fe_3O_4) and maghemite (γ-Fe_2O_3), are by far the most used materials as a constituent of MNPs in the biomedical field.[24] These oxides crystallize in the cubic spinel structure, where two cationic sites with different geometries define the two magnetic sublattices, labelled as A and B sites. The cations at A and B sublattices have different atomic magnetic moments and result in a noncancelling total magnetic moment, and this type of magnetic order is known as ferrimagnetism. The macroscopic behaviour is similar to a ferromagnetic one, with remanence (i.e. a net magnetization at zero applied field), hysteresis (i.e. magnetization dependence on the magnetic history) and magnetic ordering (Néel) temperature. As mentioned before, in MNPs with size below the critical domain size the formation of domain walls is energetically unfavourable and the magnetic state has a single-domain configuration. If the MNP's volume is small enough, the thermal fluctuations at room temperature makes the magnetic moment to relax within timescales shorter than the measuring time, yielding a null-average of the magnetic moment.[25] This state is known as superparamagnetism. The magnetic relaxation of MNPs in magnetic colloids is therefore governed by the dynamics of the magnetization vector and have been modelled by Usov et al.[26] using the stochastic Landau–Lifshitz model.

As mentioned before, the specific properties of a given magnetic material must be considered when using MNPs as pulling agents. Specifically, those aspects governing the interaction expressed by Equation 22.1 between the magnetic field **B** and the magnetic moment **m** of the material. Since the material properties of the MNPs enter Equation 22.1 through the magnetic moment **m**, it is expected that the resulting magnetic forces should be more or less independent on the physicochemical environment of the MNPs for a given application. Instead, the most relevant parameters are the saturation magnetic moment M_S of the particles and their magnetic anisotropy. Together with the average volume *V*, these parameters will define the magnetic response of the material under the magnetic field intensity *H* applied. If the MNPs have a single-domain configuration, the magnetic moment measured under a given *H* value is given by

$$M(T,H) = M_S\ \mathcal{L}(M_S VH/k_B T), \tag{22.2}$$

where $\mathcal{L}(x)$ is the Langevin function, and $k_B T$ is the thermal energy factor at temperature *T*. This expression, together with Equation 22.1, show that if the MNPs are small enough the thermal fluctuations will render the magnetization small therefore decreasing the magnetic force. On the other hand, for multidomain MNPs the magnetization is governed mainly by domain wall motion and for large *H* values also by magnetic moment rotation within domains. Therefore, the preferred materials that provide magnetic saturation at low fields (and therefore large magnetic forces) would be magnetically soft materials (i.e. low magnetic anisotropy) without crystalline defects or vacancies, to avoid domain wall pinning.

22.2.3 Instrumentation: Simulation and Application of Magnetic Forces

Experimentally, the **B** profiles required for *in vitro* experiments can be produced by a suitable configuration of permanent magnets (e.g. $FeSm_5$ or NdFeB magnets) as well as different types of electromagnets. In most biomedical applications, processing conditions require working within fluidic phases and in small volumes. These prerequisites combine well with the use of microfluidics as a complementary technique to handle highly stable microflows and to fit the small volumes of liquids like culture media for *in vitro* experiments, where the total volumes can be as small as 10^{-9} L. The downscaling of magnetic separators allow to integrate them into more complex systems like detection devices for diagnostics and clinical assays, environmental monitoring, food-contaminant analysis, etc.[27–29] Also, small sample spaces allow larger field gradients to be applied without the need of large magnetic fields.

On the other hand, for larger working spaces the use of high-power electromagnets seems to be the only workable choice. One of the possible arrangements to combine continuous sorting and large working volumes is schematized in Figure 22.1a, where a superconductor coil of cylindrical symmetry surrounds a ferromagnetic matrix immersed in the flowing medium. Here the external field provides the large intensity of **B** inside the tube, whereas the ferromagnetic network provides the local inhomogeneity (i.e. field gradient) to retain the magnetic particles. The scalability of magnetic separation by magnetic forces is technically simple, although the amounts of energy required at industrial scales make it expensive. In any case, the use of high-gradient magnetic fields is being used successfully for treatment of industrial wastewaters and removal of heavy metals.[30]

At small working volumes (i.e. *in vitro* or small *in vivo* applications) the adequate choice of the **B** source will depend on the specific details of the experimental setup, but in most cases commercially available permanent magnets can produce suitable magnetic field gradients of several thousand *T*/*m*. A simple quadrupole configuration used for magnetic separation is produced by four permanent magnets placed on the external side of a supporting tube through which the colloid is pumped

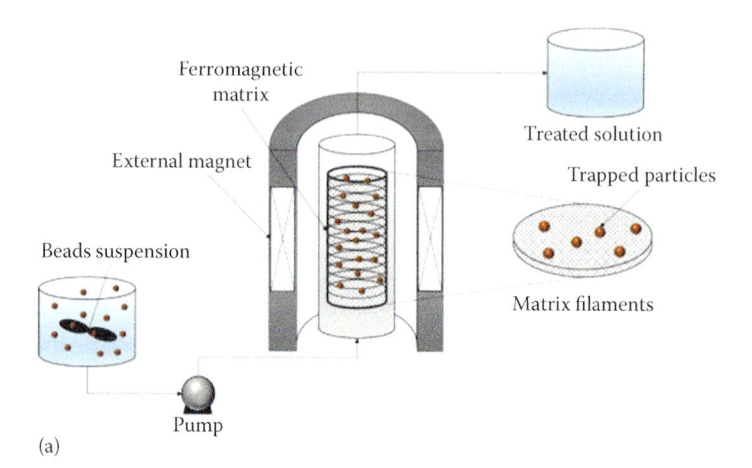

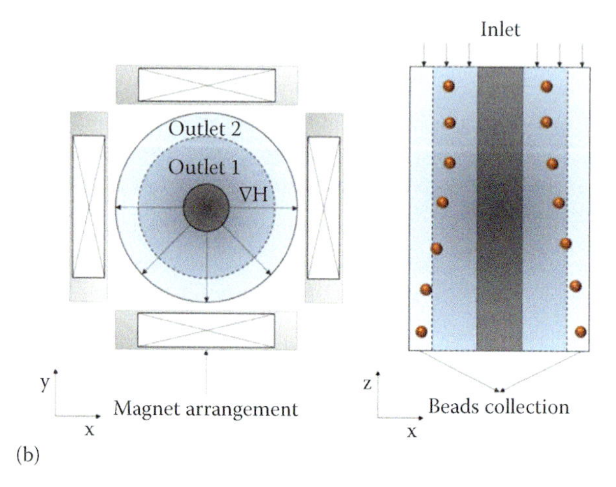

FIGURE 22.1 Schematic drawing of two approaches for magnetic separation under continuous flow sorting (a) a high gradient field separator based on superconducting magnets and (b) a magnetic quadrupole configuration, frequently used in small-volume applications. (Reprinted from *Sep. Purif. Technol.*, 172, J. Gomez-Pastora, X. Z. Xue, I. H. Karampelas, E. Bringas, E. P. Furlani, and I. Ortiz, 16, Copyright 2017, with permission from Elsevier.)

(see Figure 22.1b).[31] This configuration produces four regions with maximum field along the circular perimeter of the tube, whereas **B** = 0 at the centre.

A similar approach than the one used for magnetic separation, i.e. the use of the magnetic forces between external dc fields and the MNPs magnetic moment discussed above, is the basis for magnetic applicators designed for magnetic targeting. However, for MNPs to be concentrated at any internal body space there are additional difficulties. First, any realistic *in vivo* situation should consider not only the dynamic nature of the circulating blood but also the nonlinear character of the systemic paths that will carry the MNPs. In addition, the inherent pulling nature of the magnetic forces makes difficult to direct a net magnetic force towards an inner body volume using an external array of magnetic field sources. A potential solution to this problem was proposed through active targeting and accumulation of magnetic actuators to neural cells. This strategy has been successfully applied to control the mammalian nervous system in mice.[32]

22.3 MAGNETIC ACTUATION ON NEURAL CELLS

22.3.1 Effects of DC Magnetic Fields on Neural Cells

A substantial portion of the early research on biomagnetism was devoted to elucidating the influence of static and alternate magnetic fields at cellular and tissue levels.[33] Such investigations have disclosed many biochemical pathways that are influenced by a magnetic field. Only a small number of those investigations were related to physiological mechanisms in vertebrates under the influence of static magnetic fields, describing how reactions to magnetic stimuli were effected through the CNS.[34] The physical mechanism by which an exogenous magnetic field affects the biological pathways in eukaryotic cells is still under discussion, although there is long-standing experimental evidence that demonstrates the measurable effects on cell proliferation, migration and adhesion.[35] The existence of the earth's magnetic field (H = 39.8 A/m or 500 mG) provides examples of biological interactions that are well documented in bees, pigeons, bacteria and fish. The phenomena involving the capacity of a living organism to perceive or detect such weak magnetic field is known as magnetoreception.[36]

Also, the effects of intense static magnetic fields (i.e. up to several MA/m, or kGauss) have been studied in several different animal species, with different results, a relation between long-term application of strong static fields and biological pathways has been suggested. For example, experiments in young mice subjected to strong DC magnetic fields (i.e. H = 334 kA/m or 4200 G) have demonstrated measurable effects including growth retardation, changes in the population of bone marrow-derived monocytes, and increased rates of appearance of spontaneous cancer.[37,38]

As mentioned above, there is abundant experimental evidence that the application of static (or very low-frequency) magnetic fields on eukaryotic cells affects many biochemical pathways significantly, including cell proliferation, adhesion[35] and expression of heat-shock protein.[39] In the case of neural cells, the influence of magnetic fields could be expected on those mechanisms involving the exchange of ions through the cell membrane. Theoretical explanations[40] for these effects were proposed through perturbation effects of the magnetic field on moving charges. Since these neural communication mechanisms involve electrical signaling through ion channels at the cell membrane, it seems reasonable to expect that magnetic fields can influence the dynamics of cross-membrane ion pumping, impacting on cell differentiation and cell growth. However, there is experimental evidence excluding measurable effects on Na+ and K+ transmembrane currents down to one part in 1000.[46] On the other hand, it has been suggested that changes in nerve activity when exposed to strong DC magnetic fields (e.g. >100 kA/m) could be related to the diamagnetic anisotropy of some molecular components of the cell membrane. Under high magnetic fields, it is expected that the anisotropy axis of the membrane molecules will align along the field direction, and this realignment would suffice to modify the ion channel activity.

It is interesting to note that the two mechanisms differ on their physical basis: the action of $\vec{B}$ on moving charges $\pm q$ is the Lorentz force $\vec{F} = \pm q(\vec{v} \times \vec{B})$ applied to those charges with velocity $\vec{v}$, whereas the diamagnetic alignment of membrane molecules is the response of the closed-shell orbital atomic moments to the applied magnetic field. These differences make it in principle possible to design experiments to identify which mechanism will contribute under specific conditions. Both effects could be significant under strong fields, but the different B-thresholds at which these mechanisms start to operate and to what extent they are independent remain to be elucidated. In any case, the experimental evidence supporting the influence of static magnetic fields on neural cells is already quite solid, and explains why most reports on clinical effects of magnetic fields refer to the nervous system.[47]

Due to the complex interaction between electric and magnetic phenomena, the disentanglement of each source when a given (electrical or magnetic) experiment is performed is always challenging. The classification of 'pure' magnetic or electrical stimulation can be useful sometimes but the electromagnetic theory makes this distinction unfitting in the sense that a 'pure' static field B can modify the distribution of electrical charges existing in any material. Regarding biological materials (e.g. membranes, tissues, body fluids) it has been shown that the most influential physical parameters to be considered to affect cell functions are the electric and magnetic field amplitudes (E_0 and H_0, respectively), the intensity of induced currents, the induced voltage and the frequency.[43] In any case, general considerations indicate that the time scale of the electromagnetic stimulus must be of the same order of magnitude than the physical mechanism involved because otherwise the time average of the shorter magnetic pulses on the much larger time scales of biochemical dynamics in a cell membrane would produce a null effect out by simple time averaging any effect. For this reason both DC and extremely low-frequency magnetic fields are the usually chosen regimes to influence the response of biological systems.

Some works published in the 1980s about a 'cyclotron resonant effect' attempted to link weak electromagnetic fields to an enhanced Ca^{2+} transport through cell membrane due to resonant mechanisms.[44] However, attempts to replicate this effect were unsuccessful.[45] Moreover, theoretical considerations about the influence of viscosity and molecular collisions in fluid biological media seem to preclude any possibility of resonance associated with ion trajectories in such magnetic fields.

Iron is a relatively abundant element in most living organisms. Therefore, it is not surprising that biomineralization of iron, i.e. the biochemical processes through which an organism synthesizes hard minerals have made magnetite (Fe_3O_4) ubiquitous across both kingdoms of prokaryotes and eukaryotes including bacteria, protozoa and mammals. The occurrence of Fe_3O_4 crystals in the human brain resulting from iron biomineralization was first reported by Kirschvink,[46] who showed the presence of 20–50 nm crystals both isolated and forming linear structures similar to those typical of magnetotactic bacteria. The presence of nanostructured magnetite in the brain has been related to NDs in which disruption of normal iron homeostasis occurs.[47] The excess of iron and senile plaques found in brain tissue seem to support this idea.[48] It is interesting to mention that a recent study has suggested airborne pollution as an exogenous source of the Fe_3O_4 NPs found in brain tissue,[49] which poses the question of whether the major sources of MNPs in the brain have an internal or external origin. In any case, the idea that these magnetite MNPs within the brain could have relation with some of the biological effects related to AC magnetic fields in humans[50] merits further investigation.

22.3.2 Magnetic Forces Can Actuate on Cells

Although magnetic fields do have an influence on neural tissue, it is evident from the previous discussion that the nature of the interaction makes difficult to envisage their uses for remote tethering or actuation. Magnetic actuation is the action of influencing the behaviour of a cell by magnetic forces, generated from MNPs previously uploaded/attached to the cell.

To have the capacity of influencing axonal growth, magnetic forces must produce an effect larger than the drag forces within the cell, even at the nanometric scale. Magnetic forces originate in the interaction between the magnetic moment of MNPs and the magnetic field, as already discussed in Section 22.2.1, together with drag forces. For cell actuation a way to overcome the effects of drag forces is through the design of the MNPs. Furthermore, novel therapies that use exogenous cells (cell therapy) to gain lost functionalities in target tissues or organs have been proposed, which provide a fascinating tool for concurrent uses for MNPs. For example, stem cell-based treatments have been established as a clinical standard of care for some conditions, such as hematopoietic stem cell transplants for leukaemia and epithelial stem cell-based treatments.[51] Although the scope of potential cell-based therapies has expanded in recent years due to advances in basic research, attempts to develop a cell-based intervention into an accepted standard of medical practice are particularly difficult processes for different reasons. One of the unresolved issues relating to the clinical use of transplanted cells concerns the localization of these cells to the diseased site, since only a small percentage of the implanted/injected cells *in vivo* reach the desired location.[52]

There is enough evidence that for neural or neural precursor cells MNPs can be incorporated into the cytoplasm in large amounts. For example, the iron uptake in the oligodendroglial cell line OLN-93 has been reported[53] to increase the contents of intracellular iron up to ≈200 times the basal concentration in a concentration-dependent way. A comparative study on internalization in primary and immortalized cells showed that immortalized PC12 cells have a more intense activity than primary cells regarding MNP uptake.[54] The same study revealed that in a mixed (neuronal and glial) primary cell culture the predominant uptake of MNPs was done by microglia, whereas the number of astroglia and oligodendroglia incorporating MNPs was lower. Moreover, comparison against organotypic cocultures of spinal cord and peripheral nerve grafts yielded MNP-uptake levels similar to those of the primary cell cultures.[54]

The way by which a MNP is delivered to the cytoplasmic space can be very different depending on the type of cells or MNPs involved. Little work has been reported on the mechanisms of MNP uptake by neural cells and, more generally, about the interactions between MNPs and neural cell lines. Tay et al. reported a meticulous study on the interactions of MNPs with primary cortical neural networks in different developmental stages.[55] These authors found that chitosan-coated MNPs were internalized whereas starch-coated MNPs were not, the latter being attached to the cell membrane. By inhibiting selectively different uptake mechanisms, they concluded that the mechanisms by which chitosan-coated MNPs were incorporated was micropinocytosis and clathrin-mediated endocytosis.

The latest evolution of nanoscience into the neuroscience field has provided incipient solutions for the remote guidance of functional cells related to the above-mentioned cell therapies.[56,57] The ability to introduce MNPs into cells and magnetize them was the first step towards remote manipulation by magnetic fields to carry healing cells to the desired site, enabling the cells to colonize and differentiate into any desired cell type.[58] Also, different approaches based on magnetic forces to destroy target (cancer) cells have been reported. For example, Kim et al.[59] have succeeded in provoking cell damage using magnetic microdisks that could be forced to rotate by an external magnetic field of very low frequency (i.e. a few hertz) due to their vortex structure. The mechanical rotation was reported to compromise the integrity of the cell membrane, triggering an apoptotic mechanism. More recently, the same concept has been successfully applied *in vivo* to reduce an intracranial glioma tumour with no observed side effects.[60]

However, the rationale for remote guiding of axonal growth includes not only the successful uptake of the MNPs by the target cells. Given the large number of cells that participate in the repair after nerve injury, there is the question of whether some specific cell types could be more efficient in internalizing the MNPs injected than the target neurons. Most of the previous reports about the effects of NPs on neural cells (e.g. cell uptake, toxicity, etc.) have been conducted on immortalized cell lines (see for example Riggio et al.[61]) and only a few studies have been performed to investigate the effects of MNPs on primary cells of the nervous systems.[61,62]

The design of any magnetically guided axon regeneration therapy must consider how the external magnetic forces will act on an MNPs-loaded cell. The basis of the remotely guided neural regeneration involves (a) physical mechanisms to direct axonal regrowth along selected directions, and (b) biochemical mechanisms to stimulate axonal elongation across the nerve lesion site.[63] Also, the molecular guidance of axonal growth based on high-affinity molecules (such as growth factors and extracellular matrix proteins) can orientate growing cells,[64] although no therapeutic outcome has yet been reported.

Regarding physical guidance, autologous and heterologous tissue grafts or bioderived materials as scaffolds have been partially successful in providing growth conduits to guide the nerve during regeneration.[65,66] On the other hand, the uses of contactless magnetic forces have been much less studied. Some studies have been reported to be effective in both axon orientation and growth,[67] although these results are up to now limited to *in vitro* experiments. For *in vitro* situations, there are specific adhesion forces chemically sticking the cells to a substrate,[68] and therefore interaction of MNPs with H must be strong enough to overcome the adhesive forces, which have been reported to be in the 1–200 pN range depending on the measuring conditions.[69]

22.4 NERVE REPAIR

22.4.1 Magnetic Guidance

The concept of magnetically assisted nerve repair is based two complementary actions that can be performed when MNPs are used. The first action is related to the interactions with a remote magnetic field, which generates a pulling force from the MNPs to the growing axons. The second, i.e. the possibility of having neurotrophic factors on the MNP's surface, would make it possible to stimulate growth rates during the application. This synergistic approach for nerve repair using MNPs is schematically illustrated in Figure 22.2. It is evident that the complexity of the actual biological process, which includes a plethora of different cell types acting on the injured nerve, makes it necessary to verify some hypotheses regarding the effects and fate of MNPs once they are injected into the injured tissue. For example, Schwann cells are recognized as helpful agents promoting axonal regeneration in the PNS while astrocytes and oligodendrocytes in the CNS are not. Therefore, the successful delivery of MNPs to the target axon will depend on the relative affinity of these cell types for the MNPs.

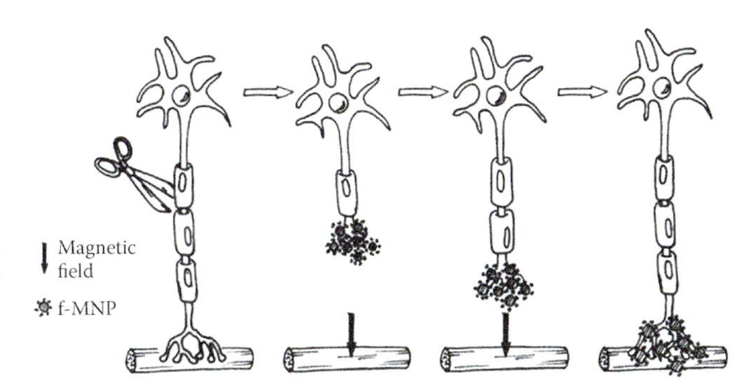

FIGURE 22.2 Schematic illustration of magnetically assisted nerve repair process. The injured nerve is targeted by the magnetic NPs, and then a remote magnetic field guides axonal growth along the field lines. The particles, in turn, could be surface-functionalized with growth-stimulating molecules to accelerate the healing.[76] (Reprinted from *Nanomed-Nanotechnol*, 10, C. Riggio et al., 1549, Copyright 2014, with permission from Elsevier.)

Although mechanisms involved in axonal growth are not completely understood, there is increasing evidence that mechanical force generation is a crucial process for both axonal guidance and lengthening.[70] The existing literature suggests that neurons and their axons possess fine sensors to sense and transduce mechanical force in axon initiation/elongation/guidance. The involvement of mechanical tension in the morphogenesis of the nervous system was clear in the late 1970s when pioneering experimental work revealed that neuronal processes *in vitro* are under tension.[66] Later, different teams demonstrated that the external application of mechanical tension alone is sufficient to initiate *de novo* axonal sprouting. There is a consensus that neurite elongation is a linear function of the applied force and its rate has been found to be similar to both PNS and CNS (about 0.1–1 $\mu m\ h^{-1}$ per pN of applied force). MNPs, which develop a strong magnetic force when an external magnetic field is applied, could be used to induce an extremely rapid regeneration of the injured axons, purely directed by mechanical forces on MNP-labelled axon tracts. Fass and colleagues used magnetic beads to precisely develop forces in the piconewton range, finding cells able to sustain mechanical-driven elongation with applied tensions between 15 and 100 pN.[71] In addition to the evidence that mechanical tension can induce elongation of neurite or process initiation, recently its influence on axonal guidance has also been investigated. It was demonstrated in a neuron-like cell line that MNPs can be used to gain control of directional movements of neurites. Specifically, by using magnetic nanobeads, it was found that the application of 0,5 pN force on cell neurites was enough to preferentially align them along the direction imposed by the mechanical force.[7] Moreover, using a model based on the effects of the applied forces acting on the receptor–ligand bond, dynamic process of bond loading, breaking and formation during cytoskeletal movements, the authors could reproduce the experimental data successfully.[63] A basic setup for this experimental approach using four parallel NdFeB magnets is depicted in Figure 22.3, where the micrograph (inset) shows the preferential growth along the

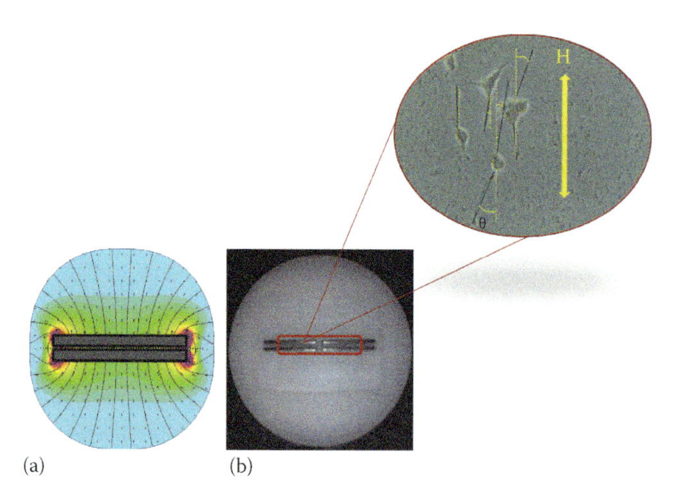

FIGURE 22.3 (a) Representation of the magnetic field applied to the neural PC12 cell cultures. The magnetic field was homogeneous in the Y and X direction (0.19–0.20 T). The maximum magnetic field gradient was 0.019 T/m. (b) Image of the support where the T-25 flasks were incorporated and an example of the images obtained by an optical microscope in the area where the cells are analyzed. The image shows the analysis of the neurite direction; each neurite is manually traced and then the angle formed between the neurite and the direction of the magnetic field (θ) is recorded.

field lines (yellow arrow), quantified through the angle θ between H and the direction of the main dendrites.

The ability to generate mechanical tensions on neurites, to promote elongation, and to guide directional movement could make MNPs a powerful strategy to address the dream of axonal re-innervation from the CNS to the desired target, e.g. the neuromuscular junction. The MNP-mediated mechanical force has also been used to manipulate neuronal compartments. MNPs have been used to bind filopodia cell membrane of retinal ganglion cell growth cone and to elicit axonal growth and guidance by exerting mechanical tension with an externally applied magnetic field.[72] MNPs functionalized with TrkB agonist antibodies have been used to target the particles to signaling endosomes, to manipulate them by focal magnetic fields, and to alter their localization in the growth cone, thus deregulating growth cones motility and neurite growth.[73] The synaptosomes of brain nerve terminal labelled with MNPs were spatially manipulated with external magnetic field without affecting the key characteristics of glutamatergic neurotransmission.[74] Recently, manipulation via MNP has also been performed at the molecular level, by influencing protein segregation during axonal development, *in vitro* and *in vivo*, to dictate axon formation.[75] In general, MNPs offer the distinct advantage of being easily functionalized with ligands for high affinity binding to specific neuronal cell types, compartments or proteins,[76] which makes particularly effective present and future strategies of neuronal manipulation via MNP-induced mechanical forces. MNPs have been used also to manipulate the extracellular environment, which plays a key role in the process of nerve regeneration. Recently, magnetic particles have been used to orientate collagen fibres under an external magnetic field, opening the possibility to develop oriented scaffolds to strongly promote the process of functional reinnervation.[77]

22.4.2 Neuroprotection

Functionalization of MNPs with neurotrophic factors to promote neuron survival/growth can also be achieved.[78] Although the free growth factors have a very short half-life (e.g. few minutes)[79] *in vitro* studies have proved that the conjugation to iron oxide NPs can prolong the biological activity of NGF, glial cell-derived neurotrophic factor (GDNF) and basic fibroblast growth factor (FGF-2).[79,80] An additional advantage of MNPs is that they can be remotely guided by magnetic forces. They have also been used as magnetically guided nanocarriers for spatially controlled drug delivery, e.g. for local release of anaesthetics for local nerve block[81] or for targeting neurotrophic factors to the blood–brain barrier (BBB).[82] The idea of improving neuroprotection using drug-loaded nanocarriers through the BBB is many years old.[83] The capability of a nanometre-sized device with a therapeutic payload to cross the BBB is appealing since about 95% of the therapeutic drugs for treating CNS diseases fail to do so in the brain.[84] This is mainly related to the impenetrability of the BBB for such molecules. Several strategies to overcome this problems using MNPs have been reported, based on the functionalization of the particles with peptides, proteins and similar small molecules.[82]

However, the actual neurotoxicity levels of MNPs *in vivo* are not yet completely known. It has been reported that MNPs entering into the body fluid system can result in adverse effects on the CNS.[85] Also, systemic administration of MNPs has been reported to induce breakdown of the BBB, an effect not only exclusive of magnetic particles but NPs in general.[86] Different interactions between MNPs and CNS in physiological vs. pathological conditions cannot be also excluded. A recent work showed that NPs can target myeloid cells in epileptogenic brain tissue, suggesting their use for detecting immune system involvement in epilepsy or for localization of epileptic foci.[87] A related, more subtle question of whether MNPs influences the physiological brain responses under pathological conditions has been addressed only rarely in the literature, but is certainly a subject that merits investigation.

22.4.3 Magnetofection

The concept of transfection can be defined as the procedure by which any type of genetic material from a foreign source is introduced into a different mammalian cell. When dealing with DNA, this process enables the expression of proteins from the original source by the host cell's machinery. The transfer of the genetic material can be done by different methods, in many cases using coadjuvant molecules to improve the transfection rates. One example is the use of cationic lipids (e.g. Lipofectamine®) with a positively charged head group that favours DNA condensation and also facilitates the fusion of the liposome/nucleic acid with the cell membrane prior to the endocytosis.[88]

The first experiments on the use of magnetic fields to enhance nucleic material delivery were reported by C. Mah et al.,[89] and soon after the term 'magnetofection' was coined by C. Plank's group.[90] Since then, this concept of MNP-mediated transfection has been customized and improved regarding the dose–response ratios and transfection rates. Today, there are many commercially available kits that provide user-ready MNPs and reagents for routine laboratory applications such as CombiMag™ (Ozbiosciences SAS, France) or Magnetofection™ (Chemicell GmbH, Germany). The basic mechanism is depicted in Figure 22.4: through the use of an external magnetic field gradient, the forces acting on the magnetic vectors increase the contact time of the genetic material and the cell membrane, increasing the uptake dynamics of the cell membrane and thus the efficiency of nucleic acid delivery. Some simple physical models have been proposed for this interaction, based on a drift-diffusion equation through the cell membrane,[91] but a complete model accounting for the different physiological pathways is still lacking. However, there is an emerging consensus that for these applications the surface chemical composition of the MNPs is a key factor determining the final efficiency, irrespective of the details of the magnetic structure of the magnetic cores.

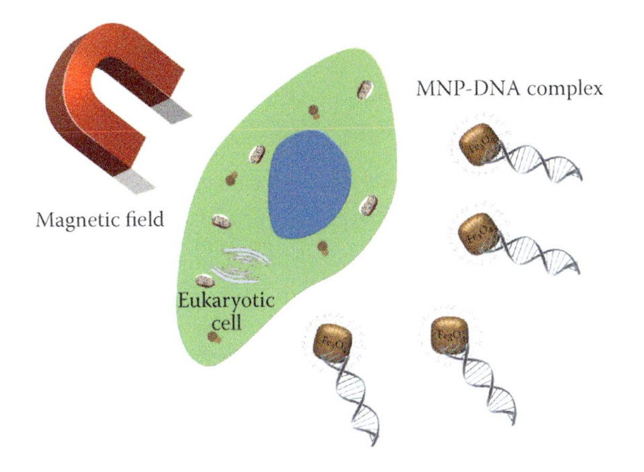

FIGURE 22.4 Schematic illustration of the magnetofection principle. The nucleic acid and the magnetic NP form the magnetic nanovector complex that is pulled towards the cell by a noncontact magnetic force from an external magnetic field gradient. The forces increase the rate of contact events between the vectors and the cell membrane, thus improving the uptake dynamics of the cell.

Recently, a method to increase transfection in neural stem cells (NSCs) using MNPs and very low frequency (4 Hz) magnetic fields demonstrated that transfection efficacy could be improved significantly, while keeping the differentiation capabilities unaffected.[92] As shown from the differentiation profiles in Figure 22.5, magnetofected NSCs show positive for all transfected markers. Moreover, the authors reported that magnetofected NSCs displayed disrupted cell membranes as compared to control cells. Although the physical mechanisms involved are not completely understood, experimental data suggest that the higher efficiency under magnetic fields is due

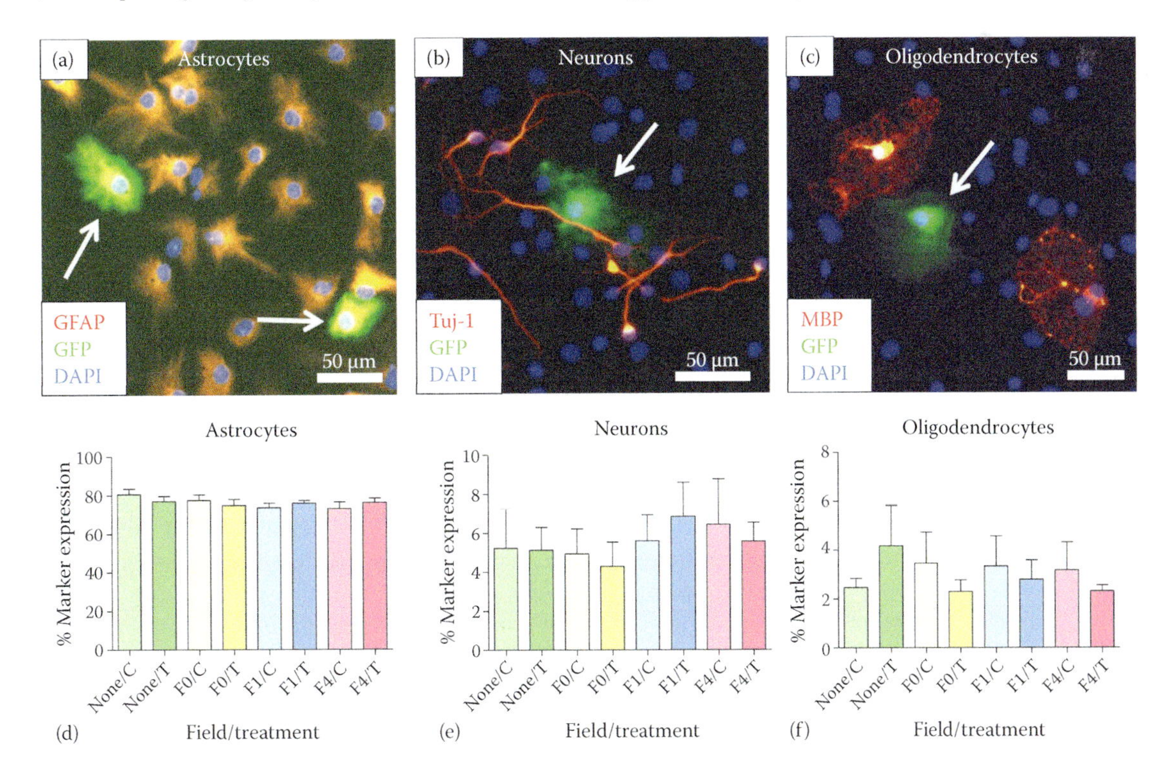

FIGURE 22.5 Triple merged images of magnetofected ($f = 4$ Hz) neural stem cells (NSCs), postdifferentiation showing cells positive for GFAP (a), Tuj-1 (b), and MBP (c). GFP expressing GFAP+ cells are seen in (a, arrows) and GFP+ cells with the morphological appearance of astrocytes in (b and c, arrows). (d–f) Bar charts showing proportions of GFAP, Tuj-1, and MBP positive cells, n = 4 cultures. None: no field, F(N): frequency of oscillation, C: control, T: transfected. (Reprinted from *Nanomedicine: Nanotechnology, Biology and Medicine*, 9, C. F., Adams, M. R., Pickard, and D. M., Chari, 737, Copyright 2013, with permission from Elsevier.)

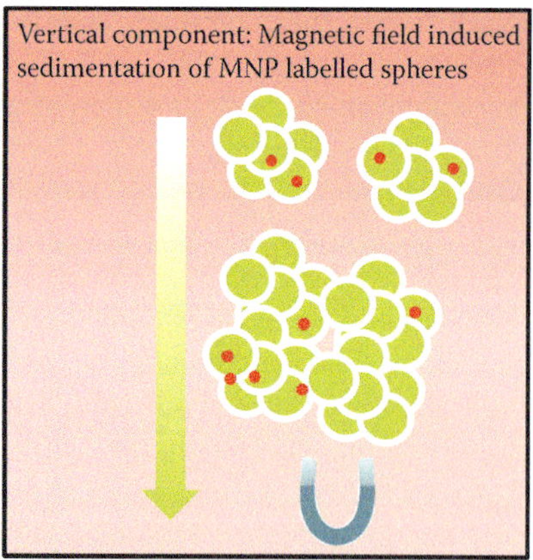

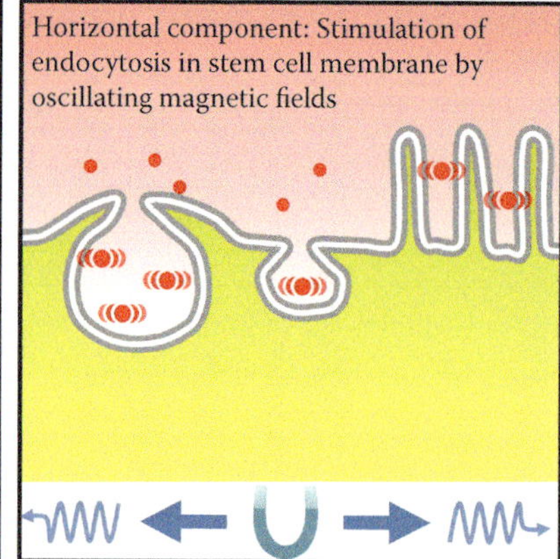

FIGURE 22.6 Proposed mechanism of transfection of neurospheres. Schematic diagram illustrating a hypothetical model to explain the mechanism of oscillating field enhancement of transfection in neurospheres.[110] (Reprinted from *Nanomedicine: Nanotechnology, Biology and Medicine*, 9, C. F., Adams, M. R., Pickard, and D. M., Chari, 737, Copyright 2013, with permission from Elsevier.)

to both an increase of the MNP-cell interaction time and a frequency-driven stimulus of the endocytic activity of the cell membrane, as depicted in Figure 22.6.[92]

22.4.4 Magnetotransduction

Similarly to the use of magnetic forces, the use of viral vectors has provided a fruitful solution to increase the low efficiency of nonviral gene vectors, a technique known as transduction. The term was coined more than 50 years ago by Zinder et al.,[93] in their genetic studies on *Salmonella typhimurium* and has been improved notably along the last decade.[94] Viral vectors have been intensively used as a tool for fighting NDs, through the delivery of neurotrophic factors that prevent degeneration and enhance recovery of target neurons. The potential of this technique for clinical uses is apparent, especially in the field of NDs. For example, two powerful neuroprotective molecules for the treatment of neurodegenerative pathologies affecting both motor and cognitive functions are GDNF and insulin-like growth factor I (IGF-I).[95] In spite of some promising results, the efficiency of viral (and nonviral) vectors for therapeutic gene delivery into the brain still remains one of the limiting factors to be overcome before clinical trials can be safely implemented. Additionally, most protocols currently in use for nucleic acid delivery[96] require some improvement of either the efficiency or specificity of nucleic acid delivery.[97]

Based on the concepts of magnetofection, i.e. the use of magnetic forces on MNPs to improve transfection efficiency, therapeutic approaches against NDs have begun to use magnetically labelled viral units to deliver genetic material. The construction of magnetic viral vectors (usually adenovirus or lentivirus) for a magnetic field-assisted viral transduction has been reported for some years now. This technique is known as *magnetotransduction*, and is often related (but not restricted) to strategies for delivering neuroprotective molecules to target cells as a therapy against ND diseases. One of the main goals of this approach is related to the enhancement in the levels of neurotrophic factors delivered, since it is accepted that an increase in the delivered concentration of these factors can prevent neural degeneration and enhance recovery of remaining neuron neuroprotective molecules at the target site.[98]

In magnetotransduction, the MNPs also work as the 'pulling' agents when conjugated with viral vectors to construct a magnetic-viral vector of higher efficacy than virus or MNPs alone. Some configurations using Fe_3O_4-based MNPs and recombinant adenoviral vector harbouring reporter genes have been already used to magneto-transduce glial and neuronal brain cells (ependymal, hypothalamic and substantia nigra) with high efficiency.[99] Some proof-of-principle experiments with MNP-AAV (adeno-associated viral) vectors showed partial success,[100] but the need for further optimization of vector formulation remains, especially if neuroprotective and neurotrophic factors (e.g. IGF-1, GDNF) are to be used for clinical applications to ND diseases. If successful, this approach could represent a major improvement towards new therapies for NDs.

22.4.5 Scavenging Strategies

When the CNS is affected, nerve injury results in a disruption of the blood–spinal cord barrier. Moreover, the damage induced in surrounding blood vessels stimulates a proliferation of Schwann cells, leucocytes, monocytes and macrophages around the nerve lesion that provokes the loss of nervous tissue. At the cellular level, axons show deteriorated myelin layers, and the resulting growth-inhibitory myelin debris is only partially removed by macrophages. Therefore, a containment/scavenging protocol is desired before actual nerve regeneration. Gathering those cells that are activated in response to pathological situations can be achieved by the use of remote magnetic forces on the injured area. The incorporation of MNPs by scavenger cells has been already observed in organotypic culture,[61] and therefore it can be expected that similar targeting can be achieved *in vivo*.[101] Indeed, the *in vitro* preloading of macrophages and the subsequent infiltration *in vivo* for magnetic resonance imaging of injured nerve has been successfully tested some years ago.[102] It is, therefore, a matter of time before similar magnetic labeling of scavenger cells can be used for improved magnetically driven nerve repair. MNPs possess themselves scavenging properties. In particular, their capacity to scavenge free radicals has been used to attenuate oxidative damage induced by H_2O_2 in SCI rats when localized by an external magnetic field.[101] Additionally, their functionalization with biomolecules can confer new scavenger capabilities, as recently demonstrated by MNP functionalization with O-methyl-β-cyclodextrin to reduce the extracellular level of L-glutamate in brain nerve terminals.[103]

22.4.6 Cell Therapies

Cellular therapies exploit the regenerative potential of cells for nerve repair.[104] They are considered promising, especially for the repair of CNS injuries and long gaps in the PNS. Several cell types such as stem cells, Schwann cells, OECs have been utilized as transplantable cells in nerve regeneration, demonstrating improved regenerative outcomes[105,106] but, similarly to any cell-based strategy, this approach suffers from drawbacks, which limit the translation from experimental to clinical stages. A great help for implementing safe and effective cell transplantation could be the development of strategies for cell homing and cell tracking, allowing for monitoring of the fate of the transplanted cells and to retain them in the injury/pathology site, maximizing the therapeutic effects while avoiding dangerous migrations to ectopic sites. Several lines of evidence suggest that MNPs hold a great potential to overcome these limitations. Recently, a clinical study* has demonstrated in healthy volunteers that MNP can be used for *in vivo* tracking of magnetically labelled human mononuclear cells using MRI scanning. Following intravenous administration, the distribution of iron-labelled cells was monitored as well as their ability to migrate to a site of inflammation. Cell labeling with MNPs can be thus easily imaged via MRI and this approach offers the distinct advantage to correlate the study outcome to the cell localization at the site, or biodistribution in the organism. MNPs have been used to label oligodendrocyte precursor cells, which showed high promise as a transplant population to remyelinate nerve fibres and promote regeneration in the CNS. Indeed, clinical trials using these types of cells have been initiated in some areas.[107] The migration of MNP labelled OPCs was followed via MRI, when injected into the spinal cord of myelin-deficient rats[108] or after transplantation into adult rat brain.[109] Magnetic manipulation is also an advantageous method for guiding cells remotely. Neural progenitor cells[110] or olfactory ensheathing cells[111] have been labelled with MNPs and magnetically localized to promote axon growth in organotypic cocultures. This approach was also used *in vivo* to remotely guide MNP labelled stem cell in the spinal cord of SCI mice, demonstrating enhanced localization and axon regeneration.

* www.clinicaltrials.gov. Study ref. NCT01169935

22.5 OUTLOOK FOR THE FUTURE

There are several nanotherapies already proposed as substitutes for (a) surgical nerve grafting after peripheral nerve injury, (b) pharmacological treatment after drug abuses and (c) neuroprotective drug delivery.[112–114] However, there are no reports to date that can show conclusive clinical improvements over the established surgical procedures. The near future will probably see new nanotherapies as coadjuvant protocols. MNPs have already opened new paths for noninvasive therapies based on the exploitation of the remotely driven mechanical forces on MNP-loaded neurons. The ability of these approaches to promote migration and axonal elongation/growth have already passed the first proof-of-concept challenges, but many fundamental questions are yet unresolved. It is also clear that a 'second generation' of enhanced MNPs is required offering minimum toxicity and better reproducibility. A major issue still not addressed, which will determine the final efficacy of these magnetic vectors, is the creation of a flexible surface for functionalization with neurotrophic/neuroprotective factors. If this flexible platform is developed, it will open boundless possibilities for novel molecular therapies, as well as the basis (together with multipotent stromal cells) for more effective cell therapies.

ACKNOWLEDGEMENTS

This work was partially supported by the Spanish Ministerio de Economia y Competitividad (MINECO) through project MAT2016-78201-P; and the Aragon Regional Government (DGA, Project No. E26) and the Wings for Life Spinal Cord Research Foundation (WFL, Project No. 163).

REFERENCES

1. Millesi H. 2007. Bridging defects: Autologous nerve grafts. *Acta Neurochirurg. Suppl.* **100,** 37–8.
2. Ruijs A C, Jaquet J B, Kalmijn S, Giele H and Hovius S E. 2005. Median and ulnar nerve injuries: a meta-analysis of predictors of motor and sensory recovery after modern microsurgical nerve repair. *Plastic Reconstruct. Surg.* **116,** 484–94; discussion 95–6.
3. Lin M Y, Manzano G and Gupta R. 2013. Nerve allografts and conduits in peripheral nerve repair. *Hand Clin.* **29,** 331–48.
4. Deumens R, Bozkurt A, Meek M F, Marcus M A, Joosten E A, Weis J and Brook G A. 2010. Repairing injured peripheral nerves: Bridging the gap. *Progr. Neurobiol.* **92,** 245–76.
5. Daly W, Yao L, Zeugolis D, Windebank A and Pandit A. 2012. A biomaterials approach to peripheral nerve regeneration: bridging the peripheral nerve gap and enhancing functional recovery. *J. R. Soc. Interface/R. Soc.* **9,** 202–21.
6. Pabari A, Lloyd-Hughes H, Seifalian A M and Mosahebi A. 2014. Nerve conduits for peripheral nerve surgery. *Plastic Reconstruct. Surg.* **133,** 1420–30.
7. Piotrowicz A and Shoichet M S. 2006. Nerve guidance channels as drug delivery vehicles. *Biomaterials.* **27,** 2018–27.
8. Villegas-Perez M P, Vidal-Sanz M, Bray G M and Aguayo A J. 1988. Influences of peripheral nerve grafts on the survival and regrowth of axotomized retinal ganglion cells in adult rats. *J, Neurosci* **8,** 265–80.

9. David S and Aguayo A J. 1981. Axonal elongation into peripheral nervous system "bridges" after central nervous system injury in adult rats. *Science.* **214,** 931–3.
10. Raisman G. 1985. Specialized neuroglial arrangement may explain the capacity of vomeronasal axons to reinnervate central neurons. *Neuroscience.* **14,** 237–54.
11. Tabakow P, Raisman G, Fortuna W, Czyz M, Huber J, Li D, Szewczyk P, Okurowski S, Miedzybrodzki R, Czapiga B, Salomon B, Halon A, Li Y, Lipiec J, Kulczyk A and Jarmundowicz W. 2014. Functional regeneration of supraspinal connections in a patient with transected spinal cord following transplantation of bulbar olfactory ensheathing cells with peripheral nerve bridging. *Cell Transplant.* **23,** 1631–55.
12. Lai C Y, Trewyn B G, Jeftinija D M, Jeftinija K, Xu S, Jeftinija S and Lin V S Y. 2003. A mesoporous silica nanosphere-based carrier system with chemically removable CdS nanoparticle caps for stimuli-responsive controlled release of neurotransmitters and drug molecules. *J. Am. Chem. Soc.* **125,** 4451–9.
13. Alexander H S. 1962. Biomagnetics—Biological effects of magnetic fields. *Am. J. Med. Electron.* **1,** 181–7.
14. Jackson J D. 1999. *Classical Electrodynamics* (New York: Wiley).
15. Ortega D., in *Magnetic nanoparticles: From fabrication to clinical applications*, edited by N.T.K. Thanh, (CRC Press. Taylor & Francis Group, Boca Raton, FL, 2012). Chapter 1, p. 3. ISBN 9781439869321. 616 Pages.
16. Cullity B D. 1972. *Introduction to Magnetic Materials* (Reading, Mass.,: Addison-Wesley Pub. Co.).
17. Seifriz W. 1924. An elastic value of protoplasm, with further observations on the viscosity of protoplasm. *J. Exp. Biol.* **2,** 1–11.
18. Dobson J. 2008. Remote control of cellular behaviour with magnetic nanoparticles. *Nat. Nano* **3,** 139–43.
19. Hallmark B, Darton N J, James T, Agrawal P and Slater N K H. 2010. Magnetic field strength requirements to capture superparamagnetic nanoparticles within capillary flow. *J. Nanopart. Res.* **12,** 2951–65.
20. Frankel R B, Blakemore R P and Wolfe R S. 1979. Magnetite in freshwater magnetotactic bacteria. *Science.* **203,** 1355–6.
21. Buongiorno J. 2005. Convective transport in nanofluids. *J. Heat Transf.* **128,** 240–50.
22. Rogers H B, Anani T, Choi Y S, Beyers R J and David A E. 2015. Exploiting size-dependent drag and magnetic forces for size-specific separation of magnetic nanoparticles. *Int. J. Mol. Sci.* **16,** 20001–19.
23. Kim H K, Hong S H, Hwang S W, Hwang J S, Ahn D, Seong S and Park T H. 2005. Magnetic capture of a single magnetic nanoparticle using nanoelectromagnets *J. Appl. Phys.* **98,** 104307.
24. Pankhurst Q A, Connolly J, Jones S K and Dobson J. 2003. Applications of magnetic nanoparticles in biomedicine. *J. Phys. D—Appl. Phys.* **36,** R167-R81.
25. Brown W F 1963 Thermal fluctuations of a single-domain particle. *Phys. Rev.* **130,** 1677–86.
26. Usov N A and Liubimov B Y. 2012. Dynamics of magnetic nanoparticle in a viscous liquid: Application to magnetic nanoparticle hyperthermia. *J. Appl. Phys.* **112,** 11.
27. Lee H, Shin T-H, Cheon J and Weissleder R. 2015. Recent developments in magnetic diagnostic systems. *Chem. Rev.* **115,** 10690–724.
28. Modak N, Datta A and Ganguly R. 2008. Cell separation in a microfluidic channel using magnetic microspheres. *Microfluid. Nanofluid.* **6,** 647.
29. Song H P, Li X G, Sun J S, Xu S M and Han X. 2008. Application of a magnetotactic bacterium, Stenotrophomonas sp to the removal of Au(III) from contaminated wastewater with a magnetic separator. *Chemosphere.* **72,** 616–21.
30. Babel S and del Mundo Dacera D. 2006. Heavy metal removal from contaminated sludge for land application: A review. *Waste Manage.* **26,** 988–1004.
31. Gomez-Pastora J, Xue X Z, Karampelas I H, Bringas E, Furlani E P and Ortiz I. 2017. Analysis of separators for magnetic beads recovery: From large systems to multifunctional microdevices. *Sep. Purif. Technol.* **172,** 16–31.
32. Wheeler M A, Smith C J, Ottolini M, Barker B S, Purohit A M, Grippo R M, Gaykema R P, Spano A J, Beenhakker M P, Kucenas S, Patel M K, Deppmann C D and Guler A D. 2016. Genetically targeted magnetic control of the nervous system. *Nat. Neurosci.* **19,** 756.
33. Tenforde T T. 1979. *Magnetic Field Effect on Biological Systems* (USA: Springer).
34. Nakhilni Zn. 1974. Biological effect of constant magnetic-fields. *Kosm. Biol. Avia. Med.* **8,** 3–15.
35. Dini L and Abbro L. 2005. Bioeffects of moderate-intensity static magnetic fields on cell cultures. *Micron.* **36,** 195–217.
36. Kirschvink J L, Jones D S and MacFadden B J. 2013. *Magnetite Biomineralization and Magnetoreception in Organisms: A New Biomagnetism,* vol 5 (Berlin, Germany: Springer Science & Business Media.).
37. Lerchl A, Nonaka K, Stokkan K-A and Reiter R. 1990. Marked rapid alterations in nocturnal pineal serotonin metabolism in mice and rats exposed to weak intermittent magnetic fields. *Biochem. Biophys. Res. Commun.* **169,** 102–8.
38. Semm P and Demaine C. 1986. Neurophysiological properties of magnetic cells in the pigeon's visual system. *J. Comp. Physiol. A.* **159,** 619–25.
39. Goodman R and Blank M. 1998. Magnetic field stress induces expression of hsp70. *Cell Stress Chaperones.* **3,** 79.
40. Lednev V V. 1991. Possible mechanism for the influence of weak magnetic fields on biological systems. *Bioelectromagnetics.* **12,** 71–5.
41. Miyakoshi J. 2005. Effects of static magnetic fields at the cellular level. *Progr. Biophys. Molec. Biol.* **87,** 213–23.
42. van Deventer E, Simunic D and Repacholi M. 2006. EMF standards for human health. In *Biological and Medical Aspects of Electromagnetic Fields*, BG Frank and S. Barnes, editors. CRC Taylor & Francis Group, Boca Ratón, FL. p. 277.
43. Blank M. 1987. Ionic processes at membrane surfaces. In *Mechanistic Approaches to Interactions of Electric and Electromagnetic Fields with Living Systems*, Blank M and Findl E, editors. Springer Science & Business Media, LLC, New York. p. 1.
44. Liboff A R, Smith S D and McLeod B R. 1987. *Mechanistic Approaches to Interactions of Electric and Electromagnetic Fields with Living Systems,* eds M Blank and E Findl (Boston, MA: Springer).
45. Parkinson W C and Sulik G L. 1992. Diatom response to extremely low-frequency magnetic fields. *Radiat. Res.* **130,** 319–30.
46. Sandweiss J. 1990. On the cyclotron resonance model of ion transport. *Bioelectromagnetics.* **11,** 203–5.
47. Dobson J. 2001. Nanoscale biogenic iron oxides and neurodegenerative disease. *FEBS Lett.* **496,** 1–5.
48. Moon W-J, Kim H-J, Roh H G, Choi J W and Han S-H. 2012. Fluid-attenuated inversion recovery hypointensity of the pulvinar nucleus of patients with Alzheimer disease: Its possible association with iron accumulation as evidenced by the T2* map. *Korean J. Radiol.* **13,** 674–83.

49. Maher B A, Ahmed I A, Karloukovski V, MacLaren D A, Foulds P G, Allsop D, Mann D M, Torres-Jardón R and Calderon-Garciduenas L. 2016. Magnetite pollution nanoparticles in the human brain. *Proc. Natl. Acad. Sci.* **113,** 10797–801.
50. Kirschvink J L, Kobayashi-Kirschvink A and Woodford B J. 1992. Magnetite biomineralization in the human brain. *Proc. Natl. Acad. Sci.* **89,** 7683–7.
51. Daley George Q. 2012. The promise and perils of stem cell therapeutics. *Cell Stem Cell.* **10,** 740–9.
52. Li X, Ling W, Khan S, Wang Y P, Pennisi A, Barlogie B, Shaughnessy J and Yaccoby S. 2010. Systemically transplanted human bone marrow mesenchymal stem cells primarily traffic to mesenteric lymph nodes. *Blood.* **116,** 1068.
53. Hohnholt M C, Geppert M and Dringen R. 2011. Treatment with iron oxide nanoparticles induces ferritin synthesis but not oxidative stress in oligodendroglial cells. *Acta Biomater.* **7,** 3946–54.
54. Pinkernelle J, Calatayud P, Goya G F, Fansa H and Keilhoff G. 2012. Magnetic nanoparticles in primary neural cell cultures are mainly taken up by microglia. *BMC Neurosci.* **13,** 32.
55. Tay A, Kunze A, Jun D, Hoek E and Di Carlo D. 2016. The age of cortical neural networks affects their interactions with magnetic nanoparticles. *Small.* **12,** 3559–67.
56. Elder J B, Liu C Y and Apuzzo M L J. 2008. Neurosurgery in the realm of 10(-9), Part 1: Stardust and nanotechnology in neuroscience. *Neurosurgery.* **62,** 1–19.
57. Kumar P, Choonara Y E, Modi G, Naidoo D and Pillay V. 2014. Nanoparticulate strategies for the five R's of traumatic spinal cord injury intervention: Restriction, repair, regeneration, restoration and reorganization. *Nanomedicine.* **9,** 331–48.
58. Goya G F, Marcos-Campos I, Fernandez-Pacheco R, Saez B, Godino J, Asin L, Lambea J, Tabuenca P, Mayordomo J I, Larrad L, Ibarra M R and Tres A. 2008. Dendritic cell uptake of iron-based magnetic nanoparticles. *Cell Biol. Int.* **32,** 1001–5.
59. Kim D H, Rozhkova E A, Ulasov I V, Bader S D, Rajh T, Lesniak M S and Novosad V. 2010. Biofunctionalized magnetic-vortex microdiscs for targeted cancer-cell destruction. *Nat. Mater.* **9,** 165–71.
60. Cheng Y, Muroski M E, Petit D, Mansell R, Vemulkar T, Morshed R A, Han Y, Balyasnikova I V, Horbinski C M, Huang X L, Zhang L J, Cowburn R P and Lesniak M S. 2016. Rotating magnetic field induced oscillation of magnetic particles for in vivo mechanical destruction of malignant glioma. *J. Control. Release.* **223,** 75–84.
61. Riggio C, Calatayud M P, Hoskins C, Pinkernelle J, Sanz B, Torres T E, Ibarra M R, Wang L, Keilhoff G, Goya G F, Raffa V and Cuschieri A. 2012. Poly-l-lysine-coated magnetic nanoparticles as intracellular actuators for neural guidance. *Int. J. Nanomed.* **7,** 3155–66.
62. Pickard M R and Chari D M. 2010. Robust uptake of magnetic nanoparticles (MNPs) by central nervous system (CNS) microglia: Implications for particle uptake in mixed neural cell populations. *Int. J. Mol. Sci.* **11,** 967–81.
63. Riggio C, Calatayud M P, Giannaccini M, Sanz B, Torres T E, Fernandez-Pacheco R, Ripoli A, Ibarra M R, Dente L, Cuschieri A, Goya G F and Raffa V. 2014. The orientation of the neuronal growth process can be directed via magnetic nanoparticles under an applied magnetic field. *Nanomed.-Nanotechnol. Biol. Med.* **10,** 1549–58.
64. Dickson B J. 2002. Molecular mechanisms of axon guidance. *Science.* **298,** 1959–64.
65. Daly W, Yao L, Zeugolis D, Windebank A and Pandit A. 2012. A biomaterials approach to peripheral nerve regeneration: Bridging the peripheral nerve gap and enhancing functional recovery. *J. R. Soc. Interface.* **9,** 202–21.
66. Bray D. 1979. Mechanical tension produced by nerve-cells in tissue-culture. *J. Cell Sci.* **37,** 391–410.
67. Pilar Calatayud M, Riggio C, Raffa V, Sanz B, Torres T E, Ricardo Ibarra M, Hoskins C, Cuschieri A, Wang L, Pinkernelle J, Keilhofff G and Goya G F. 2013. Neuronal cells loaded with PEI-coated Fe3O4 nanoparticles for magnetically guided nerve regeneration. *J. Mater. Chem. B.* **1,** 3607–16.
68. Pierrat S, Brochard-Wyart F and Nassoy P. 2004. Enforced detachment of red blood cells adhering to surfaces: Statics and dynamics. *Biophys. J.* **87,** 2855–69.
69. Merkel R, Nassoy P, Leung A, Ritchie K and Evans E. 1999. Energy landscapes of receptor-ligand bonds explored with dynamic force spectroscopy. *Nature.* **397,** 50–3.
70. Suter D M and Miller K E. 2011. The emerging role of forces in axonal elongation. *Progr. Neurobiol.* **94,** 91–101.
71. Fass J N and Odde D J. 2003. Tensile force-dependent neurite elicitation via anti-beta1 integrin antibody-coated magnetic beads. *Biophys. J.* **85,** 623–36.
72. Pita-Thomas W, Steketee M B, Moysidis S N, Thakor K, Hampton B and Goldberg J L. 2015. Promoting filopodial elongation in neurons by membrane-bound magnetic nanoparticles. *Nanomedicine.* **11,** 559–67.
73. Steketee M B, Moysidis S N, Jin X L, Weinstein J E, Pita-Thomas W, Raju H B, Iqbal S and Goldberg J L. 2011. Nanoparticle-mediated signaling endosome localization regulates growth cone motility and neurite growth. *Proc. Natl. Acad. Sci. USA.* **108,** 19042–7.
74. Borisova T, Krisanova N, Borsmall u C A, Sivko R, Ostapchenko L, Babic M and Horak D. 2014. Manipulation of isolated brain nerve terminals by an external magnetic field using D-mannose-coated gamma-Fe_2O_3 nano-sized particles and assessment of their effects on glutamate transport. *Beilstein J. Nanotechnol.* **5,** 778–88.
75. Suarato G, Lee S I, Li W, Rao S, Khan T, Meng Y and Shelly M. 2016. Micellar nanocomplexes for biomagnetic delivery of intracellular proteins to dictate axon formation during neuronal development. *Biomaterials.* **112,** 176–91.
76. Roy S, Johnston A H, Newman T A, Glueckert R, Dudas J, Bitsche M, Corbacella E, Rieger G, Martini A and Schrott-Fischer A. 2010. Cell-specific targeting in the mouse inner ear using nanoparticles conjugated with a neurotrophin-derived peptide ligand: Potential tool for drug delivery. *Int. J. Pharmaceut.* **390,** 214–24.
77. Antman-Passig M and Shefi O. 2016. Remote magnetic orientation of 3D collagen hydrogels for directed neuronal regeneration. *Nano Lett.* **16,** 2567–73.
78. Pinkernelle J, Raffa V, Calatayud M P, Goya G F, Riggio C and Keilhoff G. 2015. Growth factor choice is critical for successful functionalization of nanoparticles. *Front. Neurosci.-Switz.* **9**, 305. doi: 10.3389/fnins.2015.00305
79. Zhang S and Uludag H. 2009. Nanoparticulate systems for growth factor delivery. *Pharmaceut. Res.* **26,** 1561–80.
80. Ziv-Polat O, Shahar A, Levy I, Skaat H, Neuman S, Fregnan F, Geuna S, Grothe C, Haastert-Talini K and Margel S. 2014. The role of neurotrophic factors conjugated to iron oxide nanoparticles in peripheral nerve regeneration: in vitro studies *BioMed Res. Int.* **2014,** 267808.
81. Nadri S, Mahmoudvand H and Eatemadi A. 2016. Magnetic nanogel polymer of bupivacaine for ankle block in rats. *J. Microencapsul.* **33.7,** 656–662.
82. Pilakka-Kanthikeel S, Atluri V S R, Sagar V, Saxena S K and Nair M. 2013. Targeted brain derived neurotropic factors (BDNF) delivery across the blood-brain barrier for neuro-protection using magnetic nano carriers: An in-vitro study. *PLos One.* **8.4,** e62241.

83. Suri S S, Fenniri H and Singh B. 2007. Nanotechnology-based drug delivery systems. *J. Occup. Med. Toxicol.* **2,** 1.
84. Pardridge W M. 2003. Blood-brain barrier drug targeting: The future of brain drug development. *Molec. Interv.* **3,** 90–105, 51.
85. Sharma H S and Sharma A. 2007. Nanoparticles aggravate heat stress induced cognitive deficits, blood-brain barrier disruption, edema formation and brain pathology. *Progr. Brain Res.* **162,** 245–73.
86. Sun Z, Worden M, Wroczynskyj Y, Yathindranath V, van Lierop J, Hegmann T and Miller D W. 2014. Magnetic field enhanced convective diffusion of iron oxide nanoparticles in an osmotically disrupted cell culture model of the blood–brain barrier. *Int. J. Nanomed.* **9,** 3013–26.
87. Portnoy E, Polyak B, Inbar D, Kenan G, Rai A, Wehrli S L, Roberts T P, Bishara A, Mann A, Shmuel M, Rozovsky K, Itzhak G, Ben-Hur T, Magdassi S, Ekstein D and Eyal S. 2016. Tracking inflammation in the epileptic rat brain by bi-functional fluorescent and magnetic nanoparticles. *Nanomedicine.* **12,** 1335–45.
88. Felgner P L, Gadek T R, Holm M, Roman R, Chan H W, Wenz M, Northrop J P, Ringold G M and Danielsen M. 1987. Lipofection—A Highly Efficient, Lipid-Mediated DNA-Transfection Procedure. *Proc. Natl. Acad. Sci. USA.* **84,** 7413–7.
89. Mah C, Zolotukhin I, Fraites T, Dobson J, Batich C and Byrne B. 2000. Microsphere-mediated delivery of recombinant AAV vectors in vitro and in vivo. *Mol Ther.* **1,** S239.
90. Scherer F, Anton M, Schillinger U, Henkel J, Bergemann C, Kruger A, Gansbacher B and Plank C. 2002. Magnetofection: Enhancing and targeting gene delivery by magnetic force in vitro and in vivo. *Gene Ther.* **9,** 102–9.
91. Furlani E P and Ng K C. 2008. Nanoscale magnetic biotransport with application to magnetofection. *Phys. Rev. E,* **77.6,** 061914.
92. Adams C F, Pickard M R and Chari D M. 2013. Magnetic nanoparticle mediated transfection of neural stem cell suspension cultures is enhanced by applied oscillating magnetic fields *Nanomed. Nanotechnol. Biol. Med.* **9,** 737–41.
93. Zinder N D and Lederberg J. 1952. Genetic exchange in *Salmonella. J. Bacteriol.* **64,** 679–99.
94. Zahid M and Robbins P D. 2012. Protein transduction domains: Applications for molecular medicine. *Curr. Gene Ther.* **12,** 374–80.
95. Campos C, Rocha N B F, Lattari E, Paes F, Nardi A E and Machado S. 2016. Exercise-induced neuroprotective effects on neurodegenerative diseases: the key role of trophic factors. *Exp. Rev. Neurother.* **16,** 723–34.
96. Naldini L, Blomer U, Gallay P, Ory D, Mulligan R, Gage F H, Verma I M and Trono D. 1996. In vivo gene delivery and stable transduction of nondividing cells by a lentiviral vector. *Science.* **272,** 263–7.
97. Ellis B L, Hirsch M L, Barker J C, Connelly J P, Steininger R J and Porteus M H. 2013. A survey of ex vivo/in vitro transduction efficiency of mammalian primary cells and cell lines with Nine natural adeno-associated virus (AAV1-9) and one engineered adeno-associated virus serotype. *Virol. J.* **10,** 74.
98. Lim S, Airavaara M and Harvey B K. 2010. Viral vectors for neurotrophic factor delivery: A gene therapy approach for neurodegenerative diseases of the CNS. *Pharmacol. Res.* **61,** 14–26.
99. J I, Goya G F, Pilar Calatayud M, Herenu C B, Reggiani P C and Goya R G. 2012. Magnetic field-assisted gene delivery: Achievements and therapeutic potential. *Curr. Gene Ther.* **12,** 116–26.
100. Hwang J H, Lee S, Kim E, Kim J S, Lee C H, Ahn I S and Jang J H. 2011. Heparin-coated superparamagnetic nanoparticle-mediated adeno-associated virus delivery for enhancing cellular transduction. *Int. J. Pharmaceut.* **421,** 397–404.
101. Pal A, Singh A, Nag T C, Chattopadhyay P, Mathur R and Jain S. 2013. Iron oxide nanoparticles and magnetic field exposure promote functional recovery by attenuating free radical-induced damage in rats with spinal cord transection. *Int. J. Nanomed.* **8,** 2259–72.
102. Stoll G and Bendszus M. 2009. Imaging of inflammation in the peripheral and central nervous system by magnetic resonance imaging. *Neuroscience.* **158,** 1151–60.
103. Horak D, Benes M, Prochazkova Z, Trchova M, Borysov A, Pastukhov A, Paliienko K and Borisova T. 2016. Effect of O-methyl-beta-cyclodextrin-modified magnetic nanoparticles on the uptake and extracellular level of l-glutamate in brain nerve terminals. *Colloids Surf. B Biointerf.* **149,** 64–71.
104. Khuong H T and Midha R. 2013. Advances in nerve repair. *Curr. Neurol. Neurosci. Rep.* **13,** 322.
105. Bunge M B. 2002. Bridging the transected or contused adult rat spinal cord with Schwann cell and olfactory ensheathing glia transplants. *Progr. Brain Res.* **137,** 275–82.
106. Wakao S, Hayashi T, Kitada M, Kohama M, Matsue D, Teramoto N, Ose T, Itokazu Y, Koshino K, Watabe H, Iida H, Takamoto T, Tabata Y and Dezawa M. 2010. Long-term observation of auto-cell transplantation in non-human primate reveals safety and efficiency of bone marrow stromal cell-derived Schwann cells in peripheral nerve regeneration. *Exp. Neurol.* **223,** 537–47.
107. Jenkins S I, Yiu H H, Rosseinsky M J and Chari D M. 2014. Magnetic nanoparticles for oligodendrocyte precursor cell transplantation therapies: Progress and challenges. *Molec. Cell. Ther.* **2,** 23.
108. Bulte J W, Zhang S, van Gelderen P, Herynek V, Jordan E K, Duncan I D and Frank J A. 1999. Neurotransplantation of magnetically labeled oligodendrocyte progenitors: Magnetic resonance tracking of cell migration and myelination. *Proc. Natl. Acad. Sci. USA.* **96,** 15256–61.
109. Franklin R J, Blaschuk K L, Bearchell M C, Prestoz L L, Setzu A, Brindle K M and ffrench-Constant C. 1999. Magnetic resonance imaging of transplanted oligodendrocyte precursors in the rat brain. *Neuroreport.* **10,** 3961–5.
110. Hamasaki T, Tanaka N, Kamei N, Ishida O, Yanada S, Nakanishi K, Nishida K, Oishi Y, Kawamata S, Sakai N and Ochi M. 2007. Magnetically labeled neural progenitor cells, which are localized by magnetic force, promote axon growth in organotypic cocultures. *Spine.* **32,** 2300–9.
111. Riggio C, Nocentini S, Catalayud M P, Goya G F, Cuschieri A, Raffa V and del Rio J A. 2013. Generation of magnetized olfactory ensheathing cells for regenerative studies in the central and peripheral nervous tissue. *Int. J. Molec. Sci.* **14,** 10852–68.
112. Sagar V, Atluri V S R, Pilakka-Kanthikeel S and Nair M. 2016. Magnetic nanotherapeutics for dysregulated synaptic plasticity during neuroAIDS and drug abuse. *Molec. Brain.* **9.1,** 57.
113. Wu Y-W, Goubran H, Seghatchian J and Burnouf T. 2016. Smart blood cell and microvesicle-based Trojan horse drug delivery: Merging expertise in blood transfusion and biomedical engineering in the field of nanomedicine. *Transf. Apheresis Sci.* **54,** 309–18.
114. Zhang R P, Wang L J, He S, Xie J and Li J D. 2016. Effects of magnetically guided, SPIO-labeled, and neurotrophin-3 gene-modified bone mesenchymal stem cells in a rat model of spinal cord injury. *Stem Cells Int.* **2016,** 2018474.

Dr Gerardo F. Goya (E-mail: goya@unizar.es) is associate professor at the University of Zaragoza, Spain. He has been associate professor at the University of Sao Paulo (Brazil) and he is currently researcher at the Institute of Nanoscience of Aragón (INA), University of Zaragoza. Prof. Goya's pioneering team (http://www.unizar.es/gfgoya) on magnetic hyperthermia in Spain established that induced cell death with magnetic hyperthermia without temperature rise is possible. His team has developed engineered MNPs for neural guidance under externally applied magnetic fields. He has over 130 publications on nanomagnetism and bioapplications and holds two patents. Prof. Goya has led the design, development and building of devices for measuring power absorption for magnetic hyperthermia, which established the basis for a spin-off company, nB Nanoscale Biomagnetics, of which he is cofounder and scientific advisor.

Vittoria Raffa holds an MSc in Chemical Engineering, PhD in nanotechnology and is associate professor in molecular biology at the University of Pisa. She is the team leader of the Nanomedicine Lab at the University of Pisa (Department of Biology). Prof. Raffa's research interests include nanomedicine and its applications to molecular biology and neuroscience. She was author in the last 10 years of 60 publications in ISI journals and 5 patents on technologies related to nanomedicine.

23 Radionuclide Labeling and Imaging of Magnetic Nanoparticles

*Benjamin P. Burke, Christopher Cawthorne and Stephen J. Archibald**

CONTENTS

23.1 ROLE OF IMAGING IN THE DEVELOPMENT OF MAGNETIC NANOPARTICLES FOR DIAGNOSTIC AND THERAPEUTIC APPLICATIONS

23.1.1 BACKGROUND

Magnetic nanoparticles (MNPs) have a large surface area which can be easily functionalized, allowing the introduction of multiple agents onto the same vector, e.g. active targeting groups, therapeutic drugs and imaging labels (Figure 23.1). The particle size, morphology and surface chemistry can be readily modified to influence their behaviour *in vivo*, offering great potential as both diagnostic and therapeutic agents. However, levels of NP accumulation in tumours still remain low despite many years of research and this remains a barrier to efficacy and clinical translation.[1]

The fate of MNPs when systemically administered is governed by their absorption, distribution, metabolism and excretion (ADME). A range of factors have a significant impact on these pharmacokinetic parameters and thus the ultimate biodistribution of the NPs, including size, shape, coating, charge and interaction with plasma proteins. An understanding of how these parameters interact is central to the rational design of agents for both tumour targeting and toxicity. The major clearance route for MNPs is the mononuclear phagocytic system (MPS),[2] consisting of blood-borne monocytes as well as the various tissue-resident macrophage populations that derive from them.[3] NP interactions with macrophages are mediated in the same way as extracellular pathogens, i.e. facilitated by opsonization and recognition,[4] and the rate of degradation of MNPs after phagocytosis is largely unknown.

Hydrodynamic size, in combination with morphology, coating and charge, dictates which clearance pathway will be favoured. Broadly speaking, particles of <15 nm are cleared renally, 15–100 nm are taken up by hepatocytes, >100 nm are taken up by liver-resident Kuppfer cells, and >200 nm are captured in the red pulp of the spleen; with neutral particles circulating longer than negative and positive ones (reviewed extensively by Arami et al.).[2] In the absence of aggregation, clearance via liver and spleen resident macrophages therefore dominates for most intravenously injected NPs,[5] although, at saturating doses, accumulation by alveolar and adipose tissue-resident cells also becomes important.[6]

While liver accumulation is desirable for the visualization of liver metastases by MRI,[7] a central area of research in the field is the functionalization of NPs to avoid recognition and

* Corresponding author.

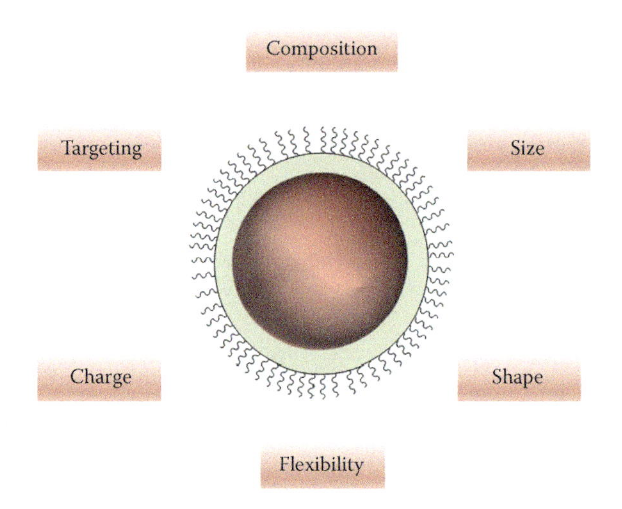

FIGURE 23.1 Characteristics of MNPs that affect their biodistribution *in vivo*.

engulfment by Kuppfer cells to improve tumour targeting.[8–10] This is commonly achieved by modification of coating (which also aids aqueous stability), with PEGylation the most widely reported method. Polyethylene glycol (PEG) shields surface charge and increases hydrophilicity, thus reducing opsonization and interaction with the MPS, reviewed by Karakoti et al.,[11] leading to the development of 'stealth' NPs which can avoid detection. However, in practice, this mainly results in an increase in the blood pool circulation time and partial changes in clearance; there is only limited evidence of a significant increase in target tissue uptake using these methods.[7]

Targeting of MNPs to malignant tissue is either passive or active. Passive targeting is thought to exploit the inherent 'leaky' vasculature and compromised lymphatic drainage of many cancer types which causes NP accumulation, known as the enhanced permeability and retention (EPR) effect,[12] although this is highly variable amongst patients, which can cause issues on clinical translation.[13] Active targeting requires the modification of the NP surface to promote interactions with molecular elements of the tumour or tumour microenvironment. A range of targeting moieties have been used to target receptors expressed on the tumour surface, including monoclonal antibodies and derived fragments, aptamers, lectins, cell penetrating peptides and small molecules such as folic acid.[14] Targeting the tumour vasculature, e.g. via integrin binding using cyclic Arginylglycylaspartic acid (RGD) motifs, has the advantage that the NPs do not need to extravasate for integrin binding. Targeting via application of an external magnetic field is also possible for iron oxide and other MNPs.[15] There has also been some interest in using MNPs for MRI cardiovascular imaging.[16,17]

NP drug delivery systems (DDS), which aim to selectively deliver chemotherapeutics to the tumour and thus reduce off-target side effects, also require functionalization to avoid capture by the MPS and promote long circulation times.[18–20,21] MNPs offer the potential for magnetically triggered drug release and magnetic fluid hyperthermia (see Chapter 23), although these systems must also optimize biocompatibility and stability/selective drug release. The use of MNPs for drug delivery has been recently reviewed by Huang et al.[22]

23.1.2 Multimodal Imaging of NPs

The assessment of biodistribution, clearance, tumour accumulation and extent of drug release are key for the successful translation of MNPs.[2] MNPs are ideal constructs for delivering contrast for detection with multiple modalities due to the large number of attachment sites that are present on the surface in comparison to a small molecule, so that multiple functional components can be attached to the surface.[23] No single molecular imaging modality has the capability to offer all of the required data to fully characterize the properties of an administered agent.[24] Optical techniques have issues with tissue penetration *in vivo*, MRI has high resolution but low sensitivity, while nuclear imaging techniques offer much improved sensitivity but relatively poor resolution. While less useful for whole organism imaging, fluorescent groups are excellent tools for *in vitro* and *ex vivo* validation in cellular or tissue samples and, in some cases, preclinical imaging in small animals, particularly with near-infrared (NIR) emitting probes.[25,26] This type of multimodal validation imaging accelerates progression along the translational pathway through to clinical trials and also provides key design/development data at the preclinical stage. This can increase the speed of development and, clinically, can offer a secondary mode of analysis in excised tissue, or be used to guide surgery.[27] However, optically active moieties can have an effect on the properties they report on and, hence, may require significant re-optimization if an established nanoconstruct is to be surface modified; they can also add to the complexity of nanoconstructs, which is a barrier to translation.

MNPs have inherent contrast properties for MRI; however, the low sensitivity of this technique requires high-dose administration. The nuclear imaging modalities positron emission tomography (PET) and single-photon emission computed tomography (SPECT) rely on the incorporation of a short-lived radioisotope (emitting positrons for PET and gamma rays for SPECT) into a molecular species, and allow for dynamic quantification of such radiolabeled agents noninvasively due to their high sensitivity (10^{-11}–10^{-12} M). A wide range of isotopes are available, with different decay properties, chemical compatibilities and half-lives (common radioisotopes are summarized in Table 23.1).

TABLE 23.1
Common Isotopes Used in PET and SPECT Imaging

Nuclide	Modality	Half-Life (min)	Production Method	Imaging Decay (%)	E_{max} (MeV)
^{18}F	PET	110	Cyclotron	97% (β^+)	0.63
^{11}C	PET	20	Cyclotron	99% (β^+)	0.96
^{68}Ga	PET	68	Generator	89% (β^+)	1.89
^{64}Cu	PET	762	Cyclotron	19% (β^+)	0.65
^{89}Zr	PET	4710	Cyclotron	23% (β^+)	0.90
^{99m}Tc	SPECT	360	Generator	99% (γ)	0.14
^{111}In	SPECT	4032	Cyclotron	94% (γ)	0.171, 0.245

The radiolabeling of MNPs for PET and SPECT can allow them to be assessed at tracer doses in 'phase 0' clinical trials, and to demonstrate that MRI is appropriate for routine clinical monitoring at higher doses. Functionalizing NPs for molecular imaging with PET and SPECT allows their tumour targeting and drug delivery properties to be determined rapidly, noninvasively and with minimal animal use.[28] Preclinical studies that make use of radiolabeled NPs are summarized below.

23.1.3 Nuclear Imaging in Preclinical Studies of NPs

Several preclinical studies have used PET and SPECT to quantify how NP surface modification alters their biodistribution, clearance and targeting. Radiometals are an appropriate choice for labeling of NPs as, in some cases, they can coordinate stably with the surface under mild conditions, which avoids the late stage covalent attachment necessary for using nonmetal radioisotopes. Chen et al. have used PET imaging with copper-64 to develop an optimized gold NP system which is rapidly and predominantly cleared by the renal system (elimination half-life of ca. 6 min).[29] The effects of different coatings on the biodistribution of gold NPs was examined by Frellsen et al. using PET, comparing a dodecane/tween mixture, a methoxy PEG and sulfonate/quaternary ammonium mixture, see Figure 23.2,[30] demonstrating a significant increase in blood circulation time for the PEG-coated NPs. Perez-Campana et al. developed a method for radiolabeling commercial Al_2O_3 NPs (see Section 23.2.2.3) and are able to generally correlate size with clearance pathway.[31,32] Similar approaches can be carried out using MNPs.

PET and SPECT are especially suited to the quantitation of NP pharmacokinetics and targeting in tumour models.[33–35] MRI contrast has also been used to investigate the MNPs

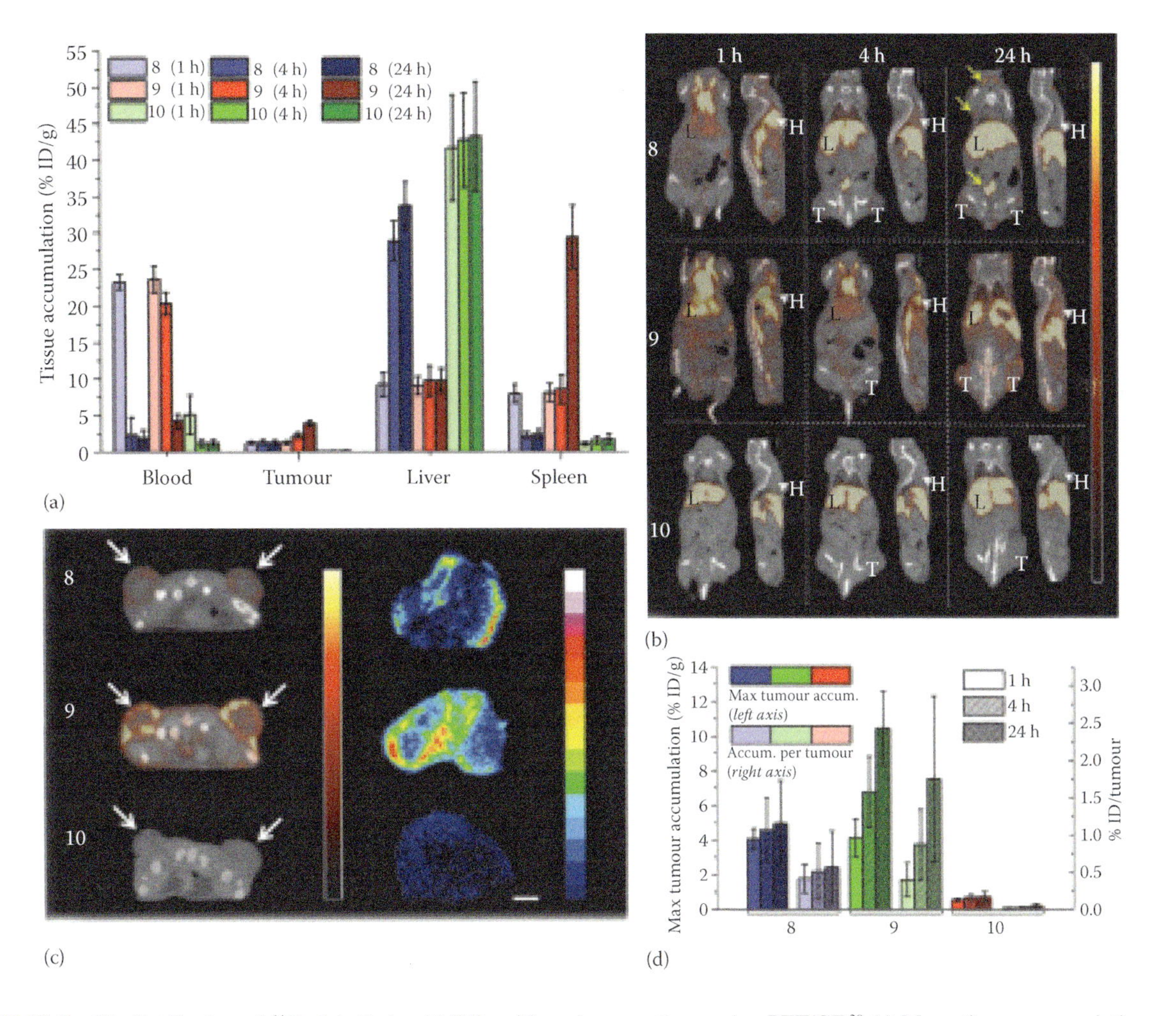

FIGURE 23.2 Biodistribution of ^{64}Cu labelled gold NPs with various coatings using PET/CT.[30] (a) Mean tissue accumulation given as %ID/g ± SD. (b) Representative PET/CT images. Organs are indicated as (H) heart and the large intracardiac blood volume, (L) liver, and (T) tumours. Colour bar indicates activity level of copper-64. (c) Left column: Representative axial PET/CT images at 24 h after intravenous injection. Tumours are indicated by white arrows (scale bar 0–16%ID/g). Right column: Autoradiography in 8 μm tumour cryosections. (d) Maximum tumour accumulation given as %ID/g ± SD. (Reprinted with permission from Frellsen, A. F., A. E. Hansen, R. I. Jølck, P. J. Kempen, G. W. Severin, P. H. Rasmussen, A. Kjær, A. T. I. Jensen and T. L. Andresen. Mouse positron emission tomography study of the biodistribution of gold nanoparticles with different surface coatings using embedded copper-64. *ACS Nano*, 10, 9887–9898. Copyright 2016 American Chemical Society.)

localization properties with quantification of uptake achieved by relaxivity mapping,[22,36,37] but this is laborious and less suited to dynamic imaging.[38] For example, Cai et al. labelled quantum dots (QDs) with copper-64 to quantify tumour localization after targeting using RGD peptides, showing uptake of 4 and 1% injected dose (ID)/g for targeted and nontargeted NPs respectively.[39] Lee et al. conducted a similar study using iron-oxide NPs, again targeting $\alpha_v\beta_3$ with RGD peptides and radiolabeling with ^{64}Cu. The study also uses a blocking experiment (which is a standard approach in the development of targeted PET imaging agents) to saturate available receptors and prevent specific NP binding, demonstrating tumour uptake can be reduced to the same level of nontargeted derivatives, from 10 to 4% ID/g respectively.[40] Similarly, Zhang et al. used ^{125}I SPECT/CT to assess tumour specificity and determine the optimal time point for applying photothermal therapy using RGD coated gold nanorods.[41]

An additional application of PET for the assessment of NP drug delivery is afforded by the ability to radiolabel many drugs,[42] and thus obtain direct information on tissue pharmacokinetics; for example carbon-11 labelled Irinotecan[43] (a topoisomerase I inhibitor used in the treatment of colon cancer) would facilitate the study of liposomal delivery of coformulations with cisplatin.[44] Detection of radiolabeled reporters incorporated into NPs has recently been shown to predict therapy response in preclinical models,[45] allowing stratification of patients for NP therapy.[46] As mentioned above, the saturability of the MPS means that there is the potential for significant differences in biodistribution, which are dependent on the concentration of NPs administered, and thus low specific activity assays should be carried out to determine this.

There are many potential non-oncology applications of MNPs where the introduction of a nuclear imaging component could have a significant impact on the optimization of design; including stroke,[47] gene therapy,[48] respiration,[49,50] and tissue engineering.[51] In a paradigmatic example, Ruiz-de-Angulo et al. engineered a vaccine displaying iron oxide NP to enhance antigen-specific immunity (see Figure 23.3), using ^{68}Ga (PET) and ^{67}Ga (SPECT) imaging to provide key information in the selection of optimal size for the specific targeting of dendritic cells in the lymph nodes.[52]

23.2 IMAGING METHODS TO LABEL NPS AND TRACK THEM *IN VIVO*

23.2.1 General Considerations for Radiolabeling and Imaging Studies

When developing a radiolabeled NP system, key design criteria include the following:

Robust and stable radiochemistry (essential) – Formation of a stable construct to allow *in vivo* biodistribution to be assessed by imaging.

Original properties of the NP are unperturbed (essential) – Care must be taken to characterize the system to ensure that any modifications or radiolabeling methodologies have not affected the inherent NP properties (magnetization, targeting etc.)

Radiolabeling achieved using the simplest method possible (preferable) – If a previously developed (or clinically approved) system that is well characterized can be radiolabeled in a simple fashion without significant redesign or modification then this will be the optimal approach. More laborious methodologies are only justified if there is significant benefit from the additional complexity.

PET preferred to SPECT – The use of PET imaging allows for higher sensitivity dynamic data, which can give valuable information about NP biodistribution.

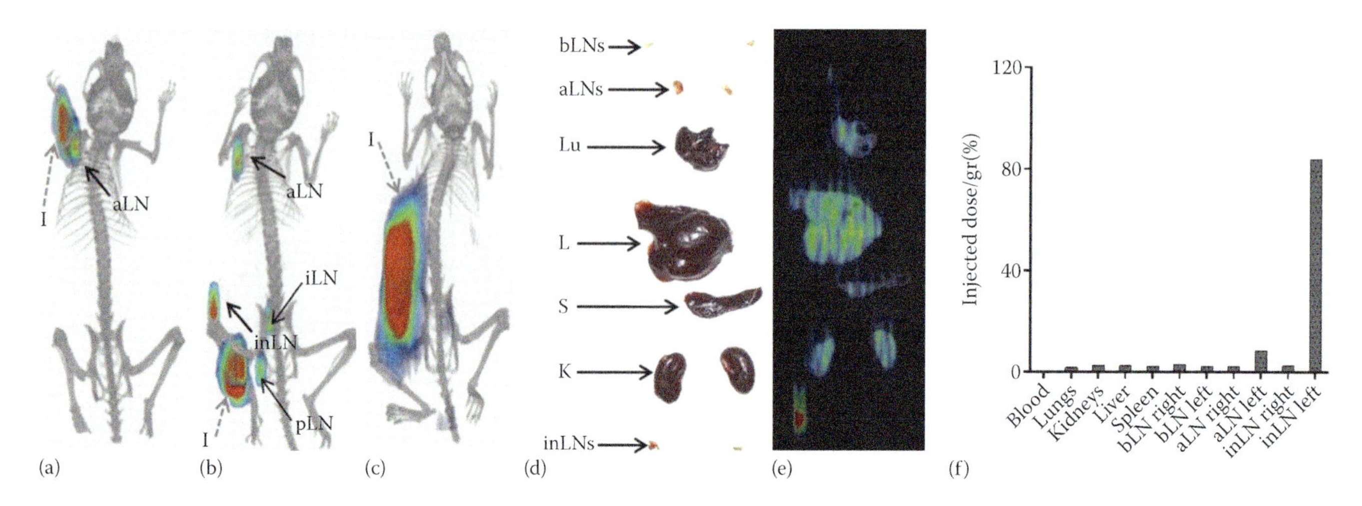

FIGURE 23.3 Radionuclide imaging to track a vaccine displaying iron oxide NP that enhances antigen-specific immunity.[52] SPECT/CT images of (a) forearm, (b) hind hock and (c) flank. Ex vivo analysis of the biodistribution after 24 h: (d) photograph of selected harvested organs, (e) SPECT/CT image of the harvested organs and (f) biodistribution expressed as percent injected dose per gram of tissue. (Reprinted with permission from Ruiz-de-Angulo, A., A. Zabaleta, V. Gomez-Vallejo, J. Llop and J. C. Mareque-Rivas. Microdosed lipid-coated Ga-67-magnetite enhances antigen-specific immunity by image tracked delivery of antigen and CpG to lymph nodes. *ACS Nano*, 10, 1602–1618. Copyright 2016 American Chemical Society.)

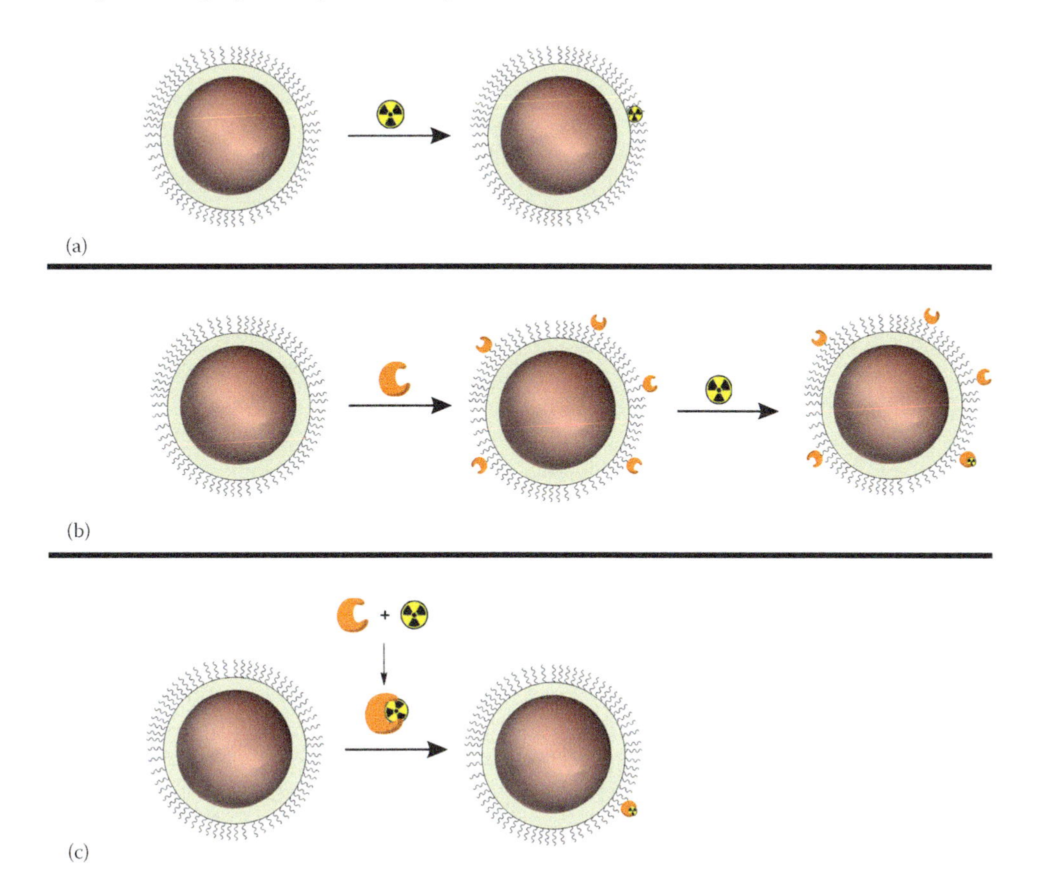

FIGURE 23.4 Common methods for the formation of radiolabeled MNPs with radiometals. (a) Chelator free, (b) final-step chelator and (c) two-step chelator.

23.2.2 Formation of a Radiolabeled MNP

There are three common methods for the formation of radiolabeled NPs (see Figure 23.4), they will be discussed in order of increasing complexity.

23.2.2.1 Chelator Free Radiolabeling Methods

The simplest way to radiolabel a NP is via the direct interaction of the radioisotope with the surface of a preformed nanoconstruct. The term 'chelator free' generally refers to the use of radioactive metals that form stable interactions directly with the surface or core of NPs. This methodology allows the direct labeling of constructs without significant alteration of surface properties, making it attractive for use when the properties of the NP have already been optimized for e.g. targeting, magnetism or drug delivery. For example, Voulgari et al. developed a PMAA-graft-PEG copolymer-derived iron oxide NP system for the delivery of cisplatin to tumours which showed improved *in vivo* characteristics (enhanced anticancer efficacy/reduced toxicity) in HT-29 tumour-bearing mice compared to free cisplatin, particularly in the presence of an external magnetic field.[53] Final step chelator-free gallium-68 radiolabeling and dynamic PET imaging gave an indication that the increased potency in the presence of a magnetic field was likely to be caused by an increase in internalization of the cisplatin-loaded carrier rather than a direct increase in NP (and therefore drug) concentration in the tumour.

If it is necessary to modify the surface to allow stable radiolabeling, then the parent construct must be reassessed to ensure that modification has not altered characteristics *in vivo*. Chelator free methodologies can also be used to directly understand clinically approved NPs. Ferumoxytol, an FDA approved NP, has been successfully radiolabeled with a range of radiometals by a heat-induced method to allow highly sensitive evaluation,[54,55] although this method is only feasible when nanoconstructs are stable with respect to heat. Loudos and co-workers have used chelator-free methods for imaging of ^{99m}Tc labelled MNPs using a planar gamma camera,[56–58] comparing the biodistribution of cobalt ferrite and magnetite NPs[56–58] and quantifying the increase in tumour uptake with actively targeted NPs.[57] Cheng et al. radiolabeled $FeSe_2$ doped Bi_2Se_3 nanosheets with ^{64}Cu to form a stable system,[59] PET imaging was used to offer dynamic information on tumoural accumulation over 24 h.

Once a nanoconstruct has been identified and a radioisotope selected with appropriate half-life and emission properties, robust radiolabeling methodology must be developed. The goal is to form a stable construct at the lowest concentration of NPs using the mildest conditions, which significantly decreases the chance of affecting the NP properties; practically this is achieved by optimizing a small number of variables: concentration of NPs; type, concentration and pH of reaction buffer; temperature and reaction time. The initial aim is to find conditions where some degree of radiolabeling can be achieved. It must be considered that with variation in conditions (high concentration/low temperature and low concentration/high temperature) different labelled analogues may be formed which will vary in their stability (*vide infra*). The analysis carried out to determine radiochemical incorporation is almost exclusively radio-TLC, which offers rapid feedback to the radiochemist on the effectiveness of the procedure.[58–62] In general, NPs will not move from the application point on silica gel TLC plate (i.e. baseline), whereas free metal ions are able to move upwards under appropriate elution conditions. Analysis of the radioactivity across the plate with a TLC plate reader allows quantification of incorporation of the radioisotope. This methodology can also be used to determine reaction kinetics as time points can be sampled without stopping the reaction and multiple TLC plates can be eluted in parallel (as opposed to HPLC which is sequential, limiting the amount of data that can be collected). TLC plates can be read rapidly (ca. 30 s) on a plate reader and results analyzed in the accompanying software package.

Typically, radiochemistry experiments will consist of multiple reactions (with variable conditions or repeats) investigated in parallel, with samples taken from each reaction every few minutes for analysis by TLC. This creates a significant amount of data from a single elution, with manpower (rather than equipment access) the rate-limiting factor.

In some cases, quantitative radiolabeling may not be possible and so purification of the stably labelled construct from unbound radioisotope is required. Gel filtration can be used for separation, with PD-10 columns frequently used for MNPs.[57,63] This method also purifies NPs from any other reagents used in the radiolabeling procedure, such as buffer salts. The biggest limitation of this approach is the dilution, which occurs during the procedure, with the purified NP now being present in around 1–1.5 mL of solution. Depending on the amount of radioactivity used, this can have an impact on dosing in preclinical studies (see Section 23.2.3.2). Spin filtration techniques use a molecular weight cut-off diaphragm and centrifugal force to allow radioactive ions to pass through the filter, while retaining the NPs. Hence, NPs can be washed multiple times to ensure purity without an increase in volume. The NPs can then be recovered from the filter in a low volume (typically 0.2–0.3 mL). However, there are some potential limitations to this approach: spin filtration is slower to carry out than size exclusion purification methods and is therefore incompatible with shorter half-life isotopes, the NPs can stick to the membrane filter thus reducing recovery, and prolonged high-speed centrifugation may affect the construct stability.

Careful assessment of the radiochemical stability of the construct is one of the most important and often overlooked (or unreported) steps in the development of a radiolabeled derivative. Inadequate assessment of stability could result in the collection of *in vivo* data that are not representative of the construct properties as the biodistribution will follow the released radionuclide in addition to any remaining intact construct. The standard *in vitro* stability test is incubation in serum to look for trans-chelation to proteins found in blood, showing that release of the radionuclide would occur after intravenous administration.[57–61] Another regularly used procedure to determine labeling stability is based on competition with a known chelator for the metal radionuclide.[58,63] This allows more extreme conditions to be investigated than would be encountered *in vivo*, which is useful for a robust comparison of the different methods and techniques for labeling. Stability measurements may show that some of the radionuclide is released relatively rapidly after which the remaining labelled construct is stable, even if further competing chelator is introduced. This can indicate that multiple coordination environments are available for attachment of the radionuclide to the NP. Hence it is important that during radiolabeling method development, multiple conditions are investigated in radiochemical stability assays, even if a 100% radiochemical yield can be achieved under mild conditions. If the stability assays show that both weakly and strongly bound radionuclide are present then a chelator washing step should be included in the protocol as a purification method to remove the radiometal from the less stable coordination site.[63] This will then require separation of the stable labelled NPs from the transchelated radioisotope. Other stability assays may also be performed that are specific to the construct being developed, e.g. magnetic field stability for hyperthermia experiments or magnetic targeting.[53,64,65]

The main advantage of chelator free methodology is the simplicity of the procedure, which allows incorporation of a radionuclide into a preformed nanoconstruct with well-developed and optimized properties. There can be issues if the required conditions are harsh, hence it is imperative that the radiolabeled construct (often after decay) is analyzed appropriately to ensure that there has been no modification of characteristics; such as aggregation, loss/modification of coating or disruption of targeting capability.

Chen et al. developed an iron oxide NP radiolabeled with a mixture of arsenic isotopes ($^{71/72/74/76}As$) which is an unusual and effective approach, although these are not routinely available isotopes.[60] Arsenic is known to bind to the magnetite surface and this method could be used for imaging a range of preformed NPs with low-density coatings, although the current lack of serum stability would require modification of the radiosynthesis process. A range of methodologies have been developed for radiolabeling other metallic/metal oxide nanostructures via chelator-free methodologies, which are likely to be directly translatable to MNPs.[29,66–69]

Determining the mechanism by which chelator-free radiolabeling of MNPs proceeds is challenging, as standard analytical techniques cannot detect metal ions at sufficiently low concentrations to emulate radiolabeling with a nonradioactive isotope. Using higher concentrations of metal ions for detection by spectroscopy, mass spectrometry, elemental analysis, etc. may result in different interactions to the picomolar amount of radiometal used for radiolabeling.[61]

Chelator-free radiolabeling processes may rely on direct interactions with the metal oxide surface, with cavities or pores in a polymer coating (e.g. silica) or via coordination interactions with donor atoms on the modified surface (e.g. PEG or PLA). These interactions are challenging to predict as stability is achieved through a combination of weaker interactions and the polymeric structure or supramolecular arrangement of the surface coating components. Despite this, chelator-free radiolabeling with ^{68}Ga is an attractive first step due to the availability of this isotope and the lack of significant effect on surface chemistry discussed above.

23.2.2.2 Chelator Conjugate Based Radiolabeling Methods

In some cases, initial investigation of a simpler chelator free method will show that a chelator is required for stable and effective radiolabeling of the construct. Burke et al. assessed the inclusion of a conjugated chelator in the ^{68}Ga radiolabeling of silica coated magnetite NPs and, in this case, determined that the presence or absence of a chelator made no difference to the radiolabeling reaction conditions or stability characteristics (both *in vitro* and *in vivo*), see Figure 23.5.[61,62]

However, in some systems, chelator-free methodologies are unsuccessful and a chelator-based approach needs to be developed. The chelator should be selected to match with the radiometal being used to ensure rapid binding kinetics to form a stable coordination complex.[70] A range of common chelators have been used to form radiolabeled MNPs, see Figure 23.6.

23.2.2.2.1 Final-Step Radiolabeling

The standard method for chelator-mediated nanoconstruct radiolabeling is to form the construct with the chelators displayed on the surface coating prior to the addition of the radiometal. Chelators are attached either during synthesis or coating, or by reaction with the formed NPs to give covalent bonds with reactive groups on the surface coating. The resulting system can then be assessed to investigate whether the presence of the chelator influences the properties (magnetization, targeting, etc.). Chelator selection should be based on matching the properties of the bifunctional chelators that are available to the coordination chemistry characteristics of the metal radionuclide.[70,71] The key desired property is rapid binding kinetics under mild conditions (to minimize chances of disruption of coating) to form a stable complex.

The majority of reported examples of final step chelator-based radiolabeling use ^{64}Cu as the radioisotope and DOTA as the chelator.[72,73] This is not an obvious choice in terms of metal complex stability and is more likely to have been based on the availability of the DOTA (bifunctional) chelator.[70] Lee et al. used standard peptide coupling reaction conditions to attach DOTA to the surface of a polyaspartic acid coated iron oxide NP system functionalized with RGD groups and, after radiolabeling with ^{64}Cu, used PET imaging in a mouse model to show specific/blockable uptake in $\alpha_v\beta_3$ expressing tumours.[40]

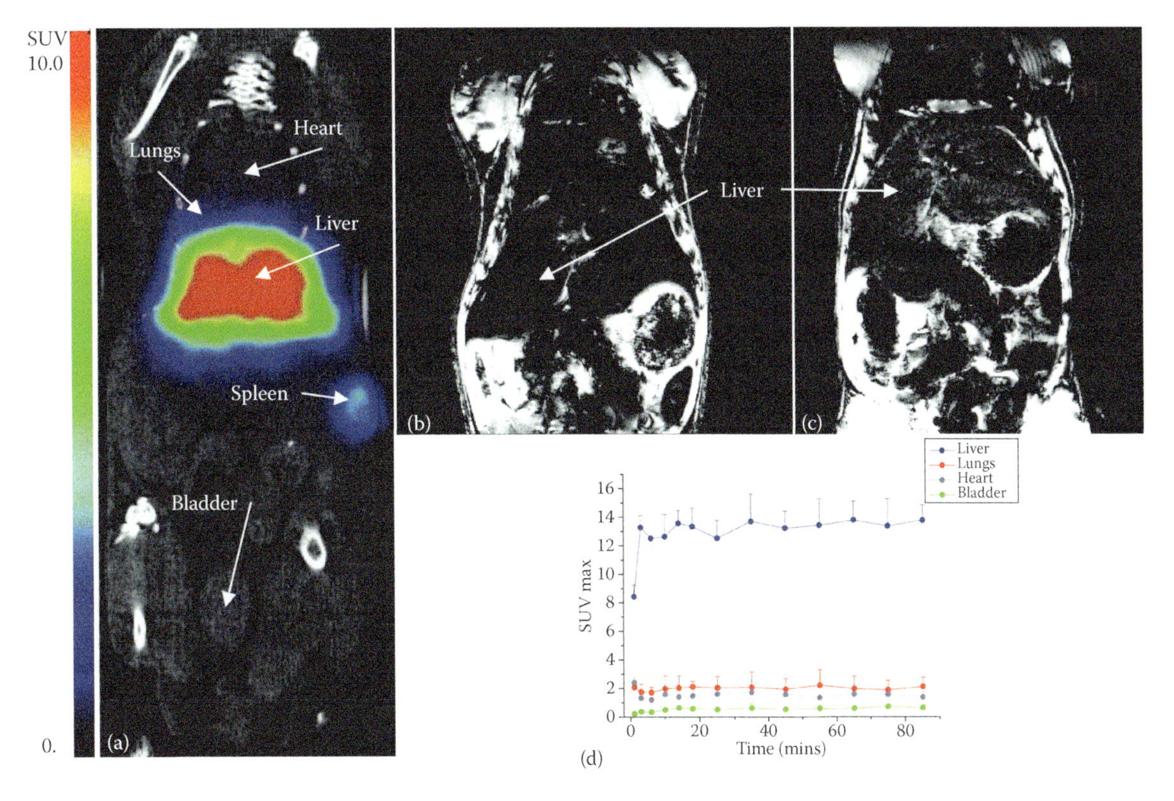

FIGURE 23.5 Biodistribution PET/CT and MRI studies of ^{68}Ga labelled iron oxide NPs.[61] In vivo mouse images of (a) fused PET-CT coronal slice image showing main organ uptake, (b) T2-weighted MR image, (c) control mouse without nanoparticle administration, T2-weighted MR image and (d) time–activity curve for major organs from PET image. (From Burke, B. P., N. Baghdadi, A. E. Kownacka, S. Nigam, G. S. Clemente, M. M. Al-Yassiry, J. Domarkas, M. Lorch, M. Pickles, P. Gibbs, R. Tripier, C. Cawthorne and S. J. Archibald. Chelator free gallium-68 radiolabelling of silica coated iron oxide nanorods via surface interactions. *Nanoscale*, 7, 14889–14896, 2015. Reproduced by permission of The Royal Society of Chemistry.)

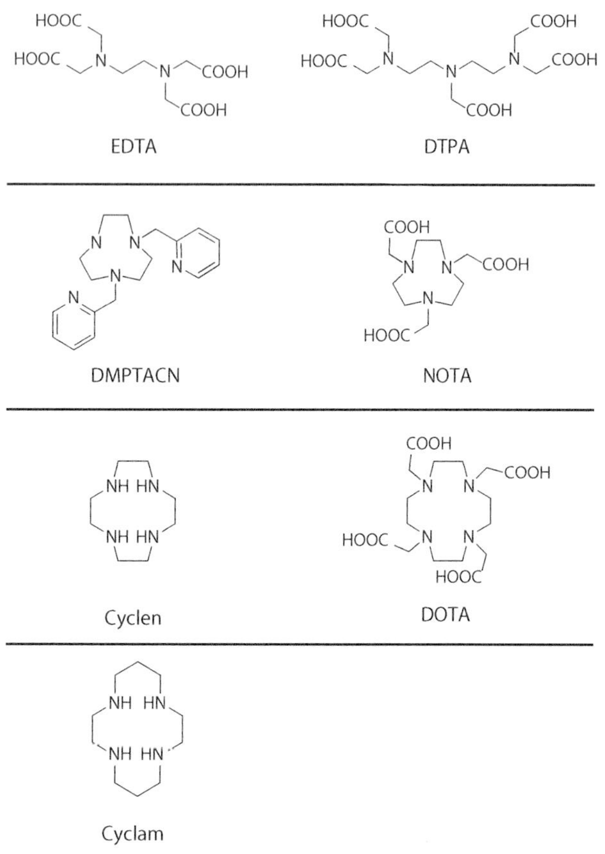

EDTA DTPA

DMPTACN NOTA

Cyclen DOTA

Cyclam

FIGURE 23.6 Chelators used for the formation of radiolabeled MNPs.

Xie et al. designed a system for PET/MRI/optical imaging which uses final step radiolabeling of iron oxide NPs functionalized with DOTA and a cy5.5 dye that are modified for active tumour targeting using human serum albumin.[74] Unexpectedly, *ex vivo* organ analysis showed little correlation between near infrared fluorescence (NIRF) imaging and the PET image. This was attributed to the high background of NIRF imaging; however, the article does not report any stability measurements and so an alternative explanation is possible. Either weakly bound isotope on the particle surface or the characterized inadequacy of DOTA as an *in vivo* chelator for copper(II)[75] may explain the high liver and kidney uptake shown for the PET signal, whereas the covalently linked dye would remain attached to the NP.

Radiochemical stabilities analogous to that of the parent chelator may be expected for nanoconstructs, however, this must be assessed. An assumption is made that all of the radiometal is bound to the chelator and none is nonspecifically bound to other sites on the NP. The processes that were described for stability analysis of chelator free radiolabeled compounds, see Section 23.2.2.1, remain valid. Care must be taken to account for the potential presence of any chelator that has been associated with the nanoconstruct but is not covalently bound, as this can complex to the radiometal but will not be stably attached to the nanoconstruct. Purification can be carried out using similar methods to chelator-free methodologies. In particular, incubation with chelating agents to remove nonchelator based interactions is common and recommended.[63]

Glaus et al. radiolabeled DOTA conjugated iron oxide NPs with ^{64}Cu, with radiochemical yields above 95% after only a 1 h reaction time.[76] Ethylenediaminetetraacetic acid (EDTA) and diethylenetriaminepentaacetic acid (DTPA) were used to remove weakly attached nonchelate bound ^{64}Cu, after which, a system which was stable in serum was produced. PET imaging determined the circulation half-life to be 143 min, which offered important design information on how surface modifications effect *in vivo* properties. Barreto et al. synthesized silica coated iron-oxide NPs with a range of macrocyclic chelators (cyclam, cyclen and DMPTACN) attached to the surface for ^{64}Cu radiolabeling.[77] Radiolabeling could be carried out very rapidly (5 min) and a stable system could be formed after introduction of a washing protocol with cyclam to remove weakly bound ^{64}Cu; however, these chelators are typically not stable enough for *in vivo* use. The *in vitro* stability assays showed only small differences between the chelators and, given our work on ^{68}Ga chelator-free radiolabeling of silica coated iron oxide NPs,[61] it is possible that stable binding is occurring to the silica surface. This requires further investigation.

23.2.2.2.2 *Two-Step Radiolabeling*

An alternative method for forming chelator-based radiolabeled NPs is the initial radiolabeling of a bifunctional chelator (BFC, chelating agent with a reactive group) followed by conjugation of the radiometal complex to the NP. The advantage of this method is that the radiometal is stably bound to the chelate before attachment to the NP, hence other surface coordination interactions are not possible. Chelate/metal stability can be assessed prior to conjugation and so only the stability of the chelator attachment would need to be assessed for the formed conjugate with the NP. The most common method of two-step radiolabeling is via attachment of the BFC complex with the radiometal to reactive groups on the surface coating, often via amide bond formation. Louie and coworkers developed one of the first radiolabeled MNP systems using this methodology.[78] Copper-64 radiolabeling of an amine functionalized DOTA derivative was performed before conjugation to a dextran sulfate NP. In a seminal study for the development of radiolabeled MNPs, the authors attempted chelator-free methodologies as well as final step radiolabeling of DOTA functionalized NPs. Only after both of these approaches were unsuccessful, was a two-step radiolabeling system developed, to form a stable construct – although radiochemical yields are significantly affected by using the more labour-intensive multistep approach. This methodology has also been successfully carried out using a NOTA based chelator for ^{68}Ga radiolabeling.[79]

An alternative approach is to use a BFC which can bind to the NP core. Torres and coworkers have developed a bisphosphonate bifunctional system which, by modification of the chelating unit, can bind to either ^{99m}Tc for SPECT imaging or ^{64}Cu for PET imaging.[80,81] These units can then coordinate directly to the metal oxide core surface via strong bisphosphonate interactions with the iron atoms. This methodology can have significant advantages as, similar to chelator-free methods, it can be used with preformed fully characterized NPs, avoiding the development of a specific system for covalent chelator attachment, and has the versatility to be used with a range of coating types (dependent on surface coating density). As with all of these methodologies, analysis should be carried out on systems exposed to radiolabeling conditions to ensure that the desired properties are unperturbed.

23.2.2.3 Other Radiolabeling Methodologies

Alternative approaches have also been used to radiolabel MNPs and although their versatility can be somewhat limited, there are situations where they can offer useful data. Radioactive ions can be incorporated directly into the core of the NP during synthesis, forming constructs which are often highly stable with respect to leaching as the ion is incorporated into the solid state structure. An obvious choice when trying to form radiolabeled iron-oxide NPs is to use a radioactive form of iron. ^{59}Fe is a gamma-emitting radioisotope, which would allow like-for-like incorporation, however the long half-life (45 days) and poor availability limit its use.[82,83] Other metallic radionuclides, which are more widely available and have more favourable characteristics for medical imaging including ^{64}Cu,[84] ^{111}In[85] and ^{68}Ga,[86] have been added during NP formation. In an interesting study, Wang et al. introduced multiple radioactive isotopes in the same construct at different positions to track the *in vivo* behaviour and metabolism of the NP.[87] ^{59}Fe was used to label the core, ^{111}In chelated to DTPA was added to the lipid (DMPE) coating along with ^{14}C-oleic acid as a stabilizer. *In vivo* biodistribution studies were used to see how the different components were processed. As expected,[83] the oleic acid used as a surfactant does not remain attached *in vivo* after only 10 min, whereas ^{59}Fe and ^{111}In largely correlate, assuming initially that the DTPA-DMPE remains attached to the NP. However, significant differences were noticed in kidney and robust control experiments (administration of ^{111}In-DTPA-DMPE alone) showed a similar biodistribution potentially caused by micelle formation, suggesting instability of chelator attachment. This type of multi-isotope study, where different components can be separately tracked, is complex to set up but can give useful and accurate information about the processing of NPs *in vivo*.

It is also worth noting that there is the potential to form radiolabeled NPs by producing the radionuclide *in situ*. The NPs can be bombarded with protons or neutrons in the target of particle accelerators, causing nuclear reactions in the NP core. Perez-Campana et al. irradiated ^{18}O enriched Al_2O_3 NPs with a proton beam to form ^{18}F labelled particles which are completely stable to radiochemical leaching.[32] They subsequently formed ^{13}N labelled NPs using a similar approach, see Figure 23.7.[31] These approaches have not been studied in depth but have significant potential for understanding the *in vivo* behaviour of preformed NPs, so long as there is no

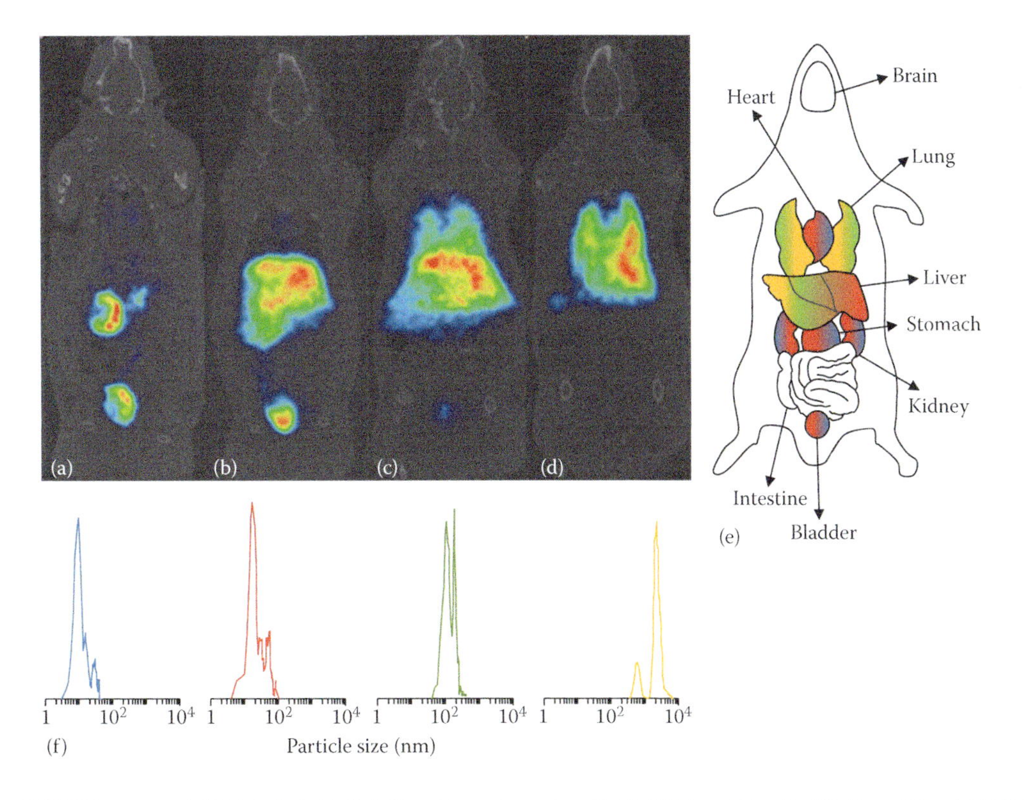

FIGURE 23.7 PET/CT images of ^{13}N-labeled Al_2O_3 NPs with various sizes.[31] (a) NS10nm NPs, (b) NS40nm NPs, (c) NS150nm NPs and (d) NS10μm NPs. (e) Organ accumulation as a function of particle size, according to colour codes depicted in (f). (f) Particle size distribution as determined by TEM (NS10nm, NS40nm, NS150nm) or DLS (NS10μm) from left to right. (Reprinted with permission from Perez-Campana, C., V. Gomez-Vallejo, M. Puigivila, A. Martin, T. Calvo-Fernandez, S. E. Moya, R. F. Ziolo, T. Reese and J. Llop. Biodistribution of different sized nanoparticles assessed by positron emission tomography: A general strategy for direct activation of metal oxide particles. *ACS Nano*, 7, 3498–3505. Copyright 2013 American Chemical Society.)

significant effect of the bombardment on the coating or other properties of the NPs.

23.2.3 *In Vivo* Imaging Evaluation Studies

After successful radiolabeling has been achieved and *in vitro/ex vivo* evaluation carried out as far as possible (*vide infra*), NPs will be assessed *in vivo*; the translation of NP agents to the clinic depends at least partly on their evaluation in robust animal models.[88] In preclinical oncology (as in preclinical imaging in general) rodent models dominate, accounting for >75% of the total number of animal procedures, although there is an increasing use of companion animals in trials.[89] The imaging of radiolabeled NPs in rodents requires careful consideration of a wide range of factors if robust and reproducible data are to be obtained. First, the choice of animal model will ultimately depend on the nature of the required assessment, e.g. biodistribution and pharmacokinetics vs. tumour or stroma targeting, assessment of tumour targeting vs. assessment of drug delivery, assessment of drug delivery vs. assessment of therapeutic effect. Pharmacokinetic parameters can be assessed in naïve animals, although historical controls should not be employed to interpret novel data due to unavoidable genetic drift in commercial stocks.[90] The majority of *in vivo* assessment is carried out on NPs administered directly to the venous circulation via injection into the tail vein; although other routes include direct intratumoural injection or delivery via the lungs,[91] how the latter is achieved has implications for the relevance of the data.[92]

Regardless of the model or experimental purpose, animal research cannot take place without explicit consideration of the '3Rs' of replacement, reduction and refinement as defined by Russell and Birch,[93] and this is reflected in national and international guidance.[94,95] It is currently difficult to replace the data obtained from *in vivo* preclinical imaging with simpler systems, although this is an active area of research (as is the development of sophisticated *in vitro* systems to assess NP interactions with tumour cells).[96] To comply with the criteria for reduction, the minimum number of animals must be used to obtain a statistically valid result. By enabling longitudinal study, *in vivo* imaging increases data quality (each animal has the potential to act as its own control)[97] and has the potential to reduce the number of animals necessary for the determination of pharmacokinetic parameters,[98] although there are a number of factors that must be taken into account to achieve this (*vide infra*). Statistical software is readily available for the determination of animal numbers,[99] with several publications indicating group sizes for typical imaging experiments.[98,100] Key to reducing animal numbers is to reduce as far as possible sources of biological variability; this includes providing environmental enrichment and avoiding single-housing of animals[101] as well as (for example) considering standard housing conditions on murine physiology;[102] these are also key requirements for the refinement of animal procedures. These considerations of course also extend to all aspects of the scanning procedure as detailed below.

23.2.3.1 Selection of Animal Models

For the assessment of pharmacokinetic parameters, naïve animals can be employed as outlined above; genetically engineered (but nontumour-bearing) mice can be used to assess the interaction of immune system components for MNP surface chemistries.[103] For assessment of tumour targeting and drug delivery, oncology models must be employed. These are commonly hosted in mice due to the lower cost and availability of immunodeficient strains (*vide infra*), however syngeneic rat models are also used as they offer higher resolution for radionuclide imaging. The small physical size of rodent models compared to man, especially mice, has a number of implications for preclinical imaging,[104] not least of which is a higher requirement for high specific activity of the radiolabeled construct when an imaging application is envisaged for low-density receptor targets.[105] The choice of model will depend on the chosen application, with the overall proviso that the model should mimic as far as possible the clinical situation. Mouse models can be classified in several ways: spontaneous (syngeneic vs. genetically engineered mouse models [GEMM]) vs. transplanted (ectoptic vs. orthotopic vs. patient-derived xenograft [PDX]); immunocompetent vs. immunodeficient; these are summarized in Table 23.2.[106,107]

A key parameter in the assessment of NP targeting is the presence of the enhanced permeability and retention effect

TABLE 23.2
Classification of Mouse Models Used in Animal Research

Description	Tumour Type	Tumour Species	Host	Tumour Location	Refs
Xenograft	Cells derived from patients, cells derived from patient blood (CDx), tumour pieces (PDx)	Human	Immunodeficient mice: Nu/Nu, SCID/Bg, NOD/SCID, NOD/SCID/IL2Rγ_{null} etc.	Ectoptic or orthotopic, subrenal capsule for PDx	108–119
Syngeneic/Allograft	Spontaneously arising or chemically induced tumour cells derived from genetically related host	Murine	Immunocompetent	Ectoptic or orthotopic	45,120–125
Genetically engineered	Tumour arises in animal as the result of programmed gene expression, conditional or inducible	Murine	Immunocompetent	Ectoptic or orthotopic	103,120,126

(EPR) as outlined in Section 23.1.1. The EPR occurs as a result of abnormal tumour vasculature, with high permeability as a result of overexpressed vascular factors such as nitric oxide, prostaglandins and bradykinin as well as poor lymphatic drainage. The magnitude of this effect depends on a number of factors including NP size, charge, circulation time and biocompatibility, all of which are NP-specific,[12] as well as the degree of EPR present in a particular preclinical model.[1,45,127] The most common model by far is the subcutaneous/ectopic implant of established human tumour cell lines into immunodeficient mice, which is considered to result in a tumour exhibiting high levels of EPR that is unrepresentative of the situation in man (which itself is highly variable).[13] Ectopic models do not capture key elements of the tumour/stroma interactions found in sporadic cancer that can contribute to therapy resistance, and the stromal targeting of NPs cannot be evaluated in these models. Xenograft models using patient-derived material have been shown to recapitulate the clinical situation with respect to histology and therapy response,[128] although they require highly immune-deficient animal hosts and generally take much longer to grow. Orthotopic models, where tumour material is implanted into the anatomically correct organ of the host, can be derived from cells, organoids or patient tumour material; these may better reflect the clinical situation[129] although they are of greater technical difficulty (involving laparotomy/craniotomy for example) and require some form of noninvasive imaging to monitor tumour burden over time.

Although such xenograft models allow the use of human tissue, a key disadvantage is the absence of adaptive immunity in the host, which is not reflective of the clinical situation. Syngeneic models include a functional host immune system, and commonly consist of murine cells that have spontaneously arisen or been induced, and are implanted into a genetically related host in an ectopic or orthotopic site. The 4T1 model (murine breast cancer cells implanted into the mammary fat pad of Balb/C mice) is often reported for NP evaluation, with fluorescent and luminescent cell variants reported for imaging of disease burden.[130] Syngeneic models also include mice that have been genetically engineered to bear cell lineages that mimic the mutational sequence undergone by particular cancer types (genetically engineered mouse models, GEMMs). There is some evidence that these models better reflect the clinical situation.[131]

Other microenvironmental variables, such as induction of cytokine expression, level of tumour associated phagocytic cells and vascular state have been shown to affect the targeting of liposomal NPs in preclinical models.[132] Although orthotopic implant of cells or the spontaneous generation of tumours in GEMMS can recapitulate the microenvironment, this does not in itself reflect the biology of metastasis. Metastatic models are either based on systemic administration of cells or inoculation in a primary site with subsequent spread via lymphatic or hematogenous spread;[133] metastatic models are reflective of metastatic behaviour clinically which is of key importance for therapy.[134] Metastatic tumours are also smaller clinically, and have a less pronounced EPR effect,[135] which is of particular importance for NP targeting. A list of primary site inoculations and resulting metastases is given in Figure 23.8 and Table 23.3.

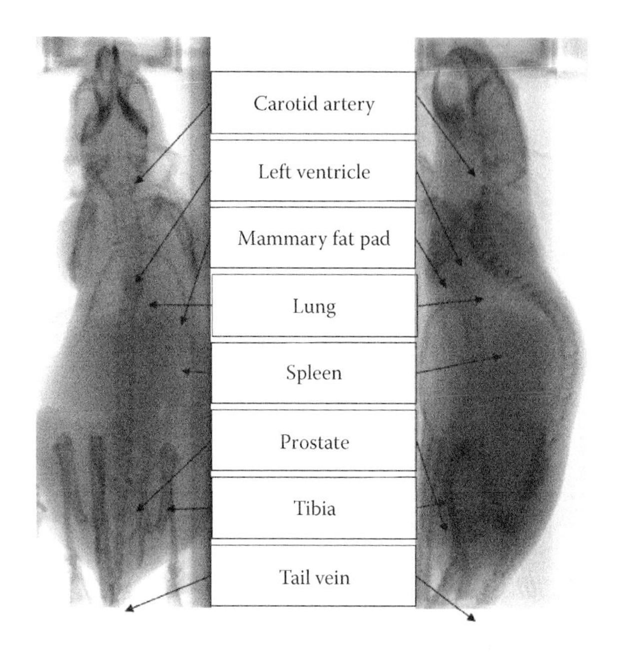

FIGURE 23.8 Common sites of cell implantation for the generation of metastatic cancer models

TABLE 23.3
Site of Implant vs. Metastatic Site for Common Metastatic Cancer Models

Implant Site	Tumour Type	Resulting Metastasis
Carotid artery	Brain, cells	Brain
Left ventricle	Various, cells	Bone, brain
Mammary fat pad	Breast, cells	Liver, lung, bone, lymph nodes
Lung	Lung, cells	Lymph nodes
Spleen	Various, cells or pieces	Liver
Prostate	Prostate, cells	Lymph nodes
Tibia	Various	Lung
Tail vein	Various	Lung

Orthotopic metastasis models, where tumour cells or pieces are implanted into a site from which cells can metastasize to the anatomically correct location, have been shown to generate tumours with different responses to therapy compared to ectopic implant and primary growth in the same host, thus representing improved models for validating imaging agents for therapy response.[136]

23.2.3.2 Scanning Protocol

There is a wide range of preclinical (rodent) scanners available for both PET and SPECT, with varying characteristics.[137–139] Preclinical scanning facilities need to implement a series of quality control procedures to ensure accurate quantitation[139,140] and, importantly, to characterize individual system responses and the effects of the image reconstruction protocol implemented.[141–143] This is especially important if multiple animals are to be imaged simultaneously, and a necessity if dose–response relationships are to be established.[144] Also important is the calibration of dose calibrators to allow accurate doses to be calculated, as well as the cross-calibration of imaging equipment, dose calibrators and gamma counters to allow comparison between (for example) imaging data and blood or tissue samples. The direct comparison of imaging quantitation and organ activity as measured after dissection requires consideration of partial volume effects, among others.[145,146]

Alongside scanner characterization and quality control, animal handling and anaesthesia regimes need to be optimized for the scanning protocol. Induction of anaesthesia should be rapid and (as far as possible) stress-free. Rodents lose the ability to thermoregulate on anaesthesia, and the high surface area relative to weight of a mouse means that the induction unit and imaging cell need to be tightly temperature-controlled. Commonly, respiration and temperature are monitored throughout the scan, with ECG if required for gating, with several proprietary systems currently available. Commonly anaesthesia is gaseous isoflurane on an oxygen carrier, although this is known to affect several physiological systems. Dynamic procedures may take upwards of 90 min, and the respiratory depression, hypercapnia and acidosis subsequent to isoflurane occurring at 10–50 min warrant careful control of anaesthetic depth.

In addition, NPs should be provided with high volumetric activity, such that the optimal dose for imaging can be administered in a volume consistent with ethical guidelines and provide high-count data (this will depend on the noise equivalent count rate of the scanner). On scaling to therapeutic doses, biodistribution may change as a result of 'saturation' of the MPS (*vide supra*). Preparations must also be free of toxic contaminants and aggregation, and be sufficiently stable in a buffered solute to prevent degradation between synthesis and injection. Successful catheterization should be assessed either by including the tail in the field of view, or confirming low tail uptake postscan.[147]

Postscan, animals should be recovered in a warmed environment, without unnecessarily exposing staff to radiation. If NPs are of a size consistent with renal excretion or the radiolabel is cleaved, special attention should be paid to handling during recovery due to the relatively large amounts of activity that may be present in the urine if micturition is induced.

Commonly, data are quantified using injected dose per gram (ID/g) or SUV; however, these measures can be affected by a number of variables especially if imaging is carried out subsequent to treatment; inclusion of tumour sizes and full experimental details is necessary for benchmarking.[148,149] True quantitation requires the application of modeling, which is technically challenging in small animals.[150] *Ex vivo* tissue analyses are necessary to obtain information about the localization of NPs at the cellular and subcellular level[7,151]; with measurement of accumulation in human tumours ultimately required for full validation.[152]

23.2.4 Translation to Human Clinical Trials

23.2.4.1 Clinically Approved NPs

The majority of the current clinically approved NP agents are administered for one of three purposes; cancer therapeutics, imaging/contrast agents or as iron replacement therapies.[153,154] Passively targeted organic NP-based drug delivery systems have shown potential for selective cancer therapy with a range of agents clinically approved.[155] Therapeutics are generally liposomal formulations of known anticancer drugs ranging from Doxil (doxorubicin), approved in 1995, to Onivyde (irinotecan), approved in 2015. In terms of contrast agents, both of the iron oxide MRI contrast agents that were approved ca. 20 years ago (Endorem and Resovist) have now been discontinued, although new applications in lymph node metastases are developing for related constructs such as the dextran coated iron oxide NPs Combidex.[153] There is also a number of microsphere and microbubble agents for ultrasound imaging; however, there is no requirement for them to be magnetic. The widest application of iron oxide NPs is in iron replacement therapies; however, this has opened the door for materials such as Ferumoxytol which was originally approved for iron deficiency but also has cancer (and other) MRI contrast imaging applications.

There are many new nanomaterials currently in clinical trials, with the main focus on liposomal formulations for combination cancer drug delivery; the ^{188}Re-BMEDA liposome is an example of the incorporation of a therapeutic isotope.[156,157] Other types of particles, such as inorganic materials, are also in trials and include silica/gold NPs for thermal ablation of lung tumours (Aurolase); radiation-enhancing hafnium oxide NPs; multimodal particles incorporating NIR dyes, ^{124}I reporters and RGD targeting peptides for brain tumour imaging; and iron oxide NPs for prostate tumour thermal ablation (Magnablate).

A new advance taken into clinical trials (July 2016) is gadolinium chelate NPs (AGuIX) formed around a silica shell created from siloxane compounds and DO3A chelates which was designed and synthesized by Tillement and coworkers.[158] The small, ca. 5 nm diameter, particles were originally investigated as an MRI contrast agent but were later repurposed for radiosensitisation applications, with the phase 1 clinical trial focusing on the treatment of multiple brain metastases.

23.2.4.2 Issues for Clinical Translation

There have been significant concerns from regulatory and public bodies regarding the use of NPs due to toxicity issues and environmental impact.[159] From a clinical perspective, data are required to mitigate these factors and alleviate the concerns for each potential application which will, in turn, facilitate clinical translation.[160] The low mass modification required for nuclear medicine techniques generally results in negligible change to the labelled particle's surface properties, which is attractive from the regulatory perspective. There are three main reasons to attach radionuclides to nanomaterials in clinical translation studies:

1. The use of NPs for imaging at the tracer level. MRI requires a high mass amount of material for contrast but nuclear imaging techniques can offer tracking with subpharmacological amounts of material that allows exploitation of the properties of NPs with reduced safety concerns (as the amount will be below the toxicity threshold).
2. *In vivo* tracking of NPs to facilitate clinical trials through patient selection, treatment times and validation of biodistribution/pharmacokinetics (PK modeling at the phase 0 stage may be one of the key factors that will increase progression of translational research to clinical application).
3. The use of NPs as constructs for drug delivery or as therapeutic radioisotope delivery platforms allows the dosimetry/dosing level to be determined. This is achieved through tracking the amount of radiolabeled particle reaching the target (PET/SPECT) or tracking a radiolabeled drug to show the release profile allows calculation of the dose that should be administered to obtain the desired response.

As described above, a number of recent studies with PET/CT in preclinical models are now helping to define particle characteristics that will quantify suitable dimensions and properties to optimize the exploitation of this effect.[44,161] Biodistribution profiles are important to develop this understanding and radionuclide labeling offers a route to access this information.

However, a general conclusion from recent progress in this area of medicine is that there are very few approvals by the regulatory bodies of nanoconstructs and NPs in comparison to the level of research activity.[1,162] One of the key challenges is in the ADME factors which will impact on the delivery and localization of the drug to target. Reliance on the EPR effect can be problematic as it is dependent on factors that have a high degree of variability; interpretation of animal data can also be problematic due to the inherent variability of preclinical models and the lack of standardized reporting criteria as mentioned in Sections 23.2.3.1 and 23.2.3.2.[1,13,44,148,163,164]

A useful method to mitigate the issues in translation of research listed above is to more rapidly proceed to phase 0 trials with few patients and fast fail at a relatively low cost.[165,166] This approach may be combined with further large animal studies (e.g. pigs or companion animals) to move away from the late stage reliance on rodent models.[89] The development of improved methods to determine nanoconstruct stability and processing in biological systems is also extremely important. There is a frequent disconnect between the breadth of methods at an *in vitro* level and those that can be combined with *in vivo* studies.

There are also issues in scaled manufacture of some of the complex/multicomponent systems. Key concerns include the presence of solvent residues, the stability of formulation and batch-to-batch variation.[162] Careful investigation of the subtle influences of parameters such as temperature, pH and ionic strength in a small scale synthesis can be very informative in the development of the large-scale process. The other issue is the availability of dedicated GMP facilities to perform the syntheses prior to clinical trials, necessitating considerable investment at this stage.

Overall *in vivo* monitoring of nanoconstructs using radionuclide labeling has been identified as a key facilitating technique for clinical trials.[28,167] An immediate readout will be given of the distribution and excretion patterns of the nanomaterial. This pharmacokinetic data will allow selection of appropriately engineered particles that show high uptake in the desired target tissue (tumour) and appropriate excretion properties.

23.3 NEW APPLICATIONS

The key advances in applications are likely to be driven by novel combinations (imaging modalities, therapy and imaging or multiple therapeutic actions) and novel targeting approaches (which could include alternate administration protocols). Radionuclide imaging is an excellent validation tool for these applications and potentially for clinical use in patient selection and stratification.

It is anticipated that the approach of validating NP biodistribution using nuclear imaging will become widespread and potentially a regulatory requirement in the future. Holland and coworkers have developed a chelator-free methodology for radiolabeling Fermoxytol with a range of metal radionuclides; PET imaging of tumour uptake using this agent may offer significant advantages in patient stratification and therapy prediction.[53] Fermoxytol has been used clinically to successfully predict therapeutic response in solid tumours and thus there is significant potential for a radiolabeled derivative to have imaging applications beyond validation.[168,169] This is appealing as a rapid transition to clinical use is possible.

Magnetic hyperthermia is an application that is growing in importance with an increasing number of studies, which are ideal for combination with an imaging component.[170] This could be used to monitor optimization of targeting and levels of tissue accumulation, which will map onto requirements for sufficient localized temperature increase to result in apoptosis and immune response.

There is also growing interest in alternate systemic administration methods to the standard IV injection, such as inhalation, which can be validated using radionuclide labeling methods.[49,91]

23.4 CONCLUSIONS AND FUTURE PERSPECTIVE

The area of nanotechnology in medicine has been one of significant global research activity in the past 10–15 years and has attracted many investigators into the field. Issues stymying the development to clinical applications include a lack of clarity in the required characteristics of nanoconstructs to make them fit for purpose; and identified standard methods for *in vivo* assessment and validation. The regulatory requirements for approval have also not been well defined in the past.

This is not unexpected in a young field, but the rapid expansion has led to a lot of research that is of some chemical and physical interest but ultimately has little or no clinical relevance. New approaches and methods with clearer validation are required, and radionuclide imaging is central to their success.

The potential for clinical applications in drug delivery, therapy enhancement and imaging is compelling and once a clearer pathway for validation and translation is established it is likely there will be an increased influx of nanomedicines into clinical use. Wilhelm et al. offered an assessment of the current state-of-the-art for NP delivery to tumours with a metrics analysis of the literature that exemplifies this lack of progress.[1] They highlight the key requirement for improved control of transport inside the body. Improved methodology for analysis and tracking should be adopted across the field to increase the likelihood of translational success and rapidly consign the unsuitable approaches to a 'fast fail'.[27]

REFERENCES

1. Wilhelm, S., A. J. Tavares, Q. Dai, S. Ohta, J. Audet, H. F. Dvorak and W. C. W. Chan. 2016. Analysis of nanoparticle delivery to tumours. *Nat Rev Mater,* 1.
2. Arami, H., A. Khandhar, D. Liggitt and K. M. Krishnan. 2015. In vivo delivery, pharmacokinetics, biodistribution and toxicity of iron oxide nanoparticles. *Chem Soc Rev,* 44, 8576–8607.
3. Murray, P. J. and T. A. Wynn. 2011. Protective and pathogenic functions of macrophage subsets. *Nat Rev Immunol,* 11, 723–737.
4. Beckmann, N., C. Cannet, A. L. Babin, F. X. Ble, S. Zurbruegg, R. Kneuer and V. Dousset. 2009. In vivo visualization of macrophage infiltration and activity in inflammation using magnetic resonance imaging. *Wiley Interdiscip Rev Nanomed Nanobiotechnol,* 1, 272–298.
5. Lee, M. J., O. Veiseh, N. Bhattarai, C. Sun, S. J. Hansen, S. Ditzler, S. Knoblaugh, D. Lee, R. Ellenbogen, M. Zhang and J. M. Olson. 2010. Rapid pharmacokinetic and biodistribution studies using cholorotoxin-conjugated iron oxide nanoparticles: A novel non-radioactive method. *Plos One,* 5, e9536.
6. Levy, M., N. Luciani, D. Alloyeau, D. Elgrabli, V. Deveaux, C. Pechoux, S. Chat, G. Wang, N. Vats, F. Gendron, C. Factor, S. Lotersztajn, A. Luciani, C. Wilhelm and F. Gazeau. 2011. Long term in vivo biotransformation of iron oxide nanoparticles. *Biomaterials,* 32, 3988–3999.
7. Cole, A. J., A. E. David, J. Wang, C. J. Galban and V. C. Yang. 2011. Magnetic brain tumor targeting and biodistribution of long-circulating PEG-modified, cross-linked starch-coated iron oxide nanoparticles. *Biomaterials,* 32, 6291–6301.
8. Xie, J., C. Xu, N. Kohler, Y. Hou and S. Sun. 2007. Controlled PEGylation of monodisperse Fe3O4 nanoparticles for reduced non-specific uptake by macrophage cells. *Adv Mater,* 19, 3163–3166.
9. Kulkarni, S. A. and S. S. Feng. 2013. Effects of particle size and surface modification on cellular uptake and biodistribution of polymeric nanoparticles for drug delivery. *Pharm Res,* 30, 2512–2522.
10. Cole, A. J., A. E. David, J. X. Wang, C. J. Galban, H. L. Hill and V. C. Yang. 2011. Polyethylene glycol modified, cross-linked starch-coated iron oxide nanoparticles for enhanced magnetic tumor targeting. *Biomaterials,* 32, 2183–2193.
11. Karakoti, A. S., S. Das, S. Thevuthasan and S. Seal. 2011. PEGylated inorganic nanoparticles. *Angew Chem Int Ed,* 50, 1980–1994.
12. Maeda, H., H. Nakamura and J. Fang. 2013. The EPR effect for macromolecular drug delivery to solid tumors: Improvement of tumor uptake, lowering of systemic toxicity, and distinct tumor imaging in vivo. *Adv Drug Deliv Rev,* 65, 71–79.
13. Prabhakar, U., H. Maeda, R. K. Jain, E. M. Sevick-Muraca, W. Zamboni, O. C. Farokhzad, S. T. Barry, A. Gabizon, P. Grodzinski and D. C. Blakey. 2013. Challenges and key considerations of the enhanced permeability and retention effect for nanomedicine drug delivery in oncology. *Cancer Res,* 73, 2412–2417.
14. Bakhtiary, Z., A. A. Saei, M. J. Hajipour, M. Raoufi, O. Vermesh and M. Mahmoudi. 2016. Targeted superparamagnetic iron oxide nanoparticles for early detection of cancer: Possibilities and challenges. *Nanomedicine,* 12, 287–307.
15. Muthana, M., A. J. Kennerley, R. Hughes, E. Fagnano, J. Richardson, M. Paul, C. Murdoch, F. Wright, C. Payne, M. F. Lythgoe, N. Farrow, J. Dobson, J. Conner, J. M. Wild and C. Lewis. 2015. Directing cell therapy to anatomic target sites in vivo with magnetic resonance targeting. *Nature Communications,* 6, 8009.
16. Bjornerud, A. and L. Johansson. 2004. The utility of superparamagnetic contrast agents in MRI: Theoretical consideration and applications in the cardiovascular system. *NMR Biomed,* 17, 465–477.
17. Prince, M. R., H. L. Zhang, S. G. Chabra, P. Jacobs and Y. Wang. 2003. A pilot investigation of new superparamagnetic iron oxide (ferumoxytol) as a contrast agent for cardiovascular MRI. *J Xray Sci Technol,* 11, 231–240.
18. Jain, T. K., M. A. Morales, S. K. Sahoo, D. L. Leslie-Pelecky and V. Labhasetwar. 2005. Iron oxide nanoparticles for sustained delivery of anticancer agents. *Mol Pharm,* 2, 194–205.
19. Dobson, J. 2006. Magnetic nanoparticles for drug delivery. *Drug Develop Res,* 67, 55–60.
20. Yang, J., C. H. Lee, H. J. Ko, J. S. Suh, H. G. Yoon, K. Lee, Y. M. Huh and S. Haam. 2007. Multifunctional magneto-polymeric nanohybrids for targeted detection and synergistic therapeutic effects on breast cancer. *Angew Chem Int Ed,* 46, 8836–8839.
21. Liang, C., L. Xu, G. Song and Z. Liu. 2016. Emerging nanomedicine approaches fighting tumor metastasis: Animal models, metastasis-targeted drug delivery, phototherapy, and immunotherapy. *Chem Soc Rev,* 45, 6250–6269.
22. Huang, J., Y. Li, A. Orza, Q. Lu, P. Guo, L. Wang, L. Yang and H. Mao. 2016. Magnetic nanoparticle facilitated drug delivery for cancer therapy with targeted and image-guided approaches. *Adv Funct Mater,* 26, 3818–3836.
23. Cai, W. B. and X. Y. Chen. 2008. Multimodality molecular imaging of tumor angiogenesis. *J Nucl Med,* 49, 113S–128S.
24. Kim, J., Y. Piao and T. Hyeon. 2009. Multifunctional nanostructured materials for multimodal imaging, and simultaneous imaging and therapy. *Chem Soc Rev,* 38, 372–390.
25. Yen, S. K., D. Janczewski, J. L. Lakshmi, S. Bin Dolmanan, S. Tripathy, V. H. B. Ho, V. Vijayaragavan, A. Hariharan, P. Padmanabhan, K. K. Bhakoo, T. Sudhaharan, S. Ahmed, Y. Zhang and S. T. Selvan. 2013. Design and synthesis of polymer-functionalized NIR fluorescent dyes-magnetic nanoparticles for bioimaging. *ACS Nano,* 7, 6796–6805.
26. John, R., R. Rezaeipoor, S. G. Adie, E. J. Chaney, A. L. Oldenburg, M. Marjanovic, J. P. Haldar, B. P. Sutton and S. A. Boppart. 2010. In vivo magnetomotive optical molecular imaging using targeted magnetic nanoprobes. *Proc Natl Acad Sci USA,* 107, 8085–8090.
27. Choi, H. S. and J. V. Frangioni. 2010. Nanoparticles for biomedical imaging: Fundamentals of clinical translation. *Molec Imaging,* 9, 291–310.

28. Choi, H., Y. S. Lee, D. W. Hwang and D. S. Lee. 2016. Translational radionanomedicine: A clinical perspective. *Eur J Nanomed,* 8, 71–84.
29. Chen, F., S. Goel, R. Hernandez, S. A. Graves, S. X. Shi, R. J. Nickles and W. B. Cai. 2016. Dynamic positron emission tomography imaging of renal clearable gold nanoparticles. *Small,* 12, 2775–2782.
30. Frellsen, A. F., A. E. Hansen, R. I. Jølck, P. J. Kempen, G. W. Severin, P. H. Rasmussen, A. Kjær, A. T. I. Jensen and T. L. Andresen. 2016. Mouse positron emission tomography study of the biodistribution of gold nanoparticles with different surface coatings using embedded copper-64. *ACS Nano,* 10, 9887–9898.
31. Perez-Campana, C., V. Gomez-Vallejo, M. Puigivila, A. Martin, T. Calvo-Fernandez, S. E. Moya, R. F. Ziolo, T. Reese and J. Llop. 2013. Biodistribution of different sized nanoparticles assessed by positron emission tomography: A general strategy for direct activation of metal oxide particles. *ACS Nano,* 7, 3498–3505.
32. Perez-Campana, C., V. Gomez-Vallejo, A. Martin, E. San Sebastian, S. E. Moya, T. Reese, R. F. Ziolo and J. Llop. 2012. Tracing nanoparticles in vivo: A new general synthesis of positron emitting metal oxide nanoparticles by proton beam activation. *Analyst,* 137, 4902–4906.
33. van der Geest, T., P. Laverman, J. M. Metselaar, G. Storm and O. C. Boerman. 2016. Radionuclide imaging of liposomal drug delivery. *Expert Opin Drug Deliv,* 13, 1231–1242.
34. Kunjachan, S., J. Ehling, G. Storm, F. Kiessling and T. Lammers. 2015. Noninvasive imaging of nanomedicines and nanotheranostics: Principles, progress, and prospects. *Chem Rev,* 115, 10907–10937.
35. Chakravarty, R., H. Hong and W. B. Cai. 2014. Positron emission tomography image-guided drug delivery: Current status and future perspectives. *Mol Pharm,* 11, 3777–3797.
36. Huang, J., X. D. Zhong, L. Y. Wang, L. L. Yang and H. Mao. 2012. Improving the magnetic resonance imaging contrast and detection methods with engineered magnetic nanoparticles. *Theranostics,* 2, 86–102.
37. Mouli, S. K., P. Tyler, J. L. McDevitt, A. C. Eifler, Y. Guo, J. Nicolai, R. J. Lewandowski, W. G. Li, D. Procissi, R. K. Ryu, Y. A. Wang, R. Salem, A. C. Larson and R. A. Omary. 2013. Image-guided local delivery strategies enhance therapeutic nanoparticle uptake in solid tumors. *ACS Nano,* 7, 7724–7733.
38. Loudos, G., G. C. Kagadis and D. Psimadas. 2011. Current status and future perspectives of in vivo small animal imaging using radiolabeled nanoparticles. *Eur J Radiol,* 78, 287–295.
39. Cai, W. B., K. Chen, Z. B. Li, S. S. Gambhir and X. Y. Chen. 2007. Dual-function probe for PET and near-infrared fluorescence imaging of tumor vasculature. *J Nucl Med,* 48, 1862–1870.
40. Lee, H. Y., Z. Li, K. Chen, A. R. Hsu, C. J. Xu, J. Xie, S. H. Sun and X. Y. Chen. 2008. PET/MRI dual-modality tumor imaging using arginine-glycine-aspartic (RGD) – Conjugated radiolabeled iron oxide nanoparticles. *J Nucl Med,* 49, 1371–1379.
41. Zhang, L., H. Su, J. Cai, D. Cheng, Y. Ma, J. Zhang, C. Zhou, S. Liu, H. Shi, Y. Zhang and C. Zhang. 2016. A multifunctional platform for tumor angiogenesis-targeted chemo-thermal therapy using polydopamine-coated gold nanorods. *ACS Nano,* 10, 10404–10417.
42. Jones, T. and P. Price. 2012. Development and experimental medicine applications of PET in oncology: A historical perspective. *Lancet Oncol,* 13, e116–125.
43. Kawamura, K., H. Hashimoto, M. Ogawa, J. Yui, H. Wakizaka, T. Yamasaki, A. Hatori, L. Xie, K. Kumata, M. Fujinaga and M. R. Zhang. 2013. Synthesis, metabolite analysis, and in vivo evaluation of C-11 irinotecan as a novel positron emission tomography (PET) probe. *Nucl Med Biol,* 40, 651–657.
44. Tardi, P. G., N. Dos Santos, T. O. Harasym, S. A. Johnstone, N. Zisman, A. W. Tsang, D. G. Bermudes and L. D. Mayer. 2009. Drug ratio-dependent antitumor activity of irinotecan and cisplatin combinations in vitro and in vivo. *Mol Cancer Ther,* 8, 2266–2275.
45. Perez-Medina, C., D. Abdel-Atti, J. Tang, Y. Zhao, Z. A. Fayad, J. S. Lewis, W. J. Mulder and T. Reiner. 2016. Nanoreporter PET predicts the efficacy of anti-cancer nanotherapy. *Nat Commun,* 7, 11838.
46. Lammers, T., L. Y. Rizzo, G. Storm and F. Kiessling. 2012. Personalized nanomedicine. *Clin Cancer Res,* 18, 4889–4894.
47. Hoehn, M., E. Kustermann, J. Blunk, D. Wiedermann, T. Trapp, S. Wecker, M. Focking, H. Arnold, J. Hescheler, B. K. Fleischmann, W. Schwindt and C. Buhrle. 2002. Monitoring of implanted stem cell migration in vivo: A highly resolved in vivo magnetic resonance imaging investigation of experimental stroke in rat. *Proc Natl Acad Sci USA,* 99, 16267–16272.
48. Scherer, F., M. Anton, U. Schillinger, J. Henkel, C. Bergemann, A. Kruger, B. Gansbacher and C. Plank. 2002. Magnetofection: Enhancing and targeting gene delivery by magnetic force in vitro and in vivo. *Gene Ther,* 9, 102–109.
49. Dames, P., B. Gleich, A. Flemmer, K. Hajek, N. Seidl, F. Wiekhorst, D. Eberbeck, I. Bittmann, C. Bergemann, T. Weyh, L. Trahms, J. Rosenecker and C. Rudolph. 2007. Targeted delivery of magnetic aerosol droplets to the lung. *Nat Nanotechnol,* 2, 495–499.
50. Choi, H. S., Y. Ashitate, J. H. Lee, S. H. Kim, A. Matsui, N. Insin, M. G. Bawendi, M. Semmler-Behnke, J. V. Frangioni and A. Tsuda. 2010. Rapid translocation of nanoparticles from the lung airspaces to the body. *Nat Biotechnol,* 28, 1300–U1113.
51. Santiesteban, D. Y., K. Kubelick, K. S. Dhada, D. Dumani, L. Suggs and S. Emelianov. 2016. Monitoring/imaging and regenerative agents for enhancing tissue engineering characterization and therapies. *Ann Biomed Eng,* 44, 750–772.
52. Ruiz-de-Angulo, A., A. Zabaleta, V. Gomez-Vallejo, J. Llop and J. C. Mareque-Rivas. 2016. Microdosed lipid-coated Ga-67-magnetite enhances antigen-specific immunity by image tracked delivery of antigen and CpG to lymph nodes. *ACS Nano,* 10, 1602–1618.
53. Voulgari, E., B. Aristides, S. Galtsidis, V. Zoumpourlis, B. P. Burke, G. S. Clemente, C. Cawthorne, S. J. Archibald, J. Tucek, R. Zboril, V. Kantarelou, A. Karydas and K. Avgoustakis. Synthesis, characterization and in vivo evaluation of a magnetic cisplatin delivery Nanosystem based on PMAA-graft-PEG copolymers. *J Control Release,* DOI: http://dx.doi.org/10.1016/j.jconrel.2016.10.021
54. Boros, E., A. M. Bowen, L. Josephson, N. Vasdev and J. P. Holland. 2015. Chelate-free metal ion binding and heat-induced radiolabeling of iron oxide nanoparticles. *Chem Sci,* 6, 225–236.
55. Normandin, M. D., H. S. Yuan, M. Q. Wilks, H. H. Chen, J. M. Kinsella, H. Cho, N. J. Guehl, N. Absi-Halabi, S. M. Hosseini, G. El Fakhri, D. E. Sosnovik and L. Josephson. 2015. Heat-induced radiolabeling of nanoparticles for monocyte tracking by PET. *Angew Chem Int Ed,* 54, 13002–13006.
56. Psimadas, D., G. Baldi, C. Ravagli, P. Bouziotis, S. Xanthopoulos, M. C. Franchini, P. Georgoulias and G. Loudos. 2012. Preliminary evaluation of a Tc-99m labeled hybrid nanoparticle bearing a cobalt ferrite core: In vivo biodistribution. *J Biomed Nanotechnol,* 8, 575–585.

57. Tsiapa, I., E. K. Efthimiadou, E. Fragogeorgi, G. Loudos, A. D. Varvarigou, P. Bouziotis, G. C. Kordas, D. Mihailidis, G. C. Nikiforidis, S. Xanthopoulos, D. Psimadas, M. Paravatou-Petsotas, L. Palamaris, J. D. Hazle and G. C. Kagadis. 2014. Tc-99m-labeled aminosilane-coated iron oxide nanoparticles for molecular imaging of alpha(v)beta(3)-mediated tumor expression and feasibility for hyperthermia treatment. *J Colloid Interface Sci,* 433, 163–175.
58. Psimadas, D., G. Baldi, C. Ravagli, M. C. Franchini, E. Locatelli, C. Innocenti, C. Sangregorio and G. Loudos. 2014. Comparison of the magnetic, radiolabeling, hyperthermic and biodistribution properties of hybrid nanoparticles bearing CoFe2O4 and Fe3O4 metal cores. *Nanotechnology,* 25.
59. Cheng, L., S. D. Shen, S. X. Shi, Y. Yi, X. Y. Wang, G. S. Song, K. Yang, G. Liu, T. E. Barnhart, W. B. Cai and Z. Liu. 2016. FeSe2-decorated Bi2Se3 nanosheets fabricated via cation exchange for chelator-free Cu-64-Labeling and multimodal image-guided photothermal-radiation therapy. *Adv Funct Mater,* 26, 2185–2197.
60. Chen, F., P. A. Ellison, C. M. Lewis, H. Hong, Y. Zhang, S. X. Shi, R. Hernandez, M. E. Meyerand, T. E. Barnhart and W. B. Cai. 2013. Chelator-free synthesis of a dual-modality PET/MRI agent. *Angew Chem Int Ed,* 52, 13319–13323.
61. Burke, B. P., N. Baghdadi, A. E. Kownacka, S. Nigam, G. S. Clemente, M. M. Al-Yassiry, J. Domarkas, M. Lorch, M. Pickles, P. Gibbs, R. Tripier, C. Cawthorne and S. J. Archibald. 2015. Chelator free gallium-68 radiolabelling of silica coated iron oxide nanorods via surface interactions. *Nanoscale,* 7, 14889–14896.
62. Burke, B. P., N. Baghdadi, G. S. Clemente, N. Camus, A. Guillou, A. E. Kownacka, J. Domarkas, Z. Halime, R. Tripier and S. J. Archibald. 2014. Final step gallium-68 radiolabelling of silica-coated iron oxide nanorods as potential PET/MR multimodal imaging agents. *Farad Discuss,* 175, 59–71.
63. Stelter, L., J. G. Pinkernelle, R. Michel, R. Schwartlander, N. Raschzok, M. H. Morgul, M. Koch, T. Denecke, J. Ruf, H. Baumler, A. Jordan, B. Hamm, I. M. Sauer and U. Teichgraber. 2010. Modification of aminosilanized superparamagnetic nanoparticles: Feasibility of multimodal detection using 3T MRI, small animal PET, and fluorescence imaging. *Mol Imaging Biol,* 12, 25–34.
64. Shin, T.-H., Y. Choi, S. Kim and J. Cheon. 2015. Recent advances in magnetic nanoparticle-based multi-modal imaging. *Chem Soc Rev,* 44, 4501–4516.
65. Yu, X. J., I. Trase, M. Q. Ren, K. Duval, X. Guo and Z. Chen. 2016. Design of nanoparticle-based carriers for targeted drug delivery. *J Nanomater,* DOI: 10.1155/2016/1087250
66. Hu, H., P. Huang, O. J. Weiss, X. F. Yan, X. Y. Yue, M. G. Zhang, Y. X. Tang, L. M. Nie, Y. Ma, G. Niu, K. C. Wu and X. Y. Chen. 2014. PET and NIR optical imaging using self-illuminating Cu-64-doped chelator-free gold nanoclusters. *Biomaterials,* 35, 9868–9876.
67. Sun, X. L., X. L. Huang, X. F. Yan, Y. Wang, J. X. Guo, O. Jacobson, D. B. Liu, L. P. Szajek, W. L. Zhu, G. Niu, D. O. Kiesewetter, S. H. Sun and X. Y. Chen. 2014. Chelator-free Cu-64-integrated gold nanomaterials for positron emission tomography imaging guided photothermal cancer therapy. *ACS Nano,* 8, 8438–8446.
68. Goel, S., F. Chen, E. B. Ehlerding and W. B. Cai. 2014. Intrinsically radiolabeled nanoparticles: An emerging paradigm. *Small,* 10, 3825–3830.
69. Sun, X., W. Cai and X. Chen. 2015. Positron emission tomography imaging using radiolabeled inorganic nanomaterials. *Acc Chem Res,* 48, 286–294.
70. Price, E. W. and C. Orvig. 2014. Matching chelators to radiometals for radiopharmaceuticals. *Chem Soc Rev,* 43, 260–290.
71. Ramogida, C. F. and C. Orvig. 2013. Tumour targeting with radiometals for diagnosis and therapy. *Chem Commun,* 49, 4720–4739.
72. Tu, C. Q., T. S. C. Ng, R. E. Jacobs and A. Y. Louie. 2014. Multimodality PET/MRI agents targeted to activated macrophages. *J Biol Inorg Chem,* 19, 247–258.
73. Xu, C., S. X. Shi, L. Z. Feng, F. Chen, S. A. Graves, E. B. Ehlerding, S. Goel, H. Y. Sun, C. G. England, R. J. Nickles, Z. Liu, T. H. Wang and W. B. Cai. 2016. Long circulating reduced graphene oxide-iron oxide nanoparticles for efficient tumor targeting and multimodality imaging. *Nanoscale,* 8, 12683–12692.
74. Xie, J., K. Chen, J. Huang, S. Lee, J. H. Wang, J. Gao, X. G. Li and X. Y. Chen. 2010. PET/NIRF/MRI triple functional iron oxide nanoparticles. *Biomaterials,* 31, 3016–3022.
75. Cooper, M. S., M. T. Ma, K. Sunassee, K. P. Shaw, J. D. Williams, R. L. Paul, P. S. Donnelly and P. J. Blower. 2012. Comparison of Cu-64-complexing bifunctional chelators for radioimmunoconjugation: Labeling efficiency, specific activity, and in vitro/in vivo stability. *Bioconjug Chem,* 23, 1029–1039.
76. Glaus, C., R. Rossin, M. J. Welch and G. Bao. 2010. In vivo evaluation of Cu-64-Labeled magnetic nanoparticles as a dual-modality PET/MR imaging agent. *Bioconjug Chem,* 21, 715–722.
77. Barreto, J. A., M. Matterna, B. Graham, H. Stephan and L. Spiccia. 2011. Synthesis, colloidal stability and Cu-64 labeling of iron oxide nanoparticles bearing different macrocyclic ligands. *N J Chem,* 35, 2705–2712.
78. Jarrett, B. R., B. Gustafsson, D. L. Kukis and A. Y. Louie. 2008. Synthesis of Cu-64-labeled magnetic nanoparticles for multimodal imaging. *Bioconjug Chem,* 19, 1496–1504.
79. Hwang, D. W., H. Y. Ko, S. K. Kim, D. Kim, D. S. Lee and S. Kim. 2009. Development of a quadruple imaging modality by using nanoparticles. *Chem Eur J,* 15, 9387–9393.
80. de Rosales, R. T. M., R. Tavare, R. L. Paul, M. Jauregui-Osoro, A. Protti, A. Glaria, G. Varma, I. Szanda and P. J. Blower. 2011. Synthesis of Cu-64(II)-Bis(dithiocarbamatebisphospho nate) and Its conjugation with superparamagnetic iron oxide nanoparticles: In vivo evaluation as dual-modality pet-mri agent. *Angew Chem Int Ed,* 50, 5509–5513.
81. de Rosales, R. T. M., R. Tavare, A. Glaria, G. Varma, A. Protti and P. J. Blower. 2011. Tc-99m-bisphosphonate-iron oxide nanoparticle conjugates for dual-modality biomedical imaging. *Bioconjug Chem,* 22, 455–465.
82. Raabe, N., E. Forberich, B. Freund, O. T. Bruns, M. Heine, M. G. Kaul, U. Tromsdorf, L. Herich, P. Nielsen, R. Reimer, H. Hohenberg, H. Weller, U. Schumacher, G. Adam and H. Ittrich. 2015. Determination of liver-specific r(2)* of a highly monodisperse USPIO by Fe-59 iron core-labeling in mice at 3 T MRI. *Contrast Media Mol Imaging,* 10, 153–162.
83. Freund, B., U. I. Tromsdorf, O. T. Bruns, M. Heine, A. Giemsa, A. Bartelt, S. C. Salmen, N. Raabe, J. Heeren, H. Ittrich, R. Reimer, H. Hohenberg, U. Schumacher, H. Weller and P. Nielsen. 2012. A simple and widely applicable method to fe-59-radiolabel monodisperse superparamagnetic iron oxide nanoparticles for in vivo quantification studies. *ACS Nano,* 6, 7318–7325.
84. Wong, R. M., D. A. Gilbert, K. Liu and A. Y. Louie. 2012. Rapid size-controlled synthesis of dextran-coated, Cu-64-doped iron oxide nanoparticles. *ACS Nano,* 6, 3461–3467.
85. Zeng, J. F., B. Jia, R. R. Qiao, C. Wang, L. H. Jing, F. Wang and M. Y. Gao. 2014. In situ In-111-doping for achieving biocompatible and non-leachable In-111-labeled Fe3O4 nanoparticles. *Chem Commun,* 50, 2170–2172.

86. Pellico, J., J. Ruiz-Cabello, M. Saiz-Alia, G. del Rosario, S. Caja, M. Montoya, L. F. de Manuel, M. P. Morales, L. Gutierrez, B. Galiana, J. A. Enriquez and F. Herranz. 2016. Fast synthesis and bioconjugation of Ga-68 core-doped extremely small iron oxide nanoparticles for PET/MR imaging. *Contrast Media Mol Imaging,* 11, 203–210.
87. Wang, H. T., R. Kumar, D. Nagesha, R. I. Duclos, S. Sridhar and S. J. Gatley. 2015. Integrity of In-111-radiolabeled superparamagnetic iron oxide nanoparticles in the mouse. *Nucl Med Biol,* 42, 65–70.
88. European Medicines Agency. 2011. Reflection paper on non-clinical studies for generic nanoparticle iron medicinal product applications.
89. Axiak-Bechtel, S. M., C. A. Maitz, K. A. Selting and J. N. Bryan. 2015. Preclinical imaging and treatment of cancer: The use of animal models beyond rodents. *Q J Nucl Med Mol Imaging,* 59, 303–316.
90. Chia, R., F. Achilli, M. F. Festing and E. M. Fisher. 2005. The origins and uses of mouse outbred stocks. *Nat Genet,* 37, 1181–1186.
91. Stocke, N. A., S. A. Meenach, S. M. Arnold, H. M. Mansour and J. Z. Hilt. 2015. Formulation and characterization of inhalable magnetic nanocomposite microparticles (MnMs) for targeted pulmonary delivery via spray drying. *Int J Pharm,* 479, 320–328.
92. Fernandes, C. A. and R. Vanbever. 2009. Preclinical models for pulmonary drug delivery. *Expert Opin Drug Deliv,* 6, 1231–1245.
93. Russell, W. M. S. and R. L. Burch, *The principles of humane experimental technique*, Methuen, London, 1959.
94. Workman, P., E. O. Aboagye, F. Balkwill, A. Balmain, G. Bruder, D. J. Chaplin, J. A. Double, J. Everitt, D. A. Farningham, M. J. Glennie, L. R. Kelland, V. Robinson, I. J. Stratford, G. M. Tozer, S. Watson, S. R. Wedge, S. A. Eccles and I. Committee of the National Cancer Research. 2010. Guidelines for the welfare and use of animals in cancer research. *Br J Cancer,* 102, 1555–1577.
95. Kilkenny, C., W. Browne, I. C. Cuthill, M. Emerson, D. G. Altman, R. National Centre for the Replacement and R. Reduction of Amimals in. 2011. Animal research: Reporting in vivo experiments — The ARRIVE guidelines. *J Cereb Blood Flow Metab,* 31, 991–993.
96. Ho, D. N. and S. Sun. 2012. The gap between cell and animal models: Nanoparticle drug-delivery development and characterization using microtissue models. *Ther Deliv,* 3, 915–917.
97. Scheibe, P. O., D. R. Vera and W. C. Eckelman. 2005. What is to be gained by imaging the same animal before and after treatment? *Nucl Med Biol,* 32, 727–732.
98. Eckelman, W. C., M. R. Kilbourn, J. L. Joyal, R. Labiris and J. F. Valliant. 2007. Justifying the number of animals for each experiment. *Nucl Med Biol,* 34, 229–232.
99. Faul, F., E. Erdfelder, A. Buchner and A. G. Lang. 2009. Statistical power analyses using G*Power 3.1: tests for correlation and regression analyses. *Behav Res Methods,* 41, 1149–1160.
100. Vanhove, C., J. P. Bankstahl, S. D. Kramer, E. Visser, N. Belcari and S. Vandenberghe. 2015. Accurate molecular imaging of small animals taking into account animal models, handling, anaesthesia, quality control and imaging system performance. *EJNMMI Phys,* 2, 31.
101. Wolfer, D. P., O. Litvin, S. Morf, R. M. Nitsch, H. P. Lipp and H. Wurbel. 2004. Laboratory animal welfare: Cage enrichment and mouse behaviour. *Nature,* 432, 821–822.
102. David, J. M., S. Knowles, D. M. Lamkin and D. B. Stout. 2013. Individually ventilated cages impose cold stress on laboratory mice: A source of systemic experimental variability. *J Am Assoc Lab Anim Sci,* 52, 738–744.
103. Inturi, S., G. Wang, F. Chen, N. K. Banda, V. M. Holers, L. Wu, S. M. Moghimi and D. Simberg. 2015. Modulatory role of surface coating of superparamagnetic iron oxide nanoworms in complement opsonization and leukocyte uptake. *ACS Nano,* 9, 10758–10768.
104. de Jong, M. and T. Maina. 2010. Of mice and humans: Are they the same? — Implications in cancer translational research. *J Nucl Med,* 51, 501–504.
105. Hume, S. P., R. N. Gunn and T. Jones. 1998. Pharmacological constraints associated with positron emission tomographic scanning of small laboratory animals. *Eur J Nucl Med,* 25, 173–176.
106. Ruggeri, B. A., F. Camp and S. Miknyoczki. 2014. Animal models of disease: Pre-clinical animal models of cancer and their applications and utility in drug discovery. *Biochem Pharmacol,* 87, 150–161.
107. Mou, H., Z. Kennedy, D. G. Anderson, H. Yin and W. Xue. 2015. Precision cancer mouse models through genome editing with CRISPR-Cas9. *Genome Med,* 7, 53.
108. Zhou, H. Y., W. P. Qian, F. M. Uckun, L. Y. Wang, Y. A. Wang, H. Y. Chen, D. Kooby, Q. Yu, M. Lipowska, C. A. Staley, H. Mao and L. Yang. 2015. IGF1 Receptor targeted theranostic nanoparticles for targeted and image-guided therapy of pancreatic cancer. *ACS Nano,* 9, 7976–7991.
109. Ghosh, D., Y. Lee, S. Thomas, A. G. Kohli, D. S. Yun, A. M. Belcher and K. A. Kelly. 2012. M13-templated magnetic nanoparticles for targeted in vivo imaging of prostate cancer. *Nat Nanotechnol,* 7, 677–682.
110. Gao, X., Y. Luo, Y. Wang, J. Pang, C. Liao, H. Lu and Y. Fang. 2012. Prostate stem cell antigen-targeted nanoparticles with dual functional properties: In vivo imaging and cancer chemotherapy. *Int J Nanomed,* 7, 4037–4051.
111. Cao, C., X. Wang, Y. Cai, L. Sun, L. Tian, H. Wu, X. He, H. Lei, W. Liu, G. Chen, R. Zhu and Y. Pan. 2014. Targeted in vivo imaging of microscopic tumors with ferritin-based nanoprobes across biological barriers. *Adv Mater,* 26, 2566–2571.
112. Leuschner, C., C. S. Kumar, W. Hansel, W. Soboyejo, J. Zhou and J. Hormes. 2006. LHRH-conjugated magnetic iron oxide nanoparticles for detection of breast cancer metastases. *Breast Cancer Res Treat,* 99, 163–176.
113. Chen, T. J., T. H. Cheng, C. Y. Chen, S. C. Hsu, T. L. Cheng, G. C. Liu and Y. M. Wang. 2009. Targeted Herceptin-dextran iron oxide nanoparticles for noninvasive imaging of HER2/neu receptors using MRI. *J Biol Inorg Chem,* 14, 253–260.
114. Huh, Y. M., Y. W. Jun, H. T. Song, S. Kim, J. S. Choi, J. H. Lee, S. Yoon, K. S. Kim, J. S. Shin, J. S. Suh and J. Cheon. 2005. In vivo magnetic resonance detection of cancer by using multifunctional magnetic nanocrystals. *J Am Chem Soc,* 127, 12387–12391.
115. Xie, J., K. Chen, H. Y. Lee, C. J. Xu, A. R. Hsu, S. Peng, X. Y. Chen and S. H. Sun. 2008. Ultrasmall c(RGDyK)-coated Fe(3)O(4) nanoparticles and their specific targeting to integrin alpha(v)beta(3)-rich tumor cells. *J Am Chem Soc,* 130, 7542–7542.
116. Jiang, T., C. Zhang, X. Zheng, X. Xu, X. Xie, H. Liu and S. Liu. 2009. Noninvasively characterizing the different alphavbeta3 expression patterns in lung cancers with RGD-USPIO using a clinical 3.0T MR scanner. *Int J Nanomed,* 4, 241–249.

117. Yan, C., Y. Wu, J. Feng, W. Chen, X. Liu, P. Hao, R. Yang, J. Zhang, B. Lin, Y. Xu and R. Liu. 2013. Anti-alphavbeta3 antibody guided three-step pretargeting approach using magnetoliposomes for molecular magnetic resonance imaging of breast cancer angiogenesis. *Int J Nanomed,* 8, 245–255.
118. Zhang, C., X. Xie, S. Liang, M. Li, Y. Liu and H. Gu. 2012. Mono-dispersed high magnetic resonance sensitive magnetite nanocluster probe for detection of nascent tumors by magnetic resonance molecular imaging. *Nanomedicine,* 8, 996–1006.
119. Suwa, T., S. Ozawa, M. Ueda, N. Ando and M. Kitajima. 1998. Magnetic resonance imaging of esophageal squamous cell carcinoma using magnetite particles coated with anti-epidermal growth factor receptor antibody. *Int J Cancer,* 75, 626–634.
120. Mahajan, U. M., S. Teller, M. Sendler, R. Palankar, C. van den Brandt, T. Schwaiger, J. P. Kuhn, S. Ribback, G. Glockl, M. Evert, W. Weitschies, N. Hosten, F. Dombrowski, M. Delcea, F. U. Weiss, M. M. Lerch and J. Mayerle. 2016. Tumour-specific delivery of siRNA-coupled superparamagnetic iron oxide nanoparticles, targeted against PLK1, stops progression of pancreatic cancer. *Gut,* 65, 1838–1849.
121. Shevtsov, M. A., B. P. Nikolaev, L. Y. Yakovleva, Y. Y. Marchenko, A. V. Dobrodumov, A. L. Mikhrina, M. G. Martynova, O. A. Bystrova, I. V. Yakovenko and A. M. Ischenko. 2014. Superparamagnetic iron oxide nanoparticles conjugated with epidermal growth factor (SPION-EGF) for targeting brain tumors. *Int J Nanomed,* 9, 273–287.
122. Jiang, L., Q. Zhou, K. Mu, H. Xie, Y. Zhu, W. Zhu, Y. Zhao, H. Xu and X. Yang. 2013. pH/temperature sensitive magnetic nanogels conjugated with Cy5.5-labled lactoferrin for MR and fluorescence imaging of glioma in rats. *Biomaterials,* 34, 7418–7428.
123. Kresse, M., S. Wagner, D. Pfefferer, R. Lawaczeck, V. Elste and W. Semmler. 1998. Targeting of ultrasmall superparamagnetic iron oxide (USPIO) particles to tumor cells in vivo by using transferrin receptor pathways. *Magn Reson Med,* 40, 236–242.
124. Lim, S. W., H. W. Kim, H. Y. Jun, S. H. Park, K. H. Yoon, H. S. Kim, S. Jon, M. K. Yu and S. K. Juhng. 2011. TCL-SPION-enhanced MRI for the detection of lymph node metastasis in murine experimental model. *Acad Radiol,* 18, 504–511.
125. Peiris, P. M., R. Toy, E. Doolittle, J. Pansky, A. Abramowski, M. Tam, P. Vicente, E. Tran, E. Hayden, A. Camann, A. Mayer, B. O. Erokwu, Z. Berman, D. Wilson, H. Baskaran, C. A. Flask, R. A. Keri and E. Karathanasis. 2012. Imaging metastasis using an integrin-targeting chain-shaped nanoparticle. *ACS Nano,* 6, 8783–8795.
126. Kievit, F. M., Z. R. Stephen, O. Veiseh, H. Arami, T. Wang, V. P. Lai, J. O. Park, R. G. Ellenbogen, M. L. Disis and M. Zhang. 2012. Targeting of primary breast cancers and metastases in a transgenic mouse model using rationally designed multifunctional SPIONs. *ACS Nano,* 6, 2591–2601.
127. Duncan, R., Y. N. Sat-Klopsch, A. M. Burger, M. C. Bibby, H. H. Fiebig and E. A. Sausville. 2013. Validation of tumour models for use in anticancer nanomedicine evaluation: The EPR effect and cathepsin B-mediated drug release rate. *Cancer Chemother Pharmacol,* 72, 417–427.
128. Tentler, J. J., A. C. Tan, C. D. Weekes, A. Jimeno, S. Leong, T. M. Pitts, J. J. Arcaroli, W. A. Messersmith and S. G. Eckhardt. 2012. Patient-derived tumour xenografts as models for oncology drug development. *Nat Rev Clin Oncol,* 9, 338–350.
129. Kerbel, R. S. 2003. Human tumor xenografts as predictive preclinical models for anticancer drug activity in humans: Better than commonly perceived-but they can be improved. *Cancer Biol Ther,* 2, S134–139.
130. Paschall, A. V. and K. Liu. 2016. An orthotopic mouse model of spontaneous breast cancer metastasis. *J Vis Exp,* DOI: 10.3791/54040
131. Singh, M., C. L. Murriel and L. Johnson. 2012. Genetically engineered mouse models: Closing the gap between preclinical data and trial outcomes. *Cancer Res,* 72, 2695–2700.
132. Song, G., D. B. Darr, C. M. Santos, M. Ross, A. Valdivia, J. L. Jordan, B. R. Midkiff, S. Cohen, N. Nikolaishvili-Feinberg, C. R. Miller, T. K. Tarrant, A. B. Rogers, A. C. Dudley, C. M. Perou and W. C. Zamboni. 2014. Effects of tumor microenvironment heterogeneity on nanoparticle disposition and efficacy in breast cancer tumor models. *Clin Cancer Res,* 20, 6083–6095.
133. Gieling, R. G., R. J. Fitzmaurice, B. A. Telfer, M. Babur and K. J. Williams. 2015. Dissemination via the lymphatic or angiogenic route impacts the pathology, microenvironment and hypoxia-related drug response of lung metastases. *Clin Exp Metastasis,* 32, 567–577.
134. Francia, G., W. Cruz-Munoz, S. Man, P. Xu and R. S. Kerbel. 2011. Mouse models of advanced spontaneous metastasis for experimental therapeutics. *Nat Rev Cancer,* 11, 135–141.
135. Schroeder, A., D. A. Heller, M. M. Winslow, J. E. Dahlman, G. W. Pratt, R. Langer, T. Jacks and D. G. Anderson. 2011. Treating metastatic cancer with nanotechnology. *Nat Rev Cancer,* 12, 39–50.
136. Cawthorne, C., N. Burrows, R. G. Gieling, C. J. Morrow, D. Forster, J. Gregory, M. Radigois, A. Smigova, M. Babur, K. Simpson, C. Hodgkinson, G. Brown, A. McMahon, C. Dive, D. Hiscock, I. Wilson and K. J. Williams. 2013. [18F]-FLT positron emission tomography can be used to image the response of sensitive tumors to PI3-kinase inhibition with the novel agent GDC-0941. *Mol Cancer Ther,* 12, 819–828.
137. Peterson, T. E. and S. Shokouhi. 2012. Advances in preclinical SPECT instrumentation. *J Nucl Med,* 53, 841–844.
138. Deleye, S., R. Van Holen, J. Verhaeghe, S. Vandenberghe, S. Stroobants and S. Staelens. 2013. Performance evaluation of small-animal multipinhole muSPECT scanners for mouse imaging. *Eur J Nucl Med Mol Imaging,* 40, 744–758.
139. Goertzen, A. L., Q. Bao, M. Bergeron, E. Blankemeyer, S. Blinder, M. Canadas, A. F. Chatziioannou, K. Dinelle, E. Elhami, H. S. Jans, E. Lage, R. Lecomte, V. Sossi, S. Surti, Y. C. Tai, J. J. Vaquero, E. Vicente, D. A. Williams and R. Laforest. 2012. NEMA NU 4-2008 comparison of preclinical PET imaging systems. *J Nucl Med,* 53, 1300–1309.
140. Bernsen, M. R., P. E. Vaissier, R. Van Holen, J. Booij, F. J. Beekman and M. de Jong. 2014. The role of preclinical SPECT in oncological and neurological research in combination with either CT or MRI. *Eur J Nucl Med Mol Imaging,* 41 Suppl 1, S36–49.
141. Harteveld, A. A., A. P. Meeuwis, J. A. Disselhorst, C. H. Slump, W. J. Oyen, O. C. Boerman and E. P. Visser. 2011. Using the NEMA NU 4 PET image quality phantom in multipinhole small-animal SPECT. *J Nucl Med,* 52, 1646–1653.
142. Disselhorst, J. A., M. Brom, P. Laverman, C. H. Slump, O. C. Boerman, W. J. Oyen, M. Gotthardt and E. P. Visser. 2010. Image-quality assessment for several positron emitters using the NEMA NU 4-2008 standards in the Siemens Inveon small-animal PET scanner. *J Nucl Med,* 51, 610–617.
143. Visser, E. P., J. A. Disselhorst, M. Brom, P. Laverman, M. Gotthardt, W. J. Oyen and O. C. Boerman. 2009. Spatial resolution and sensitivity of the Inveon small-animal PET scanner. *J Nucl Med,* 50, 139–147.

144. Aide, N., E. P. Visser, S. Lheureux, N. Heutte, I. Szanda and R. J. Hicks. 2012. The motivations and methodology for high-throughput PET imaging of small animals in cancer research. *Eur J Nucl Med Mol Imaging,* 39, 1497–1509.
145. Soret, M., S. L. Bacharach and I. Buvat. 2007. Partial-volume effect in PET tumor imaging. *J Nucl Med,* 48, 932–945.
146. Cawthorne, C., C. Prenant, A. Smigova, P. Julyan, R. Maroy, K. Herholz, N. Rothwell and H. Boutin. 2011. Biodistribution, pharmacokinetics and metabolism of interleukin-1 receptor antagonist (IL-1RA) using [(1)(8)F]-IL1RA and PET imaging in rats. *Br J Pharmacol,* 162, 659–672.
147. Abbey, C. K., A. D. Borowsky, J. P. Gregg, R. D. Cardiff and S. R. Cherry. 2006. Preclinical imaging of mammary intraepithelial neoplasia with positron emission tomography. *J Mammary Gland Biol Neoplasia,* 11, 137–149.
148. Dawidczyk, C. M., L. M. Russell and P. C. Searson. 2014. Nanomedicines for cancer therapy: State-of-the-art and limitations to pre-clinical studies that hinder future developments. *Front Chem,* 2, 69.
149. Dawidczyk, C. M., L. M. Russell and P. C. Searson. 2015. Recommendations for benchmarking preclinical studies of nanomedicines. *Cancer Res,* 75, 4016–4020.
150. Dupont, P. and J. Warwick. 2009. Kinetic modelling in small animal imaging with PET. *Methods,* 48, 98–103.
151. Zuckerman, J. E., C. H. Choi, H. Han and M. E. Davis. 2012. Polycation-siRNA nanoparticles can disassemble at the kidney glomerular basement membrane. *Proc Natl Acad Sci U S A,* 109, 3137–3142.
152. Clark, A. J., D. T. Wiley, J. E. Zuckerman, P. Webster, J. Chao, J. Lin, Y. Yen and M. E. Davis. 2016. CRLX101 nanoparticles localize in human tumors and not in adjacent, nonneoplastic tissue after intravenous dosing. *Proc Natl Acad Sci U S A,* 113, 3850–3854.
153. Anselmo, A. C. and S. Mitragotri. 2015. A review of clinical translation of inorganic nanoparticles. *Aaps J,* 17, 1041–1054.
154. Svenson, S. 2012. Clinical translation of nanomedicines. *Curr Opin Solid State Mater Sci,* 16, 287–294.
155. Shi, J., P. W. Kantoff, R. Wooster and O. C. Farokhzad. 2016. Cancer nanomedicine: Progress, challenges and opportunities. *Nat Rev Cancer,* DOI: 10.1038/nrc.2016.108
156. Bao, A., B. Goins, R. Klipper, G. Negrete and W. T. Phillips. 2003. Re-186-liposome labeling using Re-186-SNS/S complexes: In vitro stability, imaging, and biodistribution in rats. *J Nucl Med,* 44, 1992–1999.
157. Chang, Y. J., C. H. Chang, T. J. Chang, C. Y. Yu, L. C. Chen, M. L. Jan, T. Y. Luo, W. Te, Lee and G. Ting. 2007. Biodistribution, pharmacokinetics and microSPECT/CT imaging of Re-188-BMEDA-liposome in a C26 murine colon carcinoma solid tumor animal model. *Anticancer Res,* 27, 2217–2225.
158. Sancey, L., S. Kotb, C. Trulllet, F. Appaix, A. Marais, E. Thomas, B. van der Sanden, J. P. Klein, B. Laurent, M. Cottier, R. Antoine, P. Dugourd, G. Panczer, F. Lux, P. Perriat, V. Motto-Ros and O. Tillement. 2015. Long-term in vivo clearance of gadolinium-based aguix nanoparticles and their biocompatibility after systemic injection. *Acs Nano,* 9, 2477–2488.
159. Wiesing, U. and J. Clausen. 2014. The Clinical research of nanomedicine: A new ethical challenge? *Nanoethics,* 8, 19–28.
160. Fatehi, L., S. M. Wolf, J. McCullough, R. Hall, F. Lawrenz, J. P. Kahn, C. Jones, S. A. Campbell, R. S. Dresser, A. G. Erdman, C. L. Haynes, R. A. Hoerr, L. F. Hogle, M. A. Keane, G. Khushf, N. M. P. King, E. Kokkoli, G. Marchant, A. D. Maynard, M. Philbert, G. Ramachandran, R. A. Siegel and S. Wickline. 2012. Recommendations for nanomedicine human subjects research oversight: An evolutionary approach for an emerging field. *J Law Med Ethics,* 40, 716–750.
161. Sun, X. L., W. B. Cai and X. Y. Chen. 2015. Positron emission tomography imaging using radio labeled inorganic nanomaterials. *Acc Chem Res,* 48, 286–294.
162. Landesman-Milo, D. and D. Peer. 2016. Transforming nanomedicines from lab scale production to novel clinical modality. *Bioconjug Chem,* 27, 855–862.
163. Arrieta, O., L. A. Medina, E. Estrada-Lobato, L. A. Ramirez-Tirado, V. O. Mendoza-Garcia and J. de la Garza-Salazar. 2014. High liposomal doxorubicin tumour tissue distribution, as determined by radiopharmaceutical labelling with Tc-99m-LD, is associated with the response and survival of patients with unresectable pleural mesothelioma treated with a combination of liposomal doxorubicin and cisplatin. *Cancer Chemother Pharmacol,* 74, 211–215.
164. Koukourakis, M. I., S. Koukouraki, A. Giatromanolaki, S. C. Archimandritis, J. Skarlatos, K. Beroukas, J. G. Bizakis, G. Retalis, N. Karkavitsas and E. S. Helidonis. 1999. Liposomal doxorubicin and conventionally fractionated radiotherapy in the treatment of locally advanced non-small-cell lung cancer and head and neck cancer. *J Clin Oncol,* 17, 3512–3521.
165. Burt, T., K. Yoshida, G. Lappin, L. Vuong, C. John, S. N. de Wildt, Y. Sugiyama and M. Rowland. 2016. Microdosing and other phase 0 clinical trials: Facilitating translation in drug development. *Clin Transl Sci,* 9, 74–88.
166. Heuveling, D. A., R. de Bree, D. J. Vugts, M. C. Huisman, L. Giovannoni, O. S. Hoekstra, C. R. Leemans, D. Neri and G. van Dongen. 2013. Phase 0 microdosing PET study using the human mini antibody F16SIP in head and neck cancer patients. *J Nucl Med,* 54, 397–401.
167. Lee, D. S., H. J. Im and Y. S. Lee. 2015. Radionanomedicine: Widened perspectives of molecular theragnosis. *Nanomedicine,* 11, 795–810.
168. Ramanathan, R. K., R. L. Korn, J. C. Sachdev, G. J. Fetterly, K. Marceau, V. Marsh, J. M. Neil, R. G. Newbold, N. Raghunand, J. Prey, S. G. Klinz, E. Bayever and J. B. Fitzgerald. 2014. Abstract CT224: Pilot study in patients with advanced solid tumors to evaluate feasibility of ferumoxytol (FMX) as tumor imaging agent prior to MM-398, a nanoliposomal irinotecan (nal-IRI). *Cancer Res,* 74, CT224–CT224.
169. Kalra, A. V., J. Spernyak, J. Kim, A. Sengooba, S. Klinz, N. Paz, J. Cain, W. Kamoun, N. Straubinger, Y. Qu, S. Trueman, E. Bayever, U. Nielsen, D. Drummond, J. Fitzgerald and R. Straubinger. 2014. Abstract 2065: Magnetic resonance imaging with an iron oxide nanoparticle demonstrates the preclinical feasibility of predicting intratumoral uptake and activity of MM-398, a nanoliposomal irinotecan (nal-IRI). *Cancer Res,* 74, 2065–2065.
170. Singh, D., J. M. McMillan, A. V. Kabanov, M. Sokolsky-Papkov and H. E. Gendelman. 2014. Bench-to-bedside translation of magnetic nanoparticles. *Nanomedicine,* 9, 501–516.

Stephen J. Archibald (E-mail: S.J.Archibald@hull.ac.uk) is a professor of molecular imaging chemistry at the University of Hull, the director of the Positron Emission Tomography Research Centre at the University of Hull and the research director of the Molecular Imaging Centre at Castle Hill Hospital. Key research areas in the Archibald group are PET probe development, multimodal imaging, novel radiosynthesis methodology and chemokine receptor binding molecules. Recently Archibald has led a project to develop new lab-on-a-chip devices for integrated synthesis and quality control of radiopharmaceuticals. Current imaging probe development projects include chemokine receptor targeted agents, multimodal (PET/MR) imaging with coated nanoparticles, targeting fibrotic plaques, new bifunctional chelators for PET metalloradioisotopes, mitochondrial function reporting agents, PET/MR imaging of tissue engineering constructs and contrast agent encapsulation. Website: http://pet.hull.ac.uk.

Benjamin P. Burke (E-mail: B.Burke@hull.ac.uk) earned his PhD in 2013, during which he developed methodologies for chelator-free radiolabelling of SPIONs to form PET/MR imaging agents. He is currently the principle translational radiochemist between the PET Research Centre at the University of Hull and the Molecular Imaging Research Centre at Castle Hill Hospital where he is focussed on the development of novel imaging agents from bench-to-bedside. His research interests include using nuclear imaging to understand nanoparticle behaviour, chemokine receptor imaging for oncology and improving radiochemical methodologies using radiometals and fluorine-18. Website: http://pet.hull.ac.uk.

Christopher Cawthorne (E-mail: C.Cawthorne@hull.ac.uk) is a lecturer in molecular imaging and the head of preclinical imaging at the University of Hull. He was trained as a molecular pharmacologist, earning his PhD from the University of Manchester. He then carried out postdoctoral work developing preclinical PET tracers for oncology under Profs. Pat Price and Terry Jones at the Wolfson Molecular Imaging Centre. Since moving to the University of Hull in 2013, his focus has been on the validation of novel imaging biomarkers for use in oncology, the assessment of existing ones for targeted/cytotoxic/radiotherapy response and quantitation of preclinical PET via the incorporation of tracer kinetic models into image reconstruction methodologies. Website: http://pet.hull.ac.uk.

24 Red Blood Cells Constructs to Prolong the Life Span of Iron-Based Magnetic Resonance Imaging/Magnetic Particle Imaging Contrast Agents *In Vivo*

*Antonella Antonelli and Mauro Magnani**

CONTENTS

24.1 INTRODUCTION

Successful clinical applications of magnetic nanoparticles (MNPs) require performances that are not easily achieved unless size, shape and surface properties are carefully designed to meet specific needs. When MNPs are used *in vivo* as tracers or contrast agents, their circulation and life span become critical factors. In the last years, most of the intrinsic limits of nanomaterials have been overcome by developing cell-based biomimetic constructs that take advantage of the long circulation time of blood cells, especially red blood cells (RBCs), maintaining the key magnetic properties of the nanoparticles (NPs). This chapter will describe the basic properties of these cellular carriers, the methods to encapsulate selected nanomaterials, the requirements for optimal *in vivo* biocompatibility and survival. Finally, some applications as *in vivo* tracers are illustrated.

24.2 RED BLOOD CELLS

In the past few years several papers have described the use of RBCs or erythrocytes as delivery vehicles for different kinds of biologically active compounds. In fact, human RBCs, which are the most abundant cells in blood, and have a diameter of 7–8 μm, an average thickness of 2 μm (2.5 μm at the thickest point and less than 1 μm at the centre) and a large internal capacity of 90 μm^3, represent a potential natural carrier system for several compounds, including contrasting agents, due to their unique properties that include biocompatibility, membrane flexibility and long circulation time (~120 days). The average adult male has about 5×10^6 RBCs per mm^3 of blood (about 4.5 million in adult females), but this may vary according to the person's geographical location – people living at high altitudes usually have more RBCs. Under normal circumstances, the body produces enough RBCs every day to offset the removal of senescent (old) cells, but RBCs may be removed from the bloodstream at any time if they are severely damaged and non-functional. The aging of RBCs in the bloodstream is associated with a number of modifications. The most obvious changes are related to an increase in cell density, a decrease in deformability and a significant metabolic decline involving the loss of many key enzymatic activities. Aging is also associated with membrane modifications that ultimately result in RBC opsonization by autologous immunoglobulins and complement, which serve as recognition signals for the removal of senescent cells from the bloodstream by Kupffer cells in the liver, and macrophages in the spleen and bone marrow. These macrophages release iron, which is carried by transferrin to the bone marrow where it can be reused for the production of new RBCs. The remaining porphyrin portion (heme) of the haemoglobin molecule is converted and excreted as bilirubin released in bile.[1]

The number of circulating RBCs must be adequate to supply oxygen to tissues. In the case of hypoxia, the haemopoietic stem cells in the bone marrow, stimulated by erythropoietin, go through various phases of development until the mature

* Corresponding author.

RBCs can be released into the bloodstream. The mature RBCs are released from the bone marrow into the blood at the reticulocyte stage and reticulocytes become fully functional RBCs after 1–2 days. This process of developing from erythropoietic bone marrow cells to mature RBCs takes only a few days.[2] The final stage of maturation, when the cell lacks a nucleus, mitochondria and endoplasmic reticulum requires iron, vitamin B12 and folic acid. The enzymes within the RBC allow it to produce small amounts of energy. Glucose, which is the main energy source for these cells, is metabolized in the glycolytic pathway anaerobically to pyruvate or lactate, producing two molecules of adenosine triphosphate (ATP) per molecule of glucose that is used. Thus, glycolysis is the main energy pathway, and it is regulated in the hexokinase (Hk) and phosphofructokinase steps.[3] The most important and functional component of RBCs is haemoglobin, which accounts for the oxygen carrying capacity of these cells. In addition to carrying oxygen, which is the main function of RBCs, they carry out the following functions: (1) conversion of carbon dioxide in bicarbonate, thanks to the enzyme carbonic anhydrase, and its transport to the lungs where it is expelled; and (2) control of pH in the bloodstream by acting as an acid-base buffer.

Although RBCs are wider than some capillaries, their elasticity allows them to become distorted as they are squeezed through narrow passageways and then return to their original shape. All healthy mammalian RBCs are disc-shaped (discocyte) when not subjected to external stress. The RBC membrane comprises a phospholipid bilayer and an underlying two-dimensional network of spectrin molecules that, together with several other cytoskeleton molecules, account for the typical discocyte morphology of healthy RBCs (see Figure 24.1). The RBC membrane is elastic with a high surface-to-volume ratio that facilitates considerable reversible deformation of the RBC as it repeatedly passes through small capillaries during microcirculation.

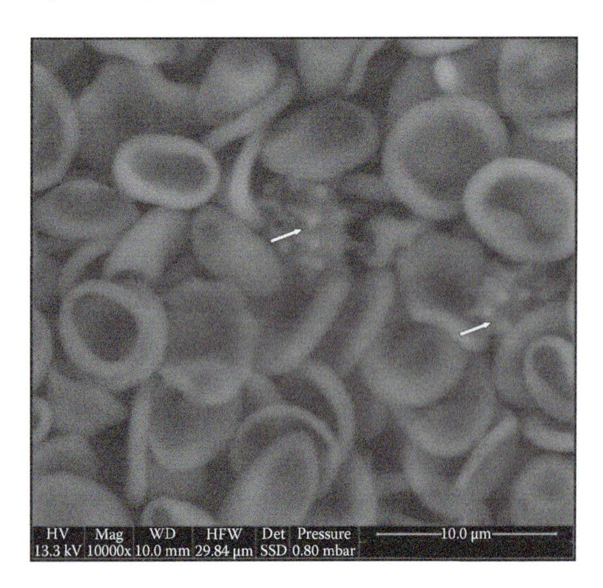

FIGURE 24.1 Image of human RBCs obtained by FEI Quanta 200 environmental scanning electron microscope (ESEM) with EDAX EDS system. Morphology: Backscattered Electrons (BSE) Detector at 15 kV of beam accelerating voltage EDS (energy dispersive x-ray spectroscopy). The image shows RBCs with typical morphology (disk-shaped or discocyte) and some occasionally echinocytes (indicated by white arrows).

The biconcave shape and corresponding deformability of the RBC is an essential feature of its biological function. Indeed, RBC deformability is key to circulation, which is necessary to transport oxygen and carbon dioxide. Usually, a rise in pH or reduced oncotic pressure results in the appearance of echinocytes (spiculated RBCs with short projections over their surface) while a lowering of pH or an excess of proteins induces the formation of stomatocytes (bowl-shaped RBCs with a single concavity). A number of agents induce the formation of echinocytes and stomatocytes. The former are likely formed by the expansion of the outer half of the phospholipid bilayer, while the latter are formed by the expansion of the inner half. A reduction in RBC life span occurs when there are molecular defects in one or more enzymes of the cell or when there are defects in the membrane proteins. RBC survival in the bloodstream may also be affected by autoantibodies, haemolytic substances or mechanical trauma.[4]

24.3 USE OF RBCs TO DELIVER SPIO- AND USPIO-BASED NPs

The growing interest in superparamagnetic NPs has led to their use as both diagnostic agents in MRI as well as drug delivery vehicles. In the last few years, a variety of functional magnetic nanostructures have been developed as a potential alternative to gadolinium (Gd) chelates in MRI and applied in the fields of medicine and biology.[5] Iron oxide NPs (IONPs), which become superparamagnetic if their core particle diameter is ~30 nm or less, have R_1 and R_2 relaxivities which are much higher than those of conventional paramagnetic gadolinium chelates.[6] In particular, biocompatible superparamagnetic IONPs (SPIONs) based on magnetite (Fe_3O_4) or maghemite (γFe_2O_3), which can be manipulated by external magnetic fields, have received the most attention for their utility as contrast agents in MRI applications,[7] as mediators for cancer magnetic hyperthermia[8] or as active constituents of drug-delivery platforms.[9] These particles can be arranged into several categories according to their hydrodynamic diameter, size, crystalline structure, coating and higher order organization[10]: standard SPIOs (with hydrodynamic diameter of 50–180 nm), ultra small superparamagnetic iron oxides (USPIOs) (10–50 nm), very small superparamagnetic iron oxide particles (VSPIOs) (<10 nm), monocrystalline iron oxide particles (MION) and cross-linked iron oxide (CLIO).

Several SPIO compounds have been registered and approved, such as Resovist® (or SHU 555A, consisting of an aqueous suspension of SPIONs that are carboxydextran-coated with a mean hydrodynamic diameter of 62 nm, Bayer HealthCare Pharmaceuticals) in Europe and Japan, and Endorem® (or AMI 25, consisting of IONPs dextran-coated with a hydrodynamic diameter of 80–150 nm, Guerbet) in Europe or Feridex® (Berlex) in the United States. In all cases, the clinical targets are the liver and spleen since SPIOs are rapidly cleared from the blood by the mononuclear phagocyte system of these organs.[11,12]

IONPs have become extremely popular due to their ability to dramatically shorten T_2^* relaxation times[13,14] in the liver, spleen and bone marrow by selective uptake and accumulation in the cells of the reticuloendothelial system (RES). For example, commercially available SPIOs such as Resovist® are extremely strong enhancers of proton relaxation and excellent MRI contrasting agents, superior to gadolinium derivatives, but have very short useful half-lives after intravenous administration, as they are rapidly cleared from the blood (within minutes) and accumulate in the RES of the liver and spleen.[15] Indeed, upon intravenous injection, the surface of the NPs are subjected to adsorption by plasma proteins during the opsonization process, which renders the particles recognizable by the RES, the body's major defence system.[16,17]

The current optimization of the properties of these magnetic particles aims to (1) provide an increase in MNP concentration in blood vessels, (2) reduce early clearance from the body, (3) minimize nonspecific cell interactions, thus minimizing side effects and (4) increase their internalization efficiency within target cells, thus reducing the total dose required.[18] However, the biological barriers that serve to protect the body against foreign bodies, including injected therapeutics and contrasting agents, restrict NP function since they are eliminated before they reach their destinations.[19,20] In fact, although several techniques, including size reduction,[20] have been employed to increase MNP blood circulation time or to enhance their target specificity, SPION applications, such as their direct use in MRI applications in the vascular system, are still limited. The majority of past research has addressed this issue via chemical modification of NPs in the form of hydrophilic coatings, and reducing adsorption of opsonins, which trigger RES clearance. Numerous surface modifiers or coating agents such as dextran, alginate and polyethylene glycol (PEG)[21] have been proposed to provide new MNPs with better survival capacity in the blood. For example PEGylated magnetoliposomes, which remain in circulation for a long time, are used as an MRI diagnostic agent and a magnetic-targeting drug agent.[22] Generally, the fate of these NPs is dependent on their physicochemical properties, namely size, morphology, charge and surface chemistry, which directly affect their pharmacokinetics and biodistribution.[23,24] Modifying the surface characteristics of SPION with biocompatible polymers and controlling their size within the desirable range can yield powerful targeted delivery vehicles. The functionalization of these NPs with targeting agents has also been used for targeted imaging via the site-specific accumulation of NPs at the site of interest.[25–27]

In this context, the development of innovative delivery systems able to prolong the residence time of iron oxide-based contrast agents in the bloodstream is required. Among the several proposed approaches, strategies based on RBCs and/or their mechanobiological features are pursued to develop carriers useful in biomedical and diagnostic applications such as MRI of the vascular system.

RBCs that are biocompatible, biodegradable and nonimmunogenic can be used as a valuable carrier system with a life span that is remarkably prolonged and controllable compared to synthetic carriers. Moreover, the possibility of loading drugs into autologous RBCs prior to their transfusion into patients has been studied in small animal models and primates, as well as in clinical studies of human patients.[28–30] Several approaches have been developed to load agents into RBCs or to attach agents onto RBCs' outer surface by either chemical or physical methods. However, possible alterations following the reengineering of RBCs are critical issues to preserve *in vivo* performance and to ensure the safety and the efficacy of the delivery system for translating to the clinical setting. Muzykantov et al.[31,32] have already extensively reported rigorous *in vitro* and *in vivo* functional assays aimed to estimate the potential damaging or sensitizing effects of the cargo attached to or encapsulated in RBCs. The sensitivity to osmotic, mechanical, oxidative and complement stress as well as RBC agglutination of these RBC carriers were studied through a set of assays and the resulting data showed great resistance of coupled RBC. These data are consistent with the minimal effect on their circulation obtained *in vivo* where the life span of labeled RBCs was not markedly affected by the surface modifications. In fact, these authors also demonstrated that the regulation of the surface density of biotin residues on the membrane of biotinylated-RBCs provide immunoerythrocytes which display only marginally altered biocompatibility and life time in the circulation.[33] Other authors such as Bax et al.[34] reported the *in vivo* survival results of human carrier RBCs demonstrated the regular mean cell life and cell half-life of the carrier RBCs within the normal range (89–131 and 19–29 days, respectively) by monitoring the disappearance of ^{51}Cr label from circulation of unloaded RBCs (cells submitted to a loading procedure without the addition of cargo). Moreover, the evidence that the encapsulation of therapeutic or diagnostic compounds only marginally alters their lifetime in the circulation and their biocompatibility was reported. For example, osmotic fragility studies demonstrated that after the drug loading process no significant variation is shown in the osmotic fragility profiles of both treated and untreated cells.[35–37] Coker et al.[38] have presented data reporting the *in vivo* survival of human RBCs loaded with dexamethasone sodium phosphate (DSP) through an automated EryDex System (EDS). Healthy volunteer consenting subjects were randomized to receive autologous RBCs prepared using EDS. DSP-loaded RBCs were radiolabeled with ^{51}Cr and followed over 49 days post-infusion. The mean ± SD RBC life span resulted 84.3 ± 8.3 days and with a mean of RBC half-life (T50) of 42.1 ± 3.1 days showing that the *in vivo* pharmacokinetic of these carrier RBCs is similar to unloaded RBCs treated in the same condition with EDS (life span, 88 ± 6.2 days; mean T50, 44.4 ± 3.1 days)[38] and within the normal range defined by the FDA for the infusion of blood-derived products.

Some authors have reported the diffusion of enzymes, nucleic acids and drugs, such as diclofenac sodium, into RBCs by using an electroporation method, thus exposing cells to a strong external electrical field to induce pore opening in the RBC membrane.[39–41] However, this method may irreversibly damage the cell membrane integrity. In fact phosphatidylserine was found to be externalized after electroporation treatment.[42,43] This is a signal for the mononuclear-phagocyte uptake system, thus shortening the blood circulation time of these RBCs.

The most common method to entrap a broad spectrum of drugs, proteins/enzymes and any other biologically active substances is based on the principle that RBCs swell in a hypotonic solution with the reversible opening of membrane pores that can then be resealed under isotonic conditions.[44–46]

Many strategies used to load therapeutic agents into RBCs are based on osmosis. Essentially, RBCs placed in a hypotonic solution swell, increasing their volume due to the reduced osmotic pressure created outside the cell, which leads to a water influx that is faster than the salt efflux because of the cell's higher water permeability. The RBC responds to the entrance of water by changing shape from biconcave to spherocytic. Additional water influx causes increased membrane stretching and ruptures begin to appear on the lipid bilayer with the formation of large pores. The cell membrane maintains its integrity to a tonicity of just under 150 mOsm/kg, after which the cell bursts. The opening of these transient pores in the RBC membrane leads to the escape of the cellular contents, which equilibrate on the inside and outside of the cell, but it also allows the entrance of external macromolecules or material. By incubation with an isotonic solution, the membrane reseals entrapping the substance(s) of interest inside, and RBCs reassume their natural shape.

Different variants of hypotonic haemolysis are known such as hypotonic dilution, hypotonic preswelling, osmotic pulse and the most commonly used hypotonic dialysis.[47,48] These techniques are all based on the principle that RBCs swell in hypotonic solution leading to the formation of pores ranging in diameter from 10 to 200 nm.[49] The increase in membrane permeability allows soluble agents to diffuse into cells driven by the concentration gradient. The pores are then resealed under isotonic conditions to form nanomaterial and/or drug-loaded RBCs. However, similarly to electroporation, osmosis-based methods may also result in varying degrees of damage to the RBC membrane and the loss of important cellular components that impair the structural integrity and physiological functions of the cell and thus increase the likelihood that it will be recognized and cleared by the body's immune system. Moreover, methods such as the osmotic pulse method have limitations that stem from the very brief opening of the transient pores, which results in a small amount of the drug or substance in question being entrapped and some haemoglobin being lost. The osmotic pulse method, based on the addition of dimethyl sulphoxide (DMSO) to RBCs to raise the osmolality of the suspension, incorporates inositol hexaphosphate, an allosteric effector of hemoglobin.[50] Isotonic haemolysis is achieved using this method by subjecting the cells to a short but intense osmotic stress. After the incubation with DMSO, which has high membrane permeability, the RBC suspension is mixed with an isotonic solution containing the drug or substance to be encapsulated. The DMSO external concentration decreases immediately resulting in a transient gradient of DMSO in the cell that determines water influx and cellular swelling. This mechanism is based on the fact that DMSO transport across the RBC's membrane is slower than water influx. Once the DMSO has left the cell, osmotic equilibrium is restored and the cell reacquires its original permeability and shape. However, the osmotic pulse method results in only a small fraction of loaded cells (a total of 9%–25%) with a small amount of drug or substance in question being entrapped and some haemoglobin being lost.[50] Covalent and noncovalent loading protocols with variable degrees of molecular specificity have also been reported. Proteins and carbohydrates on the RBC's surface contain an abundance of amine and thiol groups available for conjugation via a variety of chemistries involving modification of the RBC surface.[51] For example, chemical conjugation of biotin to the RBC surface allows avidin functionalized particles to readily bind and attach to the biotinylated RBC.[44,52,53]

The surface of RBCs offers many membrane proteins representing potential sites for carriage of therapeutics. Surface loading to RBCs can be achieved *ex vivo* by incubating either intact or modified RBCs with drugs or drug carriers. The drug of interest may react chemically with the RBC surface to form a covalent linkage or be nonspecifically adsorbed. Drugs and carriers can be conjugated or fused with antibodies, antibody fragments, peptides or other ligands that bind to RBCs' surface.[53–55] For example, avidin–biotin bridge is a common way to conjugate bioactive agents on the surface of RBCs; the biotin N-hydrosuccinimide ester (NHS-biotin) was first reacted with the amino groups on RBC membrane. Subsequently, the bioactive agents were conjugated with avidin following by incubation with biotin modified RBCs. Biopharmaceuticals, like fibrinolytic agents, HIV-1 TAT protein, bovine serum albumin, even organic/inorganic hybrid NPs could all be conjugated to RBC surfaces via the biotin–avidin bridge.[56,57] Coating particles with antibodies or other ligands of RBC proteins offers the most specific loading.[58] Recently, peptide ligands capable of mediating NP adhesion to human RBCs or binding of other cargo, were identified from a large bacterial display peptide library.[59–63] Shi et al.[64] have introduced genes into erythrocyte precursors encoding surface proteins that can be covalently and site specifically modified on the cell surface. Briefly, the authors engineered erythroid precursors to express sortase-modifiable proteins that are retained on the plasma membrane of mature RBCs. Sortase A from *Staphylococcus aureus* recognizes an LPXTG motif positioned close to C terminus of substrates such as proteins, peptides, small molecules (organic or inorganic), carbohydrates, lipids and other substances. These engineered RBCs can be labeled in a sortase-catalyzed reaction under native conditions without inflicting damage to the membrane or cell. The site specificity of the sortase reaction and its ability to accommodate a wide range of substituents offer distinct advantages, including modification of RBCs with substituents that cannot be encoded genetically.[64]

Additional strategies adopted for producing synthetic drug delivery carriers are based on recreation of the complex morphology of RBCs. In fact, in literature there are reports of both the fabrication of synthetic NP structures mimicking the natural features of RBCs and the use of the RBC itself manipulated to produce new drug/nanomaterial delivery systems.

Several authors have also described RBC carriers in comparison with synthetically engineered carriers such as polyelectrolyte multilayer capsules, inorganic particles or liposomes and micelles that confer distinct advantages/disadvantages for specific medical applications.[65,66]

24.3.1 Strategies for SPION Carriage by RBCs

There are a variety of strategies to attach NPs with different characteristics to the RBC surface or to encapsulate NPs in RBCs; however, the method used to combine NPs and RBCs must take into account both the properties of the particles and the final application of the RBC delivery systems in order to obtain both optimal attachment/inclusion and limited damage to or impaired function of the RBC. A common and simple method of NP carriage by RBCs reported by several authors involves nonspecific physical adsorption of the hydrophobic particles on the areas of the RBC membrane containing hydrophobic domains.[67] Mesoporous silica nanoparticles (MSNs) with silanol functional groups have been shown to adsorb to RBCs by interacting with the phosphatidylcholine present on the RBC membranes. The interactions between the human RBC membrane and MSNs with different particle sizes and surface properties were investigated by Zhao et al.[68] using fluorescence and electron microscopies. Small MSNs (~100 nm) were found to adsorb to the surface of RBCs without disturbing their membrane or morphology. In contrast, adsorption of larger MSNs (~600 nm) to RBCs induced a strong local membrane deformation leading to internalization of the particles, and eventual cell haemolysis. These authors have hypothesized that the internalization of MSNs could be consequent to a membrane wrapping after their contact with a flat cell membrane; driven by a local reduction in free energy, the phospholipids in the immediate neighbourhood of the site of contact are drawn to the surface of the particle leading to eventual encapsulation. Such internalization is different from phagocytosis or endocytosis because it appears to be driven by the balance of two opposing forces rather than by an active uptake of nutrients by the cell.[68] Recently, Laurencin et al.[69] also described a method of nonspecific adsorption of silica MNPs to RBCs by interactions between amorphous silica and the main constituent of the biological membrane (i.e. phosphatidylcholine) to obtain functionalized RBCs as multifunctional drug carriers with a long circulating time. The authors emphasize the efficiency of the core-shell MNP (CSMN) adsorption process on RBCs. CSMNs obtained by citrated MNPs embedded in a silica shell were also derivatized with amino groups, which provides a positive charge and increases the cell specificity for negatively charged human RBC surfaces. Moreover, CSMNs are derivatized with PEG chains to limit the haemolytic activity and because PEGylated silica shells are noncytotoxic and resistant to biodegradation. This is the first investigation in which MNPs were used for RBC membrane decoration (Figure 24.2), thus allowing the combination of relevant properties for diagnosis (MRI imaging) and therapeutic (hyperthermia) uses (Figure 24.3).

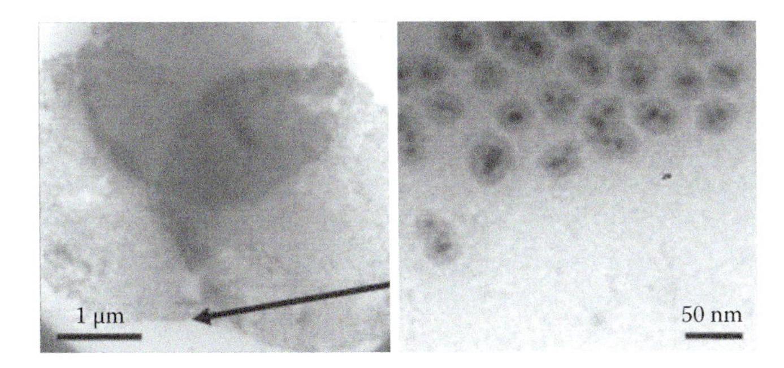

FIGURE 24.2 Human RBCs (hRBCs) obtained by a nonspecific method of CSMNs adsorption on the RBC membrane were observed by TEM analysis. Magnetic hRBCs appeared as circular packs of 7 μm diameter (TEM micrograph on the left); the focusing on the membrane allowed the visualization of CSMNs that caused the presence of an unconventional electronic contrast (TEM micrograph on the right). (Laurencin, M., Cam, N., Georgelin, T., Clément, O., Autret, G., Siaugue, J. M. and Ménager, C.: Human erythrocytes covered with magnetic core-shell nanoparticles for multimodal imaging. *Adv Healthc Mater.* 2013. 2(9). 1209–1212. Copyright Wiley-VCH Verlag GmbH & Co. KGaA. Reproduced with permission.)

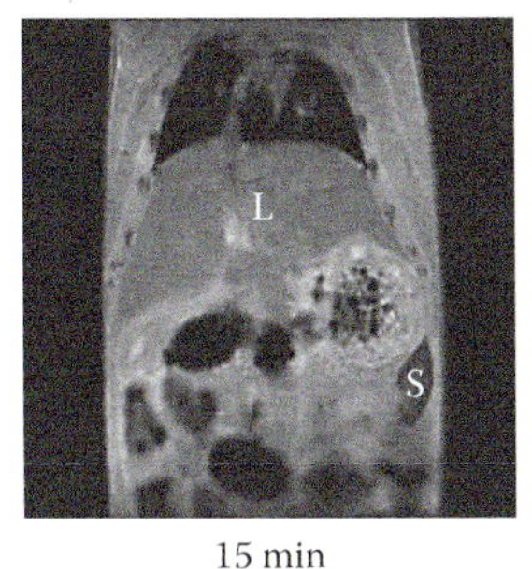

15 min

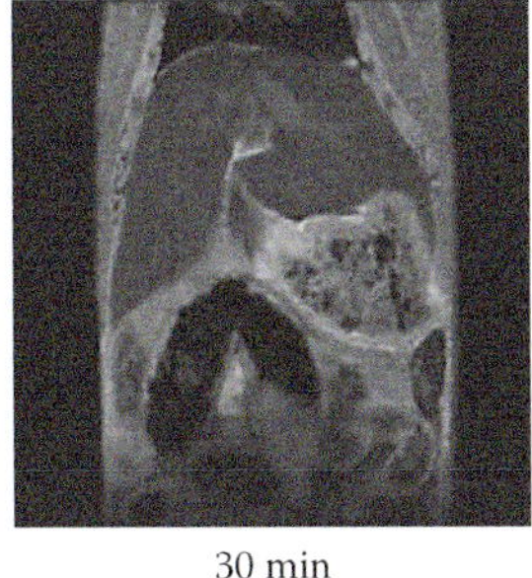
30 min

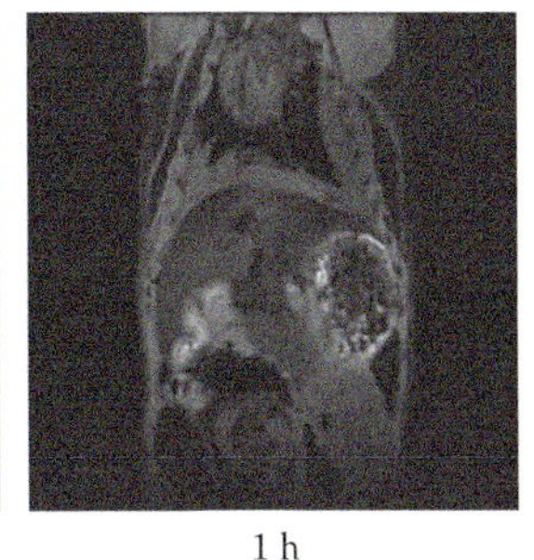
1 h

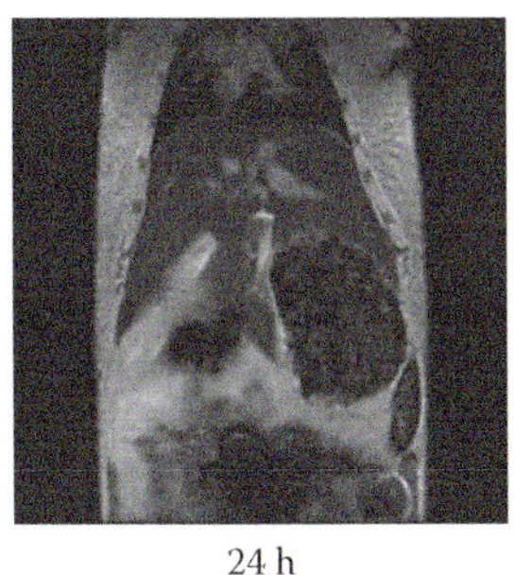
24 h

FIGURE 24.3 MR images obtained in a mouse after intravenous injection of magnetic human RBCs prepared by the adsorption of CSMNs onto RBC surface. A negative contrast enhancement of MR images of abdominal organs such as liver (L) and the spleen (S) was observed and interestingly this contrast increased during 24 h confirming the presence of magnetic hRBCs still circulating in this period of time. (Laurencin, M., Cam, N., Georgelin, T., Clément, O., Autret, G., Siaugue, J. M. and Ménager, C.: Human erythrocytes covered with magnetic core-shell nanoparticles for multimodal imaging. *Adv Healthc Mater.* 2013. 2(9). 1209–1212. Copyright Wiley-VCH Verlag GmbH & Co. KGaA. Reproduced with permission.)

The decoration of RBCs with aminated and carboxylated CSMNs has also been studied and elucidated by Mai et al.[70] These authors performed a series of experiments in which the charge of CSMNs was modulated, demonstrating that only aminated CSMNs could decorate the RBCs. Although phosphatidylcholine, sphingomyelin and cholesterol are the main components of the outer lipid membrane of RBCs, the RBC surface charge is mainly due to the carboxyl groups of N-acetylneuraminic (sialic) acid residues in the glycoproteins of the external surface. The adsorption is mainly ruled by the electrostatic attraction between the positively charged amino groups on CSMNs and the abundant sialic acid groups on the outer surface of RBCs.[70] The absorption of IONPs onto the RBC membrane also occurs by a specific binding through the interaction between ligand and receptor or by chemical conjugation. In fact, coating particles with antibodies or other binding RBC proteins offers the most specific loading.[58] Wang et al.[71] reported the attachment of IONPs coated with a photodynamic agent, chlorine e6 (Ce6) and functionalized with avidin onto the biotinylated murine RBC surface. Moreover, doxorubicin (DOX) was loaded inside the RBCs as a chemotherapy drug and the surface of the IONPs outside the RBCs was coated with PEG to prolong blood-circulation half-life and reduce accumulation in RES organs. The authors demonstrated that the RBC-based platform appears to have better *in vivo* pharmacokinetics than free nanoparticulate drug-delivery systems and a strong spleen retention of the RBCs after injection into healthy mice; however, the Ce6 fluorescent signals in the spleen and liver of tumour-bearing mice with magnetic tumour targeting were rather low, indicating that the external magnetic field may be able to attract a significant portion of those magnetic RBCs to the targeted tumour and reduce their accumulation in RES organs. In contrast, IONP-Ce6-PEG NPs not attached to RBCs show limited tumour uptake in response to the external magnetic field but dominant accumulation in the liver and spleen. The authors report significant tumour contrast as a result of highly effective magnetic field-enhanced tumour accumulation together with the enhanced r_2 relaxivity of these engineered magnetic RBCs. The use of this IONP/RBC-based drug delivery system could yield a strong tumour growth inhibition effect in a magnetic-field enhanced combined cancer therapy.

Other authors in a very recent work report a different strategy based on a rational design of RBC membrane camouflaged iron oxide magnetic clusters (MNC@RBCs) as an alternative to polymer-encapsulated Fe_3O_4 materials.[72] This investigation shows that by simply introducing an 'ultra-stealth' biomimetic coating to iron oxide magnetic nanoclusters (MNCs), the resultant MNC@RBCs have superior prolonged blood retention time. Therefore, MNC@RBCs, derived from the fusion of MNCs and RBC-derived membrane vesicles, could be more effective than other therapies since they maintain both the stealth properties of natural RBCs and the MRI and photothermal therapy capabilities of pristine iron oxide cores. When intravenously injected in a breast cancer xenograft mouse model, the MNC@RBCs show high tumour accumulation and relatively low liver biodistribution leading to greatly enhanced photothermal therapeutic efficacy by a single treatment without further magnetic force manipulation.[72]

Although these approaches seem to be promising, several papers report that the most commonly used method is IONP encapsulation based on transient opening of RBC membrane pores by exploiting the osmotic properties of RBCs. Typically the IONPs for imaging applications are encapsulated inside RBCs[73] taking advantage of the same principle of hypotonic swelling that opens up pores in the membrane, allowing NPs to diffuse into the cell before returning the cell to isotonic conditions to seal the pores.[74]

24.3.1.1 Loading Procedures for SPION Encapsulation in RBCs

The creation of a circulating carrier-RBC for sustained drug and/or nanomaterial delivery requires an optimized loading procedure resulting in a carrier bearing a strong similarity to native RBCs.

To date, few research groups have reported the encapsulation of MNPs using basically hypotonic haemolysis methods or hypotonic dialysis.

Thirty years ago, Sprandel et al.[75] showed that it was possible to entrap magnetite particles in erythrocyte ghosts paving the way for the potential use of magnetically responsive erythrocytes as carriers for *in vivo* drug targeting. The authors reported the first attempt to encapsulate ferromagnetic micromolecules, so-called ferrofluids, in human erythrocyte ghosts to promote drug targeting. Erythrocyte ghosts were prepared by a hypotonic dialysis procedure, as described previously.[76,77] However, the cytotoxic effect of the ferrofluids has been documented in morphological studies using scanning electron microscopy (SEM). These studies have also shown a relatively low entrapment ranging between 3% and 15% of the ferrofluid added to the cells during the encapsulation procedure. Moreover, only 20–30% of the erythrocyte ghosts containing ferrofluids appeared as normal biconcave discocytes. The rest were stomatocytes, echinocytes and cells with abnormal or destroyed forms.

A few years later, in 1994, Vyas and Jain reported the preparation of magnetically responsive ibuprofen-loaded erythrocytes[78] and diclofenac sodium-loaded erythrocytes[79] using the preswell dilution technique showing that it is possible to entrap drugs and magnetite in the erythrocytes of rats and rabbits. However, the *in vitro* osmotic fragility test, performed according to the method reported by Sprandel and Zollner,[80] showed that these drug- and magnetite-loaded erythrocytes have less resistance to osmotic fragility than normal and drug-loaded erythrocytes. Moreover, a higher percentage of the drug was released from drug loaded magnetic erythrocytes compared to drug-loaded erythrocytes indicating a more porous cellular membrane in the case of the magnetic cellular system, which could be related to the reduced deformity caused by magnetite. Thus, the observations were in agreement with the data previously reported by Sprandel et al.[75] Unfortunately, these authors did not perform *in vivo* experiments to test the performance of these drug-loaded magnetic erythrocytes, but since the procedure that was used led to crenated ghost cells, it is conceivable that these loaded cells have limited stability. In fact, nonviable, or dead, erythrocytes, such as crenated or ghost erythrocytes are immediately recognized by RES *in vivo*.

Others have reported the use of autologous red cells loaded with a ferromagnetic colloid compound and aspirin for intravenous administration to abort arteriothrombosis. The authors showed that it was possible to prevent a local arterial thrombosis in animal models, specifically, clot formation in the canine carotid artery and the central artery of the rabbit ear, by using aspirin-loaded magnetically charged red cells. The aspirin dosage administered to the animals did not interfere with the blood coagulation properties in a magnetic field-deprived artery despite its efficiency in the magnetically targeted vessel. Local concentration of magnetically charged red cells by means of a magnetic field could be used in vector drug delivery.[81]

More recently, the encapsulation of citrate-coated SPIONs into RBCs by a hypotonic haemolysis method was proposed by Brähler et al.[82] and Stenberg et al.[83]

In the loading procedure, RBCs and magnetite NPs were incubated under stirring in hypo-osmotic lysing buffer (Na_2HPO_4/NaH_2PO_4, 20 mOsm, pH 8) at 4°C for 1 h. This incubation was performed by a dilution of packed RBCs with 30 volumes of hypo-osmotic buffer. Then, free haemoglobin and excess Fe_3O_4 NPs were eliminated by washing the RBCs with PBS using a stirred filtration cell with a 3 μm membrane filter. The washed RBCs were resealed by incubation in phosphate-buffered saline at 37°C for 1 h. However, as demonstrated by TEM analyses, SPIONs were not only distributed inside the cells but also strongly attached on their surface (Figure 24.4a). This cell surface modification constitutes a limitation for the application of these loaded cells *in vivo* because it could activate elimination of the cells by the immune system.

The same authors subsequently attempted to reduce RBC membrane damage by applying different variations of the dilutional hypotonic haemolysis method,[83] showing that a careful handling of the cells during the whole procedure, namely dilution of the RBCs in a buffer with higher osmolarity and in a volume ratio of 1:1–2, can better preserve membrane integrity.

In 2011, Cinti et al.[84] developed a new drug delivery system based on erythro-magneto-HA virosomes that, thanks to both their magnetic and highly efficient fusion properties, can be used to specifically drive the drug at target organs or tissues releasing the therapeutic compound directly at the intracellular level of target cells, including malignant cells. The preparation method of erythro-magneto-HA virosomes consists in attaching a viral spike fusion glycoprotein (haemagglutinin, HA) to the erythrocytes' membrane and encapsulating superparamagnetic NPs and drugs into the erythrocytes. Briefly, human erythrocytes were incubated with a lysis buffer (10 mM TRIS, 0.1 mM EDTA, 1 mM $MgCl_2$ at pH 7.2) for

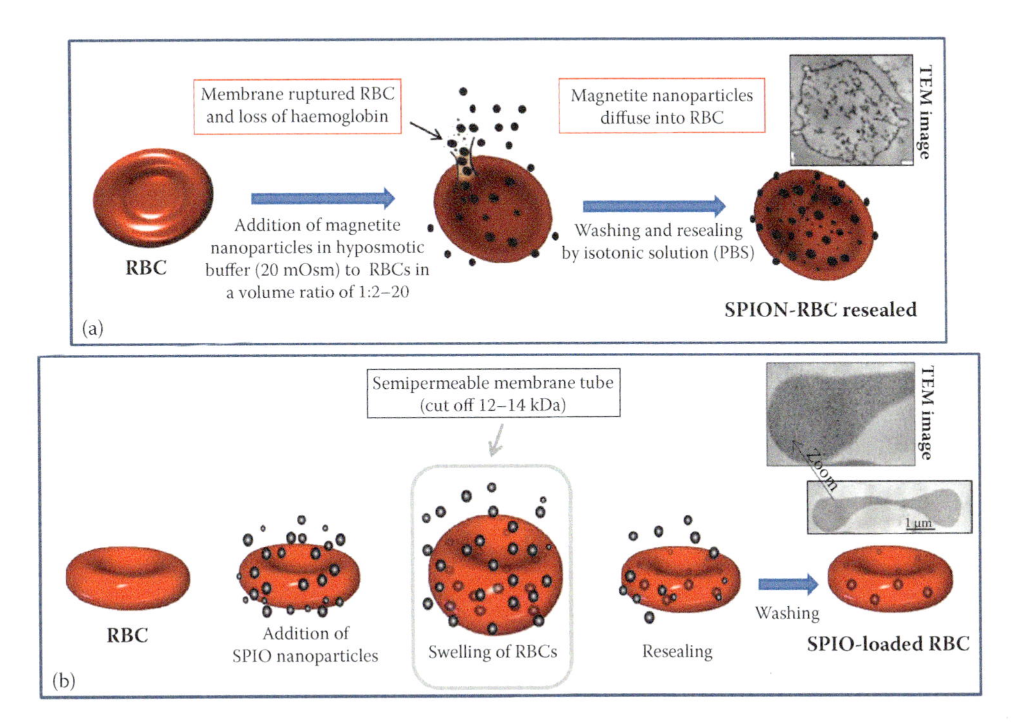

FIGURE 24.4 (a) Schematic representation of osmotic shock by the hypotonic dilution method, used by Brähler et al.[82] to induce swelling of RBCs and simultaneous diffusion of SPIONs into cells. TEM image of SPION loaded-RBC showed that magnetic NPs were located mainly inside cells forming aggregates of different size and also strongly incorporated in the membrane. (b) Schematic representation of the hypotonic dialysis method used by Antonelli et al.[85–88] The isotonic suspension of RBCs and SPIO NPs is placed in a dialysis tubing and the dialysis bag is immersed in a hypotonic solution under gentle stirring. Resealing of RBCs is obtained by adding a resealing buffer solution containing concentrated potassium chloride to achieve isotonicity. TEM image of SPIO-loaded RBC showed the presence of magnetic NPs uniformly distributed only into cell cytoplasm and not bound to the external membrane surface.

60 min at 0°C and the isotonicity was restored by a resealing buffer supplemented with 100 nm red-labeled superparamagnetic NPs, HA influenza viral spike glycoprotein and the anticancer drug 5-Aza-2-deoxycytidine. These engineered erythrocytes are able to fuse with the cytoplasmatic membrane of target cells in a very short time and release the MNPs and therapeutic compound inside these cells. To date, the hypotonic dialysis method appears to be the most promising for encapsulation of superparamagnetic IONPs into RBCs since it makes it possible to obtain a high loading efficiency, high cell recovery and viable cells that avoid RES uptake as shown by Antonelli et al.[85,86] Since 2008, these authors have evidenced that it is possible to encapsulate MNPs in human and murine RBCs with a procedure that permits a transient opening of the cell membrane pores by controlled hypotonic dialysis and the subsequent isotonic resealing and reannealing of the cells (Figure 24.4b), without changing the main features of the natural cells. The hypotonic solution induces a transient opening of pores in the RBC membrane, so that particles can enter the cells. After resealing the pores in an isotonic buffer, particles are trapped in the RBCs. Importantly, it has been noted that the MNPs should be monodispersed and sufficiently small so that they can readily pass into the RBCs when they become porous upon exposure to the hypotonic dialysis solution. The size range of the particles includes SPIOs, such as Resovist® and Endorem® and USPIOs, such as Sinerem® NPs, which can be loaded successfully into RBCs. The total preparation procedure typically results in a cell recovery of loaded RBCs ranging from 60% to 70%, similar to the recovery rate for unloaded cells. Moreover, several biological features such as mean corpuscular volume (MCV), mean haemoglobin concentration (MCH) and mean corpuscular haemoglobin (MCHC) of USPIO/SPIO-loaded RBCs resultant after the loading procedure are similar to those of untreated cells, showing that these magnetic-RBC carriers retain the same properties as those of native red cells. As shown by SEM, the cell morphology of loaded RBCs is not significantly different from that of control cells; the majority of these magnetic-RBCs appear to have a biconcave discoid shape with occasional stomatocytes and rarer echinocytes. It is notable, as shown by TEM images, that the magnetic nanomaterial is homogeneously distributed in the cytoplasm inside of the RBCs and not bound to the external membrane surface of the cells (Figure 24.5, Table 24.1). Nuclear magnetic resonance (NMR) measurements showed that SPIO loaded RBCs, such as Resovist®-, Endorem®- and Sinerem®-loaded RBCs, contain final iron concentrations ranging from 5.3 to 16.7 mM for human RBCs and from 1.4 to 3.55 mM for murine RBCs.[87] Antonelli et al.[88] have also applied the loading procedure to a new USPIO NP suspension, P904 iron oxides coated with hydrophilic derivatives of glucose (developed by Guerbet Laboratories). The results showed that this nanomaterial can be efficiently loaded into human and murine RBCs at iron concentrations ranging from 1.5 to 12 mM, Table 24.2.[88] These USPIO- or SPIO-loaded RBCs are responsive to external magnetic fields, namely the field created by a magnet, and their relaxation rate (R), which is around one order of magnitude higher than the value of control cells, is maintained for several days.[85]

24.3.1.2 Not All SPIO NP Can Be Encapsulated into RBCs

Antonelli et al.[88,89] have tested several SPIONs, both commercially available and newly synthesized, for encapsulation into RBCs, and have found that not all NPs can be successfully loaded into erythrocytes. Different parameters, such as size, coating and/or dispersant agents used to obtain NPs in monodispersed form, are very important in order to obtain efficient loading into RBCs. Dextran or carboxydextran coated SPIONs, such as Resovist®, Sinerem®, Endorem® and PMP-50, can be successfully loaded into RBCs; likewise, newly synthesized Np-1 NPs dispersed in the Disperbyk®-190 agent can be efficiently encapsulated into RBCs.[89] Other NPs

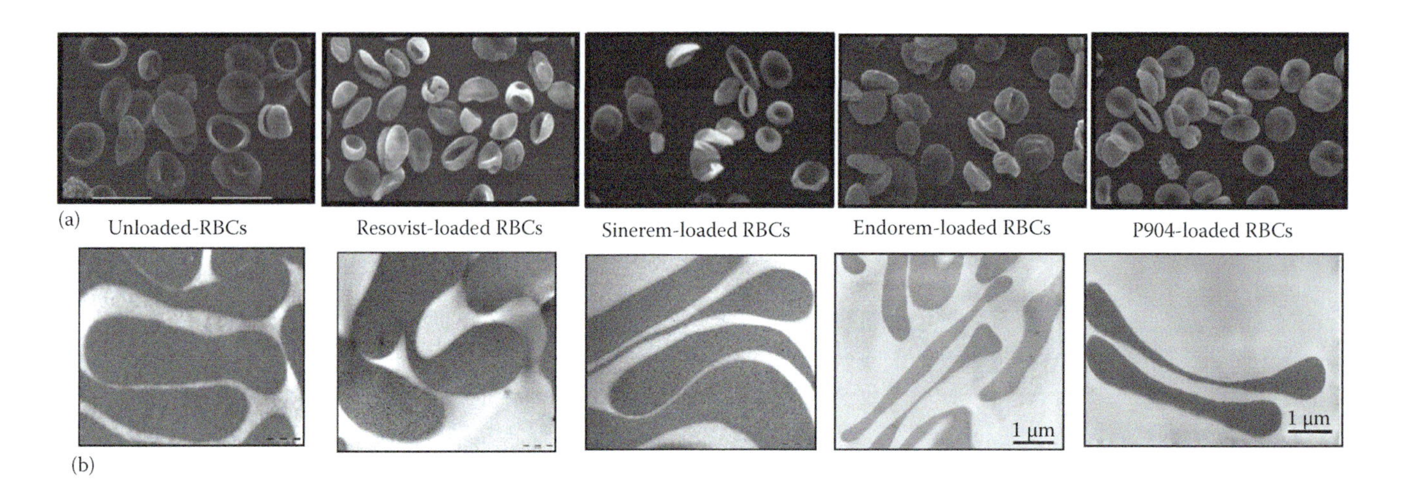

FIGURE 24.5 SEM and TEM analyses of control unloaded- and USPIO/SPIO NPs-loaded RBCs obtained with the hypotonic dialysis method reported by Antonelli et al.[85–90] (a) SEM images showed RBCs loaded with SPIO or USPIO NPs with normal cell morphology comparable to that of control cells. (b) TEM images of these USPIO/SPIO-RBCs constructs showed the effective encapsulation of magnetic NPs in cells and their homogeneous distribution in monodispersed form only into RBC cytoplasm.

TABLE 24.1
Characteristics of SPIO and USPIO NPs (Produced by Different Companies) That Were Encapsulated into Human and Murine RBCs through the Loading Method Reported by Antonelli et al.[85–90]

Short or/and Trade Name	Obtained From	[Fe] (mg/ml)	Coating	Hydrodynamic Diameter (nm)
SHU 555A (Resovist®)	Bayer Healthcare	28	Carboxydextran	62
AMI 227 (Sinerem®)	Guerbet	20	Dextran	20–40
PMP-50	Kisker	10	Dextran	50
AMI-25 (Endorem®)	Guerbet	11.2	Dextran	80–150
P904	Guerbet	28	Amino-alcohol derivative of glucose	21
Ferucarbotran	TOPASS GmbH	59.2	Carboxydextran	45–65

Sources: Antonelli, A. et al., *J Nanosci Nanotechnol*, 8, 2270–8, 2008; Magnani, M. and Antonelli, A. *International Patent Application* No. PCT/EP2007/006349, July 3, 2007; Antonelli, A. et al., *PLoS One*, 8(10), e78542, 2013; Antonelli, A. et al., *Medical Imaging 2013*: *Biomedical Applications in Molecular, Structural, and Functional Imaging*, Weaver, J. B. and Molthen, R. C. (Eds.) SPIE, Bellingham, 8672, 86721D, 2013; Antonelli, A. et al., *Nanomedicine*, 6(2), 211–223, 2011; Antonelli, A. et al. Proceeding presented at: 5th International Workshop on Magnetic Particle Imaging (IWMPI). Istanbul, Turkey, 26–28 March 2015.

TABLE 24.2
Comparison of Main *In Vitro* and *In Vivo* Properties of Three Types of Murine SPIO-Loaded RBCs

Properties	Ferucarbotran-Loaded RBCs	Resovist®-Loaded RBCs	P904-Loaded RBCs
MCV (*fL*)	40	40	39
MCH (*pg*)	14.3	14.6	14.1
MCHC (*g dL*$^{-1}$)	33.3	31	36.4
Cell recovery after loading procedure (%)	35–50	15–35	30–50
Iron encapsulated (*mM*)	1.25–12.0	1.4–3.55	1.5–10
24h survival (*% of injected dose*)	60 ± 3.4	33 ± 1.12	56 ± 2.15
Half-life in bloodstream	Exponential half-life: 4.5 days	Biphasic half-life: 1st 5 min; 2nd 4.1 days	Exponential half-life: 5.5 days

Note: Values of MCV, MCH, MCHC and cell recovery were obtained by using an automated haemocytometer on SPIO-loaded RBC samples at the end of loading procedure. The concentrations of iron encapsulated into SPIO-loaded RBCs and their half-life in the bloodstream were calculated by T_1 NMR measurements. Iron concentration encapsulated in Ferucarbotran-, Resovist®- and P904-loaded RBCs was calculated using r_1 value relaxivity obtained as reported.[87–89] The 24 h survival was obtained after calculation of Fe µmoles from T_1 NMR value, contained in blood withdrawn from mice at 24 h post SPIO-loaded RBC administration expressed as percentage of total Fe injected dose. All values correspond to the means ± SD of at least three experiments. The *in vivo* half-life ($t_{1/2}$) of SPIO-loaded RBCs was determined using $t_{1/2} = (\ln 2)/K$.

(e.g. USM, ultra small magnetite; USM14A, silica coated ultra small magnetite; USM5, 4% Disperbyk®-190 coated-ultra small magnetite; USMSiO$_2$ITween20, Tween20-silica coated ultra small magnetite; USMdextran, dextran coated ultra small magnetite) tested in the authors' laboratory do not appear suitable for encapsulation into RBCs (Figure 24.6). Moreover, some NP suspensions, namely USMSiO$_2$ITween20 and USM-5, showed a haemolytic effect on RBCs and, consequently, a very low final cell recovery (13–16%). This haemolytic effect depended on both the NP synthesis protocol (e.g. dispersant agent addition during or postsynthesis) and the amount of dispersing agent used; for example, the addition of 4% (w/v) Disperbyk®-190 during the synthesis of USM5 NPs led to a lower cell recovery than that obtained with the Np1 NPs prepared with the postsynthesis addition of the same dispersing agent (10% w/v Disperbyk®-190). In fact, Np1 NPs do not appear to affect RBC viability, permitting cell recovery at the end of the loading procedure similar to which occurs in control cells. These results show that among the coating materials used for the achievement of stable and monodispersed NP colloidal suspensions, the dextran and Disperbyk polymers appear to be the most suitable for NP encapsulation into RBCs. However, although USMdextran NPs were synthesized using dextran as a dispersing agent, they did not show the same level of performance in terms of encapsulation as the dextran-coated commercial SPIOs (e.g. Resovist® and Sinerem®).

Np-2 NP suspension (tetramethyl ammonium hydroxide coated) is another non-optimal material for encapsulation into RBCs since, at the end of the loading procedure, particles are present in aggregated forms on the extracellular RBC surface.[89]

Thus, the production of new MNPs needs to be further investigated in order to produce materials suitable for loading into erythrocytes, which so far have shown the longest *in vivo* half-life in circulation. This is a very important consideration especially with regard to the imaging of the circulatory system.

As reported above, a significant challenge associated with the biological application of IONPs is their behaviour *in vivo*. Their utility is often compromised when these IONPs, including materials used as contrast agents for bowel, liver and spleen imaging,[11,91,92] are recognized and cleared by the body's major defence system.

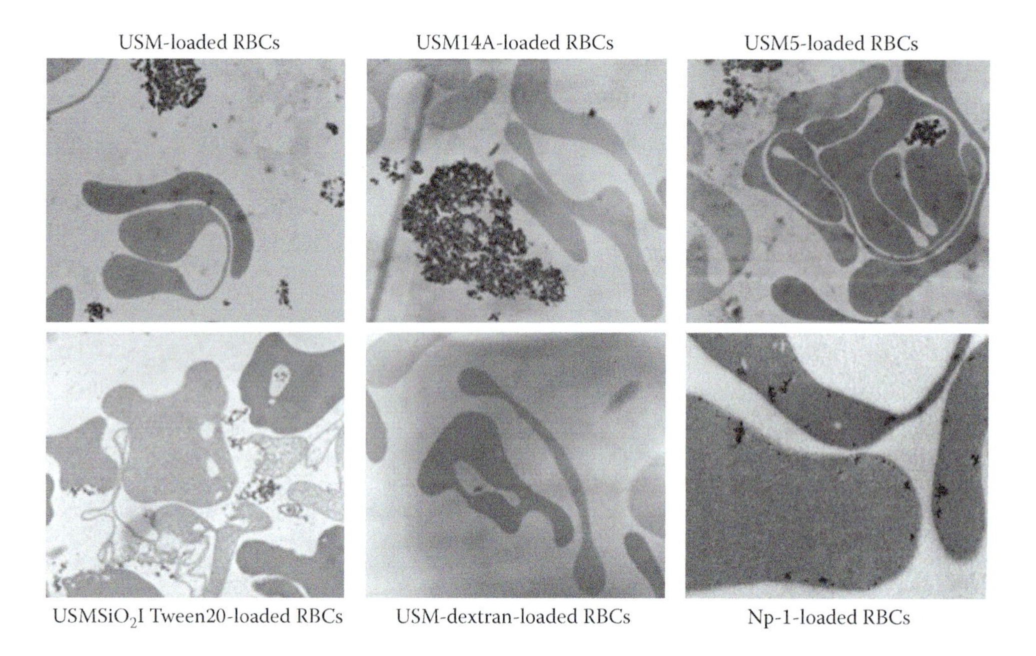

FIGURE 24.6 TEM analysis of human erythrocytes loaded with different types of newly synthesized magnetic NPs. TEM images showed the presence of aggregated magnetic NPs mainly in the extracellular space or on the membrane surface of RBC-loaded samples obtained using these nanomaterials. The results demonstrated that internalization of magnetic NPs such as USM (ultra small magnetite), USM14A (silica-coated ultra small magnetite), USM5 (4% Disperbyk®-190-coated ultra small magnetite), USMSiO2ITween20 (Tween20-silica-coated ultra small magnetite), USM-dextran (dextran-coated ultra small magnetite) into erythrocytes is not possible. On the contrary, Np-1 NP sample appears to be internalized in RBCs. (From Antonelli, A., Sfara, C., Manuali, E., Bruce, I. J. and Magnani, M. *Nanomedicine (Lond.)*, 6(2), 211–223. Copyright 2011, Future Medicine Ltd. Reproduced with permission.)

Recently, Antonelli et al.[93] reported the encapsulation of a new Ferucarbotran suspension obtained from TOPASS GmbH (Berlin, Germany). The new suspension is very similar to Resovist®, which is currently available only in Japan.

Ferucarbotran consists of a hydrophilic colloidal suspension of superparamagnetic IONPs with a hydrodynamic diameter of 45–65 nm and coated with carboxydextran. The data show that the amount of ferucarbotran NPs incorporated into human RBCs increases in proportion to the iron dose that is used. The loading procedure, performed using Ferucarbotran amounts ranging from 2.8 to 22.4 mg of iron added to 1 ml of dialyzed RBCs, led to iron concentrations in human RBCs ranging from 1.6 to 16.4 mM. At the end of the procedure, Ferucarbotran-loaded RBCs showed a cell recovery of 64–70%, which is similar to that of unloaded cells. In contrast, the loading procedure performed with murine RBCs led to a cell recovery of only 35–50%. Moreover, murine Ferucarbotran-loaded RBCs showed a T_1 value range from 486 to 58.8 ms (vs. 2370 ms for control RBCs) corresponding to an encapsulation of iron in the range of 1.25–12 mM (see Table 24.2). TEM analyses of human and murine Ferucarbotran-loaded RBCs showed that there were no iron oxides in the extracellular space or adhering to the cell surface of Ferucarbotran-loaded RBCs. In fact, the Ferucarbotran NPs were only present in the cell cytoplasm, showing no evidence of aggregation. This is an important result since the presence of NPs on the external surface of the RBC membrane would activate loaded RBC recognition by RES macrophages, leading to their rapid elimination from the bloodstream.

To test the viability of the loaded RBCs, the authors first administered human Ferucarbotran-loaded RBCs *in vitro* to human macrophages to study their interaction and therefore evaluate if these RBC-constructs were still viable cells able to avoid macrophage uptake. Second, the authors studied the *in vivo* survival of murine Ferucarbotran-loaded RBCs and iron biodistribution in a mouse model,[93] the results of which will be discussed in the next section.

24.3.2 *In Vivo* Delivery of MNPs by RBCs

Undoubtedly, the circulation time of SPIO-loaded RBCs strongly depends on the loading procedure and the properties and coating of the nanomaterials that are used. Such procedures should maintain the biochemical and physiological characteristics of native cells as well as membrane integrity. Indeed, the selection of the NPs and the optimization of the loading procedure to create biomimetic RBC carriers for drug and/or nanomaterial delivery is vitally important.[44]

The MNP-RBC constructs proposed by Antonelli et al. showed *in vivo* biocompatibility and stability since they were recognized as viable cells by the host. The authors demonstrated that iron oxide-based NPs encapsulated in murine RBCs and intravenously injected in mouse bloodstream are effective biomimetic constructs remaining in circulation much longer than free nanomaterial, extending the blood retention time toward the half-life of RBCs, which is around 11 days in mice[94] and 30 days in humans.[95] Hence, the use of SPIO-loaded RBC constructs overcame the limitations of the current contrast agents used for

MRA applications. For example, the murine Resovist®-loaded RBC blood retention time is ~12 days while free Resovist® retention time at the same iron concentration is ~1 h (Figure 24.7).[85,87] These results appear to be significant considering that, although iron oxide-based contrast agents have major advantages over gadolinium chelate, their stability in blood circulation is very short-lived, and contrast agents such as Resovist® have only been used for MRI of the liver and spleen. Resovist® has an effect on the shortening of both T_1 and T_2 relaxation times; however, due to the high r_2 relaxivity, it is more suited to T_2/T_2^*-weighted imaging.[11] In fact, the use of Resovist® for MR angiography (MRA) is not feasible since its T_1 effect does not allow the commonly requested elaboration of data into maximum intensity projections.[92] Moreover, none of the available SPIO agents have been approved for MRA; the only USPIO particle approved to date, Ferumoxytol (Feraheme; AMAG Pharmaceuticals, MA, USA), has been approved in the United States by the FDA for treating iron-deficient anemia in adults with chronic kidney disease.[24] There is an urgent need for the development of novel contrast agents that overcome low efficiency, poor biodistribution, as well as safety issues for highly efficient angiography.

As mentioned above, when applied to P904, the loading procedure used for Resovist also results in an efficient nanomaterial encapsulation both into human and murine RBCs.[88]

Boni et al.[96] have reported a relaxometric study of these P904-loaded RBC constructs, consisting of longitudinal (r_1) and transverse (r_2) relaxivity measurements over a wide range of Larmor frequencies (0.01–300 MHz) in comparison with control samples with P904 NPs dispersed in blood. The internalization of P904 into RBCs resulted in smaller r_1, and in a very high r_2/r_1 ratio (232) in the highest field (300 MHz). Moreover, a shift of the Curie peak to high fields was observed in P904-loaded RBCs, possibly the result of NP size selection caused by the internalization process as shown by Markov et al. for Resovist®- and Sinerem®-loaded RBCs.[97]

In vivo experiments performed in mice showed a higher survival rate for P904 in circulation when encapsulated into RBCs compared to free P904. The data showed a half-life (>120 h) of P904-loaded RBCs, which was longer than the half-life (<2 h) of free P904. The presence of P904-loaded RBCs in the mouse bloodstream can be detected for more than 1 month until T_1 values return to control values. An

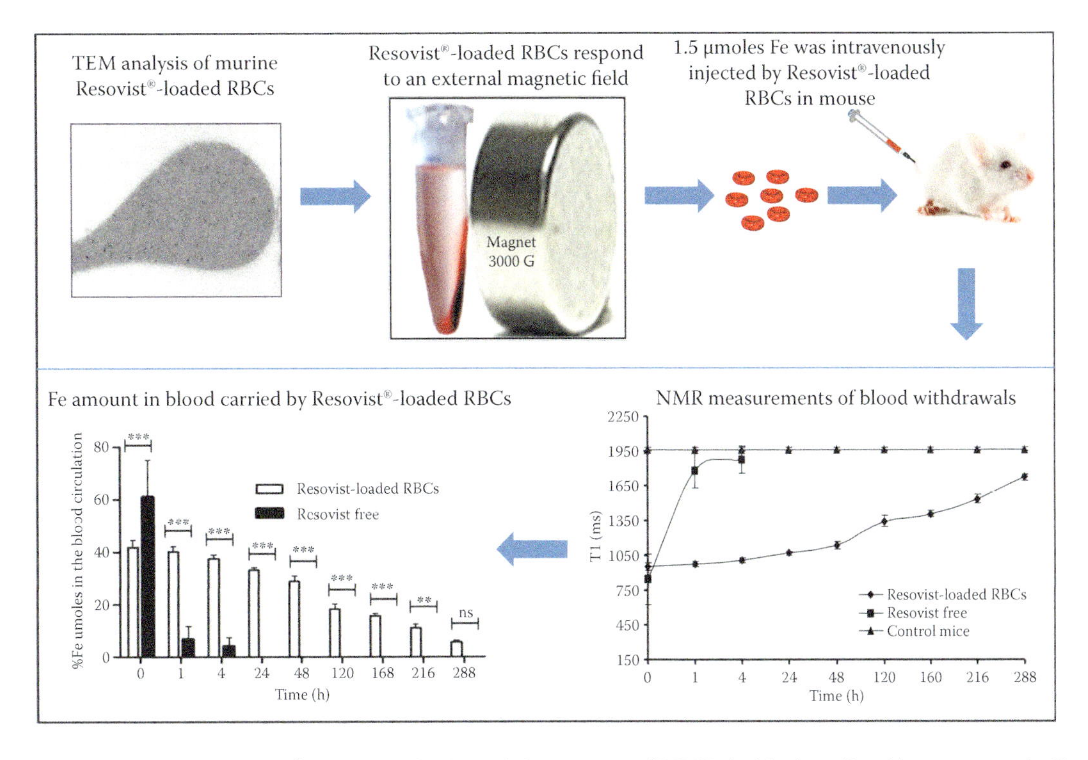

FIGURE 24.7 TEM analysis of Resovist®-loaded RBCs showed the presence of IONPs inside the cell and homogeneously distributed only in the cytoplasm and were not bound to the membrane surface. These RBCs constructs were responsive to an external magnetic field and moved toward the attached magnet. After the intravenous injection of murine Resovist®-loaded RBCs in mice, survival of these magnetic RBCs in the vascular system was investigated by T_1 NMR measurements. Three mice (32 g) received a single bolus injection in the tail vein of 500 µl of a murine Resovist®-loaded RBC suspension (3 mM Fe) at 44% haematocrit corresponding to 1.5 µmol Fe total (or 46.8 µmol Fe/kg). Three other mice received a single bolus injection of bulk Resovist® suspension at the same concentration. T_1 NMR measurements were performed on blood withdrawn from mice at different times and the concentrations of magnetic iron carried in blood circulation by Resovist®-loaded RBC were determined by using r_1 value relaxivity as reported.[87] Fe µmoles in the blood system of mice treated with Resovist®-loaded RBCs or bulk Resovist® suspension were expressed as percentage of total iron injected dose. All values correspond to the means of three experiments and the error bars show the standard deviations. Asterisks indicate statistical significance between Resovist®-loaded RBCs and bulk Resovist® treated mice. * $p < 0.05$; ** $p < 0.01$; *** $p < 0.001$. The graphs reporting NMR results were reproduced with permission.[87] (From Antonelli, A., Sfara, C., Battistelli, S., Canonico, B., Arcangeletti, M., Manuali, E., Salamida, S., Papa, S. and Magnani, M. *PLoS One*, 8, e78542. Copyright 2013, PLOS ONE.)

equivalent amount of free P904 suspension injected into mice disappeared from circulation within only a few hours. Their very long blood half-life and high r_2 relaxivity make P904-loaded RBCs a promising blood-pool negative contrast agent for MR diagnostic applications. In addition, the new Ferucarbotran-RBC constructs recently described by Antonelli et al.[93] could be suitable to improve diagnostic imaging procedures. The authors showed, for the first time, how free Ferucarbotran injected into the bloodstream of mice immediately induces a decrease in blood T_1 values (730 ms ± 200 vs. 2354 ms ± 102) that is still measurable at 1 h (1705 ms ± 171) but that disappears within 3 h (2124 ms ± 132) from injection. In contrast, Ferucarbotran-loaded RBCs similarly induce a blood T_1 value decrease after injection (605 ms ± 104), but the T_1 decrease is maintained for up to 9 days, at which time the values are still significantly lower (1753 ms ± 59) than those of control mice. The decrease in T_1 values persists for up to 12 days, after which they rise to values similar to those of control mice.[91] Ferucarbotran-loaded RBCs that showed a half-life of about 4 days (Table 24.2) and free ferucarbotran (half-life <1 h) are eliminated in the same way, but they have different *in vivo* kinetic removal. In fact, ICP-OES analyses showed that the same amount of iron sequestered at 24 h by the liver of free Ferucarbotran-treated mice is instead detectable at 13 days in the liver of mice treated with Ferucarbotran-loaded RBCs (70% ± 1). Furthermore, at 13 days the amount of iron in the liver of mice treated with Ferucarbotran-loaded RBCs is also higher than the amount measured in free ferucarbotran treated mice (14% ± 9). All of these data point to potentially very interesting applications in clinical settings to improve MRI diagnosis. The increased stability and viability of the SPIO-based contrast agents loaded into RBCs would allow the same patient to be imaged on a number of occasions over time. Clearly, MRI *in vivo* imaging of animal models treated with SPIO-loaded RBCs could contribute to the improvement of actual diagnostic and therapeutic applications including cerebral blood volume-weighted functional MRI sensitivity[98] and the monitoring of vessel maturation in angiogenesis and cardiovascular diseases.[99,100] Furthermore, it is noteworthy that automatic loading of the drug/nanomaterial in RBCs has already been achieved by companies such as EryDel SpA in Italy (www.erydel.com)[101] and Erytech Pharma in France (www.erytech.com).[102]

These companies are specialized in the development of drugs and diagnostics delivered through human RBCs via a proprietary medical device (to encapsulate the therapeutic agent) that is not yet commercially available; nevertheless, clinical trials are being carried out on volunteer patients for the treatment of chronic inflammatory[103] and neurological diseases.[104]

24.4 SPIO-LOADED RBCs AS TRACERS

The recent development of SPIO-RBC constructs could greatly enhance imaging technologies, including MRI or combined multimodality imaging, but these constructs could also be useful for magnetic particle imaging (MPI), a novel real-time imaging technique introduced by Philips in 2005.[105] This technique is different from MRI[106] because it is able to directly visualize magnetic particles rather than their effect on proton relaxation, which is the basis of MRI detection. MPI therefore allows direct and quantitative visualization of IONP concentration without confounding background signals.[107–111] MPI takes advantage of the nonlinear magnetization response of superparamagnetic NPs to applied magnetic fields and promises to deliver high spatial and temporal resolution with a sensitivity exceeding that of MRI. A comprehensive coverage of MPI can be also be found in Chapter 20, 'Magnetic Particle Imaging for Angiography, Stem Cell Tracking, Cancer Imaging and Inflammation Imaging' by P. Goodwill et al. in the book *Magnetic Nanoparticles: From Fabrication to Clinical Applications*.[112]

Commercially available contrast agents that are currently being used in preclinical MPI studies are not optimized and thus do not allow the MPI technique to attain optimal resolution and sensitivity. The properties of MNPs are strongly dependent on their size, which often exhibit a broad distribution. To this end, in many applications only a small proportion of particles contribute to the desired magnetic effect. This applies in particular to the MPI performance, which is strongly influenced by the magnetic characteristics of the tracer, particularly the core size and anisotropic contributions, i.e. shape and crystal structure of tracers.[113,114]

Using tracer material formulated for intravenous injection, the obvious MPI applications are those that can take advantage of the fact that the material remains in the bloodstream for a certain length of time. In particular, due to its high temporal resolution, MPI will be especially suited for functional cardiac diagnosis, including angiography, cardiac wall motion assessment and quantitative myocardial perfusion imaging.[112] In the future, MPI might also be a safe alternative for CT and MRI angiography in patients with chronic kidney disease who cannot tolerate iodine- or gadolinium-based contrast agents.[115–118]

The SPIO-based MRI liver contrast agent Resovist® has been successfully used as an *in vivo* MPI tracer for real-time visualization of the bolus passage through a mouse heart.[119,120] However, the properties of commercial tracer materials affect MPI image quality, and it has been shown that the MPI signal is mainly generated by particles with a 30 nm magnetic core diameter.[105,121]

However, Resovist® contains IONPs with an average nanocrystal size of 4 to 6 nm. Larger-sized iron oxide nanocrystal cores that display steeper magnetization curves would be more effective in MPI. It was estimated that only ~3% of the particles in Resovist® colloidal suspension exhibit optimal NP diameter able to contribute to significant MPI signals.[105]

In fact, from the broad spectrum of particle sizes found in Resovist®,[122] MPI mainly makes use of particles with a core diameter larger than 15 nm, which are the particles that are removed most rapidly from the bloodstream by the mononuclear phagocyte system. Thus, the time frame for bolus-based measurements is only a few minutes. Moreover, within this time frame, the particle concentration does not maintain a constant level because a substantial number of particles are already removed during the dilution of the particle bolus in the blood.

In this context, the use of SPIO-loaded RBCs produced by Antonelli et al. has been investigated to assess their potential as new MPI blood pool tracer materials. Markov et al.[97,123] have investigated for the first time the magnetic characterization of Resovist®- and Sinerem®-loaded RBC constructs as long half-life MPI blood tracer agents for applications in imaging of the circulatory system. NMR measurements performed to quantify iron oxide concentrations within loaded cells were benchmarked against the results obtained by magnetic particle spectroscopy (MPS)[124] and vibrating sample magnetometry (VSM)[125] in combination with inductively coupled plasma-optical emission spectroscopy (ICP-OES).[126] Based on magnetic moment saturation levels analyzed by VSM, tracer iron concentrations were derived using bulk Resovist® and bulk Sinerem® magnetization curves as a reference. SPIO-loaded RBCs magnetic moments normalized by tracer iron content (NMR and VSM based) were characterized as a function of frequency with MPS. The MPS signal of 50 μl of bulk tracers or Resovist- and Sinerem-loaded RBCs was acquired over 30 s upon application of an oscillating magnetic field with an amplitude of 10 mT/μ 0 at 25 kHz. After Fourier transform of the output signal, the spectrum, which is related to the nonlinear contribution in the magnetization response, shows odd multiple harmonics of the base frequency of 25 kHz.

The results show a reduced MPS signal (up to one order of magnitude) for Resovist®- and Sinerem®-loaded RBCs compared to bulk tracers. This could be attributed to a preferential encapsulation in cells of smaller NPs, which are not optimal as MPI tracer materials. In fact, the presence of NPs with a core diameter ≥30 nm in Resovist® accounts for its superior performance over Sinerem®. The change in hydrodynamic diameter particle distribution upon encapsulation explains the reduction of the MPS signal for SPIO-loaded RBCs. With typical loading parameters, particles with a hydrodynamic diameter smaller than 60 nm are preferentially entrapped. Since the hydrodynamic diameter strongly correlates with the diameter of the iron core, the peak of the core size distribution of entrapped particles is shifted to smaller diameters and also drops much more steeply toward larger diameters compared to bulk NP suspensions (see Figure 24.8).

A preferential encapsulation of smaller NPs with a mean core diameter of 4 to 7 nm, which hardly contribute to the MPS signal, could be the mechanism that accounts for MPS signal suppression. Nevertheless, this is counterbalanced by the superior *in vivo* stability of SPIO loaded-RBCs compared to free NPs, which are rapidly removed from the bloodstream. This has been validated by Rahmer et al., who reported the

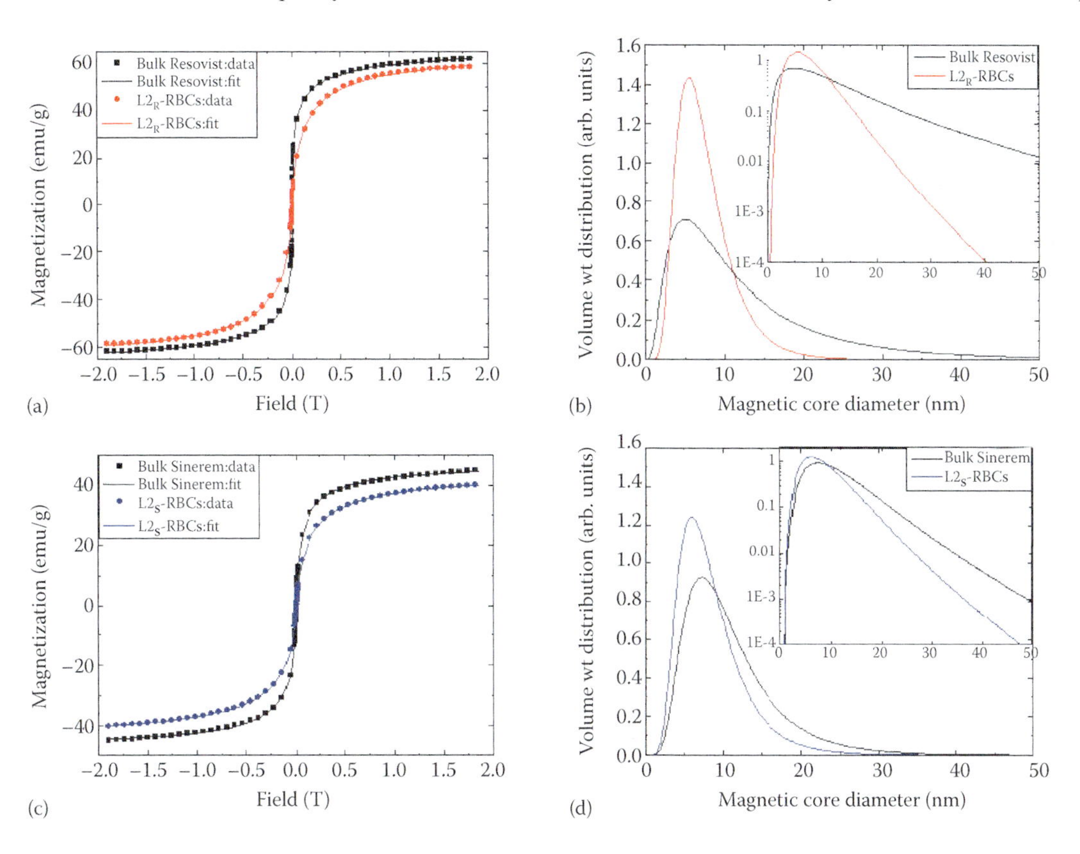

FIGURE 24.8 Quantification of relative contribution of magnetic NPs with a different core size in the magnetization curves for bulk tracers Resovist® and Sinerem and for respective tracer-loaded RBCs (Resovist®-loaded RBCs and Sinerem-loaded RBCs). A generic Langevin model with the volume weighted log normal distribution of the magnetic core diameter was used. (a). Resovist® and Resovist®-loaded RBC magnetization curves. Symbols show experimental data, solid lines – indicate simulations. (b). Magnetic core diameter distributions derived from simulations for Resovist® and Resovist®-loaded RBCs. (c). Sinerem and Sinerem-loaded RBC magnetization curves. Symbols show experimental data, solid lines indicate – simulations. (d). Magnetic core diameter distributions derived from simulations for Sinerem and Sinerem-loaded RBCs. (From Markov, D. E., Boeve, H., Gleich, B., Borgert, J., Antonelli, A., Sfara, C. and Magnani, M. *Phys Med Biol* 55, 6461–6473. Copyright 2010, Institute of Physics and Engineering in Medicine. Reproduced with permission.)

first *in vivo* MPI[127] of mice using Resovist®-loaded RBCs. Despite the reduced sensitivity and resolution resulting from the size selection effect and the difficulties related to the application of the loading procedure to mice instead of humans, the sensitivity of the experimental MPI instrumentation allowed dynamic imaging up to several hours after Resovist®-loaded RBC injection. In fact, MPI *in vivo* experiments showed the feasibility of detecting a heartbeat in mice after intravenous injection of murine Resovist®-loaded RBCs. This was demonstrated by visualization of cardiac motion with current MPI equipment and by the determination of the concentration of iron-loaded RBCs in the blood using MPS. The MPS data obtained after injection of Resovist®-loaded RBCs showed that these magnetic RBC constructs can circulate in mouse blood for many hours. In fact, MPS measurements of blood withdrawals from mice treated with Resovist®-loaded RBCs show that 50% and 30% of the administered dose are still visible at 3 h and 24 h after intravenous injection, respectively.[127,128] The achieved iron concentrations in the blood are close to the sensitivity limit of current equipment, but the real-time information delivered by the rapid 3D MPI acquisition provides enough additional information to allow assignment of signal to certain anatomies. It was only in the scan performed 3 h after injection that a clear signature of the beating heart and thus of the signal generated from loaded RBCs still flowing in the blood could be observed. A dynamic data evaluation was used to discern heart motion from the signal in the liver, which is mainly subject to breathing motion.[127] After 24 h, the signal of particles accumulated in organs like the liver dominated the images. It is assumed that during the loading procedure, a substantial number of murine RBCs were slightly damaged and subsequently rather rapidly removed by the liver and spleen. Due to their higher membrane fragility, the loading process is slightly different for murine RBCs, and cell recovery rates are typically only in the range of 25 to 30% compared to a cell recovery rate of 60 to 70% achieved with human RBCs. Figure 24.9 shows MPI images of the heart region of the living mouse acquired 3 h and 24 h after Resovist®-loaded RBC

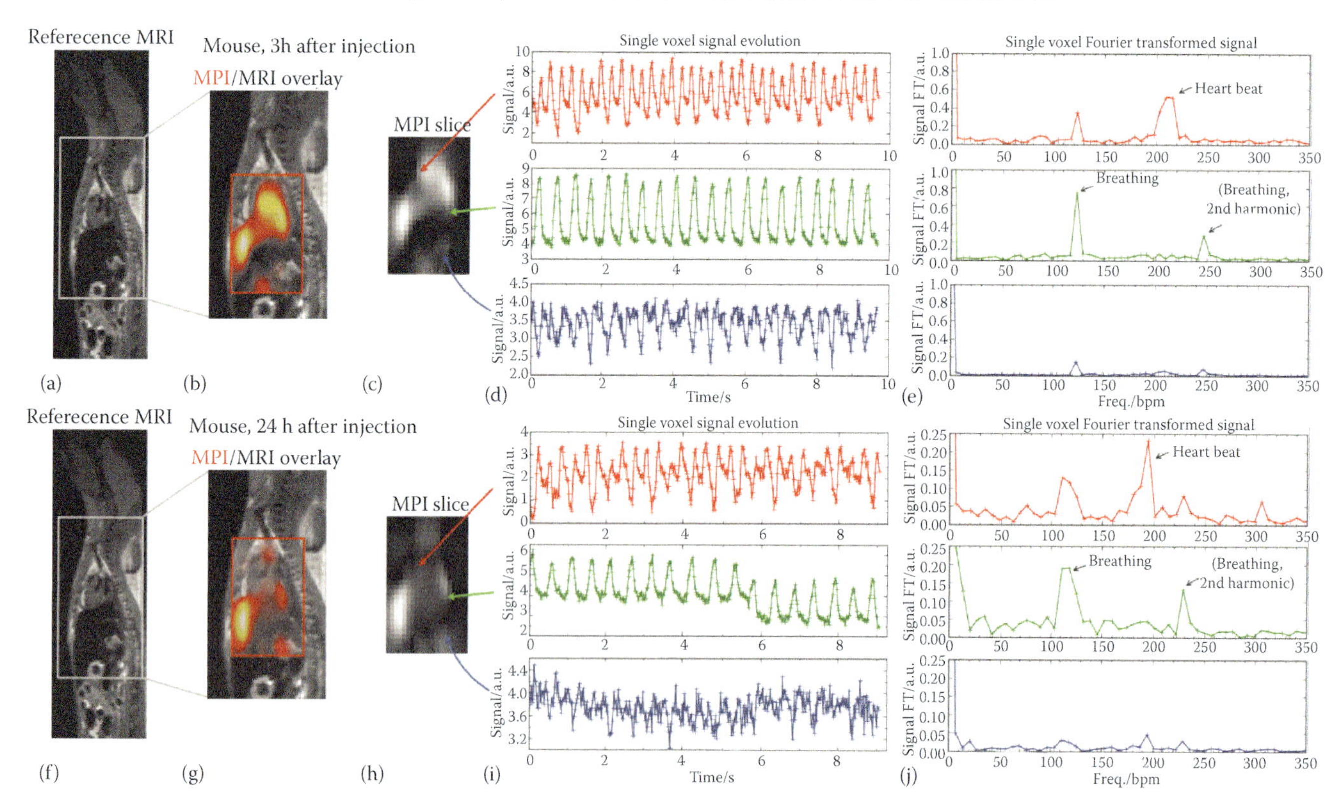

FIGURE 24.9 Intravenous administration of murine Resovist®-loaded RBCs in mice and MPI acquisition. After intravenous injection of 1 μmole Fe by SPIO-loaded RBCs, namely Resovist®-loaded RBCs, MPI of the blood pool of living mice is feasible several hours after injection with current MPI equipment, proving the efficacy of the RBC-loading method of Antonelli et al.[83–88] in increasing the blood retention time of IONPs. From the 3D MPI data acquired 3 h after injection of Resovist®-loaded RBCs (top row) and 24 h (bottom row) after injection, a sagittal slice through the position of the heart was selected for display (c), (h). The corresponding slices of the MRI data set are shown for reference (a), (f) as well as the overlays of the interpolated MPI signal (orange) on the MRI data (b), (g). The signal variations for three selected voxels were extracted from each slice (d), (i). The voxels were selected at the edge of the suspected heart (red arrows), the edge of the suspected liver (green arrows) and at a position deeper in the abdomen (blue arrows). The selected time series were Fourier transformed (e), (j) to show that distinct peaks for the heartbeat and breathing motion could be observed. (From Rahmer, J., Antonelli, A., Sfara, C., Tiemann, B., Gleich, B., Magnani, M., Weizenecker, J. and Borgert, J., *Phys Med Biol*, 58, 3965–3977. Copyright 2013, Institute of Physics and Engineering in Medicine. Reproduced with permission.)

injection. This work presented the first evidence that SPIO-loaded RBCs can be imaged in the blood pool of mice several hours after injection using MPI, thus pointing to the efficacy of this RBC loading method for increasing the blood retention time of IONPs. Prolonged blood retention time allows therapy monitoring, but also more reliable concentration and blood volume evaluations compared to the dynamic concentration variations occurring in bolus exams. Current and future improvements in MPI hardware[129–131] and optimal particle performance in MPI are expected to increase the signal levels and thus greatly enhance image quality. Based on these results, new MPI applications can be envisioned, such as permanent 3D real-time visualization of the vessel tree during surgical procedures, as well as monitoring and treatment control of cardiovascular diseases.

24.5 CHALLENGES AND OUTLOOK

The use of intravascular contrast agents to improve the sensitivity and resolution of medical imaging is of central importance in patient diagnosis, and there is a particularly urgent need to develop highly efficient blood-pool contrast agents with long blood circulation times. Such agents could improve our ability to assess the vascular system in functional cardiac diagnosis, angiography or cardiac wall motion assessment. The use of RBCs as carriers for intravascular contrast agents could be an effective strategy to increase the time frame for vascular imaging by prolonging the agents' circulation time. Moreover, the use of autologous RBCs strongly reduces the risk of triggering allergic or other immune reactions that may occur after administration of some types of contrast agents. Considering that actual procedures for encapsulating agents into the RBCs of a patient require only 1.5–2.0 h and as little as 50 ml of blood, which is processed *ex-vivo* and then reinfused into the same patient (autologous transfusion), we think that SPIO-loaded RBCs could be used in diagnostic procedures that require repeated angiography or other surgical procedures. Since the same RBCs can be loaded simultaneously with SPIONs and drugs, the cell-based constructs described in this chapter can be considered as theranostics agents able to deliver the proper drug with the proper pharmacokinetic, while also allowing assessment of the pharmacological effect when the drug is acting on the cardiovascular system. The major challenges envisaged are those related to clinical development of the product, which consists of a cell (RBCs) and a contrasting agent (NPs) and will therefore likely be evaluated as a combination product requiring expertise from different fields. Additional challenges are related to the selection of the most relevant clinical condition requiring repeated visualizations of the circulatory system and/or relevant organs (i.e. brain or heart) following trauma or blood vessel leakage. Finally, SPIONs as contrasting agents, delivered free or encapsulated into RBCs, are removed through the same organs (liver and spleen), but with very different kinetics. Free NPs are taken up in 1–2 days,[24] while cell-loaded NPs are removed from circulation over a period of weeks.[87,88,93] When the same amount of iron is administered, a rapid uptake of iron in the liver and spleen occurs when free NPs are used. This poses potential risks of iron overload for the patient, especially in the case of repeated administrations. Conversely, when iron is administered encapsulated into red cells, its slow removal from circulation, and consequent slow uptake by liver and spleen, allows the metabolism of iron without altering the physiology of these organs. The clear advantages of such a delivery system, and the ongoing work of a highly committed interdisciplinary team, should lead to the development of new safe (radiation-free) procedures able to rapidly deliver high-resolution diagnostic information.

ACKNOWLEDGEMENTS

This work was partially supported by the R24 MH109085 (Prof. Jeff W. Bulte). The authors would like to thank Prof. Pison Ulrich for providing us with the ferucarbotran suspension, (TOPASS Clinical Solution) and Prof. Pietro Gobbi for the RBC image in ESEM and Carla Sfara, PhD, for the reference formatting.

REFERENCES

1. de Back, D. Z., Kostova, E. B., van Kraaij, M., van den Berg, T. K. and van Bruggen, R. 2014. Of macrophages and red blood cells; A complex love story. *Front Physio*, 5, 9.
2. Erhabor, O. and Adias T. C. 2013. *Haematology Made Easy*. Bloomington, IN: AuthorHouse.
3. Beutler, E. 2000. Energy metabolism and maintenance of erythrocytes. In: Beutler, E., Lichtman, M. A., Coller, B. S., Kipps, T. J. and Seligsohn U. (eds.) *Williams Hematology*, 6th ed. 319–332. New York, NY: McGraw-Hill.
4. McCance, K. L. and Huether S. E. 2014. *Pathophysiology: The Biologic Basis for Disease in Adults and Children*, 7th ed. St. Louis, MO: Elsevier.
5. Neuwelt, E. A., Hamilton, B. E., Varallyay, C. G., Rooney, W. R., Edelman, R. D., Jacobs, P. M. and Watnick, S. G. 2009. Ultrasmall superparamagnetic iron oxides (USPIOs): A future alternative magnetic resonance (MR) contrast agent for patients at risk for nephrogenic systemic fibrosis (NSF)? *Kidney Int*, 75, 465–474.
6. Rümenapp, C., Gleich, B. and Haase, A. 2012. Magnetic nanoparticles in magnetic resonance imaging and diagnostics. *Pharm Res*, 29, 1165–1179.
7. Wang, Y. X. J. 2011. Superparamagnetic iron oxide based MRI contrast agents: Current status of clinical application. *Quant Imaging Med Surg*, 1, 35–40.
8. Bañobre-López, M., Teijeiro, A., and Rivas, J. 2013. Magnetic nanoparticle-based hyperthermia for cancer treatment. *Rep Pract Oncol Radiother*, 18(6), 397–400.
9. Ulbrich, K., Holá, K., Šubr, V., Bakandritsos, A., Tuček, J. and Zbořil, R. 2016. Targeted drug delivery with polymers and magnetic nanoparticles: Covalent and noncovalent approaches, release control, and clinical studies. *Chem Rev*, 116(9), 5338–5431.
10. Strijkers, G. J., Mulder, W. J., van Tilborg, G. A. and Nicolay, K. 2007. MRI contrast agents: Current status and future perspectives. *Anticancer Agents Med Chem*, 7, 291–305.
11. Wang, Y. X., Hussain, S. M. and Krestin, G. P. 2001. Superparamagnetic iron oxide contrast agents: Physicochemical characteristics and applications in MR imaging. *Eur Radiol*, 11, 2319–2331.
12. Zhou, Z., Wu, C., Liu, H., Zhu, X., Zhao, Z., Wang, L., Xu, Y., Ai, H. and Gao, J. 2015. Surface and interfacial engineering of iron oxide nanoplates for highly efficient magnetic resonance angiography. *ACS Nano*, 9, 3012–3022.

13. Rohrer, M., Bauer, H., Mintorovitch, J., Requardt, M. and Weinmann, H. J. 2005. Comparison of magnetic properties of MRI contrast media solutions at different magnetic field strengths. *Invest Radiol*, 40, 715–724.
14. Na, H. B., Song, I. C. and Hyeon, T. 2009. Inorganic nanoparticles for MRI contrast agents. *Adv Mater*, 21, 2133–2148.
15. Grazioli, L., Bondioni, M. P., Romanini, L., Frittoli, B., Gambarini, S., Donato, F., Santoro, L. and Colagrande, S. 2009. Superparamagnetic iron oxide-enhanced liver MRI with SHU 555 A (RESOVIST): New protocol infusion to improve arterial phase evaluation-a prospective study. *J Magn Reson Imaging*, 29, 607–616.
16. Berry, C. C. and Curtis, A. S. G. 2003. Functionalisation of magnetic nanoparticles for applications in biomedicine. *J Phys D: Appl Phys*, 36, R198–R206.
17. Aggarwal, P., Hall, J. B., McLeland, C. B., Dobrovolskaia, M. A. and McNeil, S. E. 2009. Nanoparticle interaction with plasma proteins as it relates to particle biodistribution, biocompatibility and therapeutic efficacy. *Adv Drug Deliv Rev*, 61, 428–437.
18. Muthiah, M., Park, I. K. and Cho, C. S. 2013. Surface modification of iron oxide nanoparticles by biocompatible polymers for tissue imaging and targeting. *Biotechnol Adv*, 31, 1224–1236.
19. Longmire, M., Choyke, P. L. and Kobayashi, H. 2008. Clearance properties of nano-sized particles and molecules as imaging agents: Considerations and caveats. *Nanomedicine*, 3, 703–717.
20. Corot, C., Robert, P., Idée, J. M. and Port, M. 2006. Recent advances in iron oxide nanocrystal technology for medical imaging. *Adv Drug Deliv Rev*, 58, 1471–1504.
21. Laurent, S., Forge, D., Port, M., Roch, A., Robic, C., Vander Elst, L. and Muller, R. N. 2008. Magnetic iron oxide nanoparticles: Synthesis, stabilization, vectorization, physicochemical characterizations, and biological applications. *Chem Rev*, 108(6), 2064–2110.
22. Martina, M. S., Fortin, J. P., Ménager, C., Clément, O., Barratt, G., Grabielle-Madelmont, C., Gazeau, F., Cabuil, V., Lesieur, S. 2005. Generation of superparamagnetic liposomes revealed as highly efficient MRI contrast agents for in vivo imaging. *J Am Chem Soc*, 127, 10676–10685.
23. Weissleder, R., Stark, D. D., Engelstad, B. L., Bacon, B. R., Compton, C. C., White, D. L., Jacobs, P. and Lewis, J. 1989. Superparamagnetic iron oxide: Pharmacokinetics and toxicity. *AJR Am J Roentgenol*, 152, 167–173.
24. Arami, H., Khandhar, A., Liggitt, D. and Krishnan, K. M. 2015. In vivo delivery, pharmacokinetics, biodistribution and toxicity of iron oxide nanoparticles. *Chem Soc Rev*, 44, 8576–8607.
25. Amstad, E., Zurcher, S., Mashaghi, A., Wong, J. Y., Textor, M. and Reimhult, E. 2009. Surface functionalization of single superparamagnetic iron oxide nanoparticles for targeted magnetic resonance imaging. *Small*, 5, 1334–1342.
26. Meng, J., Fan, J., Galiana, G., Branca, R. T., Clasen, P. L., Ma, S., Zhou, J., Leuschner, C., Kumar, C. S. S. R., Hormes, J., Otiti, T., Beye, A. C., Harmer, M. P., Kiely, C. J., Warren, W., Haataja, M. P. and Soboyejo, W. O. 2009. LHRH-functionalized superparamagnetic iron oxide nanoparticles for breast cancer targeting and contrast enhancement in MRI. *Mater Sci Eng C*, 29, 1467–1479.
27. Rosen, J. E., Chan, L., Shieh, D. B. and Gu, F. X. 2012. Iron oxide nanoparticles for targeted cancer imaging and diagnostics. *Nanomedicine*, 8, 275–290.
28. Flower, R., Peiretti, E., Magnani, M., Rossi, L., Serafini, S., Gryczynski, Z. and Gryczynski, I. 2008. Observation of erythrocyte dynamics in the retinal capillaries and choriocapillaris using ICG-loaded erythrocyte ghost cells. *Invest Ophthalmol Vis Sci*, 49, 5510–5516.
29. Leuzzi, V., Micheli, R., D'Agnano, D., Molinaro, A., Venturi, T., Plebani, A., Soresina, A., Marini, M., Ferremi Leali, P., Quinti, I., Pietrogrande, M. C., Finocchi, A., Fazzi, E., Chessa, L. and Magnani, M. 2015. Positive effect of erythrocyte-delivered dexamethasone in ataxia-telangiectasia. *Neurol Neuroimmunol Neuroinflamm*, 2, e98.
30. Godfrin, Y. and Bax, B. E. 2012. Enzyme bioreactors as drugs. *Drugs of the Future*, 37(4), 263–272.
31. Pan, D., Vargas-Morales, O., Zern, B., Anselmo, A. C., Gupta, V., Zakrewsky, M., Mitragotri, S., Muzykantov, V. 2016. The effect of polymeric nanoparticles on biocompatibility of carrier red blood cells. *PLoS One*, 11(3), e0152074.
32. Muzykantov, V., Smirnov, M., Samokhin, G. 1992. Avidin-induced lysis of biotinylated erythrocytes by homologous complement via the alternative pathway depends on avidin's ability of multipoint binding with biotinylated membrane. *Biochim Biophys Acta*, 1107(1), 119–125.
33. Muzykantov, V. R., Murciano, J. C., Taylor, R. P., Atochina, E. N., Herraez, A. 1996. Regulation of the complement-mediated elimination of red blood cells modified with biotin and streptavidin, *Anal Biochem*, 241(1), 109–119.
34. Bax, B. E., Bain, M. D., Talbot, P. J., Parker-Williams, E. J. and Chalmers, R. A. 1999. Survival of human carrier erythrocytes in vivo. *Clin Sci*, 96, 171–178.
35. Gutiérrez Millán, C., Zarzuelo Castañeda, A., González López, F., Sayalero Marinero, M. L., Lanao, J. M. 2008. Pharmacokinetics and biodistribution of amikacin encapsulated in carrier erythrocytes. *J Antimicrob Chemother*, 61(2), 375–381.
36. Kwon, Y. M., Chung, H. S., Moon, C., Yockman, J., Park, Y. J., Gitlin, S. D., David, A. E., Yang, V. C. 2009. L-Asparaginase encapsulated intact erythrocytes for treatment of acute lymphoblastic leukemia (ALL). *J Control Release*, 139(3), 182–189.
37. Bax, B. E., Bain, M. D., Fairbanks, L. D., Webster, A. D., Chalmers, R. A. 2000. In vitro and in vivo studies with human carrier erythrocytes loaded with polyethylene glycol-conjugated and native adenosine deaminase. *Br J Haematol*, 109(3), 549–554.
38. Coker, S., Szczepiorkowski, Z. M., Seigel, A. H., Ferrari, A., Benatti, L., Mambrini, G., Anand, R., Hartman R. D., Dumont L. J. 2016. The in vivo recovery/survival and pharmacokinetic properties of dexamethasone sodium phosphate encapsulated in autologous erythrocytes. *Blood*, 128, 2629.
39. Dong, Q. and Jin, W. 2001. Monitoring diclofenac sodium in single human erythrocytes introduced by electroporation using capillary zone electrophoresis with electrochemical detection. *Electrophoresis*, 22, 2786–2792.
40. Lizano, C., Sanz, S., Luque, J. and Pinilla, M. 1998. In vitro study of alcohol dehydrogenase and acetaldehyde dehydrogenase encapsulated into human erythrocytes by an electroporation procedure. *Biochim Biophys Acta Gen Subj*, 1425, 328–36.
41. Cabrales, P., Tsai, A. G. and Intaglietta M. 2008. Modulation of perfusion and oxygenation by red blood cell oxygen affinity during acute anemia. *Am J Respir Cell Molec Biol*, 38, 354–361.
42. Tsong, T. Y. 1991. Electroporation of cell membranes. *Biophys J*, 60(2), 297–306.
43. Vernier, P. T., Sun, Y., Marcu, L., Craft, C. M., Gundersen, M. A. 2004. Nanoelectropulse-induced phosphatidylserine translocation. *Biophys J*, 86(6), 4040–4048.
44. Magnani, M., Serafini, S., Fraternale, A., Antonelli, A., Biagiotti, S., Pierigè, F., Sfara, C. and Rossi, L. 2011. Red blood cell based delivery of drugs and nanomaterials for

therapeutic and diagnostic applications. In: Nalwa, H. S. (ed.) *Encyclopedia of Nanoscience and Nanotechnology*. Los Angeles: American Scientific Publishers, 22, 309–354.

45. Muzykantov, V. R. 2010. Drug delivery by red blood cells: Vascular carriers designed by Mother Nature. *Expert Opin Drug Deliv*, 7, 403–427.
46. Rossi, L., Pierigè, F., Antonelli, A., Bigini, N., Gabucci, C., Peiretti, E., Magnani, M. 2016. Engineering erythrocytes for the modulation of drugs' and contrasting agents' pharmacokinetics and biodistribution. *Adv Drug Deliv Rev*, 106, 73–87.
47. Millan, C. G., Marinero, M. L., Castaneda, A. Z. and Lanao J. M. 2004. Drug, enzyme, and peptide delivery using erythrocytes as carriers. *J Control Release*, 95, 27–49.
48. Hu, C.-M. J., Fang, R. H. and Zhang, L. 2012. Erythrocyte-inspired delivery systems. *Adv Healthc Mater*, 1, 537–547.
49. Ihler, G. M. and Tsang, H. C. 1987. Hypotonic hemolysis methods for entrapment of agents in resealed erythrocytes. *Methods Enzymol*, 149, 221–229.
50. Mosca, A., Paleari, R., Russo, V., Rosti, E., Nano, R., Boicelli, A., Villa, S. and Zanella, A. 1992. IHP entrapment into human erythrocytes: Comparison between hypotonic dialysis and DMSO osmotic pulse. *Adv Exp Med Biol*, 326, 19–26.
51. Stephan, M. T. and Irvine, D. J. 2011. Enhancing cell therapies from the outside in: Cell surface engineering using synthetic nanomaterials. *Nano Today*, 6(3), 309–325.
52. Muzykantov, V. R. and Taylor, R. P. 1994. Attachment of biotinylated antibody to red blood cells: Antigen-binding capacity of immunoerythrocytes and their susceptibility to lysis by complement. *Anal Biochem*, 223(1), 142–148.
53. Muzykantov, V. R., Zaltsman, A. B., Smirnon, M. D., Samokhin, G. P. and Morgan, B. P. 1996. Target-sensitive immunoerythrocytes: Interaction of biotinylated red blood cells with immobilized avidin induces their lysis by complement. *Anal Biochem*, 241(1), 109–119.
54. Slowing, I. I., Wu, C. W., Vivero-Escoto, J. L., Lin, V. S. 2009. Mesoporous silica nanoparticles for reducing hemolytic activity towards mammalian red blood cells. *Small*, 5(1), 57–62.
55. Tan, S., Wu, T., Zhang, D., Zhang, Z. 2015. Cell or cell membrane-based drug delivery systems. *Theranostics*, 5(8), 863–881.
56. Hamidi, M., Zarrin, A., Foroozesh, M., Mohammadi-Samani, S. 2007. Applications of carrier erythrocytes in delivery of biopharmaceuticals. *J Control Release*, 118(2), 145–160.
57. Corinti, S., Chiarantini, L., Dominici, S., Laguardia, M. E., Magnani, M., Girolomoni, G. 2002. Erythrocytes deliver Tat to interferon-gamma-treated human dendritic cells for efficient initiation of specific type 1 immune responses in vitro. *J Leukoc Biol*, 71(4), 652–658.
58. Villa, C. H., Pan, D. C., Zaitsev, S., Cines, D. B., Siegel, D. L. and Muzykantov, V. R. 2015. Delivery of drugs bound to erythrocytes: New avenues for an old intravascular carrier. *Ther Deliv*, 6, 795–826.
59. Kontos, S. and Hubbell, J. A. 2010. Improving protein pharmacokinetics by engineering erythrocyte affinity. *Mol Pharm*, 7(6), 2141–2147.
60. Hall, S., Mitragotri, S. and Daugherty, P. S. 2007. Identification of peptide ligands facilitating nanoparticle attachment to erythrocytes. *Biotechnol Prog*, 23(3), 749–754.
61. Lorentz, K. M., Kontos, S., Diaceri, G., Henry, H., and Hubbell, J. A. 2015. Engineered binding to erythrocytes induces immunological tolerance to *E. coli* asparaginase. *Sci Adv*, 1, e1500112.
62. Grimm, A. J., Kontos, S., Diaceri, G., Quaglia-Thermes, X., and Hubbell, J. A. 2015. Memory of tolerance and induction of regulatory T cells by erythrocyte-targeted antigens. *Sci Rep*, 5, 15907.
63. Cremel, M., Guérin, N., Horand, F., Banz, A., Godfrin, Y. 2013. Red blood cells as innovative antigen carrier to induce specific immune tolerance. *Int J Pharm*, 443, 39–49.
64. Shi, J., Kundrat, L., Pishesha, N., Bilate, A., Theile, C., Maruyama, T., Dougan, S. K., Ploegh, H. L., Lodish, H. F. 2014. Engineered red blood cells as carriers for systemic delivery of a wide array of functional probes. *Proc Natl Acad Sci U S A*, 111(28), 10131–10136.
65. Kolesnikova, T. A., Skirtach, A. G., and Möhwald, H. 2013. Red blood cells and polyelectrolyte multilayer capsules: Natural carriers versus polymer-based drug delivery vehicles. *Expert Opin Drug Deliv*, 10(1), 47–58.
66. Bhateria, M., Rachumallu, R., Singh, R. and Bhatta, R. S. 2014. Erythrocytes-based synthetic delivery systems: Transition from conventional to novel engineering strategies. *Expert Opin Drug Deliv*, 11(8), 1219–1236.
67. Chambers, E. and Mitragotri, S. 2007. Long circulating nanoparticles via adhesion on red blood cells: Mechanism and extended circulation. *Exp Biol Med*, 232(7), 958–966.
68. Zhao, Y., Sun, X., Zhang, G., Trewyn, B. G., Slowing, II, Lin, V. S. 2011. Interaction of mesoporous silica nanoparticles with human red blood cell membranes: Size and surface effects. *ACS Nano*, 5(2), 1366–1375.
69. Laurencin, M., Cam, N., Georgelin, T., Clément, O., Autret, G., Siaugue, J. M. and Ménager, C. 2013. Human erythrocytes covered with magnetic core-shell nanoparticles for multimodal imaging. *Adv Healthc Mater*, 2(9), 1209–1212.
70. Mai, T. D., d'Orlyé, F., Ménager, C., Varenne, A. and Siaugue J.-M. 2013. Red blood cells decorated with functionalized core-shell magnetic nanoparticles: Elucidation of the adsorption mechanism. *Chem Commun*, 49(47), 5393–5395.
71. Wang, C., Sun, X., Cheng, L., Yin, S., Yang, G., Li, Y. and Liu, Z. 2014. Multifunctional theranostic red blood cells for magnetic-field enhanced in vivo combination therapy of cancer. *Adv Mater*, 26(28), 4794–4802.
72. Ren, X., Zheng, R., Fang, X., Wang, X., Zhang, X., Yang, W. and Sha X. 2016. Red blood cell membrane camouflaged magnetic nanoclusters for imaging-guided photothermal therapy. *Biomaterials*, 92, 13–24.
73. Villa, C. H., Anselmo, A. C., Mitragotri, S. and Muzykantov, V. 2016. Red blood cells: Supercarriers for drugs, biologicals, and nanoparticles and inspiration for advanced delivery systems. *Adv Drug Deliv Rev*, 106(Pt A), 88–103.
74. Wu, Z., Li, T., Li, J., Gao, W., Xu, T., Christianson, C., Gao, W., Galarnyk, M., He, Q., Zhang, L. and Wang, J. 2014. Turning erythrocytes into functional micromotors. *ACS Nano*, 8(12), 12041–12048.
75. Sprandel, U., Lanz, D. J. and von Hörsten, W. 1987. Magnetically responsive erythrocyte ghosts. *Methods Enzymol*, 149, 301–312.
76. Sprandel, U., Hubbard, A. R. and Chalmers, R. A. 1979. In vitro studies on resealed erythrocyte ghosts as protein carriers. *Res Exp Med (Berl.)*, 175, 239–245.
77. DeLoach, J. and Ihler, G. 1977. A dialysis procedure for loading erythrocytes with enzymes and lipids. *Biochim Biophys Acta*, 496, 136–145.
78. Vyas, S. P. and Jain, S. K. 1994. Preparation and in vitro characterization of a magnetically responsive ibuprofen-loaded erythrocytes carrier. *J Microencapsul*, 11, 19–29.
79. Jain, S. K. and Vyas, S. P. 1994. Magnetically responsive diclofenac sodium-loaded erythrocytes: Preparation and in vitro characterization. *J Microencapsul*, 11, 141–151.
80. Sprandel, U. and Zöllner, N. 1985. Osmotic fragility of drug carrier erythrocytes. *Exp Med*, 185, 77–85.

81. Orekhova, N. M., Akchurin, R. S, Belyaev, A. A., Smirnov, M. D., Ragimov, S. E. and Orekhov, A. N. 1990. Local prevention of thrombosis in animal arteries by means of magnetic targeting of aspirin-loaded red cells. *Thromb Res*, 57(4), 611–616.
82. Brahler, M., Georgieva, R., Buske, N., Müller, A., Müller, S., Pinkernelle, J., Teichgräber, U., Voigt, A. and Bäumler, H. 2006. Magnetite-loaded carrier erythrocytes as contrast agents for magnetic resonance imaging. *Nano Lett*, 6, 2505–2509.
83. Stenberg, N., Georgieva, R., Duft K. and Bäumler, H. 2012. Surface modified loaded human red blood cells for targeting and delivery of drugs. *J Microencapsul*, 29, 9–20.
84. Cinti, C., Taranta, M., Naldi, I. and Grimaldi, S. 2011. Newly engineered magnetic erythrocytes for sustained and targeted delivery of anticancer therapeutic compounds. *PLoS One*, 6, e17132.
85. Antonelli, A., Sfara, C., Mosca, L., Manuali, E. and Magnani, M. 2008. New biomimetic constructs for improved in vivo circulation of superparamagnetic nanoparticles. *J Nanosci Nanotechnol*, 8, 2270–2278.
86. Magnani, M. and Antonelli, A. July 3, 2007. Delivery of contrasting agents for magnetic resonance imaging, International Patent Application No. PCT/EP2007/006349.
87. Antonelli, A., Sfara, C., Battistelli, S., Canonico, B., Arcangeletti, M., Manuali, E., Salamida, S., Papa, S. and Magnani, M. 2013. New strategies to prolong the in vivo life span of iron-based contrast agents for MRI. *PLoS One*, 8(10), e78542.
88. Antonelli, A., Sfara, C., Manuali, E., Salamida, S. Louin, G. and Magnani, M. 2013. Magnetic red blood cells as new contrast agents for MRI applications. In: *Medical Imaging 2013: Biomedical Applications in Molecular, Structural, and Functional Imaging*, Weaver, J. B. and Molthen, R. C. (Eds.) SPIE, Bellingham, 8672, 86721D.
89. Antonelli, A., Sfara, C., Manuali, E., Bruce, I. J. and Magnani, M. 2011. Encapsulation of superparamagnetic nanoparticles into red blood cells as new carriers of MRI contrast agents. *Nanomedicine (Lond.)*, 6(2), 211–223.
90. Antonelli, A., Weber, O., Sfara, C., Pison, U. and Magnani, M. 26–28 March 2015. *Encapsulation of new ferucarbotran nanoparticles into red blood cells as potential MPI contrast agent*. Proceeding presented at: 5th International Workshop on Magnetic Particle Imaging (IWMPI). Istanbul, Turkey.
91. Morana, G., Salviato, E. and Guarise, A. 2007. Contrast agents for hepatic MRI. *Cancer Imaging* 7, S24–S27.
92. Reimer, P. and Balzer, T. 2003. Ferucarbotran (Resovist): A new clinically approved RES-specific contrast agent for contrast-enhanced MRI of the liver: Properties, clinical development, and applications. *Eur Radiol*, 13, 1266–1276.
93. Antonelli, A., Sfara, C., Weber, O., Pison, U., Manuali, E., Salamida, S. and Magnani, M. 2016. Characterization of ferucarbotran-loaded RBCs as long circulating magnetic contrast agents. *Nanomedicine (Lond)*, 11(21), 2781–2795.
94. Bourgeaux, V., Aufradet, E., Campion, Y., De Souza, G., Horand, F., Bessaad, A., Chevrier, A. M., Canet-Soulas, E., Godfrin, Y., Martin, C. 2012. Efficacy of homologous inositol hexaphosphate-loaded red blood cells in sickle transgenic mice. *Br J Haematol*, 157(3), 357–369.
95. Franco, R. S. 2012. Measurement of red cell lifespan and aging. *Transf Med Hemother*, 39(5), 302–307.
96. Boni, A., Ceratti, D., Antonelli, A., Sfara, C., Magnani, M., Manuali, E., Salamida, S., Gozzi, A. and Bifone, A. 2014. USPIO-loaded red blood cells as a biomimetic MR contrast agent: A relaxometric study. *Contrast Media Mol Imaging*, 9, 229–236.
97. Markov, D. E., Boeve, H., Gleich, B., Borgert, J., Antonelli, A., Sfara, C. and Magnani, M. 2010. Human erythrocytes as nanoparticle carriers for magnetic particle imaging. *Phys Med Biol*, 55, 6461–6473.
98. Kim, S. G., Harel, N., Jin, T., Kim, T., Lee, P. and Zhao, F. 2013. Cerebral blood volume MRI with intravascular superparamagnetic iron oxide nanoparticles. *NMR Biomed*, 26, 949–962.
99. Barrett, T., Brechbiel, M., Bernardo, M. and Choyke, P. L. 2007. MRI of tumor angiogenesis. *J Magn Reson Imaging*, 26, 235–249.
100. Jaffer, F. A., Libby, P. and Weissleder, R. 2007. Molecular imaging of cardiovascular disease. *Circulation*, 116, 1052–1061.
101. Magnani, M., Rossi, L., D'Ascenzo, M., Panzani, I., Bigi, L. and Zanella, A. 1998. Erythrocyte engineering for drug delivery and targeting. *Biotechnol Appl Biochem*, 28, 1–6.
102. Godfrin, Y. February 16, 2006. Lysis/resealing process and device for incorporating an active ingredient in erythrocytes, International Patent Application No. PCT/IB2005/002323.
103. Bossa, F., Latiano, A., Rossi, L., Magnani, M., Palmieri, O., Dallapiccola, B., Serafini, S., Damonte, G., De Santo, E., Andriulli, A. and Annese, V. 2008. Erythrocyte-mediated delivery of dexamethasone in patients with mild-to-moderate ulcerative colitis, refractory to mesalamine: A randomized, controlled study. *Am J Gastroenterol*, 103, 2509–2516.
104. Chessa, L., Leuzzi, V., Plebani, A., Soresina, A., Micheli, R., D'Agnano, D., Venturi, T., Molinaro, A., Fazzi, E., Marini, M., Leali, P. F., Quinti, I., Cavaliere, F. M., Girelli, G., Pietrogrande, M. C., Finocchi, A., Tabolli, S., Abeni, D. and Magnani, M. 2014. Intra-erythrocyte infusion of dexamethasone reduces neurological symptoms in ataxia teleangiectasia patients: Results of a phase 2 trial. *Orphanet J Rare Dis*, 9, 5.
105. Gleich, B. and Weizenecker, J. 2005. Tomographic imaging using the nonlinear response of magnetic particles. *Nature*, 435, 1214–1217.
106. Lautembur, P. C. 1973. Image formation by induced local interactions: Examples employing nuclear magnetic resonance. *Nature*, 242, 190–191.
107. Weizenecker, J., Borgert, J. and Gleich, B. 2007. A simulation study on the resolution and sensitivity of magnetic particle imaging. *Phys Med Biol*, 52, 6363–6374.
108. Gleich, B., Weizenecker, J. and Borgert, J. 2008. Experimental results on fast 2D-encoded magnetic particle imaging. *Phys Med Biol*, 53, N81–N84.
109. Weizenecker, J., Gleich, B. and Borgert, J. 2008. Magnetic particle imaging using a field free line. *J Phys D: Appl Phys*, 41, 105009.
110. Borgert, J., Schmidt, J. D., Schmale, I., Rahmer, J., Bontus, C., Gleich, B., David, B., Eckart, R., Woywode, O., Weizenecker, J., Schnorr, J., Taupitz, M., Haegele, J., Vogt, F. M. and Barkhausen, J. 2012. Fundamentals and applications of magnetic particle imaging. *J Cardiovasc Comput Tomogr*, 6, 149–153.
111. Rahmer, J., Weizenecker, J., Gleich, B. and Borgert, J. 2012. Analysis of a 3-D system function measured for magnetic particle imaging. *IEEE Trans Med Imaging*, 31, 1289–1299.
112. Goodwill, P., Krishnan, K. M. and Conolly, S. M. 2012. Magnetic particle imaging for angiography, stem cell tracking, cancer imaging and inflammation imaging (Chapter 20). In: Thanh, N. T. K. (Ed.) *Magnetic Nanoparticles: From Fabrication to Clinical Applications*. Boca Raton, FL: CRC Press, Taylor & Francis Group, pp. 523–540.

113. Löwa, N., Knappe, P., Wiekhorst, F., Eberbeck, D., Thünemann, A. F., and Trahms, L. 2015. Hydrodynamic and magnetic fractionation of superparamagnetic nanoparticles for magnetic particle imaging. *J Magn Magn Mater*, 380, 266–270.
114. Eberbeck, D., Dennis, C. L., Huls, N. F., Krycka, K. L., Grüttner, C., and Westphal, F. 2013. Multicore magnetic nanoparticles for magnetic particle imaging. *IEEE Trans Magn*, 49(1), 269–274.
115. Saritas, E. U., Goodwill, P. W., Croft, L. R., Konkle, J. J., Lu, K., Zheng, B. and Conolly, S. M. 2013. Magnetic particle imaging (MPI) for NMR and MRI researchers. *J Magn Reson*, 229, 116–126.
116. Goodwill, P. W., Saritas, E. U., Croft, L. R., Kim, T. N., Krishnan, K. M., Schaffer, D. V. and Conolly, S. M. 2012. X-space MPI: Magnetic nanoparticles for safe medical imaging. *Adv Mater*, 24, 3870–3877.
117. Haegele, J., Biederer, S., Wojtczyk, H., Gräser, M., Knopp, T., Buzug, T. M., Barkhausen, J. and Vogt, F. M. 2013. Toward cardiovascular interventions guided by magnetic particle imaging: First instrument characterization. *Magn Reson Med*, 69, 1761–1767.
118. Keselman, P., Yu, E.Y., Zhou, X. Y., Goodwill, P. W., Chandrasekharan, P., Ferguson, M., Khandhar, A. P., Kemp, S. J., Krishnan, K. M., Zheng, B., Conolly, S. M. 2017. Tracking short-term biodistribution and long-term clearance of SPIO tracers in magnetic particle imaging. *Phys Med Biol*, 62(9), 3440–3453.
119. Weizenecker, J., Gleich, B., Rahmer, J., Dahnke, H. and Borgert, J. 2009. Three-dimensional real-time in vivo magnetic particle imaging. *Phys Med Biol*, 54, L1–L10.
120. Schmale, I., Rahmer, J., Gleich, B., Kanzenbach, J., Schmidt, J. D., Bontus, C., Woywode, O. and Borgert, J. 2011. First phantom and in vivo MPI images with an extended field of view. In: *Medical Imaging 2011: Biomedical Applications in Molecular, Structural, and Functional Imaging*. Weaver, J. B. and Molthen, R. C. (Eds.), SPIE, Bellingham, 7965, 796510.
121. Ferguson, R. M., Minard, K. R., Khandhar, A. P. and Krishnan, K. M. 2011. Optimizing magnetite nanoparticles for mass sensitivity in magnetic particle imaging. *Med Phys*, 38, 1619–1626.
122. Eberbeck, D., Wiekhorst, F., Wagner, S. and Trahms, L. 2011. How the size distribution of magnetic nanoparticles determines their magnetic particle imaging performance. *Appl Phys Lett*, 98, 182502.
123. Markov, D., Boeve, H., Gleich, B. Borgert, J., Antonelli, A., Sfara, C. and Magnani, M. 2010. SPIO nanoparticles encapsulation into human erythrocytes for MPI application. In: Buzug, T. M., Borgert, J., Knopp, T., Biederer, S., Sattel, T. F., Erbe, M. and Lüdtke-Buzug, K. (Eds.) *Magnetic Nanoparticles. Particles Science, Imaging Technology, and Clinical Application*. Singapore: World Scientific Publishing Co. Pte. Ltd. pp. 26–31.
124. Wawrzik, T., Schilling, M. and Ludwig, F. 2012. Magnetic particle imaging: Exploring particle mobility. In: Buzug, T. M. and Borgert, J. (Eds.) *Magnetic Particle Imaging*. SPPHY, Berlin: Springer-Verlag. pp. 140, 41–45.
125. Foner, S. 1975. Further improvements in vibrating sample magnetometer sensitivity. *Rev Sci Instrum*, 46, 1425–1426.
126. Fassel, V. A. and Kniseley, R. N. 1974. Inductively coupled plasma. Optical emission spectroscopy. *Anal Chem*, 46, 1110A–1120A.
127. Rahmer, J., Antonelli, A., Sfara, C., Tiemann, B., Gleich, B., Magnani, M., Weizenecker, J. and Borgert, J. 2013. Nanoparticle encapsulation in red blood cells enables blood-pool magnetic particle imaging hours after injection. *Phys Med Biol*, 58, 3965–3977.
128. Antonelli, A., Sfara, C., Rahmer, J., Gleich, B., Borgert, J. and Magnani, M. 2013. Red blood cells as carriers in magnetic particle imaging. *Biomed Tech (Berl.)*, 58, 517–525.
129. Pablico-Lansigan, M. H., Situ, S. F. and Samia, A. C. S. 2013. Magnetic particle imaging: Advancements and perspectives for real-time in vivo monitoring and image-guided therapy. *Nanoscale*, 5, 4040–4055.
130. Knopp, T. and Buzug T. M. (eds.), 2012. *Magnetic Particle Imaging: An Introduction to Imaging Principles and Scanner Instrumentation*. Berlin: Springer-Verlag.
131. Buzug, T. M. and Borgert J. (eds.), 2012. *Magnetic Particle Imaging: A Novel SPIO Imaging Technique*. SPPHY. Berlin: Springer-Verlag. P. 140.

Antonella Antonelli (E-mail: antonella.antonelli@uniurb.it) earned her PhD in biochemical and pharmacological methodologies in 1999 from the University of Urbino Carlo Bo, Italy. She currently works as a research assistant in the Department of Biomolecular Sciences and is involved in national and international scientific projects. She has extensive experience in animal models and preclinical development of new drug delivery approaches using a red blood cell as delivery system. At present, her primary research area includes the development of new contrast agents based on the encapsulation of superparamagnetic nanoparticles into RBCs as potential biomimetic constructs useful for biomedical and diagnostic applications such as magnetic resonance imaging (MRI) and magnetic particle imaging (MPI).

Mauro Magnani (E-mail: mauro.magnani@uniurb.it) is full professor of biochemistry and dean of the School of Biological and Biotechnological Sciences at the University of Urbino Carlo Bo, Italy. He has long-standing experience in the development of new drug- and antigen-delivery systems and has published over 500 papers in international scientific journals. Inventor in 14 patent families, he is also cofounder and board member of two start-up companies: Diatheva SrL, founded in 2002, and EryDel SpA, founded in 2007. He served as referee for many international agencies and has received several awards for scientific merit. More recently his interests were devoted to the use of nanomaterials in the development of new *in vitro* and *in vivo* diagnostic applications. The combination of the nanomaterial properties with the biological properties of red blood cells resulted in the generation of new biomimetic constructs advantageously employed in *in vivo* imaging.

25 Stimuli-Regulated Cancer Theranostics Based on Magnetic Nanoparticles

*Yanmin Ju, Shiyan Tong and Yanglong Hou**

CONTENTS

25.1 INTRODUCTION

Cancer is a major public health problem the world over. Many people have lost their lives because of cancer, which has been ranked as the second leading cause of death.[1] One in four deaths in the United States is due to cancer. There are three main clinical methods to treat cancer: surgical removal, radiation therapy and chemotherapy. However, surgery can only cure nonhaematological and accessible tumours when entirely ablated. In radiation therapy, radioactive rays cause atom ionization of single- or double-stranded DNA through photon or charged particles, thus interfering the growth and division of cells. Nevertheless, it causes significant side effects in young patients and provides attenuated efficiency in late-stage tumours. Chemotherapy usually affects cell division via cytotoxic drugs and harms normal cells, especially those with high replacement rate, owing to its poor specificities and high toxicities. Thus, lots of cancer patients still cannot be cured by the above-mentioned methods. Therefore, a new method with high efficiency to kill cancer cells and without side effect is needed. To achieve that purpose, distinguishing between tumour and normal cells is extremely important.

* Corresponding author.

In fact, tumours have unique physiopathologic characteristics, which are not observed in normal tissues. Compared with blood vessels of normal tissue, tumour blood vessels have more porous structures, with pores varying from 100 to 780 nm in diameter.[2] Therefore, tumours exhibit enhanced vascular permeability, ensuring a sufficient supply of nutrients and oxygen for rapid growth. Meanwhile, nanoparticles (NPs) less than 200 nm can gradually accumulate at tumour sites after intravenous administration by enhanced permeability and retention (EPR) effect.[3]

Based on this passive targeting, NPs can be developed as carriers and deliver drug to tumour tissues. Besides, cancer cells overexpress proteins on cell surfaces, which are at low levels in normal cells, or express proteins exclusively. Active targeting is achieved by conjugating tumour associated antigen (or receptor) ligands to a NP surface, leading to efficient accumulation in tumours and reduced side effects in normal tissues.[4] These targeting moieties include small molecules (e.g. biotin,[5] folate,[6] transferrin[7]), peptides (RGD)[8], affibodies,[9] and antibodies.[10] Tumour microenvironments are characterized as hypoxic because of the defective blood vessel system. Reductive enzymes, associated with tumour hypoxia, are overexpressed in tumour microenvironments and thus provide an important strategy for active tumour targeting.[11,12] Acidosis is a biomarker of tumour microenvironments and appears both at very early and at advanced stages of tumours. Strategies binding antitumour agents with NPs via pH-sensitive chemical bonds, and releasing the agents in the acidic tumour microenvironments have been utilized.[13] NPs can be engineered for stimuli-controlled cancer detection relying on the above-mentioned internal stimuli. Well-designed NPs can also respond to external stimuli, such as temperature,[9] magnetic fields,[14,15] and light, which can have spatial and temporal control to the target sites.

Among the numerous kinds of NPs, magnetic nanoparticles (MNPs), including metallic, bimetallic and superparamagnetic iron oxide nanoparticles (SPIONs),[16,17] have shown great promise due to their unique properties in a magnetic field without depth-penetration limit during treatment. Their responses to external magnetic fields allow MNPs to enable real-time monitoring and drug delivery with high accuracy. In addition, MNPs have adjustable size and morphology through controllable synthetic methods. They have comparable size with biological molecules, which renders them very useful for biomedical applications. Besides, MNPs can also respond to an alternating magnetic field (AMF) and increase local temperature like a heater, offering a promising therapeutic solution called hyperthermia.

MNPs have been widely used for biomedical applications.[18] Besides internal stimuli generated by tumour microenvironments (e.g. pH, redox and enzymes), external stimuli (e.g. temperature, magnetic field and light) have also been exploited.[14,15] In this chapter, we focus on recent advances in stimulus-regulated cancer theranostics based on MNPs. Internal stimuli-responsive NPs, including pH, reduction-sensitive NPs and external stimuli-inductive NPs, such as magnetic field- and light-controlled NPs will all be covered. We will conclude with an outlook on future perspectives in this area of research.

25.2 EXTERNAL STIMULI-TRIGGERED THERANOSTICS

25.2.1 Magnetic Field-Responsive Theranostics

NPs can cross biological barriers and enter into cells, where they are activated on demand owing to their minute size. Among them, MNPs are extremely promising, as they can be activated even by a remote magnetic field with reduced biodistribution, dosage, and thus no side effects to normal tissues. Besides, MNPs can be fixed at a local place and escape capture by the reticuloendothelial system (RES).[19] Therefore, they have potential in magnetic specific targeting, drug and gene delivery, hyperthermia, magnetically controlled cell fate and magnetic resonance imaging (MRI).[20]

25.2.1.1 Magnetic Specific Targeting

As mentioned above, common targeting methods are classified to passive targeting through EPR effect and active targeting through binding of tumour-associated proteins. In cancer therapy, NPs can accumulate at tumour sites more efficiently by these common targeting technique compared to non-targeting methods, thus eventually leading to reduced dosage and side effects during treatment.[7] However, passive targeting may not work in hypovascular portions of tumours, such as in pancreatic cancer, where the EPR effect is absent,[21] while active targeting can only apply to one kind of receptor, which is expressed or overexpressed in one kind of tumour. Different from these common targeting techniques, MNPs, with their ability to respond to an external magnetic field, have their own targeting method called 'magnetic targeting.' Using an external magnetic field placed near the tumour, MNPs are magnetically attracted toward this area. This physical interaction can even work in hypovascular tumours and is not confined by specific receptor expression, enhancing uptake at the target site and resulting in effective treatment at lower doses.[22]

In 1963, Meyers et al. first made use of magnetic targeting, putting a magnet on the leg skin of a dog to accumulate metallic iron particles.[23] Since then, many similar works have been reported in the following decades. However, only when the development of MNPs took place, did the magnetic targeting make huge progress. Recently, Liu et al. developed a kind of multifunctional NP with up conversion NPs as the core, and a layer of ultrasmall iron oxide nanoparticles (UIONPs) as the intermediate shell (Figure 25.1). These NPs tend to migrate toward the tumour after intravenous injection and show high tumour accumulation (>8-fold compared with nonmagnetic targeting) when placing a magnet with a surface magnetic field strength of 3000 Gs on the tumour for 2 h.[22] This unique and highly effective targeting can control NP migration processes remotely and noninvasively, which can be used as a technique for drug delivery and gene translation.

By conjugating drugs or genes on MNPs, magnetic targeting can be applied in drug and gene delivery.[24] Hou et al. reported that a multifunctional Fe_5C_2 NPs exhibit tumour targeting both *in vitro* and *in vivo* through a magnetic field (Figure 25.2). Thus far, drug delivery by magnetic targeting

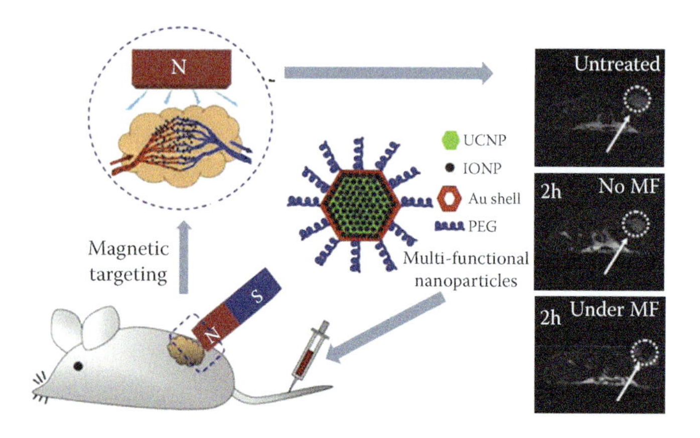

FIGURE 25.1 Schematic illustration showing the composition of the multifunctional NPs and the concept of *in vivo* magnetically targeted tumour therapy with more obvious darkening effect of tumour site 2 h after injection.[22] (Reprinted from *Biomaterials*, 33(7), Cheng, L. et al., 2215–2222, Copyright 2011, with permission from Elsevier.)

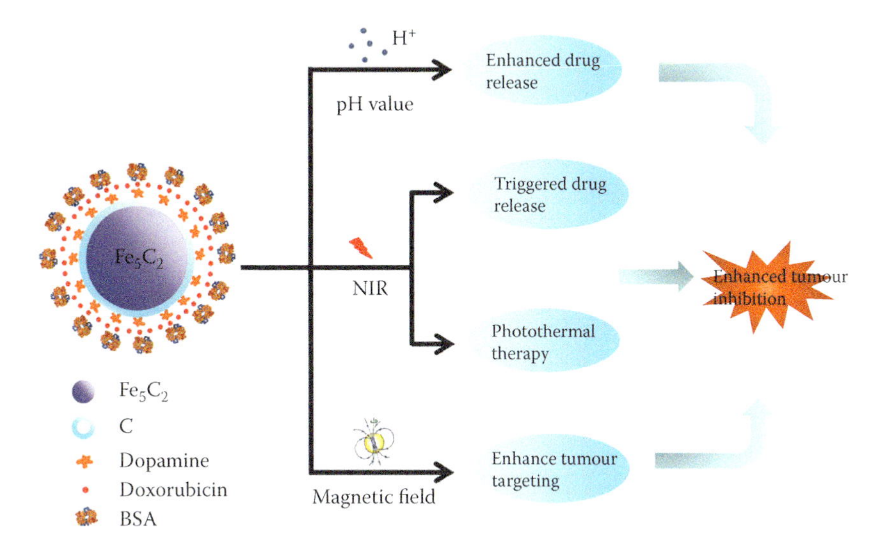

FIGURE 25.2 Schematic illustration of Fe_5C_2 NPs exhibiting tumour targeting both *in vitro* and *in vivo* through a magnetic field.[24] (Reprinted with permission from Yu, J. et al., *ACS Nano*, 10(1), p. 159–169, 2015. Copyright 2016, American Chemical Society.)

has been utilized in various cancer cells, such as in the lung, prostate, brain, breast and liver. A method termed magnetofection can transfect both DNA and RNA, by using a magnetic force acting on gene vectors that are associated with MNPs. The great advantages of this method are high transfection efficiency and a controllable transfection process. Compared with commercially transfection agents, magnetofection can obviously enhance the transfection efficiency both *in vitro* and *in vivo*.[25,26] For example, Plank et al. associated gene vectors with superparamagnetic NPs to achieve targeted gene delivery by a magnetic field, which improved the efficacy of any vector up to hundred-fold, and resulted in reduction of gene delivery duration to minutes, compared to nonviral gene vectors.[26]

Magnetic targeting shows promising therapeutic perspective. However, there are still some problems that need to be solved. For instance, external magnetic fields need adequate focusing and deep penetration into the tissues to reach the targeted sites with sufficient drive. Future efforts to identify the best magnetic targeting technologies are needed.

25.2.1.2 Magnetically Triggered Drug/Gene Delivery

MNP-based drug delivery can transport drugs to specific sites and subsequently control drug release remotely.[27,28] Hou et al. reported that hollow iron oxide NPs (IONPs) can load more antitumour drugs than solid IONPs with the same core size and coating strategy (Figure 25.3). When drug-loading hollow IONPs incubated with multidrug resistant cells, drugs were more effectively absorbed by the cells.[29]

Drug release kinetics are regulated by controlling structural features and chemical bindings within MNP conjugates. A pH sensitive linker binding drugs to MNPs can trigger drug release after MNPs accumulate at tumour sites, due to the different pH values between tumour microenvironment and normal tissues.[30] Under an alternating magnetic field (AMF), a heat sensitive linker can also promote drug release

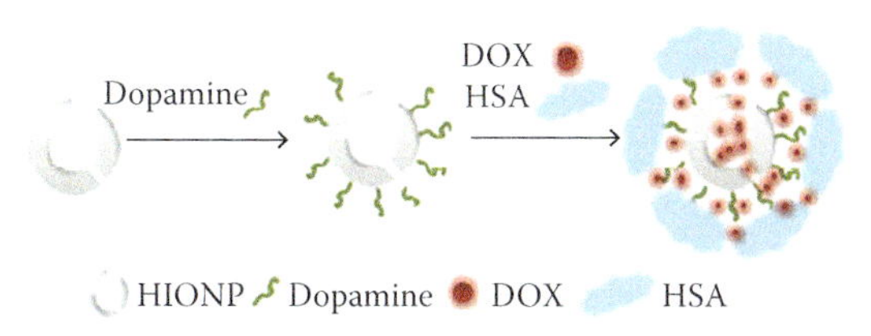

FIGURE 25.3 Schematic illustration of the formation of a drug delivery system based on hollow IONPs.[30] (With kind permission from Springer Science+Business Media: *Nano Research*, 6(1), 2013, Xing, R. et al. Copyright 2013, Tsinghua University Press.)

from the nanoplatform due to the disruption of the linker or polymer. Pellegrino et al. developed a kind of nanocube coated with a thermoresponsive polymer shell composed of poly(N-isopropylacrylamide-co-polyethylene glycolmethyl ether acrylate) that can go through phase transition under temperature variations and act as a drug carrier.[31] Such nanohybrids exhibit no drug release below 37°C, but will release doxorubicin (DOX) on demand by generating heat under an AMF (Figure 25.4). Chen et al. reported a drug-containing silica

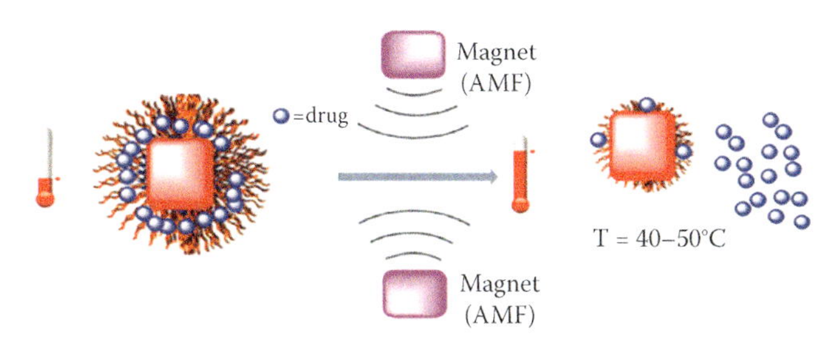

FIGURE 25.4 Schematic illustration of nanohybrids including a thermoresponsive polymer shell with a proposed mechanism for controlled release of DOX.[28] (Reprinted with permission from Kakwere, H. et al., *ACS Applied Materials & Interfaces*, 7(19), p. 10132–10145. Copyright 2015, American Chemical Society.)

core surrounded by iron oxide shell. With such a unique core/shell configuration, drugs can be protected from damage from harsh environments and be released upon exposure to a high-frequency magnetic field.[28] Nanocomposite membranes, consisting of thermosensitive nanogels and magnetite NPs designed by Kohane et al. can achieve 'on-demand' drug delivery under an oscillating magnetic field.[32] Valves can be used to control the release of drugs loaded into porous NPs. As the increasing heat can improve internal pressure, the molecular valves can move away and drug can be released subsequently. For example, drugs can be loaded in a mesoporous silica framework, incorporating with zinc-doped iron oxide nanocrystals[33] by a thermally sensitive link. When applying an AMF, heat can be generated, the link disassembles and allows drugs to be released (Figure 25.5). Temperature-sensitive polymers can wrap both drugs and MNPs together. While the local heating increases under the AMF, the polymers can shrink, crack or deform, accompanied with drug release.[34] Although MNP-based drug carriers remain in preclinical investigation, they still hold great prospect for clinical application.[35]

As mentioned above, magnetofection can protect genes and deliver genes efficiently after accumulation at target sites through magnetic force.[36] In 2002, Byrne et al. first proposed the concept that linkage of adeno-associated virus vectors to microspheres can achieve greater numbers of vector particles transfected to each cell.[37] Nowadays, magnetofection has been applied to transfecting numerous kinds of cells, including endothelial cells, keratinocytes, lung epithelial, etc.[38] Both DNA and RNA have already been successfully transfected by MNPs. Initially, researchers attached DNA with MNPs by charge or chemical interactions.[39] For example, plasmid DNA conjugated with superparamagnetic NPs coated with polyethyleneimine (PEI) polymer have the ability to protect DNA

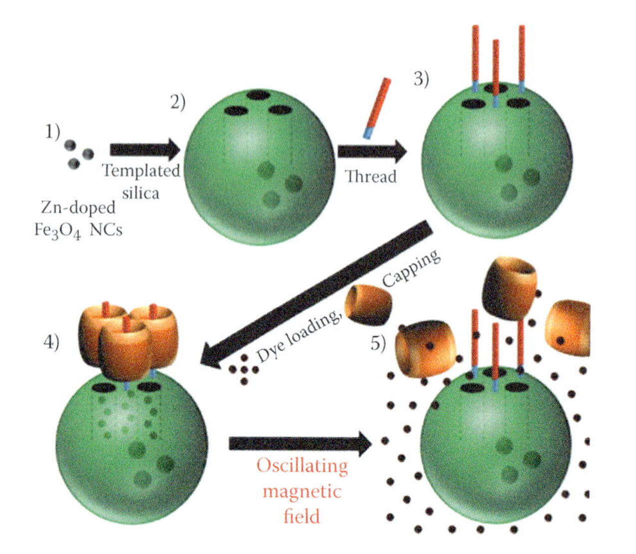

FIGURE 25.5 Schematic illustration showing the assembly and drug release of a drug delivery system.[33] (Reprinted with permission from Thomas, C.R. et al., *Journal of the American Chemical Society*, 132(31), p. 10623–10625. Copyright 2010, American Chemical Society.)

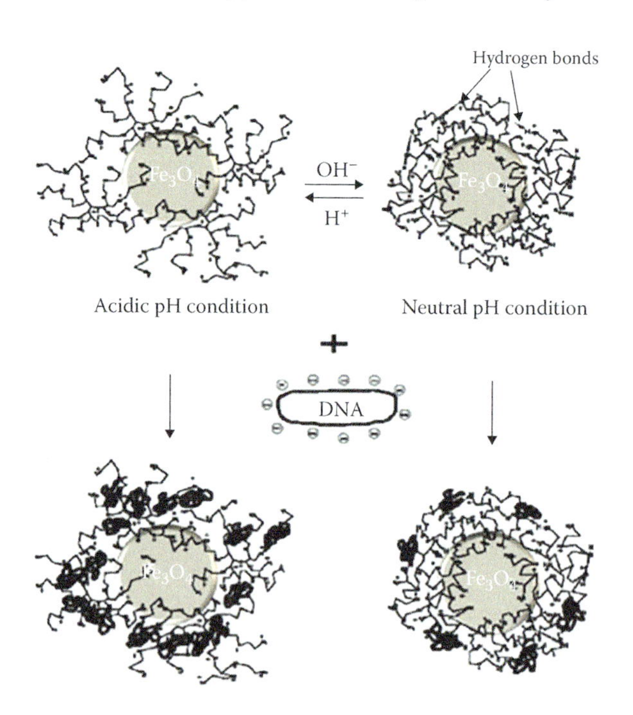

FIGURE 25.6 Schematic illustration demonstrating that PEI structural changes under acidic and neutral pH conditions enable the possibility of DNA entrapment.[39] (Reprinted with permission from Al-Deen, F.N. et al., *Langmuir* 27(7), p. 3703–3712. Copyright 2011, American Chemical Society.)

molecules due to the structure of the polymer at acidic pH and enhances transfection efficiency under external magnetic field (Figure 25.6). Later on, small interfering RNA (siRNA), which can selectively inhibit targeted genes at posttranscriptional mRNA level by a mechanism called RNA interference, was also transferred by MNPs.[40] Actually, many experiments proved that magnetically driven siRNA can suppress specific protein expression effectively both *in vitro* and *in vivo*.[41,42] Besides, short hairpin RNA have the demonstrated ability to be delivered by MNPs.[43] Interestingly, Dobson's group found that oscillating magnetic field can enhance magnetofection efficiency, which can even be levelled up 10-fold compared with static magnetic fields.[44] Although the underlying mechanisms are still unclear, magnetofection may be a prospecting method for gene delivery in the future.

25.2.1.3 Magnetic Hyperthermia

Hyperthermia is a new cancer treatment which entails exposing body tumour sites to high temperatures, thus changing the physiology of tumour cells directly and eventually causing cell apoptosis.[45] Hyperthermia treatment can be classified into three different types, according to temperature. Increasing temperatures above 46°C (up to 56°C) is generally referred to as thermoablation, which may result in direct cell death in acute treatments lasting only a few minutes and is characterized by tissue necrosis, coagulation or carbonization. Diathermia uses lower temperatures, i.e. under 41°C for treatment. Moderate hyperthermia (41°C–46°C), traditionally termed hyperthermia treatment, alters the function of intercellular proteins, leads to cellular degradation and ultimately induces cell apoptosis.[46] Interestingly, tumour cells can be killed by moderate hyperthermia without hurting the normal cells.

Early hyperthermia treatment basically transfers energy such as microwaves or ultrasounds into thermal energy. However, these energy sources are a serious threat to normal tissues. Gilchrist et al., in the 1950s,[47] first proposed magnetic materials for hyperthermia when he investigated this heating phenomenon by using 20–100 nm γ-Fe_2O_3 particles exposed to a 1.2 MHz magnetic field. After that, MNPs-based hyperthermia, which is referred to as magnetic hyperthermia, has been a hotspot in cancer treatment and used for selective heating of tumours.[48–50] Magnetic hyperthermia has some exceptional advantages: penetration depth of magnetic fields is much higher compared to other activation (light or acoustic waves), and the heating capabilities of MNPs are controllable.[51,52] To investigate the effects of AMF on the death rate of cells, Goya et al. loaded MNPs as heating agents into dendritic cells and found that AMF exposure time and amplitude, as well as the MNPs concentration, can control the viability of cells.[53]

The main mechanisms of MNPs generating heat under an oscillating or alternating magnetic field, are reported as hysteresis loss and/or Néel relaxation.[54] Hysteresis is originated from the internal energy of MNPs, especially for ferromagnetic NPs. Hysteresis loss depends on the rapid changing of magnetic movement and is proportional to the hysteresis loop area.[55] Meanwhile, relaxation behaviour, including Brown and Néel relaxation, is suitable for superparamagnetic NPs. Brownian relaxation represents the rotation of the particles, while Néel relaxation relates to the magnetic moment rotating within the magnetic core (Figure 25.7). Actually, the main contribution to heat is from Néel relaxation, as internal environment (tissues viscosity) usually hampers the motion of NPs.[56]

By using the thermal energy generated by MNPs, tumours can be cured through prompting the immune system or inducing cell death directly. Hyperthermia treatment can activate an immune response because of the heat shock, and then both primary tumours exposed to heat and metastatic tumours unexposed to heat may regress. The activity of natural killer cells was found to be increased under a temperature of about 42°C because of heat shock proteins, which can be induced by heat stress, protect cells from heat-induced apoptosis and provoke an antitumour immunity.[57] Kobayashi et al. transplanted a T-9 rat glioma tumour into a rat femur and chose one side of the tumour to be subjected to hyperthermia by magnetite liposomes. He observed that both sides of tumours disappeared after therapy. Interestingly, this phenomenon cannot be observed in immunodeficient rats, implying the contribution of immune system.[58,59] The induction of systemic antitumour immune responses by magnetic hyperthermia has also been reported in several B16 mouse melanoma models.[60] These results may indicate that magnetic hyperthermia-induced immunotherapy holds great promise for tumour therapy, especially metastatic tumours.

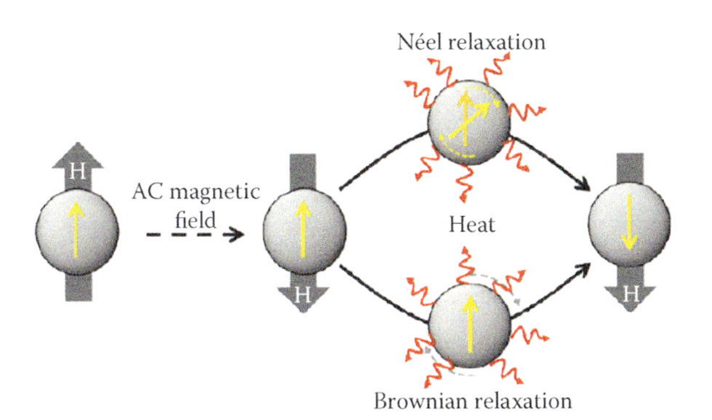

FIGURE 25.7 Schematic illustration of Néel and Brownian relaxation processes.[56] (Reprinted with permission from Yoo, D. et al., *Accounts of Chemical Research*, 44(10), p. 863–874. Copyright 2011, American Chemical Society.)

Apoptosis, which induces cell death in a programmed way, retains most cell membrane functionality and elicits no inflammation. It can be induced by magnetic hyperthermia, with characteristic morphological changes, including membrane blebbing, cell rounding and actin detachment.[61] Marano et al. demonstrated that NPs can induce reactive oxygen species (ROS) in tumour sites, disrupt organelles like mitochondria, lysosomes and nuclei, and eventually lead to cell death through apoptosis or necrosis.[62] Magnetic hyperthermia based on γ-$Mn_xFe_{2-x}O_3$ ($0 \leq x \leq 1.3$) NPs has been proved to be able to cause cancer cell apoptosis.[63] However, the mechanism by which NPs cause cell apoptosis remains unclear. The intrinsic pathway releases cytochrome c from mitochondria. Cytochrome c can bind to apoptotic protease activating factor-1, which then activates caspase-3 and caspase-9 to promote cell apoptosis. Hsieh HC et al. proved that superparamagnetic NPs can result in reactive oxidative stress, disturb the mitochondrial membrane, cause the release of pro-apoptotic proteins, cytochrome c and apoptosis-inducing factor, and finally induce cell apoptosis.[64,65]

Quite different from apoptosis, necrosis is a process of cell death triggered directly by destroying cellular structure and metabolic pathways completely. Exposed to an AMF, MNPs can disrupt the endosome membrane by high heat after being engulfed in an endosome. Then, the released endosome content may induce damage to cell membrane.[53] As shown in Figure 25.8, when MNPs are incubated with tumour cells before hyperthermia, MNPs are organized in endosomes and the cellular structure remains unbroken, but the cytoplasm and the nucleus are clearly damaged under direct[66] exposure to the magnetic field. Marcos-Campos et al. reported that iron oxide coated by dextran can induce a necrotic-like cell death, along with loss of membrane structure and shrinking of cells.[67]

Heating efficiency of MNPs is determined by the following factors: the magnitude and frequency of the applied magnetic field, and the magnetic and physical properties of MNPs. In general, heat produced in magnetic hyperthermia is proportional to the amplitude and frequency of the applied magnetic field.[68,69] However, magnetic field strength is limited in the clinical hyperthermia application, as large-amplitude,

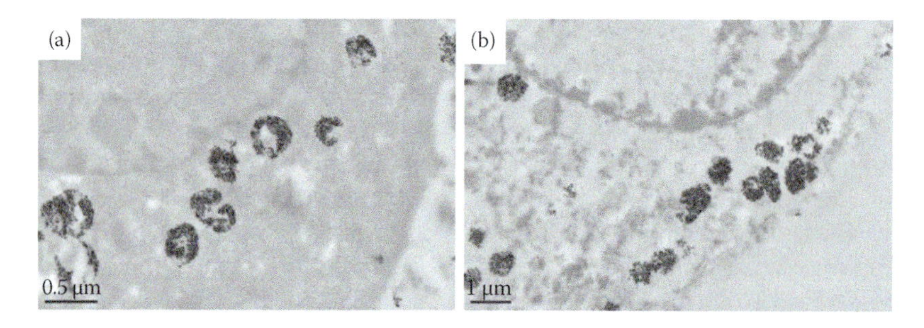

FIGURE 25.8 TEM characterization of treated cells incubated with MNPs before (a) and after (b) hyperthermia treatment.[66] (Reprinted with permission from Guardia, P. et al., *ACS Nano*, 6(4), 3080–3091. Copyright 2012, American Chemical Society.)

high-frequency magnetic fields may induce local heating in nonmagnetic tissues.[70] Properties, like anisotropy, diameter, magnetization and homogeneity of MNPs can change the heat generation efficiency. For instance, diameter and anisotropy, both of which are distinctive in different MNPs, are needed in their optimal ranges to generate heat efficiently.[71] Higher magnetization of MNPs can cause higher temperature owing to MNPs' larger area of the hysteresis loop. In comparison with polydispersed MNPs, monodispersed ones exhibit considerable high heat generation rates.[72] Moreover, surface modification of MNPs, in addition to enabling stabilization and dispersion, can also improve heat generating efficiency. Pawar et al. reported that the hyperthermia effect of Fe_3O_4 NPs coated by chitosan/glutaraldehyde could be dramatically enhanced as the formation of larger aggregates of Fe_3O_4 is hampered.[73] Based on this principle, the Hou group developed 12 nm and 25 nm IONPs stabilized by protocatechuic acid through a ligand exchange protocol. Solution with iron concentration of 1 mg/mL could be heated from 36 to 46°C within 5 min under an AMF.[74]

25.2.1.4 Magnetically Regulated Cell Fate Control

MNPs can also be used to generate mechanical forces to influence cell activity, such as differentiation, growth and death.[75] Under an external magnetic field, MNPs conjugated with antibodies or specific ligands, can move to cell membrane or between cells by binding to targeted receptors, and exert mechanical stimulations on cells, which can activate cellular signaling pathways via receptor clustering or ion channel activation and change cell fate such as cell death.[76,77]

Clusterization of receptors is usually induced by multivalent biochemical ligands. As magnetic IONPs can produce strong magnetic forces with each other and aggregate under a static magnetic field, they can induce the same clusterization effect artificially and affect cell fate. Angiogenesis is quite important both in the growth and development of blood vessels and in tumour metastasis.[78] Cheon et al. developed $TiMo_{214}$ monoclonal antibody-conjugated Zn^{2+}-doped ferrite MNPs (Ab-Zn-MNP) to target Tie2 receptors, the clusterization of which is critical to activate signaling steps and participates as well in angiogenic processes. After exposed to an external magnetic field, the Tie2 receptors on cell membrane can magnetically cluster with the aggregation of Ab-Zn-MNPs, further initiate intracellular signal pathways, and finally lead to angiogenesis.[77] The advantages of magnetically activated cellular transformation process are the ability to control the cellular activities spatially, temporally and remotely.

Switching ion channels can also control cell signal transduction, which controls cell fate. There are two activation methods explored to control ion channels. The Ingber and Dobson group demonstrated that activation of ion channels is possible by using MNPs. They developed 30 nm superparamagnetic beads, which are coated with monovalent ligands and bound to transmembrane receptors. When exposed to magnetic fields, they magnetize and aggregate together owing to bead–bead attraction and switch calcium signaling in cells when exposed to magnetic fields.[76] Efforts have been made, demonstrating that a cellular membrane receptor, FcεRI, can agglomerate to activate calcium signaling under an external magnetic field (Figure 25.9). Dobson et al. succeeded in using

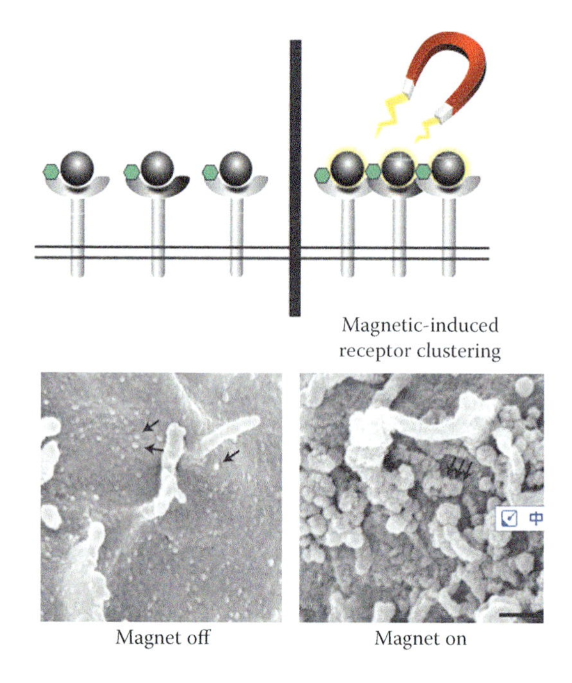

FIGURE 25.9 Monovalent ligand-coated magnetic nanobeads (dark grey circles) individually bind to IgE/FcεRI receptor complexes without inducing receptor clustering (left). When a magnetic field magnetizes the nanobeads, receptors are pulled into tight clusters (right) rapidly, parallel to electron microscopy images.[76] (Reprinted by permission from Macmillan Publishers Ltd. *Nature Nanotechnology*, Mannix, R.J. et al., 3(1), 36–40. Copyright 2008.)

MNPs coated with antibodies to activate a TREK-1 mechanosensitive ion channel, a stretch-activated potassium ion channel.[79]

25.2.1.5 MRI-Monitoring Cancer Therapy

Magnetic resonance imaging is based on the response of hydrogen spin to magnetic field and can provide high resolution imaging for monitoring tissue morphology and anatomical details, thus cancer therapeutic efficiency can be monitored by this technique.[80] Although MRI has already become the main imaging modality, magnetic contrast agents, generally in the form of T_1 positive contrast agents and T_2 negative contrast agents, are still necessarily required for enhanced contrast and amplified signal from the background. Magnetic NPs are actively used as MRI contrast agents for cancer therapy to discriminate tumours and normal tissues by conjugating with biological molecules.[81,82]

T_1 positive contrast agents are mainly paramagnetic gadolinium (Gd^{3+}), manganese (Mn^{2+}) chelates and ultrasmall IONPs, which provide brighter contrast T_1-weighted images. T_2 negative contrast agents are predominantly iron oxide particles which can be roughly classified according to their hydrodynamic sizes and produce darker contrast T_2-weighted images. The efficiency of an MRI contrast agent is expressed in terms of its relaxivity (r_1, r_2), defined as the paramagnetic relaxation rate enhancement referred to 1 mM agent.[83] The relaxivity ratio r_2/r_1 of positive contrast agents is commonly in the range of 1–2, while negative contrast agents show high r_2/r_1 ratios (at least 10).[84]

25.2.1.5.1 T_1 Positive Contrast Agents

Gd-chelates, for example, Gd-diethylenetriaminepentaacetic acid (Gd-DTPA) and Gd- *N,N′,N″,N‴*-tetracarboxymethyl-1,4,7,10-tetraazacyclododecane (Gd-DOTA) are the most widely used T_1-weighted contrast agents for clinical therapy. However, it has relatively low r_1 value and short retention time *in vivo*. Besides, the toxicity and biocompatibility after entering into the body still need to be solved.[85] In order to reduce the toxicity of Gd^{3+} and enhance the T_1-weighted MR signal, efforts have been devoted to incorporate Gd^{3+} with NPs, which can also increase the cellular uptake of Gd^{3+} by tuning the size and shape of NPs.[83,86] Gd-chelate grafted inorganic NPs have been developed and can even behave as contrast agents for multimodality imaging for cancer therapy.[87] Xia et al. have designed a kind of core-shell NP, with up conversion nanoparticles (UCNPs) as the core and SiO_2 as the shell layer, and Gd-DTPA as the surface ligand for upconversion luminescence, computed tomography (CT) and T_1-weighted MR trimodality *in vivo* imaging.[88] Inorganic NPs containing Gd^{3+}, such as gadolinium oxide (Gd_2O_3), gadolinium phosphate ($GdPO_4$) and gadolinium fluoride (GdF_3) have recently been utilized as T_1-weighted MRI contrast agents in cancer therapy.[89–91] Compared to Gd-chelate grafted NPs, these NPs have the ability to carry large payloads of active magnetic centres and increase relaxivity values. For example, ultrasmall paramagnetic Gd_2O_3 NPs, with an average diameter of 1 nm can act as T_1 positive contrast agents.[92]

Mn^{2+}, with five unpaired electrons, can be used as a new kind of MRI contrast agent. Manganese-based NPs, such as MnO, Mn_3O_4, MnO@mesoporous SiO_2, Mn_3O_4@SiO_2 and hollow MnO NPs have been synthesized as the T_1 contrast agents for cancer therapy recently.[93–95] Hou et al. developed a hollow manganese phosphate NPs drug delivery system with antitumour drugs loaded into the hollow cavities. Conjugated with folic acid, this system enables tracing drug delivery to cell lines expressing different amounts of folate receptor by T_1 MRI signal. Folic acid-positive tumour cells show higher drug release as well as a bright T_1 MRI signal through a folic acid-mediated method.[6]

The UIONPs have recently been designed as T_1 MRI contrast agents. Iron oxide particles in T_1-weighted imaging are limited in most cases due to the large r_2/r_1 ratio. Weller et al. reported adjusted size of small PEGylated IONPs, which have a r_2/r_1 ratio 2.4 at 1.41 T and a two-time higher r_1 relaxivity than Magnevist, a typical T_1 contrast agent based on gadolinium as a clinical standard.[96] F. Herranz et al. carried out the synthesis of 2.5 nm extremely small ^{68}Ga core-doped IONPs, which performed quite well as positive contrast in MRI and nanoradiotracers for ^{68}Ga-based PET in an angiogenesis murine model.[97] Compared with Gd based T_1 contrast agents, the main advantage of iron oxide particles is they provide low long-term toxicity.

25.2.1.5.2 T_2 Positive Contrast Agents

Magnetic IONPs (magnetite [Fe_3O_4] or maghemite [γ-Fe_2O_3]) have emerged as a new kind of MRI T_2 contrast agents as they can selectively shorten T_2 relaxation times in the body and lengthen blood retention time by changing their core size and surface modification.[98] SPIONs are commonly applied as T_2 contrast agents.[99] However, iron oxide contrast agents approved by FDA or EMA all show extremely low activity in current clinical trials. Feridex I.V./Endorem and Resovist/Cliavist were all discontinued, which is reflected by the lack of presence in current clinical trials.[100] Therefore, developing new types of T_2 positive contrast agents is urgently needed. Recently, numerous studies have been conducted to develop new T_2 positive contrast agents from the following three aspects: composition, surface properties and NP size.

Changing the composition of MNPs can significantly affect their r_2 values as it decides the magnetic movement at the atomic level. Thus, incorporating other metal ions into the iron oxide can regulate the magnetic movements of the IONPs. Cheon et al. doped a series of +2 cation including Mn, Fe, Co and Ni to form ferrites MFe_2O_4. Among the different cation doped MFe_2O_4 NPs, $MnFe_2O_4$ NPs show the highest mass magnetization value of 110 (emu/mass of magnetic atoms). Consistent with the magnetization results, Mn-doped $MnFe_2O_4$ shows the strongest MR contrast effect, with a relaxivity value up to 358 $mM^{-1}sec^{-1}$ (1.5 T).[101]

Surface properties can influence MRI efficiency due to the interactions between water and IONPs occurring on the surface of NPs.[102] It is reported that IONPs coated by casein protein exhibit prominent T_2 enhancing capability with a transverse relaxivity r_2 of 273 $mM^{-1}sec^{-1}$ at 3 T, which is 2.5-fold higher

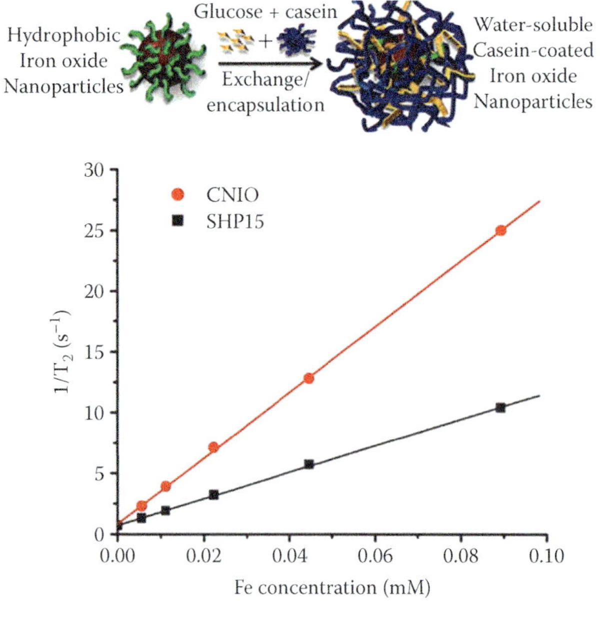

FIGURE 25.10 Scheme illustration of the exchange-encapsulation process for coating IONPs with casein (above). The transverse relaxation rates ($1/T_2$) of CNIOs is much higher than polymer-coated IONPs (SHP15) as a function of the iron concentration (mM) (below).[102] (Reprinted with permission from Huang, J. et al., *ACS Applied Materials & Interfaces*, 5(11), p. 4632–4639. Copyright 2013, American Chemical Society.)

than IONPs coated with amphiphilic polymer (Figure 25.10). Tong et al. used different weight of 1,2-distearoyl-sn-glycero-3-phosphoethanolamine-*N*-methoxypoly(ethyleneglycol) copolymer (DSPE-mPEG) to modify SPIONs and found that T_2 relaxation time augments as the consequence of increasing molecular weight of PEG.[103]

Relaxivity values also depend on the size of NPs. There are two main theories to explain the interaction. 'Motional averaging regime' (MAR), first introduced by Brooks, is suitable for smaller NPs. It claims that the proton diffusion that is much faster than resonance frequency shift determines signal decay, and r_2 values increase with the increasing size of NPs. Meanwhile, 'Static dephasing regime' (SDR), first introduced by Yablonskiy and Haacke, applies to larger NPs and suggests that proton diffusion is not the dominant factor for signal decay as the induced surrounding perturbing field is stronger. Therefore, r_2 values are independent of NP size.[104] Recently, experimental results proved that MNPs synthesized for MRI applications follow the MAR theory, which means larger MNPs have higher r_2 values.[105]

25.2.1.5.3 Dual (T1- and T2-) Weighted MRI Contrast Agents

Single modality contrast agents all have advantages and disadvantages. For example, Gd-based T_1 positive MRI contrast agents have brighter excellent images with risks of biological toxicity.[106] MNP-based T_2-weighted MRI contrast agents have low toxicity to the human body, while the resulting dark signal might mislead the clinical diagnosis as their negative contrast is easily confused with a low level MR signal tissues such as bone or vasculature.[107] Therefore, the development of new types of MR contrast agents with robust dual MRI contrast agents is urgently needed as well. The dual T_1- and T_2-weighted MRI contrast agent significantly improve detection accuracy for molecular imaging and diagnostic application.[108,109] Hu et al. synthesized 5.4 nm SPIONs of high crystallinity and size uniformity for dual contrast T_1- and T_2-weighted MRI. The SPIONs exhibit an impressive magnetization of 94 emu/g Fe_3O_4, the highest r_1 of 19.7 $mM^{-1}s^{-1}$ and the lowest r_2/r_1 ratio of 2.0 at 1.5 T. T_1- and T_2-weighted MR images showed that the SPIONs can improve surrounding water proton signals in the T_1-weighted image and induce significant signal reduction in the T_2-weighted image. *In vitro* cell experiments demonstrated that the SPIONs have little effect on the viability of tumour cell.[110]

25.2.1.5.4 Dual Agents

Besides MRI, there are still many molecular imaging methods to characterize and measure biological processes at the cellular and/or molecular level, such as optical fluorescence, ultrasound, computed tomography (CT) and positron emission tomography (PET). None of the single imaging methods can provide complete information about a subject's structure and function. So combining two or more imaging techniques together for multiple imaging modalities has been an attractive goal. Impressively, MNPs can serve as multifunctional probes for multiple imaging applications, such as MRI/CT probes.

Another useful imaging modality, PET, uses the signals emitted by positron-emitting radiotracers to construct images. A series of positron emitting elements, including ^{11}C, ^{18}F, ^{64}Cu, ^{68}Ga and ^{124}I are commonly used as radiotracer in PET imaging.[111,112] It is an extraordinarily sensitive imaging modality with low resolution, while MRI gives high spatial resolution. Highly sensitive and high-resolution images are possible by combining them. In order to combine PET with MRI in one probe, ^{68}Ga is labeled onto PEG coated SPIONs. After intravenous administration of this probe in normal mice, Sarker et al. proved that ^{68}GaPEG-SPIONs have capabilities as PET-MRI agents for detection of liver and spleen malignancies.[113]

Noble-metal NPs, such as gold and silver NPs, are widely used in CT imaging recently owing to their biocompatibility and high electron density. They have advantages of high spatial and density resolution, 3D tomography reconstruction and disadvantages of low sensitivity resulting in poor soft-tissue contrast. With the combination of CT and MR imaging techniques, drawbacks appear to be solved since MRI/CT can address multiple issues including sensitivity, resolution and tissue penetration during diagnosis. To date, many MRI/CT agents have been developed with incorporation of radiodense elements for CT and MR imaging. For instance, cysteamine-modified FePt NPs conjugated with anti-HER2 antibody can be prepared for dual modality CT/MR molecular imaging both *in vitro* and *in vivo* for FePt NPs can display both X-ray attenuation property and T_2-weighted MR contrast enhancement.[114] Shi et al. generated PEGylated Gd@Au NPs by using branched PEI as a nanoscaffold for *in vivo* dual-mode CT/MR imaging. The PEGylated Gd@Au NPs display a high X-ray attenuation intensity and uncompromised r_1 relaxivity due to the coexistence of Au and Gd elements in one nanoplatform. Hence, they can be used as a contrast agent for dual-mode CT/MR imaging of blood pool and major organs of mice without obvious adverse effects.[115]

25.2.2 Light-Active Theranostics

Certainly, magnetic fields can permit remote theranostics without physical or chemical contact; however, light can flexibly control the cancer therapy by optimizing its wavelength and intensity and achieve precise on-demand drug release in response to illumination of a specific wavelength in the ultraviolet, visible or near-infrared (NIR), λ = 700–1100 nm, regions. NIR imaging has attracted great interest because its penetration depth is optimal and its absorption in human tissues is minimal. In this section, light-active therapy based on photothermal therapy (PTT), photodynamic therapy (PDT), light-triggered delivery and imaging-guided therapy will be introduced.

25.2.2.1 Photothermal Therapy

Although MNPs-based hyperthermia has been a hotspot in cancer treatment, nanotechnology has also been exploring other alternatives. Efficient heat generation under illumination with laser radiation has been developed in recent years.[116] However, laser radiation has obvious disadvantages: (1) it is only suitable for superficial tumours as human tissues also have strong extinction coefficients in the visible range of optical spectrum[117]; (2) it is absorbed by both healthy and cancerous tissues, which leads to damage to adjacent healthy tissue and reduction of therapy efficiency. Interestingly, PTT, by incorporating light-activated heating NPs into tumours can achieve high temperature in tumour sites at lower laser light intensity with no damage to surrounding healthy tissue. Besides, using specific laser wavelengths, which are non-absorbable by healthy tissues, can further improve the photothermal efficiency. Therefore, NPs to be used in PTT should have properties including large absorption cross sections of optical wavelengths with low absorption by healthy tissues, low toxicity and biocompatibility. At present, there are four types of NIR laser-driven photothermal NPs, including gold nanostructures such as gold nanorods[118] and gold nanoshells,[119] copper chalcogenide semiconductors,[120] carbon-based nanomaterials such as carbon nanotubes[121] and graphene,[122] and organic NPs such as indocyanine green.[123]

MNP-based photothermal agents are often incorporated in the above-mentioned nanomaterials. Shi et al. developed a core-shell structured nanocomposite with superparamagnetic Fe_3O_4 NPs as the core, an organic–inorganic hybrid as the mediate layer and an outer gold nanoshell. The nanocomposites can kill cancer cells effectively when exposed to a 808 nm laser, while both the nanocomposite and the laser alone have no significant influence on cell viability.[124] Ultrasmall (<10 nm) $Fe_3O_4@Cu_{2-x}S$ core-shell NPs offer both high

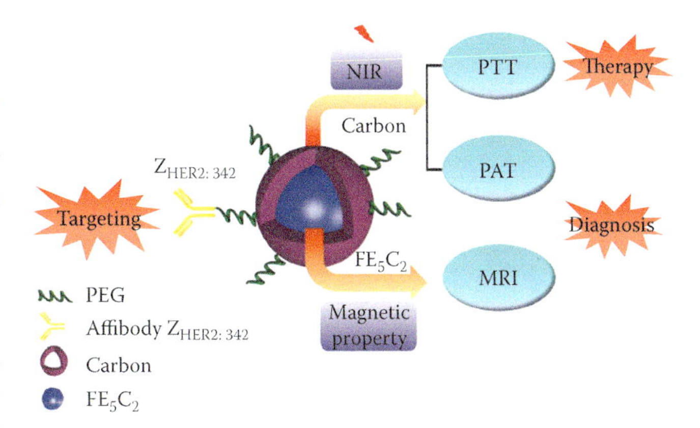

FIGURE 25.11 Schematic illustration for the design of Fe5C2 NPs as a targeted theranostic platform.[9] (Yu, J. et al.: *Advanced Materials*. 2014. 26(24). 4114–4120, 2014. Copyright Wiley-VCH Verlag GmbH & Co. KGaA. Reproduced with permission.)

photothermal stability and superparamagnetic properties.[125] Indocyanine green (an NIR imaging agent)-based superparamagnetic $Fe_3O_4@mSiO_2$ core-shell NPs can perfectly absorb NIR light for laser-mediated PTT.[126] Hou et al. developed a kind of magnetic iron carbide NPs (Fe_5C_2 NPs) with a thin carbon shell coating, which provide a unique opportunity to possess photothermal property. *In vivo* experiments, as shown in Figure 25.11, proved that Fe_5C_2 probes can achieve efficient tumour ablation with no observed side effects.[9]

25.2.2.2 Photodynamic Therapy

Compared to current tumour treatments including surgery, radiation therapy and chemotherapy, PDT is less invasive to surrounding healthy tissues and has emerged as an important area in preclinical research and clinical practice. Based on photoexcitation-triggered photosensitizers (PSs) to generate ROS, PDT can treat certain kinds of cancers and influence precancerous conditions.[127] Most PSs are highly hydrophobic, so the PDT agent is required to deliver PSs to tumour sites very precisely and improve PDT efficiency. NPs are ideal carriers for PSs because of their large surface area, controlled PSs release, biocompatibility and efficient uptake by cells. Konan et al. divided NPs for the delivery of PSs into passive and active group.[128] The former ones are just carriers without interaction with light, while the latter ones can directly be used as PSs or as energy donors to transduce energy to PSs in addition to serving as carrier. Traditional passive NPs for PDT are usually polymer-based materials, typically poly(lactic-*co*-glycolic acid), poly(lactic acid) or their copolymers. Active PS agents include quantum dots, up conversion NPs, gold-based NPs, etc.

MNPs can be covalently modified with PSs that have been proved to be capable of generating $^1O_2$ reduced from $^3O_2$ under laser radiation. Magnetofluorescent NP with iron oxide core binding with Alexa Fluor 750, conjugated to a PS, 5-(4-carboxyphenyl)-10,15,20-triphenyl-2,3-dihydroxychlorin, forms a magneto-fluorescent NP that allows for highly efficient PDT to murine and human macrophages under 650 nm laser light *in vitro*.[129] Magnetite, Fe_3O_4 coated by meso-tetra(4-sulfonatophenyl)porphyrin dihydrochloride, a photosensitive drug used in PDT, is capable of generating singlet oxygen on the surface of MNPs when irradiated at 400 nm.[130] However, when MNPs are conjugated with dye molecules or quantum dots, these two moieties often result in significant luminescence quenching[131–133] and perhaps a consequent decrease of $^1O_2$ production because of the direct connection. It was reported that when Fe_3O_4 MNPs are mixed with CdSe NPs, both magnetic and luminescent properties are retained.[134] Unfortunately, the PDT functionality is lacked due to the unnoticeable amount of $^1O_2$ production.[135]

One problem remaining in PDT is that most PSs used currently are excited by visible or ultraviolet (UV) light, which can be absorbed by human tissues and is limited to superficial treatment. Near IR has deeper penetration for PDT but efficiency of PSs excitation dramatically reduces when exposed to NIR. Up conversion NPs, particularly lanthanide-doped rare-earth nanocrystals, which are able to emit high-energy photons under NIR light, have shown potential prospects as energy donors. Thus, loading PSs to UCNPs with appropriate methods can improve the efficiency of resonance energy transfer from UCNPs to PSs.[136] Yan et al. presented a core-shell structured nanomaterial, with $NaGdF_4$: Yb,Er@CaF_2 UCNPs as the core, and PS-grafted mesoporous silica as the shell. These nanomaterials exhibit excellent PDT efficiency *in vitro* through an energy transfer process between the core and PS molecules under 980 nm NIR laser's radiation. Moreover, the Gd^{3+} ions in the core with paramagnetic properties can provide T_1-weighted MR images.[137]

25.2.2.3 Light-Triggered Delivery

As light can serve as a spatiotemporal trigger in controlled release, photoresponsive systems have been developed in the past few years to achieve on-demand drug release under the illumination of specific wavelength light by photosensitiveness-induced structural modifications of the nanocarriers. Azobenzene and its derivatives can change their structures from *trans* to *cis* under illumination of 300–380 nm wavelength light, and from *cis* to *trans* by irradiating light in the visible region, which can elicit photo-regulated release of drug. Many systems have been developed for the on-demand drug delivery by incorporating azobenzene to nanocarriers, including liposomes,[138] micelles,[139] or polymers.[140] Suzuki et al. proposed micellar MRI contrast agents with photochromic molecule azobenzene derivative incorporated into the hydrophobic chain of DTPA-Gd derivative. These bifunctional micelles can release the included compound upon photoirradiation within 10 min, indicating that the platform is a potential MRI-traceable drug carrier that can be trigged by light.[141]

Photodimerization cleavage under irradiation can also be utilized for light-triggered delivery. o-Nitrobenzyl and its derivatives is a typical group which can cleave the bond between drugs and particles irreversibly by transforming into o-nitrobenzaldehyde.[142] Lin et al. developed a kind of core-shell NP, with magnetic iron oxide core and mesoporous silica shell, which is loaded with anticancer drugs. The entrances of the mesopores are blocked with 2-nitro-5-mercaptobenzyl alcohol functionalized CdS NPs through a photocleavable linkage. When irradiated with UV light, drug release from this magnetic drug delivery system is successfully triggered, which confirms its potential application for selective targeted treatment of cancer.[143] Coumarin is another group for the on-demand drug delivery based on the light-controlled dimerization/cleavage of the dimer. Coumarin moieties can form coumarin photodimers under illumination with wavelengths of 320–400 nm, undergo cleavage and regenerate the coumarin moieties after irradiation by 200–280 nm UV light.[144] Singh et al. first developed photoresponsive MNPs using coumarin-based phototrigger and Fe/Si MNPs for controlled delivery of anticancer drug chlorambucil. *In vitro* experiments revealed that coumarin tethered Fe/Si MNPs efficiently delivered chlorambucil into cancer cells and decreased the viability of cancer cells upon irradiation.[145]

25.2.2.4 Image-Guided Therapy

Using imaging agents to visualize the biodistribution of drugs and monitor tumours is very significant for therapeutic efficiency. There are various kinds of imaging methods, among which, optical imaging has emerged as a promising diagnostic biomedical tool for *in situ* imaging of biological tissues and materials.[146,147] Since the image contrasts mainly rely on the scattering and absorption of light by tissues, several types of light-responsive NPs or dyes based on optical scattering or absorption have been demonstrated for optical imaging.[148] For instance, MNPs can be used to quantitatively determine their long-term biodistribution and tumour localization with or without an external magnetic field in mice with xenograft breast tumours with optical imaging by using NIR dyes.[149]

Photoacoustic tomography (PAT) is a new imaging technique and can overcome a high degree of scattering of optical photons in human tissue by using a photoacoustic effect through transforming light into heat and then launch ultrasonic waves.[150] PAT is often combined with photothermal therapy for imaging-guided therapy. For example, Nie et al. developed hybrid Janus-like vesicles with controlled distribution of 50 nm gold NPs and 15 nm MNPs in the vesicular membrane which show a strong absorption in the NIR window and enhance the transverse relaxation (T_2) contrast effect. The vesicles are applied as imaging agents for *in vivo* bimodal PAT/MR imaging of tumours by intravenous injection.[151]

25.3 INTERNAL STIMULI-TRIGGERED THERANOSTICS

25.3.1 pH-Responsive Theranostics

As a common endogenous stimulus, pH variation has been extensively applied in cancer therapy for targeting drug delivery and controlled drug release. As opposed to the alkalescence in blood (pH_b, 7.35–7.45) and normal tissues (7.0–7.4), the pH value in extracellular tumour environment (pH_e) is mildly acidic (6.85–6.95).[152] However, the intracellular pH (pH_i) of tumours exhibits similar or relatively basic level to

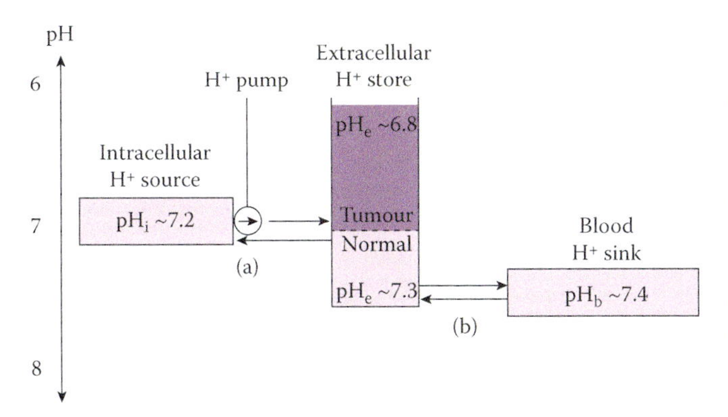

FIGURE 25.12 Fluctuation of pH caused by H^+ flow in normal and tumour tissues. Generally, H^+ produced in metabolism is transported from an intracellular compartment to an extracellular matrix via H^+-pump on cell membrane, and diffused into blood afterwards, regulating pH variation under a normal level. In tumour tissues, however, the increase of H^+ pumping and resistance at interstitial/vascular interface lead to the decrease of pH^e.[153] (Reprinted from *Molecular Medicine Today*, 6(1), Stubbs, M. et al., p. 15, Copyright 2000, with permission from Elsevier.)

that of normal tissues.[154] Such phenomenon primarily derives from the genetic disorder of tumour cells that determines their metabolic phenotype (oxidative phosphorylation and glycolysis). During such metabolic process, acidic metabolites including lactate and CO_2 are produced, and further exported (in the form of H^+) and accumulated in extracellular interstitial fluid, due to the absence of prompt exportation to blood. The clearance of intracellular H^+ sustains normal pH level of cytoplasm, enabling fast proliferation of tumour cells in competition with normal cells (Figure 25.12). Meanwhile, inordinate angiogenesis during rapid tumour growth leads to deficiency in both nutrients and oxygen at the tumour site, which subsequently consolidates glycolytic reaction to increase extracellular acidity.[153] Other than the slightly acidic condition of pH_e and near-neutral condition of pH_i, the intracellular organelles, late endosomes and lysosomes, have a greater pH decrease (5.0–5.5 and 4.5–5.0, respectively). By utilizing subtle pH variations induced by pathological changes, many nanomaterials with various structures and surface modifications have been constructed to respond to certain pH ranges and to release loaded drugs efficiently to desired tumour sites.

Strategies to properly design such pH-responsive nanocarrier system have been exploited as follows: (1) use ionizable chemical groups that undergo conformational switch along with physical/chemical property changes under different pH values to release therapeutic agents; (2) introduce acid-labile chemical bonds sensitive to pH changes whose cleavage disintegrates the delivery system and triggers drug release; and (3) embed carbon oxide-generating precursors with drug molecules into nanodevices, so that CO_2 bubble production will rupture the shell by increasing internal pressure to unload the encapsulated drug.[155]

25.3.1.1 Ionizable Chemical Groups

Ionizable chemical groups, such as amine, phosphoric acid and carboxylic acid, can accept or donate protons, which leads to changes in conformation and solubility. Material integrated with these groups can be classified into three categories according to their chemical composition: organic, inorganic and hybrid.

25.3.1.1.1 Organic Material

Investigated pH-sensitive organic material includes polymers, peptides, lipids and DNA. Polymers with great biocompatibility and pH-sensitivity have been widely used in drug delivery system, such as micelles,[156] liposomes,[157] polymersomes,[158] nanospheres,[159] dendrimers,[160] hydrogel[161] and film.[162] Polymers go through pH-dependent conformational changes via dissociation, destabilization or distribution coefficient shift between drug and carrier. Anionic polymers are incorporated with carboxylic acid groups[163] or sulphonamide groups,[164] while cationic polymers with tertiary amine groups,[165] pyridine groups,[166] or imidazole groups.[167] Peptides utilized in pH-responsive drug delivery system can promote membrane fusion by destabilizing lipid bilayers to accomplish endosomal escape.[168,169] Lipids, which are usually used to generate liposomes, maintain stable and intact structure in neutral pH and undergo protonation and production of nonbilayer structure, an intermediate state before membrane fusion, at acidic pH.[170] DNA with particular sequences can transform from random coil structure to four-stranded motif swiftly and reversibly through protonation when environmental pH decreases, enabling fast drug release.[171] Through combination with such organic material, MNPs can be utilized to construct pH-responsive drug delivery systems. For instance, the magnetic resonance agent, manganese ferrite ($MnFe_2O_4$) and the anticancer drug DOX are encapsulated into anti HER2/neu (HER: Herceptin) antibody-modified α-pyrenyl-ω-carboxyl poly(ethylene glycol) (Py-PEG-COOH or pyrenyl-PEG) via nanoemulsion. The doxorubicin and pyrene groups conjugate with each other through a π–π interaction under physiological pH, and disintegrate under acidic endosomal pH to trigger fast drug release due to the decrease of π–π interaction resulting from protonation of DOX. Thereby, targeted drug delivery, pH-responsive drug release as well as MR imaging can be achieved in a single particle.[172] Poly (beta-amino ester) (PBAE) copolymers are obtained by integrating di(ethylene glycol) diacrylate (DEGDA), amine containing PEG, dopamine (DA) and DOX, in which the tertiary amines and ester groups separately endow the pH-sensitivity and biodegradability onto the backbone. The drug-loaded PBAE polymers are then attached to a SPION core through DA (Figure 25.13). Such polymers have good drug loading capacity, pH-sensitive drug release efficiency and strong MRI contrast.[173]

25.3.1.1.2 Inorganic Material

Application of inorganic material can be divided into two parts: material integrated with pH-induced ionizable chemical bonds, and material itself that can dissolve in acidic environment.[174,175] For instance, a versatile core-shell PB@MIL-100(Fe)(PB: Prussian blue, MIL: materials of Institut Lavoisier) dual metal-organic-framework (d-MOF) nanoparticle is synthesized as a combinatorial therapeutic agent

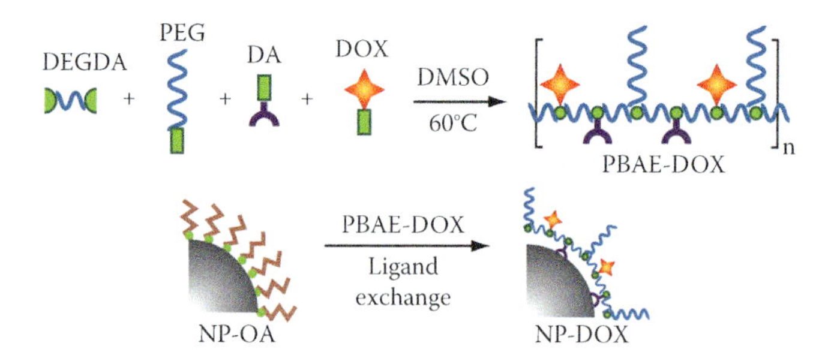

FIGURE 25.13 Schematic illustration of synthesis of PBAE and nanoparticle surface modification. PEG, DA and DOX are conjugated to the backbone of PBAE covalently through linkage between amines and DEGDA. The PBAE-DOX molecules are anchored to the surface of oleic acid (OA) coated SPION by strong chemical bonds formed by DA and metal oxide through ligand exchange.[173] (Reprinted from *Journal of Controlled Release*, 162(1), Fang, C. et al., 233–241, Copyright 2012, with permission from Elsevier.)

along with tri-modal imaging. Metal–organic framework is a porous crystalline inorganic–organic framework that is pH sensitive due to its degradation under acidic conditions. The inner PB MOFs can serve as T_1/T_2 MRI contrast agent, fluorescence optical imaging (FOI) agent and photothermal therapy (PTT) agent under 808 nm NIR laser's radiation.[176] The outer MIL-100(Fe) MOFs are responsible for efficient drug loading (i.e. artemisinin, ART), pH-responsive drug release and T_2-weighted MRI (Figure 25.14).

25.3.1.1.3 Hybrid Material

With complementary physical, chemical and biological properties, organic and inorganic material can be combined to construct promising drug delivery systems, which orchestrate their distinctive function as well as retain their favourable features.[177–179] For example, graphene oxide (GO) is magnetically

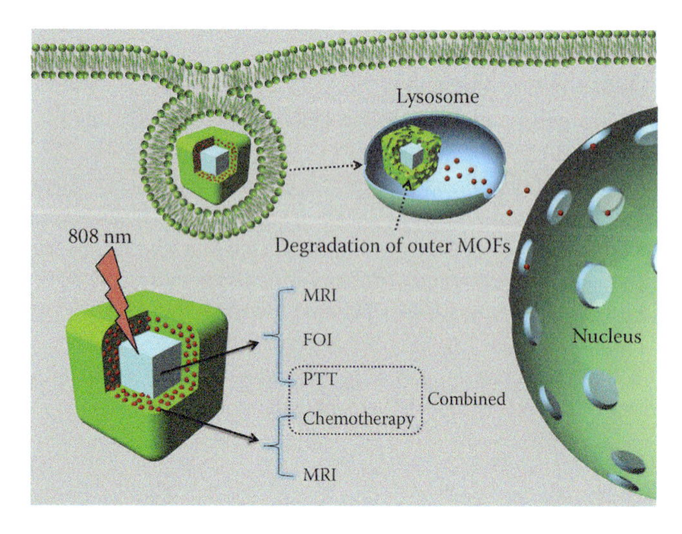

FIGURE 25.14 Schematic illustration of d-MOFs targeting tumours for combined therapy. The drug loading d-MOFs diffuse into tumours through EPR effect and degrade outer MOFs to release ART and trigger dual-modal FOI and MRI-guided cancer therapy.[176] (Reprinted from *Biomaterials*, 100, Wang, D. et al. 27, Copyright 2016, with permission from Elsevier.)

functionalized by integrating Fe_3O_4 MNPs onto GO nanoplatelets through chemical coprecipitation for targeted delivery. As further modification, chitosan and PEG are covalently attached to magnetic graphene oxide. The chemotherapeutic agent, irinotecan or DOX, is loaded to the nanocarrier system through π-π interaction, endowing the system with pH-triggered drug release property.[180]

25.3.1.2 Acid-Labile Chemical Bonds

Acid-labile chemical bonds, as the name implies, are stable in basic/neutral situations, but are degraded or hydrolyzed when encountered with acid. These chemical bonds, including acetal,[181] orthoester,[182] hydrazone,[183] imine,[184] and cis-aconityl,[185] either are incorporated within the nanostructure maintaining their pH-sensitivity to trigger collapse of the whole nanostructure and release of the cargo,[183,186] or attach the desired drug molecules covalently to nanocarriers, forming inactive prodrugs whose activation and release correspond with the cleavage of the chemical bonds.[187] A multifunctional and water-soluble drug delivery system with bimodal imaging is established by conjugating the anticancer drug DOX to SPIONs through a pH-sensitive hydrazone bond. The tumour targeting ligands, cyclic arginine-glycine-aspartic acid (cRGD) peptide and positron emission tomography (PET) ^{64}Cu chelators, macrocyclic 1,4,7-triazacyclononane-*N,N,N″*-triacetic acid, are conjugated onto SPIONs through PEG.[188] Thus, these nanodevices can target integrin $\alpha_v\beta_3$-expressing tumour cells and trigger pH-responsive drug release, combined with PET/MRI dual-modal imaging (Figure 25.15). Superparamagnetic Fe_3O_4 NPs are grafted to pore openings of MSNs via an acid-labile linker 1,3,5-triazaadamantane (TAA) group, which serve as a nanogate, entrapping therapeutic drug dexamethasone into mesopores under physiological pH.[189] On entering tumour cells, the TAA linkers are hydrolyzed into tri(amino methyl)ethane under acidic endosomal pH, removing Fe_3O_4 caps and leaking encapsulated drugs (Figure 25.16).

25.3.1.3 Gas-Generating Precursors

By utilizing the reaction between HCO_3^- and H^+, which yields CO_2 gas and water, resulting in disintegration of the delivery

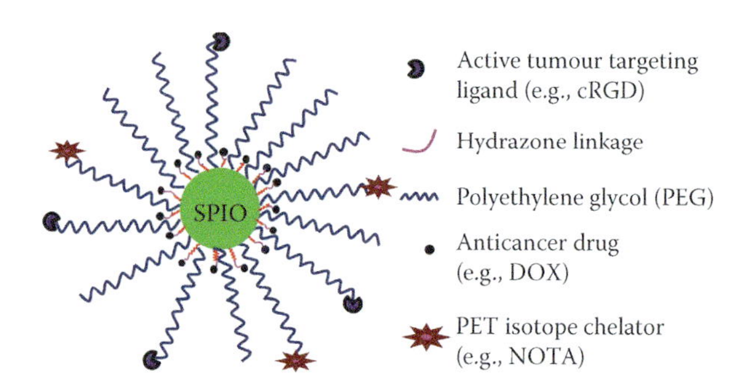

FIGURE 25.15 Schematic illustration of the multifunctional cRGD-conjugated SPIO nanocarriers for combined tumour-targeting drug delivery and PET/MR imaging.[188] (Reprinted from *Biomaterials*, 32(17), Yang, X. et al. 4151, Copyright 2011, with permission from Elsevier.)

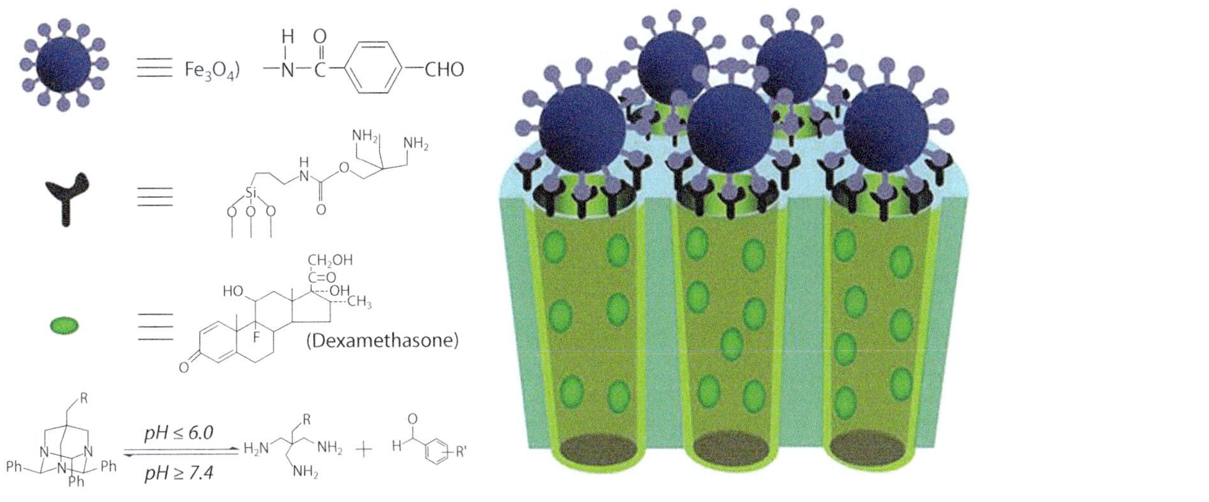

FIGURE 25.16 Schematic illustration of the reversible pH-responsive Fe_3O_4-capped mesoporous silica ensembles with an acid-labile linker.[189] (From Gan, Q. et al., *Journal of Materials Chemistry*, 22(31), p. 15960–15968, 2012. Reproduced by permission of The Royal Society of Chemistry.)

system and release of loaded cargos, another type of nanocarriers has been developed. Commonly used CO_2-generating agents include sodium bicarbonate[190] and ammonium bicarbonate.[191] Nevertheless, this method is barely used in MNP-based drug delivery system.

25.3.2 Reduction-Responsive Theranostics

ROS, including peroxide, superoxide, hydroxyl radical and singlet oxygen, can regulate many signaling pathways like cell proliferation and differentiation, enzyme activity, inflammation and pathogen elimination. Cells can maintain ROS homeostasis by controlling ROS generation as well as eliminating them via ROS-scavenging systems. Studies have demonstrated that a variety of cancer cell types have a higher level of ROS than normal cells, which leads to oxidative stress.[192] Intrinsic factors such as oncogene activation, aberrant metabolism, mitochondrial dysfunction and antioxidant deficit (e.g. loss of p53 protein)[193,194] and extrinsic factors such as inflammatory cytokines, nutrient imbalance and hypoxic environment,[195] can induce more ROS production, affecting the intracellular redox homeostasis of cancer cells. A moderate increase of ROS is usually associated with cell proliferation and differentiation in normal cells.[196] In cancer cells, however, the consistent high level of ROS can cause oxidative damage to lipids, proteins and DNAs, resulting in gene mutation, metabolic disturbance, p53 malfunction and compromised DNA repair capability,[197] which leads to genomic instability. The genomic instability can in turn stimulate intrinsic factors to promote ROS generation.[198] Nevertheless, when the increased ROS accumulates to a certain threshold that is intolerant to cancer cells, the excessive amount of ROS will elicit cytotoxicity and apoptosis, thus inhibiting cancer development.[199] Therefore, adaptive mechanism is generated in cancer cells to prevent such conditions. Redox adaptation can cope with oxidative stress by activating redox-sensitive transcription factors, involving increased expression of ROS-scavenging enzymes, such as superoxide dismutase, catalase, thioredoxin and glutathione (GSH),[200] elevation of cell survival molecules, such as BCL2 and MCL1,[201] and suppression of cell death factors, such as caspases.[202]

By utilizing the different concentration of antioxidant molecules such as GSH between cancer cells and normal tissues, as well as extracellular (2–10 μM) and intracellular (2–10 mM) environment, redox-responsive system is established for specific drug or nucleic acid delivery. Disulphide bonds, whose cleavage can be induced by GSH, are widely used in this system.[203]

A magnetic iron oxide core is coated with a redox-responsive chitosan-PEG copolymer shell that is modified with tumour targeting ligand chlorotoxin.[204] The chemotherapeutic agents O^6-benzylguanine (BG) are covalently conjugated to chitosan. Through convection-enhanced delivery, these drug-loading NPs diffuse into cancer cells and release BG under reductive intracellular conditions in a more effective but less toxic way (Figure 25.17). Cationic disulphide-containing

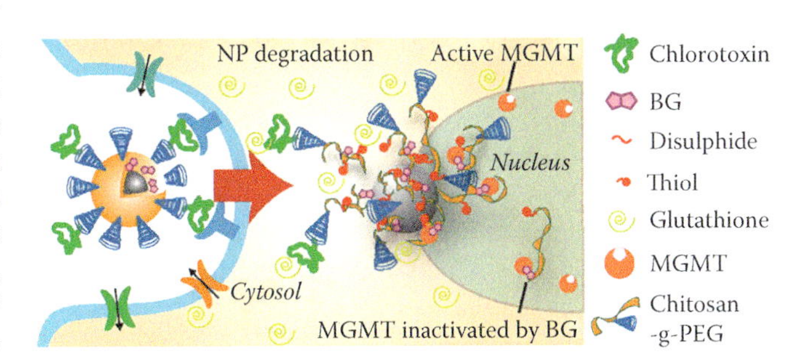

FIGURE 25.17 Due to the upregulation of the DNA repair protein O^6-methylguanine-DNA methyltransferase (MGMT), glioblastoma multiforme (GBM) exhibits resistance to temozolomide-based chemotherapy. As a MGMT inhibitor, BG is transported to GBM through magnetic nanocarriers. The cleavage of disulphide bridges induced by glutathione in cancer cells leads to degradation of the NPs and reduction-sensitive drug release.[204] (Reproduced from Stephen, Z.R. et al., *ACS Nano*, 8(10), p. 10383–95, 2014. With permission.)

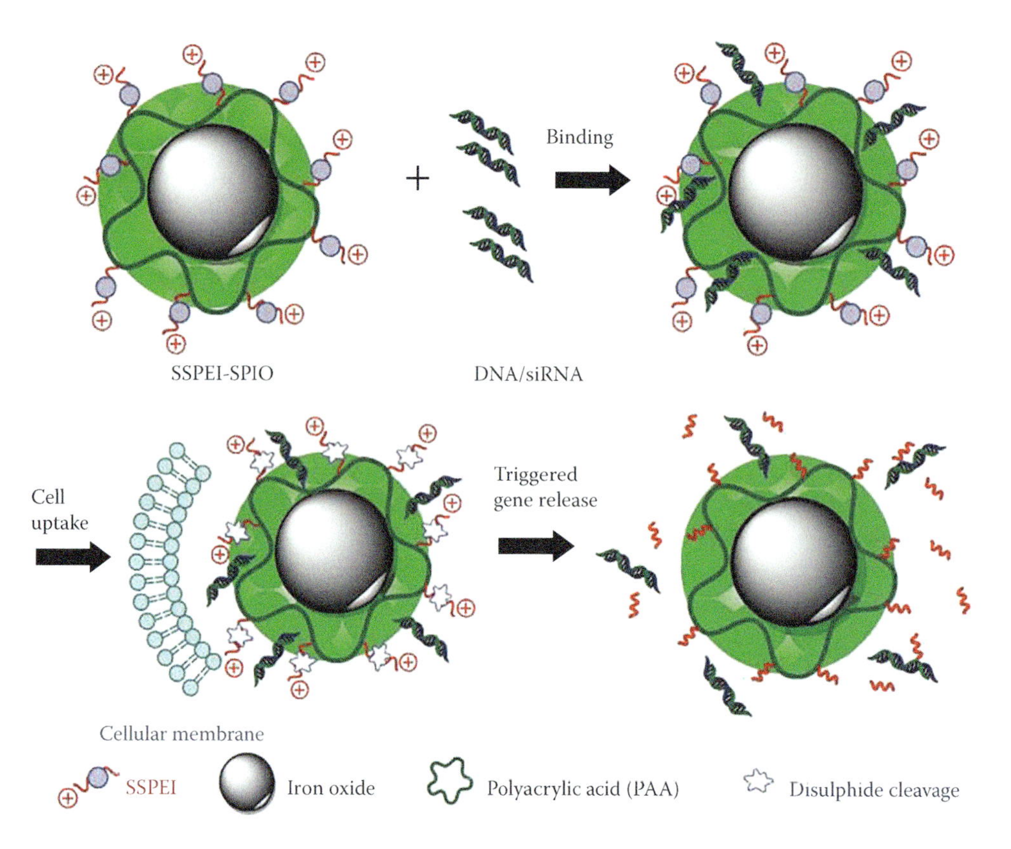

FIGURE 25.18 SSPEI-SPIO nanoparticles can efficiently bind genes and mediate subsequent gene release induced by cleavage of disulphide bonds.[205] (Reproduced from Li, D. et al., *International Journal of Nanomedicine*, 2014(Issue 1), p. 3347–3361. 2014. With permission.)

polymers polyethyleneimine (SSPEI) are integrated with SPIONs through a weak anionic polyelectrolyte polyacrylic acid, forming SSPEI-SPIO nanoparticles.[205] These NPs are capable of binding DNA and siRNA, and mediating localized gene release in intracellular reductive environment (Figure 25.18).

Another approach investigated in redox-responsive system is oxidation responsiveness by targeting high level of ROS in cancer cells, via introducing ROS-sensitive chemicals into drug delivery systems.[206] A core-shell nanoplatform is developed with pH- and H_2O_2 responsiveness and O_2 generation. The inner SiO_2-methylene blue core has a high payload of photosensitizer for PDT. The outer MnO_2 shell can avoid undesirable leakage of the cargo loaded in core. Simultaneously, O_2 is produced via the reduction of Mn(IV) to Mn(II) in acidic H_2O_2 environment within tumour tissues, which prevents hypoxia in tumour cells and enables the nanosystem MRI capability.[207]

25.4 MULTIMODALITY THERANOSTICS

Since a single type of stimulus can merely respond to a restricted proportion of tumour environment, the therapeutic efficacy is far from optimal. Besides, each therapeutic modality has its own limitations. For instance, internal stimuli-dependent therapies exhibit lower flexibility than external stimuli-responsive modalities. Light-mediated therapies suffer shallow penetration and magnetic-triggered therapies can only sustain function under high concentrations of MNPs. Moreover, the complexity of biological system sometimes requires more than one stimulus to trigger enzymatic function. Therefore, NPs need to be optimized to achieve higher sensitivity and better modulation. Thus, dual-responsive NPs, such as pH- and redox-,[208] glucose- and pH-,[209] magneto- and pH-,[210] thermo- and pH-[178] and enzyme- and pH-,[211] in addition to multistimuli active NPs,[212,213] have been exploited. A kind of core–shell composite microsphere is synthesized by encapsulating DOX into magnetic MSN core coated with poly(N-isopropyl acrylamide-co-methacrylic acid)(P(NIPAM-co-MAA)) shell via precipitation polymerization. Such composites present thermal sensitivity and pH-dependent drug release.[214] A controllable drug release nanocontainer is synthesized by construction of a core-shell structure and further dissolution of the core to create a cavity for drug loading. When encountered with a reducing environment in combination with pH changes, the nanocontainer collapses, unloading the encapsulated drug.[215] The magnetic response can further induce local heating via hyperthermia under AMF (Figure 25.19).

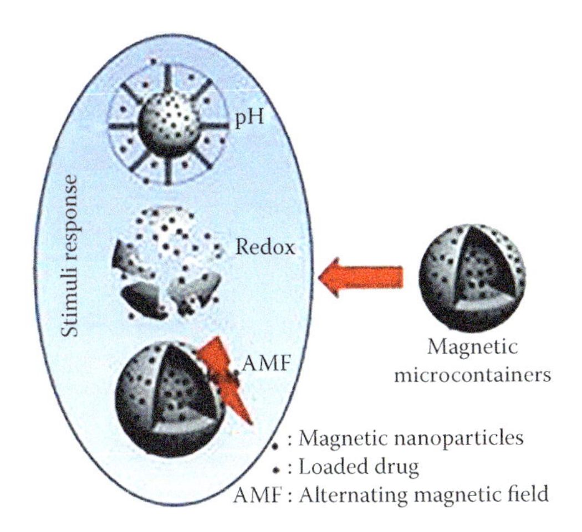

FIGURE 25.19 Schematic illustration of magnetic, pH and redox-sensitive nanocontainer[215] (From Bilalis, P. et al., *Journal of Materials Chemistry*, 22(27), 13451–13454, 2012. Reproduced by permission of The Royal Society of Chemistry.)

25.5 CONCLUSIONS AND PERSPECTIVES

Tremendous efforts have been made to overcome the side effects of cancer treatment in the past few decades. In this context, MNPs sensitive to external or internal stimuli represent an attractive alternative for cancer theranostics. The different kinds of stimuli can spatiotemporally trigger drug release from nanocarriers to tumour sites. Meanwhile, various responsive MNPs can be designed with different size, shape or even distinct incorporation with other functional molecules, allowing great flexibility of stimuli-responsive systems. Besides, MNPs themselves can be utilized as contrast agents for monitoring therapeutic efficiency and release signals, such as heat and mechanical forces to kill cancer cells when the MNPs are triggered by certain stimuli.

However, despite considerable achievements having been reported in stimulus-responsive MNP-based systems for cancer therapy, only a few have attained the preclinical stage, and very few have reached the clinical stage. Magnetic NPs can sometimes be harmful to the human body when the composition, physicochemical properties, administration route and dosage are altered. For example, iron-containing MNPs may release Fe^{2+} and/or ferric Fe^{3+} ions, and electron interaction between these two ions may break the body's homeostasis and lead to abnormal cellular responses, such as oxidative stress and inflammatory responses. Besides, many available stimuli-regulated systems may be too impractical to be put into clinical practice, due to limitations such as defective degradability, insufficient biocompatibility, complexity of the systems and the demand for synthetic procedures. Most systems previously discussed can only interact with a fraction of tumour cells overexpressing specific proteins, lacking universal therapeutic effect. It is difficult to pinpoint which stimuli-responsive nanosystems have the best chances to reach the clinical stage because every stimuli-regulated system has its drawbacks. Externally applied stimuli are limited by the penetration depth and concentration of contrast agents, while internal stimuli-responsive therapies are only sensitive to a certain kind of stimuli and cannot be achieved straightforwardly. Moreover, the body's environment is very complex and pathological change is a dynamic process. Multiple sensitive systems are needed to constitute intelligent therapy applicable to environmental changes. Perhaps more attention should now be addressed to clinically acceptable systems. In the future, we expect to have more stimulus-responsive MNP-based systems that demonstrate easy excretion and good biocompatibility for effective clinical cancer therapy.

REFERENCES

1. Siegel, R., D. Naishadham, and A. Jemal, *Cancer statistics*, 2013. **63**(1): p. 11–30.
2. Jang, S.H. et al., Drug delivery and transport to solid tumors. *Pharmaceutical research*, 2003. **20**(9): p. 1337–50.
3. Kim, K. et al., Tumor-homing multifunctional nanoparticles for cancer theragnosis: Simultaneous diagnosis, drug delivery, and therapeutic monitoring. *Journal of controlled release*, 2010. **146**(2): p. 219–27.
4. Park, J.H. et al., Polymeric nanomedicine for cancer therapy. *Progress in polymer science*, 2008. **33**(1): p. 113–37.
5. Heo, D.N. et al., Gold nanoparticles surface-functionalized with paclitaxel drug and biotin receptor as theranostic agents for cancer therapy. *Biomaterials*, 2012. **33**(3): p. 856–66.
6. Yu, J. et al., Hollow manganese phosphate nanoparticles as smart multifunctional probes for cancer cell targeted magnetic resonance imaging and drug delivery. *Nano research*, 2012: p. 1–16.
7. Yhee, J.Y. et al., Tumor-targeting transferrin nanoparticles for systemic polymerized siRNA delivery in tumor-bearing mice. *Bioconjugate chemistry*, 2013. **24**(11): p. 1850–60.
8. Danhier, F., A. Le Breton, and V.r. Préat, RGD-based strategies to target alpha (v) beta (3) integrin in cancer therapy and diagnosis. *Molecular pharmaceutics*, 2012. **9**(11): p. 2961–73.
9. Yu, J. et al., Multifunctional Fe_5C_2 nanoparticles: A targeted theranostic platform for magnetic resonance imaging and photoacoustic tomography – guided photothermal therapy. *Advanced materials*, 2014. **26**(24): p. 4114–20.
10. Barua, S. et al., Particle shape enhances specificity of antibody-displaying nanoparticles. *Proceedings of the National Academy of Sciences*, 2013. **110**(9): p. 3270–75.
11. Cho, H. et al., Redox-sensitive polymeric nanoparticles for drug delivery. *Chemical communications*, 2012. **48**(48): p. 6043–45.
12. Zhang, J. et al., Multifunctional envelope-type mesoporous silica nanoparticles for tumor-triggered targeting drug delivery. *Journal of the American Chemical Society*, 2013. **135**(13): p. 5068–73.
13. Han, L. et al., pH – controlled delivery of nanoparticles into tumor cells. *Advanced healthcare materials*, 2013. **2**(11): p. 1435–9.
14. Yu, M.K. et al., Drug – loaded superparamagnetic iron oxide nanoparticles for combined cancer imaging and therapy in vivo. *Angewandte chemie*, 2008. **120**(29): p. 5442–5.
15. Gupta, A.K. and M. Gupta, Synthesis and surface engineering of iron oxide nanoparticles for biomedical applications. *Biomaterials*, 2005. **26**(18): p. 3995–4021.

16. Sun, C., J.S. Lee, and M. Zhang, Magnetic nanoparticles in MR imaging and drug delivery. *Advanced drug delivery reviews*, 2008. **60**(11): p. 1252–65.
17. McCarthy, J.R. and R. Weissleder, Multifunctional magnetic nanoparticles for targeted imaging and therapy. *Advanced drug delivery reviews*, 2008. **60**(11): p. 1241–51.
18. Pankhurst, Q. et al., Progress in applications of magnetic nanoparticles in biomedicine. *Journal of physics D: Applied physics*, 2009. **42**(22): p. 224001.
19. Widder, K.J. et al., Selective targeting of magnetic albumin microspheres to the Yoshida sarcoma: Ultrastructural evaluation of microsphere disposition. *European journal of cancer and clinical oncology*, 1983. **19**(1): p. 141–47.
20. Jing, Y. et al., Magnetic nanoparticle-based cancer therapy. *Chinese physics B*, 2013. **22**(2): p. 027506.
21. Park, D.H. et al., Biodegradable inorganic nanovector: Passive versus active tumor targeting in siRNA transportation. *Angewandte chemie international edition*, 2016. **55**(14): p. 4582–86.
22. Cheng, L. et al., Multifunctional nanoparticles for upconversion luminescence/MR multimodal imaging and magnetically targeted photothermal therapy. *Biomaterials*, 2012. **33**(7): p. 2215–22.
23. Meyers, P.H., F. Cronic, and C. Nice Jr, Experimental approach in the use and magnetic control of metallic iron particles in the lymphatic and vascular system of dogs as a contrast and isotropic agent. *American Journal of roentgenology radium therapy and nuclear medicine*, 1963. **90**: p. 1068.
24. Yu, J. et al., Multistimuli-regulated photochemothermal cancer therapy remotely controlled via Fe5C2 nanoparticles. *ACS nano*, 2015. **10**(1): p. 159–169.
25. Pereyra, A.S. et al., Magnetofection enhances adenoviral vector-based gene delivery in skeletal muscle cells. *Journal of nanomedicine & nanotechnology*, 2016. **7**(2): p. 364.
26. Scherer, F. et al., Magnetofection: Enhancing and targeting gene delivery by magnetic force in vitro and in vivo. *Gene therapy*, 2002. **9**(2): p. 102.
27. Veiseh, O., J.W. Gunn, and M. Zhang, Design and fabrication of magnetic nanoparticles for targeted drug delivery and imaging. *Advanced drug delivery reviews*, 2010. **62**(3): p. 284–304.
28. Hu, S.H. et al., Core/single-crystal-shell nanospheres for controlled drug release via a magnetically triggered rupturing mechanism. *Advanced materials*, 2008. **20**(14): p. 2690–2695.
29. Xing, R. et al., Hollow iron oxide nanoparticles as multidrug resistant drug delivery and imaging vehicles. *Nano research*, 2013. **6**(1): p. 1–9.
30. Hervault, A. et al., Doxorubicin loaded dual pH-and thermo-responsive magnetic nanocarrier for combined magnetic hyperthermia and targeted controlled drug delivery applications. *Nanoscale*, 2016. **8**(24): p. 12152–61.
31. Kakwere, H. et al., Functionalization of strongly interacting magnetic nanocubes with (thermo) responsive coating and their application in hyperthermia and heat-triggered drug delivery. *ACS applied materials & interfaces*, 2015. **7**(19): p. 10132–45.
32. Hoare, T. et al., A magnetically triggered composite membrane for on-demand drug delivery. *Nano letters*, 2009. **9**(10): p. 3651–57.
33. Thomas, C.R. et al., Noninvasive remote-controlled release of drug molecules in vitro using magnetic actuation of mechanized nanoparticles. *Journal of the american chemical society*, 2010. **132**(31): p. 10623–25.
34. Zhang, J. and R. Misra, Magnetic drug-targeting carrier encapsulated with thermosensitive smart polymer: Core–shell nanoparticle carrier and drug release response. *Acta biomaterialia*, 2007. **3**(6): p. 838–850.
35. Prijic, S. and G. Sersa, Magnetic nanoparticles as targeted delivery systems in oncology. *Radiology and oncology*, 2011. **45**(1): p. 1–16.
36. Czugala, M. et al., Efficient and safe gene delivery to human corneal endothelium using magnetic nanoparticles. *Nanomedicine*, 2016. **11**(14): p. 1787–800.
37. Mah, C. et al., Improved method of recombinant AAV2 delivery for systemic targeted gene therapy. *Molecular therapy*, 2002. **6**(1): p. 106–12.
38. Dobson, J., Gene therapy progress and prospects: Magnetic nanoparticle-based gene delivery. *Gene therapy*, 2006. **13**(4): p. 283.
39. Al-Deen, F.N. et al., Superparamagnetic nanoparticles for effective delivery of malaria DNA vaccine. *Langmuir*, 2011. **27**(7): p. 3703–12.
40. Lee, J.H. et al., All-in-one target-cell-specific magnetic nanoparticles for simultaneous molecular imaging and siRNA delivery. *Angewandte chemie*, 2009. **121**(23): p. 4238–43.
41. Uthaman, S. et al., Fabrication and development of magnetic particles for gene therapy. *Polymers and nanomaterials for gene therapy*, 2016: p. 215.
42. Lee, J.Y. et al., Ultrasound-enhanced siRNA delivery using magnetic nanoparticle – loaded chitosan – deoxycholic acid nanodroplets. *Advanced healthcare materials*, 2017. **6**(8).
43. Pizzimenti, S. et al., Challenges and opportunities of nanoparticle–based theranostics in skin cancer. *Nanoscience in dermatology*, 2016: p. 177.
44. McBain, S.C., H.H. Yiu, and J. Dobson, Magnetic nanoparticles for gene and drug delivery. *International journal of nanomedicine*, 2008. **3**(2): p. 169.
45. Yadollahpour, A. and S.A. Hosseini, Magnetic nanoparticle based hyperthermia: A review of the physiochemical properties and synthesis methods. *International journal of pharmaceutical research and allied sciences*, 2016. **5**(2): p. 242–246.
46. Sellins, K.S. and J.J. Cohen, Hyperthermia induces apoptosis in thymocytes. *Radiation research*, 1991. **126**(1): p. 88–95.
47. Gilchrist, R. et al., Selective inductive heating of lymph nodes. *Annals of surgery*, 1957. **146**(4): p. 596.
48. Gobbo, O.L. et al., Magnetic nanoparticles in cancer theranostics. *Theranostics*, 2015. **5**(11): p. 1249.
49. Kossatz, S. et al., Efficient treatment of breast cancer xenografts with multifunctionalized iron oxide nanoparticles combining magnetic hyperthermia and anti-cancer drug delivery. *Breast cancer research*, 2015. **17**(1): p. 66.
50. Estelrich, J. et al., Iron oxide nanoparticles for magnetically-guided and magnetically-responsive drug delivery. *International journal of molecular sciences*, 2015. **16**(4): p. 8070–101.
51. Jang, J.t. et al., Critical enhancements of MRI contrast and hyperthermic effects by dopant – controlled magnetic nanoparticles. *Angewandte chemie*, 2009. **121**(7): p. 1260–1264.
52. Johannsen, M. et al., Clinical hyperthermia of prostate cancer using magnetic nanoparticles: Presentation of a new interstitial technique. *International journal of hyperthermia*, 2005. **21**(7): p. 637–47.
53. Asín, L. et al., Controlled cell death by magnetic hyperthermia: Effects of exposure time, field amplitude, and nanoparticle concentration. *Pharmaceutical research*, 2012. **29**(5): p. 1319–27.

54. Tartaj, P., S. Veintemillas-Verdaguer, and C.J. Serna, The preparation of magnetic nanoparticles for applications in biomedicine. *Journal of physics D: Applied physics*, 2003. **36**(13): p. R182.
55. Kita, E. et al., Ferromagnetic nanoparticles for magnetic hyperthermia and thermoablation therapy. *Journal of physics D: Applied physics*, 2010. **43**(47): p. 474011.
56. Yoo, D. et al., Theranostic magnetic nanoparticles. *Accounts of chemical research*, 2011. **44**(10): p. 863–74.
57. Kobayashi, T. et al., Antitumor immunity by magnetic nanoparticle-mediated hyperthermia. *Nanomedicine*, 2014. **9**(11): p. 1715–26.
58. Mosser, D.D. et al., The chaperone function of hsp70 is required for protection against stress-induced apoptosis. *Molecular and cellular biology*, 2000. **20**(19): p. 7146–59.
59. Yanase, M. et al., Antitumor immunity induction by intracellular hyperthermia using magnetite cationic liposomes. *Cancer science*, 1998. **89**(7): p. 775–82.
60. Jimbow, K. et al., Melanoma-targeted chemothermotherapy and in situ peptide immunotherapy through HSP production by using melanogenesis substrate, NPrCAP, and magnetite nanoparticles. *Journal of skin cancer*, 2013. **2013**: p. 742925.
61. Nedelcu, G., Magnetic nanoparticles impact on tumoral cells in the treatment by magnetic fluid hyperthermia. *Digest journal of nanomaterials and biostructures*, 2008. **3**(3): p. 103–107.
62. Marano, F. et al., Nanoparticles: Molecular targets and cell signalling. *Archives of toxicology*, 2011. **85**(7): p. 733–741.
63. Prasad, N. et al., Mechanism of cell death induced by magnetic hyperthermia with nanoparticles of γ-$Mn_xFe_{2-x}O_3$ synthesized by a single step process. *Journal of materials chemistry*, 2007. **17**(48): p. 5042–5051.
64. Ozben, T., Oxidative stress and apoptosis: Impact on cancer therapy. *Journal of pharmaceutical sciences*, 2007. **96**(9): p. 2181–2196.
65. Hsieh, H.-C. et al., ROS-induced toxicity: Exposure of 3T3, RAW264. 7, and MCF7 cells to superparamagnetic iron oxide nanoparticles results in cell death by mitochondria-dependent apoptosis. *Journal of nanoparticle research*, 2015. **17**(2): p. 71.
66. Guardia, P. et al., Water-soluble iron oxide nanocubes with high values of specific absorption rate for cancer cell hyperthermia treatment. *ACS nano*, 2012. **6**(4): p. 3080–3091.
67. Marcos-Campos, I. et al., Cell death induced by the application of alternating magnetic fields to nanoparticle-loaded dendritic cells. *Nanotechnology*, 2011. **22**(20): p. 205101.
68. Mehdaoui, B. et al., Large specific absorption rates in the magnetic hyperthermia properties of metallic iron nanocubes. *Journal of magnetism and magnetic materials*, 2010. **322**(19): p. L49–L52.
69. Dennis, C. et al., Nearly complete regression of tumors via collective behavior of magnetic nanoparticles in hyperthermia. *Nanotechnology*, 2009. **20**(39): p. 395103.
70. Deatsch, A.E. and B.A. Evans, Heating efficiency in magnetic nanoparticle hyperthermia. *Journal of magnetism and magnetic materials*, 2014. **354**: p. 163–172.
71. Lee, J.-H. et al., Exchange-coupled magnetic nanoparticles for efficient heat induction. *Nature nanotechnology*, 2011. **6**(7): p. 418–422.
72. Rosensweig, R.E., Heating magnetic fluid with alternating magnetic field. *Journal of magnetism and magnetic materials*, 2002. **252**: p. 370–374.
73. Patil, R. et al., Superparamagnetic iron oxide/chitosan core/shells for hyperthermia application: Improved colloidal stability and biocompatibility. *Journal of magnetism and magnetic materials*, 2014. **355**: p. 22–30.
74. Hao, R. et al., Developing Fe_3O_4 nanoparticles into an efficient multimodality imaging and therapeutic probe. *Nanoscale*, 2013. **5**(23): p. 11954–11963.
75. Golovin, Y.I. et al., Towards nanomedicines of the future: Remote magneto-mechanical actuation of nanomedicines by alternating magnetic fields. *Journal of controlled release*, 2015. **219**: p. 43–60.
76. Mannix, R.J. et al., Nanomagnetic actuation of receptor-mediated signal transduction. *Nature nanotechnology*, 2008. **3**(1): p. 36–40.
77. Lee, J.H. et al., Artificial control of cell signaling and growth by magnetic nanoparticles. *Angewandte chemie international edition*, 2010. **49**(33): p. 5698–5702.
78. Coultas, L., K. Chawengsaksophak, and J. Rossant, Endothelial cells and VEGF in vascular development. *Nature*, 2005. **438**(7070): p. 937.
79. Hughes, S. et al., Selective activation of mechanosensitive ion channels using magnetic particles. *Journal of the Royal Society interface*, 2008. **5**(25): p. 855–863.
80. Vandenberghe, S. and P.K. Marsden, PET-MRI: A review of challenges and solutions in the development of integrated multimodality imaging. *Physics in medicine and biology*, 2015. **60**(4): p. R115.
81. MacDonald, M.E. and R. Frayne, Cerebrovascular MRI: A review of state-of-the-art approaches, methods and techniques. *NMR in biomedicine*, 2015. **28**(7): p. 767–791.
82. Wu, M. et al., Nanocluster of superparamagnetic iron oxide nanoparticles coated with poly (dopamine) for magnetic field-targeting, highly sensitive MRI and photothermal cancer therapy. *Nanotechnology*, 2015. **26**(11): p. 115102.
83. Louie, A., Multimodality imaging probes: Design and challenges. *Chemical reviews*, 2010. **110**(5): p. 3146–3195.
84. Aime, S. et al., Lanthanide (III) chelates for NMR biomedical applications. *Chemical society reviews*, 1998. **27**(1): p. 19–29.
85. Estelrich, J., M.J. Sánchez-Martín, and M.A. Busquets, Nanoparticles in magnetic resonance imaging: From simple to dual contrast agents. *International journal of nanomedicine*, 2015. **10**: p. 1727.
86. Xu, W. et al., Paramagnetic nanoparticle T_1 and T_2 MRI contrast agents. *Physical chemistry chemical physics*, 2012. **14**(37): p. 12687–12700.
87. Lin, W.-I. et al., High payload Gd (III) encapsulated in hollow silica nanospheres for high resolution magnetic resonance imaging. *Journal of materials chemistry* B, 2013. **1**(5): p. 639–645.
88. Xia, A. et al., Gd 3+ complex-modified NaLuF 4-based upconversion nanophosphors for trimodality imaging of NIR-to-NIR upconversion luminescence, x-ray computed tomography and magnetic resonance. *Biomaterials*, 2012. **33**(21): p. 5394–5405.
89. He, M. et al., Dual phase – controlled synthesis of uniform lanthanide-doped $NaGdF_4$ upconversion nanocrystals via an OA/ionic liquid two-phase system for in vivo dual-modality imaging. *Advanced functional materials*, 2011. **21**(23): p. 4470–4477.
90. Wu, Y. et al., A new type of silica-coated $Gd_2(CO_3)_3$: Tb nanoparticle as a bifunctional agent for magnetic resonance imaging and fluorescent imaging. *Nanotechnology*, 2012. **23**(20): p. 205103.

91. Liu, Z. et al., Long-circulating Gd_2O_3: Yb^{3+}, Er^{3+} up-conversion nanoprobes as high-performance contrast agents for multi-modality imaging. *Biomaterials,* 2013. **34**(6): p. 1712–1721.
92. Park, J.Y. et al., Paramagnetic ultrasmall gadolinium oxide nanoparticles as advanced T_1 MRI contrast agent: Account for large longitudinal relaxivity, optimal particle diameter, and in vivo T_1 MR images. *ACS nano*, 2009. **3**(11): p. 3663–3669.
93. Chen, Y. et al., Manganese oxide-based multifunctionalized mesoporous silica nanoparticles for pH-responsive MRI, ultrasonography and circumvention of MDR in cancer cells. *Biomaterials*, 2012. **33**(29): p. 7126–7137.
94. Abbasi, A.Z. et al., Manganese oxide and docetaxel co-loaded fluorescent polymer nanoparticles for dual modal imaging and chemotherapy of breast cancer. *Journal of controlled release*, 2015. **209**: p. 186–196.
95. Hsu, B.Y.W. et al., A Hybrid silica nanoreactor framework for encapsulation of hollow manganese oxide nanoparticles of superior T_1 magnetic resonance relaxivity. *Advanced functional materials*, 2015. **25**(33): p. 5269–5276.
96. Tromsdorf, U.I. et al., A highly effective, nontoxic T_1 MR contrast agent based on ultrasmall PEGylated iron oxide nanoparticles. *Nano letters*, 2009. **9**(12): p. 4434–4440.
97. Pellico, J. et al., Fast synthesis and bioconjugation of ^{68}Ga core-doped extremely small iron oxide nanoparticles for PET/MR imaging. *Contrast media & molecular imaging*, 2016. **11**(3): p. 203–210.
98. Jain, T.K. et al., Iron oxide nanoparticles for sustained delivery of anticancer agents. *Molecular pharmaceutics*, 2005. **2**(3): p. 194–205.
99. Bulte, J.W. and D.L. Kraitchman, Iron oxide MR contrast agents for molecular and cellular imaging. *NMR in biomedicine*, 2004. **17**(7): p. 484–499.
100. Anselmo, A.C. and S. Mitragotri, Nanoparticles in the clinic. *Bioengineering & translational medicine*, 2016. **1**(1): p. 10–29.
101. Jae-Hyun, L. et al., Artificially engineered magnetic nanoparticles for ultra-sensitive molecular imaging. *Nature medicine*, 2007. **13**(1): p. 95.
102. Huang, J. et al., Casein-coated iron oxide nanoparticles for high MRI contrast enhancement and efficient cell targeting. *ACS applied materials & interfaces*, 2013. **5**(11): p. 4632–4639.
103. Tong, S. et al., Coating optimization of superparamagnetic iron oxide nanoparticles for high T_2 relaxivity. *Nano letters*, 2010. **10**(11): p. 4607–4613.
104. Yablonskiy, D.A. and E.M. Haacke, Theory of NMR signal behavior in magnetically inhomogeneous tissues: The static dephasing regime. *Magnetic resonance in medicine*, 1994. **32**(6): p. 749.
105. Berret, J.F. et al., Controlled clustering of superparamagnetic nanoparticles using block copolymers: Design of new contrast agents for magnetic resonance imaging. *Journal of the American Chemical Society*, 2006. **128**(5): p. 1755.
106. Kuo, P.H. et al., Gadolinium-based MR contrast agents and nephrogenic systemic fibrosis. *Radiology*, 2007. **242**(3): p. 647.
107. Iron oxide MR contrast agents for molecular and cellular imaging. *NMR in biomedicine*, 2004, 17(7): 484–499.
108. Li, Z. et al., Ultrasmall water-soluble and biocompatible magnetic iron oxide nanoparticles as positive and negative dual contrast agents. *Advanced functional materials*, 2012. **22**(11): p. 2387–2393.
109. Zhou, Z. et al., Engineered iron-oxide-based nanoparticles as enhanced T_1 contrast agents for efficient tumor imaging. *ACS nano*, 2013. **7**(4): p. 3287–3296.
110. Hu, F. et al., Facile synthesis of ultrasmall PEGylated iron oxide nanoparticles for dual-contrast T_1- and T_2-weighted magnetic resonance imaging. *Nanotechnology*, 2011. **22**(24): p. 245604.
111. Judenhofer, M.S. et al., Simultaneous PET-MRI: A new approach for functional and morphological imaging. *Nature medicine*, 2008. **14**(4): p. 459–465.
112. Eiber, M. et al., Simultaneous (68)Ga-PSMA HBED-CC PET/MRI improves the localization of primary prostate cancer. *European urology*, 2016. **70**(5): p. 829–836.
113. Lahooti, A. et al., PEGylated superparamagnetic iron oxide nanoparticles labeled with ^{68}Ga as a PET/MRI contrast agent: A biodistribution study. *Journal of radioanalytical & nuclear chemistry*, 2017: p. 1–6.
114. Chou, S.W. et al., In vitro and in vivo studies of FePT nanoparticles for dual modal CT/MRI molecular imaging. *Journal of the american chemical society*, 2010. **132**(38): p. 13270.
115. Zhou, B. et al., PEGylated polyethylenimine-entrapped gold nanoparticles loaded with gadolinium for dual-mode CT/MR imaging applications. *Nanomedicine*, 2016. **11**(13): p. 1639.
116. Pissuwan, D., S.M. Valenzuela, and M.B. Cortie, Therapeutic possibilities of plasmonically heated gold nanoparticles. *Trends in biotechnology*, 2006. **24**(2): p. 62–67.
117. Huang, X. et al., Cancer cell imaging and photothermal therapy in the near-infrared region by using gold nanorods. *Journal of the American Chemical Society*, 2006. **128**(6): p. 2115–2120.
118. Choi, W.I. et al., Tumor regression in vivo by photothermal therapy based on gold-nanorod-loaded, functional nanocarriers. *ACS nano*, 2011. **5**(3): p. 1995–2003.
119. Liu, H. et al., Multifunctional gold nanoshells on silica nanorattles: A platform for the combination of photothermal therapy and chemotherapy with low systemic toxicity. *Angewandte chemie*, 2011. **50**(4): p. 891–895.
120. Tian, Q. et al., Hydrophilic flower-like CuS superstructures as an efficient 980 nm laser-driven photothermal agent for ablation of cancer cells. *Advanced materials*, 2011. **23**(31): p. 3542.
121. Liu, X. et al., Optimization of surface chemistry on single-walled carbon nanotubes for in vivo photothermal ablation of tumors. *Biomaterials*, 2011. **32**(1): p. 144–151.
122. Kai, Y. et al., Multimodal imaging guided photothermal therapy using functionalized graphene nanosheets anchored with magnetic nanoparticles. *Advanced materials*, 2012. **24**(14): p. 1868.
123. Zheng, M. et al., Single-step assembly of DOX/ICG loaded lipid–polymer nanoparticles for highly effective chemo-photothermal combination therapy. *ACS nano*, 2013. **7**(3): p. 2056.
124. Dong, W. et al., Facile synthesis of monodisperse superparamagnetic Fe_3O_4 Core@hybrid@Au shell nanocomposite for bimodal imaging and photothermal therapy. *Advanced materials*, 2011. **23**(45): p. 5392–5397.
125. Tian, Q. et al., Sub-10 nm Fe_3O_4@$Cu_{(2-x)}S$ core-shell nanoparticles for dual-modal imaging and photothermal therapy. *Journal of the American Chemical Society*, 2013. **135**(23): p. 8571.
126. Li, J. et al., Multifunctional uniform core-shell Fe_3O_4 @$mSiO_2$ mesoporous nanoparticles for bimodal imaging and photothermal therapy. *Chemistry—An Asian journal*, 2013. **8**(2): p. 385–391.
127. Derosa, M.C. and R.J. Crutchley, Photosensitized singlet oxygen and its applications. *Coordination chemistry reviews*, 2002. s **233–234**(02): p. 351–371.

128. Konan, Y.N., R. Gurny, and E. Allémann, State of the art in the delivery of photosensitizers for photodynamic therapy. *Journal of photochemistry & photobiology B biology*, 2002. **66**(2): p. 89–106.
129. Mccarthy, J.R., F.A. Jaffer, and R. Weissleder, A macrophage-targeted theranostic nanoparticle for biomedical applications. *Small*, 2010. **2**(8–9): p. 983–987.
130. Ding, J. et al., In vivo photodynamic therapy and magnetic resonance imaging of cancer by TSPP-coated Fe_3O_4 nanoconjugates. *Journal of biomedical nanotechnology*, 2010. **6**(6): p. 683–686.
131. Josephson, L. et al., Near-infrared fluorescent nanoparticles as combined MR/optical imaging probes. *Bioconjugate chemistry*, 2002. **13**(3): p. 554.
132. Dubertret, B., M. Calame, and A.J. Libchaber, Single-mismatch detection using gold-quenched fluorescent oligonucleotides. *Nature biotechnology*, 2001. **19**(4): p. 365.
133. Yi, D.K. et al., Silica-coated nanocomposites of magnetic nanoparticles and quantum dots. *Journal of the american chemical society*, 2005. **127**(14): p. 4990–4991.
134. Selvan, S.T. et al., Synthesis of silica-coated semiconductor and magnetic quantum dots and their use in the imaging of live cells. *Angewandte chemie*, 2007. **46**(14): p. 2448–2452.
135. Samia, A.C., X. Chen, and C. Burda, Semiconductor quantum dots for photodynamic therapy. *Journal of the american chemical society*, 2003. **125**(51): p. 15736–15737.
136. Chao, W., C. Liang, and L. Zhuang, Upconversion nanoparticles for photodynamic therapy and other cancer therapeutics. *Theranostics*, 2013. **3**(5): p. 317–330.
137. Qiao, X.F. et al., Triple-functional core-shell structured upconversion luminescent nanoparticles covalently grafted with photosensitizer for luminescent, magnetic resonance imaging and photodynamic therapy in vitro. *Nanoscale*, 2012. **4**(15): p. 4611.
138. Bisby, R.H., C. Mead, and C.G. Morgan, Wavelength-programmed solute release from photosensitive liposomes. *Biochemical & biophysical research communications*, 2000. **276**(1): p. 169–173.
139. Wang, G., X. Tong, and Z. Yue, Preparation of azobenzene-containing amphiphilic diblock copolymers for light-responsive micellar aggregates. *Macromolecules*, 2004. **37**(24): p. 8911–8917.
140. Wang, Y. et al., Photocontrolled self-assembly and disassembly of block ionomer complex vesicles: A facile approach toward supramolecular polymer nanocontainers. *Langmuir the ACS journal of surfaces & colloids*, 2010. **26**(2): p. 709.
141. Heta, Y. et al., Gadolinium containing photochromic micelles as potential magnetic resonance imaging traceable drug carriers. *Photochemistry & photobiology*, 2012. **88**(4): p. 876–883.
142. Park, C., K. Lee, and C. Kim, Photoresponsive cyclodextrin-covered nanocontainers and their sol-gel transition induced by molecular recognition. *Angewandte chemie*, 2009. **48**(7): p. 1275.
143. Knežević, N.Ž. and V.S. Lin, A magnetic mesoporous silica nanoparticle-based drug delivery system for photosensitive cooperative treatment of cancer with a mesopore-capping agent and mesopore-loaded drug. *Nanoscale*, 2013. **5**(4): p. 1544–1551.
144. Mal, N.K., M. Fujiwara, and Y. Tanaka, Photocontrolled reversible release of guest molecules from coumarin-modified mesoporous silica. *Nature*, 2003. **421**(6921): p. 350–353.
145. Karthik, S. et al., Photoresponsive coumarin-tethered multifunctional magnetic nanoparticles for release of anti-cancer drug. *Applied materials & interfaces*, 2013. **5**(11): p. 5232–5238.
146. Srinivasarao, M., C.V. Galliford, and P.S. Low, Principles in the design of ligand-targeted cancer therapeutics and imaging agents. *Nature reviews drug discovery*, 2015. **14**(3): p. 203.
147. Prodi, L. et al., Imaging agents based on lanthanide doped nanoparticles. *Chemical society reviews*, 2015. **44**(14): p. 4922.
148. Huang, Y. et al., Biomedical nanomaterials for imaging-guided cancer therapy. *Nanoscale*, 2012. **4**(20): p. 6135.
149. Foy, S.P. et al., Optical imaging and magnetic field targeting of magnetic nanoparticles in tumors. *ACS nano*, 2010. **4**(9): p. 5217–5224.
150. Wang, L.V. and S. Hu, Photoacoustic tomography: In vivo imaging from organelles to organs. *Science*, 2012. **335**(6075): p. 1458–1462.
151. Liu, Y. et al., Magneto-plasmonic Janus vesicles for magnetic field-enhanced photoacoustic and magnetic resonance imaging of tumors. *Angewandte chemie*, 2016. **55**(49): p. 15297.
152. Engin, K. et al., Extracellular pH distribution in human tumours. *International journal of hyperthermia*, 1995. **11**(2): p. 211–216.
153. Stubbs, M. et al., Causes and consequences of tumour acidity and implications for treatment. *Molecular medicine today*, 2000. **6**(1): p. 15.
154. Gerweck, L.E. and K. Seetharaman, Cellular pH gradient in tumor versus normal tissue: Potential exploitation for the treatment of cancer. *Cancer research*, 1996. **56**(6): p. 1194–1198.
155. Liu, J. et al., pH-sensitive nano-systems for drug delivery in cancer therapy. *Biotechnology advances*, 2014. **32**(4): p. 693–710.
156. Chen, J. et al., Polyion complex micelles with gradient pH-sensitivity for adjustable intracellular drug delivery. *Polymer chemistry*, 2014. **6**(3): p. 397–405.
157. Yuba, E. et al., Gene delivery to dendritic cells mediated by complexes of lipoplexes and pH-sensitive fusogenic polymer-modified liposomes. *Journal of controlled release*, 2008. **130**(1): p. 77–83.
158. Chen, W. et al., pH-sensitive degradable polymersomes for triggered release of anticancer drugs: A comparative study with micelles. *Journal of controlled release*, 2010. **142**(1): p. 40–46.
159. Makhlof, A., Y. Tozuka, and H. Takeuchi, pH-sensitive nanospheres for colon-specific drug delivery in experimentally induced colitis rat model. *European journal of pharmaceutics & biopharmaceutics*, 2009. **72**(1): p. 1–8.
160. Shen, M. et al., Multifunctional drug delivery system for targeting tumor and its acidic microenvironment. *Journal of controlled release*, 2012. **161**(3): p. 884–892.
161. Hu, X. et al., Preparation and characterization of a novel pH-sensitive Salecan-g-poly (acrylic acid) hydrogel for controlled release of doxorubicin. *Journal of materials chemistry B*, 2015. **3**(13): p. 2685–2697.
162. Kavitha, T., S.I.H. Abdi, and S.Y. Park, pH-sensitive nanocargo based on smart polymer functionalized graphene oxide for site-specific drug delivery. *Physical chemistry chemical physics*, 2013. **15**(14): p. 5176–5185.
163. Chen, H. et al., Monitoring pH-triggered drug release from radioluminescent nanocapsules with X-ray excited optical luminescence. *ACS nano*, 2013. **7**(2): p. 1178–1187.
164. Kang, S.I. and Y.H. Bae, A sulfonamide based glucose-responsive hydrogel with covalently immobilized glucose oxidase and catalase. *Journal of controlled release*, 2003. **86**(1): p. 115–121.
165. Song, W. et al., Tunable pH-sensitive poly(β-amino ester)s synthesized from primary amines and diacrylates for intracellular drug delivery. *Macromolecular bioscience*, 2012. **12**(10): p. 1375–1383.

166. Risbud, M.V. et al., pH-sensitive freeze-dried chitosan-polyvinyl pyrrolidone hydrogels as controlled release system for antibiotic delivery. *Journal of controlled release*, 2000. **68**(1): p. 23.
167. Wu, H., L. Zhu, and V.P. Torchilin, pH-sensitive poly(histidine)-PEG/DSPE-PEG co-polymer micelles for cytosolic drug delivery. *Biomaterials*, 2013. **34**(4): p. 1213–1222.
168. Collins, L. et al., Self-assembly of peptides into spherical nanoparticles for delivery of hydrophilic moieties to the cytosol. *ACS nano*, 2010. **4**(5): p. 2856–2864.
169. Guo, X.D. et al., Oligomerized alpha-helical KALA peptides with pendant arms bearing cell-adhesion, DNA-binding and endosome-buffering domains as efficient gene transfection vectors. *Biomaterials*, 2012. **33**(26): p. 6284–6291.
170. Drummond, D.C., M. Zignani, and J. Leroux, Current status of pH-sensitive liposomes in drug delivery. *Progress in lipid research*, 2000. **39**(5): p. 409.
171. Shieh, Y.A. et al., Aptamer-based tumor-targeted drug delivery for photodynamic therapy. *ACS nano*, 2010. **4**(3): p. 1433.
172. Lim, E.K. et al., pH-triggered drug-releasing magnetic nanoparticles for cancer therapy guided by molecular imaging by MRI. *Advanced materials*, 2011. **23**(21): p. 2436–2442.
173. Fang, C. et al., Fabrication of magnetic nanoparticles with controllable drug loading and release through a simple assembly approach. *Journal of controlled release*, 2012. **162**(1): p. 233–241.
174. Yuan, Q., S. Hein, and R.D. Misra, New generation of chitosan-encapsulated ZnO quantum dots loaded with drug: Synthesis, characterization and in vitro drug delivery response. *Acta biomaterialia*, 2010. **6**(7): p. 2732–2739.
175. Muhammad, F. et al., Acid degradable ZnO quantum dots as a platform for targeted delivery of an anticancer drug. *Journal of materials chemistry*, 2011. **21**(35): p. 13406–13412.
176. Wang, D. et al., Controllable synthesis of dual-MOFs nanostructures for pH-responsive artemisinin delivery, magnetic resonance and optical dual-model imaging-guided chemo/photothermal combinational cancer therapy. *Biomaterials*, 2016. **100**: p. 27.
177. Han, L. et al., Enhanced siRNA delivery and silencing gold-chitosan nanosystem with surface charge-reversal polymer assembly and good biocompatibility. *ACS nano*, 2012. **6**(8): p. 7340.
178. Hu, X. et al., Multifunctional hybrid silica nanoparticles for controlled doxorubicin loading and release with thermal and pH dually response. *Journal of materials chemistry B*, 2013. **1**(8): p. 1109.
179. Zhang, Z.Y. et al., Biodegradable ZnO@polymer core-shell nanocarriers: pH-triggered release of doxorubicin in vitro. *Angewandte chemie*, 2013. **52**(15): p. 4127.
180. Huang, Y.S., Y.J. Lu, and J.P. Chen, Magnetic graphene oxide as a carrier for targeted delivery of chemotherapy drugs in cancer therapy. *Journal of magnetism & magnetic materials*, 2016. **427**: p. 34–40.
181. Knorr, V. et al., Acetal linked oligoethylenimines for use as pH-sensitive gene carriers. *Bioconjugate chemistry*, 2008. **19**(8): p. 1625–1634.
182. Tang, R. et al., Block copolymer micelles with acid-labile ortho ester side-chains: Synthesis, characterization, and enhanced drug delivery to human glioma cells. *Journal of controlled release*, 2011. **151**(1): p. 18.
183. Ding, M. et al., Toward the next-generation nanomedicines: Design of multifunctional multiblock polyurethanes for effective cancer treatment. *ACS nano*, 2013. **7**(3): p. 1918–1928.
184. Kim, Y.H. et al., Polyethylenimine with acid-labile linkages as a biodegradable gene carrier. *Journal of controlled release*, 2005. **103**(1): p. 209–219.
185. Kakinoki, A. et al., Synthesis of poly(vinyl alcohol)-doxorubicin conjugates containing cis-aconityl acid-cleavable bond and its isomer dependent doxorubicin release. *Biological & pharmaceutical bulletin*, 2008. **31**(1): p. 103–110.
186. Min, S.S. and Y.J. Kwon, Ketalized poly(amino ester) for stimuli-responsive and biocompatible gene delivery. *Polymer chemistry*, 2012. **3**(9): p. 2570–2577.
187. Du, C. et al., A pH-sensitive doxorubicin prodrug based on folate-conjugated BSA for tumor-targeted drug delivery. *Biomaterials*, 2013. **34**(12): p. 3087–3097.
188. Yang, X. et al., cRGD-functionalized, DOX-conjugated, and Cu-labeled superparamagnetic iron oxide nanoparticles for targeted anticancer drug delivery and PET/MR imaging. *Biomaterials*, 2011. **32**(17): p. 4151.
189. Gan, Q. et al., Endosomal pH-activatable magnetic nanoparticle-capped mesoporous silica for intracellular controlled release. *Journal of materials chemistry*, 2012. **22**(31): p. 15960–15968.
190. Ke, C.J. et al., Real-time visualization of pH-responsive PLGA hollow particles containing a gas-generating agent targeted for acidic organelles for overcoming multi-drug resistance. *Biomaterials*, 2013. **34**(1): p. 1.
191. Liu, J. et al., CO_2 gas induced drug release from pH-sensitive liposome to circumvent doxorubicin resistant cells. *Chemical communications (Cambridge, England)*, 2012. **48**(40): p. 4869.
192. Kawanishi, S. et al., Oxidative and nitrative DNA damage in animals and patients with inflammatory diseases in relation to inflammation-related carcinogenesis. *Biological chemistry*, 2006. **387**(4): p. 365–372.
193. Brandon, M., P. Baldi, and D.C. Wallace, Mitochondrial mutations in cancer. *Oncogene*, 2006. **25**(34): p. 4647.
194. Rodrigues, M.S., M.M. Reddy, and M. Sattler, Cell cycle regulation by oncogenic tyrosine kinases in myeloid neoplasias: From molecular redox mechanisms to health implications. *Antioxidants & redox signaling*, 2008. **10**(10): p. 1813–1848.
195. Azad, N., Y. Rojanasakul, and V. Vallyathan, Inflammation and lung cancer: Roles of reactive oxygen/nitrogen species. *Journal of toxicology & environmental health part B*, 2008. **11**(1): p. 1.
196. Boonstra, J. and J.A. Post, Molecular events associated with reactive oxygen species and cell cycle progression in mammalian cells. *Gene*, 2004. **337**(35): p. 1–13.
197. Houten, B.V., V. Woshner, and J.H. Santos, Role of mitochondrial DNA in toxic responses to oxidative stress. *DNA repair*, 2006. **5**(2): p. 145–152.
198. Schneider, B.L. and M. Kuleszmartin, Destructive cycles: The role of genomic instability and adaptation in carcinogenesis. *Carcinogenesis*, 2004. **25**(11): p. 2033–2044.
199. Fruehauf, J.P. and M.F. Jr, Reactive oxygen species: A breath of life or death? *Clinical cancer research*, 2007. **13**(3): p. 789.
200. Sullivan, R. and C.H. Graham, Chemosensitization of cancer by nitric oxide. *Current pharmaceutical design*, 2008. **14**(11): p. 1113–1123.
201. Trachootham, D. et al., Redox regulation of cell survival. *Antioxidants & redox signaling*, 2008. **10**(8): p. 1343.
202. Chen, E.I. et al., Adaptation of energy metabolism in breast cancer brain metastases. *Cancer research*, 2007. **67**(4): p. 1472.
203. Yang, P. et al., Biodegradable yolk-shell microspheres for ultrasound/MR dual-modality imaging and controlled drug delivery. *Colloids & surfaces B biointerfaces*, 2017. **151**: p. 333–343.

204. Stephen, Z.R. et al., Redox-responsive magnetic nanoparticle for targeted convection-enhanced delivery of O^6-benzylguanine to brain tumors. *ACS nano*, 2014. **8**(10): p. 10383–10395.
205. Li, D. et al., Theranostic nanoparticles based on bioreducible polyethylenimine-coated iron oxide for reduction-responsive gene delivery and magnetic resonance imaging. *International journal of nanomedicine*, 2014. **2014**(Issue 1): p. 3347–3361.
206. Wilson, D.S. et al., Orally delivered thioketal nanoparticles loaded with TNF-α–siRNA target inflammation and inhibit gene expression in the intestines. *Nature materials*, 2010. **138**(11): p. 923–928.
207. Ma, Z. et al., MnO_2 Gatekeeper: An intelligent and O_2-evolving shell for preventing premature release of high cargo payload core, overcoming tumor hypoxia, and acidic H_2O_2-sensitive MRI. *Advanced functional materials*, 2016: p. 1604258.
208. Noh, J. et al., Amplification of oxidative stress by a dual stimuli-responsive hybrid drug enhances cancer cell death. *Nature communications*, 2015. **6**: p. 6907.
209. Wu, S., X. Huang, and X. Du, Glucose-and pH-responsive controlled release of cargo from protein-gated carbohydrate-functionalized mesoporous silica nanocontainers. *Angewandte chemie*, 2013. **52**(21): p. 5580–5584.
210. Curcio, A. et al., Magnetic pH-responsive nanogels as multifunctional delivery tools for small interfering RNA (siRNA) molecules and iron oxide nanoparticles (IONPs). *Chemical communications*, 2012. **48**(18): p. 2400–2402.
211. Pramod, P.S., R. Shah, and M. Jayakannan, Dual stimuli polysaccharide nanovesicles for conjugated and physically loaded doxorubicin delivery in breast cancer cells. *Nanoscale*, 2015. **7**(15): p. 6636–6652.
212. Zhou, S. et al., Multi-responsive and logic controlled release of DNA-gated mesoporous silica vehicles functionalized with intercalators for multiple delivery. *Small*, 2014. **10**(5): p. 980–988.
213. An, X. et al., Rational design of multi-stimuli-responsive nanoparticles for precise cancer therapy. *ACS nano*, 2016. **10**(6): p. 5947.
214. Chang, B. et al., Thermo and pH dual responsive, polymer shell coated, magnetic mesoporous silica nanoparticles for controlled drug release. *Journal of materials chemistry*, 2011. **21**(25): p. 9239–9247.
215. Bilalis, P. et al., Nanodesigned magnetic polymer containers for dual stimuli actuated drug controlled release and magnetic hyperthermia mediation. *Journal of materials chemistry*, 2012. **22**(27): p. 13451–13454.

Yanglong Hou (E-mail: hou@pku.edu.cn) obtained his PhD in Materials Science from Harbin Institute of Technology (China) in 2000. After research activities at Peking University, the University of Tokyo and Brown University from 2000 to 2007, he joined the Peking University in 2007 and is now the Chang Jiang Chair Professor of Materials Science. His research interests include the design and chemical synthesis of functional nanoparticles and graphene-based nanocomposites, and their biomedical and energy related applications

Yanmin Ju earned her BS degree in veterinary medicine in 2014 from Nanjing Agricultural University (China). She is now a PhD candidate at Peking University; her research interest is focused on the biomedicine applications of magnetic nanoparticles.

Shiyan Tong earned her BS degree in biological science at Ocean University of China in 2012. She worked as a technician at Peking University until 2015. She was admitted to Peking University as a PhD candidate in 2015. Her research interest is focused on the biomedical applications of magnetic nanoparticles.

Section V

Good Manufacturing Practice

26 Good Manufacturing Practices (GMP) of Magnetic Nanoparticles

Nazende Günday Türeli and Akif Emre Türeli*

CONTENTS

26.1 INTRODUCTION: BACKGROUND AND DRIVING FORCES

Good manufacturing practice (GMP) defines a set of regulations for the pharmaceutical industry that is EU- and U.S.-wide accepted, in order to ensure that quality is guaranteed and sustained throughout the whole supply chain. GMP covers not only requirements that are related directly to the manufacturing process in a pharmaceutical company. It also regulates all the related areas, premises and other processes, including equipment validations, hygiene, training and staff. It is designed to minimize the risks involved in any step of pharmaceutical production which cannot be identified through testing the final product.

Even though the concept and the principles do not show major differences, the structure and organization of GMP regulations and guidelines differ in the EU and the United States.

European GMP contains three parts and annexes. Part I provides guidelines on the manufacturing process for medicinal products, Part II refers to GMP of active substances, whereas annexes 1–20 supports the requirements for related side topics that play a crucial role in control management strategies such as 'annex-8: Sampling of starting and packaging materials, annex-11: Computerized systems, annex-15: Qualification and validation, annex-16: Certification by qualified person and batch release, annex-19: Reference and retention samples and annex-20: Quality risk management'. Since 2011, GMP guidelines have been updated and part III has been introduced for 'GMP related documents' to enlighten the regulatory expectations, i.e. Site Master File.

In the United States, the Food and Drug Administration (FDA) provides regulatory guidance. The FDA GMP, also known as current GMP (cGMP), is defined under the Part 210/211 of 'Code of Federal Regulation (CFR) – Title 21 Food and Drugs' and the related topics are introduced as paragraphs under these two parts.

In addition to complying with GMP regulations and guidelines[†] each country ensures the quality over pharmaceutical products by the laws regulating finished pharmaceutical or medicinal products and active pharmaceutical ingredient (API) manufacturing. Additionally, European Pharmacopeia and U.S. Pharmacopeia help to ensure the quality and safety by setting standards. Thus, it is very important to understand the specific regulatory expectations for each country where a pharmaceutical product is intended to be manufactured and marketed for human and veterinary use.

Quality assurance (QA) in a united form of GMP guidelines and regulations, quality control (QC) and product quality review (PQR) builds the quality management (QM) system in a pharmaceutical company.

The GMP serves as an integrated part of QA to warrant that the product quality is consistent and is ensured by producing and controlling with quality standards of the intended product specifications and regulatory requirements. 'Quality' is expected to be produced as an output of core rules and regulations under standardized conditions. Controls over the process in pharmaceutical companies serve as tools for quality. The starting materials, API and inactive substances, must be manufactured under GMP; manufacturing and control processes must be clearly specified; responsibilities must be clearly defined, and all the necessary controls and validations must be performed. In parallel, QC deals with sampling, specifications and controls, as well as organization, documentation and release processes, as

* Corresponding author.

[†] That is, International Conference on Harmonisation (ICH) of Technical Requirements for Registration of Pharmaceuticals for Human and Pharmaceutical Inspection Convention and Pharmaceutical Inspection Co-operation Scheme (http://www.ich.org/products/guidelines.html).

a part of GMP. The PQR supports the QC by periodic quality testing of already released pharmaceuticals, and verifies the suitability of specifications, checks for trends and identifies the potential improvements of products and procedures.

The GMP manufacturing starts by acquisition of raw materials and their identification, processing raw materials to finished products, preparation of master batch records and release testing. Release testing ensures quality by verifying that the product meets the specifications derived from the critical quality attributes (CQA). Under the umbrella of the QM system, GMP covers not only production and QC aspects, but also regulates other aspects including purchase, supply, warehouse, regulatory affairs, IT and personnel, as well as contracted manufacturers and suppliers.

26.2 GMP REQUIREMENTS

GMP is based on a very simple but essential rule: all processes that have an impact on quality must be planned, controlled and monitored.

All those organizations, documentation, controls and monitors are documented in a series of harmonized documents such as the site master file (SMF). The SMF defines specific aspects of GMP carried out at the premises for production and control of pharmaceutical products. This information includes but is not limited to

- Brief information on the company
- Availability and use of appropriate resources:
  - Premises, space, equipment and services
  - Materials, containers, labels
  - Clear, written instructions and procedures
  - Laboratories and in-process control
  - Qualified and trained staff
  - Trained operators
  - Procedures for storage, transport
- Documentation system and specifications
- Qualification and validation
- Change control (CC) management, corrective and preventive action (CAPA) setting
- Records of actions, deviations and investigations
- Records for manufacture and distribution
- Proper storage and distribution
- Systems for complaints and recalls

These quality commitments require documentation of all the processes, and continuous review and improvement of all the existing documents. As seen in Figure 26.1, QM documentation structure can be divided into three hierarchical groups.

The 'QM Manual' is a master document describing the regulations that the pharmaceutical company follows, ensures that products and services meet the demands, defines the implementation of all elements of the QM system with applicable policies and responsibilities and shows the processes and their interaction(s). Standard operating procedures (SOPs) give step-by-step instructions for performing operational tasks or activities and define the responsibilities. Every other

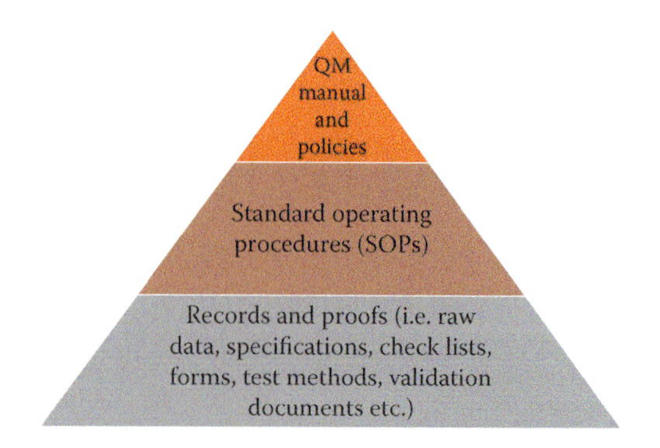

FIGURE 26.1 Structure of QM-documentation.

document can be listed under records and proofs. These type of documents show that QM-Manual, policies and SOPs have been followed. Records enable traceability of actions taken on a specific process. Raw data include all worksheets, notes or original copies that are generated during the activities. Raw data are necessary to comprehend and evaluate a work, a qualification, a validation, a process or a report. Raw data can be available as screenshots, hard copies or soft copies, as long as they are signed and controlled. Quality records are used to demonstrate the effectiveness of the QM system as well as the conformity of the products and processes. These are particularly important in cases of liability. Document management SOPs set out the measures for labeling, preparation, storage, protection, retrieval, retention and availability of records.

Overall, GMP requirements and regulations ensures the quality of the pharmaceutical products by tracking, controlling and improving all the aspects that are involved in the pharmaceutical manufacturing process starting from the very beginning of the chain.

26.3 FROM RESEARCH AND DEVELOPMENT TO GMP ENVIRONMENT

From a regulatory point of view a 'nanotechnology product' is a product that contains or is manufactured using materials in the nanoscale range, as well as products that contain or are manufactured using certain materials that otherwise exhibit related dimension-dependent properties or phenomena.[1] For example, chemical, biological or magnetic properties being altered in the nanoscale range in comparison to coarse particles of same chemical composition.[2]

Prerequisites of NP manufacturing can be listed as the size, size distribution and type of ingredients that are used for the (nano)pharmaceuticals, since they directly determine the solubility and the dissolution rate of the pharmaceuticals. On the other hand, costs of goods and economics determine if the nanoparharmaceutical will make it to the market or not, regardless of the success brought to the field and the high quality offered. Finally, the complexity, yield and batch sizes that can be achieved, as well as the applicability of the employed manufacturing method to the existing resources and/or manufacturing lines determine the nanopharmaceuticals' penetration into the market.

Since the mid-1990s, following the very first FDA approved nanopharmaceutical, Doxil®, there has been an increasing interest in nanoreformulations of existing pharmaceutical products, since nanotechnology offers novel solutions to solubility, stability, dose-dependent toxicity and many other problems.

Development of a clinically successful liposomal formulation, Doxil, encouraged the scientific and industrial community. Meanwhile, inorganic NPs such as MRI contrast agent and imaging agents (i.e. Feridex® and GastroMARK® developed by AMAG Pharmaceuticals, Inc.) prepared by batch manufacturing processes made the first penetration into the market.

The following nanopharmaceutical products were liposomal formulations and a second generation was initiated for micellar nanoformulations (i.e. Taxotere®). Both bottom-up manufacturing methods were easily realized by utilizing the existing production equipment, as batch size was not a limiting factor, the postmanufacturing removal of side products and/or other components did not require extreme measures, and the ingredients used were already pharmaceutically approved, safe excipients. However, these types of nanoformulations suffered from low drug loading. In order to circumvent this, other approaches were established. Right after the millennium, the nanoparticulate formulations prepared by top-down NP preparation methods (wet milling and media milling) (i.e. NanoCrystal® technology from Elan Pharmaceutical Technologies) were commercialized. Milling is realized by mechanical energy input to physically reduce the size of coarse drug particles.[3] Drug particles that are milled can be rarely used as it is due to their poor flow properties. Thus, addition and/or mixing excipients during milling was required for the majority of the nanoformulations in the market in the first decade of 2000. Additionally, milling processes require energy input and produce heat, requiring extra measures due to the high risk of dust explosion. Through a sequence of process generations, manufacturers overcame these problems; however, agglomeration tendency of the small particles during the milling process and handling difficulties still remain a challenge. Table 26.1 summarizes the successfully established nanoformulations during the decade following approval of the first nanopharmaceutical.[4]

After 2009, there was a long silent period for market penetration of nanoformulations. Even other manufacturing technologies, such as Nanopure® technology by Pharmasol GmbH, DissoCubes™ (high pressure homogenization) from SykePharm PLC or Nanoedge™ (high pressure homogenization combined with precipitation) technology, were made available, none of formulations manufactured by those technologies could make it to the market, mainly due to their batch size limitations.

The production of innovative nanopharmaceuticals in quantities and qualities required for them to enter clinical trials remains a challenge. Not only manufacturing those nanopharmaceuticals at larger scales, but also complying with regulatory requirements, is a struggle for developers. For conventional pharmaceutical development already standardized and/or pharmacopeia defined tests such as assays, impurity levels, microbial contamination, suitable container closure systems, stability and efficacy are still challenging for innovative nanopharmaceuticals. This challenge

TABLE 26.1
Approved Pharmaceutical Nanoformulations

	Trade Name	Indication	Approved Date
Liposome	Abelcet	Fungal Infections	20/11/1995
	AmBisome	Fungal Infections	11/8/1997
	Amphotec	Fungal Infections	22/11/1996
	DaunoXome	Antineoplastic	8/4/1996
	DepoCyt	Lymphomatous Meningitis	1/4/1999
	Doxil	Antineoplastic	17/11/1995
	Visudyne (Verteporfin for injection)	Photodynamic Therapy for Age-Related Macular Degeneration	12/4/2000
Micelle	Amphotec	Antifungal	22/11/1996
	Estrasorb	Vasomotor Symptoms Associated with Menopause	9/10/2003
	Taxotere	Antineoplastic	14/5/1996
Nanocrystal	Emend	Antiemetic	27/3/2003
	Tricor	Hypercholesterolemia and hypertriglyceridemia	5/11/2004
	Triglide	Hypercholesterolemia and hypertriglyceridemia	7/5/2005
	Megace ES	Anorexia, Cachexia or an Unexplained Significant Weight Loss in AIDS Patients	5/7/2005
	Rapamune	Immunosuppressant; The Prophylaxis of Organ Rejection in Patients Receiving Renal Transplants	25/8/2000
Nanoparticle	Abraxane	Metastatic Breast Cancer	7/1/2005
	Anthelios 20	Sunscreen	5/10/2006
	Helioblock SX Sunscreen Cream	Sunscreen	31/3/2008
Nanotube	Somatuline Depot	Acromegaly	30/8/2007
Superparamagnetic Iron Oxide	Feraheme Injection	Treatment of Iron Deficiency Anemia in Patients with Kidney Disease (CKD)	30/6/2009
	Feridex	MRI Contrast Agent	30/8/1996
	GastroMARK	Imaging of Abdominal Structures	6/12/1996

Source: Pena, C., A FDA perspective on nanomedicine, current initiatives in the US. *EMA First International Workshop on Nanomedicine*, http://www.ema.europa.eu/docs/en_GB/document_library/Presentation/2010/09/WC500096201.pdf, 2010.

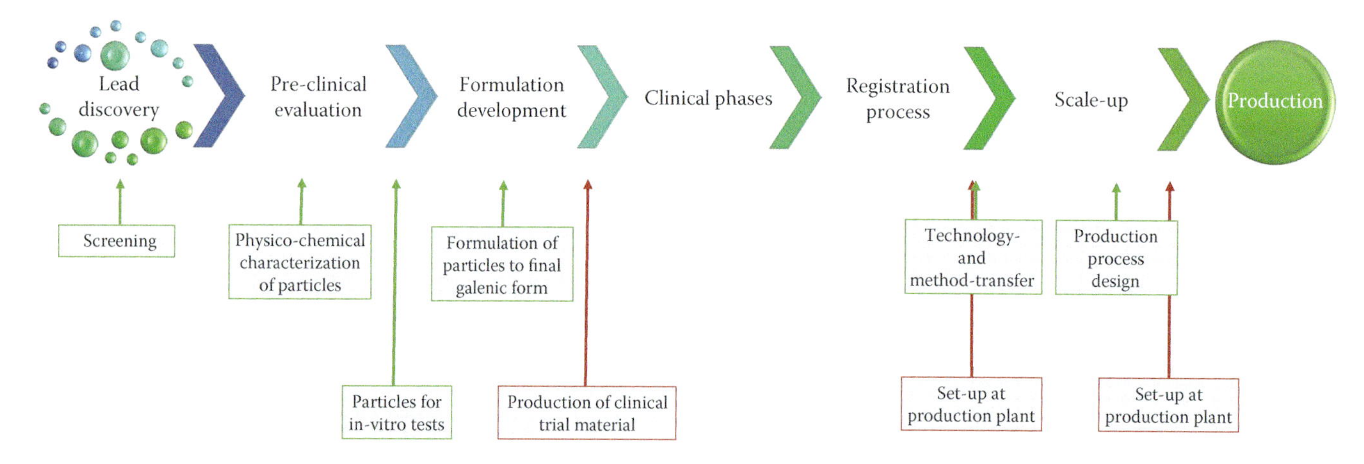

FIGURE 26.2 Pharmaceutical development chain of NPs.

is not only limited to quality ensuring tests but also proactive quality management approaches such as manufacturing design space, in process controls (IPCs), identification of risks and risk management, and batch-to-batch reproducibility.

26.3.1 Scale-Up

During the scaling-up process, preclinical optimal parameters, such as NPs' efficacy, which depend on their average size, size distribution, surface charge and surface chemistry should be preserved. However, most NP manufacturing methods are not easy to implement in existing manufacturing plants and characterization methods are not yet standardized and/or not easy to establish. Moreover, small companies, often spin-offs and academic institutions, who are the major players of innovative nanopharmaceutical developments, lack resources to scale-up and implement GMP manufacturing to meet the demands for the quality control.

To ensure industrial sustainability, from early phases of the pharmaceutical development chain, translation of manufacturing and characterization methods should be considered important aspects. As shown in Figure 26.2, once lead candidates are chosen, particles chosen for galenic formulation development should also be evaluated for translatability.

Inactive materials (excipients) are used to control the *in vivo* characteristics and performance of the nanoformulations such as maximizing the probability of absorption which increases the bioavailability. Not only physicochemical characterization methods, to ensure efficacy and safety, but also manufacturing processes should be applicable to GMP conditions. The technology and method transfers to the GMP sites should be implemented in the process as early as possible (labeled in red). Considering that batch sizes produced during the development and preclinical characterization phases are most of the time not applicable to industrial manufacturing-process demands, it would not be incorrect to label the scale-up of developed manufacturing methods as the first translational challenge on the hard road from bench to bedside.

In the scale-up process, the physicochemical properties of NPs should be maintained.[5] Thus, the effect of manufacturing processes on physicochemical properties of NPs should be well understood and regulatory requirements should be taken into consideration. During the scale-up phase, if possible, suitable statistical methods should be employed to improve the understanding of the process, and quality-by-design concepts (QbD), as required by International Council of Harmonization (ICH) quality guidelines and FDA guidelines, should be implemented from the very early phases of the development process.

Scale-up possibilities include conventional technologies (i.e. high pressure homogenization, batch processes in existing reactors/mixers[6–9]) or innovative technologies (i.e. microchannel reactors,[10] micromixers and continuous mixers[11–14]). As listed in Table 26.2, both technologies have pros and cons, which need to be evaluated before choosing a scale-up approach.

26.3.2 Quality by Design

ICH quality guidelines‡ provide the essential information on the minimum requirements for ensuring product quality. The ICH Q8: 'Pharmaceutical Development' describes the principles of QbD and shows how concepts and tools (e.g. design space) could be put into practice for all dosage forms. Application of QbD and quality risk management according to guideline ICH Q9: 'Quality Risk Management', linked to an appropriate pharmaceutical quality system according to ICH Q10, provides opportunities to enhance science- and risk-based regulatory approaches. Science and risk-based regulatory approaches can be enhanced by linking the pharmaceutical quality system (conforming ICH Q10 guideline) to QbD applications and quality risk management, as described in guideline ICH Q9: 'Quality Risk Management'.

QbD is defined in ICH guidelines as 'a systematic approach to development that begins with predefined objectives and emphasizes product and process understanding and process control.'

QbD implementation should provide a higher level of product quality together with cost and energy saving. The knowledge gained over the product's lifecycle should not only serve for design controls and testing, based on scientific understanding

‡ http://www.ich.org/products/guidelines/quality/article/quality-guidelines.html

TABLE 26.2
Conventional and Innovative Scale-Up Possibilities, Pros and Cons

	Pro	Cons
Conventional methods	– No expensive equipment investment – The equipment used are well-known by the industry – Calibration methods and/or qualifications are already established in-house or available in pharmacopeia – Easy to integrate and/or operate in existing GMP lines	– Not flexible in equipment design – Suitable for small to medium batch sizes – Often high energy input required (i.e. high pressure, sonication) – Possible metal contamination – High temperature processes – Potential product degradation due to high temperature
Innovative methods	– Robust set-up – No contamination of product – Easy to clean – Easy to realize in process control – Continuous production up to tons – Low process costs – No energy input required – Often no stress on product	– Calibration methods and/or qualifications are not established in-house or available in pharmacopeia

at pharmaceutical development stage, but also for continuous improvement of the product, as shown in Figure 26.3.

The Design of Experiments (DoE) approach enables systematic problem-solving with minimum numbers of experiments by gaining information about the process parameters, defining CQAs and finding the process parameter effect on CQAs through fully planned experiments. Assigned design spaces serve as multidimensional combination and interaction of input variables and process parameters that are used to provide assurance of quality. Using the statistical DoE it is possible to accurately determine the relationship between factors (i.e. independent variables) and targets (i.e. dependent variables) with few experiments. Advantages are relatively low experimental efforts, identification of a regression function for the entire experimental space, and evaluation of each influencing factor on the response. Results of DoE-analysis are very useful for setting the IPC of the manufacturing process.[15–17] Finally, control strategies, which are a planned set of controls derived from the current product and process understanding, ensure process performance and product quality.

Early integration of these concepts will pioneer establishment of quality systems for regulatory compliance. However, it is worth mentioning that QbD is product-specific and requires a heavy workload, including but not limited to multivariate analysis, design space(s) description(s), quality risk management and control strategy establishment. Even though it aims to achieve efficiency and lower manufacturing cost over the entire lifecycle in the long term, it means an additional investment during the development phase.

26.4 PREREQUISITES OF GMP CONFORMED MANUFACTURING

In order to comply with regulatory requirements, a number of actions are required for the production of NPs. The regulatory requirements extracted from ICH guidelines can be summarized as follows;

a. Setting the specifications of API and every other ingredient in the nanoformulation.
b. Development and validation of analytical methods for the release of the starting materials such as API and excipients and solvents.

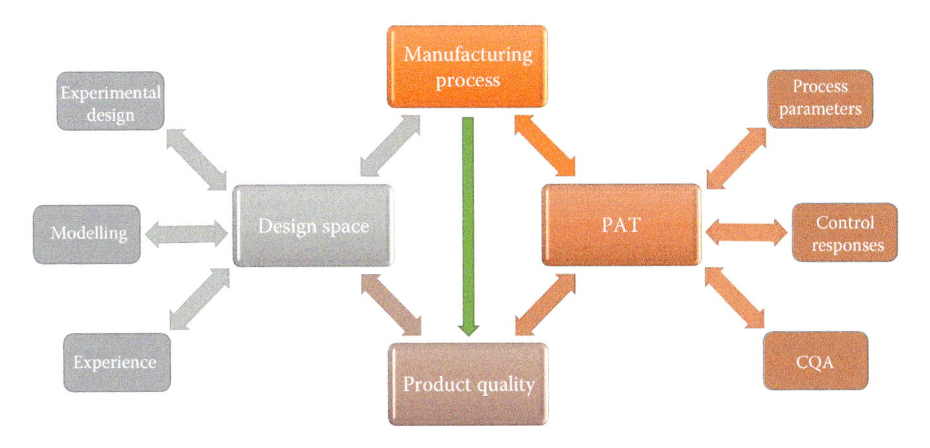

FIGURE 26.3 Ensuring product quality via quality-by-design.

c. Setting the specifications and in-process control parameters for the production of NPs.
d. Development and validation of the analytical methods for the in-process control of the NP production.
e. Setting the specification for the intermediate product(s), i.e. NP suspension, freeze dried powder.
f. Development and validation of analytical methods for the intermediate product(s).
g. Setting the specifications for primary packaging.
h. Development and validation of analytical methods for the primary packaging.
i. Batch release.

These requirements, applicable from starting materials to end products, can be met by establishing Process Analytical Technology (PAT). PAT serves as 'a system for designing, analyzing and controlling manufacturing through timely measurements (i.e. during processing) of CQA and performance attributes of raw and in-process materials and processes with the goal of ensuring final product quality.'

26.4.1 Quality Control Analytics

NPs for imaging, diagnostic and theranostics purposes, as well as some of the already marketed drug delivery systems have sizes <100 nm, while many other drug delivery systems fall in the range of 100–1000 nm. Conventional particle range defined in regulatory guidelines (>100 nm, absence of particles >10 μm) does not cover state-of-the-art NP technology. There is an immediate need for establishment of relevant methods and equipment for NP characterization under GMP. Additionally, quality systems must be modified for integration of NP-specific precautions and controls. Site-specific policy must be determined and these major policies must be enriched by NP manufacturing-process/product specific standard operating procedures, test methods and batch release methods.

Thorough multidimensional characterization of particles' long-term stability must be evaluated (i.e. for agglomeration, disintegration and drug leakage, etc.), since these particle characteristics have great impact on product performance, bioavailability, efficacy, safety and other properties. Nanoscale materials are mostly still treated in the same way as conventional formulations in a regulatory context, although several scientific opinions, guidelines and EU regulations specifically address nano(bio)materials. However, to date, no guideline or pharmacopeia chapter has been dedicated to minimum requirements for characterization of nanotechnology products. In one of the latest communications, FDA announced that existing guidelines are expected to apply for nanotechnology products, and they reflect the FDA's current opinion on all the aspects of GMP manufacturing of nanopharmaceuticals. These guidelines listed in Table 26.3 are open to interpretation for their implementation.

European and United States pharmacopoeias define a particle as 'the smallest entity of a particulate system, which can be liquid or semi-solid droplets, single or poly-crystals,

TABLE 26.3
Existing Guidance Documents for Pharmaceutical Nanotechnology Products

Topic Area	Guidance Title	Publication Date
CMC	Analytical Procedures and Method Validation: Chemistry, Manufacturing and Controls Documentation	2000
	Comparability Protocols – Chemistry, Manufacturing and Control Information	2003
	Current Good Manufacturing Practice for Combination Products	2004
	Residual Solvent in Drug Products Marketed in the United States	2009
	Guideline on General Principles of Process Validation	1987
	Good Laboratory Practice Regulations: Questions and Answers	1998
	Liposome Drug Products: Chemistry, Manufacturing and Controls: Human Pharmokinetics and Bioavailability, and Labeling Documentation	2002
	Process Validation: General Principles and Practices	2008
Imaging	Developing Medical Imaging Drug and Biological Products Part I: Conducting Safety Assessment; Part II: Clinical Indications; Part III: Design, Analysis and Interpretation of Clinical Studies	2004
Procedural	Content and Format of Investigational New Drug Applications (INDs) for Phase I Studies for Drugs, Including Well-Characterized, Therapeutic, Biotechnology-Derived Products	1995
	Early Development Considerations for Innovative Combination Products	2006
	Guidance for Reviewers: Pharmacology/ Toxicology Review Format	2001
	Guidelines for Submitting Documentation for Manufacture of and Control for Drug Products	1987
	Guidelines for Submitting Supporting Documentation in Drug Applications for the Manufacture of Drug Substances	
PK/ADME	Drug Interaction Studies-Study Design, Data Analysis and Implications for Dosing and Labeling	2006
	Drug Metabolism/Drug Interaction Studies in the Drug Development Process: Studies *In Vitro*	1997
	Nonclinical Safety Evaluation of Reformulated Drug Products and Products Intended for Administration by an Alternative Route	2008
	Nonclinical Studies for Safety Evaluation of Pharmaceutical Excipients	2005
	Safety Testing of Drug Metabolites	2008
	Single Dose Acute Toxicity Testing for Pharmaceuticals	1996
	Statistical Aspects of the Design, Analysis and Interpretation for Chronic Rodent Carcinogenicity Studies of Pharmaceuticals	2001

amorphous structures or agglomerates.' NPs can be considered as undissolved active substances in solid or liquid drug formulations. Particle size has an effect on active ingredient release, bioavailability or stability. Thus, particle size and size distribution, and appropriate measurement methods should be carefully designed to meet regulatory requirements.

Particle size is directly correlated with the shape of the particle. The diameter of the sphere is used for spherical particles, whereas for nonspherical particles the particle size measurement becomes more complicated. Microscopic or screening (sieving) methods provide real particle diameter while indirect methods such as photon correlation spectroscopy determine the hydrodynamic diameter. The hydrodynamic diameter is the actual particle including the solvated shell moving with the particle. Although photon correlation spectroscopy is not included in the Pharmacopoeias and regulated by ISO standard 22412, it is still the most extensively used method in GMP environment for the measurement of particles.[18]

In general, methods for particle characterization, which are used in the QC of medicinal products, should meet the requirements of the established guidelines (i.e. ICH Q2-R1 Validation of analytical procedures: text and methods, FDA Guidance for Industry, Analytical Procedures and Methods Validation). The requirements of validation of analytical procedures are predominately applied for chromatographic methods such as High Pressure Liquid Chromatography (HPLC) but can also be adapted for particle size measurement methods. For this purpose, an extensive technical risk analysis must be carried out and critical parameters of the characterization method and assessment of the influence on product safety and efficacy must be evaluated.

Additionally, a sterilization process should be applied to comply with regulatory requirements in case of parenteral products. Different sterilization methods should be evaluated to find the most suitable method, where physicochemical properties and particle functionality are preserved. Sterilization by filtration can be the method of choice, as long as proper filtration method development is performed. Type of filter, pore sizes, housing materials and diameters must be carefully designed to prevent 'traffic jams'. Such incident might cause alteration of the particles due to high pressure and the mechanical stress they will be exposed to. If this technique is not evaluated as suitable, other options include terminal sterilization by moist heat and/or ionizing radiation. In addition, the level of bacterial endotoxin must be determined and the bacterial endotoxin levels must be kept within the required limits. A sterilization process validation must be performed to ensure the suitability of the process. After the sterilization process, the physicochemical properties of formulations must be evaluated and preservation of those properties must be shown.

Release from NPs must also be evaluated as part of quality control. There are a variety of methods currently available for nanoformulation release studies,[19] including but not limited to dialysis bag methods,[20] cross flow filtration based methods, filtration methods,[21] solid phase extractions,[22] and microcentrifugation based single-time-point measurements. Unfortunately, none of these methods have made their way into the compendial methods, thus, there are no strategies for successfully standardizing and qualifying those methods. This might lead to batch-dependent result deviations, simply because of method treatment-based deviations regardless of the quality and/or repeatability of the nanoformulation.

The biggest issue to consider for innovative nanopharmaceuticals is the appropriateness of the equipment used for the intended nanoproduct and the need for 'self-developed' process validation and equipment qualification concepts.

As summarized in Table 26.4, ICH guidelines require that process development studies should provide the basis for process improvement, validation (or verification) and any IPC to be integrated.

The usual analytical method validation parameters are:

- Specificity
- Linearity and range
- Accuracy (comparison with reference material)
- Precision (repeatability and intermediate precision)
- Detection limit (determination limit for particles obsolete)
- Robustness

A very important, but often neglected, validation parameter is the robustness. In this case, the influence of minor changes on the results are checked for a change in the performance of the measurement method.

TABLE 26.4
Application of Process Performance and Product Quality Monitoring System Throughout the Product Lifecycle

Pharmaceutical Development	Technology Transfer	Commercial Manufacturing
Process and product knowledge generated, and process and product monitoring conducted throughout development can be used to establish a control strategy for manufacturing.	Monitoring during scale-up activities can provide a preliminary indication of process performance and successful integration into manufacturing. Knowledge obtained during transfer and scale-up activities can be useful in further developing the control strategy.	A well-defined system for process performance and product quality monitoring should be applied to assure performance within a state of control and to identify improvement areas.

Source: Adapted from ICH Q10: Pharmaceutical quality system.

In the case of methods for direct work with particles, the important influence of sample preparation must be considered. Typical test parameters are:

- Homogenization time and parameters (i.e. stirring time and speed)
- Service life
- Instrument parameters (e.g. pump rates, air pressure)
- Sample stability
- Measurement time

Ultimately, a variety of particle characterization methods are available, even though not all of these methods are fully established in the GMP environment, and not all are suitable for each type of particle, or must be combined with one or more alternative methods to enable qualitative data acquisition.

Furthermore, orthogonal measurement methods (i.e. optical or microscopic methods such as SEM/TEM/AFM) must be available to verify the primary method.

In order to remedy these shortcomings and reach a consensus in the pharmaceutical industry, further studies must be conducted to determine the connection between the certain manufacturing process parameters and CQAs included in the batch release process.

26.5 CONCLUSION

In recent years, nanotechnology has provided an intriguing tool to solve some of the unmet needs of the pharmaceutical industry. However, tremendous efforts are still required to fill the gap between research laboratories and the market, and sustain translational research on nanotechnology products with clinical potential. Market penetration success of a pharmaceutical nanotechnology product is dictated by its successful development through careful design of processes during the very early stages of research, its applicability to GMP and its compliance with regulation policies.

REFERENCES

1. FDA, Guidance for Industry: Considering whether an FDA-regulated product involves the application of nanotechnology, U.S. Department of Health and Human Service, Editor. 2014.
2. FDA, Nanotechnology. *A Report of the U.S. Food and Drug Administration Nanotechnology Task Force*. 2007.
3. Loh, Z.H., A.K. Samanta, and P.W. Sia Heng, Overview of milling techniques for improving the solubility of poorly water-soluble drugs. *Asian Journal of Pharmaceutical Sciences*, 2015. **10**(4): p. 255–274.
4. Pena, C., A FDA perspective on nanomedicine, current initiatives in the US. *EMA First International Workshop on Nanomedicine*, http://www.ema.europa.eu/docs/en_GB/document_library/Presentation/2010/09/WC500096201.pdf, 2010.
5. Wicki, A. et al., Large-scale manufacturing of GMP-compliant anti-EGFR targeted nanocarriers: Production of doxorubicin-loaded anti-EGFR-immunoliposomes for a first-in-man clinical trial. *International Journal of Pharmaceutics*, 2015. **484**(1–2): p. 8–15.
6. Bohm, B.H. and R.H. Muller, Lab-scale production unit design for nanosuspensions of sparingly soluble cytotoxic drugs. *Pharmaceutical Science & Technology Today*, 1999. **2**(8): p. 336–339.
7. Chang, S.F. et al., Nonionic polymeric micelles for oral gene delivery in vivo. *Human Gene Therapy*, 2004. **15**(5): p. 481–93.
8. Danhier, F. et al., PLGA-based nanoparticles: An overview of biomedical applications. *Journal of Controlled Release*, 2012. **161**(2): p. 505–22.
9. Galindo-Rodriguez, S. et al., Physicochemical parameters associated with nanoparticle formation in the salting-out, emulsification-diffusion, and nanoprecipitation methods. *Pharmaceutical Research*, 2004. **21**(8): p. 1428–39.
10. Ali, H.S., P. York, and N. Blagden, Preparation of hydrocortisone nanosuspension through a bottom-up nanoprecipitation technique using microfluidic reactors. *International Journal of Pharmaceutics*, 2009. **375**(1–2): p. 107–13.
11. Johnson, B.K., Prud, and R.K. homme, Flash nanoprecipitation of organic actives and block copolymers using a confined impinging jets mixer. *Australian Journal of Chemistry*, 2003. **56**(10): p. 1021–1024.
12. Xie, H. and J.W. Smith, Fabrication of PLGA nanoparticles with a fluidic nanoprecipitation system. *Journal of Nanobiotechnology*, 2010. **8**: p. 18.
13. Zhao, C.-X. et al., Nanoparticle synthesis in microreactors. *Chemical Engineering Science*, 2011. **66**(7): p. 1463–1479.
14. Zhu, Z. et al., Formation of block copolymer-protected nanoparticles via reactive impingement mixing. *Langmuir*, 2007. **23**(21): p. 10499–504.
15. Beyer, S. et al., Bridging laboratory and large scale production: Preparation and in vitro-evaluation of photosensitizer-loaded nanocarrier devices for targeted drug delivery. *Pharmaceutical Research*, 2015. **32**(5): p. 1714–26.
16. Günday Türeli, N., A.E. Türeli, and M. Schneider, Optimization of ciprofloxacin complex loaded PLGA nanoparticles for pulmonary treatment of cystic fibrosis infections: Design of experiments approach. *International Journal of Pharmaceutics*, 2016. **515**(1–2): p. 343–351.
17. Draheim, C. et al., A design of experiment study of nanoprecipitation and nano spray drying as processes to prepare PLGA nano- and microparticles with defined sizes and size distributions. *Pharmaceutical Research*, 2015. **32**(8): p. 2609–2624.
18. Schichtel, J., A.E. Tuereli, and M. Limberger, Partikelmessung von F&E bis GMP in Partikel in der Pharmaproduktion, S.F., Editor. 2015, Editio Cantor Verlag Aulendorf.
19. Shen, J. and D.J. Burgess, In vitro dissolution testing strategies for nanoparticulate drug delivery systems: Recent developments and challenges. *Drug Delivery and Translational Research*, 2013. **3**(5): p. 409–415.
20. Bhardwaj, U. and D.J. Burgess, A novel USP apparatus 4 based release testing method for dispersed systems. *International Journal of Pharmaceutics*, 2010. **388**(1–2): p. 287–94.
21. Juenemann, D. et al., Biorelevant in vitro dissolution testing of products containing micronized or nanosized fenofibrate with a view to predicting plasma profiles. *European Journal of Pharmaceutics and Biopharmaceutics*, 2011. **77**(2): p. 257–64.
22. Guillot, A. et al., Solid phase extraction as an innovative separation method for measuring free and entrapped drug in lipid nanoparticles. *Pharmaceutical Research*, 2015. **32**(12): p. 3999–4009.

Nazende Günday Türeli (E-mail: N.Guenday@mjr-pharmjet.com) possesses over 10 years of international research and development experience in international pharmaceutical companies and contract research organizations. She is experienced on Quality-by-Design (QbD) approaches and GMP manufacturing, as well as EU regulatory requirements (submissions for marketing authorization in EU and their applications). Her academic expertise is on pharmaceutical nano(bio)technology, and her research interests focus on nanoparticulate drug delivery systems for oral and pulmonary administration. Her research has been rewarded, and recognized through several international and national organizations, as well as peer-reviewed journals. She is the winner of the 2014 Global CPhI Pharma Awards for Best Innovation in Formulation category.

Dr Akif Emre Türeli is the chief scientific officer and co-founder of MJR PharmJet, GmbH. He has more than 10 years of international research and development experience in the pharmaceutical industry. He has held research and development positions in various pharmaceutical companies and in contract research organizations, where he has gained experience with different stages of drug and formulation development. He is the developer of the microjet reactor technology for nanopharmaceutical applications. His expertise is on drug delivery, nanoparticle production and formulation. His focus is on the development of innovative formulations using the MJR nanotechnology for different industry applications. He holds a PhD degree in pharmaceutical technology from Johannes Gutenberg University in Mainz, Germany.

Index

A

B

C

D

E

F

N

O

T

U

V

W

X

Z

Clinical Applications of Magnetic Nanoparticles